W9-CEI-124

eighth edition

Environmental Science

A Study of Interrelationships

Eldon D. Enger
Delta College

Bradley F. Smith
Western Washington University

Boston Burr Ridge, IL Dubuque, IA Madison, WI New York San Francisco St. Louis
Bangkok Bogotá Caracas Kuala Lumpur Lisbon London Madrid Mexico City
Milan Montreal New Delhi Santiago Seoul Singapore Sydney Taipei Toronto

McGraw-Hill Higher Education

ENVIRONMENTAL SCIENCE: A STUDY OF INTERRELATIONSHIPS
EIGHTH EDITION

Published by McGraw-Hill, a business unit of The McGraw-Hill Companies, Inc., 1221 Avenue of the Americas, New York, NY 10020.

Some ancillaries, including electronic and print components, may not be available to customers outside the United States.

 This book is printed on recycled, acid-free paper containing 10% postconsumer waste.

2 3 4 5 6 7 8 9 0 VNH/VNH 0 9 8 7 6 5 4 3 2 1

ISBN 0–07–231547–4

Executive editor: *Margaret J. Kemp*
Senior developmental editor: *Kathleen R. Loewenberg*
Marketing manager: *Heather K. Wagner*
Lead project manager: *Peggy J. Selle*
Production supervisor: *Enboge Chong*
Design manager: *Stuart D. Paterson*
Cover/interior designer: *Jamie A. O'Neal*
Cover images: *Sandra Nykerk/John McColgan*
Senior photo research coordinator: *Lori Hancock*
Photo research: *LouAnn K. Wilson*
Supplement producer: *Brenda A. Ernzen*
Executive producer: *Linda Meehan Avenarius*
Compositor: *GAC–Indianapolis*
Typeface: *10/12 Times Roman*
Printer: *Von Hoffmann Press, Inc.*

The credits section for this book begins on page 473 and is considered an extension of the copyright page.

Library of Congress Cataloging-in-Publication Data

Enger, Eldon D.
Environmental science : a study of interrelationships / Eldon D. Enger, Bradley F. Smith. — 8th ed.
p. cm.
Includes index.
ISBN 0–07–231547–4 (acid-free paper)
1. Environmental sciences. I. Smith, Bradley Fraser. II. Title.

GE105 .E54 2002
363.7—dc21 2001030344
CIP

www.mhhe.com

To Judy, my wife and friend,
for sharing life's adventures.
Eldon Enger

To my wife, Daria, for her love, support,
patience and understanding.
Brad Smith

Brief Contents

PART ONE

Interrelatedness 3

CHAPTER 1 Environmental Interrelationships 4
CHAPTER 2 Environmental Ethics 19
CHAPTER 3 Risk and Cost: Elements of Decision Making 38

PART TWO

Ecological Principles and Their Application 65

CHAPTER 4 Interrelated Scientific Principles: Matter, Energy, and Environment 66
CHAPTER 5 Interactions: Environment and Organisms 80
CHAPTER 6 Kinds of Ecosystems and Communities 105
CHAPTER 7 Population Principles 130
CHAPTER 8 Human Population Issues 146

PART THREE

Energy 167

CHAPTER 9 Energy and Civilization: Patterns of Consumption 168
CHAPTER 10 Energy Sources 187
CHAPTER 11 Nuclear Energy: Benefits and Risks 215

PART FOUR

Human Influences on Ecosystems 237

CHAPTER 12 Human Impact on Resources and Ecosystems 238
CHAPTER 13 Land-Use Planning 269
CHAPTER 14 Soil and Its Uses 291
CHAPTER 15 Agricultural Methods and Pest Management 315
CHAPTER 16 Water Management 339

PART FIVE

Pollution and Policy 371

CHAPTER 17 Air Quality Issues 372
CHAPTER 18 Solid Waste Management and Disposal 401
CHAPTER 19 Regulating Hazardous Materials 419
CHAPTER 20 Environmental Policy and Decision Making 438

APPENDIX 1 Critical Thinking 459
APPENDIX 2 Metric Unit Conversion Tables 460
APPENDIX 3 The Periodic Table of the Elements 462
APPENDIX 4 What You Can Do to Make the World a Better Place in Which to Live 463
APPENDIX 5 How to Write to Your Elected Officials 464

Glossary 465
Credits 473
Index 476

Contents

Preface xi
Guided Tour xvi
About the Authors xxiv

Part One
Interrelatedness 3

CHAPTER 1
Environmental Interrelationships 4
The Field of Environmental Science 5
The Interrelated Nature of Environmental Problems 5
• **Environmental Close-Up:** Science Versus Policy 6
An Ecosystem Approach 7
Regional Environmental Concerns 7
• **Global Perspective:** Fish, Seals, and Jobs 8
The Wilderness North 8
• **Environmental Close-Up:** The Greater Yellowstone Ecosystem 9
• **Environmental Close-Up:** Headwaters Forest 10
The Agricultural Middle 11
The Dry West 11
The Forested West 12
The Great Lakes and Industrial Northeast 13
The Diverse South 14

CHAPTER 2
Environmental Ethics 19
Views of Nature 20
• **Environmental Close-Up:** What Is Ethical? 21
Environmental Ethics 21
Environmental Attitudes 22
• **Environmental Close-Up:** Naturalist Philosophers 23
• **Environmental Close-Up:** Environmental Philosophy 24
Societal Environmental Ethics 24
• **Environmental Close-Up:** A Corporate Perspective 25
Corporate Environmental Ethics 26
• **Global Perspective:** Chico Mendes and Extractive Reserves 28
Environmental Justice 28
Individual Environmental Ethics 29
• **Global Perspective:** International Trade in Endangered Species 30
Global Environmental Ethics 30
• **Global Perspective:** Earth Summit on Environment and Development 31
• **Issues & Analysis:** Antarctica—Resource or Refuge? 33
• **Global Perspective:** The Kyoto Protocol on Greenhouse Gases 34

CHAPTER 3
Risk and Cost: Elements of Decision Making 38
Measuring Risk 39
Risk Assessment 39
• **Environmental Close-Up:** What's in a Number? 41
Risk Management 41
True and Perceived Risks 42
Economics and the Environment 43
Economic Concepts 43
Market-Based Instruments 45
• **Global Perspective:** Wombats and the Australian Stock Exchange 47
• **Environmental Close-Up:** Georgia-Pacific Corporation: Recycled Urban Wood—A Case Study in Extended Product Responsibility 48
Extended Product Responsibility 48
Cost-Benefit Analysis 49
Concerns about the Use of Cost-Benefit Analysis 50
Economics and Sustainable Development 50
• **Environmental Close-Up:** "Green" Advertising Claims—Points to Consider 51
External Costs 53
Common Property Resource Problems 53
• **Global Perspective:** Pollution Prevention Pays! 54
Economic Decision Making and the Biophysical World 55
• **Environmental Close-Up:** Placing a Value on Ecosystem Services 56
Economics, Environment, and Developing Nations 56
• **Global Perspective:** Costa Rican Forests Yield Tourists and Medicines 57
The Tragedy of the Commons 58
Lightening the Load 58
• **Issues & Analysis:** Shrimp, Turtles, and Turtle Excluder Devices 59

Part Two
Ecological Principles and Their Application 65

CHAPTER 4
Interrelated Scientific Principles: Matter, Energy, and Environment 66
Scientific Thinking 67
• **Environmental Close-Up:** Typical Household Chemicals 68
Limitations of Science 69
The Structure of Matter 69
Atomic Structure 69
Molecules and Mixtures 70
Acids, Bases, and pH 70
Inorganic and Organic Matter 71
Chemical Reactions 71
Chemical Reactions in Living Things 72
Energy Principles 72
Kinds of Energy 72
States of Matter 73
First and Second Laws of Thermodynamics 73
Environmental Implications of Energy Flow 74
• **Issues & Analysis:** Improvements in Lighting Efficiency 76

CHAPTER 5
Interactions: Environment and Organisms 80
Ecological Concepts 81
Environment 81
Limiting Factors 82
Habitat and Niche 82
The Role of Natural Selection and Evolution 84
Species Definition 84
• **Environmental Close-Up:** Habitat Conservation Plans: Tool or Token? 85
Natural Selection 85
Evolutionary Patterns 86
Kinds of Organism Interactions 87
Predation 87
Competition 88
Symbiotic Relationships 88
Some Relationships Are Difficult to Categorize 90
• **Environmental Close-Up:** Human Interaction–A Different Look 91
Community and Ecosystem Interactions 91
Major Roles of Organisms in Ecosystems 92
Keystone Species 92
Energy Flow Through Ecosystems 93
Food Chains and Food Webs 94
• **Environmental Close-Up:** Contaminants in the Food Chain of Fish from the Great Lakes 96
Nutrient Cycles in Ecosystems 96
Human Impact on Nutrient Cycles 100
• **Issues & Analysis:** Reintroducing Wolves to the Yellowstone Ecosystem 101

CHAPTER 6
Kinds of Ecosystems and Communities 105
Succession 106
Primary Succession 106
Secondary Succession 109
The Changing Nature of the Climax Concept 109
Biomass: Major Types of Terrestrial Climax Communities 110
The Effect of Elevation on Climate and Vegetation 111
Desert 111
Grassland 113
• **Environmental Close-Up:** Grassland Succession 114
Savanna 114
• **Global Perspective:** Tropical Rainforests: A Special Case? 116
Tropical Rainforest 117
Temperate Deciduous Forest 117
• **Environmental Close-Up:** Forest Canopy Studies 119
Taiga, Northern Coniferous Forest, or Boreal Forest 119
Tundra 120
Major Aquatic Ecosystems 120
Marine Ecosystems 121
Freshwater Ecosystems 125
• **Issues & Analysis:** Protecting Old-Growth Temperate Rainforests of the Pacific Northwest 127

CHAPTER 7
Population Principles 130
Population Characteristics 131
Natality and Mortality 131
Sex Ratio and Age Distribution 132
Population Density and Spatial Distribution 133
Summary of Factors That Influence Population Growth Rates 133
A Population Growth Curve 134
Carrying Capacity 135
• **Environmental Close-Up:** Population Growth of Invading Species 137
Reproductive Strategies and Population Fluctuations 138
Human Population Growth 138
• **Global Perspective:** Managing Elephant Populations—Harvest or Birth Control? 139
Social Factors Influence Human Population Growth 141
Ultimate Size Limitation 141
• **Issues & Analysis:** Wolves and Moose on Isle Royale 143

CHAPTER 8
Human Population Issues 146
Human Population Trends and Implications 147

• **Global Perspective:** Thomas Malthus and His Essay on Population 148
Factors That Influence Population Growth 148
Biological Factors 148
Social Factors 149
• **Environmental Close-Up:** Control of Births 150
Political Factors 151
Population Growth and Standard of Living 152
Population and Poverty—A Vicious Cycle? 153
Hunger, Food Production, and Environmental Degradation 154
The Demographic Transition Concept 156
• **Global Perspective:** The Urbanization of the World's Population 157
The U.S. Population Picture 157
Anticipated Changes with Continued Population Growth 158
• **Global Perspective:** North America—Population Comparisons 160
• **Issues & Analysis:** The Impact of AIDS on Populations 161

Part Three
Energy 167

CHAPTER 9
Energy and Civilization: Patterns of Consumption 168
History of Energy Consumption 169
Biological Energy Sources 169
Increased Use of Wood 169
Fossil Fuels and the Industrial Revolution 170
Energy and Economics 171
Economic Growth and Energy Consumption 171
The Role of the Automobile 172
Gasoline Prices and Government Policy 172
• **Global Perspective:** Five Ways to Curb Traffic 173
How Energy is Used 173
Residential and Commercial Energy Use 173
Industrial Energy Use 174
Transportation Energy Use 174
The Variability of Gasoline Prices 175
Electrical Energy 175
• **Environmental Close-Up:** Hybrid Electric Vehicles 176
Energy Consumption Trends 177
• **Global Perspective:** OPEC 178
• **Environmental Close-Up:** Alternative-Fuel Vehicles 181
• **Global Perspective:** Energy Development in China 182
• **Global Perspective:** Potential World Petroleum Resources 183

CHAPTER 10
Energy Sources 187
Energy Sources 188
Resources and Reserves 188
Fossil-Fuel Formation 190
Coal Formation 190
Oil and Natural Gas Formation 191
Issues Related to the Use of Fossil Fuels 192
Coal Use Issues 192
Oil Use Issues 194
Natural Gas Use Issues 196
Renewable Sources of Energy 197
Hydroelectric Power 197
• **Global Perspective:** Hydroelectric Sites 198
Tidal Power 200
• **Global Perspective:** The Three Gorges Dam 201
Geothermal Power 201
Wind Power 202
Solar Energy 203
Biomass Conversion 205
• **Global Perspective:** Electricity from the Ground Up 208
Fuelwood 208
Solid Waste 209
Energy Conservation 210
• **Issues & Analysis:** The Arctic National Wildlife Refuge and Oil 212

CHAPTER 11
Nuclear Energy: Benefits and Risks 215
The Nature of Nuclear Energy 216
The History of Nuclear Energy Development 217
Nuclear Reactors 217
Plans for New Reactors Worldwide 219
Plant Life Extension 220
Breeder Reactors 221
Nuclear Fusion 222
The Nuclear Fuel Cycle 223
Nuclear Material and Weapons Production 223
Nuclear Power Concerns 224
Reactor Safety: The Effects of Three Mile Island and Chernobyl 225
Exposure to Radiation 227
Thermal Pollution 228
Decommissioning Costs 229
Radioactive Waste Disposal 230
• **Global Perspective:** The Nuclear Legacy of the Soviet Union 231

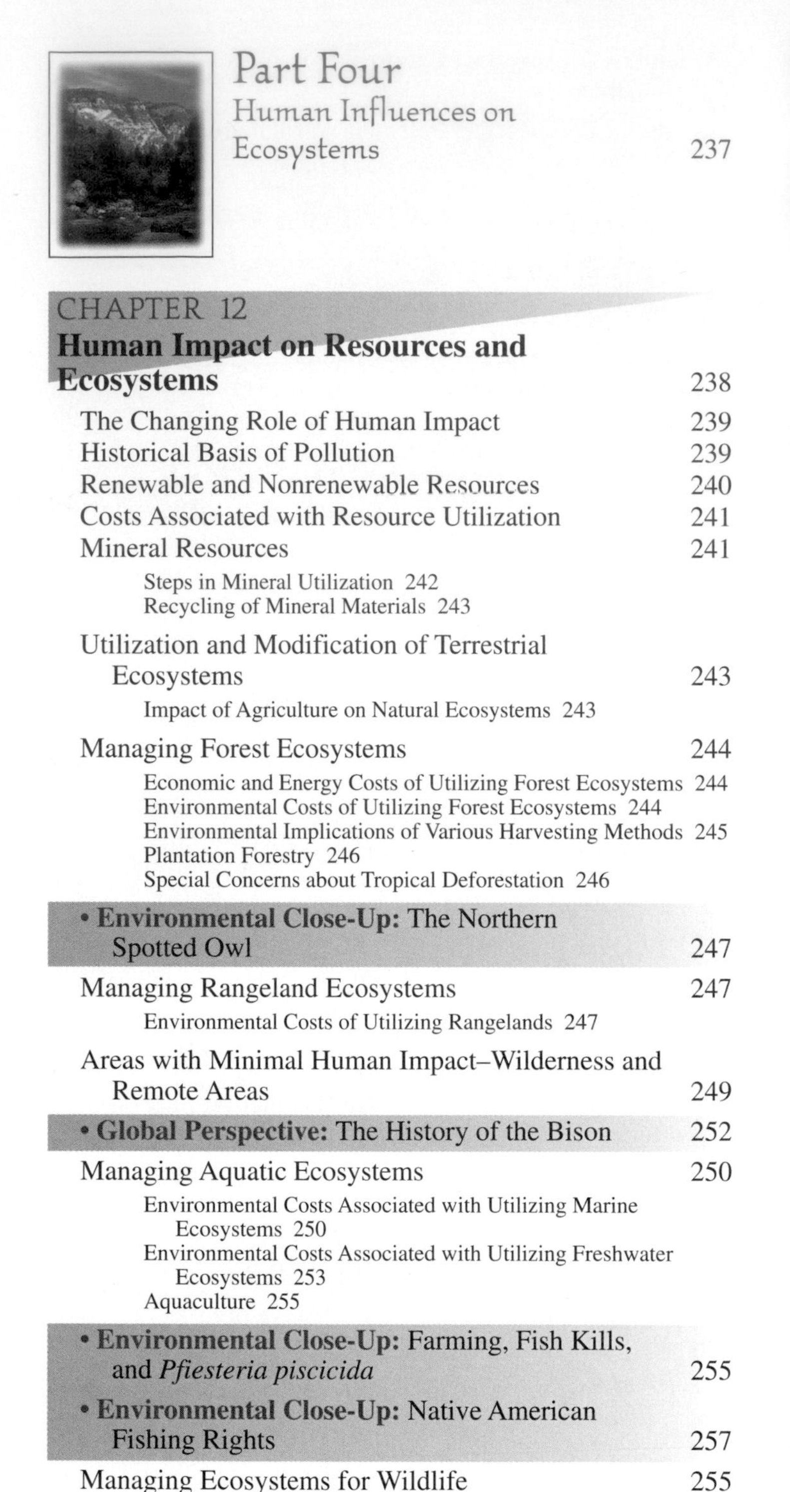

Part Four
Human Influences on Ecosystems 237

CHAPTER 12
Human Impact on Resources and Ecosystems 238

The Changing Role of Human Impact 239
Historical Basis of Pollution 239
Renewable and Nonrenewable Resources 240
Costs Associated with Resource Utilization 241
Mineral Resources 241
Steps in Mineral Utilization 242
Recycling of Mineral Materials 243
Utilization and Modification of Terrestrial Ecosystems 243
Impact of Agriculture on Natural Ecosystems 243
Managing Forest Ecosystems 244
Economic and Energy Costs of Utilizing Forest Ecosystems 244
Environmental Costs of Utilizing Forest Ecosystems 244
Environmental Implications of Various Harvesting Methods 245
Plantation Forestry 246
Special Concerns about Tropical Deforestation 246
• **Environmental Close-Up:** The Northern Spotted Owl 247
Managing Rangeland Ecosystems 247
Environmental Costs of Utilizing Rangelands 247
Areas with Minimal Human Impact–Wilderness and Remote Areas 249
• **Global Perspective:** The History of the Bison 252
Managing Aquatic Ecosystems 250
Environmental Costs Associated with Utilizing Marine Ecosystems 250
Environmental Costs Associated with Utilizing Freshwater Ecosystems 253
Aquaculture 255
• **Environmental Close-Up:** Farming, Fish Kills, and *Pfiesteria piscicida* 255
• **Environmental Close-Up:** Native American Fishing Rights 257
Managing Ecosystems for Wildlife 255
Habitat Analysis and Management 255
Population Assessment and Management 256
Predator and Competitor Control 258
Special Issues with Migratory Waterfowl Management 259
Extinction and Loss of Biodiversity 260
Human-Accelerated Extinction 260
Why Worry about Extinction? 261
What Is Being Done to Prevent Extinction and Protect Biodiversity? 262
• **Environmental Close-Up:** The California Condor 263
• **Issues & Analysis:** Fire As a Forest Management Tool 266

CHAPTER 13
Land-Use Planning 269

The Need for Planning 270
Historical Forces That Shaped Land Use in North America 270
The Importance of Waterways 270
The Rural-to-Urban Shift 270
• **Global Perspective:** Urbanization in the Developing World 272
Migration from the Central City to the Suburbs 272
Factors That Contribute to Sprawl 273
Lifestyle Factors 274
Economic Factors 274
Planning and Policy Factors 274
Problems Associated with Unplanned Urban Growth 275
Transportation Problems 275
Air Pollution 275
Low Energy Efficiency 275
Loss of Sense of Community 275
Death of the Central City 275
Higher Infrastructure Costs 276
Loss of Open Space 276
Loss of Farmland 276
Water Pollution Problems 277
Floodplain Problems 277
Wetlands Misuse 277
• **Environmental Close-Up:** Wetlands Loss in Louisiana 278
Other Land-Use Considerations 279
Land-Use Planning Principles 279
Mechanisms for Implementing Land-Use Plans 280
Establishing State or Regional Planning Agencies 280
Purchasing Land or Use Rights 281
Regulating Use 281
• **Environmental Close-Up:** Land-Use Planning and Aesthetic Pollution 282
Special Urban Planning Issues 282
Urban Transportation Planning 283
Urban Recreational Planning 283
Redevelopment of Inner-City Areas 285
Federal Government Land-Use Issues 285
• **Issues & Analysis:** Decision Making in Land-Use Planning: The Malling of America 287

CHAPTER 14
Soil and Its Uses 291

Geologic Processes 292
Soil and Land 294
Soil Formation 294
Soil Properties 296
Soil Profile 297
Soil Erosion 299
Soil Conservation Practices 301

Contour Farming 303
Strip Farming 304
Terracing 304
Waterways 304
Windbreaks 304

Conventions Versus Conservation Tillage 305
Protecting Soil on Nonfarm Land 307

• **Global Perspective:** Worldwide Soil Degradation 308
• **Environmental Close-Up:** Land Capability Classes 310
• **Issues & Analysis:** Soil Erosion in Virginia 312

CHAPTER 15
Agricultural Methods and Pest Management 315

Different Approaches to Agriculture 316
Fossil Fuel Versus Muscle Power 318
The Impact of Fertilizer 318
Agricultural Chemical Use 319

Insecticides 320

• **Environmental Close-Up:** Regulation of Pesticides 321

Herbicides 322

• **Environmental Close-Up:** A New Generation of Insecticides 323

Fungicides and Rodenticides 323
Other Agricultural Chemicals 324

Problems with Pesticide Use 324

Persistence 324
Bioaccumulation and Biomagnification 324

• **Environmental Close-Up:** Politics and the Control of Ethylene Dibromide (EDB) 325

Pesticide Resistance 325
Effects on Nontarget Organisms 327
Human Health Concerns 327

• **Global Perspective:** China's Ravenous Appetite 328
• **Global Perspective:** Contaminated Soils in the Former Soviet Union 329

Why Are Pesticides So Widely Used? 329
Alternatives to Conventional Agriculture 329

Techniques for Protecting Soil and Water Resources 330

• **Environmental Close-Up:** Food Additives 331

Integrated Pest Management 331

• **Issues & Analysis:** Herring Gulls As Indicators of Contamination in the Great Lakes 335

CHAPTER 16
Water Management 339

The Water Issue 340
The Hydrologic Cycle 341
Human Influences on the Hydrologic Cycle 342
Kinds of Water Use 343

Domestic Use of Water 343
Agricultural Use of Water 344
Industrial Use of Water 346
In-Stream Use of Water 347

• **Environmental Close-Up:** Is It Safe to Drink the Water? 351

Kinds and Sources of Water Pollution 349

Municipal Water Pollution 351
Agricultural Water Pollution 352
Industrial Water Pollution 353
Thermal Pollution 353

• **Global Perspective:** The Cleanup of the Holy Ganges 352
• **Global Perspective:** Comparing Water Use and Pollution in Industrialized and Developing Countries 354

Marine Oil Pollution 355
Groundwater Pollution 355

Water-Use Planning Issues 356

Water Diversion 357
Wastewater Treatment 358

• **Environmental Close-Up:** Restoring the Everglades 360
• **Global Perspective:** Death of a Sea 363

Salinization 361
Groundwater Mining 361
Preserving Scenic Water Areas and Wildlife Habitats 362

• **Global Perspective:** ECOPARQUE 366
• **Issues & Analysis:** The California Water Plan 367

Part Five
Pollution and Policy 371

CHAPTER 17
Air Quality Issues 372

The Atmosphere 373
Primary Air Pollutants 374

Carbon Monoxide (CO) 375
Hydrocarbons (HC) 375

• **Global Perspective:** Air Pollution in Mexico City 376

Particulates 376
Sulfur Dioxide (SO_2) 376
Oxides of Nitrogen (N and NO_2) 377

Photochemical Smog 377
Other Significant Air Pollutants 378
Control of Air Pollution 379

Clean Air Act 380

Acid Deposition 381

• **Environmental Close-Up:** Secondhand Smoke 384

Global Warming and Climate Change 385

Worsening Health Effects 389

Rising Sea Level 390
Disruption of the Water Cycle 390
Changing Forests and Natural Areas 391
Challenges to Agriculture and Food Supply 391

Addressing Climate Change 392
Ozone Depletion 393
• **Environmental Close-Up:** Radon 394
Indoor Air Pollution 396
• **Environmental Close-Up:** Noise Pollution 397
• **Issues & Analysis:** International Air Pollution 398

CHAPTER 18
Solid Waste Management and Disposal 401
Introduction 402
The Disposable Decades 402
The Nature of the Problem 402
Methods of Waste Disposal 404
Landfilling 405
• **Environmental Close-Up:** Resins Used in Consumer Packaging 406
Incineration 407
Composting 408
Source Reduction 409
Recycling 411
• **Environmental Close-Up:** What You Can Do to Reduce Waste and Save Money 413
• **Environmental Close-Up:** Recycling Is Big Business 413
• **Environmental Close-Up:** Recyclables Market Basket 415
• **Issues & Analysis:** Corporate Response to Environmental Concerns 416

CHAPTER 19
Regulating Hazardous Materials 419
Hazardous and Toxic Materials in Our Environment 420
Hazardous and Toxic Substances—Some Definitions 420
Defining Hazardous Waste 421
Issues Involved in Setting Regulations 421
Identification of Hazardous and Toxic Materials 422
• **Environmental Close-Up:** Exposure to Toxins 423
Setting Exposure Limits 423
Acute and Chronic Toxicity 423
Synergism 423
Persistent and Nonpersistent Pollutants 424
Environmental Problems Caused by Hazardous Wastes 424
Health Risks Associated with Hazardous Wastes 425
Hazardous-Waste Dumps—A Legacy of Abuse 426
• **Global Perspective:** Lead and Mercury Poisoning 427
• **Environmental Close-Up:** Computers—A Hazardous Waste 429
Toxic Chemical Release 429
Managing Hazardous Wastes 429
Pollution Prevention 430
Waste Minimization 430
Recycling of Waste 431
Treatment of Waste 431
Land Disposal 431
Hazardous-Waste Management Choices 432
International Trade in Hazardous Wastes 432
• **Global Perspective:** Hazardous Wastes and Toxic Materials in China 433
Hazardous-Waste Program Evaluation 434
• **Issues & Analysis:** Love Canal 435

CHAPTER 20
Environmental Policy and Decision Making 438
New Challenges for a New Century 439
Learning from the Past 440
Thinking about the Future 441
Defining the Future 442
The Development of Environmental Policy in the United States 442
Environmental Backlash—The Wise Use Movement 444
The Changing Nature of Environmental Policy 445
Environmental Policy and Regulation 446
The Greening of Geopolitics 447
• **Environmental Close-Up:** Changing the Nature of Environmental Regulation–The Safe Drinking Water Act 449
• **Global Perspective:** Eco-Terrorism 451
International Environmental Policy 451
• **Global Perspective:** Environmental Policy and the European Union 452
• **Global Perspective:** Overview of an International Organization–The International Whaling Commission (IWC) 453
• **Global Perspective:** Eco-Labels 454
New International Instruments 455
It All Comes Back to You 455

Appendix 1:
Critical Thinking 459
Appendix 2:
Metric Unit Conversion Tables 460
Appendix 3:
The Periodic Table of the Elements 462
Appendix 4:
What *You* Can Do to Make the World a Better Place In Which to Live 463
Appendix 5:
How to Write to Your Elected Officials 464

Glossary 465
Credits 473
Index 476

Preface

Environmental science is an interdisciplinary field. Because environmental disharmonies occur as a result of the interaction between humans and the natural world, we must include both when seeking solutions to environmental problems. It is important to have a historical perspective, appreciate economic and political realities, recognize the role of different social experiences and ethical backgrounds, and integrate these with the science that describes the natural world and how we affect it. *Environmental Science: A Study of Interrelationships* incorporates all of these sources of information when discussing any environmental issue. Furthermore, the authors have endeavored to present a balanced view of issues, diligently avoiding personal biases and fashionable philosophies.

Environmental Science: A Study of Interrelationships is intended as a text for a one-semester, introductory course for students with a wide variety of career goals. They will find it interesting and informative. The central theme is interrelatedness. No text of this nature can cover all issues in depth. What we have done is to identify major issues and give appropriate examples that illustrate the complex interactions that are characteristic of all environmental problems. There are many facts—presented in charts, graphs, and figures—that help to illustrate the scope of environmental issues. However, this is not the core of the text, since the facts will change.

Organization and Content

This book is divided into five parts and twenty chapters. It is organized to provide an even, logical flow of concepts and to provide clear illustrations of the major environmental issues of today.

Part 1 establishes the theme of the book—in chapter 1—by looking at the kinds of environmental issues typical of different regions of North America. In each region, the specific issues selected involve scientific, social, political, and economic components typical of environmental problems. Chapter 2 focuses on the philosophical base needed to examine environmental issues by discussing various ethical and moral stands that shape how people approach environmental issues. Chapter 3 introduces economic issues and the concept of risk analysis. Both of these topics will be brought up at several points later in the text.

Part 2 provides an understanding of the ecological principles that are basic to organism interactions and the flow of matter and energy in ecosystems. The nature of food chains and how they affect the flow of matter and energy are discussed. Other topics included are the efficiency of energy flow through ecosystems, the intricacies of organism-to-organism interaction, and the creative role of natural selection in shaping ecological relationships. Principles of population structure and organization are also developed in this section, with particular attention to the implications of these principles to growth and impact of human populations.

Part 3 focuses on energy. A major emphasis is on the historically important, nonrenewable fossil fuels that have stimulated economic success of the developed economies of the world. Renewable sources of energy are discussed, but with the recognition that they currently are a small part of the world energy picture. Weapons production and nuclear power plants use enormous amounts of energy that can be released from the nucleus of the atom. Both of these uses have caused fear among the public related to the dangers of radiation and the adequacy of waste disposal. These issues are discussed in this section.

Part 4 emphasizes the impact of human activity on natural ecosystems. As human populations grow, and technology changes, the magnitude of human actions becomes more apparent. The natural ecosystems on land and water are modified to meet human needs. The heavy use of pesticides in agriculture is discussed in this section.

Part 5 deals with the major types of pollution. Pollution affects the health and welfare of humans and other organisms. Air pollution, solid waste, and hazardous and toxic substances are discussed in this section. The cost of pollution cannot always be measured in financial terms but may be reflected in the mental and physical health of the populace. Ultimately, governments must address environmental concerns and develop policy to address the concerns. Increasingly, the concerns are international in scope and require negotiations between governments with very different economic conditions and concerns.

New to this Edition

1. The text has been edited throughout and rewritten where needed to include the most recent data and ways of thinking about environmental issues.
2. Many new illustrations were developed and many others were modified to improve their ability to convey information.
3. Several chapters have been substantially revised:
   - Chapter 3, Risk and Cost: Elements of Decision Making, has been moved near the front of the text. Since economics and risk are integral parts of many kinds of environmental discussions, several reviewers have

suggested that this discussion should appear early in the text. This chapter includes expanded coverage of market-based instruments for addressing environmental issues and additional material on the concept of sustainability.

- Chapter 4, Interrelated Scientific Principles, has been substantially rewritten. The section on the scientific method was rewritten, the concepts of entropy and pH were expanded, and a new table was added that describes the various subunits of matter.
- Chapter 5, Interactions: Environment and Organisms, was rewritten to include significantly more material on the concepts of evolution and natural selection and how evolution relates to what is seen in ecosystems. The material on nutrient cycles was extensively rewritten and several boxed readings have been incorporated into the text to provide a better flow of ideas.
- Chapter 12, Human Impact on Resources and Ecosystems, was substantially reorganized with more meaningful headings. The material on minerals has been reduced and the topic of speciation has been moved to chapter 5.
- Chapter 13, Land-Use Planning, was completely rewritten with expanded sections on land-use planning principles, the causes and problems associated with urban sprawl, and redevelopment of inner cities.
- Chapter 17, Air Quality Issues, was substantially rewritten with expanded coverage of climate change and its effects, as well as steps that can be taken to reduce human impact on the global climate.

4. Many new topics or boxed readings have also been added, or have replaced previous readings:
   - Chapter 1 has new material on the harp seal hunting, and a new Environmental Close-Up on forest management.
   - Chapter 2 has expanded coverage of the topic of environmental justice and a new Environmental Close-up on illegal trade in rare species.
   - Chapter 3 has a new Global Perspective—Wombats and the Australian Stock Exchange and much new information on market approaches to managing environmental problems.
   - Chapter 4 has new material on subunits of matter, and an expanded discussion of pH and of latent heat and sensible heat.
   - Chapter 5 has a new section on evolution and natural selection that includes discussion of evolutionary patterns and coevolution, and includes several new examples. A new section on keystone species was added.
   - Chapter 8 has new material on total fertility rate and on the importance of breastfeeding in population control. Additional material was also added on India as a major force in human population growth.
   - Chapter 9 has expanded coverage of OPEC, energy development in China, and a new Environmental Close-Up on hybrid vehicles.
   - Chapter 10 has new material on the forces that cause rising fuel prices, the potential for energy conservation, and expanded wind energy and biomass conversion technologies. The status of the Three Gorges Dam in China has also been updated.
   - Chapter 12 contains a new Issues and Analysis case study dealing with the use of fire as a forest management tool, and a new figure on the impact of technology on natural systems.
   - Chapter 13 has major new sections on land-use planning principles, the cause and consequences of urban sprawl, and a new figure related to flooding.
   - Chapter 15 has added material on the efforts of the World Wildlife Fund to ban DDT use worldwide. The topics of precision agriculture and the controversy surrounding the use of genetically modified crops are also introduced.
   - Chapter 17 has increased coverage of climate change and its impacts, and a new section that deals with strategies for addressing climate change.
   - Chapter 19 has a new Environmental Close-Up on computers a hazardous waste problem. There is also expanded coverage of pollution prevention and international awareness of hazardous wastes as a problem.
   - Chapter 20 has a new section on the cyanide poisoning incident on the Danube River. In addition, there is expanded coverage of current environmental policy and environmental security.

Useful Ancillaries

1. An **Instructor's Manual** accompanies the text and includes chapter outlines, objectives, key terms, a range of test and discussion questions, suggestions for demonstrations, and suggestions for audiovisual materials and other teaching aids.
2. A set of one hundred **transparencies** is also available to users of the text. The transparencies duplicate text figures that clarify essential ecological, political, economic, social, and historical concepts.
3. **Computerized Testing Software** allows for easy test generation using the questions found in the printed test bank.
4. The **Environmental Science Visual Resource Library (VRL)** is a dual platform CD-ROM that allows the user to search with key words or terms and access hundreds of images to illustrate classroom lectures, with just the click

of a mouse. It contains images from four McGraw-Hill textbooks and over 400 additional photographs.

5. Visit our comprehensive **online learning center** at http://www.mhhe.com/environmentalscience/ and discover a variety of valuable resources for both instructor and student. Examples include chapter-by-chapter Internet links (updated regularly) that correspond to each chapter, laboratory exercises, case studies, classroom activities, concept mapping exercises, current global environmental events in the news, practice quizzing, career information, and more.
6. Available on CD-Rom, or accessed via the Online Learning Center, the Environmental Science **Essential Study Partner** is a complete, interactive study tool offering animations and learning activities to help students understand complex environmental science concepts. This valuable resource also includes self-quizzing to help students review each topic and provides hyperlinks to tutorial sections for further review.
7. **BioCourse.com** is an electronic meeting place for students and instructors. Its breadth and depth goes beyond our Online Learning Centers to offer six major areas of up-to-date and relevant information: Faculty Club, Student Center, News Briefing Room, BioLabs, Lifelong Learning Warehouse, and R & D Center.

Related Titles

Field and Laboratory Exercises in Environmental Science

ISBN = 0-07-0290913-7

This lab manual provides hands-on experiences that are relevant, easy to understand, and applicable to students' lives. The experiments are designed to be concise, unique, inexpensive, and easily tailored to any course.

Online *Taking Sides: Clashing Views on Controversial Environmental Issues*

ISBN = 0-07-243097-4 • www.dushkin.com/online

This debate-style reader is designed to introduce one to controversies in environmental policy and science, and reflects a variety of viewpoints staged as "pro" and "con" debates. Issues are organized around four core areas: general philosophical and political issues, the environment and technology, disposing of wastes, and the environment and the future.

Annual Editions: Environment 01/02

ISBN = 0-07-243359-0 • www.dushkin.com/online

A compilation of current articles from such sources as *World Watch, Audubon, The Atlantic Monthly*, and *Scientific American,* this text explores the global environment, the world's population, energy, the biosphere, natural resources, and pollution.

Sources: Notable Selections in Environmental Studies

ISBN = 0-07-303186-0 • www.dushkin.com/online

This volume brings together primary source selections of enduring intellectual value—classic articles, book excerpts, and research studies that have shaped environmental studies and our contemporary understanding of it.

You Can Make a Difference: Be Environmentally Responsible

ISBN = 0-07-292416-0

This book is organized around the three parts of the biosphere: land, water, and air.

Acknowledgements

The production of a textbook requires a dedicated team of professionals who provide guidance, criticism, and encouragement. It is also important to have open communication and dialog in order to deal with the many issues that arise during the development and production of a text. Therefore we would like to thank Marge Kemp, Kathy Loewenberg, Peggy Selle, Jane Stembridge, Jamie O'Neal, and LouAnn Wilson for their critiques and kindnesses. We would also like to acknowledge our many colleagues who have reviewed all or part of *Environmental Science: A Study of Interrelationships.* Their valuable input has contributed significantly to the quality of this textbook.

Ghulam Aasef
Kaskaskia College

Thomas J. Algeo
University of Cincinnati

John Vincent Aliff
Georgia Perimeter College

Julius Alker
Central Florida Community College

Margaret M. Avard
Southeastern Oklahoma State University

Robert M. Barry
Palm Beach Community College

R. P. Benard
American International College

William B. N. Berry
University of California, Berkeley

Patricia J. Beyer
Bloomsburg University

Deborah Bird
Pima Community College

Andrew David Bixler
Chaffey College

Del Blackburn
Clark College

Dorothy Boorse
Gordon College

Richard A. Boutwell
Missouri Western State College

Steven G. Brumbaugh
Tacoma Community College

David Byres
Florida Community College, Jacksonville

Catherine W. Carter
Georgia Perimeter College

David A. Charlet
Community College Southern Nevada

Terence H. Cooper
University of Minnesota

William C. Culver
Saint Petersburg Junior College

William T. Davin
Berry College

James N. DeVries
Lancaster Bible College

Darren Divine
University of Nevada, Las Vegas

Patricia Dooris
Saint Leo College

Vernon P. Dorweiler
Michigan Technological University

Terese Dudek
Kishwaukee College

Tom Dudley
Angelina College

Andrew Evans, Jr.
James Madison University

Sharon Flanagan
Nunez Community College

Stephen Fleckenstein
Sullivan County Community College

Patrick B. Fulks
Bakersfield College

Lesley Garner
University of West Alabama

John Gault
Missouri Valley College

J. Phil Gibson
Agnes Scott College

Nancy L. Goodyear
Bainbridge College

Andrew P. Goyke
Northland College

Thomas F. Grittinger
University of Wisconsin, Sheboygan

Carl W. Grobe
Westfield State College

Elizabeth A. Guthrie
University of Tennessee, Chattanooga

Dallas Hanks
Utah Valley State College

Steve Hardin
Ozarks Technical Community College

Keith R. Hench
Kirkwood Community College

Stephen R. Herr
Oral Roberts University

Graham C. Hickman
Texas A & M University, Corpus Christi

Sue Holt
Cabrillo College

Alan R. Holyoak
Manchester College

Huey-Min Hwang
Jackson State University

Al Iglar
East Tennessee State University

Joseph Jay Jacquot
Grand Valley State University

Wendel J. Johnson
University of Wisconsin, Marinette

Gina Johnston
California State University, Chico

Eric N. Jones
Mary Baldwin College

Martin G. Kelly
State University of New York, Buffalo

Suzanne Kempke
Armstrong Atlantic State University

Barry Kilch
Maine Maritime Academy

Richard J. Kirker
Dutchess Community College

Peter A. Kish
Southwest Oklahoma State University

Ned J. Knight
Linfield College

Bob Koningsor, Jr.
Grossmont College

Erica Kosal
North Carolina Wesleyan College

Edward S. Kubersky
Felician College

John Lange
Gannon University

Ronald L. Laughlin
Elizabethtown College

John F. Logue
University of South Carolina, Sumter

David A. Lovejoy
Westfield State College

Robert Lovely
Marycrest International University

Paul E. Lutz
Lenoir-Rhyne College

Heidi Marcum
Baylor University

David C. Martin
Centralia College

Michael McCarthy
Eastern Arizona College

Mark McConnaughhay
Dutchess Community College

G. Wayne McGraw
Louisiana College

Katheryne McKenzie
Oregon Coast Community College

Jean B. McManus
North Greenville College

Michael T. Mengak
Ferrum College

Mark Mitch
New England College

David A. Munn
The Ohio State University, Agricultural Technical Institute

Thomas E. Murray
Elizabethtown College

Herman S. Muskatt
Utica College of Syracuse University

Muthena Naseri
Moorpark College

William A. Niering
Connecticut College

Chuks Ogbonnaya
Mountain Empire College

Niamh O'Leary
Wells College

Roger G. Olson
Saint Joseph's College

Joyce Ownbey
Sacramento City College

Carl S. Pavetto
Wesley College

Charles R. Peebles
Michigan State University

Chris E. Petersen
College of DuPage

Ervand M. Peterson
Sonoma State University

Jon K. Piper
Bethel College

Rosann Poltrone
Arapahoe Community College

William G. Raschi
Bucknell University

Joyce O. Rasdall
Western Kentucky University

C. Lee Rockett
Bowling Green State University

Steven J. Ropski
Gannon University

Eleanor Saboski
University of New England

Robert Sanford
University of Southern Maine

Bradley A. Sarchet
Colby—Sawyer College

Christine L. Schadler
University of New Hampshire

Fred Schindler
Indian Hills Community College

Bruce A. Schulte
Providence College

Alan M. Schwartz
Saint Lawrence University

Harold F. Sears
University of South Carolina, Union

Janet Anne Sherman
Pennsylvania College of Technology

Leslie A. Sherman
Providence College

Gary Silverman
Bowling Green State University

Stephanie Smith
Northeast Mississippi Community College

Rolf Sohn
Brevard Community College

John R. Spear
Colorado School of Mines

Douglas J. Spieles
Southwest State University

Joseph D. Stogner
Ferrum College

Max R. Terman
Tabor College

Jamey Thompson
Maysville Community College

Darwin R. Thorpe
Compton Community College

Olli H. Tuovinen
Ohio State University

Daniel A. Underwood
Peninsula College

Dale A. Utt, Jr.
Oklahoma Baptist University

W. M. von Zharen
Texas A & M University, Galveston

Marjorie Welch
Southwestern Community College

Arlene A. Westhoven
Ferris State University

Thomas M. Wolf
Washburn University

Richard J. Wright
Valencia Community College

Len Yannielli
Naugatuck Valley Community Technical College

Metin Yersel
Lyndon State College

John M. Zamora
Middle Tennessee State University

Astatkie Zikarge
Texas Southern University

GUIDED TOUR

The features of this book are Unique

The organization and principle features of this book were planned with the students' wholistic learning and comprehension in mind:

WORLD MAP

A **world map** with political boundaries can be found immediately following the preface. This will help the reader to more fully understand and appreciate global environmental issues.

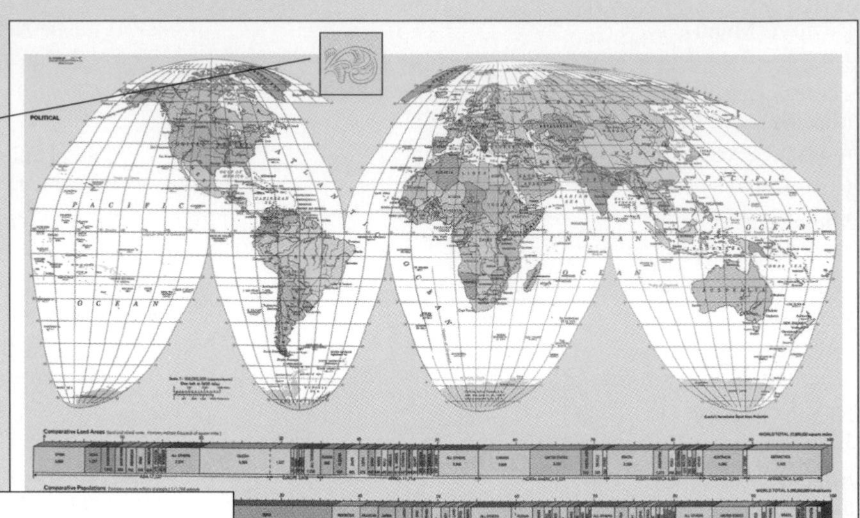

1 CHAPTER

Environmental Interrelationships

Objectives

After reading this chapter, you should be able to:

- Understand why environmental problems are complex and interrelated.
- Realize that environmental problems involve social, ethical, political, and economic issues, not just scientific issues.
- Understand that acceptable solutions to environmental problems are not often easy to achieve.
- Understand that all organisms have an impact on their surroundings.
- Understand what is meant by an ecosystem approach to environmental problem solving.
- Recognize that different geographic regions have somewhat different environmental problems, but the process for resolving them is the same and involves compromise.

Chapter Outline

The Field of Environmental Science
The Interrelated Nature of Environmental Problems
Environmental Close-Up: *Science Versus Policy*
Global Perspective: *Fish, Seals, and Jobs*
An Ecosystem Approach
Regional Environmental Concerns
Environmental Close-Up: *The Greater Yellowstone Ecosystem*
Environmental Close-Up: *Headwaters Forest*
The Wilderness North
The Agricultural Middle
The Dry West
The Forested West
The Great Lakes and Industrial Northeast
The Diverse South

INTRODUCTION

The five parts of the text each present an **introduction** that places the upcoming chapters in context for the reader by recalling previously discussed material and by describing the organization of the chapters to come.

LEARNING OBJECTIVES, OUTLINE, CONCEPTUAL DIAGRAM

Each chapter begins with a set of **learning objectives,** an **outline,** and a **conceptual diagram,** all of which give the student a broad overview of the interrelated forces that are involved in the material to be discussed. Students are encouraged to refer to these resources while reading and reviewing the chapter.

TABLES, CHARTS, GRAPHS, MAPS, DRAWINGS, OR PHOTOGRAPHS

To dramatize and clarify text material, each chapter includes a number of **tables, charts, graphs, maps, drawings,** or **photographs.** Every illustration has been carefully chosen to provide a pictorial image or an organized format for showing detailed information, which helps the reader comprehend the chapter material.

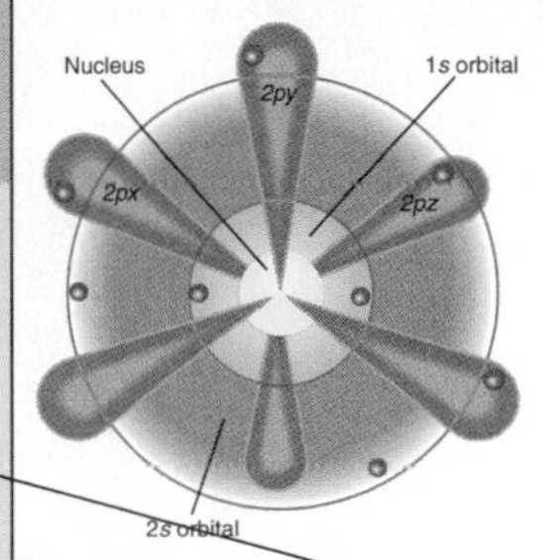

figure 4.2 **Diagrammatic Oxygen Atom** Most oxygen atoms are composed of a nucleus containing eight positively charged protons and eight neutrons without charges. Eight negatively charged electrons spin around the nucleus.

Table 4.1 Relationships Between the Kinds of Subunits Found in Matter

Category of Matter	Subunits	Characteristics
Subatomic Particles	protons	Positively charged
		Located in nucleus of the atom
	neutrons	Have no charge
		Located in nucleus of the atom
	electrons	Negatively charged
		Located outside the nucleus of the atom
Elements	atoms	Atoms of an element are composed of specific arrangements of protons, neutrons, and electrons.
		Atoms of different elements differ in the number of protons, neutrons, and electrons present
Compounds	molecules	Molecules of compounds are composed of two or more atoms chemically bonded together
		Molecules of different compounds contain different atoms or different proportions of atoms.
Mixtures	molecules	Molecules of mixtures are not chemically bonded to each other.
		The number of each kind of molecule present is variable.

many plants such as tobacco, poison ivy, and rhubarb leaves naturally contain toxic materials, while the use of chemical fertilizers has contributed to the health of major portions of the world since their use accounts for about one-third of the food grown in the world. However, it is appropriate to question if the use of agricultural chemicals is always necessary or if trace amounts of specific agricultural chemicals in food are dangerous. It is often easy to jump to conclusions or confuse fact with hypothesis, particularly when we generalize.

The Structure of Matter

stant motion. Although different kinds of matter have different properties, they are similar in one fundamental way. They are all made up of one or more kinds of smaller subunits called atoms.

Atomic Structure

Atoms are the fundamental subunits of matter. They in turn are made up of protons, neutrons, and electrons. There are 92 kinds of atoms found in nature. Each kind forms a specific type of matter known as an **element.** Gold (Au), oxygen (O), and mercury (Hg) are examples of elements. All atoms are composed of a central region known as a **nucleus,** which is composed of two kinds of relatively heavy particles: positively charged particles called **protons** and uncharged particles called **neutrons.** Surrounding the nucleus are clouds of relatively lightweight, fast-moving, negatively charged particles called **electrons.** As mentioned earlier, each kind of element is composed of a specific kind of atom. The atoms of different kinds of elements differ from one another in the number of protons, neutrons, and electrons present. For example, a typical atom of mercury

figure 4.2.) (Appendix 3 contains a periodic table of the elements.) All atoms of an element always have the same number of protons and electrons, but the number of neutrons may vary from one atom to the next. Atoms of the same element that differ from one another in the number of neutrons they contain are called **isotopes.**

Molecules and Mixtures

Atoms can be attached to one another into stable units called **molecules.** When two or more different kinds of atoms are attached to one another, the kind of matter formed is called a **compound.** While only 92 kinds of atoms are commonly found, there are millions of ways atoms can be combined to form compounds. Water (H_2O), sugar ($C_6H_{12}O_6$), salt (NaCl), and methane gas (CH_4) are examples of compounds.

Many other kinds of matter are **mixtures,** variable combinations of atoms or molecules. Honey is a mixture of several sugars and water; concrete is a mixture of cement, sand, gravel, and reinforcing rods; and air is a mixture of several gases of which the most common are nitrogen

ENVIRONMENTAL CLOSE-UP

Exposure to Toxins

We are all exposed to materials that are potentially harmful. The question is, at what levels is such exposure harmful or toxic? One measure of toxicity is **LD_{50},** the dosage of a substance that will kill (lethal dose) 50 percent of a test population. Toxicity is measured in units of poisonous substance per kilogram of body weight. For example, the deadly chemical that causes botulism, a form of food poisoning, has an LD_{50} in adult human males of 0.0014 milligrams per kilogram. This means that if each of 100 human adult males weighing 100 kilograms consumed a dose of only 0.14 milligrams—about the equivalent of a few grains of table salt—approximately 50 of them will die.

Lethal doses are not the only danger from toxic substances. During the past decade, concern has been growing over minimum harmful dosages, or threshold dosages, of poisons, as well as their sublethal effects.

The length of exposure further complicates the determination of toxicity values. Acute exposure refers to a single exposure lasting from a few seconds to a few days. Chronic exposure refers to continuous or repeated exposure for several days, months, or even years. Acute exposure usually is the result of a sudden accident, such as the tragedy at Bhopal, India, mentioned at the beginning of

the chapter. Acute exposures often make disaster headlines in the press, but chronic exposure to sublethal quantities of toxic materials presents a much greater hazard to public health. For example, millions of urban residents are continually exposed to low levels of a wide variety of pollutants. Many deaths attributed to heart failure or such diseases as emphysema may actually be brought on by a lifetime of exposure to sublethal amounts of pollutants in the air.

other fish-eating carnivores failed to reproduce. Once these substances are identified as toxic, their use is regulated. Countries contemplating regulation of hazardous and toxic materials and wastes must consider not only how toxic each one is but also how flammable, corrosive, and explosive it is, and whether it will produce mutations or cause cancer.

Setting Exposure Limits

Even after a material is identified as hazardous or toxic, there are problems in determining appropriate exposure limits. Nearly all substances are toxic in sufficiently high doses. The question is, When does a chemical cross over from safe to toxic? There is no easy way to establish acceptable levels. For any new compounds that are to be brought on the market, extensive toxicology studies must be done to establish their ability to do harm. Usually these are tests on animals. (See Environmental Close-Up:

of exposure at which none of the test animals is affected (**threshold level**) and then set the human exposure level lower to allow for a safety margin. This safety margin is important because it is known that threshold levels vary significantly among species, as well as among members of the same species. Even when concentrations are set, they may vary considerably from country to country. For example, in the Netherlands, 50 milligrams of cyanide per kilogram of waste is considered hazardous; in neighboring Belgium, the toxicity standard is fixed at 250 milligrams per kilogram.

Acute and Chronic Toxicity

Regulatory agencies must look at both the effects of one massive dose of a substance (**acute toxicity**) and the effects of exposure to small doses over long periods (**chronic toxicity**). Acute toxicity is readily apparent because organisms respond to the toxin shortly after being exposed. Chronic toxicity is much more

an acute exposure may make an organism ill but not kill it, while chronic exposure to a toxic material may cause death. A good example of this effect is alcohol toxicity. Consuming extremely high amounts of alcohol can result in death (acute toxicity and death). Consuming moderate amounts may result in illness (acute toxicity and full recovery). Consuming moderate amounts over a number of years may result in liver damage and death (chronic toxicity and death).

Another example of chronic toxicity involves lead. Lead has been used in paints, in gasoline, and in pottery glazes for many years, but researchers discovered that it has harmful effects. The chronic effects on the nervous system are most noticeable in children, particularly when children eat paint chips.

Synergism

Another problem in regulating hazardous materials is assessing the effects of mixtures of chemicals. Most toxico-

BOX READINGS

Each chapter also includes **boxed readings.** These provide an in-depth consideration of a specific situation that is relevant to the content, an alternative viewpoint, or a wider worldview of the issues discussed in the chapter.

WEB INTEGRATION

Chapters conclude with an **"Issues and Analysis"** case study, a **summary,** a list of **key terms, review questions,** and **critical thinking questions,** and a list of **articles** and **animations** available on the accompanying website. The case studies, articles, and animations have been specifically selected to allow the reader to apply the chapter concepts to actual situations. All material in light blue reflects a direct connection to much more information available via the **online learning center.**

Interactive Exploration

Check out the website at
http://www.mhhe.com/environmentalscience
and click on the cover of this textbook for interactive versions of the following:

KNOW THE BASICS

activated-sludge sewage treatment *359*
aquiclude *341*
aquifer *341*
aquitard *341*
artesian well *342*
biochemical oxygen demand (BOD) *349*
confined aquifer *341*
domestic water *344*
eutrophication *350*
evapotranspiration *341*
fecal coliform bacteria *351*
groundwater *341*
groundwater mining *361*
hydrologic cycle *341*
industrial water use *346*
in-stream water use *347*
irrigation *345*
limiting factor *350*
nonpoint source *350*
point source *350*
porosity *342*
potable waters *340*
primary sewage treatment *358*
runoff *341*
salinization *361*
secondary sewage treatment *358*
sewage sludge *359*
storm-water runoff *356*
tertiary sewage treatment *360*
thermal pollution *353*
trickling filter system *358*
unconfined aquifer *341*
vadose zone *341*
water diversion *357*
water table *341*

On-line Flashcards
Electronic Glossary

IN THE REAL WORLD

Supposedly, the importance of wetlands is understood, but the ... ion continue. Check out the history ... ds protection. Stronger Wetland ... d and The Prairie Wetlands of ...

... drinking water comes from? Is your ... ce water? A detailed explanation of ... essee's water supply is provided in ... lls, Not from the River. ... cting groundwater from contamina- ... rotecting Groundwater Resources ...

... ning of groundwater, and politically ... es when water is a scarce resource. ... reement on Water Allotments in ... Drought Focuses on Long-standing Water Disputes in the Middle East to learn more about conflicts that result from the lack of water. Although water is not scarce in the Great Lakes region, see why Low Water Levels in the Great Lakes are causing concerns in this area.

Restoration projects, protection, and funding for cleanup activities are just a few examples. Check out Everglades Restoration: Greatest Restoration Yet, or Just More of the Same? for an example of a restoration project.

A variety of ways to clean up and improve water quality can be found in the following stories: International Accord to Clean Up the Rhine River, and Food Web Control of Primary Production in Lakes.

Another way to improve water quality is to support local conservation measures. Look over American Heritage River System Created for details on how this program aims to help coordinate efforts to improve water quality.

PUT IT IN MOTION

What is the link between fertilizer runoff and the green soupy lake in an adjacent area? Now is your chance to study it in more detail. Check out the Deoxygenation of Lakes animation for a better understanding of the process.

Normal rainfall is neutral. . . right? What causes acid rain? When is the effect of acid rain most apparent? Study the Acid Rain animation for answers to these questions and to see how acid rain affects aquatic ecosystems such as lakes.

TEST PREPARATION

Review Questions

1. Describe the hydrologic cycle.
2. Distinguish between withdrawal and consumption of water.
3. What are the similarities between domestic and industrial water use? How are they different from in-stream use?
4. How is land use related to water quality and quantity? Can you provide local examples?
5. What is biochemical oxygen demand? How is it related to water quality?
6. How can the addition of nutrients such as nitrates and phosphates result in a reduction of the amount of dissolved oxygen in the water?
7. Differentiate between point and nonpoint sources of water pollution.
8. How are most industrial wastes disposed of? How has this changed over the past 25 years?
9. What is thermal pollution? How can it be controlled?
10. Describe primary, secondary, and tertiary sewage treatment.
11. What are the types of wastes associated with agriculture?
12. Why is storm-water management more of a problem in an urban area than in a rural area?
13. Define groundwater mining.
14. How does irrigation increase salinity?

Critical Thinking Questions

1. Leakage from freshwater distribution systems accounts for significant losses. Is water so valuable that governments should require systems that minimize leakage in order to preserve the resource? Under what conditions would you change your evaluation?
2. Do non-farmers have an interest in how water is used for irrigation? Under what conditions should the general public be involved in making these decisions along with the farmers who are directly involved?
3. Should the United States allow Mexico to have water from the Rio Grande and the Colorado Rivers, both of which originate in the United States and flow to Mexico?
4. Do you believe that large scale hydroelectric power plants should be promoted as a renewable alternative to power plants that burn fossil fuels? What criteria do you use for this decision?
5. After reading the Issues and Analysis concerning the California Water Plan, do you believe water should be diverted from northern California to southern California?

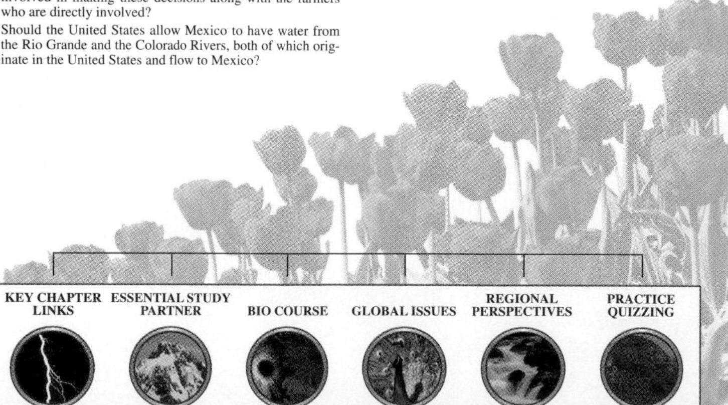

APPENDICES

The text concludes with several **appendices** that deal with critical thinking. the metric system, the periodic table of the elements, some thoughts on what you can do to make the world a better place in which to live, and how to write to public officials. In addition, there is a complete **glossary** and **index.**

Appendix 1

CRITICAL THINKING

We live in an age of information. Computers, e-mail, the Internet, CD-ROMs, instant news, and fax machines bring us information more quickly than ever before. A simple search of the Internet will provide huge amounts of information. Some of the information has been subjected to scrutiny and is quite valid, some is well-informed opinion, some is naive misinformation, and some is even designed to mislead. How do we critically evaluate the information we get?

Critical thinking involves a set of skills that help us to evaluate information, arguments, and opinions in a systematic and thoughtful way. Critical thinking also can help us better understand our own opinions as well as the points of view of others. It can help us evaluate the quality of evidence, recognize bias, characterize the assumptions behind arguments, identify the implications of decisions, and avoid jumping to conclusions.

Characteristics of Critical Thinking

Critical thinking involves skills that al- … a meaning- … or useless … izing that … ey compo- …

… *context.* All … ain assump- … tions. It is important to recognize what those assumptions are. Critical thinking involves looking closely at an argument or opinion by identifying the historical, social, political, economic, and scientific context in which the argument is being made. It is also important to understand the kinds of bias contained in the argument and the level of knowledge the presenter has.

Consider alternative views. A critical thinker must be able to understand and evaluate different points of view. Often these points of view may be quite varied. It is important to keep an open mind and to look at all the information objectively and try to see the value in alternative points of view. Often people miss obvious solutions to problems because they focus on a certain avenue of thinking and unconsciously dismiss valid alternative solutions.

Expect and accept mistakes. Good critical thinking is exploratory and speculative, tempered by honesty and a recognition that we may be wrong. It takes courage to develop an argument, engage in debate with others, and admit that your thinking contains errors or illogical components. By the same token, be willing to point out what you perceive to be shortcomings in the arguments of others. It is always best to do this with good grace and good humor.

Have clear goals. When analyzing an argument or information, keep your goals clearly in mind. It is often easy to get sidetracked. A clear goal will allow you to quickly sort information into that which is pertinent and that which may be interesting but not germane to the particular issue you are exploring.

Evaluate the validity of evidence. Information comes in many forms and has differing degrees of validity. When evaluating information, it is important to understand that not all the information from a source may be of equal quality. Often content about a topic is a mix of solid information interspersed with less certain speculations or assumptions. Apply a strong critical attitude to each separate piece of information. Often what appears to be a minor, insignificant error or misunderstanding can cause an entire argument to unravel.

Critical thinking requires practice. As with most skills, you become better if you practice. At the end of each chapter in the text, there are a series of questions that allow you to practice critical thinking skills. Some of these questions are straightforward and simply ask you to recall information from the chapter. Others ask you to apply the information from the chapter to other similar contexts. Still others ask you to develop arguments that require you to superimpose the knowledge you have gained from the chapter on quite different social, economic, or political contexts from your own.

Practice, Practice, Practice.

Glossary

A

abiotic factors Nonliving factors that influence the life and activities of an organism.

abyssal ecosystem The collection of organisms and the conditions that exist in the deep portions of the ocean.

acid Any substance that, when dissolved in water, releases hydrogen ions.

acid deposition The accumulation of potential acid-forming particles on a surface.

acid mine drainage A kind of pollution, associated with coal mines, in which bacteria convert the sulfur in coal into compounds that form sulfuric acid.

acid rain (acid precipitation) The deposition of wet acidic solutions or dry acidic particles from air.

activated sludge sewage treatment Method of treating sewage in which some of the sludge is returned to aeration tanks, where it is mixed with incoming wastewater to encourage degradation of the wastes in the sewage.

activation energy The initial energy input required to start a reaction.

active solar system A system that traps sunlight energy as heat energy and uses mechanical means to move it to another location.

acute toxicity A serious effect, such as a burn, illness, or death, that occurs shortly after exposure to a hazardous substance.

age distribution The comparative percentages of different age groups within a population.

agricultural products Any output from farming: milk, grain, meat, etc.

agricultural runoff Surface water that carries soil particles, nutrients, such as phosphate, nitrates, and other agricultural chemicals, as it runs off agricultural land to lakes and streams.

air stripping The process of pumping air through water to remove volatile materials dissolved in the water.

alpha radiation A type of radiation consisting of a particle with two neutrons and two protons.

aquiclude An impervious confining layer of an aquifer.

aquifer A porous layer of earth material that becomes saturated with water.

aquitard A partially permeable layer in an aquifer.

artesian well The result of a pressurized aquifer being penetrated by a pipe or conduit, within which water rises without being pumped.

atom The basic subunit of elements, composed of protons, neutrons, and electrons.

auxin A plant hormone that stimulates growth.

B

base Any substance that, when dissolved in water, removes hydrogen ions from solution; forms a salt when combined with an acid.

benthic Describes organisms that live on the bottom of marine and freshwater ecosystems.

benthic ecosystems A type of marine or freshwater ecosystem consisting of organisms that live on the bottom.

beta radiation A type of radiation consisting of electrons released from the nuclei of many fissionable atoms.

bioaccumulation The buildup of a material in the body of an organism.

biocentric Life-centered, a theory of moral responsibility that states that all forms of life have an inherent right to exist.

biochemical oxygen demand (BOD) The amount of oxygen required by microbes to degrade organic molecules in aquatic ecosystems.

biocide A kind of chemical that kills many different types of living things.

biodegradable Able to be broken down by natural biological processes.

biodiversity A measure of the variety of kinds of organisms present in an ecosystem.

biomagnification The increases in the amount of a material in the bodies of organisms at successively higher trophic levels.

biotic factors Living portions of the environment.

biotic potential The inherent reproductive capacity.

birthrate The number of individuals born per thousand individuals in the population per year.

black lung disease A respiratory condition resulting from the accumulation of large amounts of fine coal dust particles in miners' lungs.

boiling-water reactor (BWR) A type of light water reactor in which steam is formed directly in the reactor, which is used to generate electricity.

boreal forest A broad band of mixed coniferous and deciduous trees that stretches across northern North America (and also Europe and Asia); its northernmost edge is integrated with the arctic tundra.

brownfields Buildings and land that have been abandoned because they are contaminated and the cost of cleaning up the site is high.

brownfields development The concept that abandoned contaminated sites can be cleaned up sufficiently to allow some specified uses without totally removing all of the contaminants

C

carbamate A class of soft pesticides that work by interfering with normal nerve impulses.

carbon absorption The use of carbon particles to treat chemicals by having the chemicals attach to the carbon particles.

carbon cycle The cyclic flow of carbon from the atmosphere to living organisms and back to the atmospheric reservoir.

carbon dioxide (CO_2) A normal component of the Earth's atmosphere that in elevated concentrations may interfere with the Earth's heat budget.

carbon monoxide (CO) A primary air pollutant produced when organic materials, such as gasoline, coal, wood, and trash, are incompletely burned.

carcinogen A substance that causes cancer.

Visual Resource Library CD-ROM

A VRL is an electronic library of educational presentation resources that instructors can use to enhance their lectures. View, sort, search, and print catalog images, play chapter-specific slideshows using PowerPoint, or create customized presentations when you:

- Find and sort thumbnail image records by name, type, location, and user-defined keywords
- Search using keywords or terms
- View images at the same time with the Small Gallery View
- Select and view images at full size
- Display all the important file information for easy file identification
- Drag and place or copy and paste into virtually any graphics, desktop publishing, presentation, or multimedia application

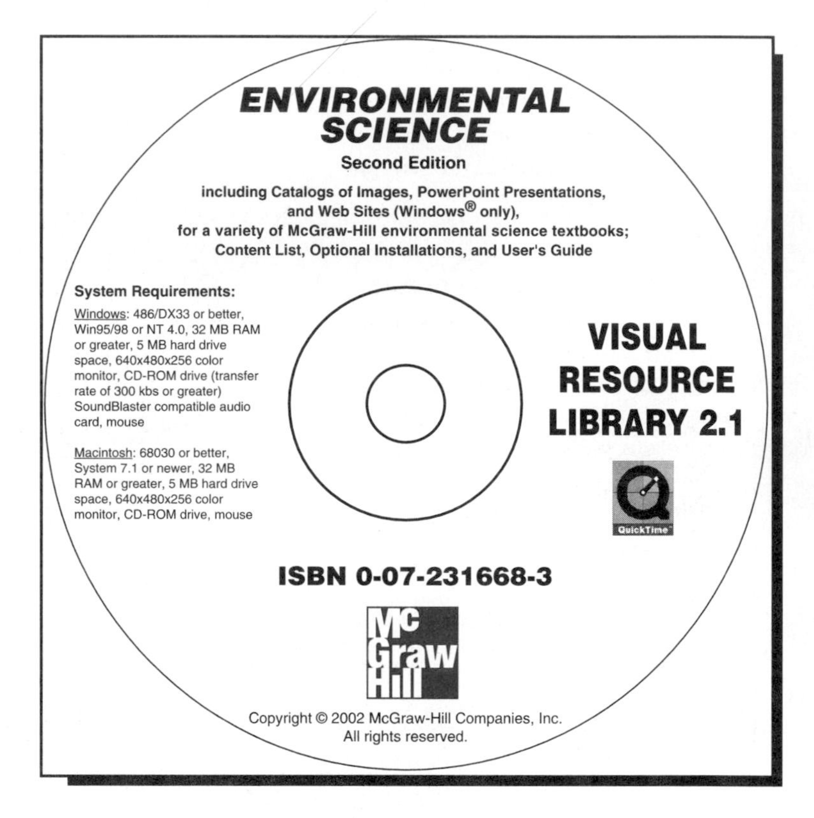

The Environmental Science Visual Resource Library CD-ROM

This helpful CD-ROM contains illustrations from this book, plus <u>four other texts</u> in environmental science and ecology. Hundreds of exclusive photographs are also included. You'll be able to create interesting multimedia presentations with the use of these images, and students will have the ability to easily access the same images in their texts to later review the content covered in class

Contact your McGraw-Hill sales representative for more information or visit *www.mhhe.com*.

The Online Learning Center with Essential Study Partner 2.0

www.mhhe.com/environmentalscience

Imagine the advantages of having so many learning and teaching tools all in one place—all at your fingertips.

Students, you'll appreciate extensive self-quizzing opportunities; interactive activities; case studies; and related web links in addition to the new Essential Study Partner 2.0—a web-based student tutorial of major environmental science topics—hosted on this site.

Instructors, you'll want to take advantage of our electronic illustrations from the text; classroom activities; corporate annual environmental science reports, Global Environmental Issues, Case Studies and access to the PageOut: Course Website Development Center—All available anytime you want them.

Online Learning Center

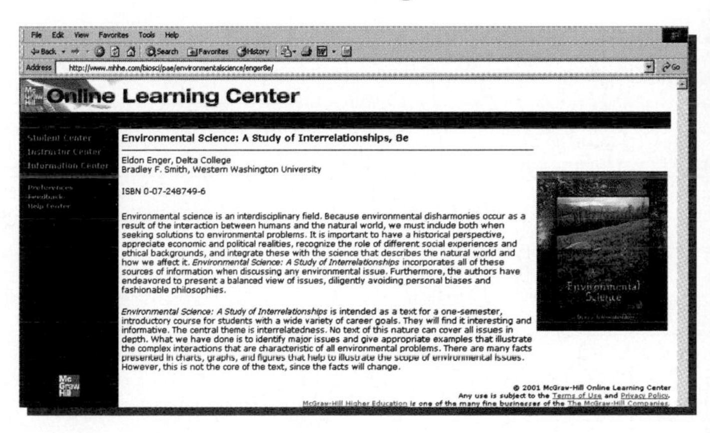

Environmental Science
Essential Study Partner 2.0

Contact your McGraw-Hill sales representative for more information or visit *www.mhhe.com*

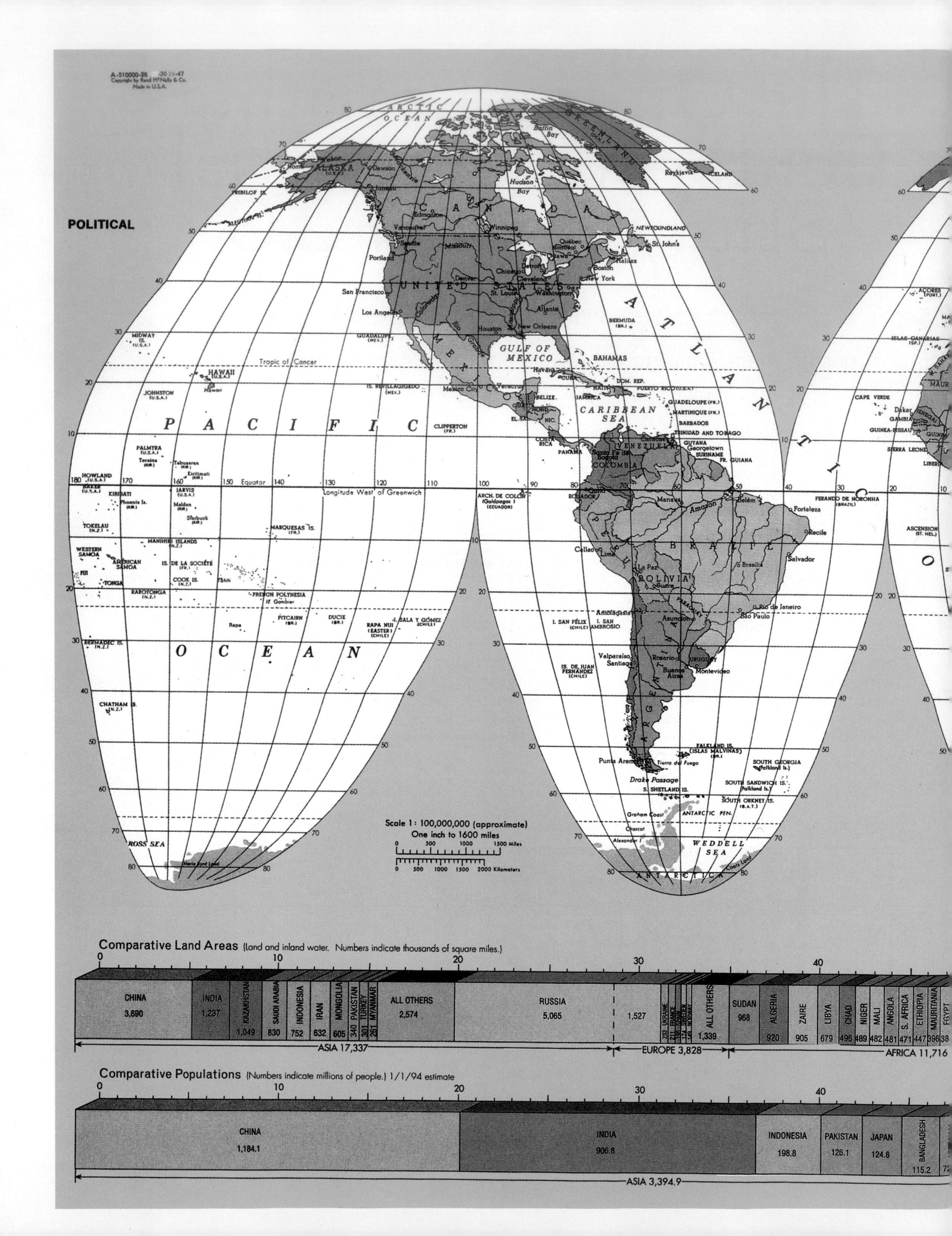

A-510000-26 -30 29-47
Copyright by Rand McNally & Co.
Made in U.S.A.
POLITICAL
ARCTIC OCEAN
ALASKA (U.S.A.)
CANADA
UNITED STATES
GREENLAND
Baffin Bay
Hudson Bay
ICELAND
Reykjavik
NEWFOUNDLAND
St. John's
Halifax
Boston
New York
Washington
Atlanta
New Orleans
Houston
Chicago
Cleveland
Detroit
St. Louis
Denver
Winnipeg
Edmonton
Vancouver
Seattle
Portland
San Francisco
Los Angeles
Québec
Montréal
Ottawa
Juneau
Dawson
Nome
PRIBILOF IS.
ALEUTIAN IS.
GULF OF MEXICO
MEXICO
Mexico City
Veracruz
Havana
CUBA
BAHAMAS
BERMUDA (BR.)
DOM. REP.
PUERTO RICO (U.S.A.)
HAITI
JAMAICA
BELIZE
HOND.
NIC.
EL SAL.
COSTA RICA
PANAMA
GUADELOUPE (FR.)
MARTINIQUE (FR.)
BARBADOS
TRINIDAD AND TOBAGO
CARIBBEAN SEA
ATLANTIC OCEAN
PACIFIC OCEAN
Tropic of Cancer
Equator
Longitude West of Greenwich
MIDWAY IS. (U.S.A.)
HAWAII (U.S.A.)
Hawaii
JOHNSTON (U.S.A.)
PALMYRA (U.S.A.)
Teraina
Tabuaeran
Kiritimati
HOWLAND (U.S.A.)
BAKER (U.S.A.)
KIRIBATI
Phoenix Is.
JARVIS (U.S.A.)
Malden
Starbuck
TOKELAU (N.Z.)
MANIHIKI ISLANDS
WESTERN SAMOA
AMERICAN SAMOA
FIJI
TONGA
IS. DE LA SOCIÉTÉ (FR.)
COOK IS. (N.Z.)
Tahiti
RAROTONGA (N.Z.)
MARQUESAS IS. (FR.)
FRENCH POLYNESIA
Is. Gambier
Rapa
PITCAIRN (BR.)
DUCIE (BR.)
RAPA NUI (EASTER) (CHILE)
I. SALA Y GÓMEZ (CHILE)
KERMADEC IS. (N.Z.)
CHATHAM IS. (N.Z.)
CLIPPERTON (FR.)
ARCH. DE COLÓN (Galápagos I.) (ECUADOR)
ROSS SEA
Marie Byrd Land
VENEZUELA
Caracas
GUYANA
Georgetown
SURINAME
FR. GUIANA
COLOMBIA
Santa Fe de Bogotá
ECUADOR
Quito
PERU
Lima
Callao
BRAZIL
Manaus
Amazon
Belém
Fortaleza
Recife
Salvador
Brasília
Rio de Janeiro
São Paulo
BOLIVIA
La Paz
Sucre
PARAGUAY
Asunción
Antofagasta
CHILE
ARGENTINA
URUGUAY
Montevideo
Rosario
Buenos Aires
Valparaíso
Santiago
I. SAN FÉLIX (CHILE)
I. SAN AMBROSIO
IS. DE JUAN FERNÁNDEZ (CHILE)
FALKLAND IS. (ISLAS MALVINAS) (BR.)
Punta Arenas
Tierra del Fuego
Drake Passage
S. SHETLAND IS.
SOUTH GEORGIA (Falkland Is.)
SOUTH SANDWICH IS. (Falkland Is.)
SOUTH ORKNEY IS. (B.A.T.)
ANTARCTIC PEN.
Graham Coast
Charcot
Alexander I
WEDDELL SEA
Coats Land
ANTARCTICA
FERNANDO DE NORONHA (BRAZIL)
ASCENSION (ST. HEL.)
AÇORES (PORT.)
ISLAS CANARIAS (SP.)
CAPE VERDE
Dakar
SENEGAL
GAMBIA
GUINEA-BISSAU
SIERRA LEONE
Scale 1: 100,000,000 (approximate)
One inch to 1600 miles
0 500 1000 1500 Miles
0 500 1000 1500 2000 Kilometers
Comparative Land Areas (Land and inland water. Numbers indicate thousands of square miles.)
0 10 20 30 40
CHINA 3,690
INDIA 1,237
KAZAKHSTAN 1,049
SAUDI ARABIA 830
INDONESIA 752
IRAN 632
MONGOLIA 605
PAKISTAN 340
TURKEY 301
MYANMAR 261
ALL OTHERS 2,574
RUSSIA 5,065
1,527
233 UKRAINE
211 FRANCE
ALL OTHERS 1,339
SUDAN 968
ALGERIA 920
ZAIRE 905
LIBYA 679
CHAD 496
NIGER 489
MALI 482
ANGOLA 481
S. AFRICA 471
ETHIOPIA 447
MAURITANIA 396
EGYPT
ASIA 17,337
EUROPE 3,828
AFRICA 11,716
Comparative Populations (Numbers indicate millions of people.) 1/1/94 estimate
0 10 20 30 40
CHINA 1,184.1
INDIA 906.8
INDONESIA 198.8
PAKISTAN 126.1
JAPAN 124.8
BANGLADESH 115.2
ASIA 3,394.9

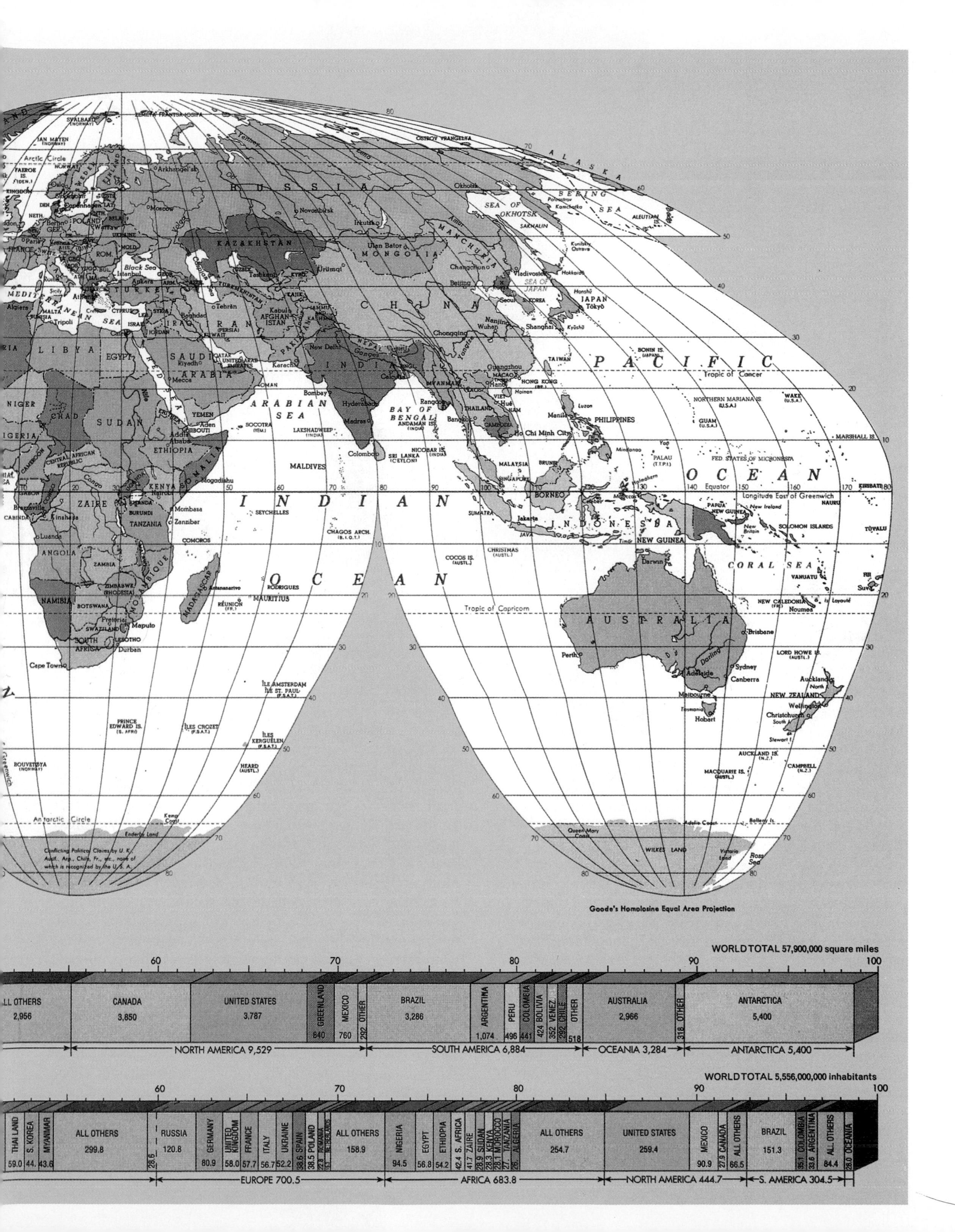

RUSSIA
KAZAKHSTAN
MONGOLIA
CHINA
JAPAN
INDIA
IRAN
SAUDI ARABIA
LIBYA
EGYPT
CHAD
SUDAN
ETHIOPIA
KENYA
ZAIRE
TANZANIA
ANGOLA
NAMIBIA
SOUTH AFRICA
MADAGASCAR
TURKEY
PHILIPPINES
BORNEO
INDONESIA
NEW GUINEA
AUSTRALIA
NEW ZEALAND
PACIFIC OCEAN
INDIAN OCEAN
ARABIAN SEA
BAY OF BENGAL
SEA OF OKHOTSK
BERING SEA
CORAL SEA
ALASKA
Tropic of Cancer
Tropic of Capricorn
Equator
Arctic Circle
Antarctic Circle
Longitude East of Greenwich
WILKES LAND
Conflicting Political Claims by U. K., Austl., Arg., Chile, Fr., etc., none of which is recognized by the U. S. A.
Goode's Homolosine Equal Area Projection
WORLD TOTAL 57,900,000 square miles
ALL OTHERS 2,956
CANADA 3,850
UNITED STATES 3,787
GREENLAND 840
MEXICO 760
OTHER 292
BRAZIL 3,286
ARGENTINA 1,074
PERU 496
COLOMBIA 441
BOLIVIA 424
VENEZ. 352
CHILE 292
OTHER 518
AUSTRALIA 2,966
OTHER 318
ANTARCTICA 5,400
NORTH AMERICA 9,529
SOUTH AMERICA 6,884
OCEANIA 3,284
ANTARCTICA 5,400
WORLD TOTAL 5,556,000,000 inhabitants
THAILAND 59.0
S. KOREA 44.
MYANMAR 43.6
ALL OTHERS 299.8
28.6
RUSSIA 120.8
GERMANY 80.9
UNITED KINGDOM 58.0
FRANCE 57.7
ITALY 56.7
UKRAINE 52.2
SPAIN 38.6
POLAND 38.5
ALL OTHERS 158.9
NIGERIA 94.5
EGYPT 56.8
ETHIOPIA 54.2
S. AFRICA 42.4
ZAIRE 41.7
SUDAN 28.9
KENYA 28.3
MOROCCO 28.1
TANZANIA 27
ALGERIA 26.
ALL OTHERS 254.7
UNITED STATES 259.4
MEXICO 90.9
CANADA 27.9
ALL OTHERS 66.5
BRAZIL 151.3
COLOMBIA 35.1
ARGENTINA 33.6
ALL OTHERS 84.4
OCEANIA 28.0
EUROPE 700.5
AFRICA 683.8
NORTH AMERICA 444.7
S. AMERICA 304.5

About the Authors

Eldon D. Enger

Eldon D. Enger is a professor emeritus of biology at Delta College, a community college near Saginaw, Michigan. He received his B.A. and M.S. degrees from the University of Michigan. Professor Enger has over 30 years of teaching experience, during which he has taught biology, zoology, environmental science, and several other courses. He has been very active in curriculum and course development. Recent activities include the development of a learning community course in stream ecology, which involves students in two weekend activities including canoeing and camping, and a plant identification course that incorporates weekend field activities with backpacking and camping. In addition, he was involved in the development of an environmental regulations course and an environmental technician curriculum.

Professor Enger is an advocate for variety in teaching methodology. He feels that if students are provided with varied experiences, they are more likely to learn. In addition to the standard textbook assignments, lectures, and laboratory activities, his classes are likely to include writing assignments, student presentation of lecture material, debates by students on controversial issues, field experiences, individual student projects, and discussions of local examples and relevant current events. Textbooks are very valuable for presenting content, especially if they contain accurate, informative drawings and visual examples. Lectures are best used to help students see themes and make connections, and laboratory activities provide important hands-on activities.

Professor Enger has been a Fulbright Exchange Teacher to Australia and Scotland, received the Bergstein Award for Teaching Excellence and the Scholarly Achievement Award from Delta College, and participated as a volunteer in an Earthwatch Research Program in Costa Rica, the Virgin Islands, and Western Australia. He has also visited New Zealand, New Guinea, Fiji, Puerto Rico, Mexico, Canada, Morocco, many areas in Europe, and much of the United States. During these travels he has spent considerable time visiting coral reefs, ocean coasts, mangrove swamps, alpine tundra, prairies, tropical rainforests, cloud forests, deserts, temperate rainforests, coniferous forests, deciduous forests, and many other special ecosystems. This extensive experience provides the background to look at environmental issues from a broad perspective.

Professor Enger is married, has two college-aged sons, and enjoys a variety of outdoor pursuits such as cross-country skiing, hiking, hunting, fishing, camping and gardening. Other interests include reading a wide variety of periodicals, beekeeping, singing in a church choir, and preserving garden produce.

Bradley F. Smith

Bradley F. Smith is the dean of Huxley College of Environmental Studies at Western Washington University in Bellingham, Washington. Prior to assuming the position as dean in 1994, he served from 1991 to 1994 as the first director of the Office of Environmental Education for the U.S. Environmental Protection Agency in Washington, D.C. Dean Smith also served as the acting president of the National Environmental Education and Training Foundation in Washington, D.C. and as a special assistant to the EPA administrator.

Before moving to Washington, D.C., Dean Smith was a professor of political science and environmental studies for fifteen years, and the executive director of an environmental education center and nature refuge for five years.

Dean Smith has considerable international experience. He was a Fulbright exchange teacher to England and worked as a research associate for Environment Canada in New Brunswick, Canada. He is a frequent speaker on environmental issues worldwide and serves on the International Scholars Program for the U.S. Information Agency. He also served as a U.S. representative on the Tri-Lateral Commission on Environmental Education with Canada and Mexico. In 1995, he was awarded a NATO fellowship to study the environmental problems associated with the closure of former Soviet military bases in Eastern Europe. Dean Smith is an adjunct professor at Far Eastern State University in Vladivostok, Russia, and is a member of the Russian Academy of Transport. He also serves as a commissioner for the International Union for the Conservation of Nature (IUCN). He is a frequent speaker at universities in China.

Nationally, Dean Smith serves as a member/advisor for many environmental organizations' board of directors, advisory councils, and executive committees, including the President's Council for Sustainable Development (Education Task Force), and the Science Advisory Boards for MOTE Marine Laboratory in Sarasota, Florida, and for the Center for Sustainable Futures in Vermont.

Dean Smith holds B.A. and M.A. degrees in political science and public administration and a Ph.D. from the School of Natural Resources and Environment at the University of Michigan.

Dean Smith lives with his wife Daria, daughter Morgan, son Ian, and English setter Skye, along Puget Sound south of Bellingham. He is an avid outdoor enthusiast.

Environmental Science

PART ONE

Interrelatedness

Environmental science is an interdisciplinary study that describes environmental contributions to quality of human life and the problems caused by human use of the natural world. It also seeks remedies for these problems. To learn about this complex field of study, it helps to understand three things: First, it is important to understand the natural processes (both physical and biological) that operate in the world. Second, it is important to appreciate the role that technology plays in our society and its capacity to alter natural processes as well as solve problems caused by human impact. Third, it helps to understand the complex social processes that characterize human populations. When we integrate that understanding with a knowledge of technology and natural processes, we can fully appreciate our role in the natural world.

Chapter 1 introduces the central theme of interrelatedness by analyzing some environmental issues in North America region by region. **Chapter 2** discusses the differences that can exist among individuals in a society and the different behaviors people exhibit, depending on whether they are acting as individuals, as part of a corporation, or as part of government.

Chapter 3 introduces the interrelationship between economics and the environment. The theme of the chapter is that it is not a question of the economy OR the environment; rather, it is a question of both. You need a healthy environment in order to have a healthy economy and, in turn, you need a healthy economy in order to have a healthy environment.

CHAPTER

Environmental Interrelationships

Objectives

After reading this chapter, you should be able to:

- Understand why environmental problems are complex and interrelated.
- Realize that environmental problems involve social, ethical, political, and economic issues, not just scientific issues.
- Understand that acceptable solutions to environmental problems are not often easy to achieve.
- Understand that all organisms have an impact on their surroundings.
- Understand what is meant by an ecosystem approach to environmental problem solving.
- Recognize that different geographic regions have somewhat different environmental problems, but the process for resolving them is the same and involves compromise.

Chapter Outline

The Field of Environmental Science

The Interrelated Nature of Environmental Problems

Environmental Close-Up: *Science Versus Policy*

An Ecosystem Approach

Regional Environmental Concerns

Global Perspective: *Fish, Seals, and Jobs*

The Wilderness North

Environmental Close-Up: *The Greater Yellowstone Ecosystem*

Environmental Close-Up: *Headwaters Forest*

The Agricultural Middle
The Dry West
The Forested West
The Great Lakes and Industrial Northeast
The Diverse South

The Field of Environmental Science

Environmental science is an interdisciplinary area of study that includes both applied and theoretical aspects of human impact on the world. Since humans are generally organized into groups, environmental science must deal with politics, social organization, economics, ethics, and philosophy. Thus, environmental science is a mixture of traditional science, individual and societal values, and political awareness. (See figure 1.1.)

Although environmental science as a field of study is evolving, it is rooted in the early history of civilization. Many ancient cultures expressed a reverence for the plants, animals, and geographic features that provided them with food, water, and transportation. These features are still appreciated by many modern people. Although the following quote from Henry David Thoreau (1817–1862) is over a century old it is consistent with current environmental philosophy:

> I wish to speak a word for Nature, for absolute freedom and wildness, as contrasted with a freedom and culture merely civil . . . to regard man as an inhabitant, or a part and parcel of Nature, rather than a member of society.

The current interest in the state of the environment began with philosophers like Thoreau and received emphasis from the organization of the first Earth Day on April 22, 1970. Subsequent Earth Days reaffirmed this commitment. As a result of this continuing interest in the state of the world and how people both affect it and are affected by it, environmental science is now a standard course or program at many colleges. It is also included in the curriculum of high schools. Most of the concepts covered by environmental science courses had previously been taught in ecology, conservation, or geography courses. Environmental science incorporates the scientific aspects of these courses with input from the social sciences, such as economics, sociology, and political science, creating a new interdisciplinary field.

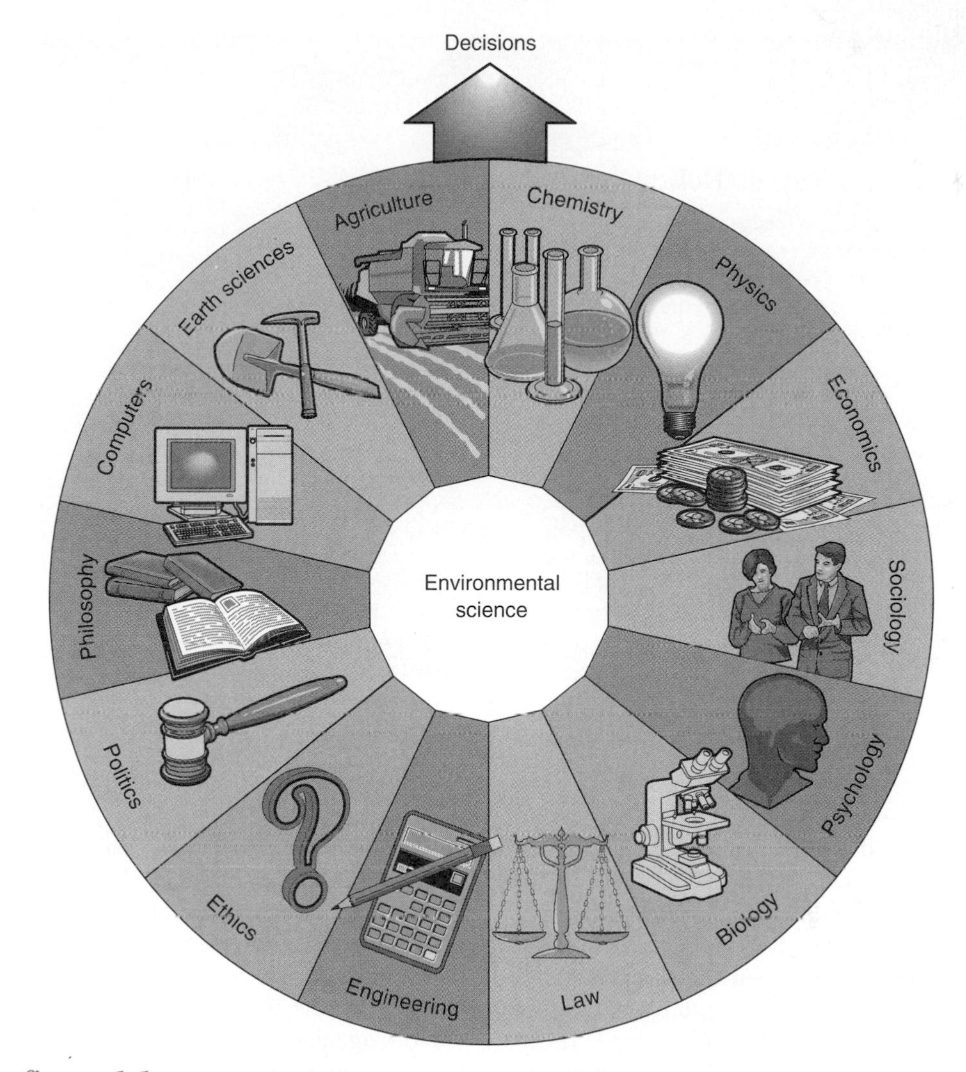

figure 1.1 **Environmental Science** The field of environmental science involves an understanding of scientific principles, economic influences, and political action. Environmental decisions often involve compromise. A decision that may be supportable from a scientific or economic point of view may not be supportable from a political point of view without modification. Often political decisions relating to the environment may not be supported by economic analysis.

The Interrelated Nature of Environmental Problems

Environmental science is by nature an interdisciplinary field. The word *environmental* is usually understood to mean the surrounding conditions that affect people and other organisms. In a broader definition, **environment** is everything that affects an organism during its lifetime. In turn, all organisms including people affect many components in their environment. (See figure 1.1.) From a human perspective, environmental issues involve concerns about science, nature, health, employment, profits, politics, ethics, and economics.

Most social and political decisions are made with respect to political jurisdictions, but environmental problems do not necessarily coincide with these artificial political boundaries. For example, air pollution may involve several local units of government, several states or

ENVIRONMENTAL CLOSE-UP

Science Versus Policy

Scientific knowledge and government policy do not always agree. The scientific community can advise governments but cannot insist that certain policies be adopted. Governments may halt some scientific research because they control funding sources, or they may introduce regulations that make continuing the research difficult. For example, much federal money was spent on alternative energy research during the Carter presidency, but many of these projects were not in favor during the Reagan and Bush presidencies. Funding was reduced and much of the research into alternative fuels stopped. Conversely, the passage of the 1990 Clean Air Act during the Bush presidency mandated that alternative-fuel automobiles be used in some cities with severe air pollution problems.

Government policy may be contrary to prevailing scientific opinion for economic or political reasons. For many years during the Reagan administration, most scientists in the United States and Canada agreed that the burning of high-sulfur coal and other acid-producing fuels was responsible for acid rain, which was leading to the deaths of lakes in parts of Canada and the northeastern United States. The administration continued to insist that the information was not conclusive and that the problem should be studied in greater detail.

The Clinton administration faced a similar dilemma regarding the debate over global warming. While the scientific community strongly supported the position that global warming is for real and that it is due in large part to human causes, Clinton was cautious in moving too fast to reduce emissions in the United States. This is in part due to the potential economic consequences involved. It is difficult to separate scientific knowledge, governmental policy, and economic policy. Can you identify similar examples of this debate in your community? Can you explain how industry, government, consumers, and environmental issues are interrelated?

provinces, and even different nations. The forest fires that raged in Mexico in 1998 had a severe impact on air quality in Texas. On a more local level, the air pollution problems in Juarez, Mexico, are also problems in El Paso, Texas. But the issue is more than air quality and human health. Lower wage rates and less strict environmental laws have influenced some U.S. industries to move to Mexico for economic advantages. Mexico and many other developing nations are struggling to improve their environmental image and need the money generated by foreign investment to improve the conditions and the environment in which their people live.

Air pollutants produced in the major industrial regions of the United States drift across the border into Canada, where acid rain damages lakes and forests. A long-standing dispute exists between the United States and Canada over this issue. Canada claims that the United States should be doing more to reduce emissions that cause acid rain, and the United States claims it is doing as much as it can. In another example, farmers who use water from the Colorado River for irrigation reduce the quality and quantity of water entering Mexico. This causes political friction between Mexico and the United States.

The issue of declining salmon stocks in the Pacific Northwest of the United States and British Columbia, Canada, is another example of political friction over a shared natural resource. It has been calculated that on the U.S. side of the salmon issue alone, there are five federal Cabinet level departments, two federal agencies, five federal laws in question as well as numerous Tribal treaties, commissions, and court decisions. All of this is in addition to many state-level departments, commissions, and rulings. If all of this were not sufficient, international bodies such as the United Nations and international treaties impact the fate of the salmon. Considering all this complexity, it is not surprising that the plight of the salmon is in such a dangerous status. (See figure 1.2.)

Because of all these political, economic, ethical, and scientific links, solving environmental problems is complicated. Environmental problems seldom have simple solutions. However, international organizations, such as the International Joint Commission, have had major bearing on the quality of the environment over broad regions of the world.

The International Joint Commission was established in 1909 when the Boundary Waters Treaty was signed between the United States and Canada. The treaty was established in part to provide that the "boundary waters and waters flowing across the boundary shall not be polluted on either side to the injury of health or property of the other." The commission has been instrumental in identifying areas of concern and encouraging the cleanup of polluted sites that affect the quality of the Great Lakes and other boundary waters. In general, the two governments have listened to the commission's advice and have responded by initiating cleanup activities.

The first worldwide meeting of heads of state directed to concern for the environment took place at the Earth Summit, formally known as the United Nations Conference on Environment and Development (UNCED) in Rio de Janeiro in 1992. Most countries have also signed agreements on **sustainable development** and biodiversity. In 1997, representatives from 125 nations met in Kyoto, Japan, for the Third Conference of the United Nations Framework Convention on Climate Change. This conference, commonly referred to as the Kyoto Conference on Climate Change, resulted in commitments from the participating nations to reduce their overall emissions of six greenhouse gases (linked to global warming) by at least 5 percent below 1990 levels and to do so

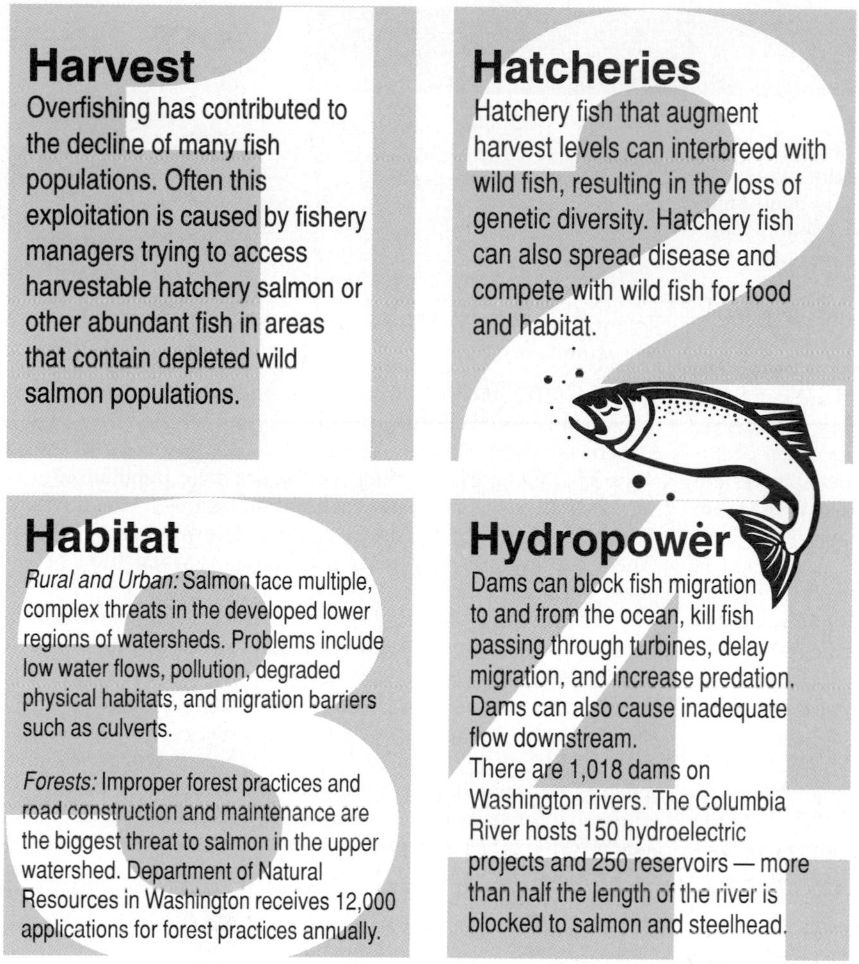

Harvest

Overfishing has contributed to the decline of many fish populations. Often this exploitation is caused by fishery managers trying to access harvestable hatchery salmon or other abundant fish in areas that contain depleted wild salmon populations.

Hatcheries

Hatchery fish that augment harvest levels can interbreed with wild fish, resulting in the loss of genetic diversity. Hatchery fish can also spread disease and compete with wild fish for food and habitat.

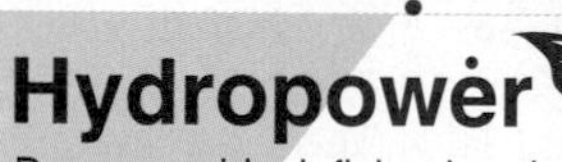

Habitat

Rural and Urban: Salmon face multiple, complex threats in the developed lower regions of watersheds. Problems include low water flows, pollution, degraded physical habitats, and migration barriers such as culverts.

Forests: Improper forest practices and road construction and maintenance are the biggest threat to salmon in the upper watershed. Department of Natural Resources in Washington receives 12,000 applications for forest practices annually.

Hydropower

Dams can block fish migration to and from the ocean, kill fish passing through turbines, delay migration, and increase predation. Dams can also cause inadequate flow downstream.

There are 1,018 dams on Washington rivers. The Columbia River hosts 150 hydroelectric projects and 250 reservoirs — more than half the length of the river is blocked to salmon and steelhead.

figure 1.2 **The Four H's: Human Activities that Affect Wild Salmon Survival** The interrelated nature of environmental problems is evident in the diminishing numbers of wild salmon in the Pacific Northwest of the United States and in British Columbia, Canada. This diagram portrays the plight of wild salmon in the State of Washington but the issue is region wide.

Source: Washington Department of Natural Resources Newsletter, Winter 1998, Department of Natural Resources, Olympia, Washington.

between the years 2008 and 2012. (See figure 1.3.) The Kyoto Protocol, as the agreement was called, was viewed by many as one of the most important steps to date in environmental protection and international diplomacy. It may be years before we will know if all countries that signed these agreements will meet their commitments to environmental improvement, but they have at least stated their intention to do so.

The United Nations, through the United Nations Educational, Scientific, and Cultural Organization (UNESCO) and the United Nations Environment Programme (UNEP), has supported many environmental programs. A recent undertaking is the International Environmental Education Programme (IEEP). This program recognizes the need for both formal environmental education in schools and the informal education that occurs through the media and groups of interested citizens. Conferences on environmental education were first held during the 1970s and continue to the present.

An Ecosystem Approach

The natural world is organized into interrelated units called ecosystems. An **ecosystem** is a region in which the organisms and the physical environment form an interacting unit. Weather affects plants, plants use minerals in the soil and affect animals, animals spread plant seeds, plants secure the soil, and plants evaporate water, which affects weather.

Ecosystems sometimes have fairly discrete boundaries, as is the case with a lake, island, or biosphere. Sometimes the boundaries are indistinct, as in the transition from grassland to desert. Grassland gradually becomes desert, depending on the historical pattern of rainfall in an area.

An ecosystem approach requires a look at the way the natural world is organized. Where do the rivers flow? What are the prevailing wind patterns? What are the typical plants and animals in the area? How does human activity affect nature? The task of an environmental scientist is to recognize and understand the natural interactions that take place and to integrate these with the uses humans must make of the natural world.

To illustrate the interrelated nature of environmental issues, we will look at several regions of North America and highlight some of the key features and issues of each.

Regional Environmental Concerns

No region is free of environmental concerns. Most regions tend to focus on specific, local environmental issues that apply directly to them. For example, protecting endangered species is a concern in many parts of the world. In the Pacific Northwest, for example, an endangered species known as the northern spotted owl depends on undisturbed mature forests for its survival. Development and logging may conflict with the survival of the owl. In most metropolitan areas the problem of endangered species is purely historical, since the construction of cities has destroyed the previously existing ecosystem. Here we present a number of regional vignettes

GLOBAL PERSPECTIVE

Fish, Seals, and Jobs

In 1995, the Canadian government announced a moratorium on cod fishing along the east coast of Canada. The cod industry contributes $700 million a year and 31,000 jobs to the Canadian economy, primarily in Newfoundland. At the same time, the government announced that it would begin a program to encourage the harvesting of harp seals by helping develop markets for seal products. Environmental groups opposed the harvesting of the seals.

How do all these pieces fit together? It is thought that the low numbers of cod in the North Atlantic are partially the result of an increasing population of harp seals. While overfishing, larger nets, and other factors have also contributed to the decline of the cod, it is true that harp seals feed on fish that could have been harvested. The current harp seal population along the Atlantic Coast has more than doubled since the 1970s to 4.8 million in 1997 and is projected to reach 6 million by 2000 if there are no hunts. The increase in harp seals is at least partly the result of actions during the 1970s by environmental groups that sought to stop the killing of seals because they considered the harvesting method inhumane. The traditional method involves clubbing the young seals to death. In 1996, the Canadian government increased the seal hunt quota from 186,000 to 275,000.

There is also a growing concern about the health of the harp seal population. The Northwest Atlantic harp seal population migrates annually between Greenland and Canada. It is hunted during the summer months in Greenland and, in the spring, along Canada's east coast. In setting its Total Allowable Catch or TAC (275,000 in 1997, 1998, 1999, and 2000), it has been argued that the Canadian government did not completely account for the number of animals harvested in the increasing and largely unregulated Greenland summer hunt which now takes up to 80,000 harp seals per year.

The following two web sites can provide additional information to this complicated and interrelated issue.

Fisheries and Oceans, Canada—
www.ncr.dfo.ca/home_e.htm

International Marine Mammal Association—
www.imma.org

In 1997, scientists from seven countries met in St. John's, Newfoundland, to consider the interactions between harp seals and fisheries in the Northwest Atlantic. The conclusions of the meeting supported the earlier findings of the dramatic population growth amongst the seals and also noted that the animals were now growing more slowly and the pregnancy rate is lower than in the 1980s. These are the effects you would likely see when food becomes more difficult to find.

The St. John's meeting reinforced the fact that the diet of harp seals in the nearshore waters of the Labrador-Newfoundland shelf is dominated by Arctic cod. It was not, however, determined whether or not harp seals were affecting commercial fishing stocks—and Atlantic cod, in particular—on the Labrador-Newfoundland shelf. This was because there is a need for an estimate of the amount of juvenile cod in both inshore and offshore areas, and for an assessment of the amounts of cod which are being taken by the other important predators such as Greenland halibut, whales, and seabirds.

It appears that what might be seen initially as a number of isolated and unrelated factors are really issues interrelated in a way that affects the economy of an entire region.

to illustrate the complexity and interrelatedness of environmental issues. (See figure 1.4.)

The Wilderness North

Much of Alaska and Northern Canada can be characterized as **wilderness**—areas with minimal human influence. Much of this land is owned by governments, not by individuals, so government policies have a large effect on what happens in these regions. These areas have important economic values in their trees, animals, scenery, and other natural resources. Exploitation of the region's natural resources involves significant trade-offs. Usually, a portion of the natural world is altered permanently, but the area altered is so small that many people consider it insignificant. Because of the severe climate, northern wilderness areas

figure 1.3 The Kyoto Protocol to the United Nations Framework Convention on Climate Change was the first time in history that such a protocol contained legally binding reduction targets for all major greenhouse gases. The 1997 protocol represented a major step forward in international efforts to avert the threat of climate change.

ENVIRONMENTAL CLOSE-UP

The Greater Yellowstone Ecosystem

In 1872, the U.S. government established Yellowstone National Park as the world's first national park. It was an expansive area that protected unique natural features such as geysers, hot springs, rivers, lakes, and mountains. It was also a preserve for many kinds of wildlife such as grizzly bears, elk, moose, and bison. At the time it was established, the park was thought to be of adequate size to protect the scenic resources and the wildlife. Since that time, the lands surrounding the park have been converted to a variety of uses, including cattle grazing, timber production, hunting, and mining.

Fortunately, most of the lands surrounding Yellowstone National Park and the adjacent Grant Teton National Park are still under government control as national forests, national wildlife refuges, and other state, local, or federal bodies. Some of the park wildlife, particularly the grizzly bear and bison, often wander across the park boundaries. The grizzly in particular needs large regions of wilderness to survive as a species.

Many people assert that it is essential that these lands be integrated into a Greater Yellowstone Ecosystem management plan encompassing about 7.3 million hectares (18 million acres). The plan is based on more natural boundaries than the original boundaries established in 1872. This would require changes in the way much of the land surrounding Yellowstone is currently being used. The trade-offs are significant. Logging, mining, hunting, and grazing would be stopped or significantly reduced. This would result in a loss of jobs in those industries. Proponents argue that additional jobs would be created in the tourist and related service industries. The advantages, they argue, would be equal to or greater than the economic losses caused by stopping current uses. Individual and group decisions result in organizational policies and consumer behavior which support or weaken the ecosystem.

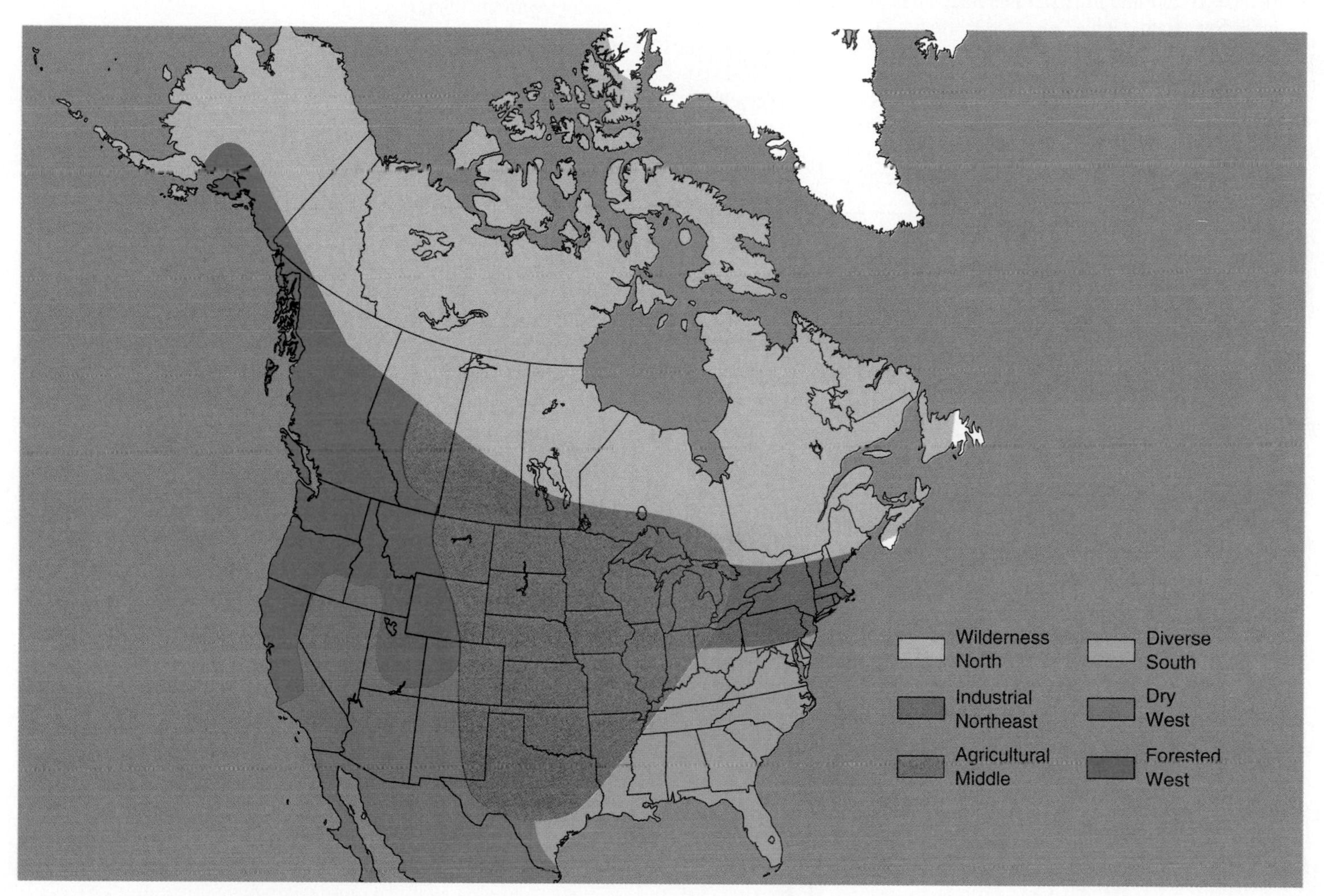

figure 1.4 **Regions of North America** Because of natural features of the land and the uses people make of the land, different regions of North America face different kinds of environmental issues. Certainly within each region people face a large number of specific issues, but certain kinds of issues are more important in some regions than others.

ENVIRONMENTAL CLOSE-UP

Headwaters Forest

The Headwaters Forest in Humbolt County, California, is the last ancient redwood forest remaining in private ownership. The old-growth redwoods are centuries-old trees, distinguished from second-growth trees, which have regenerated after logging. The term "old-growth" can apply to groves of trees and/or individual trees. The owner of this forest until 1999 was the Pacific Lumber Company (PL). Given the unique nature of the old growth and the local ecosystem, a great deal of public attention was focused on the fate of lands that were in private hands. Pacific Lumber Company purchased the forest to lumber it, not to preserve it. The fate of the forest was never certain during a decade of almost constant and bitter controversy between those who wanted to log the forest and those who wanted to preserve it.

In 1996, a federal and state agreement was entered into with Pacific Lumber. This agreement committed the federal and state government to provide $380 million for the purchase of the largest grove of old-growth redwood still in private hands anywhere in the world—the Headwaters Grove on PL lands. In addition, the agreement specified that PL would develop a Habitat Conservation Plan and a Sustained Yield Plan for the remainder of its lands, approximately 81,000 hectares (200,000 acres). In 1999, with a final cost of $450 million, the Pacific Lumber Company and federal and state government agencies signed the Headwaters Forest Agreement. The controversy, however, did not end with the signing.

On the one hand, the agreement was hailed as a landmark. It was proclaimed as a compromise that every "reasonable" person should be able to accept. It protects every extensive tract of old-growth redwood remaining in the possession of Pacific Lumber Company, which means every tract of biological importance. It places heavier protections on salmon-spawning streams running through PL property than are applied to any private timberland in California. It also mandates more extensive precautions against stream siltation and landsliding than are currently in effect anywhere in the state. The agreement includes a covenant that guarantees these protections will run with the land—PL cannot void its habitat conservation plan by selling its land to another company. At the same time, the agreement provides PL with a predictable annual level of timber harvest on which to base its economic calculations.

Not all, however, were totally pleased with the agreement. Opponents to the agreement argued that the Habit Conservation Plan gives PL too much latitude in logging sensitive habitat. All habitat conservation plans are little more than an end run around the Endangered Species Act and threaten the health of the forest, critics of the agreement stated. It was also argued that the agreement left little buffer between PL's ongoing logging operations and the public land. Apparently not all the "reasonable" people were in total agreement with the plan.

What do you think?

Should public tax dollars be spent to acquire properties such as the Headwaters Forest?

Is compromise possible on such a divisive issue?

Do you or your family have redwood furniture or a redwood deck? What is the connection? Is there one?

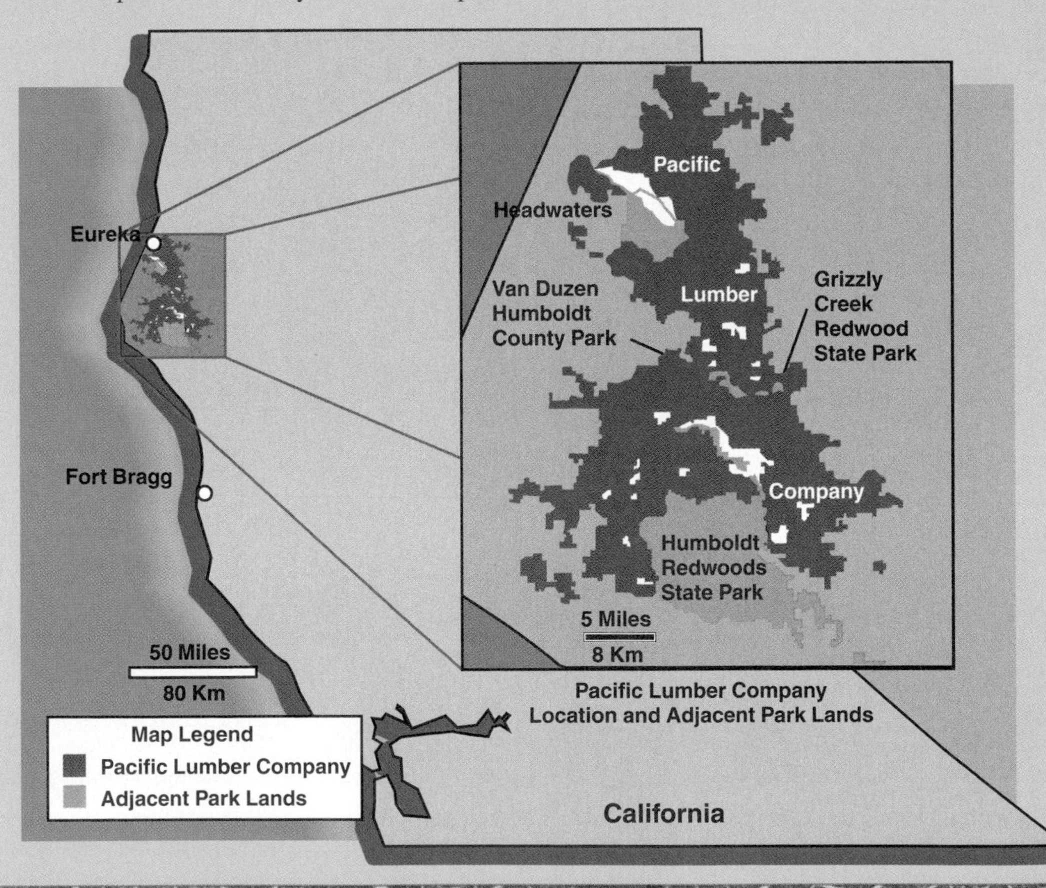

tend to be very sensitive to insults and take a long time to repair damage done by unwise exploitation. Mining, oil exploration, development of hydroelectric projects, and harvesting of timber all require roads and other human artifacts, involve the insertion of new technologies into native culture, and generate economic benefits.

In the past, many short-term political and economic decisions failed to look at long-term environmental implications. Today, however, people are concerned about these remaining wilderness areas. Politicians are more willing to look at the scientific and recreational values of wilderness as well as the economic value of exploitation.

Native people, who consider much of this region to be their land, have become increasingly sophisticated in negotiating with state, provincial, and federal governments to protect rights they feel they were granted in treaties. They are sensitive to changes in land use or government policy that would force changes in their traditional way of life.

Concerned citizens, business interests, and environmental activists have become increasingly sophisticated in influencing decisions made by government. The process of compromise is often difficult and does not always assure wise decisions, but most governments now realize they must listen to the concerns of their citizens and balance economic benefits with social and cultural benefits. (See figure 1.5.)

The Agricultural Middle

The middle of the North American continent is dominated by intensive agriculture. This means that the original, natural ecosystems have been replaced by managed agricultural enterprise. It is important to understand that this area was at one time wilderness. Today, you would need to search very hard to find regions of true wilderness in Iowa, Indiana, or southern Manitoba. Some special areas have been set aside to preserve fragments of the original natural plant and animal associations, but most of the land has been converted to agriculture wherever practical.

The economic value generated by this use of a rich soil resource is tremendous, and most of the land is privately owned. Governments cannot easily control what happens on these privately held lands. But governments indirectly encourage certain activities through departments of agriculture that encourage agricultural research, grant special subsidies to farmers in the form of guaranteed prices for their products and other special payments, and develop markets for products. Yet because the economic risks involved in farming are great, the number of farmers constantly declines. There are a number of reasons for farm failures, including drought, disease, lack of markets, increasing labor shortages, and fuel and equipment costs.

One of the major, nonpoint pollution sources (pollution that does not have an easily identified point of origin) is agriculture. Air pollution in the form of dust is an inevitable result of tilling the land. Soil erosion occurs when soil is exposed to wind and moving water and leads to siltation of rivers, impoundments, and lakes. Fertilizers and other agricultural chemicals blow or are washed from the areas where they are applied. Nutrients washed from the land enter rivers and lakes where they encourage the growth of algae, lowering water quality. The use of pesticides causes concern about human exposure, effects on wild animals that are accidentally exposed, and residues in foods produced.

Since many communities in this region rely on groundwater for drinking water, the use of fertilizers and pesticides, and their potential for entering the groundwater as a result of unwise or irresponsible use, is a consumer issue. In addition, many farmers use groundwater for irrigation, which lowers the water table and leaves less groundwater for other purposes.

In an effort to stay in business and preserve their way of life, farmers must use modern technology. Careful use of these tools can reduce their impact; irresponsible use causes increased erosion, water pollution, and risk to humans. (See figure 1.6.)

The Dry West

Where rainfall is inadequate to support agriculture, ranching and raising livestock are possible. This is true in much

Walrus harvesting

A clear-cut forest

Grizzly bear fishing for salmon

figure 1.5 **The Wilderness North** Protection of wilderness is a major issue of this region. The major points of conflict involve the government role in managing these lands and wildlife, the protection of the rights and beliefs of native people, and the desire of many to exploit the mineral and other resources of the region.

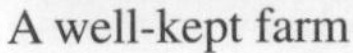

A well-kept farm

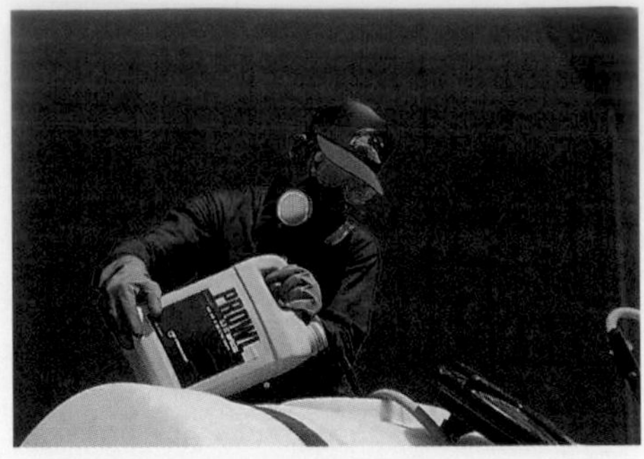

Agricultural chemicals

Barges loaded with grain

figure 1.6 **The Agricultural Middle** The rich soil resource of this region has been converted to managed agricultural activity. The use of pesticides and fertilizer and exposure of the land to erosion cause concern about pollution of surface and groundwater. Most farmers still maintain that these practices are essential in modern agriculture and that they can be used safely and with minimal pollution.

of the drier portions of western North America. Because much of the land is of low economic value, most is still the property of government, which encourages its use by providing water for livestock and irrigation at minimal cost, offering low rates for grazing rights, and encouraging mining and other development.

Many people believe that government agencies have seriously mismanaged these lands. They assert that the agencies are controlled by special interest groups and powerful politicians sensitive to the demands of ranchers, that they subsidize ranchers by charging too little for grazing rights, and that they allow destructive overgrazing because of the economic needs of ranchers. Ranchers argue that they require access to government-owned land, cannot afford significantly increased grazing fees, and that changing government policies would destroy a way of life that is important to the regional economy.

Water is an extremely valuable resource in this region. It is needed for municipal use and for agriculture. Many areas, particularly the river valleys, have fertile soils that can be used for intensive agriculture. Cash crops such as cotton, fruits, and vegetables can be grown if water is available for irrigation. Because water tends to evaporate from the soil rapidly, long-term use of irrigated lands often results in the buildup of salts in the soil, thus reducing fertility. Irrigation water flowing from fields is polluted by agricultural chemicals that make it unsuitable for other uses such as drinking. As cities in the region grow, an increasing conflict arises between urban dwellers who need water for drinking and other purposes, and ranchers and farmers who need the water for livestock and agriculture. Increased demand for water will result in shortages, and decisions will have to be made about who will ultimately get the water and at what price. If the urban areas get the water they want, some farmers and ranchers will go out of business. If the agricultural interests get the water, urban growth and development will have to be limited and expensive changes will have to be made to conserve domestic water use.

Because population density is low in most of this region, much of the land has a wilderness character. Increasingly, a conflict has developed between the economic management of the land for livestock production and the desire on the part of many to preserve the "wilderness." Designating an area as wilderness means that certain uses are no longer permitted. This offends individuals and groups who have traditionally used the area for grazing, hunting, and other pursuits. A long history of use and abuse of this land by overgrazing, modification to encourage plants valuable for livestock, and the introduction of grasses for livestock has significantly altered the region so that it cannot truly be called wilderness. The low population density does, however, provide a remoteness and natural character that many seek to preserve. (See figure 1.7.)

The Forested West

The coastal areas and mountain ranges of the western United States and Canada receive sufficient rainfall for coniferous forests to dominate as vegetation. Since most of these areas are not suitable for farmland, they have been maintained as forests with some grazing activity in the more open forests. Governments and large commercial timber companies own large sections of these lands. Government forest managers (U.S. Forest Service, Bureau of Land Management, Environment Canada, and various state and provincial departments) historically have sold timber-cutting rights at a loss and are thought by many to be too interested in the production of forest products at the expense of other, less tangible values. In 1993, the U.S. Forest Service was directed to stop below-cost timber sales.

This policy change has become a major issue in the old-growth forests of the Pacific Northwest where timber interests maintain that they must have access to government-owned forests in order to remain in business. Many of these areas have significant wilderness, scenic, and recreational value. Environmental interests point out that it makes no sense to complain about the destruction of tropical rainforests in South America while North America makes

Overgrazed land

Irrigation water and electrical generation from Glen Canyon Dam

Bryce Canyon

Wilderness area

figure 1.7 **The Dry West** Water is a key issue in this region. Both city dwellers and rural ranchers and farmers need water, and conflict results when there is not enough water to satisfy the desires of all. In addition, much of the land in this region is owned by the government. This raises concerns about how the government manages the land and how government policy affects the people of the region.

plans to cut large areas of previously uncut, temperate rainforest. Are the intangible values of preserving an ancient forest ecosystem as important as the economic values provided by timber and jobs?

Environmental interests are concerned about the consequences logging would have on organisms that require mature, old-growth forests for their survival. Grizzly bear habitat in Alaska and British Columbia could be altered significantly by logging; the northern spotted owl has become a symbol of the conflict between logging and preservation in Oregon and Washington; and preservation of coastal redwood forests has become an issue in northern California. (See figure 1.8.)

The Great Lakes and Industrial Northeast

While much of the West and Central region of North America is characterized by low population densities and small towns, major portions of the Great Lakes and Northeast are dominated by large metropolitan complexes that generate social and resource needs that are difficult to satisfy. Many of these older cities were formed around industrial centers that have declined, leaving behind poverty, environmental problems in abandoned industrial sites, and difficulties with solid waste disposal, air quality, and land-use priorities. Interspersed among the major metropolitan areas are small towns, farmland, and forests.

One of the major resources of the region is water transport. The Great Lakes and eastern seacoast are extremely important to commerce; ships can travel throughout the area by way of the St. Lawrence Seaway and the Great Lakes through a series of locks and canals that bypass natural barriers. Because of the importance of shipping in this region, harbors have been constructed and waterways have been deepened by dredging. The waterways are maintained at considerable government expense.

One of the greatest problems associated with the industrial uses of the Great Lakes and East Coast is contamination of the water with toxic materials. In some cases, unthinking or unethical individuals have dumped toxins directly into the water. In other cases, small, accidental spills or leaks over long periods of time have contaminated the sediments in harbors and bays.

A major concern about these pollutants is that they bioaccumulate (see chapter 15) in the food chain. The concentrations of some chemicals in the fat tissue of top predators, such as lake trout and fish-eating birds, can be a million times higher than the concentration in the water. Because of this, government agencies have issued consumption advisories for some fish and shellfish in contaminated areas. Since many kinds of fish can swim great distances, advisories for the Great Lakes warn against eating certain fish taken anywhere within the lakes, not just from the site of contamination. Similarly, Chesapeake Bay has been subjected to years of thoughtless pollution, resulting in reduced fish and shellfish populations and advisories against consuming some organisms taken from the bay.

Cut logs being hauled

Native elk

U.S. Forest Service ownership

figure 1.8 **The Forested West** The cutting of forested areas for timber production destroys the previous ecosystem. Some see the trees as a valuable resource that provides jobs and building materials. Others see the forest ecosystem as a natural resource that should be preserved. In addition, government ownership of much of this land has generated considerable political debate about what the appropriate use of the land should be.

Water always generates considerable recreational value. Consequently, conflicts arise between those who want to use the water for industrial and shipping purposes and those who wish to use it for recreation. Due to the fact that so much of the North American population is concentrated in this region, the economic value of recreational use is extremely high. Consumer pressure is great to clean up contaminated sites and prevent the pollution of new ones. Contaminated areas do not enhance tourism or quality of life.

Most of these older, large cities had no plan to shape their growth. As a result, open space for people is limited and urban dwellers have few opportunities to interact with the natural world. Children who grow up in these cities often do not know that milk comes from a cow—they have never seen, smelled, or touched a cow. Consequently, urban people have difficulty understanding the feeling rural people have for the land. These urban dwellers may never have an opportunity to experience wilderness. Their major environmental priorities are cleaning up contaminated sites, providing more parks and recreation facilities, reducing air and water pollution, and improving transportation. (See figure 1.9.)

The Diverse South

In many ways, the South is a microcosm of all the regions previously discussed. The petrochemical industry dominates the economies of Texas and Louisiana, and forestry and agriculture are significant elements of the economy in other parts of the region. Major metropolitan areas thrive, and much of the area is linked to the coast either directly or by the Mississippi River and its tributaries. The environmental issues faced in the South are as diverse as those in the other regions.

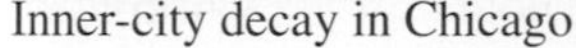

Inner-city decay in Chicago

Harbor in Duluth, Minnesota

Central Park in New York City

figure 1.9 **The Great Lakes and the Industrial Northeast** Industry, waterways, and population centers are the defining elements of this region. The historically extensive use of the Great Lakes and coastal areas of the Northeast for industry, because of the ease of providing water transportation, has resulted in many older cities with poor land-use practices. Rebuilding cities, providing recreational opportunities for urban dwellers, and repairing previous environmental damage are important issues. The water resources of the region provide transportation, recreation, and industrial opportunities.

Some areas of the South (particularly Florida) have had extremely rapid population growth, which has led to groundwater problems, transportation problems, and concerns about regulating the rate of growth. Growth means money to developers and investors, but it requires municipal services, which are the responsibility of local governments. Too many people and too much development also threaten remaining natural ecosystems.

Poverty has been a problem in many areas of the South. This creates a climate that encourages state and local governments to accept industrial development at the expense of other values. Often, jobs are more important than the environmental consequences of the jobs; low-paying jobs are better than no jobs.

The use of the coastline is of major concern in many parts of the South. The coast is a desirable place to live, which may encourage unwise development on barrier islands and in areas that are subject to flooding during severe weather. In addition, industrial activity along the coast has resulted in the loss of wetlands. (See figure 1.10.)

Miami metropolitan area

Everglades

Chemical plant on lower Mississippi

figure 1.10 **The Diverse South** Poverty has been a historically important problem in the region. Often the creation of jobs was considered more important than the environmental consequences of those jobs. The use of coastal areas for industry has resulted in pollution of coastal waters. The heavy use of the Mississippi River for transportation and industry has caused pollution problems. In addition, the desirable climate in the South has resulted in intense pressure to develop new housing for those who want to move to the region. Unwise development of housing on fragile coastal sites has resulted in damage to buildings by storms and the actions of the oceans. This causes intense debate on land use.

Summary

Artificial political boundaries create difficulties in managing environmental problems because most environmental units, or ecosystems, do not coincide with political boundaries. Therefore, a regional approach to solving environmental problems, one that incorporates natural geographic units, is ideal. Each region of the world has certain environmental issues that are of primary concern because of the mix of population, resource use patterns, and culture.

Environmental problems become issues when there is disagreement. This inevitably leads to a confrontation between groups that have different views on the consequences of an environmental problem. Many social, economic, ethical, and scientific issues shape a person's opinions. The process of environmental decision making must account for all of these issues when viewing an acceptable compromise.

Environmental problems are people problems. They occur because the uses of natural resources, which some people feel are justified, result in a diminished environment for others in the region. Environmental problems are defined by the person who perceives the problem. When perceptions differ, conflict occurs. Environmental decisions inevitably involve economic consequences because someone is receiving value from the resources being used or someone perceives an economic loss because a use has been withdrawn.

- Some argue that economic consequences should not be important when making environmental decisions; others argue that economic considerations can resolve all environmental issues.
- Some argue that regulation is necessary to protect resources; others argue that regulation hinders valuable use of resources.
- Some consider nonhuman organisms as important as humans; others feel that humans have a primary place in nature.
- Some are against change; others recognize that change must occur if negative consequences are to be prevented.
- Some believe that environmental responsibility rests on each decision maker, whether at home, in the workplace, or in the community. Each hour and dollar the consumer spends involves environmental consequences. How do you feel about this statement?

With all these differing opinions, compromise is the only way to resolve the conflicts. The social institution of government must play a role. Economic evaluation is important. Recognition of the validity of opposing points of view is essential. The field of environmental science seeks to find that middle ground.

Interactive Exploration

Check out the website at

http://www.mhhe.com/environmentalscience
and click on the cover of this textbook for interactive versions of the following:

KNOW THE BASICS

ecosystem *7*
environmental science *5*
environment *5*
sustainable development *6*
wilderness *8*

- On-line Flashcards
- Electronic Glossary

IN THE REAL WORLD

What happens when bison do not recognize Yellowstone's legal boundaries, but Montana's officials of livestock management do? Once bison leave the park, they are under the jurisdiction of state officials, and state policy dictates that bison must not roam freely outside the park. To further understand this controversial management policy, read Should Bison Leaving Yellowstone National Park Be Shot?

For further stories that emphasize the sometimes controversial interactions between people and wildlife, and the interrelated nature of environmental problems, read: Judge Rules on Excessive Fishing in Alaska, Canadian Government Takes Steps to Conserve Salmon, Environmental Concerns in the "Battle for Seattle," at World Trade Organization Talks, Aging Dams Being Removed for Environmental Benefits, and China Slated to Join the WTO.

Regional environmental concerns, from the poaching of mushrooms in Pacific Northwest forests to water levels in the Great Lakes, can be explored in the following case studies and environmental stories: Matsutake Mushroom Mania, The Prairie Wetlands of Southwest Minnesota, PCB Contaminants in the Fox River, Protecting Groundwater Resources, BP Petroleum/Amoco Admits to Dumping Toxic Waste on Alaska's North Slope, Historic Decision to Allow Drilling in Remote Alaska Oil Reserve, Major Fires Expected in Popular Wilderness Area, and Low Water Levels in the Great Lakes.

PUT IT IN MOTION

Bioaccumulation is mentioned as a major concern in the Great Lakes and in the industrial Northeast. This concept will be further discussed in chapter 15, but now you can see an animation that brings the concept to life at Biomagnification.

TEST PREPARATION

Review Questions

1. Describe why finding solutions to environmental problems is so difficult. Do you think it has always been as complicated?
2. Describe what is meant by an ecosystem approach to environmental problem solving. Is this the right approach?
3. List two key environmental issues for each of the following regions: the wilderness North, the agricultural middle, the forested West, the dry West, the Great Lakes and industrial Northeast, and the South. How are the issues changing?
4. Define environment and ecosystem and provide examples of these terms from your region.
5. Describe how environmental conflicts are resolved.
6. Select a local environmental issue and write a short essay presenting all sides of the question. Is there a solution to this problem?

Critical Thinking Questions

1. Imagine you are a United States congressional representative from a western state and a new wilderness area is being proposed for your district. Who might contact you to influence your decision? What course of action would you take? Why?
2. How do you weigh in on the issue of jobs or the environment? What limits do you set on economic growth? Environmental protection?
3. Imagine you are an environmentalist in your area that is interested in local environmental issues. What kinds of issues might these be?
4. Imagine that you lived in the urban East and that you were an advocate of wilderness preservation. What disagreements might you have with residents of the wilderness North or the arid West. How would you justify your interest in wilderness preservation to these residents?
5. You are the superintendent of Yellowstone National Park and want to move to an ecosystem approach to managing the park. How might an ecosystem approach change the current park? How would you present your ideas to surrounding landowners?
6. Look at the issue of global warming from several different disciplinary perspectives—economics, climatology, sociology, political science, agronomy. What might be some questions that each discipline could contribute to our understanding of global warming?

CHAPTER 2 Environmental Ethics

Objectives

After reading this chapter, you should be able to:

- Differentiate between ethics and morals.
- Define personal ethics.
- Explain the connection between material wealth and resource exploitation.
- Describe how industry exploits resources and consumes energy to produce goods.
- Explain how corporate behavior is determined.
- Describe the influential power that corporations wield because of their size.
- Explain why governmental action was necessary to force all companies to meet environmental standards.
- Describe the factors associated with environmental justice.
- Describe what has been the general attitude of consumers and business toward the environment.
- Explain the relationship between economic growth and environmental degradation.
- List three conflicting attitudes toward nature.

Chapter Outline

Views of Nature

Environmental Close-Up: *What Is Ethical?*

Environmental Ethics

Environmental Attitudes

Environmental Close-Up: *Naturalist Philosophers*

Environmental Close-Up: *Environmental Philosophy*

Societal Environmental Ethics

Environmental Close-Up: *A Corporate Perspective*

Corporate Environmental Ethics

Global Perspective: *Chico Mendes and Extractive Reserves*

Environmental Justice

Individual Environmental Ethics

Global Perspective: *International Trade in Endangered Species*

Global Environmental Ethics

Global Perspective: *Earth Summit on Environment and Development*

Issues & Analysis: *Antarctica—Resource or Refuge?*

Global Perspective: *The Kyoto Protocol on Greenhouse Gases*

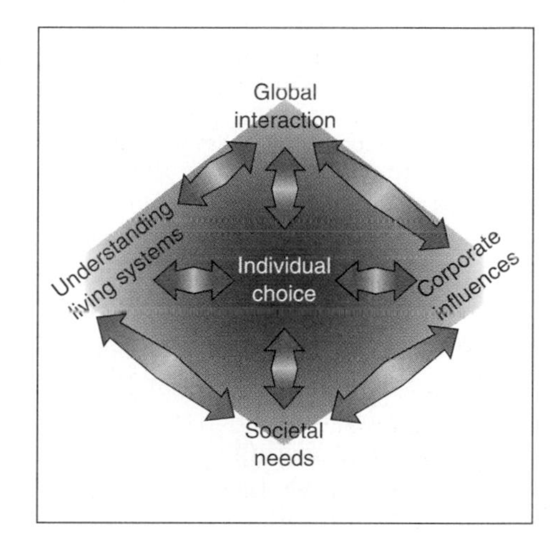

Views of Nature

The most beautiful object I have ever seen in a photograph in all my life, is the planet Earth seen from the distance of the moon, hanging in space, obviously alive. Although it seems at first glance to be made up of innumerable separate species of living things, on closer examination every one of its things, working parts, including us, is interdependently connected to all the other working parts. It is, to put it one way, the only truly closed ecosystem any of us know about.

—Lewis Thomas

There are no passengers on Spaceship Earth. We are all crew members.

—Buckminster Fuller

figure 2.1 **The Earth as Seen from Space** Political, geographical, and nationalistic differences among humans do not seem so important from this perspective. In reality, we all share the same "home."

One of the marvels of recent technology is that we can see the Earth from the perspective of space, a blue sphere unique among all the planets in our solar system. (See figure 2.1.) Looking at ourselves from space, it becomes obvious, says ecologist William Clark of Harvard University, that only as a global species, "pooling our knowledge, coordinating our actions, and sharing what the planet has to offer—do we have any prospect for managing the planet's transformation along pathways of sustainable development."

Many people see little value in an undeveloped river and feel it is unreasonable to leave it flowing in a natural state. It could be argued that rivers throughout the world have been "controlled" to provide power, irrigation, and navigation at the expense of the natural world. It could also be argued that to not use these resources would be wasteful.

In the U.S. Pacific Northwest, there is a conflict over the value of old-growth forests. Economic interests want to use the forests for timber production and feel that to not do so would cause economic hardship. They argue that the trees are going to die anyway and they might as well be used for the betterment of the human community. Others feel that all the living things that make up the forest have a value we do not yet appreciate. Removing the trees would destroy something that took hundreds of years to develop and may never be replaced.

Interactions between people and their environment are as old as human civilization. The problem of managing those interactions, however, has been transformed today by unprecedented increases in the rate, scale, and complexity of the interactions. At one time, pollution was viewed as a local, temporary event. Today, pollution may involve several countries—as with the concern over acid deposition in Europe and in North America—and will affect multiple generations. The debates over chemical and radioactive waste disposal are examples of the increasingly international nature of pollution. For example, many European countries are concerned about the transportation of radioactive and toxic wastes across their borders. What were once straightforward confrontations between ecological preservation and economic growth now involve multiple linkages that blur the distinction between right and wrong. For example, the enhanced greenhouse effect is thought to result from energy consumption, agricultural practices, and climatic change.

Many people believe that we have entered an era characterized by global change that stems from the interdependence between human development and the environment. They argue that self-conscious, intelligent management of the earth is one of the greatest challenges

ENVIRONMENTAL CLOSE-UP

What Is Ethical?

Ethics is one branch of philosophy. Ethics seeks to define fundamentally what is right and what is wrong, regardless of cultural differences. For example, most cultures have a reverence for life and hold that all humans have a right to live. It is considered unethical to deprive an individual of life.

Morals differ somewhat from ethics because morals reflect the predominant feelings of a culture about ethical issues. For example, in almost all cultures, it is certainly unethical to kill someone; however, when a country declares war, most of its people accept the necessity of killing the enemy. Therefore, it is a moral thing to do even though ethics says that killing is wrong. No nation has ever declared an immoral war.

Environmental issues require a consideration of ethics and morals. For example, because there is currently enough food in the world to feed everyone adequately, it is unethical to allow some people to starve while others have more than enough. However, the predominant mood of those in the developed world is one of indifference. They don't feel morally bound to share what they have with others. In reality, this indifference says that it is permissible to allow people to starve. This moral stand is not consistent with a purely ethical one.

As we can see, ethics and morals are not always the same; thus, it is often difficult to clearly define what is right and what is wrong. Some individuals view the world's energy situation as serious and have reduced their consumption. Others do not believe there is a problem and so have not modified their energy use. Still others do not care what the situation is. They will use energy as long as it is available. Other issues are population and pollution. Is it ethical to have more than two children when the world faces overpopulation? Should an industry persuade the public to vote "no" on a particular legislative bill because it might reduce profits, even though its passage would improve the environment? The stand we take on such issues often depends on our position. An industrial leader, for example, would probably not look upon pollution as negatively as someone who participates in outdoor activities. In fact, many business leaders view the behavior of active preservationists as immoral because it restricts growth and, in some cases, causes unemployment.

Most ethical questions are very complex. Ethical issues dealing with the environment are no different. It is important to explore environmental issues from several points of view before taking a stand. One point to consider is the difference between the short-term and long-term effects of a course of action.

When we take an ethical stand, we become open to attack from those who disagree with our stand. Often, individuals are portrayed as villains for pursuing a course of action they consider righteous.

facing humanity as we begin the twenty-first century. To meet this challenge, they believe, a new environmental ethic must evolve.

Environmental Ethics

Ethical issues dealing with the environment are different from other kinds of ethical problems. Depending on your perspective, an environmental ethic could encompass differing principles and beliefs.

Environmental ethics is a topic of applied ethics which examines the moral basis of environmental responsibility. In these environmentally conscious times, most people agree that we need to be environmentally responsible. Toxic waste contaminates groundwater, oil spills destroy shore lines, and fossil fuels produce carbon dioxide thus adding to global warming. The goal of environmental ethics, then, is not to convince us that we should be concerned about the environment—many already are. Instead, environmental ethics focuses on the moral foundation of environmental responsibility, and how far this responsibility extends. There are three primary theories of moral responsibility regarding the environment. Although each supports environmental responsibility, their approaches are different.

The first of these theories is **anthropocentric,** or human-centered. Environmental anthropocentrism is the view that all environmental responsibility is derived from human interests alone. The assumption here is that only human beings are morally significant organisms and have a direct moral standing. Since the environment is crucial to human well-being and human survival, we have a duty toward the environment, that is, a duty which is derived from human interests. This involves the duty to assure that the earth remains environmentally hospitable for supporting human life, and that its beauty and resources are preserved so that human life on earth continues to be pleasant. Some have argued that our environmental duties are derived both from the immediate benefit which living people receive from the environment, and the benefit that future generations of people will receive. But, critics have maintained that since future generations of people do not yet exist, then, strictly speaking, they cannot have rights any more than a dead person can have rights. Nevertheless, both parties to this dispute acknowledge that environmental concern derives solely from human interests.

A second theory of moral responsibility to the environment is **biocentric.** According to the broadest form of the life-centered theory, all forms of life have an inherent right to exist. Some biocentric thinkers give species a hierarchy of values. Some, for example, believe we have greater responsibility to protect animal species than plant species. Others determine the rights of various species depending on the harm they do to humans. For example, they see nothing wrong in killing pest species such as rats or mosquitoes. Some go further and believe that each individual organism, not just each species, has a basic right to survive. Individuals who support the animal rights movement tend to place more value on the individuals of animal species than on plant species. Trying to decide what types of species or individuals should be protected from early extinction or death resulting from human activities is an ethical dilemma. It is hard to know where to draw the line and be ethically consistent.

The third approach to environmental responsibility, called **ecocentrism,** maintains that the environment deserves direct moral consideration, and not one which is merely derived from human (and animal) interests. In ecocentrism it is suggested that the environment has direct rights, that it qualifies for moral personhood, that it is deserving of a direct duty, and that it has inherent worth. The environment, by itself, is considered to be on a moral par with humans.

The position of ecocentrism is the view advocated by the ecologist and writer Aldo Leopold in his essay "The Land Ethic" from *A Sand County Almanac and Sketches Here and There* (1949):

> "All ethics so far evolved rest upon a single premise: that the individual is a member of a community of interdependent parts. The land ethic simply enlarges the boundaries of the community to include soils, waters, plants, and animals, or collectively the land . . . a land ethic changes the role of Homo sapiens from conqueror of the land-community to plain member and citizen of it . . . It implies respect for his fellow-members, and also respect for the community as such."

What Leopold put forth in "The Land Ethic" was viewed by many as a radical shift in how humans perceive themselves in relation to the environment. Originally we saw ourselves as conquerors of the land. Now according to Leopold we need to see ourselves as members of a community which also includes the land and the water.

Preservation

Development

Conservation

Recreation

figure 2.2 **The Views of Nature** Individuals envision the same resources used differently.

Leopold also wrote that "a thing is right when it tends to preserve the integrity, stability, and beauty of the biotic community. It is wrong when it tends otherwise. . . . We abuse land because we regard it as a commodity belonging to us. When we see land as a community to which we belong, we may begin to use it with love and respect."

As traditional political and nationalistic boundaries begin to fade or shift globally, there are also new variations of environmental thought and ethics evolving. Some of the new thoughts on environmental ethics are founded on an awareness that humanity is part of nature and that nature's many parts are interdependent. In any natural community, the well-being of the individual and of each species is tied to the well-being of the whole. In a world increasingly without environmental borders, nations, like individuals, should have a fundamental ethical responsibility to respect nature and to care for the Earth, protecting its life-support systems, biodiversity, and beauty and caring for the needs of other countries and future generations.

Environmental ethicists argue that to consider environmental protection as a "right" of the planet is a natural extension of the concept of human rights. Many also argue that an environmental ethic considers one's actions toward the environment as a matter of right and wrong, rather than one of self-interest.

Environmental Attitudes

There are many different attitudes about the environment, most of which fall under one of three headings: (*a*) the development ethic, (*b*) the preservation ethic, and (*c*) the conservation ethic. Each of these ethical positions has its own code of conduct against which ecological mortality may be measured. (See figure 2.2.)

The **development ethic** is based on individualism or egocentrism. It assumes that the human race is and should be the master of nature and that the Earth and its resources exist for our benefit and pleasure. This view is reinforced by the

ENVIRONMENTAL CLOSE-UP

Naturalist Philosophers

The philosophy behind the environmental movement had its roots in the last century. Among many notable conservationist philosophers, several stand out: Ralph Waldo Emerson, Henry David Thoreau, John Muir, Aldo Leopold, and Rachel Carson.

In Emerson's first essay, *Nature,* published in 1836, he claimed that "behind nature, throughout nature, spirit is present." Emerson was an early critic of rampant economic development, and he sought to correct what he considered to be the social and spiritual errors of his time. In his *Journals,* published in 1840, Emerson stated that "a question which well deserves examination now is the Dangers of Commerce. This invasion of Nature by Trade with its Money, its Credit, its Steam, its Railroads, threatens to upset the balance of Man and Nature."

Ralph Waldo Emerson

Henry David Thoreau

Henry David Thoreau was a naturalist who held beliefs similar to Emerson's. Thoreau's bias fell on the side of "truth in nature and wilderness over the deceits of urban civilization." The countryside around Concord, Massachusetts, fascinated and exhilarated him as much as the commercialism of the city depressed him. It was near Concord that Thoreau wrote his classic, *Walden,* which describes a year in which he lived in the country to have direct contact with nature's "essential facts of Life." In his later writings and journals, Thoreau summarized his feelings toward nature with prophetic vision:

> But most men, it seems to me, do not care for Nature and would sell their share in all her beauty, as long as they may live, for a stated sum—many for a glass of rum. Thank God, man cannot as yet fly, and lay waste the sky as well as the earth! We are safe on that side for the present. It is for the very reason that some do not care for these things that we need to continue to protect all from the vandalism of a few. (1861)

John Muir

Aldo Leopold

John Muir combined the intellectual ponderings of a philosopher with the hard-core, pragmatic characteristics of a leader. Muir believed that "wilderness mirrors divinity, nourishes humanity, and vivifies the spirit." Muir tried to convince people to leave the cities for a while to enjoy the wilderness. However, he felt that the wilderness was threatened. In the 1876 article entitled, "God's First Temples: How Shall We Preserve Our Forests?" published in the Sacramento *Record Union,* Muir argued that only government control could save California's finest sequoia groves from the "ravages of fools." In the early 1890s, Muir organized the Sierra Club to "explore, enjoy, and render accessible the mountain regions of the Pacific Coast" and to enlist the support of the government in preserving these areas. His actions in the West convinced the federal government to restrict development in the Yosemite Valley, which preserved its beauty for generations to come.

Aldo Leopold was another thinker as well as a doer in the early conservation field. As a philosopher, Leopold summed up his feelings in *A Sand County Almanac:*

> Wilderness is the raw material out of which man has hammered the artifact called civilization. No living man will see again the long grass prairie, where a sea of prairie flowers lapped at the stirrups of the pioneer. No living man will see again the virgin pineries of the Lake States, or the flatwoods of the coastal plain, or the giant hardwoods.

Leopold founded the field of game management. In the 1920s, while serving in the Forest Service, he worked for the development of a wilderness policy and pioneered his concepts of game management. He wrote extensively in the *Bulletin* of the American Game Association and stated that the amount of space and the type of forage of a wildlife habitat determine the number of animals that can be supported in an area. Furthermore, he said that regulated hunting can maintain a proper balance of wildlife.

Rachel Carson

While most people talk about what's wrong with the way things are, few actually go ahead and change it. Rachel Carson ranks among those few. A distinguished naturalist and best-selling nature writer, Rachel Carson published in the *New Yorker* in 1960 a series of articles which generated widespread discussion about pesticides. In 1962, she published *Silent Spring,* which dramatized the potential dangers of pesticides to food, wildlife, and humans and eventually led to changes in pesticide use in the United States.

Although some technical details of her book have been shown to be in error by later research, her basic thesis that pesticides can contaminate and cause widespread damage to the ecosystem has been established. Unfortunately, Carson's early death from cancer came before her book was recognized as one of the most important events in the history of environmental awareness and action in the 20th century.

ENVIRONMENTAL CLOSE-UP

Environmental Philosophy

Nature, growth, and progress are concepts that we all use, but which we seldom define either in discussion or to ourselves. We speak about environmental ethics, environmental philosophy, ecophilosophy, and so on, but what do we put into these concepts? We say that we have a responsibility for future generations, and that this is a question of morals, but how should questions about morals be decided?

We all have some kind of reasons for our opinions on these matters—whether we believe that nature shall serve humanity, that humanity shall serve nature, or something in between—but we seldom make them explicit or draw the conclusions from them. What are these reasons? Are they reasonable, rational, defensible, scientifically grounded, emotional, religious? And what do we mean by saying that a reason is "rational"? What value does it have that something is scientifically grounded? What weight should emotions carry in this context? Can we reason rationally about them and come to an agreement about them, or do we have to put them aside and just stick to facts? How can we know what is right?

These are all philosophical questions, and as such they may seem far removed from real, concrete environmental problems such as ozone holes and forest death. Can it really be meaningful to spend energy on a debate of these questions, when if not disaster then at least crisis stands at the doorway? Is it at all meaningful to philosophize about the environmental questions? Isn't it action that is needed?

Trying to answer the philosophical questions does not, of course, in itself solve any environmental problems, but on the other hand is it questionable whether we can solve these problems without discussing them on a philosophical level? Because whether we discuss them or not, we have ideas and conceptions which guide our way of thinking, what we see as a problem, what we see as causes of problems and what we see as possible, desirable, or necessary solutions. And, quite seriously, the problems which stand before us today are hardly founded on our having thought too much.

What are your thoughts on this question? Is there already too much talk or is it too little listening?

Source: Dept. of Philosophy, University of Gothenburg, Sweden. No copyright.

work ethic, which dictates that humans should be busy creating continual change and that things that are bigger, better, and faster represent "progress," which itself is good. This philosophy is strengthened by the idea that, "if it can be done, it should be done," or that our actions and energies are best harnessed in creative work.

Examples of the development ethic abound. The notion that bigger is better is certainly not new to us, nor is the belief that if something can be done or built, it should be. The dream of upward mobility is embodied in this ethic. In some circles, questioning growth is considered almost unpatriotic. In the development ethic, nature has only instrumental value; that is, the environment has value only insofar as human beings economically utilize it. Only in the past fifty to one hundred years have the byproducts and waste associated with development been considered.

The **preservation ethic** considers nature special in itself. Nature, it is argued, has intrinsic value or inherent worth apart from human appropriation. Preservationists have diverse reasons for wanting to preserve nature. Some hold an almost religious belief regarding nature. They have a reverence for life and respect the right of all creatures to live, no matter what the social and economic costs. Some preservationists' interest in nature is primarily aesthetic or recreational. They believe that nature is beautiful and refreshing and should be available for picnics, hiking, camping, fishing, or just peace and quiet.

In addition to the religious and recreational preservationists, there are also preservationists whose reasons are essentially scientific. They argue that the human species depends on and has much to learn from nature. Rare and endangered species and ecosystems, as well as the more common ones, must be preserved because of their known or assumed long-range, practical utility. In this view, natural diversity, variety, complexity, and wilderness are thought to be superior to humanized uniformity, simplicity, and domesticity. Scientific preservationists want to lock up not all the land but only what they consider important to future generations.

The third environmental ethic is referred to as the **conservation or management ethic.** It is related to the scientific preservationist view but extends the rational consideration to the entire Earth and for all time. It recognizes the desirability of decent living standards, but it works toward a balance of resource use and resource availability. The conservation ethic stresses a balance between total development and absolute preservation. It stresses that rapid and uncontrolled growth in population and economics is self-defeating in the long run. The goal of the conservation ethic is one people living together in one world, indefinitely.

Societal Environmental Ethics

Society is composed of a great variety of people with diverse viewpoints. This variety can be distilled into a set of ideas that reflect the prevailing attitudes of society. The collective attitudes can be analyzed from an ethical point of view. Western, developed societies have long acted as if the earth has unlimited reserves of natural resources, an unlimited

ENVIRONMENTAL CLOSE-UP

A Corporate Perspective

One of the most important and yet difficult lessons learned during the past quarter century is that the environment and the economy need not be viewed as opposites. On the contrary, we are beginning to understand that in order to have a healthy environment you must also have a healthy economy. The opposite is also equally true in that a healthy economy is dependent on a healthy environment. Many of the leading multinational corporations are adopting this new attitude as a philosophy of business. This is especially true in the area of sustainability. General Motors Corporation (GM) is one such company that has publicly stated its commitment to sustainability.

GM has stated that sustainability is a management framework that drives continuous improvement in the corporation's daily business and allows it to view challenging issues as new business opportunities. Sustainability, GM states, does not require a shift in business priorities, but rather a broader interpretation of those priorities, to meet the increasing expectation of business by society.

As societal risks become interconnected and the economic success of developing countries emerges, civil society, government, and business have to work together to balance economic, environmental, and social objectives. Likewise, each sector must also do its part to integrate these objectives into its respective decisions. For business, this means companies must determine how to operate in ways that promote economic growth and comprehend the needs of the environment and society, without compromising the needs of those in the future. GM states that it is a company's ability to balance environmental, social, and economic considerations that is crucial in the future.

In 1991, the General Motors Board of Directors adopted the Environmental Principles (see below). GM has also publicly committed to integrate economic, environmental, and social issues into its business decisions. GM publishes an annual report on its progress in these issues as a form of public accountability and has done so since 1994. GM is a member of the Global Reporting Initiative (GRI) Steering Committee. The GRI is collaboration among the Coalition of Environmental Responsible Economies (CERES), the United Nations Environment Program (UNEP), and numerous organizations, united to develop a common framework for global sustainability reporting. Issues reported using these guidelines are not limited to environmental performance, but also include social and economic indicators. See www.gm.com/environment to view these reports and obtain more information on GM environmental commitment and sustainability.

The following environmental principles provide guidance to General Motors personnel worldwide in the conduct of their daily business practices.

1. We are committed to actions to restore and preserve the environment.
2. We are committed to reducing waste and pollutants, conserving resources, and recycling materials at every stage of the product's life cycle.
3. We will continue to participate actively in educating the public regarding environmental conservation.
4. We will continue to pursue vigorously the development and implementation of technologies for minimizing pollutant emissions.
5. We will continue to work with all governmental entities for the development of technically sound and financially responsible environmental laws and regulations.
6. We will continually assess the impact of our plants and products on the environment and the communities in which we live and operate with a goal of continuous improvement.

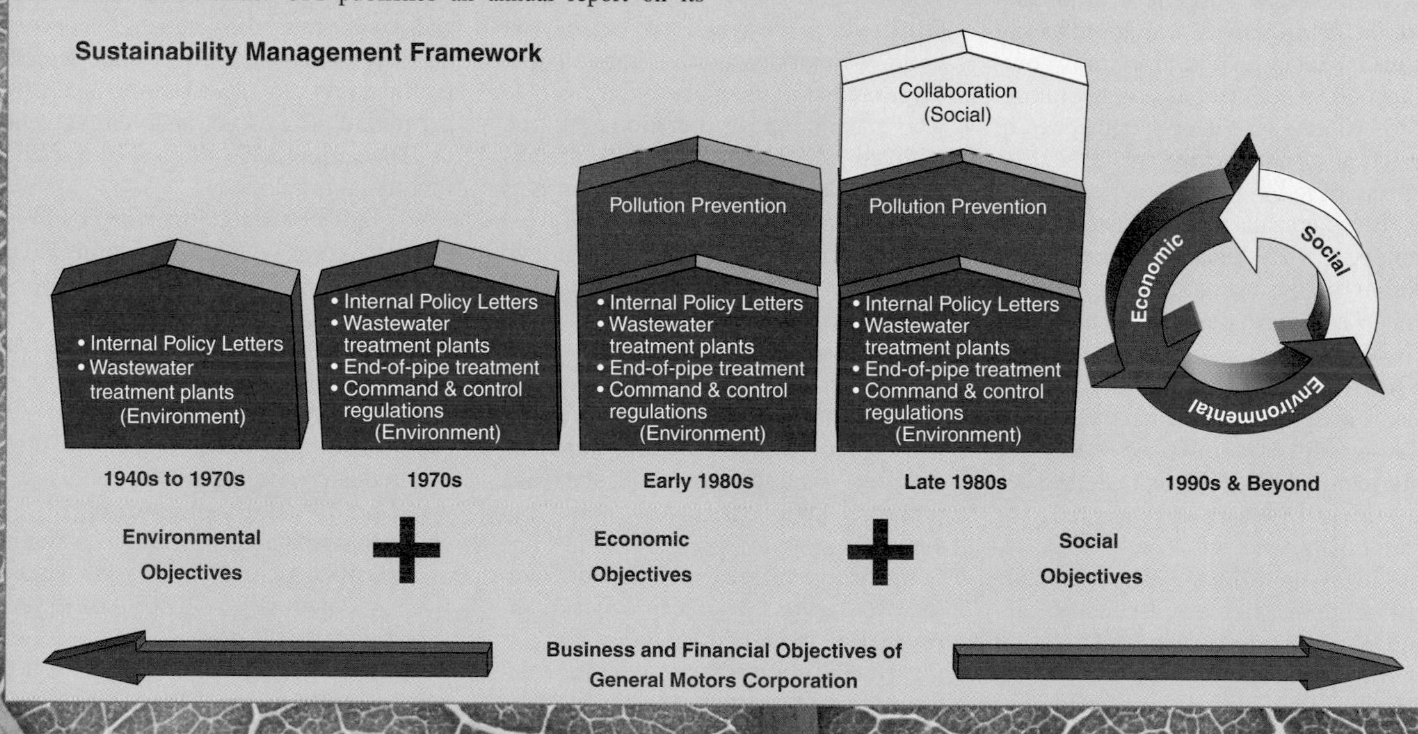

ability to assimilate wastes, and a limitless ability to accommodate unchecked growth.

The economic direction and rationale of developed nations have been that of continual growth. Unfortunately, this growth has not always been carefully planned or even desired. This "growth mania" has resulted in the use of our nonrenewable resources for comfortable homes, well-equipped hospitals, convenient transportation, fast-food outlets, VCRs, home computers, and battery-operated toys, among other things. In economic statistics, such "growth" measures out as "productivity." But the question arises, "What is enough?" Poor societies have too little, but rich societies never say, "Halt! We have enough." The Indian philosopher and statesman Mahatma Ghandi said, "The earth provides enough to satisfy every person's need, but not every person's greed."

Growth, expansion, and domination remain the central sociocultural objectives of most advanced societies. **Economic growth** and **resource exploitation** are attitudes shared by developing societies. We continue to consume natural resources as if the supplies were never ending. All of this is reflected in our increasingly unstable relationship with the environment, which grows out of our tendency to take from the "common good" without regard for the future.

This attitude is deeply embedded in the fabric of our society. Since the first settlers arrived in North America, nature has been considered an enemy. Frequently, the colonists expressed their relation to the wilderness in military terms. They viewed nature as an enemy to be "conquered," "subdued," or "vanquished" by a pioneer "army." Any qualms the pioneers may have felt about invading and exploiting the wilderness were justified by religious beliefs. They were driven by what they perceived to be a "moral imperative." This attitude toward nature is still popular today. Many view wilderness solely as underdeveloped land and see value in land only if it is farmed, built upon, or in some way developed. The notion that land and wilderness should be preserved is incomprehensible to some. The thought of purposely opting to not develop a resource is considered almost a sin.

Corporate Environmental Ethics

Many tasks of industry, such as procuring raw materials, manufacturing and marketing, and disposing of wastes, are in large part responsible for pollution. This is not because any industry or company has adopted pollution as a corporate policy. Industry is naturally dirty because it consumes energy and resources. When raw materials are processed, some waste (useless material) is inevitable. It is usually not possible to completely control the dispersal of all by-products of a manufacturing process. Also, some of the waste material may simply be useless.

For example, the food-service industry uses energy to prepare meals. Much of this energy is lost as waste heat. Smoke and odors are released into the atmosphere, and discolored food items must be discarded.

The cost of controlling waste can be very important in determining a company's profit margin. **Corporations** are legal entities designed to operate at a profit, which is not in itself harmful. The corporation has no ethics, but the people who make up the corporation are faced with ethical decisions. Ethics are involved when a corporation cuts corners in production quality or waste disposal to maximize profit. The cheaper it is to produce an item, the greater the possible profit. It is cheaper in the short run to dump wastes into a river than to install a wastewater treatment facility, and it is cheaper in the short run to release wastes into the air than it is to trap them in filters. Many people consider such pollution unethical and immoral, but some corporations think of it as just one of the factors that determines **profitability.** (See figure 2.3.) Because stockholders expect an immediate return on their investment, corporations often make decisions based on short-term profitability rather than long-term benefit to society.

The amount of profit a corporation realizes determines how much it can expand. To expand continually, a corporation increases the demand for its products through advertising. The more it expands, the more power it attains. The more power it has, the greater its influence over decision makers who can create conditions favorable to its expansion plans. The process becomes a seemingly never-ending spiral.

Nations of the world must confront the problem of corporate irresponsibility toward the environment. In business, incorporation allows for the organization and concentration of wealth and power far surpassing that of individuals or partnerships. Some of the most important decisions affecting our environment are made not by governments or the public but by executives who wield massive corporate power. Often, these executives make only minimal concessions to the public interest, while they make every effort to maximize profits.

Business decisions and technological developments have increased the exploitation of natural resources. In addition, many political and legal institutions have generally supported the development of private enterprise. They have also defended and promoted private property rights rather than social and environmental concerns. Businesses and individuals typically use loopholes, political pressure, and the time-consuming nature of legal action to circumvent or delay compliance with social or environmental regulations.

Is industry becoming more environmentally concerned? Corporations have certainly made more frequent references to worldwide environmental issues over the past several years. Is such concern only rhetoric and social marketing, or is it the beginning of a new corporate ethic? The "Valdez Principles," a tool that industry and the public can use to evaluate corporate environmental responsibility, were developed as a result of the 1989 oil spill in Alaska. These were later named the CERES Principles. (See figure 2.4.)

The Oil Protection Act of 1990, passed in the wake of the *Valdez* spill, was supposed to regulate supertankers

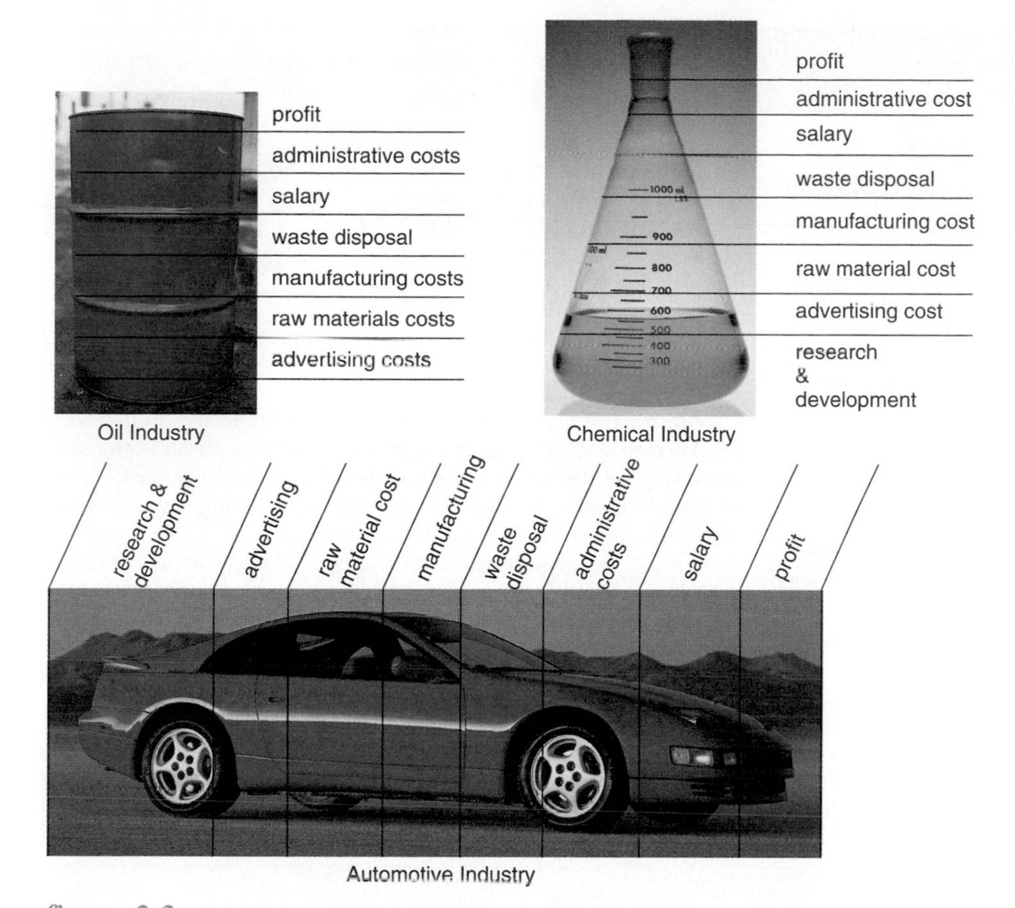

figure 2.3 **Corporate Decision Making** Corporations must make a profit. When they look at pollution control, they view its cost like any other cost: any reductions in cost increase profits.

figure 2.4 **CERES Principles** The 1989 oil spill in Alaska led to the development of the CERES Principles. (a) These waterfowl were victims of the spill. (b) The *Exxon Valdez.*

and reduce the chances of supertanker oil spills. However, to get around the law many oil carriers have shifted their oil transport operations to lightly regulated oil barges pulled by tugboats. This reduction in oil spill safety has led to several barge oil spills, including one in January 1996 in Rhode Island's Moonstone Bay and another in March 1997 in Texas's Galveston Bay.

In 1992, the Valdez Principles were proposed and adopted by the CERES Organization (Coalition for Environmentally Responsible Economics). This organization is an independent group of environmentalists and social investors.

The CERES Principles are a set of codes that businesses may adopt voluntarily. The codes provide environmental standards against which all companies can be assessed and compared. The ten principles encompass a wide range of goals that include minimizing pollutants, making sustainable use of renewable resources, reducing health and safety risks for employees and communities, and representing environmental interests on corporate boards. The CERES Principles are looked upon as a guide for corporate environmentalism. The goal, some argue, should be to make compliance with the CERES Principles a prerequisite for doing business.

Practicing an environmental ethic should not interfere with corporate and other social responsibilities or obligations, though this is not always the case. It must be integrated into overall systems of belief and coordinated with economic systems. Environmental advocates, in turn, need to consider others' objectives just as they demand that others consider environmental consequences in decision making. It makes little sense to preserve the environment if that objective produces national economic collapse. Nor does it make sense to maintain stable industrial productivity at the cost of breathable air, drinkable water, wildlife species, parks, and wilderness. But to maintain profitability, influence, and freedom, businesses must be sensitive to their impact on current and future citizens, not just in terms of the price and quality of the goods they produce but also in terms of public approval of their social and political influence. A 1997 Harris Poll, for example, found that 70 percent of Americans wanted increased government spending, either national or state, to address environmental problems, even if they had to pay higher taxes. In another poll, eight of ten Americans said they would be willing to pay extra for a product packaged with recyclable materials.

GLOBAL PERSPECTIVE

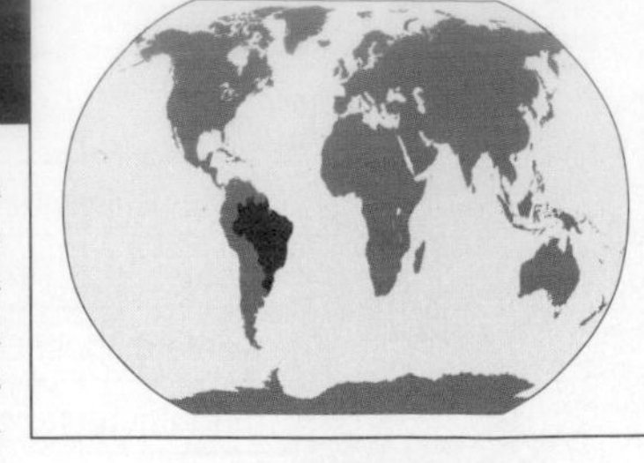

Chico Mendes and Extractive Reserves

Francisco "Chico" Alves Mendes Filho was born in 1944 in the western Brazilian Amazon. A second-generation rubber tapper, Chico was active in the rubber tapper's union for over fifteen years. He and many other peasants made a living by extracting latex from rubber trees and selling it. Rubber tappers also collect and sell other natural products of the forest, such as Brazil nuts, fruits, and native medicines. Mendes was interested in preserving the portion of the Brazilian rainforest that provided their livelihood, and he supported the concept of "extractive reserves."

Extractive reserves involves setting aside land for rubber tappers, who would continue their traditional lifestyles and use the rainforest in its natural state for generations to come. This idea put Mendes in conflict with powerful people interested in clearing the rainforest to raise cattle. Most cattle-ranching operations show short-term economic gains but ultimately become uneconomical when land fertility declines from overgrazing.

In 1987, Mendes received two international environmental awards for his efforts in establishing extractive reserves. One award was the "Global 500" award by the United Nations Environment Programme (UNEP), the other was Ted Turner's Better World Society Environment Award.

In 1988, Chico led the Xapuri Rural Workers Union in a winning effort to stop cattle rancher Darli Alves from deforesting an area the rubber tappers wanted to make into a reserve. On December 22, 1988, as he walked from his home, Chico Mendes was shot by members of a vigilante group who supported local ranchers. In 1990, Darli Alves da Silva and his son were convicted of the murder. Alves da Silva was sentenced to 19 years in prison but escaped in 1993 leading to a three-year exhaustive manhunt. He is now in a high-security prison in Brasilia.

Before his death, Mendes had said, "I want to live to defend the Amazon." His life and death appear to have made a difference in the way the Brazilian rainforest is being used. In 1990, the Chico Mendes Extractive Reserve was established covering about 6 percent of the state of Acre in northwest Brazil.

Due to the circumstances that surrounded his death and due to his role as a leader in the rubber tappers' union, his murder received international notice and caused many people to ask if the natural rainforest perhaps has as much to offer as ranches do.

While confrontations between the rubber tappers and ranchers have decreased, there have been new problems relating to mismanagement of funds in the Chico Mendes Foundation and mismanagement of the rubber tappers' cooperative that runs the Brazil-nut factory. The future of both the cooperative and the foundation remains in doubt.

In the middle 1990s a concept emerged called **industrial ecology** that reflects the link between the economy and the environment. This concept argues that good ecology is also good economics and that alternatives exist for corporations to provide goods and services in ways that do not destroy the environment.

One of the most important elements of industrial ecology is that, as in biological systems, it accounts for waste. Dictionaries define waste as useless or worthless material. In nature, however, nothing is eternally discarded; in various ways, all materials are reused. In our industrial world, discarding materials taken from the Earth at great cost is also generally unwise. Perhaps materials and products that are no longer in use should be termed *residues* rather than *wastes;* wastes are merely residues that our economy has not yet learned to use efficiently. A simpler way of saying this is to view a pollutant as a resource out of place. Such a statement forces us to view pollution and waste in a new way.

Environmental Justice

Environmental justice is a movement promoting the fair treatment of people of all races, income, and culture with respect to the development, implementation, and enforcement of environmental laws, regulations, and policies. Fair treatment implies that no person or group of people should shoulder a disproportionate share of the negative environmental impacts resulting from the execution of domestic and foreign policy programs. (The environmental justice movement is also occasionally referred to as Environmental Equity—which EPA defines as equal protection of all individuals, groups, or communities regardless of race, ethnicity, or economic status, from environmental hazards). Although environmental justice has many facets (e.g., legal, economic, and political), it may be approached appropriately in a variety of ways by the public and private sectors. In addition, the health community should naturally focus on the health aspect of environmental justice.

The environmental justice movement is generally acknowledged to have emerged in the early 1980s in response to large demonstrations opposing the siting of a PCB-landfill in a predominantly black community in Warren County, North Carolina. Subsequent studies and public attention raised concerns of the fairness and protection afforded under existing environmental programs, concerns that have received the increased attention of government as well as the private community.

At its core, **environmental justice** means fairness. It speaks to the impartiality that should guide the application of laws designed to protect the health of human beings and the productivity of ecological systems on which all human activity, economic activity included, depends. It is emerging as an issue because studies show that certain groups of North Americans and citizens of other nations may suffer disproportionately from the effects of pollution.

Governments have established numerous laws, mandates, and directives to eliminate discrimination in housing, education, and employment, but few attempts have been made to address discriminatory environmental practices. In the United States, people of color have borne a disproportionate burden in the location of municipal landfills, incinerators, and hazardous-waste treatment, storage, and disposal facilities.

Hazardous waste sites and incinerators are not randomly located. While waste generation is correlated directly with per capita income, few toxic waste sites are located in affluent suburbs. Waste facilities are often located in communities that have high percentages of poor, elderly, young, and minority residents. Often such facilities are deliberately sited in these communities because they are seen as providing the path of least resistance.

Questions of environmental justice extend beyond the location of toxic waste sites. Exposure to harmful pesticides and other toxic agricultural substances is a major health issue among hired farm workers, the majority of whom are people of color. There is also concern that because some Native American communities consume much greater amounts of fish from certain areas such as the Great Lakes than does the general population, they are at greater risk for dietary exposure to toxic chemicals.

Historically, the environmental movement has been a concern of middle-class whites, but there is a growing level of activism by people of color. Minority participation has broadened the debate to include many issues that were being ignored. It has also forced a dialogue about race, class, discrimination, and equity. Minorities have pushed the plight of their communities to the forefront. They have also brought a new perspective to the environmental movement and will be a part of any future environmental agenda.

Environmental Justice Highlights

1979
Houston community group sues over landfill: While the black community group from Houston eventually lost its court case, its work produced some of the first research on environmental justice showing that most incinerators and landfills built in the city were in black neighborhoods.

1982
More than 500 arrested in landfill fight: In mostly black Warren County, N.C., a battle over a landfill intended to hold PCB-contaminated dirt led to the arrest of hundreds including the Washington, D.C. delegate to Congress.

1983
Government report finds hazardous waste bias: An analysis by the Government Accounting Office found that three of the four largest hazardous waste sites in the Southeast were in black communities.

1987
Church finds environmental racism a national problem: The United Church of Christ published a report that showed that communities with hazardous waste facilities had higher percentages of minorities than those that had had no such facilities.

1989
EPA takes up environmental justice: Bush administration EPA Administrator William Reilly establishes the Environmental Equity Work Group, marking the first official EPA response to the problem.

1993
EPA accepts first civil rights complaint: The Louisiana case was filed by the Tulane (University) Environmental Law Clinic. The legal group later would file another case over the Shintech Inc. PVC plant in rural Louisiana, which became a nationally important battle. At about the same time, the EPA formed the National Environmental Justice Advisory Council, a group of activists, local officials, and industry experts intended to serve as advisers to the EPA.

1994
President Clinton signs environmental justice executive order: The order requires all federal agencies to begin taking the issue into account. "Each Federal agency shall make achieving environmental justice part of its mission by identifying and addressing, as appropriate, disproportionately high and adverse human health or environmental effects of its programs, policies, and activities on minority populations and low-income populations."

1998
EPA releases environmental justice "guidance": The new agency rules, produced with no input from states, cities, or industry, create an uproar.

Individual Environmental Ethics

The environmental movement has effectively influenced public opinion and moved the business community toward an environmental ethic. The result of this changing view of business's responsibilities will complicate business decision making into the next century. More complex environmental and safety demands by the public and a broadening of horizons on the part of business will be a dominant theme of corporate life during the next decade. As human populations and economic activity continue to grow, we are facing a number of environmental problems that threaten not only human health and the productivity of ecosystems, but in some cases the very habitability of the globe.

If we are to respond to those problems successfully, our environmental ethic must express itself in broader and more fundamental ways. We have to

GLOBAL PERSPECTIVE

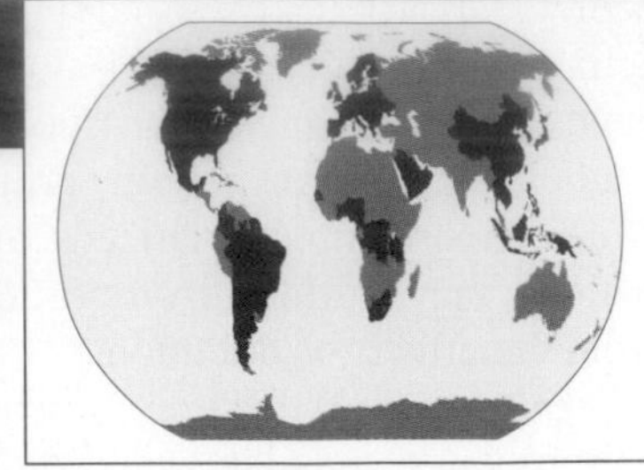

International Trade in Endangered Species

Illegal trading in rare or disappearing species as an international business turns over more than 1.6 billion dollars a year, second only to smuggling in drugs or arms. It directly affects the populations of more than 37,000 animal and plant species and represents a severe threat to their survival.

The international groups that conduct this trading make a fortune. They buy a whole range of cheap animal and plant products, which in certain markets then sell for many times the buying price. It is estimated that each year one and a half million caiman skins leave Brazil, Bolivia, and Paraguay, which as well as a tragedy for the species means a loss of millions of dollars for these countries. The most important markets are to be found in the wealthy countries; the chief buyers for these wildlife products are in Japan, the United States, and the European Union. In the United States, legal trading accounts for more than 200 million dollars, while illegal trading accounts for more than 300 million. This unregulated traffic squanders natural resources and is one of the most dangerous forms of wildlife and biosphere destruction. It does serious damage to Southern societies rich in biodiversity, as it yields no lasting profits for them and only enriches the intermediaries. It pays no taxes or customs duty. The returns for the poachers are limited, because what they receive for illegally capturing or gathering macaws, tigers, crocodiles, or orchids is very little compared to what is paid for them in the wealthy countries. Such trading threatens the sustainability of the ecosystems where the species live and perpetuates inequalities between the wealthy consumer countries and the poorer producer countries, which hardly gain anything from the trade.

The consequences of this trading is dramatic. Animals such as the rhinoceros, the tiger, the leopard, the otter, the South American caiman, the macaw, some primates, butterflies, frogs, tortoises, orchids, cacti, carnivorous plants, trees bearing precious wood like the mahogany, are just a few examples affected. The list goes on.

Many animals are sacrificed for one specific product. But in the traffic of live specimens the mortality is very high, both at the moment of capture and during shipping; many animals die so that a few can reach their destination alive. Traditional, more respectful forms of hunting and gathering are abandoned, poachers often cut down trees to reach the highest nests, taking males and females indiscriminately.

In 1973, after much debating and pressure by scientific organizations and non-governmental organizations (NGOs), the Convention on International Trade in Endangered Animal and Plant Species (CITES) was signed in Washington, D.C. by 21 Western countries. Today there are 125 member countries. The chief aim of CITES is to prevent illegal international trading in endangered species that are divided into three categories:

- Appendix 1 lists species in which trading is not allowed due to their imminent danger of extinction.
- Appendix 2 lists species in which trading is allowed under rigid scientific control, including those above born in captivity.
- Appendix 3 lists species for which there are no general restrictions on trade but which have endangered populations in certain specific countries.

The convention foresees that each country should pass its own particular legislation to support and enforce the treaty's final provisions. This means that protection can vary from one member country to another. As with any law enforcement, it is not easy. Illegal trade continues to grow because the demand is there. Have you ever encountered illegally traded plants or wildlife? Have you ever asked when purchasing fish for an aquarium or perhaps a pet bird where they came from? Would you purchase a plant, fish, or perhaps jewelry with coral in it if you thought that it was traded illegally?

recognize that each of us is individually responsible for the quality of the environment we live in and that our personal actions affect environmental quality, for better or worse. The recognition of individual responsibility must then lead to changes in individual behavior. In other words, our environmental ethic must begin to express itself not only in national laws, but also in subtle but profound changes in the ways we all live our daily lives.

Various public opinion polls conducted over the past decade have indicated that Americans think environmental problems can often be given a quick technological fix. The Roper polling organization has stated that, "They believe that cars, not drivers, pollute, so business should invent pollution-free autos. Coal utilities, not electricity consumers, pollute, so less environmentally dangerous generation methods should be found." It appears that many individuals want the environment cleaned up, but they do not want to make major lifestyle changes to make that happen.

Decisions and actions by individuals faced with ethical choices collectively determine the hopes and quality of life for everyone. As ecological knowledge and awareness begin to catch up with good intentions, people in all walks of life will need to live by an environmental ethic.

Global Environmental Ethics

In 1990, Noel Brown, the director of the United Nations North American Environmental Programme, stated:

> Suddenly and rather uniquely the world appears to be saying the same thing. We are approaching what I have termed a consensual moment in history, where suddenly from most quarters we get a sense that the world community is now agreeing that the environment has become a matter of global priority and action.

GLOBAL PERSPECTIVE

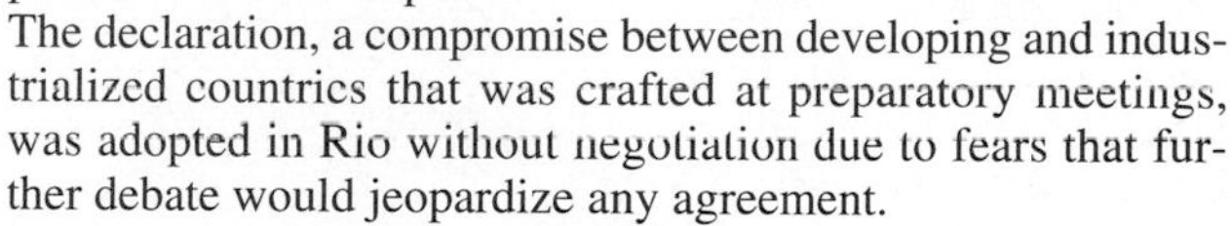

Earth Summit on Environment and Development

In June 1992, representatives from 178 countries, including 115 heads of state, met in Rio de Janeiro, Brazil, at the Earth Summit. Officially, the meeting was titled the United Nations Conference on Environment and Development (UNCED), and it was the largest gathering of world leaders ever held. The first Earth Summit had been held 20 years earlier in Stockholm, Sweden. At that time, the planet was divided into rival East and West blocs and was preoccupied with the perils of the nuclear arms race. With the collapse of the East bloc and the thawing of the cold war, a fundamental shift in the global base of power had occurred.

Today, the more important diversion, especially on environmental issues, is not between East and West but between "North" (Europe, North America, and Japan) and "South" (most of Asia, Africa, and Latin America). And, though the immediate threat of nuclear destruction has lifted, the planet is still at risk.

The idea behind the Earth Summit was that the relaxation of cold war tensions, combined with the growing awareness of ecological crises, offered a rare opportunity to persuade countries to look beyond their national interests and agree to some basic changes in the way they treat the environment. The major issues are clear: The developed countries of the North have grown accustomed to lifestyles that are consuming a disproportionate share of natural resources and generating the bulk of global pollution. Many of the developing countries of the South are consuming irreplaceable global resources to provide for their growing populations.

The Earth Summit was intended to promote better integration of nations' environmental goals with their economic aspirations. Although the hopes of some developing nations for large commitments of new foreign assistance did not fully materialize, much was accomplished during the Summit.

- The *Rio Declaration on Environment and Development* sets out 27 principles to guide the behavior of nations toward more environmentally sustainable patterns of development. The declaration, a compromise between developing and industrialized countries that was crafted at preparatory meetings, was adopted in Rio without negotiation due to fears that further debate would jeopardize any agreement.
- States at UNCED also adopted a voluntary action plan called *Agenda 21,* named because it is intended to provide an agenda for local, national, regional, and global action into the 21st century. UNCED Secretary General Maurice Strong called Agenda 21 "the most comprehensive, the most far-reaching and, if implemented, the most effective program of international action ever sanctioned by the international community." Agenda 21 includes hundreds of pages of recommended actions to address environmental problems and promote sustainable development. It also represents a process of building consensus on a "global work-plan" for the economic, social, and environmental tasks of the United Nations as they evolve over time.
- The third official product of UNCED was a *"non-legally binding authoritative statement of principles for a global consensus on the management, conservation, and sustainable development of all types of forests."* Negotiations on the forest statement, begun as negotiations for a legally binding convention on forests, were among the most difficult of the UNCED process. Many states and experts, dissatisfied with the end result, came away from UNCED seeking further negotiations toward agreement on a framework convention on forests.

This new sense of urgency and common cause about the environment is leading to unprecedented cooperation in some areas. Despite their political differences, Arab, Israeli, Russian, and American environmental professionals have been working together for several years. Ecological degradation in any nation almost inevitably impinges on the quality of life in others. For years, acid rain has been a major irritant in relations between the United States and Canada. Drought in Africa and deforestation in Haiti have resulted in waves of refugees. From the Nile to the Rio Grande, conflicts flare over water rights. The growing megacities of the Third World are time bombs of civil unrest.

Much of the current environmental crisis is rooted in and exacerbated by the widening gap between rich and poor nations. Industrialized countries contain only 20 percent of the world's population, yet they control 80 percent of the world's goods and create most of its pollution. The developing countries are hardest hit by overpopulation, malnutrition, and disease. As these nations struggle to catch up with the developed world and improve the quality of life for their people, a vicious circle begins: Their efforts at rapid industrialization poison their cities, while their attempts to boost agricultural production often result in the destruction of their forests and the depletion of their soils, which lead to greater poverty. (See figure 2.5.)

Perhaps one of the most important questions for the future is, "Will the nations of the world be able to set aside their political differences to work toward a global environmental course of action?" The United Nations Conference on Human Environment held in Stockholm, Sweden, in 1972 was a step in the right direction. Out of that international conference was born the U.N. Environment Programme, a separate department of the United Nations that deals with environmental issues. A second world environmental conference was held in 1992 in Brazil. It followed up the Stockholm conference with many new international initiatives. A major world conference on climate change was held in Kyoto, Japan, in 1997. (See Global Perspective: Earth Summit, and Global Perspective: The Kyoto Protocol.) Through organizations and conferences such as these, nations can work together to solve common environmental problems.

figure 2.5 **Lifestyle and Environmental Impact** Significant differences in lifestyles and their environmental impact exist between the rich and poor nations of the world. What would be the environmental impact on the Earth if the citizens of China and India and other less developed countries enjoyed the standard of living of North Americans? Can we deny them that opportunity?

Antarctica—Resource or Refuge?

ISSUES & ANALYSIS

Few places on earth have not been exploited by humans. One such place is Antarctica. It is as close to an unpolluted environment as there is on earth, but it is not without its problems.

Seals and whales were the earliest exploited resource in Antarctica. There was money to be made, and this "opportunity" resulted in the near extinction of the southern fur seal, the elephant seal, and the blue whale.

By the 1950s, aboveground nuclear testing had spread radioactive particles around the planet, including Antarctica. Pesticides like DDT were turning up in the tissues and blood of certain Antarctic bird and marine mammal species. A growing hole in the ozone layer above the Antarctic continent is caused by the use of chlorofluorocarbons throughout the world. Fossil-fuel combustion contributes to the greenhouse effect, which in turn threatens to melt the ice in Antarctica's Western Peninsula.

During the past several decades, Antarctica has been the site of extensive scientific exploration. Much of this exploration has been economically motivated. For example, government scientists, with the aid of satellites, are advising oil and mineral prospectors. Much of the so-called scientific research is conducted with geopolitical or military objectives in mind.

Antarctica is also being proposed as a tourist attraction. Australia has suggested building a hotel, while Argentina is considering chartering a vessel to transport six hundred tourists from South America seven times a year. Several sites are also being viewed as potential ski resorts. Numbers of tourists to Antarctica are increasing yearly. Over 10,000 tourists visited in 1999. This was a 50 percent increase over 1998. Antarctica is being promoted as a tourist destination for those "in search of a new frontier."

From an ecological perspective, Antarctica is fragile. The thin layer on the surface of the ocean, nourished by the sun, supports the tiny shrimp-like krill, which sustain fish, whales, seals, and penguins. These short, simple food chains are extremely sensitive to environmental insults.

In the mid-1970s, New Zealand proposed designating the continent an Antarctic World Park. This would turn Antarctica into an international wilderness area, a region on earth where we recognize that humanity does not belong.

In 1991, 24 countries signed an agreement to ban mineral and oil exploration in Antarctica for 50 years. The agreement, which was hailed as historic by governments and environmental groups, includes new regulations for wildlife protection, waste disposal, marine pollution, and continued monitoring of the Antarctic, which covers nearly one-tenth of the world's land surface. The signing of the agreement in Madrid, Spain, was the result of two years of negotiations. The protocol protects Antarctica's delicate flora and fauna and sets procedures to assess environmental effects of all human activities on the continent.

- Should we turn a continent into a world park?
- Would humanity be better served by developing the natural resources of Antarctica, such as oil and minerals?
- Should the natural beauty of Antarctica be opened up to tourism so it can be enjoyed by many?
- Is it possible to strike a balance between preservation and development in a fragile ecosystem? Can you give examples?

Source: United Nations Publications.

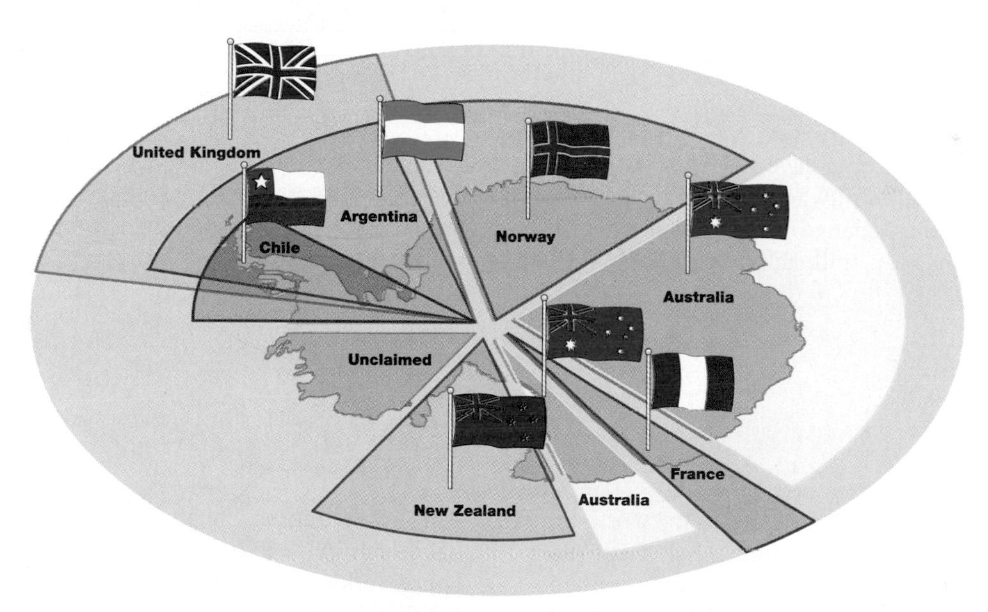

Territorial Claims

No country owns Antarctica, but seven countries have made claims to parts of it. (The United States, which has the largest research population on the continent, has not.) Overlapping claims led to the Antarctic Treaty, signed by 41 countries in 1991, which declares that "Antarctica shall be open to all nations to conduct scientific or other peaceful activities there."

Source: Composite drawing from United Nations' publications

GLOBAL PERSPECTIVE

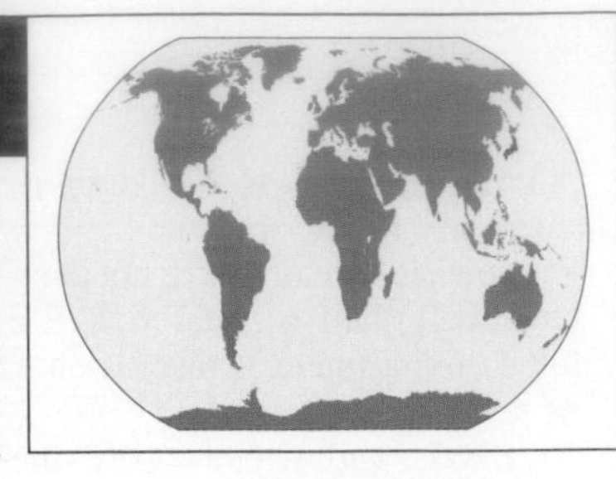

The Kyoto Protocol on Greenhouse Gases

On December 10, 1997, 160 nations reached agreement in Kyoto, Japan, on limiting emissions of carbon dioxide and other greenhouse gases. The Kyoto Protocol was a significant agreement in focusing the attention of world leaders on the issues of climate change. The Kyoto Protocol called for the industrialized nations—the so-called Annex 1 countries—to reduce their average national emissions over the period 2008–2012 to about 5 percent below 1990 levels. The United States pledge is 7 percent below the 1990 level, slightly less than the European Union's pledge and slightly more than Japan's. None of the developing countries was required to set any limits.

The protocol initially covered only three greenhouse gases: carbon dioxide, methane, and nitrous oxide. Three more compounds—hydrofluorocarbons, perfluorocarbons, and sulfur hexafluoride—were to be added in subsequent years. The protocol also contains the elements of a program for international trading of greenhouse gas emissions. Such trading would employ market incentives to help ensure that the lowest-cost opportunities for reductions are pursued.

While the Kyoto climate agreement was important in bringing a new level of international attention to greenhouse gas emissions, there remain many important issues to be resolved, issues pertaining to the agreement. These include:

1. The rules and institutions that are to govern international trading of greenhouse gas emissions among Annex 1 countries must be better established.
2. The criteria used to judge compliance, and any penalties for noncompliance, must be clearly articulated.
3. To make longer-term objectives more credible, moderate but specific near-term goals should be set for Annex 1 countries and these countries should be able to use early emissions reductions to meet longer-term requirements.

One thing is for sure, if the goals of the protocol are to be met by 2008 the developed nations of the world will be forced to focus attention on their energy policies and this in turn will affect all of us.

Key dates in the global warming story

1898—Swedish scientist Svante Ahrrenius warns that Industrial Revolution's carbon dioxide emissions, from coal and oil, could accumulate in atmosphere and lead to global warming.

1961—New observatory atop Hawaii's Mauna Loa volcano detects rise in atmospheric carbon dioxide.

1980s—Computer models of world climate project temperature rises.

1988—U.N. establishes authoritative network of climate scientists, Intergovernmental Panel on Climate Change.

1990—IPCC certifies scientific basis for "greenhouse effect" and global warming predictions.

1992—Climate change treaty signed, setting voluntary goals for industrial nations to lower greenhouse gas emissions to 1990 levels by 2000. Almost 170 nations eventually ratify.

October 1997—Negotiators end two years of preliminary tasks with major issues, including level of binding targets, unresolved and prepare for final conference in Kyoto.

Source: Data from "Cool Facts About Global Warming" USEPA. July 1998, Office of Policy, Planning and Evaluation.

At the individual level, people have begun to respond to increased awareness of global environmental change by altering their values, beliefs, and actions. Changes in individual behavior are necessary but are not enough. As a global species, we are changing the planet. By pooling our knowledge, coordinating our actions, and sharing what the planet has to offer, we can achieve a global environmental ethic.

Summary

People of different cultures view their place in the world from different perspectives. Among the things that shape their views are religious understandings, economic pressures, geographic location, and fundamental knowledge of nature. Because of this diversity of backgrounds, different cultures put different values on the natural world and the individual organisms that compose it.

Three prevailing attitudes toward nature are the development ethic, which assumes that nature is for people to use for their own purposes; the preservation ethic, which assumes that nature has value in itself and should not be disturbed; and the conservation ethic, which recognizes that we will use nature but that it should be used in a sustainable manner.

Ethical issues can be examined at several levels. Growth and exploitation have been the prevailing priorities of our society and individual consumers for generations. This does not mean that everyone in society has the same opinions, but the general attitude has been one of development rather than preservation. Most individual environmental decisions have actually been economic decisions, and the rationale has been: If a resource is available for use, it should be used.

Corporate ethics are even more strongly influenced by economics. Corporations exist to make a profit. Any way that they can reduce costs makes them more profitable. Unfortunately, pollution and exploitation of rare resources may be costly to individuals or society while being profitable to corporations. In addition, corporations wield tremendous economic power and can sway public opinion and political will. Many corporations have begun to openly acknowledge their responsibilities to carefully examine their impact on the natural world.

Society and corporations are composed of individuals. An increasing sensitivity of individual citizens to environmental concerns can change the political and economic climate for society and corporations. However, people often do not have a clear idea of what should be done and often do not act in a way that supports their stated beliefs.

Global environmental concerns have become more important. The world is getting "smaller" and more interrelated. As more people are added to the world's population each year, there is increasing competition for the resources needed to live a decent life. An environmental disaster is no longer a local problem but affects us globally. The increasing economic difference between rich and poor nations affects the global environment, since the poor aspire to have what the rich take for granted. All peoples and nations need to work together to solve environmental problems.

Interactive Exploration

Check out the website at

http://www.mhhe.com/environmentalscience

and click on the cover of this textbook for interactive versions of the following:

KNOW THE BASICS

anthropocentric *21*
biocentric *21*
conservation or management ethic *24*
corporation *26*
development ethic *22*
ecocentrism *22*
economic growth *26*
environmental justice *28*
ethics *21*
industrial ecology *28*
morals *21*
preservation ethic *24*
profitability *26*
resource exploitation *26*

On-line Flashcards

Electronic Glossary

IN THE REAL WORLD

Is it the Titantic all over again? Are ships in danger of the 'ultimate' disaster? Huge ice chunks are breaking off from the Antarctic ice sheet. Why is this happening? What are the threats? See **More Large Icebergs Calve from Antarctica's Ice Shelves** and **Portion of Antarctic Ice Sheet Weakening** for answers to these questions.

How do you view the use of land? How do your classmates and your friends view the use of land? As the book mentions, there are many contrasts in how different people and groups view the use of land. **Record Conservation Land Purchase Saves Forests in Adirondack Park** tells the story of a diverse group that came together for a common vision to purchase nearly 300,000 acres in the northeast.

Elephants, turtles and gorillas to name a few . . . international trade in endangered species is a lucrative business. For an update on how these species are faring, read **Ban on Wild Elephant Products Is Maintained,** **Turtles and Tortoises Disappearing into Soup Pots of China** and **Uganda's Mountain Gorillas Threatened by Tourist Massacre.**

TEST PREPARATION

Review Questions

1. How does personal wealth relate to ethics? Can you provide personal examples?
2. Why do industries pollute?
3. Why would normal economic forces work against pollution control? Do you feel that this is changing?
4. Is it reasonable to expect a totally unpolluted environment? Why or why not?
5. What has been the dominant societal attitude toward resource use?
6. Describe the differences between development, preservation, and conservation ethics. Must there always be conflict among these ethics?
7. What is a major motivating force of corporate management?
8. Why do decision makers view the actions of corporations differently from the way they view the actions of individuals?

Critical Thinking Questions

1. Using the definitions of moral and ethical judgment as presented in the text, identify at least two moral and ethical responses each to the issue of global climate change. What values, beliefs, and perspectives are at the root of these judgments?
2. What are our responsibilities to future generations regarding the environment? What values, beliefs, and perspectives lead you to think and act the way you do with regard to the environment?
3. Compare and contrast the three approaches to environmental ethics outlined in the text. Which is closest to your own? Why? How does, and how could, your ethical stance influence your actions?
4. The text makes the point that up until recently humans have believed, almost universally, in unchecked growth as a positive good. Now, at the beginning of the 21st century, some are beginning to question this belief. What values, beliefs, and perspectives might these critics have? Describe some ways these critics might be received in a developing country. Why?
5. Imagine you are a business executive who wants to pursue an environmental policy for your company that limits pollution and uses fewer raw materials, but would cost more. What might be the discussion at your next board of directors meeting? How would you respond to your board of directors and shareholders? Why?
6. In 1997, Ojibwa Indians in northern Wisconsin sat on railroad tracks to block from crossing their reservation a shipment of sulfuric acid that was headed for a controversial injection copper mine in northern Michigan. Try and put yourself in their position. What values, beliefs, and perspectives might have contributed to this action? Now put yourself in the position of the copper miners in northern Michigan. How might these copper miners have responded? What values, beliefs, and perspectives contribute to their action?
7. Read the Environmental Close-Up about environmental philosophy. Do you feel there is too much talk about environmental problems and not enough action? Or too little talk? Or some other problem? Please describe your position on this and your reasons for thinking the way you do.
8. Imagine yourself in the position of a person who lives on a poor Native American reservation that is contemplating building a storage facility for nuclear waste. What preconceptions, values, beliefs, or contextual perspectives might you bring to the issue? What might you propose as a course of action for yourself and for others? Why?
9. Consider environmental ethics issues in the year 2025. At the rate consumers, corporations, and governments are responding to environmental concerns, what quality-of-life consequences do you project for the year 2025? How will your health, lifestyle, income, employment, and community be affected?

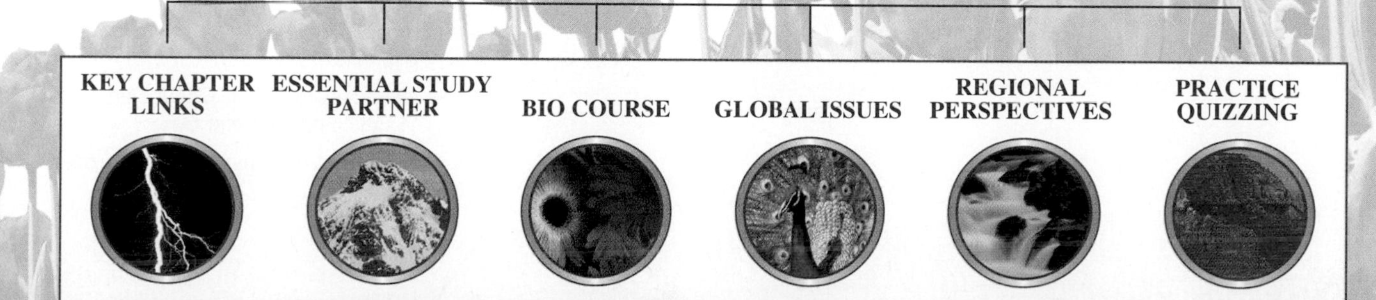

CHAPTER

Risk and Cost: Elements of Decision Making

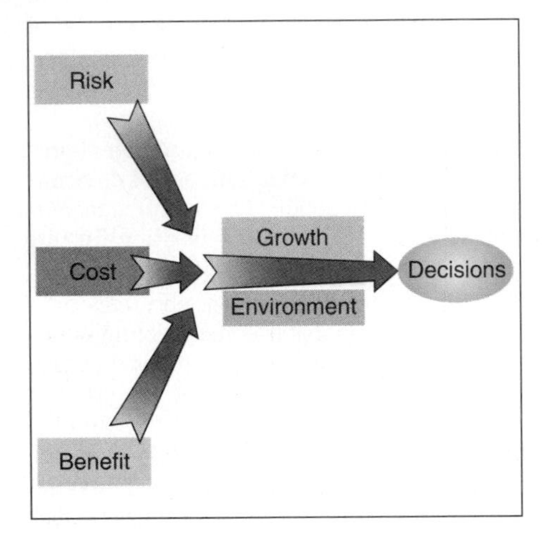

Objectives

After reading this chapter, you should be able to:

- Describe why the analysis of risk has become an important tool in environmental decision making.
- Understand the difference between risk assessment and risk management.
- Describe the issues involved in risk management.
- Understand the difference between true and perceived risks.
- Define what an economic good or service is.
- Understand the relationship between the available supply of a commodity or service and its price.
- Understand how and why cost-benefit analysis is used.
- Understand the concept of sustainable development.
- Understand environmental external costs and the economics of pollution prevention.
- Understand the market approach to curbing pollution.

Chapter Outline

Measuring Risk
Risk Assessment

Environmental Close-Up: *What's in a Number?*

Risk Management
True and Perceived Risks

Economics and the Environment
Economic Concepts
Market-Based Instruments

Global Perspective: *Wombats and the Australian Stock Exchange*

Environmental Close-Up: *Georgia-Pacific Corporation: Recycled Urban Wood—A Case Study in Extended Product Responsibility*

Extended Product Responsibility
Cost-Benefit Analysis
Concerns about the Use of Cost-Benefit Analysis
Economics and Sustainable Development

Environmental Close-Up: *"Green" Advertising Claims—Points to Consider*

External Costs
Common Property Resource Problems

Global Perspective: *Pollution Prevention Pays!*

Economic Decision Making and the Biophysical World

Environmental Close-Up: *Placing a Value on Ecosystem Services*

Economics, Environment, and Developing Nations

Global Perspective: *Costa Rican Forests Yield Tourists and Medicines*

The Tragedy of the Commons
Lightening the Load

Issues & Analysis: *Shrimp, Turtles, and Turtle Excluder Devices*

Measuring Risk

Two factors are primary in many decisions in life: risk and cost. We commonly ask such questions as "How likely is it that someone will be hurt?" and "What is the cost of this course of action?" Environmental decision making is no different. If a new air-pollution regulation is contemplated, industry will be sure to point out that it will cost a considerable amount of money to put these controls in place and will reduce profitability. Citizens will point out that their tax money will have to support another governmental bureaucracy. On the other side, advocates will point out the reduced risk of illnesses and the reduced cost of health care for people who live in areas of heavy air pollution.

Risk analysis has become an important decision-making tool at all levels of society. In the area of environmental concerns, assessing and managing risks help us determine what environmental policies are appropriate. The analysis of risk generally involves a probability statement. **Probability** is a mathematical statement about how likely it is that something will happen. Probability is often stated in terms like "The probability of developing a particular illness is 1 in 10,000," or "The likelihood of winning the lottery is 1 in 5,000,000." It is important to make a distinction between *probability* and *possibility*. When we say something is *possible*, we are just saying that it could occur. It is a very inexact term. *Probability* defines how likely *possible* events are.

Another important consideration is the consequences of an event. If a disease is likely to make 50 percent of the population ill (the probability of becoming ill is 50 percent) but no one dies, that is very different from analyzing the safety of a dam, which if it failed would cause the deaths of thousands of people downstream. We would certainly not accept a 50 percent probability that the dam would fail. Even a 1 percent probability in that case would be unacceptable. The assessment and management of risk involve an understanding of probability and the consequences of decisions. (See figure 3.1.)

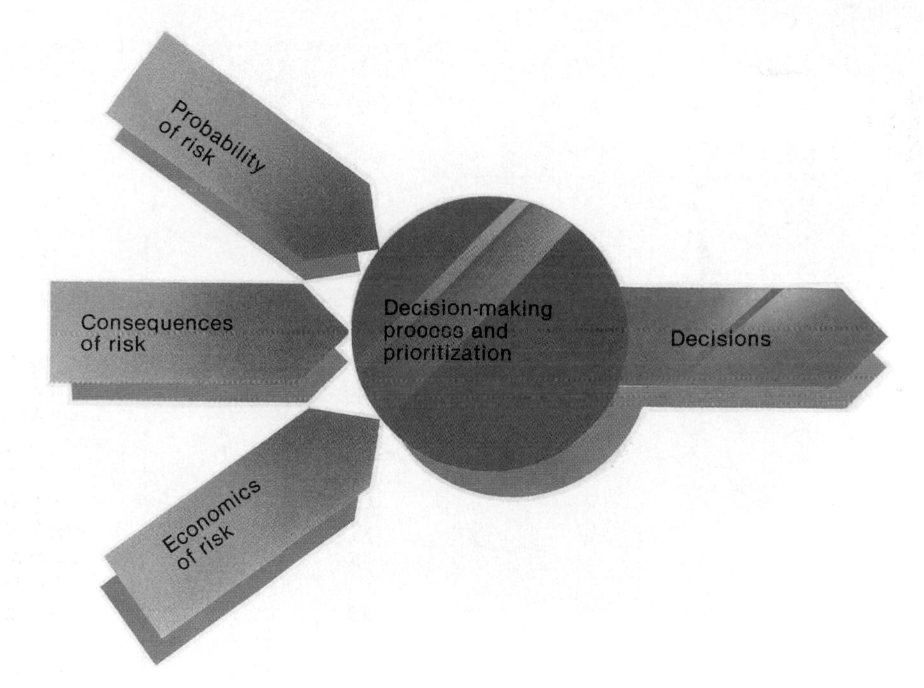

figure 3.1 **Decision-Making Process** The assessment, cost, and consequences of risks are all important to the decision-making process.

Risk assessment involves analyzing a risk to determine the probability of an adverse effect. Risk management is a much broader task that includes assessment and the consequences of risks in the decision-making process. We will look at both these tasks in the next two sections.

Risk Assessment

Environmental **risk assessment** is the use of facts and assumptions to estimate the probability of harm to human health or the environment that may result from exposures to pollutants, toxic agents, or management decisions. What risk assessment provides for environmental decision makers is an orderly, clearly stated, and consistent way to deal with scientific issues when evaluating whether a hazard exists and what the magnitude of the hazard may be.

Calculating the hazardous risk to humans of a particular activity, chemical, or technology is difficult. If a technology is well known, scientists use probabilities based on past experience to estimate risks. For example, the risk of developing black lung disease from coal dust in mines is well established. To predict the risks associated with new technology, much less accurate statistical probabilities, based on models rather than real-life experiences, must be used.

Risks associated with new chemicals are difficult to quantify. While animal tests are widely accepted in predicting whether or not a chemical will cause cancer in humans, their use in predicting how many cancers will be caused in a group of exposed people is still very controversial. Most risk assessments are *estimates* of the probability that a person will develop cancer or other negative effects. (See figure 3.2.)

Such estimates typically are based on broad assumptions to ensure that a lack of complete knowledge does not result in an underestimation of the risk. For example, people may be more or less sensitive to the effects of certain chemicals than the laboratory animals studied. Also, people vary in their sensitivity to cancer-causing compounds. Thus, what may present no risk to one person may be a high risk to others. Persons with breathing difficulties are more likely to be adversely affected by high levels of air pollutants than are healthy individuals. In addition, the estimate of human risk is based on extrapolation from animal tests in which high, chronic

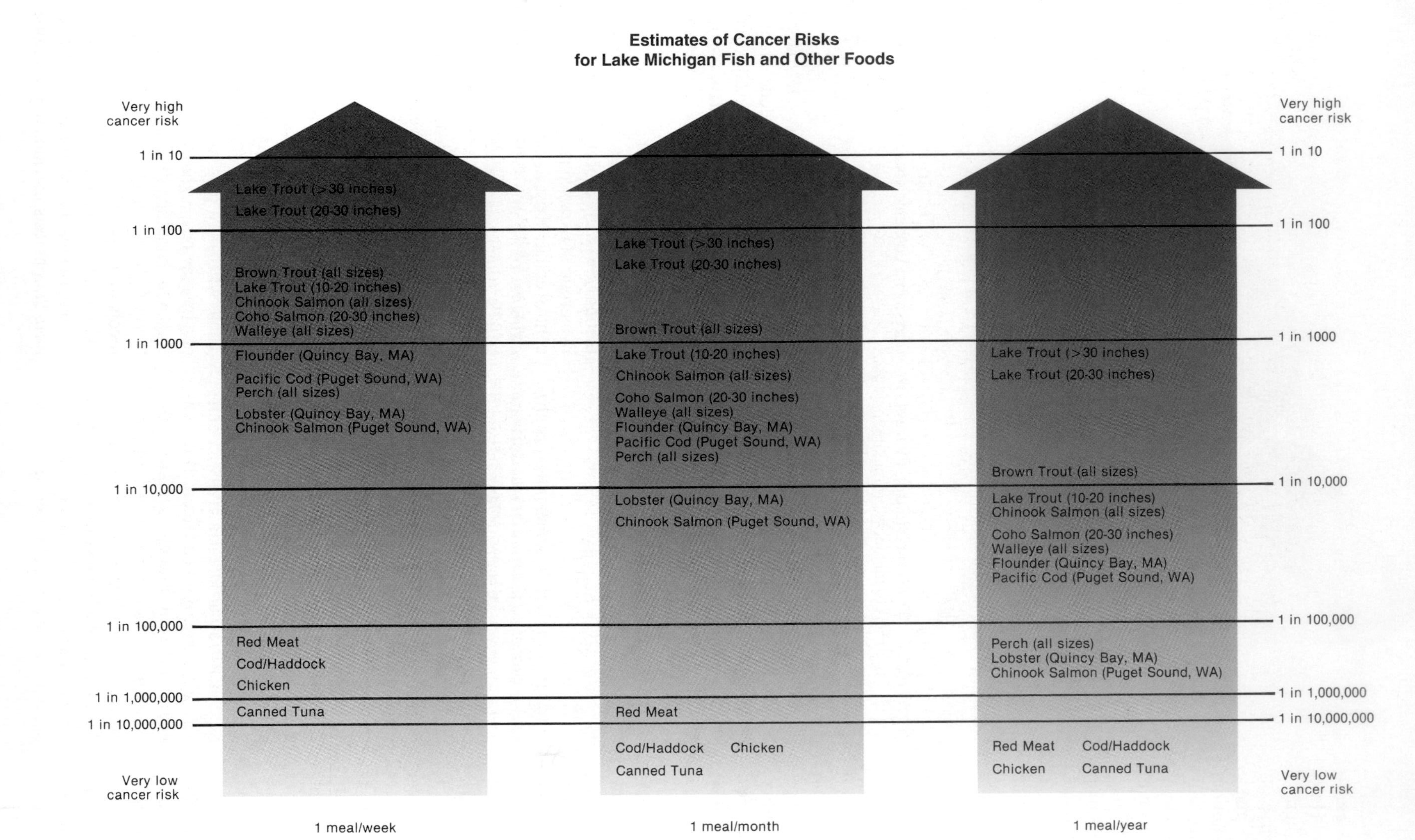

figure 3.2 **Cancer Risk and Fish Consumption** A study by the National Wildlife Federation provides estimates for cancer risks associated with the consumption of sport fish.

ENVIRONMENTAL CLOSE-UP

What's in a Number?

Risk values are often stated, shorthand-fashion, as a number. When the risk concern is cancer, the risk number represents a probability of occurrence of additional cancer cases. For example, such an estimate for Pollutant X might be expressed as 1×10^{-6}, or simply 10^{-6}. This number can also be written as 0.000001, or one in a million—meaning one additional case of cancer projected in a population of one million people exposed to a certain level of Pollutant X over their lifetimes. Similarly, 5×10^{-7}, or 0.0000005, or five in *100 million,* indicates a potential risk of five additional cancer cases in a population of 100 million people exposed to a certain level of the pollutant. These numbers signify incremental cases above the background cancer incidence in the general population. American Cancer Society statistics indicate that the background cancer incidence in the general population is one in three over a lifetime.

If the effect associated with Pollutant X is not cancer but another health effect, perhaps neurotoxicity (nerve damage) or birth defects, then numbers are not typically given as probability of occurrence, but rather as levels of exposure estimated to be without harm. This often takes the form of a reference dose (RfD). An RfD is typically expressed in terms of milligrams (of pollutant) per kilogram of body weight per day, e.g., 0.004 mg/kg/day. Simply described, an RfD is a rough estimate of daily exposure to the human population (including sensitive subgroups) that is likely to be without appreciable risk of deleterious noncancerous effects during a lifetime. The uncertainty in an RfD may be one or several orders of magnitude (i.e., multiples of 10).

What's in a number? The important point to remember is that the numbers by themselves don't tell the whole story. For instance, even though the numbers are identical, a cancer risk value of 10^{-6} for the "average exposed person" (perhaps someone exposed through the food supply) is not the same thing as a cancer risk of 10^{-6} for a "most exposed individual" (perhaps someone exposed from living or working in a highly contaminated area). It's important to know the difference. Omitting the qualifier "average" or "most exposed" incompletely describes the risk and would mean a failure in risk communication.

A numerical estimate is only as good as the data it is based on. Just as important as the *quantitative* aspect of risk characterization (the risk numbers), then, are the *qualitative* aspects. How extensive is the data base supporting the risk assessment? Does it include human epidemiological data as well as experimental data? Does the laboratory data base include less data on more than one species? If multiple species were tested, did they all respond similarly to the test substance? What are the "data gaps," the missing pieces of the puzzle? What are the scientific uncertainties? What science policy decisions were made to address these uncertainties? What working assumptions underlie the risk assessment? What is the overall confidence level in the risk assessment? All of these qualitative considerations are essential to deciding what reliance to place on a number and to characterizing a potential risk.

Source: Data from *EPA Journal.*

doses are used. Human exposure is likely to be lower or infrequent. Because of all these uncertainties, government regulators have decided to err on the side of safety to protect the public health. That approach has been criticized by those who say it carries protection to the extreme, usually at the expense of industry.

Over the past decade, risk assessment has had its largest impact in regulatory practices involving cancer-causing chemicals called **carcinogens.** In the United States, for example, the decisions to continue registration of pesticides, to list substances as hazardous air pollutants under the Clean Air Act, and to regulate water contaminants under the Safe Drinking Water Act depend to a large degree on the risk assessments for the substances in question.

Risk assessment analysis is also being used to help set regulatory priorities and support regulatory action. Those chemicals or technologies that have the highest potential to cause damage to health or the environment receive attention first, while those perceived as having minor impacts receive less immediate attention. Medical waste is perceived as having high risk, and laws have been enacted to minimize the risk, while the risk associated with the use of fertilizer on lawns is considered minimal and is not regulated.

The science supporting environmental regulatory decisions is complex and rapidly evolving. Many of the most important threats to human health and the environment are highly uncertain. Risk assessment quantifies risk and states the uncertainty that surrounds many environmental issues. This can help institutions research and plan in a way that is consistent with scientific and public concern for environmental protection.

Risk Management

Risk management is a decision-making process that involves risk assessment, technological feasibility, economic impacts, public concerns, and legal requirements. Risk management includes:

1. Deciding which risks should be given the highest priority
2. Deciding how much money will be needed to reduce each risk to an acceptable level
3. Deciding where the greatest benefit would be realized by spending limited funds
4. Deciding how much risk is acceptable
5. Deciding how the plan will be enforced and monitored

Risk management raises several issues. With environmental concerns such

Table 3.1 Risks of Death

Deaths per million hours of exposure	
Mountain climbing	40,000
Canoeing	10,000
Cigarette smoking	3000
Swimming	2560
Automobile travel	1200
Hunting	1000
Air travel	500
Being struck by lightning	100
Vaccination	1.5
Living beside a nuclear power plant	0.5

as acid rain, ozone depletion, and hazardous waste, the scientific basis for regulatory decisions is often controversial. As was previously mentioned, hazardous substances can be tested, but only on animals. Are animal tests appropriate for determining impacts on humans? There is not an easy answer to this question. Dealing with global warming, ozone depletion, and acid rain require projecting into the future and estimating the magnitude of future effects. Will the sea level rise? How many lakes will become acidified? How many additional skin cancers will be caused by depletion of the ozone layer? Estimates from equally reputable sources vary widely. Which ones do we believe?

The politics of risk management frequently focus on the adequacy of the scientific evidence. The scientific basis can be thought of as a kind of problem definition. Science determines that some threat or hazard exists, but because scientific facts are open to interpretation, there is controversy. For example, it is a fact that dioxin is a highly toxic material known to cause cancer in laboratory animals. It is also very difficult to prove that human exposure to dioxin has led to the development of cancer, although high exposures have resulted in acne in exposed workers. Acne is a common result of exposure to molecules like dioxin.

This is why problem definition is so important. Defining the problem helps to determine the rest of the policy process (making rules, passing laws, or issuing statements). If a substance poses little or no risk, then policy action is unnecessary. For example, some observers believe chemicals pose many threats that need to be addressed. Others believe chemicals pose little threat; instead, they see scare tactics and government regulations as unnecessary attacks on businesses. Logging forests poses risks of soil erosion and the loss of resident animal species. The timber industry sees these risks as minimal and as a threat to its economic well-being, while many environmentalists consider the risks unacceptable. These and similar disagreements are often serious public-relations problems for both government and business because most of the public has a poor understanding of the risks they accept daily.

True and Perceived Risks

People often overestimate the frequency and seriousness of dramatic, sensational, well-publicized causes of death and underestimate the risks from more familiar causes that claim lives one by one. Risk estimates by "experts" and by the "public" on many environmental problems differ significantly. This discrepancy and the reasons for it are extremely important because the public generally does not trust experts to make important risk decisions alone.

While public health and environmental risks can be minimized, eliminating all risks is impossible. Almost every daily activity—driving, walking, working—involves some element of risk. (See table 3.1.)

From a risk management standpoint, whether one is dealing with a site-specific situation or a national standard, the deciding question ultimately is: What degree of risk is acceptable? In general, we are not talking about a "zero risk" standard, but rather a concept of **negligible risk:** At what point is there really no significant health or environmental risk? At what point is there an adequate safety margin to protect public health and the environment?

Risk management involves comparing the estimated true risk of harm from a particular technology or product with the risk of harm perceived by the general public. The public generally perceives involuntary risks, such as nuclear power plants or nuclear weapons, as greater than voluntary risks, such as drinking alcohol or smoking. In addition, the public perceives newer technologies, such as genetic engineering or toxic-waste incinerators, as greater risks than more familiar technologies, such as automobiles and dams. Many people are afraid of flying for fear of crashing; however, automobile accidents account for a far greater number of deaths—about 45,000 in the United States each year, compared to less than a thousand annually from plane crashes. (See figure 3.3.)

A fundamental problem facing governments today is how to satisfy people who are concerned about a problem that experts state presents less hazard than another less visible problem, especially when economic resources to deal with the problems are limited. Debate continues over the health concerns raised by asbestos, dioxin and radon.

Some researchers argue that the public is frequently misled by the politics of public health and environmental safety. This is understandable since many prominent people become involved in such issues and use their public image to encourage people to look at issues from a particular point of view.

Whatever the issue, it is hard to ignore the will of the people, particularly when sentiments are firmly held and not easily changed. A fundamental issue surfaces concerning the proper role of a democratic government and other organizations in a democracy when it comes

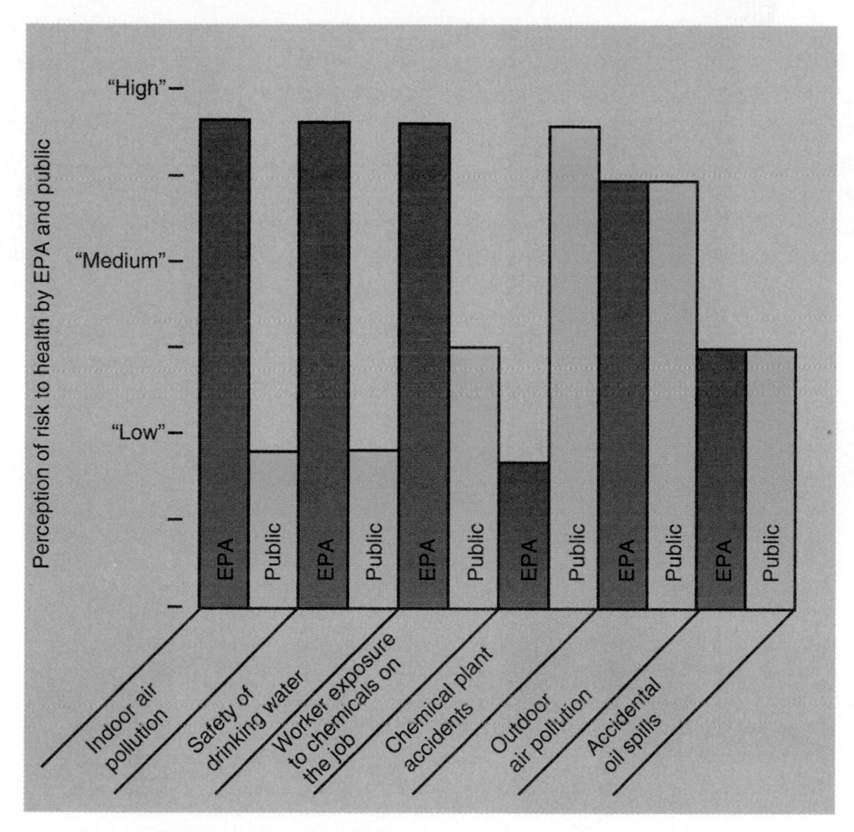

figure 3.3 **Perception of Risk** Professional regulators and the public do not always agree on what risks are.

to matters of risk. Should the government focus available resources and technology where they can have the greatest tangible impact on human and ecological well-being, or should it focus them on problems about which the public is most upset? What is the proper balance? For example, adequate prenatal health care for all pregnant women would have a greater effect on the health of children than would removing asbestos from all school buildings.

Obviously, there are no clear answers to these questions. However, experts and the public are both beginning to realize that they each have something to offer concerning how we view risk. Many risk experts who have been accustomed to looking at numbers and probabilities are now conceding that there is rationale for looking at risk in broader terms. At the same time, the public is being supplied with more data to enable them to make more informed judgments.

Throughout this discussion of risk assessment and management we have made numerous references to costs and economics. It is not economically possible to eliminate all risk. As risk is eliminated, the cost of the product or service increases. Many environmental issues are difficult to evaluate from a purely economic point of view, but economics is one of the tools useful to analyzing any environmental problems.

Economics and the Environment

Environmental problems are primarily economic problems. While this may be an overstatement, it often is difficult to separate economics from environmental issues or concerns. Basically, economics deals with resource allocation. It is a description of how we value goods and services. We are willing to pay for things or services we value highly and are unwilling to pay for things we think there is plenty of. For example, we will readily pay for a warm, safe place to live but would be offended if someone suggested that we pay for the air we breathe.

Our goal as a society is to seek long-term economic growth that creates jobs while improving and sustaining the environment. Achieving this goal requires an environmental strategy that repairs past environmental damage; helps us shift from waste management to pollution prevention; and uses valuable resources more efficiently.

In the three decades since the first Earth Day, many nations have made considerable progress in responding to threats to public health and the environment. Yet major challenges remain. We can put in place a set of policies and programs that will establish a new course for the development and use of environmental technologies into the next century. We must, however, broaden our environmental tool kit, replacing those instruments that are no longer effective with a new set of tools designed to meet today's challenges and tomorrow's needs.

Economic Concepts

An economic good or service can be defined as anything that is scarce. Scarcity exists whenever the demand for anything exceeds its supply. We live in a world of general scarcity. **Resources** are anything that contributes to making desired goods and services available for consumption. Resources are limited, relative to the desires of humans to consume. The **supply** is the amount of a good or service available to be purchased. **Demand** is the amount of a product that consumers are willing and able to buy at various prices. In economic terms, supply depends on:

1. The raw materials available to produce a good or service using present technology
2. The amounts of those materials available
3. The costs of extracting, shipping, and processing the raw materials

4. The degree of competition for those materials among users
5. The feasibility and cost of recycling already used material
6. The social and institutional arrangements that might have an impact. (See figure 3.4.)

The relationship between available supply of a commodity or service and its price is known as a **supply/demand curve.** (See figure 3.5.) The price of a product or service reflects the strength of the demand for and the availability of the commodity. When demand exceeds supply, the price rises. Cost increases cause people to seek alternatives or to decide not to use a product or service, which results in lower quantity demanded.

For example, food production depends heavily on petroleum for the energy to plant, harvest, and transport food crops. In addition, petrochemicals are used to make fertilizer and chemical

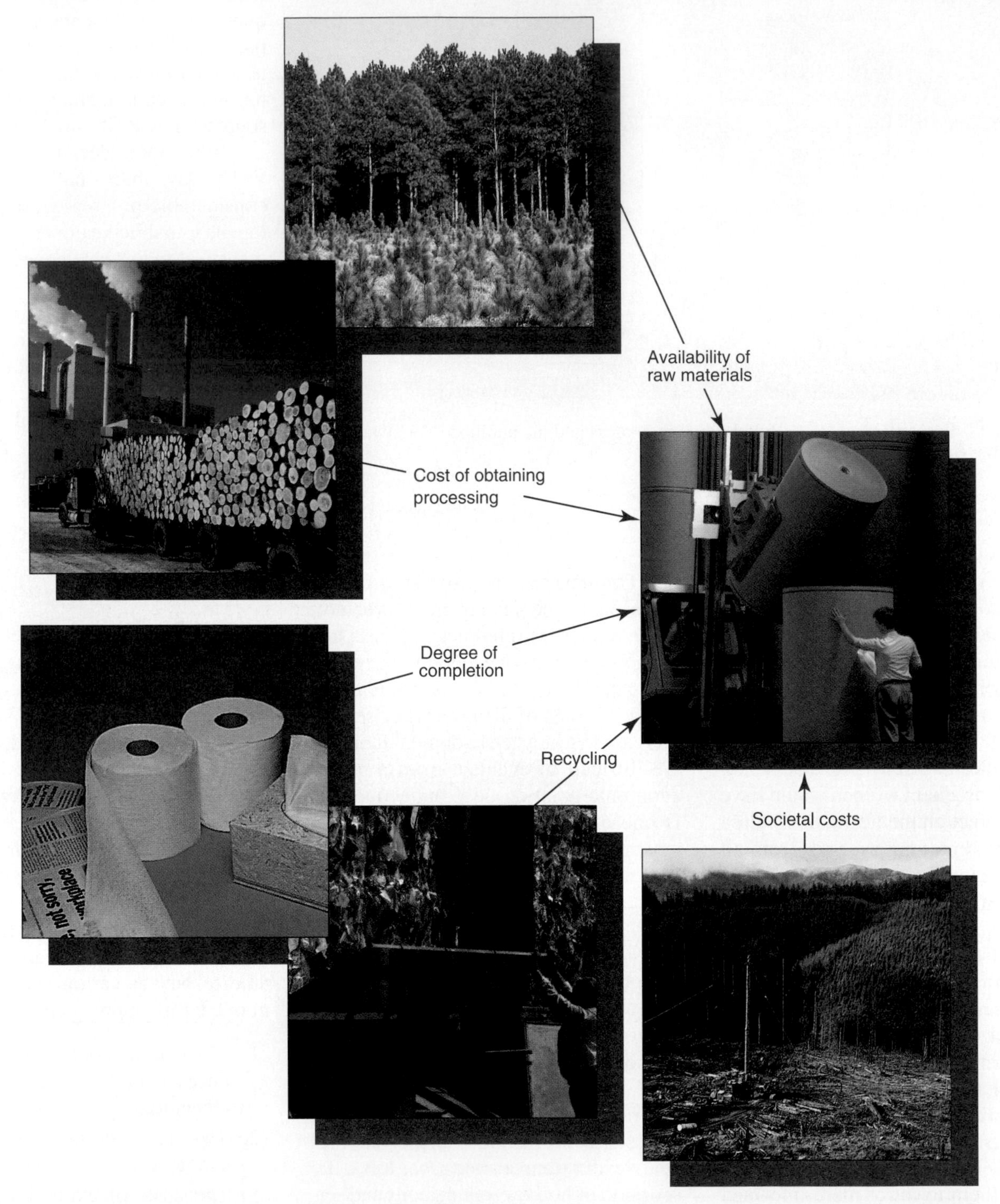

figure 3.4 **Factors that Determine Supply** The supply of a good or service is dependent on the several factors shown here.

pest-control agents. As petroleum prices rise, farmers reduce their petroleum use. Perhaps they farm less land or use less fertilizer or pesticide. Regardless, the price of food must rise as the price of petroleum rises. As the prices of certain foods rise, consumers seek less costly forms of food.

When the supply of a commodity exceeds the demand, producers must lower their prices to get rid of the product, and eventually, some of the producers go out of business. Ironically, this happens to farmers when they have a series of good years. Production is high, prices fall, and some farmers go out of business.

Market-Based Instruments

With the growing interest in environmental protection during the past decade, policy makers are examining new methods to reduce harm to the environment. One area of growing interest is market-based instruments (MBIs). MBIs provide an alternative to the common command-and-control legislation because they use economic forces and the ingenuity of entrepreneurs to achieve a high degree of environmental protection at a low cost. Instead of dictating how industry should conduct its activities, MBIs provide incentives by imposing costs on pollution-causing activities. This approach allows companies to decide for themselves how best to achieve the required level of environmental protection. To date, most of these market-based policies have been implemented in developed nations and in some rapidly growing developing nations. In virtually all cases, they have been introduced as supplements to, not substitutes for, traditional government regulations.

Implementing market-based policies is not easy in any nation. To succeed, nations optimally require open, dynamic market systems, sound macroeconomic conditions, political and institutional stability, and full development of human rights.

In developing nations with large informal sectors, other policies—capacity-building, overall policy reform, community participation, investments in education and health care—initially may be more effective than market-based mechanisms.

Nevertheless, market-based policies may offer valuable opportunities to achieve environmental goals more efficiently and at lower cost to governments and entrepreneurs. They also offer opportunities for governments to ride the momentum of economic growth, while leading it to a more sustainable footing.

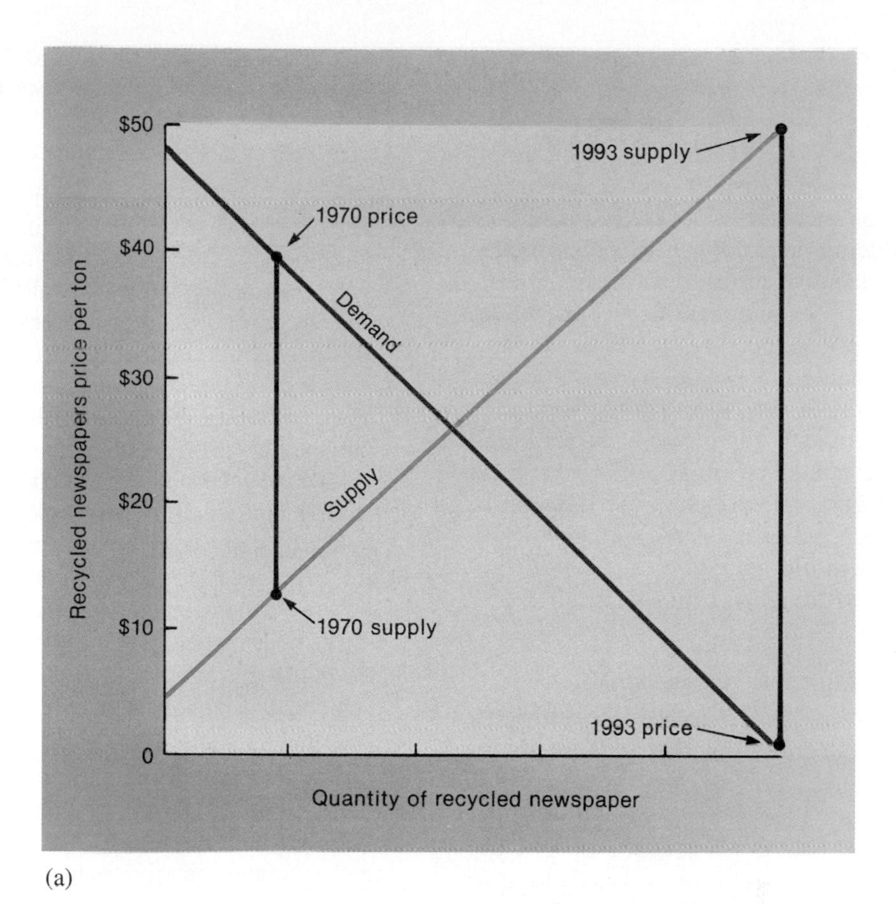

(a)

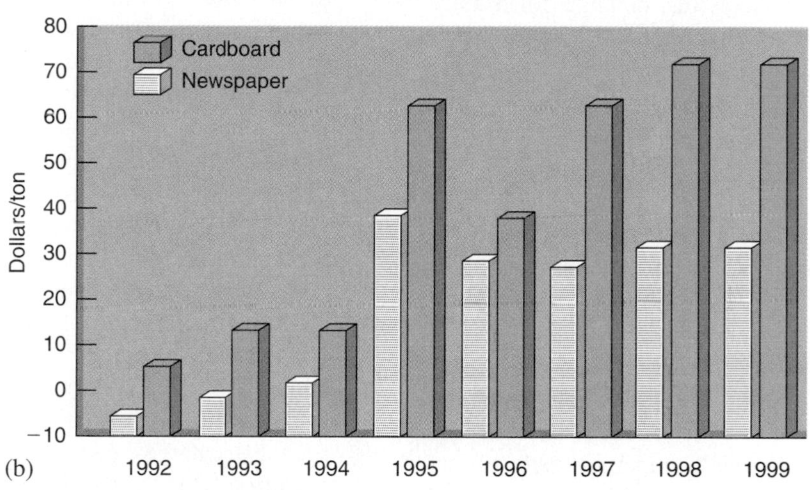

(b)

figure 3.5 **Supply and Demand** *(a)* As a result of increased interest in recycling, the supply of recycled newspapers has increased substantially between 1970 and 1993. The demand changed very little; therefore, the price in 1993 was extremely low. *(b)* Beginning in 1994, the demand for recycled newspaper and cardboard increased, thus the price rose significantly. Prices were more stable through 1999.

Source: *(b)* Data from *Recycling Times.*

Market-based instruments acknowledge the fact that environmental resources are often underpriced. First, subsidies—on water, electricity, fossil fuels, road transport, and agriculture—are an incentive to overuse a resource; reducing or removing subsidies creates an incentive to use resource more efficiently. Second, market prices reflect only private cost, not the external damages caused by pollution or resource extraction. Instead of inflexible, top-down government directives, market-based policies take advantage of price signals and give entrepreneurs the freedom to choose the solution most economically efficient for them.

MBIs can be grouped into five basic categories:

1. **Information programs**

   These programs rely on informed consumers' market choices to reduce environmental problems. Information about the environmental or risk consequences of choices make clear to consumers that it is in their personal interest to change their decisions or behavior. Examples include radon or lead testing or labeling pesticide products. Another type of program, such as the Toxic Release Inventory in the United States, discloses information on environmental releases by polluters. This provides corporations with incentives to improve their environmental performance to enhance their public image.

2. **Tradable emissions permits**

   These permits give companies the right to emit specified quantities of pollutants. Companies that emit less than the specified amounts can sell their permits to other firms or "bank" them for future use. Businesses responsible for pollution have an incentive to internalize the external cost they were previously imposing on society: If they clean up their pollution sources, they can realize a profit by selling their permit to pollute. Once a business recognizes the possibility of selling its permit, it sees that pollution is not costless. The creation of new markets (the permits) to reduce pollution reduces the external aspects to waste disposal and makes them internal costs, just like costs of labor and capital.

3. **Emission fees, taxes, and charges**

   These fees provide incentives for environmental improvement by making environmentally damaging activity or products more expensive. Businesses and individuals reduce their level of pollution wherever it is cheaper to abate the pollution than to pay the charges. Emissions fees can be useful when pollution is coming from many small sources, such as vehicular emissions or agricultural runoff, where direct regulation or trading schemes are impractical. Taxes contribute to government revenue; charges are used to fund environmental cleanup programs.

   In China, a pollution tax system is intended to raise revenue for investment in industrial pollution control, help pay for regulatory activities, and encourage enterprises to comply with emission and effluent standards. The system imposes noncompliance fees on discharges that exceed standards, and fines and other charges assessed on violations of regulations.

   In the Netherlands, a system of effluent charges on industrial wastewater has been viewed as successful. Especially among larger companies, the tax worked as an incentive to reduce pollution. In a survey of 150 larger companies, about two-thirds said the tax was the main factor in their decision to reduce discharges. As the volume of pollution from industrial sources dropped, rates were increased to cover the fixed costs of sewage water treatment plants. Rising rates are providing a further incentive for more companies to start purifying their sewage water.

4. **Performance Bond/Deposit-refund programs**

   These programs place a surcharge on the price of a product, which is refunded when the used product is returned for reuse or recycling.

   Some nations—including Indonesia, Malaysia, and Costa Rica—use performance bonds to ensure that reforestation takes place after timber harvesting.

   The United States also has used this kind of approach to ensure that strip-mined lands are reclaimed. Before a mining permit can be granted, a company must post a performance bond sufficient to cover the cost of reclaiming the site in the event the company does not complete reclamation. The bond is not fully released until all performance standards have been met and full reclamation of the site, including permanent revegatation, is successful—a five-year period in the East and Midwest and 10 years in the arid West. The bond can be partially released as various phases of reclamation are successfully completed.

   Deposit/refund schemes have been widely used to encourage recycling. In Japan, deposits are made for the return of bottles. The deposits are passed on from manufacturers to shops and ultimately consumers, who get the deposit refunded when the used packages and bottles are returned. Under this system, Japan recycled 92 percent of its beer bottles, 50 percent of waste paper, 43 percent of aluminum cans, and 48 percent of glass bottles.

5. **Subsidies**

   Subsidies may include consumer rebates for purchases of environmentally friendly goods, soft loans for businesses planning to implement environmental products, and other monetary incentives designed to reduce the costs of improving environmental performance.

A **subsidy** is a gift from government to a private enterprise that is considered important to the public interest. Agriculture, transportation, space technology, and communication are frequently subsidized by governments. These gifts, whether loans, favorable tax situations, or direct grants, are all paid for by taxes on the public.

GLOBAL PERSPECTIVE

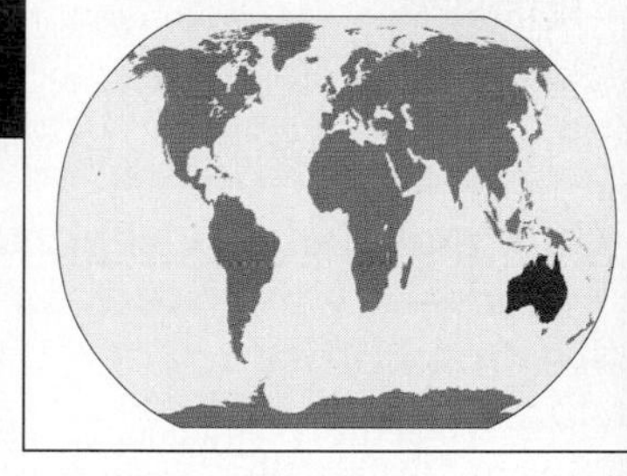

Wombats and the Australian Stock Exchange

The global New Economy can soon claim platypuses, wombats, numbats, and wallabies among its most charismatic recent recruits.

These native Australian mammals, at risk from the steady loss of their habitat, are working assets of Earth Sanctuaries Ltd., which is taking a controversial new approach to environmental activism. In 2000, Earth Sanctuaries was listed on the Australian Stock Exchange, making it the world's first conservation company to go public.

The venture, which makes money mostly from tourists attracted by access to rare animals at three Australian sanctuaries, is considered an extreme example of a trend intriguing environmentalists and investors around the world. Many believe calls to altruism have failed to reverse the rapid loss of species. It's time, they say, to focus on the bottom line.

There is however a growing debate to this new approach. On the one side, researchers see this approach as a beautiful alignment of incentives. They would argue that people care first of all about themselves and their families, and it just has not worked to try to make them choose. Do you save to give your daughter a good education or help rescue a marsupial? They argue that this way you can have both.

Yet some worry about the fund-raising approach. Their argument would be that there would always be a tension between conservation and maximizing profits. If shareholders pressure the company to make higher returns, will the company compromise its values by building megalodges or overstocking the reserves? Some would raise the question, is this approach really conserving natural ecosystems or just creating large zoos?

Is this argument purely idealistic or accurate? Does the market place have a role in habitat preservation? Are there other applications of this approach that you can identify?

Subsidies are costly in two ways: First, the bureaucracy necessary to administer a subsidy costs money, and the subsidy is an indirect way of keeping the market price of a product low. The actual cost is higher because these subsidy costs must be added to the market price to arrive at the product's true cost. Second, in many cases, subsidies encourage activities that in the long term may be detrimental to the environment. For example, the transportation subsidies for highway construction encourage use of inefficient individual automobiles. Higher taxes on automobile use to cover the cost of building and repairing highways would encourage the use of more energy-efficient public transport. Most nations are moving toward reduction of subsidies in many areas, bringing the cost of delivering resources such as water and electricity much closer to market cost. Nevertheless, estimated subsidies for energy, roads, water, and agriculture in developing and transition economies still totaled about $250 billion annually by the late 1990s.

Once established, resource subsidies are understandably difficult to dislodge. But experience suggests that subsidies can be reduced or removed without disrupting rural economic development. In China, subsidy rates for coal declined from an estimated 61 percent in 1985 to 10 percent in 1997. Private mines now account for about half of all production and some 80 percent of the coal is now sold at international prices. These reforms have had numerous benefits. Energy intensity in China has fallen by about 50 percent since 1980, operating losses at state-owned mines dropped from $1.4 billion to $230 million over the 1990–95 period, and the government's total subsidy for fossil fuels fell from about $25 billion in 1990/91 to $10 billion in 1997/98.

Land use subsidies also can create perverse incentives. In France, since the mid-nineteenth century, a tax on undeveloped land had encouraged conversion of environmentally sensitive woodlands and wetlands. Under reforms introduced in 1992, this tax has been reduced and the economic incentive to convert less productive natural areas into productive lands decreased.

While research is still needed to determine how MBIs should be structured and used, there has been sufficient experience to justify their further deployment as a means of attaining environmental objectives at least possible cost. Successful programs so far include a sulfur dioxide trading market, lead phase-down banking and trading, and hundreds of pay-by-the-bag trash collection programs in many countries.

When used correctly, market-based approaches will allow us to reach a level of environmental protection at lower total cost than would be possible with the traditional means of command and control. Market forces tend to drive decisions toward least-cost solutions. Offer the right incentives, and business will develop and adopt better pollution-control technology, rather than stagnating at "commanded" technology.

Market-oriented policies, however, will not always work. It could also be stated that no single one of these economic incentive approaches will be a panacea for all problems. It could be a mistake to start with a policy instrument, then go in search of applications. But this kind of flexible approach, using several solutions, is gaining favor among the regulated and the regulators.

Although the various economic instruments discussed have their own niches, they can be used effectively in combination. For example, trading or pricing approaches work better if supported by information programs: communities that adopted pay-by-the-bag systems of trash disposal had fewer problems if households were given adequate information well in advance. Environmental tax systems can incorporate trading features (e.g., taxes can be levied on net emissions after trades). It will be

ENVIRONMENTAL CLOSE-UP

Georgia-Pacific Corporation: Recycled Urban Wood—A Case Study in Extended Product Responsibility

Georgia-Pacific manufactures particleboard from multi-species wood recovered from commercial disposal or general urban solid waste. The company has agreements with five recycling and processing companies that accept or collect wood at various sites. The wood is cleaned of contaminants and sent to a Georgia-Pacific particleboard manufacturing plant in Martell, California, or to other end users.

The project involves five stakeholder groups: (1) wood waste producers (e.g., operations involved with construction and demolition debris, cut-to-size lumber, commercial wood waste from furniture), (2) collection agents, (3) processors of wood waste, which make the waste into a product that can be reused, (4) transportation contractors, shippers, and haulers, and (5) end users (e.g., Georgia-Pacific's Martell plant). The project has a variety of goals, including increasing the availability of the wood supply for particleboard production, contributing to Georgia-Pacific's goals of product stewardship, and contributing to California's mandated reduction in solid waste (e.g., 50 percent reduction by 2002).

Business factors driving the project include the shortage of fiber for the particleboard plant, rising costs of landfilling, and mandated solid waste reductions. Benefits include an expanded fiber supply in the Northwest United States. Contamination is one of the most significant barriers to the wood recovery program. Often the collected wood is mixed in with metal, plastic, and paper and must be cleared of these contaminants to be usable. The captured paper, plastic, and nonferrous metals are sent to a landfill. Wood by-products that cannot be used in particleboard processing are sold for use as animal bedding, playground cover, soil additive, and lawn or garden mulch. Currently virgin fiber, a by-product from sawmills is often less expensive than recovered fiber for use in particleboard. As wood becomes more scarce, however, the economics will reverse.

increasingly important to use the various MBI methods together. The challenge is to design the most appropriate instruments to deal with environmental problems, bearing in mind the relevant policy objectives: steady progress in reducing risks, cost-effectiveness, encouragement of technological innovation, fairness, and administrative simplicity.

Extended Product Responsibility

Extended product responsibility (EPR) is an emerging principle for a new generation of pollution prevention policies that focus on product systems instead of production facilities. It relies for its implementation on life cycle analysis to identify opportunities to prevent pollution and reduce resource and energy use in each stage of the product chain through changes in product design and process technology. All factors along the product chain share responsibility for the life cycle environmental impacts of products, from the upstream impacts inherent in selection of materials and impacts from the manufacturing process itself to downstream impacts from the use and disposal of the products. (See figure 3.6.)

EPR had its origins in Western Europe. The term *extended product responsibility* is derived from the term *extended producer responsibility,* which is often applied to the German packaging ordinance and similar policies in other Western European countries. The second term is something of a misnomer, however, because most extended responsibility schemes allocate the burden of environmental protection all along the product chain rather than placing it entirely on producers. Under the German packaging ordinance, for instance, consumers, retailers, and packaging manufacturers all share this responsibility, with the financial burden of waste management falling on the last two. In the United States, extended product responsibility has gained greater currency because it highlights systems of shared responsibility.

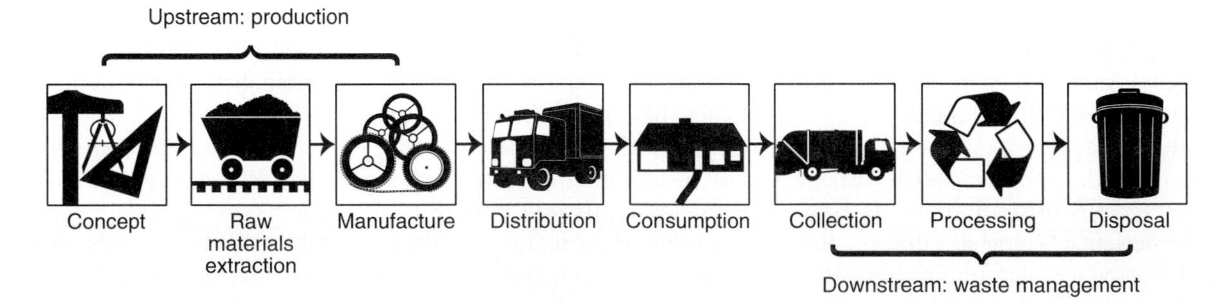

figure 3.6 **The Life Cycle of a Typical Product**

Source: *Environment,* Vol. 39, No. 7, September 1997.

Specific benefits of EPR include:

Cost savings, particularly through the process of producers taking back used products, allow manufacturers to recover valuable materials, reuse them, and save money.

Companies are looking at designing for recycling and disassembly—innovations necessitated by end-of-life management of products, which in many cases helps companies realize how to assemble products more efficiently.

There are more efficient environmental protections, since product-based environmental strategies often are a more cost-effective method of complying with environmental regulations and avoiding environmental liabilities than existing facility-based programs.

Despite these benefits, however, obstacles to EPR exist. These include:

Cost of EPR

Lack of information and tools to assess overall product system impacts

Difficulty in building relationships among actors in different life cycle stages

Hazardous waste regulations that require hazardous waste permits for collection and take-back of certain products

Antitrust laws that make it difficult for companies to cooperate

These are the kinds of questions that further explorations of EPR and, most important, that real-world experience with EPR programs can help to understand and address. Ultimately, EPR is an opportunity to explore new models of environmental policy that are less costly and more flexible.

Cost-Benefit Analysis

People use **cost-benefit analysis** to determine whether a policy generates more social costs than social benefits, and if benefits outweigh costs, how much activity would obtain optimal results. Steps in cost-benefit analysis include:

1. Identification of the project to be evaluated
2. Determination of all impacts, favorable and unfavorable, present and future, on all of society
3. Determination of the value of those impacts, either directly through market values or indirectly through price estimates
4. Calculation of the net benefit, which is the total value of positive impacts less the total value of negative impacts

For example, the cost of reducing the amount of lead in drinking water in the United States to acceptable limits is estimated to be about $125 million a year. The benefits to the nation's health from such a program are estimated at nearly $1 billion per year. Thus under a cost-benefit analysis, the program is economically sound. Recycling of solid waste, which was once cost-prohibitive, is now cost-effective in many communities because the costs of landfills have risen dramatically. Table 3.2 gives examples of the kinds of costs and benefits

Table 3.2 Costs and Benefits of Improving Air Quality

Costs	Benefits
Installation and maintenance of new technology	Reduced deaths and disease
1. Scrubbers on smokestacks	Fewer respiratory problems
2. Automobile emissions control	Reduced plant and animal damage
Redesign of industries and machines	Lower cleaning costs for industry and public
Additional energy costs to industry and public	More clear, sunny days; better visibility
Retraining of employees to use new technology	Less eye irritation
Costs associated with monitoring and enforcement	Fewer odor problems

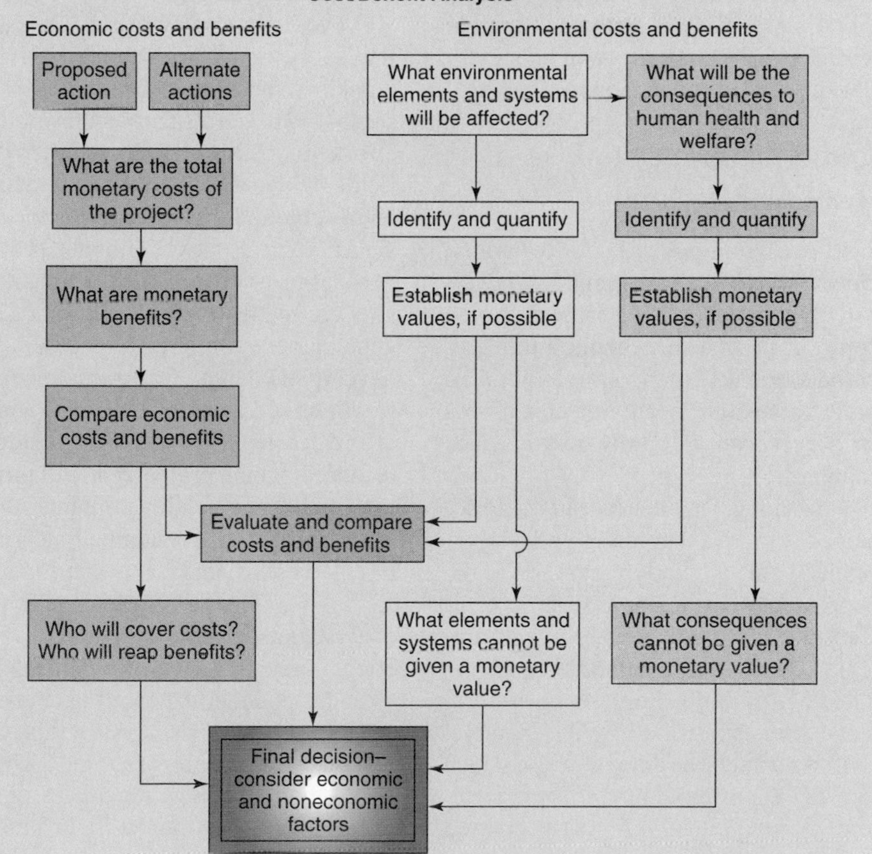

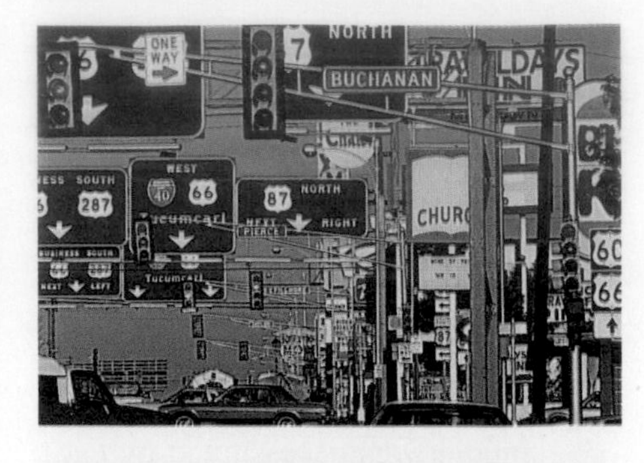

figure 3.7 **Does Everything Have an Economic Value?** The use of water and land is often based on the economic benefits obtained. Anything that humans value creates economic benefits—they just are not all easy to measure.

involved in improving air quality. Although not a complete list, the table indicates the kinds of considerations that go into a cost-benefit analysis. Some of these are easy to measure in monetary terms; others are not.

Concerns about the Use of Cost-Benefit Analysis

Does everything have an economic value? Critics of cost-benefit analysis raise this point, among others. Some people argue that, if economic thinking pervades society, many simple noneconomic values like beauty or cleanliness can survive only if they prove to be "economic." (See figure 3.7.)

It has long been the case in many developed countries that major projects, especially those undertaken by the government, require some form of cost-benefit analysis with respect to environmental impacts and regulations. In the United States, for example, such requirements were established by the National Environmental Policy Act of 1969, which requires environmental impact statements for major government-supported projects. Increasingly, similar analyses are required for projects supported by national and international lending institutions such as the World Bank.

There are clearly benefits to requiring such analysis. Although environmental issues must be considered at some point during project evaluation, efforts to do so are hampered by the difficulty of assigning specific value to environmental resources. In cases of Third World development projects, these already difficult environmental issues are made more difficult by cultural and socioeconomic differences. A less-developed country, for example, may be less inclined to insist on or be able to afford expensive emissions-treatment technology on a project that will provide jobs and economic development.

One particularly compelling critique of cost-benefit analysis is that for analysis to be applied to a specific policy, the analyst must decide which preferences count—that is, which preferences have "standing" in cost-benefit analysis. In theory, cost-benefit analysis should count all benefits and costs associated with the policy under review, regardless of who benefits or bears the costs. In practice, however, this is not always done. For example, if a cost is spread thinly over a great many people, it may not be recognized as a cost at all. The cost of air pollution in many parts of the world could fall into such a category. Debates over how to count benefits and costs for future generations, inanimate objects such as rivers, and nonhumans, such as endangered species, are also common.

Economics and Sustainable Development

The most commonly used definition of the term **sustainable development** is one that originated with the 1987 report, *Our Common Future,* by the World Commission on Environment and Development (known as the Bruntland Commission). By that formulation, sustainable development is "development that meets the needs of the present without compromising the ability of future generations to meet their own needs."

Since the release of the Bruntland Commission report, the phrase has been broadened and modified. The term "sustainable" has gained usage because of increasing concern over exploitation of natural resources and economic devel-

ENVIRONMENTAL CLOSE-UP

"Green" Advertising Claims—Points to Consider

Like many consumers, you may be interested in buying products that are less harmful to the environment. You have probably seen products with such "green" claims as "environmentally safe," "recyclable," "degradable," or "ozone friendly." But what do these claims really mean? How can you tell which products really are less harmful to the environment? Here are some pointers to help you decide.

1. Look for environmental claims that are specific. Read product labels to determine whether they have specific information about the product or its packaging. For example, if the label says "recycled," check how much of the product or packaging is recycled. Labels with "recyclable" claims mean that these products can be collected and made into useful products. This is relevant to you, however, only if this material is collected for recycling in your community or if you can find a way to send the material for recycling.
2. Be wary of overly broad or vague environmental claims. These claims provide little information to help you make purchasing decisions. Labels with unqualified claims that a product is "environmentally friendly," "eco-safe," or "environmentally safe" have little meaning, for two reasons. First, all products have some environmental impact, though some may have less impact than others. Second, these phrases alone do not provide the specific information needed to compare products and packaging on their environmental merits.
3. Some products claim to be "degradable." Degradable materials will not help save landfill space. Biodegradable materials, like food and leaves, break down and decompose into elements found in nature when exposed to air, moisture, and bacteria or other organisms. Photodegradable materials, usually plastics, disintegrate into smaller pieces when exposed to enough sunlight. Either way, however, degradation of any material occurs very slowly in landfills, where most solid waste is sent. That is because modern landfills are designed to minimize the entry of sunlight, air, and moisture. Even organic materials like paper and food may take decades to decompose in a landfill.

Three types of recycling symbols are commonly used in the United States. *(a)* This symbol simply means that the object is potentially recyclable, not that it has been or will be recycled. *(b)* This symbol indicates that a product contains recycled material, but it does not indicate how much recycled material is in the product (it could be only a very small amount). *(c)* This symbol states explicitly the percentage of recycled content found in the product.

opment at the expense of environmental quality. Although disagreement exists as to the precise meaning of the term beyond respect for the quality of life of future generations, most definitions refer to the viability of natural resources and ecosystems over time, and to maintenance of human living standards and economic growth. The popularity of the term stems from the melding of the dual objectives of environmental protection and economic growth. A sustainable agricultural system, for example, can be defined as one that can indefinitely meet the demands for food and fiber at socially acceptable economic costs and environmental impacts.

Finally, as pointed out by Tan Sri Razali, former chairman of the United Nation's Commission on Sustainable Development, the transfer of modern, environmentally sound technology to developing nations is the "key global action to sustainable development."

The past president of the Japan Economic Research Center, Saburo Okita, once stated that a slowdown of economic growth is needed to prevent further deterioration of the environment. Whether or not a slowdown is necessary provokes sharp differences of opinion.

One school of thought argues that economic growth is essential to finance the investments necessary to prevent pollution and to improve the environment by a better allocation of resources. A good school of thought, which is also progrowth, stresses the great potential of science and technology to solve problems and advocates relying on technological advances to solve environmental problems. Neither of these schools of thought sees any need for fundamental changes in the nature and foundation of economic policy. Environmental issues are viewed mainly as a matter of setting priorities in the allocation of resources.

A newer school of economic thought believes that economic and environmental well-being are mutually reinforcing goals that must be pursued simultaneously if either one is to be reached. Economic growth will create its own ruin if it continues to undermine the healthy functioning of Earth's natural systems or to exhaust natural resources. It is also true that healthy economies are most likely to provide the necessary financial investments in environmental protection. For this reason, one of the principal objectives of environmental policy must be to ensure a decent standard of living for all. The solution, at least in the broad scope, would be for a society to manage its economic growth in such a way as to do no irreparable damage to its environment. The term *sustainable development* has been criticized as ambiguous and open to a wide range of interpretations, many of which are contradictory. The confusion arises because "sustainable growth" and "sustainable use" have been used interchangeably, as if their meanings were the same. They are not. "Sustainable growth" is a contradiction in terms: Nothing physical can grow indefinitely. "Sustainable use" is applicable only to renewable resources: it means using them at rates within their capacity for renewal.

By balancing economic requirements with ecological concerns, the needs of the people are satisfied without jeopardizing the prospects of future generations. While this concept may seem to be common sense, the history of the world shows that it has not been a common practice. A major obstacle to sustainable development in many countries is a social structure that gives most of the nation's wealth to a tiny minority of its people. It has been said that a person who is worrying about his next meal is not going to listen to lectures on protecting the environment. What to residents in the Northern Hemisphere seem like some of the worst environmental outrages—cutting rain forests to make charcoal for sale as cooking fuel, for example—are often committed by people who have no other form of income.

The disparities that mark individual countries are also reflected in the planet as a whole. Most of the wealth is concentrated in the Northern Hemisphere. From the Southern Hemisphere's point of view, it is the rich world's growing consumption patterns—big cars, refrigerators, and climate-controlled shopping malls—that are the problem. The problem for the long term is that people in developing countries now want those consumer items that make life in the industrial world so comfortable—and these are the items that are environmentally so costly. If the standard of living in China and India were to rise to that of Germany or the United States, the environmental impact on the planet would be significant.

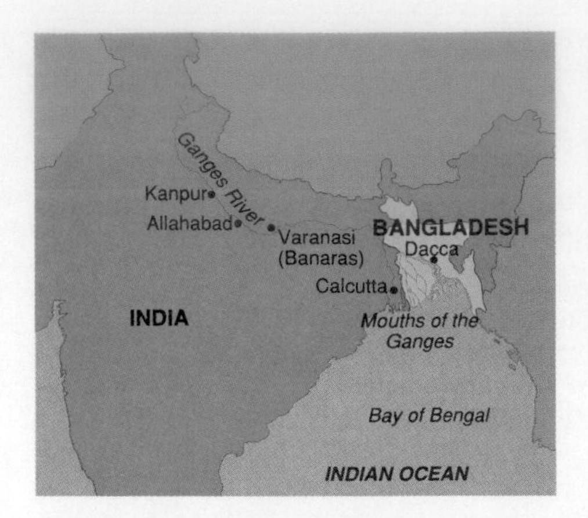

figure 3.8 **Indian Deforestation Causes Floods in Bangladesh** Because the Ganges River drains much of India and the country of Bangladesh is at the mouth of the river, deforestation and poor land use in India can result in devastating floods in Bangladesh.

If sustainable development is to become feasible, it will be necessary to transform our approach to economic policy. Steps in that direction would include changing the definition of gross national product (GNP) to include environmental improvement or decline. The concept of sustainable development may seem simple, but implementing it will be a very complex process.

Historically, rapid exploitation of resources has provided only short-term economic growth, and the environmental consequences in some cases have been incurable. For example, 40 years ago, forests covered 30 percent of Ethiopia. Today, forest covers only 1 percent, and deserts are expanding. One-half of India once was covered by trees; today, only 14 percent of the land is in forests. As the Indian trees and topsoil disappear, the citizens of Bangladesh drown in India's runoff. (See figure 3.8.)

Sustainable development requires choices based on values. Both depend upon information and education, especially regarding the economics of decisions that affect the environment. A. W. Clausen, in his final address as president of the World Bank, noted the

> increasing awareness that environmental precautions are essential for continued economic development over the long run. Conservation, in its broadest sense, is not a luxury for people rich enough to vacation in scenic parks. It is not just a motherhood issue. Rather, the goal of economic growth itself dictates a serious and abiding concern for resource management.

High-income developed nations, such as the United States, Japan, and much of Europe, are in a position to promote sustainable development. They have the resources to invest in research and the technologies to implement research findings. Some believe that the world should not impose environmental protection standards upon poorer nations without also helping them move into the economic mainstream.

Gaylord Nelson, the founder of the first Earth Day, lists five characteristics that define sustainability:

1. *Renewability:* A community must use renewable resources, like water, topsoil, and energy sources, no faster than they can replace themselves. The rate of consumption of renewable resources cannot exceed the rate of regeneration.
2. *Substitution:* Whenever possible, a community should use renewable resources instead of nonrenewable resources. This can be difficult, because there are barriers to substitution. To be sustainable, a community has to make the transition before the nonrenewable resources become prohibitively scarce.

Table 3.3 Economic Solutions to Pollution and Resource Waste

Solution	Internalizes External Costs	Innovation	International Competitiveness	Administrative Costs	Increases Government Revenue
Regulation	Partially	Can encourage	Decreased*	High	No
Subsidies	No	Can encourage	Increased	Low	No
Withdrawing harmful subsidies	Yes	Can encourage	Decreased*	Low	Yes
Tradable rights	Yes	Encourages	Decreased*	Low	Yes
Green taxes	Yes	Encourages	Decreased*	Low	Yes
User fees	Yes	Can encourage	Decreased*	Low	Yes
Pollution-prevention bonds	Yes	Encourages	Decreased*	Low	No

*Unless more cost-effective and productive technologies are developed.

3. *Interdependence:* A sustainable community recognizes that it is a part of a larger system and that it cannot be sustainable unless the larger system is also sustainable. A sustainable community does not import resources in a way that impoverishes other communities, nor does it export its wastes in a way that pollutes other communities.
4. *Adaptability:* A sustainable community can absorb shocks and can adapt to take advantage of new opportunities. This requires a diversified economy, educated citizens, and a spirit of solidarity. A sustainable community invests in, and uses, research and development.
5. *Institutional commitment:* A sustainable community adopts laws and political processes that mandate sustainability. Its economic system supports sustainable production and consumption. Its educational systems teach people to value and practice sustainable behavior.

External Costs

Many of the important environmental problems facing the world today arise because modern production techniques and consumption patterns transfer waste disposal, pollution, and health costs to society. Such expenses, whether they are measured in monetary terms or in diminished environmental quality, are borne by someone other than the individuals who use a resource. They are referred to as **external costs.** For example, consider an individual who attempts to spend a day of leisure fishing in a lake. Transportation, food, fishing equipment, and bait for the day's activities can be purchased in readily accessible markets. But suppose that upon arrival at the lake, the individual finds a lifeless, polluted body of water. Let us further suppose that a chemical plant situated on the lakeshore is responsible for degrading the water quality on the lake to the point that all or most of the fish are destroyed. In this example, fishers collectively bear the external costs of chemical production in the form of lost recreational opportunities.

Pollution-control costs include pollution-prevention costs and pollution costs. **Pollution-prevention costs** are those incurred either in the private sector or by government to prevent, either entirely or partially, the pollution that would otherwise result from some production or consumption activity. The cost incurred by local government to treat its sewage before dumping it into a river is a pollution-prevention cost; so is the cost incurred by a utility to prevent air pollution by installing new equipment.

Pollution-prevention costs can often be factored into a life cycle analysis. Life cycle analysis can help us understand the full cost, potential, and impact on new products and their associated technologies. As a systems approach, life cycle analysis examines the entire set of environmental consequences of a product, including those that result from its manufacture, use, and disposal. Because the relationships among industrial processes are complex, life cycle analysis requires understanding of material flows, resource reuse, and product substitution. Shifting to an approach that considers all resources, products, and waste as an interdependent system will take time, but governments can facilitate the shift by encouraging the transition to a systems approach.

Pollution costs can be broken down into two categories:

1. The private or public expenditures to avoid pollution damage once pollution has already occurred
2. The increased health costs and loss of the use of public resources because of pollution

The large cost of cleaning up spills, such as that from the *Exxon Valdez* and from the war in the Persian Gulf, is an example of a pollution cost, as is the increased health risk to humans from eating seafood contaminated from the oil. By the middle 1990s there were several new economic solutions being utilized to both internalize external costs and to prevent pollution. (See table 3.3.)

Common Property Resource Problems

Economists have stated that, when everybody shares ownership of a resource, there is a strong tendency to overexploit and misuse that resource. Thus, common public ownership could

GLOBAL PERSPECTIVE

Pollution Prevention Pays!

Extended product responsibility (EPR) is an emerging principle of resource conservation and pollution prevention. EPR advocates using a life cycle perspective to identify pollution prevention and resource conservation opportunities that maximize eco-efficiency. Under this principle, there is assumed responsibility for the environmental impacts of a product throughout its life cycle, including impacts on the selection of materials for the product, impacts from the manufacturer's production process, and downstream impacts from the use, recycling, or disposal of the product.

While this concept is relatively new, successful examples of it are in operation. For example, several years ago the Minnesota Mining and Manufacturing (3M) Company's European chemical plant in Belgium switched from a polluting solvent to a safer but more expensive water-based substance to make the adhesive for its Scotch™ Brand Magic™ Tape. The switch was not made to satisfy any environmental law in Belgium or the 12-nation European Community. 3M managers were complying with company policy to adopt the strictest pollution-control regulations that any of its subsidiaries is subject to—even in countries that have no pollution laws at all.

Part of the policy is founded on corporate public relations, a response to growing customer demand for "green" products and environmentally responsible companies. But as many North American multinationals with similar global environmental policies are discovering, cleaning up waste, whether voluntarily or as required by law, can cut costs dramatically.

Since 1975, 3M's "Pollution Prevention Pays" program—or 3P—has cut the company's air, water, and waste pollution around the world in half and, at the same time, has saved nearly $600 million in the last 18 years on changes in its manufacturing process, including $100 million overseas. Less waste has meant less spending to comply with pollution-control laws. But, in many cases, 3M actually has made money selling wastes it formerly hauled away. And, because of recycling prompted by the 3P program, it has saved money by not having to buy as many raw materials.

AT&T followed a similar path. In 1990, it set voluntary goals for the company's 40 manufacturing and 2,500 nonmanufacturing sites worldwide. According to its latest estimates, AT&T has (1) reduced toxic air emissions, many caused by solvents used in the manufacture of computer circuit boards, by 73 percent; (2) reduced emissions of chlorofluorocarbons—gases blamed for destroying the ozone layer in the Earth's atmosphere—by 76 percent; and (3) reduced manufacturing waste 39 percent.

Xerox Corporation had focused on recycling materials in its global environment efforts. It provides buyers of its copiers with free United Parcel Service pickup of used copier cartridges, which contain metal-alloy parts that otherwise would wind up in landfills. The cartridges and other parts are now cleaned and used to make new ones.

Control and Prevention Technologies—Some Examples

Control/Treatment/Disposal

- Sewage treatment
- Industrial wastewater treatment
- Refuse collection
- Incineration
- Off-site recovery and recycling of wastes
- Landfilling
- Catalytic conversion and oxidation
- Particulate controls
- Flue-gas desulfurization
- Nitrogen oxides control technology
- Volatile organic compound control and destruction
- Contaminated site remediation

Prevention

- Improved process control to use energy and materials more efficiently
- Improved catalysis or reactor design to reduce by-products, increase yield, and save energy in chemical processes
- Alternative processes (e.g., low or no-chlorine pulping)
- In-process material recovery (e.g., vapor recovery, water reuse, and heavy metals recovery)
- Alternatives to chlorofluorocarbons and other organic solvents
- High-efficiency paint and coating application
- Substitutes for heavy metals and other toxic substances
- Cleaner or alternative fuels and renewable energy
- Energy-efficient motors, lighting, heat exchangers, etc.
- Water conservation
- Improved "housekeeping" and maintenance in industry

Source: Data from EPA Journal.

Other recycled Xerox copier parts include power supplies, motors, paper transport systems, printed wiring boards, and metal rollers. In all, 1 million parts per year are remanufactured. The initial design and equipment investment was $10 million. Annual savings total $200 million.

The philosophy of pollution prevention is that pollution should be prevented or reduced at the source whenever feasible. It is increasingly being shown that preventing pollution can cut business costs and thus increase profits. Pollution prevention, then, does make cents!

be better described as effectively having no owner.

For example, common ownership of the air makes it virtually costless for any industry or individual to dispose of wastes by burning them. The air-pollution cost is not reflected in the economics of the polluter but becomes an external cost to society. Common ownership of the ocean makes it inexpensive for cities to use the ocean as a dump for their wastes. (See figure 3.9.)

Similarly, nobody owns the right to harvest whales. If any one country delays in getting its share of the available supply of whales, other countries may beat that country to the supply. Thus, there is a strong incentive to overharvest

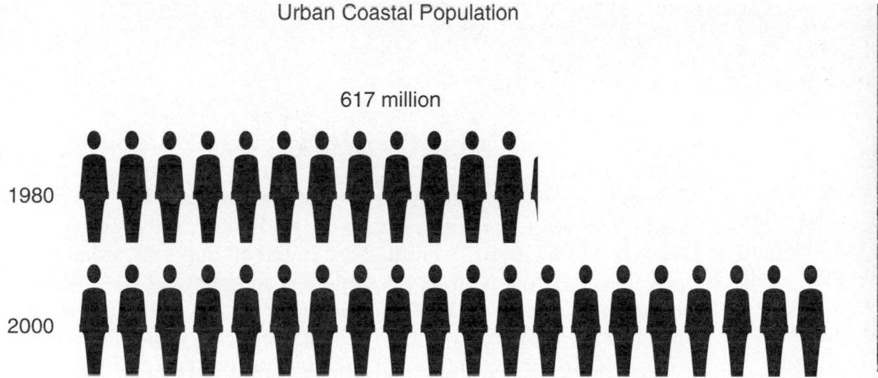

figure 3.9 **Abuse of the Ocean** Since the oceans of the world are a shared resource that nobody owns, there is a tendency to use the resource unwisely. *(a)* Growing population in coastal areas lead to more marine pollution and destruction of coastal habitats. *(b)* Many countries use the ocean as a dump for unwanted trash. Some 6.5 million tons of litter finds its way into the sea each year.

Source: (a) Data from UN Environment Programme; World Resources Institute.

the whale population which creates a consequent threat to the survival of the species. What is true on an international scale for whales is true also for other species within countries. Note that endangered species are wild and undomesticated; the survival of privately owned livestock is not a concern.

Finally, common ownership of land resources, such as parks and streets, is the source of other environmental problems. People who litter in public parks do not generally dump trash on their own property. The lack of enforceable property rights to commonly owned resources explains much of what economist John Kenneth Galbraith has termed "public squalor amid private affluence."

Economic Decision Making and the Biophysical World

For most natural scientists, current crises like biodiversity loss, climate change, and many other environmental problems are symptoms of an imbalance between the socioeconomic system and the natural world. While it is true that humans have always changed the natural world, it is also clear that this imprint is currently much greater than anything experienced in the past. One reason for the profound effect of human activity on the natural world is the fact that there are so many of us.

One of the most serious consequences of the growing human impact on the natural world is the loss of biodiversity. Biological diversity is thought to affect the stability of ecosystems and the ability to cope with crises. In the 570-million-year history of complex life on earth, there have been several major extinction events. The loss of biodiversity after major extinction episodes ranged from 20 percent to more than 90 percent. The loss in biodiversity caused by human activity since the Industrial Revolution alone is somewhere between 10 and 20 percent. If current trends continue, losses are likely to reach 50 percent by the end of the next century. After each extinction event, it took between 20 million and 100 million years for biodiversity to recover to previous levels, a length of time between 100 and 500 times longer than the 200,000-year history of *Homo sapiens.*

The greatest single cause of the loss of biodiversity is habitat destruction, that is, the destruction of the web of organisms and functions that support individual species. The recognition that individual species are supported by others within the ecosystem is foreign to the way markets view the world.

The example of biodiversity illustrates the conflicting frameworks of economics and ecology. Market decisions fail to account for the context of a species or the interconnections between resource quality and ecosystem functions. For example, the value of land used for beef production is measured according to its contribution to output. Yet long before output and the use value of land decreases, the diversity of grass varieties, microorganisms in the soil, or groundwater quality may be affected by intensive beef production. As long as yields are maintained, these changes go unnoticed by markets and are unimportant to land-use decisions.

Another obvious difference between economics and ecology is the relevant time frame in markets and ecosystems. The biophysical world operates in tens of thousands and even millions of years. The time frame for market decisions is short. Particularly where economic policy is concerned, two- to four-year election cycles are the frame of reference; for investors and dividend earners, performance time frames of three months to one year are the rule.

Space or place is another issue. For ecosystems, place is critical. Take groundwater as an example. Soil quality, hydrogeological conditions, regional precipitation rates, plants that live in the region, and losses from evaporation, transpiration, and groundwater flow all contribute to the size and location of groundwater reservoirs. These capacities are not simply transferable from one location to another. For economic activities, place is increasingly irrelevant. Topography, location and function within a bioregion, or local ecological features do not enter into economic calculations except as simple functions of

ENVIRONMENTAL CLOSE-UP

Placing a Value on Ecosystem Services

There are many services provided by functioning ecosystems that are taken for granted. Protection of watersheds by forested land has long been known to be of great value. New York City found that it could provide water to its residents less expensively by protecting the watershed from which the water comes rather than by building expensive water purification plants to clean water from local rivers. Ducks Unlimited, an organization that supports waterfowl hunting, uses money provided by its members to protect nesting habitat for ducks and geese. Many countries have planted trees to help remove carbon dioxide from the air. All of these services can be converted into monetary terms, since it takes money to purify water, purchase land, and purchase and plant trees.

Since choices between competing uses for ecosystems often are determined by financial values assignable to ecosystems, it is important to have some kind of idea about the value of the "free" services provided by functioning ecosystems. Many environmental thinkers have begun to try to put a value on the many services provided by intact, functioning ecosystems. Obviously this is not an easy task and many will belittle these initial attempts to put monetary values on ecosystem services, but it is an important first step in forcing people to consider the importance of ecosystem services when making economic decisions about how ecosystems should be used. The following table represents approximate values for ecosystems services assigned by a panel of experts including ecologists, geographers, and economists. The total of $33 trillion per year is an estimate which many consider to be low. The current world GNP is about $18 trillion per year. Therefore the "free" services of ecosystems must not be overlooked when decisions are made about land use and how natural resources should be managed.

Categories of Services	Examples of Services	Estimated Yearly Value (trillion 1994 U.S. dollars)
Soil formation	Weathering, organic matter	17.1
Recreation	Outdoor recreation, eco-tourism, sport fishing	3.0
Nutrient cycling and waste treatment	Nitrogen, phosphorus, organic matter	2.4
Water services	Irrigation, industry, transportation, watersheds, reservoirs, aquifers	2.3
Climate regulation	Greenhouse gas regulation	1.8
Refuges	Nursery areas for animals, stopping places during migration, overwintering areas	1.4
Disturbance regulation and erosion control	Protection from storms and floods, drought recovery	1.2
Food and raw materials	Hunting and gathering, fishing, lumber, fuel wood, food for animals	0.8
Genetic resources	Medicines, reservoir of genes for domesticated plants and animals	0.8
Atmospheric gas balance	Carbon dioxide, oxygen, ozone, sulfur oxides	0.7
Pollination and pest control	Increased production of fruits	0.5
Total		33*

*Disparity is due to rounding.

Source: Data from R. Costanza, et al. "The Value of the World's Ecosystem Services and Natural Capital," in *Nature* Vol. 387 (1997).

transportation costs or comparative advantage. Production is transferable, and the preferred location is anywhere production costs are the lowest.

Another difference between economics and ecology is that they are measured in different units. The unifying measure of market economics is money. Progress is measured in monetary units that everyone uses and understands to some degree. Biophysical systems are measured in physical units such as calories of energy, CO_2 absorption, centimeters of rainfall, or parts per million of nitrate contamination. Focusing only on the economic value of resources while ignoring biophysical health may mask serious changes in environmental quality or function.

Economics, Environment, and Developing Nations

As previously mentioned, the earth's "natural capital," on which humankind depends for food, security, medicines, and for many industrialized products, is its biological diversity. The majority of this diversity is in the developing world and much of it is threatened by exploitation and development. In order to pay for development projects, many economically poorer nations are forced to borrow money from banks in the developed world.

So great is the burden of external debt that many developing nations see little option but to overexploit their natural resource base. By 2000, the debt in

GLOBAL PERSPECTIVE

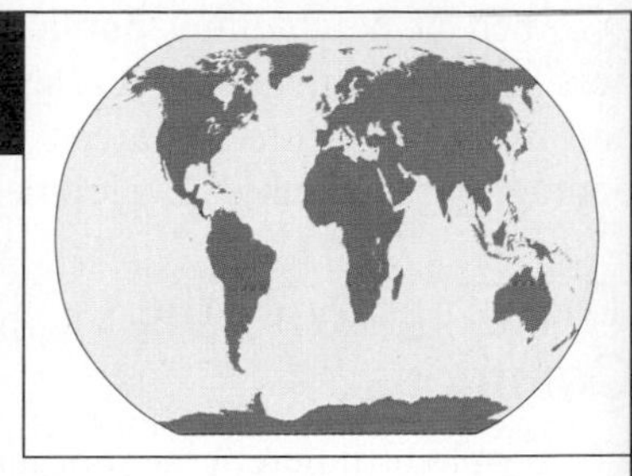

Costa Rican Forests Yield Tourists and Medicines

The Central American republic of Costa Rica has been very successful in protecting a significant amount of its remaining forests. It has a rich variety of different kinds of tropical forests and other natural resources. Mangrove swamps, cloud forests, rainforests, dry tropical forests, volcanoes, and beaches on both the Pacific Ocean and the Caribbean Sea are all part of the mix found in an area about the size of West Virginia.

Nearly 20 percent of the land in Costa Rica is protected as parks, reserves, and refuges. This is the result of several factors. The government is committed to preserving natural areas. A major part of the parks program involves educating the local people about the values of the parks, including the biological value of the large number of species of plants and animals and the economic value of the parks as a tourist attraction. The job of educating the people is made easier by the fact that 93 percent of the people are literate.

Many jobs in Costa Rica are in the developing ecotourism market. People who wish to visit natural areas require guides, transportation, food, and lodging. The jobs created in these industries encourage local people to preserve their natural resources because their livelihood depends on it. Furthermore, when people are employed, they are less likely to try to convert forested land into farmland.

In addition to using the natural resources of forests for tourists, the government of Costa Rica has an agreement between its Instituto Nacional de Biodiversidad (INBio) and Merck and Company, Inc. to prospect for possible drugs from many tropical plants and animals found in the forests. In return, Merck will pay $1 million to the Costa Rican government. Specially trained local parataxonomists (technicians trained in the identification of plants and animals) are involved in this search, resulting in additional jobs.

- What conditions have contributed to the apparent success of Costa Rica in protecting its forests?
- What are possible sources of failure in other countries?
- Who benefits from ecotourism?

the developing nations had risen to over US $1,840 billion, a figure almost half their collective gross national product. The debt burdens have led what investment there is in many developing countries to projects with safe, short-term returns and programs absolutely necessary for immediate survival. Environmental impacts are often neglected, the view being that severely indebted countries cannot afford to pay attention to environmental costs until other problems are resolved. This strategy suggests that environmental problems can be "corrected" once a country has reached a higher income level, but it ignores the growing realization that environmental impacts frequently cause international problems. Many countries under pressure from their debt crisis feel forced to overexploit their natural resources, rather than manage them sustainably.

One new method of helping manage a nation's debt crises is referred to as debt-for-nature exchange. Debt-for-nature exchanges are an innovative mechanism for addressing the debt issue while encouraging investment in conservation and sustainable development. The exchanges, or swaps, allow debt to be bought at discount but redeemed at a premium, in local currency, for use in conservation and sustainable development projects. Debt-for-nature originated in 1987, when a nonprofit organization, Conservation International, bought $650,000 of Bolivia's foreign debt in exchange for Bolivia's promise to establish a national park. By 1998, at least 16 debtor countries—in the Caribbean, Africa, Eastern Europe, and Latin America—had made similar deals with official and nongovernmental organizations. By 1998, nearly US $125 million of debt around the world had been purchased at a cost of some US $24 million but redeemed for the equivalent of US $68 million. This money was used to establish biosphere reserves and national parks, develop watershed protection programs, build inventories of endangered species, and develop environmental education.

In debt-for-nature exchanges, debtor countries benefit from the reduction of their foreign-currency debt obligations and add to their expenditure invested at home. The conservation investor receives a premium on the investment. This can be used for conservation and to establish sustainable development projects. Creditor banks gain by converting their nonpaying debts. Although they receive only part of their initial loan, some return is better than a total loss.

The primary goal of debt-for-nature exchanges has not been debt reduction but the funding of natural-resource-management investment. The contribution made by exchanges could increase, as in the case of the Dominican Republic, where 10 percent of the country's outstanding foreign commercial debt is to be redeemed by exchanges. Although eliminating the debt crisis alone is no guarantee of investment in environmentally sound projects, instruments like debt-for-nature exchanges can, on a small scale, reduce the mismanagement of natural resources and encourage sustainable development.

Attitudes of banks in the industrialized nations also seem to be changing. For example, the World Bank, which lends money for Third World development projects, has long been criticized by environmental groups for backing large, ecologically unsound programs, such as a cattle-raising project in Botswana that led to overgrazing. During the past few years, however, the World Bank has been factoring environmental concerns into its programs. One product of this new approach is an environmental action plan for Madagascar. The 20-year plan, which has been drawn up jointly with the World Wildlife Fund, is aimed at heightening public awareness of environmental issues, setting up and managing protected areas, and encouraging sustainable development.

The Tragedy of the Commons

The problems inherent in common ownership of resources were outlined by biologist Garrett Hardin in a now classic essay entitled "The Tragedy of the Commons" (1968). The original "commons" were areas of pastureland in England that were provided free by the king to anyone who wished to graze cattle.

There are no problems on the commons as long as the number of animals is small in relation to the size of the pasture. From the point of view of each herder, however, the optimal strategy is to enlarge his or her herd as much as possible: If my animals do not eat the grass, someone else's will. Thus, the size of each herd grows, and the density of stock increases until the commons becomes overgrazed. The result is that everyone eventually loses as the animals die of starvation. The tragedy is that, even though the eventual result should be perfectly clear, no one acts to avert disaster. In a democratic society, there are few remedies to keep the size of herds in line.

The ecosphere is one big commons stocked with air, water, and irreplaceable mineral resources—a "people's pasture," but a pasture with very real limits. Each nation attempts to extract as much from the commons as possible while enough remains to sustain the herd. Thus, the United States and other industrial nations consume far more than their share of the total world resource harvest each year, much of it imported from less-developed nations. The nations of the world compete frantically for all the fish that can be taken from the sea before the fisheries are destroyed. Each nation freely uses the commons to dispose of its wastes, ignoring the dangers inherent in overtaxing the waste-absorbing capacity of rivers, oceans, and the atmosphere.

The tragedy of the commons also operates on an individual level. Most people are aware of air pollution, but they continue to drive their automobiles. Many families claim to need a second or third car. It is not that these people are antisocial; most would be willing to drive smaller or fewer cars if everyone else did, and they could get along with only one small car if public transport were adequate. But people frequently get "locked into" harmful situations, waiting for others to take the first step, and many unwittingly contribute to tragedies of the commons. After all, what harm can be done by the birth of one more child, the careless disposal of one more beer can, or the installation of one more air conditioner?

Lightening the Load

Ship captains pay careful attention to a marking on their vessels called the Plimsoll line. If the water level rises above the Plimsoll line, the boat is too heavy and is in danger of sinking. When the line is submerged, rearranging items on the ship will not help much. The problem is the total weight, which has surpassed the carrying capacity of the ship.

This analogy points out that human activity can reach a scale that the earth's natural systems can no longer support. In 1992, more than 1,600 scientists, including 102 Nobel laureates, underscored this point by collectively signing a "Warning to Humanity." Their warning stated in part that "a new ethic is required, a new attitude towards discharging our responsibility for caring for ourselves and for the earth. . . .This ethic must motivate a great movement, convincing reluctant leaders and reluctant governments and reluctant peoples themselves to effect the needed changes."

Such a new successful global effort to lighten humanity's load on the earth would need to directly address three major driving forces of environmental decline: the inequitable distribution of income, resource consumptive economic growth, and rapid population growth. It would redirect technology and trade to buy time for this great change to occur. Although there is much to say about each of these challenges, some key points bear noting.

Wealth inequality may be the most difficult problem, since it has existed for centuries. The difference today, however, is that the future of rich and poor alike depends on reducing poverty and thereby eliminating this driving force of global environmental decline. In this way, self-interest joins ethics as a motive for redistributing wealth, and raises the chances that it might be done.

Important actions to narrow the income gap must include reducing Third World debt. This was talked about a great deal in the 1980s, but little was accomplished. In addition, the developed nations must focus more foreign aid, trade, and international lending policies directly on improving the living standards of the world's poor.

A key description for reducing the kinds of economic growth that harm the environment is the same as that for making technology and trade more sustainable: internalizing environmental costs. If this is done through the adoption of environmental taxes, such as taxing based on pollution emitted, governments could avoid imposing heavier taxes overall by lowering income taxes accordingly. In addition, establishing better measures of economic accounting is critical. Since the calculations used to produce the gross national product do not account for the destruction or depletion of natural resources, this popular economic measure is extremely misleading. It tells us we are making progress even as our ecological foundations are being diminished. A better guide toward a sustainable path is essential. The United Nations and several governments have been working to develop better accounting methods, and while the progress has been slow, there is growing hope in the heightened awareness that a change is necessary.

As our discussion has shown, the economics of environmental problems is complex and difficult. The single most difficult problem to overcome is the assignment of an appropriate economic value to resources that have not previously been examined from an economic perspective. When air, water, scenery, and wildlife are assigned an economic value, they are looked at from an entirely different point of view.

ISSUES & ANALYSIS

Shrimp, Turtles, and Turtle Excluder Devices

The nets used for trawling for shrimp unfortunately do not catch only shrimp. One historic victim of such nets has been turtles, including the endangered Kemp's Ridley sea turtle as well as threatened loggerhead and endangered green sea turtles. Until the mid-1990s, shrimp boats traveling the South Atlantic and Gulf of Mexico could accidentally catch an estimated 45,000 sea turtles in their nets, of which some 12,000 would drown.

The U.S. National Marine Fisheries Service (NMFS), looking for a technological innovation to stop the accidental netting and killing of the turtles, developed a device to keep turtles out of shrimp nets. The turtle excluder device, or TED, attaches to standard shrimp nets. A TED is a grid of bars with an opening either at the top or bottom. The grid is fitted into the neck of a shrimp trawl. Small animals like shrimp slip through the bars and are caught in the bag end of the trawl. Large animals such as turtles and sharks, when caught at the mouth of the trawl, strike the grid bars and are ejected through the opening. Data compiled by the NMFS show that TEDs can reduce turtle captures in shrimp nets by 97 percent with minimal loss of shrimp.

TEDs were first introduced to the shrimp industry on a voluntary basis in 1982. TEDs, however, were not widely accepted and many fishermen claimed that it was not fair for only U.S. fishermen to use them while their counterparts in Mexico did not. The U.S. fishermen argued that they were operating under an unfair economic policy. The shrimpers who opposed the TEDs also claimed that the devices were expensive and dangerous, and cut down on their catches.

TEDs were redesigned with input from the fishermen and there soon followed legislation replacing the voluntary program with a mandated one. The NMFS ensured that the TED requirements were phased in gradually to minimize the impact on the fishery. By 1998, inspectors from the NMFS uncovered only 98 TED violations in 2724 visits aboard shrimp boats.

In 1993, Mexico also mandated the use of TEDs. By the late 1990s, the use of TEDs had spread to many parts of the world. The NMFS and the U.S. Department of State have worked closely with Mexico and the other shrimp-supplying nations in Latin America to help them develop comparable TED programs. TEDs have also been successfully implemented in Thailand, Malaysia, and the Philippines. After actual demonstration and dissemination of results of experiments with TEDs, shrimp fishermen did not resist their use and were understanding of the necessity for using them.

- Can you think of other examples where the development of a new technology has ensured economic growth without endangering one or more species?
- Is it possible to always have a "technological fix"?

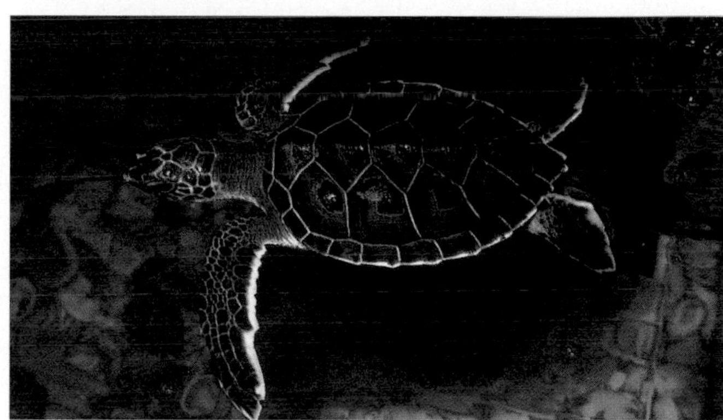

Trawling for shrimp also kills several endangered and threatened species of turtles in the South Atlantic and Gulf of Mexico.

Summary

Risk assessment is the use of facts and assumptions to estimate the probability of harm to human health or the environment that may result from exposures to pollutants, toxic agents, or management decisions. While it is difficult to calculate risks, risk assessment is used in risk management, which analyzes factors in decision making. The politics of risk management focus on the adequacy of scientific evidence, which is often open to divergent interpretations. In assessing risk, people frequently overestimate new and unfamiliar risks, while underestimating familiar ones.

To a large degree, environmental problems can be viewed as economic problems. Economic policies and concepts, such as supply/demand and subsidies, play important roles in environmental decision making. Another important economic tool is cost-benefit analysis. Cost-benefit analysis is concerned with whether a policy generates more social benefits than social costs. Criticism of cost-benefit analysis is based on the question of whether everything has an economic value. It has been argued that if economic thinking dominates society, then even noneconomic values, like beauty, can survive only if a monetary value is assigned to them.

A newer school of economic thought is referred to as sustainable development. Sustainable development has been defined as actions that address the needs of the present without compromising the ability of future generations to meet their own needs. Sustainable development requires choices based on values.

Pollution is extremely costly. When the costs are imposed on society, they are referred to as external costs. Costs of pollution control include pollution-prevention costs and pollution costs. Prevention costs are less costly, especially from a societal perspective.

Economists have stated that when everyone shares ownership of a resource, there is a strong tendency to overexploit and misuse the resource. This concept was developed by Garrett Hardin in "The Tragedy of the Commons."

Recently, a market approach to curbing pollution has been proposed that would assign a value to not polluting, thereby introducing a profit motive to pollution reduction.

Economic concepts are also being applied to the debt-laden developing countries. One such approach is the debt-for-nature swap. This program, which involves transferring loan payments for land that is later turned into parks and wildlife preserves, is gaining popularity.

Interactive Exploration

Check out the website at

http://www.mhhe.com/environmentalscience

and click on the cover of this textbook for interactive versions of the following:

KNOW THE BASICS

carcinogens *41*
cost-benefit analysis *49*
demand *43*
external costs *53*
negligible risk *42*
pollution costs *53*
pollution-prevention costs *53*
probability *39*
resources *43*
risk assessment *39*
risk management *41*
subsidy *46*
supply *43*
supply/demand curve *44*
sustainable development *50*

On-line Flashcards
Electronic Glossary

IN THE REAL WORLD

Is your shrimp dinner causing turtle deaths? Is the use of Turtle Exclusion Devices (TEDs) an example of unfair trade restrictions? Although TEDs are used frequently in shrimp trawling and seem to be a technological solution to the problem of drowning sea turtles, all is not solved in the business. Some trawlers are not using the devices, and internationally there are complaints about the use of TEDs. Gulf Shrimp Trawler Fined for Disabling Turtle Exclusion Device tells the story of a Louisiana shrimp trawler who is obviously not supportive of TEDs. You can also see how the TED works in the accompanying illustration and explanation.

The World Trade Organization overturned U.S. legislation to ban the import of shrimp from countries that refuse to use TEDs. For more information on how the WTO affects U.S. environmental laws, check out "Environmental Concerns in the Battle For Seattle," at World Trade Organization Talks.

Additional information regarding the WTO, economics, and the environment is found in China Slated to Join the WTO.

More news on the international scene includes Sweden who leads the way in changing their approach to economic policy by acknowledging the real costs of environmental degradation. Read Swedish Parliament Plans Budget for Environmental Indicators to see how this country is attempting to reduce external costs.

Are the conveniences of using plastics and pesticides enough to outweigh the possibility of declining sperm counts and increased breast cancer? Is this just a scare tactic? Investigate these questions with Endocrine Disrupters on the Gulf Coast, a case study that discusses the costs and benefits of using pesticides and plastic.

The high demand for plastics and pesticides can lead to potential problems, but there are also problems associated with the high demand for natural products. Is the quest for natural products a modern-day "gold rush" or a Tragedy of the Commons? Check out Matsutake Mushroom Mania for the drama that unfolds when there is high demand for a natural product.

TEST PREPARATION

Review Questions

1. How is risk assessment used in environmental decision making?
2. What is incorporated in a cost-benefit analysis? Develop a cost-benefit analysis for a local issue.
3. What are some of the concerns about the use of cost-benefit analysis in environmental decision making?
4. What concerns are associated with sustainable development?
5. What are some examples of environmental external costs?
6. Define what is meant by pollution-prevention costs.
7. Define the problem in common property resource ownership. Provide some examples.
8. Describe the concept of debt-for-nature.

Critical Thinking Questions

1. If you were a regulatory official, what kind of information would you require in order to make a decision about whether a certain chemical was "safe" or not? What level of risk would you deem acceptable for society? For yourself and your family?
2. Why do you suppose some carcinogenic agents, like those in cigarettes, are so difficult to regulate?
3. Imagine you were assessing the risk of a new chemical plant being built along the Mississippi River in Louisiana. Identify some of the risks that you would want to assess. What kinds of data would you need to assess whether the risk was acceptable, or not? Do you think that some risks are harder to quantify than others? Why?
4. Granting polluting industries or countries the right to buy and sell emissions permits is a controversial idea. Some argue that the market is the best way to limit pollution. Others argue that trade in permits allows polluting industries to continue to pollute and concentrates that pollution. What do you think?
5. Imagine you are an independent economist who is conducting a cost-benefit analysis of a hydroelectric project. What might be the costs of this project? The benefits? How would you quantify the costs of the project? The benefits? What kinds of costs and benefits might be hard to quantify, or might be too tangential to the project to figure into the official estimates?
6. Do you think environmentalists should stretch traditional cost-benefit analysis to include how development impacts the environment or shouldn't they? What are the benefits to this? The risks?
7. Looking at your own life, what kinds of risks do you take? What kinds would you be unwilling to take? What criteria do you use to make a decision about acceptable and unacceptable risk?
8. Is current worldwide growth and development sustainable? If there were less growth, what would be the effect on developing countries? How could we achieve a just distribution of resources and still limit growth?
9. Should our policies reflect an interest in preserving resources for future generations? If so, what level of resources should be preserved? What would you be willing to do without in order to save for the future?

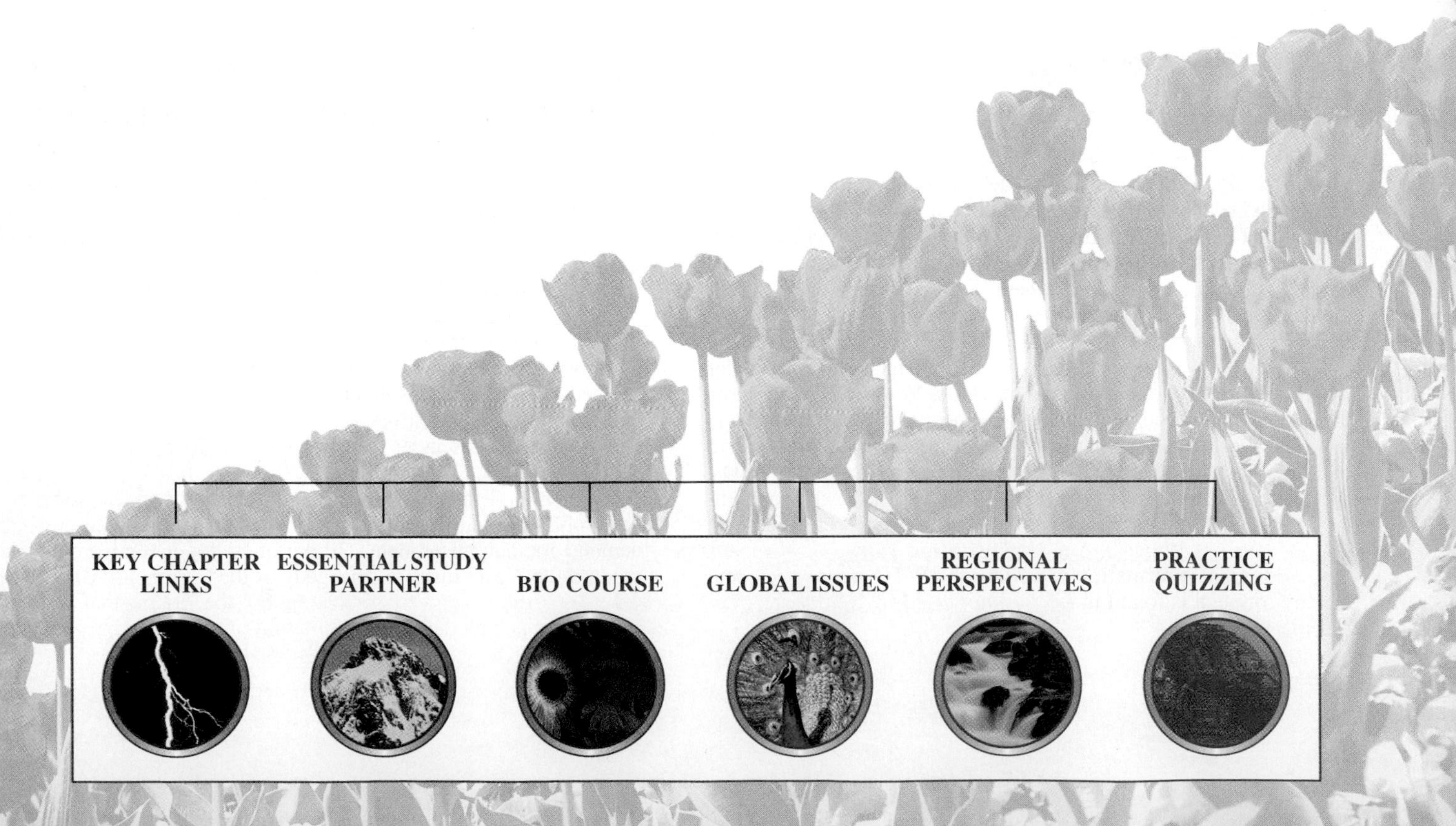

PART TWO

Ecological Principles and Their Application

The science of ecology deals with understanding and describing interrelationships among organisms and between organisms and their surroundings. Part Two discusses basic ecological principles, which are an important foundation for understanding the broader social, political, and economic contexts of environmental science. **Chapter 4** introduces concepts of matter and energy and how changes in the nature of matter and in the use of energy affect living things and human societies. **Chapter 5** explores various ways in which organisms interact with one another and how these interactions are related to the flow of nutrients through food chains. The energy relationships among organisms are also presented. **Chapter 6** discusses the interrelationships among organisms, their physical surroundings, and climate by describing various kinds of ecosystems and the ways they change. **Chapters 7 and 8** explore the dynamics of population biology by describing the primary factors that cause populations to change. Chapter 8 pays particular attention to human population issues.

CHAPTER 4

Interrelated Scientific Principles: Matter, Energy, and Environment

Matter and Energy Interactions
Nature of Matter
Nature of Energy
Human Use of Matter and Energy
Environmental Consequences

Objectives

After reading this chapter, you should be able to:

- Understand that science is usually reliable because information is gathered in a manner that requires impartial evaluation and continuous revision.
- Understand that matter is made up of atoms that have a specific subatomic structure of protons, neutrons, and electrons.
- Recognize that atoms of different elements have different atomic structures and that isotopes of the same element may differ in the number of neutrons present.
- Recognize that atoms may be combined and held together by chemical bonds to produce molecules.
- Understand that rearranging chemical bonds results in chemical reactions and that these reactions are associated with energy changes.
- Recognize that matter may be solid, liquid, or gas, depending on the amount of kinetic energy contained in the molecules.
- Realize that energy can be neither created nor destroyed, but when energy is converted from one form to another, some energy is converted into a less useful form.
- Understand that energy can be of different qualities.

Chapter Outline

Scientific Thinking

Environmental Close-Up: *Typical Household Chemicals*

Limitations of Science

The Structure of Matter
- Atomic Structure
- Molecules and Mixtures
- Acids, Bases, and pH
- Inorganic and Organic Matter
- Chemical Reactions
- Chemical Reactions in Living Things

Energy Principles
- Kinds of Energy
- States of Matter
- First and Second Laws of Thermodynamics
- Environmental Implications of Energy Flow

Issues & Analysis: *Improvements in Lighting Efficiency*

Scientific Thinking

Since environmental science involves the analysis of data, it is useful to understand how scientists gather and evaluate information. It is also important to understand some chemical and physical principles as a background for evaluating environmental issues. An understanding of these scientific principles will also help you appreciate the ecological concepts in the chapters that follow.

The word *science* creates a variety of images in the mind. Some people feel that it is a powerful word and are threatened by it. Others are baffled by scientific topics and have developed an unrealistic belief that scientists are brilliant individuals who can solve any problem. For example, there are those who believe that the conservation of fossil fuels is unnecessary because scientists will soon "find" a replacement energy source. Similarly, many are convinced that if government really "wanted to" it would allocate sufficient funds to allow scientists to find a cure for AIDS. Such images do not accurately portray what science is really like. **Science** is a body of knowledge characterized by the requirement that information be gathered and evaluated by impartial testing of hypotheses and that it be shared so that it can be evaluated by others. It should also be remembered that science is based on a collection of information. The **scientific method** of gathering information generally involves observation, asking questions about the observations, forming hypotheses, testing hypotheses, critically evaluating the results, and publishing information so that others can evaluate the process and its conclusions. (See figure 4.1.) Underlying all of these activities is constant attention to accuracy and freedom from bias.

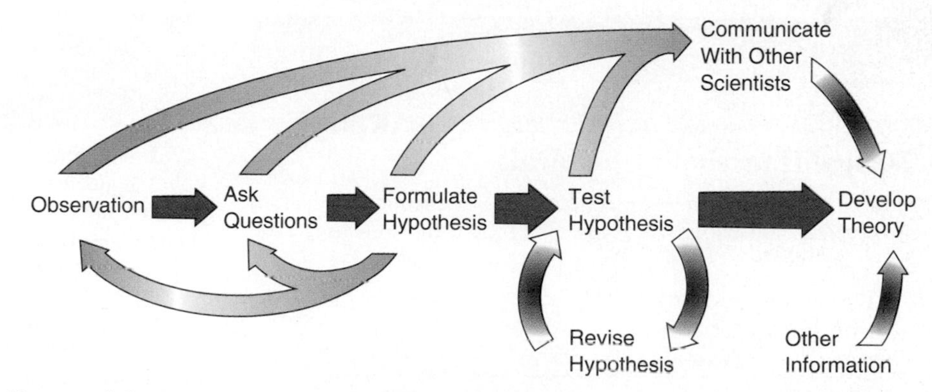

figure 4.1 **Elements of the Scientific Method** The scientific method consists of several kinds of activities. Observation of a natural phenomenon is usually the first step. Observation often leads people to ask questions about the observation they have made or to try to determine why the event occurred. This questioning is typically followed by the construction of a hypothesis that explains why the phenomenon occurred. The hypothesis is then tested to see if it is supported. Often this involves experimentation. If the hypothesis is not substantiated, it is modified and tested in its new form. It is important at all times that others in the scientific community be informed by publishing observations of unusual events, their probable cause, and the results of experiments that test hypotheses. Occasionally, this method of inquiry leads to the development of theories that tie together many bits of information into broad statements that serve to guide future thinking about a specific area of science.

Observation simply means the ability to notice something. Sometimes, the observation is made with the unaided senses—we see, feel, or smell something. Often, machines such as microscopes, chemical analyzers, or radiation detectors may be used to extend our senses. Because these machines are complicated, we might get the feeling that science is incredibly complex, when in reality the questions being asked are relatively easy to understand.

A microscope has several knobs to turn and a specially designed light source. It requires considerable skill to use properly, but it is essentially a fancy magnifying glass that allows small objects to be seen more clearly. The microscope has enabled scientists to answer some relatively simple questions such as; Are there living things in pond water? and Are living things made up of smaller subunits? Similarly, a meter stick allows us to measure distance or a pH meter to measure a chemical property of a solution. Both are simple activities, but if we are not familiar with the units of measure, we might consider the processes hard to understand.

Observations often lead one to ask questions about the observations. Why did this event happen? Will it happen again in the same circumstances? Is it related to something else? Some questions may be simple speculation but others may inspire you to further investigation. When you have formulated a question that needs scientific investigation, your first step is to form a hypothesis. A **hypothesis** is a logical statement that potentially explains an event or answers a question. A good hypothesis should be as simple as possible, while taking all of the known facts into account. Furthermore, a hypothesis must be testable. In other words, you must be able to support it or prove it incorrect. The construction and testing of hypotheses is one of the most difficult (creative) aspects of the scientific method. Often, artificial situations must be constructed to test hypotheses. These are called **experiments.** A standard kind of experiment is one called a **controlled experiment.** Two groups are created that are identical in all respects except one. The control group has nothing out of the ordinary done to it. The experimental group has one thing different. If the experimental group gives different results from the control group, those results must be caused by the single difference (variable) between the two groups.

The results of a well-designed experiment should be able to support or disprove a hypothesis. However, this does not always occur. Sometimes the results of an experiment are inconclusive. This means that a new experiment has to be conducted or that more information has to be collected. Often, it is

ENVIRONMENTAL CLOSE-UP

Typical Household Chemicals

Modern society uses many different kinds of chemicals. A survey of a typical household would probably yield the following inorganic chemicals:

Common Name	Chemical Name	Use
Table salt	Sodium chloride, $NaCl$	Flavor
Saltpeter	Potassium nitrate, KNO_3	Preservative
Baking soda	Sodium bicarbonate, $NaHCO_3$	Leavening agent
Ammonia	Ammonia, NH_3	Disinfectant
Bleach	Sodium hypochlorite, $NaHClO$	Bleaching
Lye	Sodium hydroxide, $NaOH$	Drain cleaner

Other products we use contain mixtures of inorganic chemicals. Fertilizers are good examples. They usually contain a nitrate such as ammonium nitrate (NH_4NO_3), a phosphate compound (PO_4^{3+}), and potash, which is potassium oxide (K_2O).

In addition, we use a vast array of organic chemicals: ethyl alcohol in alcoholic beverages, acetic acid in vinegar, methyl alcohol for fuel, and cream of tartar (tartaric acid) for flavoring. We also use many complex mixtures of organic molecules in flavorings, pesticides, cleaners, and other applications.

Most of us know very little about the activities of the molecules we use. Many of them can be dangerous if used improperly. Fertilizer is poisonous, caustic soda can cause severe burns, and bleach or ammonia in high enough concentrations can damage skin or other tissues. The disposal of unused or unwanted household chemicals is a problem. Many of them should not just be dumped down the sink but should be disposed of in such a way that the material is converted to a harmless product or stored in a secure place. Unfortunately, most people do not know how to dispose of unwanted chemicals. For this reason, many manufacturers of household chemicals that have a potential to cause harm print statements on the containers explaining how to properly dispose of the unused product and the container. In addition, many communities have regular cleanup efforts for household hazardous waste, in which volunteers who know the contents of such products help determine how to dispose of them properly.

CARPET BEETLES—Thoroughly apply as a spot treatment. Spray along baseboards and edges of carpeting, under carpeting, rugs and furniture, in closets and on shelving, or wherever these insects are seen or suspected. **FLEAS, BROWN DOG TICKS**—Remove soiled bed bedding and clean thoroughly or destroy. Spray sleeping quarters of pets, along baseboards, windows, door frames, cracks and crevices, carpets, rugs, floors where these pests may be found. Put fresh bedding in pet quarters after spray has dried. **DO NOT SPRAY ANIMALS.** Pets should be treated with FLEA-B-GON® Flea Killer (aerosol) or ORTHO Pet Flea & Tick Spray Formula II (pump spray).
STORAGE: To store, rotate nozzle to closed position. Keep pesticide in original container. Do not put concentrate or dilute into food or drink containers. Avoid contamination of feed and foodstuffs. Store in a cool, dry place, preferably in a locked storage area.
DISPOSAL: PRODUCT—Partially filled bottle may be disposed of by securely wrapping original container in several layers of newspaper and discard in trash. **CONTAINER**—Do not reuse empty bottle. Rinse thoroughly before discarding in trash.

NOTICE: Buyer assumes all responsibility for safety and use not in accordance with directions.

Chevron Chemical Company © 1984
Ortho Consumer Products Division
P.O. Box 5047 San Ramon CA 94583-0947
Form 10152-N Product 5466 Made in U.S.A.
EPA Reg. No. 239-2490-AA
EPA Est. 239-IA-3

0 71549 01980 8 C

necessary to have large amounts of information before a decision can be made about the validity of a hypothesis. The public often finds it difficult to understand why it is necessary to perform experiments on so many subjects, or why it is necessary to repeat experiments again and again.

The concept of **repeatability** is important to the scientific method. Because it is often not easy for scientists to eliminate unconscious bias, independent investigators must repeat the experiment to see if they get the same results. To do this, they must have a complete and accurate written document to work from. That means the scientists must publish the results of their experiment. This process of publishing results for others to examine and criticize is one of the most important steps in the process of scientific discovery. If a hypothesis is supported by many experiments and by different investigators, it is considered reliable.

A hypothesis that has survived repeated examination by many investigators over a long time and that has central importance to an area of science may become known as a **law.** For example, the **law of conservation of matter** states that matter cannot be created or destroyed. This has been tested repeatedly over hundreds of years and there have been no exceptions. Broadly written statements that cover large bodies of scientific knowledge are often called **theories.** A theory is well thought out and has lots of evidence for its support. For example, the **theory of evolution** holds that the characteristics of plants and animals and their kinds change over time. A theory is generally accepted by scientists to be true, but it cannot be proved true in *every* case because it is impossible to test *every* case. It is important to recognize that the word *theory* is often used in a much less restrictive sense. Often it is used incorrectly to describe a vague idea or a hunch. This is not a theory in the scientific sense. So when you see or hear the word *theory* you must look at the context to see if the speaker or writer is referring to a theory in the scientific sense.

Now that we have some idea of how the scientific method works, let's look at an example. In many rivers in industrial parts of the world, it is possible to notice tumors of the skin and liver in the fish that live in the rivers (*observation*). This raises the question of what causes the tumors. Many people feel that the tumors are caused by the toxic chemicals that have been released into the rivers by industrial plants (*hypothesis*). Now, how could an experiment be conducted to test the hypothesis? If an industrial plant is suspected of releasing toxic chemicals that cause tumors, resident species of fish that do not migrate can be collected upstream and downstream from the plant's wastewater discharge pipes (outfall). Fish collected above the outfall constitute the control group, and those collected below the outfall constitute the experimental group. Large numbers of fish would have to be collected and examined. If the fish below the outfall have significantly more tumors than those above the outfall, it is because of where they live in the river and so the toxic chemicals from the industrial plants are a probable cause of the tumors. This is particularly true if the chemicals are already known to cause tumors. After the data were evaluated, the results of the experiment would be published. Certainly, the owners of the industrial plants would want to look at the data and might want to repeat the experiment to see if they get the same results.

Limitations of Science

Science is a powerful tool for developing an understanding of the natural world, but it cannot analyze international politics, decide if family-planning programs should be instituted, or evaluate the significance of a beautiful landscape. These tasks are beyond the scope of scientific investigation. This does not mean that scientists cannot comment on such issues. They often do. But they should not be regarded as more knowledgeable on these issues just because they are scientists. Scientists may know more about the scientific aspects of these issues, but they struggle with the same moral and ethical questions that face all people, and their judgments on these matters can be just as faulty as anyone else's.

It is important to differentiate between the scientific data collected and the opinions scientists have about what the data mean. Scientists form and state opinions that may not always be supported by fact, just as other people do. Equally reputable scientists commonly state opinions that are in direct contradiction. This is especially true in environmental science, where predictions about the future must be based on inadequate or fragmentary data. The issue of climate change (covered in chapter 18) is an example of this.

It is important to recognize that some scientific knowledge can be used to support both valid and invalid conclusions. For example, the following statements are all factual.

1. Many of the kinds of chemicals used in modern agriculture are toxic to humans and other animals.
2. Agricultural chemicals have been detected in small amounts in some agricultural products.
3. Low levels of some toxic materials have been strongly linked to a variety of human illnesses.

This does not mean that all foods grown with the use of chemicals are less nutritious or are dangerous to health or that "organically grown" foods are necessarily more nutritious or more healthful because they have been grown without agricultural chemicals. The idea that something that is artificial is necessarily bad and something natural is necessarily good is an oversimplification. After all, many plants such as tobacco, poison ivy, and rhubarb leaves naturally contain toxic materials, while the use of chemical fertilizers has contributed to the health of major portions of the world since their use accounts for about one-third of the food grown in the world. However, it is appropriate to question if the use of agricultural chemicals is always necessary or if trace amounts of specific agricultural chemicals in food are dangerous. It is often easy to jump to conclusions or confuse fact with hypothesis, particularly when we generalize.

The Structure of Matter

Now that we have an appreciation for the methods of science, it is time to explore some basic information and theories about the structure and function of various kinds of matter. **Matter** is anything that takes up space and has mass. Air, water, trees, cement, and gold are all examples of matter. A central theory that describes the structure and activity of matter is the **kinetic molecular theory.** This theory states that all matter is made up of tiny objects that are in constant motion. Although different kinds of matter have different properties, they are similar in one fundamental way. They are all made up of one or more kinds of smaller subunits called atoms.

Atomic Structure

Atoms are the fundamental subunits of matter. They in turn are made up of protons, neutrons, and electrons. There are 92 kinds of atoms found in nature. Each kind forms a specific type of matter known as an **element.** Gold (Au), oxygen (O), and mercury (Hg) are examples of elements. All atoms are composed of a central region known as a **nucleus,** which is composed of two kinds of relatively heavy particles: positively charged particles called **protons** and uncharged particles called **neutrons.** Surrounding the nucleus are clouds of relatively lightweight, fast-moving, negatively charged particles called **electrons.** As mentioned earlier, each kind of element is

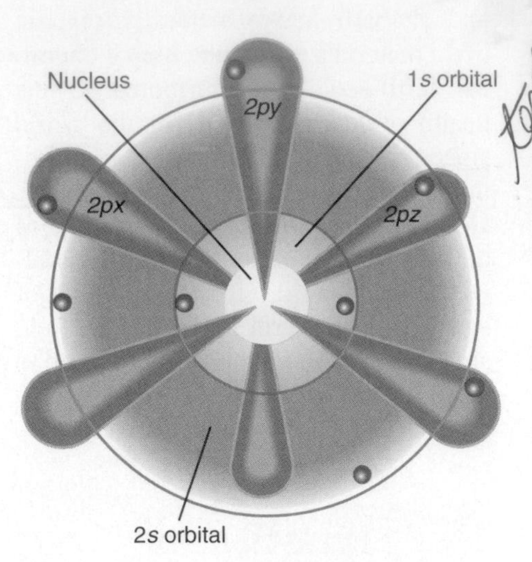

figure 4.2 **Diagrammatic Oxygen Atom** Most oxygen atoms are composed of a nucleus containing eight positively charged protons and eight neutrons without charges. Eight negatively charged electrons spin around the nucleus.

Table 4.1 Relationships Between the Kinds of Subunits Found in Matter

Category of Matter	Subunits	Characteristics
Subatomic Particles	protons	Positively charged Located in nucleus of the atom
	neutrons	Have no charge Located in nucleus of the atom
	electrons	Negatively charged Located outside the nucleus of the atom
Elements	atoms	Atoms of an element are composed of specific arrangements of protons, neutrons, and electrons. Atoms of different elements differ in the number of protons, neutrons, and electrons present
Compounds	molecules	Molecules of compounds are composed of two or more atoms chemically bonded together Molecules of different compounds contain different atoms or different proportions of atoms.
Mixtures	molecules	Molecules of mixtures are not chemically bonded to each other. The number of each kind of molecule present is variable.

composed of a specific kind of atom. The atoms of different kinds of elements differ from one another in the number of protons, neutrons, and electrons present. For example, a typical atom of mercury contains 80 protons and 80 electrons; gold has 79, and oxygen only eight. (See figure 4.2.) (Appendix 3 contains a periodic table of the elements.) All atoms of an element always have the same number of protons and electrons, but the number of neutrons may vary from one atom to the next. Atoms of the same element that differ from one another in the number of neutrons they contain are called **isotopes.**

Molecules and Mixtures

Atoms can be attached to one another into stable units called **molecules.** When two or more different kinds of atoms are attached to one another, the kind of matter formed is called a **compound.** While only 92 kinds of atoms are commonly found, there are millions of ways atoms can be combined to form compounds. Water (H_2O), sugar ($C_6H_{12}O_6$), salt (NaCl), and methane gas (CH_4) are examples of compounds.

Many other kinds of matter are **mixtures,** variable combinations of atoms or molecules. Honey is a mixture of several sugars and water; concrete is a mixture of cement, sand, gravel, and reinforcing rods; and air is a mixture of several gases of which the most common are nitrogen and oxygen. Table 4.1 summarizes the various kinds of matter and the subunits of which they are composed.

Acids, Bases, and pH

Acids and bases are two classes of compounds that are of special interest. Their characteristics are determined by the nature of their chemical bonds. When acids are dissolved in water, hydrogen ions (H^+) are set free. An **ion** is an atom or molecule that has gained or lost one or more electrons and, therefore, has either a positive charge or a negative charge. A *hydrogen ion* is positive because it has lost its electron and now has only the positive charge of its proton. Therefore, a hydrogen ion is a proton. An **acid** is any ionic compound that releases hydrogen ions (protons) in a solution. We can also think of acids as compounds that act like a hydrogen ion: they attract negatively charged particles. An example of a common acid is the sulfuric acid (H_2SO_4) in our automobile batteries.

A **base** is the opposite of an acid in that it is an ionic compound that releases a group known as a **hydroxide ion,** or OH^- ion. This ion is composed of an oxygen atom and a hydrogen atom bonded together but with an additional electron. The hydroxide ion is negatively charged. It is a base because it is able to donate electrons to the solution. A base can also be thought of as any substance able to attract positively charged particles such as hydrogen ions. A very strong base used in oven cleaners is NaOH, or sodium hydroxide.

The strength of an acid or base is represented by a number called its **pH** number. The pH scale is a measure of hydrogen ion concentration. However, the pH scale is different from what you might expect. First it is a reciprocal scale which means that the smaller the number, the higher the number of hydrogen ions present. Second, the scale is a logarithmic scale which means that a difference between two pH numbers is really a difference of ten times. For example, a pH of 7 indicates that the solution is neutral and has an equal number of H^+ ions and OH^- ions, but a pH of 6 means that the solution is 10 times more acid than it would be at a pH of 7 and that the number of hydrogen ions has increased by 10 times while the number of hydroxide ions has decreased by 10 times. As the number of hydrogen ions in the solution increases, the pH number gets smaller. The lower the number, the stronger the acid. A number higher than seven indicates that the solution has more OH^- than H^+. As the number of hydroxide ions increases, the pH number gets larger. The higher the number, the stronger the base. (See figure 4.3.)

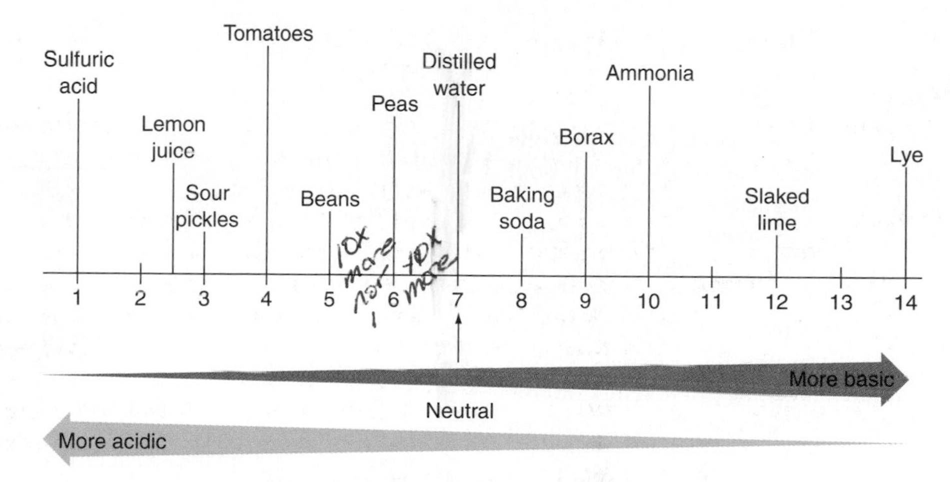

figure 4.3 **The pH Scale** The concentration of acid is greatest when the pH number is lowest. As the pH number increases, the concentration of base increases. At a pH of 7.0, the concentrations of H^+ and OH^- are equal. We usually say as the pH number gets smaller, the solution becomes more acid. As the pH number gets larger, the solution becomes more basic or alkaline.

Inorganic and Organic Matter

Inorganic and organic matter are usually distinguished from one another by one fact: organic matter consists of molecules that contain carbon atoms bonded to form chains or rings. Consequently, organic molecules are usually large. Many different kinds of organic molecules are possible. Inorganic molecules are generally small and are of relatively few kinds. All living things contain organic molecules. They must either be able to manufacture organic molecules from inorganic molecules or be able to modify organic molecules they obtain from eating organic material. Typically, organic molecules contain a large amount of chemical energy that can be released when they are broken down to inorganic molecules. Salt, water, metals, sand, and oxygen are examples of inorganic matter. Sugars, proteins, and fats are examples of organic molecules that are produced and used by living things. Natural gas, oil, and coal are all examples of organic matter that was originally produced by living things but has been modified by geologic processes.

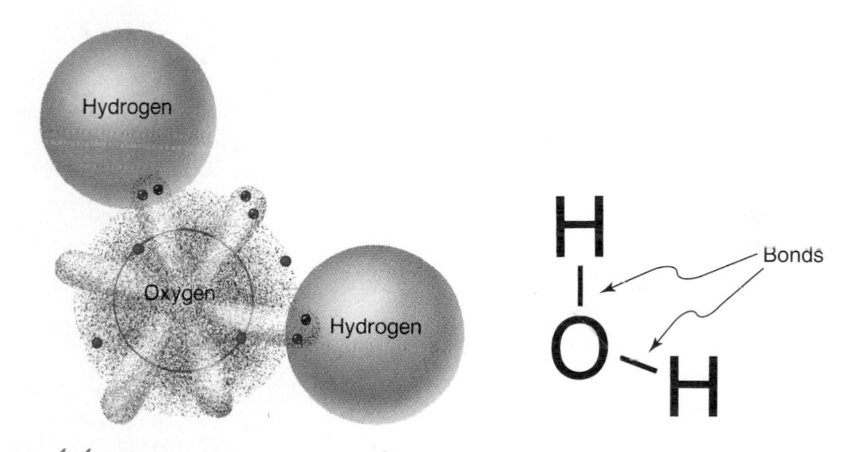

figure 4.4 **Water Molecule** Water molecules are an atom of oxygen bonded to two atoms of hydrogen, with the hydrogen atoms located on one side of the oxygen.

Chemical Reactions

The atoms within a molecule are held together by chemical bonds. (See figure 4.4.) **Chemical bonds** are physical attractions between atoms resulting from the interaction of their electrons. When chemical bonds are broken or formed, a chemical reaction occurs. When it does, the amount of energy within the chemical bonds changes, and some of the energy may be released as heat and light. A common example is the burning of natural gas. The primary ingredient in natural gas is the compound methane. When methane and oxygen are mixed together and a small amount of energy is used to start the reaction, the chemical bonds in the methane and oxygen (reactants) are rearranged to form two different compounds, carbon dioxide and water (products). In this kind of reaction, some chemical-bond energy is left over; it is released as light and heat. (See figure 4.5.) In every reaction, the amount of energy in the reactants and in the products can be compared and the differences accounted for by energy loss or gain. Even energy-yielding reactions usually need an input of energy to get the reaction started. This initial input of energy is called **activation energy.** In certain cases, the amount of activation energy required to start the reaction can be reduced by the use of a catalyst. A **catalyst** is a substance that alters the rate of a reaction, but the catalyst itself is not altered in the process. Catalysts are used in catalytic converters, which are attached to automobile exhaust systems. The purpose of the catalytic converter is to bring about more complete burning of the fuel, thus resulting in less air pollution. Most of the materials that are not completely burned by the engine require high temperatures to react further; with the presence of catalysts, these reactions can occur at lower temperatures.

Chemical Reactions in Living Things

Living things are constructed of cells that are themselves made up of both inorganic and organic matter in very specific arrangements. The chemical reactions that occur in living things are regulated by protein molecules called **enzymes** that reduce the activation energy needed to start the reactions. This is important since the high temperatures required to start these reactions without enzymes would destroy living organisms. Many of these enzymes are arranged in such a way that they cooperate in controlling a chain of reactions, as in photosynthesis and respiration.

Photosynthesis is the process plants use to convert inorganic material into organic matter, using the assistance of light energy. Light energy enables the smaller inorganic molecules (water and carbon dioxide) to be converted into organic sugar molecules. In the process, oxygen is released. Photosynthesis takes place in the green portions of the plant, usually the leaves. (See figure 4.6.) The organic molecules produced as a result of photosynthesis can be used as a source of energy by the plants and by organisms that eat the plants.

Respiration involves the use of oxygen to break down large, organic molecules (sugars, fats, and proteins) into smaller, inorganic molecules (carbon dioxide and water). This process releases energy the organisms can use. (See figure 4.7.) All organisms must carry on some form of respiration, since all need a source of energy to maintain life.

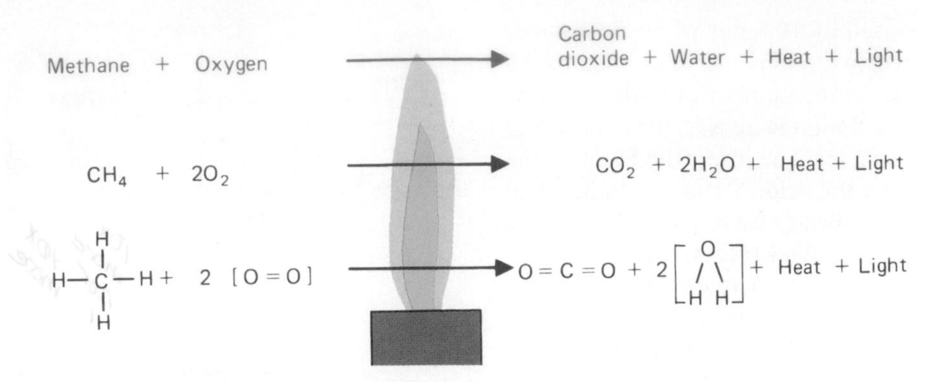

figure 4.5 **A Chemical Reaction** When methane is burned, chemical bonds are changed, and the excess chemical bond energy is released as light and heat. The same atoms are present, but they are bonded in different ways, resulting in different molecules.

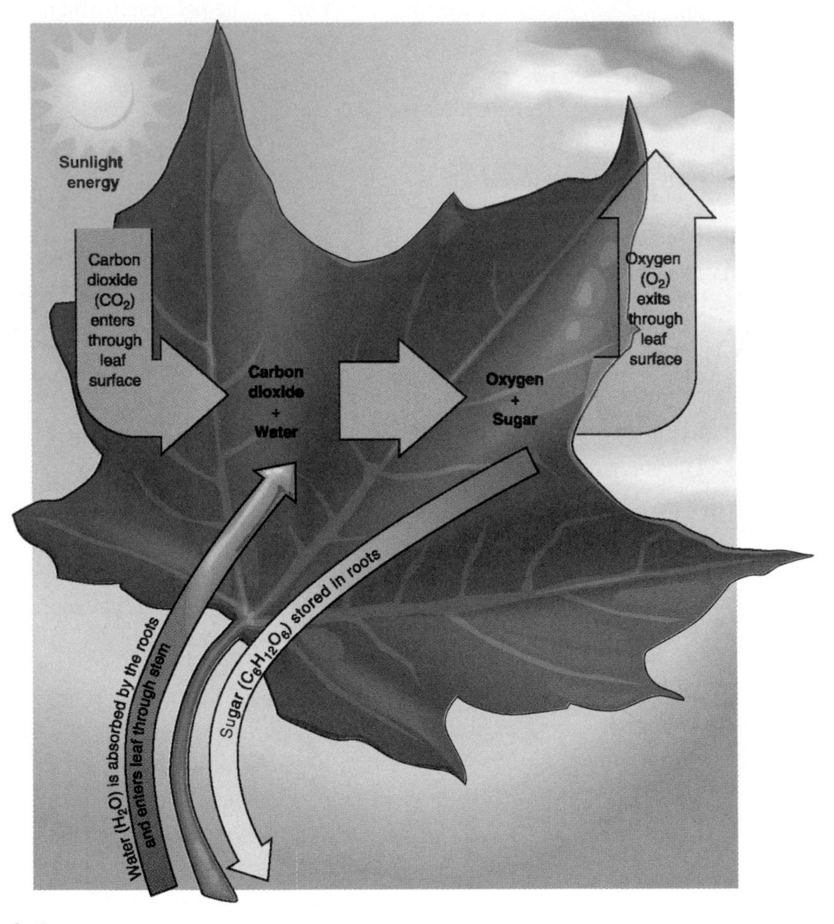

figure 4.6 **Photosynthesis** This reaction is an example of one that requires an input of energy (sunlight) to combine low-energy molecules (CO_2 and H_2O) to form sugar with a greater amount of chemical bond energy. Oxygen is also produced.

Energy Principles

The previous section started out with a description of matter, yet it used the concepts of energy to describe chemical bonds, chemical reactions, and the movement of atoms and molecules. That is because energy and matter are inseparable. It is difficult to describe one without the other. **Energy** is the ability to do work, which typically results in matter being moved over a distance. This occurs even at the molecular level.

Kinds of Energy

There are several kinds of energy. Heat, light, electricity, and chemical energy are common forms. The energy contained by moving objects is called **kinetic energy.** The moving molecules in air have kinetic energy, as does water running downhill or a dog chasing a ball. **Potential energy** is in a special category; it is the energy matter has because of its position. The water behind a dam has potential energy by virtue of its elevated position. (See figure 4.8.) An

figure 4.7 **Respiration** Respiration involves the release of energy from organic molecules when they react with oxygen. In addition to providing energy in a usable form, respiration produces carbon dioxide and water.

electron moved to a position farther from the nucleus has increased potential energy due to the increased distance between the electron and the nucleus.

States of Matter

Depending on the amount of energy present, matter can occur in three states: solid, liquid, or gas. The physical nature of matter changes when the amount of kinetic energy its molecules contain changes, but the chemical nature of matter and the kinds of chemical reactions it will undergo remain the same. For example, water vapor, liquid water, and ice all have the same chemical composition but differ in the arrangement and activity of their molecules. The amount of kinetic energy molecules have determines how rapidly they move. (See figure 4.9.) In solids, the molecules have low amounts of energy, and they vibrate in place very close to one another. In liquids, the higher-energy molecules are farther apart and will roll, tumble, and flow over each other. The molecules of gases move very rapidly and are very far apart. All that is necessary to change the physical nature of a type of matter is an energy change. Heat energy must be added or removed.

When two forms of matter have different temperatures, heat energy will flow from the one with the higher temperature to the one with the lower temperature and the temperature of the cooler matter increases while that of the warmer matter decreases. You experience this whenever you touch a cold or hot object. This is referred to as a **sensible heat** transfer. When heat energy is used to change the state of matter from solid to liquid or liquid to gas, heat is transferred but the temperature of the matter does not change. This is called a **latent heat** transfer. You have experienced this effect when water evaporates from your skin. Your body supplies the heat necessary to convert liquid water to water vapor. While the temperature of the water did not change, the physical state of the water did change and you transferred heat to the water to allow the evaporation to take place. When materials change from gas to liquid or liquid to solid there is a corresponding release of heat energy without a change in temperature.

First and Second Laws of Thermodynamics

Energy can exist in several different forms, and it is possible to convert one kind of energy into another. However, the total amount of energy remains constant. The **first law of thermodynamics** states that energy can neither be created nor destroyed; it can only be changed from one form into another. From a human perspective, some forms of energy are more useful than others. We tend to make extensive use of electrical energy for a variety of purposes, but there is very little electrical energy present in nature. Therefore, we convert other forms of energy into electrical energy. When converting energy from one form to another, some of the useful energy is lost. This is the **second law of thermodynamics.** The energy that cannot be used to do useful work is called **entropy.** Therefore, another way to state the second law of thermodynamics is to say that when energy is converted from one form to another entropy increases. An alternative way to look at the idea of entropy is to say that entropy is a measure of disorder. It is important to understand that when energy is converted from one form to another there is no loss of *total* energy, but there is a loss of *useful* energy. For example, coal, which contains chemical energy, can be burned

figure 4.8 **Kinetic and Potential Energy** Kinetic and potential energy are interconvertible. The potential energy possessed by the water behind a dam is converted to kinetic energy as the water flows to a lower level.

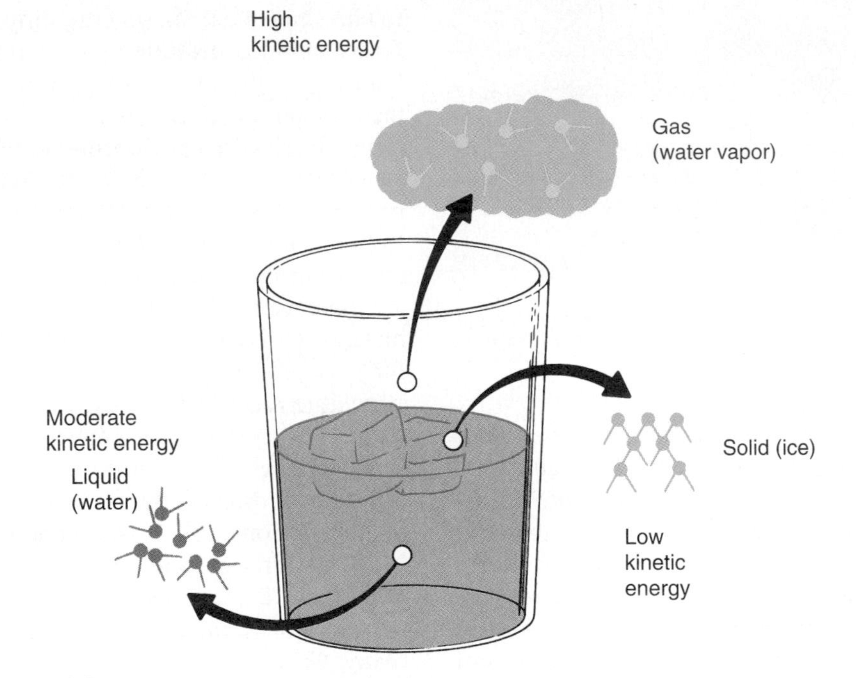

figure 4.9 **States of Matter** Matter exists in one of three states, depending on the amount of kinetic energy the molecules have. The higher the amount of energy, the greater the distance between molecules and the greater their degree of freedom of movement.

in a power plant to produce electrical energy. The heat from the burning coal is used to heat water to form steam, which turns turbines that generate electricity. At each step in the process, some heat energy is lost from the system. Therefore, the amount of useful energy (electricity) coming from the plant is much less than the total amount of chemical energy present in the coal that was burned. (See figure 4.10.)

Energy is being converted from one form to another continuously within the universe. Stars are converting nuclear energy into heat and light. Animals are converting the chemical potential energy found in food into the kinetic energy of motion. Plants are converting sunlight energy into the chemical bond energy of sugar molecules. In each of these cases, some energy is produced that is not able to do useful work. This is generally in the form of heat.

Environmental Implications of Energy Flow

The heat produced when energy conversions occur is dissipated throughout the universe. This is a common experience. Valuable things always disintegrate unless we work to maintain them. Houses fall into ruin, automobiles rust, and appliances wear out. In reality, all of these phenomena involve the loss of heat. The organisms that decompose the wood in our houses release heat. The chemical reaction that causes rust releases heat. Friction, caused by the movement of parts of a machine against each other, generates heat and causes the parts to wear.

Orderly arrangements of matter, such as clothing, automobiles, or living organisms, always tend to become disordered. There is an increase in entropy. Eventually, nonliving objects wear out and living things die and decompose. This process of becoming more disordered coincides with the constant flow of energy toward a dilute form of heat. This dissipated, low-quality heat has little value to us.

It is important to understand that different energy forms are of different quality. Some are of high quality, such as electrical energy, which can be easily used to perform a variety of useful actions. Some are of low quality, such as the heat in the water of the ocean. Although the total *quantity* of heat energy in the ocean is much greater than the total amount of electrical energy in the world, little useful work can be done with the heat energy in the ocean because it is of low *quality.* Therefore, it is not as valuable as other forms of energy that can be used to do work for us.

The reason the heat of the ocean is of little value involves the small temperature difference between two sources of heat. When two objects differ in temperature, heat will flow from the warmer to the cooler object. The greater the temperature difference, the more useful the work that can be done. For example, fossil-fuel power plants burn fuel to heat water and convert it to steam. High temperature steam enters the turbine, while cold cooling water condenses the steam as it leaves the turbine. This steep temperature gradient also provides a steep pressure gradient as heat energy flows from the steam to the cold water, which causes a turbine to turn, which generates electricity. Because the average temperature of the ocean is not high, and it is difficult to find another object that has a greatly lower temperature than the ocean, it is difficult to use the huge heat content of the ocean to do useful work for us.

These quantitative and qualitative factors are also evident in the energy expended by a stream as the water runs downhill. The steeper the slope, the greater the amount of energy expended per kilometer of its length. If there is no point along the stream where the slope is very steep, the stream has low-quality energy, because the energy is dissipated along the entire length of the stream. To make this a high-quality (concentrated) source of energy, the water must be dammed so that it will drop a long distance at one point. This means that it will give up much of its energy over a short distance. With damming, the *quantity* of energy has not changed but the *quality* has.

Organisms such as plants do photosynthesis and are able to convert low-quality light energy to high-quality chemical energy in the organic molecules they produce. Eventually, they will use this stored energy for their needs or it will be used by some other organism that has eaten the plant. In accordance with the second law of thermodynamics, all organisms, including humans, are in the process of converting high-quality energy into low-quality energy. Waste heat is produced when the chemical-bond energy in food is converted into the energy needed to move, grow, or respond. The process of releasing chemical-bond energy from food by organisms is known as cellular respiration. From an energy point of view, it is similar to the process of **combustion,** which is the burning of fuel to obtain heat, light, or some other form of useful energy. The efficiency of cellular respiration is relatively high. About 40 percent of the energy contained in food is released in a useful form. The rest is dissipated as low-quality heat. Table 4.2 lists the efficiencies of many common energy conversion systems.

figure 4.10 **Second Law of Thermodynamics** Whenever energy is converted from one form to another, some of the useful energy is lost, usually in the form of heat. The conversion of fuel to electricity produces heat, which is lost to the atmosphere. As the electricity moves through the wire, resistance generates some additional heat. When the electricity is converted to light in a light bulb, heat is produced as well. All of these steps produce useless heat in accordance with the second law of thermodynamics.

Table 4.2 The Efficiency of Some Energy Conversion Systems

Energy Conversion System	% Efficiency*
Electric motor	93
Hydroelectric power plant	85
Home oil furnace	65
Fluorescent lamp	65
Steam-power plant	47
High-intensity lamp	32
Automobile engine	25
Incandescent lamp	4

*Efficiency with which the energy of the power or fuel source is converted to a useful form.
Source: Data from U.S. Department of Energy, 1998.

An unfortunate consequence of energy conversion is pollution. The heat lost from most energy conversions is a pollutant. The wear of the brakes used to stop cars results in pollution. The emissions from power plants pollute. All of these are examples of the effect of the second law of thermodynamics. If each individual on earth used less energy, there would be less waste heat and other forms of pollution that result from energy conversion. The amount of energy in the universe is limited. Only a small portion of that energy is of high quality. The use of high-quality energy decreases the amount of useful energy available, as more low-quality heat is generated. All life and all activities are subject to these important physical principles known as the first and second laws of thermodynamics.

ISSUES & ANALYSIS

Improvements in Lighting Efficiency

In the United States, 36 percent of the energy consumed is electricity. Of that, nearly 40 percent is used for lighting. Improvements in the efficiency of lighting would significantly reduce the demand for electrical energy. All forms of lighting involve the conversion of electricity into light, but some systems are much more efficient than others. The incandescent light bulb is extremely inefficient yet is still used in many homes and commercial buildings. Standard incandescent light bulbs are 5 to 10 percent efficient. That means that 90 to 95 percent of the energy entering an incandescent light bulb is being released as heat. Compact fluorescent lights are four times more efficient. They use about 25 percent of the energy of an incandescent bulb to produce the same amount of light and can be used in the standard incandescent light bulb socket. Modern compact fluorescent bulbs have color qualities similar to incandescent bulbs. In many situations such as commercial buildings, newer high-efficiency fluorescent lights with electronic ballasts can reduce energy consumption by about 15 percent over standard fluorescent bulbs which are already four times more efficient than incandescent bulbs.

In some cases, fluorescent lighting is not practical. It does not work well in the cold and in most situations cannot be used with dimmer switches. Other kinds of higher efficiency lighting are available, however. Halogen lights are incandescent lights with about 10 percent better efficiency than standard incandescent lights. Sodium vapor, mercury vapor, and metal halide lights are very efficient but produce a light of a different color than normal daylight. These lights also require several minutes to come up to full lighting power. They are used in places where color is not important and where they are not turned on and off repeatedly, such as exterior lighting in parking lots. The U.S. Department of Energy has helped develop a new sulfur lamp that is even more efficient than fluorescent lighting and has better color than other high-efficiency lamps.

Any improvement in the efficiency with which electricity is converted to light also reduces the amount of waste heat produced, which reduces the amount of electricity needed to cool buildings. Often the cost of replacing inefficient incandescent light bulbs is offset by subsidies from local electric utilities, since increased lighting efficiency reduces the demand for electricity and allows utilities to put off building expensive new power plants.

- How many incandescent light bulbs do you have in your home?
- Why haven't they been replaced?

Summary

Science is a method of gathering and organizing information. It involves observation, asking questions, hypothesis formation, the testing of hypotheses, and publication of the results for others to evaluate. A hypothesis is a logical prediction about how things work that must account for all the known information and be testable. The process of science attempts to be careful, unbiased, and reliable in the way information is collected and evaluated. This often involves conducting experiments to test the validity of a hypothesis. If a hypothesis is continually supported by the addition of new facts, it may be incorporated into a theory. A theory is a broadly written, widely accepted generalization that ties together large bodies of information.

The fundamental unit of matter is the atom, which is made up of protons and neutrons in the nucleus and electrons circling the nucleus. The number of protons for any one type of atom is constant, but the number of neutrons in different atoms of the same type may vary. The number of electrons is equal to the number of protons. Protons have a positive charge, neutrons lack a charge, and electrons have a negative charge.

When two or more atoms combine with one another, they form stable units known as molecules. Chemical bonds are physical attractions between atoms resulting from the interaction of their electrons. When chemical bonds are broken or formed, a chemical reaction occurs, and the amount of energy within the chemical bonds is changed. Chemical reactions require activation energy to get the reaction started.

Matter that is composed of only one kind of atom is known as an element. Matter that is composed of molecules containing atoms bonded in specific ratios is known as a compound. Atoms or molecules that have gained or lost electrons so that they have an electric charge are known as ions.

Matter can occur in three states: solid, liquid, and gas. These three differ in the amount of energy the molecules contain and the distance between the molecules. Kinetic energy is the energy contained by moving objects. Potential energy is the energy an object has because of its position.

The first law of thermodynamics states that the amount of energy in the universe is constant, that energy can neither be created nor destroyed. The second law of thermodynamics states that when energy is converted from one form to another, some of the useful energy is lost (entropy increases). Some forms of energy are more useful than others. The quality of the energy determines how much useful work can be accomplished by expending the energy. Low-temperature heat sources are of poor quality, since they cannot be used to do useful work.

Interactive Exploration

Check out the website at

http://www.mhhe.com/environmentalscience

and click on the cover of this textbook for interactive versions of the following:

KNOW THE BASICS

acid *70*
activation energy *71*
atom *69*
base *70*
catalyst *71*
chemical bond *71*
combustion *75*
compound *70*
controlled experiment *67*
electron *69*
element *69*
energy *72*
entropy *73*
enzyme *72*
experiment *67*
first law of thermodynamics *73*
hydroxide ion *70*
hypothesis *67*
ion *70*
isotope *70*
kinetic energy *72*
kinetic molecular theory *69*
latent heat *73*
law *68*
law of conservation of matter *68*
matter *69*
mixture *70*
molecule *70*
neutron *69*
nucleus *69*
observation *67*
pH *70*
photosynthesis *72*
potential energy *72*
proton *69*
repeatability *68*
respiration *72*
science *67*
scientific method *67*
second law of thermodynamics *73*
sensible heat *73*
theory *68*
theory of evolution *68*

- On-line Flashcards
- Electronic Glossary

IN THE REAL WORLD

We flip that switch and expect the light to turn on. What factors affect our supply of electricity? One thing that is constantly working 'against' us is the second law of thermodynamics. Read Heat Wave Threatens Nuclear Power Plants to see why electricity generation is commonly reduced during heat waves (it is *not* due to increased demand).

The quest for high quality energy continues to provide convenience in most of our lives and a Historic Decision to Allow Drilling in Remote Alaska Oil Reserve explains how we are using a petroleum reserve in Alaska.

PUT IT IN MOTION

You've probably studied and learned about photosynthesis before. But do you remember what happens? Try out the Photosynthesis animation to refresh your memory of this process and how it relates to potential, kinetic, and chemical energy.

For a refresher of atoms, electrons, protons, and neutrons, check out the Atom animation.

What is the difference between atoms and molecules, and what is a covalent bond? See the Covalent Bond animation for a review.

TEST PREPARATION

Review Questions

1. How do scientific disciplines differ from nonscientific disciplines?
2. What is a hypothesis? Why is it an important part of the way scientists think?
3. Why are events that happen only once difficult to analyze from a scientific point of view?
4. What is the scientific method, and what processes does it involve?
5. How are the second law of thermodynamics and pollution related?
6. Diagram an atom of oxygen and label its parts.
7. What happens to atoms during a chemical reaction?
8. State the first and second laws of thermodynamics.
9. How do solids, liquids, and gases differ from one another at the molecular level?
10. List five kinds of energy.
11. Are all kinds of energy equal in their capacity to bring about changes? Why or why not?

Critical Thinking Questions

1. You observe that a high percentage of frogs, especially sensitive to environmental poisons, in small ponds in your agricultural region have birth defects. Suspecting agricultural chemicals present in runoff to be the culprit, state the hypothesis in your own words. Next devise an experiment that might help you support or reject your hypothesis.
2. Given the experiment you proposed in Critical Thinking Question 1, imagine some results that would support that hypothesis. Now imagine you are a different scientist, one who is very skeptical of the initial hypothesis. How convincing do you find these data? What other possible explanations (hypotheses) might there be to explain the results? Devise a different experiment to test this new hypothesis.
3. Increasingly, environmental issues like global climate change are moving to the forefront of world concern. What role should science play in public policy decisions? How should we decide between competing scientific explanations about an environmental concern like global climate change? What might be some of the criteria for deciding what is "good science" and what is "bad science"?
4. How important are the first and second laws of thermodynamics to explain environmental issues? Using the concepts in these laws of thermodynamics, try to explain a particular environmental issue. How does an understanding of thermodynamics change your conceptual framework regarding this issue?
5. The text points out that incandescent light bulbs are only 4 percent efficient at using energy to accomplish their task, while new, initially more expensive, compact fluorescent lighting uses significantly less electricity to provide the same quantity of light. Examine the contextual framework of those who advocate for new lighting methods and the contextual framework of those who continue to design and build using the old methods. What are the major differences in perspective? What could you suggest be done to help bring these different perspectives closer together?
6. Some scientists argue that living organisms constantly battle against the principles of the second law of thermodynamics using the principles of the first law of thermodynamics. What might they mean by this? Do you think this is accurate? What might be some of the implications of this for living organisms?

KEY CHAPTER LINKS | ESSENTIAL STUDY PARTNER | BIO COURSE | GLOBAL ISSUES | REGIONAL PERSPECTIVES | PRACTICE QUIZZING

CHAPTER 5

Interactions: Environment and Organisms

Objectives

After reading this chapter, you should be able to:

- Identify and list abiotic and biotic factors in an ecosystem.
- Define niche.
- Describe the process of natural selection as it operates to refine the fit between organism, habitat, and niche.
- Describe predator–prey, parasite–host, competitive, mutualistic, and commensalistic relationships.
- Differentiate between a community and an ecosystem.
- Define the roles of producer, herbivore, carnivore, omnivore, scavenger, parasite, and decomposer.
- Describe energy flow through an ecosystem.
- Relate the concepts of food webs and food chains to trophic levels.
- Explain the cycling of nutrients such as nitrogen, carbon, and phosphorus through an ecosystem.

Chapter Outline

Ecological Concepts
  Environment
  Limiting Factors
  Habitat and Niche
The Role of Natural Selection and Evolution
  Species Definition
Environmental Close-up: *Habitat Conservation Plans: Tool or Token?*
  Natural Selection
  Evolutionary Patterns
Kinds of Organism Interactions
  Predation
  Competition
  Symbiotic Relationships
  Some Relationships Are Difficult to Categorize
Environmental Close-up: *Human Interaction—A Different Look*
Community and Ecosystem Interactions
  Major Roles of Organisms in Ecosystems
  Keystone Species
  Energy Flow Through Ecosystems
  Food Chains and Food Webs
Environmental Close-Up: *Contaminants in the Food Chain of Fish from the Great Lakes*
  Nutrient Cycles in Ecosystems
  Human Impact on Nutrient Cycles
Issues & Analysis: *Reintroducing Wolves to the Yellowstone Ecosystem*

Ecological Concepts

The science of **ecology** is the study of the way organisms interact with each other and with their nonliving surroundings. Ecology deals with the ways in which organisms are molded by their surroundings, how they make use of these surroundings, and how an area is altered by the presence and activities of organisms. These interactions involve energy and matter. Living things require a constant flow of energy and matter to assure their survival. If the flow of energy and matter ceases, the organisms die.

All organisms are dependent on other organisms in some way. One organism may eat another and use it for energy and raw materials. One organism may temporarily use another without harming it. One organism may provide a service for another, such as when animals distribute plant seeds or bacteria break down dead organic matter for reuse. The study of ecology can be divided into many specialties and be looked at from several levels of organization. (See figure 5.1.) Before we can explore the field of ecology in greater depth, we must become familiar with some of the standard vocabulary of this field.

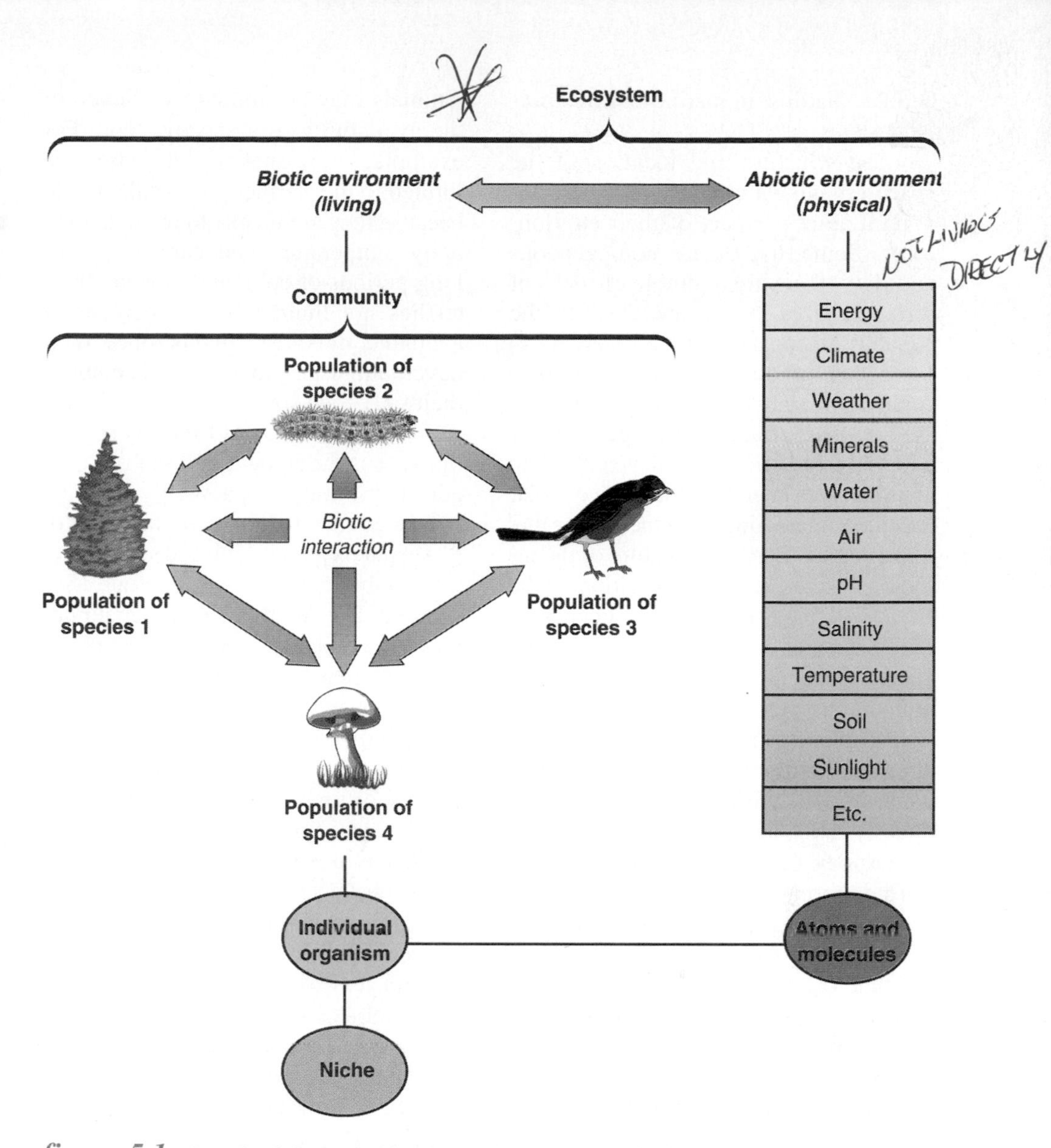

figure 5.1 **Levels of Organization in Ecology** Ecology is the science that deals with the interactions between organisms and their environment. This study can take place at several different levels, from the broad ecosystem level through community interactions to populations studies and the study of the niche of individual organisms. Ecology also involves study of the physical environment and the atoms and molecules that make up both the living and nonliving parts of an ecosystem.

Environment

Everything that affects an organism during its lifetime is collectively known as its **environment.** Environment is a very broad concept. For example, during its lifetime an animal such as a raccoon is likely to interact with millions of other organisms (bacteria, food organisms, parasites, mates, predators), drink copious amounts of water, breathe huge quantities of air, and respond to daily changes in temperature and humidity. This list only begins to describe the various components that make up the raccoon's environment. Because of this complexity, it is useful to subdivide the concept of environment into **abiotic** (nonliving) and **biotic** (living) **factors.**

Abiotic factors can be organized into several broad categories: energy, nonliving matter, and processes that involve the interactions of nonliving matter and energy. All organisms require a source of energy to maintain themselves. The ultimate source of energy for almost all organisms is the sun; in the case of plants, the sun directly supplies the energy necessary for them to maintain themselves. Animals obtain their energy by eating plants or other animals that eat plants. Ultimately, the amount of living material that can exist in an area is determined by the amount of energy plants, algae, and bacteria can trap.

All forms of life require atoms such as carbon, nitrogen, and phosphorus, and molecules such as water, to construct and maintain themselves. Organisms constantly obtain these materials from their environment by taking them up through the process of photosynthesis or obtaining them from the food they eat. The atoms become part of an organism's body structure for a short time period, and eventually all of them are returned to the environment through respiration, excretion, or death and decay. Aquatic organisms are influenced by the materials dissolved in the water. The pH, salinity, dissolved oxygen, and inorganic molecules dissolved are all important in the success of bacteria, algae, plants, and animals that live in marine and freshwater systems.

The structure and location of the space organisms inhabit is also an important abiotic aspect of their environment. Some spaces are homogeneous and flat; others are a jumble of rocks of different sizes. Some are close to the equator; others are near the poles.

Many important ecological processes involve interactions of matter and energy. The climate (average weather patterns over a number of years) of an area involves energy in the form of solar radiation interacting with the matter that makes up the Earth. The nature of the interaction is influenced by position on the Earth. The intensity and duration of sunlight in an area causes daily and seasonal changes in temperature. Differences in temperature generate wind. Solar radiation is also responsible for generating ocean currents and the evaporation of water into the atmosphere which subsequently falls as precipitation patterns which may be rain, snow, hail, or fog. Furthermore, there may be seasonal precipitation patterns. Soil building processes are influenced by prevailing weather patterns and local topography and produce soils that range from sandy, dry, infertile soils to fertile, moist soils with fine particles.

The biotic factors influencing an organism include all forms of life with which it interacts. Plants that carry on photosynthesis; animals that eat other organisms; bacteria and fungi that cause decay; bacteria, viruses, and other parasitic organisms that cause disease; and other individuals of its own species are all part of an organism's biotic environment.

Limiting Factors

Although organisms interact with their surroundings in many ways, certain factors may be critical to a particular species' success. A shortage or absence of this factor restricts the success of the species; thus, the factor is known as a **limiting factor.** Limiting factors may be either abiotic or biotic and can be quite different from one species to another. Many plants are limited by scarcity of water, light, or specific soil nutrients. Animals may be limited by climate or the availability of a specific food. For example, many snakes and lizards are limited to the warmer parts of the world because they are unable to regulate their body temperature and cannot survive long periods of cold, and monarch butterflies are limited by the number of available milkweed plants since their developing caterpillars use this plant as their only food source.

The limiting factor for many species of fishes is the amount of dissolved oxygen in the water. In a swiftly flowing, tree-lined mountain stream, the level of dissolved oxygen is high and so provides a favorable environment for trout. (See figure 5.2.) As the stream continues down the mountain, the steepness of the slope decreases, which results in fewer rapids where the water tumbles over rocks and becomes oxygenated. In addition, as the stream becomes wider the canopy of trees over the stream usually is thinner, allowing more sunlight to reach the stream and warm the water. Warm water cannot hold as much dissolved oxygen as cool water. Therefore, slower-flowing, warm-water streams contain less oxygen than rapidly moving, cool streams. Fishes such as black bass and walleye are adapted to such areas, since they are able to tolerate lower oxygen concentrations and higher water temperatures. Trout are not able to survive under such conditions and are not found in warm, less well-oxygenated water. Each of these species has a specific **range of tolerance** to oxygen concentration and water temperature. Thus, low levels of oxygen and high water temperatures are limiting factors for the distribution of trout.

Other factors, such as the abundance of silt, may influence the ability of water to support certain species of fishes. Silt reduces visibility, making it difficult for fish to find food, and covers gravel beds needed for spawning. Under these conditions, the bass and walleye may be replaced by such species as carp and catfish, which have an even greater ability to withstand high temperatures and low oxygen concentrations and are better able to survive in water with a high amount of silt.

Habitat and Niche

As we have just seen, it is impossible to understand an organism apart from its environment. The environment influences the organism, and organisms affect the environment. To focus attention on specific elements of this interaction, ecologists have developed two concepts that need to be clearly understood: habitat and niche.

The **habitat** of an organism is the space that the organism inhabits, the place where it lives (its address). We tend to characterize an organism's habitat by highlighting some prominent physical or biological feature of their environment such as: soil type, availability of water, climatic conditions, or predominant plant species that exist in the area. For example, mosses are small plants that must be covered by a thin film of water in order to reproduce. In addition, many kinds dry out and die if they are exposed to sunlight, wind, and drought. Therefore, the typical habitat of moss is likely to be cool, moist, and shady. (See figure 5.3.) Likewise, a rapidly flowing, cool, well-oxygenated stream with many bottom-dwelling insects is good trout habitat, while open prairie with lots of grass is preferred by bison, prairie dogs, and many kinds of hawks and falcons. Elm bark beetles will reside only in areas where elm trees are found. The particular biological requirements of an organism determine the kind of habitat in which it is likely to be found.

The **niche** of an organism is the functional role it has in its surroundings (its profession). A description of an organism's niche includes all the ways it affects the organisms with which it interacts as well as how it modifies its physical surroundings. In addition, the description of a niche includes all of the things that happen to the organism. For example, beavers frequently flood areas by building dams of mud and sticks across streams. (See figure 5.4.) The flooding has several effects. It provides beavers with a larger area of deep water, which they need for protection; it provides a pond habitat for many other species of animals like ducks and fish; and it kills trees that cannot live in satu-

figure 5.2 **Limiting Factors** In aquatic habitats, the amount of oxygen dissolved in the water is often a limiting factor for many species of fish. Cool, highly oxygenated water, which is typical of the rapidly flowing upper sections of river system, supports trout, but warmer, less oxygenated water is unsuited for trout. Other fish, which are more tolerant of low levels of oxygen, such as bass, catfish, bullheads, and carp, occupy the lower sections of the river, where the water is warmer, there is less oxygen, and the river contains much silt and other soil particles.

rated soil. The animals attracted to the pond and the beavers often fall to predators. After the beavers have eaten all the suitable food, such as aspen, they abandon the pond and migrate to other areas along the stream and begin the whole process over again.

In this recitation of beaver characteristics, we have listed several effects that the animal has. It changes the physical environment by flooding, it kills trees, it enhances the environment for other animals, and it is a food source for predators. This is only a superficial glimpse of the many aspects of the beaver's interaction with its environment. A complete catalog of all aspects of its niche would make up a separate book.

Another familiar organism is the dandelion. (See figure 5.5.) It is an opportunistic plant that rapidly becomes established in sunny, disturbed sites. In a few days, it can produce thousands of parachutelike seeds that are easily carried by the wind over long distances. (You have probably helped this process by blowing on the fluffy, white collections of seeds of a mature dandelion fruit.) Furthermore, it often produces several sets of flowers per year. Since there are so many seeds and they are so easily distributed, the plant can easily establish itself in any sunny, disturbed site, including lawns. Since it is a plant, one major aspect of its niche is the ability to carry on photosynthesis and grow. Dandelions need direct sunlight to grow successfully. Mowing lawns helps provide just the right conditions for dandelions, since the vegetation is never allowed to get so tall that dandelions are shaded. Many kinds of animals, including some humans, use the plant for food. The young leaves may be eaten in a salad, and the blossoms can be used to make dandelion wine. Bees visit the flowers regularly to obtain nectar and pollen.

figure 5.3 **Moss Habitat** The habitat of mosses is typically cool, moist, and shady, since many mosses die if they are exposed to drying. In addition, mosses must have a thin layer of water present in order to reproduce sexually.

figure 5.4 **Ecological Niche** The ecological niche of an organism is a complex set of interactions between an organism and its surroundings, which includes all of the ways an organism influences its surroundings as well as all of the ways the organism is affected by its environment. A beaver's ecological niche includes streams with aspen trees nearby, building dams and flooding forested areas, killing trees, providing habitat for ducks and other animals, and many other effects.

figure 5.5 **The Niche of a Dandelion** A dandelion is a common plant that does well in disturbed sites with lots of sunlight. It is able to invade these areas easily because it produces many seeds that are blown easily to new areas.

The Role of Natural Selection and Evolution

Since organisms generally are well adapted to their surroundings and fill a particular niche, it is important that we develop an understanding of the processes that led to this situation. Furthermore, since the mechanisms that result in adaptation occur within a species, we need to understand the nature of a species.

Species Definition

A **species** is a category of organisms in which the individuals within the group are potentially able to interbreed and produce fertile offspring. Individual organisms are not a species but are members of a species. This definition of species contains two points that require explanation. Obviously, there are individuals in any population that never reproduce, and many pairs of individuals that will never meet one another. However, they still have the potential to interbreed and are considered members of the same species. The second point is the ability to produce fertile offspring. In some instances, two kinds of organisms may interbreed and produce offspring, but the offspring are sterile and never reproduce. For example, horses and donkeys can breed and produce offspring called mules, but since the mules are sterile, the horse and donkey are considered separate species.

Some species are easy to recognize. We easily recognize humans as a distinct species. Most people recognize a dandelion when they see it and do not confuse it with other kinds of plants that have yellow flowers. Other species are not as easy to recognize. Most of us cannot tell one species of mosquito from another or identify different species of grasses. Because of this we tend to lump organisms into large categories and do not recognize the many subtle niche differences that exist among the similar-appearing species. However, different species of mosquitoes are quite distinct from one another. Only certain ones carry and transmit the human disease malaria. Other species transmit the dog heartworm parasite. Each mosquito species is active during certain portions of the day or night. And each species requires specific conditions to reproduce.

ENVIRONMENTAL CLOSE-UP

Habitat Conservation Plans: Tool or Token?

The Endangered Species Act places strict regulations on the destruction of the habitat of an endangered species. Since endangered species typically have narrow niches, they are restricted to specific habitats and often have very local distribution. Therefore, preservation of specific patches of habitat is critical to their survival. The Endangered Species Act states that persons cannot "incidentally take" (accidentally kill) members of the endangered species. Since many economically important land uses (farming, development, mining) would alter the habitat and result in the incidental taking of members of the endangered species, many landowners feel that the presence of an endangered species on their land unfairly deprives them of the use of their land. Many landowners have argued that, since they have lost the use of their land due to the presence of an endangered species, they should be compensated by the government for the value they have lost. The cost to the government would be enormous. However, the Endangered Species Act allows for habitat conservation plans to be put in place which would allow the landowner some limited use of the land while assuring the protection of the endangered species.

The process of developing a habitat conservation plan results in a negotiated settlement between the federal government and the landowner. Often these plans allow the landowner to use part of the land while setting other portions aside as protected areas. Sometimes travel corridors must be maintained or critical nesting sites must be protected. Many conservationists argue that the process is totally inadequate because it is impossible to determine all the critical habitat features of poorly understood endangered species. Therefore, habitat conservation plans are very likely to be inadequate to protect the species from extinction. An additional problem is that once a habitat conservation plan is established it becomes a binding document and allows landowners to continue to operate under the habitat conservation plan even if new scientific information becomes available that demonstrates that the plan is inadequate. Landowners feel that a binding plan is necessary. Without a binding plan they could be held responsible for unforeseen consequences that would put their investments at risk. Critics feel that most plans are political compromises that do not protect the endangered species and that the plans make no provisions for the recovery of the species. And since modifications to the habitat are often allowed, the ultimate fate of most species the plans are supposed to protect is likely to be extinction.

From a purely scientific point of view, we know that each species has a specific niche and has critical habitat requirements necessary for its survival. If a species is endangered and it is to be preserved, its habitat must be protected. However, the economic and political forces of human populations are also important. Habitat conservation plans are compromises between total protection and total conversion of habitat to human use. Some well-constructed plans will be successful while others will temporarily delay the inevitable extinction of vulnerable species.

Natural Selection

As we have seen, each species of organism is specifically adapted to a particular habitat in which it has a very specific role (niche). But how is it that each species of plant, animal, fungus, or bacterium fits into its environment in such a precise way? Most of the structural, physiological, and behavioral characteristics organisms display are determined by the genes they possess. These genes are passed from one generation to the next when individuals reproduce. The process that leads to this close fit between the characteristics organisms display and demands of their environment is known as natural selection. **Natural selection** is the process that determines which individuals within a species will reproduce and pass their genes to the next generation. There are several conditions and steps involved in the process of natural selection.

1. *Individuals within a species show variation; some of the variations are useful and others are not.* For example, individual animals that are part of the same species show color variations. Some colors make the animal more conspicuous while others make it less conspicuous.
2. *Organisms within a species typically produce huge numbers of offspring. Most of them die.* One apple tree may produce hundreds of apples with several seeds in each apple, or a pair of rabbits may have three to four litters of offspring each summer, with several young in each litter. Few of the seeds or baby rabbits become reproducing adults.
3. *The excess number of individuals results in a shortage of specific resources.* Individuals within a species must compete with each other for food, space, mates, or other requirements that are in limited supply. If you plant 100 bean seeds in a pot, many of them will begin to grow, but eventually some will become taller and get the majority of the sunlight while the remaining plants are shaded. Great horned owls typically produce two young at a time, but if food is in short supply, the larger of the two young will get the majority of the food.
4. *Because of variation among individuals, some have a greater chance of obtaining needed resources and, therefore, have a greater likelihood of surviving and reproducing than others.* Individuals that have genes that allow them to obtain needed resources and avoid threats to their survival will be more likely to survive and reproduce. Even if less well-adapted individuals survive, they may mature more slowly and

not be able to reproduce as many times as the more well-adapted members of the species. Often the degree to which organisms are adapted to their environment is referred to as fitness. It is important to recognize that fitness does not necessarily mean strong or vigorous. In this context, it means how well does the organism fit in with all the aspects of its surroundings.

5. *As time passes and each generation is subjected to the same process of natural selection, the percentage of individuals showing favorable variations will increase and those having unfavorable variations will decrease.* Those that reproduce more successfully pass on to the next generation the genes for the characteristics that made them successful in their environment, and the genes that made them successful become more common in future generations. Thus, each species of organism is continually refined to be adapted to the environment in which it exists.

Evolutionary Patterns

When we look at the effects of natural selection over long periods of time (thousands to millions of years), we can see considerable change in the characteristics of a species and kinds of species present. These changes in the kinds of organisms that exist and in their characteristics are together called **evolution.** Natural selection involves the processes that bring about change in species and the end result of the natural selection process observable in organisms is called evolution.

Scientists have continuously shown that this theory of natural selection can explain the development of most aspects of the structure, function, and behavior of organisms. It is the central idea that helps explain how species adapt to their surroundings. When we discuss environmental problems, it is helpful to understand that species change, and that as the environment is changed either naturally or by human action, some species will adapt to the new conditions while others will not.

There are many examples that demonstrate the validity of the process of natural selection and the evolutionary changes that result from natural selection. In recent times, we have become aware that many species of insects have become resistant to insecticides that formerly were effective against them. When an insecticide is first used against an insect pest, it kills most of them. However, in many cases there are some individual insects within the species that happen to have genes that allow them to resist the effects of the pesticide. These individuals are better adapted to survive in the presence of the insecticide and have a higher likelihood of surviving. When they reproduce, they pass on to their offspring the same genes that contributed to their survival. After several generations of such selection, a majority of the individuals in the species will contain genes for resistance to the pesticide and the pesticide is no longer effective against the pest.

When we look at the evolutionary history of organisms in the fossil record over long time periods, it becomes obvious that new species come into being while other species disappear. The production of new species from previously existing species is known as **speciation** and is thought to occur as a result of a species dividing into two isolated subpopulations. If the two subpopulations contain some genetic differences and their environments are somewhat different, natural selection will work on the two groups differently and they will begin to diverge from each other. Eventually, the differences may be so great that the two subpopulations are not able to interbreed. At this point they are two different species.

The environment in which organisms exist does not remain constant over long time periods. Those species that lack the genetic resources to cope with a changing environment go extinct. **Extinction** is the loss of an entire species and is a common feature of the evolution of organisms. Of the estimated 500 million species of organisms that are believed to have ever existed on Earth since life began, perhaps 5 million to 10 million are currently active. This represents an extinction rate of 98 to 99 percent. Obviously, these numbers are estimates, but the fact remains that extinction has been the fate of most species of organisms. In fact, studies of recent fossils and other geologic features show that only thousands of years ago, huge glaciers covered much of Europe and the northern parts of North America. Humans coexisted with mammoths, saber-toothed tigers, and giant cave bears. As the climate became warmer and the glaciers receded, and humans continued to prey on these animals, new pressures affected the organisms in the area. Some, including the mammoths, saber-toothed tigers, and giant cave bears, did not adapt and became extinct. Others, such as humans, horses, and many kinds of plants, adapted to the new conditions and so survive to the present.

It is also possible to have the extinction of specific populations of a species. Most species consist of many different populations that may differ from one another in significant ways. Often some of these populations have small local populations which can easily be driven to extinction. While these local extinctions are not the same as the extinction of an entire species, local extinctions often result in the loss of specific gene combinations. Many of the organisms listed on the endangered species list are really local populations of a more widely distributed species.

Natural selection is constantly at work shaping organisms to fit a changing environment. It is clear that humans have had a significant impact on the extinction of many kinds of species. Wherever humans have modified the environment for their purposes (farming, forestry, cities, hunting, and introducing exotic organisms), species are typically displaced from the area. If large areas are modified, entire species may have been displaced. Ultimately, humans are also subject to evolution and the possibility of extinction as well.

Coevolution is the concept that two or more species of organisms can reciprocally influence the evolutionary direction of the other. In other words, organisms affect the evolution of other organisms. Since all organisms are influenced by other organisms, this is a common pattern. For example, grazing animals and the grasses they consume have coevolved. Grasses which are eaten by grazing animals grow from the base of the plant near the ground rather than from the tips of the branches as many plants do. Furthermore, grasses have hard materials in their cell walls that make it difficult for animals to crush the cell walls and digest them. Grazing animals have different kinds of adaptations that overcome these deterrents. Many grazers have teeth that are very long or grow continuously to compensate for the wear associated with grinding hard cell walls. Others, such as cattle, have complicated digestive tracts that allow microorganisms to do most of the work of digestion. Similarly, the red color and production of nectar by many kinds of flowers is attractive to hummingbirds which pollinate the flowers at the same time as they consume nectar from the flower. The next section will explore in more detail the ways that organisms interact and the results of long periods of coevolution.

Kinds of Organism Interactions

Ecologists look at organisms and how they interact with their surroundings. Perhaps the most important interactions occur between organisms. Ecologists have identified several general types of organism-to-organism interactions that are common in all ecosystems. When we closely examine how organisms interact, we see that each organism has specific characteristics that make it well suited to its role. An understanding of the concept of natural selection allows us to see how interactions between different species of organisms can result in species that are finely tuned to a specific role. As you read this section, notice how each species has special characteristics that equip it for its specific role (niche). Because these interactions involve two kinds of organisms interacting, we should expect to see examples of coevolution. If the interaction between two species is the result of a long period of interaction, we should expect to see that each species has characteristics that specifically adapt it to be successful in its role.

figure 5.6 **Predator-Prey Relationship** Lions are predators on zebras. The quicker lions are more likely to get food, and the slower, sickly, or weaker zebras are more likely to become prey.

Predation

One common kind of interaction called **predation** occurs when one organism, known as a **predator,** kills and eats another, know as the **prey.** (See figure 5.6.) The predator benefits from killing and eating the prey and the prey is harmed. Some examples of predator-prey relationships are lions and zebras, birds and worms, wolves and moose, and frogs and insects. There are even a few plants that show predatory behavior. The Venus flytrap has specially modified leaves that can quickly fold together and trap insects which are then digested. To succeed, predators employ several strategies. Some strong and speedy predators (lions, sharks) chase and overpower their prey; other species (lizards, hawks) lie in wait and quickly strike prey that happen to come near them; and some (spiders) use snares to help them catch prey. At the same time, prey species have many characteristics that help them avoid predation. Many have keen senses that allow them to detect predators, others are camouflaged so they are not conspicuous, and many can avoid detection by remaining motionless when predators are in the area. An adaptation common to many prey species is a high reproductive rate. For example, field mice may have 10 to 20 offspring per year, while hawks typically have two to three. Because of this high reproductive rate, prey species can endure a high mortality rate and still maintain a viable population. Certainly, the *individual* organism that is killed and eaten is harmed, but the prey *species* is not, since the prey individuals that die are likely to be the old, the slow, the sick, and the less well-adapted members of the population. The healthier, quicker, and better-adapted individuals are more likely to survive. When these survivors reproduce, their offspring are more likely to have characteristics that help them survive; they are better adapted to their environment. At the same time, a similar process is taking place in the predator population. Since poorly adapted individuals are less likely to capture prey, they are less likely to survive and reproduce. The predator-and prey-species are both participants in the natural selection process. This dynamic relationship between predator and prey species is a complex one that continues to intrigue ecologists.

figure 5.7 **Competition** Whenever a needed resource is in limited supply, organisms compete for it. This competition may be between members of the same species and is called intraspecific competition, or it may be between different species and is called interspecific competition. This photograph shows several vultures competing for a food source.

Competition

A second type of interaction between species is **competition,** in which two organisms strive to obtain the same limited resource. In the process, both organisms are harmed to some extent. (See figure 5.7.) However, this does not mean that there is no winner. If a large number of lodgepole pine trees begin growing close to one another, they will compete for water, minerals, and sunlight. None of the trees grows as rapidly as it could because their access to resources is restricted by the presence of the other trees. Eventually, some of the pines will grow faster and will get a greater share of the resources. The taller trees will get more sunlight and the shorter trees will receive less. In time, some of the smaller trees die. Similarly, when two robins are competing for the same worm, only one gets it. Both organisms were harmed because they had to expend energy in fighting for the worm, but one got some food and was harmed less than the one that fought and got nothing. These examples of competition, in which members of the same species compete for resources, is known as **intraspecific competition.** Other examples of intraspecific competition include corn plants in a field competing for water and nutrients, male elk competing with one another for the right to mate with the females, and certain species of woodpeckers competing for the holes in dead trees to use for nesting sites.

Competition among members of the same species is a major force in shaping the evolution of a species. When resources are limited, less well-adapted individuals are more likely to die or be denied mating privileges. Consequently, each succeeding generation will contain more of the genetic characteristics that are favorable for survival of the species in that particular environment. Since individuals of the same species have similar needs, competition among them is usually very intense. A slight advantage on the part of one individual may mean the difference between survival and death.

Competition between organisms of different species is called **interspecific competition.** Many species of predators (hawks, owls, foxes, coyotes) may use the same prey species (mice, rabbits) as a food source. If the supply of food is inadequate, intense competition for food will occur and certain predator species may be more successful than others. In grasslands, the same kind of competition for limited resources occurs. Rapidly growing, taller grasses get more of the water, minerals, and sunlight, while shorter species are less successful. Often the shorter species are found to be more abundant when the taller species are removed by grazers, fire, or other activities.

As with intraspecific competition, one of the effects of interspecific competition is that the species that has the larger number of successful individuals emerges from the interaction better adapted to its environment than its less successful rivals. The more similar two species are, the more intense will be the competition between them. If one of the two competing species is better adapted to live in the area than the other, the less-fit species must evolve into a slightly different niche, migrate to a different geographic area, or become extinct. This concept is often formally called the **competitive exclusion principle,** which states that no two species can occupy the same ecological niche in the same place at the same time. When the niche requirements of two similar species are examined closely, we usually find significant differences between the niches of the two species. The difference in niche requirements reduces the intensity of the competition between the two species. For example, there are many small forest birds that eat insects. However, they may obtain them in different ways; a flycatcher sits on a branch and makes short flights to snatch insects from the air, a woodpecker excavates openings to obtain insects in rotting wood, and many warblers flit about in the foliage capturing insects. Even among these categories there are specialists. Different species of warblers look in different parts of trees for their insect food.

Symbiotic Relationships

Symbiosis is a close, long-lasting, physical relationship between two different species. In other words, the two species are usually in physical contact and at least one of them derives some sort of benefit from this contact. There are three different categories of symbiotic relationships: parasitism, commensalism, and mutualism.

Parasitism

Parasitism is a relationship in which one organism, known as the **parasite,** lives in or on another organism, known as the **host,** from which it derives nourishment. Generally, the parasite is much smaller than the host. Although the host is harmed by the interaction, it is generally not killed immediately by the parasite, and some host individuals may live a long time and be relatively little affected by their parasites. Some parasites are much more destructive than others, however. Newly established parasite/host relationships are likely to be more destructive than those that have a long evolutionary history. With a long-standing interaction between the parasite and the host, the two species generally evolve in such a way that they can accommodate one another. It is not in the parasite's best interest to kill its host. If it does, it must find another. Likewise, the host evolves defenses against the parasite, often reducing the harm done by the parasite to a level the host can tolerate.

Many parasites have complex life histories that involve two or more host species for different stages in the parasite's life cycle. Many worm parasites have their adult, reproductive stage in a carnivore (the definitive host), but they have an immature stage that reproduces asexually in another animal (the intermediate host) that the carnivore uses as food. Thus, a common dog tapeworm is found in its immature form in certain internal organs of rabbits. Other parasite life cycles involve animals that carry the parasite from one host to another. These carriers are known as **vectors.** For example, many biting insects and mites can transmit parasites when they obtain a blood meal. Malaria, lyme disease, and sleeping sickness are transmitted by vectors.

Parasites that live on the surface of their hosts are known as **ectoparasites.** Fleas, lice, and some molds and mildews are examples of ectoparasites. (See figure 5.8.) Many other parasites, like tapeworms, malaria parasites, many kinds of bacteria, and some fungi, are called **endoparasites,** because they live inside the bodies of their hosts. A tapeworm lives in the intestines of its host where it is able to resist being digested and makes use of the nutrients in the intestine. If a host has only one or two tapeworms, it can live for some time with little discomfort, supporting itself and its parasites. If the number of parasites is large, the host may die.

Flea (external parasite)

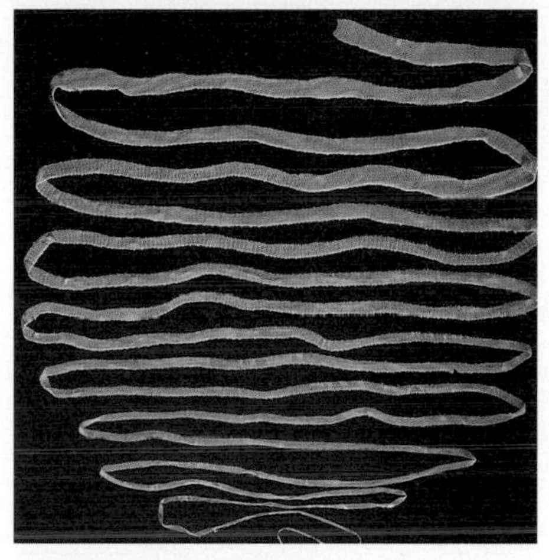

Tapeworm (internal parasite)

figure 5.8 **Parasitism** Fleas are small insects that live in the feathers of birds or the fur of mammals, where they bite their hosts to obtain blood. Since they live on the outside of their hosts, they are called ectoparasites. Tapeworms live inside the intestines of their hosts, where they absorb food from their hosts' intestines. Since they live inside their hosts, they are called endoparasites.

Even plants can be parasites. Mistletoe is a flowering plant that is parasitic on trees. It establishes itself on the surface of a tree when a bird transfers the seed to the tree. It then grows down into the water-conducting tissues of the tree and uses the water and minerals it obtains from these tissues to support its own growth.

Parasitism is a very common life strategy. If we were to categorize all the organisms in the world, we would find many more parasitic species than nonparasitic species. Each organism, including you, has many others that use it as a host.

figure 5.9 **Commensalism** Remoras hitchhike a ride on sharks and feed on the scraps of food lost by the sharks. This is a benefit to the remoras. The sharks do not appear to be affected by the presence of the remoras.

Commensalism

If the relationship between organisms is one in which one organism benefits while the other is not affected, it is called **commensalism.** It is possible to visualize a parasitic relationship evolving into a commensal one. Since parasites generally evolve to do as little harm to their host as possible and the host is combating the negative effects of the parasite, they might eventually evolve to the point where the host is not harmed at all. There are many examples of commensal relationships. Many orchids use trees as a surface upon which to grow. The tree is not harmed or helped, but the orchid needs a surface upon which to establish itself and also benefits by being close to the top of the tree, where it can get more sunlight and rain. Some mosses, ferns, and many vines also make use of the surfaces of trees in this way.

In the ocean, many sharks have a smaller fish known as a remora attached to them. Remoras have a sucker on the top of their heads that they can use to attach to the shark. In this way, they can hitchhike a ride as the shark swims along. When the shark feeds, the remora frees itself and obtains small bits of food that the shark misses. Then, the remora reattaches. The shark does not appear to be positively or negatively affected by remoras. (See figure 5.9.) Many commensal relationships are rather opportunistic and may not involve long-term physical contact. For example, many birds rely on trees of many different species for places to build their nests, but do not use the same tree year after year. Similarly, in the spring bumble bees typically build nests in underground mouse nests that are no longer in use.

Mutualism

Mutualism is another kind of symbiotic relationship and is actually beneficial to both species involved. In many mutualistic relationships, the relationship is obligatory; the species cannot live without each other. In others, the species can exist separately but are more successful when they are involved in a mutualistic relationship. Some species of *Acacia,* a thorny tree, provide food in the form of sugar solutions in little structures on their stems. Certain species of ants feed on the solutions and live in the tree, which they will protect from other animals by attacking any animal that begins to feed on the tree. Both organisms benefit; the ants receive food and a place to live, and the tree is protected from animals that would use it as food.

One soil nutrient that is usually a limiting factor for plant growth is nitrogen. Many kinds of plants, such as beans, clover, and alder trees, have bacteria that live in their roots in little nodules. The roots form these nodules when they are infected with certain kinds of bacteria. The bacteria do not cause disease but provide the plants with nitrogen-containing molecules that the plants can use for growth. The nitrogen-fixing bacteria benefit from the living site and nutrients that the plants provide, and the plants benefit from the nitrogen they receive. (See figure 5.10.) Similarly, there are many kinds of fungi that form an association with the roots of plants. The root-fungus associations are called **mycorrhizae.** The fungus obtains organic molecules from the roots of the plant, and the branched nature of the fungus assists the plant in obtaining nutrients such as phosphates and nitrates. In many cases, it is clear that the relationship is obligatory.

figure 5.10 **Mutualism** The growths on the roots of this plant contain beneficial bacteria that make nitrogen available to the plant. The relationship is also beneficial to the bacteria, since the bacteria obtain necessary raw materials from the plant. It is a mutually beneficial relationship.

figure 5.11 **Nest Parasitism** This red-eyed vireo is feeding the nestling of a cowbird. A female cowbird laid its egg in the vireo's nest. The vireo is harmed because it is not raising its own young, and the cowbird benefits because it did not need to expend energy to build and defend a nest or collect food for its own young.

Some Relationships Are Difficult to Categorize

Sometimes it is not easy to categorize the relationships that organisms have with each other. For example, it is not always easy to say whether a relationship is a predator-prey relationship or a host-parasite relationship. How would you classify a mosquito or a tick? Both of these animals require blood meals to live and reproduce. They don't kill and eat their prey. Neither do they live in or on a host for a long period of time. This question points out the difficulty encountered when we try to place all kinds of organism interactions into a few categories. However, we can eliminate this problem if we call them temporary parasites or blood predators.

Another relationship that doesn't fit well is the relationship that certain birds like cowbirds and European cuckoos have with other birds. Cowbirds and European cuckoos do not build nests but lay their eggs in the nests of other species of birds, who are left to care for a foster nestling at the expense of their own nestlings, who generally die. This situation is usually called nest parasitism or brood parasitism. (See figure 5.11.)

What about grazing animals? Are they predators or parasites on the plants that they eat? Sometimes they kill the plant they eat, while at other times they simply remove part of the plant and the rest continues to grow. In either case, the plant has been harmed by the interaction and the grazer has benefitted.

There are also mutualistic relationships that do not require permanent contact between the participants in the relationship. Bees and the flowering plants they pollinate both benefit from their interactions. The bees get pollen and nectar for food and the plants are pollinated. But the active part of the relationship involves only a part of the life of any plant and the bees are not restricted to any one species of plant for the food. They must actually switch to different flowers at different times of the year.

ENVIRONMENTAL CLOSE-UP

Human Interaction—A Different Look

Humans are the dominant organisms on Earth. Our niche is very broad and we interact in many ways with the organisms with which we share the planet. If we examine our activities, we can see that we have complicated interactions with other organisms, and these interactions can be placed into the same categories we use to describe relationships between nonhuman organisms.

Predator—Humans throughout the world use animals as food. Some actually kill animals themselves while others rely on employees of slaughterhouses to do the killing for them.

Herbivore—Humans rely on many kinds of plants as their primary source of food.

Grazing and browsing involve consuming part of a living plant without killing the entire plant. We graze or browse parts of plants such as asparagus, rhubarb, lettuce, broccoli, and many other kinds of plants.

Foraging involves searching for food that is available from nature. Hunter/gatherer peoples spend a great deal of time foraging for edible plant materials such as roots, fruits, and seeds. Even individuals from sophisticated cultures engage in foraging activities when they pick wild berries, mushrooms, or asparagus.

Scavenger—Scavenging involves finding and consuming animals that are already dead. Our distant ancestors were probably actively engaged in scavenging by seizing the kills of more efficient carnivores. Even today in places where protein-rich food is in short supply, any recently dead animal is a valued food source. And many states in the United States have laws that allow people to take animals killed by collisions with automobiles.

Commensal—Humans find themselves on both sides of the commensal relationship. Many kinds of organisms use our homes as places to live without affecting us. Birds may nest on our buildings, spiders build webs in our windows, and rats may live under our decks. Other animals benefit from the animals we accidentally kill along our highways. We also derive benefit from organisms without affecting them, such as when we are able to get out of the hot sun by sitting under the shade of a tree or when we rely on decomposers to decay wastes.

Parasite—Although humans do not live in or on other living things we do engage in relationships that are parasitic in nature. In some African cultures, blood is drawn from cows and mixed with milk to serve as food. Maple syrup is made by "bleeding" maple trees. Similarly, humans tap rubber trees to obtain sap that is used to make rubber. Many human activities regularly rob animals. Honeybees are robbed of their honey and chickens are robbed of their eggs.

Mutualism—Humans have many mutualistic relationships with plants and animals. Our domesticated plants and animals rely on us for support and nutrition and we extract from them payment in the form of companionship, food, or other valuable resources.

Competition—Humans are in competition with all other organisms on Earth. As we convert land and aquatic resources to our uses, we deprive other organisms of what they need to survive. When humans hunt and kill large grazing animals such as bison or gazelles, we are in direct competition with other predators such as wolves or leopards. Because of our technological superiority and our huge population, we usually win in the game of competition.

Community and Ecosystem Interactions

Thus far, we have discussed specific ways in which individual organisms interact with one another and with their physical surroundings. However, often it is useful to look at ecological relationships from a broader perspective. Two concepts that focus on relationships that involve many different kinds of interactions are community and ecosystem. A **community** is an assemblage of all the interacting species of organisms in an area. Some species play minor roles while others play major roles, but all are part of the community. For example, the grasses of the prairie have a major role since they carry on photosynthesis and provide food and shelter for the animals that live in the area. Grasshoppers, prairie dogs, and bison are important consumers of grass. However, a meadowlark, though a conspicuous and colorful part of the prairie scene, has a relatively minor role and has little to do with maintaining a prairie community.

Communities consist of interacting species, but these species interact with their physical world as well. An **ecosystem** is a defined space in which interactions take place between a community, with all its complex interrelationships, and the physical environment. The physical world has a major impact on what kinds of plants and animals can live in an area. We do not expect to see a banana tree in the arctic or a walrus in the Mississippi River. Banana trees are adapted to warm, moist, tropical areas, and walruses require cold ocean waters. Some ecosystems, such as grasslands and certain kinds of forests, are shaped by periodic fires. The kind of soil and the amount of moisture also influence the kinds of organisms found in an area.

While it is easy to see that the physical environment places limitations on the kinds of organisms that can live in an area, it is also important to recognize that organisms impact their physical surroundings. Trees break the force of the wind, grazing animals form paths, and earthworms create holes that aerate the soil. While the concepts of community and ecosystem are closely related, an ecosystem is a broader concept because it involves physical as well as biological processes.

Every system has parts that are related to one another in specific ways. A bicycle has wheels, a frame, handlebars, brakes, pedals, and a seat. These parts must be organized in a certain way or the system known as a bicycle will not function. Similarly, ecosystems have parts that must be organized in specific ways or the systems will not operate. In order to more fully develop the concept of ecosystem, we will look at ecosystems from three points of view: the major roles played by organisms, the way energy is utilized within ecosystems, and the way atoms are cycled from one organism to another.

Major Roles of Organisms in Ecosystems

Several categories of organisms are found in any ecosystem. **Producers** are organisms that are able to use sources of energy to make complex, organic molecules from the simple inorganic substances in their environment. In nearly all ecosystems, energy is supplied by the sun, and organisms such as, plants, algae, and tiny aquatic organisms called phytoplankton use light energy to carry on photosynthesis. Since producers are the only organisms in an ecosystem that can trap energy and make new organic material from inorganic material, all other organisms rely on producers as a source of food, either directly or indirectly. These other organisms are called **consumers** because they consume organic matter to provide themselves with energy and the organic molecules necessary to build their own bodies. An important part of their role is the process of respiration in which they break down organic matter to inorganic matter.

However, some consumers have significantly different roles from others. **Primary consumers,** also known as **herbivores,** are animals that eat producers (plants or phytoplankton) as a source of food. Herbivores, such as leaf-eating insects and seed-eating birds, are usually quite numerous in ecosystems where they serve as food for the next organisms in the chain. **Secondary consumers** or **carnivores** are animals that eat other animals. Secondary consumers can be further subdivided into categories based on what kind of prey they capture and eat. Some carnivores, like ladybird beetles, primarily eat herbivores, like aphids; others, such as eagles, primarily eat fish that are themselves carnivores. While these are interesting conceptual distinctions, most carnivores will eat any animal they can capture and kill. In addition, there are many animals, called **omnivores,** that include both plants and animals in their diet. Even animals that are considered to be carnivores (foxes, bears) regularly include large amounts of plant material in their diets. Conversely, animals often thought of as herbivores (mice, squirrels, seed-eating birds) regularly consume animals as a source of food.

A final category of consumer is the decomposer. **Decomposers** are organisms of many kinds that use nonliving organic matter as a source of energy and raw materials to build their bodies. Whenever an organism sheds a part of itself, excretes waste products, or dies, there is a source of food for decomposers. Since decomposers carry on respiration, they are extremely important in recycling matter by converting organic matter to inorganic material. Many small animals, fungi, and bacteria fill this niche. (See table 5.1.)

Table 5.1 Roles in an Ecosystem

Category	Major Role or Action	Examples
Producer	Converts simple inorganic molecules into organic molecules by the process of photosynthesis	Trees, flowers, grasses, ferns, mosses, algae
Consumer	Uses organic matter as a source of food	Animals, fungi, bacteria
Herbivore	Eats plants directly	Grasshopper, elk, human vegetarian
Carnivore	Kills and eats animals	Wolf, pike, dragonfly
Omnivore	Eats both plants and animals	Rats, raccoons, most humans
Scavenger	Eats meat, but often gets it from animals that died by accident or illness, or that were killed by other animals	Coyote, vulture, blowflies
Parasite	Lives in or on another living organism and gets food from it	Tapeworm, many bacteria, some insects
Decomposer	Returns organic material to inorganic material; completes recycling of atoms	Fungi, bacteria, some insects and worms

Keystone Species

Each ecosystem has many species of organisms interacting in many ways. However, some species have more central roles than others. In recognition of this idea, ecologists have developed the concept of keystone species. A **keystone species** is one that has a critical role to play in the maintenance of specific ecosystems. In prairie ecosystems, grazing animals are extremely important in maintaining the mix of species typical of a grassland. Without the many influences of the grazers, the nature of the prairie changes. A study of the American tallgrass prairie indicated that when bison are present they increase the biodiversity of the site. Bison typically eat grasses and, therefore, allow smaller plant species that would normally be shaded by tall grasses to be successful. In ungrazed plots, the tall grasses become the dominant vegetation and biodiversity decreases. Bison dig depressions in the soil, called wallows, to provide themselves with dust or mud with which they can coat themselves. These wallows retain many species of plants that typically live in disturbed areas. Their urine has also been shown to be an important source of nitrogen for the plants.

The activities of bison even affect the extent and impact of fire, another important feature of grassland ecosystems. Since bison prefer to feed on recently burned sites and revisit these sites several times throughout the year, they

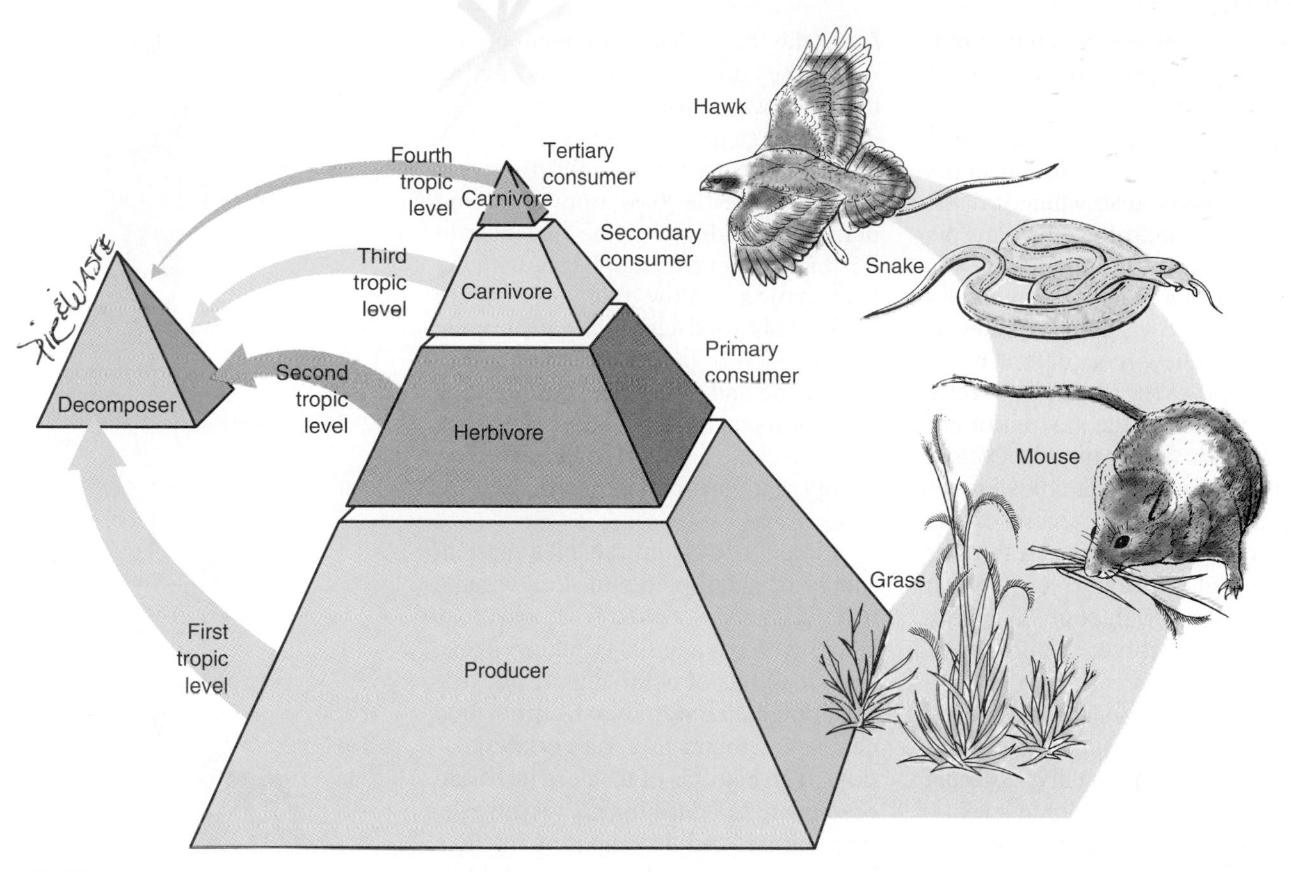

figure 5.12 **Energy Flow Through an Ecosystem** As energy flows through an ecosystem, it passes through several levels known as trophic levels. Each time energy moves to a new trophic level, approximately 90 percent of the useful energy is lost. Therefore, in most ecosystems, higher trophic levels contain less energy and fewer organisms.

tend to create a patchwork of grazed and ungrazed areas. The grazed areas are less likely to be able to sustain a fire and it is likely that fires will be more prevalent in ungrazed patches.

The concept of keystone species has also been applied to marine ecosystems. The relationship between sea urchins, sea otters, and kelp forests suggests that sea otters are a keystone species. Sea otters eat sea urchins, which eat kelp. A reduction in the number of otters results in an increase in the number of sea urchins. Increased numbers of sea urchins leads to heavy grazing of the kelp by sea urchins. When the amount of kelp is severely reduced, fish and many other animals that live within the kelp beds lose their habitat and biodiversity is significantly reduced.

The concept of keystone species is useful to ecologists and resource managers because it helps them to realize that all species cannot be treated equally. Some species have pivotal roles and their elimination or severe reduction can significantly alter ecosystems. In some cases, the loss of a keystone species can result in the permanent modification of an ecosystem into something considerably different from the original mix of species.

Energy Flow Through Ecosystems

An ecosystem is a stable, self-regulating unit. To maintain itself, it must have a continuous input of energy. The only significant source of energy for most ecosystems is sunlight energy. Producers are the only organisms that are capable of trapping solar energy through the process of photosynthesis and making it available to the ecosystem. The energy is stored in the form of chemical bonds in large organic molecules such as carbohydrates (sugars, starches), fats, and proteins. The energy stored in the molecules of producers is transferred to other organisms when the producers are eaten. Each step in the flow of energy through an ecosystem is known as a **trophic level.** Producers (plants, algae, phytoplankton) constitute the first trophic level, and herbivores constitute the second trophic level. Carnivores that eat herbivores are the third trophic level, and carnivores that eat other carnivores are the fourth trophic level. Omnivores, parasites, and scavengers occupy different trophic levels, depending on what they happen to be eating at the time. If we eat a piece of steak, we are at the third trophic level; if we eat celery, we are at the second trophic level. (See figure 5.12.)

The second law of thermodynamics states that whenever energy is converted from one form to another, some of the energy is converted to a non-useful form. Thus, there is always less useful energy following an energy conversion. Therefore, when energy passes from one trophic level to the next, there is less useful energy left with each successive trophic level. Much of this loss is in the form of low-quality heat, which is dissipated to the surroundings and warms the air, water, or soil. In addition to this loss of heat, organisms must expend energy to maintain their own life processes. It takes energy to chew food, defend nests, walk to waterholes, or produce and raise offspring. Therefore, the amount of energy contained in higher trophic levels is considerably less than that at lower levels. Approximately 90 percent of the useful energy is lost with each transfer to the next highest trophic level. So in any ecosystem, the amount of energy contained in the herbivore trophic level is only about 10 percent of the energy contained in the producer trophic level. The amount of energy at the third trophic level is approximately 1 percent of that found in the first trophic level.

Because it is difficult to actually measure the amount of energy contained in each trophic level, ecologists often use other measures to approximate the relationship between the amounts of energy at each level. One of these is the biomass. The **biomass** is the weight of living material in a trophic level. It is often possible in a simple ecosystem to collect and weigh all the producers, herbivores, and carnivores. The weights often show the same 90 percent loss from one trophic level to the next as happens with the amount of energy.

Food Chains and Food Webs

The passage of energy from one trophic level to the next as a result of one organism consuming another is known as a **food chain.** For example, willow trees grow well in very moist soil, perhaps near a pond. The trees' leaves capture sunlight and convert carbon dioxide and water into sugars and other organic molecules. The leaves serve as a food source for insects, such as caterpillars and leaf beetles, that have chewing mouth parts and a digestive system adapted to plant food. Some of these insects are eaten by spiders, which fall from the trees into the pond below, where they are consumed by a frog. As the frog swims from one lily pad to another, a large bass consumes the frog. A human may use an artificial frog as a lure to entice the bass from its hiding place. A fish dinner is the final step in this chain of events that began with the leaves of a willow tree. (See figure 5.13.) This food chain has six trophic levels. Each organism occupies a specific niche and has special abilities that fit it for its niche, and each organism in the food chain is involved in converting energy and matter from one form to another.

Some food chains rely on a constant supply of small pieces of dead organic material being supplied from situations where photosynthesis is taking place. The small bits of nonliving organic material are called **detritus.** Detritus food chains are found in a variety of situations. The bottoms of the deep lakes and oceans are too dark for photosynthesis. The animals and decomposers that live there rely on a steady rain of small bits of organic matter from the upper layers of the water where photosynthesis does take place. Similarly, in most streams, leaves and other organic debris serve as the major source of organic material and energy. A sewage treatment plant is also a detritus food chain in which particles and dissolved organic matter are constantly supplied to a series of bacteria and protozoa that use this material for food.

In another example, the soil on a forest floor receives leaves, which fuel a detritus food chain. In detritus food chains, a mixture of insects, crustaceans, worms, bacteria, and fungi cooperate in the breakdown of the large pieces of organic matter, while at the same time feeding on one another. When a leaf dies and falls to the forest floor, it is colonized by bacteria and fungi, which begin the breakdown process. An earthworm will also feed on the leaf and at the same time consume the bacteria and fungi. If that earthworm is eaten by a bird, it becomes part of a larger food chain that includes material from both a detritus food chain and a photosynthesis-driven food chain. When several food chains overlap and intersect, they make up a **food web.** (See figure 5.14.)

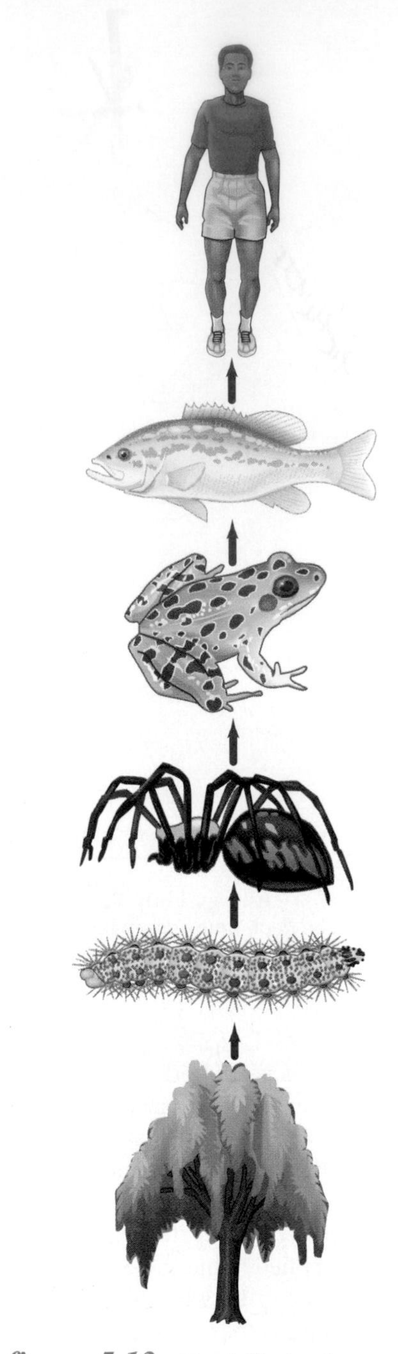

figure 5.13 **Food Chain** As one organism feeds on another organism, energy flows through the series. This is called a food chain.

Notice in the upper-left-hand corner of figure 5.14 that the Cooper's and sharp-shinned hawks use many different kinds of birds as a source of food. These hawks fit into several food chains. If one source of prey is in short supply, they can switch to something else without too

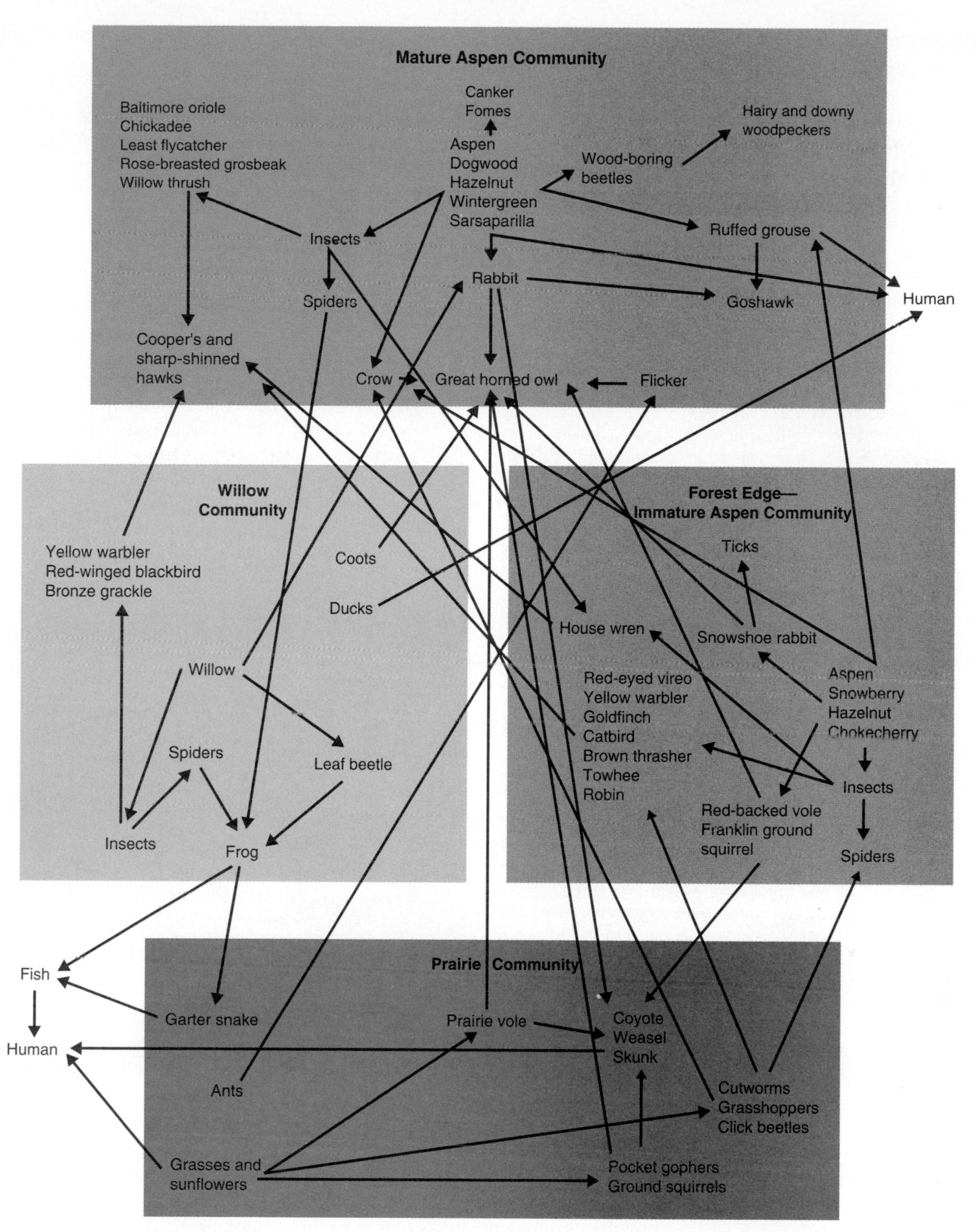

figure 5.14 **Food Web** The many kinds of interactions among organisms in an ecosystem constitute a food web. In this network of interactions, several organisms would be affected if one key organism were reduced in number. Look at the rabbit in the mature aspen community and note how many organisms use it as food.

Source: R.D. Bird, "Biotic Communities of the Aspen Parkland of Central Canada" in *Ecology,* 11:410, April 1930.

ENVIRONMENTAL CLOSE-UP

Contaminants in the Food Chain of Fish from the Great Lakes

As organisms eat, molecules move through the food chain from one organism to another. Most molecules are simply recycled through nutrient pathways because they can be broken down by decay organisms. However, humans have invented a large number of organic molecules that are not easily decomposed and these tend to remain in food chains and accumulate in higher concentrations in organisms that are at higher trophic levels.

The Great Lakes area developed as an industrial center because the lakes provided an efficient way to move raw materials and products and because water was important for manufacturing processes. In the past, many of these industries released heavy metals and organic molecules into the water as an accidental byproduct of the manufacturing process, or because it was a cheap way to get rid of unwanted material. Many of the organic molecules are products of modern organic synthesis and, therefore, are not something that bacteria and fungi are able to decay. Abnormally high concentrations of inorganic materials in the environment can also lead to abnormally high concentrations in organisms.

Approximately five hundred different organic compounds that scientists think are contaminants have been identified in the bodies of fish from the Great Lakes. Most are present in extremely small amounts and probably do not represent a serious hazard, but others are present in high enough concentrations to cause public health officials to be concerned. Since these materials do not break down, fish tend to accumulate more of these toxic materials in their bodies as they get older. Furthermore, carnivorous fish that are feeding at higher trophic levels accumulate more of these compounds. It just so happens that most of the fish that people catch and eat are carnivores, which tend to have the highest concentrations of contaminants.

If people place themselves in this food chain by eating the contaminated fish, they will tend to accumulate the contaminants in their bodies, which could have health effects. It is not possible to check every fish caught to see if it is fit to eat. The cost of doing so would be on the order of several hundred to several thousands of dollars per individual fish, depending on which contaminants are looked for. Therefore, the states and provinces surrounding the Great Lakes have developed advisory statements to help people avoid unsafe fish. While there are some minor differences among state and provincial governments, the following advisories are typical.

1. Some species of fish, such as carp and catfish, feed on the bottom and tend to accumulate contaminants that are in the bottom sediments. In many areas of the Great Lakes, people are advised not to eat these fish.
2. Larger fish have generally consumed more food and have had an opportunity to accumulate more contaminants. Therefore, people are advised to eat only smaller fish.
3. Since many of the organic contaminants are fat soluble, removal of the fat or cooking in such a way that the fat is allowed to separate from the flesh is also advised in many areas.
4. Because the amount of contamination a person is exposed to is directly related to the number of fish a person eats, people are advised to limit the number of fish consumed, and women of childbearing age and young children are often advised to not eat the fish at all.

The lip cancer on this bullhead indicates carcinogens in Wisconsin's Fox River. In some tributaries of the Great Lakes, fish cancer rates may reach 84 percent.

much trouble. These kinds of complex food webs tend to be more stable than simple food chains with few cross-links.

Nutrient Cycles in Ecosystems

All matter is made up of atoms. These atoms are cycled between the living and nonliving portions of an ecosystem. Some atoms are more common in living things than are others. Carbon, nitrogen, oxygen, hydrogen, and phosphorus are found in important organic molecules like proteins, DNA, carbohydrates, and fats, which are found in all kinds of living things. Organic molecules contain large numbers of carbon atoms attached to one another. These organic molecules are initially manufactured from inorganic molecules by the activities of producers and are transferred from one living organism to another in food chains. The processes of respiration and decay ultimately break down the complex organic molecules of organisms and convert them to simpler, inorganic constituents which are returned to the abiotic environment. In this section, we will look at the flow of three kinds of atoms within communities and between the abiotic and abiotic portions of an ecosystem: carbon, nitrogen, and phosphorus.

Carbon Cycle

All living things are composed of organic molecules that contain the atom carbon. The **carbon cycle** includes the

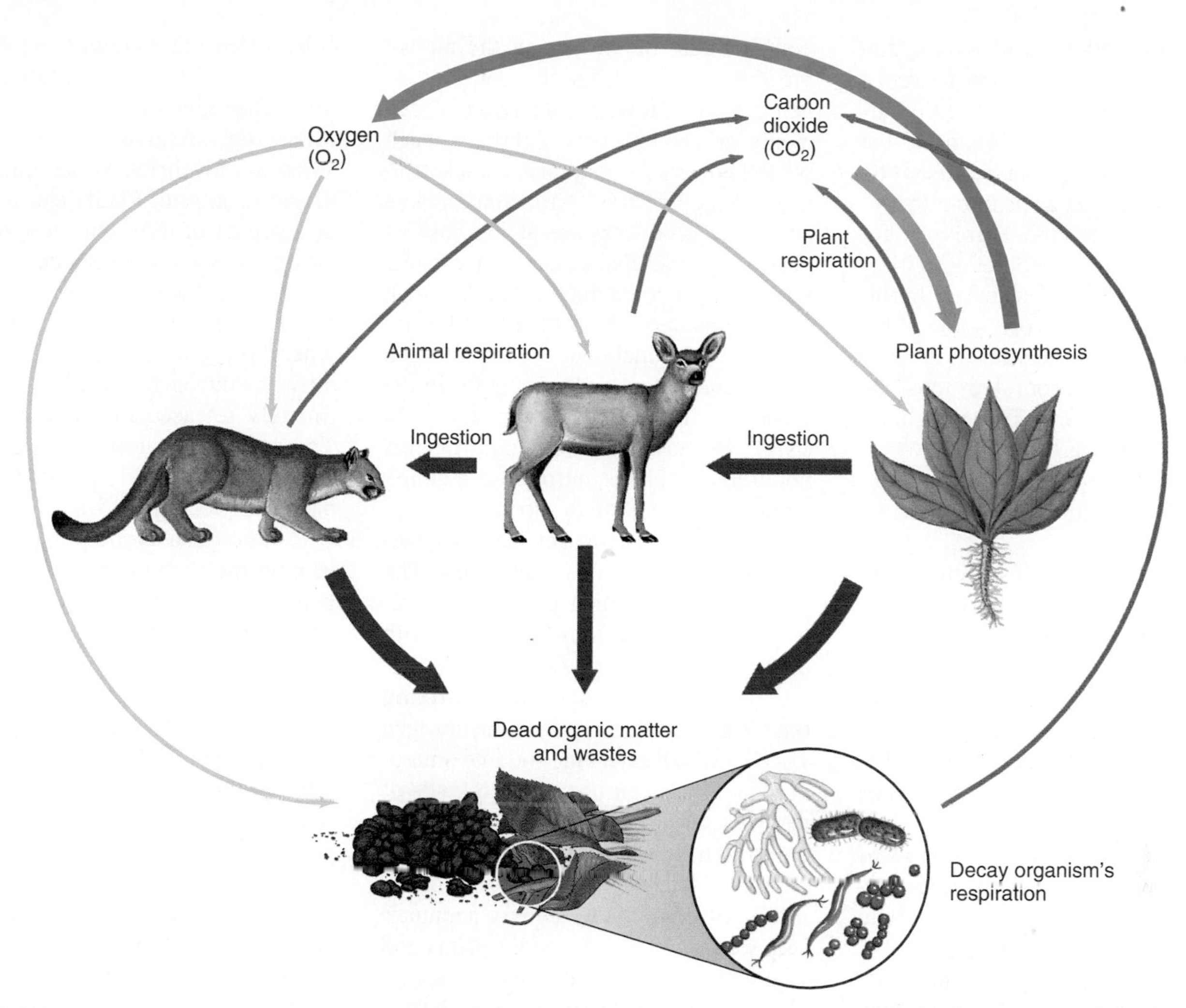

figure 5.15 **Carbon Cycle** Carbon atoms are cycled through ecosystems. Plants can incorporate carbon atoms from carbon dioxide into organic molecules when they carry on photosynthesis. The carbon-containing organic molecules are passed to animals when they eat plants or other animals. Organic wastes or dead organisms are consumed by decay organisms. All organisms, plants, animals, and decomposers return carbon atoms to the atmosphere when they carry on respiration. Oxygen atoms are being cycled at the same time that carbon atoms are being cycled.

processes and pathways involved in capturing inorganic carbon-containing molecules, converting them into organic molecules that are used by organisms, and the ultimate release of inorganic carbon molecules back to the abiotic environment. (See figure 5.15.) Carbon and oxygen combine to form the molecule carbon dioxide (CO_2), which is present in small quantities as a gas in the atmosphere and dissolved in water. During photosynthesis, carbon dioxide from the atmosphere is taken into the leaves of plants where it is combined with hydrogen from water molecules, (H_2O), which are absorbed from the soil by the roots and transported to the leaves. Many kinds of aquatic organisms such as algae and some bacteria also perform photosynthesis but absorb carbon dioxide and water molecules from the water in which they live. The energy needed to perform photosynthesis is provided by sunlight. As a result of photosynthesis, complex organic molecules such as carbohydrates (sugars) are formed. Oxygen molecules (O_2) are released into the atmosphere or water during the process of photosynthesis because water molecules are split to provide hydrogen atoms necessary to manufacture carbohydrate molecules. The remaining oxygen is released as a waste product of photosynthesis. In this process, light energy is converted to chemical-bond energy in organic molecules, such as sugar. Plants and other producer organisms use these sugars for growth and to provide energy for other necessary processes.

Herbivores can use these complex organic molecules as food. When an herbivore eats plants or algae, it breaks down the complex organic molecules into simpler organic molecular building blocks, which can be reassembled into the specific organic molecules that are part of its chemical structure. The carbon atom, which was once part of an organic molecule in a producer, is now part of an organic molecule in an herbivore. All organisms also carry on the process of respiration, in which oxygen from the atmosphere is used to break down large organic molecules into carbon dioxide and water. Much of the chemical-bond energy is released by respiration and is lost as heat, but the remainder is used by the herbivore for movement, growth, and other activities.

In similar fashion, when an herbivore is eaten by a carnivore, some of the carbon-containing molecules of the herbivore become incorporated into the body of the carnivore. The remaining organic molecules are broken down in the process of respiration to obtain energy and carbon dioxide and water are released.

The organic molecules contained in animal waste products and dead organisms are acted upon by decomposers that use these organic materials as a source of food. The decay process of decomposers involves respiration and releases carbon dioxide and water so that naturally occurring organic molecules are typically recycled.

In the carbon cycle, all organisms require organic molecules for their survival and must either manufacture them or consume them. Photosynthetic organisms (producers) capture inorganic carbon in the form of carbon dioxide molecules and manufacture organic molecules. Nearly all organisms, including plants, carry on respiration, in which organic molecules are broken down to provide energy and inorganic carbon dioxide is released. The same carbon atoms are used over and over again. In fact, you are not exactly the same person today that you were yesterday. Some of your carbon atoms are different. Furthermore, those carbon atoms have been involved in many other kinds of living things over the past several billion years. Some of them were temporary residents in dinosaurs, extinct trees, or insects, but at this instant they are part of you.

Nitrogen Cycle

Another very important nutrient cycle, the **nitrogen cycle,** involves the cycling of nitrogen atoms between the abiotic and biotic components and among the organisms in an ecosystem. Seventy-eight percent of the gas in the air we breathe is made up of molecules of nitrogen gas (N_2). However, the two nitrogen atoms are bound very tightly to each other, and very few organisms are able to use nitrogen in this form. Since plants and other producers are at the base of nearly all food chains, they must make new nitrogen-containing molecules, such as proteins and DNA. Plants and other producers are unable to use the nitrogen in the atmosphere and must get it in the form of nitrate (NO_3^-) or ammonia (NH_3). Because atmospheric nitrogen is not usable by plants, nitrogen-containing compounds are often in short supply and the availability of nitrogen is often a factor that limits the growth of plants. The primary way in which plants obtain nitrogen compounds they can use is with the help of bacteria that live in the soil.

Bacteria, called **nitrogen-fixing bacteria,** are able to convert the nitrogen gas (N_2) which enters the soil into ammonia that plants can use. Certain kinds of these bacteria live freely in the soil and are called **free-living nitrogen-fixing bacteria.** Others, known as **symbiotic nitrogen-fixing bacteria,** have a mutualistic relationship with certain plants and where they live in nodules in the roots of plants known as legumes (peas, beans, and clover) and certain trees such as alders. Some grasses and evergreen trees appear to have a similar relationship with certain root fungi that seem to improve the nitrogen-fixing capacity of the plant.

Once plants and other producers have nitrogen available in a form they can use, they can construct proteins, DNA, and other important nitrogen-containing organic molecules. When herbivores eat plants, the plant protein molecules are broken down to smaller building blocks called amino acids. These amino acids are then reassembled to form proteins typical for the herbivore. This same process is repeated throughout the food chain.

Bacteria and other types of decay organisms are involved in the nitrogen cycle also. Dead organisms and their waste products contain molecules, such as proteins, urea, and uric acid, that contain nitrogen. Decomposers break down these nitrogen-containing organic molecules, releasing ammonia, which can be used directly by many kinds of plants. Still other kinds of soil bacteria called **nitrifying bacteria** are able to convert ammonia to nitrite, which can be converted to nitrate. Plants can use nitrate as a source of nitrogen for synthesis of nitrogen-containing organic molecules.

Finally, bacteria, known as **denitrifying bacteria,** are, under conditions where oxygen is absent, able to convert nitrite to nitrogen gas (N_2), which is ultimately released into the atmosphere. These nitrogen atoms can re-enter the cycle with the aid of nitrogen-fixing bacteria. (See figure 5.16.)

Although a cyclic pattern is present in both the carbon cycle and the nitrogen cycle, the nitrogen cycle shows two significant differences. First, most of the difficult chemical conversions are made by bacteria and other microorganisms. Without the activities of bacteria, little nitrogen would be available and the world would be a very different place. Second, although nitrogen enters organisms by way of nitrogen-fixing bacteria and returns to the atmosphere through the actions of denitrifying bacteria, there is a secondary loop in the cycle that recycles nitrogen compounds directly from dead organisms and wastes directly back to producers.

In naturally occurring soil, nitrogen is often a limiting factor of plant growth. To increase yields, farmers provide extra sources of nitrogen in several ways. Inorganic fertilizers are a primary method of increasing the nitrogen available. These fertilizers may contain ammonia, nitrate, or both.

Since the manufacture of nitrogen fertilizer requires a large amount of energy, fertilizer is expensive. Therefore, farmers use alternative methods to supply nitrogen and reduce their cost of production. Several different techniques are effective. Farmers can alternate nitrogen-yielding crops like soybeans with nitrogen-demanding crops like corn. Since soybeans are legumes that have symbiotic nitrogen-fixing bacteria in their roots, if soybeans are planted one year, the excess nitrogen left in the soil can be used by the corn plants grown the next year. Some farmers even plant

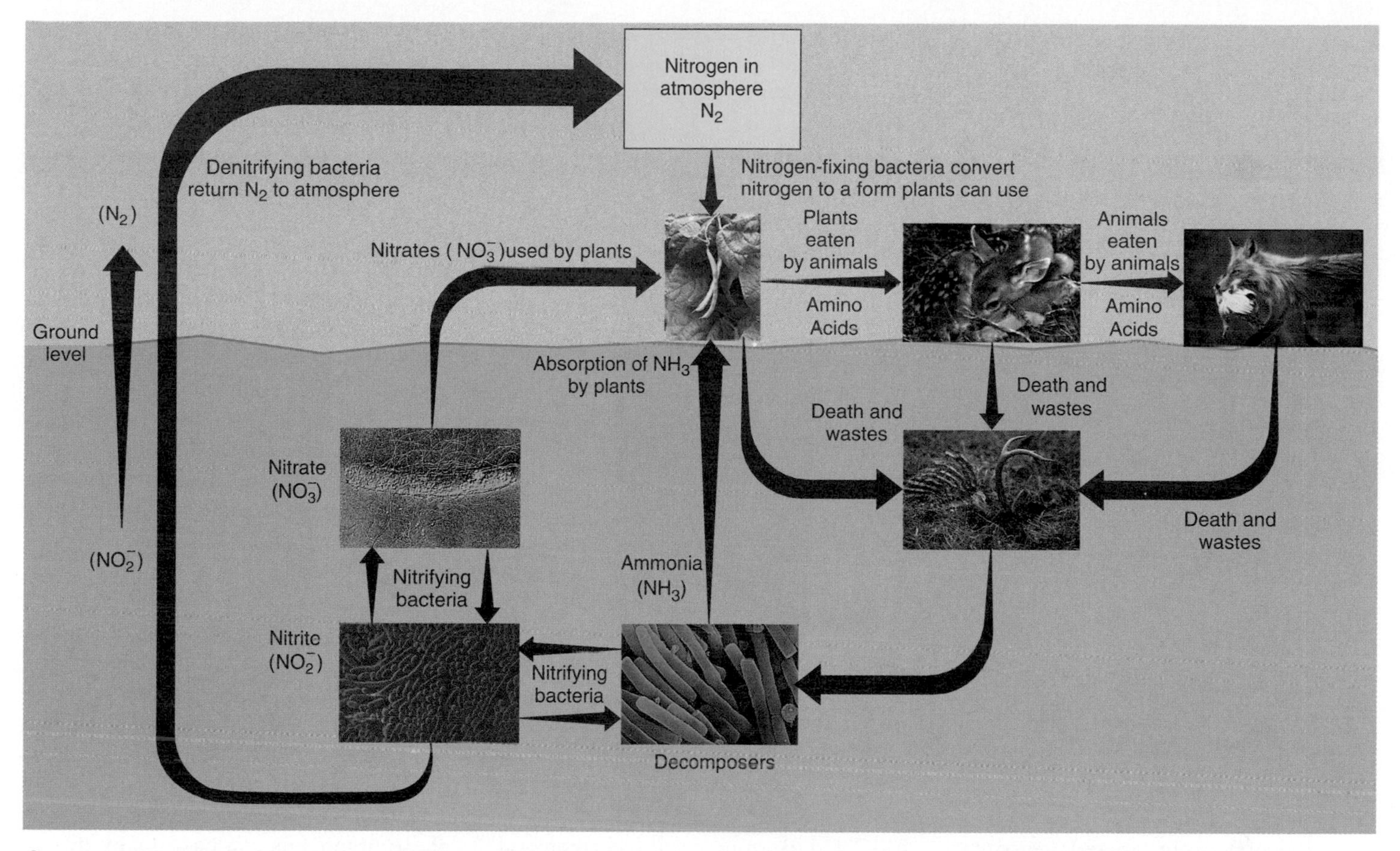

figure 5.16 **Nitrogen Cycle** Nitrogen atoms are cycled in ecosystems. Atmospheric nitrogen is converted by nitrogen-fixing bacteria to a form that plants can use to make protein and other compounds. Proteins are passed to other organisms when one organism is eaten by another. Dead organisms and waste products are acted on by decay organisms to form ammonia, which may be reused by plants or converted to other nitrogen compounds by other kinds of bacteria. Denitrifying bacteria are able to convert inorganic nitrogen compounds into atmospheric nitrogen.

alternating strips of soybeans and corn in the same field. A slightly different technique involves growing a nitrogen-fixing crop for a short period of time and then plowing the crop into the soil and letting the organic matter decompose. The ammonia released by decomposition serves as fertilizer to the crop that follows. This is often referred to as green manure. Farmers can also add nitrogen to the soil by spreading manure from animal production operations or dairy farms on the field and relying on the soil bacteria to decompose the organic matter and release the nitrogen for plant use.

Phosphorus Cycle

Phosphorus is another kind of atom common in the structure of living things. It is present in many important biological molecules such as DNA and in the membrane structure of cells. In addition, the bones and teeth of animals contain significant quantities of phosphorus. The ultimate source of phosphorus atoms is rock. In nature, new phosphorus compounds are released by the erosion of rock and become dissolved in water. Plants use the dissolved phosphorus compounds to construct the molecules they need. Animals obtain the phosphorus they need when they consume plants or other animals. When an organism dies or excretes waste products, decomposer organisms recycle the phosphorus compounds back into the soil. Phosphorus compounds that are dissolved in water are ultimately precipitated as deposits. Geologic processes elevate these deposits and expose them to erosion, thus making these deposits available to organisms. Waste products of animals often have significant amounts of phosphorus. In places where large numbers of seabirds or bats congregate for hundreds of years, the thickness of their droppings (called guano) can be a significant source of phosphorus for fertilizer. (See figure 5.17.) In many soils, phosphorus is in short supply and must be provided to crop plants to get maximum yields. Phosphorus is also in short supply in aquatic ecosystems.

Fertilizers usually contain nitrogen, phosphorus, and potassium compounds. The numbers on a fertilizer bag indicate the percentage of each in the fertilizer. For example, a 6–24–24 fertilizer has 6 percent nitrogen, 24 percent phosphorus, and 24 percent potassium compounds. In addition to carbon, nitrogen, and phosphorus, potassium and other elements are cycled within ecosystems. In an agriculture ecosystem, these elements are removed when the crop is harvested. Therefore, farmers must not only return

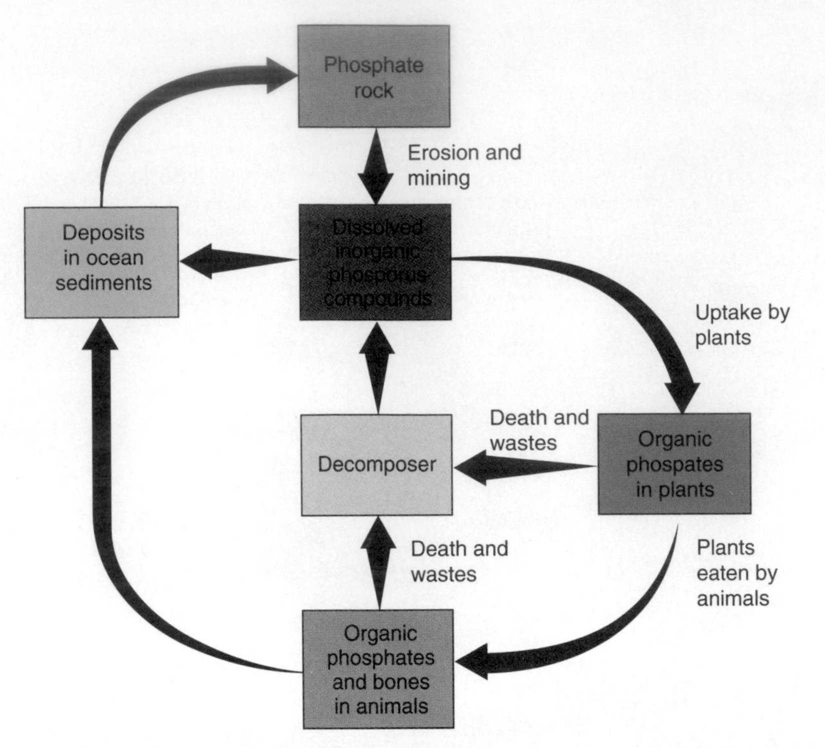

figure 5.17 **Phosphorus Cycle** The source of phosphorus is rock that, when dissolved, provides a source of phosphate used by plants and animals.

the nitrogen, phosphorus, and potassium, they must also analyze for other less prominent elements and add them to their fertilizer mixture as well. Aquatic ecosystems are also sensitive to nutrient levels. High levels of nitrates or phosphorus compounds often result in rapid growth of aquatic producers. In aquaculture, such as that used to raise catfish, fertilizer is added to the body of water to stimulate the production of algae which is the base of many aquatic food chains.

Human Impact on Nutrient Cycles

In order to appreciate how ecosystems function, it is important to have an understanding of how nutrients such as carbon, nitrogen, and phosphorus flow through them. When we look at these cycles from a global perspective, it is apparent that humans have significantly altered them in many ways. Two activities have caused significant changes in the carbon cycle: burning fossil fuels and converting forests to agricultural land. Fossil fuels are carbon-containing molecules produced when organisms were fossilized. Burning fossil fuels (coal, oil, and natural gas) releases large amounts of carbon dioxide into the atmosphere. The conversion of forest ecosystems that tend to store carbon for long periods to agricultural ecosystems that store carbon only temporarily has also disrupted the natural carbon cycle. Less carbon is being stored in the bodies of large, long-lived plants such as trees. One consequence of these actions is that the amount of carbon dioxide in the atmosphere has been increasing steadily since humans began to use fossil fuels extensively. It has become clear that increasing carbon dioxide is causing changes in the climate of the world, and many nations are seeking to reduce energy use and prevent deforestation. This topic will be discussed in more detail in chapter 17 which deals with air pollution.

The burning of fossil fuels has also altered the nitrogen cycle. When fossil fuels are burned, the oxygen and nitrogen in the air are heated to high temperatures and a variety of nitrogen-containing compounds are produced. These compounds are used by plants as nutrients for growth. Many people suggest that these sources of nitrogen, along with that provided by fertilizers, have doubled the amount of nitrogen available today as compared to pre-industrial times.

Fertilizer is used in agriculture to increase the growth of crops. These nutrients are intended to become incorporated into the bodies of the plants and animals that we raise for food. However, if too much nitrogen or phosphorus is applied as fertilizer or if they are applied at the wrong time, much of this fertilizer is carried into aquatic ecosystems. In addition, raising large numbers of animals for food in concentrated settings results in huge amounts of animal waste that contain nitrogen and phosphorus compounds that often enter local water sources. These additions of nitrogen and phosphorus to aquatic ecosystems is particularly significant since aquatic ecosystems normally are starved for these nutrients. The presence of large amounts of these nutrients in either freshwater or saltwater results in increased rates of growth of bacteria, algae, and aquatic plants. Increases in these organisms can have many different effects. Many algae are toxic and when their numbers increase significantly, fish kills and human poisoning result. An increase in the number of plants and algae in aquatic ecosystems also can lead to low oxygen concentrations in the water. When these organisms die, decomposers use oxygen from the water as they break down the dead organic matter. This lowers the oxygen concentration and many organisms die. For example, each summer a major "dead zone" of about 18,000 km^2 (7,000 mi^2) develops in the Gulf of Mexico off the mouth of the Mississippi River. This "dead zone" contains few fish and bottom-dwelling organisms. It is caused by low oxygen levels brought about by the rapid growth of algae and bacteria in the nutrient-rich waters. The nutrients can be traced to the extensive use of fertilizer in the major farming areas of the central United States, farming areas drained by the Mississippi River and its tributaries. Thus, fertilizer use in the agricultural center of the United States results in the death of fish in the Gulf of Mexico. (See figure 5.18).

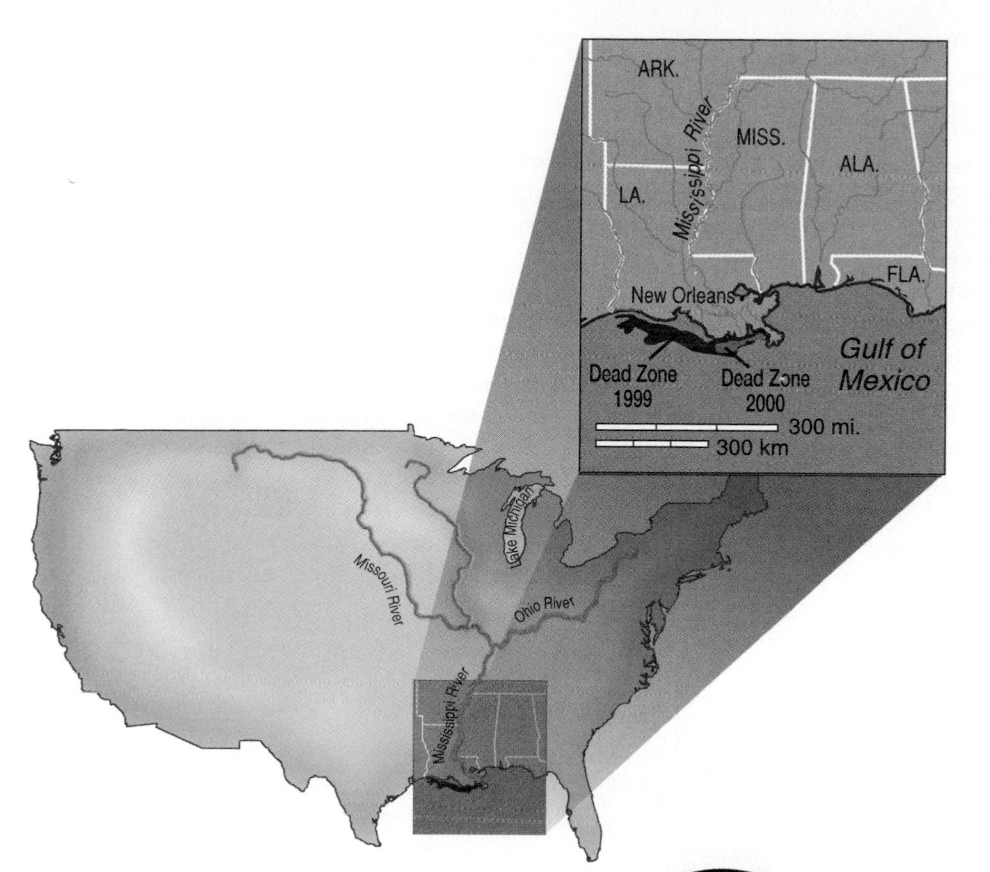

figure 5.18 **Nutrient Impact on Aquatic Organisms** The Mississippi River drainage system carries water from the agricultural center of the United States to the Gulf of Mexico. Extensive use of fertilizer for agriculture results in nitrogen and phosphorus compounds being carried to the Gulf of Mexico. When the organisms that grew as a result of the fertilizer die, a major region of the Gulf of Mexico has oxygen concentrations too low to support most kinds of life.

ISSUES & ANALYSIS

Reintroducing Wolves to the Yellowstone Ecosystem

At one time, large predators such as wolves, mountain lions, and grizzly bears were an important part of the community of organisms in the region of Yellowstone National Park. For generations, ranchers in the area killed these large predators because they were a threat to livestock. By the early 1900s, the wolf was extinct in the U.S. Rocky Mountains and, in 1967, was listed as an endangered species. Without large predators, the populations of large grazing animals in the park (elk, moose, and bison) increased to levels that were not natural. In the early 1990s, the U.S. Fish and Wildlife Service reintroduced wolves into the park to restore a more natural mix of species and to control the burgeoning grazer populations.

By 2000, 11 packs of wolves and several lone wolves were active in and around Yellowstone National Park. The total population is now over 100. The wolves are successful in reducing the elk population. During the winter, when elk are weaker, a typical pack of wolves will kill an elk every two to three days. Another effect of the wolf introduction has been the reduction in the coyote population, since wolves kill coyotes. Historically, coyotes were not important in Yellowstone, but when wolves were eliminated, the coyote population rose significantly. The reintroduction of wolves has restored a more natural balance between these two predators.

However, many of the ranchers in the area were not pleased with the reintroduction of wolves and fought the plan in the courts. It is highly likely that as the population of wolves increases, the wolves will migrate from the park and prey on livestock. This has already happened in a few instances. The losses of livestock to wolves was anticipated and a fund was set up to reimburse ranchers for livestock they lost to wolf predation. However, many ranchers were not satisfied with this mechanism and filed a lawsuit to have the wolves removed from the park. The lawsuit was based on a technicality of the Endangered Species Act. The wolves in Yellowstone received a "nonessential experimental" designation under the Endangered Species Act. Because the wolves were migrating from Canada into the area, the reintroduction program simply accelerated the process. The special designation would allow ranchers to kill reintroduced wolves that attack their livestock; however, wolves that migrate naturally from Canada were not to be killed. The ranchers argued that they would not be able to tell reintroduced "nonessential experimental" wolves from those that arrived by migration from other areas.

In December of 1997, a U.S. District Judge ordered the removal of the wolves from Yellowstone. The ruling was appealed. Finally, in April 2000, the American Farm Bureau, which had been challenging the reintroduction program on behalf of local ranchers, announced that it would not pursue further legal action. Therefore, wolves are again a legal as well as an important biological component of the Yellowstone ecosystem.

- Is it important to return species to areas that were part of their former range?
- Should the economic interests of ranchers be more important than the biological interests of managing Yellowstone National Park as an example of a natural ecosystem?
- Should special interests be allowed to use "technicalities" to circumvent the generally accepted policy?

Summary

Everything that affects an organism during its lifetime is collectively known as its environment. The environment of an organism can be divided into biotic (living) and abiotic (nonliving) components.

The space an organism occupies is known as its habitat, and the role it plays in its environment is known as its niche. The niche of a species is the result of natural selection directing the adaptation of the species to a specific set of environmental conditions.

Organisms interact with one another in a variety of ways. Predators kill and eat prey. Organisms that have the same needs compete with one another and do mutual harm, but one is usually harmed less and survives. Symbiotic relationships are those in which organisms live in physical contact with one another. Parasites live in or on another organism and derive benefit from the relationship, harming the host in the process. Commensal organisms derive benefit from another organism but do not harm the host. Mutualistic organisms both derive benefit from their relationship.

A community is the biotic portion of an ecosystem which is a set of interacting groups of organisms. Those organisms and their abiotic environment constitute an ecosystem. In an ecosystem, energy is trapped by producers and flows from producers through various trophic levels of consumers (herbivores, carnivores, omnivores, and decomposers). About 90 percent of the energy is lost as it passes from one trophic level to the next. This means that the amount of biomass at higher trophic levels is usually much less than that at lower trophic levels. The sequence of organisms through which energy flows is known as a food chain. Several interconnecting food chains constitute a food web.

The flow of atoms through an ecosystem involves all the organisms in the community. The carbon, nitrogen, and phosphorus cycles are examples of how these materials are cycled in ecosystems.

Interactive Exploration

Check out the website at

http://www.mhhe.com/environmentalscience

and click on the cover of this textbook for interactive versions of the following:

KNOW THE BASICS

abiotic factors *81*
biomass *94*
biotic factors *81*
carbon cycle *96*
carnivore *92*
coevolution *86*
commensalism *89*
community *91*
competition *88*
competitive exclusion principle *88*
consumer *92*
decomposer *92*
denitrifying bacteria *98*
detritus *94*
ecology *81*
ecosystem *91*
ectoparasite *89*
endoparasite *89*
environment *81*
evolution *86*
extinction *86*
food chain *94*
food web *94*
free-living nitrogen-fixing bacteria *98*
habitat *82*
herbivore *92*
host *88*
interspecific competition *88*
intraspecific competition *88*
keystone species *92*
limiting factor *82*
mutualism *90*
mycorrhizae *90*
natural selection *85*
niche *82*
nitrogen cycle *98*
nitrifying bacteria *98*
nitrogen-fixing bacteria *98*
omnivore *92*
parasite *88*
parasitism *88*
predation *87*
predator *87*
prey *87*
primary consumer *92*
producer *92*
range of tolerance *82*
secondary consumer *92*
speciation *86*
species *84*
symbiosis *88*
symbiotic nitrogen-fixing bacteria *98*
trophic level *93*
vector *89*

On-line Flashcards

Electronic Glossary

IN THE REAL WORLD

A change in the predator/prey relationship between orcas and otters is apparently affecting the sea otter population. This is especially important since sea otters are considered keystone species in the marine ecosystem. For more information on this relationship and how humans are playing the role of competitor in the interaction, read Mysterious Disappearance of Sea Otters in Alaska. For a related story, check out Judge Rules on Excessive Fishing in Alaska.

Another keystone species, bison, are also affected by human interactions. Once bison leave Yellowstone Park, they are no longer protected and are actually targeted because bison are not supposed to roam freely outside the park. Why would we purposely destroy a keystone species? To further understand the reasons behind this controversial management policy, read Should Bison Leaving Yellowstone National Park Be Shot?

What's large, brown, floats, and provides critical habitat and nutrients for hundreds of fish and invertebrates off the southern U.S. coast? Sargassum is a brown algae that grows in huge floating mats in the Atlantic Ocean. The U.S. agricultural community recognizes the nutritional value of the algae and uses it to feed hogs and cattle. This has led to competition and conflict. The controversial harvest of this habitat is further explained in Controversy over Algae Harvest and Habitat Conservation in the Atlantic.

We know human activities impact nutrient cycles, but can people change food webs to benefit the ecosystem? Look at the Food Web Control of Primary Production in Lakes case study for answers to these questions and more.

Historically, we have manipulated populations and interactions among organisms for our benefit. Using drugs could be considered an example of how humans manipulate parasite populations. Drug-resistant Strain of Sleeping Sickness Appears in Ethiopia explains how this parasite population is "striking back" through natural selection.

The dead zone in the Gulf of Mexico off the mouth of the Mississippi River is not alone. Another dead zone off the coast of North Carolina is the result of runoff from hog farms. Read N.C. Aquatic Dead Zone from Floods after Hurricane Floyd for this story.

The Great Lakes are also suffering from runoff problems. In this chapter you read about organic contaminants in Great Lakes' fish and how these contaminants can influence human health. Now see the story of a contaminated river that empties into Lake Michigan. PCB Contamination in the Fox River case study.

PUT IT IN MOTION

Evolution . . . Natural selection . . . Speciation . . . you may have seen these terms before, but how are new species produced and how does natural selection occur? For a great animation that clearly illustrates the speciation process, check out Allopatric Speciation, and for an *excellent* animation that clearly describes the natural selection process is Pesticide Resistance.

If you would like to check your understanding of the major roles of organisms in an ecosystem and how the roles correspond to the energy flow through an ecosystem, study the, Energy Flow animation.

Two more animations worth looking at are the Carbon Cycle and the Nitrogen Cycle. These animations illustrate how carbon and nitrogen cycle through ecosystems and how humans impact the flow of these materials.

TEST PREPARATION

Review Questions

1. Define environment.
2. Describe, in detail, the niche of a human.
3. How is natural selection related to the concept of niche?
4. List five predators and their prey organisms.
5. How is an ecosystem different from a community?
6. Humans raising cattle for food is what kind of relationship?
7. Give examples of organisms that are herbivores, carnivores, and omnivores.
8. What are some different trophic levels in an ecosystem?
9. Describe the carbon cycle, the nitrogen cycle, and the phosphorus cycle.
10. Analyze an aquarium as an ecosystem. Identify the major abiotic and biotic factors. List members of the producer, primary consumer, secondary consumer, and decomposer trophic levels.

Critical Thinking Questions

1. Ecologists and political scientists look at habitat destruction differently. Consider the Environmental Close-Up about conservation plans and the political/economic and scientific issues that surround conservation plans. Identify some perspectives each discipline has to contribute to our understanding of habitat destruction. What values does each place on the ideas of the other discipline? What do you think about the issue of creating conservation plans? Protecting habitat from destruction? Why?
2. Even before humans entered the scene, many species of plants and animals were extinct and new ones had developed. Why are we even concerned about endangered species, given the fact that species have always come and gone?
3. Humans have had a major impact on many ecosystems. Consider the questions in the Issues and Analysis section about reintroducing wolves to Yellowstone Park. Is this the appropriate way for humans to try to repair the environmental damage they have caused? Identify some of the values and beliefs that make up these different perspectives: ranchers and conservationists. Which do you find the most compelling? Why?
4. Concentrations of industrial chemicals are high in some species of fish, high enough to call for an advisory to limit the number of fish a person should eat within a given period of time. Many of these chemicals are thought to cause cancer, but cancer is an effect that is often not felt for decades after exposure. How do scientists decide how many fish can be safely eaten? Is there any "safe" level? What evidence would convince you that there is danger? How could you tell?
5. You notice that after using pesticides on your farm field that the number of insects declines for a year. The next year, though, they come back and you need to reapply the pesticide. This time, though, there is less of an effect on the insect population. A third application in another year has even less of an effect. What is your hypothesis about what is happening here? Design an experiment that tests your hypothesis.

KEY CHAPTER LINKS | ESSENTIAL STUDY PARTNER | BIO COURSE | GLOBAL ISSUES | REGIONAL PERSPECTIVES | PRACTICE QUIZZING

CHAPTER

Kinds of Ecosystems and Communities

Objectives

After reading this chapter, you should be able to:

- Recognize the difference between primary and secondary succession.
- Describe the process of succession from pioneer to climax community in both terrestrial and aquatic situations.
- Associate typical plants and animals with the various terrestrial biomes.
- Recognize the physical environmental factors that determine the kind of climax community that will develop.
- Differentiate the forest biomes that develop based on temperature and rainfall.
- Describe the various kinds of aquatic ecosystems and the factors that determine their characteristics.

Chapter Outline

Succession
- Primary Succession
- Secondary Succession
- The Changing Nature of the Climax Concept

Biomes: Major Types of Terrestrial Climax Communities
- The Effect of Elevation on Climate and Vegetation
- Desert
- Grassland

Environmental Close-Up: *Grassland Succession*

- Savanna

Global Perspective: *Tropical Rainforests—A Special Case?*

- Tropical Rainforest
- Temperate Deciduous Forest

Environmental Close-Up: *Forest Canopy Studies*

- Taiga, Northern Coniferous Forest, or Boreal Forest
- Tundra

Major Aquatic Ecosystems
- Marine Ecosystems
- Freshwater Ecosystems

Issues & Analysis: *Protecting Old-Growth Temperate Rainforests of the Pacific Northwest*

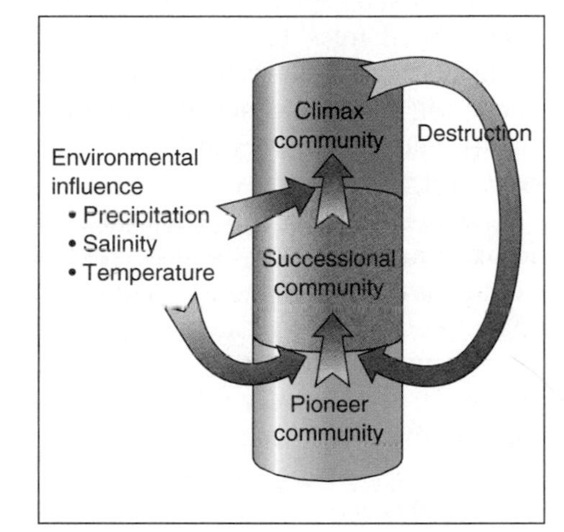

Succession

Ecosystems are dynamic, changing units. On a daily basis, plants grow and die, animals feed on plants and on one another, and decomposers recycle the chemical elements that make up the biotic portion of any ecosystem. Abiotic factors (such as temperature, rainfall, intensity of sunlight, and seasonality) also have a major influence on the kind of community that will be established. Since all organisms are linked together in a community, any change in the community affects many organisms within it. Certain conditions within a community are keys to the kinds of organisms that are found associated with one another. Grasshoppers need grass for food, robins need trees to build nests, and herons need shallow water to find food. Each organism has specific requirements that must be met in the community, or it will not survive.

Over long time periods, it is possible to see trends in the way the structure of a community changes and to recognize that climate greatly influences the kind of community that becomes established in an area. Generally, this series of changes eventually results in a relatively long-lasting, stable combination of species that is self-perpetuating. The concept that communities proceed through a series of regular, predictable changes in structure over time is called **succession.** The stable, long-lasting community that is the result of succession is called a **climax community.** The kind of climax community that will develop is primarily determined by climate. Some will be forests, while others will be grassland or deserts. Succession occurs because the activities of organisms cause changes in their surroundings that make the environment suitable for other kinds of organisms. When new species become established, they compete with the original inhabitants and often replace them completely. Slowly, over time, species are replaced by invaders, and it is recognized that a significantly different community has become established. There are several factors that determine the pace and direction of the successional process. Ecologists recognize two different kinds of succession. **Primary succession** is a successional progression, which begins with a total lack of organisms and bare mineral surfaces or water. Primary succession often takes an extremely long time, since there is no soil and there are few readily available nutrients for plants to use for growth. **Secondary succession** is much more commonly observed and generally proceeds more rapidly, because it begins with the destruction or disturbance of an existing ecosystem. Therefore, there is at least some soil and often there are seeds or roots from which plants can begin growing almost immediately.

Primary Succession

Primary succession can begin on a bare rock surface, pure sand, or standing water. Since succession on rock and sand is somewhat different from that which occurs with watery situations, we deal with them separately. We discuss terrestrial succession first.

Terrestrial Primary Succession

There are several factors that determine the rate of succession and the kind of climax community that will develop in an area. The kind of substrate (rock, sand, clay) will greatly affect the kind of soil that will develop. The kinds of spores, seeds, or other reproductive structures will determine the species available to colonize the area. The climate will determine the species that will live in an area and how rapidly they will grow. The rate of growth will determine how quickly organic matter will accumulate in the soil. The kind of substrate, climate, and amount of organic matter will influence the amount of water available for plant growth. Let's look at a specific example of how these factors are interrelated in an example of primary succession from bare mineral surfaces.

Bare rock or sand is a very inhospitable place for organisms to live. The temperature changes drastically, there is no soil, there is little moisture, the organisms are exposed to the damaging effect of the wind, few nutrients are available, and few places are available for organisms to attach themselves or hide. However, a few kinds of organisms can survive in even this inhospitable environment. This collection of organisms is known as the **pioneer community** because it is the first to colonize bare rock. (See figure 6.1.)

figure 6.1 **Pioneer Organism** The lichen growing on this rock is able to accumulate bits of debris, carry on photosynthesis, and aid in breaking down the rock. All of these activities contribute to the formation of a thin layer of soil, which is necessary for plant growth in the early stages of succession.

The dominant organism in this initial community is something called a lichen. Lichens are actually mutualistic relationships between two kinds of organisms: algae or bacteria that carry on photosynthesis and fungi that attach to the rock surface and retain water. The growth and development of lichens is often a slow process. It may take lichens one hundred years to grow as large as a dinner plate. Lichens are the producers in this simple ecosystem, and many tiny consumer organisms may be found associated with lichens. Some feed on the lichen and many use it as a place of shelter, since even a drizzle is like a torrential rain for a microscopic animal. Since lichens are firmly attached to rock surfaces, they also tend to accumulate bits of airborne debris and store small amounts of water that would otherwise blow away or run off the rock surface. Acids produced by the lichen tend to cause the breakdown of the rock substrate into smaller particles. This fragmentation of rock, aided by physical and

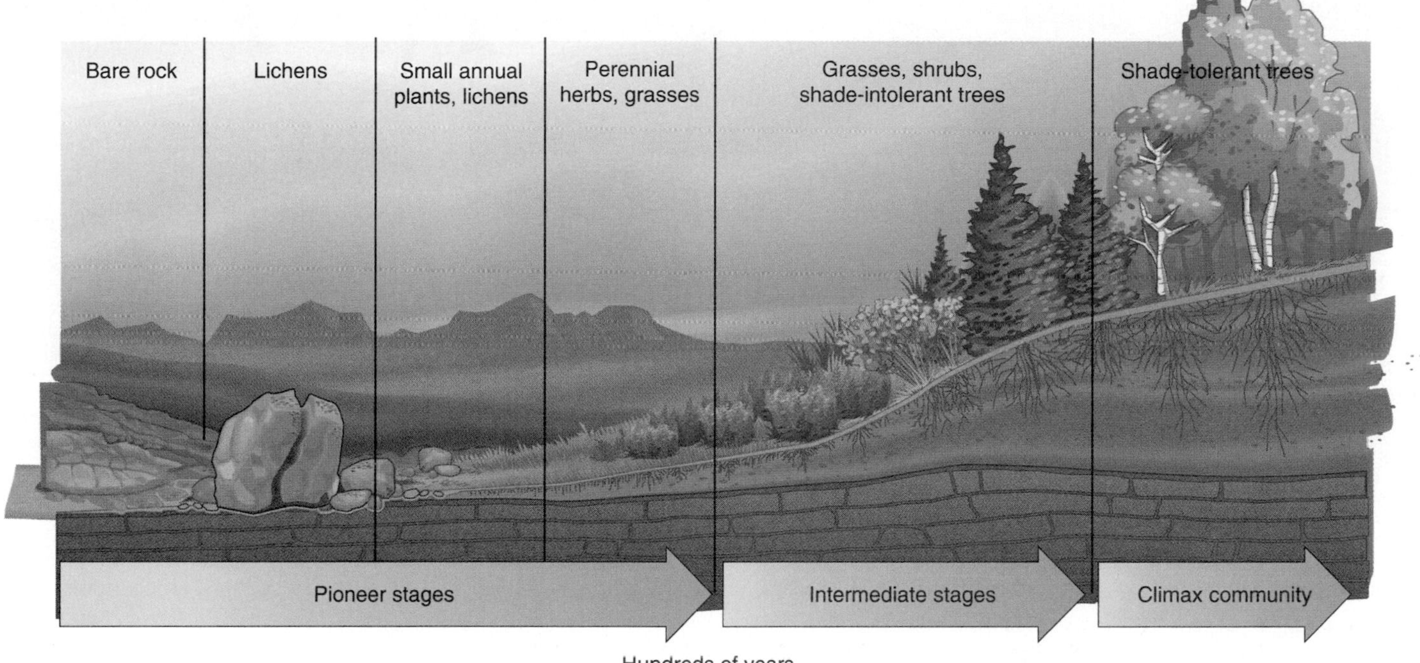

figure 6.2 **Primary Succession on Land** The formation of soil is a major step in primary succession. Until soil is formed, the area is unable to support large amounts of vegetation, which modify the harsh environment. Once soil formation begins, the site proceeds through an orderly series of stages toward a climax community.

chemical weathering processes, along with the trapping of debris and the contribution of organic matter by the death of lichens and other organisms, ultimately leads to the accumulation of a very thin layer of soil.

This thin layer of soil is the key to the next stage in the successional process. The layer can retain some water and support some fungi, certain small worms, insects, bacteria, protozoa, and perhaps a few tiny annual plants that live for only one year but produce flowers and seeds that fall to the soil and germinate the following growing season. Many of these initial organisms or their reproductive structures are very tiny and will arrive as a result of wind and rain. As these organisms grow, reproduce, and die, they contribute additional organic material for the soil-building process, and the soil layer increases in thickness and is better able to retain water. This stage, which is dominated by annual plants, eliminates the lichen community because the plants are taller and shade the lichens, depriving them of sunlight. This stage is itself replaced by a community of small perennial grasses and herbs that live for several years. The short, nonwoody grasses and herbs are often replaced by larger perennial woody shrubs, which are often replaced by larger trees that require lots of sunlight, which are replaced by trees that can tolerate shade. Sun-loving (shade-intolerant) trees are replaced by shade-tolerate trees because the seedlings of shade-intolerant trees cannot grow in the shade of their parents while seedlings of shade-tolerant trees can. Eventually, a relatively stable, long-lasting, complex, and interrelated climax community of plants, animals, fungi, and bacteria is produced. Each step in this process from pioneer community to climax community is called a **successional stage,** or **seral** stage, and the entire sequence of stages—from pioneer community to climax community—is called a **sere.** (See figure 6.2.)

Although in this example we have described a successional process that began with a lichen pioneer community and ended with a climax forest, it is important to recognize that the process of succession can stop at any point along this continuum. In certain extreme climates, lichen communities may last for hundreds of years and must be considered climax communities. Others reach a grass-herb stage and proceed no further. The specific kind of climax community produced depends on such things as climate and soil type, which are discussed in greater detail later in this chapter. However, when successional communities are compared to climax communities, climax communities show certain characteristics.

1. Climax communities maintain their mix of species for a long period of time.
2. They are in energy balance, while successional communities show changes in species and tend to accumulate large amounts of new material (gain energy).
3. They tend to have many more kinds of organisms and kinds of interactions among organisms than does a successional community.

The general trend in succession is toward increasing complexity and energy efficiency, compared to the successional communities that preceded them.

figure 6.3 **Primary Succession from a Pond to a Wet Meadow** A shallow pond will fill slowly with organic matter from producers in the pond. Eventually, a wet soil will form and grasses will become established. In many areas, this will be succeeded by a climax forest.

Aquatic Primary Succession

The principal concepts of land succession can be applied to aquatic ecosystems. Except for the oceans, most aquatic ecosystems are considered temporary. Certainly, some are going to be around for thousands of years, but eventually they will disappear and be replaced by terrestrial ecosystems as a result of normal successional processes. All aquatic ecosystems receive a continuous input of soil particles and organic matter from surrounding land, which results in the gradual filling in of shallow bodies of water like ponds and lakes.

In deep portions of lakes and ponds, only floating plants and algae can exist, but as the amount of sediment accumulates and the water depth becomes less, it becomes possible for certain species of submerged plants to establish their roots in the sediments of the bottom of shallow bodies of water. They carry on photosynthesis, which results in a further accumulation of organic matter. These plants also tend to trap sediments that flow into the pond or lake from streams or rivers, resulting in a further decrease in water depth. Eventually, as the water becomes shallower, emergent plants become established. They have leaves that float on the surface of the water or project into the air. The network of roots and stems below the surface of the water results in the accumulation of more material, and the water depth decreases as material accumulates on the bottom. As the process continues, a wet soil is formed and grasses and other plants that can live in wet soil become established. This successional stage is often called a wet meadow. The activities of plants tend to draw moisture from the soil, and, as more organic matter is added to the top layer of the soil, it becomes somewhat drier. Once this occurs, the stage is set for a typical terrestrial successional series of changes, eventually resulting in a climax community typical for the climate of the area. (See figure 6.3.)

Since the shallower portions of most lakes and ponds are at the shore, it is often possible to see the various stages in aquatic succession from the shore. In the central, deeper portions of the lake, there are only floating plants and algae. As we approach the shore, we first find submerged plants like *Elodea* and algal mats, then emergent vegetation like water lilies and cattails, then grasses and sedges that can tolerate wet soil, and on the shore the beginnings of a typical terrestrial succession resulting in the climax community typical for the area.

In many northern ponds and lakes, sphagnum moss forms thick, floating mats. These mats may allow certain plants that can tolerate wet soil conditions to become established. The roots of the plants bind the mat together and establish a floating bog, which may contain small trees and shrubs as well as many other smaller, flowering plants. (See figure 6.4.) You can recognize that the entire system is floating only when you jump on it and the trees sway or when you step through a weak zone in the mat and sink to your hips in water. Eventually, these bogs will become increasingly dry and the normal climax vegetation for the area will succeed the more temporary bog stage.

figure 6.4 **Floating Bog** In many northern regions, sphagnum moss forms a floating mat that can support the growth of other plants. If a person were to walk on this mat, it would bounce up and down, because it is floating on water.

Mature oak/hickory forest destroyed	Farmland abandoned	Annual plants	Grasses and biennial herbs	Perennial herbs and shrubs begin to replace grasses and biennials	Pines begin to replace shrubs	Young oak and hickory trees begin to grow	Pines die and are replaced by mature oak and hickory trees	Mature oak/hickory forest
		1–2 years	**3–4 years**	**4–15 years**	**5-15 years**	**10–30 years**	**50–75 years**	

figure 6.5 **Secondary Succession on Land** A plowed field in the southeastern United States shows a parade of changes over time, involving plant and animal associations. The general pattern is for annual weeds to be replaced by grasses and other perennial herbs, which are replaced by shrubs, which are replaced by trees. As the plant species change, so do the animal species.

Secondary Succession

The same processes and activities that drive primary succession result in secondary succession. The major difference is that secondary succession occurs when an existing community is destroyed in some way. A forest fire, a flood, or the conversion of a natural ecosystem to agriculture may be the cause. Usually, the destroyed ecosystem is not completely returned to bare rock. Much of the soil may remain, and many of the nutrients necessary for plant growth may be available for the reestablishment of the previously existing ecosystem. In addition, some plants and other organisms may survive the disturbance and continue to grow, while others will survive as roots or seeds and quickly re-establish themselves in the area. Furthermore, undamaged communities adjacent to the disturbed area can serve as sources of seeds and animals that migrate into the disturbed area. Consequently, secondary succession tends to be more rapid than primary succession. Figure 6.5 shows the typical secondary succession found on abandoned farmland in the southeastern United States.

Similarly, when beavers flood an area, the existing terrestrial community is replaced by an aquatic ecosystem. As the area behind the dam fills in with sediment and organic matter, it goes through a series of changes that may include floating plants, submerged plants, emergent plants, and wet meadow stages, but it eventually returns to the typical climax community for the area. (See figure 6.6.)

There are many kinds of communities that exist only as successional stages and are continually re-established following disturbances. Many kinds of woodlands along rivers exist only where floods remove vegetation, allowing specific species to become established on the disturbed flood plain. Some kinds of forest and shrub communities exist only if fire occasionally destroys the mature forest. Windstorms such as hurricanes are also important in causing openings in forests that allow the establishment of certain kinds of plant communities.

The Changing Nature of the Climax Concept

When European explorers traveled across the North American continent, they saw huge expanses of land dominated by specific types of communities: hardwood forests in the East, evergreen forests in the North, grasslands in central North America, and deserts in the Southwest. These regional communities came to be considered the steady-state or normal situation for those parts of the world. When ecologists began to explore the way in which ecosystems developed over time, they began to think of these ecosystems as the end point or climax of a long journey, beginning with the formation of soil and its colonization by a variety of plants and other organisms.

As settlers removed the original forests or grasslands and converted the land to farming, the original "climax" community was destroyed. Eventually, as poor farming practices destroyed the soil, many farms were abandoned and the land was allowed to return to its "original" condition. This secondary succession often resulted in forests that resembled those that had been destroyed. However, in most cases these successional forests contained fewer species and in some cases were entirely different kinds of communities from the originals. These new stable communities were also called climax communities, but they were not the same as the originals.

In addition, the introduction of species from Europe and other parts of the world changed the mix of organisms that might colonize an area. Many of grasses and herbs that were introduced either on purpose or accidentally have become well established. Today, some communities are dominated by these introduced species. Even diseases have altered the nature of climax communities. Chestnut blight and Dutch elm disease

figure 6.6 **Secondary Succession from a Beaver Pond** A colony of beavers can dam up streams and kill trees by the flooding that occurs and by using trees for food. Once the site is abandoned, it will slowly return to the original forest community by a process of succession.

have removed tree species that were at one time dominant species in certain plant communities.

Ecologists began to recognize that there was not a fixed, predetermined community for each part of the world and began to modify the way they looked at the concept of climax communities. The concept today is a more plastic one. It is still used to talk about a stable stage following a period of change, but ecologists no longer feel that land will eventually return to a "preordained" climax condition. They have also recognized in recent years that the type of climax community that develops depends on many factors other than simply climate. One of these is the availability of seeds to colonize new areas. Two areas with very similar climate and soil characteristics may develop very different successional and "climax" communities because of the seeds that were present in the area when the lands were released from agriculture. Furthermore, we need to recognize that the only thing that differentiates a "climax" community from a successional one is the time scale over which change occurs. "Climax" communities do not change as rapidly as successional ones. However, all communities are eventually replaced, as were the swamps that produced coal deposits, the preglacial forests of Europe and North America, and the pine forests of the northeastern United States.

So what should we do with this concept? Although the climax concept embraces a false notion that there is a specific end point to succession, it is still important to recognize that there is a predictable pattern of change during succession and that later stages in succession are more stable and longer lasting than early stages. Whether we call a specific community of organisms a climax community is not really important.

Biomes: Major Types of Terrestrial Climax Communities

Biomes are terrestrial climax communities with wide geographic distribution. (See figure 6.7.) Although the concept of biomes is useful for discussing general patterns and processes, it is important to recognize that when different communities within a particular biome are examined there will be variations in the exact species present. However, in broad terms the general structure of the ecosystem and the kinds of niches present are similar. Two primary nonbiological factors have major impacts on the kind of climax community that develops in any part of the world: precipitation and temperature. Several aspects of precipitation are important: the total amount of precipitation per year, the form in which it arrives (rain, snow, sleet), and its seasonal distribution. Precipitation may be evenly spaced throughout the year or it may be concentrated at particular times so that there are wet and dry seasons.

The temperature patterns are also important and can vary considerably in different parts of the world. Tropical areas have warm, relatively unchanging temperatures throughout the year. Areas near the poles have long winters with extremely cold temperatures and relatively short, cool summers. Other areas

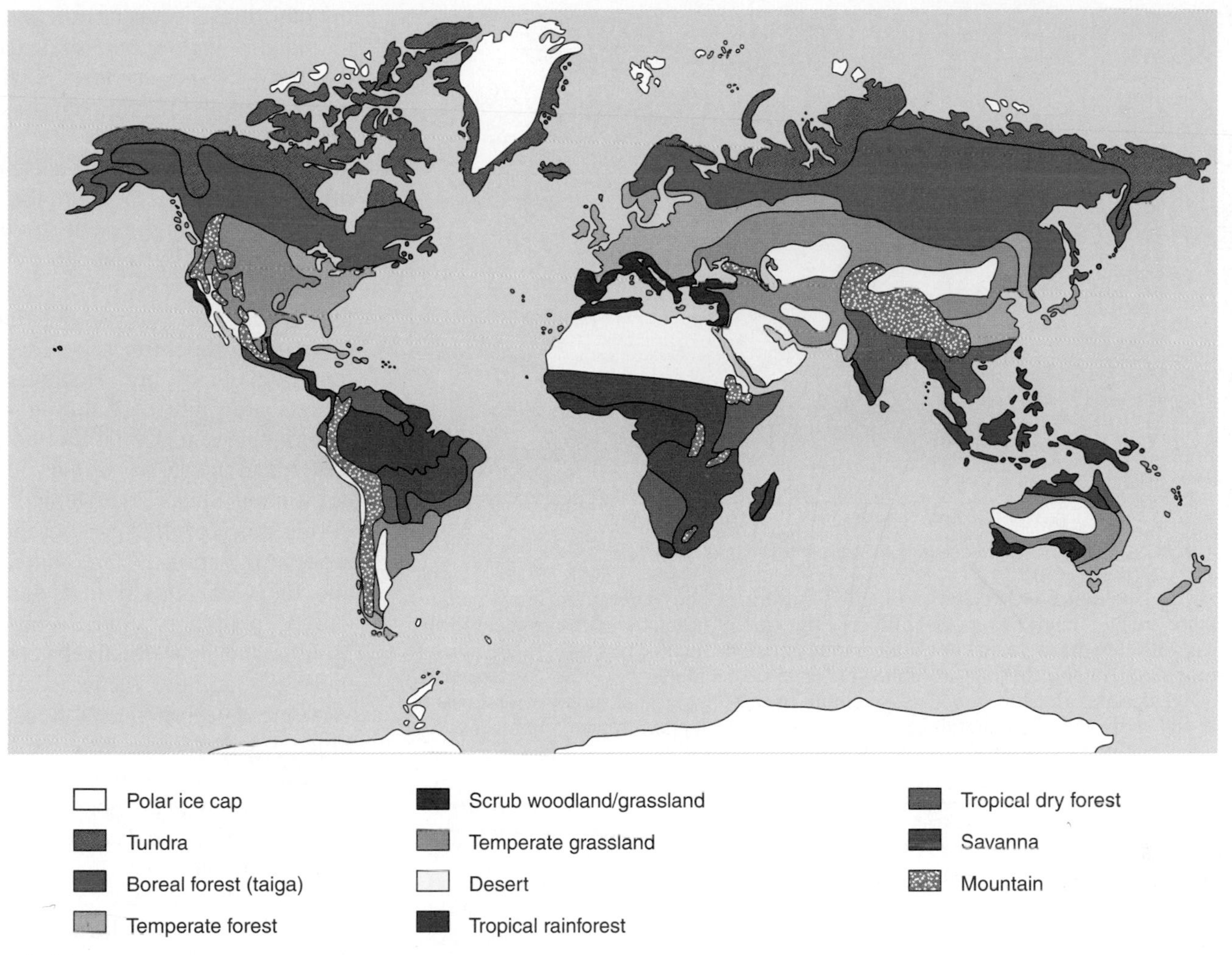

figure 6.7 **Biomes of the World** Although most biomes are named for a major type of vegetation, each includes a specialized group of animals adapted to the plants and the biome's climatic conditions.

are more evenly divided between cold and warm periods of the year. (See figure 6.8.)

Although temperature and precipitation are of primary importance, there are several other factors that may influence the kind of climax community present. Periodic fires are important in maintaining some grassland and shrub climax communities because the fires prevent the establishment of larger, woody species. Some parts of the world have frequent, strong winds that prevent the establishment of trees and cause rapid drying of the soil. The type of soil present is also very important. Sandy soils tend to dry out quickly and may not allow the establishment of more water-demanding species like trees, while extremely wet soils may allow only certain species of trees to grow. Obviously, the kinds of organisms currently living in the area are also important, since their offspring will be the ones available to colonize a new area.

The Effect of Elevation on Climate and Vegetation

The distribution of terrestrial ecosystems is primarily related to precipitation and temperature. The temperature is warmest near the equator and becomes cooler toward the poles. Similarly, as the height above sea level increases, the average temperature decreases. This means that even at the equator it is possible to have cold temperatures on the peaks of tall mountains. As one proceeds from sea level to the tops of mountains, it is possible to pass through a series of biomes that are similar to what would be encountered as one traveled from the equator to the North Pole. (See figure 6.9.)

In the next sections, we will look at the major biomes of the world and highlight the abiotic and biotic features typical of each biome.

Desert

Deserts are areas that generally average less than 25 centimeters (10 inches) of precipitation per year. A lack of water is the primary factor that determines that an area will be a desert. (See figure 6.10.) When and how precipitation arrives is quite variable in different deserts. Some deserts receive most of

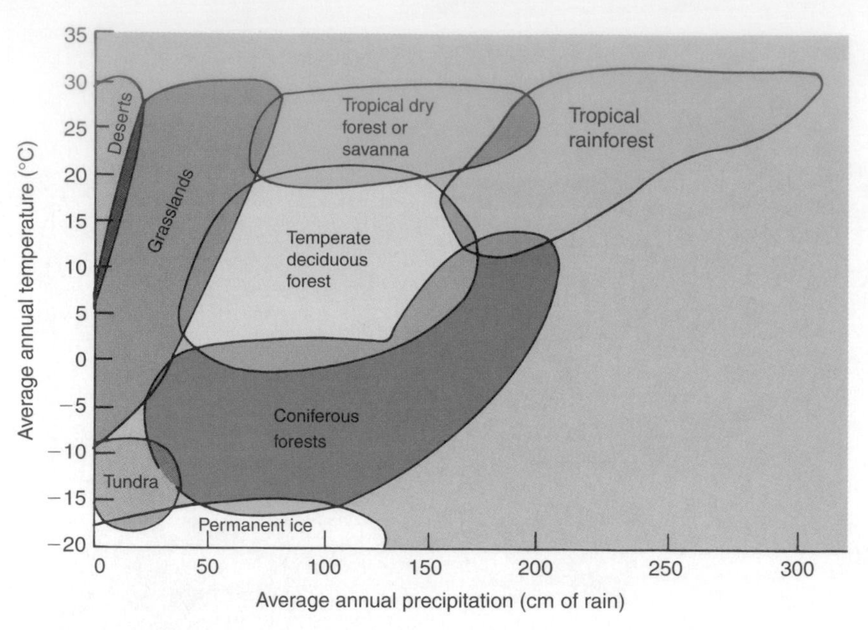

figure 6.8 **Influence of Precipitation and Temperature on Vegetation** Temperature and moisture are two major factors that influence the kind of vegetation that can occur in an area. Areas with low moisture and low temperatures produce tundra; areas with high moisture and freezing temperatures during part of the year produce deciduous or coniferous forests; dry areas produce deserts; moderate amounts of rainfall or seasonal rainfall support grasslands or savannas; and areas with high rainfall and high temperatures support tropical rainforests.

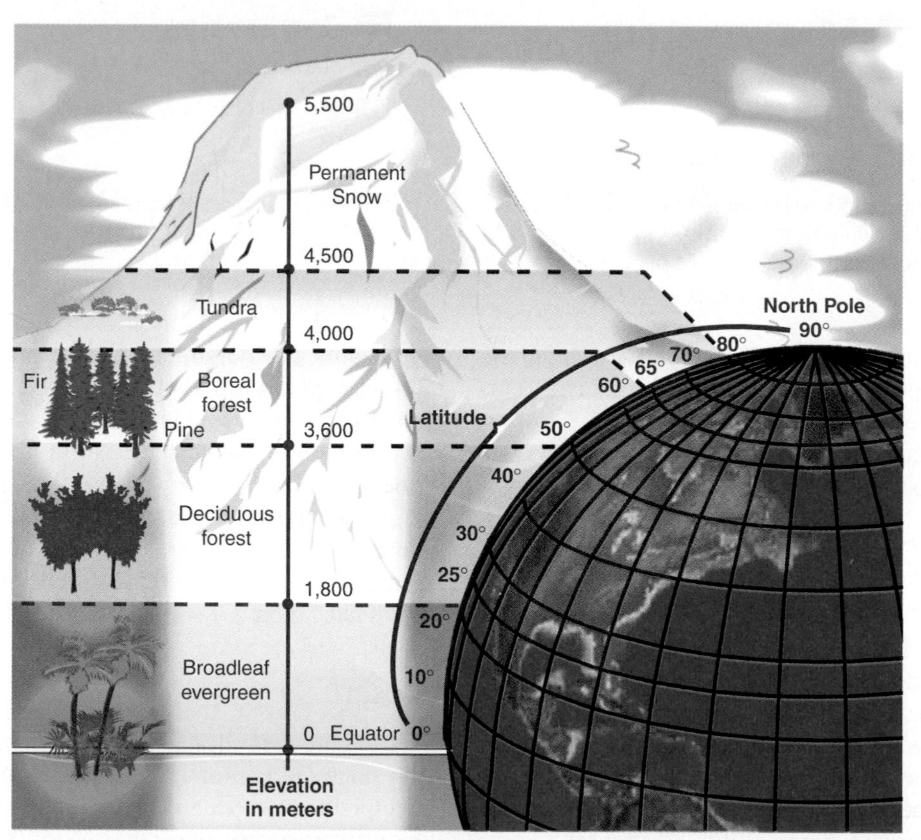

figure 6.9 **Relationship Between Height above Sea Level, Latitude, and Vegetation** As one travels up a mountain, the climate changes. The higher the elevation, the cooler the climate. Even in the tropics, tall mountains can have snow on the top. Thus, it is possible to experience the same change in vegetation by traveling up a mountain as one would experience traveling from the equator to the North Pole.

the moisture as snow or rain in the winter months, while in others rain comes in the form of thundershowers at infrequent intervals. If rain comes as heavy thunder showers, much of the water does not sink into the ground but runs off into gullies. Also, since the rate of evaporation is high, plant growth and flowering usually coincide with the periods when moisture is available. Deserts are also likely to be windy. We often think of deserts as hot, dry wastelands devoid of life. However, many deserts are quite cool during a major part of the year. Certainly, the Sahara Desert and the deserts of the southwestern United States and Mexico are hot during much of the year, but the desert areas of the northwestern United States and the Gobi Desert in Central Asia can be extremely cold during winter months and have relatively cool summers. Furthermore, the temperature can vary greatly during a 24-hour period. Since deserts receive little rainfall, it is logical that most will have infrequent cloud cover. With no clouds to block out the sun, during the day the soil surface and the air above it tend to heat up rapidly. After the sun has set, the absence of an insulating layer of clouds allows heat energy to be reradiated from the earth, and the area cools off rapidly. Cool to cold nights are typical even in "hot" deserts, especially during the winter months.

Another misconception about deserts is that few species of organisms live in the desert. There are many species, but they typically have low numbers of individuals. For example, a conspicuous feature of deserts is the dispersed nature of the plants. There is a significant amount of space between them Similarly, animals do not have large, dense populations. However, those species that are present are specially adapted to survive in dry, often hot environments. For example, water evaporates from the surfaces of leaves. As an adaptation to this condition many desert plants have very small leaves that allow them to conserve water. Some even lose their leaves entirely during the driest part of the year. Some, like cactus, have the ability to store water in their

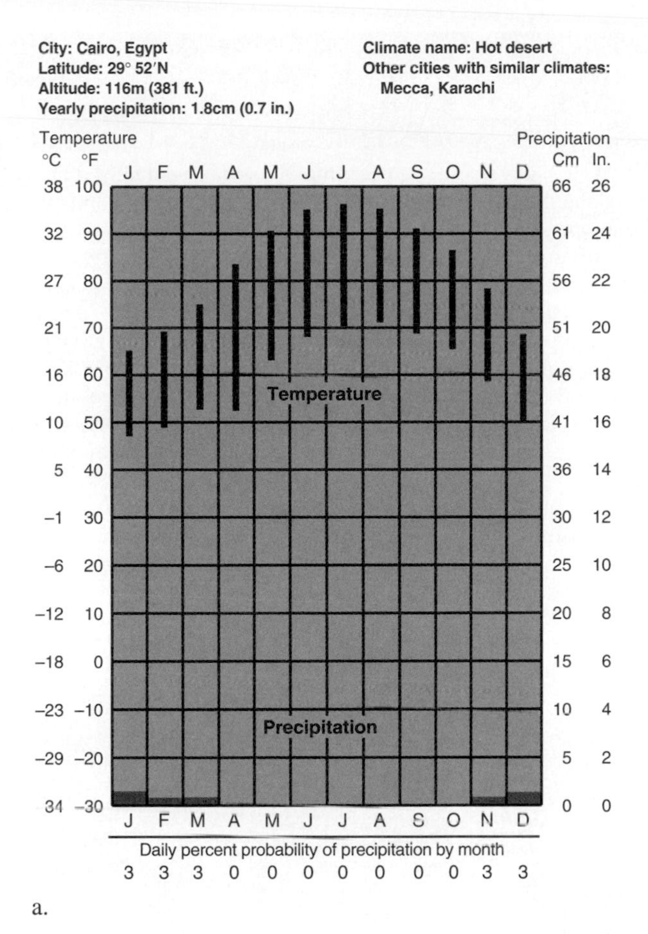

a.

b.

figure 6.10 **Desert** (*a*) Climagraph for Cairo, Egypt. (*b*) The desert receives less than 25 centimeters (10 inches) of precipitation per year, yet it teems with life. Cactus, sagebrush, lichens, snakes, small mammals, birds, and insects inhabit the desert. Because daytime temperatures are often high, most animals are active only at night, when the air temperature drops significantly. Cool deserts also exist in many parts of the world, where rainfall is low but temperatures are not high.

spongy bodies or their roots for use during drier periods. Other plants have parts or seeds that lie dormant until the rains come. Then they grow rapidly, reproduce, and die, or become dormant until the next rains. Even the perennial plants are tied to the infrequent rains. During these times, the plants are most likely to produce flowers and reproduce. Many desert plants are spiny. The spines discourage large animals from eating the leaves and young twigs.

The desert has many kinds of animals. However, they are often overlooked because their populations are low, numerous species are of small size, and many are inactive during the hot part of the day. They also aren't seen in large, conspicuous groups. Many insects, lizards, snakes, small mammals, grazing mammals, carnivorous mammals, and birds are common in desert areas. All of the animals that live in deserts are able to survive with a minimal amount of water. Some get nearly all of their water from the moisture in the food they eat. They generally have an outer skin or cuticle that resists water loss, so they lose little water by evaporation. They often limit their activities to the cooler part of the day (the evening) and may spend considerable amounts of time in underground burrows during the day, which allows them to avoid extreme temperatures and to conserve water.

Grassland

Grasslands, also known as **prairies** or **steppes,** are widely distributed over temperate parts of the world. As with deserts, the major factor that contributes to the establishment of a grassland is the amount of available moisture. Grasslands generally receive between 25 and 75 centimeters (10 to 30 inches) of precipitation per year. These areas are windy with hot summers and cold to mild winters. In many grasslands, fire is an important force in preventing the invasion of trees and releasing nutrients from dead plants to the soil. Grasses make up 60 to 90 percent of the vegetation. Many other kinds of flowering plants are interspersed with the grasses. (See figure 6.11.) Typically, the grasses and other plants are very close together and their roots form a network that binds the soil together. Trees, which generally require greater amounts of water, are rare in these areas except along watercourses.

The primary consumers are animals that eat the grasses, such as large herds of migratory, grazing mammals like bison, wildebeests, wild horses, and various kinds of sheep, cattle, and goats. While the grazers are important as consumers of the grasses, they also supply fertilizer from their dung and discourage invasion by woody species of plants because they eat the young shoots.

In addition to grazing mammals, many kinds of insects, including grasshoppers and other herbivorous insects,

ENVIRONMENTAL CLOSE-UP

Grassland Succession

Because there are many kinds of grasslands, it is difficult to generalize about how succession takes place in these areas. Most grasslands in North America have been heavily influenced by agriculture and the grazing of domesticated animals. The grasslands reestablished in these areas may be quite different from the original ecosystem. However, there appear to be several stages typically involved in grassland succession.

After land is abandoned from cultivation, a short period of one to three years elapses in which the field is dominated by annual weeds. In this respect, grassland succession is like deciduous forest succession. The next stage varies in length (10 or more years) and is dominated by annual grasses. Usually, in these early stages, the soil is in poor condition, lacking organic matter and nutrients. After several years, the soil fertility increases as organic material accumulates from the death and decay of annual grasses. This leads to the next stage in development, perennial grasses. Eventually, a mature grassland develops as prairie flowers invade the area and become interspersed with the grasses. In general, throughout this sequence, the soil becomes more fertile and of higher quality.

Because so much of the original North American grassland has been used for agriculture, when the land is allowed to return to a prairie, there may not be seeds of all of the original plants native to the area. Thus, the grassland that results from secondary succession may not be exactly like the original; some species may be missing. Consequently, in many managed restorations of prairies, seeds that are no longer available in the local soils are introduced from other sources.

The low amount of rainfall and the fires typical of grasslands generally cause the successional process to stop at this point. However, if more water becomes available or if fire is prevented, woody trees may invade moist sites.

Actively farmed

Recently abandoned

Several years of succession

dung beetles (which feed on the dung of grazing animals), and several kinds of flies are common. Some of these flies bite to obtain blood. Others lay their eggs in the dung of large mammals. Some feed on dead animals and lay their eggs in carcasses. Small herbivorous mammals, such as mice and ground squirrels, are also common. Birds are often associated with grazing mammals. They eat the insects stirred up by the mammals or feed on the insects that bite them. Other birds feed on seeds and other plant parts. Reptiles (snakes and lizards) and other carnivores like coyotes, foxes and hawks feed on small mammals and insects.

Most of the moist grasslands of the world have been converted to agriculture, since the rich, deep soil that developed as a result of the activities of centuries of soil building is useful for growing cultivated grasses like corn (maize) and wheat. The drier grasslands have been converted to the raising of domesticated grazers like cattle, sheep, and goats. Therefore, there is little undisturbed grassland left and those fragments that remain need to be preserved as refuges for the grassland species that once occupied huge portions of the globe.

Savanna

Tropical parts of Africa, South America, and Australia have extensive grasslands spotted with occasional trees or patches of trees. (See figure 6.12.) This kind of a biome is often called a **savanna.** Although savannas receive 50 to 150 centimeters (20 to 60 inches) of rain per year, the rain is not distributed evenly throughout the year. Typically, a period of heavy rainfall is followed by a prolonged drought. This results in a very seasonally structured ecosystem. The plants and animals time their reproductive activities to coincide with the rainy period, when limiting factors are least confining. The predominant plants are grasses, but many drought-resistant, flat-topped, thorny trees are common. As with grasslands, fire is a common feature of the savanna and the trees present are resistant to fire damage. Many of these trees are particularly important because they are legumes that are involved in nitrogen fixation. They also provide shade and nesting sites for animals. As with grasslands, the predominant mammals are the grazers. Wallabies in Australia, wildebeests, zebras, elephants and various species of antelope in Africa, and capybaras (rodents) in South America are examples. Many kinds of rodents,

City: Tehran, Iran
Latitude: 35° 41′N
Altitude: 1220m (4002 ft.)
Yearly precipitation: 26cm (10.1 in.)

Climate name: Midaltitude dryland
Other cities with similar climates:
Salt Lake City, Ankara

Temperature
°C °F
J F M A M J J A S O N D
Precipitation
Cm In.
38 100
32 90
27 80
21 70
16 60
10 50
5 40
−1 30
−6 20
−12 10
−18 0
−23 −10
−29 −20
−34 −30
66 26
61 24
56 22
51 20
46 18
41 16
36 14
30 12
25 10
20 8
15 6
10 4
5 2
0 0
Temperature
Precipitation
J F M A M J J A S O N D
Daily percent probability of precipitation by month
13 14 16 10 6 3 3 0 0 3 10 13

a.

b.

figure 6.11 **Grassland** (*a*) Climagraph for Tehran, Iran. (*b*) Grasses are better able to withstand low water levels than are trees. Therefore, in areas that have moderate rainfall, grasses are the dominant plants.

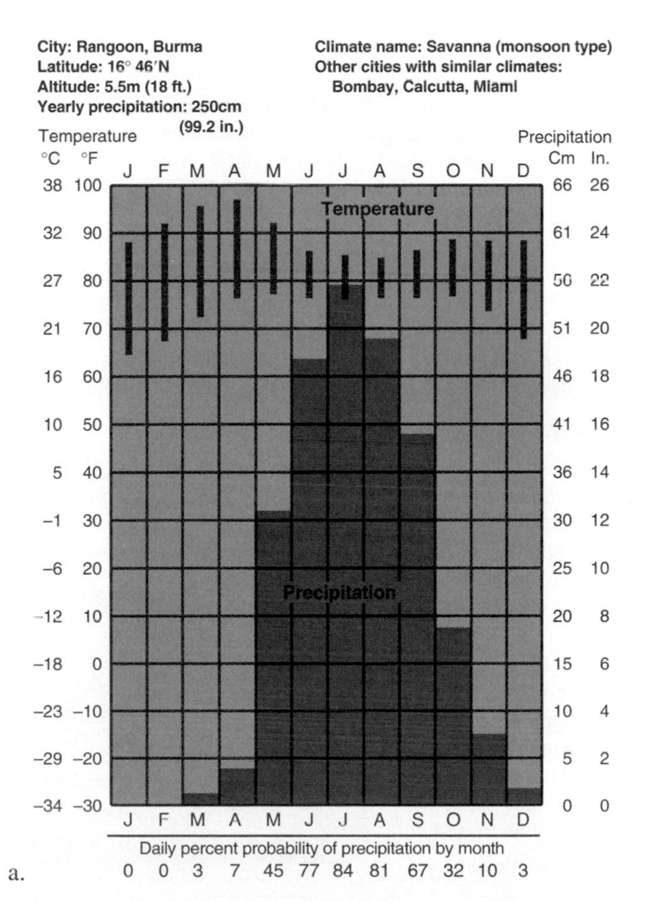

a.

b.

figure 6.12 **Savanna** (*a*) Climagraph for Rangoon, Burma. (*b*) Savannas develop in tropical areas that have seasonal rainfall. They typically have grasses as the dominant vegetation with drought and fire-resistant trees scattered through the area.

GLOBAL PERSPECTIVE

Tropical Rainforests—A Special Case?

Today, there is considerable political and economic interest in how tropical rainforests are used. Some would preserve them in their current state while others would use the trees and other forest resources for economic gain. However, since tropical rainforests are located in countries in which there are large numbers of poor people, there are strong pressures to exploit forests for economic benefit. Most economic uses of the rainforest result in its destruction or reduce its biodiversity.

Two hundred years ago, tropical rainforests covered about 1500 million hectares (3700 million acres), an area the size of Europe, but today only 900 million hectares (2200 million acres) remain. Modern technology makes short work of clearing the rainforest, taking less than an hour to clear one hectare (2.47 acres). Exactly how much rainforest is disappearing is not known, but it seems likely that over 20 million hectares (50 million acres) are destroyed each year. At this rate, there will be no rainforests remaining in 50 years. The causes of deforestation are easy to identify.

Logging

Rainforests were spared from exploitation in earlier years because of their inaccessibility, the relative low value of most of the trees for timber purposes, and the limited world demand. Recently, this situation has changed and a wide variety of tree species previously considered worthless are now used for pulp or as cellulose for the production of plastics. With new machines and better transportation, it has become profitable to remove trees from previously remote areas. Often logging companies are interested only in one or two hardwood species, such as teak or mahogany. As there may be only three suitable trees in a hectare, to remove them may not seem a threat. However, heavy machinery is needed to clear a path, and when the tree is felled and dragged through the forest, the amount of damage to other trees can be enormous. Faced with a high demand for their forest products, most countries with rainforests have been willing to sign over timber rights to foreign companies, hoping thereby to increase their national incomes. Unfortunately, most of these timber contracts contain few or no provisions for conservation.

Farming

Poor people seeking farmland make use of roads built by logging and mining companies to gain access to previously remote areas. Government policies often encourage people to settle in logged areas, however inappropriate these are for growing crops. The settlers establish a shifting form of agriculture in which the trees are cut down and burned. They grow vegetables in the clearing for two to three years until the soil is depleted of its nutrients, then abandon the clearing and move on to another place in the rainforest.

Ranching

Large areas of rainforest are being burned down and converted into ranchland even though this is the worst possible use of the land. Central America has already lost two-thirds of its rainforest to cattle ranching. Where tropical rainforests have been cleared for pasture and cattle ranches established, the production of meat hardly reaches 50 kilograms of meat/hectare/year (45 lb/acre/ year) whereas North American farms produce more than 600 kilograms of meat/hectare/ year (535 lb/acre/year). The soil quickly loses its scarce nutrients and becomes useless. When this happens, the land is abandoned and more rainforest is cleared.

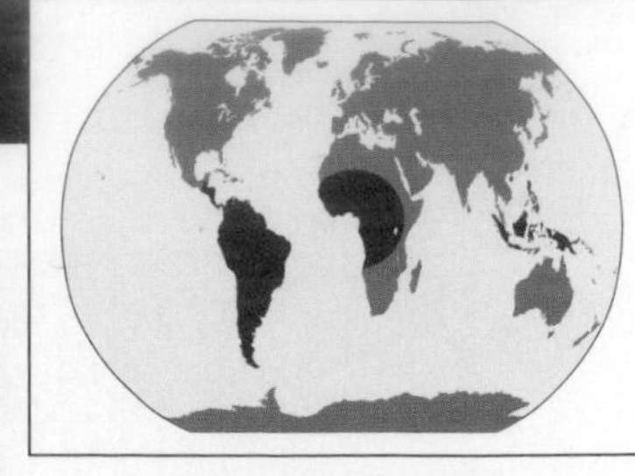

Mining

Many rainforests are rich in oil deposits and mineral reserves such as bauxite, coal, copper, diamonds, gold, iron ore, nickel, tin, and uranium. While mining is a minor cause of deforestation, still land is cleared for access and the mines are often in areas which are the only habitats of certain plant and animal species. The small amount of deforestation can cause a disproportionate amount of damage and the mining often releases toxic wastes into rivers, destroying the animal and plant life.

Biodiversity Resources

Tropical rainforests contain an amazing array of different kinds of organisms. Many kinds of foods and other useful products are derived from rainforest plants. Every time we drink coffee, eat chocolate, bananas, or nuts, or use anything made of rubber, we are using products originally discovered in the rainforests.

The following list points out the value of the diverse kinds of plants that live in the tropics.

Fruit: pineapples, bananas, oranges, lemons, limes, grapefruit, and tangerines

Vegetables: tomatoes, avocados, and many types of beans

Nuts: Brazil nuts, cashew nuts, peanuts, sesame seeds, and coconuts

Spices: chilies, pepper, cloves, nutmeg, vanilla, cinnamon, and turmeric

Drinks: tea, coffee, cocoa, chocolate, kola nut extract used in Coca-Cola

Gums and resins: rubber for household and industrial goods, copals for paints and varnishes, chicle gum used for chewing gum, balata used in golf balls

One of the greatest potential benefits of rainforests is the possibility of discovering new drugs. Plants contain many substances known as phytochemicals which help them deter insects from eating their leaves or have other value to the plant. These same chemicals often have use as drugs. There is a one-in-four chance that the next time you enter a pharmacy for a prescription, you will leave with a product derived from the rainforest. Seventy percent of plants identified as having anticancer properties come from the rainforests. In fact, almost 50 percent of our medications are derived from plants, yet less than 1 percent of tropical rainforest species have been examined for their possible value in medicine.

Global Rainforest Services

Many people value rainforests for the services they provide. Large expanses of rainforest alter weather conditions, protect soil from erosion, and store large amounts of carbon. This carbon-storing function is particularly important as we recognize we are adding large amounts of carbon to the atmosphere as we burn fossil fuels. While these may be valuable services, it is hard to put a monetary value on them. Therefore, in most cases, the immediate economic benefits of logging and agriculture tend to outweigh the long-term biodiversity and service values provided by rainforests.

birds, insects, and reptiles are associated with this biome. Among the insects, mound-building termites are particularly common.

Tropical Rainforest

Tropical rainforests are located near the equator in Central and South America, Africa, Southeast Asia, and some islands in the Caribbean Sea and Pacific Ocean. (See figure 6.13.) The temperature is normally warm and relatively constant. There is no frost, and it rains nearly every day. Most areas receive in excess of 200 centimeters (80 inches) of rain per year. Some receive 500 centimeters (200 inches) or more. Because of the warm temperatures and abundant rainfall, most plants grow very rapidly; however, soils are usually poor in nutrients because water tends to carry away any nutrients not immediately taken up by plants. Many of the trees have extensive root networks, associated with fungi (mycorrhiza), near the surface of the soil that allow them to capture nutrients from decaying vegetation before the nutrients can be carried away. Because most of the nutrients in a tropical rainforest are tied up in the biomass, not in the soil, these areas do not make good farmland. Rainfall is a source of new nutrients since atmospheric particles and gases dissolve as the rain falls. The canopy contains many kinds of epiphytic plants that trap many of these nutrients in the canopy before they can reach the soil.

Tropical rainforests have a greater diversity of species than any other biome. More species are found in the tropical rainforests of the world than in the rest of the world combined. A small area of a few square kilometers is likely to have hundreds of species of trees. Furthermore, it is typical to have distances of a kilometer or more between two individuals of the same species. Balsa, teakwood, and many other ornamental woods are from tropical trees. Each of those trees is home to a set of animals and plants that use it as food, shelter, or support. The canopy, which forms a solid wall of leaves between the sun and the forest floor, consists of two or three levels. A few trees, called emergent trees, protrude above the canopy. Below the canopy is a layer of understory tree species. Recently, biologists discovered a whole new community of organisms that live in the canopy of these forests. Since most of the sunlight is captured by the trees, only shade-tolerant plants live beneath the trees' canopy.

Epiphytes (plants that live on the surface of other plants) are common in the rainforest. Many understory species are vines that attach themselves to the tall trees as they grow toward the sun. When they reach the canopy they can compete effectively with their supporting tree for available sunlight. In addition to supporting various vines, each tree serves as a surface for the growth of ferns, mosses, and orchids.

Associated with this variety of plants is an equally large variety of animals. Insects, such as ants, termites, moths, butterflies, and beetles, are particularly abundant. Birds also are extremely common, as are many climbing mammals, lizards, and tree frogs. The insects are food to many of these species. Since flowers and fruits are available throughout the year, there are many kinds of nectar- and fruit-feeding birds and mammals. Their activities are important in spreading seeds throughout the forest. Because of the low light levels and the difficulty of maintaining visual contact with one another, many of the animals communicate by making noise.

Tropical rainforests are under intense pressure from logging and agriculture. Many of the countries where tropical rainforests are present are poor and seek to obtain jobs and money by exploiting this resource. Generally, agriculture has not been successful because the soils are poor and cannot withstand constant agricultural activity. Forestry can be a sustainable activity, but in many cases it is not. The forests are being cut down with no effort to protect them for long-term productivity.

Temperate Deciduous Forest

Forests in temperate areas of the world that have a winter–summer change of seasons typically have trees that lose their leaves during the winter and replace them the following spring. This kind of forest is called a **temperate deciduous forest** and is typical of the eastern half of the United States, parts of south central and southeastern Canada, southern Africa, and many areas of Europe and Asia.

These areas generally receive 75 to 100 centimeters (30 to 60 inches) of relatively evenly distributed precipitation per year. The winters are relatively mild and plants are actively growing for about half the year. Each area of the world has certain species of trees that are the major producers for the biome. (See figure 6.14.) In contrast to tropical rainforests, where individuals of a tree species are scattered throughout the forest, temperate deciduous forests generally have many fewer species, and many

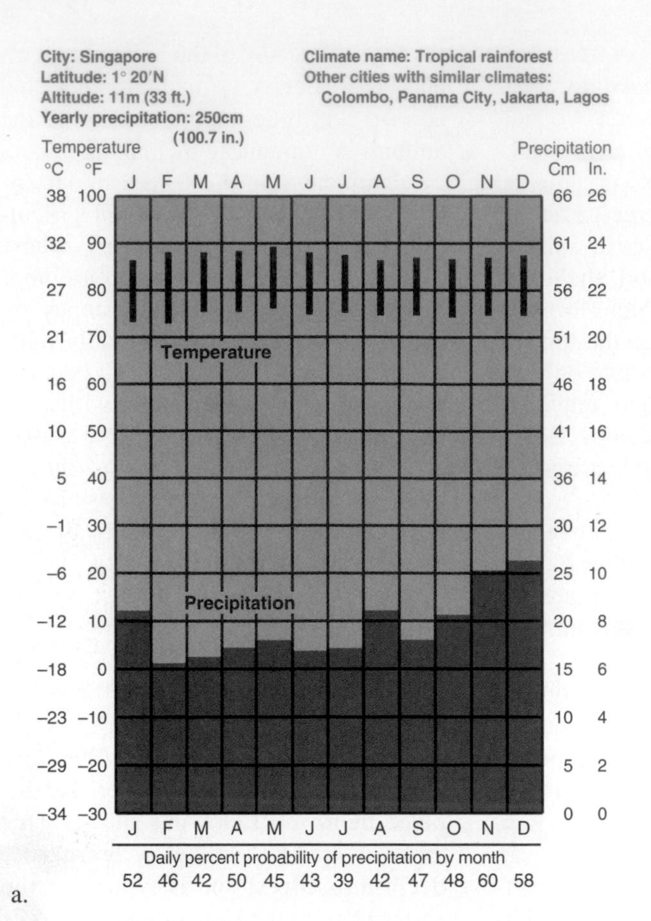

a.

b.

figure 6.13 **Tropical Rainforest** (*a*) Climagraph for Singapore. (*b*) Tropical rainforests develop in areas with high rainfall and warm temperatures. They have an extremely diverse mixture of plants and animals.

City: Chicago, Illinois
Latitude: 41° 52′N
Altitude: 181m (595 ft.)
Yearly precipitation: 85cm (33.3 in.)

Climate name: Humid continental (warm summer)
Other cities with similar climates:
New York, Berlin, Warsaw

Temperature
°C °F
Precipitation
Cm In.
J F M A M J J A S O N D
38 100
32 90
27 80
21 70
16 60
10 50
5 40
−1 30
−6 20
−12 10
−18 0
−23 −10
−29 −20
−34 −30
66 26
61 24
56 22
51 20
46 18
41 16
36 14
30 12
25 10
20 8
15 6
10 4
5 2
0 0
Temperature
Precipitation
J F M A M J J A S O N D
Daily percent probability of precipitation by month
35 36 39 37 39 37 29 29 30 29 33 35

a.

b.

figure 6.14 **Temperate Deciduous Forest** (*a*) Climagraph for Chicago, Illinois. (*b*) A temperate deciduous forest develops in areas that have significant amounts of moisture throughout the year, but where the temperature falls below freezing for parts of the year. During this time the trees lose their leaves. This kind of forest once dominated the eastern half of the United States and southeastern Canada.

ENVIRONMENTAL CLOSE-UP

Forest Canopy Studies

In the past few years, a new frontier of ecological study has developed. Scientists have traditionally looked at forest ecosystems at ground level. Trees were identified, understory species were categorized, and the animals that live on or near the forest floor were studied. Gradually, scientists began to realize that many kinds of plants and animals that are important parts of forest ecosystems rarely descend from the tops of trees to the forest floor. They began to devise methods for studying these canopy-dwelling organisms. Initially, they relied on techniques and gear used by mountain climbers, but that approach was labor intensive and dangerous. Recently, ecologists have established several sites where large construction cranes have been built in forests. These cranes allow researchers to study the chemistry, climate, and organisms found in the canopy.

Several surprising discoveries have been made, including the discovery of new species of insects. A canopy study in an old-growth temperate rainforest on Vancouver Island, British Columbia, identified more than 60 new species of insects. Researchers estimate that hundreds of new species will eventually be identified. Similar studies in tropical forests have identified hundreds more new species of insects. Studies in Panama recognized that vines (lianas) can constitute up to 70 percent of the forest canopy and that these vines compete for sunlight with the trees that support them.

The way in which animals use the canopy is another interesting area of research. Birds, bats, monkeys, squirrels, and insects use specific parts of the forest canopy for food, nesting, and travel routes. Researchers are learning much as they spend time in the forest canopy with the animals.

forests may consist of two or three dominant tree species. In deciduous forests of North America and Europe, common species are maples, aspen, birch, beech, oaks, and hickories. These tall trees shade the forest floor, where many small flowering plants bloom in the spring. These spring wildflowers store food in underground structures. In the spring, before the leaves come out on the trees, the wildflowers can capture sunlight and reproduce before they are shaded. Many smaller shrubs also are found in the understory of these forests.

These forests are home to a great variety of insects, many of which use the leaves and wood of trees as food. Beetles, moth larvae, wasps, and ants are examples. The birds that live in these forests are primarily migrants that arrive in the spring of the year, raise their young during the summer, and leave in the fall. Many of these birds rely on the large summer insect population for their food. Others use the fruits and seeds that are produced during the summer months. A few kinds of birds, including woodpeckers, grouse, turkeys, and some of the finches, are year-round residents. Amphibians (frogs, toads, salamanders) and reptiles (snakes and lizards) prey on insects and other small animals. Several kinds of small and large mammals inhabit these areas. Mice, squirrels, deer, shrews, moles, and opossums are common examples. Major predators on these mammals are foxes, badgers, weasels, coyotes, and birds of prey.

Taiga, Northern Coniferous Forest, or Boreal Forest

Throughout the southern half of Canada, parts of northern Europe, and much of Russia, there is an evergreen coniferous forest known as the **taiga, northern coniferous forest,** or **boreal forest.** (See figure 6.15.) The climate is one of short, cool summers and long winters with abundant snowfall. The winters are extremely harsh and can last as long as six months. Typically, the soil freezes during the winter. Precipitation ranges between 25 and 100 centimeters (10 to 40 inches) per year, but the climate is humid because the generally low temperatures during all parts of the year reduce evaporation. The landscape is typically dotted with lakes, ponds, and bogs.

Spruces, firs, and larches are trees most common in these areas. These trees are specifically adapted to winter conditions. Winter is relatively dry as far as the trees are concerned because the moisture falls as snow and stays above the soil until it melts in the spring. The needle-shaped leaves are adapted to prevent water loss; in addition the larches lose their needles in the fall. The branches of these trees are flexible, allowing them to bend under a load of snow so that the snow slides off the pyramid-shaped trees without greatly damaging them. As with the temperate deciduous forest, many of the inhabitants of this biome are temporarily active during the summer. Most birds are migratory and feed on the abundant summer insect population, which is not available during the long, cold winter. A few birds, such as woodpeckers, owls, and grouse, are permanent residents. Typical mammals are deer, caribou, moose, wolves,

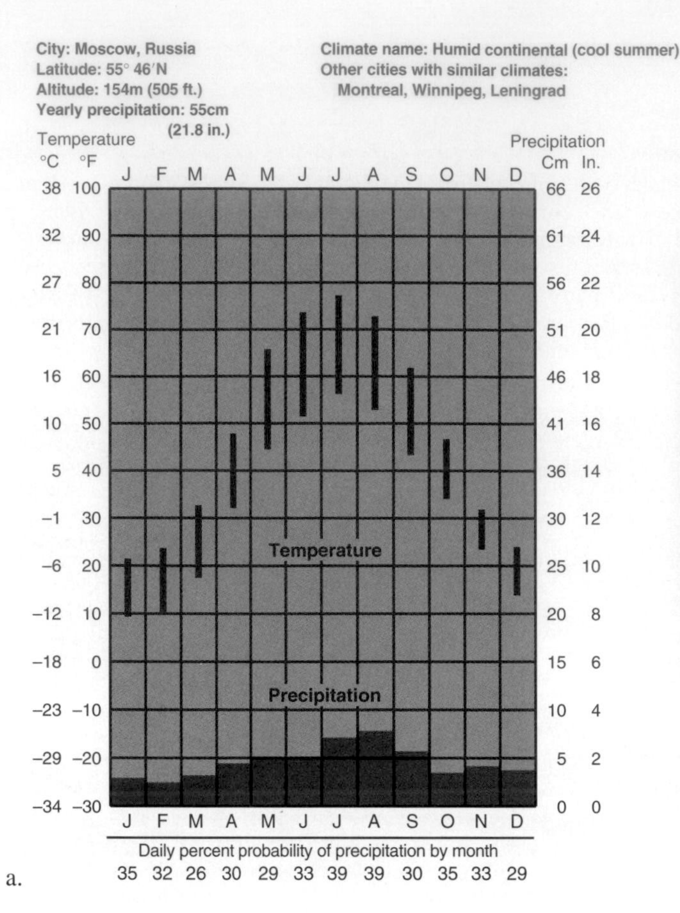

figure 6.15 **Taiga, Northern Coniferous Forest, or Boreal Forest** (*a*) Climagraph for Moscow. (*b*) The taiga, northern coniferous forest, or boreal forest occurs in areas with long winters and heavy snowfall. The trees have adapted to these conditions and provide food and shelter for the animals that live there.

weasels, mice, snowshoe hares, and squirrels. Because of the cold, few reptiles and amphibians live in this biome.

Tundra

North of the taiga is the **tundra,** a biome that lacks trees and has a permanently frozen soil. This frozen soil layer is known as **permafrost.** (See figure 6.16.) Because of the permanently frozen soil and extremely cold, windy climate (up to 10 months of winter), no trees can live in the area. Although the amount of precipitation is similar to that in some deserts—less than 25 centimeters (10 inches) per year—the short summer is generally wet because the winter snows melt in the spring and summer temperatures are usually less than 10°C (50°F) which reduces the evaporation rate. Since the permafrost does not let the water sink into the soil, waterlogged soils and many shallow ponds and pools are present. Many waterfowl like ducks and geese migrate to the tundra in the spring; there, they mate and raise their young during the summer before migrating south in the fall. When the top few centimeters (inches) of the soil thaw, many plants (grasses, dwarf birch, dwarf willow) and lichens, such as reindeer moss, grow. The plants are short, usually less than 20 centimeters (8 inches).

Clouds of insects are common during the summer and serve as food for migratory birds. Permanent resident birds are the ptarmigan and snowy owl. No reptiles or amphibians survive in this extreme climate. A few hardy mammals like musk oxen, caribou (reindeer), arctic hare, and lemmings can survive by feeding on the grasses and other plants that grow during the short, cool summer. The arctic fox, wolves, and owls are the primary predators in this region. Because of the very short growing season, damage to this kind of ecosystem is slow to heal, so the land must be handled with a great deal of care.

Scattered patches of tundralike communities also are found on mountaintops throughout the world. These are known as **alpine tundra.** Although the general appearance of the alpine tundra is similar to true tundra, many of the species of plants and animals are different. The animals present often migrate up to the alpine tundra during the summer and return to lower elevations as the weather turns cold.

Major Aquatic Ecosystems

Terrestrial biomes are determined by the amount and kind of precipitation and by temperatures. Other factors, such as soil type and wind, also play a part. Aquatic ecosystems also are shaped by primary determiners. Four such factors are the ability of the sun's rays to penetrate the water, the nature of the bottom substrate, the water temperature, and the amount of dissolved materials.

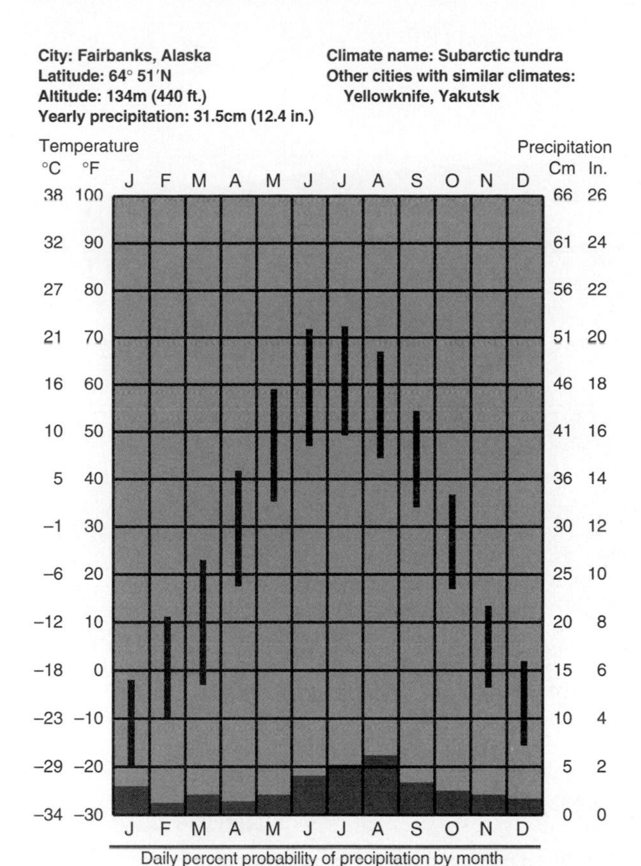

a.

b.

figure 6.16 **Tundra** (*a*) Climagraph for Fairbanks, Alaska. (*b*) In the northern latitudes and on the tops of some mountains, the growing season is short and plants grow very slowly. Trees are unable to live in these extremely cold areas, in part because there is a permanently frozen layer of soil beneath the surface, known as the permafrost. Because growth is so slow, damage to the tundra can still be seen generations later.

Marine Ecosystems

An important determiner of the nature of aquatic ecosystems is the amount of salt dissolved in the water. Those that have little dissolved salt are called **freshwater ecosystems** and those that have a high salt content are called **marine ecosystems.**

Pelagic Marine Ecosystems

In the open ocean, many kinds of organisms float or swim actively. Crustaceans, fish, and whales swim actively as they pursue food. These kinds of organisms that are not attached are called **pelagic** organisms, and the ecosystem they are a part of is called a **pelagic ecosystem.** As with all ecosystems, the organisms at the bottom of the energy pyramid carry on photosynthesis. The term **plankton** is used to describe aquatic organisms that are so small and weakly swimming that they are simply carried by currents. The planktonic organisms that carry on photosynthesis are called **phytoplankton.** In the open ocean, a majority of these organisms are small, microscopic, floating algae and bacteria. The upper layer of the ocean, where the sun's rays penetrate, is known as the **euphotic zone.** It is in this euphotic zone where phytoplankton are most common. The thickness of the euphotic zone varies with the degree of clarity of the water but in clear water can be up to 150 meters (500 feet) in depth. Small, weakly swimming animals of many kinds, known as **zooplankton,** feed on the phytoplankton. Zooplankton are often located at a greater depth in the ocean than the phytoplankton but migrate upward at night and feed on the large population of phytoplankton. The zooplankton are in turn eaten by larger animals like fish and larger shrimp, which are eaten by larger fish like salmon, tuna, sharks, and mackerel. (See figure 6.17.)

A major factor that influences the nature of a marine community is the kind and amount of material dissolved in the water. Probably more important is the amount of nutrients available to the organisms carrying on photosynthesis. Phosphorus, nitrogen, and carbon are all required for the construction of new living material. In water, these are often in short supply. Therefore, the most productive aquatic ecosystems are those in which these essential nutrients are most common. These areas include places in oceans where currents bring up nutrients that have settled to the bottom and areas where rivers deposit their load of suspended and dissolved materials.

Benthic Marine Ecosystems

Organisms that live on the ocean bottom, whether attached or not, are known as **benthic** organisms, and the ecosystem of which they are a part is called a **benthic ecosystem.** Some fish, clams, oysters, various crustaceans, sponges, sea anemones, and many other kinds of organisms live on the bottom. In shallow water, sunlight can penetrate to the bottom, and

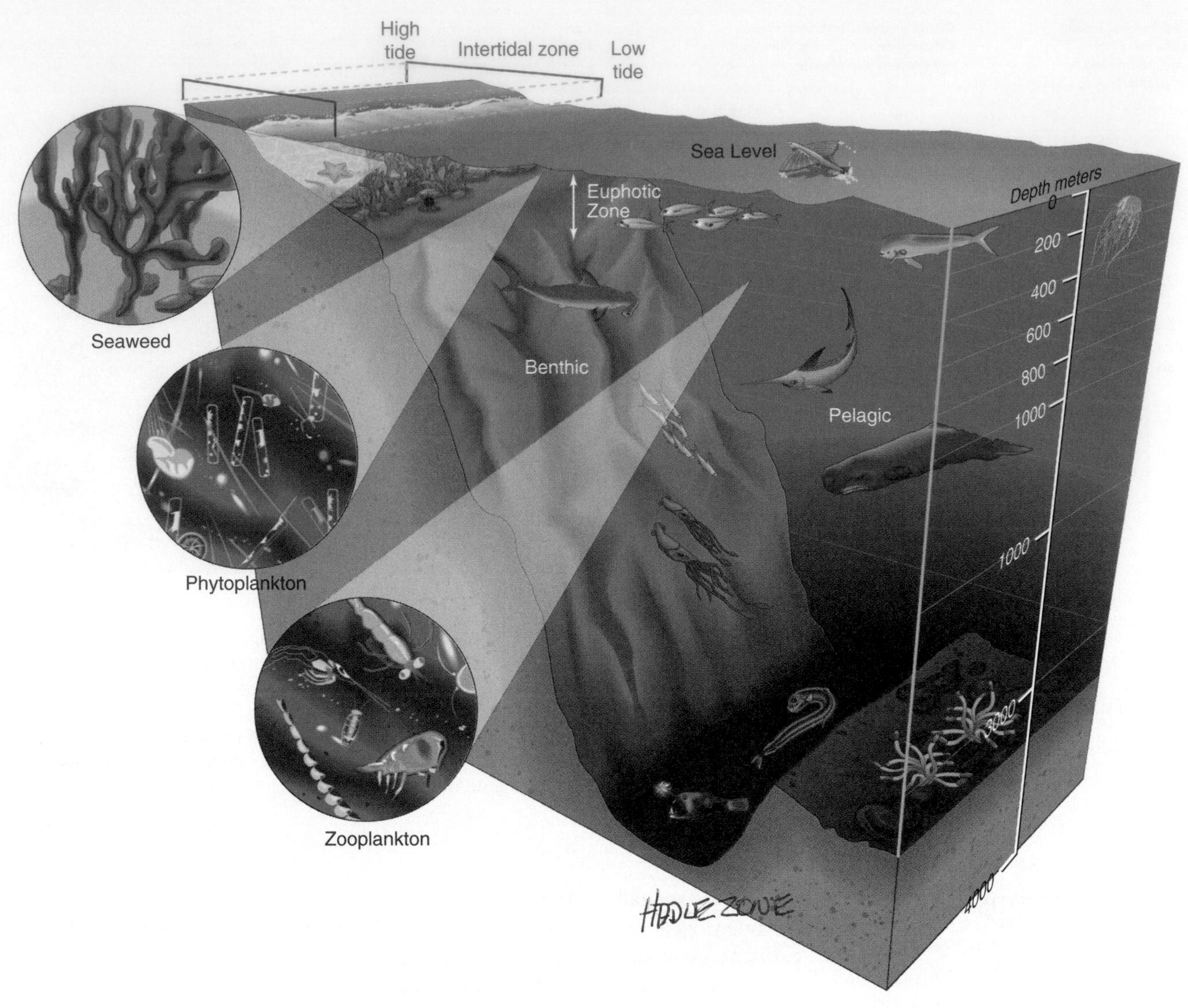

figure 6.17 **Marine Ecosystems** All of the photosynthetic activity of the ocean occurs in shallow water called the euphotic zone, either by attached algae near the shore or by minute phytoplankton in the upper levels of the open ocean. Consumers are either free-swimming pelagic organisms or benthic organisms that live on the bottom. Small animals that feed on phytoplankton are known as zooplankton.

a variety of attached photosynthetic organisms like kelp (commonly called seaweeds) are common. Since they are attached and can grow to very large size, many other bottom-dwelling organisms, such as sea urchins, worms, and fish, are associated with them.

The substrate is very important in determining the kind of benthic community that develops. Sand tends to shift and move, making it difficult for large plants or algae to become established, although some clams, burrowing worms, and small crustaceans find sand to be a suitable habitat. Clams filter water for plankton and detritus, or burrow through the sand, feeding on other inhabitants. Mud may provide suitable habitats for some kinds of rooted plants, such as mangrove trees or sea grasses. Although mud usually contains little oxygen, it still may be inhabited by a variety of burrowing organisms that feed by filtering the water above them or feed on other animals in the mud. Rocky surfaces in the ocean provide a good substrate for many kinds of large algae. Associated with this profuse growth of algae is a large variety of animals. (See figure 6.18.)

Temperature also has an impact on the kind of benthic community established. Some communities, such as coral reefs or mangrove swamps, are found only in areas where the water is warm. **Coral reef ecosystems** are the result of large numbers of small animals that build cup-shaped external skeletons around themselves. They are able to protrude from their skeletons to capture food and expose themselves to the sun. This is important because corals contain single-celled algae within their bodies. These algae carry on photosynthesis and provide both themselves and the coral with the nutrients necessary for growth. Because they require warm water, coral ecosystems are found only near the equator. Coral ecosystems also require shallow, clear water since the algae must have ample sunlight to carry on

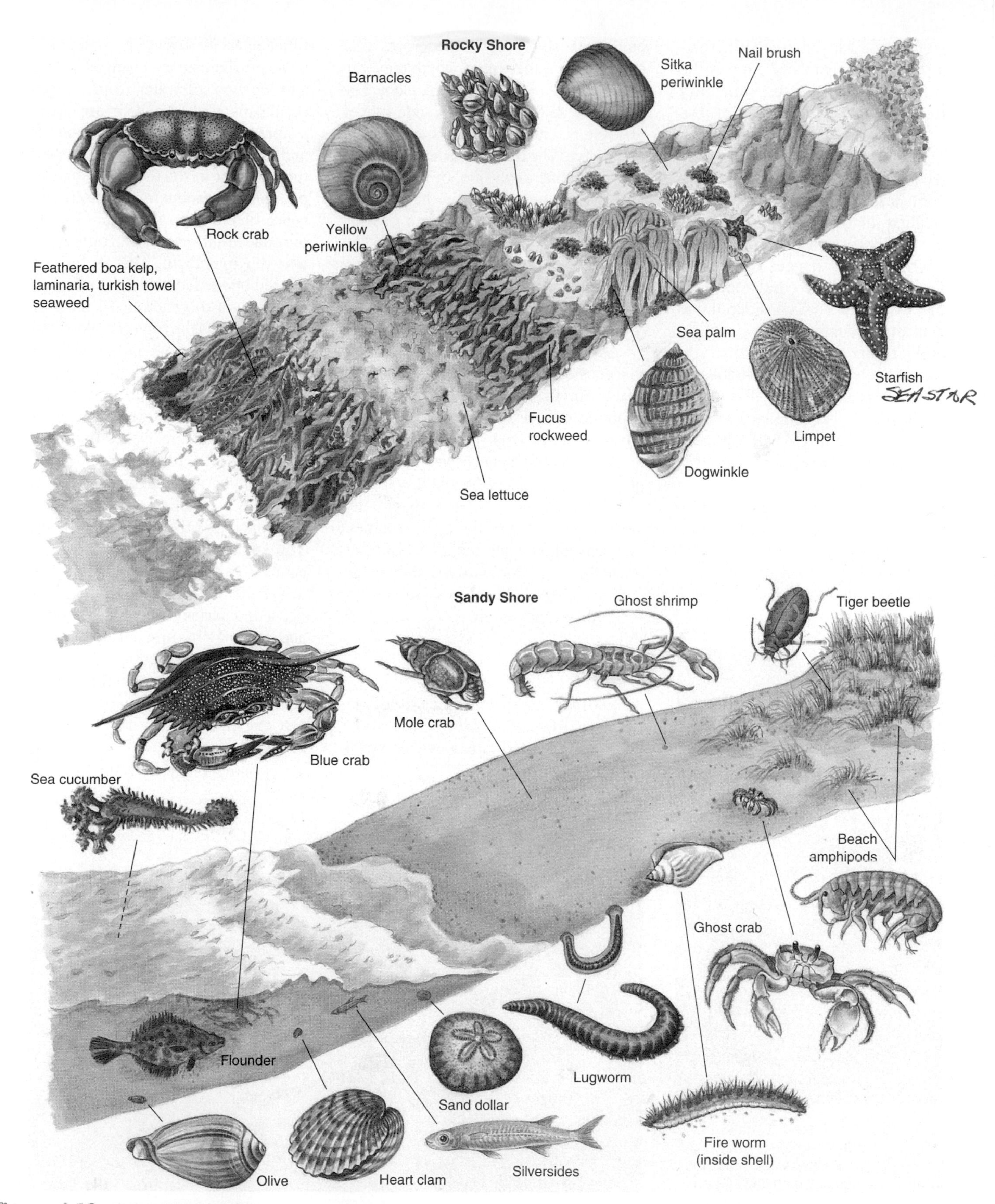

figure 6.18 **Types of Shores** The kind of substrate determines the kind of organisms that can live near the shore. Rocks provide areas for attachment that sands do not, since sands are constantly shifting. Muds usually have little oxygen in them; therefore, the organisms that live there must be adapted to those kinds of conditions.

photosynthesis. This mutualistic relationship between algae and coral is the basis for a very productive community of organisms. The skeletons of the corals provide a surface upon which many other kinds of animals live. Some of these animals feed on corals directly, while others feed on small plankton and bits of algae that establish themselves among the coral organisms. Many kinds of fish, crustaceans, sponges, clams, and snails are members of coral reef ecosystems. Coral reefs are considered one of the most productive ecosystems on earth. (See figure 6.19.)

Mangrove swamp ecosystems occupy a region near the shore. The dominant organisms are special kinds of trees that are able to tolerate the high salt content of the ocean. In areas where the water is shallow and wave action is not too great, the trees can become established. They have long seeds that float in the water. When the seeds become trapped in mud, they take root. The trees can excrete salt from their leaves. They also have extensively developed roots that extend above the water, where they can obtain oxygen and prop up the plant. The trees trap sediment and provide places for oysters, crabs, jellyfish, sponges, and fish to live. The trapping of sediment and the continual extension of mangroves into shallow areas result in the development of a terrestrial ecosystem in what was once shallow ocean. Mangroves are found in South Florida, the Caribbean, Southeast Asia, Africa, and other parts of the world where tropical mudflats occur. (See figure 6.20.)

At great depths in the ocean is a benthic ecosystem that must rely on a continuous rain of organic matter from the euphotic zone. These areas are known as abyssal areas, and the ecosystem is known as an **abyssal ecosystem.** No light penetrates to this region and the amount of food available is limited. Essentially, all of the organisms in this environment are scavengers that feed on whatever drifts their way. Many of the animals are small and have light sources that they use for finding or attracting food.

figure 6.19 **Coral Reef** Corals are small sea animals that secrete external skeletons. They have a mutualistic relationship with certain algae, which allows both kinds of organisms to be very successful. The skeletal material serves as a substrate upon which many other kinds of organisms live.

Estuaries

An **estuary** is a special category of marine ecosystem, that consists of shallow, partially enclosed areas where freshwater enters the ocean. The saltiness of the water in the estuary changes with tides and the flow of water from rivers. The organisms that live here are specially adapted to this set of physical conditions, and the number of species is less than in the ocean or in freshwater. Estuaries are particularly productive ecosystems because of the large amounts of nutrients introduced into the basin from the rivers that run into them. This is further enhanced by the fact that the shallow water allows light to penetrate to most of the water in the basin. Phytoplankton and attached algae and plants are able to use the sunlight and the nutrients for rapid growth. This photosynthetic activity supports many kinds of organisms in the estuary. Estuaries are especially important as nursery sites for fish and crustaceans like flounder and shrimp. The adults enter these productive, sheltered areas to reproduce and then return to the ocean. The young spend their early life in the estuary and eventually leave as they get larger and are more able to survive in the ocean. Estuaries also trap sediment. This activity tends to prevent many kinds of pollutants from reaching the ocean and also results in the gradual filling in of the estuary, which may eventually become a salt marsh and then part of a terrestrial ecosystem.

figure 6.20 **Mangrove Swamp** Mangroves are tropical trees that are able to live in very wet, salty muds found along the ocean shore. Since they are able to trap additional sediment, they tend to extend farther seaward as they reproduce.

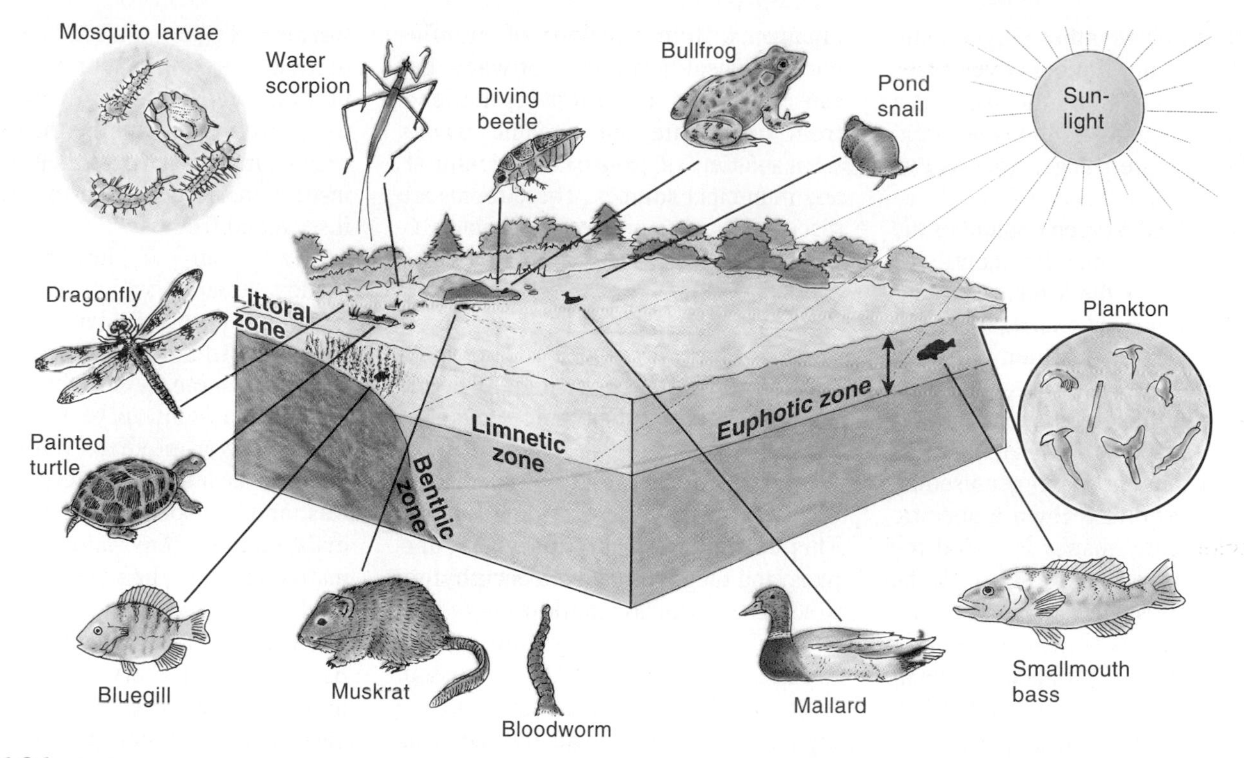

figure 6.21 **Lake Ecosystem** Lakes are similar in structure to oceans except that the species are different because most marine organisms cannot live in freshwater. Insects are common organisms in freshwater lakes, as are many kinds of fish, zooplankton, and phytoplankton.

Freshwater Ecosystems

Freshwater ecosystems differ from marine ecosystems in several ways. The amount of salt present is much less, the temperature of the water can change greatly, the water is in the process of moving downhill, oxygen can often be in short supply, and the organisms that inhabit freshwater systems are different.

Freshwater ecosystems can be divided into two categories: those in which the water is relatively stationary, such as lakes, ponds, and reservoirs, and those in which the water is running downhill, such as streams and rivers.

Lakes and Ponds

Large lakes have many of the same characteristics as the ocean. If the lake is deep, there is a euphotic zone at the top, with many kinds of phytoplankton, and zooplankton that feed on the phytoplankton. Small fish feed on the zooplankton, which are in turn eaten by larger fish. The species of organisms found in freshwater lakes are different from those found in the ocean, but the roles played are similar, so the same terminology is used.

Along the shore and in the shallower parts of lakes, many kinds of flowering plants are rooted in the bottom. Some have leaves that float on the surface or protrude above the water and are called **emergent plants.** Cattails, bulrushes, arrowhead plants, and water lilies are examples. Rooted plants that stay submerged below the surface of the water are called **submerged plants.** *Elodea* and *Chara* are examples.

Many kinds of freshwater algae also grow in the shallow water, where they may appear as mats on the bottom or attached to vegetation and other objects in the water. Associated with the plants and algae are a large number of different kinds of animals. Fish, crayfish, clams, and many kinds of aquatic insects are common inhabitants of this mixture of plants and algae. This region, with rooted vegetation, is known as the **littoral zone,** and the portion of the lake that does not have rooted vegetation is called the **limnetic zone.** (See figure 6.21.)

The productivity of the lake is determined by several factors. Temperature is important, since cold temperatures tend to reduce the amount of photosynthesis. Water depth is important because shallow lakes will have light penetrating to the lake bottom and, therefore, photosynthesis can occur throughout the entire water column. Shallow lakes also tend to be warmer as a result of the warming effects of the sun's rays. A third factor that influences the productivity of lakes is the amount of nutrients present. This is primarily determined by the rivers and streams that carry nutrients to the lake. River systems that run through areas that donate many nutrients will carry the nutrients to the lakes. Exposed soil and farmland tend to release nutrients, as do other human activities like depositing sewage into streams and lakes. Deep, cold, nutrient-poor lakes are low in productivity and are called **oligotrophic lakes.** Shallow, warm, nutrient-rich lakes are called **eutrophic lakes.**

Although the water molecule (H_2O) has oxygen as part of its structure, this oxygen is not available to organisms. The oxygen that they need is dissolved molecular oxygen (O_2), which enters water from the air or when it is released as a result of photosynthesis by aquatic plants. When water tumbles over rocks in a stream or crashes on the shore as a result of wave action, air and water mix, which allows more oxygen to dissolve in the water.

The dissolved oxygen content of the water is important since the quantity of oxygen determines the kinds of organisms that can inhabit the lake. When organic molecules enter water, they are broken down by bacteria and fungi. These decomposer organisms use oxygen from the water as they perform respiration. The amount of oxygen used by decomposers to break down a specific amount of organic matter is called the **biochemical oxygen demand** or **BOD.** Organic materials enter aquatic ecosystems in several ways. The organisms that live in the water produce the metabolic wastes. When organisms that live in or near water die or shed parts, their organic matter is contributed to the water. The amount of nutrients entering the water is also important, since the algae and plants whose growth is stimulated will eventually die and their decomposition will reduce oxygen concentration. Many bodies of water experience a reduced oxygen level during the winter when producers die. The amount and kinds of organic matter determine, in part, how much oxygen is left to be used by other organisms, such as fish, crustaceans, and snails. Many lakes may experience periods when oxygen is low, resulting in the death of fish and other organisms. Human activity often influences the health of bodies of water, because we tend to introduce nutrients from agriculture and organic wastes from a variety of industrial, agricultural, and municipal sources. These topics are discussed in greater depth in chapter 16.

Streams and Rivers

Streams and rivers are a second category of freshwater ecosystem. Since the water is moving, planktonic organisms are less important than are attached organisms. Most algae grow attached to rocks and other objects on the bottom. This collection of attached algae, animals, and fungi is called the **periphyton.** Since the water is shallow, light can penetrate easily to the bottom (except for large or extremely muddy rivers). Even so, it is difficult for photosynthetic organisms to accumulate the nutrients necessary for growth, and most streams are not very productive. As a matter of fact, the major input of nutrients is from organic matter that falls into the stream from terrestrial sources. These are primarily the leaves from trees and other vegetation, as well as the bodies of living and dead insects. Within the stream is a community of organisms that are specifically adapted to use the debris as a source of food. Bacteria and fungi colonize the organic matter, and many kinds of insects shred the material as they eat it along with the fungi and bacteria living on it. The feces (intestinal wastes) of these insects and the tiny particles produced during the eating process become food for other insects that build nets to capture the tiny bits of organic matter that drift their way. These insects are in turn eaten by carnivorous insects and fish.

Organisms in larger rivers and muddy streams, which have less light penetration, rely in large part on the food that drifts their way from the many streams that empty into the river. These larger rivers tend to be warmer and to have slower-moving water. Consequently, the amount of oxygen is usually less, and the species of plants and animals change. Any additional organic matter added to the river system adds to the BOD, further reducing the oxygen in the water. Plants may become established along the river bank and contribute to the ecosystem by carrying on photosynthesis and providing hiding places for animals.

Just as estuaries are a bridge between freshwater and marine ecosystems, swamps and marshes are a transition between aquatic and terrestrial ecosystems. **Swamps** are wetlands that contain trees that are able to live in places that are either permanently flooded or flooded for a major part of the year. **Marshes** are wetlands that are dominated by grasses and reeds. Many swamps and marshes are successional states that eventually become totally terrestrial communities.

Summary

Ecosystems change as one kind of organism replaces another in a process called succession. Ultimately, a relatively stable stage is reached, called the climax community. Succession may begin with bare rock or water, in which case it is called primary succession, or may occur when the original ecosystem is destroyed, in which case it is called secondary succession. The stages that lead to the climax are called successional stages.

Major regional terrestrial climax communities are called biomes. The primary determiners of the kinds of biomes that develop are the amount and yearly distribution of rainfall and the yearly temperature cycle. Major biomes are desert, grassland, savanna, tropical rainforest, temperate deciduous forest, taiga, and tundra. Each has a particular set of organisms that is adapted to the climatic conditions typical for the area. As one proceeds up a mountainside, it is possible to witness the same kind of change in biomes that occurs if one were to travel from the equator to the North Pole.

Aquatic ecosystems can be divided into marine (saltwater) and freshwater ecosystems. In the ocean, some organisms live in open water and are called pelagic organisms. Light penetrates only the upper layer of water; therefore, this region is called the euphotic zone. Tiny photosynthetic organisms that float near the surface are called phytoplankton. They are eaten by small animals known as zooplankton, which in turn are eaten by fish and other larger organisms.

The kind of material that makes up the shore determines the mixture of organisms that lives there. Rocky shores provide surfaces for organisms to attach; sandy shores do not. Muddy shores are often poor in oxygen, but marshes and swamps may develop in these areas. Coral reefs are tropical marine ecosystems dominated by coral animals. Mangrove swamps are tropical marine shoreline ecosystems dominated by trees. Estuaries occur where freshwater streams enter the ocean. They are usually shallow, very productive areas. Many marine organisms use estuaries for reproduction.

Insects are common in freshwater and are absent in marine systems. Lakes show similar structure to the ocean, but the species are different. Deep, cold-water lakes with poor productivity are called oligotrophic, while shallow, warm-water, highly productive lakes are called eutrophic. Streams differ from lakes in that most of the organic matter present in streams falls into it from the surrounding land. Thus, organisms in streams are highly sensitive to the land uses that occur near the streams.

ISSUES & ANALYSIS

Protecting Old-Growth Temperate Rainforests of the Pacific Northwest

The coastal areas of northern California, Oregon, Washington, British Columbia, and southern Alaska have an unusual set of environmental conditions that support a special kind of forest, a temperate rainforest. The prevailing winds from the west bring moisture-laden air to the coast. As this air meets the coastal mountains and is forced to rise, it cools and the moisture falls as rain or snow. Most of these areas receive 200 or more centimeters (80 or more inches) of precipitation per year. This abundance of water, along with fertile soil and mild temperatures, results in a luxuriant growth of plants.

Sitka spruce, Douglas fir, and western hemlock are typical evergreen coniferous trees. Undisturbed (old-growth) forests of this region have trees as old as 800 years that are almost as tall as the length of a football field. Deciduous trees of various kinds (red alder, bigleaf maple, black cottonwood) grow in places where they can get enough light. All trees are covered with mosses, ferns, and other plants that grow on their surface. The dominant color is green, since most surfaces have something growing on them.

When a tree dies and falls to the ground, it rots in place and often serves as a site for the establishment of new trees. This is such a common feature of the forest that the fallen, rotting trees are called nurse trees. The fallen trees also serve as a food source for a variety of insects, which are food for a variety of other animals. Several endangered or threatened animals, such as the northern spotted owl, the marbled murrelet (a seabird), and the Roosevelt elk, are dependent on undisturbed forest for their survival.

Because of the rich resource of trees, 90 percent of the original temperate rainforest has been logged. What remains has become a source of controversy. Some maintain that it should be protected as a remnant of the original forest of the region, just as small patches of prairie and eastern woodland have been preserved in other parts of North America. Others point out that the intact old-growth forest must be preserved to protect endangered species. Some maintain that since the trees are old and will die in the near future, they should be harvested for their timber value and the jobs the logging will provide. Others counter that dying and dead trees are important to the maintenance of this unique ecosystem. The fate of this unusual ecosystem is not likely to be resolved without legislation or legal action.

- List three components you would include in a law that would protect this ecosystem.
- What compromises can you offer to those who feel the forest should be used for timber?
- If there were no endangered species in the region would you feel differently?

Interactive Exploration

Check out the website at

http://www.mhhe.com/environmentalscience

and click on the cover of this textbook for interactive versions of the following:

KNOW THE BASICS

abyssal ecosystem *124*
alpine tundra *120*
benthic *121*
benthic ecosystem *121*
biochemical oxygen demand (BOD) *126*
biome *110*
boreal forest *119*
climax community *106*
coral reef ecosystem *122*
desert *111*
emergent plants *125*
estuary *124*
euphotic zone *121*
eutrophic lake *125*
freshwater ecosystem *121*
grassland *113*
limnetic zone *125*
littoral zone *125*
mangrove swamp ecosystem *124*
marine ecosystem *121*
marsh *126*
northern coniferous forest *119*
oligotrophic lake *125*
pelagic *121*
pelagic ecosystem *121*
periphyton *126*
permafrost *120*
phytoplankton *121*
pioneer community *106*
plankton *121*
prairie *113*
primary succession *106*
savanna *114*
secondary succession *106*
seral stage *107*
sere *107*
steppe *113*
submerged plants *125*
succession *106*
successional stage *107*
swamp *126*
taiga *119*
temperate deciduous forest *117*
tropical rainforest *116*
tundra *120*
zooplankton *121*

- On-line Flashcards
- Electronic Glossary

IN THE REAL WORLD

Can geese actually threaten a biome? Take a look at Snow Goose Population Threatens Arctic Tundra Habitat and see how an animal that is *normally* found in an area can become a problem.

A biome that has changed considerably in the past two hundred years is the grasslands of the Great Plains region in North America. What has caused the major changes? The story of this region's freshwater marshes is told in The Prairie Wetlands of Southwest Minnesota.

Temperate rainforests have had a lot of publicity with the controversy over logging and spotted owls. Now is it mushrooms? Check out the Matsutake Mushroom Mania case study for an update on this craze.

Tropical rainforests have also received publicity for a high rate of deforestation and loss of species. Estimating the rate of deforestation is difficult for many reasons but read the latest in New Study Raises Estimates of Deforestation in Amazon.

Speaking of tropical rainforests, what can you do to help protect forest species? Every morning (especially during finals week) you could have an impact. Find out how by reading Why You Should Buy Organic Coffee: It Helps Migratory Birds and Other Forest Species.

Green, soupy, and no fun for swimming. A lake in the summer with too many nutrients and an algae bloom can cause problems for fish and humans alike. Read how some ecologists are proposing to manipulate the food web for better water quality in Food Web Control of Primary Production in Lakes.

Speaking of human manipulation, the restoration of ecosystems is an issue that has some people in south Florida worrying about their water supply. Read Everglades Restoration: Greatest Restoration Yet, or Just More of the Same? for a better understanding of the issues and concerns related to this ecosystem restoration.

PUT IT IN MOTION

What can cause the oxygen of a lake to decrease? Chapter 16 will investigate this process in more depth but you can get a sneak preview by checking out the Deoxygenation of Lakes animation.

Some animations can make a concept clearer, even if you have heard of it since elementary school. The animations that deal with climate are worth checking out since they make the concepts clear and let you *see* the process and hear it at the same time. Study Global Air Circulation for an excellent explanation of why there are tropical rainforests near the equator and desert belts at 30 degrees north and south latitude.

The rainshadow effect is another process that you've probably studied in the past. Check out the Rainshadow Effect animation for a clear and concise explanation of this phenomenon.

Did you know that when some college graduates were asked to explain the four seasons, they were not able to do so? Study the Four Seasons animation to make sure you can describe the seasons without embarassment.

TEST PREPARATION

Review Questions

1. Describe the process of succession. How does primary succession differ from secondary succession?
2. How does a climax community differ from a successional community?
3. List three characteristics typical of each of the following biomes: tropical rainforest, desert, tundra, taiga, savanna, grassland, and temperate deciduous forest.
4. What two primary factors determine the kind of terrestrial biome that will develop in an area?
5. How does height above sea level affect the kind of biome present?
6. What areas of the ocean are the most productive?
7. How does the nature of the substrate affect the kinds of organisms found at the shore?
8. What is the role of each of the following organisms in a marine ecosystem: phytoplankton, zooplankton, algae, coral organisms, and fish?
9. List three differences between freshwater and marine ecosystems.
10. What is an estuary? Why are estuaries important?

Critical Thinking Questions

1. Does the concept of a "climax community" make sense? Why or why not?
2. What do you think about restoring ecosystems that have been degraded by human activity? Should it be done or not? Why? Who should pay for this reconstruction?
3. Identify the biome in which you live. What environmental factors are instrumental in maintaining this biome? What is the current health of your biome? What are the current threats to its health? How might your biome have looked 100, 1000, 10,000 years ago?
4. Imagine you are a conservation biologist who is being asked by local residents what the likely environmental outcomes of development would be in the tropical rainforest in which they live. What would you tell them? Why do you give them this evaluation? What evidence can you cite for your claims?
5. The text says that 90 percent of the old-growth temperate rainforest in the Pacific Northwest has been logged. What to do with the remaining 10 percent is still a question. Some say it should be logged and others say it should be preserved. What values, beliefs, and perspectives are held by each side? What is your ethic regarding logging old-growth in this area? What values, beliefs, and perspectives do you hold regarding this issue?
6. Much of the old-growth forest in the United States has been logged, economic gains have been realized, and second-growth forests have become established. This is not the case in the tropical rainforests, although they are being lost at alarming rates. Should developed countries, who have already "cashed in" on their resources, have anything to say about what is happening in developing countries? Why do you think the way you do?

KEY CHAPTER LINKS | ESSENTIAL STUDY PARTNER | BIO COURSE | GLOBAL ISSUES | REGIONAL PERSPECTIVES | PRACTICE QUIZZING

CHAPTER

Population Principles

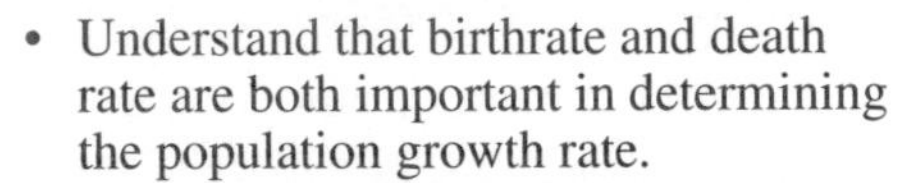

Objectives

After reading this chapter, you should be able to:

- Understand that birthrate and death rate are both important in determining the population growth rate.
- Define the following characteristics of a population: natality, mortality, sex ratio, age distribution, biotic potential, and spatial distribution.
- Explain the significance of biotic potential to the rate of population growth.
- Describe the lag, exponential growth, and stable equilibrium phases of a population growth curve. Explain why each of these stages occurs.
- Describe how the limiting factors determine the carrying capacity for a population.
- List the four categories of limiting factors.
- Describe a death phase that is typical of some kinds of populations.
- Recognize that humans are subject to the same forces of environmental resistance as are other organisms.
- Understand the implications of overreproduction.
- Understand that the human population is still growing rapidly.
- Explain how human population growth is influenced by social, theological, philosophical, and political thinking.

Chapter Outline

Population Characteristics
- Natality and Mortality
- Sex Ratio and Age Distribution
- Population Density and Spatial Distribution
- Summary of Factors That Influence Population Growth Rates

A Population Growth Curve

Carrying Capacity

Environmental Close-Up: *Population Growth of Invading Species*

Reproductive Strategies and Population Fluctuations

Human Population Growth

Global Perspective: *Managing Elephant Populations—Harvest or Birth Control?*
- Social Factors Influence Human Population
- Ultimate Size Limitation

Issues & Analysis: *Wolves and Moose on Isle Royale*

Population Characteristics

A population can be defined as a group of individuals of the same species inhabiting an area. Just as individuals within a population are recognizable, different populations of the same species have specific characteristics that distinguish them from one another. Some important ways in which populations differ include natality (birthrate), mortality (death rate), sex ratio, age distribution, growth rates, density, and spatial distribution.

Natality and Mortality

Natality refers to the number of individuals added to the population through reproduction over a particular time period. There are two ways in which new individual organisms are produced—asexual reproduction and sexual reproduction. Bacteria and other tiny organisms reproduce primarily asexually when they divide to form new individuals that are identical to the original organism. Even plants and many kinds of animals reproduce asexually by dividing into two parts or by budding off small portions of themselves that become independent individuals. However, most organisms have some stage in their life cycle in which they reproduce sexually. In plant populations, sexual reproduction results in the production of numerous seeds, but the seeds must land in appropriate soil conditions before they will germinate to produce a new individual. Animal species also typically produce large numbers of offspring as a result of sexual reproduction. In human populations, natality is usually described in terms of the **birthrate,** the number of individuals born per one thousand individuals per year. For example, if a population of 2000 individuals produced 20 offspring during one year, the birthrate would be 10 per thousand per year. The natality for most species is typically quite high. Most organisms produce many more offspring than are needed to replace the parents.

It is important to recognize that the growth of a population is not determined by the birthrate (natality) alone. **Mortality,** the number of deaths in a population over a particular time period, is also important. For most organisms, mortality rates are very high. Of all the seeds that plants produce, very few will result in a mature plant that itself will produce offspring. Many seeds are eaten by animals, some never find proper soil conditions, and those that germinate must compete with other organisms for nutrients and sunlight. In animals, most immature organisms die before they have an opportunity to reproduce. In human population studies, mortality is usually discussed in terms of the **death rate,** the number of people who die per one thousand individuals per year. Compared to the high mortality of the young of most species, the infant death rate of long-lived animals like humans is relatively low. In order for the size of a population to grow, the number of individuals added by reproduction must be greater than the number leaving it by dying. (See figure 7.1)

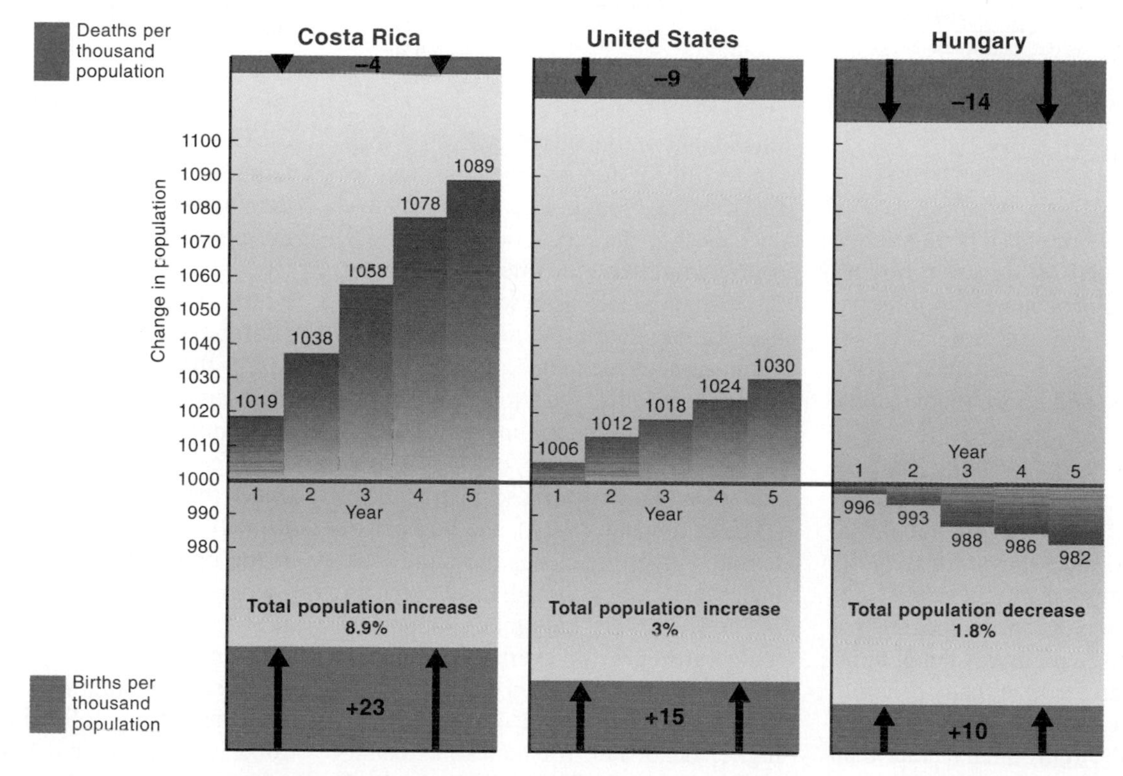

figure 7.1 **Effect of Birthrate and Death Rate on Population Size** For a population to grow, the birthrate must exceed the death rate for a period of time. These three human populations illustrate how the combined effects of births and deaths would change population size if birthrates and death rates were maintained for a five-year period.

Source: Data from World Population Data Sheet 2000, Population Reference Bureau, Inc., Washington, D.C.

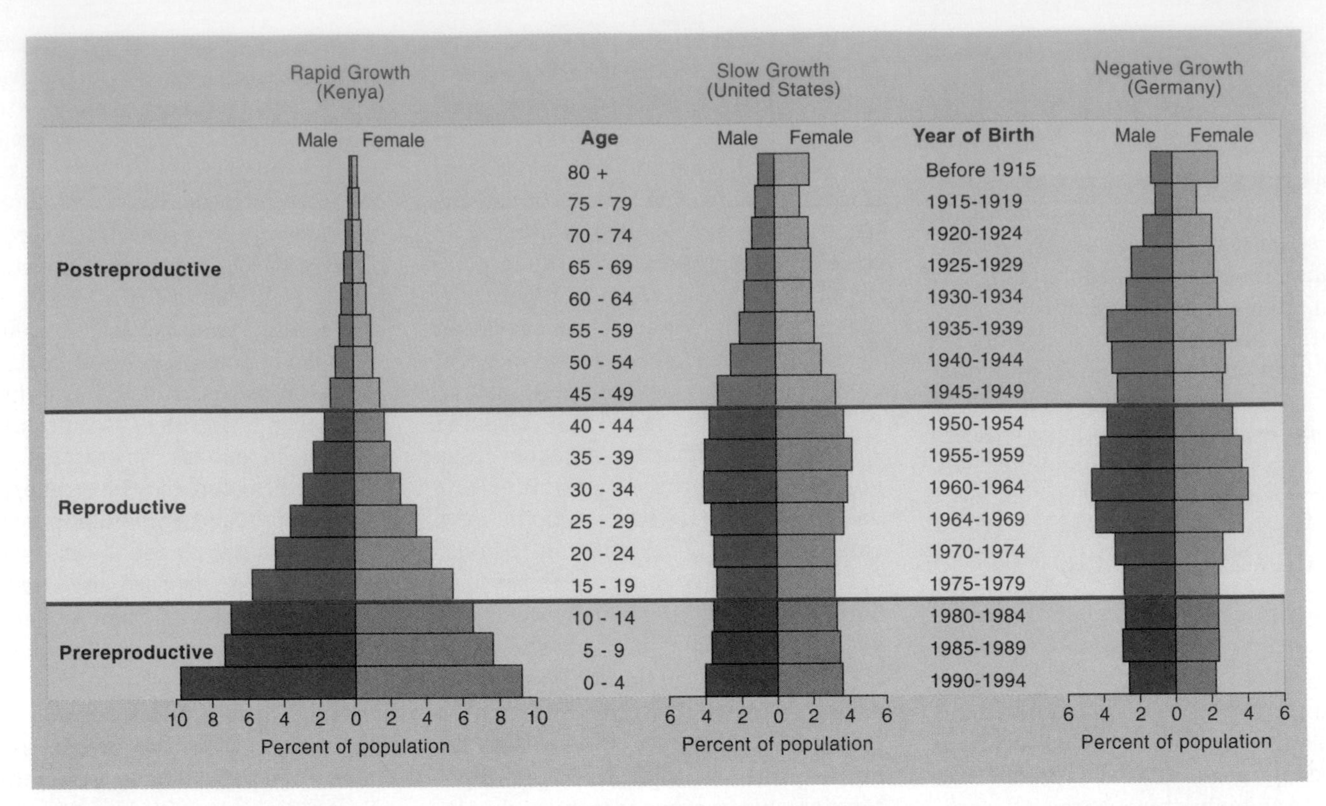

figure 7.2 **Age Distribution in Human Populations** The relative numbers of individuals in each of the three categories (prereproductive, reproductive, and postreproductive) are good clues to the future growth of a population. Kenya has a large number of young individuals who will become reproducing adults. Therefore, this population is likely to grow rapidly. The United States has a large proportion of reproductive individuals and moderate number of prereproductive individuals. Therefore, it is likely to grow slowly. Germany has a declining number of reproductive individuals and very small number of prereproductive individuals. Therefore, it has begun to decline.

Source: Data from Population Reference Bureau.

Sex Ratio and Age Distribution

The reproductive rate of a population is greatly influenced by its sex ratio and age distribution. The **sex ratio** refers to the relative numbers of males and females. (Many kinds of organisms, such as earthworms and most plants, have both kinds of sex organs in the same body; sex ratio has no meaning for these species.) In organisms that have two different sexes, the number of females is very important, since they ultimately determine the number of offspring produced in the population. In species that are polygamous, a relatively small number of males can fertilize many different females, so the number of males is less important to the reproductive rate than the number of females. In monogamous species in which mated pairs raise the young, unpaired females may not be fertilized and produce young. Even if they are fertilized, they will be less successful in raising young. It is typical in most species that the sex ratio is about 1:1 (one female to one male). However, there are populations in which this is not true. In populations of many species of game animals the males are shot (have a higher mortality) and the females are not. This results in an uneven sex ratio in which the females outnumber the males. In many social insect populations (bees, ants, and wasps), the number of females greatly exceeds the number of males at all times, though most of the females are sterile. In humans, about 106 males are born for every 100 females. However, in the United States, by the time people reach their mid-twenties, a higher death rate for males has equalized the sex ratio. The higher male death rate continues into old age, when women outnumber men.

As you can see in figure 7.2, populations can differ in **age distribution,** the number of individuals of each age in the population. Some are prereproductive juveniles, some are reproducing adults, and some are postreproductive adults. Many kinds of organisms, particularly those that have short life spans, have age distributions that change significantly during the course of a year.

The age distribution greatly influences the reproductive rate of a population. If the majority of a population is made up of reproducing adults, a rapid increase in the number of young will probably result. Most species of plants and animals typically produce a large number of young at a specific time of the year (spring, following rain, or when food is plentiful) but have a very high mortality among the young. Thus the number of young declines sharply as time passes. A good example of a

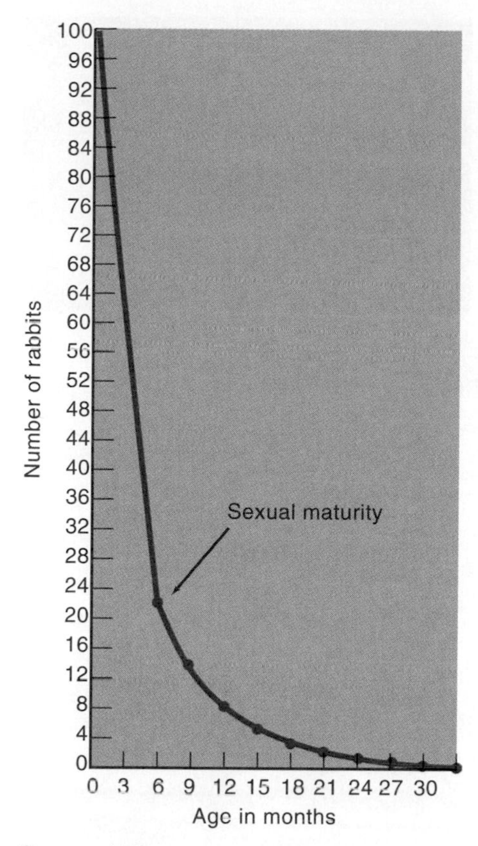

figure 7.3 **Survivorship Curve of Cottontail Rabbit** In most natural populations, mortality rates are so high that very few individuals reach sexual maturity, and even fewer reach old age.

species of this type is the cottontail rabbit. Many young are produced in the spring, but most of them die, a few reach sexual maturity, and very few live to old age. (See figure 7.3.)

In species that live a long time, it is possible for a population to have a balanced age distribution in which the number of individuals in these three categories is relatively constant. For most long-lived plant and animal populations, this means more prereproductive individuals than reproductive individuals, and more reproductive individuals than postreproductive individuals. If the majority of a population is postreproductive, the population declines. This age structure develops in many insect populations in the fall after eggs have been laid.

Human populations exhibit several types of age distribution. (See figure 7.2.) Kenya's population has a large prereproductive and reproductive component. This means that it will continue to increase rapidly for some time. Germany has an age distribution with high postreproductive and low prereproductive portions of the population. With low numbers of prereproductive individuals entering their reproductive years, the population of Germany has begun to decline. The United States has a very large reproductive component with a declining number of prereproductive individuals. Eventually, the U.S. population will begin to decline if current trends in birthrates and death rates continue.

Population Density and Spatial Distribution

Because of such factors as soil type, quality of habitat, and availability of water, organisms normally are distributed unevenly. Some populations have many individuals clustered into a small space while other populations of the same species may be widely dispersed. **Population density** is the number of organisms per unit area. For example, fruitfly populations are very dense around a source of rotting fruit, while they are rare in other places. Similarly, humans are often clustered into dense concentrations we call cities, with lower densities in rural areas. When the population density is too great, all individuals within the population are injured because they compete severely with each other for necessary resources. Plants may compete for water, soil nutrients, or sunlight. Animals may compete for food, shelter, or nesting sites. In animal populations, overcrowding might cause some individuals to explore and migrate into new areas. This movement from densely populated locations to new areas is called **dispersal.** It relieves the overcrowded conditions in the home area and, at the same time, increases the population in the places to which they migrate. Often, it is juvenile individuals that relieve overcrowding by leaving. The pressure for out-migration (**emigration**) may be a result of seasonal reproduction leading to a rapid increase in population size, or environmental changes that intensify competition among members of the same species. For example, as water holes dry up, competition for water increases, and many desert birds emigrate to areas where water is still available.

The organisms that leave one population often become members of a different population. This in-migration (**immigration**) may introduce characteristics that were not in the population originally. When Europeans immigrated to North America, they brought genetic and cultural characteristics that had a tremendous impact on the existing Native American population. Among other things, Europeans brought diseases that were foreign to the Native Americans. These diseases increased the death rate and lowered the birthrate of Native Americans, resulting in a sharp decrease in the size of their populations.

Droughts, wars, and political persecution have caused people to emigrate from their native lands to other countries. In 1999, the United Nations High Commissioner for Refugees estimated that there were about 11.5 million refugees worldwide. Often, the receiving countries are unable to cope with the large influx of new inhabitants.

Summary of Factors That Influence Population Growth Rates

Populations have an inherent tendency to increase in size. However, as we have just seen, many factors influence the rate at which a population can grow. At the simplest level, the rate of increase is determined by subtracting the number of individuals leaving the population from the number entering. Individuals leave the population either by death or emigration. Individuals enter the population by birth or immigration. Birthrates and death rates are influenced by several factors, including the number of females in the population and their age. In addition, the density of a population may encourage individuals to leave because of intense competition for a limited supply of resources.

Apples

Geese

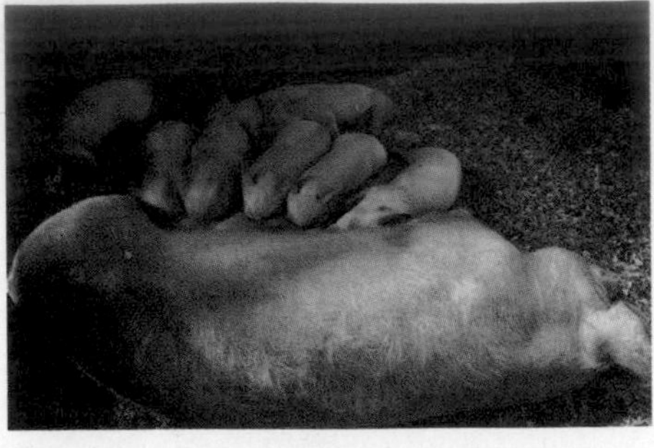
Pigs

figure 7.4 **Biotic Potential** The ability of a species to reproduce greatly exceeds the number necessary to replace those who die. Here are some examples of the prodigous reproductive abilities of some species.

A Population Growth Curve

Sex ratios and age distributions directly influence the rate of reproduction within a population. However, each species also has an inherent reproductive capacity, or **biotic potential,** which is its biological ability to produce offspring. Some species, like apple trees, may produce thousands of offspring (seeds) per year, while others, like pigs or geese, may produce 10 to 12 young per year. (See figure 7.4.) Some large animals, like bears or elephants, may produce one young every two to three years. Although there are large differences among species, generally, adults produce many more offspring during their lifetimes than are needed to replace themselves when they die. Furthermore, most of the young die; only a few survive to become reproductive adults themselves.

Because most species have a high biotic potential, there is a natural tendency for populations to increase. For example, if two mice produce four offspring and the offspring live, they will produce offspring of their own, while their parents continue to reproduce as well. Under these conditions, the population will grow exponentially (2, 4, 8, 16, 32, etc.).

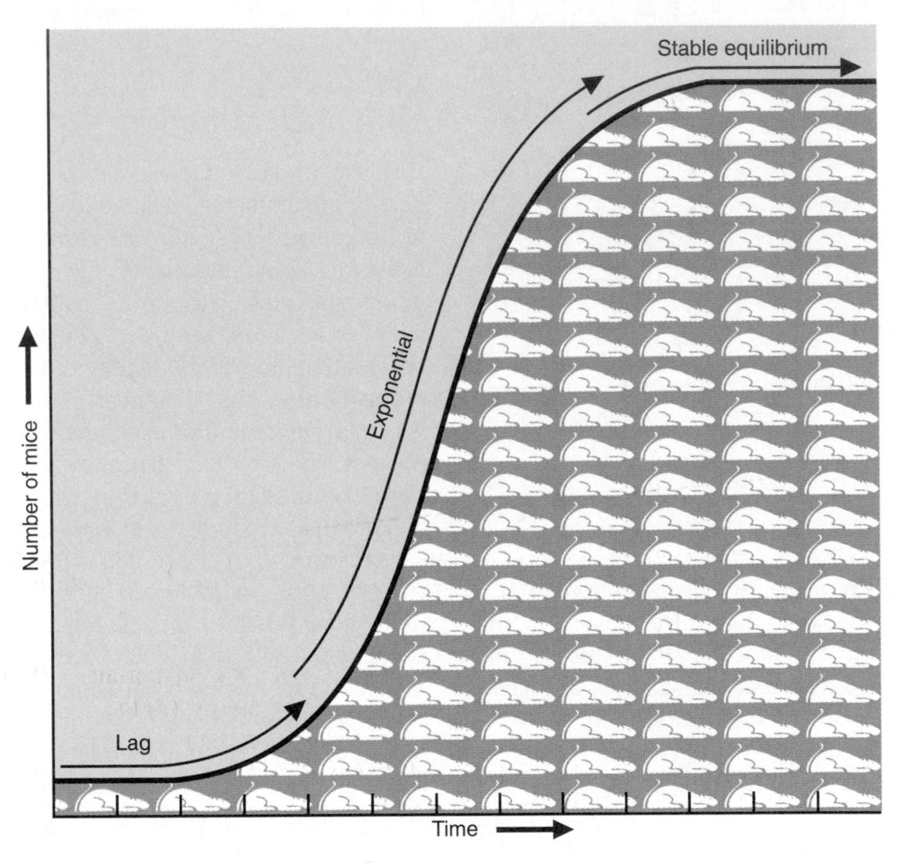

figure 7.5 **A Typical Population Growth Curve** In this mouse population, there is little growth during the lag phase. During the exponential growth phase, the population increases rapidly as increasing numbers of individuals reach reproductive age. Eventually, the population reaches a stable equilibrium phase, during which the birthrate equals the death rate.

Population growth often follows a particular pattern, consisting of a lag phase, an exponential growth phase, and a stable equilibrium phase. Figure 7.5 shows a typical population growth curve. During the first portion of the curve, known as the **lag phase,** the population grows very slowly because there are few births since the process of reproduction and growth of offspring takes time. Organisms must mature into adults before they can reproduce. When the offspring begin to mate and have young, the parents may be producing a second set of offspring. Since more organisms now are reproducing, the population begins to increase at an accelerating rate. This stage is known as the **exponential growth phase (log phase).** This growth will continue for as long as the birthrate exceeds the death rate. Eventually, however, the death rate and the birthrate will come to equal one another, and the population will stop growing and reach a relatively stable population size. This stage is known as the **stable equilibrium phase.**

It is important to recognize that although the size of the population may not be changing, the individuals are changing. As new individuals enter by birth or immigration, others leave by death or emigration. For most organisms, the first indication that a population is entering a stable equilibrium phase is an increase in the death rate. A decline in the birthrate may also contribute to the stabilizing of population size. Usually, this occurs after an increase in the death rate. To understand why populations cannot grow continuously, it is necessary to discuss the concept of carrying capacity.

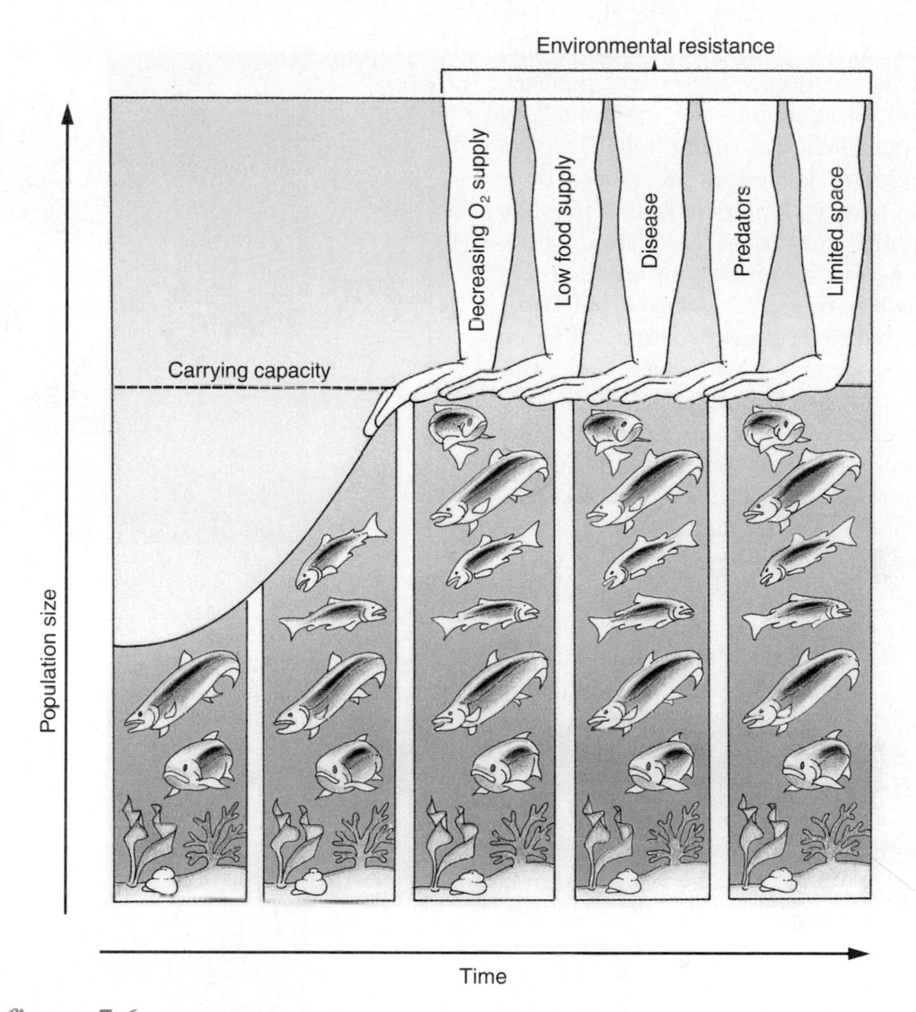

figure 7.6 **Carrying Capacity** A number of factors in the environment, such as food, oxygen supply, diseases, predators, and space, determine the number of organisms that can survive in a given area—the carrying capacity of that area. The environmental factors that limit populations are known collectively as environmental resistance.

Carrying Capacity

The **carrying capacity** of an area is the number of individuals of a species that can survive in that area over time. The concept of carrying capacity is usually applied to relatively long-lasting habitats and is helpful when we examine why populations stabilize. However, nothing is permanent and as the habitats change, because of disturbance or succession, the carrying capacity for a species changes also. There are also seasonal changes that influence the numbers of individuals that can be supported in an area. The number of individuals that can be supported during the summer may be much larger than can be supported in winter months. Indeed, some animals (such as birds) use certain habitats only during the summer and migrate in the fall when living conditions become difficult. The combination of factors that sets the carrying capacity for an area is called **environmental resistance.**

When a particular condition or factor can be identified as a key component that limits the size of a population, it is identified as a **limiting factor.** For most populations, four categories of limiting factors are recognized as components of environmental resistance that set the carrying capacity. These are: (1) the availability of raw materials, (2) the availability of energy, (3) the accumulation of waste products and their means of disposal, and (4) interactions among organisms.

In some cases, these limiting factors are easy to identify. Lack of food, lack of oxygen, competition with other species, or disease are examples. In other cases, the limiting factors may be less obvious. (See figure 7.6.) For example, in grass plants, nitrogen and magnesium in the soil are necessary raw materials for the manufacture of chlorophyll. If these minerals are not present in sufficient quantities, the grass population cannot increase. The application of fertilizers containing these minerals removes this limiting factor, and the individual grass plants grow and reproduce, resulting in a larger population. In effect, the carrying capacity has been increased because this limiting factor has been removed. There is still a carrying capacity but it is set at a new level and some new primary limiting factor will emerge. Perhaps it will be the amount of water, the number of insects that feed on the grass, or competition for sunlight. Because plants require energy in the form of sunlight for photosynthesis, the amount of light can be a limiting factor for many plants. When small plants are in the shade of trees, they often do not grow well and have small populations.

Accumulation of waste products is not normally a limiting factor for plants, since they produce few wastes, but it can be for other kinds of organisms. Bacteria, other tiny organisms, and many kinds of aquatic organisms that live in small ecosystems like puddles, pools, or aquariums may be limited by wastes. When a small number of a species of bacterium are placed on a petri plate with nutrient agar (a jellylike material containing food substances), the population growth follows a curve shown in figure 7.7. As expected, it begins with a lag phase, continues through an exponential growth phase, and eventually levels off in a stable equilibrium phase. However, in this small, enclosed space, there is no way to get rid of the toxic waste products, which accumulate, eventually killing the bacteria. This decline in population size is known as the **death phase.** When a population decreases rapidly, it is said to crash.

Interactions among organisms are also important in controlling population size. Since many birds eat grass seeds, birds have a limiting effect on the size of grass populations. Decomposer organisms may allow for increased populations because they prevent the buildup of toxic wastes. Parasites and predators may cause the premature death of individuals, thus limiting the size of the population. A good example of predator–prey interaction is the relationship between the cat known as the Canada lynx and a member of the rabbit family known as the varying hare. The varying hare produces large numbers of young. In peak reproductive years a female varying hare can produce 16-18 young. As with many animals a primary cause of death is predation. The varying hare population is a good food source for a variety of predators including the lynx. When the population of varying hares increases it provides an abundant source of food for the lynx and the size of the lynx population rises and when the population of hares decreases so does that of the lynx. This pattern repeats itself in a 10 year cycle (figure 7.8).

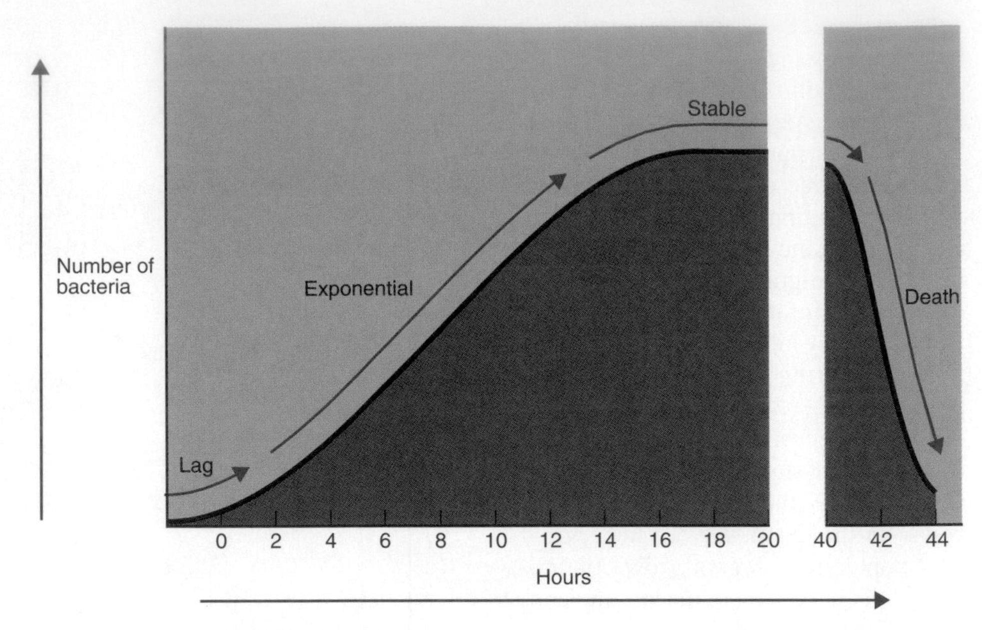

figure 7.7 **A Bacterial Growth Curve** The initial change in population size follows a typical population growth curve until waste products become lethal. The buildup of waste products lowers the carrying capacity. When a population begins to decline, it enters the death phase.

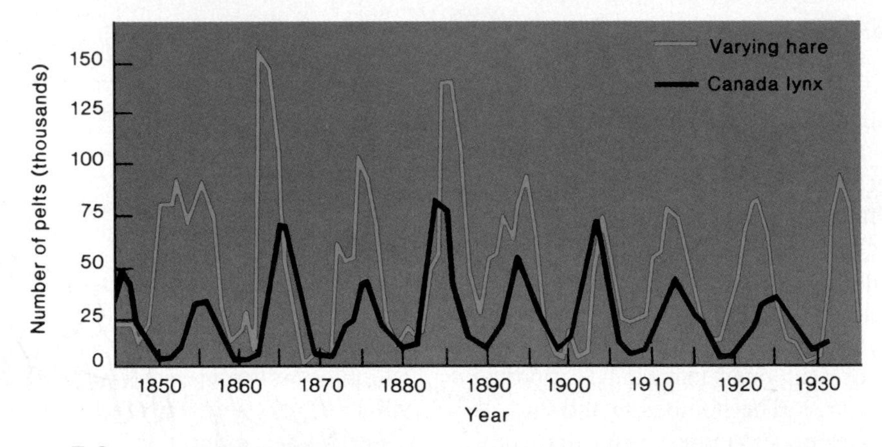

figure 7.8 **Interaction of Predator–Prey Populations** Interaction between predator and prey species is complex and often difficult to interpret. These data were collected from the records of the number of pelts purchased by the Hudson Bay Company. They show that the two populations fluctuate with about 10 years between successive high populations. The change in the lynx population usually following changes in the varying hare population.

Source: Data from D.A. MacLulich, "Fluctuations in the Numbers of the Varying Hare" (*Lepus americanus*), in *University of Toronto Studies, Biology Series #43 1937* (reprinted 1974), University of Toronto Press, Toronto.

Recent studies indicate that one of the causes of the decline in varying hare populations is a reduction in their reproductive rate. The causes of this reduction rate may be related to a variety of factors including: reduced quality of food and higher levels of stress resulting from greater difficulty in finding food and avoiding predators. With reduced reproduction and continued high predation the varying hare population drops. With reduced numbers of hares, lynx populations drop. Eventually the reproductive rate of hares increases and the population rebounds followed by a rebound in the lynx population as well. It appears that both food availability and predation are important limiting factors that determine the size of the varying hare population and the number of varying hares is a primary limiting factor for the lynx.

Some studies indicate that populations can be controlled by interaction among individuals within the population. A study of laboratory rats shows

ENVIRONMENTAL CLOSE-UP

Population Growth of Invading Species

When a new species is introduced into an area suitable for its survival, it often has great potential to increase its population size, since it may not have natural enemies to keep mortality high. New species may also be able to compete favorably for available resources and lower the populations of native species. A typical population growth curve starts with a few individuals being released. Once established, the population will increase in number and expand its range. The zebra mussel, for example, is thought to have entered the Great Lakes about 1985. Today, it is found throughout all the Great Lakes and has been introduced into the Mississippi River and its tributaries, where it has been discovered as far south as New Orleans. Similarly, the gypsy moth has spread from its place of original release to much of the forested land of the Midwest, and is still expanding its range. Another invading species is the kudzu vine, which has become a pest in many areas of the southern United States.

Eventually, the invading species occupies all the habitat suitable to it and the population stabilizes. Dandelions and starlings are introduced species that are no longer expanding their range. They have simply become a normal part of the biology of North America.

Dandelion

Zebra mussels

Kudzu vine

Starlings

Gypsy moths

that crowding causes a breakdown in normal social behavior, which leads to fewer births and increased deaths. The changes observed include abnormal mating behavior, decreased litter size, fewer litters per year, lack of maternal care, and increased aggression in some rats or withdrawal in others. Thus, limiting factors can reduce birthrates as well as increase death rates.

Reproductive Strategies and Population Fluctuations

So far, we have talked about population growth as if all organisms reach a stable population when they reach the carrying capacity. That is an appropriate way to begin to understand population changes, but the real world is much more complicated. Species can be divided into two broad categories based on their reproductive strategies. **K-strategists** are usually large organisms that have relatively long lives, produce few offspring, and provide care for their offspring. Their populations typically stabilize at a carrying capacity. Their reproductive strategy is to invest a great deal of energy in producing a few offspring that have a good chance of living to reproduce. Deer, lions, and swans are examples of this kind of organism. Humans generally produce single offspring, and even in countries with high infant mortality, 80 percent of the children survive beyond one year of age, and the majority of these will reach adulthood. Generally, populations of K-strategists are controlled by density-dependent limiting factors. **Density-dependent limiting factors** are those that become more severe as the size of the population increases. For example, as size of a hawk population increases, the competition among hawks for available food, such as mice, snakes, and small birds, becomes more severe. When food is in short supply, many of the young in the nests die, and population growth slows. They have reached the carrying capacity.

The **r-strategist** is typically a small organism that has a short life, produces many offspring, and does not reach a carrying capacity. Examples are grasshoppers, gypsy moths, and some mice. The reproductive strategy of r-strategists is to expend large amounts of energy producing many offspring but to provide limited care (often none) for them. Consequently, there is high mortality among the young. For example, one female oyster may produce a million eggs, but few of them ever find suitable places to attach themselves and grow. Typically, these populations are limited by **density-independent limiting factors** in which the size of the population has nothing to do with the limiting factor. Typical density-independent limiting factors are changing weather conditions that kill large numbers of organisms, the drying up of a small pond, or the death of entire populations due to the destruction of their food source. The population size of r-strategists is likely to fluctuate wildly. They reproduce rapidly, population size increases until some factor causes the population to crash, and then they begin the cycle all over again.

Since humans are K-strategists, it may be difficult for us to appreciate that the r-strategy can be viable from an evolutionary point of view. There is no carrying capacity for temporary resources. Resources that are present only for a short time can be exploited most effectively if many individuals of one species monopolize the resource, while denying other species access to it. Rapid reproduction can place a species in a position to compete against other species that are not able to increase numbers as rapidly. Obviously, most of the individuals will die, but not before they have left some offspring or resistant stages that will be capable of exploiting the resource should it become available again.

Even K-strategists, however, have population fluctuations for a variety of reasons. One reason is that the environment is not constant from year to year. Floods, droughts, fires, extreme cold, and similar events may affect the carrying capacity of an area, thus causing fluctuations in population size. Epidemic disease or increased predation may also lead to populations that vary from year to year. Figure 7.8 shows rather substantial changes in the populations of lynx and varying hare. The size of the lynx population seems to be tied to the size of the hare population, which is logical since lynx eat hares. However, the causes of fluctuations in the varying hare population are unclear. One possibility is that periodic epidemic disease may cause dense populations of organisms like hares to crash, leading to the crash of their predators' populations.

Although local human populations often show fluctuations, the worldwide human population has increased continually for the past several hundred years. Humans have been able to reduce environmental resistance by eliminating competing organisms, increasing food production, and controlling disease organisms.

Human Population Growth

The human population growth curve has a long lag phase followed by a sharply rising exponential growth phase that is still rapidly increasing. (See figure 7.9.) A major reason for the continuing increase is that the human species has lowered its death rate. When various countries reduce environmental resistance by increasing food production or controlling disease, they share this technology throughout the world. Developed countries send health care personnel to all parts of the globe to improve the quality of life for people in less-developed countries. Physicians offer advice on nutrition, and engineers develop wastewater treatment systems. Improved sanitary facilities in India and Indonesia, for example, decreased deaths caused by cholera. These advancements tend to reduce mortality while birthrates remain high. Thus the size of the human population increases rapidly.

Let us examine the human population situation from a different perspective. The world population is currently increasing at an annual rate of 1.4 per-

GLOBAL PERSPECTIVE

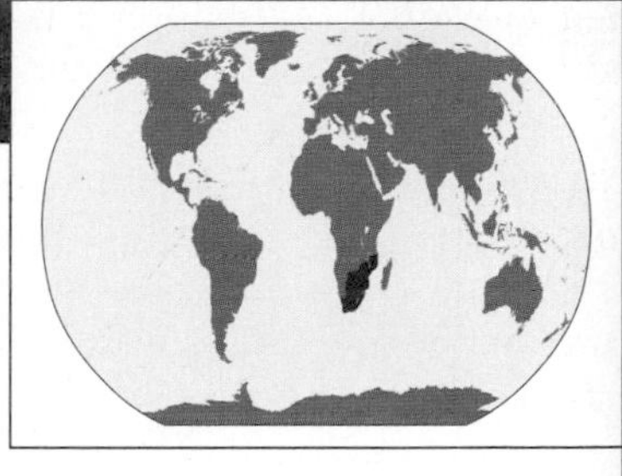

Managing Elephant Populations—Harvest or Birth Control?

During the 1980s, uncontrolled hunting of elephants for ivory and food had reduced the African elephant, *Loxodonta africana,* population from 1.3 million to 650,000. Many conservation organizations became alarmed by the rapid decline in the numbers of elephants and supported a ban on the export of ivory. In 1989, the Convention on International Trade in Endangered Species of Wild Fauna and Flora (CITES) banned all international trade in elephant products, of which ivory is the most important. The ban worked. The price of ivory fell and poaching became unprofitable. Since then, some southern African countries have experienced an elephant population explosion that threatens to destroy the limited habitat available to elephants and increasingly results in conflicts between farmers and elephants.

The African elephant requires huge amounts of food [150–250 kg/day (330–550 lb/day)] to sustain its large body. The animals strip bark from trees, uproot the trees, and eat large quantities of grass. Where humans and elephants share the same habitat, elephants can do great damage to crops. Because elephants have been increasingly confined to national parks and nature preserves, two natural options for relieving population pressure (migration or starvation) have not been available. As a consequence, by 2000 the population had fallen to 400,000. Migration from the park results in increased agricultural damage and risk of injury to farmers, and increasing populations can cause irreparable damage to the protected habitat they currently occupy as they seek food. If they are allowed to starve, there will be enormous public pressure from wildlife groups, and the park managers will be condemned. If the populations are to remain healthy, not destroy their habitat, and not infringe on farming communities, the size of the population must be controlled. There are only two ways to do that—increase the death rate (culling) or decrease the birthrate (sterilization or birth control).

However, several southern African countries, such as Botswana, Namibia, Zimbabwe, and South Africa, have growing elephant populations. They have argued that they should be allowed to manage their elephant populations as a natural resource in the same way that other large game animals (deer, elk, caribou) are managed. The sustainable harvest of surplus elephants would be wise use of a natural resource that would provide income to the local people as well as a much needed source of protein in this region. Furthermore, the income from the harvesting of elephants could be used to provide additional funding for park management.

In 1997, CITES approved the sale of 60 tonnes (66 tons) of ivory from Botswana, Namibia, and Zimbabwe to Japan. Following the sale, conservation organizations reported an increase in poaching of elephants and unlawful sale of ivory. In 2000, several countries again petitioned CITES to approve additional sales of ivory. Their request was not approved because of concerns about poaching and monitoring of illegal sales.

The Humane Society of the United States has advocated the use of birth control to reduce the birthrate and solve the elephant population problem. Several methods have been tested and they can work. However, critics of this approach suggest that preventing female elephants from conceiving for long periods of time may disrupt the normal social structure of the herd. Under most circumstances, a female that is pregnant will not come into heat again for two years. If females do not conceive, they will come into heat approximately every 15 weeks, and there will be more females in heat than is normal throughout the year. Since the mating activity of elephants is disruptive of their usual pattern of activity, some people are concerned that the normal social structure of the herd will be disrupted. Furthermore, a herd usually includes various ages of immature individuals. The younger individuals learn from older siblings or cousins. The use of birth control could disrupt the typical spacing of pregnancies and interfere with the normal process of "educating" young elephants. What looks like a simple problem to solve has resulted in two diametrically opposed camps: those that advocate increased deaths by harvesting and those that advocate decreased births by manipulating births. Neither is "natural," but something must be done or the survival of elephants could again become a problem.

cent. That may not seem like much, but even at 1.4 percent the population is growing rapidly. It can be difficult to comprehend the impact of a 1 or 2 percent annual increase. Remember that a growth rate in any population compounds itself, since the additional individuals eventually reproduce, thus adding more individuals. One way to look at this growth is to determine how much time is needed to double the population. This is a valuable method

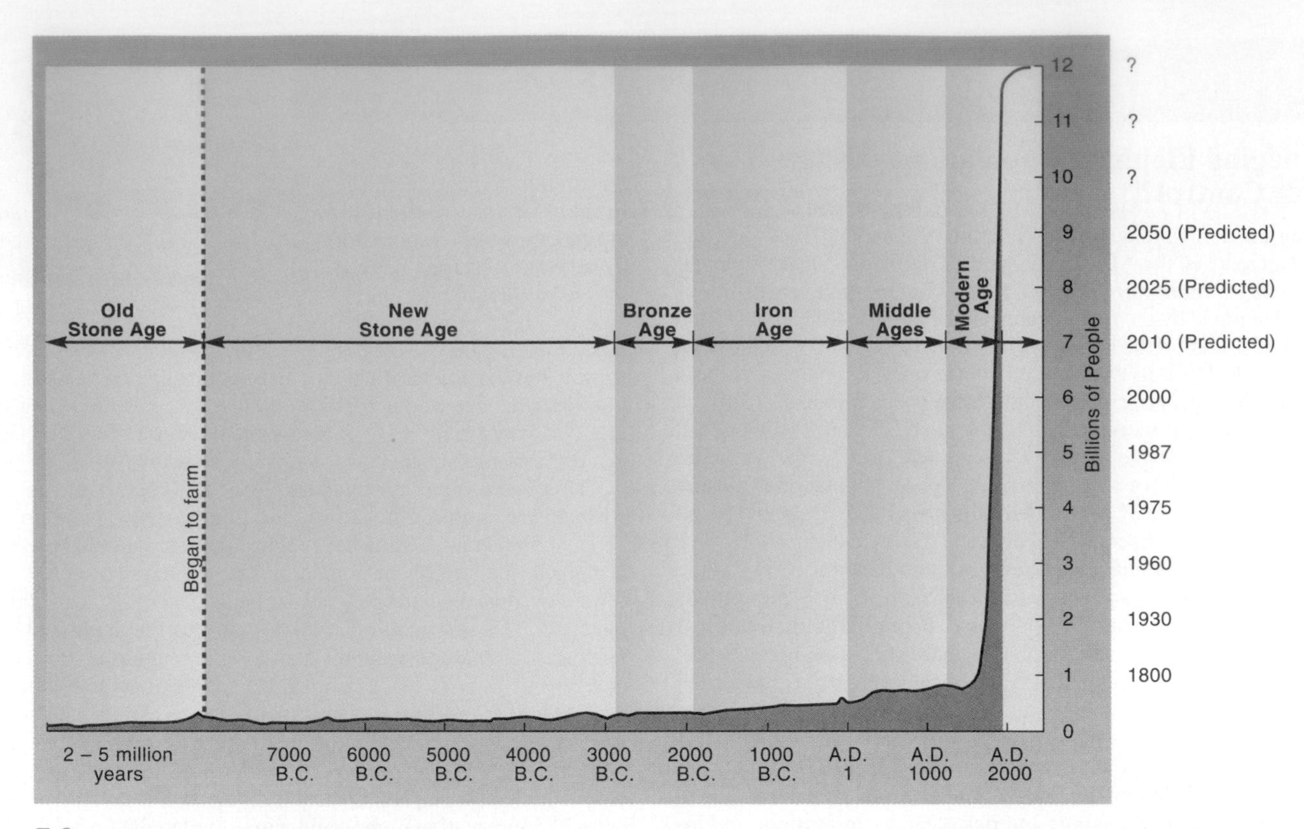

figure 7.9 **The Historical Human Population Curve** From A.D. 1800 to A.D. 1930, the number of humans doubled (from one billion to two billion) and then doubled again by 1975 (four billion) and could double again (eight billion) by the year 2025. How long can this pattern continue before the Earth's ultimate carrying capacity is reached?

Source: Data from Jean Van Det Tak, et al., "Our Population Predicament: A New Look" in *Population Bulletin,* Vol. 34, No. 5, December 1979, Population Reference Bureau, Washington, D.C.; and more recent data taken from Population Reference Bureau.

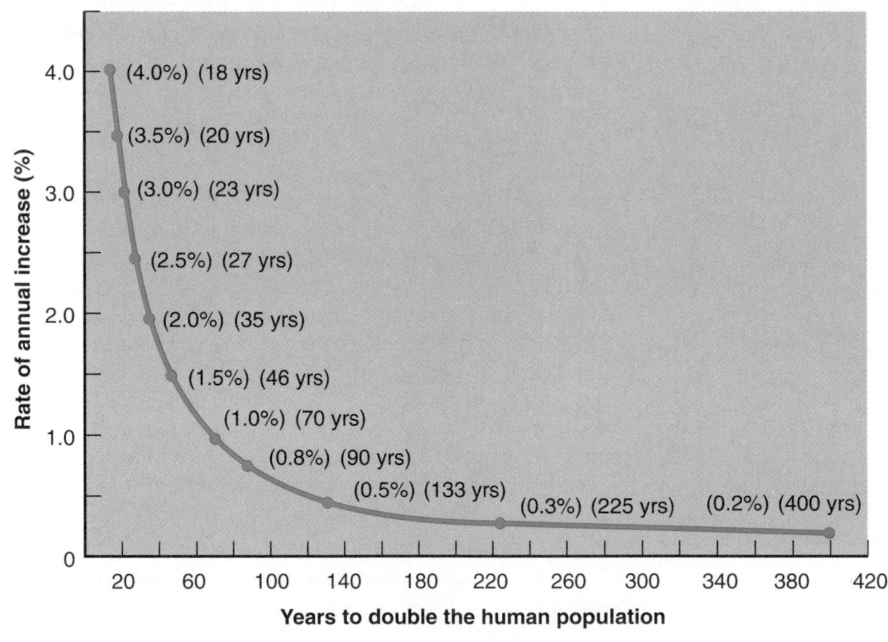

figure 7.10 **Doubling Time for the Human Population** This graph shows the relationship between the rate of annual increase in percent and doubling time. A population growth rate of 1 percent per year would result in the doubling of the population in about seventy years. A population growth rate of 3 percent per year would result in a population doubling in about twenty-three years.

because most of us can appreciate what life would be like if the number of people in our locality were doubled, particularly if the number were to occur within our lifetime.

Figure 7.10 shows the relationship between the rate of annual increase for the human population and the number of years it would take to double the population if that rate were to continue. At a 1 percent rate of annual increase, the population will double in approximately 70 years. At a 2 percent rate of annual increase, the population will double in about 35 years. The current worldwide rate of annual increase of about 1.4 percent will double the world population in about 51 years.

What does this very rapid rate of growth mean to the human species? As a species, humans are subject to the same limiting factors as all other species. We cannot increase beyond our ability to acquire raw materials and energy and safely dispose of our wastes. We also must remember that interactions with other species and with other humans will help determine our carrying capacity.

Let us look at these four factors in more detail. Many of us think of raw materials simply as the amount of food available. However, we have become increasingly dependent on technology, and our lifestyles are directly tied to our use of other kinds of resources, such as irrigation water, genetic research, and antibiotics. Food production is becoming a limiting factor for some segments of the world's human population. Malnutrition is a serious problem in many parts of the world because sufficient food is not available. Currently, about 1 billion people (1/6 of the world's population) suffer from a lack of adequate food. Chapter 8 deals in greater detail with the problems of food production and distribution and their relationship to human population growth.

The second factor, available energy, involves problems similar to those of raw materials. Essentially, all species on earth are ultimately dependent on sun light for their energy. New, less disruptive methods of harnessing this energy must be developed to support an increasing population. Currently, the world population depends on fossil fuels to raise food, modify the environment, and move from place to place. When energy prices increase, much of the world's population is placed in jeopardy because incomes are not sufficient to pay the increased costs for energy and other essentials.

Waste disposal is the third factor determining the carrying capacity for humans. Most pollution is, in reality, the waste product of human activity. Lack of adequate sewage treatment and safe drinking water causes large numbers of deaths each year. Some people are convinced that disregard for the quality of our environment will be a major limiting factor. In any case, it makes good sense to control pollution and to work toward cleaning our environment.

The fourth factor that determines the carrying capacity of a species is interaction with other organisms. We need to become aware that we are not the only species of importance. When we convert land to meet our needs, we displace other species from their habitats. Many of these displaced organisms are not able to compete with us successfully and must migrate or become extinct. Unfortunately, as humans expand their domain, the areas available to these displaced organisms become more rare. Parks and natural areas have become tiny refuges for the plants and animals that once occupied vast expanses of land. If these refuges fall to the developer's bulldozer or are converted to agricultural use, many organisms will become extinct. What today seems like an unimportant organism, one that we could easily do without, may someday be seen as an important link to our very survival.

Social Factors Influence Human Population

Human survival depends upon interaction and cooperation with other humans. Current technology and medical knowledge are available to control human population growth and to improve the health of the people of the world. Why then does the population continue to increase, and why do large numbers of people continue to live in poverty, suffer from preventable diseases, and endure malnutrition? Humans are social animals who have freedom of choice and frequently do not do what is considered "best" from an unemotional, uninvolved, biological point of view. People make decisions based on history, social situations, ethical and religious considerations, and personal desires. The biggest obstacles to controlling human population are not biological but are the province of philosophers, theologians, politicians, and sociologists. People in all fields need to understand that the cause of the population problem has both biological and social components if they are to successfully develop strategies for addressing it.

Ultimate Size Limitation

The human population is subject to the same biological constraints as other species of organisms. We can say with certainty that our population will ultimately reach its carrying capacity and stabilize. There is disagreement about how many people can exist when the carrying capacity is reached. Some people suggest we are already approaching the carrying capacity, while others maintain that we could more than double the population before the carrying capacity is reached. Furthermore, there is uncertainty about what the primary limiting factors will be, and about the quality of life the inhabitants of a more populous world would have. If the human population continues to reproduce at its current rate, the population will double from its current 6 billion to 12 billion people by 2050.

If the reproduction rate falls so that each woman produces only two children during her lifetime, the world will contain 9 billion people by the year 2050. As with all K-strategist species, when the population increases, density-dependent limiting factors will become more forceful. Some people suggest that a lack of food, a lack of water, or increased waste heat will ultimately control the size of the human population. Still others suggest that, in the future, social controls will limit population growth. These social controls could be either voluntary or involuntary. In the economically developed portions of the world, families have voluntarily lowered their birthrates to less than two children per woman. Most of the poorer countries of the world have higher birthrates. What kinds of measures are needed to encourage them to limit their populations? Will voluntary compliance with stated national goals be enough or will enforced sterilization and economic penalties become the norm? Others are concerned that countries will launch wars to gain control of limited resources or to simply eliminate people who compete for the use of those resources.

It is also important to consider the age structure of the world population. In most of the world, there are many reproductive and prereproductive individuals. Since most of these individuals are currently reproducing or will reproduce in the near future, even if they reduce their rate of reproduction, there will be sharp increase in the number of people in the world in the next few years.

No one knows what the ultimate human population size will be or what the most potent limiting factors will be, but most agree that we are approaching the maximum sustainable human population. If the human population continues to increase, eventually the amount of agricultural land available will not be able to satisfy the demand for food.

Summary

A population is a group of organisms of the same species that inhabits an area. The birthrate (natality) is the number of individuals entering the population by reproduction during a certain period. The death rate (mortality) measures the number of individuals that die in a population during a certain period. Population growth is determined by the combined effects of the birthrate and death rate.

The sex ratio of a population is a way of stating the relative number of males and females. Age distribution and the sex ratio have a profound impact on population growth. Most organisms have a biotic potential much greater than that needed to replace dying organisms.

Interactions among individuals in a population, such as competition, predation, and parasitism, are also important in determining population size. Organisms may migrate into (immigrate) or migrate out of (emigrate) an area as a result of competitive pressure.

A typical population growth curve shows a lag phase followed by an exponential growth phase and a stable equilibrium phase at the carrying capacity. The carrying capacity is determined by many limiting factors that are collectively known as environmental resistance. The four major categories of environmental resistance are available raw materials, available energy, disposal of wastes, and interactions among organisms. Some populations experience a death phase following the stable equilibrium phase.

K-strategists typically are large, long-lived organisms that reach a stable population at the carrying capacity. Their population size is usually controlled by density-dependent limiting factors. Organisms that are r-strategists are generally small, short-lived organisms that reproduce very quickly. Their populations do not generally reach a carrying capacity but crash because of some density-independent limiting factor.

The human population is increasing at a rapid rate. The Earth's ultimate carrying capacity for humans is not known. The causes for human population growth are not just biological but also social, political, philosophical, and theological.

Wolves and Moose on Isle Royale

ISSUES & ANALYSIS

Isle Royale in Lake Superior has been the site of a long-term study of the relationship between moose and wolf populations. Wolves (probably a single pair) reached Isle Royale by crossing the ice from Canada during the winter of 1947–48 and had reached a population of about 20 by 1958 when a succession of biologists began a long-term study of the wolves and their relationships with moose and other organisms on the island. In over 50 years, they have learned a great deal about the food habits of wolves, pack behavior, the relationships among various wolf packs, and the relationship between the predator (wolf) and prey (moose) populations. Although there are several factors that affect population size for wolves and moose, an examination of the fluctuations in moose and wolf populations suggests that they have an effect on one another. When wolf populations decline, moose populations rise and when wolf populations rise, moose populations fall. The major changes in moose and wolf populations since 1980 appear to have had triggering events that resulted in major reductions in one of the two populations that had a subsequent effect on the other. For example, the major decline in wolves in the early 1980s from about 50 to 13 wolves was probably initiated by the introduction of a viral disease of domesticated dogs (parvovirus). The reduction in the number of wolves allowed a greater number of new young moose to survive and the moose population increased rapidly.

The rapid decline in the moose population during the mid-1990s was probably initiated by a combination of factors. The rapid growth in the years prior to 1995 had reduced the amount of food available and resulted in malnutrition, slow growth, and poor reproduction. A very severe winter and a tick infestation probably contributed to the die-off as well. During the winter of 1995–96, the moose population suffered a 50 percent reduction as it fell to about 1,200. This decline continued as the population fell to about 500 individuals in 1997, an 80 percent reduction from its maximum. However, it rebounded to about 700 individuals in 1998. What was a bad time for moose was a good time for wolves, since weak and dying moose provided an abundant food supply, and the wolf population rose to 24 individuals, its highest population since the early 1980s. However, this abundant food supply was short-lived and the remaining moose were younger, healthy individuals that are harder to take as prey. In 1998, the wolf population fell to 14, raising concern that the population may not be able to sustain itself. On the positive side, a smaller wolf population reduces competition for food. In addition, most of the remaining wolves are young and have the potential to produce many sets of pups over the next several years. On the negative side, recent survival of pups has been low. This could be due to several factors. Perhaps the canine parvovirus is affecting the survival of the pups. Lack of genetic variety may also be contributing to the low survival, since the entire population is thought to be descended from a single mated pair.

Since 1998, both populations have increased. The moose population increased to 750 in 1999 and to 850 in 2000. The wolf population has increased even more dramatically—from 14 in 1998 to 25 in 1999 and 29 in 2000. Observations over the next few years may identify causes for the fluctuations in the wolf population. For example, blood samples taken from captured wolves could clarify questions about the presence of parvovirus and concerns about genetic variety. It is also possible that the wide fluctuations seen in these two populations may be normal for isolated, island populations in which prey and predator populations have significant influences on one another.

- When populations of animals become endangered, efforts are often made to preserve them. Should special efforts be made to preserve this wolf population?
- If lack of genetic variety is shown to be a contributing factor to the decline of the wolf population, should efforts be made to introduce genetic variety?
- In what ways is this wolf population different from that of populations of California condors or whooping cranes?

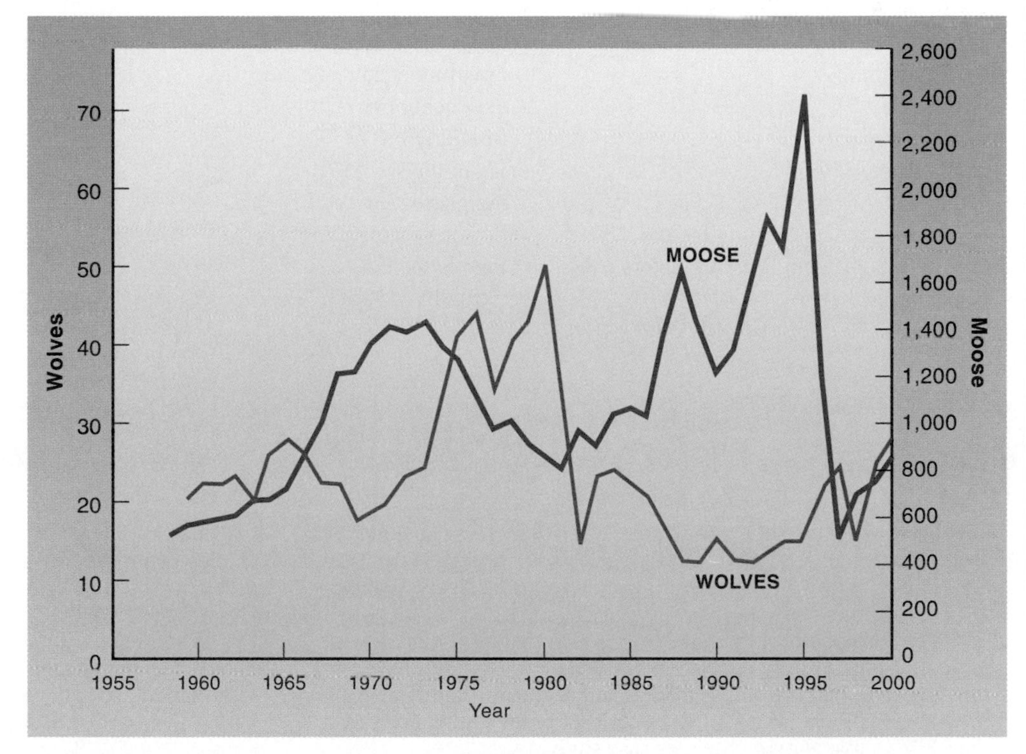

Source: Data from Rolf O. Peterson, "Ecological Studies of Wolves on Isle Royale," Annual Report 1995–96 and subsequent reports.

Interactive Exploration

Check out the website at

http://www.mhhe.com/environmentalscience

and click on the cover of this textbook for interactive versions of the following:

KNOW THE BASICS

age distribution *132*
biotic potential *134*
birthrate *131*
carrying capacity *135*
death phase *136*
death rate *131*
density-dependent limiting factors *138*
density-independent limiting factors *138*
dispersal *133*
emigration *133*
environmental resistance *135*
exponential growth phase (log phase) *134*
immigration *133*
K-strategists *138*
lag phase *134*
limiting factor *135*
log phase *134*
mortality *131*
natality *131*
population density *133*
r-strategists *138*
sex ratio *132*
stable equilibrium phase *135*

On-line Flashcards
Electronic Glossary

IN THE REAL WORLD

What is causing the snow goose population to increase so quickly? What are the consequences of this rapid population growth? Take a look at Snow Goose Population Threatens Arctic Tundra Habitat for this story.

The European green shore crab is also making its mark with increased population growth. This non-native (invading) species has been compared to the European zebra mussel and is causing problems for West Coast (U.S.) fisheries. For the entire story, read the Invasive Marine Species case study.

Increased population is not always gloom and doom. Two examples of population growth resulting from the restoration and protection of species are found in Pennsylvania's Elk Herd Growing and Positive Turn in Kenya's Rhino Population.

What are your thoughts about managing elephant populations? Should they be managed through harvest or birth control? For more information about the legality of using elephant products, look at Ban on Wild Elephant Products Is Maintained.

PUT IT IN MOTION

Why is it getting so crowded? When a population is growing exponentially, it happens *fast!* See a bacteria population go through the stages of growth in Exponential Population Growth.

TEST PREPARATION

Review Questions

1. How is biotic potential related to the rate at which a population will grow?
2. List three characteristics populations might have.
3. Why do some populations grow? What factors help to determine the rate of this growth?
4. Under what conditions might a death phase occur?
5. List four factors that could determine the carrying capacity of an animal species.
6. How do the concepts of birthrate and population growth differ?
7. How does the population growth curve of humans compare with that of bacteria on a petri dish?
8. How do r-strategists and K-strategists differ?
9. As the human population continues to increase, what might happen to other species?
10. All successful organisms overproduce. What advantage does this provide for the species? What disadvantages may occur?

Critical Thinking Questions

1. Why do you suppose some organisms display high natality and others display lower natality? For example, why do cottontail rabbits show high natality and wolves relatively low natality? Why wouldn't all organisms display high natality?
2. Do you think African elephants should be managed by southern African countries like deer are managed in the United States? What values, beliefs, and perspectives lead you to your conclusion?
3. Where do you stand on the issue of birth control for elephants or culling of the herd? Why do you think the way you do? What is the position of the Humane Society in the United States? Why do you think they hold the position they do? Do you think conservationists in the United States should have a say over what is done with the elephants in Africa? Should Africans have a say in what is done in the United States? Why?
4. Consider the differences between r-strategists and K-strategists. What costs are incurred by adopting either strategy? What evolutionary benefits does each strategy enjoy?
5. Why do invading species, which can survive in a new environment, often show exponential growth rates soon after they are introduced to that environment?
6. If humans see that a population of organisms, like the population of wolves on Isle Royale, is declining for "natural" reasons, should humans intervene and try to preserve these populations? Why?

CHAPTER

Human Population Issues

Objectives

After reading this chapter, you should be able to:

- Apply some of the principles discussed in chapter 7 to the human population.
- Differentiate between birthrate and population growth rate.
- Describe the current population situation in the United States.
- Explain why the age distribution and the status and role of women affects population growth projections.
- Recognize that countries in the developed world are experiencing an increase in the average age of their populations.
- Recognize that most countries of the world have a rapidly growing population.
- Describe the implications of the demographic transition concept.
- Understand how an increasing world population will alter the worldwide ecosystem.
- Recognize that rapid population growth and poverty are linked.
- Explain why less-developed nations have high birthrates and why they will continue to have a low standard of living.
- Recognize that the developed nations of the world will be under greater pressure to share their abundance.

Chapter Outline

Human Population Trends and Implications

Global Perspective: *Thomas Malthus and His Essay on Population*

Factors That Influence Population Growth

- Biological Factors
- Social Factors

Environmental Close-Up: *Control of Births*

- Political Factors

Population Growth and Standard of Living

Population and Poverty—A Vicious Cycle?

Hunger, Food Production, and Environmental Degradation

The Demographic Transition Concept

Global Perspective: *The Urbanization of the World's Population*

The U.S. Population Picture

Anticipated Changes with Continued Population Growth

Global Perspective: *North America—Population Comparisons*

Issues & Analysis: *The Impact of AIDS on Populations*

Human Population Trends and Implications

The human population dilemma is very complex. In order to appreciate it, we must understand current population trends and how they are related to social, political, and economic conditions. Currently, the world population is over 6 billion people. By the year 2025, this is expected to increase to about 7.8 billion. Much of this increase is expected to occur in Africa, Asia, and Latin America, which already have over 82 percent of the world population. (See figure 8.1.) If these trends continue, the total population of Africa, Asia, and Latin America will increase from the current 5 billion to over 6.6 billion by 2025, when these continents will contain over 85 percent of the world's people.

There are many ways to show that world population growth is a contributing factor in nearly all environmental problems. Current population growth has led to famine in areas where food production cannot keep pace with increasing numbers of people; political unrest in areas with great disparities in availability of resources (jobs, goods, food); environmental degradation (erosion, desertification) by poor agricultural practices; water pollution by human and industrial waste; air pollution caused by the human need to use energy for personal use and for industrial applications; extinctions caused by people converting natural ecosystems to managed agricultural ecosystems; and destructive effects of exploitation of natural resources (strip mining, oil spills, groundwater mining).

Several factors interact to determine the impact of a society on the resources of its country. These include the size of the population, the land area the people occupy, and their degree of technological development. The larger the size of a population, the greater the demand on the resources of the country. However, population size is not the only important factor. **Population density,** the number of people per unit of land area, is also important. A million people spread out over the huge area of the Amazon have much less impact on resources than that same million people in a small island country, because the impact is distributed over a greater land surface.

The degree of technological development and affluence is also important. The environmental impact of the developed world is often underestimated because the population in these countries is relatively stable and local environmental conditions are good. However, the people in highly developed countries consume huge amounts of resources. Citizens of these countries eat more food, particularly animal protein, which requires larger agricultural inputs than does a vegetarian diet. They have more material possessions and consume vast amounts of energy. These developed countries purchase goods and services from other parts of the world, often degrading environmental conditions in less-developed countries. Thus, the

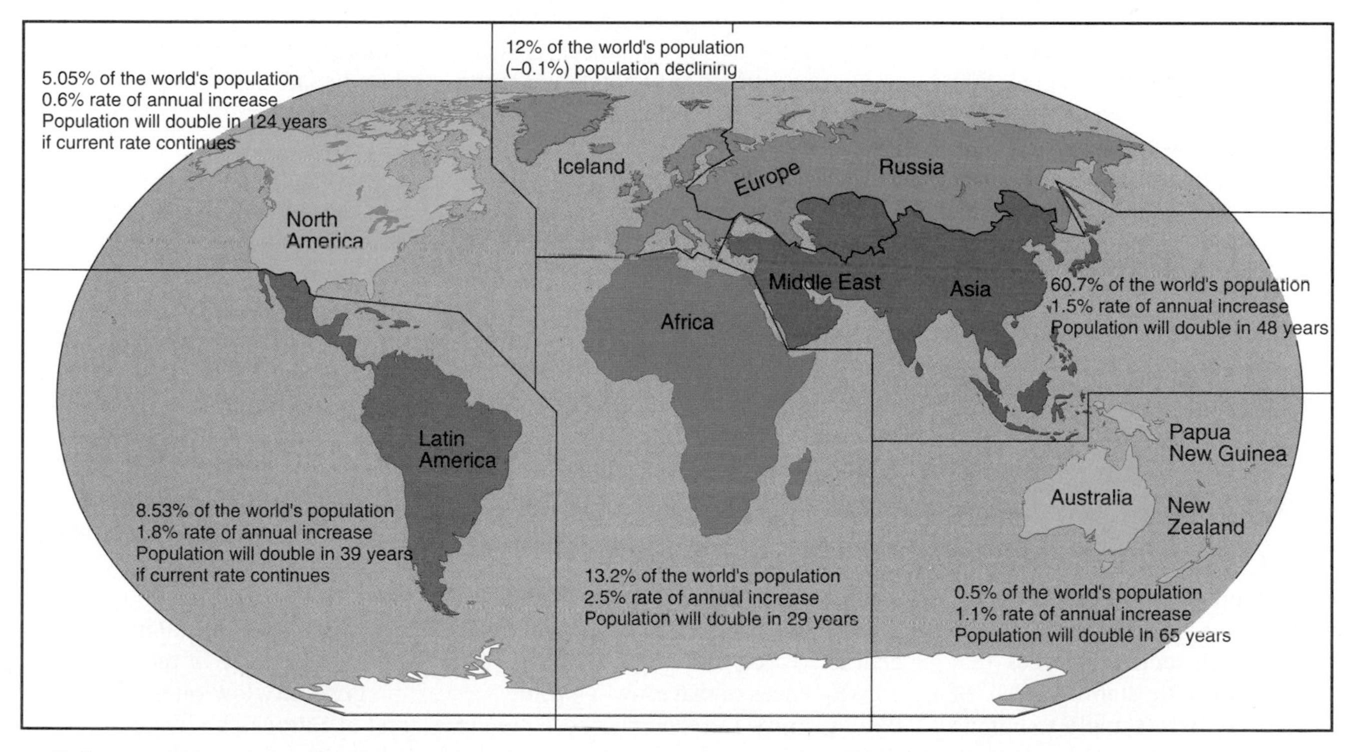

figure 8.1 **Population Growth in the World (2000)** The population of the world is not evenly distributed. Currently over 82 percent of the world's population is in Latin America, Africa, and Asia. These areas also have the highest rates of increase and are generally considered less developed. Because of the high birthrates, they are likely to remain less developed and will constitute over 85 percent of the world's population by the year 2025.

GLOBAL PERSPECTIVE

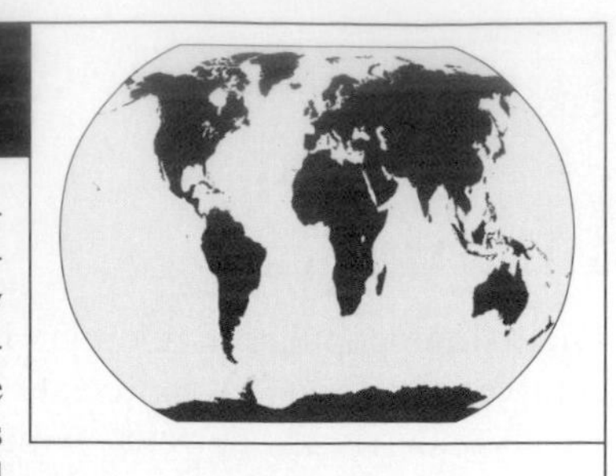

Thomas Malthus and His Essay on Population

In 1798 Thomas Robert Malthus, an Englishman, published an essay on human population. In it, he presented an idea that was contrary to popular opinion. His basic thesis was that human population increased in a geometric or exponential manner (2, 4, 8, 16, 32, 64, etc.), while the ability to produce food increased only in an arithmetic manner (1, 2, 3, 4, 5, 6, etc.). The ultimate outcome of these different rates would be that population would outgrow the ability of the land to produce food. He concluded that wars, famines, plagues, and natural disasters would be the means of controlling the size of the human population. His predictions were hotly debated by the intellectual community of his day. His assumptions and conclusions were attacked as erroneous and against the best interest of society. At the time he wrote the essay, the popular opinion was that human knowledge and "moral constraint" would be able to create a world that would supply all human needs in abundance. One of Malthus's basic postulates was that "commerce between the sexes" (sexual intercourse) would continue unchanged, while other philosophers of the day believed that sexual behavior would take less procreative forms and human population would be limited. Only within the past 50 years, however, have really effective conception-control mechanisms become widely accepted and used, and they are used primarily in developed countries.

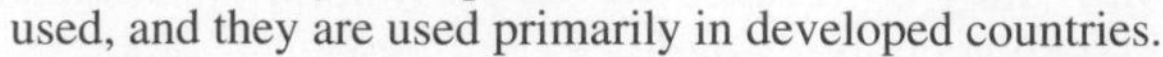

Malthus did not foresee the use of contraception, major changes in agricultural production techniques, or the exporting of excess people to colonies in the Americas, Australia, and other parts of the world. These factors, as well as high death rates, prevented the most devastating of his predictions from coming true. However, in many parts of the world today, people are experiencing the forms of population control (famine, epidemic disease, wars, and natural disasters) predicted by Malthus in 1798. Many people feel that his original predictions were valid—only his time scale was not correct—and that we are seeing his predictions come true today.

environmental impact of highly developed regions like North America, Japan, and Europe is often felt in distant places, while the impact on resources in the home country may be minimal.

While controlling world population growth would not eliminate all environmental problems, it could reduce the rate at which environmental degradation is occurring. It is also generally believed that the quality of life for many people in the world would improve if their populations grew less rapidly. Why, then, does the human population continue to grow at such a rapid rate?

Factors That Influence Population Growth

In chapter 7, we examined populations from a biological point of view. We looked at their characteristics, the causes of growth, and the forces that cause populations to stabilize. All of these biological factors apply to human as well as non-human populations. There is an ultimate carrying capacity for the human population. Eventually, limiting factors will cause human populations to stabilize. However, unlike other kinds of organisms, humans are also influenced by social, political, economic, and ethical factors. We have accumulated knowledge that allows us to predict the future. We can make conscious decisions based on the likely course of events and adjust our lives accordingly. Part of our knowledge is the certainty that as populations continue to increase, death rates and birthrates will become equal. This can happen by allowing the death rate to rise or by choosing to limit the birthrate. Controlling human population would seem to be a simple process. Once people understand that lowering the birthrate is more humane than allowing the death rate to rise, they should make the "correct" decision and control their birthrates; however, it is not quite that simple.

Biological Factors

The scientific study of human populations, their characteristics, how these characteristics affect growth, and the consequences of that growth is known as **demography.** Demographers can predict the future growth of a population by looking at several biological indicators. Currently, in almost all countries of the world, the birthrate exceeds the death rate. Therefore, the size of the population must increase. (See table 8.1.) Some countries that have high birthrates and high death rates—with birthrates greatly exceeding the death rates—will grow rapidly (Afghanistan and Ethiopia). Such countries usually have an extremely high mortality rate among children because of disease and malnutrition.

Some countries have high birthrates and low death rates and will grow extremely rapidly (Mexico and Syria). Infant mortality rates are moderately high in these countries. Other countries have low birthrates, and death rates that closely match the birthrates; they will grow slowly (Japan and the United Kingdom). These and other more-developed countries typically have very low infant mortality rates. The disruption caused by the political upheaval in the former Soviet Union and Eastern Europe has resulted in several countries (e.g., Russia and Germany) with death rates that are equal to or exceed birthrates, causing their populations to decline. Because of these countries and the generally low rates of growth in the rest of Europe, the European region as a whole has a declining population.

The most important determinant of the rate at which human populations grow is related to how many women in

Table 8.1 Population Characteristics of Selected Countries (2000)

Country	Current Population (millions)	Births per 1000 Individuals	Deaths per 1000 Individuals	Infant Mortality Rate (deaths per 1000 live births)	Total Fertility Rate (children per woman per lifetime)	Rate of Natural Increase (annual %)	Time Needed to Double Population (years)
World	*6,067.3*	*22.0*	*9.0*	*57.0*	*2.9*	*1.4*	*51*
Russia	145.2	8.4	14.6	16.5	1.2	(−0.63)	
Germany	82.1	9.0	10.0	5.0	1.3	(−0.1)	—
Sweden	8.9	10.0	11.0	4.0	1.5	(−0.08)	—
Belgium	10.2	11.0	10.0	6.0	1.6	0.1	770
United Kingdom	59.8	12.0	11.0	6.0	1.7	0.1	546
Japan	126.9	9.0	8.0	4.0	1.3	0.15	462
Canada	30.8	11.0	7.0	6.0	1.5	0.4	178
United States	275.6	15.0	9.0	7.0	2.1	0.6	120
China	1,264.5	15.2	6.5	31.4	1.8	0.9	79
Zimbabwe	11.3	30.1	20.1	80.0	4.0	1.0	69
Argentina	37.0	19.0	8.0	19.0	2.6	1.1	62
Turkey	65.3	21.8	6.8	37.9	2.5	1.5	46
Uzbekistan	24.8	23.0	5.8	21.9	2.8	1.72	40
India	1,002.1	27.0	9.0	72.0	3.3	1.8	39
Mexico	99.6	23.9	4.4	31.5	2.7	1.95	36
Ethiopia	64.1	45.1	21.1	116.0	6.7	2.4	29
Afghanistan	26.7	43.0	18.2	149.8	6.1	2.49	28
Syria	16.5	33.2	5.6	24.6	4.7	2.76	25
Togo	5.0	41.8	11.1	79.7	6.1	3.07	23

Source: Data from *World Population Data Sheet 2000,* Population Reference Bureau, Washington, D.C.

the population are having children and the number of children each woman will have. The **total fertility rate** of a population is the number of children born per woman per lifetime. A total fertility rate of 2.1 is known as **replacement fertility,** since parents produce 2 children who will replace the parents when they die. In the long run, if the total fertility rate is 2.1, population growth will stabilize. A rate of 2.1 is used rather than 2.0 because some children do not live very long after birth and therefore will not contribute to the population for very long. When population is not growing, and the number of births equals the number of deaths, it is said to exhibit **zero population growth.**

A total fertility rate of 2.1 will not necessarily immediately result in a stable population with zero growth, for several reasons. First, the death rate may fall as living conditions improve and people live longer. If the death rate falls faster than the birthrate, there will still be an increase in the population even though it is reproducing at the replacement rate.

The **age distribution,** the number of people of each age in the population, also has a great deal to do with the rate of population growth. If a population has many young people who are raising families or who will be raising families in the near future, the population will continue to increase even if the families limit themselves to two children. Depending on the number of young people in a population, it may take 20 years to a century for the population of a country to stabilize so that there is no net growth.

Social Factors

It is clear that populations in economically developed countries of the world have low fertility rates and low rates of population growth and that the less-developed countries have high fertility rates and high population growth rates. It also appears obvious that reducing fertility rates would be to everyone's advantage; however, not everyone in the world feels that way. Several factors influence a person's desired family size. Some are religious, some are traditional, some are social, and some are economic.

The major social factors that determine family size are the status and desires of women in the culture. In many male-dominated cultures, the traditional role of women is to marry and raise children. Often this role is coupled with strong religious input as well. Typically, little value is placed on educating women, and early marriage is encouraged. In these cultures, women are totally dependent on their husbands and children in old age. Because early marriage is encouraged, fertility rates are high, since women are exposed to the probability of pregnancy for more of

ENVIRONMENTAL CLOSE-UP

Control of Births

The use of technology to control disease and famine has greatly reduced the death rate of the human population. Technological developments can also be used to control the birthrate. Birth control refers to anything that reduces the number of births. Some processes prevent conception and can be called contraceptives. A variety of contraceptive methods are available to help people regulate their fertility. Research is continuing to develop more effective, more acceptable, and less expensive methods of controlling conception. Because of cultural and religious differences, some forms of contraception may be more acceptable to one segment of the world's population than to another.

The most common methods of contraception are oral contraceptive pills, diaphragms and spermicidal jelly, spermicidal vaginal foam, condoms, vasectomy, and tubal ligation. The range of effectiveness of these methods, shown in the table, is the result of individual fertility differences and the degree of care employed in the use of each method.

Some birth control devices such as intrauterine contraceptive devices (IUDs) are thought to prevent embryos from attaching to the uterine wall and completing development and, therefore, control births but do not prevent conception. Abortion is the physical intervention in the development of the embryo in the uterus which is used to terminate unwanted pregnancies. Most countries with low birthrates, such as the United States, Japan, and many European countries, have ready access to contraceptive methods and allow abortions.

Effectiveness of Various Methods of Contraception

	Percent of women experiencing an unintended pregnancy within the first year of use	
Method	**Typical Use**	**Perfect Use**
No contraceptive method used	85	85
Spermicidal foams, creams, gels, suppositories, and vaginal films	26	6
Cervical cap		
Women who have had children	40	26
Women who have not had children	20	9
Sponge		
Women who have had children	40	20
Women who have not had children	20	9
Female condom	21	5
Diaphragm with spermicide	20	6
Withdrawal	19	4
Male condom	14	3
Periodic abstinence (natural family planning)		
Calendar method		9
Ovulation method		3
Temperature method		2
Post-ovulation method		1
Intrauterine device (IUD)	2	1.5
Female sterilization (tubal ligation)	0.5	0.5
Contraceptive pill		0.5
Contraceptive injection (Depo-Provera)	0.3	0.3
Male sterilization (vasectomy)	0.15	0.10
Contraceptive implant (Norplant)	0.05	0.05

Source: Data from J. Trussel, "Contraceptive Efficacy" in R. A. Hatcher, et al., *Contraceptive Technology,* 17th Revised Edition, 1998, Irvington Publishers.

their fertile years. Lack of education reduces options for women in these cultures. They do not have the option to not marry or to delay marriage and thus reduce the number of children they will bear. By contrast, in much of the developed world women are educated, delay marriage, and have fewer children. It has been said that the single most important activity needed to reduce the world population growth rate is to educate women. Whenever the educational level of women increases, fertility rates fall.

Data on the age of mothers giving birth indicate that 12% of births in Africa are to women in the 15- to 19-year-old range. This is true of 6 percent of births in Asia (excluding China) and 8 percent of births in Latin America. The total fertility rates of these areas are 5.0, 3.0, and 3.3 children per woman per lifetime, respectively. In the developed world, the average age of first marriage is much higher, between age 25 and 27; early marriages are rare; about 3 percent of births are to mothers who are between 15 to 19 years of age, and the total fertility rate is less than replacement fertility at about 1.5 children per woman per lifetime.

Even childrearing practices have an influence on population growth rates. In countries where breast feeding is practiced, several benefits accrue. The breast milk is an excellent source of nutrients for the infant as well as a source of antibodies against some diseases. Furthermore, since many women do not return to a normal reproductive cycle until after they have stopped nursing, during the months a woman is breast feeding her child, she is less likely to become pregnant again. Since in many cultures, breast feeding may continue for one to two years, it serves to increase the time between successive births. Increased time between births results in a lower mortality among women of childbearing age.

As women become better educated and obtain higher paying jobs, they become financially independent and can afford to marry later and consequently have fewer children. Better-educated women are also more likely to have access to and use birth control. In economically advanced countries, about 70 percent of women typically use contraception. In the less-developed countries, contraceptive use is much lower—less than 20 percent in Africa, about 45 percent in Asia (except China), and about 60 percent in Latin America.

It is important to recognize that access to birth control alone will not solve the population problem, What is most important is the desire of women to limit the size of their families. In developed countries, use of birth control is extremely important in regulating the birthrate. This is true regardless of religion and previous historical birthrates. Italy and Spain, both traditionally Catholic countries, both have a total fertility rate of 1.2. The average for the developed countries of the world is 1.5. Obviously, women in these countries make use of birth control to help them regulate the size of their families. (See Environmental Close-Up: Control of Births.) By contrast, Mexico, which is also a traditionally Catholic country, has a total fertility rate of 2.7, which is typical of birthrates in the less-developed world regardless of religious tradition.

Women in the less-developed world typically have more children than they think is ideal and the number of children they have is higher than the replacement fertility rate of 2.1 children. Access to birth control will allow them to limit the number of children they actually have to their desired number and will allow them to space their children at more convenient intervals, but they still desire more children than the 2.1 needed for replacement. Why do they desire large families? There are several reasons. In areas where infant mortality is high, it is traditional to have large families since several of a woman's children may die before they reach adulthood. This is particularly important in the less-developed world where there is no government program of social security. Parents are more secure in old age if they have several children to contribute to their needs when they can no longer work.

In less-developed countries, the economic benefits of children are extremely important. Even young children can be given jobs that contribute to the family economy. They can protect livestock from predators, gather firewood, cook, or carry water. In the developed world, large numbers of children are an economic drain. They are prevented by law from working, they must be sent to school at great expense, and they consume large amounts of the family income. Parents in the developed world make an economic decision about having children in the same way they buy a house or car: "We are not having children right away. We are going to wait until we are better off financially."

Political Factors

Two other factors that influence the population growth rate of a country are government policies on population growth and immigration. Many countries in Europe have official policies that state that their population growth rates are too low. As their populations age and there are few births, they are concerned about a lack of working-age people in the future and have instituted programs that are meant to encourage people to have children. For example, Hungary, Sweden, and several other European countries provide paid maternity leave for mothers during the early months of a child's life and the guarantee of a job when the mother returns to work. Many countries provide childcare facilities and other services that make it possible for both parents to work. This removes some of the economic barriers that tend to reduce the birthrate. The tax system in many countries, such as the United States, provides an indirect payment for children by allowing a deduction for each child. Canada pays a bonus to couples on the birth of a child.

By contrast, most countries in the developing world publicly state that their population growth rates are too high. In order to reduce the birthrate, they have programs that provide information on maternal and child health and on birth control. The provision of free or low-cost access to contraceptives is usually a part of their population-control effort as well.

China and India are the two most populous countries in the world, each with over a billion people. China has taken steps to control its population and now has a total fertility rate of 1.8 children per woman while India has a total fertility rate of 3.4. This difference between these two countries is the result of different policy decisions over the last 50 years. The history of China's population policy is an interesting study of how government policy affects reproductive activity among its citizens. When the People's Republic of China was established in 1949, the official policy of the government was to encourage births, because more Chinese would be able to produce more goods and services, and production was the key to economic prosperity. The population grew from 540 million to 614 million between 1949 and 1955, while economic progress was slow. Consequently, the government changed its policy and began to promote population control.

The first family-planning program in China began in 1955, as a means of improving maternal and child health. Birthrates fell. (See figure 8.2.) In addition, other social changes resulted in widespread famine, increased death rates, and low birthrates in the late 1950s and early 1960s.

The present family-planning policy began in 1971 with the launching of the *wan xi shao* campaign. Translated, the phrase means "later" (marriages), "longer" (intervals between births), and "fewer" (children). This program raised the legal ages for marriage. For women and men in rural areas, the ages were raised to 23 and 25, respectively; for women and men in urban areas, the ages were raised to 25 and 28, respectively. These policies resulted in a reduction of birthrates by nearly 50 percent between 1970 and 1979.

An even more restrictive, one-child campaign was begun in 1978–79. The program offered incentives for couples to restrict their family size to one child. Couples enrolled in the program would receive free medical care, cash bonuses for their work, special housing treatment, and extra old-age benefits. Those who broke their pledge were penalized by the

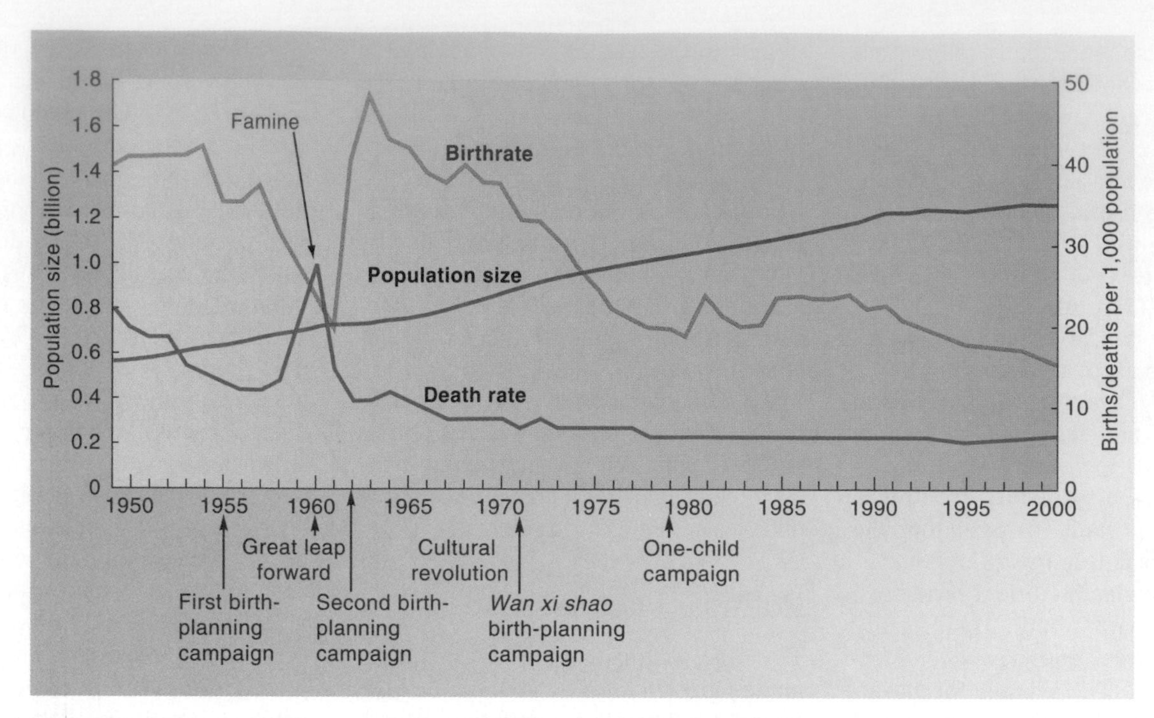

figure 8.2 **Population Changes in China** China has had a long history of actively promoting population control. This graph shows the changes that have occurred in birthrates, death rates, and total population as a result of significant policy initiatives.

Source: Data from H. Yuan Tien, "China's Demographic Dilemmas" in *Population Bulletin, 1992,* Population Reference Bureau, Inc., Washington, D.C. and National Family Planning Commission of China; and more recent data taken from Population Reference Bureau.

loss of these benefits as well as other economic penalties. By the mid-1980s, less than 20 percent of the eligible couples were signing up for the program. Rural couples particularly desired more than one child. In fact, in a country where about 70 percent of the population is rural, the rural total fertility rate was 2.5 children per woman. In 1988, a second child was sanctioned for rural couples if their first child was a girl, which legalized what had been happening anyway.

The current total fertility rate is 1.8 children per woman. Over 80 percent of couples use contraception; the most commonly used forms are male and female sterilization and the intrauterine device. Abortion is also an important aspect of this program, with a ratio of over 600 abortions per 1,000 live births.

By contrast, during the same 50 years, India has had little success in controlling its population. In 2000, a new plan was unveiled which has the goal of bringing the total fertility rate from its current 3.4 children per woman to 2 (replacement rate) by 2010. In the past , the emphasis of government programs was on meeting goals of sterilization and contraceptive use, but this has not been successful. Today, about 40 percent of couples use contraceptives. This new plan will emphasize improvements in the quality of life of the people. The major thrusts will be to reduce infant and maternal death, immunize children against preventable disease, and encourage girls to attend school. It is hoped that improved health will remove the perceived need for large numbers of births. Currently, less than 50 percent of the women in India can read and write. The emphasis on improving the educational status of women is related to the experiences of other developing countries. In many other countries, it has been shown that an increase in the education level of women is linked to lower fertility rates.

The immigration policies of a country also have a significant impact on the rate at which the population grows. Birthrates are currently so low in several European countries, Japan, and China that these countries will likely have a shortage of those of working age in the near future. One way to solve this problem is to encourage immigration from other parts of the world.

The developed countries are under tremendous pressure to accept immigrants. The standard of living in these countries is a tremendous magnet for refugees or people who seek a better life than is possible where they currently live. The continuing economic and political reorganization in central Europe is causing significant increases in the number of immigrants and placing considerable strain on the social systems of Germany, Austria, and other countries of the region. In the United States, approximately one-third of the population increase experienced each year is the result of immigration. Canada encourages immigrants and has set a goal of accepting 300,000 new immigrants each year. This is 1 percent of its current population.

Population Growth and Standard of Living

There appears to be an inverse relationship between the rate at which the population of a country is growing and its standard of living. The **standard of living** is an abstract concept that attempts to

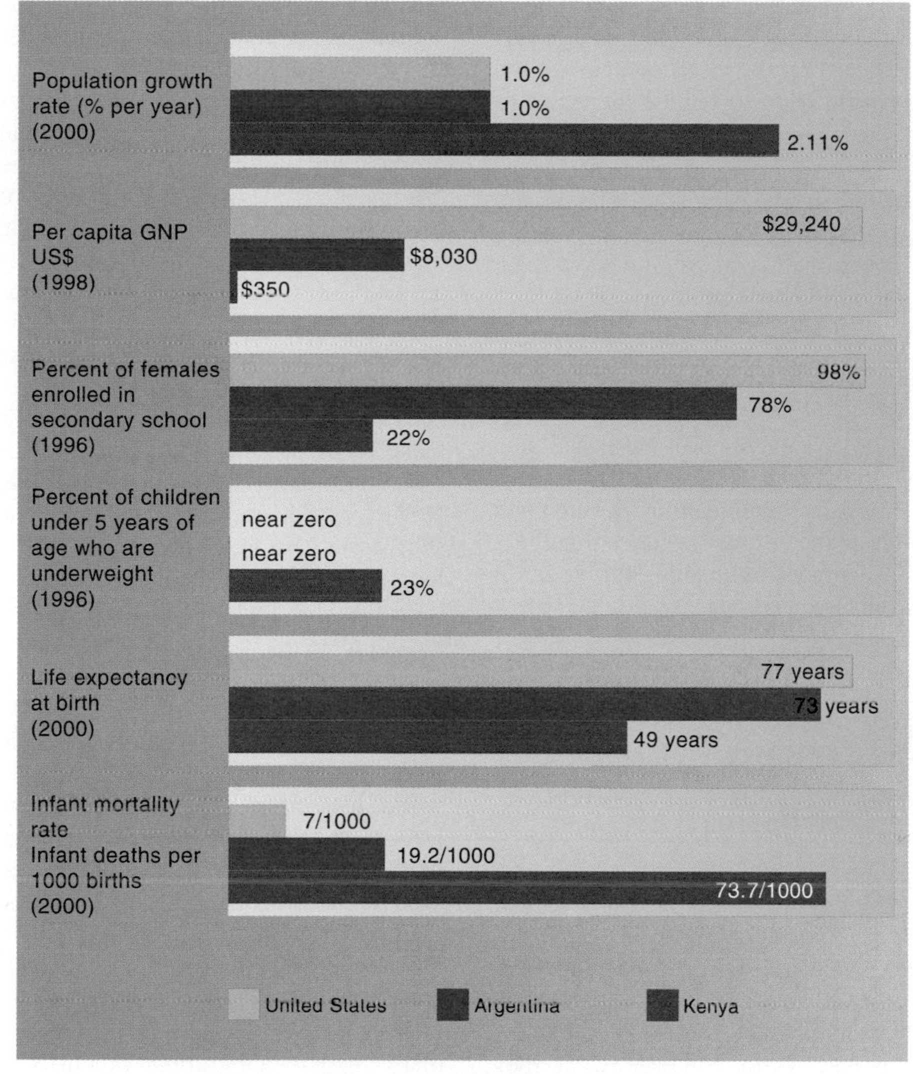

figure 8.3 **Standard of Living and Population Growth in Three Countries** Standard of living is a measure of how well one lives. It is not possible to get a precise definition, but when we compare the United States, Argentina, and Kenya, it is obvious that there are great differences in how the people in these countries live. Kenya has a high population growth rate, a low life expectancy, a high infant mortality rate, and many people without adequate food. Furthermore, their incomes are low and they are poorly educated. The United States has a low population growth rate, a high life expectancy, a low infant mortality rate, and many people who eat too much. People in the United States have high educational levels and high incomes. Argentina is intermediate in all of these characteristics.

quantify the quality of life of people. Standard of living is a difficult concept to quantify since various cultures have different attitudes and feelings about what is desirable. However, several factors can be included in an analysis of standard of living: economic well-being, health conditions, and the ability to changes one's status in the society. Figure 8.3 lists several factors that are important in determining standard of living and contrasts three countries with very different standards of living (the United States, Argentina, and Kenya). One important economic measure of standard of living is the per capita **gross national product (GNP).** The GNP is an index that measures the total goods and services generated within a country. As you can see from figure 8.3, a wide economic gap exists between economically advanced countries and those that are less developed. Yet the people of less-developed countries aspire to the same standard of living enjoyed by people in the developed world.

Health criteria reflect many aspects of standard of living. Access to such things as health care, safe drinking water, and adequate food are reflected in life expectancy, infant mortality, and growth rates of children. The United States and Argentina have similar life expectancies (over 70 years) and adequate nutrition. Kenya has a low life expectancy (49 years), many undernourished children (23 percent are underweight), and a high infant mortality rate (73.7/1000). The United States has a low infant mortality rate (7/1000). Argentina has an intermediate infant mortality rate—19.2/1000.

Finally, the educational status of people determines the kinds of jobs that are available and the likelihood of being able to improve one's status. In general, men are more likely to receive an education than women, but the educational status of women has a direct bearing on the number of children they will have and, therefore, on the economic well-being of the family, Nearly all girls in the United States attend high school, while 78 percent do so in Argentina, but only 22 percent do so in Kenya. Obviously, tremendous differences exist in the standard of living among these three countries. What the average U.S. citizen would consider poverty level would be considered a luxurious life for the average person in Kenya.

Population and Poverty—A Vicious Cycle?

It is clear that the areas of the world where the human population is growing most rapidly are those that have the lowest standard of living. The developed countries of the world (Europe, North America, Japan, Australia, and New Zealand) have abundant wealth and have relatively slowly growing populations. The less-developed countries of Africa, Latin America, and Asia are generally poor and have high population growth rates. Although not all cases are the same, poverty, high birthrates, poor health, and lack of education seem to be interrelated.

1. Poor people cannot afford birth control and, since they are often poorly educated, may not be able to read the directions on how to use various birth control mechanisms correctly. Therefore, they have more children than they may wish to have.
2. Poor people need to obtain income in many ways. Often this includes taking children out of school so that they are able to work on the farm or in other jobs to provide income for the family. Poorly educated people cannot get high paying jobs and so remain in poverty.
3. Poor people have little access to health care and are therefore more likely to suffer preventable illnesses that lower their ability to earn an income.
4. Women in poor countries are usually poorly educated and do not have disposable income. Therefore, they are dependent on their husbands or the family unit for their livelihood. Women who do not have independent income are more likely to have children they do not want, because they cannot afford birth control.
5. High infant mortality rates result from poor health, but children (particularly sons) are desired by parents because the sons will provide for the parents when the parents are old. High infant mortality rates make it likely that parents will desire larger numbers of offspring, since some offspring will die as children.

At the United Nations International Conference on Population and Development held in Cairo, Egypt, in September 1994, attention was focused on breaking this cycle of poverty and high population growth rates. Several important conclusions were reached that have the potential to break this cycle.

1. There was recognition that economic well-being is tied to solving the population problem. However, the huge, rapidly growing population of the poor countries of the world cannot hope to consume at the rate the rich countries do. Furthermore, the rich countries of the world need to reduce their rate of consumption.
2. Improving the educational status of women was promoted. This would lead to improved financial standing for women, which could allow them to have fewer children.
3. Access to birth control and health care would reduce infant and maternal deaths.

In fact, several countries have instituted population control programs that focus on improving the health of women and children, increasing the educational levels of women, and making birth control universally available.

Hunger, Food Production, and Environmental Degradation

As the human population increases, there is an increased demand for food. People must either grow food themselves or purchase it. Most people in the developed world purchase what they need and have more than enough food to eat. Most people in the less-developed world must grow their own food and have very little money to purchase additional food. Typically, these farmers have very little surplus. If crops fail, people starve. Even in countries with the highest population (China and India), the majority of the people live on the land and farm.

The human population can increase only if the populations of other kinds of plants and animals decrease. Each ecosystem has a maximum biomass that can exist within it. There can be shifts within ecosystems to allow an increase in the population of one species, but this always adversely affects certain other populations because they are competing for the same basic resources.

When humans need food, they convert natural ecosystems to artificially maintained agricultural ecosystems. The natural mix of plants and animals is destroyed and replaced with species useful to humans. If these agricultural ecosystems are mismanaged, the region's total productivity may fall below that of the original ecosystem. The dust bowl of North America, desertification in Africa, and destruction of tropical rainforests are well-known examples. In countries where food is in short supply and the population is growing, there is intense pressure to convert remaining natural ecosystems to agriculture. Typically, these areas are the least desirable for agriculture and will not be productive. However, to a starving population, the short-term gain is all that matters. The long-term health of the environment is sacrificed for the immediate needs of the population.

A consequence of the basic need for food is that people in less-developed countries generally feed at lower trophic levels than do those in the developed world. (See figure 8.4.) Converting the less-concentrated carbohydrates of plants into more nutritionally valuable animal protein and fat is an expensive process. During the process of feeding plants to animals and harvesting animal products, approximately 90 percent of the energy in the original plants is lost. Although many modern agricultural practices in the developed world obtain better efficiencies than this, most of the people in the developing world are not able to use such sophisticated systems. Thus, these people approach the 90 percent loss characteristic of natural ecosystems. Therefore, in terms of economics and energy, people in less-developed countries must consume the plants themselves rather than feed the plants to animals and then consume the animals. In most cases, if the plants were fed to animals, many people would starve to death. On the other hand, a lack of protein in diets that consist primarily of plants can lead to malnutrition. Many people in the less-developed world suffer from a lack of adequate protein, which stunts their physical and mental development.

In contrast, in most of the developed world, meat and other animal protein sources are important parts of the

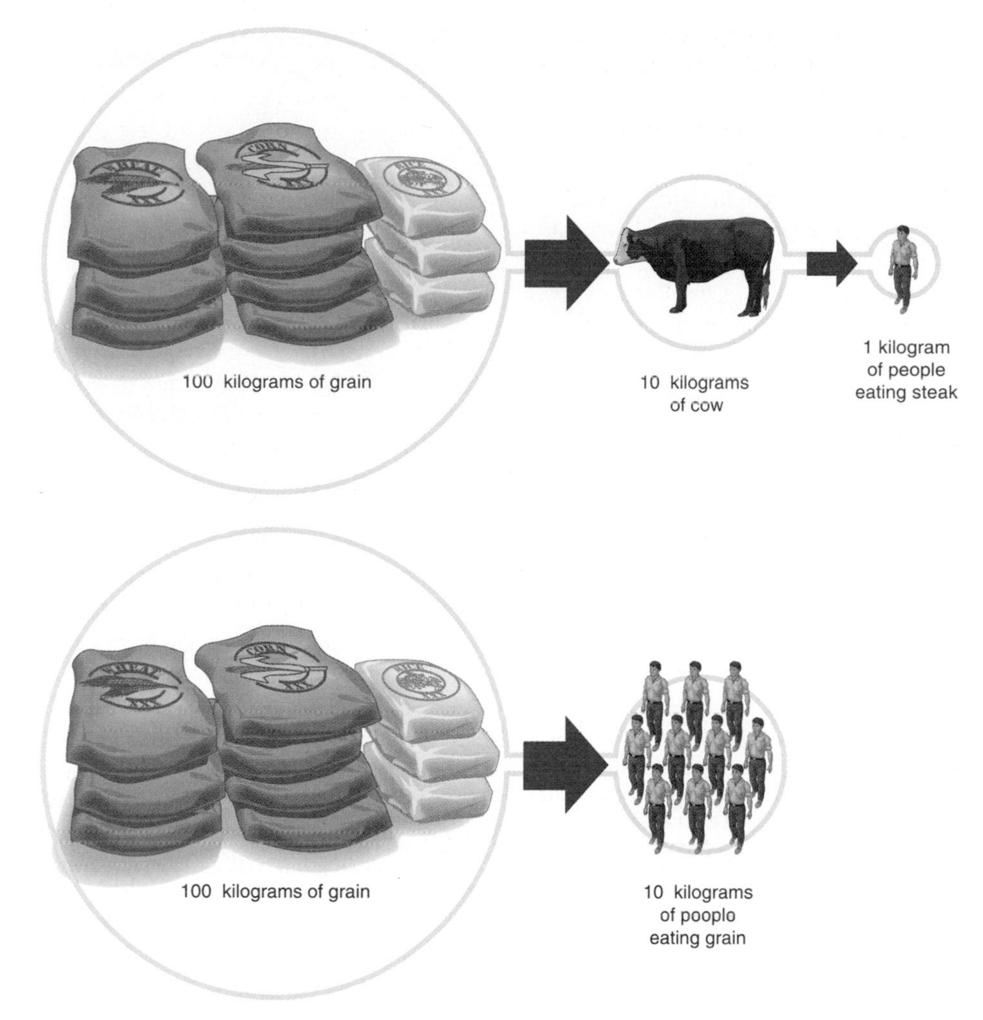

figure 8.4 **Population and Trophic Levels** The larger a population, the more energy it takes to sustain the population. Every time one organism is eaten by another organism, approximately 90 percent of the energy is lost. Therefore, when countries are densely populated, they usually feed at the herbivore trophic level because they cannot afford the 90 percent energy loss that occurs when plants are fed to animals. The same amount of grain can support ten times more people at the herbivore level than at the carnivore level.

diet. Many people suffer from overnutrition (they eat too much); they are "malnourished" in a different sense. Fifty-five percent of North Americans are overweight and 25 percent are obese. The ecological impact of one person eating at the carnivore level is about ten times that of a person eating at the herbivore level. If people in the developed world were to reduce their animal protein intake, they would significantly reduce their demands on world resources. Almost all of the corn and soybeans grown in the United States is used as animal feed. If these grains were used to feed people rather than animals, less grain would have to be grown and the impact on farmland would be less.

In countries where food is in short supply, agricultural land is already being exploited to its limit, and there is still a need for more food. This makes the United States, Canada, Australia, Argentina, New Zealand, and the European Economic Community net food exporters. Many countries, like India and China, are able to grow enough food for their people but do not have any left for export. Others, including many nations of the former Soviet Union, are not able to grow enough to meet their own needs and, therefore, must import food.

A country that is a net food importer is not necessarily destitute. Japan and some European countries are net food importers but have enough economic assets to purchase what they need. Hunger occurs when countries do not produce enough food to feed their people and cannot obtain food through purchase or humanitarian aid.

The current situation with respect to world food production and hunger is very complicated. It involves the resources needed to produce food, such as arable land, labor, and machines; appropriate crop selection; and economic incentives. It also involves the maldistribution of food within countries. This is often an economic problem, since the poorest in most countries have difficulty finding the basic necessities of life, while the rich have an excess of food and other resources. In addition, political activities often determine food availability. War, payment of foreign debt, corruption, and poor management often contribute to hunger and malnutrition.

Improved plant varieties, irrigation, and improved agricultural methods have dramatically increased food production in some parts of the world. In recent years, India, China, and much of southern Asia have moved from being food importers to being self-sufficient, and in some cases, food exporters.

The areas of greatest need are in sub-Saharan Africa. Africa is the only major region of the world where per capita grain production has decreased over the past few decades. People in these regions are trying to use marginal lands for food production, as forests, scrubland, and grasslands are converted to agriculture. Often, this land is not able to support continued agricultural production. This leads to erosion and desertification.

What should be done about countries that are unable to raise enough food for their people and are unable to buy the food they need? This is not an easy question. A simple humanitarian solution to the problem is for the developed countries to supply food. Many religious and humanitarian organizations do an excellent service by taking food to those who need it, and save many lives. However, the aim should always be to provide temporary help and insist that the people of the country develop mechanisms for solving their own problem.

Often, emergency food programs result in large numbers of people migrating from their rural (agricultural) areas to cities, where they are unable to support themselves. They become dependent on the food aid and stop working to raise their own food, not because they do not want to work, but because they need to leave their fields to go to the food distribution centers. Many humanitarian organizations now recognize the futility of trying to feed people with gifts from the developed world. The emphasis must be on self-sufficiency.

The Demographic Transition Concept

The relationship between the standard of living and the population growth rate seems to be that countries with the highest standard of living have the lowest population growth rate, and those with the lowest standard of living have the highest population growth rate. This has led many people to suggest that countries naturally go through a series of stages called **demographic transition.** This model is based on the historical, social, and economic development of Europe and North America. In a demographic transition, the following four stages occur (see figure 8.5):

1. Initially, countries have a stable population with a high birthrate and a high death rate. Death rates often vary because of famine and epidemic disease.
2. Improved economic and social conditions (control of disease and increased food availability) bring about a period of rapid population growth as death rates fall. Birthrates remain high.
3. As countries become industrialized, the birthrates begin to drop because people desire smaller families and make use of contraceptives.
4. Eventually, birthrates and death rates again become balanced, with low birthrates and low death rates.

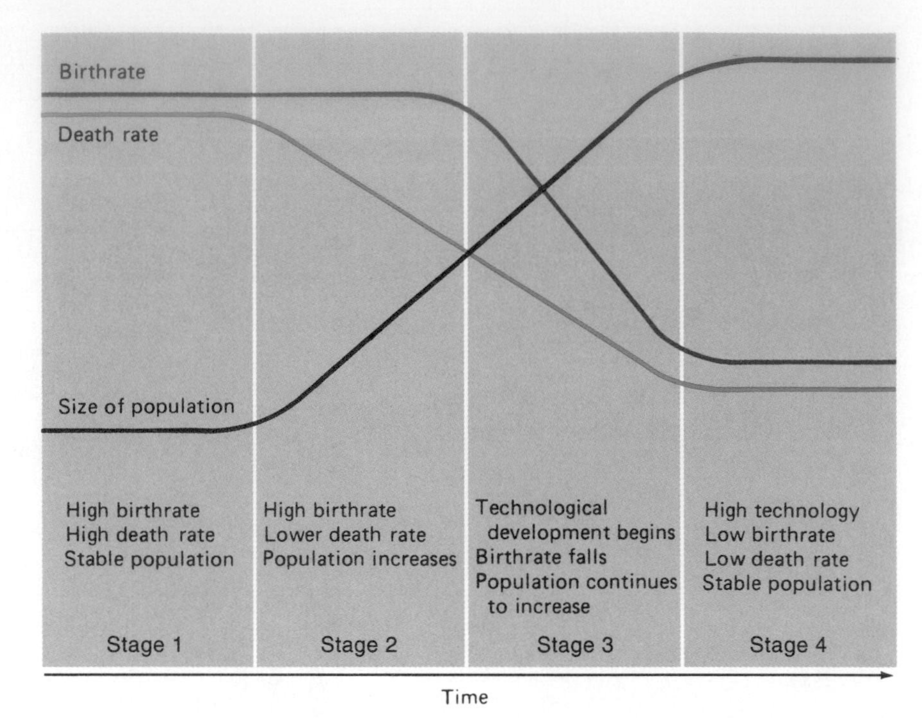

figure 8.5 **Demographic Transition** The demographic transition model suggests that as a country develops technologically, it automatically experiences a drop in the birthrate. This certainly has been the experience of the developed countries of the world. However, the developed countries make up about 20 percent of the world's population. It is doubtful whether the less-developed countries can achieve the kind of technological advances experienced in the developed world.

This is a very comfortable model because it suggests that if a country can become industrialized, then social, political, and economic processes will naturally cause its population to stabilize.

However, the model leads to some serious questions. Can the historical pattern exhibited by Europe and North America be repeated in the less-developed countries of today? Europe, North America, Japan, and Australia passed through this transition period when world population was lower and when energy and natural resources were still abundant. It is doubtful whether these supplies are adequate to allow for the industrialization of the major portion of the world currently classified as less developed.

A second concern is the time element. With the world population increasing as rapidly as it is, industrialization probably cannot occur fast enough to have a significant impact on population growth. As long as people in less-developed countries are poor, there is a strong incentive to have large numbers of children. Children are a form of social security because they take care of their elderly parents. Only people in developed countries can save money for their old age. They can choose to have children, who are expensive to raise, or to invest money in some other way.

When the countries of Europe and North America passed through the demographic transition, they had access to large expanses of unexploited lands, either within their boundaries or in their colonies. This provided a safety valve for expanding populations during the early stages of the transition. Without this safety valve, it would have been impossible to deal adequately with the population while simultaneously encouraging economic development. Today, less-developed countries may be unable to accumulate the necessary capital to develop economically, since an ever-increasing population is a severe economic drain.

GLOBAL PERSPECTIVE

The Urbanization of the World's Population

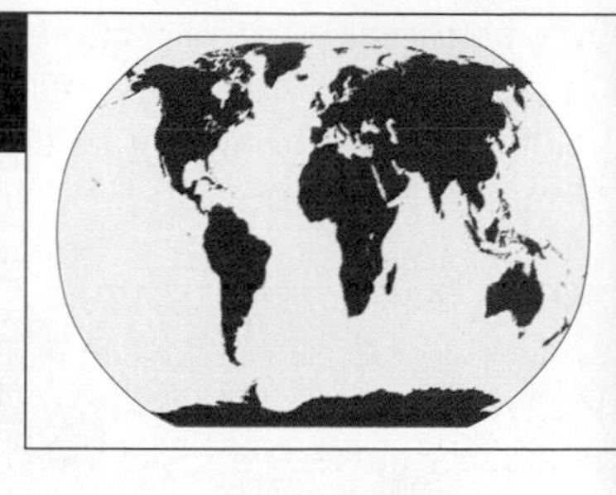

The United Nations estimates that currently about 45 percent of the world's population lives in cities. This is expected to increase to over 50 percent by 2005 and reach 60 percent by 2025. The economically developed world is currently about 75 percent urban, and the less-developed world is about 40 percent urban. Therefore, the increase in the urban population will occur primarily in the less-developed world where resources to deal with urban problems are least available.

Urbanization is not necessarily bad. The economic activity of the developed world takes place primarily in cities. Cities can be planned and managed to be healthy, interesting places to live. Cities offer jobs, health care, schools, and other services that are usually lacking in rural areas. Often the rural economy of a country is considered to be of low status, and young people wish to move to cities where higher status jobs and desirable cultural amenities are available.

Dense populations of people have a concentrated impact on local resources. Often water must be transported long distances, wastes are difficult to get rid of, and air quality drops as each additional person places a burden on the local environment. Unfortunately, in the less-developed world it is impossible to plan for growth and provide basic resources such as drinking water, sewers, transportation, and housing as fast as the population is growing. New migrants to the cities often lack education or training for the high wage jobs available, have lost the opportunity to raise their own food, and lack the support of their family units. Consequently, when these poor people migrate to the cities, they often live in shantytowns on the edge of the city or form squatter settlements on hillsides and other unused land. These centers of poverty lack basic services since they are often outside the bureaucratic structure of the city. The rich in such cities live a short distance from the poor and may be relatively isolated, but the negative environmental impact of the entire urban region (air and water pollution, transportation problems, etc.) affects all the residents. Even the long-established urban centers of the developed world have major problems. Many contain pockets of poverty that have led to problems of social unrest and crime.

The U.S. Population Picture

As a result of the 1990 census in the United States, several changes have occurred in how demographers view the structure of and future trends in the U.S. population. In many ways, the U.S. population is similar to those of other developed countries of the world with low birthrates and slow population growth. However, the U.S. population includes a **postwar baby boom** component, which has significantly affected population trends. These baby boomers were born during an approximately 15-year period (1947–1961) following World War II, when birthrates were much higher than today, and constitute a bulge in the age distribution profile. (See figure 8.6.) As members of this group have raised families, they have had a significant influence on how the U.S. population has grown. Some of the older persons in this group are beginning to retire, and, as more of them do, the population will gradually age. By 2030, about 20 percent of the population will be 65 years of age or older.

A changing age structure will lead to social changes as well. The baby boom of the late 1940s and the 1950s encouraged growth in service industries needed by young families. Maternity wards had to be expanded, schools could not be built fast enough, baby-care companies saw unprecedented sales, and the toy industry flourished. Today, these "babies" are in their forties and fifties. They are buying homes, cars, and appliances, but are raising fewer children than did their parents. However, because baby boomers are such a large segment of the population, they have contributed a large number of children to the population. The children of the baby boomers are now having children and, in many parts of the country, schools are seeing increased numbers of children in the early grades. At the same time, there are more elderly, because people are living longer. That creates a need for additional services for the elderly. This trend toward an aging population will be accentuated as the baby boomers retire. What will the social needs be in the year 2020 when many of these baby boomers will have retired?

Even with the current lower total fertility rate of 2.1 children per woman, the population is still growing by about 1.1 percent per year. About 0.6 percent is the result of natural increases owing to the difference between birthrates and death rates. The remainder is the result of immigration into the United States, The U.S. Census Bureau projects that immigration will increase significantly and account for 50 percent of population growth by the year 2050.

Current immigration policy in the United States is difficult to characterize. Strong measures are being taken to reduce illegal immigration across the southern border. This is in part due to pressures placed on Congress by states that receive large numbers of illegal immigrants. Illegal immigrants add to the education and health care that states must fund. At the same time, some

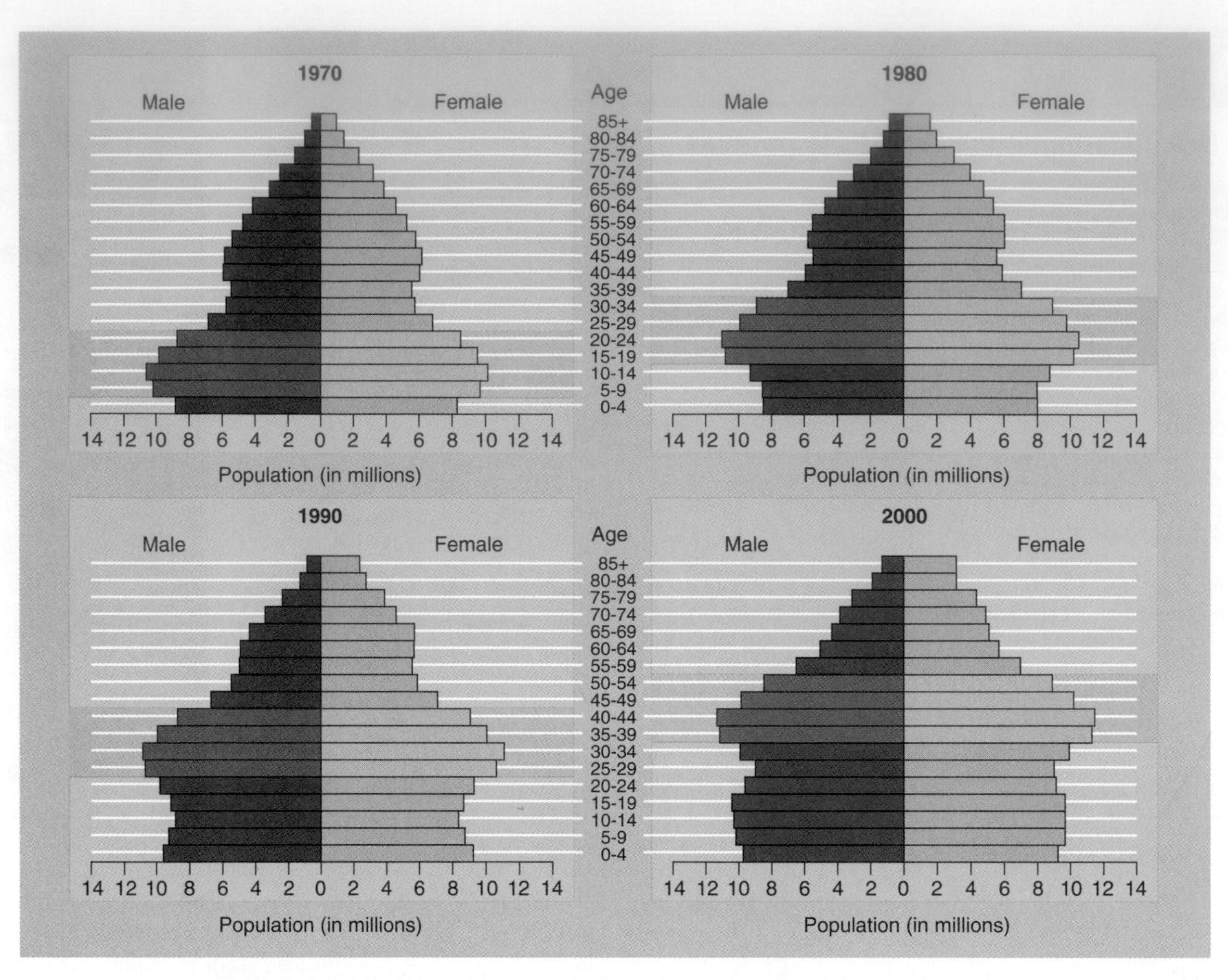

figure 8.6 **Changing Age Distribution of U.S. Population (1970–2000)** These graphs show the number of people in the United States at each age level. Notice that in the year 2000 a bulge begins to form at age 35 to 39 and ends at about age 50 to 54. These people represent the "baby boom" that followed World War II. As you compare the age distribution for 1970, 1980, and 1990 with 2000, you can see that this group of people moves through the population. As this portion of the population has aged, it has had a large impact on the nature of the U.S. population. In the 1970s and 1980s, baby boomers were in school. In the 1990s, they were in their middle working years. In 2000, some of them are beginning to retire and many more will do so throughout the decade.

Source: Data from the U.S. Department of Commerce, Bureau of the Census.

segments of the U.S. economy (agriculture, tourism) maintain that they are unable to find workers to do certain kinds of work. Consequently, special guest workers are allowed to enter the country for limited periods to serve the needs of these segments of the economy. There is also a consistent policy of allowing immigration that reunites families of U.S. residents. Obviously, the families that fall into this category are likely to be U.S. citizens who were recent immigrants themselves. It is obvious that most immigration policy is the result of political decisions rather than decisions that relate to population policy or a concern about the rate at which the U.S. population is growing.

Projections based on the 1990 census indicate that the population will grow more rapidly than previously projected, and it does not seem to be moving toward zero growth. Indeed, it appears that the total fertility rate, which is currently about 2 children per woman per lifetime, will increase to 2.12 by the year 2050. This would result in an increase from about 275 million people in 2000 to about 383 million by 2050. (See figure 8.7.)

These increases are partly the result of higher fertility rates among some minority populations. The Hispanic and Asian-American portions of the population will grow rapidly, and the Caucasian portion will decline. Thus, the U.S. population will be much more ethnically diverse in the future.

Anticipated Changes with Continued Population Growth

As the world human population continues to increase, pressure for the necessities of life will become greater. Differences in the standard of living

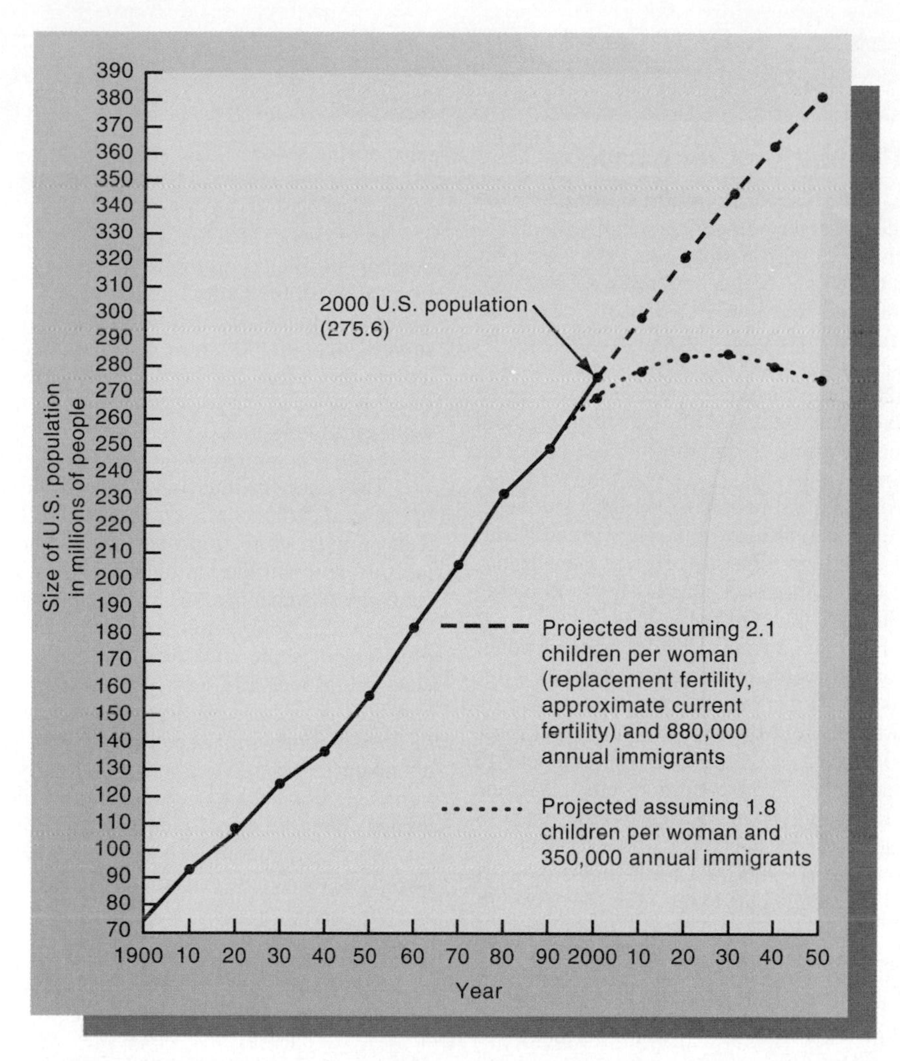

figure 8.7 **United States Population Growth** The population of the United States has grown continuously since colonization. The graph indicates that the size of the population was about 230 million people in 1980. In 1980, the U.S. Census Bureau made projections based on different birthrates and immigration rates. The ultimate size of the U.S. population differs considerably, depending on the estimates used. The current total fertility rate is 2.1 and immigration is about 1 million people per year. It appears that the upper line is the course the current population is following.

Source: Data from the U.S. Department of Commerce, Bureau of the Census.

between developed and less-developed countries will remain significant because population will increase most in less-developed countries. The supply of fuel and other resources is dwindling. Pressure for these resources will intensify as industrialized countries seek to maintain their current standard of living. People in less-developed countries will seek more land to raise crops and feed themselves unless major increases in food production per hectare occur. Since most of these people live in tropical areas, tropical forests will be cleared for farmland. The resulting erosion or alteration of the soil will make it no longer suitable for forest or crops. This conversion of natural ecosystems to agricultural ecosystems could cause profound changes in the world ecosystem.

Developed countries may have to choose between two alternatives: helping the less-developed countries, thus maintaining their friendship, or isolating themselves from the problems of the less-developed nations. Neither of these policies will be able to prevent a change in lifestyle as world population increases. The resources of the world are finite. Even if industrialized countries continue to use a disproportionate share of the world's resources, the amount available per person will decline as population rises. As world population increases, the less-developed areas will probably maintain their low standard of living. It is difficult to see how that standard of living could get much lower, since some people in these countries are already starving to death. Lifestyles in developed nations will probably change very little but may become less

GLOBAL PERSPECTIVE

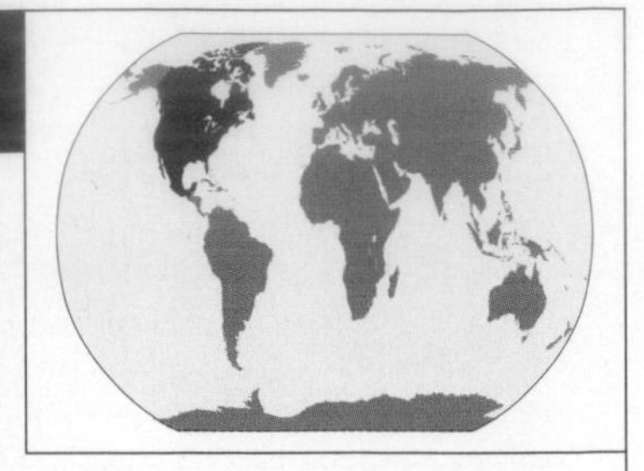

North America—Population Comparisons

The three countries that make up North America (Canada, the United States, and Mexico) interact politically, socially, and economically. The characteristics of their populations determine how they interact. Canada and the United States are both wealthy countries with similar age structures and low total fertility rates. The United States has a total fertility rate of 2.1 and Canada has a total fertility rate of 1.5. Both countries have a relatively small number of young people (about 20 percent of the population is under 15 years of age) and a relatively large number of older people (about 13 percent are 65 or older). Without immigration, these countries would have stable or falling populations. Therefore, they must rely on immigration to supply additional people to fill the workplace. This is particularly true for jobs with low pay. The United States currently receives about 1 million immigrants per year. Canada currently receives about 175,000 immigrants per year but is planning to increase that to about 300,000 in the future.

Mexico, on the other hand, has a young, rapidly growing population. About 37 percent of the population is under 15 years of age and the total fertility rate is 2.7. At that rate, the population will double in about 36 years. Furthermore, Mexico has a per capita gross national product about 20 percent of that in Canada and about 13 percent of that in the United States. These conditions create a strong incentive for individuals to migrate from Mexico to other parts of the world. The United States is the usual country of entry. The number is hard to evaluate, since many enter the United States illegally or as seasonal workers and eventually plan to return to Mexico. Current estimates are that about 250,000 people enter the United States from Mexico each year.

In response to the large number of illegal immigrants from Mexico, the United States has erected fences and increased surveillance. In 2000, over a million people were apprehended attempting to cross the border from Mexico to the United States. Thus, instead of crossing at normal points of entry, illegal immigrants are likely to cross the border in remote desert areas, which has resulted in numerous deaths due to exposure.

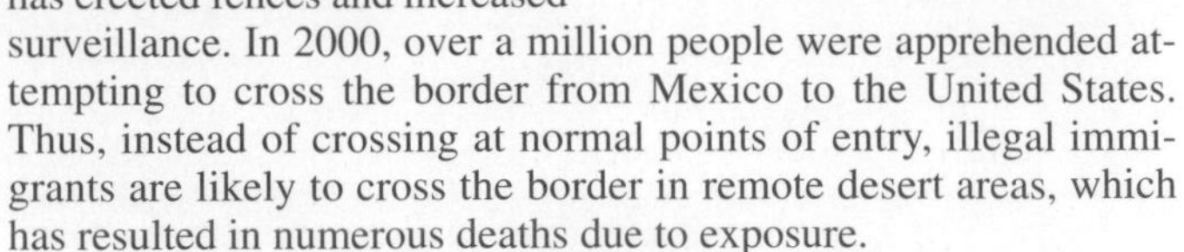

The economic interplay between Mexico and the United States has several components. Working in the United States allows Mexican immigrants to improve their economic status. Furthermore, their presence in the United States has a significant effect on the economy of Mexico, since many immigrants send much of their income to Mexico to support family members. The North American Free Trade Agreement (NAFTA) allows for relatively free exchange of goods and services between Canada, the United States, and Mexico. Consequently, many Canadian, U.S., and European businesses have built assembly plants in Mexico to make use of the abundant inexpensive labor, particularly along the U.S./Mexican border. Labor leaders in Canada and the United States are concerned about the effect access to low-cost labor will have on their membership and complain that many high-paying jobs have been moved to Mexico where labor costs are lower.

	Population Size 2000 (millions)	Birthrate per 1000	Death Rate per 1000	Rate of Natural Increase	Total Fertility Rate (children/ woman)	Life Expectancy (years)	Infant Mortality Rate (deaths/ 1000 births)	Per Capita GNP 1998 U.S. dollars	Percent Under 15 Years of Age	Percent over 65 Years of Age	Annual Immigration/ Emigration (estimate)*
Canada	30.8	11.2	7.3	0.39	1.5	79	5.5	19,170	19	12	+175,000
United States	275.6	14.5	8.7	0.58	2.1	77	7.0	29,240	21	13	+1,000,000
Mexico	99.6	23.9	4.4	1.95	2.7	72	31.5	3,840	37	5	−250,000

Source: Population Reference Bureau "2000 Population Data Sheet"

*Immigration/emigration estimates from United States and Canadian government sources

consumption-oriented. What some people currently view as necessities (meals in restaurants, vacations to remote sites, and two cars per family) will probably become luxuries. Many people who enjoy the freedom of mobility associated with the automobile will have their travel limited as public transportation replaces private transportation. Recreation may cease to involve expensive, energy-demanding machines (powerboats, motorcycles, and electric-powered toys) and emphasize instead such activities as hiking, bicycling, and reading. These changes will not come quickly, unless some catastrophic political or economic force causes major worldwide adjustments. Most likely, changes will occur only as economic pressures affect families. Many economists and political thinkers feel that as the economics of the world become more linked, and jobs flow more freely from country to country, wealth will be redistributed from the former rich countries to emerging economies in developing countries. The possibility of such a redistribution has become a major political issue in many of the developed countries of the world.

The Impact of AIDS on Populations

ISSUES & ANALYSIS

The AIDS (acquired immunodeficiency syndrome) virus has caused a worldwide epidemic, which can be called a pandemic because it continues to spread to all countries of the world. The disease is spread through direct transfer of body fluids containing the virus into the bloodstream of another person. Sharing of contaminated needles among intravenous drug users and sexual contact are the most likely methods of passage. In the United States, the disease was once considered a problem only for the homosexual community and those who use intravenous drugs. This perception is rapidly changing. Many of the new cases of AIDS are being found in women infected by male sex partners and in children born to infected mothers.

The World Health Organization (WHO) estimated that by 2000 about 47 million people were infected with the AIDS virus worldwide. WHO estimates that 2.6 million people died of AIDS in 1999 and that over 16 million have died since the pandemic began. The distribution of the virus is lowest in the economically developed countries and highest in the developing countries. The figure shows estimates by the World Health Organization of the numbers of HIV-infected people in various parts of the world at the end of 1999. In the less-developed world, there is little medical care to treat AIDS and a lack of resources to identify those who have HIV. Many people do not know they are infected and will continue to pass the disease to others.

Sub-Saharan Africa has been hit hardest by this disease. In Africa, AIDS has always been primarily a sexually transmitted disease spread through heterosexual contact. Many people believe that in the poor countries of central Africa, permissive sexual behavior and prostitution have created conditions for a rapid spread of the disease. This is particularly evident along major transportation routes. The World Health Organization estimates that nearly 70 percent of all persons infected with the AIDS virus (23.3 million) live in this region of the world. In some large urban hospitals in this part of Africa, over 50 percent of the hospital beds are taken up by AIDS patients, and AIDS is the leading cause of death from disease. Since those who are dying are young, there is a change in the age structure of the population of heavily infected countries. This has resulted in large numbers of AIDS orphans—about 10 million AIDS orphans in sub-Saharan Africa by the end of 1999. With the death of young infected adults, villages are composed primarily of older people and children.

The economic burden on these countries is tremendous. Those with AIDS symptoms are unable to work and need medical care and medication. Because of poverty, there is often little medical care available. Also, the millions of orphaned children have resulted in an additional economic burden on relatives.

In the developed world, drugs are available that can extend life and perhaps prevent transmission between infected mothers and their infants. These drugs are expensive and therefore are unavailable to most of the people in the world who need them.

- Should the developed countries of the world provide funding and drugs to combat the AIDS epidemic in Africa?
- What actions should the governments of these countries institute to combat AIDS?
- Should economic aid be given to countries that have large numbers of AIDS patients?

Source: Data from World Health Organization of the United Nations.

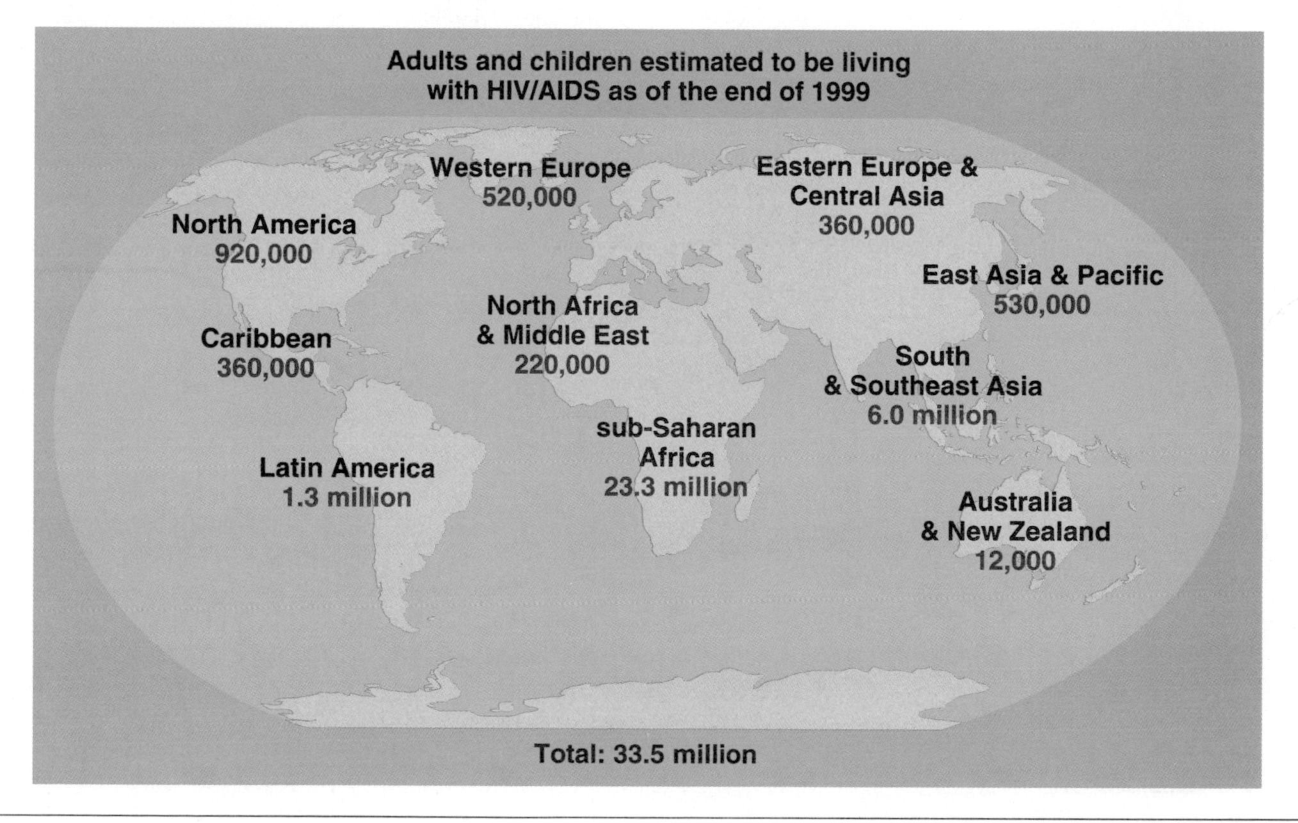

Summary

Many of the problems of the world are caused by or made worse by an increasing human population. Currently, the world's population is growing very rapidly. Most of the growth is occurring in the less-developed areas of the world (Africa, Asia, and Latin America) where people have a low standard of living. The developed regions of the world, with their high standard of living, have relatively slow population growth and, in some instances, have declining populations.

Demography is the study of human populations and the things that affect them. Demographers study the sex ratio and age distribution within a population to predict future growth. Population growth rates are determined by biological factors such as birthrate, which is determined by the number of women in the population and the age of the women, and death rate. Sociological and economic conditions are also important, since they affect the number of children desired by women, which helps set the population growth rate. In developed countries, women usually have access to jobs. Couples marry later, and they make decisions about the number of children they will have, based on the economic cost of raising children. In the less-developed world, women marry earlier, and children have economic value as additional workers, as future caregivers for the parents, and as status for either or both parents.

The current U.S. population will continue to grow. The average age of the population will increase and the racial mix of the population will change significantly.

The demographic transition model suggests that as a country becomes industrialized, its population begins to stabilize. However, there is little hope that the earth can support the entire world in the style of the industrialized nations. It is doubtful whether there are enough energy resources and other natural resources to develop the less-developed countries and/or whether there is enough time to change trends of population growth. Highly developed nations should anticipate increased pressure in the future to share their wealth with less-developed countries.

Interactive Exploration

Check out the website at

http://www.mhhe.com/environmentalscience

and click on the cover of this textbook for interactive versions of the following:

KNOW THE BASICS

age distribution *149*
demographic transition *156*
demography *148*
gross national product (GNP) *153*
population density *147*
postwar baby boom *157*
replacement fertility *149*
standard of living *152*
total fertility rate *149*
zero population growth *149*
On-line Flashcards
Electronic Glossary

IN THE REAL WORLD

China has the greatest population, but India is gaining! In October 1999, the world's population reached six billion and the largest contributor to population growth is India. Examine the social, political, and economic factors that are leading to this growth and how it contrasts with China in **India's Population Passes 1 Billion.**

Speaking of China's population, examine the implications of continued economic growth in the country with the world's greatest population in **China Slated to Join the WTO.**

PUT IT IN MOTION

You've read about the demographic transition, now see it in action. Check out the Stages of Population Growth animation for a better understanding of the concept.

TEST PREPARATION

Review Questions

1. What is demography?
2. What is demographic transition? What is it based upon?
3. What is a baby boom?
4. What does age distribution of a population mean?
5. List 10 differences between your standard of living and that of someone in a less-developed country.
6. Why do people who live in overpopulated countries use plants as their main source of food?
7. Although predicting the future is difficult, describe what you think your life will be like in 10 years. Why?
8. List five changes you might anticipate if world population were to double in the next 50 years.
9. Which three areas of the world have the highest population growth rate? Which three areas of the world have the lowest standard of living?
10. How many children per woman would lead to a stable U.S. population?
11. What role does the status of women play in determining population growth rates?
12. Describe three reasons why women in the less-developed world might desire more than two children.

Critical Thinking Questions

1. Do you think it is appropriate for developed countries to persuade less-developed countries to limit their population growth? What would be appropriate and inappropriate interventions, according to your ethics? Now imagine you are a citizen of a less developed country. What might be your reply to those who live in more developed countries? Why?
2. The Chinese government has been very involved in regulating population growth in China. What do you think about this kind of government intervention in China's population problem? What values, beliefs, and perspectives do you hold that lead you to think the way you do?
3. Population growth causes many environmental problems. Identify some of these problems. What role do you think technology will play in solving these problems? Are you optimistic or pessimistic about these problems being solved through technology? Why?
4. Do you think that demographic transition will be a viable option for world development? What evidence leads you to your conclusions? What role should the developed countries play in the current demographic transition of developing countries? Why?
5. The United States Census Bureau projects that by 2050 immigration will account for 50 percent of the population growth in the United States. What values and perspectives should guide our immigration policy? Why?
6. Imagine a debate between an American and a Sudanese about human population and the scarcity of resources. What perspectives do you think the American might bring to the debate? What perspectives do you think the Sudanese would bring? What might be their points of common ground? On what might they differ?
7. Many people in developing countries hope to achieve the standard of living of those in the developed world. What might be the effect of this pressure on the environment in developing countries? The political relationship between developing countries and already developed countries? What ethical perspective do you think should guide this changing relationship?
8. The demographic changes occurring in Mexico have an influence on the United States. What problems does Mexico face regarding its demographics? Should the United States be involved in Mexican population policy?

KEY CHAPTER LINKS

ESSENTIAL STUDY PARTNER

BIO COURSE

GLOBAL ISSUES

REGIONAL PERSPECTIVES

PRACTICE QUIZZING

PART THREE

Energy

All living systems can be described by the flow of energy through them: Energy enables simple forms of matter to be changed into more complex forms, and energy is needed to maintain this complexity. Energy sustains the complex technical and social units typical of human populations.

All living things (including humans) rely on the sun as a source of energy. Coal, oil, and natural gas are energy sources available today because organisms in the past captured sunlight energy and stored it in the complex organic molecules that made up their bodies, which were then compressed and concentrated.

Technological development and fossil-fuel exploitations are directly related to one another and have allowed us an increasingly higher standard of living. **Chapter 9** traces the development of energy consumption, its interrelationships with economic development and lifestyles, and current uses and demands. **Chapter 10** discusses the sources of energy currently being used, as well as the significance of each source and its impact. **Chapter 11** discusses nuclear energy, other applications of radioactive substances, and the concerns their use generates.

CHAPTER

Energy and Civilization: Patterns of Consumption

Objectives

After reading this chapter, you should be able to:

- Explain why all organisms require a constant input of energy.
- Describe how per capita energy consumption increased as civilization developed from hunting and gathering to primitive agriculture to advanced cultures.
- Describe how advanced modern civilizations developed as new fuels were used to run machines.
- Recognize that coal deposits are not uniformly distributed throughout the world.
- Correlate the Industrial Revolution with social and economic changes.
- Explain how cheap oil and natural gas led to a consumption-oriented society.
- Explain how the automobile changed people's lifestyles.
- Explain why overall energy use in the United States declined during the 1970s and 1980s.

Chapter Outline

History of Energy Consumption
- Biological Energy Sources
- Increased Use of Wood
- Fossil Fuels and the Industrial Revolution

Energy and Economics
- Economic Growth and Energy Consumption
- The Role of the Automobile
- Gasoline Prices and Government Policy

Global Perspective: *Five Ways to Curb Traffic*

How Energy Is Used
- Residential and Commercial Energy Use
- Industrial Energy Use
- Transportation Energy Use
- The Variability of Gasoline Prices

Electrical Energy

Environmental Close-Up: *Hybrid Electric Vehicles*

Energy Consumption Trends

Global Perspective: *OPEC*

Environmental Close-Up: *Alternative-Fuel Vehicles*

Global Perspective: *Energy Development in China*

Global Perspective: *Potential World Petroleum Resources*

History of Energy Consumption

Every form of life and all societies require a constant input of energy. If the flow of energy through organisms or societies ceases, they stop functioning and begin to disintegrate. Some organisms and societies are more energy efficient than others. In general, history shows that complex industrial societies use the most energy. If societies are to survive, they must continue to expend energy. However, they may need to change their pattern of energy consumption as traditional sources become limited.

figure 9.1 **Hunter-Gatherer Society** In this type of society, people obtain nearly all of their energy from the collection of wild plants and the hunting of animals. These societies do not make large demands on fossil fuels.

Biological Energy Sources

Energy is essential to maintain life. In every ecosystem, the sun provides that energy. (See chapter 5.) The first transfer of energy occurs during photosynthesis, when plants convert light energy into chemical energy in the production of food. Herbivorous animals utilize the food energy in the plants. The herbivores, in turn, are a source of energy for carnivores. Because nearly all of their energy requirements were supplied by food, primitive humans were no different from other animals in their ecosystems. In such hunter-gatherer cultures, nearly all human energy needs were met by using plants and animals as food, tools, and fuel. (See figure 9.1.)

Early in human history, people began to use additional sources of energy to make their lives more comfortable. They domesticated plants and animals to provide a more dependable supply of food. They no longer needed to depend solely upon gathering wild plants and hunting wild animals for sustenance. Domesticated animals also furnished a source of energy for transportation, farming, and other tasks. (See figure 9.2.) Wood provided a source of fuel for heating and cooking. Eventually, this biomass energy was used in simple technologies, such as shaping tools and extracting metals.

figure 9.2 **Animal Power** This bas-relief panel from an Egyptian tomb depicts an important accomplishment in the development of human civilization. With the use of domesticated animals, people had a source of power other than their own muscles.

Increased Use of Wood

Early civilizations, such as the Aztecs, Greeks, Egyptians, Romans, and Chinese, were culturally advanced, but their societies used human muscle, animal muscle, and fire as sources of energy. Except for limited use of some wind-powered and water-powered devices such as ships and canoes, the controlled use of fire was the first use of energy in a form other than food. Wood was the primary fuel. (Wood was also used for building materials and other cultural uses.) The energy provided by wood enabled people to cook their food, heat their dwellings, and develop a primitive form of metallurgy. Such advances separated humans from other animals. When dense populations of humans made heavy use of wood for fuel and building materials, they eventually used up the readily available sources and had to import wood or seek alternative forms of fuel.

Because of a long history of high population density, India and some other parts of the world experienced a wood shortage hundreds of years before Europe and North America did. In many of these areas, animal dung replaced wood as a fuel source. It is still used today in some parts of the world.

Western Europe and North America were able to use wood as a fuel for a longer period of time. The forests of Europe supplied sufficient fuel until the thirteenth century. In North America, vast expanses of virgin forests supplied adequate fuel until the late nineteenth century. Fortunately, when local supplies of wood declined in Europe and North America, coal, formed from fossilized plant remains, was available as an alternative energy source. By 1880, coal had replaced wood as the primary energy source.

figure 9.3 **Carboniferous Period** Approximately 300 million years ago, this kind of ecosystem was common throughout the world. Plant material accumulated in these swamps and was ultimately converted to coal.

Fossil Fuels and the Industrial Revolution

Fossil fuels are the remains of plants, animals, and microorganisms that lived millions of years ago. (The energy in these fuels is stored sunlight, just as the biomass of wood represents stored sunlight.) During the Carboniferous period, 275 to 350 million years ago, conditions in the world were conducive to the formation of large deposits of fossil fuels. (See figure 9.3.) Ever since machines replaced muscle power, the major energy sources for the world have been fossil remains from the distant past.

Historically, the first fossil fuel to be used extensively was coal. In the early eighteenth century, regions of the world that had readily available coal deposits were able to switch to this new fuel and participate in a major cultural change known as the **Industrial Revolution.** The Industrial Revolution began in England and spread to much of Europe and North America. It involved the invention of machines that replaced human and animal labor in manufacturing and transporting goods. Central to this change was the invention of the steam engine, which could convert heat energy into the energy of motion. The steam engine made possible the large-scale mining of coal. Before steam engines, coal mines flooded and thus were not economically mined. The source of energy for steam engines was either wood or coal; wood was quickly replaced by coal in most cases. Nations without a source of coal, or those possessing coal reserves that were not easily exploited, did not participate in the Industrial Revolution.

Prior to the Industrial Revolution, Europe and North America were predominately rural. Goods were manufactured on a small scale in the home. As machines and the coal to power them became increasingly available, the factory system of manufacturing products replaced the small home-based operation. Because expanding factories required a constantly increasing labor supply, people left the farms and congregated in areas surrounding the factories. Villages became towns, and towns became cities. Widespread use of coal in cities resulted in increased air pollution. In spite of these changes, the Industrial Revolution was viewed as progress. Energy consumption increased, economies grew, and people prospered. Within a span of two hundred years, the daily per capita energy consumption of industrialized nations increased eightfold. This energy was furnished primarily by coal, but a new source of energy was about to be discovered: oil.

The Chinese used some gas and oil as early as 1000 B.C., yet these resources

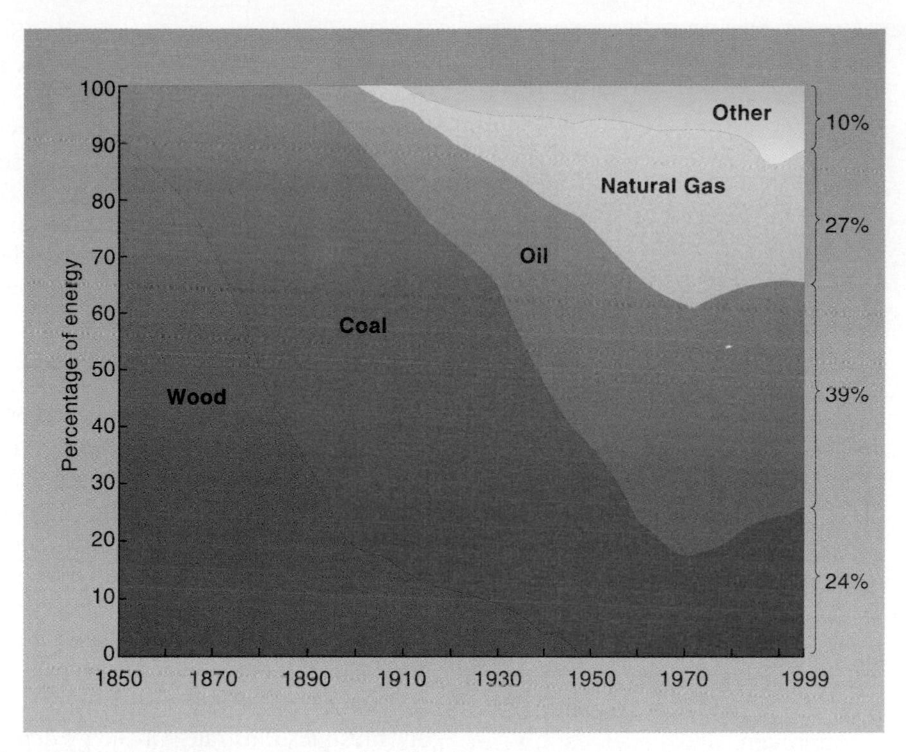

figure 9.4 **Oil Replaces Coal** Just as wood was replaced by coal, coal was later replaced by oil. This graph represents the production of energy by various sources in the United States. Oil has remained a dominant energy source for the past 40 years; coal has decreased slightly; natural gas has increased; and other sources such as nuclear, hydroelectric, wind, and solar power have been increasing.

Source: Data from Annual Energy Review, 1999, Energy Information Administration, and more recent data from *BP Statistical Review of World Energy,* 1999 and pervious editions.

remained virtually untapped until fairly recently. The oil well that Edwin L. Drake, an early oil prospector, drilled in Pennsylvania in 1859 was not the world's first oil well, but it was the beginning of the modern petroleum era. By 1870, oil production in the United States had reached over four million barrels a year and supplied 1 percent of the nation's energy requirements. It grew to nearly 50 percent by 1970 and currently contributes just over 40 percent. (See figure 9.4.)

For the first 60 years of production, the principal use of oil was to make kerosene, a fuel for lamps. The gasoline produced was discarded as a waste product. During this time, oil was abundant relative to its demand and thus it had a low price. The low price led the oil industry to have the federal government control the price of oil. However, the automobile dramatically increased the demand. In 1900, the United States had only 8000 automobiles. By 1920, it had 8 million cars, and by 1999, 162 million. More oil was needed to make automobile fuel and lubricants.

The use of natural gas did not increase as rapidly as the use of oil. It was used primarily for home heating. In the early 1900s, 90 percent of the natural gas was "flared," that is, burned as a waste product at oil wells.

A series of events involving the U.S. government and the need for more oil was ultimately responsible for increased use of natural gas in the United States. World War II greatly increased the energy demand for manufacturing and transportation. In 1943, a federally financed pipeline was constructed to transport oil within the United States. This 2,000-kilometer (1200 mile) pipeline transported oil more efficiently from wells in Texas, Louisiana, and Oklahoma to refineries and factories in the eastern section of the country. In 1944, a longer [2,400 kilometers (1500 mile)] federally financed pipeline was built to increase the flow of oil to the country's eastern and midwestern regions.

After the war, the federal government sold these pipelines to private corporations. The corporations converted the pipelines to transport natural gas. Thus, a direct link was established between the natural gas fields in the Southwest and the markets in the Midwest and East. By 1971, there were 400,000 kilometers (250,000 mile) of long-range transmission to pipelines and 986,000 (612,000 mile) kilometers of distribution pipelines in the United States. Approximately 1,613,000 kilometers (1 million miles) of natural gas distribution pipelines are used today. Natural gas is used in many parts of the developed world both for home heating and for industrial purposes.

Energy and Economics

Most industrial societies want to ensure a continuous supply of affordable energy. The higher the price of energy, the more expensive goods and services become. To keep costs down, many countries have subsidized their energy industries and maintained energy prices artificially low. International trade in fossil fuels has a major influence on the world economy and politics. The emphasis on low-priced fuels has encouraged high rates of consumption.

Economic Growth and Energy Consumption

There is a direct link between economic growth and the availability of inexpensive energy. The replacement of human and animal energy with fossil fuels began with the Industrial Revolution and was greatly accelerated by the supply of cheap, easy-to-handle, and highly efficient fuels. Because the use of inexpensive fossil fuels allows each worker to produce more goods and services, productivity increased. The result was unprecedented economic growth in Europe, North America, and the rest of the industrialized world.

figure 9.5 **Energy-Demanding Lifestyle** Building private homes on large individual lots some distance from shopping areas and places of employment is directly related to the heavy use of the automobile as a mode of transportation. Heating and cooling a large enclosed shopping mall, along with the gasoline consumed in driving to the shopping center, increase the demand for energy.

In North America, World War II was a prime factor in ending the economic depression of the 1930s. Military activities created millions of defense jobs. Almost everyone was employed, but there was a scarcity of consumer goods. After World War II, consumer goods that had been unavailable during the war were in great demand. Industries set up to produce military goods turned to the production of consumer goods. High employment, a rapidly expanding population, and a supply of inexpensive energy encouraged a period of rapid economic growth.

In Europe and Japan, where much of the industrial base was destroyed by the war, the recovery was slow. However, foreign aid and an intense rebuilding effort on the part of the people resulted in a reindustrialization of these countries. Because they had to build new facilities, often their technology was better than that of North America where few factories were affected by the war.

The Role of the Automobile

The cheap, abundant energy that fueled industries produced an ever-increasing amount and array of consumer goods. One product was the automobile. The growth of the automobile industry, first in the United States and then in other industrialized countries, led to roadway construction, which required energy. Thus, the energy costs of driving a car were greater than just the fuel consumed in travel. As roads improved, higher speeds were possible. People demanded faster cars, and automobile companies were quick to build them. Bigger and faster cars required more fuel and even better roads. So roads were continually being improved, and better cars were being produced. A cycle of *more chasing more* had begun. In North America and much of Europe, the convenience of the automobile encouraged two-car families, which created a demand for more energy. It requires energy to mine ore, process it into metals, form the metals into automobile components, and transport all the materials. As the economy grew, so did energy needs.

More cars meant more jobs in the automobile industry, the steel industry, the glass industry, and hundreds of other industries. Constructing thousands of kilometers of roads created additional jobs. From their beginnings as suppliers of lamp oil, the oil companies grew into one of the largest industries in the world. Thus, the automobile industry played a major role in the economic development of the industrialized world. All this wealth gave people more money for cars and other necessities of life. The car, originally a luxury, was now considered a necessity.

The car not only created new jobs but also altered people's lifestyles. Vacationers could travel greater distances. New resorts and chains of motels, restaurants, and other businesses developed to serve the motoring public, creating thousands of new jobs. Because people could live farther from work, they began to move to the suburbs. (See figure 9.5.) Large shopping centers in suburban areas hastened the decline of central business districts. Today, fewer than 50 percent of retail sales are made in central business districts of North American cities, which has resulted in a loss of jobs in these areas. In Philadelphia, 79 percent of all retail jobs were located in the city in 1930; by 1970, they had declined to 43 percent.

As people moved to the suburbs, they also changed their buying habits. Labor-saving, energy-consuming devices became essential in the home. The vacuum cleaner, dishwasher, garbage disposal, and automatic garage door opener are only a few of the ways human power has been replaced with electrical power. Eleven percent of the electrical energy in North America is used to operate home appliances. Other aspects of our lifestyles illustrate our energy dependence. The small, horse-powered farm of yesterday has grown into the huge, diesel-powered farm of today. Regardless of where we live, we expect Central American bananas, Florida oranges, California lettuce, Texas beef, Hawaiian pineapples, Ontario fruit, and Nova Scotia lobsters to be readily available at all seasons. What we often fail to consider is the amount of energy required to process, refrigerate, and transport these items. The car, the modern home, the farm, and the variety of items on our grocery shelves are only a few indications of how our lifestyles are based on cheap, abundant energy.

Gasoline Prices and Government Policy

The price of a liter of gasoline is determined by two major factors: (1) the cost of purchasing and processing crude oil into gasoline and (2) various taxes. Most of the differences in gasoline prices among countries are a result of taxes and reflect differences in government policy toward motor vehicle transportation.

A major objective of governments is to collect money to build and repair roads. Governments often charge road users by taxing the fuel their cars or trucks run on. Governments can also

GLOBAL PERSPECTIVE

Five Ways to Curb Traffic

Here is how six cities around the world are discouraging traffic in already congested areas.

Hong Kong

Electronic sensors on cars record highway travel and time of day. Drivers are issued a monthly bill (commuter hours are the most expensive).

Singapore

Automobiles entering downtown Singapore during rush hour are required to display a $30-a-month sticker, but cars carrying four or more passengers may pass without charge.

Gothenburg, Sweden

To encourage pedestrian traffic, the central business district has been divided like a pie into zones, with cars prohibited from moving directly from one zone to another. Autos move from zone to zone by way of a peripheral ring road.

Rome and Florence

All traffic except buses, taxis, delivery vehicles, and cars belonging to area residents have been banned between 7:30 A.M. and 7:30 P.M.

Tokyo

Before closing the sale, the buyer of a standard-size vehicle must show evidence that a permanent parking space is available for the car. To comply with the law, some drivers have constructed home garages with lifts to permit double parking!

discourage the use of automobiles by increasing the cost of fuel. An increase in fuel costs also creates a demand for increased fuel efficiency in all forms of motor transport.

Many European countries raise more money from fuel taxes than they spend on building and repairing roads. The United States, on the other hand, raises approximately 60 percent of the monies needed for roads from fuel taxes. The relatively low cost of fuel in the United States encourages more travel, which increases road repair costs. The cost of taxes to the U.S. consumer is about 28 percent of the retail gasoline price, while in Japan and many European countries, the cost is 47 to 77 percent.

Table 9.1 Energy Consumption 1999

Region	Energy Consumption per Capita per Year (in tonnes of oil equivalent)
Africa	0.32
Latin America	0.67
Japan	3.72
France	4.05
Germany	4.11
Canada	7.63
United States	7.86

Source: Data from *BP Statistical Review of World Energy,* June 2000 and Population Reference Bureau *2000 World Population Data Sheet.*

How Energy Is Used

The amount of energy consumed by countries of the world varies widely. (See table 9.1.) The highly industrialized countries use most of the world's energy; less-developed countries use much less. Even countries with the same level of development vary in the amount of energy they use as well as how they use it. To maintain their style of living, individuals in Canada and the United States use about twice as much energy as people in France or Japan and about 25 times as much energy as people in Africa.

Countries also use energy in different ways. Industrialized nations use energy about equally for three purposes: (1) residential and commercial uses, (2) industrial uses, and (3) transportation. Less-developed nations with little industry use most of their energy for residential purposes (cooking and heating). Developing countries use much of their energy to develop their industrial base.

Residential and Commercial Energy Use

The amount of energy required for residential and commercial use varies greatly throughout the world. Although a country with a high gross domestic product (GDP) uses a large amount of energy, it uses a lower percentage of its energy per capita for residential and commercial needs than does a less-developed country. For example, about 30 percent of the energy used in North America is for residential and commercial purposes, while in India, 90 percent of the energy is for residential uses. The ways residential and commercial energy is used also vary widely. In North America, 75 percent is used for air conditioning, refrigeration, water heating, and space heating. In India, almost all of the energy is used in the home for cooking since the scarcity and high cost of fuel precludes other uses.

As countries industrialize, their consumption of energy increases. Energy consumption and industrialization are interrelated.

figure 9.6 **Open-Fire Cooking** About half of the energy demand in Africa is for cooking. Using a solar stove instead of an open fire could greatly reduce this energy need.

The current pattern of residential and commercial energy use in each region of the world determines what conservation methods will be effective. In Canada, which has a cold climate, 40 percent of the residential energy is used for heating. Proper conservation practices could reduce this by 50 percent. In Africa, almost half of the energy used in the home is for cooking. (See figure 9.6.) Using fuel-efficient stoves instead of open fires could reduce these energy requirements by 50 percent.

Table 9.2 Per Capita Energy Use for Transportation, 2000

Country	Energy Use in Gigajoules/Capita
India	2
Zimbabwe	4
Mexico	17
Argentina	18
Russia	26
Japan	28
Netherlands	41
Denmark	43
Australia	86
United States	105

Industrial Energy Use

The amount of energy countries use for industrial processes varies considerably. Nonindustrial countries use little energy for industry. Countries that are developing new industries dedicate a high percentage of their energy use to them. They divert energy to the developing industries at the expense of other sectors of their economy. Highly industrialized countries use a significant amount of their energy in industry, but their energy use is high in other sectors as well. In the United States, industry claims about 30 percent of the energy used.

The amount of energy required in a country's industrial sector depends on the types of industrial processes used. Many countries use inefficient processes and could reduce their energy consumption by converting to more energy-efficient ones. However, they need capital investment to upgrade their industries and reduce energy consumption. Some countries cannot afford the upgrade. For example, India, a nation with few coal deposits, still uses outdated open-hearth furnaces to produce steel. These furnaces require nearly double the worldwide energy average to produce a metric ton of steel. The high cost of converting forces India to continue to use this energy-expensive method rather than convert to the more efficient electric-hearth method.

Transportation Energy Use

As with residential, commercial, and industrial uses, the amount of energy used for transportation varies widely throughout the world. In some of the less-developed nations, transportation uses are very small. Per capita energy use for transportation is larger in developing countries, and highest in highly developed countries. (See table 9.2.)

Once a country's state of development has been taken into account, the mix of bus, rail, water, and private automobiles is the main factor in determining a country's energy use for transportation. In Europe, Latin America, and

figure 9.7 **Public Transportation** In regions of the world where energy is expensive, such as in Germany, people make maximum use of cheaper public transportation.

many other parts of the world, rail and bus transport are widely used because they are more efficient than private automobile travel, governments support these transportation methods, or a large part of the populace is unable to afford an automobile. In countries with high population densities, rail and bus transport is particularly efficient. (See figure 9.7.) In these countries, automobiles require about four times more energy per passenger kilometer than bus or rail transport require. In addition, most of these countries have high taxes on fuel, which raise the cost to the consumer and encourage the use of public transport. In North America, the situation is different. Government policy has kept the cost of energy low and supported the automobile industry while removing support for bus and rail transport. Consequently, the automobile plays a dominant role, and public transport is primarily used only in metropolitan areas. Rail and bus transport are about twice as energy efficient as private automobiles. Private automobiles in North America consume over 15 percent of the world's oil production, while the rest of the automobiles in the world consume 7 percent. Air travel is relatively expensive in terms of energy, although it is slightly more efficient than private automobiles. Passengers, however, are paying for the convenience of rapid travel over long distances.

The Variability of Gasoline Prices

The price of gasoline at your local station is driven by the market forces of supply and demand. There are many external factors that influence the price you pay for gasoline, including the world price of crude oil, the cost of refining and delivering the gasoline to your local station, and the international gasoline market. Each of these factors can and does vary from day to day.

The market forces that help determine the price of gasoline vary from the state of the local economy to the marketing strategies of the companies that produce the gasoline. Customer-buying preferences and the number of stations in any given area are also factors. In addition to the above, the wholesale price of gasoline at any given time is also a factor. These factors can all affect the price of the gasoline you buy.

You have also no doubt witnessed what are referred to as price wars over gasoline at your local stations. Price wars can begin when a single dealer may use gasoline as the "loss leader" to attract you to the station to purchase other items such as car parts or groceries. The evolution from traditional gas stations to "mini-marts" has increased price wars and "loss leader" marketing.

Gasoline is a very competitive market and the visible pricing of the cost of a gallon or liter will attract customers. Pricing, however, in the long run will always stabilize in order to provide the station owner a reasonable profit after accounting for wholesale delivery and operating costs.

Governmental bodies at both the federal and state level closely monitor gasoline price swings. Such swings are indicative of market competition and often benefit the consumer by offering lower prices.

Electrical Energy

Electrical energy is such a large proportion of energy consumed in most countries that it deserves special comment. Electricity is both a way that energy is consumed and a way that it is supplied. Almost all electric energy is produced as a result of burning fossil fuels. Thus, we can look at electrical energy as a use to which fossil fuel energy is put. In the same way we use natural gas to heat homes, we can use natural gas to produce electricity. Because the transportation of electrical energy is so simple and the uses to which it can be put are so varied, electricity is a major energy source for many people of the world.

As with other forms of energy use, electrical consumption in different regions of the world varies widely. The amount of electricity used by all of the less-developed nations of the world, which have about 80 percent of the world's population, is only 66 percent of that used by the United States alone. The per capita use of electricity in North America is 25 times greater than average per capita use in the less-developed countries. In Nepal, the annual per capita use of electricity is 23 kilowatt hours, which is enough to light a 100-watt lightbulb for one week. The per capita consumption of electricity in North America is 270 times greater than in Nepal. There is also a

ENVIRONMENTAL CLOSE-UP

Hybrid Electric Vehicles

In many metropolitan areas, air pollution is a significant problem. Emissions from automobiles are a major contributor to this air-quality problem. An increase in the efficiency with which the chemical energy of fuel is converted to the motion of automobiles would greatly reduce air pollution. Several conceptually simple modifications to automobiles can significantly reduce the energy needed to move an automobile. Reducing the weight by using lighter-weight materials to build automobiles and streamlining the shape are techniques that have been used for many years. To further reduce energy consumption (increase energy efficiency) more innovative techniques are required.

A great deal of energy is needed to get a vehicle to begin moving from a stop (accelerate), and it takes an equal amount of energy to stop a vehicle once it is moving (decelerate). Internal combustion engines operate most efficiently when they are running at a specific speed (rpm) and work at less than peak efficiency when the vehicle is accelerating or decelerating. Thus, using an internal combustion engine to accelerate contributes significantly to air pollution. When the brakes are applied to stop the vehicle, the kinetic energy possessed by the moving vehicle is converted to heat in the braking system. So the energy which has just been used to accelerate the vehicle is lost as heat when it is brought to a stop. Furthermore, in metropolitan areas, an automobile is sitting in traffic with its engine running a significant amount of the time. Thus, the stop-and-go traffic common in city driving provides conditions that significantly reduce the efficient transfer of the chemical energy of fuel to the kinetic energy of turning wheels.

Hybrid electric vehicles (HEVs) combine the internal combustion engine of a conventional vehicle with the battery and electric motor of an electric vehicle. This combination offers the extended range and rapid refueling that consumers expect from a conventional vehicle, with a significant portion of the energy and environmental benefits of an electric vehicle. The practical benefits of HEVs include improved fuel economy and lower emissions compared to conventional vehicles. The flexibility of HEVs will allow them to be used in a wide range of applications, from personal transportation to commercial hauling.

HEVs have several advantages over conventional vehicles:

- Regenerative braking capability helps minimize energy loss and recover the energy used to slow down or stop a vehicle.
- Engines can be sized to accommodate average load, not peak load, which reduces the engine's weight.
- Fuel efficiency is greatly increased (hybrids consume significantly less fuel than vehicles powered by gasoline alone).
- Emissions are greatly decreased.
- HEVs can reduce dependency on fossil fuels because they can run on alternative fuels.

Hybrids are growing in importance to automakers and government agencies because they offer greater fuel efficiency than conventional vehicles and are more appealing to consumers than pure electric vehicles. A Toyota official stated that Toyota's single year 1999 hybrid sales exceeded the sales of all pure electric vehicles from 1970 through 1999.

The GM Precept, introduced at the 2000 North American International Auto Show, is an example of the new type of HEV. The Precept is an aerodynamic five-passenger sedan designed to achieve 80 mpg (gasoline equivalent). The Precept was developed by GM as its contribution to the Partnership for a New Generation of Vehicles (PNGV). The PNGV is a collaboration that began in 1993 between the U.S. government and the domestic auto industry. The specific aims of this partnership are lower emissions and up to three times the fuel efficiency of conventional cars without compromising safety, performance affordability, or utility. GM's Precept has nearly 130 technology innovations and 44 records of invention.

In October 2000, southern California was among the first regions in the United States to receive hybrid electric transit buses, from New Flyer of America, that feature series hybrid propulsion systems developed by GM's Allison Transmission Division. The hybrid technology, which uses a combination of electric motors, batteries, and an internal combustion engine, significantly reduces emissions, increases fuel economy by up to 50 percent, and increases acceleration by up to 50 percent. The system will cut nitrous oxide emissions by up to 50 percent, while cutting particulate, hydrocarbon, and carbon monoxide emissions by up to 90 percent, when fueled with low-sulfur fuel.

General Motors will produce a full-size pickup truck featuring a hybrid power train beginning in 2004. GM's full-size hybrid pickup trucks, versions of the Chevrolet Silverado and the GMC Sierra, will deliver nearly 15 percent better fuel economy. GM will test its hybrid pickups in demonstration fleets in several U.S. cities. The fleet of 10 trucks will feature a conventional power train and driveline with an electric motor integrated between the engine and transmission. Low-voltage lead-acid batteries will power the motor.

wide variation in electrical consumption among developed countries. For example, the per capita use in Europe is about half that in North America.

The production and distribution of electricity is a major step in the economic development of a country. In developed nations, about a quarter of the electricity is used by industry. The remainder is used primarily for residential and commercial purposes. In nations that are developing their industrial base, over half of the electricity is used by industry. For example, industries use 55 percent of the electricity used in Mexico and 70 percent of that used in South Korea.

Energy Consumption Trends

From a historical point of view, it is possible to plot changes in energy consumption. Economics, politics, public attitudes, and many other factors must be incorporated into an analysis of energy use trends. (See tables 9.3, p.179 and 9.4, p.180)

In 1999, world energy consumption was around 24 million metric tons of oil equivalent per day, an increase of 18 percent since 1985. Of this total, conventional fossil fuels—oil, natural gas, and coal—accounted for about 90 percent.

Over half of world energy is consumed by the 25 countries that are members of the Organization for Economic Cooperation and Development (OECD). These countries (Australia, New Zealand, Japan, Canada, Mexico, the United States, and the countries of Europe) are the developed nations of the world. In the last decade, per capita energy consumption in OECD countries has risen moderately while economic growth has continued. There has also been a shift toward service-based economies, with energy-intensive industries moving to non-OECD countries. In contrast, in countries that are becoming more economically advanced, energy consumption is increasing at a faster rate.

Oil remains the world's major source of energy, accounting for about 39 percent of primary energy demand. Coal accounts for 24 percent, and natural gas for 27 percent; the remainder is supplied mainly by nuclear energy and hydropower.

Since 1973, the year of the first "oil shock," demand for natural gas has increased faster than the demand for other fossil fuels. From 1973 to 1995, consumption of natural gas increased by about 80 percent, coal consumption increased by about 40 percent, and oil consumption by about 15 percent.

Since 1950, North American energy consumption has steadily increased. However, there were two periods of decline, one beginning in 1973 and the other in 1979. Both of these episodes were the result of political turmoil in the Middle East and the increasing influence of OPEC. (See Global Perspective: OPEC.) (See figure 9.8.) The same trend occurred in Western Europe, Japan, and Australia, resulting in a slowing of the world's growth in energy consumption. (See figure 9.9. p.180) Several factors contributed to these declines in energy consumption. Increased prices for oil and all other forms of energy forced businesses and individuals to become more energy conscious and to expand efforts to conserve energy. From 1970 to 1983, the amount of energy used for heat per dwelling in the United States declined by 20 percent. Comparable reductions were made in Denmark, West Germany, and Sweden.

However, since 1970, energy consumption has increased in North America and Europe. Like the stabilization of energy use in 1973 and 1979, the increase since 1980 is due to the price of oil. In 1979, oil was selling for about forty dollars a barrel. Beginning in 1980, the price began to drop, and in 1998, oil was selling for less than 20 dollars a barrel. When Iraq invaded

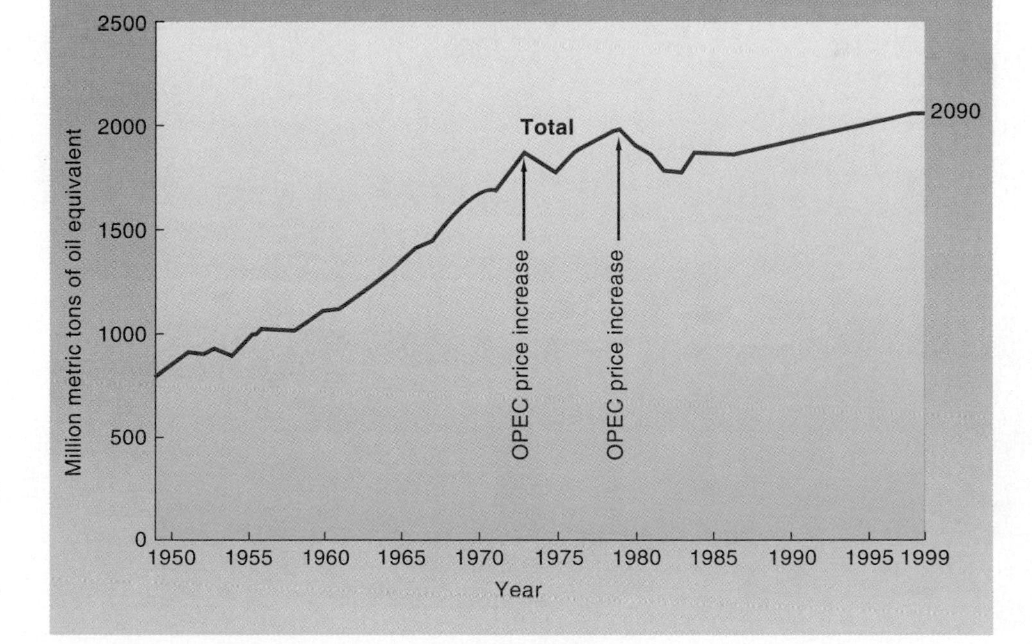

figure 9.8 **Changes in U.S. Energy Consumption** Energy consumption in the United States experienced a gradual increase until 1973 when the OPEC countries increased the price of oil. The increase in price reduced consumption. A similar action in 1979 resulted in another decrease. However, since 1983, consumption has steadily increased.

Source: Data from *BP Statistical Review of World Energy,* 1999.

GLOBAL PERSPECTIVE

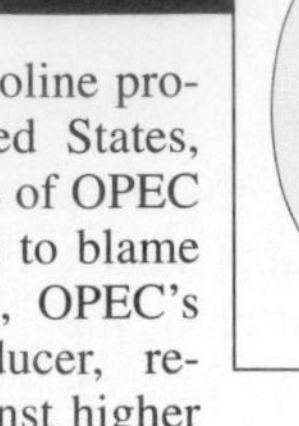

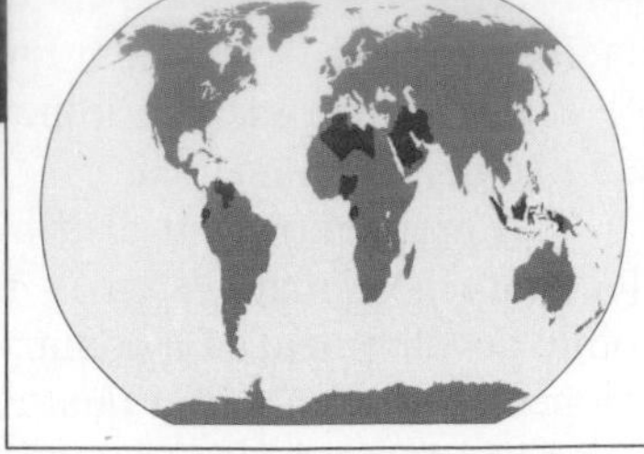

OPEC

OPEC, the Organization of the Petroleum Exporting Countries, began in September 1960 when the governments of five of the world's leading oil-exporting countries agreed to form a cartel. Three of the original members—Saudi Arabia, Iraq, and Kuwait—were Arab countries, while Venezuela and Iran were not. Today, 12 countries belong to OPEC. These include seven Arab states—Saudi Arabia, Kuwait, Libya, Algeria, Iraq, Qatar, and United Arab Emirates—and five non-Arab members—Iran, Indonesia, Nigeria, Gabon, and Venezuela. (Ecuador was a member but withdrew from OPEC in 1992. See map.) OPEC nations control over 75 percent of the world's estimated oil reserves of 1200 billion barrels of oil. Middle Eastern OPEC countries control over 60 percent of this total, which makes OPEC and the Middle East important world influences.

During the Arab-Israeli War of 1973, OPEC was supplying 55 percent of the world's oil. To protest the war, seven Arab members of OPEC reduced their production of oil. This resulted in a worldwide oil shortage, which caused all oil companies, in OPEC and non-OPEC countries, to increase their prices. In the United States, the price of oil skyrocketed from \$3.39 per barrel to \$13.93 per barrel, resulting in oil shortages, long lines at gas stations, and increased domestic exploration. The unity of OPEC, however, was not to last into the 1980s.

During the 1980s, OPEC members had important differences concerning oil pricing and production rules. These differences weakened the cartel. Then the 1990 invasion of Kuwait by Iraq deeply divided many of the OPEC countries.

Owing in part to friction caused by the Kuwait conflict and the decline in world oil prices during the middle and late 1990s, OPEC's power has continued to slide. The price of oil began to rise in 1999 and, by the middle of 2000, had reached over \$31 per barrel. Gasoline prices in the United States began to exceed \$2 per gallon. The United States exerted diplomatic pressure on OPEC to increase production levels to lower the cost of gasoline in the U.S. OPEC stated that gasoline production in the United States, rather than a shortage of OPEC crude oil output, was to blame for high prices. Iran, OPEC's second largest producer, remained strongly against higher oil production, stating that there was no shortage of crude.

By the end of the 1990s and into 2000, the weakened OPEC of the mid-1990s appears to have regained much of its earlier global power.

OPEC Oil Production, 1990–2010

(Million Barrels per Day)

Year	Ref. Case	Sensitivity Range	
History			
1990	**25.1**	**—**	**—**
1995	**29.5**	**29.3**	**30.3**
Projections			
2000	**34.8**	**33.4**	**38.9**
2005	**41.7**	**39.7**	**48.3**
2010	**46.2**	**43.5**	**55.0**

Note: Includes the production of crude oil, natural gas plant liquids, refinery gain, and other liquid fuels.

Sources: History data from Energy Information Administration (EIA); data from California Energy Commission, 1999.

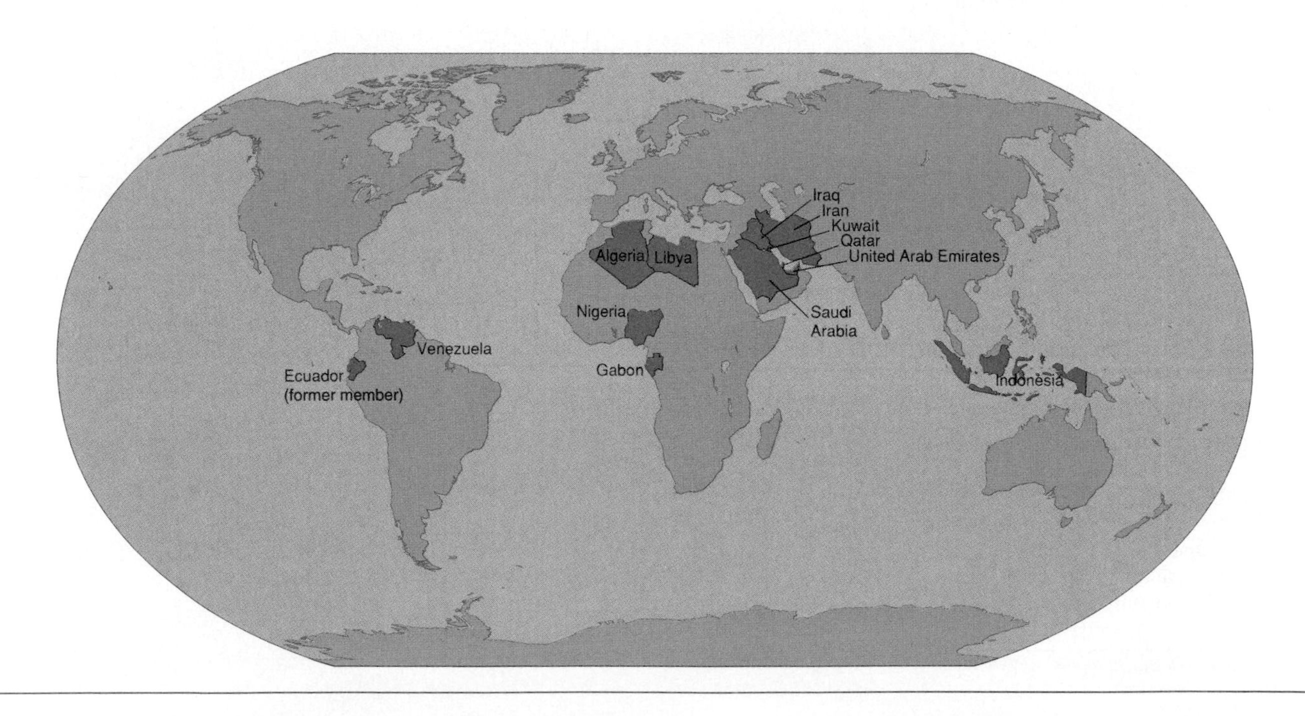

Table 9.3 World Total Energy Consumption by Region and Fuel, 1990–2010

(quadrillion Btu)

	History		Projections	
Region/Fuel	**1995**	**2000**	**2005**	**2010**
OECD	195.1	214.4	227.4	239.0
Oil	83.2	89.6	94.7	98.2
Natural gas	40.0	45.6	49.5	53.9
Coal	37.0	40.5	42.4	44.0
Nuclear	17.8	18.8	19.2	18.7
Renewables	16.8	19.8	21.5	24.0
EE/FSU	62.1	63.7	69.4	74.7
Oil	16.4	12.8	15.7	19.2
Natural gas	26.0	29.1	31.6	33.3
Coal	17.1	15.4	15.4	15.0
Nuclear	2.7	3.4	3.3	3.3
Renewables	2.6	3.1	3.4	3.8
Non-OECD Asia	59.2	79.4	92.1	104.2
Oil	19.1	29.5	32.7	34.6
Natural gas	3.8	4.7	5.9	7.2
Coal	30.1	39.1	45.7	52.8
Nuclear	0.9	1.3	1.8	1.9
Renewables	3.4	4.8	6.0	7.5
Middle East	12.4	15.2	16.9	18.4
Oil	8.1	9.4	10.4	11.4
Natural gas	4.8	5.4	6.0	6.3
Coal	0.2	0.2	0.2	0.3
Nuclear	0.0	0.0	0.0	0.0
Renewables	0.1	0.3	0.4	0.4
Africa	11.1	12.6	13.6	14.6
Oil	4.9	5.8	6.3	6.9
Natural gas	1.7	1.9	2.1	2.4
Coal	3.9	4.1	4.2	4.4
Nuclear	0.1	0.1	0.1	0.1
Renewables	0.5	0.7	0.8	0.9
Central and South America	15.1	17.2	19.1	20.8
Oil	7.6	9.4	10.2	10.9
Natural gas	2.3	2.5	3.1	3.7
Coal	0.9	1.1	1.4	1.6
Nuclear	0.1	0.2	0.3	0.3
Renewables	4.1	4.0	4.2	4.4
World total	355.0	402.6	438.6	471.7
Oil	139.3	156.5	170.0	181.3
Natural gas	78.6	89.2	98.1	106.8
Coal	89.1	100.5	109.4	118.0
Nuclear	21.6	23.7	24.6	24.4
Renewables	27.5	28.1	36.4	41.1

Notes: OECD = Organization for Economic Cooperation and Development. EE/FSU = Eastern Europe/Former Soviet Union.

Sources: History data from Energy Information Administration (EIA), *International Energy Annual,* DOE/EIA-0219(92), Washington, D.C.; projections from EIA, World Energy Projections 1998.

Table 9.4 International Petroleum Supply and Demand

(Million barrels per day)

	Year				Year		
	1999	2000	2001		1999	2000	2001
Demand				**Supply**			
OECD				OECD			
U.S. (50 States)	**19.5**	19.6	*20.0*	U.S. (50 States)	**9.0**	9.0	*9.0*
U.S. Territories	**0.3**	0.4	*0.4*	Canada	**2.6**	2.7	*2.8*
Canada	**1.9**	1.9	*1.9*	North Sea	**6.3**	6.6	*6.7*
Europe	**14.5**	14.7	*14.9*	Other OECD	**1.5**	**1.7**	*1.7*
Japan	**5.6**	5.5	*5.6*				
Australia and New Zealand	**1.0**	1.0	*1.0*				
Total OECD	**42.8**	43.2	*43.8*	Total OECD	**19.5**	20.1	*20.1*
Non-OECD				Non-OECD			
Former Soviet Union	**3.6**	3.7	*3.7*	OPEC	**29.3**	30.4	*31.9*
Europe	**1.6**	1.6	*1.7*	Former Soviet Union	**7.4**	7.6	*7.7*
China	**4.3**	4.5	*4.8*	China	**3.2**	3.3	*3.3*
Other Asia	**8.8**	9.2	*9.7*	Mexico	**3.4**	3.6	*3.6*
Other Non-OECD	**13.5**	13.9	*14.3*	Other Non-OECD	**11.2**	11.3	*11.5*
Total Non-OECD	**31.9**	32.9	*34.2*	Total Non-OECD	**54.5**	56.2	*58.0*
Total World Demand	**74.7**	76.1	*78.0*	Total World Supply	**74.0**	76.3	*78.1*

OECD: Organization for Economic Cooperation and Development: Australia, Austria, Belgium, Canada, Denmark, Finland, France, Germany, Greece, Iceland, Ireland, Italy, Japan, Luxembourg, the Netherlands, New Zealand, Norway, Portugal, Spain, Sweden, Switzerland, Turkey, the United Kingdom, and the United States. The Czech Republic, Hungary, Mexico, Poland, and South Korea are all members of OECD, but are not yet included in our OECD estimates.

OPEC: Organization of Petroleum Exporting Countries: Algeria, Indonesia, Iran, Iraq, Kuwait, Libya, Nigeria, Qatar, Saudi Arabia, the United Arab Emirates, and Venezuela.

Former Soviet Union: Armenia, Azerbaijan, Belarus, Estonia, Georgia, Kazakhstan, Kyrgyzstan, Latvia, Lithuania, Moldova, Russia, Tajikistan, Turkmenistan, Ukraine, and Uzbekistan.

Sources: Energy Information Administration: latest data available from EIA databases supporting the following reports: *International Petroleum Statistics Report,* DOE/EIA-0520; Organization for Economic Cooperation and Development, Annual and Monthly Oil.

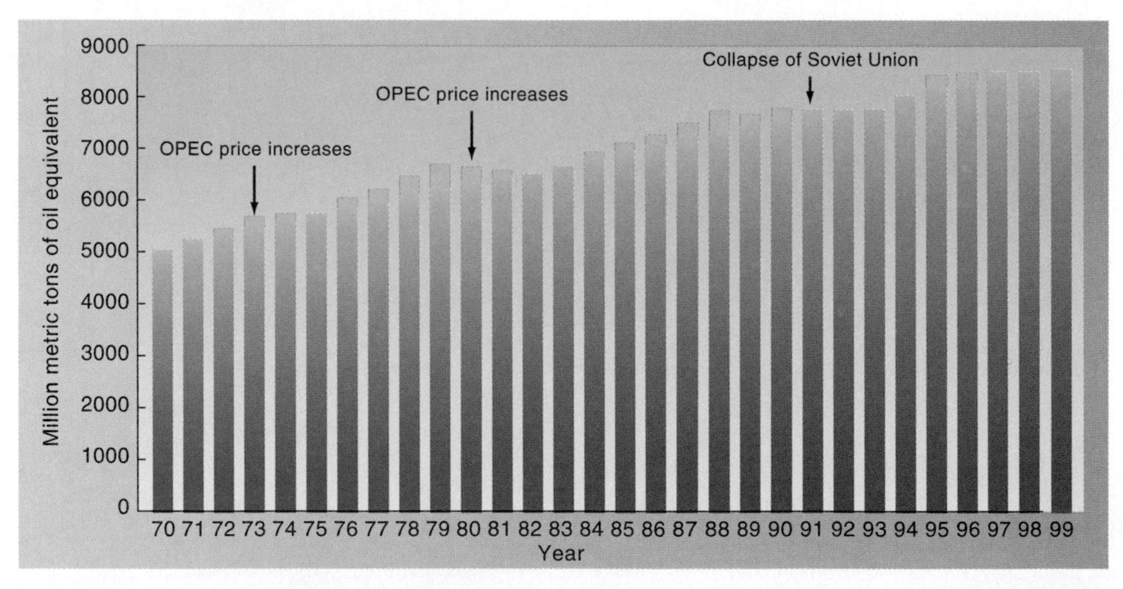

figure 9.9 **World Energy Consumption** Energy consumption has increased steadily for the past 25 years. The slowdown in growth beginning in 1973 and the decline in the early 1980s is the result of price increases instituted by OPEC. The slight decline noted in 1991 and 1992 is the result of decreased energy consumption in the former Soviet Union and Eastern Europe, and is related to economic problems in those countries.

Source: Data from *BP Statistical Review of World Energy,* June 1999.

ENVIRONMENTAL CLOSE-UP

Alternative-Fuel Vehicles

Electricity is the only zero-emission option now available for vehicles, but many experts believe that other fuels will prove viable in *reduced*-emission vehicles, although emissions are still produced in the generation of electricity at the power plant. Here are some of the options.

Compressed Natural Gas (CNG)

Status:
Fuels thousands of vehicles already on U.S. highways. Mass production is still a couple of years away, at the earliest.

Advantages:

- Substantial U.S. supply of natural gas.
- Easy—though expensive—to convert conventional gasoline cars to CNG.
- Cheaper: the equivalent of buying gasoline at 70 cents a gallon.

Disadvantages:

- Requires heavy, bulky fuel tanks.
- Public refueling stations expensive to build.
- Lower performance and shorter range than gasoline vehicles.

Emissions Benefits:

- Burns 80 percent cleaner than gasoline.

Methanol/Flex-Fuel

Status:
Some heavy vehicles—including 354 buses operated by Southern California's Metropolitan Transportation Authority—run on pure methanol. More than 10,000 flex-fuel vehicles—running on methanol-gasoline combinations—now on U.S. roads.

Advantages:

- Can be made from natural gas (98 percent of current production) and a range of renewable sources.
- Requires only modest changes in gasoline engines and fueling infrastructure.
- Higher octane than gasoline, giving 5 percent better performance.

Disadvantages:

- Highly corrosive.
- Lower energy content than gasoline, so requires bigger fuel tanks.
- Fuel cost slightly higher than gasoline.

Emissions Benefits:

- An 85 percent methanol mix burns 30 to 50 percent cleaner than pure gasoline; pure methanol vehicles potentially could be much cleaner.

Hydrogen

Status:
Still largely experimental. Vehicles could be powered by hydrogen fuel cells or by engines that burn the gas. Practical, wide-scale use is at least a decade away.

Advantages:

- A high-energy fuel, boosting vehicle range.
- Potentially unlimited fuel supply if produced from water; currently made from natural gas.
- Produces no carbon dioxide, the gas blamed for possible global warming.

Disadvantages:

- Highly explosive, though ultimately could be less dangerous than gasoline.
- Costly and difficult to produce.
- Costly and difficult to store.

Emissions Benefits:

- If produced from water using solar energy—and then used in a fuel cell—would be a zero-emissions technology.

Propane

Status:
Broadly used in transportation worldwide.

Advantages:

- A proven, low-cost fuel.
- Public refueling infrastructure in place.
- Nontoxic, so storage tanks are exempt from environmental regulations.

Disadvantages:

- Supplies are limited, compared with other alternatives. At most, could replace 10 percent of gasoline.
- Highly flammable.

Emissions Benefits:

- At least 50 percent cleaner than conventional gasoline.

Sources: Data from California Air Resources Board, California Energy Commission, South Coast Air Quality Management District, Ward's Communications, Ford Motor Co., General Motors Corp., Chrysler Corp., and LP Gas Coalition.

GLOBAL PERSPECTIVE

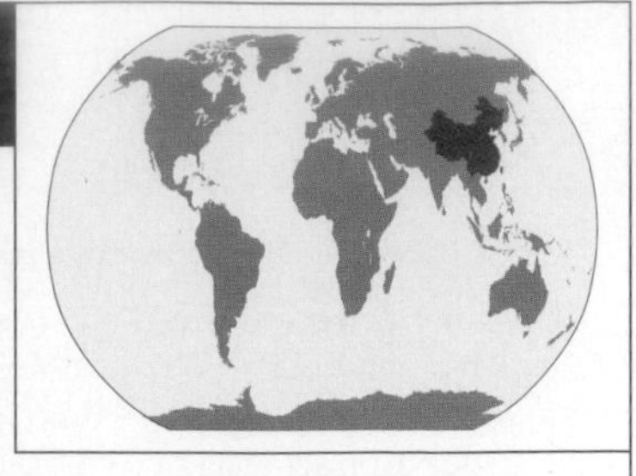

Energy Development in China

China is currently the second largest energy consumer in the world, following the United States (about 39 quadrillion Btu in 1999 versus 90 quadrillion Btu in the United States). China's rapidly growing economy will drive energy demand growth of about four to five percent annually through 2015 (compared with growth of about 1 percent in the industrialized countries). China currently consumes about 10 percent of the world's energy, and also accounts for about 10 percent of the world energy production.

China became a net importer of energy in 1995, and is expected to become increasingly dependent on imports; however, it is expected to remain a net exporter of coal. China has been a net importer of oil since 1993. Most production is from onshore. Offshore and some onshore areas are now open to foreign investment. At current rates, China's oil imports could exceed 1 million barrels daily by 2002.

In 2000, coal accounted for 70 percent of China's primary energy production. In the same year, petroleum accounted for 19 percent, hydroelectricity 11 percent, natural gas 2 percent, and nuclear power 0.5 percent. China's electricity is generated overwhelmingly by coal (about 70 percent) followed by hydroelectricity at about 20 percent. Because of China's extensive domestic coal resources and its wish to minimize dependence upon foreign energy sources, it is expected that coal will remain the main energy source for electricity generation in China for the foreseeable future. Since 1979, the government has pursued a policy of replacing oil consumption with coal in the generation of electric power. The proportion of coal-fired electric-power generation has grown from 60 percent in 1980 to 70 percent in 1999.

High coal use has serious environmental repercussions. China plans to diversify by doubling its 1990 level of hydropower production by 2002 and increasing nuclear capacity. Three Gorge's hydro project, when completed, will contribute significantly to the country's hydropower production.

One of the biggest concerns regarding energy consumption in China is that of carbon emissions and the threat of increasing global warming. China's carbon emissions are expected to increase 4 percent annually through 2015, driven by rapid economic growth and a rapid increase in coal use. The country's total carbon emissions should exceed 2 billion tonnes (2.2 U.S. tons) shortly after 2015, over three times the 1986 emissions and more than double the 1995 emissions.

The growing demand for power has driven reforms in China's energy-sector process and structure. The traditional, centralized regulatory regime is moving toward an increasingly decentralized and market-oriented regime.

Until recently, the Ministry of Electric Power (MOEP) regulated the electric power industry of China. The Ninth People's Congress held in 1998 introduced new reform measures to streamline the central government. This resulted in the abolition of a number of industrial ministries, including MOEP. Much of the former MOEP's function has now been taken over by the State Power Corporation of China (SPCC). SPCC's main duties in this respect are:

- Formulating China's electric-power development strategies, legislation, and policies, including investment policy, technical policy, and major energy production and consumption policies.
- Formulating unified energy-industry planning in collaboration with the State Development Planning Commission and other governmental agencies.
- Supervising the implementation of related national policies decrees and plans.
- Providing services to regional and provincial electric power enterprises.

Furthermore, SPCC owns all of China's state-owned power-generating assets, other than those owned by companies not directly managed by SPCC, and a few smaller units directly owned by local governments.

Source: Data from EIA, U.S. Department of Energy, International Energy Outlook, 2000.

a. China's energy production and consumption

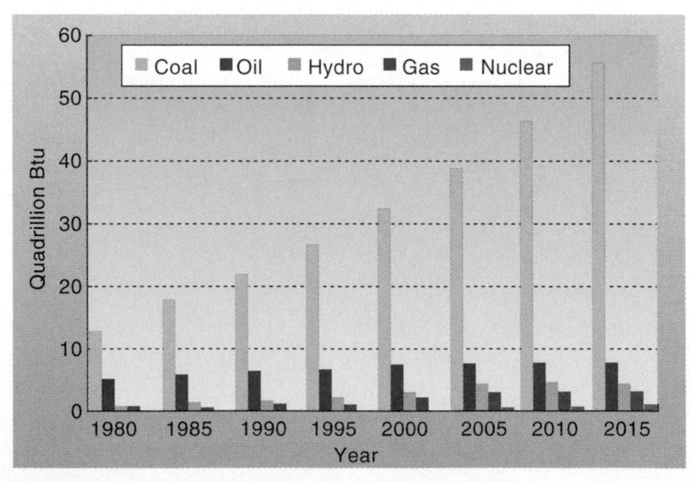

b. Energy production by fuel type, 1980–2015

Source: History data from Energy Information Administration (EIA), International Petroleum Statistics Report, DOE/EIA–520 (92/08), August 1992, Washington, D.C. Projections from EIA, National Energy Modeling System, International Energy Module, 2000.

GLOBAL PERSPECTIVE

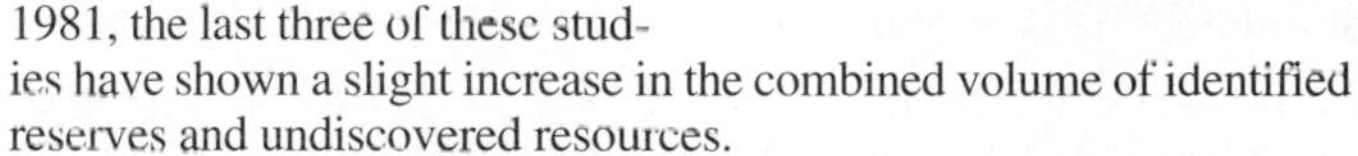

Potential World Petroleum Resources

Oil is a finite, nonrenewable resource. It will eventually be exhausted. The primary questions about when we will run out of oil revolve around price, rates of use, and rates at which new discoveries are made. In 2000, the U.S. Geological Survey issued an assessment of potential world petroleum resources. The report stated that undiscovered oil and gas resources of the world had increased, with a 20 percent increase in undiscovered oil and a slight decrease in undiscovered natural gas. This assessment estimates the volume of oil and gas, exclusive of the U.S., that may be added to the world's reserves in the next 30 years.

The assessment indicates that there is more oil and gas in the Middle East and in the offshore areas of western Africa and eastern South America than previously reported, less oil and gas in Canada and Mexico, and significantly lower volumes of natural gas in the former Soviet Union.

With the evolution of technology and new understandings of petroleum systems, the USGS World Petroleum Assessment 2000 is the first of its kind to provide a rigorous geologic foundation for estimating undiscovered energy resources for the world. The results have important implications for energy prices, policy, security, and the global resource balance.

The assessment provides a snapshot of current information about the location and abundance of undiscovered oil and gas resources. Such an overview provides exploration geologists, economists, and investors a general picture of where oil and gas resources are likely to be developed in the future. The USGS periodically estimates the amount of oil and gas remaining to be found. Since 1981, the last three of these studies have shown a slight increase in the combined volume of identified reserves and undiscovered resources.

In the USGS 2000 report, the world was divided into approximately one thousand petroleum provinces, based primarily on geologic factors, and grouped into eight regions. Estimates of reserve growth were based on the following:

- As drilling and production within discovered fields progresses, new pools or reservoirs are found that were not previously known.
- Advances in exploration technology make it possible to identify new targets within existing fields.
- Advances in drilling technology make it possible to recover oil and gas not previously considered recoverable in the initial reserve estimates.

To view the entire report, go to the USGS home page at www.usgs.gov.

Estimation of undiscovered oil and natural gas

Region	Undiscovered Oil (billion barrels)	Percent of world total	Undiscovered natural gas (trillion cubic feet)	Percent of world total
1. Former Soviet Union	116	17.9%	1611	34.5%
2. Middle East and North Africa	230	35.4%	1370	29.3%
3. Asia-Pacific	30	4.6%	379	8.1%
4. Europe	22	3.4%	312	6.7%
5. North America*	70	10.9%	154	3.3%
6. Central and South America	105	16.2%	487	10.4%
7. Sub-Saharan Africa and Antarctica	72	11.0%	235	5.0%
8. South Asia	4	0.6%	120	2.6%
WORLD TOTALS*	**649**		**4669**	

*Exclusive of the United States

Source: USGS World Petroleum Assessment 2000.

Kuwait in August 1990, prices rose dramatically, but dropped back at the end of hostilities.

During the 1980s, with energy costs declining, people in North America and Europe became less concerned about their energy consumption. They used more energy to heat and cool their homes and buildings, used more home appliances, and bought bigger cars. Obviously, the two primary factors that determine energy use are political stability in parts of the world that supply oil, and the price of that oil. Insecurity leads to a price rise, but governments can also manipulate prices by changing taxes, granting subsidies, and using other means. The energy consumption behavior of most people is motivated by economics rather than by a desire to wisely use energy resources.

Summary

A constant supply of energy is required by all living things. Energy has a major influence on society. A direct correlation exists between the amount of energy used and the complexity of civilizations.

Wood furnished most of the energy and construction materials for early civilizations. Heavy use of wood in densely populated areas eventually resulted in shortages, so fossil fuels replaced wood as a prime source of energy. Fossil fuels were formed from the remains of plants, animals, and microorganisms that lived about 300 million years ago. Fossil-fuel consumption in conjunction with the invention of labor-saving machines resulted in the Industrial Revolution, which led to the development of technology-oriented societies today in the developed world.

Throughout the world, residential and commercial uses, industries, transportation, and electrical utilities require energy. Because of financial, political, and other factors, nations vary in the amount of energy they use as well as in how they use it. Analysts expect the worldwide demand for energy to increase steadily.

Interactive Exploration

Check out the website at

http://www.mhhe.com/environmentalscience

and click on the cover of this textbook for interactive versions of the following:

KNOW THE BASICS

fossil fuels *170*

Industrial Revolution *170*

- On-line Flashcards
- Electronic Glossary

IN THE REAL WORLD

With oil as the world's major energy source, it is nearly impossible to prevent oil spills when oil is transported by oil tankers or used as fuel in ships. Read about the status of—and what we have learned from—a famous ten-year-old spill and more recent spills in these stories: The Exxon Valdez Oil Spill, Ten Years Later; Major Oil Spill; New Carissa Oil Spill on the Oregon Coast.

TEST PREPARATION

Review Questions

1. Why was the sun able to provide all energy requirements for human needs before the Industrial Revolution?
2. In addition to food, what energy requirements does a civilization have?
3. Why were some countries unable to use the technologies developed during the Industrial Revolution?
4. What factors caused a shift from wood to coal as a source of energy?
5. How were energy needs in World War II responsible for the subsequent increased consumption of natural gas?
6. What part does government regulation play in changing the consumption of natural gas and oil?
7. Why was much of the natural gas that was first produced wasted?
8. What was the initial use of oil? What single factor was responsible for a rapid increase in oil consumption?
9. List the three purposes for which a civilization uses energy.
10. Why is OPEC important in the world's economy?

Critical Thinking Questions

1. Imagine you are an historian writing about the Industrial Revolution. Imagine that you also have your new knowledge of environmental science and its perspective. What kind of a story would you tell about the development of industry in Europe and the United States? Would it be a story of triumph or tragedy, or some other story? Why?
2. What might be some of the effects of raising gasoline taxes in the United States to the rate that most Europeans pay for gasoline? Why? What do you think about this possibility?
3. Some argue that the price of gasoline in the United States is artificially low because it does not take into account all of the costs of producing and using gasoline. If you were to figure out the "true" cost of gasoline, what kinds of factors would you want to take into account?
4. How has the ubiquitous nature of automobiles changed the United States? Do you feel these changes are, in balance, positive or negative? What should the future look like regarding automobile use in the United States? How can this be accomplished?
5. Some energy experts predict that the Organization of Petroleum Exporting Countries (OPEC) will control about 65 percent of the known oil reserves by the year 2010. What political and economic effects do you think this will this have? Will this have any effect on energy use?
6. How do you think projected energy consumption will affect world politics and economics, given current concerns about global warming?

Energy Sources

Objectives

After reading this chapter, you should be able to:

- Differentiate between resources and reserves.
- Identify peat, lignite, bituminous coal, and anthracite coal as steps in the process of coal formation.
- Recognize that natural gas and oil are formed from ancient marine deposits.
- Explain how various methods of coal mining can have negative environmental impacts.
- Explain why surface mining of coal is used in some areas and underground mining in other areas.
- Explain why it is more expensive to find and produce oil today than it was in the past.
- Recognize that secondary recovery methods have been developed to increase the proportion of oil and natural gas obtained from deposits.
- Recognize that transport of natural gas is still a problem in some areas of the world.
- Explain why the amount of energy supplied by hydroelectric power is limited.
- Describe how wind, geothermal, and tidal energy are used to produce electricity.
- Recognize that wind, geothermal, and tidal energy can be developed only on areas with the proper geologic or geographical features.
- Describe the use of solar energy in passive heating systems, active heating systems, and the generation of electricity.
- Recognize that fuelwood is a major source of energy in many parts of the less-developed world and that fuelwood shortages are common.
- Describe the potential and limitations of biomass conversion and waste incineration as a source of energy.
- Recognize that energy conservation can significantly reduce our need for additional energy sources.

Chapter Outline

Energy Sources

Resources and Reserves

Fossil-Fuel Formation
- Coal Formation
- Oil and Natural Gas Formation

Issues Related to the Use of Fossil Fuels
- Coal Use Issues
- Oil Use Issues
- Natural Gas Use Issues

Renewable Sources of Energy
- Hydroelectric Power

Global Perspective: *Hydroelectric Sites*
- Tidal Power

Global Perspective: *The Three Gorges Dam*
- Geothermal Power
- Wind Power
- Solar Energy
- Biomass Conversion

Global Perspective: *Electricity from the Ground Up*
- Fuelwood
- Solid Waste

Energy Conservation

Issues & Analysis: *The Arctic National Wildlife Refuge and Oil*

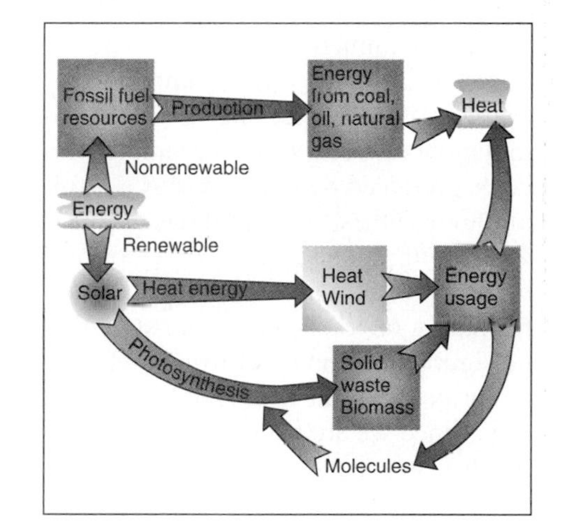

Energy Sources

Chapter 9 outlined the historical development of energy consumption and how advances in civilizations were closely linked to the availability and exploitation of energy. New manufacturing processes relied on dependable sources of energy. Technology accelerated in the twentieth century. Between 1900 and 2000, world energy consumption increased by a factor of fourteen, the quantity of products manufactured increased by nearly fortyfold, but population increased only threefold. (See table 10.1.)

The energy sources most commonly used by industrialized nations are the fossil fuels: oil, coal, and natural gas, which supply about 90 percent of the world's commercially traded energy. Fossil fuels were formed hundreds of millions of years ago. They are the accumulation of energy-rich organic molecules produced by organisms as a result of photosynthesis over millions of years. We can think of fossil fuels as concentrated, stored solar energy. The rate of formation of fossil fuels is so slow that no significant amount of fossil fuels will be formed over the course of human history. Since we are using these resources much faster than they can be produced and the amount of these materials is finite, they are known as **nonrenewable energy sources.** (See figure 10.1.) Eventually, human demands will exhaust the supplies of coal, oil, and natural gas.

Table 10.1 World Population, Economic Output, and Fossil-Fuel Consumption

	Population (Billions)	Gross World Product (Trillion 1995 Dollars)	Fossil-fuel Consumption (Billion Tons Coal Equivalent)
1900	1.6	0.6	1
1950	2.5	2.9	3
1999	6.0	34.0	15

Source: Data from *CIA World Fact Book.*

In addition to nonrenewable fossil fuels, there are several renewable energy sources. **Renewable energy sources** replenish themselves or are continuously present as a feature of the solar system. For example, in plants, photosynthesis converts light energy into chemical energy.

Carbon dioxide + water + Sunlight Energy → Biomass (Chemical Energy) + Oxygen

This energy is stored in the organic molecules of the plant as wood, starch, oils, or other compounds. Any form of biomass—plant, animal, alga, or fungus—can be traced back to the energy of the sun. Since biomass is constantly being produced, it is a form of renewable energy. Solar, geothermal, and tidal energy are renewable energy sources because they are continuously available. Anyone who has ever lain in the sun, seen a geyser or hot springs, or been swimming in the surf has experienced these forms of energy. However, many technical problems must be solved before these renewable energy forms can contribute significantly to meeting humans' energy demands.

Resources and Reserves

When discussing deposits of nonrenewable resources, such as fossil fuels, we must differentiate between deposits that can be extracted and those that cannot. From a technical point of view, a **resource** is a naturally occurring substance of use to humans that can *potentially* be extracted using current technology. **Reserves** are known deposits from which materials *can* be extracted profitably with existing technology *under certain economic conditions.* It is important to

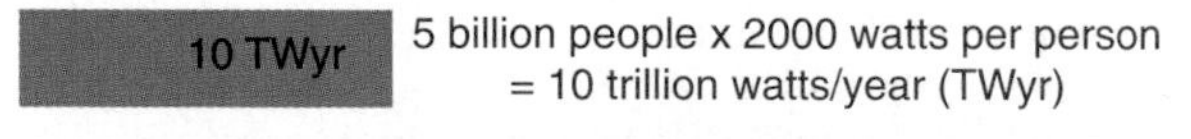

figure 10.1 **Predicting Energy Needs**

Source: Princeton Plasma Physics Laboratory, U.S. Department of Energy, Washington, D.C., 1997.

recognize that *reserves* is an economic concept and is only loosely tied to the total quantity of a material present in the world. Therefore, reserves are smaller than resources. (See figure 10.2.) Both terms are used when discussing the amount of mineral or fossil-fuel deposits a country has at its disposal. This can cause considerable confusion if the difference between these concepts is not understood. The total amount of a resource such as coal or oil changes only by the amount used each year. The amount of a reserve changes as technology advances, new deposits are discovered, and economic conditions vary. There can be large changes in the amount of reserves, while the resource remains almost constant.

When we read about the availability of fossil fuels, we must remember that energy is needed to extract the wanted material from the earth. If the material is dispersed or hard to reach, it is expensive to extract. If the cost of removing and processing a fuel is greater than the fuel's market value, no one is going to produce it. Also, if the amount of energy used to produce, refine, and transport a fuel is greater than its potential energy, the fuel will not be produced. A net useful energy yield is necessary to exploit the resource. However, in the future, new technology or changing prices may permit the profitable removal of some fossil fuels that currently are not profitable. If so, those resources will be reclassified as reserves.

To further illustrate the concept of reserves and how technology and economics influence their magnitude, let us look at the history of oil. The ancient Chinese are said to have been the first to use oil as a fuel. The only oil available to them was the small amount that naturally seeped out of the ground. These seepages represented the known oil reserves as well as the known oil resources at that time.

Nearly two thousand years passed before oil reserves increased significantly. When the first oil well in North America was drilled in Pennsylvania in 1859, it greatly expanded the estimate of the amount of oil in the earth. There was a sudden increase in the known oil reserves. In the years that followed, new deposits were discovered. Better drilling techniques led to the discovery of deeper oil deposits, and offshore drilling established the location of oil under the ocean floor. At the time of their initial discovery, these deep deposits and the offshore deposits added to the estimated size of the world's oil resources. But they did not necessarily add to the reserves because it was not always profitable to extract the oil. With advances in drilling and pumping methods and increases in oil prices, it eventually became profitable to obtain oil from many of these deposits. As it became economical to extract them, they were reclassified as reserves.

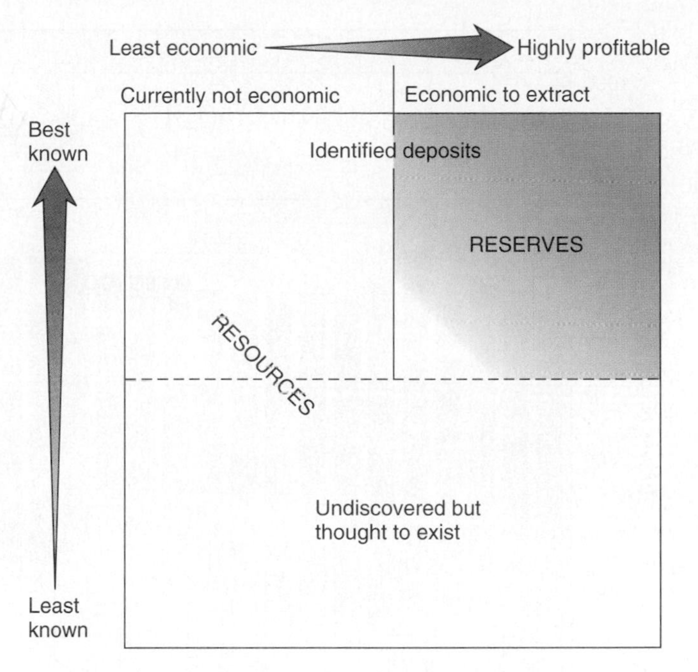

figure 10.2 **Resources and Reserves** Each term describes the amount of a natural resource present. Reserves are those known deposits that can be profitably obtained using current technology under current economic conditions. Reserves are shown in the box in the upper right-hand corner in this diagram. The darker the color, the more valuable the reserve. Resources are much larger quantities that include undiscovered deposits and deposits that currently cannot be profitably used, although it might be feasible to do so if technology or market conditions change.

Source: Adapted from the U.S. Bureau of Mines.

The amount of fossil fuel reserves is in a constant state of flux. For example, prior to 1973, many oil wells were capped because it was not profitable to remove the oil. The oil still in the ground at these wells was not economic to produce and was not included in the reserves category. During the oil embargoes of 1973–1974 and 1979–1980, when some of the OPEC countries reduced oil production, the price of oil increased. With the increase in oil prices, these wells became profitable, and the oil was reclassified as reserves. Then, in the 1980s, an oil surplus developed because the high oil prices resulting from the embargo caused increased production worldwide. Prices fell. Many of these same wells were capped again, and the oil was no longer classified as part of the reserve. With the exception of 1990 and 1991 when Iraq invaded Kuwait, the price of oil in the 1990s had been relatively stable. (See figure 10.3.)

By the end of the decade, however, oil prices were once again heading higher. Beginning in 1999, the price of oil began to climb, and, by the first half of 2001, oil had risen to over $40 per barrel up from $12 a barrel only two years earlier. The United States met frequently with OPEC leaders in the hope of having OPEC increase its production of oil and thus reduce the price at the

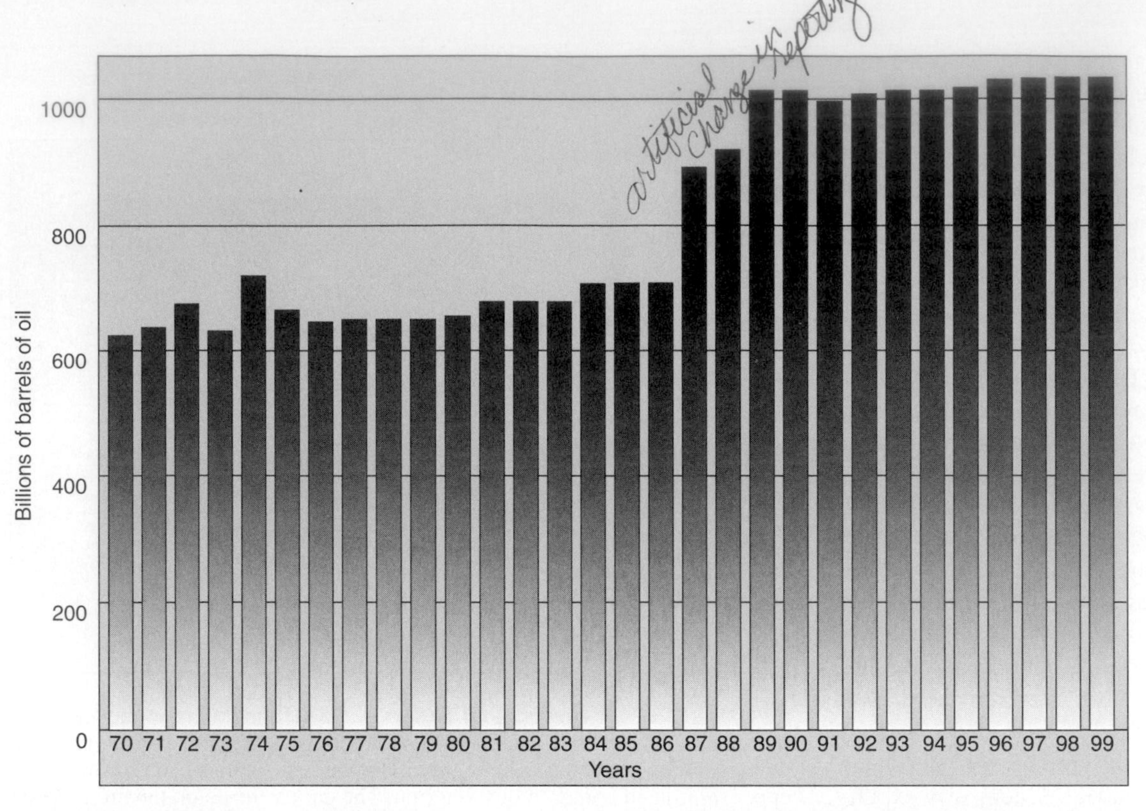

figure 10.3 **Changes in Oil Reserves** The figure shows the changes in oil reserves over a 28-year period. The changes that occurred in 1987 and 1989 are the result of changes in reporting, not the result of new discoveries. Since 1989, new discoveries and revisions have matched consumption. Thus, reserves have remained nearly constant.

Source: Data from *BP Statistical Review of World Energy,* June 1999.

pump in the United States. By the summer of 2000, the price of a gallon of gasoline exceeded $2 in many parts of the United States—double what it was a year earlier.

Many reasons were given for this sharp increase. Gasoline and petroleum stocks were low. With little crude oil or gasoline in inventory, refiners began to purchase crude oil in a market short on supplies, and that translated to higher prices. At the same time, refineries, squeezed by the high crude prices, were not operating at maximum capacity. By the spring of 2000, refineries on average were running at about 85 percent capacity, compared with nearly 92 percent at the same time a year earlier. In addition to the supply issues, the oil companies claimed that new environmental rules on the manufacturing of gasoline contributed to the increased cost of gasoline at the pump. Perhaps the only certainty of gasoline pricing is its continuing uncertainty.

Fossil-Fuel Formation

Fossil fuels are the remains of once living organisms that were preserved and altered as a result of geologic forces. There are significant differences in the formation of coal from that of oil and natural gas.

Coal Formation

Tropical freshwater swamps covered many regions of the earth 300 million years ago. Conditions in these swamps favored extremely rapid plant growth, resulting in large accumulations of plant material. Because this plant material collected under water, decay was inhibited, and a spongy mass of organic material formed, called peat. Peat moss (peat) deposits are 90 percent water, 5 percent carbon, and 5 percent volatile materials. In some parts of the world, peat is cut, dried, and used as fuel. However, because of its high water content, it is regarded as a low-grade fuel.

Due to geological changes in the earth, some of the swamps containing peat were submerged by seas. The plant material that had collected in the swamps was then covered by sediment. The weight of the plant material plus the weight of the sediment on top of it compressed it into a harder form of low-grade coal known as lignite, which contains less water and a higher proportion of burnable materials.

If the weight of the sediment was great enough, the heat from the earth high enough, and the length of time long enough, the lignite would have changed into bituminous (soft) coal. The major change from lignite to bituminous coal is a reduction in the water content from over 40 percent to about 3 percent. If the heat and pressure continued over time, some of the bituminous coal could have changed to anthracite (hard) coal, which is about 96 percent carbon. Through this

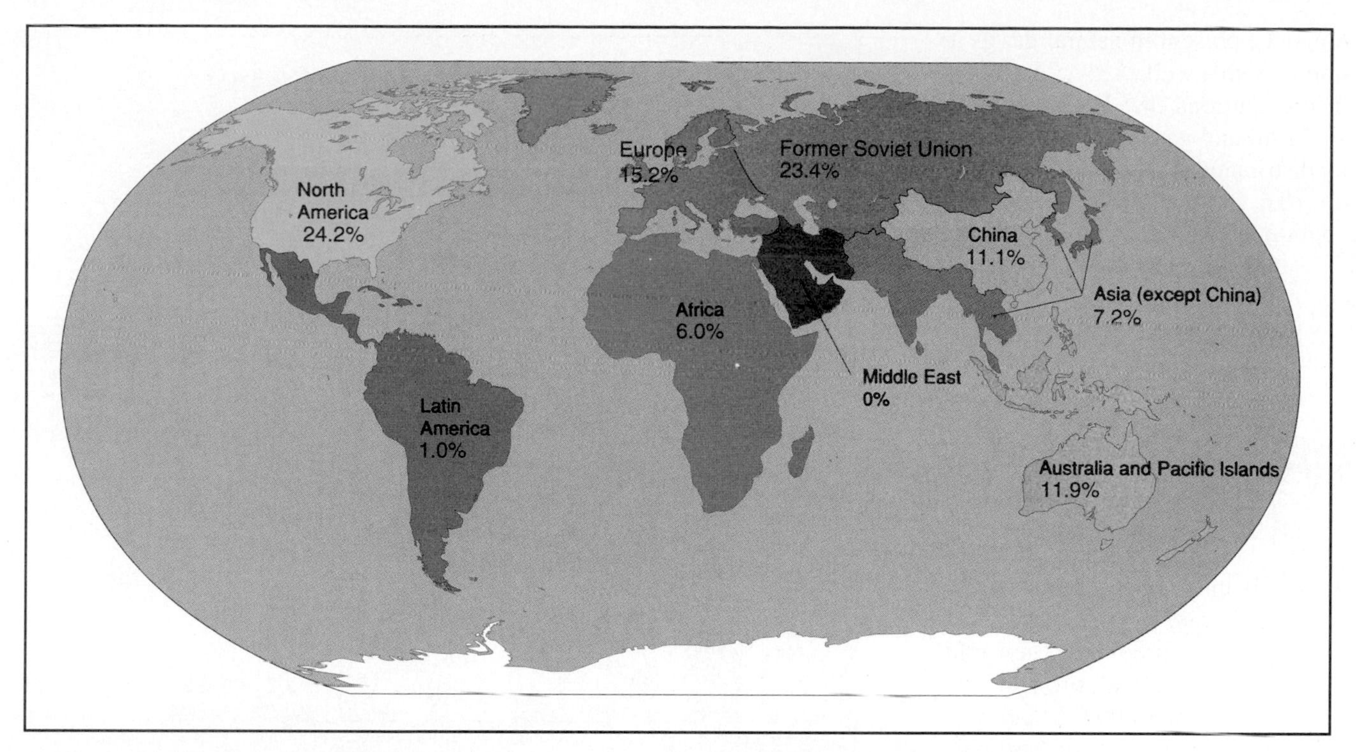

figure 10.4 **Recoverable Coal Reserves of the World 1999** The percentage indicates the coal reserves in different parts of the world. This coal can be recovered under present local economic conditions using available technology.

Source: Data from *BP Statistical Review of the World Energy,* June 2000.

combination of events, which occurred over hundreds of millions of years, present-day coal deposits were created. Most parts of the world have coal deposits. (See figure 10.4.)

Oil and Natural Gas Formation

Oil and natural gas, like coal, are products from the past. They probably originated from microscopic marine organisms. When these organisms died and accumulated on the ocean bottom and were buried by sediments, their breakdown released oil droplets. Gradually, the muddy sediment formed rock called shale, which contained dispersed oil droplets. Although shale is common and contains a great deal of oil, extraction from shale is difficult because the oil is not concentrated. If a layer of sandstone formed on the top of the oil-containing rock, and an impermeable layer of rock formed on top of the sandstone, conditions might be suitable for oil pools to form. Usually, the trapped oil does not exist as a liquid mass but rather as a concentration of oil within sandstone pores, where it accumulates because water and gas pressure force it out of the shale. (See figure 10.5.) These accumulations of oil are more likely to occur if the rock layers were folded by geological forces.

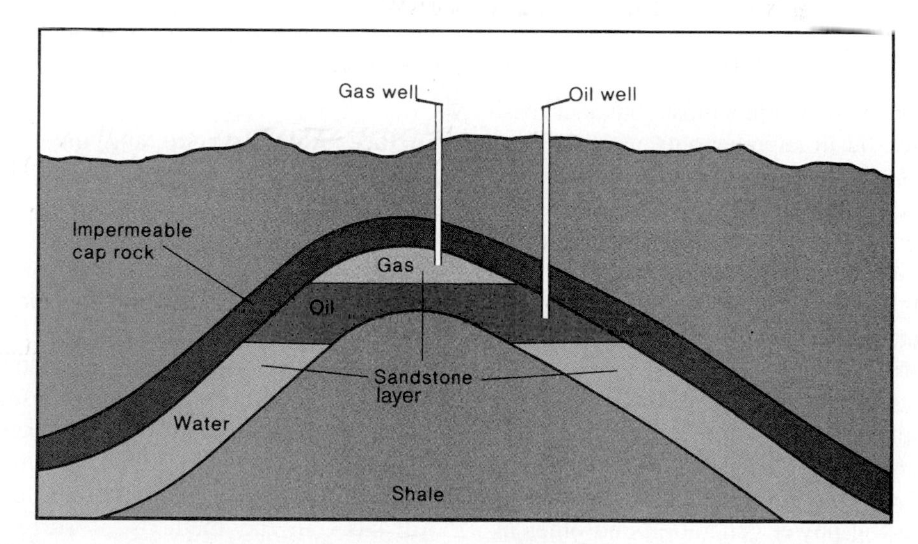

figure 10.5 **Crude Oil and Natural Gas Pool** Water and gas pressure force oil and gas out of the shale and into sandstone beneath the impermeable rock.

Natural gas, like coal and oil, forms from fossil remains. In fact, the geological conditions favorable for oil formation are the same as those for natural gas, and the two fuels are often found together. However, in the formation of natural gas, the organic material changed to lighter, more volatile (easily evaporated) hydrocarbons than those found in oil. The most common hydrocarbon in natural gas is the gas methane (CH_4). Water, liquid hydrocarbons, and other gases may be present in natural gas as it is pumped from a well.

The conditions that led to the formation of oil and gas deposits were not evenly distributed throughout the world. Figure 10.6 illustrates the geographic distribution of oil reserves and figure 10.7 illustrates the geographic distribution of natural gas reserves. Some of these deposits are easy to extract, while others are not.

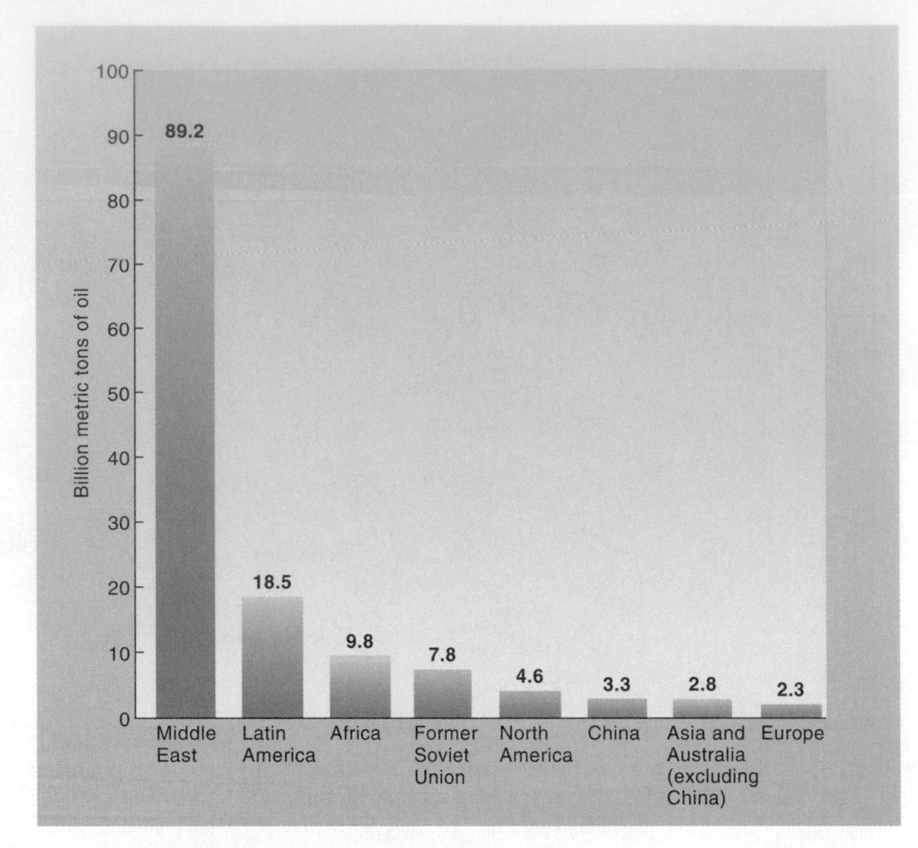

figure 10.6 **World's Oil Reserves 1999** The world's supply of oil is not distributed equally. Certain areas of the world enjoy an economic advantage because they control vast amounts of oil. Reserves are given in billions of metric tons of oil.

Source: Data from *BP Statistical Review of World Energy,* June 2000.

Issues Related to the Use of Fossil Fuels

As previously mentioned, of the world's commercial energy, over 90 percent is furnished by the three nonrenewable fossil-fuel resources: coal, oil, and natural gas. Coal supplies about 27 percent, oil supplies about 40 percent, and natural gas supplies about 23 percent. Each fuel has advantages and disadvantages and requires special techniques for its production and use.

Coal Use Issues

Coal is the world's most abundant fossil fuel, but it supplies only about 27 percent of the energy used in the world. It varies in quality and is generally classified in three categories: lignite, bituminous, and anthracite. Lignite coal has a high moisture content and is crumbly in nature, which makes it the least desirable form. Bituminous coal is the most widely used because it is the easiest to mine and the most abundant. It supplies about 20 percent of the world's energy requirements. Coal is primarily used for electric power generation and other industrial uses. For most uses, anthracite coal is the most desirable because it furnishes more energy than the other grades and is the cleanest burning. Anthracite is not as common, however, and is usually more expensive because it is found at great depths and is difficult to obtain.

Because coal was formed as a result of plant material being buried under layers of sediment, it must be mined. There are two methods of extracting coal: surface mining and underground mining. **Surface mining** (strip mining) involves removing the material on top of a vein of coal, called **overburden,** to get at the coal beneath. (See figure 10.8.) Coal is usually surface mined when the overburden is less than 100 meters thick. This type of mining operation is efficient because it removes most of the coal in a vein and can be profitably used for a seam of coal as thin as half a meter. For these reasons, surface mining results in the best utilization of coal reserves. Advances in the methods of surface mining and the development of better equipment have increased surface mining activity in the United States from 30 percent of the coal production in 1970 to more than 60 percent today. This trend toward increased surface mining has also occurred in Canada, Australia, and the former Soviet Union.

If the overburden is thick, surface mining becomes too expensive, and the coal is extracted through **underground mining.** The deeply buried coal seam can be reached in two ways: in flat country, where the vein of coal lies buried beneath a thick overburden, the coal is reached by a vertical shaft. (See figure 10.9a.) In hilly areas, where the coal seam often comes to the surface along the side of a hill, the coal is reached from a drift-mine opening. (See figure 10.9b.)

The mining, transportation, and use of coal as an energy source present several significant problems. Surface mining disrupts the landscape, as the topsoil and overburden are moved to get at the coal. It is possible to minimize this disturbance by reclaiming the area after mining operations are completed. (See figure 10.10.) However, reclamation rarely, if ever, returns the land to its previous level of productivity. The cost of reclamation is passed on to the consumer in the form of higher coal prices. Underground mining methods do not disrupt the surface environment as much as surface mining does, but subsidence

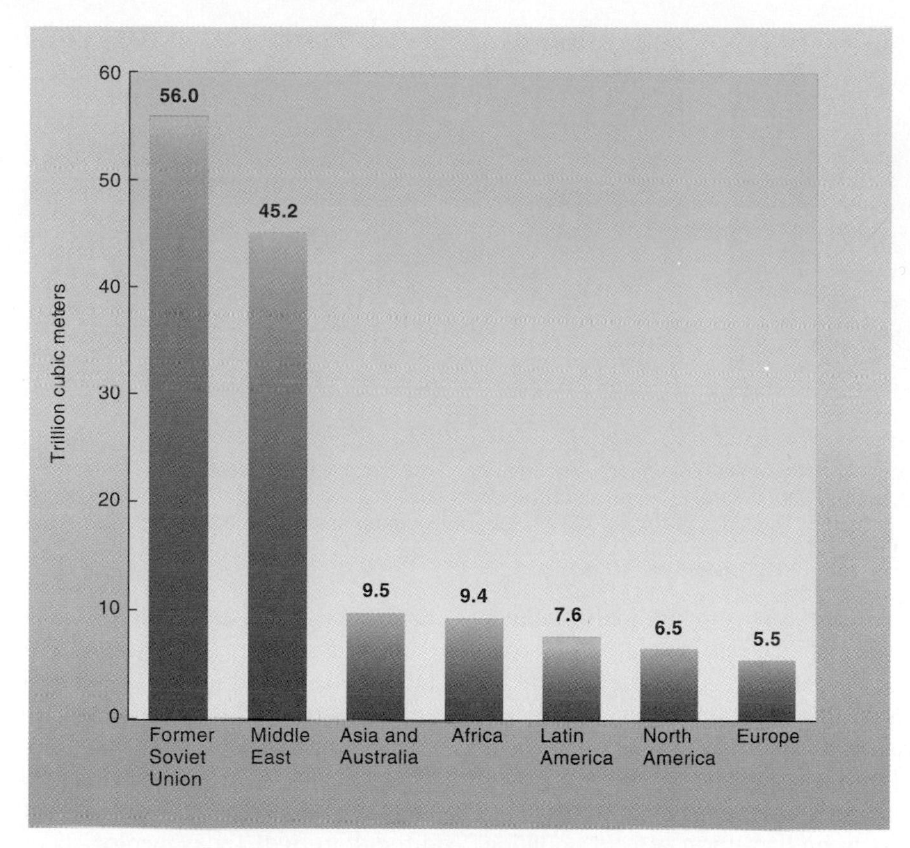

figure 10.7 **Natural Gas Reserves 1999** Natural gas reserves, like oil and coal, are concentrated in certain regions of the world. Figures are trillion cubic meters.

Source: Data from *BP Statistical Review of World Energy,* June 2000.

figure 10.8 **Surface Mining** Large power draglines are used to remove the overburden, which is piled to the side. The coal can then be loaded into trucks. When the coal has been removed, the overburden is placed back in the trench.

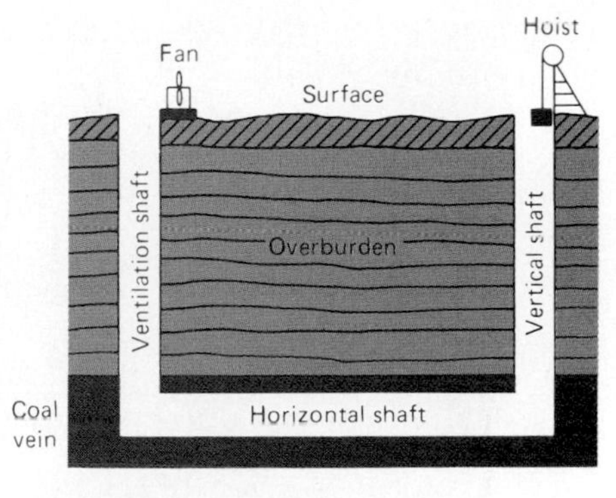

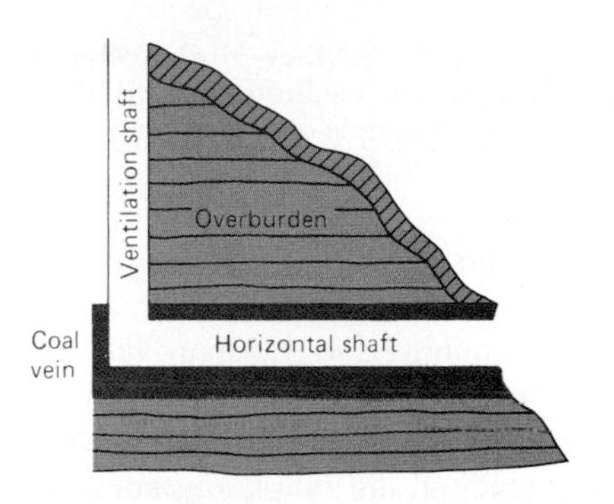

figure 10.9 **Underground Mining** If the overburden is too thick to allow surface mining, underground mining must be used. (*a*) If the coal vein is not exposed, a vertical shaft is sunk to reach the coal. (*b*) In hilly areas, if the vein is exposed, a drift mine is used in which miners enter from the side of the hill.

(sinking of the land) occurs if the mine collapses. In addition, large waste heaps are produced around the mine entrance from the debris that must be removed and separated from the coal.

Health and safety are important concerns related to coal mining, which is one of the most dangerous jobs in the world. This is particularly true with underground mining. Many miners suffer from **black lung disease,** a respiratory condition that results from the accumulation of fine coal-dust particles in the miners' lungs. The coal particles inhibit the exchange of gases between the lungs and the blood. The health care costs and death benefits related to black lung disease are an indirect cost of coal mining. Since these costs are partially paid by the federal government, their full price is not reflected in the price of coal but is paid by taxpayers in the form of federal taxes and higher health premiums.

Because coal is bulky, shipping presents a problem. Generally, the coal can be used most economically near where it is produced. Rail shipment is the most economic way of transporting coal from the mine. Rail shipment costs include the expense of constructing and maintaining the tracks, as well as the cost of the energy required to move the long strings of

a.

b.

figure 10.10 **Surface-Mine Reclamation** (*a*) This photograph shows a large area that has been surface mined with little effort to reclaim the land. The windrows created by past mining activity are clearly evident, and little effort has been made to reforest the land. By contrast, (*b*) is an example of proper surface-mining reclamation. The sides of the cut have been graded and planted with trees. The topsoil has been returned, and the level land is now productive farmland.

railroad cars. In some areas, the coal is transferred from trains to ships.

Coal mining and transport generate a great deal of dust. The large amounts of coal dust released into the atmosphere at the loading and unloading sites can cause local air-pollution problems. If a boat or railroad car is used to transport coal, there is the expense of cleaning it before other types of goods can be shipped. In some cases, the coal can be ground and mixed with water to form a slurry that can be pumped through pipelines. This helps to alleviate some of the air-pollution problems without causing significant water-pollution problems.

Air pollution from coal burning releases millions of metric tons of material into the atmosphere and is responsible for millions of dollars of damage to the environment. The burning of coal for electric generation is the prime source of this type of pollution.

Since coal is a fossil fuel formed from plant remains, it contains sulfur, which was present in the proteins of the original plants. Sulfur is associated with **acid mine drainage** and air pollution. Acid mine drainage occurs when the combined action of oxygen, water, and certain bacteria causes the sulfur in coal to form sulfuric acid. Sulfuric acid can seep out of a vein of coal even before the coal is mined. However, the problem becomes worse when the coal is mined and the overburden is disturbed, allowing rains to wash the sulfuric acid into streams. Streams may become so acidic that they can support only certain species of bacteria and algae. Today, many countries regulate the amount of runoff allowed from mines, but underground and surface mines abandoned before these regulations were enacted continue to contaminate the water.

Currently, a form of acid pollution called acid deposition is becoming a serious problem. Acid deposition occurs when coal is burned and sulfur oxides are released into the atmosphere, causing acid-forming particles to accumulate. Each year, over 150 million metric tons of sulfur dioxide are released into the atmosphere worldwide. This problem is discussed in greater detail in chapter 17.

The release of carbon dioxide from the burning of coal has become a major issue in recent years. Increasing amounts of carbon dioxide in the atmosphere are said to contribute to global warming. Environmentalists have suggested that the amount of coal used be decreased, since the other fossil fuels (oil and natural gas) produce less carbon dioxide for an equivalent amount of energy.

Because coal is difficult to transport and often has a high sulfur content resulting in air pollution, people seek alternative sources of fuel. The most common alternatives for coal are oil and natural gas.

Oil Use Issues

Oil has several characteristics that make it superior to coal as a source of energy. Its extraction causes less environmental damage than does coal mining. It is a more concentrated source of energy than coal, it burns with less pollution, and it can be moved easily through pipes. These characteristics make it an ideal fuel for automobiles. However, it is often difficult to find. Today, geologists use a series of tests to locate underground formations that may contain oil. When a likely area is identified, a test well is drilled to determine if oil is actually present. The many easy-to-reach oil fields have already been tapped. Drilling now focuses on smaller amounts of oil in less accessible sites, which means that the cost of oil from most recent discoveries is higher than that of the large, easy-to-locate sources of the past. As oil deposits on land have become more difficult to find, geologists have widened the search to include the ocean floor. Building an offshore drilling platform can cost millions of dollars. To reduce the cost, as many as 70 wells may be sunk from a single platform. (See figure 10.11.) Total proven world reserves of oil in 1998 were estimated at 1000 billion barrels. Of this, more than three-quarters are in OPEC countries and more than half lie in four Middle Eastern countries—Saudi Arabia, Iraq, Kuwait, and Iran.

One of the problems of extracting oil is removing it from the ground. If the water or gas pressure associated with a pool of oil is great enough, the oil is forced to the surface when a well is drilled. When the natural pressure is not

a.

b.

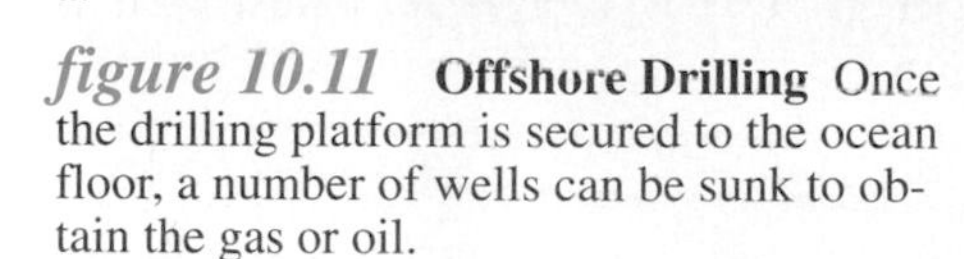

figure 10.11 **Offshore Drilling** Once the drilling platform is secured to the ocean floor, a number of wells can be sunk to obtain the gas or oil.

Source: (*b*) American Petroleum Institute.

great enough, the oil must be pumped to the surface. Present technology allows only about one-third of the oil in the ground to be removed. This means that two barrels of oil are left in the ground for every barrel produced. In most oil fields, **secondary recovery** is used to recover more of the oil. Secondary recovery methods include pumping water or gas into the well to drive the oil out or even starting a fire in the oil-soaked rock to liquefy thick oil. As oil prices increase, more expensive secondary recovery methods will need to be used.

Processing crude oil to provide useful products generates a variety of problems. Oil, as it comes from the ground, is not in a form suitable for use. It must be refined. The various components of crude oil can be separated and collected by heating the oil in a distillation tower. (See figure 10.12.) After distillation, the products may be further refined by "cracking." In this process, heat, pressure, and catalysts are used to produce a higher percentage of volatile chemicals, such as gasoline, from less volatile liquids, such as diesel fuel and furnace oils. It is possible, within limits, to obtain many products from one barrel of oil. In addition, petrochemicals from oil serve as raw materials for a variety of synthetic compounds. (See figure 10.13.) All of these processing activities are opportunities for accidental or routine releases that may cause air or water pollution. The petrochemical industry is a major contributor to air pollution.

The environmental impacts of producing, transporting, and using oil are somewhat different from those of coal. Oil spills in the oceans have been widely reported by the news media. (See figure 10.14.) However, these accidental spills are responsible for only about one-third of the oil pollution resulting from shipping. Nearly 60 percent of the oil pollution in the oceans is the result of routine shipping operations. Oil spills on land can contaminate soil and underground water. The evaporation of oil products

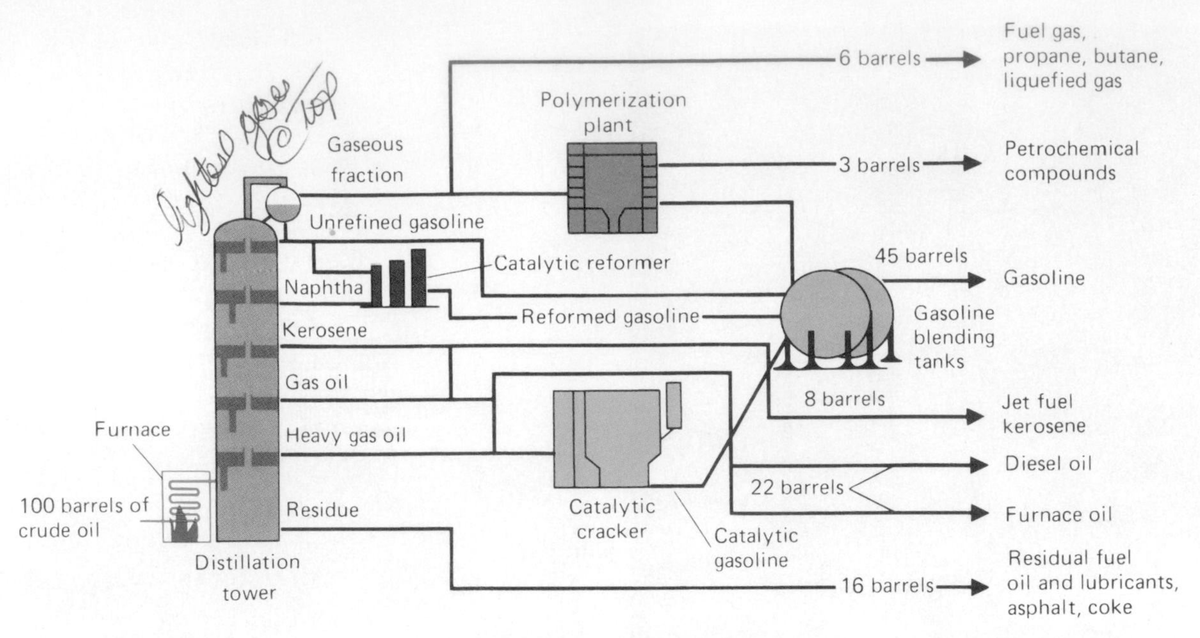

figure 10.12 **Uses of Crude Oil** A great variety of products can be obtained from distilling and refining crude oil. A barrel of crude oil produces slightly less than half a barrel of gasoline. This figure shows the many steps in the refining process and the variety of products that can be obtained from crude oil.

and the incomplete burning of oil fuels contribute to air pollution. These problems are discussed in chapter 17.

Natural Gas Use Issues

Natural gas, the third major source of fossil-fuel energy, supplies 23 percent of the world's energy. The drilling operations to obtain natural gas are similar to those used for oil. In fact, a well may yield both oil and natural gas. As with oil, secondary recovery methods that pump air or water into a well are used to obtain the maximum amount of natural gas from a deposit. After processing, the gas is piped to the consumer for use.

Transport of natural gas still presents a problem in some parts of the world. In the Middle East, Mexico, Venezuela, and Nigeria, wells are too far from consumers to make pipelines practical, so much of the natural gas is burned as a waste product at the wells. However, new methods of transporting natural gas and converting it into other products are being explored. At −162°C (-126°F), natural gas becomes a liquid and has only 1/600 of the volume of its gaseous form. Tankers have been designed to transport **liquefied natural gas** from the area of production to an area of demand. In 1999, over 75 billion cubic meters (2,600 billion cubic feet) of natural gas were shipped between countries as liquefied natural gas. This is nearly 4.5 percent of the natural gas consumed in the world. Of that amount, Japan alone imported 58 billion cubic meters (2,000 billion cubic feet). As the demand for natural gas increases, the amount of it wasted will decrease and new methods of transportation will be employed. Higher prices will make it profitable to transport natural gas greater distances from the wells to the consumers.

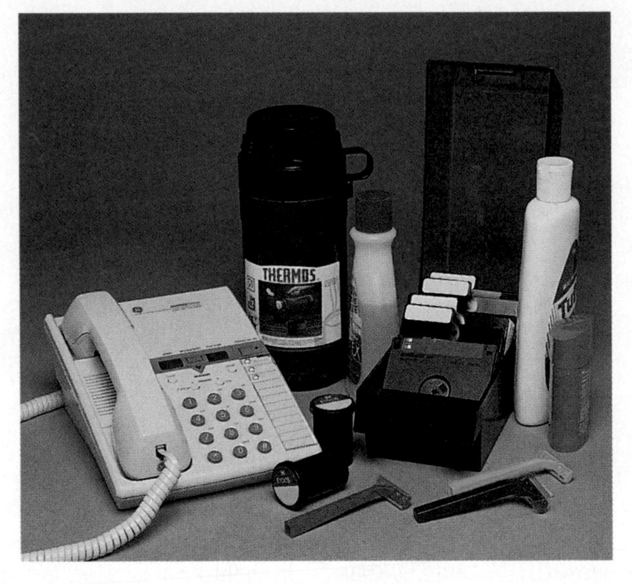

figure 10.13 **Oil-Based Synthetic Materials** These common household items are produced from chemicals derived from oil. Although petrochemicals represent only about 3 percent of each barrel of oil, they are extremely profitable for the oil companies.

Year	Number of tankers afloat	Accidental oil spills	Oil lost (metric tons)
1970	3750	36	84,485
1975	4140	45	188,042
1980	3898	32	135,635
1985	3285	9	15,000
1990	3090	7	4,700
1995	3001	9	7,200
1998	2947	10	6,400

figure 10.14 **Oil Spills** Accidents involving oil tankers are sources of water pollution.

Source: Data from Tanker Advisory Center Reference Data for 1973-1998.

One concern about transporting liquefied natural gas is accidents that might cause tankers to explode. Another, safer process converts natural gas to methanol, a liquid alcohol, and transports it in that form.

Of the three fossil fuels, natural gas is the least disruptive to the environment. A natural gas well does not produce any unsightly waste, although there may be local odor problems. Except for the danger of an explosion or fire, natural gas poses no harm to the environment during transport. Since it is clean burning, it causes almost no air pollution. The products of its combustion are carbon dioxide and water. The burning of natural gas does produce carbon dioxide which contributes to global warming. Global warming is discussed in Chapter 17.

Although natural gas is used primarily for heat energy, it does have other uses, such as the manufacture of petrochemicals and fertilizer. Methane contains hydrogen atoms that are combined with nitrogen from the air to form ammonia, which can be used as fertilizer.

More than two-thirds of the natural gas reserves are located in Russia and the Middle East. In fact, more than a third of the total reserves are located in ten giant fields; six are located in Russia and the remainder are in Qatar, Iran, Algeria, and the Netherlands.

From 1985 to 1999, world production of natural gas rose by more than 25 percent. In Russia, which has the largest natural gas reserves, production increased by 40 percent from 1985 through 1991 but has since fallen as a result of the economic disruption associated with the collapse of the Soviet Union.

Because oil reserves have been located beneath the North Sea, several European countries (the United Kingdom, Norway, and the Netherlands) have sizable reserves of natural gas. During oil shortages in the 1970s, they increased their use of natural gas to compensate for the increased cost of oil. This trend has continued; natural gas consumption increased by 30 percent from 1985 to 1999.

Renewable Sources of Energy

The three nonrenewable fossil-fuel sources of energy—coal, oil, and natural gas—furnish about 90 percent of the world's commercially traded energy. Nuclear power provides over 7 percent of the world's energy. The remainder is supplied by renewable energy sources, including hydroelectric power, tidal power, geothermal power, wind power, solar energy, fuelwood, biomass conversion, and solid waste. Hydroelectric power accounts for nearly 98 percent of electric utilities generated by renewables. (See figure 10.15.) These data do not include the use of wood, animal dung, and other locally produced energy sources typically used in the less-developed world.

Hydroelectric Power

People have long used water to power a variety of machines. Some early uses of water power were to mill grain, saw wood, and run machinery for the textile industry. Today, water power is used almost exclusively to generate electricity. As the water flows from higher to lower levels, it supplies the energy to turn a generator and produce electricity. Hydroelectric power plants are commonly located on human-made reservoirs. (See figure 10.16.) The impounded water represents a potential energy source. In some areas of the world where the streams have steep gradients and a constant flow of water, hydroelectricity may be generated without a reservoir. Such sites are usually found in mountainous regions and can support only small power-generating stations. At present, hydroelectricity produces about 2.5 percent of the world's commercially traded energy.

Hydroelectric power potential is distributed among the continents in rough proportion to land area; China alone possesses one-tenth of the world's potential. Mountainous regions and large river valleys are the most promising. Besides the United States, the eastern area of the former Soviet Union, and southern Canada, the regions that have done the most to harness hydroelectric energy are Europe and Japan. Europe has exploited almost 60 percent of its potential. Although it has only one-fourth of Asia's resources, it generates nearly twice as much hydroelectric power. In contrast, Africa has developed only 5 percent of its potential, half of which comes from only three dams: Kariba in East Africa, Aswan on the Nile, and Akosombo in Ghana.

GLOBAL PERSPECTIVE

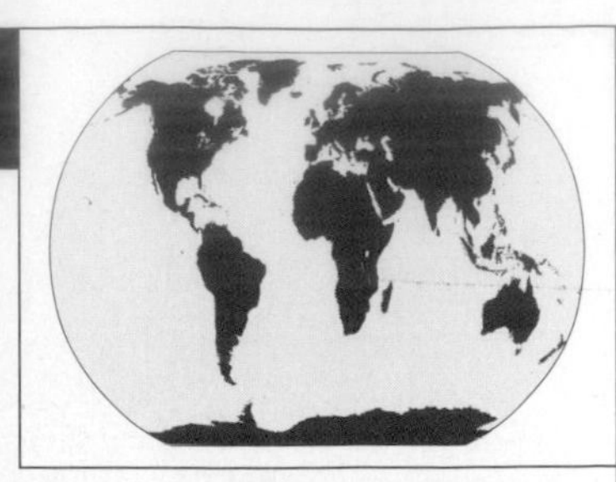

Hydroelectric Sites

Approximately 17 percent of the potential hydroelectric sites of the world have been developed. (See the tables.) The World Energy Conference estimates that the electricity produced by hydropower will increase six times by the year 2020. The less-developed countries, which have developed about 10 percent of their hydropower, will experience most of this growth.

The projected increase will come mainly from the development of plants on large reservoirs. However, construction of "mini-hydro" (less than 10 megawatts) and "micro-hydro" (less than 1 megawatt) plants is also increasing. Such plants can be built in remote places and supply electricity to small areas. China has built over 80,000 such small stations, and the United States has nearly 1,500.

About 50 percent of the U.S. hydroelectric capacity has been developed. However, this statistic is somewhat misleading. The Wild and Scenic Rivers Act (1968) prevents the construction of dams on designated streams. Presently, 37 potential hydroelectric sites are on streams protected by this act. The hydroelectric generating potential often quoted for the United States includes these areas, even though, at present, construction of plants at these sites is not possible. In fact, some dams on rivers in states such as Maine and Washington are scheduled to be destroyed in order to improve native fisheries. The U.S. political climate, which now favors the protection of certain rivers, might change if the demand for more energy becomes acute.

Developed Hydroelectric Sites, 2000

Region	Percent of Hydropower Developed
Asia	17
South America	18
Africa	10
North America	62
Russia	9
Europe	90
Oceania	34

Source: Data from *Survey of Energy Resources,* World Energy Conference.

World Consumption of Hydroelectricity and Other Renewable Energy by Region (Quadrillion BTU)

			Projections	
Region/Country	**Historical to 1995**	**2000**	**2005**	**2010**
OECD	16.8	19.8	21.5	24.0
U.S.*	6.6	7.8	8.3	9.4
Canada	3.1	4.1	4.7	5.3
Mexico	0.3	0.3	0.4	0.5
Japan	1.0	1.8	1.9	2.0
OECD Europe	5.0	5.3	5.7	6.2
U.K.	0.1	0.2	0.2	0.2
France	0.7	0.7	0.7	0.7
Germany	0.3	0.3	0.4	0.4
Italy	0.6	0.6	0.7	0.7
Netherlands	0.0	0.1	0.1	0.2
Other Eur.	3.3	3.5	3.6	3.9
Other OECD	0.4	0.5	0.5	0.7
EE/FSU	2.7	3.1	3.4	3.8
F.S.U.	2.4	2.6	2.8	2.9
E. Europe	0.5	0.5	0.6	1.0
Non-OECD Asia	3.2	4.8	6.0	7.5
China	1.6	2.6	3.5	4.5
Other Asia	1.8	2.1	2.6	3.0
Middle East	0.1	0.3	0.4	0.4
Africa	0.5	0.7	0.8	0.9
Central and South America	4.0	4.0	4.2	4.4
Total world	26.9	32.6	36.4	41.1

*Includes the 50 states, the District of Columbia, and U.S. territories.

Notes: OECD = Organization for Economic Cooperation and Development. EE/FSU = Eastern Europe/Former Soviet Union.

Sources: History data from Energy Information Administration (EIA), *International Energy Annual,* DOE/EIA−0219(92), Washington, D.C.; Projections from EIA, *Annual Energy Outlook,* DOE/EIA−0383(95), Table B1, Jan. 1995, Washington, D.C.; and World Energy Projections 2000.

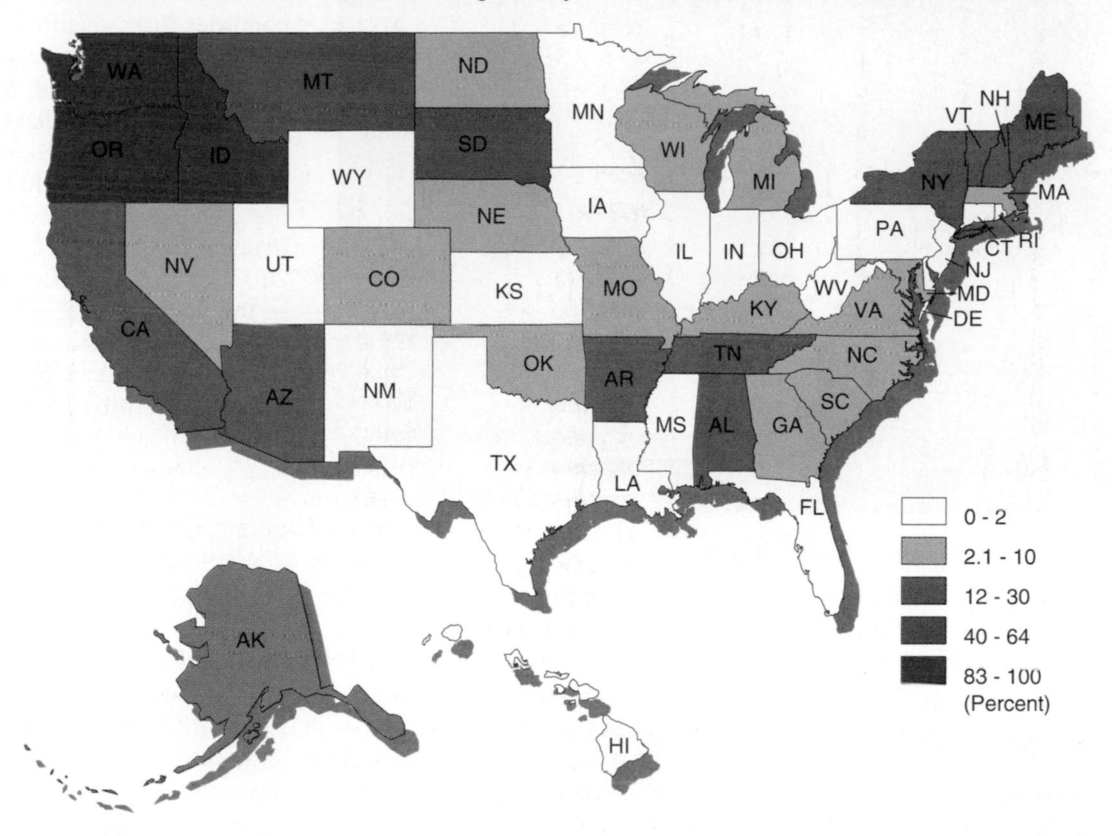

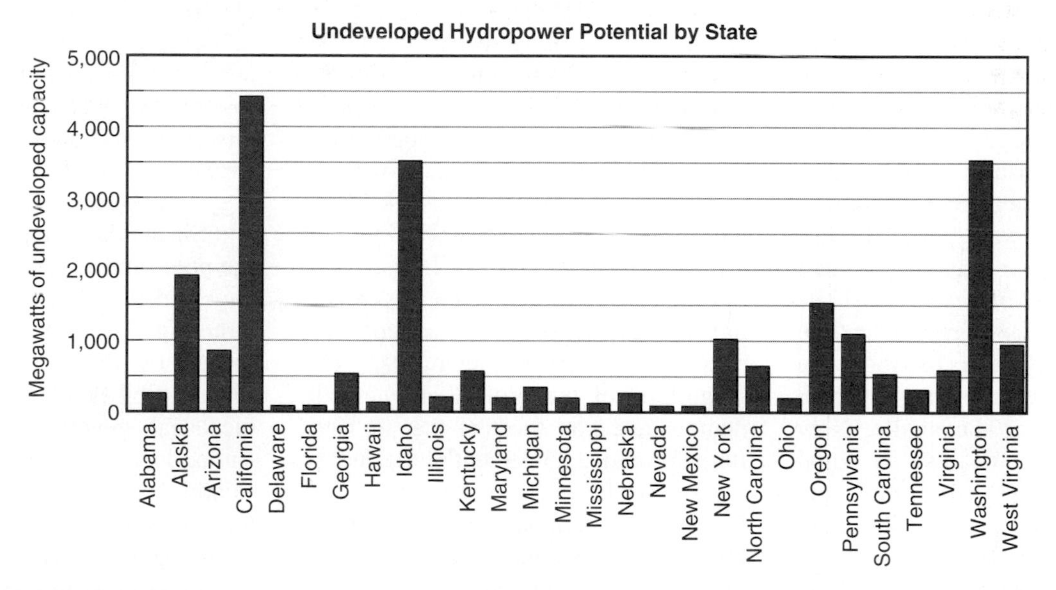

figure 10.15 **Hydroelectric Power and Potential** The top map indicates the percent of hydroelectric power in the United States. The bottom graph indicates the current undeveloped hydropower in the United States.

Source: USGS and the Foundation for Water and Energy Education: 2000.

In some areas of the world, hydroelectric power is the main source of electricity. More than 35 nations already obtain more than two-thirds of their electricity from falling water. In South America, 73 percent of the electricity used comes from hydroelectric power, compared to 44 percent in the developing world as a whole. Norway gets 99 percent of its electricity and 50 percent of all its energy from falling water.

It is important to recognize that the construction of a reservoir for a hydroelectric plant causes environmental and social problems, including loss of fertile farmland, destruction of the natural aquatic ecosystem, relocation of entire

figure 10.16 **Hydroelectric Power Plant** The water impounded in this reservoir is used to produce electricity. In addition, this reservoir serves as a means of flood control and provides an area for recreation.

communities, and a reduction in the amount of nutrient-rich silt deposited on downriver agricultural lands. The building of the Tellico Dam in Tennessee was delayed for several years because it might have caused the extinction of the snail darter, a fish that lived only in streams that would be flooded. The construction of the Aswan Dam in Egypt resulted in the displacement of eighty thousand people. It created an environment that increased schistosomiasis, a waterborne disease caused by flatworm parasites that spend part of their life cycle in snails that live in slowly moving water. The irrigation canals built to distribute the water from the Aswan Dam provide ideal conditions for the snails. Now many of the people who use the canal water for cooking, drinking and bathing are infected with the flatworm parasites.

Even though hydroelectric projects cause problems, new sites continue to be developed. From 1984 to 1999, the energy furnished by hydroelectricity for world use increased by about 27 percent. The most significant change took place in South America, where hydroelectric use increased about 40 percent over that time.

China is currently constructing a huge hydroelectric dam, known as the Three Gorges Dam, on the Yangtze River. If it is completed, it will be the largest hydroelectric dam in the world. (See Global Perspectives: The Three Gorges Dam.)

Today's large dams rank among humanity's greatest engineering feats. The Itaipu hydroelectric power plant is the largest development of its kind in operation in the world (until the completion of the Three Gorges Dam in China). Built from 1975 to 1991, in a development on the Parana River, Itaipu is a joint project between Brazil and Paraguay. The power plant's 18 generating units add up to a total production capacity of 12,600 megawatts. The magnitude of the project also can be demonstrated by the fact that in 1997 Itaipu provided 25 percent of the energy supply in Brazil and 80 percent in Paraguay. By means of comparison, Itaipu produces enough power to meet most of California's annual need. The lakes created by major dams also number among the planet's largest freshwater bodies. Ghana's Lake Volta covers 8500 square kilometers (3280 sq mi), an area the size of Lebanon. Hydroelectric dam projects figure prominently in the economic and investment plans of many developing countries. Egypt electrified virtually all of its villages with power from Aswan. Since the 1950s, large-scale hydropower development in these countries has occurred primarily because energy-intensive industries need cheap electricity and because global lending institutions have been willing to advance multibillion-dollar loans. According to the World Bank, developing countries will need to raise an estimated $100 billion (U.S.) by the year 2000 for hydroelectric plants currently in the planning stage. Given the environmental concerns of large dam construction, the trend for the future will probably be to build small-scale hydro plants that can generate electricity for a single community.

Tidal Power

Tidal flow is another source of energy related to local geological conditions. The gravitational pull of the sun and the moon, along with the earth's rotation, causes tides. The tidal movement of water represents a great deal of energy. For years, engineers have suggested that this moving water could be used to produce electricity. The principle is the same as that employed in a hydroelectric plant. As water flows from a higher level to a lower one, it can be used to generate electricity. The greater the difference between high and low tides, the more energy can be extracted.

Since tidal changes are greatest near the poles and are accentuated in narrow bays and estuaries, suitable sites of constructing power plants are limited. In the 1930s, the United States explored the possibility of constructing a tidal electrical generating facility at Passamaquoddy, Maine, on the Bay of Fundy. After spending considerable time and money on a feasibility study, it abandoned the idea.

In 1966, France constructed a commercial tidal generating station. (See figure 10.17.) Located on the Rance River Estuary on the Brittany coast, it is the world's only large tidal generating station. It was built to generate 240 megawatts, but because of the tides, it usually produces only 62 megawatts. This satisfies the electric power needs of about 100,000 people (less than 0.2 percent of the population of France).

GLOBAL PERSPECTIVE

The Three Gorges Dam

A project that began in late 1997 is already being compared to the building of the pyramids and the Great Wall of China. China's Three Gorges Dam across the world's third longest river, the Yangtze, is a project that when completed will be visible from the moon and will be the largest dam in the world. The Three Gorges Dam will stretch 2 kilometers (1.3 mi) across the Yangtze River, tower 185 meters into the air (610 ft), and create a 600-kilometer (385-mi) reservoir behind it. The dam is expected to cost in excess of $40 billion before completion in 2009. The government of China contends that the dam is necessary for several reasons. The primary reason is the 18,000 megawatts of electric power that the dam will generate. China's rapidly developing economy needs the energy for its expanding industrial centers. It is also argued that the project will transform the Yangtze River, especially the upper regions, into a more navigable and hence, economic, waterway. The final major argument for the dam relates to flood control in the middle and lower reaches of the river. These parts of the river have been prone to frequent and disastrous floods. The economic arguments, however, are not the only issues being debated in relation to the Three Gorges Dam.

Scientists in China and from other nations have warned that the project could threaten migratory fish, concentrate water pollution, and endanger to the point of extinction the Chinese alligator, river dolphins, the Siberian white crane, and the Chinese sturgeon. Environmental concerns relating to the project were instrumental in the 1993 decision of the United States to withdraw technical assistance. The impact on the regional Chinese population is also a critical issue. Upon completion, the lake formed by the dam will inundate 153 towns, 4,500 villages, numerous archeological sites, and the scenic canyons of the Three Gorges that have inspired poets and painters for centuries.

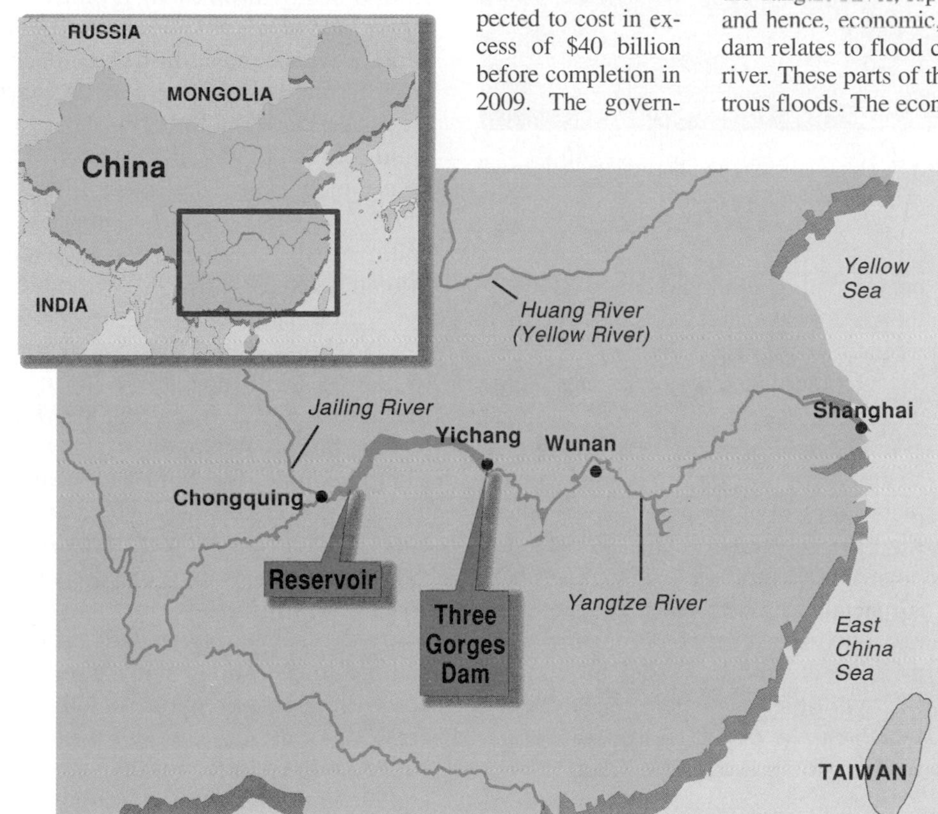

Source: Magellan Geographix/ABCNEWS.com. Data from International Rivers Network.

- Do you feel that the benefits of this massive project exceed its economic and environmental costs?
- Would you have supported the project in its developmental phase?

There is also a 16-megawatt plant in Nova Scotia, Canada.

The British government studied the feasibility of constructing a 7200-megawatt tidal station in the Severn Estuary in southwestern England. It would have thirty times the capacity of the French station. The facility could be built for a price comparable to that of a coal-fired plant of similar generating capacity, but it would disrupt the normal estuary flow and would concentrate pollutants in the area. The plant has not been built.

Geothermal Power

In areas where steam is trapped underground, **geothermal energy** is tapped by drilling wells to obtain the steam. The steam is then used to power electrical generators. At present, geothermal energy is practical only in areas where this hot mass is near the surface. (See figure 10.18.) Geothermal energy is not considered a true renewable energy source but rather an alternative source.

The United States has about half of the world's geothermal electrical gener-

figure 10.17 **Tidal Generating Station** The Rance River Estuary Power Plant in France is the world's largest tidal electrical generating station.

figure 10.18 **Geothermal Power Plant** Steam obtained from geothermal wells is used in the production of electricity.

ating capacity. More than 130 generating plants are operating in twelve other countries. The Philippines, Italy, Mexico, Japan, New Zealand, and Iceland produce sizable amounts of electricity by geothermal methods.

California alone produces 40 percent of the world's geothermal electricity, about 1,900 megawatts. One megawatt provides energy for about one thousand households. The Pacific Gas and Electric Company (PG&E) has been producing electricity from geothermal energy since 1960. PG&E's complex of generating units located north of San Francisco is the largest in the world and provides 700 megawatts of power, enough for 700,000 households or 2.9 million people.

The state of Hawaii has considered building two 25-megawatt geothermal plants on the island of Hawaii. Most of the electricity currently generated for the state comes from oil-fired power plants; geothermal-generated electricity would be less costly and reduce the risk of accidental oil spills. However, the plants would be built in a rainforest area that environmentalists believe would be destroyed by the construction.

In addition to producing electricity, geothermal energy is used directly for heating. In Iceland, half of the geothermal energy is used to produce electricity and half is used for heating. In the capital, Reykjavik, all of the buildings are heated with geothermal energy at a cost that is less than 25 percent of what it would be if oil were used.

Geothermal energy creates some environmental problems. The steam contains hydrogen sulfide gas, which has the odor of rotten eggs and is an unpleasant form of air pollution. (The sulfides from geothermal sources can, however, be removed.) The minerals in the steam corrode pipes and equipment, causing maintenance problems. The minerals are also toxic to fish.

All objects contain heat energy, which can be extracted from and transferred to other locations. A refrigerator extracts heat from the interior and exports it to the coils on the back of the unit. Similarly, pipes placed within the earth can extract heat and transfer it to a home. It is important to recognize that such an application of geothermal energy requires the expenditure of energy through an electric-powered device, so the amount of heat energy extracted costs. In some instances, obtaining heat in this manner is less expensive than other traditional sources, such as oil or natural gas, but it is not a "free" source of heat.

Wind Power

As the sun's radiant energy strikes the Earth, that energy is converted into heat, which warms the atmosphere. The earth is unequally heated because various portions receive different amounts of sunlight. Since warm air is less dense and rises, cooler, denser air flows in to take its place. This flow of air is wind. For centuries, wind has been used to move ships, grind grains, pump water, and do other forms of work. In more recent times, wind has been used to generate electricity. (See figure 10.19.)

In 1999, California was using 14,600 wind turbines to produce 300 megawatts of electricity at a cost competitive with newly constructed coal or nuclear plants. By 2010, according to the American Wind Energy Association, wind could power 10 million U.S. households. The U.S. Department of Energy has stated that the Great Plains could supply 48 states with 75 percent of their electricity. In 1999, wind facilities produced enough power for 425,000 homes. That figure is expected to increase with new wind "farms" in Minnesota, Iowa, Wyoming, and Texas coming online. Wind power is now considered competitive with new coal and natural gas plants and cheaper than nuclear plants. Since the 1980s, costs associated with wind power have decreased considerably due in part to evolving technology and government actions. In addition, many consumers are willing to pay extra for clean energy. In Colorado, through a program called Windsource, 10,000 residents and businesses pay an additional $2.50 a month for every 100-kilowatt block of wind power.

According to the U.S. Department of Energy, wind power was the world's fastest growing energy source in the 1990s. Technological improvements in the last 20 years cut the cost from 40 cents per kilowatt hour (kWh) to between 4 and 5 cents kWh. The cost of electricity produced by wind is expected to fall more as the technology advances. Meanwhile, a push for energy deregulation and concerns about smog, acid rain, and global warming are driving policy makers to require utilities to sell electricity from renewable sources. Eight states—Texas, Wisconsin, Massachusetts, Connecticut, New Jersey, Minnesota, Nevada, and Pennsylvania—

require utilities to provide some "green" electricity. At least 36 utilities include wind energy as a component of their green power programs.

Most U.S. wind sources remain untapped. Texas and the Dakotas alone have enough wind to power the nation, but that's not likely to happen. Variable wind speeds make it unreliable as a primary energy source; energy companies and regulators view it as supplementary to fossil fuels. While places like the Dakotas have the strongest winds, they are far from energy-using population centers and lack suitable transmission grids.

India, China, Germany, and Spain have plans for major increases in wind-generated electricity through the year 2005. Since 1998, the world's wind-energy capacity has grown more than 35 percent, topping 10,000 megawatts (MW). Germany contributed a third of the recent growth, largely by guaranteeing wind farms access to the power grid at a competitive price for the power they generate. And as U.S. companies begin building 1 or 2 MW wind turbines, European firms are exploring 5 MW machines. In Inner Mongolia, nomadic herders carry with them small, portable, wind-driven generators that provide electricity for light, television, and movies in their tents as well as for electric fences to contain their animals.

A steady and dependable source of wind makes the use of wind power more productive in some regions than others. Wide, open areas, such as the Great Plains in North America, are better suited for wind power than are heavily wooded areas. Wind-generated electricity is usually used in conjunction with other sources of electricity that take over when the wind does not blow. Wind generators do have some negative effects. The moving blades are a hazard to birds and produce a noise that some find annoying. Vibrations from the generators can also cause structural problems. In addition, some people consider the sight of a large number of wind generators, such as are found in California, to be visual pollution.

figure 10.19 **Wind Energy** Fields of wind-powered generators such as these in California can produce large amounts of electricity.

Solar Energy

The sun is often mentioned as the ultimate answer to the world's energy problems. It provides a continuous supply of energy that far exceeds the world's demands. In fact, the amount of energy received from the sun each day is six hundred times greater than the amount of energy produced each day by all other energy sources combined. The major problem with solar energy is its intermittent nature. It is available only during the day and when it is sunny. All systems that use solar energy must store energy or use supplementary sources of energy when sunlight is not available. Because of differences in the availability of sunlight, some parts of the world are more suited to the use of solar energy than others.

Solar energy is utilized in three ways:

1. In a passive heating system, the sun's energy is converted directly into heat for use at the site where it is collected.
2. In an active heating system, the sun's energy is converted into heat, but the heat must be transferred from the collection area to the place of use.
3. The sun's energy also can be used to generate electricity, which may be used to operate solar batteries or may be transmitted along normal transmission lines.

Passive Solar Systems

Anyone who has walked barefoot on a sidewalk or a blacktopped surface on a sunny day has experienced the effects of passive solar heating. In a **passive solar system,** light energy is transformed to heat energy when it is absorbed by a surface. Some of the earliest uses of passive solar energy were to dry food and clothes and to evaporate seawater to produce salt. Homes and buildings may now be designed to use passive solar energy for heating. (See figure 10.20.) Such systems require a large window through which sunlight may enter and a large mass that collects and stores the heat. This large mass may be a thick wall or floor. The type of construction material, the color of the roofing, and the landscaping all are important factors in passive solar heating. Since there are no moving parts, a passive solar system is maintenance free. None of the energy is used to transfer heat within the system, and there are no operating costs. However, passive solar design is usually practical only in new construction.

Active Solar Systems

An **active solar system** requires a solar collector, a pump, and a system of pipes to transfer the heat from the site of production to the area to be heated. (See figure 10.21.) Active solar systems are most easily installed in new buildings, but in some cases can be installed in existing structures. A major consideration in the use of an active solar system is the initial cost of installation. An active system requires a specially designed collector, consisting of a series of liquid-filled tubes; a pump; and pipes to transfer warm liquid from the collector to the space to be heated. Because an active system has moving parts, it also has operation and maintenance costs like any other heating or cooling system.

Rock, water, or specially produced products are used to store heat. The hot liquid in the pipes heats the storage medium, which releases its heat when the sun is not shining.

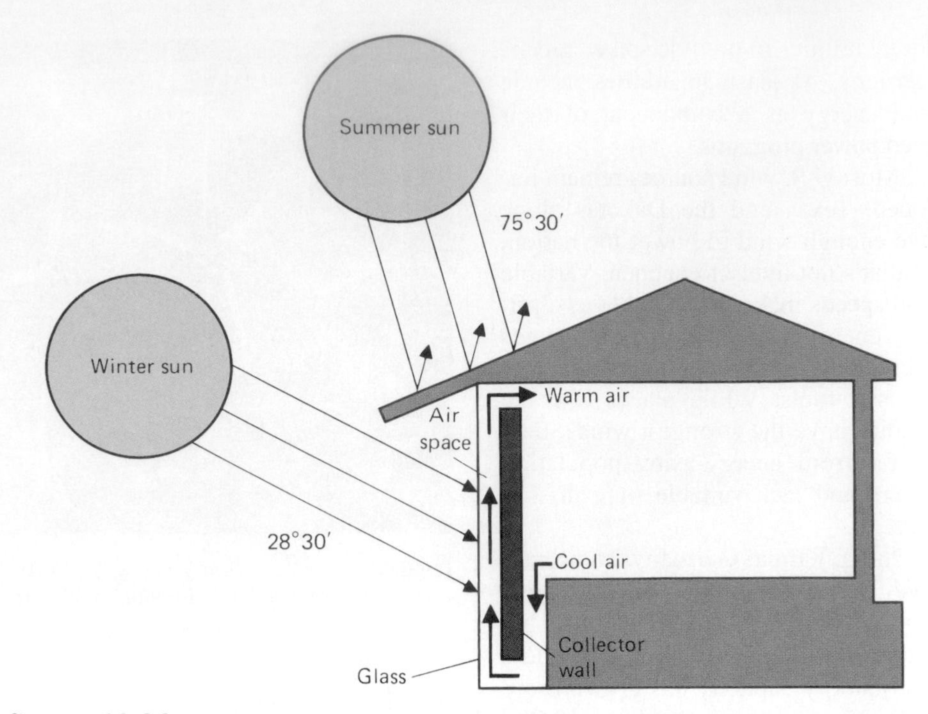

figure 10.20 **Passive Solar Heating** The length of overhang in this home is designed for solar heating at the latitude of St. Louis, Missouri (38°N). In this design, a wall 30 to 40 centimeters thick is used to collect and store heat. The collector wall is located behind a glass wall and faces south. During a midwinter day, when the sun's angle is 28 degrees, light energy is collected by the wall and stored as heat. At night, the heat stored in the wall is used to warm the house. Natural convection causes the air to circulate past the wall, and the house is heated. During a midsummer day, when the sun's angle is 75 degrees, the overhang shades the collector wall from the sun.

Solar-Generated Electricity

When the first **photovoltaic cell,** a bimetallic unit that allows the direct conversion of sunlight to electricity, was developed by Bell Laboratories in 1954, it was regarded as an expensive novelty. However, as more efficient batteries were developed and production costs were reduced, practical uses were found for photovoltaic cells. By the mid-1980s, more than 60 million solar calculators were being produced annually. These calculators used over 10 percent of the photovoltaic cells manufactured.

Photovoltaic cells appear to be emerging as sources of small amounts of electricity for special uses like running equipment in remote regions. The normal system of generating electricity in large, centrally located plants and distributing it by high tension lines is costly, and is practical only in highly populated areas. Photovoltaic cells are more practical in remote regions of the world.

Although sales of solar cells for items like highway signs, roofs, and radios increased 20 percent from 1998 to 1999, most of that growth was outside the United States. Over 70 percent of solar cells made in the United States are sent abroad, often to remote spots, like rural India, that are not connected to the grid. Since 1997, the U.S. share of the global market for solar cell products has dropped from 44 percent to 35 percent. It is expected that Japan, where electricity is relatively expensive, will move ahead to lead solar PV sales worldwide. Not coincidentally, Japan spends $240 million annually on solar power versus $72 million in 1999 for the U.S. Over ten thousand solar-electric homes have been built in parts of Alaska and the Australian outback. (See figure 10.22.) The French government has subsidized the installation of over two thousand solar electric units on eighteen islands in the Pacific. These units provide electricity for a thousand homes and five hospitals. Many of the developing countries of the world will introduce electricity to villages through the use of photovoltaic cells rather than the use of generators that require fuel and distribution lines.

It has to be argued that to compete with fossil fuels, solar engineers will have to think bigger. A novel Dutch effort is setting out to do just that. Near Amersfoort, the Netherlands, the NV REMU power company is leading a $13 million project to build 500 houses with roofs covered with photovoltaic (PV) panels. By the time the homes are finished, they should be drawing 1.3 MW of energy from the sun, enough to supply about 60 percent of the community's energy needs. The goal of the Amersfoort project is to demonstrate the construction of a solar energy system at the level of an entire community.

The price of photovoltaic cells has been falling as better technology is developed. Eventually, the cells may become competitive with other energy sources, particularly as the cost of fossil fuels rises.

Solar energy is also being used to generate electricity in a more conventional way. A company named Luz International in California has built solar collector troughs that can heat oil in pipes to 390°C (734°F) (See figure 10.23.). This heat can be transferred to water, which is turned into steam that is used to run conventional electricity-generating turbines. As with photovoltaic cells, the cost of producing

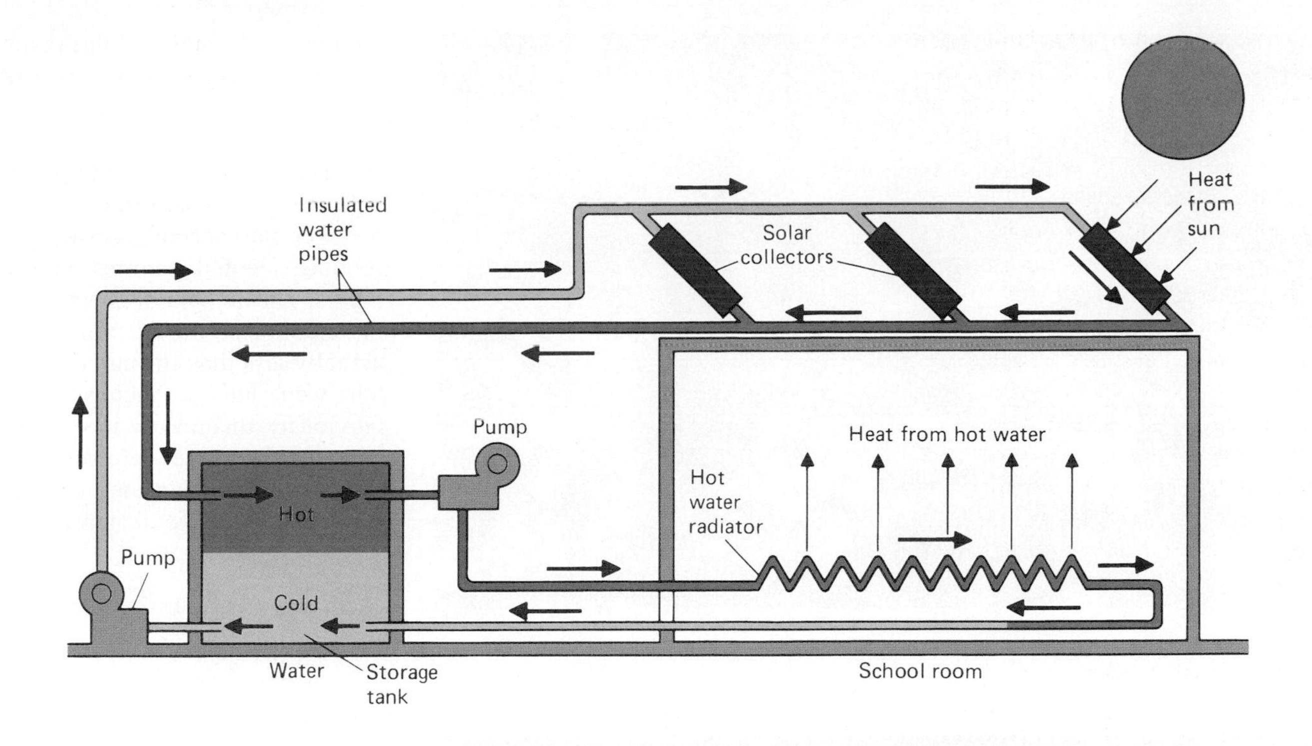

electricity in this manner is falling and is becoming competitive with conventional sources.

Limitations of Solar Energy

Solar energy provides less than 1 percent of the world's energy for several reasons. The most obvious is that it works only during the day, which means that some type of heat or electrical storage mechanism is needed for night use, which adds to the expense of relying on solar energy. The fact that solar heating is most practical in new construction also limits its use. In colder climates, solar heat is inadequate as the sole source of heat, and some type of a conventional heating system is required for backup. Climate is also a problem, since many areas have extensive cloudy periods, which reduce the amount of energy that can be collected. Although the price of collectors and related equipment has decreased in recent years, many collector systems are still expensive. For example, electricity from photovoltaic cells is still more expensive than conventional electric generating systems. Prices of photovoltaic cells continue to fall, however, and the cost of generating electricity from conventional fuels is rising.

figure 10.21 **Active Solar System** Collectors provide 65 percent of the hot water in this dental school building in California.

Biomass Conversion

Biomass is any accumulation of organic material produced by living things. The most commonly used biomass sources are fuelwood, agricultural residue from the harvesting of crops, crops grown for their energy content, and animal waste. (Because of its major impact on world energy resources, fuelwood will be discussed separately in the next section.) These traditional, often noncommercial sources of fuel provide more than 10 percent of the world's energy, but are not reported in most statistics about global energy. In many developing countries, these sources of fuel are a large proportion of the energy available.

Biomass conversion is the process of obtaining energy from the chemical

figure 10.22 **Solar Energy** In some remote areas, solar energy is an economical method of electricity production.

figure 10.23 **Solar Generation of Electricity** This solar-powered electricity generating plant is capable of generating electricity at a cost that is competitive with other methods of generating electricity.

energy stored in biomass. It is not a new idea; burning wood is a form of biomass conversion that has been used for thousands of years. Biomass can be burned directly as a source of heat for cooking, burned to produce electricity, converted to alcohol, or used to generate methane. (See figure 10.24.) The People's Republic of China has 500,000 small methane digesters in homes and on farms; India has 100.000; and Korea has 50,000. Brazil is the largest producer of alcohol from biomass. The low price of sugar coupled with the high price of oil have prompted Brazil to use its large crop of sugar cane as a source of energy. Alcohol provides 50 percent of Brazil's automobile fuel.

Biomass conversion raises some environmental and economic concerns. Countries that use large amounts of biomass for energy are usually those that have food shortages. Biomass conversion means that fewer nutrients are being returned to the soil, and this compounds the food shortage. If the price of food rises or the price of oil falls, there could be less biomass conversion.

The energy required to produce usable energy stocks from biomass must be taken into account. Growing corn to produce alcohol requires large energy inputs. The amount of energy present in the alcohol produced from the corn is actually less than the amount of energy that went into producing the alcohol. Obviously, this makes no sense from an energy point of view. However, the convenience of a liquid fuel may be worth paying for in economic terms.

Biomass Conversion Technologies

There currently exist several technologies capable of converting biomass into energy. These include anaerobic digestion, pelletising, direct combustion and cogeneration, pyrolysis, gasification, and ethanol production.

Anaerobic Digestion

Anaerobic digestion is the decomposition of wet and green biomass, through bacterial action in the absence of oxygen, to produce a mixed gas output of methane and carbon dioxide known as biogas. The anaerobic digestion of municipal solid waste buried in landfill sites produces a gas known as landfill gas. This process occurs naturally as the bacterial decomposition of the organic matter continues over time. The methane gas produced in landfill sites eventually escapes into the atmosphere. However, the landfill gas can be extracted from existing landfill sites by inserting perforated pipes into the landfill. In this way, the gas will travel through the pipes, under natural pressure, to be used as an energy source, rather than simply escaping into the atmosphere to contribute to greenhouse gas emissions.

Pelletising

Pelletising involves the compaction of biomass at high temperatures and very high pressures. The biomass particles are compressed in a die to produce briquettes

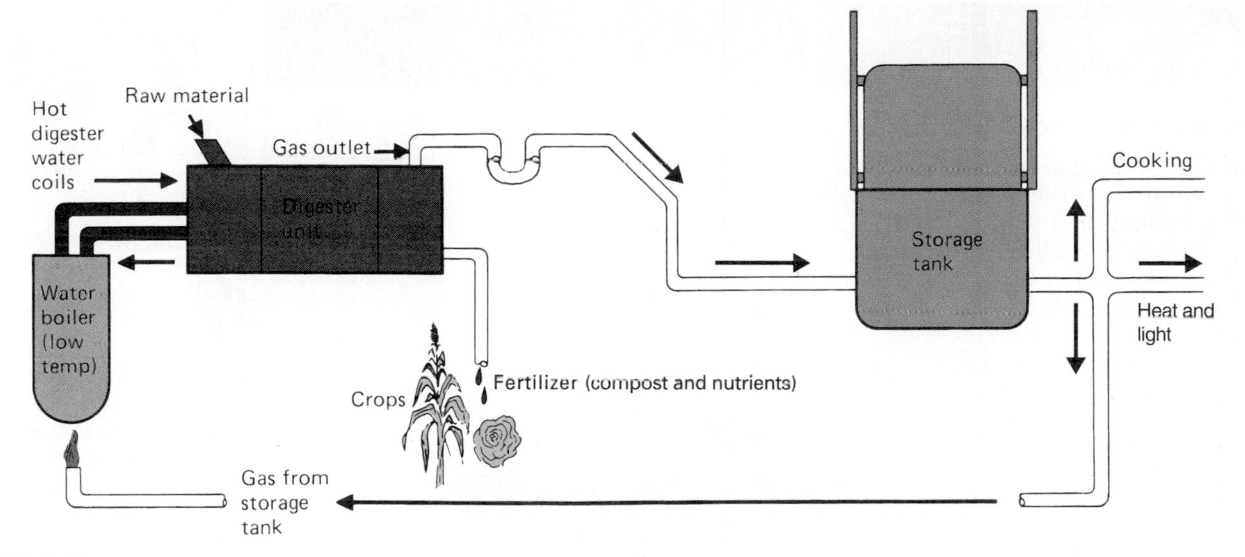

figure 10.24 **Methane Digester** In the digester unit, anaerobic bacteria convert animal waste into methane gas. This gas is then used as a source of fuel. The sludge from this process serves as a fertilizer. In many less-developed countries, this type of digester has the advantages of providing a source of energy and a supply of fertilizer and managing animal wastes, which helps reduce disease.

or pellets. These products have significantly smaller volume than the original biomass and thus have a higher volumetric energy density (VED), making them a more compact source of energy. They are also easier to transport and store than natural biomass. The pellets can be used directly on a large scale as direct combustion feed, or on a small scale in domestic stoves or wood heaters.

Direct Combustion and Cogeneration

Direct combustion is the main process adopted for utilizing biomass energy. The energy produced can be used to provide heat and/or steam for cooking, space heating, and industrial processes, or for electricity generation. Large biomass power-generation systems can have comparable efficiencies to fossil-fuel systems, but this comes at a higher cost due to the design of the burner to handle the higher moisture content of biomass. However, by using the biomass in a combined heat- and electricity-production system (or cogeneration system), the economics are significantly improved.

Pyrolysis

Pyrolysis is the basic thermochemical process for converting solid biomass to a more useful liquid fuel. Biomass is heated in the absence of oxygen, or partially combusted in a limited oxygen supply, to produce a hydrocarbon-rich gas mixture, an oil-like liquid, and a carbon-rich solid residue. Traditionally, in developing countries, the solid residue produced is charcoal, which has a higher energy density than the original fuel. The traditional charcoal kilns are simply mounds of wood covered with earth, or pits in the ground. However, the process of carbonization is very slow and inefficient in these kilns, and more sophisticated kilns are replacing the traditional ones. The pyrolitic or "bio-oil" produced can be easily transported and refined into a series of products. The process is similar to refining crude oil.

Gasification

Gasification is a form of pyrolysis, carried out with more air, and at high temperatures, in order to optimize the gas production. The resulting gas, known as producer gas, is a mixture of carbon monoxide, hydrogen and methane, together with carbon dioxide and nitrogen. The gas is more versatile than the original solid biomass, and it can be used as a source of heat or used in internal combustion engines or gas turbines to produce electricity. During the Second World War, countries such as Australia and Germany even used it to power vehicles.

Ethanol Production

Ethanol can be produced from certain biomass materials that contain sugars, starch, or cellulose. The best-known feedstock for ethanol production is sugar cane, but other materials can be used, including wheat, corn, and other cereals, and sugar beets. Starch-based biomass is usually cheaper than sugar-based materials but requires additional processing. Similarly, cellulose materials, such as wood and straw, are readily available but require expensive preparation.

Ethanol is produced by a process known as fermentation. Typically, sugar is extracted from the biomass crop by crushing and mixing with water and yeast, and then being kept warm in large tanks called fermenters. The yeast breaks down the sugar and converts it to ethanol. A distillation process is required to remove the water and other impurities from the dilute alcohol product. Brazil has a successful industrial-scale ethanol project, which produces ethanol from sugar cane for blending with gasoline. In the U.S., corn is used for ethanol production and then blended with gasoline to produce "gasahol."

GLOBAL PERSPECTIVE

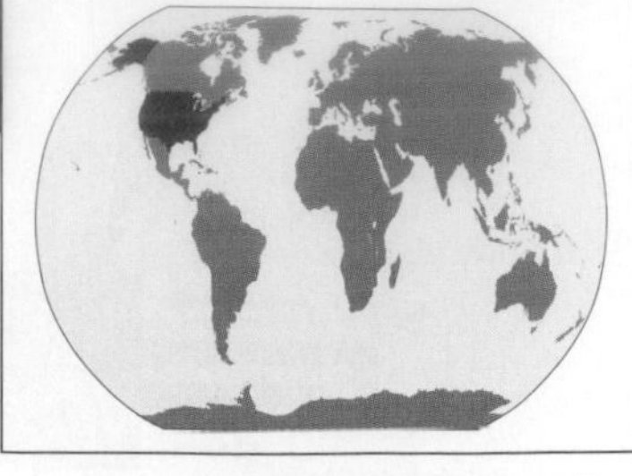

Electricity from the Ground Up

The naturally occurring heat beneath the planet's surface powers volcanoes, hot springs, and geysers such as Yellowstone National Park's Old Faithful. In a few places in the world, such as Iceland and parts of California, natural heat and natural water come together in sufficient quantities to provide energy. In most places where there is abundant subsurface water, useful heat is five kilometers (3 miles) or more below the surface, beyond the reach of economical drilling technology. In places where heat is within reach, there is often no subsurface water.

Searching for cheap, nonpolluting energy, scientists at the Los Alamos National Laboratory in New Mexico have begun a process to "mine heat." Mining heat is a new effort to use geothermal energy. In this process, water is pumped at high pressure three kilometers (1.9 miles) down into the Earth through a well. When the water comes back up through a parallel well, it has been heated beyond the boiling point by the Earth's natural heat. The hot dry rock project is an attempt to bring heat and water together as a geothermal energy source.

On average, temperatures below the Earth's surface increase about 27°C (80°F per mile) per kilometer. In much of the western United States, however, residual heat from ancient volcanoes increases temperatures by more than 69°C/km (200°F per mile). At the Los Alamos site, the temperature at the base of the well is about 240°C (430°F).

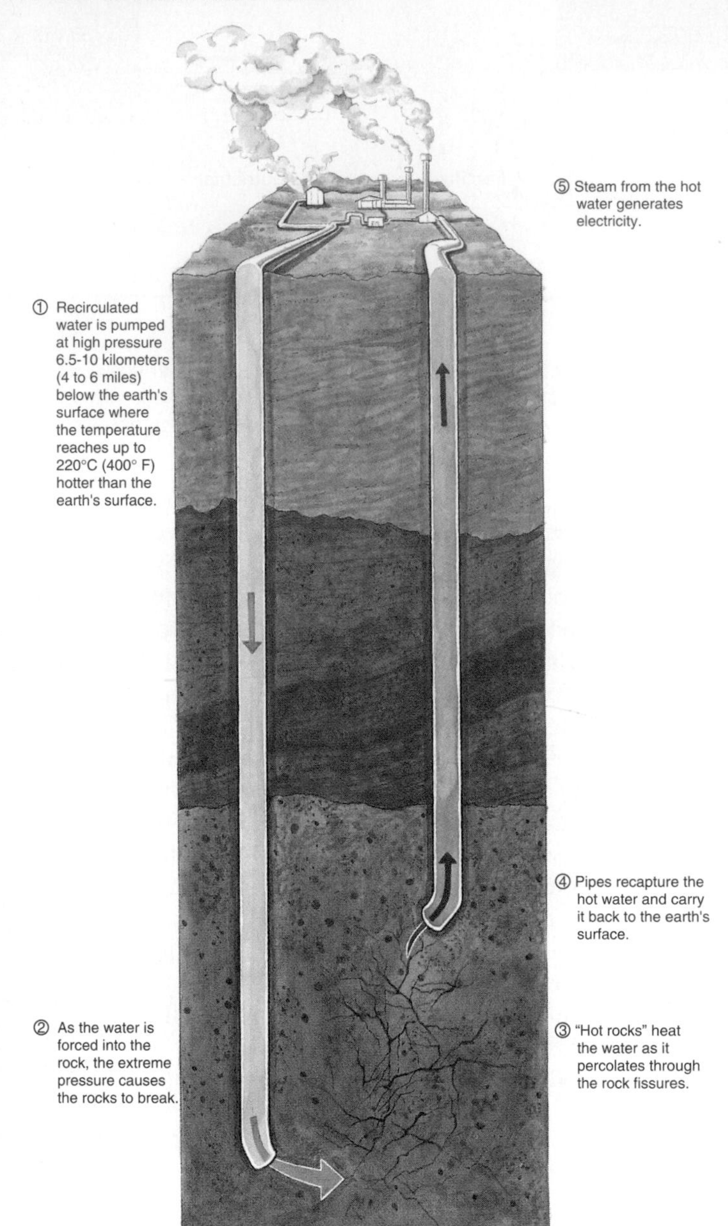

According to researchers at Los Alamos, a fairly conservative estimate is that there are at least 500,000 quads (quadrillion British thermal units, or Btus) of useful heat in hot dry rock at accessible drilling depths beneath the United States. This is about 6,000 times the total amount of energy used in the country in one year.

The fundamental question now being asked is if the energy can be developed economically for commercial use. At this point, the answer is no; however, if new technology is developed, the answer could change. Other questions also must be addressed: Will the heat in a particular well last long enough to justify the cost? Can the system be sealed to minimize water leakage? Can a constant flow of water be sustained?

Even if it proves commercially competitive, hot dry rock mining might encounter obstacles involving site access, hookups to distant power grids, and legal disputes over water rights. If such problems can be overcome, maybe one day the energy used to run your computer will come from water heated kilometers beneath the surface of the Earth.

Fuelwood

In less-developed countries, wood has been the major source of fuel for centuries. In fact, wood is the primary source of energy for nearly half of the world's population. In these regions, the primary use of wood is for cooking.

The use of wood as a prime energy source, a rapid population increase, and the high cost of other types of fuel have combined to create some serious environmental problems in many areas of the world. It is estimated that 1.3 billion people are not able to obtain enough wood or must harvest wood at a rate that exceeds its growth. This has resulted in the destruction of much forest land in Asia and Africa and has hastened the rate of desertification in these regions. (See figure 10.25.)

Because of its bulk and low level of energy compared to equal amounts of coal or oil, wood is not practical to transport over a long distance, so most of it is used locally. In the United States, Norway, and Sweden, wood furnishes 10 percent of the energy for home heating. Canada obtains 3 percent of its total energy, not just home-heating energy, from wood. Most of this energy is used in forest product industries, such as lumbering and paper mills.

Burning wood is also a source of air pollution. Studies indicate that more than 75 organic compounds are released when wood is burned, 22 of which are hydrocarbons known or suspected to be carcinogens. Often, woodstoves are not operated in the most efficient manner, and high amounts of particulate matter and other products of incomplete combustion, such as carbon monoxide, are released, contributing to ill health and death.

In areas with a high population density, the heavy use of wood releases large amounts of fly ash into the air. In Missoula, Montana, in recent years, 55 percent of the particles in the air during the summer were from burning wood. In the winter, wood was responsible for 75 percent of the particles. A number of steps have been taken to reduce air pollution resulting from burning wood. Some cities, such as London, England, have a total ban on burning wood. Vail, Colorado, permits only one wood-burning stove per dwelling. Many areas require woodstoves to have special pollution controls that reduce the amount of particulates and other pollutants released.

figure 10.25 **Desertification** The demand for fuelwood in many regions has resulted in the destruction of forests. This is a major cause of desertification.

Table 10.2 Amount of Solid Waste Produced per Capita

High-income Cities	Kilograms per Day
New York, United States	1.8
Singapore	1.6
Tokyo, Japan	0.94
Rome, Italy	0.72
Low-income Cities	
Tunis, Tunisia	0.56
Medellin, Colombia	0.54
Calcutta, India	0.51
Kano, Nigeria	0.46

Solid Waste

Residents of New York City discard in excess of 25,000 metric tonnes (27,500 tons) of waste each day, which is 1.8 kilograms (4 pounds) per person. In fact, New Yorkers lead the world in the production of municipal trash. High-income cities such as New York usually produce more waste than low-income cities. (See table 10.2.) About 80 percent of this waste is combustible and, therefore, represents a potential energy source. (See figure 10.26.)

"Trash power," the use of municipal waste as a source of energy, requires several steps. First, the waste must be sorted so that the burnable organic material is separated from the inorganic material. The sorting is accomplished most economically by the person who produces the waste. That means the producer must separate trash before putting it out for pickup. It must be separated into garbage, burnable materials, glass, and metals, and picked up by compartmentalized collection trucks. Second, the efficient use of waste as a source of fuel requires a large volume and a dependable supply. If a community constructs a facility to burn 200 metric tonnes (220 tons) of waste a day, the community must generate and collect 200 metric tonnes (220 tons) of waste each day.

Communities have been burning trash as a means of reducing the volume of waste for a number of years. The first consolidated incineration of waste was done in Nottingham, England, in 1874. Burning the municipal waste of Munich, Germany, not only reduces the volume of the waste but supplies the energy for 12 percent of the city's electricity. Rotterdam, the Netherlands, operates a 55-megawatt power plant from its garbage. In the United States, only 3 percent of household waste is burned. This is low compared to the 26 percent burned in Japan, 51 percent in Sweden, and 75 percent in Switzerland.

Burning trash is not profitable. It is a way to decrease the cost of trash disposal because it reduces the need for landfill sites. Baltimore produces gas from 1000 metric tonnes (1,100 tons) of waste per day. This gas generates steam for the

figure 10.26 **Waste to Energy** Municipal trash can be burned to produce heat and electricity. This refuse pit is used to feed hoppers of high-temperature furnaces.

Baltimore Gas and Electric Company. The daily waste also yields 80 metric tonnes (88 tons) of a charcoal-like material (char), 70 tonnes (77 tons) of ferrous metals, and 170 metric tonnes (190 tons) of glass. These products all have potential uses. The char can be burned as an additional source of energy. The ferrous metals can be sold, reducing the cost of operating the plant, as well as conserving mineral resources. The glass can be sold for recycling, further reducing the economic costs of the plant.

Although the burning of trash reduces the trash volume and furnishes energy, it poses environmental concerns, one of which is air pollution. Many of the older incinerating plants do not comply with today's air-quality standards. Also, much of the waste material, such as bleached paper and plastics, have chlorine-containing organic compounds. When burned, these compounds can form dioxins, which are highly toxic and suspected carcinogens. Another problem associated with waste-to-energy systems is the popularity of recycling. Many of the items that are now recycled, such as plastics and wood, have high heat content. The reduction in the amount of these items in the waste stream reduces its value as an energy source.

Energy Conservation

Many observers have pointed out that demanding more energy while failing to conserve is like demanding more water to fill a bathtub while leaving the drain open. To be sure, conservation and efficiency strategies by themselves will not eliminate demands for energy, but they can make the demands much easier to meet, regardless of what options are chosen to provide the primary energy.

Much of the energy we consume is wasted. This statement is not meant as a reminder to simply turn off lights and lower furnace thermostats; it is a technological challenge. Our use of energy is so inefficient that most potential energy in fuel is lost as waste heat, becoming a form of environmental pollution.

Many conservation techniques are relatively simple and highly cost effective. More efficient and less energy intensive industry, transportation, and domestic practices—could save large amounts of energy. Improved automobile efficiency, better mass transit, and increased railroad use for passenger and freight traffic are simple and readily available means of conserving transportation energy. In response to the 1970s' oil price shocks, automobile gas-mileage averages in the United States more than doubled, from 5.55 kilometers/liter (13 mpg) in 1975 to 12.3 kilometers/liter (28.8 mpg) in 1988. Unfortunately, the oil glut and falling fuel prices of the late 1980s discouraged further conservation. Between 1990 and 1997, the average slipped to only 11.8 kilometers/liter (27.6 mpg). It remains to be seen if the sharp increase of gasoline early in 2000 will translate into increased miles per gallon in new car design.

Conservation is not a way of generating energy, but it is a way of reducing the need for additional energy consumption and saves money for the consumer. Some conservation technologies are sophisticated, while others are quite simple. For example, if a small, inexpensive wood-burning stove were developed and used to replace open fires in the less-developed world, energy consumption in these regions could be reduced by 50 percent.

Several technologies that reduce energy consumption are now available. (See figure 10.27.) Highly efficient fluorescent light bulbs that can be used in regular incandescent fixtures give the same amount of light for 25 percent of the energy, and they produce less heat. Since lighting and air conditioning (which removes the heat from inefficient incandescent lighting) account for 25 percent of U.S. electricity consumption, widespread use of these lights could significantly reduce energy consumption. Low-emissive glass for windows can reduce the amount of heat entering a building while allowing light to enter. The use of this glass in new construction and replacement windows could have a major impact on the energy picture. Many other technologies, such as automatic dimming devices or automatic light-shutoff devices, are being used in new construction.

The shift to more efficient use of energy needs encouragement. Often, poorly designed, energy-inefficient buildings and machines can be produced inexpensively. The short-term cost is low, but the long-term cost is high. The public needs to be educated to look at

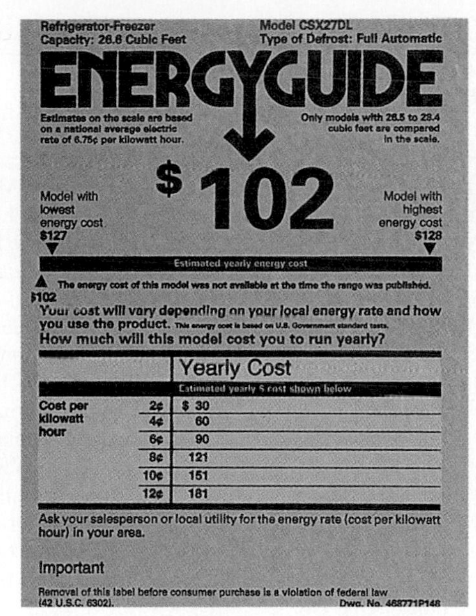

figure 10.27 **Energy Conservation** The use of fluorescent light bulbs, energy-efficient appliances, and low-emissive glass could reduce energy consumption significantly.

the long-term economic and energy costs of purchasing poorly designed buildings and appliances.

Electric utilities have recently become part of the energy conservation picture. In some states, they have been allowed to make money on conservation efforts; previously, they could make money only by building more power plants. This encourages them to become involved in energy conservation education, because teaching their customers how to use energy more efficiently allows them to serve more people without building new power plants.

Summary

A resource is a naturally occurring substance of use to humans, a substance that can potentially be extracted using current technology. Reserves are known deposits from which materials can be extracted profitably with existing technology under present economic conditions.

Coal is the world's most abundant fossil fuel. Coal is obtained by either surface mining or underground mining. Problems associated with coal extractions are disruption of the landscape due to surface mining and subsidence due to underground mining. Black lung disease, waste heaps, water and air pollution, and acid mine drainage are additional problems. Oil was originally chosen as an alternative to coal because it was more convenient and less expensive. However, the supply of oil is limited. As oil becomes less readily available, multiple offshore wells, secondary recovery methods, and increased oil exploration will become more common. Natural gas is another major source of fossil-fuel energy. The primary problem associated with natural gas is transport of the gas to consumers.

Fossil fuels are nonrenewable: The amounts of these fuels are finite. When the fossil fuels are exhausted, they will have to be replaced with other forms of energy, probably renewable forms. Hydroelectric power can be increased significantly, but its development must flood areas and in so doing may require the displacement of people. The use of geothermal and tidal energy is limited by geographic locations. Wind power may be used to generate electricity but may require wide, open areas and a large number of wind generators. Solar energy can be collected and used in either passive or active systems and can also be used to generate electricity. Lack of a constant supply of sunlight is solar energy's primary limitation. Fuelwood is a minor source of energy in industrialized countries but is the major source of fuel in many less-developed nations. Biomass can be burned to provide heat for cooking or to produce electricity, or it can be converted to alcohol or used to generate methane. In some communities, solid waste is burned to reduce the volume of the waste and also to supply energy.

Energy conservation can reduce energy demands without noticeably changing standards of living.

The Arctic National Wildlife Refuge and Oil

ISSUES & ANALYSIS

The Arctic National Wildlife Refuge (ANWR) has been a source of controversy for many years. The major players are environmentalists who seek to preserve this region as wilderness; the state of Alaska, which funds a major portion of its activities with dividends from oil production; Alaska residents, who receive a dividend payment from oil revenues; oil companies that want to drill in the refuge; and members of Congress who see the oil reserves in the region as important economic and political issues.

In 1960, 3.6 million hectares (8.9 million acres) were set aside as the Arctic National Wildlife Range. Passage of the Alaskan National Interest Lands Conservation Act in 1980 expanded the range to 8 million hectares (19.8 million acres) and established 3.5 million hectares (8.6 million acres) as wilderness. The act also renamed the area the Arctic National Wildlife Refuge. There are international implications to this act. The refuge borders the Northern Yukon National Park. Many animals, particularly members of the Porcupine caribou herd, travel across the border on a regular yearly migration. The United States is obligated by treaty to protect these migration routes.

The act requires specific authorization from Congress before oil drilling or other development activities can take place on the coastal plain in the refuge. The coastal plain has the greatest concentration of wildlife, is the calving ground for the Porcupine caribou, and has the greatest potential for oil production. In each case, the potential authorization caused a collision of three forces: environmental protection, economic development, and political benefit. Furthermore, there are great differences of opinion within each of the competing interest groups. Some Alaskan citizens support drilling; others oppose it. Members of Congress are similarly split. Even members of the Department of the Interior have provided conflicting testimony about the risks and benefits of drilling for oil in the refuge.

In 1998, the Secretary of the Interior, acting under the recommendation of President Clinton, cleared the way for oil development on Alaska's North Slope. Under the plan, about a third of a 4.6-million-acre study area in the northeastern corner of the federal reserve would be off limits to drilling. Oil leases would be sold on 4 million acres of the government's National Petroleum Reserve west of the Prudhoe Bay oil fields. Some of the leases would allow only slant drilling because the surface is to be protected. In 2000, the Energy Information Administration (EIA) released a report on the potential oil production from the coastal plain of ANWR. The coastal plain region, which comprises approximately 8 percent of the 7.7 million hectares (19 million acres) ANWR, is the largest unexplored, potentially productive geologic onshore basin in the United States. A decision on permitting the exploration and development is up to the U.S. Congress. The EIA report estimated a 95 percent probability that at least 5.7 billion barrels of technically recoverable undiscovered oil are in the ANWR coastal plain. There is a 5 percent probability that at least 16 billion barrels of oil are recoverable. The report states that once oil has been discovered, more than 80 percent of the technically recoverable oil is commercially developable at an oil price of $25 per barrel (oil was $32 per barrel in July of 2000). The value of the oil in 2000 dollars could be between $125 and $350 billion. Oil companies have repeatedly stated that the oil can be recovered without endangering wildlife or the fragile arctic ecosystem. Conservationists have argued that none of the reserve should be developed when improvements in energy could reduce demand. They argue that drilling in the reserve will harm the habitat of millions of migratory birds, caribou, and polar bears. Only time will tell which side, if either, is correct.

- What do you think of the decision to drill?
- Was the decision a compromise between two opposite sides or a politically motivated answer to a controversial issue?
- What would your decision have been?

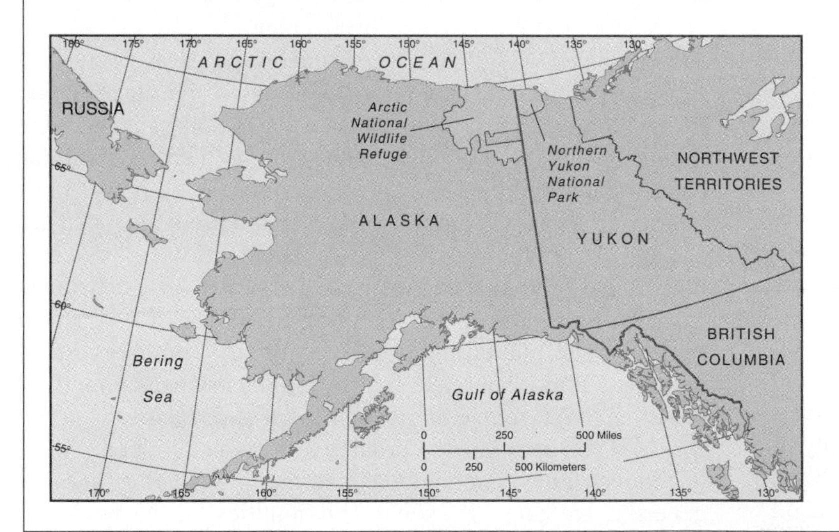

Migrating caribou in the Arctic National Wildlife Refuge.

Interactive Exploration

Check out the website at

http://www.mhhe.com/environmentalscience

and click on the cover of this textbook for interactive versions of the following:

KNOW THE BASICS

acid mine drainage *194*
active solar system *203*
biomass *205*
black lung disease *193*
geothermal energy *201*
liquefied natural gas *196*
nonrenewable energy sources *188*
overburden *192*
passive solar system *203*
photovoltaic cell *204*
renewable energy sources *188*
reserves *188*
resources *188*
secondary recovery *195*
surface mining *192*
underground mining *192*

On-line Flashcards

Electronic Glossary

IN THE REAL WORLD

Would you commit suicide if your culture and natural environment were about to be destroyed? The native people of a remote cloud forest are threatening mass suicide if oil exploration and drilling begins. For this controversy, read Ecuadorian Native Group Threatens Mass Suicide in Opposition to Drilling.

Another controversial decision regarding oil exploration and drilling is in a remote coastal Alaska reserve. It is hoped that this decision by the United States government will relieve pressure to open the Arctic National Wildlife Refuge. For the full story, check out Historic Decision to Allow Drilling in Remote Alaska Oil Reserve.

One of the arguments for opening the Arctic National Wildlife Refuge to oil exploration is that there would be minimum environmental impact. Unfortunately, some past practices by oil companies do not support these claims. What would a company do to maximize profits? Take a look at British Petroleum/Amoco Admits to Dumping Toxic Waste on Alaska's North Slope to see what happened.

Another environmental consequence of oil production, use, and transport is the probabilities of spills. For the update of an oil spill that received heavy media coverage ten years ago, read The Exxon Valdez Oil Spill, Ten Years Later.

Two other stories that received recent attention include New Carissa Oil Spill on the Oregon Coast and Major Oil Spill.

What's tall, huge, and abundant in the Gulf of Mexico? More than 4000 oil and gas production platforms loom in the seascape of the Gulf of Mexico. What does one do when they are no longer used? The Oilrigs As Artificial Reefs case study details a relatively inexpensive method that has the potential to also increase biodiversity.

Building a dam . . . Environmental catastrophe, cheap electricity, or protection against future flooding? For a better understanding of this issue, read Floods Devastate Coastal Mozambique.

TEST PREPARATION

Review Questions

1. Why are fossil fuels important?
2. Distinguish between reserves and resources.
3. What are the advantages of surface mining of coal compared to underground mining? What are the disadvantages of surface mining?
4. Compare the environmental impacts of the use of coal and the use of oil.
5. What are some limiting factors in the development of new hydroelectric generating sites?
6. What factors limit the development of tidal power as a source of electricity?
7. In what parts of the world and why is geothermal energy available?
8. Why can wind be considered a form of solar energy?
9. Compare a passive solar-heating system with an active solar-heating system.
10. What problems are associated with the use of solid waste as a source of energy?
11. List three energy conservation techniques.

Critical Thinking Questions

1. Given what you know about the economic and environmental costs of different energy sources, would you recommend that your local utility company use hydroelectricity or coal to supplement electric production? What criteria would you use to make your recommendation?
2. Coal-burning electric power plants in the Midwest have contributed to acid rain in the eastern United States. Other energy sources would most likely be costlier than coal, thereby raising electricity rates. Should citizens of another state be able to pressure these utility companies to change the method of generating electricity? What mechanisms might be available to make these changes? How effective are these mechanisms?
3. Imagine you are an official with the Department of Energy and are in the budgeting process for alternative energy research. Where would you put the money? Why?
4. Given your choices from question 3, what do you think the political repercussions of your decision would be? Why?
5. Do you believe that large dam projects like the Three Gorges Dam project in China are, on the whole, beneficial or not? What alternatives would you recommend? Why?
6. Energy conservation is one way to decrease dependence on fossil fuels. What are some things you can do at home, work, or school that would reduce fossil-fuel use and save you money?
7. What alternative energy resources that the text has outlined are most useful in your area? How might these be implemented?

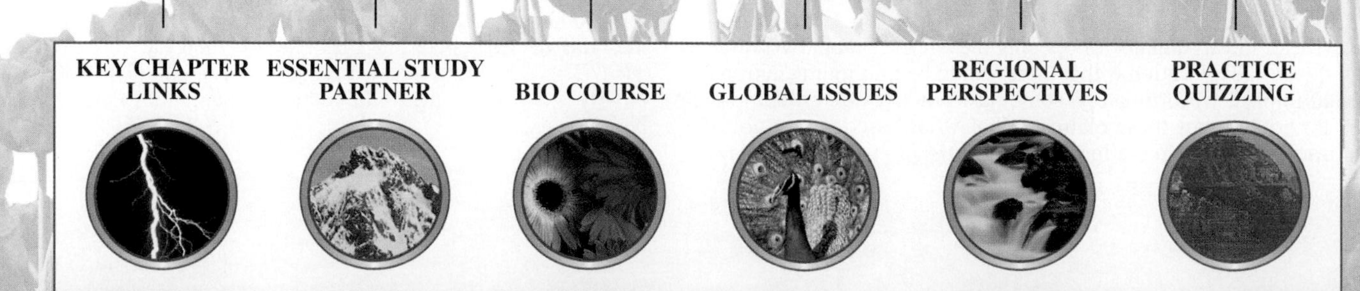

CHAPTER 11

Nuclear Energy: Benefits and Risks

Objectives

After reading this chapter, you should be able to:

- Explain how nuclear fission has the potential to provide large amounts of energy.
- Describe how a nuclear reactor produces electricity.
- Describe the basic types of nuclear reactors.
- Explain the steps involved in the nuclear fuel cycle.
- List concerns regarding the use of nuclear power.
- Explain the problem of decommissioning a nuclear plant.
- Describe how high-level radiation waste is stored.
- Describe the accident at Chernobyl.
- Explain how a breeder reactor differs from other nuclear reactors.
- List the technical problems associated with the design and operation of a liquid metal fast-breeder reactor.
- Explain the process of fusion.

Chapter Outline

The Nature of Nuclear Energy

The History of Nuclear Energy Development

Nuclear Reactors
- Plans for New Reactors Worldwide
- Plant Life Extension

Breeder Reactors

Nuclear Fusion

The Nuclear Fuel Cycle

Nuclear Material and Weapons Production

Nuclear Power Concerns
- Reactor Safety: The Effects of Three Mile Island and Chernobyl
- Exposure to Radiation
- Thermal Pollution
- Decommissioning Costs
- Radioactive Waste Disposal

Global Perspective: *The Nuclear Legacy of the Soviet Union*

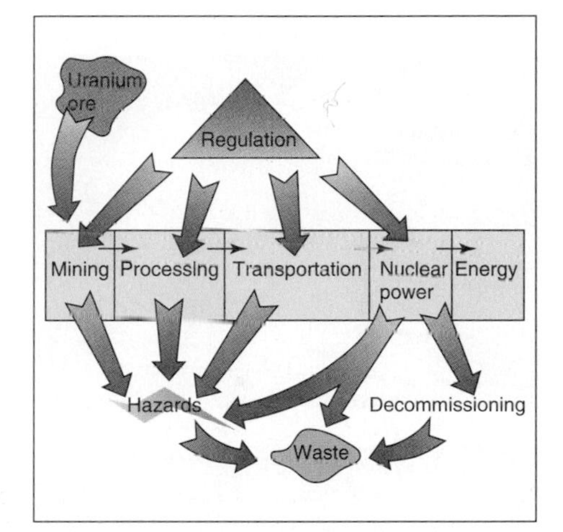

The Nature of Nuclear Energy

Energy from disintegrating atomic nuclei has a tremendous potential to do good for the people of the world. We routinely use X rays to examine bones for fractures, treat cancer with radiation, and diagnose disease with the use of radioactive isotopes. About 17 percent of the electrical energy generated in the world comes from nuclear power plants. Engineers in the former Soviet Union used nuclear explosions to move large amounts of earth and rock to construct dams, canals, and underground storage facilities. The U.S. government briefly considered the possibility of using nuclear explosions to build a new Panama Canal. (See figure 11.1.)

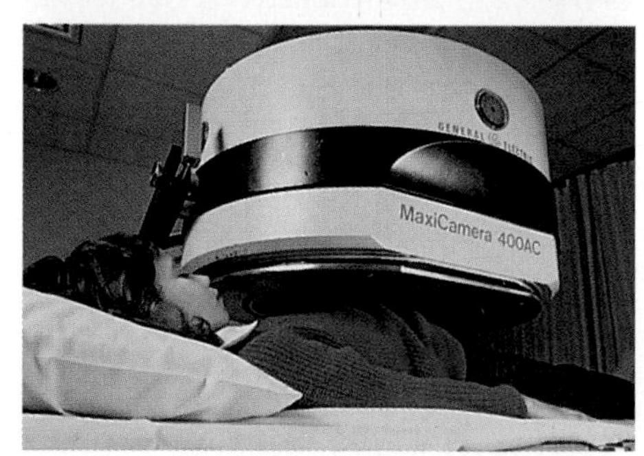

figure 11.1 **Uses of Nuclear Energy**
Each of these photographs illustrates the uses of the energy available from the splitting of atoms.

On the other hand, nuclear energy has the potential to do great harm. The Japanese cities of Hiroshima and Nagasaki were destroyed by nuclear bombs. Military uses of nuclear energy have left a legacy of radioactive wastes. In many cases, these wastes have been mismanaged or carelessly disposed of.

In order to understand where nuclear energy comes from, it is necessary to review some of the aspects of atomic structure presented in chapter 3. All atoms are composed of a central region called the nucleus, which contains positively charged protons, and neutrons that have no charge. Moving around the nucleus are smaller, negatively charged electrons. Since the positively charged particles in the nucleus repel one another, energy is needed to hold the protons and neutrons together. However, some isotopes of atoms are **radioactive;** that is, the nuclei of these atoms are unstable and spontaneously decompose. Neutrons, electrons, protons, and other larger particles are released during nuclear disintegration, and a great deal of energy is released as well. The rate of decomposition is consistent for any given isotope. It is measured and expressed as **radioactive half-life,** which is the time it takes for one-half of the radioactive material to spontaneously decompose. Table 11.1 lists the half-lives of several radioactive isotopes.

Nuclear disintegration releases energy from the nucleus as **radiation,** of which there are three major types: **Alpha radiation** consists of a moving particle composed of two neutrons and two protons. Alpha radiation usually travels through air for less than a meter and can be stopped by a sheet of paper or the outer layer of the skin. **Beta radiation** consists of electrons released from nuclei. Beta particles travel more rapidly than alpha particles and will travel through air for a couple of meters. They are stopped by a layer of clothing, glass, or aluminum. **Gamma radiation** is a type of electromagnetic radiation, like X rays, light, and radio waves. It can pass through several centimeters of concrete. If the radiation reaches living tissue, equivalent doses of beta and gamma radiation cause equal amounts of biological damage. Alpha radiation can cause up to twenty times more damage than beta or gamma radiation because it is able to cause more changes in tissue as it passes through. The ingestion or inhalation of radioactive materials places the source of the radiation in direct contact with cells and therefore this kind of exposure is much more dangerous than exposure from distant sources.

The release of neutrons is particularly important in obtaining energy from nuclear disintegration. In addition to

Table 11.1 The Half-Life of Some Radioactive Isotopes

Radioactive Isotope	Half-Life
Iodine 132	2.4 hours
Technetium 99	6.0 hours
Rhodium 105	36.0 hours
Xenon 133	5.3 days
Barium 140	12.8 days
Cerium 144	284.0 days
Cesium 137	30 years
Carbon 14	5730 years
Uranium 234	250,000 years
Chlorine 36	300,000 years
Beryllium 10	4.5 million years
Potassium 40	1.3 billion years
Helium 4	12.5 billion years

releasing alpha, beta, and gamma radiation when they disintegrate, the nuclei of atoms release neutrons. When moving neutrons hit the nuclei of certain other atoms, they can cause those nuclei to split as well. This process is known as **nuclear fission.** If these splitting nuclei also release neutrons, they can strike the nuclei of other atoms, which also disintegrate, resulting in a continuous process called a **nuclear chain reaction.** Only certain kinds of atoms are suitable for the development of a nuclear chain reaction. The two materials commonly used in nuclear reactions are uranium-235 and plutonium-239. In addition, there must be a certain quantity of nuclear fuel (a critical mass) in order for a nuclear chain reaction to occur. It is this process that results in the large amounts of energy released from bombs or nuclear reactors.

The History of Nuclear Energy Development

The first controlled fission of an atom occurred in Germany in 1938, but the United States was the first country to develop an atomic bomb. In 1945, the U.S. military dropped atomic bombs on the Japanese cities of Hiroshima and Nagasaki. The incredible devastation of these two cities demonstrated the potential of nuclear energy for destruction. For many years, most atomic research involved military applications of nuclear energy as bombs and as power sources for ships. During the 50 years following World War II, the two major military powers of the world—the United States and the former Soviet Union—conducted secret nuclear research projects related to the building and testing of bombs. This continued to be a primary focus of nuclear research until the recent changes in the former Soviet Union, which led to a world in which nuclear war is much less of a concern—although the explosion of nuclear devices in June of 1998 by India and Pakistan has heightened concern about nuclear war somewhat. A legacy of this military research is a great deal of soil, water, and air contaminated with radioactive material. Many of these contaminated sites have come to light recently and require major cleanup efforts. The U.S. Department of Energy has begun to clean up the pollution created by its weapons production activities and will begin shipping transuranic waste to a storage site in New Mexico in the near future.

After World War II, people began to see the potential for using nuclear energy for peaceful purposes rather than as weapons. The world's first electricity-generating reactor was constructed in the United States in 1951, and the Soviet Union built its first reactor in 1954. In December 1953, President Dwight D. Eisenhower, in his "Atoms for Peace" speech, made the following prediction:

> "Nuclear reactors will produce electricity so cheaply that it will not be necessary to meter it. The users will pay an annual fee and use as much electricity as they want. Atoms will provide a safe, clean, and dependable source of electricity."

More than 45 years have passed since Eisenhower's predictions. Although nuclear power currently is being used throughout the world as a reliable source of electricity, it has not fulfilled such overly optimistic promises. Several serious accidents have caused worldwide concern about safety, and construction of most new nuclear power projects has stopped. At the same time, many energy experts predict a rebirth of the nuclear power industry as energy demands increase and a new generation of safer nuclear power plants is designed. These experts believe the public will favor nuclear power plants because they do not generate carbon dioxide, which contributes to global warming.

Nuclear Reactors

A **nuclear reactor** is a device that permits a controlled fission chain reaction. In the reactor, neutrons are used to cause a controlled fission of heavy atoms, such as uranium. **Uranium-235 (U-235)** is a uranium isotope used to fuel nuclear fission reactors. When the nucleus of a U-235 atom is struck by a slowly moving neutron from another atom, the nucleus of the atom of U-235 is split into several smaller particles. Because the nucleus will split, it is said to be **fissionable.** When the nucleus is split, two to three rapidly moving neutrons are released, along with large amounts of energy. This energy is an important product of nuclear fission reactions. The neutrons released strike the nuclei of other atoms of U-235 and also cause them to undergo fission, which, in turn, releases more energy and more neutrons, thus resulting in a chain reaction. (See figure 11.2.) Once begun, this chain reaction continues to release energy until the fuel is spent or the neutrons are prevented from striking other nuclei.

In addition to fuel rods containing uranium, reactors contain control rods of cadmium, boron, graphite, or other nonfissionable materials used to control the rate of fission by absorbing neutrons. When control rods are lowered into a reactor, they absorb the neutrons produced by fissioning uranium. There are fewer neutrons to continue the chain reaction, and the rate of fission decreases. If the control rods are withdrawn, more fission occurs, and more particles, radiation, and heat are produced.

The fuel rods housed in a reactor are surrounded by water or some other type of moderator. A **moderator** absorbs energy, which slows neutrons, enabling them to split the nuclei of other atoms more effectively. Fast-moving neutrons are less effective at splitting atoms than slow-moving neutrons. As U-235 undergoes fission, the energy of the fast-moving neutrons is transferred to water; the neutrons slow down, and the water is heated.

In the production of electricity, a nuclear-powered reactor serves the same function as any fossil-fueled boiler. It produces heat, which converts water to steam to operate a turbine that generates electricity. After passing through the turbine, the steam must be cooled, and the water is returned to the reactor to be heated again. Various types of reactors have been constructed to furnish heat for the production of steam. They differ in the moderator used, in how the reactor core is cooled, and in how the heat from the core is used to generate steam. Water is the most commonly used reactor-core coolant and also serves as a neutron moderator. **Light-water reactors (LWR),** which make up 90 percent of reactors operating today, use ordinary water, which contains the lightest, most common isotope of hydrogen, having an atomic mass of one. The two types of LWRs are **boiling-water reactors (BWR)** and **pressurized-water reactors (PWR).**

In a BWR (the type of construction used in about 20 percent of the nuclear reactors in the world), the water functions as both a moderator and reactor-core coolant. (See figure 11.3.) Steam is formed within the reactor and transferred directly to the turbine, which generates electricity. A disadvantage of the BWR is that the steam passing to the turbine must be treated to remove any radiation. Even then, some radioactive material is left in the steam; therefore, the generating building must be shielded.

In a PWR (the type of construction used in over 70 percent of the nuclear reactors in the world), the water is kept under high pressure so that steam is not allowed to form in the reactor. (See figure 11.4.) A secondary loop transfers the heat from the pressurized water in the reactor to a steam generator. The steam is used to turn the turbine and generate electricity. Such an arrangement reduces the risk of radiation in the steam but adds to the cost of construction by requiring a secondary loop for the steam generator.

A third type of reactor that uses water as a coolant is a **heavy-water reactor (HWR),** developed by Canadians. It uses water that contains the hydrogen isotope deuterium in its molecular structure as the reactor-core coolant and moderator. Since the deuterium atom is twice as heavy as the more common hydrogen isotope, the water that contains deuterium weighs slightly more than ordinary water. An HWR also uses a steam generator to convert regular water to steam in a secondary loop. Thus, it is similar in structure to a PWR. The major advantage of an HWR is that naturally occurring uranium isotopic mixtures serve as a suitable fuel. This is possible because heavy water is a better neutron moderator than is regular water, while other reactors require that the amount of U-235 be enriched to get a suitable fuel. Since it does not require enriched fuel, the operating costs of an HWR are less than that of an LWR.

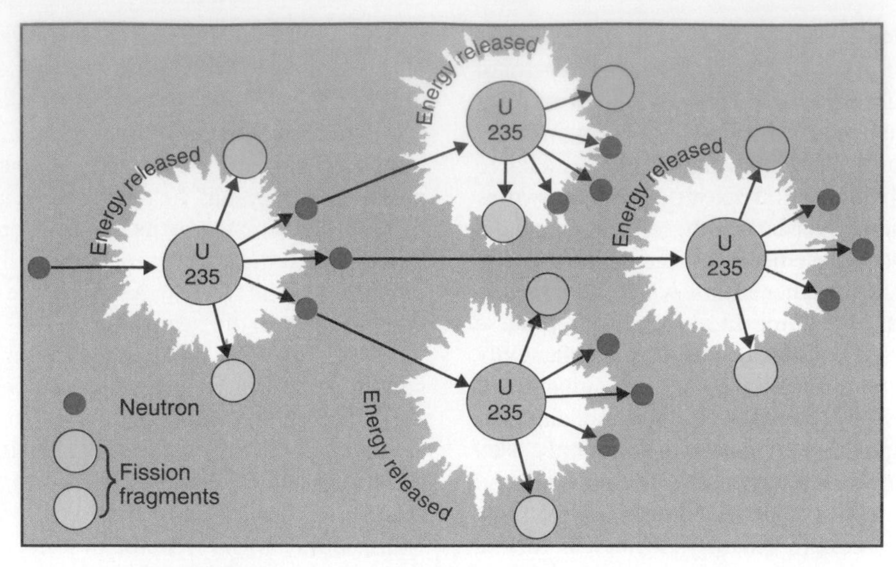

figure 11.2 **Nuclear Fission Chain Reaction** When a neutron strikes a nucleus of U-235, energy is released, and several fission fragments and neutrons are produced. These new neutrons may strike other atoms of U-235, causing their nuclei to split. This series of events is called a chain reaction.

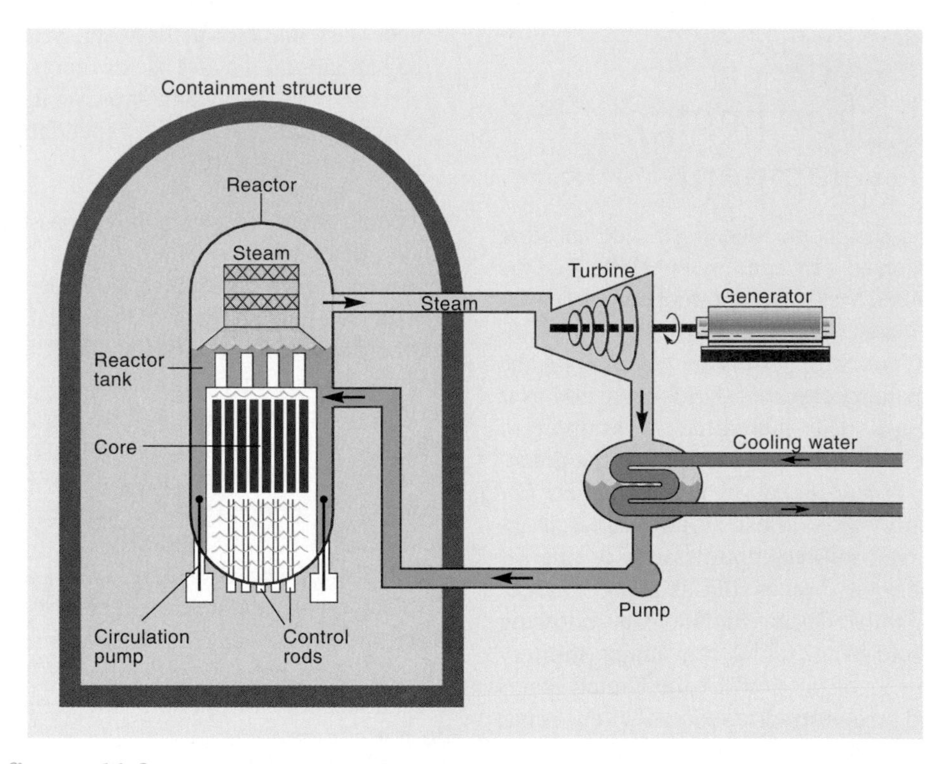

figure 11.3 **Boiling-Water Reactor (BWR)** A boiling-water reactor is a type of light-water reactor that produces steam to directly power the turbine and produce electricity. Water is used as a moderator and as a reactor-core coolant.

The **gas-cooled reactor (GCR)** was developed by atomic scientists in the United Kingdom. Carbon dioxide serves as a coolant for a graphite-moderated core. As in the HWR, natural isotopic mixtures of uranium are used as a fuel. (See figure 11.5.)

The various types of reactors represent differing approaches to building safe, economical plants. Each method has its advantages and disadvantages, and its supporters and critics. Modifications will continue to be made to these basic types. Figure 11.6 shows the distribution of nuclear power plants in North America and table 11.2 shows numbers of reactors present in each country.

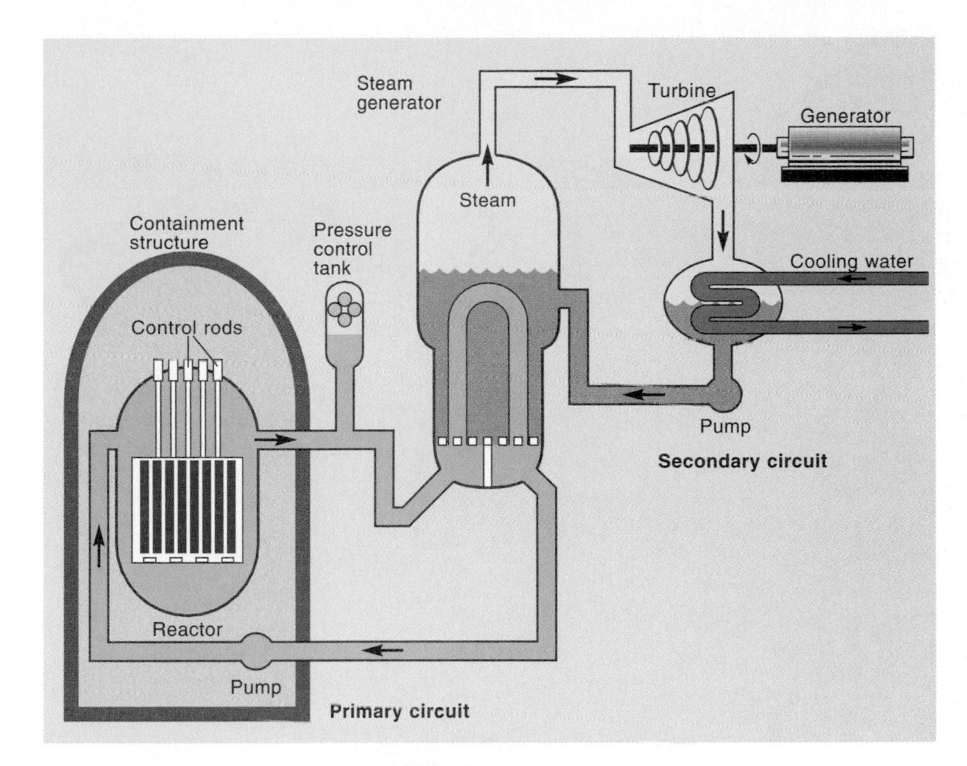

figure 11.4 **Pressurized-Water Reactor (PWR)** A pressurized-water reactor is a type of light water reactor that uses a steam generator to form steam and a secondary loop to transfer this steam to the turbine.

Plans for New Reactors Worldwide

Currently, there are 438 nuclear power reactors in 31 countries, with a combined capacity of 352 Gwe. In 2000, these provided 2401 billion kWh, over 16 percent of the world's electricity. Although some countries, notably Japan, China, and the Republic of Korea, intend to continue major nuclear power construction programs, the rate of growth of installed nuclear-generating capacity over the next ten years is expected to be low.

Some 35 power reactors are currently being constructed in 13 countries, notably China, the Republic of Korea, and Japan. Construction is well advanced on many of them and, based on reported progress and allowing for delays in countries, 13 with a total net capacity of over 9000 MWe are expected to be in operation before 2004. This excludes two Russian and two Ukrainian reactors for which funding is uncertain.

After about 2005, forecasts of installed nuclear capacity become much less certain. When the present construction programs are completed, most significant nuclear power growth is expected to continue only in the Asian region. The International Atomic Energy Agency (IAEA) forecasts that the total installed nuclear capacity in 2015 will be little more than that in 2000, with the nuclear share of world electricity output decreased from 17 percent in 1997 to 13 percent in 2015.

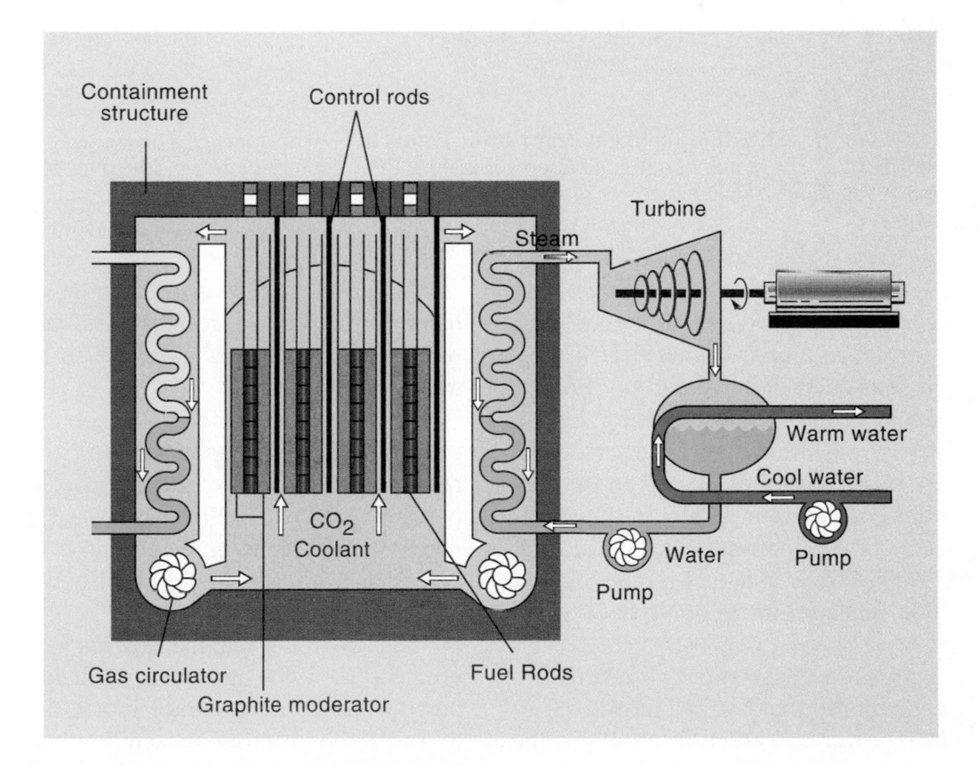

figure 11.5 **Advanced Gas-Cooled Reactor** This type of reactor uses graphite as a moderator and the gas carbon dioxide as the reactor-core coolant. A steam generator forms steam and a secondary loop is used to transfer steam to the turbine.

figure 11.6 **Distribution of Nuclear Power Plants in North America** The United States has the largest number of nuclear power plants of any country in the world—over 100. Canada has about 20 nuclear power plants and Mexico has two.

At least six countries with existing nuclear power programs (Russia, China, India, Japan, South Korea, and Taiwan) and four countries without any present nuclear capacity (Indonesia, Egypt, Turkey, and Iran) have plans to build new power reactors beyond those now under construction. In all, 59 power reactors with a total net capacity of over 57,000 MWe are planned. None of these are in western Europe or North or South America.

In France, the only western European country that has had an active nuclear power construction program, the national utility announced in 1994 that it would order no new generating capacity, nuclear or conventional, until about 2002. Sites have, however, been designated for new power reactors, and new reactor construction is expected to resume by 2004.

In a break with the rest of western Europe, Germany announced in 2000 that it would close all of their 19 nuclear power plants by 2020. Germany, Europe's largest economy, currently receives one-third of its electrical power from nuclear sources.

In eastern Europe, the Russian government in 1997 approved a nuclear power construction program. In addition to the three reactors presently under construction, a further 9, taking total capacity to about 29,200 MWe, are planned to be operating by 2010. Several of the oldest Russian reactors are expected to be retired by 2010, and it is Russia's announced intention to replace retired nuclear capacity by new construction at the same site, to optimize the use of established infrastructure and personnel.

Most planned reactors are in the Asian region, with fast-growing economies and rapidly rising electricity demand. Nuclear power will continue to play a major role in the future electricity supply mix in both South Korea and Japan. South Korea plans to bring a further 12 reactors, with a total capacity of 13,100 MWe, into operation by the year 2015. Japan has plans and, in most cases, designated sites and announced timetables for a further 20 power reactors, totaling over 25,000 MWe, and some of these are negotiating the governmental approval process. Fulfilling the necessary conditions for approval can take over a decade.

China, with three operating reactors, has begun the next phase of its nuclear power program. Construction has started on Qinshan 2 & 3 (1200 MWe), two French 900 MWe units for Lingao, Guangdong, two 700 MWe Canadian-designed CANDU reactors at Qinshan, and two Russian 950 MWe at Lianyungang. These are expected to start up from 2002 to 2005, and to add some 6200 MWe to the existing 2167 MWe nuclear capacity. China plans to increase its nuclear capacity to 20,000 MWe by 2010.

Plant Life Extension

Most nuclear power plants originally had a nominal design lifetime of up to 40 years, but engineering assessments of many plants over the last decade has established that many can operate longer. In the United States, most reactors now have confirmed life spans of 40 to 60 years, and in Japan, 40 to 70 years. In the United States, the first two reactors have been granted license renewals, which extends their operating lives to 60 years.

When the oldest commercial nuclear power stations in the world, Calder Hall and Chapelcross in the United Kingdom, were built in the 1950s, it was assumed that they would have a useful

Table 11.2 World Nuclear Power Reactors 1999–2000 and Uranium Requirements

Country	Reactors Operating		Reactors Building		On Order or Planned		Uranium Required
	No.	Megawatt	No.	Megawatt	No.	Megawatt	tonnes U
Argentina	2	935	1	692	0	0	149
Armenia	1	376	0	0	0	0	72
Belgium	7	5680	0	0	0	0	1050
Brazil	1	626	1	1245	0	0	652
Bulgaria	6	3538	0	0	0	0	574
Canada	18	12058*	0	0	0	0	1325
China	3	2079	8	6320	2	1800	717
Czech Republic	4	1648	2	1824	0	0	750
Egypt	0	0	0	0	1	600	0
Finland	4	2656	0	0	0	0	569
France	59	63203	0	0	0	0	10528
Germany	20	22326	0	0	0	0	3615
Hungary	4	1742	0	0	0	0	336
India	12	2144	4	1304	10	4480	262
Indonesia	0	0	0	0	1	600	0
Iran	0	0	1	950	3	2850	0
Japan	53	43505	1	796	14	18288	7882
Korea DPR (North)	0	0	0	0	2	1900	0
Korea RO (South)	16	12970	4	3800	10	11200	2393
Lithuania	2	2370	0	0	0	0	360
Mexico	2	1308	0	0	0	0	224
Netherlands	1	452	0	0	0	0	92
Pakistan	2	425	0	0	0	0	53
Romania	1	650	1	620	0	0	94
Russia	29	19843	3	2825	9	7450	3948
Slovakia	6	2472	0	0	0	0	730
Slovenia	1	620	0	0	0	0	129
South Africa	2	1842	0	0	0	0	358
Spain	9	7345	0	0	0	0	1515
Sweden	11	9445	0	0	0	0	1580
Switzerland	5	3170	0	0	0	0	578
Taiwan	6	4884	2	2600	0	0	940
Turkey	0	0	0	0	1	1400	0
Ukraine	14	12120	2	1900	0	0	1898
United Kingdom	33	12518	0	0	0	0	2481
USA	104	98015	0	0	0	0	18739
WORLD	**438**	**352,965**	**30**	**24,876**	**53**	**50,568**	**64,593**

Reprinted by permission of Uranium Information Centre Ltd., Melbourne, Australia

lifetime of 20 years. As of 2001, they were still authorized to operate.

Sweden's oldest reactor, which started up in 1971, has been fully rebuilt at a cost equivalent to 8 percent of a replacement unit, and all Sweden's reactors are maintained so that a further 20 years of life is in prospect.

It should be noted, however, that economic, regulatory, and political considerations have led to the premature closure of some power reactors. In the United States, reactor numbers have fallen from 110 to 104. Germany, as previously mentioned, will close all of its plants by 2020.

Breeder Reactors

During the early stages of the development of nuclear power plants, breeder reactor construction was seen as the logical step after nuclear fission development. A regular fission reactor produces heat to generate electricity but does not

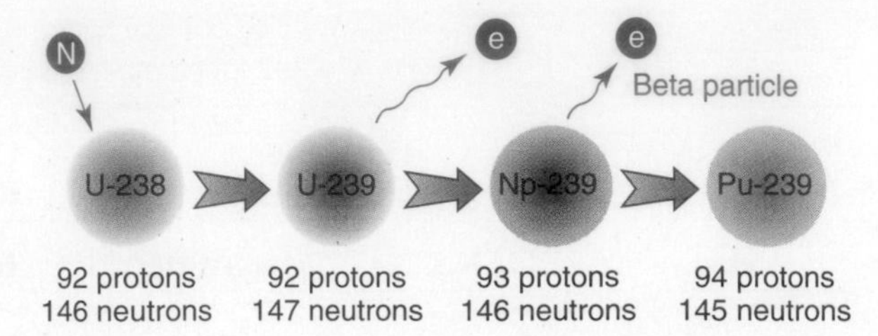

figure 11.7 **Formation of Pu-239 in a Breeder Reactor** When a fast-moving neutron (N) is absorbed by the nucleus of a U-238 atom, a series of reactions results in the formation of Pu-239 from U-238. Two intermediate atoms are U-239 and Np-239, which release beta particles (electrons) from their nuclei. (A neutron in the nucleus can release a beta particle and become a proton.) This is an important reaction because, while the U-238 does not disintegrate readily and therefore is not a nuclear fuel, the Pu-239 is fissionable and can serve as a nuclear fuel.

form radioactive products useful as a fuel. A **nuclear breeder reactor** is a nuclear fission reactor that produces heat to be converted to steam to generate electricity and also forms a new supply of radioactive isotopes. If a fast-moving neutron hits a uranium-238 (U-238) nucleus and is absorbed, an atom of fissionable **plutonium-239 (Pu-239)** is produced. (See figure 11.7.) In a breeder reactor, water is not used as a moderator because water slows the neutrons too much and Pu-239 is not produced. Breeder reactors need a moderator that allows the neutrons to move more rapidly and that also has good heat transfer properties.

The **liquid metal fast-breeder reactor (LMFBR)** appears to be the most promising model. In this type of reactor, the fuel rods in the core are surrounded by rods of U-238 and liquid sodium. The energy of the neutrons released from U-235 in the fuel rods heats the sodium to a temperature of 620°C. In addition to furnishing heat, these fast-moving neutrons are absorbed by the rods containing U-238, and some of these atoms are converted into Pu-239. After approximately 10 years of operation, during which electricity is produced, the LMFBR will have also produced enough radioactive material to operate a second reactor.

There are some serious drawbacks to the LMFBR. Sodium reacts violently if it comes into contact with water or air. Therefore, costly, highly specialized equipment is required to contain and pump the sodium. If these systems fail, the sodium boils, which allows the chain reaction to proceed at a faster rate and could damage the reactor, leading to a nuclear accident.

There are also problems in the startup of an LMFBR. When a breeder reactor is starting up or going on-line after a shutdown, the solid sodium cannot be moved by the pumps until it becomes a liquid. This presents a technical problem in developing LMFBRs.

Another problem is that reaction rates are extremely rapid and very difficult to regulate. The instrumentation needed to monitor an LMFBR must be of extremely high quality because control of the reaction involves precise adjustments over very short periods of time.

Finally, the product of a breeder reactor, plutonium-239, is extremely hazardous to humans who come in contact with it. And, because plutonium-239 can be made into nuclear weapons, it must be transported, processed, or produced under very close security. The more breeder reactors in use, the more difficult the security problems and the more likely the chance that the small amount of plutonium needed to manufacture a bomb could be stolen.

Because of these problems, no breeder reactors are scheduled for commercial use in the United States. In fact, the only experimental U.S. breeder reactor is scheduled to be shut down. The Clinch River Fast-Breeder Reactor Plant near Knoxville, Tennessee, which was to be operational in the 1970s, was never completed.

In Europe, as in the United States, the development of breeder reactors has slowed. In 1981, after running smoothly for eight years, a 250-megawatt French prototype plant developed a leak in the cooling system that resulted in a sodium fire. This raised some serious questions about the safety of breeder reactors. In 1982, the French government announced that it was scaling down the planned construction of breeder reactors from five to one. Today, it is doubtful that even that one will be built. There are a total of five LMFBRs in operation in the world today, in France, Japan, Kazakhstan, and Russia.

Nuclear Fusion

Another aspect of nuclear power that may have promise for the future involves the even more advanced technology of nuclear fusion. When two lightweight atomic nuclei combine to form a heavier nucleus, a large amount of energy is released. This process is known as **nuclear fusion.** The energy produced by the sun is the result of fusion. Most studies of fusion have involved small atoms like hydrogen. Most hydrogen atoms have one proton and no neutrons in the nucleus. The hydrogen isotope deuterium (H^2) has a neutron and a proton. Tritium (H^3) has a proton and two neutrons. When deuterium and tritium isotopes combine to form heavier atoms, large amounts of energy are released. (See figure 11.8.) The energy that would be released by combining the deuterium in 1 cubic kilometer of ocean water would be greater than that contained in the world's entire supply of fossil fuels.

Although fusion could solve the world's energy problems, technology must answer several questions before we can actually use fusion power. Three conditions must be met simultaneously if fusion is to occur: high temperature, adequate density, and confinement. If heat is used to provide the energy necessary for fusion, the temperature must approach that of the center of the sun. At the same time, the walls of the vessel confining the atoms must be protected from the heat, or they will vaporize.

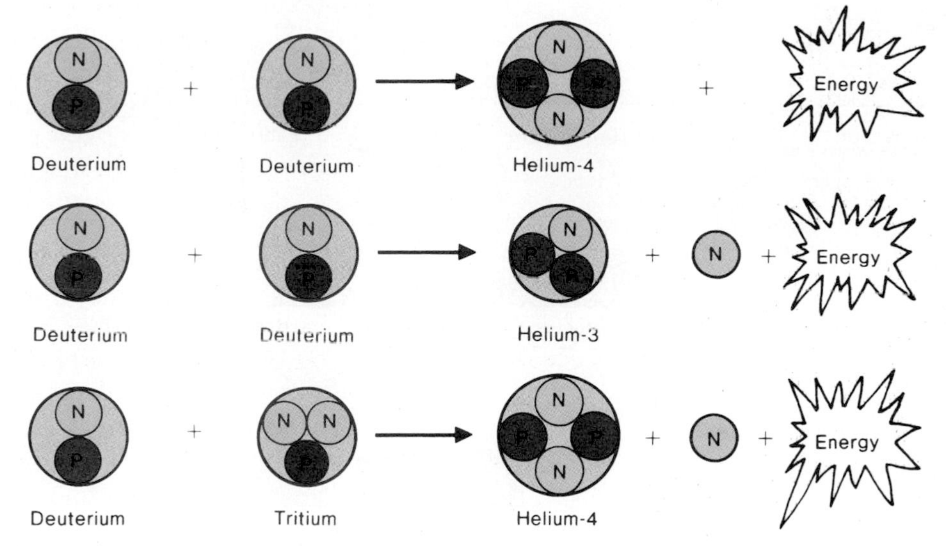

figure 11.8 **Nuclear Fusion** In nuclear fusion, small atomic nuclei are combined to form heavier nuclei. Large amounts of energy are released when this occurs. Different isotopes of hydrogen can be used in the process of fusion. There are three possible types of fusion: (1) two deuterium isotopes can combine to form helium-4 and energy; (2) two deuterium isotopes can combine to form helium-3, a free neutron, and energy; and (3) a deuterium and a tritium isotope can combine to form helium-4, a free neutron, and energy.

However, the main problem is containment of the nuclei. Because they have a positive electric charge, the nuclei repel one another.

Even though, in theory, fusion promises to furnish a large amount of energy, technical difficulties appear to prevent its commercial use in the near future. Even the governments of nuclear nations are budgeting only modest amounts of money for fusion research. And, as with nuclear fission and the breeder reactor, economic costs and fear of accidents may continue to delay the development of fusion reactors.

The Nuclear Fuel Cycle

In order to appreciate the consequences of using nuclear fuels to generate energy, it is important to understand how the fuel is processed. The nuclear fuel cycle begins with the mining operation. (See figure 11.9.) Low-grade uranium ore is obtained by underground or surface mining. The ore contains about 0.2 percent uranium by weight. After it is mined, the ore goes through a milling process. It is crushed and treated with a solvent to concentrate the uranium. Milling produces yellowcake, a material containing 70 to 90 percent uranium oxide.

Naturally occurring uranium contains about 99.3 percent nonfissionable U-238 and only 0.7 percent fissionable U-235. This concentration of U-235 is not high enough for most types of reactors, so the amount of U-235 must be increased by enrichment. Since the masses of the isotopes U-235 and U-238 vary only slightly, and there is no chemical difference, enrichment is a difficult and expensive process. However, it increases the U-235 content from 0.7 percent to 3 percent.

Fuel fabrication converts the enriched material into a powder, which is then compacted into pellets about the size of a pencil eraser. These pellets are sealed in metal fuel rods about 4 meters in length, which are then loaded into the reactor.

As fission occurs, the concentration of U-235 atoms decreases. After about three years, a fuel rod does not have enough radioactive material to sustain a chain reaction, and the spent fuel rods must be replaced by new ones. The spent rods are still very radioactive, containing about 1 percent U-235 and 1 percent plutonium. These rods are the major source of radioactive waste material produced by a nuclear reactor.

When nuclear reactors were first being built, scientists proposed that spent fuel rods could be reprocessed. The remaining U-235 could be enriched and used to manufacture new fuel rods. Since plutonium is fissionable, it could also be fabricated into fuel rods. Besides providing new fuel, reprocessing would reduce the amount of nuclear waste. However, the cost of producing fuel rods by reprocessing was found to be greater than the cost of producing fuel rods from ore, and the United States closed its reprocessing facilities. At present, India, Japan, Russia, France, and the United Kingdom operate reprocessing plants that reprocess spent fuel rods as an alternative to storing them as a nuclear waste.

Each step in the nuclear fuel cycle involves the transport of radioactive materials. The uranium mines are some distance from the processing plants. The fuel rods must be transported to the power plants, and the spent rods must be moved to a reprocessing plant or storage area. Each of these links in the fuel cycle presents the possibility of an accident or mishandling that could release radioactive material. Therefore, the methods of transport are extremely carefully designed and tested before they are used. Many people are convinced that the transport of radioactive materials is hazardous, while others are satisfied that the utmost care is being taken and that the risks are extremely small.

Ultimately, both high-level and low-level radioactive wastes must be stored. Thus, each step in the nuclear fuel cycle, from the mining of uranium to the storage of nuclear waste, poses health and environmental concerns.

Nuclear Material and Weapons Production

Producing nuclear materials for weapons and other military uses involves many of the same steps used to produce nuclear fuel for power reactors.

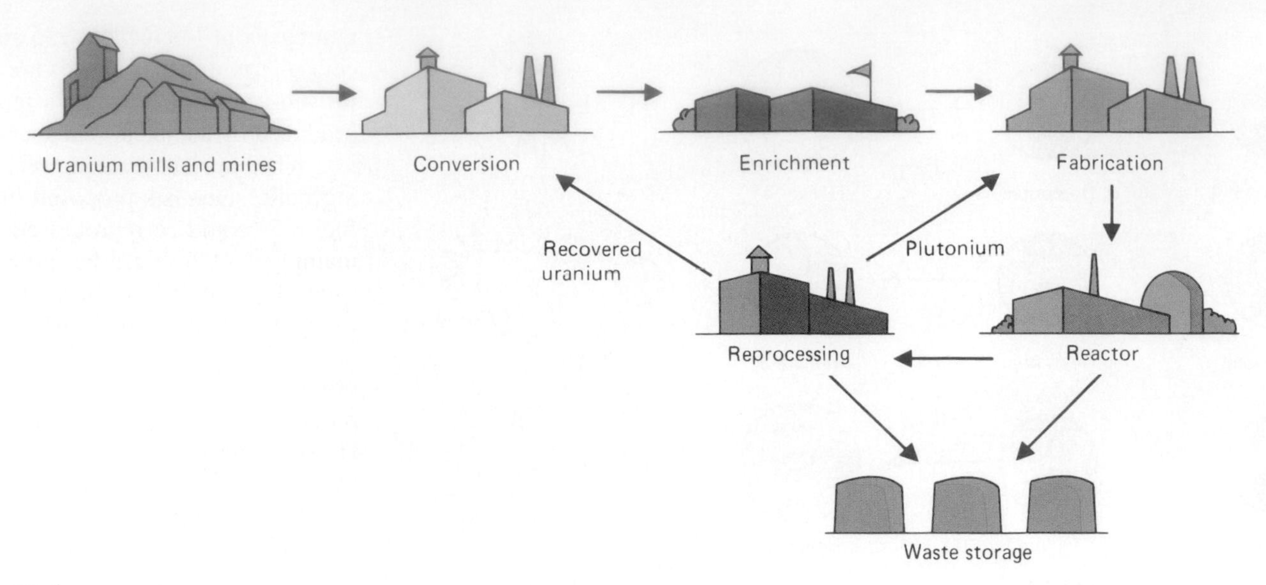

figure 11.9 **Steps in the Nuclear Fuel Cycle** The process of obtaining nuclear fuel involves mining, extracting the uranium from the ore, concentrating the U-235, fabricating the fuel rods, installing and using the fuel in a reactor, and disposing of the waste. Some countries reprocess the spent fuel as a way of reducing the amount of waste they must deal with.

Source: From U.S. Department of Energy.

In fact, the nuclear power industry is an outgrowth of the weapons industry. In the United States, the Department of Energy currently is responsible for nuclear research for both weapons and peaceful uses and stewardship of the facilities used for research and weapons production. Some facilities are used for both processing nuclear fuel and providing materials for weapons. In both cases, uranium must be mined, concentrated, and transported to sites of use. Furthermore, military uses involve the production and concentration of plutonium. The production and storage facilities invariably become contaminated, as does the surrounding land.

Research and production facilities have typically dealt with hazardous chemicals and low-level radioactive wastes by burying them, pumping them into the ground, storing them in ponds, or releasing them into rivers. Despite environmental regulations that prevented such activities, the Department of Energy (formerly the Atomic Energy Commission) maintained that it was exempt from such federal environmental legislation. As a result, the DOE has become the steward of a large number of sites that are contaminated with both hazardous chemicals and radioactive materials. The magnitude of the problem is huge. There are 3365 square miles of DOE properties that are or have been involved in weapons development or production. These include:

- 3700 contaminated sites,
- 330 underground storage tanks with high-level radioactive waste,
- more than a million 55-gallon drums of radioactive, hazardous, or mixed waste in storage,
- 5700 sites where wastes are moving through the soil, and
- millions of cubic meters of low-level and high-level radioactive wastes.

The Department of Energy has pledged to clean up these sites by 2019. Environmental cleanup is now the largest single item in its budget. Several U.S. sites are currently being cleaned up, but things are not going smoothly. Local residents and the states that are hosts to these facilities distrust the DOE and are insisting that the sites be cleaned completely. This may not be technologically or economically possible. Furthermore, environmental cleanup is a new mission for the DOE and the department is having difficulty adjusting. The cleanup process will take many years and require the expenditure of tens of billions of dollars. Figure 11.10 shows the location of waste sites the U.S. Department of Energy has responsibility for cleaning up.

An additional problem has arisen as a result of the reduced importance of nuclear weapons. The political disintegration of the Soviet Union and Eastern Europe has made large numbers of nuclear weapons, both in the East and West, unnecessary. This has probably made the world a safer place, but how are nations of the world to dispose of their nuclear weapons? Some nuclear material can be diverted to fuel use in nuclear reactors, and some reactors can be modified to accept enriched uranium or plutonium. The security of these materials, particularly in the former Soviet Union, is a concern. Investigators have uncovered several incidents in which plutonium has been sold to other countries.

Nuclear Power Concerns

Nuclear power has provided a significant amount of electricity for the people of the world. Currently, over 7 percent of the energy consumed worldwide and 17 percent of the electricity consumed worldwide comes from nuclear power. However, several accidents have raised

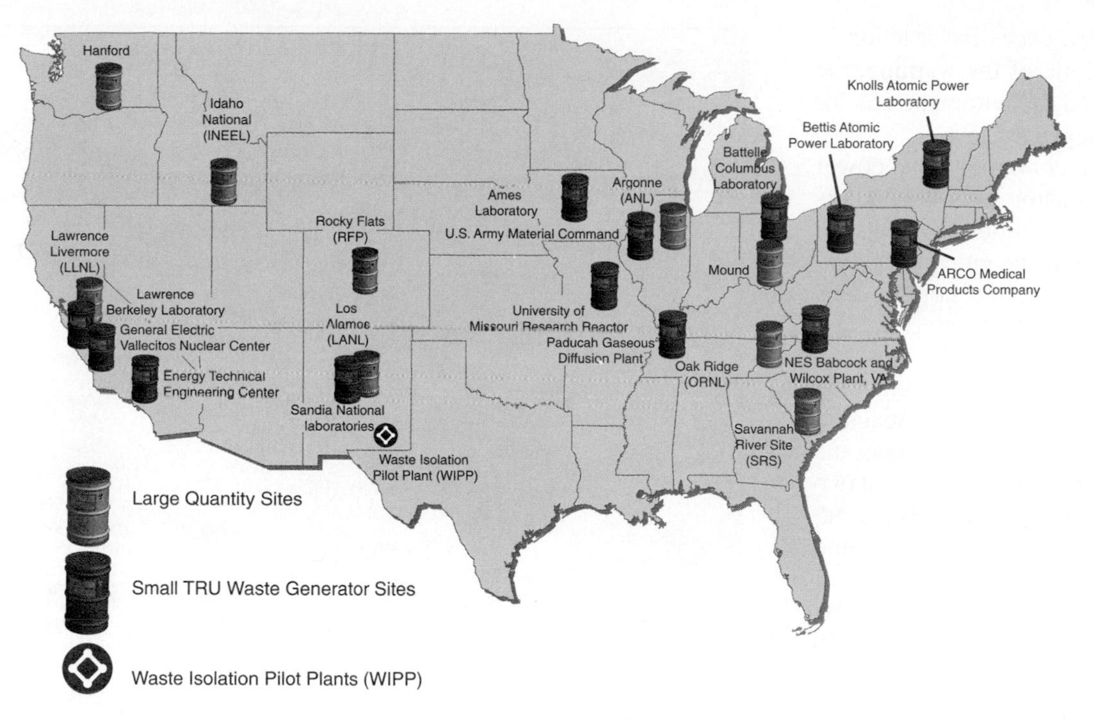

figure 11.10 **The U.S. Department of Energy Waste Sites** As a result of many activities related to research in nuclear energy and to the production of nuclear weapons, the U.S. Department of Energy has responsibility for cleaning up many contaminated sites.

questions about safety, radiation has been released into the air and water, the production and transportation of fuel have caused contamination, and the disposal of wastes is a continuing problem.

Reactor Safety: The Effects of Three Mile Island and Chernobyl

Although there have been accidents at nuclear power plants since they were first used, two relatively recent accidents have had the greatest effect on people's attitudes toward nuclear power plant safety: Three Mile Island in 1979 and Chernobyl in 1986.

On March 14, 1979, the valves closed to three auxiliary pumps at one of the reactors at the Three Mile Island nuclear plant in Pennsylvania. On March 28, 1979, the main pump to the reactor broke down. The auxiliary pumps failed to operate because of the closed valves, and the electrical-generating turbine stopped. At this point, an emergency coolant should have flooded the reactor and stabilized the temperature. The coolant did start to flow, but a faulty gauge indicated that the reactor was already flooded. Relying on this faulty reading, an operator overrode the automatic emergency cooling system and stopped it. Without the emergency coolant, the reactor temperature rose rapidly. The control rods eventually stopped fission, but a partial core meltdown had occurred. Later that day, radioactive steam was vented into the atmosphere.

The crippled reactor was eventually defueled in 1990 at a cost of about $1 billion. It has been placed in monitored storage until the companion reactor, which is still operating, reaches the end of its useful life. At that time, both reactors will be decommissioned.

Chernobyl is a small city in Ukraine near the border with Belarus, north of Kiev. As is true of many small cities in the world, most people had never heard of it. (See figure 11.11.) However, in the spring of 1986, the world's largest nuclear accident catapulted Chernobyl into the news.

At 1:00 A.M. on April 25, 1986, at Chernobyl Nuclear Power Station-4, a test was begun to measure the amount of electricity that the still-spinning turbine would produce if the steam were shut off. This was important information since the emergency core cooling system required energy for its operation and the coasting turbine could provide some of that energy until another source became available. The amount of steam being produced was reduced by lowering the control rods into the reactor. But the test was delayed because of a demand for electricity and a new shift of workers came on duty. The operators failed to program the computer to maintain power at 700 megawatts, and output dropped to 30 megawatts. This presented an immediate need to rapidly increase the power, and many of the control rods were withdrawn. Meanwhile, an inert gas (xenon) had accumulated on the fuel rods. The gas absorbed the neutrons and slowed the rate of power increase. In an attempt to obtain more power, operators withdrew all the control rods. This was a second serious safety violation.

At 1:00 A.M. on April 26, the operators shut off most emergency warning signals and turned on all eight pumps to provide adequate cooling for the reactor following the completion of the test. Just as final stages of the test were beginning, a signal indicated excessive reaction in the reactor. In spite of the warning, the operators blocked the automatic reactor shutdown and began the test.

As the test continued, the power output of the reactor rose beyond its normal level and continued to rise. The operators activated the emergency system designed to put the control rods back into the reactor and stop the fission. But it was too late. The core had already been deformed, and the rods would not fit properly; the reaction could not be stopped. In 4.5 seconds, the energy level of the reactor increased two thousand times. The fuel rods ruptured, the cooling water turned into steam, and a steam explosion occurred. The lack of cooling water allowed the reactor to explode. The explosion blew the 1000 metric ton concrete roof from the reactor and the reactor caught fire.

In less than 10 seconds, Chernobyl became the scene of the world's worst nuclear accident. A core meltdown had occurred. (See figure 11.12.) It took 10 days to bring the runaway reaction under control. By November, the damaged reactor was entombed in a concrete covering, but the hastily built structure, known as the sarcophagus, may have structural flaws. The Ukrainian government is planning to have a second containment structure built around the current structure. The immediate consequences were 31 fatalities; 500 persons hospitalized, including 237 with acute radiation sickness; and 116,000 people evacuated. Of the evacuees, 24,000 received high doses of radiation. The delayed effects are more difficult to assess. Many people suffer from illnesses they feel are related to their exposure to the fallout from Chernobyl. In 1996, it became clear that at least one delayed effect was strongly correlated with exposure. Children or fetuses exposed to fallout are showing increased frequency of thyroid cancer. The thyroid gland accumulates iodine, and radioactive iodine 131 was released from Chernobyl. It is still too early to tell if there are other delayed health effects.

figure 11.11 **Chernobyl** The town of Chernobyl became infamous as the location of the world's worst nuclear power plant accident.

figure 11.12 **The Accident at Chernobyl** An uncontrolled chain of reactions in the reactor of unit four resulted in a series of explosions and fires. See the circled area in the photograph.

More than a year after the disaster at Chernobyl, the decontamination of 27 cities and villages within 40 kilometers of Chernobyl was considered finished. This does not mean that the area was cleaned up, only that all practical measures were completed. Some areas were simply abandoned. The largest city to be affected was Pripyat, which had a population of 50,000 and was only 4 kilometers from the reactor. A new town was built to accommodate those displaced by the accident, and Pripyat remains a ghost town. Sixteen other communities within the 40-kilometer radius have been cleaned up, and the original inhabitants have returned, although there is still controversy about the safety of these towns.

One important impact of Chernobyl is that it deepened public concern about the safety of nuclear reactors. Even before Chernobyl, between 1980 and 1986, the governments of Australia, Denmark, Greece, Luxembourg, and New Zealand had officially adopted a "no nuclear" policy. Since 1980, 10 countries have canceled nuclear plant orders or mothballed plants under construction. Argentina canceled 4 plants; Brazil, 8; Mexico, 18; and the United States, 54. There have been no orders for new plants in the United States since 1974. Sweden, Austria, Germany, and the Philippines have decided to phase out and dismantle their nuclear power plants.

Before Chernobyl, 65 percent of the people in the United Kingdom were opposed to nuclear power plants; after Chernobyl, 83 percent were against them. In Germany, opposition increased from 46 percent to 83 percent. (See figure 11.13.) Opposition in the United States rose from 67 to 78 percent. Even the French, who have the greatest commitment to nuclear power, were against it by 52 percent. The number of new nuclear plants being constructed has been reduced. In addition, many plants have been shut down prematurely, resulting in the prediction that the amount of energy furnished by nuclear fission reactors will actually decline in the future.

At the same time some people are predicting the death of nuclear power as a viable energy source, others are examining ways to make nuclear power facilities safer. They expect nuclear power to be an important energy source in the future. At both Three Mile Island and Chernobyl, operator error caused or contributed to the accidents. Operators manually stopped normal safety actions from taking place. However, a contributing factor was the design: active mechanical processes had to work properly to shut the reactors down. Many of the new designs for reactors include passive mechanisms that will shut down the reactor, special catching basins for the reactor core should it melt down, and better containment buildings.

figure 11.13 **Anti-nuclear Demonstration** This anti-nuclear demonstration was held in 1998 in Germany as members of the German public protested the transport of nuclear waste across the country to a storage site.

Nuclear power will continue to be a part of the energy mix, particularly in countries that lack fossil-fuel reserves. Furthermore, as fossil fuels are used up, the pressure for energy will probably create a market for nuclear power technology.

Exposure to Radiation

Although nuclear accidents are spectacular, frightening, and can cause immediate death because of the incredible amount of energy involved, the long-term problems resulting from exposure to radiation are even more worrisome. Radiation is converted to other forms of energy when it is absorbed by matter. When organisms are irradiated, this energy conversion causes damage at a cellular, tissue, organ, or organism level. The degree and kind of damage vary with the kind of radiation, the amount of radiation, the duration of the exposure, and the types of cells irradiated.

Radiation can also cause mutations, which are changes in the genetic messages within cells. Mutations can cause two quite different kinds of problems. Mutations that occur in the ovaries or testes can form mutated eggs or sperm, which can lead to abnormal offspring. Care is usually taken to shield these organs from unnecessary radiation. Mutations that occur in other tissues of the body may manifest themselves as abnormal tissue growths known as cancer. Two common cancers that are strongly linked to increased radiation exposure are leukemia and breast cancer. Because mutations are essentially permanent, they may accumulate over time. Therefore, the accumulated effects of radiation over many years may result in the development of cancer later in life.

Human exposure to radiation is usually measured in **rems** (*r*oentgen *e*quivalent *m*an), a measure of the biological damage to tissue. The effects of large doses (1000 to 1,000,000 rems) are easily seen and can be quantified, because there is a high incidence of death at these levels, but demonstrating known harmful biological effects from smaller doses is much more difficult. (See table 11.3.) Moderate doses (10 to 1000 rems) are known to increase the likelihood of cancer and birth defects. The higher the dose, the higher the incidence of abnormality. Lower doses may cause temporary cellular changes, but it is difficult to demonstrate long-term effects. Thus, the effects of low-level, chronic radiation generate much controversy. Some people feel that all radiation is harmful, that there is no safe level, and that special

Table 11.3 Radiation Effects

Source	Dose	Biological Effect
Nuclear bomb blast or exposure in a nuclear facility	100,000 rems/incident	Immediate death
	10,000 rems/incident	Coma, death within one to two days
X rays for cancer patients	1000 rems/incident	Nausea, lining of intestine damaged, death in one to two weeks
	100 rems/incident	Increased probability of leukemia
	10 rems/ incident	Early embryos may show abnormalities
Upper limit for occupationally exposed people	5 rems/year	Effects difficult to demonstrate
X ray of the intestine	1 rem/procedure	Effects difficult to demonstrate
Upper limit for release from nuclear installations (except nuclear power plants)	0.5 rem/year	Effects difficult to demonstrate
Natural background radiation	0.2–0.3 rem/year	Effects difficult to demonstrate
Upper limit for release by nuclear power plants	0.005 rem/year	Effects difficult to demonstrate

figure 11.14 **Protective Equipment** Persons working in an area subjected to radiation must take steps to protect themselves. These workers are wearing protective clothing and filtering the air they breathe.

care must be taken to prevent exposure. (See figure 11.14.) Others feel that the increased risk of low-level radiation is extremely small and that current radiation standards are adequate to protect the public, especially in light of the benefits of radiation, such as medical diagnoses and electrical energy. Current research is trying to assess the risks associated with repeated exposure to low-level radiation.

Each step in the nuclear fuel cycle poses a radiation exposure problem, beginning with the mining of uranium. Although the radiation level in the ore is low, miners' prolonged exposure to low-level radiation increases their rates of certain cancers, such as lung cancer. Uranium miners who smoke have an even higher lung cancer rate. After mining, the ore must be crushed in a milling process. This releases radioactive dust into the atmosphere, so workers are chronically exposed to low levels of radioactivity. The crushed rock is left on the surface of the ground as mine tailings. There are over 150 million metric tons of low-level radioactive mine tailings in the United States, and at least as much in the rest of the world. These tailings constitute a hazard because they are dispersed into the environment. (See figure 11.15.)

In the enrichment and fabrication processes, the main dangers are exposure to radiation and accidental release of radioactive material into the environment. Transport also involves exposure risks. Most radioactive material is transported by highways or railroads. If a transporting vehicle were involved in an accident, radioactive material could be released into the environment. Even after fuel rods have been loaded into the reactor, people who work in the area of the reactor risk radiation exposure. Countries that reprocess spent fuel rods must be concerned about the exposure of workers and the possible theft of rods or reprocessed material by terrorist organizations that would use the radioactive material to construct an atomic bomb.

Thermal Pollution

Thermal pollution is the addition of waste heat to the environment. This is a problem particularly in aquatic environments, since many aquatic organisms are very sensitive to changes in temperature. All industrial processes release waste heat, so the problem of thermal pollution is not unique to nuclear power plants. In both fossil-fuel and nuclear plants, generating steam to produce electricity results in a great deal of waste heat. In a fossil-fuel plant, half of the heat energy produces electricity, and half is lost as waste heat. In a nuclear power plant, only one-third of the heat generates electricity, and two-thirds is waste heat. Therefore, the less efficient nuclear power plant increases the amount of thermal pollution more than does an equivalent fossil-fuel power plant. To reduce the effects of this waste heat, utilities build costly cooling facilities. Cooling processes usually involve water, so nuclear power plants often are constructed next to a water source. In some cases, water is drawn directly from lakes, rivers, or oceans and returned. In other cases, it is supplied by giant cooling towers. (See figure 11.16.)

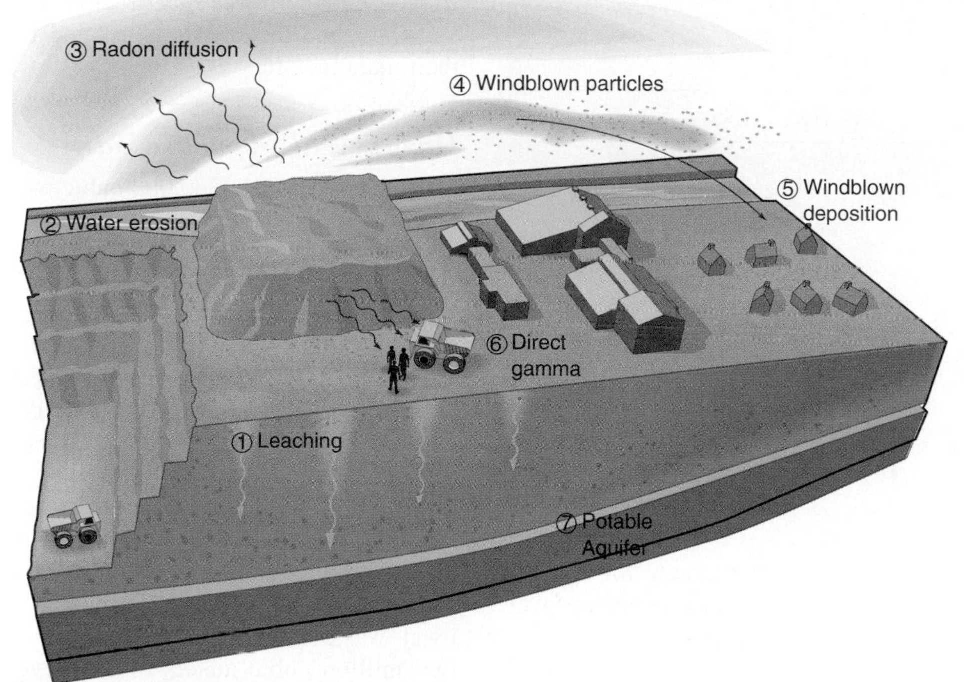

figure 11.15 **Uranium Mine Tailings** Even though the amount of radiation in the tailings is low, the radiation still represents a threat to human health. Radioactivity may be dispersed throughout the environment and come in contact with humans in several ways: (1) radioactive materials may leach into groundwater; (2) radioactive materials may enter surface water through erosion; (3) radon gas may diffuse from the tailings and enter the air; (4) persons living near the tailings may receive particles in the air, (5) come in contact with objects that are coated with particulate, or (6) receive direct gamma radiation.

Source: From U.S. Environmental Protection Agency.

figure 11.16 **Cooling Towers** Cooling towers draw air over wet surfaces. The evaporation of the water cools the surface, removing heat from the process and releasing the heat into the air.

Decommissioning Costs

All industrial facilities have a life expectancy, that is, the number of years they can be profitably operated. The life expectancy for an electrical generating plant, whether fossil-fuel or nuclear, is about 30 to 40 years, after which time the plant is demolished. With a fossil-fuel plant, the demolition is relatively simple and quick. A wrecking ball and bulldozers reduce the plant to rubble, which is trucked off to a landfill. The only harm to the environment is usually the dust raised by the demolition.

Demolition of a nuclear plant is not so simple. In fact, nuclear plants are not demolished, they are decommissioned. **Decommissioning** involves removing the fuel, cleaning surfaces, and permanently preventing people from coming into contact with the contaminated buildings or equipment. Today, over 70 nuclear plants in the world are shut down and waiting to be decommissioned, and the number of plants scheduled for decommissioning will grow. By 2005, 68 of the 104 nuclear plants in the United States will be 20 years old or older. The Nuclear Regulatory Commission authorizes their operation for 40 years but is considering extending authorization for an additional 20 years. This would put off the need for decommissioning for a few years, but eventually it will be necessary. In the United States, many small experimental facilities have been decommissioned and fourteen U.S. plants have been removed from service and are already awaiting decommissioning. Canada has decommissioned three plants and has one scheduled for decommissioning.

The decommissioning of a plant is a two-step process. In stage 1, the plant is shut down and all the fuel rods are removed; all of the water used as a reactor-core coolant, as a moderator, or to produce steam is drained; and the reactor and generator pipes are cleaned and flushed. The spent fuel rods, the drained water, and the material used to clean the pipes are all radioactive and must be safely stored or disposed of. This removes 99 percent of the radioactivity. Stage 2 consists of dismantling all the parts of the plant except the reactor and safely containing the reactor.

Utilities have three decommissioning options: (1) decontaminate and dismantle the plant as soon as it is shut down; (2) shut the plant down for 20 to 100 years to allow radioactive materials that have a short half-life to disintegrate, and then dismantle the plant; (3) entomb the plant by covering the reactor with reinforced concrete and placing a barrier around the plant.

Originally, entombment was thought to be the best method. But because of the long half-life of some of the radioactive material and the danger of groundwater contamination, this method now appears to be the least favorable. Most U.S. utilities plan to allow about 60 years to elapse between stage 1 and stage 2. Japan, which has greater needs for land, plans to mothball its plants for 5 to 10 years before dismantling them. The costs associated with decommissioning include dismantling the plant and then packaging, transporting, and burying the wastes. Because of contamination and activated material, ordinary dismantling methods cannot be used to demolish the buildings. Special remote-control machinery must be developed. Techniques that do not release dust into the atmosphere must be

Table 11.4 Low-Level Radioactive Contaminants from Decommissioning an 1100-Megawatt Pressurized-Water Reactor

Material	Volume (Cubic Meters)
Radioactive	618
Activated	
Metal	484
Concrete	707
Contaminated	
Metal	5,465
Concrete	10,613
Total	17,887

Source: Data from *Worldwatch Paper 69*, "Decommissioning: Nuclear Power's Missing Link," and U.S. Nuclear Regulatory Commission.

used, and workers must wear protective clothing. Waste disposal is estimated to be 40 percent of the decommissioning costs. Table 11.4 shows the volume of material that would be handled from an 1100-megawatt power plant.

Recent experience with the cost of decommissioning indicates that the cost for decommissioning a large plant will be between $200–400 million, about 5 percent of the cost of generating electricity. Although the mechanisms vary with countries, the money for decommissioning is generally collected over the useful life of the plant.

Radioactive Waste Disposal

When the world entered the atomic age, the problem of the disposal of nuclear waste was not fully appreciated. Low-level radioactive waste is generated by nuclear power plants, military facilities, hospitals, and research institutions. High-level radioactive waste results from spent fuel rods, obsolete nuclear weapons, and the wastes generated by their manufacture.

High-Level Radioactive Waste

In the United States, 380,000 cubic meters of highly radioactive military waste are temporarily stored at several U.S. sites. (See figure 11.17.) A high-level waste site has been constructed near Carlsbad, New Mexico. Known as the Waste Isolation Pilot Plant, it was built to house the high-level radioactive waste from the U.S. weapons program. The wastes are commonly referred to as **transuranic wastes**, which consist primarily of various isotopes of plutonium. This facility began accepting waste in March 1999. In addition to the high-level waste from weapons programs, two million cubic meters of low-level radioactive military and commercial waste are buried at various sites. In addition, about 30,000 metric tons of high-level radioactive waste from spent fuel rods are being stored in special storage ponds at nuclear reactor sites. Many plants are running out of storage space and have been authorized by the

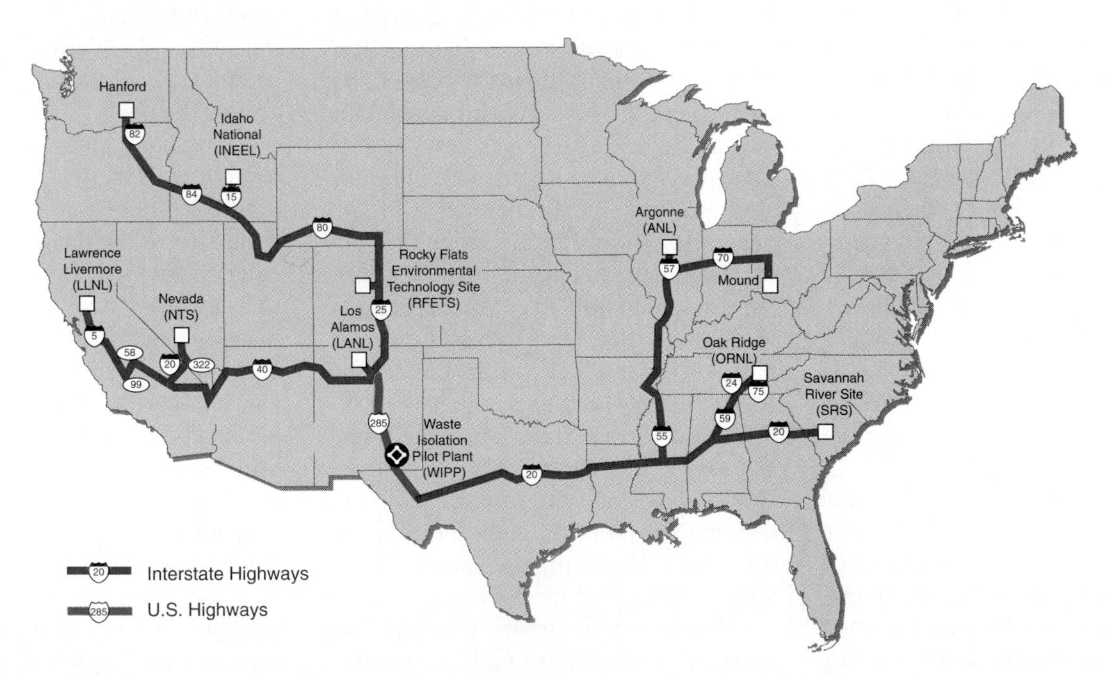

figure 11.17 **Department of Energy High-Level Radioactive Transuranic Waste Sites** The high-level transuranic radioactive waste generated by the U.S. Department of Energy will be shipped to the Waste Isolation Pilot Plant near Carlsbad, New Mexico, for storage.

Source: Nuclear Regulatory Commission.

GLOBAL PERSPECTIVE

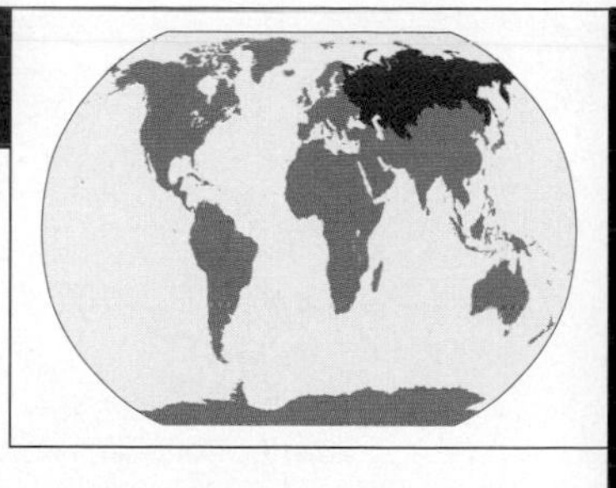

The Nuclear Legacy of the Soviet Union

The tremendous political changes that have occurred in the former Soviet Union and Eastern Europe have brought to light the magnitude of nuclear contamination caused by those nations' unwise use of nuclear energy. The Soviet Union used nuclear energy for several purposes: generating power for electricity and nuclear-powered ships, testing and producing weapons, and blasting to move earth. The accident at Chernobyl was only the most recent and most public of many problems associated with the use of nuclear energy in the Soviet Union. About 13 nuclear reactors of the Chernobyl type are still operating in the former Soviet Union. These are considered unsafe by most experts, including former Soviet scientists.

The production of nuclear fuels and weapons and the reprocessing of nuclear fuels, all of which produced nuclear wastes, occurred at three major sites: Chelyabinsk, Tomsk, and Krasnoyarsk. These facilities were secret until the recent breakup of the Soviet Union. At Chelyabinsk, radioactive waste was dumped into the Techa River and Lake Karachai. Other nuclear wastes were secretly dumped into the ocean or were buried in shallow dumps. It is estimated that there are 600 secret nuclear dump sites in and around Moscow alone.

More than 100 nuclear bombs were exploded to move earth for mining purposes, to create underground storage caverns, or to "dig" canals. In addition, 467 nuclear blasts were conducted to test weapons at a site near Semipalatinsk in Kazakhstan. As a result, nuclear fallout contaminated farmland to the northeast.

The Barents Sea and the Kara Sea have been polluted by nuclear tests and the dumping of nuclear waste. At least 15 nuclear reactors from obsolete ships were dumped into the Kara Sea. Seals that live in the area have high rates of cancer.

None of these problems is unique to the former Soviet Union. Similar secrecy surrounded the early development of nuclear power in the United States and the rest of the world. What makes the situation in the former Soviet Union different is its magnitude, caused by decades of secrecy and a commitment to nuclear weapons as the primary deterrent against enemies. It can be argued that the threat posed to the Soviet Union by its former enemies contributed to the reckless development of nuclear power.

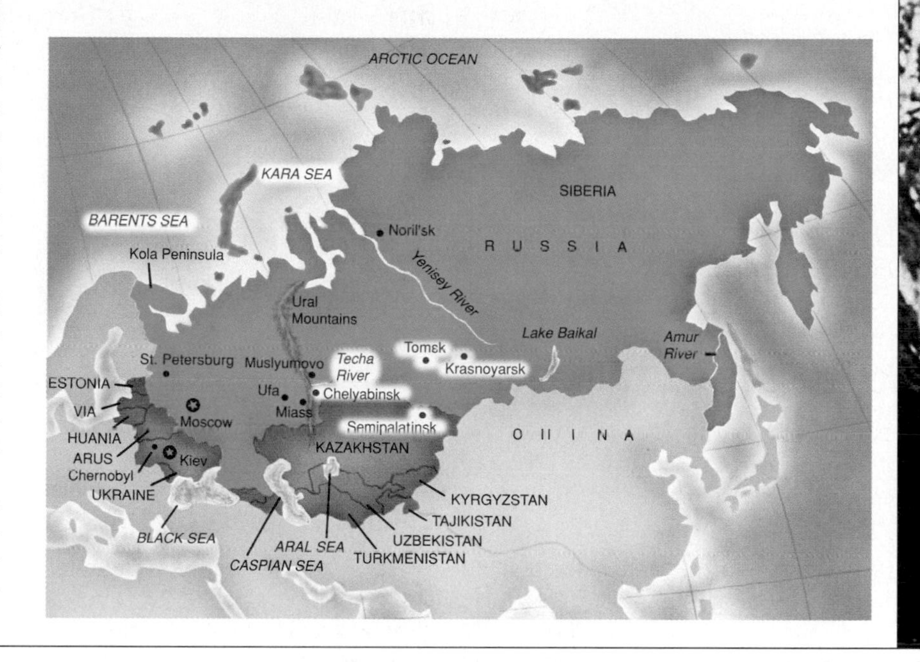

Nuclear Regulatory Commission to store radioactive waste in aboveground casks, because the storage ponds will not accommodate additional waste.

In other countries, spent fuel rods are either stored in special ponds or sent to reprocessing plants. Even though reprocessing is more expensive than manufacturing fuel rods from ore, some countries reprocess as an alternative to waste storage. The reprocessing plants in France and the United Kingdom accept domestic and foreign fuel rods, and any fuel rods fabricated by the former Soviet Union can be returned to reprocessing plants in Russia.

At this time, no country has a permanent storage solution for the disposal of high-level radioactive waste. Most experts feel the best solution is to bury it in a stable geologic formation. Several countries are investigating storage in salt deposits. The waste would be placed in borosilicate glass containers. Each container would then be sealed in a thick-walled, stainless steel canister. Due to the high temperature of radioactive waste, the containers would be stored aboveground for 10 years. At the end of this time, the temperature of the container would be reduced, and the material could be buried in a salt deposit 600 meters below the surface.

Sweden intends to store its waste 500 meters underground in granite. Site construction is not planned until 2010, when Sweden's 12 reactors are scheduled to be decommissioned. (Sweden plans to discontinue nuclear power generation of electricity, but some now doubt that this will occur.) France plans to use its reprocessing plants and "temporary" storage indefinitely.

The politics concerning the disposal of high-level radioactive waste are probably as critical as developing a suitable method. No communities want a radioactive disposal site in their area. In December 1982, the U.S. Congress passed legislation calling for a high-level radioactive disposal site for the storage of spent fuel rods from nuclear power plants to be selected by March 1987 and to be completed by 1998. In 1984, the Department of Energy stated that it was already three years behind schedule. Final site selection occurred in

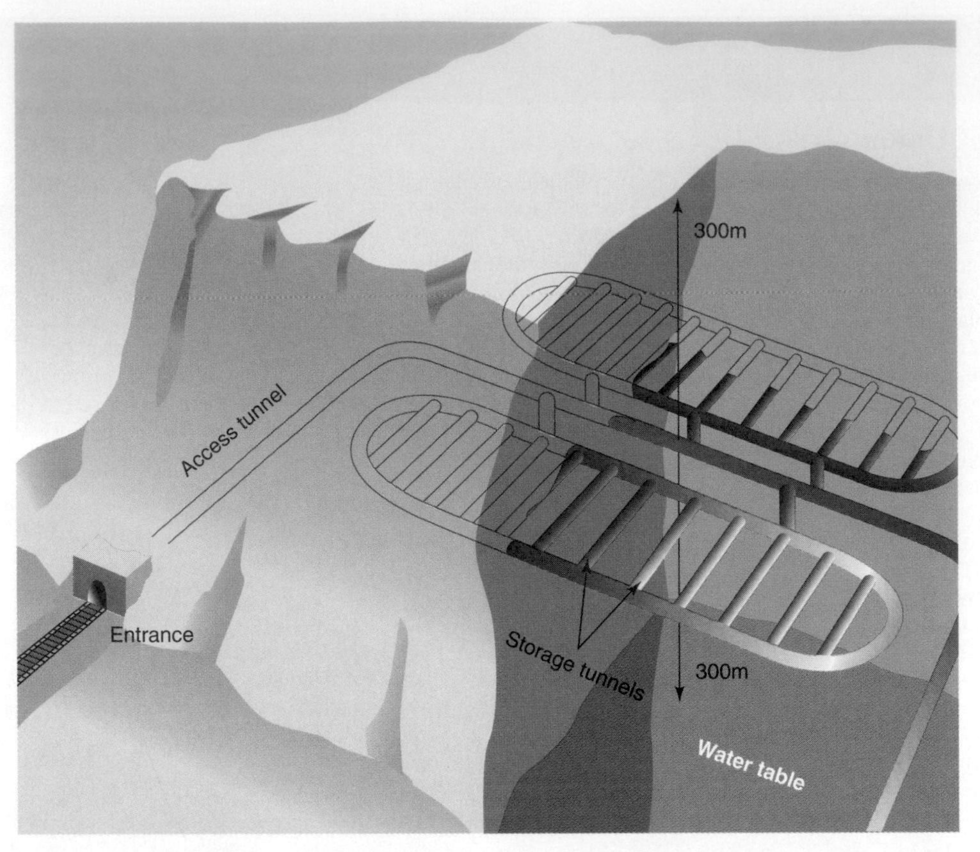

figure 11.18 **High-Level Nuclear Waste Disposal** Current plans for the disposal of high-level nuclear waste involve placing materials in tunnels underground at Yucca Mountain, Nevada. The material could be stored about 300 meters below the surface and about 300 meters above the water table. Because of the dry climate, it is considered unlikely that water infiltrating the soil would move nuclear material downward to the water table.

1989, nearly three years later than the date set by legislation. The location is Yucca Mountain, Nevada. The site was chosen for several reasons. It is in an unpopulated area near the Nevada Test Site where several nuclear devices were exploded. It is a very dry area and the water table is about 600 meters below the mountain, so groundwater will not likely be contaminated. It was also considered to be geologically stable, however, the site has witnessed seismic activity since being chosen as a site.

Work has begun on the series of tunnels that would serve as storage places for the waste. (See figure 11.18.) However, current work is primarily exploratory and is seeking to characterize the likelihood of earthquake damage and the movement of water through sediments. If completed, the facility would hold about 70,000 metric tons of spent fuel rods and other highly radioactive material. It will not be completed before 2015, and by that time the total amount of waste produced by nuclear power plants will exceed the storage capacity of the site. Furthermore, the amount of waste from nuclear weapons production greatly exceeds that produced by nuclear power plants. In 1998, a geologic study suggested that the site may not be as stable as first thought and this is likely to further delay the completion of the project. Local protests against the site are continuing. Nevada can refuse the site (its refusal could be overridden by Congress), and Congress has not appropriated adequate funds to proceed rapidly. Therefore, it is still uncertain if or when it will be open to receive waste.

Low-Level Radioactive Waste

Low-level radioactive waste includes the cooling water from nuclear reactors, material from decommissioned reactors, radioactive materials used in the medical field, protective clothing worn by persons working with radioactive materials, and materials from many other modern uses of radioactive isotopes. Disposal of this type of radioactive waste is very difficult to control. Estimates indicate that much of it is not disposed of properly.

In 1970, a U.S. moratorium halted the dumping of radioactive waste in the oceans. Prior to this, the United States placed some 90,000 barrels of radioactive waste on the ocean floor. European countries also have dumped both high-level and low-level radioactive material into the Atlantic Ocean. Before 1983, when ocean dumping was halted, these countries disposed of 90,000 metric tons of radioactive waste in the ocean.

The United States produces about 800,000 cubic meters of low-level radioactive waste per year. This is presently being buried in disposal sites in Nevada, South Carolina, and Washington. However, these states balked at accepting all the country's low-level waste. In 1980, Congress set a deadline of 1986 (later extended to 1993) for each state to provide for its own low-level radioactive waste storage site. Later, a change allowed several states to cooperate and form regional coalitions, called compacts. Under this arrangement, one state would provide a disposal site for the entire compact. Several states did not join compacts and several others have since withdrawn from the compacts they were originally a part of. Figure 11.19 shows the current compacts and the status of their disposal sites. Several of the proposed sites are under severe pressure from activists who do not want such sites in their locality. As a result, most states still do not have a permanent place to deposit low-level radioactive wastes and are relying on temporary storage facilities in Richland, Washington or Barnwell, South Carolina. As a further example of the creative politics created by the need to have low-level radioactive waste disposal sites and the unwillingness to host such a site, the states of Texas, Maine, and Vermont have applied to be a compact with Texas serving as the host for the site.

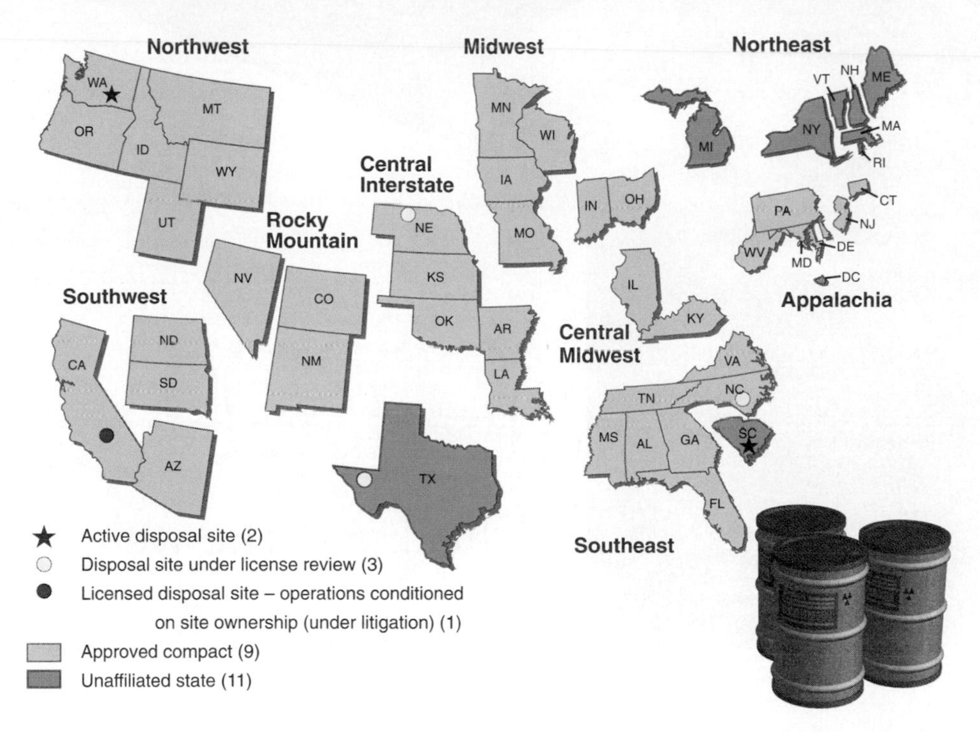

figure 11.19 **Low-Level Radioactive Waste Sites** Each state is responsible for the disposal of its low-level radioactive waste. Many states have formed compacts and selected a state to host the disposal site. None of the proposed sites is currently accepting waste. Many other states do not have an acceptable site and are temporarily relying on a site in South Carolina to accept their waste. Eventually, they will need to find other acceptable sites. The map shows the current compacts and proposed disposal sites.

Source: Nuclear Regulatory Commission.

Summary

Nuclear fission is the splitting of the nucleus of the atom. The resulting energy can be used for a variety of purposes. The splitting of U-235 in a nuclear reactor can be used to heat water to produce steam that generates electricity. Various kinds of nuclear reactors have been constructed, including boiling-water reactors, pressurized-water reactors, heavy-water reactors, gas-cooled reactors, and experimental breeder reactors. Scientists are also conducting research on the possibilities of using fusion to generate electricity. All reactors contain a core with fuel, a moderator to control the rate of the reaction, and a cooling mechanism to prevent the reactor from overheating.

The nuclear fuel cycle involves mining and enriching the original uranium ore, fabricating it into fuel rods, using the fuel in reactors, and reprocessing or storing the spent fuel rods. The fuel and wastes must also be transported. At each step in the cycle, there is danger of exposure. During the entire cycle, great care must be taken to prevent accidental releases of nuclear material.

The use of nuclear materials for military purposes has resulted in the same kinds of problems caused by nuclear power generation. The reduced risk of nuclear warfare has resulted in a need to destroy nuclear weapons and clean up the sites contaminated by their construction.

Public acceptance of nuclear power plants has been declining. Initial promises of cheap electricity have not been fulfilled because of expensive construction, cleanup, and decommissioning costs. The accidents at Three Mile Island in the United States and Chernobyl in Ukraine have accelerated concerns about the safety of nuclear power plants.

Other concerns are the unknown dangers associated with exposure to low-level radiation, the difficulty of agreeing to proper long-term storage of high-level and low-level radioactive waste, and thermal pollution.

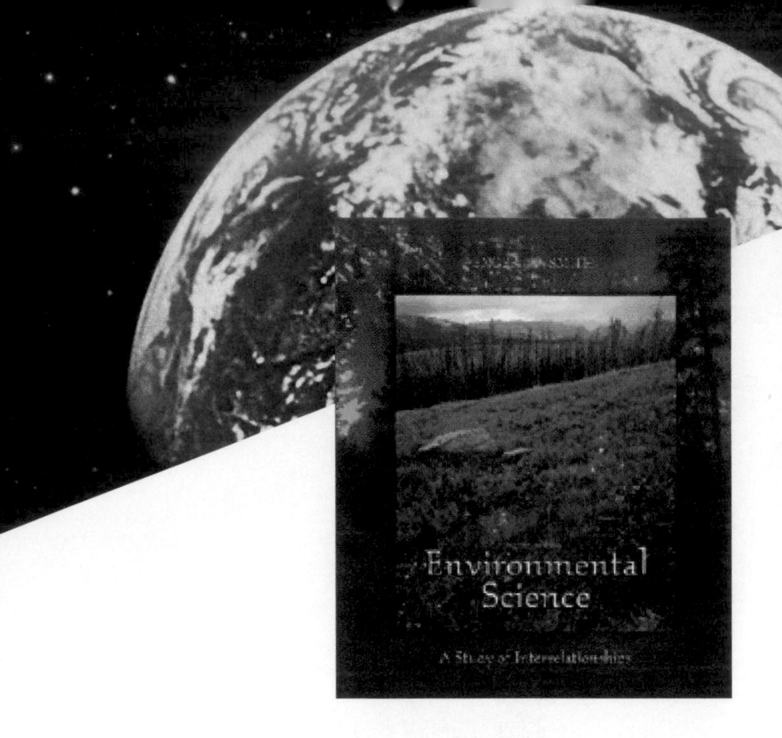

Interactive Exploration

Check out the website at
http://www.mhhe.com/environmentalscience
and click on the cover of this textbook for interactive versions of the following:

KNOW THE BASICS

alpha radiation *216*
beta radiation *216*
boiling-water reactor (BWR) *218*
decommissioning *229*
fissionable *217*
gamma radiation *216*
gas-cooled reactor (GCR) *219*
heavy-water reactor (HWR) *218*
light-water reactor (LWR) *218*
liquid metal fast-breeder reactor (LMFBR) *222*
moderator *217*
nuclear breeder reactor *222*
nuclear chain reaction *217*
nuclear fission *217*
nuclear fusion *222*
nuclear reactor *217*
plutonium-239 (Pu-239) *222*
pressurized-water reactor (PWR) *218*
radiation *216*
radioactive *216*
radioactive half-life *216*
rem *227*
thermal pollution *228*
transuranic waste *230*
uranium-235 (U-235) *217*

- On-line Flashcards
- Electronic Glossary

IN THE REAL WORLD

Is it an answer to the worldwide nuclear waste disposal problem? Moscow Proposes to Import Waste Plutonium from other countries to earn much-needed revenue. Critics are quick to point out the safety issues of this option. Check out this recent story for the pros and cons of the proposal.

The safety of operating nuclear power plants is an environmental issue that has led to a worldwide decrease in the use of nuclear reactor plants. See why hot summers are a potential safety hazard in Heat Wave Threatens Nuclear Power Plants.

Safety is also an issue with decommissioned and obsolete military machinery. Right now Russia is faced with decommissioning old nuclear-powered submarines because they are deteriorating and are a potential source of radioactive accidents. Currently, Russia does not have the money to complete this expensive job. Take a look at Deteriorating Submarine Fleet Threatens Nuclear Contamination in Arctic Seas for the entire story and why other neighboring countries are willing to help with the cleanup.

One positive spin-off to the world's worst nuclear accident is the development of a wildlife refuge in the Chernobyl region. Check out Wildlife Take Refuge in Chernobyl's Wasteland to see why endangered species are returning and establishing populations in this region.

PUT IT IN MOTION

Just in case you didn't see the refresher of atoms, electrons, protons, and neutrons, check out the Atom animation.

TEST PREPARATION

Review Questions

1. How does a nuclear power plant generate electricity?
2. Name the steps in the nuclear fuel cycle.
3. What is a rem?
4. What is a nuclear chain reaction?
5. How will nuclear fuel supplies, the cost of decommissioning facilities, and storage and ultimate disposal of nuclear wastes influence the nuclear power industry?
6. What happened at Chernobyl, and why did it happen?
7. Describe a boiling-water reactor, and explain how it works.
8. How is plutonium-239 produced in a breeder reactor?
9. Why is plutonium-239 considered dangerous?
10. Why is fusion not currently being used as a source of energy?
11. What are the major environmental problems associated with the use of nuclear power?
12. List three environmental problems associated with the construction and subsequent de-emphasis of nuclear weapons.

Critical Thinking Questions

1. Recent concerns about global warming have begun to revive the nuclear industry in the United States. Do you think nuclear power should be used instead of coal for generating electricity? Why?
2. Why has nuclear power not become the power source that is "too cheap to meter," as foreseen by President Eisenhower?
3. Should there be international regulations concerning nuclear power plants, their operation, and decommissioning? Who should be responsible for developing these regulations? What kinds of regulations, if any, would you like to see?
4. Disposal of radioactive wastes is a big problem for the nuclear energy industry. What are some of the things that need to be evaluated when considering nuclear waste disposal? What criteria would you use to judge whether a storage proposal is adequate or not?
5. Nuclear weapons testing has released nuclear radiation into the environment. These tests have always been justified as necessary for national security. Do you agree or not? What are the risks? What are the benefits?
6. If your electric utility were to build a nuclear power reactor in your area, what kind would you advise them to build? Why?
7. Some states allow consumers to choose an electric supplier. Would you choose an alternative to nuclear or coal even if it cost more?

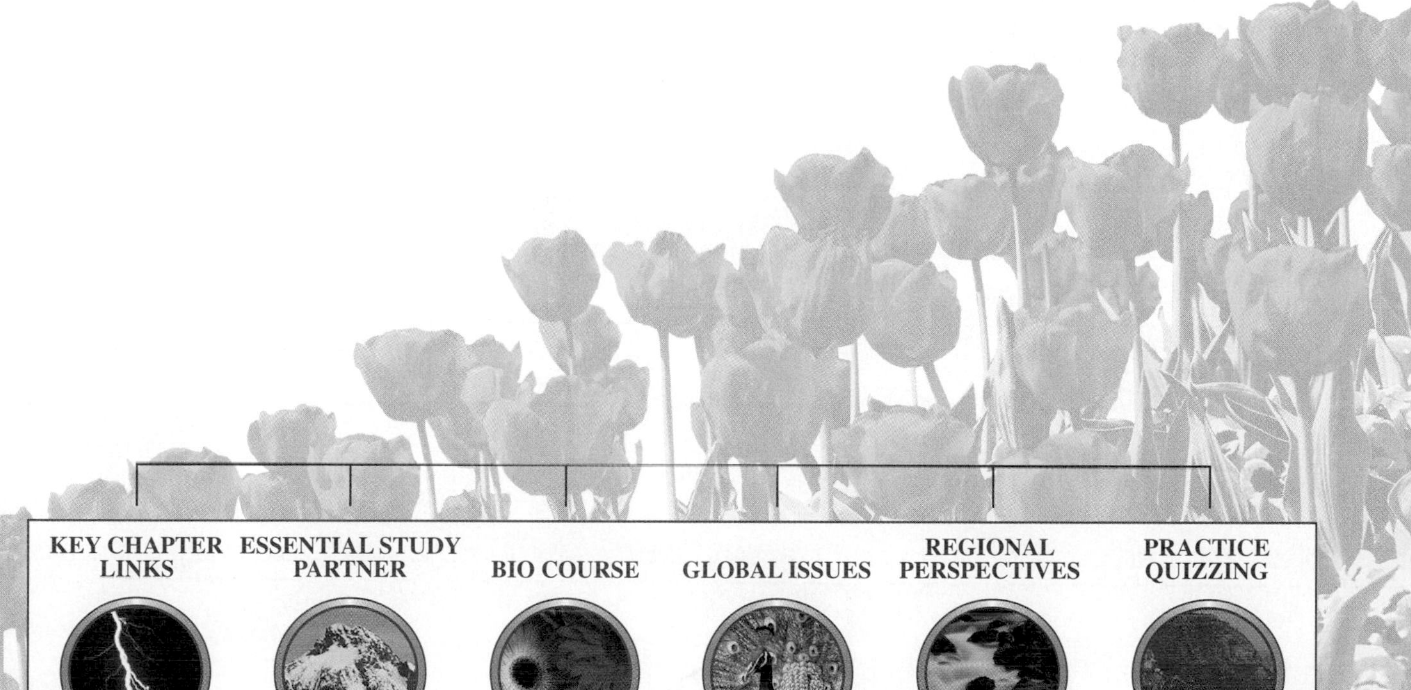

PART FOUR

Human Influences on Ecosystems

The success of the human species is a result of our ability to change our surroundings so that we have tools, food, water, and shelter. There is little of the world that has not been explored and influenced by human activity. Much of the surface of the Earth has been substantially modified by human activities. We derive food from agricultural land, wood products from forests, fish from the ocean, and minerals from the earth. Each of these activities involves economic considerations and leads to changes in the natural systems present. Decisions about how resources will be used are determined by the economic and political wants and needs of the population of a country balanced against its desire to preserve natural ecosystems. Our use of natural resources is important to many aspects of modern society. **Chapter 12** discusses how humans exploit mineral resources and use natural ecosystems. Often, this means that existing natural ecosystems are altered significantly by extracting resources, or natural ecosystems are replaced by agricultural ecosystems. In some cases, organisms are driven to extinction by human activities.

Chapter 13 emphasizes the use of the land surface, while **chapter 14** discusses the nature and wise use of soil. Land and soil are not the same. Land is the part of the world that is not covered by water, while soil is a thin layer on the land's surface that supports plant growth. Land-use decisions must take into account the nature of the soil.

Agriculture allows us to support about 6 billion people on the earth. The use of the soil, discussed in chapter 14, and the use of pesticides, discussed in **chapter 15,** are both important to food production. Misuse of the soil or pesticides can lead to the loss of usable farmland or to the poisoning of people.

Water is necessary for human life. It is important as a nutrient, but also for agriculture, industrial processes, transportation, and many other functions. **Chapter 16** deals with a wide range of water-use issues and conservation practices.

C H A P T E R

Human Impact on Resources and Ecosystems

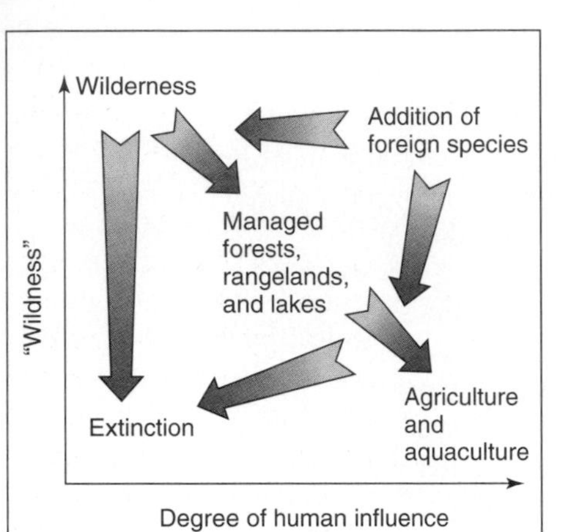

Objectives

After reading this chapter, you should be able to:

- Recognize that humans have an increasing impact on natural ecosystems.
- Define pollution.
- Differentiate between renewable and nonrenewable resources.
- Recognize that mineral resources are unevenly distributed, which creates international trade in these commodities.
- List three types of costs associated with mineral exploitation.
- Understand that some wilderness areas still have minimal human influence.
- Appreciate the ways humans modify forests.
- Identify causes of desertification.
- Recognize that aquatic systems are modified by terrestrial changes.
- Identify changes that occur to aquatic systems as a result of human activity.
- Recognize that wildlife management focuses on specific species.
- Appreciate that waterfowl management is an international problem.
- Recognize that extinction is a natural process.
- Recognize that humans increase the rate of extinction.
- Identify ways that humans cause extinctions.
- Recognize that many extinctions can be prevented if societies are willing to preserve crucial habitats and prevent the hunting of endangered species.

Chapter Outline

The Changing Role of Human Impact

Historical Basis of Pollution

Renewable and Nonrenewable Resources

Costs Associated with Resource Utilization

Mineral Resources
- Steps in Mineral Utilization
- Recycling of Mineral Materials

Utilization and Modification of Terrestrial Ecosystems
- Impact of Agriculture on Natural Ecosystems

Managing Forest Ecosystems
- Economic and Energy Costs of Utilizing Forest Ecosystems
- Environmental Costs of Utilizing Forest Ecosystems
- Environmental Implications of Various Harvesting Methods
- Plantation Forestry
- Special Concerns about Tropical Deforestation

Environmental Close-Up: *The Northern Spotted Owl*

Managing Rangeland Ecosystems
- Environmental Costs of Utilizing Rangelands

Areas with Minimal Human Impact—Wilderness and Remote Areas

Managing Aquatic Ecosystems
- Environmental Costs Associated with Utilizing Marine Ecosystems

Global Perspective: *The History of the Bison*

- Environmental Costs Associated with Utilizing Freshwater Ecosystems

Environmental Close-Up: *Farming, Fish Kills, and* Pfiesteria piscicida
- Aquaculture

Managing Ecosystems for Wildlife
- Habitat Analysis and Management
- Population Assessment and Management

Environmental Close-Up: *Native American Fishing Rights*

- Predator and Competitor Control
- Special Issues with Migratory Waterfowl Management

Extinction and Loss of Biodiversity
- Human-Accelerated Extinction
- Why Worry about Extinction?

What is Being Done to Prevent Extinction and Protect Biodiversity?

Environmental Close-Up: *The California Condor*

Issues & Analysis: *Fire As a Forest Management Tool*

The Changing Role of Human Impact

At one time, a human was just another consumer somewhere in the food chain. Humans fell prey to predators and died as a result of disease and accident just like other animals. The simple tools they used would not allow major changes in their surroundings, so these people did not have a long-term effect on their surroundings. They only minimally exploited mineral and energy resources. They mined chert, obsidian, and raw copper for the manufacture of tools, and used certain other mineral materials, such as salt, clay, and ocher, as nutrients, for making pots, or as pigments. Aside from occasional use of surface seams of coal or surface oil seeps, early humans met most of their energy needs by muscle power or the burning of biomass.

As human populations grew, and as their tools and methods of using them became more advanced, the impact that a single human could have on his or her surroundings increased tremendously. The purposeful use of fire was one of the first events that marked the capability of humans to change ecosystems. Although in many ecosystems fires were natural events, the use of fire by humans to capture game and to clear land for gardens could destroy climax communities and return them to earlier successional stages more frequently than normal.

As technology advanced, wood was needed for fuel and building materials, land was cleared for farming, streams were dammed to provide water power, and various mineral resources were exploited to provide energy and build machines. These modifications allowed larger human populations to survive, but always at the expense of previously existing ecosystems (see figure 12.1).

Today, with over 6 billion people on the Earth, nearly all of the surface of the Earth has been affected in some way by human activity. Even the polar ice caps show the effects of human activity. Various kinds of organic pollutants and lead residues from the burning of leaded gasoline can be identified in the layers of ice that build up from the continuous accumulation of snow in these areas. (The amount of lead is currently decreasing because some countries have been making the transition from burning leaded fuel to burning unleaded fuel in automobiles.)

Historical Basis of Pollution

Pollution is any addition of matter or energy that degrades the environment for humans and other organisms. However, when we think about pollution we usually mean something that people produce in large enough quantities that it interferes with our health or well-being. Two primary factors that affect the amount of damage done by pollution are the size of the population and the development of technology that "invents" new forms of pollution.

When the human population was small and people lived in a simple manner, the wastes produced were biological and so dilute that they usually did not constitute a pollution problem. People used what was naturally available and did not manufacture many products. Humans, like any other animal, fit into their natural ecosystems. Their waste products were **biodegradable** materials that were broken down into simpler chemicals, such as water and carbon dioxide by the action of decomposer organisms.

Human-initiated pollution became a problem when human populations became so concentrated that their waste materials could not be broken down as fast as they were produced. As the population increased, people began to congregate and establish cities. The release of large amounts of smoke and other forms of waste into the air caused an unhealthy condition because the pollutants were released faster than they could be absorbed and dispersed by the atmosphere.

Throughout history, humans have made numerous attempts to eliminate the misery caused by hunger and disease. In general, we rely on science and technology to improve our quality of life. However, technological progress often offers short-term solutions that in the process of solving one problem can create new ones.

The development of the steam engine allowed machines to replace animal power and human labor but increased the amount of smoke and other pollutants in the air and increased the need for fuel. The modern chemical industry has

figure 12.1 **Changes in the Ability of Humans to Modify Their World** As technology has advanced, the ability of people to modify their surroundings has increased significantly. When humans lacked technology, they had only minor impacts on the natural world. The agricultural revolution resulted in many of the suitable parts of the Earth being converted to agriculture. Modern technology allows major portions of the Earth to be moved and rearranged.

produced many extremely valuable synthetic materials (plastics, pesticides, medicines) but has also produced toxic pollutants.

Even identifying pollution is not always easy. To some, the smell of a little wood smoke in the air is pleasant; others do not like the odor. A business may consider advertising signs valuable and necessary; others consider them to be visual pollution. (See figure 12.2.) Even the presence of chemicals in drinking water can be difficult to classify as a clear-cut example of pollution. Toxic heavy metals such as arsenic certainly should be considered hazardous, but we do not know how much arsenic contamination is allowable before harm occurs. Are the small quantities of arsenic in groundwater great enough that, over a period of time, they will accumulate and cause damage to individuals? Is the arsenic a normal part of the groundwater, or is it the result of past heavy spraying of arsenic-containing pesticides on apple trees?

Certainly, if the arsenic is the result of human activity and is causing health problems, it is a pollutant. If it is a natural part of the groundwater, it may still be a health hazard but technically many would not consider it to be a pollutant.

Fish kill—an indication of water contamination

Smog—an indication that thermal inversion has kept the air contamination in the valley

Smoke from stack—contains particulate material, which could cause lung problems

Traffic—fumes (HC, NO_x, PAN, etc.) from internal combustion engines cause eye irritation

Feed lot—odor pollution as well as a source of water contamination from surface runoff

Nuclear power plant cooling towers—possible thermal pollution and radiation hazards

Strip—visual pollution, which is an annoyance but not a health hazard

Litter—an indication that we need to become more aware of how we dispose of materials

Health hazard

Annoyance

figure 12.2 **Forms of Pollution** Pollution is produced in many forms. Some are major health concerns, whereas others merely annoy.

Renewable and Nonrenewable Resources

Modern technologies have allowed us to exploit natural resources to a much greater extent than our ancestors were able to achieve. **Natural resources** are structures and processes that humans can use for their own purposes but cannot create. If the supply of a resource is very large and the demand for it is low, the resource may be thought of as free. Sunlight, oceans, and air are often not even thought of as natural resources because their supply is so large. If a resource has always been rare or it has been severely reduced by consumption, it is expensive. Pearls and precious metals are expensive because they have always been rare. In the past, land and its covering of soil was considered a limitless natural resource, but as the population grew and the demand for food, lodging, and transportation increased, we began to realize that land is a finite, nonrenewable resource. The economic value of land is highest in metropolitan areas where open land is unavailable. Unplanned, unwise, or inappropriate use can result in severe damage to the land and its soil. (See figure 12.3.)

figure 12.3 **Mismanagement of a Renewable Resource** Although soil is a renewable resource, extensive use can permanently damage it. Many of the world's deserts were formed or extended by unwise use of farmland. This photograph shows a once-productive farm, now abandoned to the wind and sand because the soil was mistreated and allowed to erode.

The landscape is also a natural resource, as we see in countries with a combination of mountainous terrain and high rainfall that can be used to generate hydroelectric power. Rivers, forests, scenery, climates, and wildlife populations are additional examples of natural resources.

Natural resources are usually categorized as either renewable or nonrenewable. **Renewable resources** can be formed or regenerated by natural processes. Soil, vegetation, animal life, air, and water are renewable primarily because they naturally undergo processes that repair, regenerate, or cleanse them when their quality or quantity is reduced. However, just because a resource is renewable does not mean that it is inexhaustible. Overuse of renewable resources can result in their irreversible degradation. **Nonrenewable resources** are not replaced by natural processes, or the rate of replacement is so slow as to be ineffective. For example, iron ore, fossil fuels, and mountainous landscapes are nonrenewable on human timescales. Therefore, when nonrenewable resources are used up, they are gone, and a substitute must be found or we must do without.

Costs Associated with Resource Utilization

Costs are always associated with the exploitation of any natural resource. These costs fall into three categories. The first category is the **economic costs** necessary to exploit the resource. Money is needed to lease or buy land, modify the land, construct roads, pay for labor, build equipment, and buy energy to run equipment. A second category is the **energy cost** of exploiting the resource. It takes energy to modify the landscape (build dams, construct roads, level the land) to allow for the use of most resources. It also takes energy to locate, extract, and transport materials such as minerals, forest products, or fish to processing sites. Since energy requires the expenditure of money, energy costs are ultimately converted to economic costs. When energy is inexpensive, inefficient processes may be profitable; however, when the cost of energy rises, energy-inefficient processes will be eliminated. For example, when fuel for machines is inexpensive it may be economically viable to transport raw materials long distances to manufacturing sites. When the price of fuel rises, distant resources will not be exploited as heavily.

A third way to look at costs is in terms of environmental effects. Air pollution, water pollution, plant and animal extinctions, and loss of scenic quality are all possible **environmental costs** of resource exploitation. Often environmental costs are difficult to assess, since they are not easily converted to monetary values. In addition, the costs may not even be recognized immediately, so they are likely to be deferred costs. Environmental costs may also be lost opportunities or lost values because the resource could not be used for another purpose. Recently, in many parts of the developed world, people are recognizing environmental costs, so these are being converted to economic costs as more strict controls on pollution and environmental degradation are enforced. It takes money to clean up polluted water and air, or to reclaim land that has been degraded by extracting or using resources.

Mineral Resources

Mineral resources are one of the major kinds of nonrenewable resources. The distribution of these resources is not uniform throughout the world. The geological and biological processes that formed mineral resources in certain places on the earth occurred many millions of years ago. Since that time, landmasses have been divided into political entities. Some have a greater wealth of mineral resources than others. Because no country has within its territorial limits all of the mineral resources it needs, an international exchange has developed. In particular, the industrially developed countries import many of the minerals they need from countries that have the resources but no economic ability to develop them.

Not only are mineral resources unevenly distributed, but those that are easiest to use and the least costly to extract have been exploited. As we continue to use mineral resources, they will be harder to find and more costly to develop. As with energy, North America is one of the primary consumers of the world's mineral resources. Reasonable estimates are that each year North America consumes over 30 percent of the minerals produced in the world, which is a disproportionate share given that the combined population of the United States and Canada is about 5 percent of the world's population.

Steps in Mineral Utilization

The extraction and use of minerals involves several steps: exploration, mining, refining, transportation, and manufacturing. Each of these activities has costs (economic, energy, and environmental) associated with it. The costs associated with locating new sources of minerals (exploration) are primarily economic because exploration takes time and new technology. There are also some energy costs and some very small environmental costs. As the better sources of mineral resources are used up, it will be necessary to look for minerals in areas that are more difficult to explore, such as under the oceans. Therefore, both economic and energy costs will increase. Some areas, such as national parks and preserves, have been off-limits to mineral exploration. As current reserves of mineral resources are depleted, pressures will build to explore in these protected areas, resulting in increased environmental costs. Conversely, as a mineral becomes scarce, its price rises and there is economic incentive to extract the mineral from less-concentrated ores at greater economic and environmental costs.

Once a mineral resource has been located and the decision made to exploit it, the resource must be taken from the earth, and concentrated before it is transported to where it will be used. These activities involve large expenditures of money to pay for labor and the construction of machines and equipment. The energy costs of mining and refining are large because they basically involve extracting a small amount of desired material from a large amount of unwanted material. Table 12.1 shows the percentage of desired material present in several ores. As energy costs increase and the more concentrated ores are depleted, the economics of producing some materials may change so drastically that seeking substitute materials that require less energy expenditure is the best alternative.

Table 12.1 Metals in Ores

Metal	Percent Metal Needed in the Ore for Profitable Extraction	Price Range per Kilogram
Iron	30.0	
Chromium	30.0	Less than a dollar
Aluminum	20.0	
Nickel	1.5	
Tin	1.0	Several dollars
Copper	0.5	
Uranium	0.1	Tens of dollars

figure 12.4 **A Strip-Mining Operation** It is easy to see the important impact a mine of this type has on the local environment. Unfortunately, many mining operations are located in areas that are also known for their scenic beauty.

In addition to the economic and energy costs, there are significant environmental costs. Mining and refining affect the environment in several ways. All mining operations involve the separation of the valuable mineral from the surrounding rock. The surrounding rock must then be disposed of in some way. These pieces of rock are usually piled on the surface of the earth, where they are known as mine tailings and present an eyesore. It is also difficult to get vegetation to grow on these deposits. Some mine tailings contain materials (such as asbestos, arsenic, lead, and radioactive materials) that can be harmful to humans and other living things.

Many types of mining operations require vast quantities of water for the extraction process. The quality of this water is degraded, so it is unsuitable for drinking, irrigation, or recreation. Since mining disturbs the natural vegetation in an area, water may carry soil particles into streams and cause erosion and siltation. Some mining operations, such as strip mining, rearrange the top layers of the soil, which lessens or eliminates its productivity for a long time. (See figure 12.4.) Strip mining has disturbed approximately 75,000 square kilometers

Table 12.2 Costs Associated with Mineral Exploitation

Steps in Exploitation	Economic Costs	Energy Costs	Environmental Costs
Exploration	High because personnel and technology are expensive	Small	Small
Mining	High because of construction of plants and the labor to run them	High because of the need to move large amounts of material to separate valuable ores from other material	High because of changes in the landscape and large amounts of wastes produced Many mining operations use large amounts of water.
Refining	High because of the labor, equipment, and energy needed to concentrate ores	High because of the need to concentrate ores	Air and water pollution from refining facilities can be a problem.
Transportation	Needed at all stages of the process to move products from the extraction site to the manufacturing site	Energy is involved in all aspects of transportation.	Air pollution can be a problem. The construction of new transportation routes may influence scenic and other ecosystem values.
Manufacturing	Manufacturing plants and labor are needed.	Large amounts of energy are used in all manufacturing processes.	Air pollution and water pollution can be a problem, but most manufacturing sites are strictly regulated.

(30,000 square miles) of U.S. land, an area equivalent to the state of Maine.

Transportation and manufacturing are the final components of the overall cost of extracting minerals from the earth. Transportation is involved in the actual mining process, getting the ore to the refinery, moving the concentrated mineral to the site where it will be made into a finished product, and distributing the product to where it will be sold. The costs of transportation and manufacturing are primarily in the form of money and energy. Table 12.2 summarizes the steps involved in the use of minerals and the principal costs involved.

Recycling of Mineral Materials

To fully appreciate mineral resource use, we must consider the concept of recycling. Many minerals are not actually consumed or used up; they are just temporarily within a structure or a process. When the material is reclaimed and used again for another structure or process, it is recycled. (See chapter 18.) In order for recycling to be economic and effective, several conditions must be met. The material must be able to be collected relatively easily and the material must have a high economic value.

Empty aluminum beverage cans and waste oil are no longer useful to the consumer. However, the aluminum atoms can be reprocessed into new cans or other aluminum products and the oil can be reprocessed or burned as fuel. For recycling to be economically practical, the obsolete items must be recaptured before they have dispersed into the environment. For example, waste oil is relatively easy to recapture at the gasoline station where engine oil is replaced, but difficult to recapture if the oil is dumped on the ground or into a sewer.

In many industries the cost of purchasing recycled raw materials is higher than the cost of purchasing virgin materials and, therefore, it is more costly to make products from recycled material than from virgin materials. For example, disassembling a building and reusing its parts is more costly than using new construction materials. Conversely, it is less costly to remanufacture recycled aluminum, iron, and copper into other products than to produce the metal from ores. Therefore, there is strong incentive to recycle these metals.

One reason recycling is not more common is that, historically, the monetary costs for energy have been extremely low. As the cost of energy rises, more attention will be given to recycling minerals rather than to mining new ones. Manufacturing products in an environmentally harmonious way will become economically advantageous in the future.

Utilization and Modification of Terrestrial Ecosystems

Whenever humans make intensive use of natural ecosystems, the ecosystems are modified. Natural ecosystems provide many valuable goods and services, such as pollination, removal of carbon dioxide, genetic resources, medicines, foods, and other useful materials. These values are the result of the activities of the organisms that are present. **Biodiversity** is a measure of the variety of kinds of organisms present in an ecosystem. Natural ecosystems have much greater biodiversity than human-managed ecosystems. Therefore, when we look at intensive human use of ecosystems, we need to recognize that there will be loss of biodiversity.

Impact of Agriculture on Natural Ecosystems

Throughout the world, forests and grasslands have been converted to agricultural land and little of the original vegetation remains, since nearly all agricultural practices involve the removal of the original vegetation and substitution of exotic domesticated crops and animals for the original inhabitants. About 40 percent of the world's land surface has been converted to cropland and permanent pasture. Typically, the most productive natural ecosystems (forests and grasslands) are the first to be modified by human use and the most intensely managed. As the human population grows, it needs more space to grow food. The pressures to modify the environment are greatest in areas that have high population density. Often the changes brought about by intense agricultural use can degrade the ecosystem and permanently alter the biotic nature of the area. For example, much of the Mediterranean and Middle East once supported extensive forests that were converted to agriculture but now consist of dry scrubland. Today, agricultural land is being pushed to feed more people and its wise use is essential to the health and welfare of the people of the world. Chapters 14 and 15 will take a close look at patterns of use of agricultural land.

Table 12.3 Changes in the Extent of Forested or Wooded Land 1981–1990

Region	Percent Change 1981–1990
All tropical countries	**−3.8**
Tropical Africa	−2.2
Tropical Asia and Oceania	−4.3
Tropical Latin America and Caribbean	−4.5
All temperate countries	**+0.1**
Temperate Africa	−7.2
Temperate Asia and Oceania	+5.3
Temperate Latin America	−5.3
Temperate North America and Europe	0.0
World	**−1.9**

Source: Food and Agriculture Organization of the United Nations.

Managing Forest Ecosystems

The forests of the world are known quantities. The economic worth of the standing timber can be assessed, and the importance of forests for wildlife and watershed protection can be given a value. The history and current status of the world's forests are well known. Originally, almost half of the United States, three-fourths of Canada, almost all of Europe, and significant portions of the rest of the world were forested. The forests were removed for fuel, for building materials, to clear land for farming, and just because they were in the way. This activity returned the forests to an earlier successional stage, which resulted in the loss of certain animal and plant species (biodiversity) that required mature forests for their habitat. In general, all continents still contain significant amounts of forested land. Most of these forest ecosystems have been extensively modified by human activity. Table 12.3 shows that temperate parts of the world (particularly North America and Europe which have large forested areas) are maintaining their forests, while tropical forests are being lost.

Economic and Energy Costs of Utilizing Forest Ecosystems

The major economic costs of utilizing forests involve the purchase or leasing of land, paying for equipment and labor, and building roads to allow for transportation of forest products to where they will be converted to lumber, paper, or other uses. The major energy costs are involved in harvesting activities and transportation. Therefore, efficient methods of harvest and transportation are important to reduce the economic cost of using forest resources. Many harvest methods that may have reduced environmental impact are rejected because of the high energy costs associated with harvest and transportation.

Environmental Costs of Utilizing Forest Ecosystems

It is often difficult to put into monetary terms the environmental value of a resource. Therefore, environmentalists must rely on ethical or biological arguments to make their points. Modern forest-management practices in many parts of the world involve a compromise that allows economic exploitation while maintaining some of the environmental value of the forest. However, the very act of harvesting trees must alter the original "wilderness" character of the forest. Logging removes the trees and, therefore, the habitat for many kinds of animals that require mature stands of timber. The pine martin, grizzly bear, and cougar all require forested habitat that is relatively untouched by human activity. The removal of trees alters plant and animal biodiversity. Today in Australia, a small squirrel-sized marsupial known as a numbat is totally dependent on old forests that have termite- and ant-infested trees. Termites are numbats' major source of food, and the hollow trees and limbs caused by the insects' activities provide numbats with places to hide. Clearly, tree harvesting will have a negative impact on this species, which is already in danger of extinction. (See figure 12.5.)

In addition to serving as habitats for many species of plants and animals, forests provide many other environmental services. Forested areas modify the climate, protect soil from erosion, reduce the rate of water runoff, and provide recreational opportunities. If the trees are removed, the climate often becomes hotter and dryer. Since there is less vegetation to hold the water, flooding and soil

figure 12.5 **A Specialized Marsupial—the Numbat** This small marsupial mammal requires termite- and ant-infested trees for its survival. Termites serve as food and the hollow limbs and logs provide hiding places. Loss of old-growth forests with diseased trees will lead to the numbat's extinction.

figure 12.6 **Extensive Clear-cutting** Large, clear-cut plots of land can lead to a loss of species and accelerated soil erosion.

erosion are more common. Because the soil's water-holding ability is related to the amount of organic material and roots it has, denuded land allows water to run off rather than sink into the soil. Soil particles can wash into streams, where they cause siltation. The loss of soil particles reduces the soil's fertility. The particles that enter streams may cover spawning sites and eliminate fish populations. Since much valuable water is lost as runoff, less water sinks into the soil to recharge groundwater resources. If the trees along the stream are removed, the water will be warmed by increased exposure to sunlight. This may also have a negative effect on the fish population. The process of cutting the trees and transporting the logs also disturbs the wildlife in the area. The roads necessary for moving equipment and removing logs are a special problem. Constant travel over these roads removes vegetation and exposes the bare soil to more rapid erosion. When the roads are not properly located and constructed, they eventually become gullies and serve as channels for the flow of water. Most of these environmental problems can be minimized by using properly engineered roads and appropriate harvesting methods.

In many parts of the world, the construction of logging roads increases access to the forest and results in colonization by peasant "squatters" who seek to clear the forest for agriculture. The roads also permit poachers to have greater access to wildlife in the forest. Finally, the "wilderness" nature of the area is destroyed, which results in a loss of value for many who like to visit mature forests for recreation. An area that has recently been logged is not very scenic, and the roads and other changes often irreversibly alter the area's wilderness character. Obviously, wilderness and logging cannot coexist. Therefore, it becomes necessary to designate specific forests as wilderness or harvestable resources. In 2000, the United States government officially recognized the role of roads in altering the wilderness nature of forests by pronouncing that no new roads will be constructed in federally held forests.

Environmental Implications of Various Harvesting Methods

One of the most controversial logging practices is **clear-cutting.** (See figure 12.6.) As the name implies, all of the trees in a large area are removed. This is a very economical method of harvesting, but it exposes the soil to significant erosive forces. If large blocks of land are cut at one time, it may slow reestablishment of forest and have significant effects on wildlife. On some sites with gentle slopes, clear-cutting is a reasonable method of harvesting trees and environmental damage is limited. This is especially true if a border of undisturbed forest is left along the banks of any streams in the area. The roots of the trees help to stabilize the stream banks and retard siltation. Also, the shade provided by the trees helps to prevent warming of the water, which might be detrimental to some fish species.

Clear-cutting can be very destructive on sites with steep slopes or where regrowth is slow. Under these circumstances, it may be possible to use **patchwork clear-cutting.** With this method, smaller areas are clear-cut among patches of untouched forest. This reduces many of the problems associated with clear-cutting and can also improve conditions for species of game animals that flourish in successional forests but not in mature forests. For example, deer, grouse, and rabbits benefit from a mixture of mature forest and early-stage successional forest.

Clear-cut sites where natural reseeding or regrowth is slow may need to be replanted with trees, a process called **reforestation.** Reforestation is especially important for many of the conifer species, which often require bare soil to become established. Many of the deciduous trees will resprout from stumps or grow quickly from the seeds that litter the forest floor, so reforestation is not as important in deciduous forests.

Selective harvesting of some species of trees is also possible but is not as efficient or as economical as other methods, from the point of view of the harvesters. It allows them, however, to take individual, mature, high-value trees without causing much change to the forest ecosystem. In many tropical forests, high-value trees, like mahogany, are often harvested selectively. However, there may still be extensive damage to the forest by the construction of roads and the damage to noncommercial trees by the felling of the selected species.

Plantation Forestry

Many lumber companies maintain forest plantations as crops and manage them in the same way farmers manage crops. They plant single species, even-aged forests of fast-growing hybrid trees that have been developed in the same way as high-yielding agricultural crops. Competing species are controlled by the use of fire in some forests, and insects are controlled by aerial spraying. In these intensively managed forests, some single-species plantations mature to harvestable size in 20 years, rather than in the approximately 100 years typical for naturally reproducing mixed forests. However, the quality of the lumber products is reduced. Such forests have low species diversity and are not as valuable for wildlife and other uses as are more natural mixed-species forests. Furthermore, the trees planted in many managed forests may be exotic species. *Eucalyptus* trees from Australia have been planted in South America, Africa, and other parts of the world; and most of the forests in northern England and Scotland have a mixture of native pines and imported species from the European mainland and North America.

The forest products industry would prefer even-aged stands of a single species of tree, but wildlife managers prefer a variety of tree species at different stages of maturity to encourage many species of wildlife. A single-species, even-aged forest does not support as wide a variety of wildlife as does a mixed-age forest. However, with proper planning, trees can be harvested in small enough patches to encourage wildlife but in large enough sections to be economical. Some older trees, or some patches of old-stand timber, can be left as refuges for species that require them as a part of their niche. For example, in the southern pine forests of the United States, the red-cockaded woodpecker requires old diseased pine trees in which to build its nests. (See figure 12.7.) A well-managed forest plantation does not provide these sites, but a willingness to retain some old diseased trees can allow harvesting of the timber while protecting the habitat of the woodpecker.

figure 12.7 **Habitat Needs of the Red-Cockaded Woodpecker** As the place for its nest, the red-cockaded woodpecker requires old living pines that have a disease known as red heart. Old diseased trees are rare in intensely managed forests. Therefore, the birds are endangered.

Special Concerns about Tropical Deforestation

Tropical forests have a greater species diversity than any other ecosystem. The diverse mixture of tree species requires different harvesting techniques from those traditionally used in northern temperate forests. Also, because of soil characteristics, tropical forests are not as likely to regenerate after logging as are many of the temperate forests. If they do not regenerate, they must be considered nonrenewable resources. If these forests are to be used, a new set of forestry principles will be needed to establish a renewable tropical forest industry. Currently, few of these forests are being managed for long-term productivity; they are being harvested on a short-term economic basis only, as if they were nonrenewable resources. Tropical forests in Africa and Latin America are being deforested at a rate of 0.6 percent per year and those in Asia are being deforested at about 1 percent per year.

There are several concerns raised by tropical deforestation. First, the deforestation of large tracts of tropical forest is significantly reducing the species diversity of the world. Second, because tropical forests very effectively trap rainfall and prevent rapid runoff and the large amount of water transpired from the leaves of trees tends to increase the humidity of the air, the destruction of these forests also can significantly alter climate, generally resulting in a hotter, more arid climate. Finally, people have

ENVIRONMENTAL CLOSE-UP

The Northern Spotted Owl

In June 1990, the northern spotted owl (*Strix occidentalis caurina*) was listed as a threatened species under the Endangered Species Act. Threatened species are those that are likely to become endangered if current conditions do not change. The northern spotted owl lives in old-growth coniferous forests in the Pacific Northwest. A primary reason for the listing is that the owl requires mature forests as its habitat. There are several characteristics of the owl's biology that make old-growth forests important. It prefers a relatively closed canopy with rather open spaces under the canopy. These conditions allow it to fly freely as it searches for food. One of its primary food sources is the northern flying squirrel, which feeds primarily on fungi that are common in old-growth forests that have many dead trees and much down timber on the forest floor. Dead and diseased trees are also important because they provide cavities and platforms for the owls to use as nesting sites. As with most carnivores at the top of the food chain, the northern spotted owl requires large areas of forest for hunting. Therefore, relatively large areas of forest are needed for nesting pairs to be successful.

Logging in the Pacific Northwest has already used up most of the trees on private land. Most of the remaining old-growth forest, which is about 10 percent of its original area, is on land managed by the U.S. Forest Service. Many of the trees in these forests are several centuries old, and it takes 150 to 200 years for regenerating forests to become suitable habitat for the northern spotted owl. This timescale would not allow for the continued existence of the northern spotted owl if major sections of the remaining old-growth forest were logged. The reduction in logging has had significant economic impact on companies that relied on harvesting timber from federally owned lands, and as the companies are hurt economically, so too are the communities where these companies function.

Listing the owl as threatened required that the U.S. Forest Service develop and implement a plan to protect it. Since the owl is found only in mature old-growth forests, and since it requires large areas for hunting, the plan set aside large tracts of forested land and protected them from logging. The plan to protect the owl resulted in several actions. Most federally owned lands that were suitable habitat for the owl have indeed been protected from logging, although some salvage logging has been permitted when forests are damaged by fire or other events. Furthermore, private land that is known to have nesting owls was protected as well. As a result of this, individuals and companies have filed lawsuits claiming that since they cannot cut the trees, they have been deprived of the use of their land. Therefore, they have asked for compensation from the government. One lumber company was awarded $1.8 million in compensation for 22.7 hectares (56 acres) of forest that was "taken" for use as owl habitat. Court cases continue in the area. In 2000, a federal district judge ruled that cutting down trees does not necessarily harm northern spotted owls and that a logging company could proceed to cut a stand of timber. He ruled that in order to prevent logging of privately held land, the federal government must prove that cutting the trees would harm the owl. It is clear that, at this point, the current arguments about the northern spotted owl are legal maneuvers rather than serious discussions about the protection of a threatened species.

become concerned about preserving the potential of forests to trap carbon dioxide. As they carry on photosynthesis, trees trap large amounts of carbon dioxide. This may help to prevent increased carbon dioxide levels that contribute to global warming. (See chapter 17 for a discussion of global warming and climate change.)

Another complicating factor is that human population is growing rapidly in tropical regions of the world. More people need more food, which means that forestland will be converted to agriculture and the value of forest for timber, fuel, watershed protection, wildlife habitat, biodiversity, and carbon dioxide storage will be lost.

Managing Rangeland Ecosystems

Rangelands consist of the many arid and semiarid lands of the world that support grasses or a mixture of grasses and drought-resistant shrubs. These lands are too dry to support crops but are often used to raise low-density populations of domesticated or semidomesticated animals in permanent open ranges or nomadic herds. Usually, the animals are introduced species not native to the region. Sheep, cattle, and goats that are native to Europe and Asia have been introduced into the Americas, Australia, New Zealand, and many areas of Africa.

Few of these lands are privately owned, so they may be used free, or a small fee may be payable to the government. Consequently, there are few economic costs of using these lands. Furthermore, the low-intensity use of the lands does not require much energy expenditure. The primary costs of using these lands are environmental.

Environmental Costs of Utilizing Rangelands

The grazing of domesticated animals on rangelands has major impacts on biodiversity. In an effort to increase the productivity of rangelands, management techniques may specifically eliminate

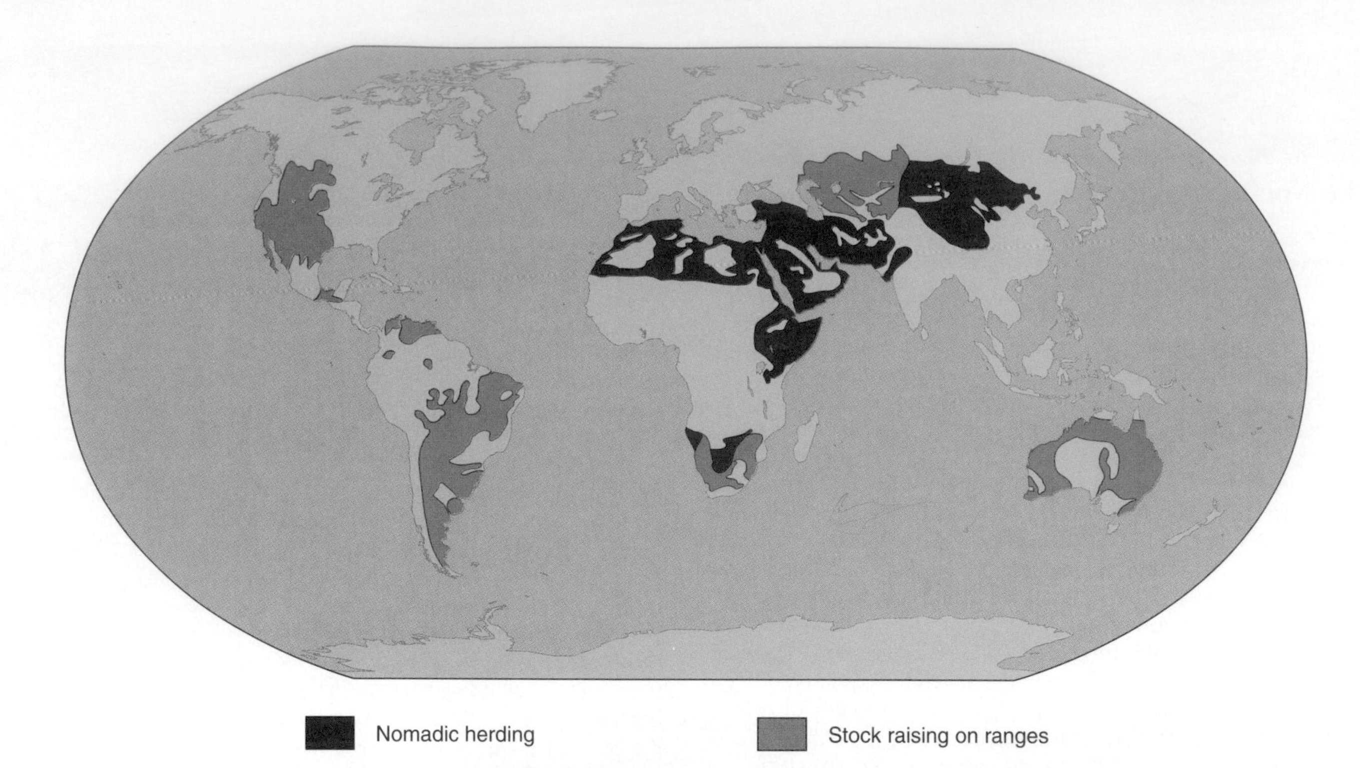

figure 12.8 **Use of Rangelands** The arid and semiarid regions of the world will not support farming without irrigation. In many of these areas, livestock can be raised. Permanent ranges occur where rainfall is low but regular. Nomadic herders can utilize areas that have irregular, sparse rainfall.

certain species of plants that are poisonous or not useful as food for the grazing animals, or specific grasses may be planted that are not native to the area. In some cases, native animals are reduced if they are a threat to livestock because they are predators or because they may spread disease to the livestock. In addition, the selective eating habits of livestock tend to reduce certain species of native plants and encourage others.

Because rainfall is low and often unpredictable, it is important to regulate the number of livestock on the range. In areas where the animals are on permanent pastures, the numbers of animals can be adjusted to meet the capacity of the range to provide forage. In many parts of the world, nomadic herders simply move their animals from areas where forage is poor to areas that have better forage. This is often a seasonal activity that involves movement of animals to higher elevations in the summer or to areas where rain has fallen recently. (See figure 12.8.)

In many parts of the world, where human population pressures are great, overgrazing is a severe problem. As populations increase, desperate people attempt to graze too many animals on the land. They also cut down the trees for firewood. If overgrazed, many plants die, and the loss of plant cover subjects the soil to wind erosion, resulting in a loss of fertility, which further reduces the land's ability to support vegetation. The cutting of trees for firewood has a similar effect, but it is especially damaging because many of these trees are legumes, which are important in nitrogen fixation. Their removal further reduces soil fertility. This severe overuse of the land results in conversion of the land to a more desertlike ecosystem. This process of converting arid and semiarid land to desert because of improper use by humans is called **desertification.** Desertification can be found throughout the world but is particularly prevalent in northern Africa and parts of Asia, where rainfall is irregular and unpredictable, and where many people are subsistence farmers or nomadic herders who are under considerable pressure to provide food for their families. (See figure 12.9.)

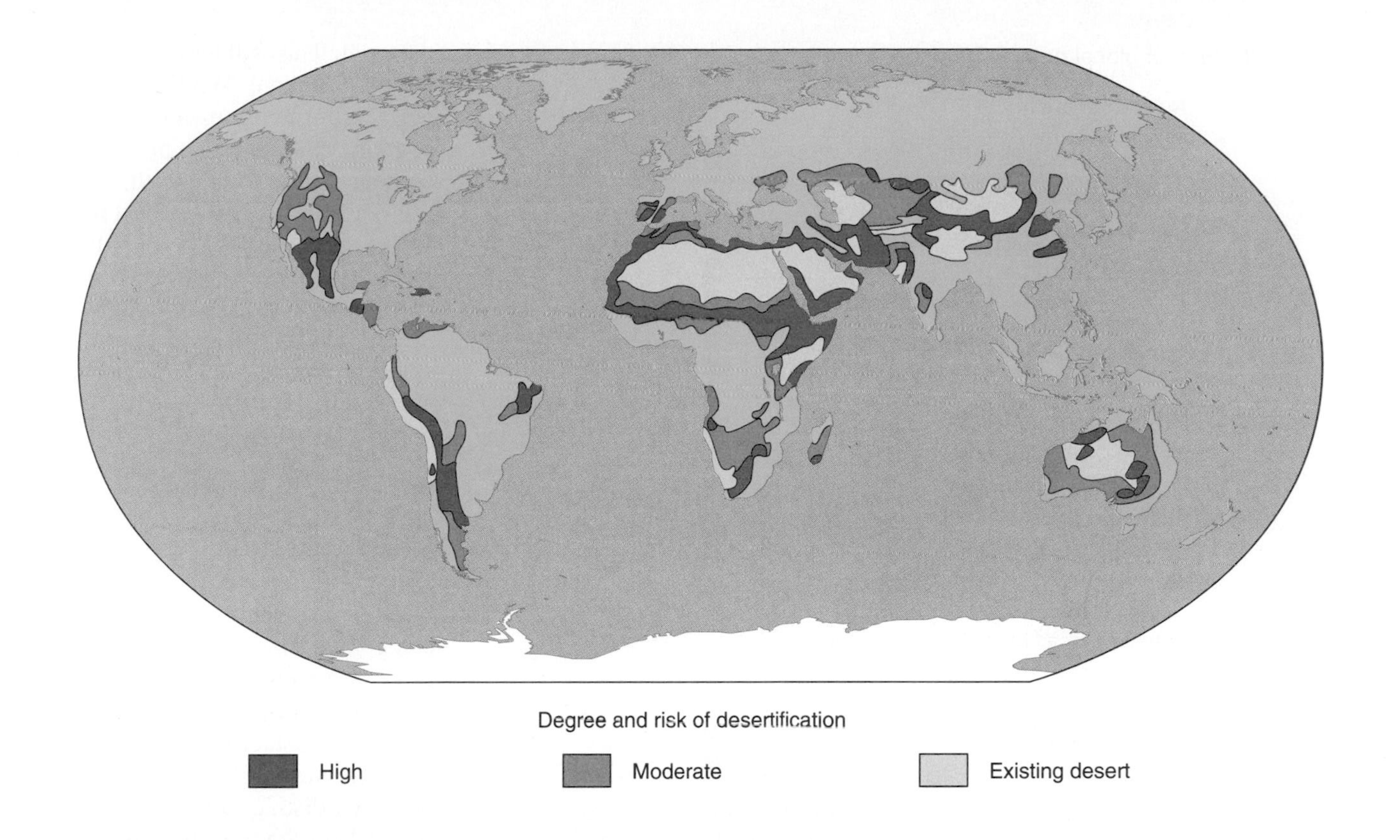

figure 12.9 **Desertification** Arid and semiarid areas can be converted to deserts by overgrazing or unsuccessful farming practices. The loss of vegetation increases erosion by wind and water, increases the evaporation rate, and reduces the amount of water that infiltrates the soil. All of these conditions encourage the development of desertlike areas.

Sources: *United Nations Map of World Desertification,* United Nations Food and Agriculture Organization, United Nations Educational, Scientific, and Cultural Organization, and the World Meteorological Organization for the United Nations Conference on Desertification, 1977, Nairobi, Kenya.

Areas with Minimal Human Impact—Wilderness and Remote Areas

There are still many areas of the world that have had minimal human impact. Some of these are remote areas with harsh environmental conditions, such as the continent of Antarctica, northern arctic areas, the tops of some tall mountains, or extremely arid areas, such as desert areas of Africa, central Australia, and central Asia. Most of these have been explored and found to be too harsh to sustain agriculture, and, therefore, dense human habitation is impossible.

Many other areas of the world, such as tropical rainforests, currently support ecosystems that have small human populations but are under threat because these ecosystems are capable of supporting agriculture or other human uses. The primary factor that will determine the survival of these natural areas is population pressure.

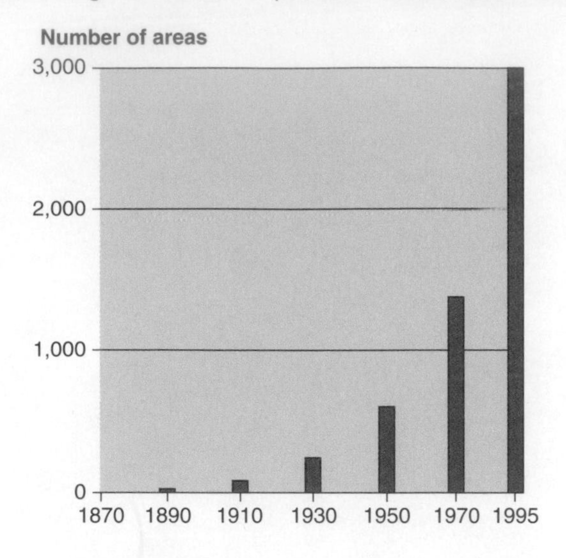

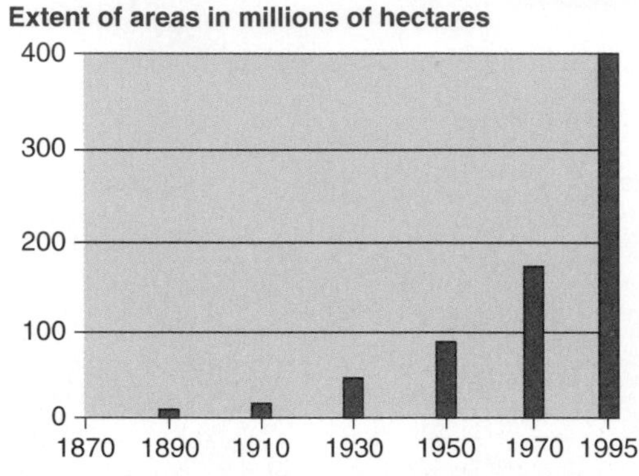

figure 12.10 **Growth in Major Park and Wilderness Areas of the World** Although there has been a steady increase in the number of protected areas and the number of hectares protected, many of the areas are in more-developed countries that can afford to set aside land for nonproductive uses.

Source: Data from International Union for Conservation of Nature and Natural Resources (IUCN), the United Nations List of National Parks and Protected Areas, 1985, Gland, Switzerland, and more recent 1995 data.

Many countries have established parks and other special designations of land use to protect areas of natural beauty or communities of organisms thought worthy of protection. This has been most noticeable in Africa, Central and South America, North America, and Australia. Until recently, these continents had large amounts of land that were relatively untouched. (See figure 12.10.) Parks and protected lands allow a variety of uses, depending on the country and its laws. Some are used primarily as tourist attractions, which generate funds for the country. Others are set up primarily to protect a species or community of organisms, while others simply seek to restrict use so that certain environmental standards are met. For example, harvesting in a forest might be restricted to a certain number of trees each year, and the number of people who visit a particular scenic site might be limited. A particularly sensitive issue is the designation of certain areas as wilderness. Although definitions of wilderness vary from country to country, the U.S. Congress, in the Wilderness Act of 1964, defined **wilderness** as "an area where the earth and its community of life are untrampled by man, where man himself is a visitor who does not remain."

Recently, there has been intense pressure on members of Congress from oil-producing states such as Alaska to allow oil exploration in areas currently designated as wilderness, such as the Arctic National Wildlife Refuge. As population increases and resources become more scarce, this pressure will become greater throughout the world.

The oceans of the world represent a major area that has been modified very little by human activities. Some areas, particularly those near the shore, receive intense use, but many areas of the open ocean are poorly understood and little affected by humans.

Managing Aquatic Ecosystems

Aquatic ecosystems are divided into two categories: marine and freshwater. The primary renewable resources obtained from these ecosystems are fish, shellfish, and crustaceans (lobsters, shrimp). Both marine and freshwater ecosystems present a distinct set of management issues. These are typically related to ownership issues, harvesting methods, and allowable harvest quotas. This section will deal primarily with the food production ability of aquatic ecosystems. Other issues related to water resources will be covered in chapter 16.

The economic and energy costs of utilizing aquatic ecosystems for food production are related to the techniques used to search for and capture aquatic organisms. Since the oceans and most freshwater systems are not owned by individuals, the cost of acquiring the right to fish is usually small. Although many countries may exercise control over specific areas along their coasts, they generally do not charge fees for the right to fish. There may be licenses that are required by fishers, but the cost of these is small compared to the cost of boats, equipment, and the energy needed to search for and capture fish and other aquatic organisms.

Environmental Costs Associated with Utilizing Marine Ecosystems

Environmental costs related to utilizing marine ecosystems fall into two broad

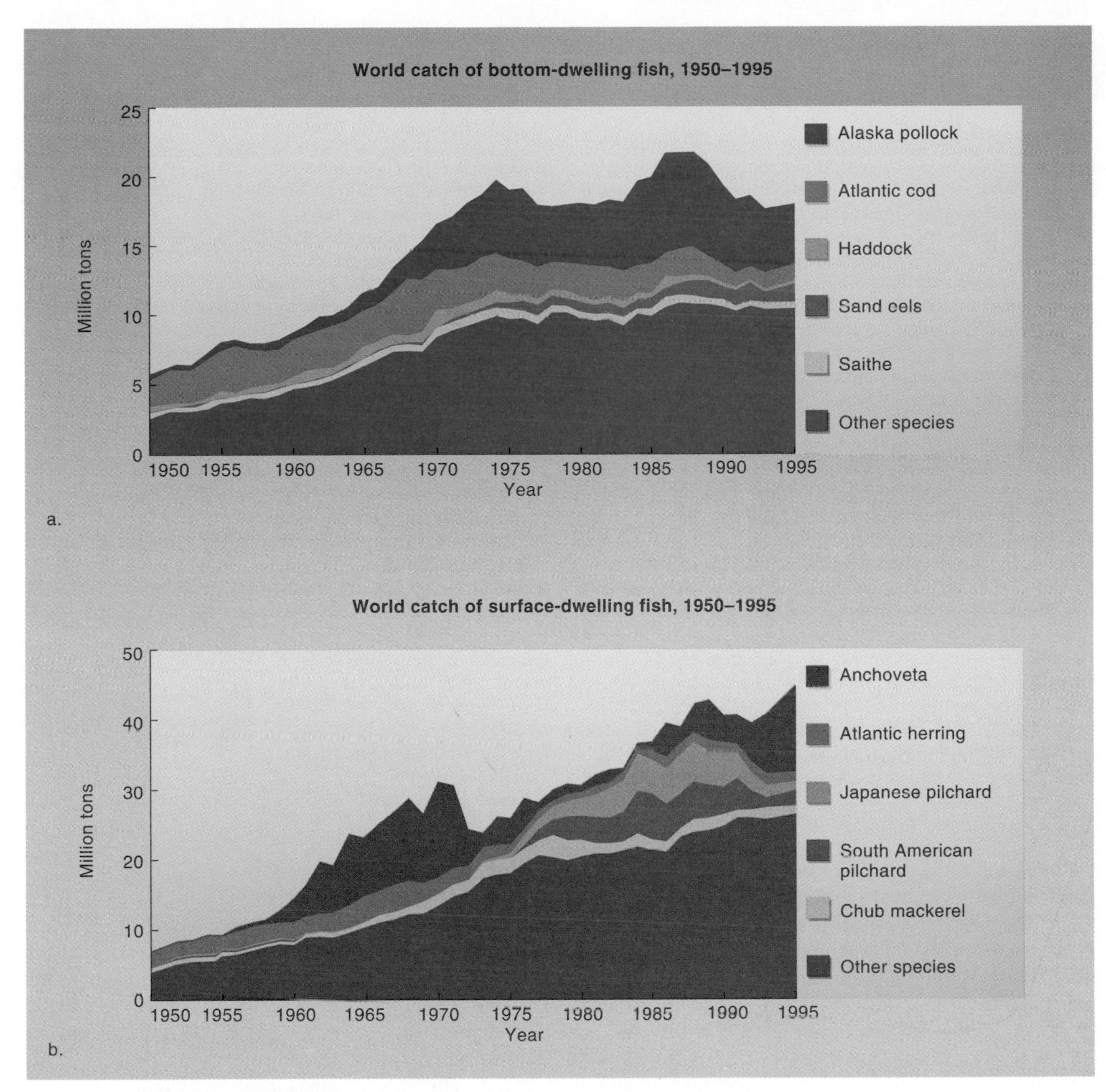

figure 12.11 **Changes in World Marine Fish Harvests 1950–1995** Increased fishing pressure has resulted in the significant reduction in many of the traditional fishing stocks. As traditional species have been overfished, other, less desirable species have been replacing them. This is true for both the bottom-dwelling species (*a*) and those that live in the open ocean (*b*).

Source: Food and Agriculture Organization of the United Nations.

categories: overfishing and the damaging effect of harvesting practices.

The United Nations estimates that 70 percent of the world's marine fisheries are being overexploited or are in danger of overexploitation as the number of fishers increases. In order to preserve the fishery, total fishing capacity should be reduced by about 30 percent. The total landings of fish peaked at just over 85 million tonnes (94 million U.S. tons) in 1989 and has been steady since then. However, the nature of the catch has changed as many of the more desirable species have declined. Figure 12.11 shows the change in species composition for both bottom-dwelling species and surface-dwelling open-ocean fish. The commercial fishing industry has been attempting to market fish species that previously were regarded as unacceptable to the consumer. These activities are the result of reduced catches of desired species. Examples of "newly discovered" fish in this category are monkfish and orange roughy.

For several reasons, the most productive areas of the ocean are those close to land. In shallow water, the entire depth (water column) is exposed to

GLOBAL PERSPECTIVE

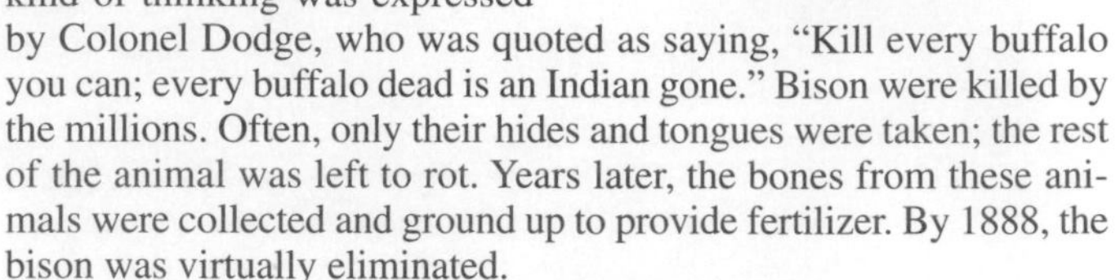

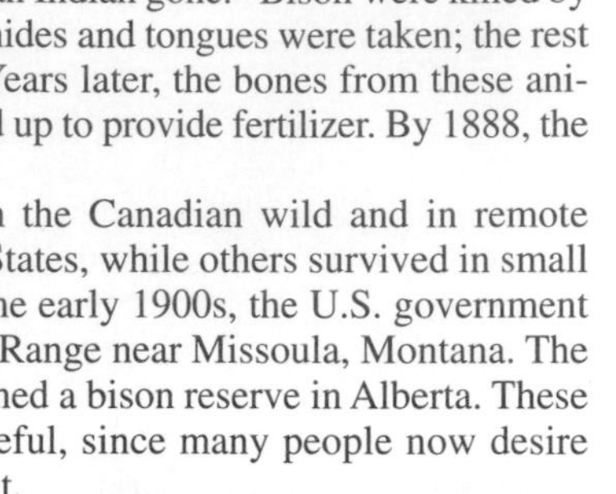

The History of the Bison

The original ecosystem in the central portions of North America was a prairie dominated by a few species of grasses. The eastern prairie, where moisture was greater, had grasses up to 2 meters (6 feet) tall, while the dryer western grasslands were populated with shorter grasses. Many kinds of animals lived in this area, including prairie dogs, grasshoppers, many kinds of birds, and bison. The bison was the dominant organism. Millions of these animals roamed the prairies of North America with few predators other than the Native Americans, who used the bison for food, their hides for shelter, and their horns for tools and ornaments. The relationship between the bison and Native Americans was a predator-prey relationship in which the humans did not significantly reduce bison numbers.

When Europeans came to North America, they changed this relationship drastically. European-born Americans sought to convert the prairie to agriculture and ranching. However, two things stood in their way: the Native Americans, who resented the intrusion of the "white man" into their territory, and the bison. Since many of the Native American tribes had horses and a history of warlike encounters with other tribes, they attempted to protect their land from this intrusion. The bison was a competitor in the eyes of the settlers, since it was impossible to use the land for agriculture or ranching with the millions of bison occupying so much area. The U.S. government established a policy of controlling the bison and the Native Americans: Since bison were the primary food source of Native Americans in many areas, eliminating them would result in the starvation of many Native Americans, which would eliminate them as a problem for the frontier settler. In 1874, the Secretary of the Interior stated that "the civilization of the Indian was impossible while the buffalo remained on the plains." Another example of this kind of thinking was expressed by Colonel Dodge, who was quoted as saying, "Kill every buffalo you can; every buffalo dead is an Indian gone." Bison were killed by the millions. Often, only their hides and tongues were taken; the rest of the animal was left to rot. Years later, the bones from these animals were collected and ground up to provide fertilizer. By 1888, the bison was virtually eliminated.

A few bison were left in the Canadian wild and in remote mountain areas of the United States, while others survived in small captive herds. Eventually, in the early 1900s, the U.S. government established the national Bison Range near Missoula, Montana. The Canadian government established a bison reserve in Alberta. These animals have proven to be useful, since many people now desire meat with a lower fat content, which bison have. Crossbreeding cattle with bison has led to a new breed of cattle, with reduced fat content, known as a beefalo. The animal that was barely saved from extinction by a few thoughtful individuals may contribute to better health for all.

sunlight. Plants and algae can carry on photosynthesis, and biological productivity is high. The nutrients washed from the land also tend to make these waters more fertile. Furthermore, land masses modify currents that bring nutrients up from the ocean bottom. Many of the commercially important fish and other seafood species are bottom dwellers, but fishing for them at great depths is not practical. Therefore, fishing pressure is concentrated on areas where the water is shallow and relatively nutrient rich. Countries have established control over the waters near them by establishing a 200-nautical-mile limit (approximately 300 kilometers) within which they control the fisheries. This has not solved conflicts, however, as neighboring countries dispute the fishing practices of waters they both claim. Canada and the United States have had continuing arguments over the management of the Pacific salmon fishery and argued as well over the North Atlantic cod fishery until its collapse in the mid-1990s.

Open-ocean species of fish are also subject to overexploitation. During the 1960s, anchovy fishing off the coast of Peru was a major industry. From 1971 to 1972, the catch dropped dramatically. Overfishing was believed to be one of the major contributing factors. This was aggravated by an increase in the area's water temperature, which prevented nutrient-rich layers from rising to the euphotic zone. Thus, productivity at all trophic levels decreased. Currently, the Peruvian anchovy fishery has rebounded but is considered to be overexploited and not sustainable. (See figure 12.11b.)

One of the major problems associated with the management of marine fisheries resources is the difficulty in achieving agreement on limits to the harvest. The oceans do not belong to any country and, therefore, each country feels it has the right to exploit the resource wherever it wants. Since each country seeks to exploit the resource for its advantage without regard for the sustainable use of the resource, international agreements are required to manage the resource. Such agreements are difficult to achieve because of political differences and disregard for the nature of the resource. (See The Tragedy of the Commons, page 58.) Figure 12.12 shows the major countries involved in harvesting marine fishes and their approximate catch. Many other countries maintain small, low-technology fishing operations that primarily supply fish to the local population.

Another environmental problem associated with these shallow-water, nearshore fisheries is the method used to harvest the fish. The bottom-dwelling fish are generally harvested with trawls—nets that are dragged along the bottom. These nets capture many different species, many of which are not commercially valuable. Typically 25 percent of the catch consists of species that have

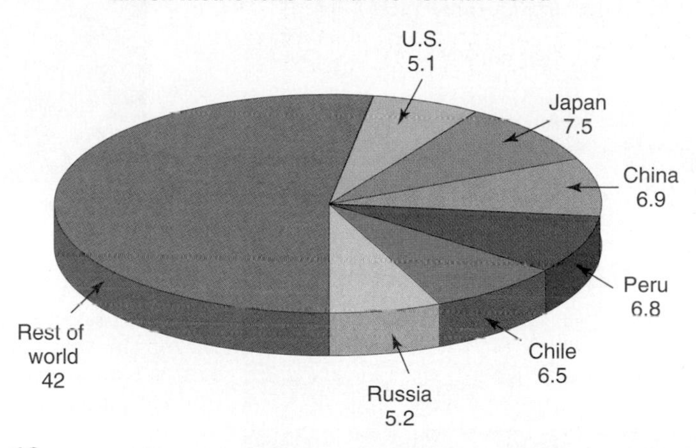

figure 12.12 **Major Harvesters of Fish** Six countries account for about half of the approximately 80 million tonnes (88 million U.S. tons) of marine fish harvested each year.

Source: Food and Agriculture Organization of the United Nations.

no commercial value. These are discarded by being thrown overboard. However, they are usually dead and their removal further alters the ecological nature of the seafloor. In addition, the trawls disturb the seafloor and create conditions that make it more difficult for the fish populations to recover. Some people have even advocated that the trawl should be banned as a fishing technique because of the damage done to the ocean bottom.

Environmental Costs Associated with Utilizing Freshwater Ecosystems

Because freshwater ecosystems are small and more intimately associated with human activities, few have not been considerably altered. Changes in water quality and the introduction of exotic species are two primary alterations. The warming of water due to thermal pollution and the addition of nutrients have made it impossible for some native species to survive. The effects of water pollution are discussed in chapter 16.

Although many places in the world harvest freshwater fish for food, the quantities are relatively small compared to those produced by marine fisheries. Furthermore, the management of the freshwater fish resource is much more intense since the bodies of water are smaller and human populations have greater access to the resource. Freshwater fish management in much of the world includes managing for both recreational and commercial food purposes. The fisheries resource manager must try to satisfy two interest groups. Both sports fishers and those who harvest for commercial purposes must adhere to regulations. However, the regulations usually allow the commercial fisher to use different harvesting methods, such as nets. In much of North America, freshwater fisheries are primarily managed for sport fishing. The management of freshwater fish populations is similar to that of other wild animals. Fish require cover, such as logs, stumps, rocks, and weed beds, so they can escape from predators. They also need special areas for spawning and raising young. These might be a gravel bed in a stream for salmon, a sandy area in a lake for bluegills or bass, or a marshy area for pike. A freshwater fisheries biologist tries to manipulate some of the features of the habitat to enhance them for the desired species of game fish. This might take the form of providing artificial spawning areas or cover. Regulation of the fishing season so that the fish have an opportunity to breed is also important.

In addition to these basic concerns, the fisheries biologist pays special attention to water quality. Whenever people use water or disturb land near the water, water quality is affected. For example, toxic substances kill fish directly, and organic matter in the water may reduce the oxygen. The use of water by industry or the removal of trees lining a stream warms the water and makes it unsuitable for certain species. Poor watershed management results in siltation, which covers spawning areas, clogs the gills of young fish, and changes the bottom so food organisms cannot live there. The fisheries biologist is probably as concerned about what happens outside the lake or stream as what happens within it.

The introduction of exotic fish species has greatly affected naturally occurring freshwater ecosystems. The Great Lakes, for example, have been altered considerably by the accidental and purposeful introduction of fish species. The sea lamprey, smelt, carp, alewife, brown trout, and several species of salmon are all new to this ecosystem. (See figure 12.13.) The sea lamprey is parasitic on lake trout and other species and nearly eliminated the native lake trout population. Controlling the lamprey problem requires use of a very specific larvicide that kills the immature lamprey in the streams. This technique works because mature lamprey migrate upstream to spawn, and the larvae spend several years in the stream before migrating downstream into the lake. With the partial control of lamprey, lake trout populations have been increasing. (See figure 12.14.) Recovery of the lake trout is particularly desirable, since at one time it was an important commercial species.

Another accidental introduction to the Great Lakes, the alewife, a small fish of little commercial or sport value, became a problem during the 1960s, when alewife populations were so great that they died in large numbers and littered beaches. Various species of salmon were introduced about this time in an attempt to control the alewife and replace the lake trout population, which had been depressed by the lamprey. While this salmon introduction has been an economic success and has generated millions of dollars for the sport fishing industry, it may have had a negative

Lake trout

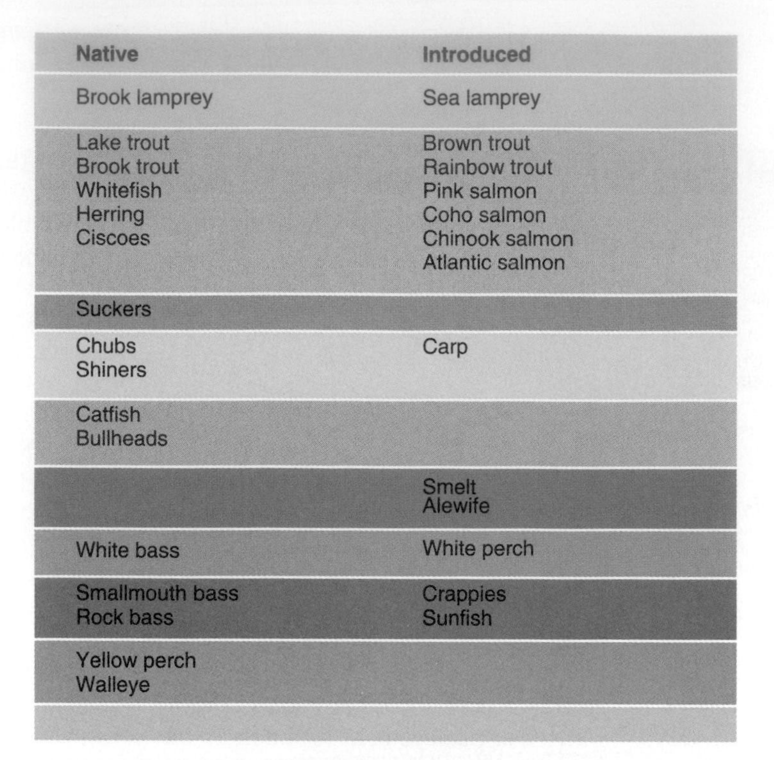

Native	Introduced
Brook lamprey	Sea lamprey
Lake trout Brook trout Whitefish Herring Ciscoes	Brown trout Rainbow trout Pink salmon Coho salmon Chinook salmon Atlantic salmon
Suckers	
Chubs Shiners	Carp
Catfish Bullheads	
	Smelt Alewife
White bass	White perch
Smallmouth bass Rock bass	Crappies Sunfish
Yellow perch Walleye	

Brown trout

Yellow perch

Smelt

figure 12.13 **Native and Introduced Fish Species in the Great Lakes** The Great Lakes have been altered considerably by the introduction of many non-native fish species. Some were introduced accidentally (lamprey, alewife, and carp), and others were introduced on purpose (salmon, brown trout, rainbow trout).

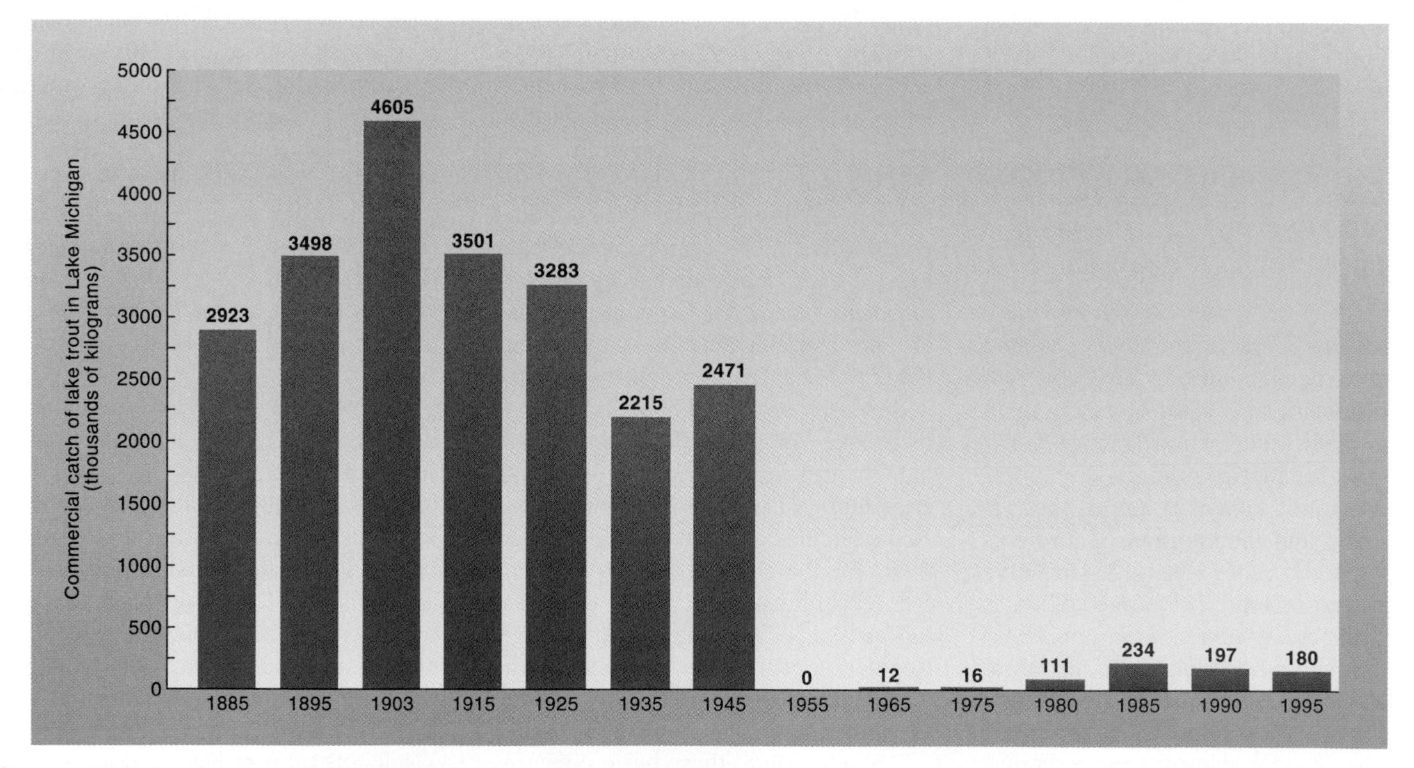

figure 12.14 **The Impact of the Lamprey on Commercial Fishing** The lamprey entered the Great Lakes in 1932. Because it is an external parasite on lake trout, it had a drastic effect on the population of lake trout in the Great Lakes. As a result of programs to prevent the lamprey from reproducing, the number of lamprey has been reduced somewhat, and the lake trout population is recovering with the aid of stocking programs.

ENVIRONMENTAL CLOSE-UP

Farming, fish kills, and *Pfiesteria piscicida*

During the summer of 1997, several instances of fish kills by *Pfiesteria piscicida* were suspected along the East Coast of the United States. The *Pfiesteria piscicida* is a dinoflagellate that has over 20 stages in its life cycle. Four of the stages release a toxin that causes fish to become lethargic, develop open sores, and often die. *Pfiesteria piscicida* appear to be triggered to transform into the toxic forms when they detect the presence of large numbers of fish. Fish are often caused to concentrate when nutrient-rich waters become depleted of oxygen. The fish move from oxygen-depleted water to the areas that have the best oxygen concentrations. It appears that *Pfiesteria piscicida* detect some component of the excrement of these concentrated populations of fish and develop into its toxic forms. The *Pfiesteria piscicida* organism then feeds on the tissues of the sick and dying fish. Humans working with *Pfiesteria* have been affected by the organism as well. Symptoms include sores, lightheadedness, and asthmalike symptoms, among others.

It appears that the problem occurs in areas of the ocean that are shallow and nutrient rich. Human and swine effluent overflows have been implicated in some cases, as has fertilizer runoff from farming operations.

impact on some native fish like the lake trout, with which the salmon probably compete. Salmon also migrate upstream, where they disrupt the spawning of native fish. Most species of salmon die after spawning, which causes a local odor problem.

Other more recent arrivals in the Great Lakes include a clam, the zebra mussel, and two fishes, the round gobi and river ruff. All of these probably arrived in ballast water from fresh or brackish water bodies in Europe. These exotic introductions are increasing rapidly and changing the mix of species present in Great Lakes' waters. (See Environmental Close-up Population Growth of Invading Species, page 137)

Streams and rivers are modified for navigation, irrigation, flood control, or power production purposes, all of which may alter the natural ecosystem and change the numbers or kinds of fish present. These topics are discussed in greater detail in chapter 16. In the Pacific Northwest, the extensive development of dams to provide power and aid navigation has made it nearly impossible for adult salmon to migrate upstream to spawn and difficult for young fish to migrate downstream to the ocean. Fish ladders and other techniques have not been successful in allowing the fish to pass the dams. As a result, many populations of Pacific salmon are nearly extinct. The only solution to the problem is to remove or greatly modify several of the dams.

Aquaculture

Fish farming (aquaculture) is becoming increasingly important as a source of marine fish production. Salmon farming has been particularly successful. This involves raising the salmon in "pens" in the ocean, which allows the introduction of food and other management techniques to achieve rapid growth of the fish. In 1995, British Columbia produced salmon worth $167 million from 80 farms. This compared to $75 million from wild-caught fish. Many other countries, including Norway, Scotland, and the United States, also farm salmon. However, even this method of producing fish is not without its environmental effects, since large numbers of fish release waste into the water and this fertilizer can increase the amount of algae in the water.

Aquaculture in freshwater systems typically involves the construction of ponds which allow the close management of the fish. Very little raising of fish occurs in open waters, although the immature stages of many species may be raised in special nursery facilities (fish hatcheries) prior to their release into the wild.

Managing Ecosystems for Wildlife

Many kinds of terrestrial wild animals are managed as game animals, to protect them from extinction, or for other purposes, and efforts are made to improve conditions for these species. Several techniques, some of which result in ecosystem modifications, are used to enhance certain wildlife populations. These techniques include habitat analysis and management, population assessment and management, predator and competitor control, and establishing refuges.

The economic costs associated with these activities include the salaries of management personnel, the purchase and leasing of land, and the cost of habitat modification activities. The energy costs are primarily involved in the modification of habitat. Often the environmental costs are considered to be minimal because the purpose of management is to enhance the populations of specific desired species. However, it needs to be understood that when one species is benefited, other species are displaced.

Habitat Analysis and Management

Managing a particular species requires an understanding of the habitat needs of that species. An animal's habitat must provide the following: food, water, and cover. **Cover** refers to any set of physical features that conceals or protects animals from the elements or enemies. Several kinds of cover are important and include safe places to rest and raise young, places where the animal can escape from enemies, and places where the animal is protected from the elements.

figure 12.15 **Cover Requirements for Quail** Quail need several kinds of cover to be successful in an area. When raising young, they need areas of tall grass and weeds to provide protection from predators. In the winter, they need thickets to provide cover from weather and predators.

Animals are highly specific in their habitat requirements. For example, bob-white quail must have a winter supply of food that protrudes above the snow. They require a supply of small rocks called grit, which the birds use in their gizzards to grind food. During most of the year, they need a field of tall grass and weeds as cover from natural predators. However, such protection is not available during the winter, so a thicket of brush is necessary. The thicket serves as cover from predators and shelter from the cold winter weather. Most areas provide suitable cover for resting, but for sleeping, quail prefer an open, elevated location, which allows the birds to quickly take flight if attacked at night. When raising young, they require grassy areas with some patches of bare ground where the young can sun themselves and dry out if they get wet. All of these requirements must be available within a radius of 400 meters (1,300 feet), because this is the extent of a quail's normal daily travels. (See figure 12.15.)

Once the critical habitat requirements of a species are understood, steps can be taken to alter the habitat to improve the success of the species. The habitat modifications made to enhance the success of a species are known as **habitat management.** The endangered Kirtland's warbler builds nests near the ground in dense stands of young jack pine. The density is important since their nests are vulnerable to predators and are better hidden in dense cover. Jack pine is a fire-adapted species that releases seeds following forest fires and naturally reestablishes dense stands following fire. Therefore, planned fires to regenerate new dense jack pine stands is a technique that has been used to provide appropriate habitat for this endangered bird.

Habitat management also may take the form of encouraging some species of plants that are the preferred food of the game species. For example, habitat management for deer may involve encouraging the growth of many young trees, saplings, and low-growing shrubs by cutting the timber in the area and allowing the natural regrowth to supply the food and cover the deer need. Both forest management and deer management may have to be integrated in this case because some other species of animals, like squirrels, will be excluded if mature trees are cut, since they rely on the seeds of trees as a major food source.

Population Assessment and Management

Population management is another important activity that requires planning. Species must be managed so that they do not exceed the carrying capacity of their habitat. Wildlife managers use several techniques to establish and maintain populations at an appropriate level. Population censuses are used to check the population regularly to see if it is within acceptable limits. These census activities include keeping records of the number of animals killed by hunters, recording the number of singing birds during the breeding season, counting the number of fecal pellets, direct counting of large animals from aircraft, and a variety of other techniques.

Given suitable habitats and protection, most wild animals can maintain a sizable population. In general, organisms produce more offspring than can survive. Figure 12.16 illustrates the reproductive potential of both quail and white-tailed deer. High reproductive potential, protection from hunting, and the management and restoration of suitable habitats have resulted in large populations of once rare animals. In Pennsylvania, where deer were once extinct, the number is now about 1.5 million. The wild turkey population in the United States has increased from about 20,000 birds in 1890 to 5 million birds today. In Zimbabwe, unlike in most African countries, the number of elephants increased from about 200 in 1900 to 35,000 in 2000. This is probably not a sustainable population, and the government would like to reduce the number to about 30,000.

Since wildlife management often involves the harvesting of animals by hunting for sport and for meat, regulation of hunting activity is an important population management technique. Seasons are usually regulated to assure adequate reproduction and provide the

ENVIRONMENTAL CLOSE-UP

Native American Fishing Rights

Throughout many parts of the United States, particularly in the Pacific Northwest and the Great Lakes states, there is a continuing controversy over fishing rights of Native Americans. This conflict is unique because it involves treaties that were made 100 to 150 years ago between the U.S. government and Native American nations. It has become a major political, economic, social, and legal issue in some states, such as Washington, Michigan, Wisconsin, and Minnesota, and has involved the entire court system, from local courts to the U.S. Supreme Court. The controversy revolves around the interpretation of treaty language. It has, on several occasions, turned to violence and has divided many communities.

According to the wording of many treaties entered into in the 1800s, the rights of Native Americans to fish would not be infringed upon by the states. Native Americans claim the treaties give them the legal authority to engage in commercial fishing enterprises even when such fishing may be restricted or banned altogether for the general public.

On the other side of the argument are many state officials and sport fishers who believe that Native Americans are seriously endangering populations of such fish as salmon and trout by their uncontrolled harvesting for commercial purposes. They further argue that many species being taken by Native Americans belong to the entire state and not only to a certain group, because the fish are stocked or planted by the state. Another concern is the fishing techniques used by Native Americans. In the 1800s, when the treaties were signed, commercial fishing technology was limited. Today, however, Native American commercial fishers use nylon nets, power boats, depth finders, and other technological aids that enable them to catch much larger quantities of fish than they could with traditional fishing practices.

In the early 1970s, when sport fishers complained that stocks of fish in Lake Michigan were being depleted because of Native Americans' gill-net fishing, the state of Michigan tried to regulate Native American fishing. The issue ended up in court, and, in 1978, a U.S. district court judge in Grand Rapids upheld fishing rights granted under 1836 and 1855 treaties between two Chippewa tribes and the U.S. government. The federal judge ruled that the state had no authority in the matter because the issue in question involved a federal treaty, and the state did not have the right to make regulations contrary to the treaty. Only Congress had such power. Subsequent to the decision, the tribes and the state negotiated changes in the areas in which the tribes could fish in an effort to reduce the potential for conflict between Native American fishers and sport or other commercial fishers.

A similar case was decided by a U.S. district judge in Tacoma, Washington, who interpreted treaties signed in 1854 and 1855 and ruled that Native Americans could catch up to 50 percent of the salmon that passed through their tribal land on the way to other parts of the state. In the face of protests by the non-native commercial fishing industry, the federal judge took over the regulation of salmon fishing in the state. Commercial fishers, who outnumbered Native American fishers by a ratio of approximately eight to one, feared for their livelihood. The case eventually went to the U.S. Supreme Court. In 1979, the Court ruled, in a six-to-three decision, to uphold the federal treaties that entitled Native Americans to half the salmon caught in the area of Puget Sound in Washington state. It further held that the 50 percent figure had to include the salmon that Native Americans caught on their lands for home consumption and religious ceremonies as well as the fish they caught for commercial purposes. The high court also voted to uphold the right of the district court to continue supervising the fishing industry because of the state's resistance to the interpretation of the treaties.

Clearly, the issues surrounding Native American fishing rights are complex and broad in scope. There are purely biological questions involving a resource and its wise use, economic issues involving families' livelihoods, cultural and religious concerns pertaining to Native Americans' use of a resource, legal questions involving federal and state conflicts, and moral questions relating to the unfair treatment of Native Americans in the past. In such conflicts, there is seldom one right answer. While the courts have ruled on the cases in Michigan and Washington, and the states and Native Americans are trying to use the fishing resource wisely, the problem still exists and is likely to continue for years to come.

largest possible healthy population during the hunting season. Hunting seasons usually occur in the fall so that surplus animals are taken before the challenges of winter. Winter taxes the animal's ability to stay warm and is also a time of low food supplies in most temperate regions. A well-managed wildlife resource allows for a large number of animals to be harvested in the fall and still leaves a healthy population to survive the winter and reproduce during the following spring. (See figure 12.17.)

In some cases, when a population of a species is below the desired number, organisms may be artificially introduced. Introductions are most often made when a species is being reintroduced to an area where it had become extinct. Wolves have been successfully reintroduced to Yellowstone National Park and large tracts of wooded land in the northern Midwest (Michigan, Wisconsin, Minnesota). Similarly an exchange between the state of Michigan and the province of Ontario reintroduced moose to Michigan and turkeys to parts of Ontario.

Some game species are non-native species that have been introduced for sport hunting purposes. The ring-necked pheasant was originally from Asia but has been introduced into Europe, Great Britain, and North America. All of the large game animals of New Zealand are introduced species, since there were none in the original biota of these islands. Several species of deer have been introduced into Europe from Asia. Many of these are raised in deer parks, where the animals are similar to free-ranging domestic cattle. They may be hunted for sport or slaughtered to provide food. Introductions of exotic species was once common. However, today it is recognized

Quail

	Adults	Young	Total
1st year	2	14	16
2nd year	16	112	128
3rd year	128	896	1024

Deer

	Adults	Yearlings	Fawns	Total
1st year	2	0	2	4
2nd year	2	2	2	6
3rd year	4	2	4	10
4th year	6	4	6	16
5th year	10	6	10	26

figure 12.16 **Reproductive Potential** If we assume no mortality, animals have a reproductive capacity far above what is required to just keep the population stable. In the real world, mortality generally keeps populations from growing beyond the ability of their habitat to sustain them.

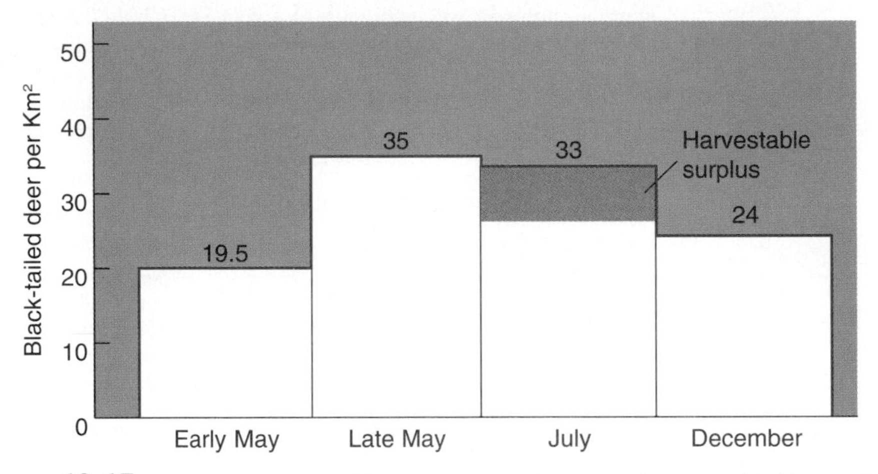

figure 12.17 **Managing a Wildlife Population** The seasonal changes in this population of black-tailed deer are typical of many game species. The hunting season is usually timed to occur in the fall so that surplus animals will be harvested before winter, when the carrying capacity is lower.

Source: Data from R.D. Taber and R.F. Dasmann, "The Dynamics of Three Natural Populations of the Deer *Odocoileus hemionus columbianus,*" in *Ecology* 38(2):233–46, 1957.

that these introductions often result in the decreased viability of native species, or in the introduced species becoming pests.

Predator and Competitor Control

At one time, it was thought that populations of game species could be increased substantially if predators were controlled. In Alaska, for example, the salmon-canning industry claimed that bald eagles were reducing the salmon population, and a bounty was placed on the bald eagle. From 1917 to 1952 in Alaska, 128,000 eagles were killed for bounty money. This theory of predator control to increase populations of game species has not proven to be valid in most cases, however, since the predators do not normally take the prime animals anyway. They are more likely to capture sick or injured individuals not suitable for game hunting.

Although predator and competitor control activities have become less common recently, they are still used in some special situations. In intensely managed European systems, gamekeepers are often charged with the responsibility of killing predators or unwanted competitors.

For some species, such as ground nesting birds, it may make some sense to control predators that eat eggs or capture the young, but, in most cases, humans have a greater impact through habitat modification and hunting than do the natural predators. Although at one time they were thought to be helpful, bounties and other forms of predator management have been largely eliminated in North America. In fact, the pendulum has swung the other way and predator control is not considered to be cost effective in most cases. One exception to this general trend is the hunting and trapping of wolves in Alaska and Canada. One rationale for the taking of wolves is that they kill moose and caribou. Since many of the people who live in these areas rely on wild game as a significant part of their food source, controlling wolf populations is politically popular. However, the number of wolves taken is regulated. Since, in many parts of the world, the major predators of game species are humans, regulation of hunting is a form of predator control.

The control of cowbird populations has been used to enhance the breeding success of the Kirtland's warbler. Cowbirds lay their eggs in the nests of other birds, including Kirtland's warblers. When the cowbird egg hatches, the hatchling pushes the warbler chicks out of the nest. Thus, trapping and killing of cowbirds in the vicinity of nesting Kirtland's warblers has been used to enhance the reproduction of this endangered species.

Many species may require refuges where they are protected from competing introduced species or human interference. Many native rangeland species of wildlife benefit from the exclusion of introduced grazing animals like cattle and sheep because the absence of grazing livestock allows a more natural grassland community to be reestablished. (See figure 12.18.)

Special Issues with Migratory Waterfowl Management

Waterfowl (ducks, geese, swans, rails, etc.) present some special management problems because they are migratory. **Migratory birds** can fly thousands of kilometers and, therefore, can travel north in the spring to reproduce during the summer months and return to the south when cold weather freezes the ponds, lakes, and streams that serve as their summer homes. (See figure 12.19.) Because many waterfowl nest in Canada and the northern United States and winter in the southern United States and Central America, an international agreement between Canada, the United States, and Mexico is necessary to manage and prevent the destruction of this wildlife resource. Habitat management has taken several forms. In Canada, where much of the breeding occurs, government and private organizations such as Ducks Unlimited have worked

figure 12.18 **Habitat Protection** The area on the left side of the fence has been protected from cattle grazing. This area provides a haven for many native species of plants and animals that cannot survive in heavily grazed areas.

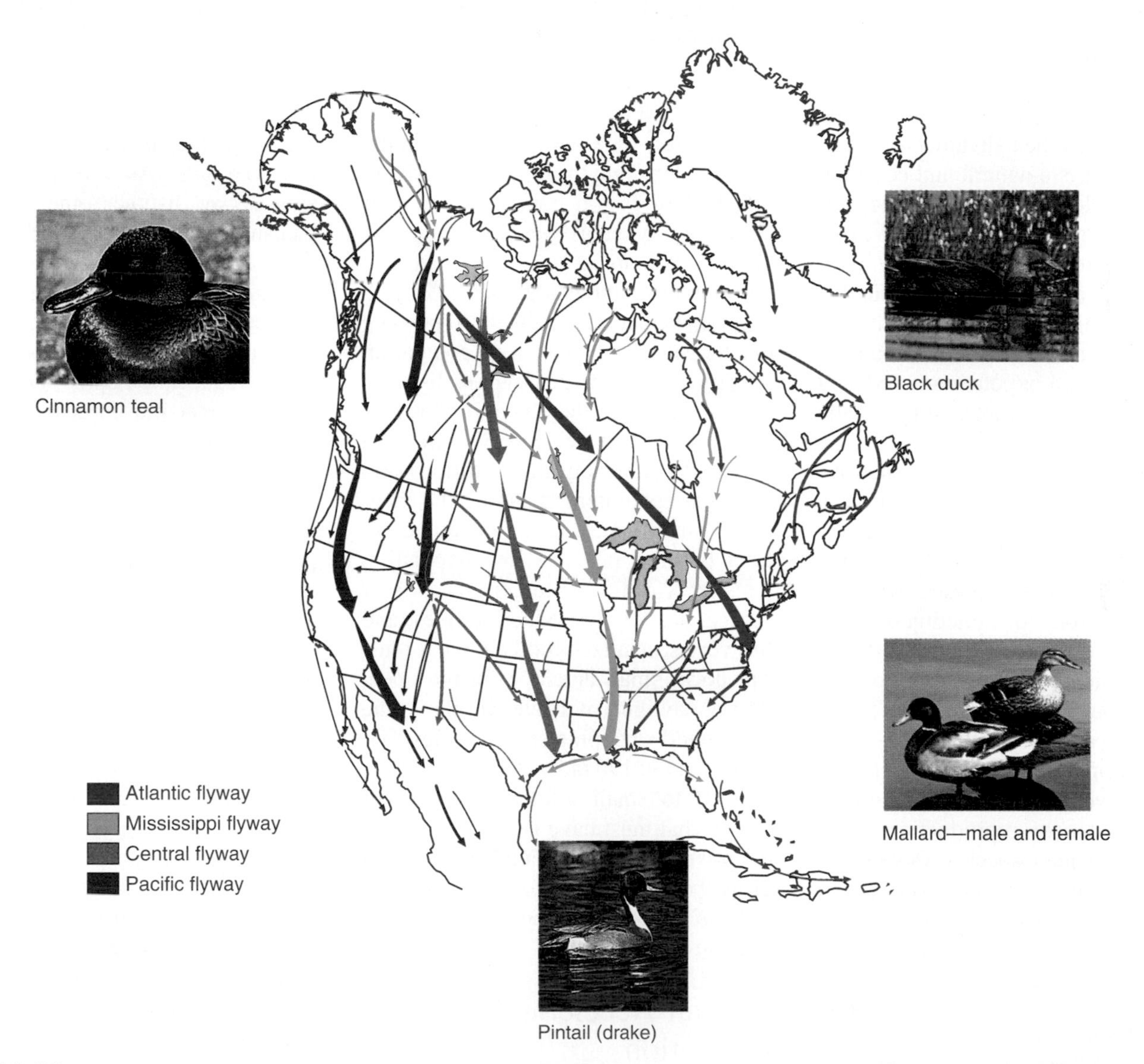

figure 12.19 **Migration Routes for North American Waterfowl** Migratory waterfowl follow traditional routes when they migrate. These have become known as the Atlantic, Mississippi, Central, and Pacific flyways. Many of these waterfowl are hatched in Canada, migrate through the United States, and winter in the southern United States or Mexico.

to prevent the draining of small ponds and lakes that provide nesting areas for the birds. In addition, new impoundments have been created where it is practical. Because birds migrate southward during the fall hunting season, a series of wildlife refuges provide resting places, food, and protection from hunting. In addition, these refuges may be used to raise local populations of birds. During the winter, many of these birds congregate in the southern United States. Refuges in these areas are important overwintering areas where the waterfowl can find food and shelter.

Table 12.4 Probability of Becoming Extinct

Most Likely to Become Extinct	Least Likely to Become Extinct
Low population density	High population density
Found in small area	Found over large area
Specialized niche	Generalized niche
Low reproductive rates	High reproductive rates

Extinction and Loss of Biodiversity

Extinction is the death of a species, the elimination of all the individuals of a particular kind. It is a natural and common event in the long history of biological evolution. In addition to complete extinction, we commonly observe local extinctions of populations. While not as final, a local extinction is an indication that the future of the species is not encouraging. Furthermore, as a population is reduced in size, some of the genetic variety in the population is likely to be lost.

Studies of modern local extinctions suggest that certain kinds of species are more likely than others to become extinct. (See table 12.4.) Species that have small populations of dispersed individuals are more prone to extinction because successful breeding is more difficult than in species that have large populations of relatively high density. Some kinds of organisms, such as carnivores at higher trophic levels in food chains, typically have low populations but also have low rates of reproduction compared to their prey species. Organisms in small restricted areas are also prone to extinction because an environmental change in their locale can eliminate the entire species at once. Organisms scattered over large areas are much less likely to be negatively affected by one event. Specialized organisms are also more likely to become extinct than are generalized ones. Since specialized organisms rely on a few key factors in the environment, anything that negatively affects these factors could result in their extinction, whereas generalists can use alternate resources.

Rabbits and rats are good examples of animals that are not likely to become extinct soon. They have high population density and a wide geographic distribution. In addition, they have high reproductive rates and are generalists that can live under a variety of conditions and use a variety of items as food. The cheetah is much more likely to become extinct because it has a low population density, is restricted to certain parts of Africa, has low reproductive rates, and has very specialized food habits. It must run down small antelope, in the open, during daylight, by itself. Similarly, the entire wild whooping crane species consists of about 190 individuals that are restricted to small winter and summer ranges that must have isolated marshes. (Captive and experimental populations bring the total number to about 250 individuals.) In addition, their rate of reproduction is low.

Human-Accelerated Extinction

Humans are among the most successful organisms on the face of the earth. We are adaptable, intelligent animals with few enemies. As our population increases, we displace other kinds of organisms. This has resulted in an accelerated rate of extinction. Wherever humans become the dominant organism, extinctions occur. Sometimes, we use other animals directly as food. In doing so, we reduce the population of our prey species. Since our population is so large and because we have an advanced technology, catching or killing other animals for food is relatively easy. In some cases, this has led to extinctions. The passenger pigeon in North America, the moas (giant birds) of New Zealand, and the bison and wild cattle of Europe were certainly helped on their way to extinction by people who hunted them for food.

We use organisms for a variety of purposes in addition to food. Many plants and animals are used as ornaments. Flowers are picked, animals skins are worn, and animal parts are used for their purported aphrodisiac qualities. In the United States, many species of cactus are being severely reduced because people like to have them in their front yards. In other parts of the world, rhinoceros horn is used to make dagger handles or is powdered and sold as an aphrodisiac. Because some people are willing to pay huge amounts of money for these products, unscrupulous

Table 12.5 Typical Species Prices (1990 Rates)

International Species	Price ($U.S.)	North American Species	Price ($U.S.)
Olive python	1500	Bald eagle	2500
Rhinoceros horn	12,500/pound	Golden eagle	200
Tiger skin (Siberian)	3500	Gila monster	200
Tiger meat	130/pound	Peregrine falcon	10,000
Cockatoo	2000	Grizzly bear	5000
Leopard	8500	Grizzly bear claw necklace	2500
Snow leopard	14,000	Polar bear	6000
Elephant tusk	250/pound	Black bear paw pad	150
Walrus tusk	50/pound	Reindeer antlers	35/pound
Mountain gorilla	150,000	Mountain lion	500
Giant panda	3700	Mountain goat	3500
Ocelot	40,000/coat	Saguaro cactus	15,000
Imperial Amazon macaw	30,000		

Source: Data from U.S. Fish and Wildlife Service.

people are willing to take the chance of poaching these animals for the quick profit they can realize. Table 12.5 shows some of the organisms, or their parts, that are highly prized by buyers. The World Wildlife Fund estimates that illegal trade in wild animals globally produces $2 billion to $30.5 billion per year. These activities have already resulted in local extinctions of some plants and animals and may be a contributing factor to the future extinction of some species.

Some organisms are extinct because they were regarded as pests. Many large predators have been locally exterminated because they preyed on the domestic animals that humans use for food. Mountain lions and grizzly bears in North America have been reduced to small, isolated populations, in part because they were hunted to reduce livestock loss. Tigers in Asia and the lion and wolf in Europe were reduced or eliminated for similar reasons. Even though commercial hunters killed thousands of passenger pigeons, their ultimate extinction was caused primarily by the increased agricultural use of the forests. Passenger pigeons ate the acorns of oaks and the beechnuts from beech, and also relied on the forests for communal nesting sites. When the forests were cleared, the pigeons became pests to farmers, who shot them to protect their crops from being eaten by the birds. Some extinctions of pest species are considered desirable. Most people would not be disturbed by the extinction of black widow spiders, mosquitoes, rats, or fleas. In fact, people work hard to drive some species to extinction. For example, in the *Morbidity and Mortality Weekly Report* (October 26, 1979), the U.S. Centers for Disease Control triumphantly announced that the virus that causes smallpox was extinct in the human population after many years of continuous effort to eliminate it.

The most important cause of extinctions related to human activity is habitat alteration. (See figure 12.20.) Whenever humans populate an area, they change it by converting the original ecosystem into something that supports themselves. Forests and grasslands have been converted to agricultural and grazing lands. In addition, humans have introduced new species to grow or graze on these lands. These new plants and animals compete with the native organisms for nutrients and living space, and, often, the native organisms lose. (See Global Perspective: The History of the Bison.) The African elephant and various species of rhinoceros are in danger because they are hunted for ivory or horn, but it was the original fragmentation of their habitat for agricultural purposes that first placed the animals in conflict with humans.

In much of the tropical world today, rainforests are being cleared to provide grazing land or agricultural land for an expanding human population. This activity destroys existing forests and fragments them into small islands. Scientists studying the effects of this activity have noticed that, as a forest is reduced to small patches, many species of birds disappear from the area. The same kinds of activities certainly happened in Europe, where little of the original forest is left. In North America, the eastern deciduous forests were reduced, which probably resulted in the extinctions of some animals and plants. Almost all of the original prairie in the United States has been replaced with agricultural land, resulting in the loss of some species.

Humans also cause extinction in less direct ways. The building of dams changes the character of rivers, making them less suitable for some species. Air and water pollution may kill all life in an area. For example, acid precipitation has lowered the pH of some lakes so much that all life has been eliminated. In other cases, indirect human activity may selectively eliminate species that are less tolerant to the pollutants being released, or the introduction of exotic species may eliminate existing species. The accidental introduction of the chestnut blight fungus into the United States resulted in the loss of the American chestnut tree over much of the Appalachian region. Some species that were dependent on these trees probably were also eliminated.

Why Worry About Extinction?

If extinction is a natural event, and if the species that become extinct are those that cannot effectively cope with the activities of humans, why should we worry about them? There are several answers to that question. First of all, strictly from a selfish point of view, many species we know little about may be useful to us.

turn to its original size. The people who depended on these fisheries now must find other ways to make a living.

A third answer, according to many people, is that all species have an intrinsic value and a fundamental right to exist without being needlessly eliminated by the unthinking activity of the human species. This is an ethical position that is unrelated to social or economic considerations. According to those who support this philosophical position, extinction by itself is not bad, but human-initiated extinction is. This contrasts with the philosophical position that humans are simply organisms that have achieved a preeminent position on Earth and, therefore, extinctions we cause are no different from extinctions caused by other forces. Regardless, there have been significant efforts to prevent the human-initiated extinction of species.

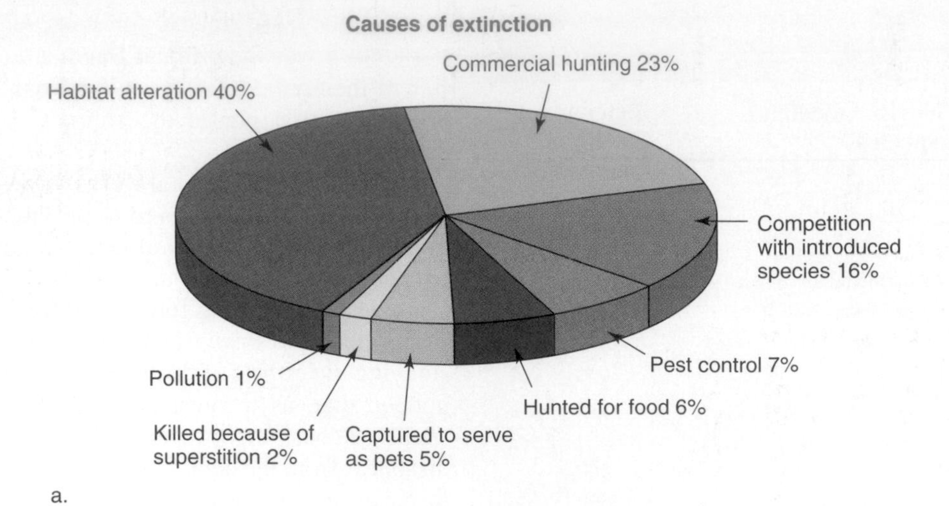

a.

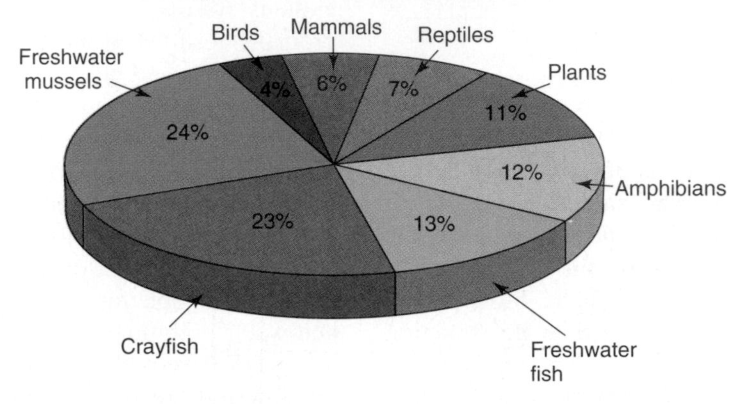

b.

figure 12.20 **Causes of Extinction** Many organisms have become extinct as a result of human activities. The most important of these activities is the indirect destruction of habitat by humans modifying the environment for their own purposes. Part (*b*) shows the species at risk in North America (percentage by group).

(*b*) Source: Data from the Nature Conservancy and the Network of Natural Heritage Programs and Conservation Data Centers, 1993.

Many plants have chemicals in them that can be used as medicines. If we drive them into extinction, we may be eliminating a potentially useful product. Also, as the human population increases, the need for different kinds of food plants and animals will increase. Once they are eliminated, we have lost the opportunity to use them for our own ends. Most of the wild ancestors of our most important food grains, like maize (corn), wheat, and rice, are thought to be extinct. What would our world be like if these plants had not been domesticated before they went extinct as wild populations? Because of this concern, many agricultural institutes and universities maintain "gene banks" of wild and primitive stocks of crop plants.

Another interesting answer to the question is that certain organisms in some ecosystems appear to play pivotal roles. Accidental extinction of one of these species could be devastating to the ecosystem and the humans that use it. For example, the sardine fishery off the coast of southern California and the anchoveta fishery off the coast of Peru appear to have been fundamentally altered by overfishing. Although the details are not known, it is thought that, once the population was significantly reduced, other organisms filled their niche, making it difficult for their population to re-

What Is Being Done to Prevent Extinction and Protect Biodiversity?

Efforts to prevent human-caused extinctions are difficult to assess. Some countries have enacted legislation to protect species that are in danger of becoming extinct. Species that receive special protection usually are given some sort of designation, such as endangered or threatened. **Endangered species** are those that have such small numbers that they are in immediate jeopardy of becoming extinct. **Threatened species** could become extinct if a critical factor in their environment were changed.

Most of the interest in preventing human-caused extinctions comes from developed countries. There, however, the problem is less acute because vulnerable species have already been eliminated. Extinction is a greater potential problem in tropical, less-developed countries. Many biologists estimate that there may be as many species in the tropical rainforests of the world as in the rest of the world combined. Unfortunately, extinction prevention is not a major issue in

ENVIRONMENTAL CLOSE-UP

The California Condor

The California condor (*Gymnogyps californianus*) is thought to have been adapted to feed on the carcasses of large mammals found in North America during the Ice Age. With the extinction of the large, Ice Age mammals, the condors' major food source disappeared. By the 1940s, their range had shrunk to a small area near Los Angeles, California. Further fragmentation of their habitat caused by human activity, death by shooting, and death by eating animals containing lead shot reduced the wild population to about 17 animals by 1986.

A low reproductive potential makes it difficult for the species to increase in numbers. They do not become sexually mature until six years of age, and females typically lay one egg every two years. Because of concerns about the survival of the species, in 1987, all the remaining wild condors were captured to serve as breeding populations. The total population was 27 individuals. The plan was to raise young condors in captivity, leading to a large population that would ultimately allow for the release of animals back into the wild. Captive breeding is a very involved activity. Extra eggs can be obtained by removing the first egg laid and incubating it artificially. The female will lay a second egg if the first is removed. Raising the young requires careful planning. Although the young are fed by humans, they must not associate food with humans. This would result in inappropriate behaviors in animals eventually released to the wild. Therefore, puppets that resemble the parent birds are used to feed the young birds.

These efforts increased the number of offspring produced per female and resulted in a captive population of 54 individuals by 1991, when two condors were released into the wild north of Los Angeles. Six more were released in 1992. In 1996, a second population of condors was introduced in Arizona north of the Grand Canyon near the Utah border. The goal is to have two populations of up to 150 individuals each. In addition to natural mortality from predation, there has been unanticipated mortality from contact with power lines and lead poisoning from ingesting lead shot with food. To help reduce these mortality problems, the birds that are currently being released go through a period of training which teaches them to avoid power lines. In addition, several condors have been recaptured to be treated for lead poisoning. These will eventually be released back to the wild.

In 2000, there were 171 California condors in existence. Seventeen were in the wild in Arizona. Thirty-one were in the wild in southern and central California, and the rest were in captive breeding programs at three different locations. The behavior of the animals seems natural and there is hope that they may eventually reproduce in the wild.

many less-developed countries. This difference in level of interest is understandable since the developed world has surplus food, higher disposable income, and higher education levels, while people in many of the less-developed countries, where population growth is high, are most concerned with immediate needs for food and shelter, not with long-term issues like extinction.

Nevertheless, many of the governments of less-developed countries have responded to pressures and suggestions from outside sources and have established preserves and parks to protect species in danger of extinction. This does not solve the problem, however, since the areas must be protected from poachers and unauthorized agricultural activity of people who trespass on these protected areas. These people are responding to basic biological and economic pressures to provide food for their families. In many countries, hunting for bush meat is an important source of income for people, and the meat provides much needed protein for the people of the area. Furthermore, protected lands generally do not generate any income for the government. Hiring security forces to patrol such areas is expensive for many of these countries, so protection is often inadequate to prevent trespass and poaching. Providing tourism opportunities that generate income for local people and the government is one way that some countries try to offset the costs of protection.

Even in countries where interest in extinction prevention is relatively high, there is a cultural bias in favor of protecting certain kinds of organisms. Most endangered or threatened species are birds, mammals, some insects (particularly butterflies), a few mollusks and fish, and certain categories of plants. Bacteria, fungi, most insects, and many other inconspicuous organisms rarely show up on endangered species lists, even though they play vital roles in the nitrogen cycle, in the carbon cycle, and as decomposers. For example, the United States Fish and Wildlife Service lists 233 vertebrates (birds, mammals, amphibians, reptiles, and fish) and 565 flowering plants, as endangered in the United States, but only 30 insects, and no fungi or bacteria.

Several international organizations work to prevent the extinction of organisms. The World Conservation Union (IUCN) estimates that, by the year 2000, at least 500,000 species of plants and animals may have been exterminated. (The International Union for the Conservation of Nature changed its name to the World Conservation Union but kept IUCN as its acronym.) The IUCN classifies species in danger of extinction into four categories: endangered, vulnerable, rare, and indeterminate. Endangered species are those whose survival is unlikely if the conditions threatening their extinction continue. These organisms need action by people to preserve them, or they will become extinct. Vulnerable species are those that have decreasing populations and will become endangered unless causal factors, such as habitat destruction, are stopped. Rare species are primarily those that have small worldwide populations and that could be at risk in the future. Indeterminate species are those that are thought to be extinct, vulnerable, or rare, but so little is known about them that they are impossible to classify.

Although the IUCN is a highly visible international conservation organization, it has very little power to effect change. It generally seeks to protect species in danger by encouraging countries to complete inventories of plants and animals within their borders. It also encourages the training of plant and animal biologists within the countries involved. (There is currently a critical shortage of plant and animal biologists who are familiar with the organisms of the tropics.) The IUCN also encourages the establishment of preserves to protect species in danger of extinction.

The U.S. Endangered Species Act was passed in 1973. This legislation gave the federal government jurisdiction over any species that were designated as endangered. About 300 U.S. species and subspecies have been so designated by the Office of Endangered Species of the Department of the Interior. (See figure 12.21.) The Endangered Species Act directs that no activity by a governmental agency should lead to the extinction of an endangered species and that all governmental agencies must use whatever measures are necessary to preserve these species.

The key to preventing extinctions is preservation of the habitat required by the endangered species. Consequently, many U.S. governmental agencies and private organizations have purchased sensitive habitats or have managed areas to preserve suitable habitats for endangered species. Setting aside certain land areas or bodies of water forces government and private enterprise to confront the issue of endangered species. The question eventually becomes one of assigning a value to the endangered species. This is not an easy task; often, politics become involved, and the endangered species does not always win.

A case in point is the controversy that surrounded Tennessee's Tellico Dam project in 1978. The U.S. Supreme Court declared that completion of the $116 million federal project would result in a violation of the Endangered Species Act because the dam would threaten the survival of an endangered species called the snail darter, a tiny fish about 8 centimeters (3 inches) long that lived in the stream that would be altered by the dam's construction. Developers, however, were not deterred. They lobbied in Congress to have all federally funded projects exempted from the act. Conservationists lobbied for the preservation of the act as it was originally written. Eventually, in 1978, Congress amended the Endangered Species Act so that exemptions to the act could be granted for federally declared major disaster areas or for national defense, or by a seven-member Endangered Species Review Committee. Because this group has the power to sanction the extinction of an organism, it has been nicknamed the "God Squad." If the committee found that the economic benefits of a project outweighed the harmful ecological effects, it would exempt a project from the Endangered Species Act. At their first meeting, the review board denied the request to exempt the Tellico Dam project on the grounds that the project was economically unsound. This should have stopped the Tellico Dam project. However, nine months later, as a result of several political maneuvers, Congress appropriated money to complete the dam. It is now complete and full of water. The snail darters that once dwelled on the site were transplanted to nearby rivers and have since been removed from the endangered classification and put in the less critical threatened category.

The amendments to the Endangered Species Act also weakened the ability of the U.S. government to add new species to the endangered and threatened lists. Before a species can be listed, it is now necessary to determine the boundaries of its critical habitat, prepare an economic impact study, and hold public hearings—all within two years of the proposal of the listing. The political aspects of endangered species continues to be important in the U.S. Congress. Intense lobbying by environmental and business interests seeks to shape the reauthorization of the act to suit their interests.

Blackfooted ferret

Whooping crane

Mission blue butterfly

Galápagos tortoise

Giant panda

Persistent trillium

figure 12.21 **Endangered Species** Many species have been placed on the endangered species list. These are plants and animals that are present in such low numbers that they are in immediate danger of becoming extinct.

ISSUES & ANALYSIS

Fire As a Forest Management Tool

Fire has been used as a forest management tool for many years. When fires are absent from ecosystems for long periods, the amount of fuel present builds up so that the intensity of any fire that occurs is more destructive than when fires are more frequent. Large destructive fires damage timber and, therefore, reduce the value of the forest for lumber. Frequent small fires that primarily burn the litter on the ground are not as hot, consume the fuel before it reaches dangerous levels, and can be used in such a way that they do little damage to trees. Many private and public organizations that manage forest lands periodically use fire to remove debris and reduce the likelihood of major damaging fires.

In many wilderness areas, natural fires are allowed to burn as long as they are not "too large" and do not threaten human habitation. Yellowstone National Park was burned extensively in 1988. At the time, a "let burn" policy was in effect for wilderness areas. However, the fires eventually threatened buildings within the park and control measures were initiated. Subsequent studies of the effects of the fires have substantiated that the park recovered and that a natural succession is taking place.

Hundreds of fires are set purposely each year as a normal part of forest management activities. Most of these fires go unnoticed since they do exactly what they are supposed to do. However, in 2000, a purposely set fire (prescribed burn) in a national forest got out of control and burned beyond the specified area, destroyed hundreds of homes, and threatened the Los Alamos National Laboratory. The U.S. Forest Service maintains that managed use of fire is still an important forest management technique, but there are members of Congress who believe the practice should be stopped.

- Should fires that are caused by lightning in wilderness areas be treated differently from fires started accidentally by humans?
- Should fire be used as a management tool?
- Should Congress or forestry professionals make decisions about what should be appropriate management practices in national forests?

Summary

Pollution is the result of technological advancements and increased population density. It is defined as any addition of matter or energy that degrades the environment for humans and other organisms.

Natural resources are structures and processes that can be used by people but cannot be created by them. Renewable resources can be regenerated or repaired; nonrenewable resources are consumed.

Mineral resources must be extracted from ores, a process that requires energy and money, and also changes the environment. The major steps of mineral exploitation are exploration, mining, processing the ore, transportation, and manufacturing the finished product. Ultimately, all costs of mineral exploitation are reduced to monetary costs. Recycling reduces the demand for new sources of mineral deposits but does not necessarily save money.

As the human population increases, we change natural ecosystems by replacing them with agricultural ecosystems, we alter species mixtures by introducing plants and animals, and we reduce populations by harvesting trees and animals for our use. Forests must be cut in a manner that allows for regrowth so that the soil is not exposed to the erosional effects of wind and water. Cutting small areas and reforestation help to prevent these problems. However, some remote areas have been changed very little by humans. Tropical rainforests represent some of the last large wilderness areas, but they are being rapidly converted to other uses by the constant pressure of growing populations. Many of these wilderness areas should be protected because of the rich diversity of species they contain.

Grazing of arid and semiarid lands can be a valuable way to provide food for people. However, the land is often overgrazed and then may be degraded to a desert that is not capable of supporting the animals needed to feed growing populations.

Aquatic ecosystems are modified by pollution and activities that occur on the land adjacent to the water. Land that is devoid of vegetation erodes and fills streams and lakes with sediment. Warming of the water is also likely to occur if trees are removed from the stream side. Many exotic species of fish have been introduced into the freshwater ecosystems of the world. This alters normal food chains and reduces the populations of native species.

Most marine fisheries are being over-fished or are at capacity. One of the problems associated with marine fisheries management is the enforcement of regulations. Managing for wildlife involves careful planning and habitat manipulation to provide the best possible population for hunting. Some areas are intensely managed, as in many European game parks, while in other parts of the world, the game animals lead a more normal wild life. Waterfowl present a unique problem because they migrate across international boundaries.

Extinction is a normal consequence of not being able to adapt to changes in the environment. However, since humans have such a great influence on nearly every ecosystem in the world, they have been the cause of increased rates of extinction. Many people recognize the value of species, both as possible helpers of humans and for their own intrinsic worth, and are trying to preserve sensitive habitats so that species will not be driven to extinction because of the appropriation of their habitats for other uses.

Interactive Exploration

Check out the website at

http://www.mhhe.com/environmentalscience

and click on the cover of this textbook for interactive versions of the following:

KNOW THE BASICS

biodegradable *239*
biodiversity *243*
clear-cutting *245*
cover *255*
desertification *248*
economic costs *241*
endangered species *262*
energy cost *241*
environmental costs *241*
extinction *260*
habitat management *256*
migratory birds *259*
natural resources *240*
nonrenewable resources *241*
patchwork clear-cutting *246*
pollution *239*
reforestation *246*
renewable resources *241*
selective harvesting *246*
threatened species *262*
wilderness *250*

- **On-line Flashcards**
- **Electronic Glossary**

IN THE REAL WORLD

With satellite technology, we have mastered the ability to see how land is being used. . . right? Take a look at **New Study Raises Estimates of Deforestation in Amazon** to see if satellite imagery is the only tool that should be used when estimating deforestation.

No matter what the estimates of deforestation are in China, officials have stated that it is too much. See what has led to this official statement in **China Officially Recognizes Link between Deforestation and Summer Flooding.**

Roads are not allowed, but fires are? How do these practices support healthy forest ecosystems? Some critics of land use in wilderness areas find that banning roads and setting intentional fires are ludicrous management practices. Read **Cerro Grande Fire Forces Evacuation,** **Major Fires Expected in Popular Wilderness Area,** and **Major Initiative Proposed to Protect National Forests** to evaluate the reasons why these practices are used.

Migrating populations offer special challenges to managers attempting to protect the species. Read how biologists are trying to restore a migratory population of waterfowl in **Trumpeter Swans Return to Historic Range and Migratory Route.**

You may not be able to help trumpeter swans return to their birthplace, but you *can* help migrating songbirds. Check out **Why You Should Buy Organic Coffee: It Helps Migratory Birds and Other Forest Species** to see how drinking organic coffee can make a difference.

How can you make a difference when technology is the culprit? Huge numbers of migratory birds are colliding into digital telephone and TV towers. Do you think better publicity would eventually lead to a technical solution for this problem? Aircraft have collision avoidance features; should the communication industry be required to design and implement collision avoidance features on towers? Take a look at Digital Telephone and TV Towers Kill Migratory Birds by the Millions for more information about how the use of modern communications is adversely affecting migrating populations.

TEST PREPARATION

Review Questions

1. Name three ways humans directly alter ecosystems.
2. Why is the impact of humans greater today than at any time in the past?
3. Define pollution. Has the definition changed with time?
4. What are three kinds of costs associated with resource exploitation?
5. Why is recycling usually more energy efficient than mining new raw materials?
6. List three problems associated with forest exploitation.
7. What is desertification? What causes it?
8. What effects do increased temperature and increased organic matter have on aquatic ecosystems?
9. List six techniques utilized by wildlife managers.
10. What special problems are associated with waterfowl management?
11. What is extinction, and why does it occur?
12. Why should humans worry about extinction? Do you?
13. List three actions that can be taken to prevent extinctions.

Critical Thinking Questions

1. Imagine you are a mining company executive trying to determine the costs of your mining operation. Create a list of your costs. How important are environmental costs to your company? Why?
2. Logging, mineral extraction, and grazing on federal lands have been subsidized in order to stimulate economic growth and support rural communities. What place should ecological value have in determining the way land is used? Should subsidies be continued? Justify your stand on this issue.
3. Perhaps 98–99 percent of all species that have ever existed are extinct. Nearly all went extinct long before humans arrived on the scene. Why should we be concerned about extinction of organisms today?
4. Would you support clearing of forests and plowing of grasslands that have significant ecological importance in order to support agriculture in countries that have significant hunger? Where do you draw the line between preserving ecosystems and human interest?
5. A Pacific Northwest Native American tribe has been permitted to hunt an endangered whale species to preserve its traditional culture and to uphold its treaty rights. Now there is fear that other countries will hunt the whales, too. Do you think the tribe should be denied its rights? Why, or why not?
6. Pharmaceutical companies are helping some developing countries preserve their rainforests so these companies can look for organisms with possible pharmacological value. How do you feel about these arrangements? What limits would you place on the pharmaceutical companies, if any? Why?

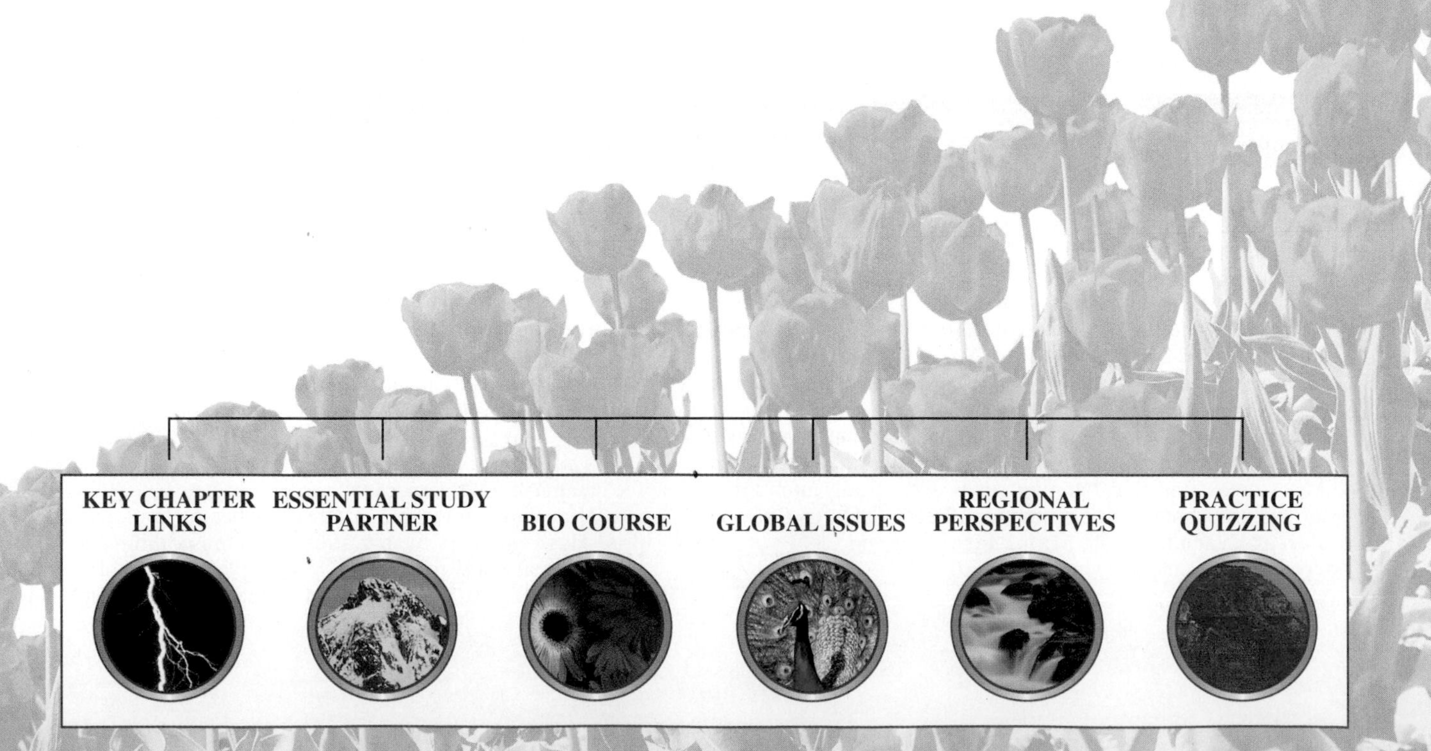

CHAPTER 13

Land-Use Planning

Objectives

After reading this chapter, you should be able to:

- Explain why most major cities are located on rivers, lakes, or the ocean.
- Describe the forces that result in farmland adjacent to cities being converted to urban uses.
- Explain why floodplains and wetlands are often mismanaged.
- Describe the economic and social values involved in planning for outdoor recreation opportunities.
- Explain why some land must be designated for particular recreational uses, such as wilderness areas, and why that decision sometimes invites disagreement from those who do not desire to use the land in the designated way.
- List the steps in the development and implementation of a land-use plan.
- Describe methods of enforcing compliance with land-use plans.
- Describe the advantages and disadvantages of both local and regional land-use planning.

Chapter Outline

The Need For Planning

Historical Forces That Shaped Land Use in North America
- The Importance of Waterways
- The Rural-to-Urban Shift

Global Perspective: *Urbanization in the Developing World*

Migration from the Central City to the Suburbs

Factors That Contribute to Sprawl
- Lifestyle Factors
- Economic Factors
- Planning and Policy Factors

Problems Associated with Unplanned Urban Growth
- Transportation Problems
- Air Pollution
- Low Energy Efficiency
- Loss of Sense of Community
- Death of the Central City
- Higher Infrastructure Costs
- Loss of Open Space
- Loss of Farmland
- Water Pollution Problems
- Floodplain Problems
- Wetlands Misuse

Environmental Close-Up: *Wetlands Loss in Louisiana*

- Other Land-Use Considerations

Land-Use Planning Principles

Mechanisms for Implementing Land-Use Plans
- Establishing State or Regional Planning Agencies
- Purchasing Land or Use Rights
- Regulating Use

Environmental Close-Up: *Land-Use Planning and Aesthetic Pollution*

Special Urban Planning Issues
- Urban Transportation Planning
- Urban Recreation Planning
- Redevelopment of Inner-City Areas

Federal Government Land-Use Issues

Issues & Analysis: *Decision Making in Land-Use Planning—The Malling of America*

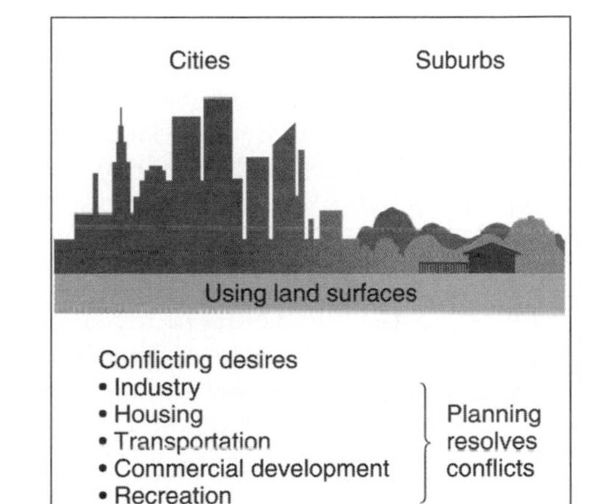

The Need for Planning

A large proportion of the land surface of the world (about one-third to one-half) has been changed by human activity. Most of this change occurred as people converted the land to agriculture and grazing, but, in our modern world, significant amounts have been covered with buildings, streets, highways, and other products of society. In many cases, cities grew without evaluating and determining the most logical use for the land. Consequently, some cities are in the wrong place. Los Angeles and Mexico City have severe air pollution problems because of their geographic location, Venice and New Orleans are threatened by high sea levels, and San Francisco and Tokyo are subject to earthquakes. Currently, most land-use decisions are still based primarily on economic considerations or the short-term needs of a growing population rather than on careful analysis of the capabilities and unique values of the land and landscape. Each piece of land has specific qualities based on its location and physical make-up. Some is valued for the unique species that inhabit it, some is valued for its scenic beauty, and some has outstanding potential for agriculture or urban uses. Since land and the resources it supports (soil, vegetation, elevation, nearness to water, watersheds) is not being created today (except volcanos, river deltas, etc.), it should be considered a nonrenewable resource.

Once land has been converted from natural ecosystems or agriculture to intensive human use, it is generally unavailable for other purposes. As the population of the world grows, there will be increased competition for the use of the land, and systematic land-use planning will become more important. Furthermore, as the population of the world becomes more urbanized and cities grow, urban planning becomes critical.

Historical Forces That Shaped Land Use in North America

Today, most of the North American continent has been significantly modified by human activity. In the United States, about 47 percent of the land is used for crops and livestock, about 45 percent is forests and natural areas, and nearly 5 percent is used intensively by people in urban centers and as transportation corridors. Canada is 54 percent forested and wooded and uses only 8 percent of its land for crops and livestock. Less than 1 percent of the land is in urban centers and transportation corridors. A large percentage of its remaining land is wilderness in the north.

This pattern of land use differs greatly from the original conditions experienced by the early European colonists who immigrated to the New World. The first colonists converted only small portions of the original landscape to farming, manufacturing, and housing, but, as the population increased, more land was converted to agriculture, and settlements and villages developed into towns and cities. Although most of this early development was not consciously planned, it was not haphazard. Several factors influenced where development took place and the form it would take.

The Importance of Waterways

Waterways were the primary method of transportation, which allowed exploration and the development of commerce in early North America. Thus, early towns were usually built near rivers, lakes, and oceans. Typically, cities developed as far inland as rivers were navigable. Where abrupt changes in elevation caused waterfalls or rapids, goods being transported by boat or barge needed to be offloaded, transported around the obstruction, and loaded onto other boats. Cities often developed at these points. Buffalo, New York, and Sault Sainte Marie, Ontario, are examples of such cities. In addition to transportation, bodies of water provided drinking water, power, and waste disposal for growing villages and towns. Those towns and villages with access to waterways that provided easy transportation could readily receive raw materials and distribute manufactured goods. Some of these grew into major industrial or trade centers. Without access to water, St. Louis, Montreal, Chicago, Detroit, Vancouver, and other cities would not have developed. (See figure 13.1.) The availability of other natural resources, such as minerals, good farm land, or forests was also important in determining where villages and towns were established. Industrial development began on the waterfront since water supplied transportation, waste disposal, and power. As villages grew into towns and cities, large factories replaced small gristmills, sawmills, and blacksmith shops. The waterfront became a center of intense industrial activity. As industrial activity increased in the cities, people began to move from rural to urban centers for the job opportunities these centers presented.

The Rural-to-Urban Shift

North America remained essentially rural until industrial growth began in the last third of the 1800s and the population began a trend toward greater urbanization. (See figure 13.2.) There were several forces that led to this rural-to-urban transformation. First, the Industrial Revolution led to improvements in agriculture that required less farm labor at the same time industrial jobs became available in the city. Thus, people migrated from the farm to the city. The average person was no longer a farmer but, rather, a factory worker, shopkeeper, or clerk living in a tenement or tiny apartment near where they worked. This pattern of rural-to-urban migration

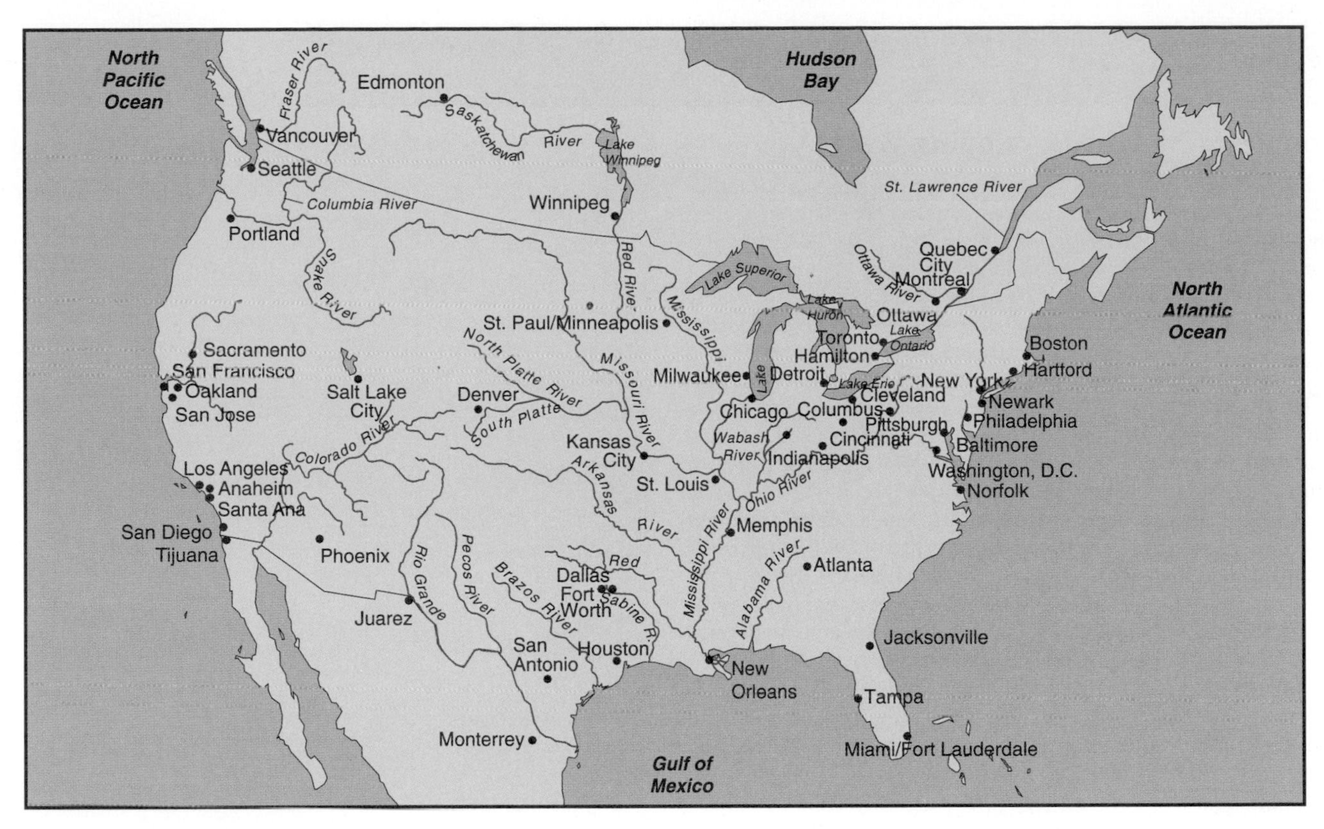

figure 13.1 **Water and Urban Centers** Note that most of the large urban centers are located on water. Water is an important means of transportation and was a major determining factor in the growth of cities. The cities shown have populations of 1,000,000 or more (except Edmonton, Winnipeg, and Quebec City).

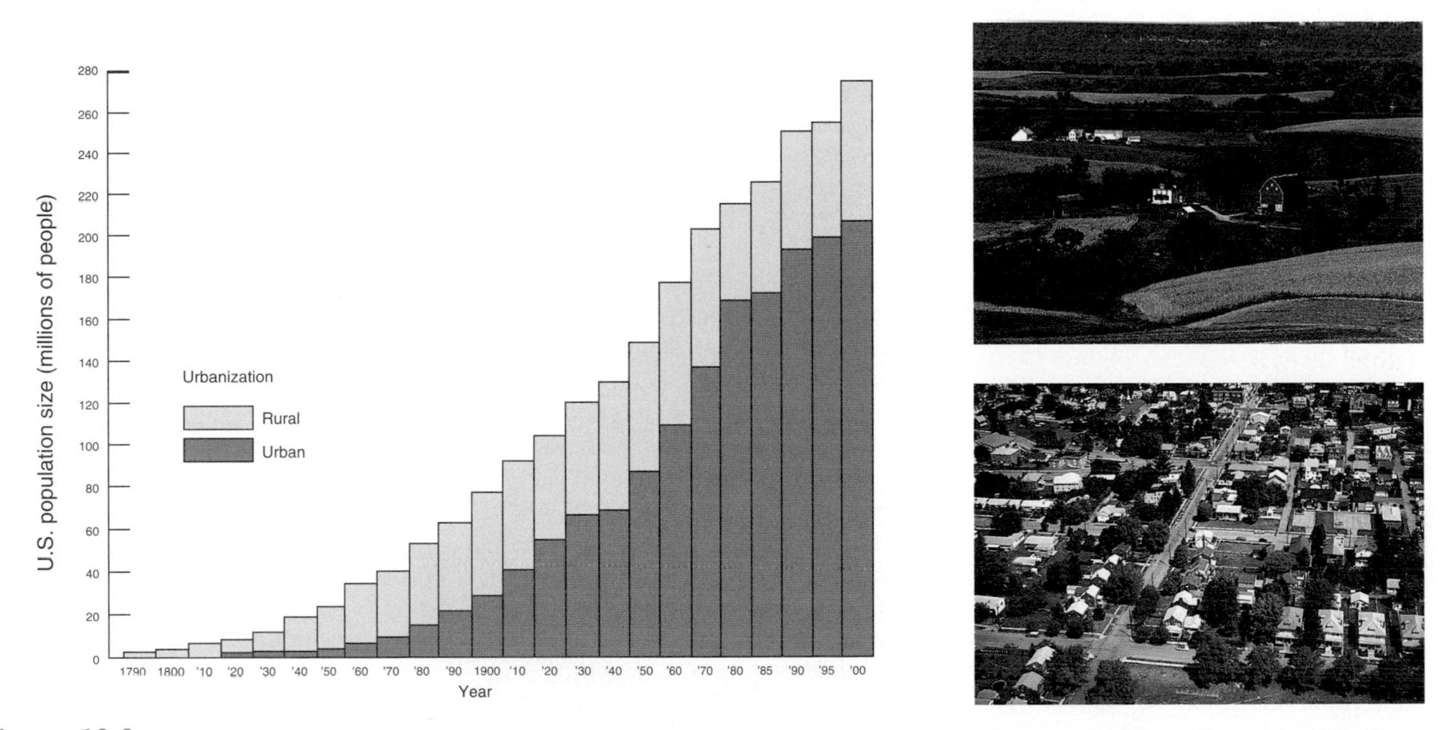

figure 13.2 **Rural-to-Urban Population Shift** In 1800, the United States was essentially a rural country. Industrialization in the late 1800s began the shift to an increasing urban population. In 2000, 75 percent of the U.S. population was urban.

Sources: (Graph) Data from the *Statistical Abstract of the United States,* 2000 data from Population Reference Bureau, Inc., Washington, D.C.

GLOBAL PERSPECTIVE

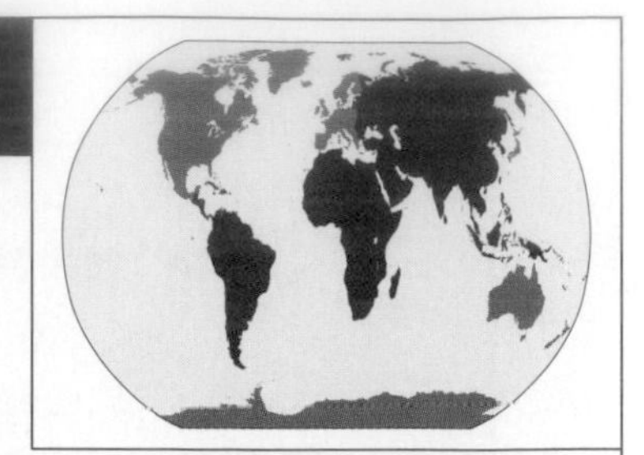

Urbanization in the Developing World

Traditionally, most of the population of the developing world has been rural. However, in recent years, there has been a rapid growth in the number of people migrating to the cities. By 2025, about 5 billion people are expected to be living in urban centers. Most of the increase in urban areas will be in the developing world. Many migrate to the cities because they feel they will have greater access to social services and other cultural benefits than are available in rural areas. Many also feel there are more employment opportunities. However, the increase in the urban population is occurring so rapidly that it is very difficult to provide the services needed by the population, and jobs are not being created as fast as the urban population is growing. Thus, many of the people live in poverty on the fringes of the city in shantytowns that lack water, sewer, and other services. Often these shantytowns are constructed without permission only a short distance from affluent urban dwellers. Because the poor lack safe drinking water and sewer services, they pollute the local water sources and disease is common. Because they burn wood and other poor quality fuels in inefficient stoves, air pollution is common. The additional people also create traffic problems of staggering proportions. The chart shows the seven cities with the largest rate of population increase. All are in the developing world.

Cities Expected to Grow by More Than 50 Percent between 1995 and 2015

Name of City	1995 Population (millions)	Projected 2015 Population (millions)
Bombay, India	15.14	26.22
Lagos, Nigeria	10.29	24.61
Delhi, India	9.95	16.86
Karachi, Pakistan	9.73	19.38
Metro Manila, Philippines	9.29	14.66
Jakarta, Indonesia	8.62	13.92
Dhaka, Bangladesh	8.55	19.49

Source: Data from *World Resources 2000–2001.*

occurred throughout North America, and it is still occurring in developing countries today. A second factor that affected the growth of cities was the influx of immigrants from Europe. Although some became farmers, many of these new citizens settled in towns and cities, where jobs were available. A third reason for the growth of cities was that they offered a greater variety of cultural, social, and artistic opportunities than did rural communities. Thus, cities were attractive for cultural as well as economic reasons.

Migration from the Central City to the Suburbs

During the early stages of industrial development, there was little control of industry activities, so the waterfront typically became a polluted, unhealthy, undesirable place to live. As roads and rail transport became available, anyone who could afford to do so moved away from the original, industrial city center. The more affluent moved to the outskirts of the city, and the development of suburban metropolitan regions began. Thus, the agricultural land surrounding the towns was converted to housing. Most cities originally had good farmland near them, since the floodplain near rivers typically has a deep, rich soil and agricultural land adjacent to the city was one of the factors that determined whether the city grew or not. This was true because until land transportation systems became well developed, farms needed to be close to the city so that farmers could transport their produce to the markets in the city. This rich farmland adjacent to the city was ideal for the expansion of the city. As the population of the city grew, demand for land increased. As the price of land in the city rose, people and businesses began to look for cheaper land farther away from the city. Developers and real estate agents were quick to respond and to help people acquire and convert agricultural land to residential or commercial uses. Land was viewed as a commodity to be bought and sold for a profit, rather than as a nonrenewable resource to be managed. As long as money could be made by converting agricultural land to other purposes, it was impossible to prevent such conversion. There were no counteracting forces strong enough to prevent it.

The conversion of land around cities in North America to urban uses destroyed many natural areas that people had long enjoyed. The Sunday drive from the city to the countryside became more difficult as people had to drive farther to escape the ever growing suburbs. The unique character of neighborhoods and communities was changed by the erection of shopping malls, apartment complexes, and expressways. Most of these alterations occurred without considering how they would affect the biological community or the lives of the people who lived in the area.

As cities continued to grow, certain sections within each city began to deteriorate. Industrial activity continued to be concentrated near water in the city's center. Industrial pollution and urban crowding turned the core of many cities into undesirable living areas. In the early

a.

b.

c.

figure 13.3 **Types of Urban Sprawl** Note the three different types of growth depicted in these photos. (*a*) The wealthy suburbs with large lots are adjacent to the city. (*b*) Ribbon sprawl develops as a commercial strip along highways. (*c*) Tract development results in neighborhoods consisting of large numbers of similar houses on small lots.

1900s, people who could afford to leave began to move to the outskirts. This trend continued after World War II, in the 1940s, 1950s, and 1960s, as a strong economy and government policies that favored new home purchases (tax deductions and low-interest loans) allowed more people to buy homes. In 1950, about 60 percent of the urban population lived in the central city; by 1990, this number was reduced to about 30 percent. Most of these new single family homes were in attractive suburbs, away from the pollution and congestion of the central city. These collections of houses were built on large lots that provided outdoor space for family activities. The blocks of homes were added to the periphery of the city in a decentralized pattern in which single family homes were separated from multifamily structures and both were some distance from places of work, shopping, and other service needs. This pattern also made it very difficult to establish efficient public transportation networks, which decreased energy efficiency, and increased the cost of supplying utility services. However, public transport was not considered important because rising automobile ownership and improved highway systems would accommodate the transportation needs of the suburban population. The convenience of a personal automobile escalated decentralized housing patterns, which, in turn, required better highways, which led to further decentralization.

By 1960, unplanned suburban growth had become known as urban sprawl. **Urban sprawl** is a pattern of unplanned, low-density housing and commercial development outside of cities that usually takes place on previously undeveloped land. In addition, blocks of housing are separated from commercial development, and the streets typically form branching patterns and often include cul-de-sacs. These large housing tracts surrounded cities, which made it difficult for people to find open space. A city dweller could no longer take a bus to the city limits and enjoy the open space of the countryside. Urban sprawl occurs in three ways. (See figure 13.3.) One type of growth involves the development of exclusive, wealthy suburbs adjacent to the city. These homes are usually on large individual lots in the more pleasing geographic areas of the city. Often they are located along water, on elevated sites, or in wooded settings. A second development pattern is tract development. **Tract development** is the construction of similar residential units over large areas. Initially, these tracts are often separated from each other by farmland. New roads are constructed to link new housing to the central city and other suburbs which stimulates the development of a third form of urban sprawl along transportation routes. This is referred to as **ribbon sprawl** and usually consists of commercial and industrial buildings that line each side of the highway that connects housing areas to the central city and shopping and service areas. Ribbon sprawl results in high costs for the extension of utilities and other public services. It also makes the extent of urbanization seem much larger than it actually is, since the driver cannot see the undeveloped land hidden by the storefronts that face the highway.

As suburbs continued to grow, cities (once separated by farmland) began to merge, and it became difficult to tell where one city ended and another began. This type of growth led to the development of regional cities. Although these cities maintain their individual names, they are really just part of one large urban area called a **megalopolis.** (See figure 13.4.) The eastern seaboard of the United States, from Boston, Massachusetts, to Washington, D.C., is an example of a continuous city. Other examples are London to Dover in England, the Toronto-Mississauga region of Canada, and the southern Florida coast from Miami northward.

In some areas throughout North America, the growth of suburbs has been slowed due to the increased cost of housing and transportation. People have migrated back to some cities on a limited scale because of the lower cost of urban houses and the fact that public transportation is generally more efficient in the city than in the suburbs, thus freeing urban residents from the cost of daily commuting. This reverse migration, however, is still greatly offset by the continual growth of the suburban communities.

Factors That Contribute to Sprawl

As we enter the twenty first century, many, including government policy

figure 13.4 **Regional Cities in the United States and Canada** The lights from this satellite image show population concentrations. More than 30 major regional cities have developed, with each having more than 1,000,000 people. Many of these cities merge with their neighbors to form huge regional cities. Major urban regions are the northeast coast of the United States (Boston to Washington, D.C.), the region south of the Great Lakes (Chicago to Pittsburgh), south Florida (Jacksonville to Miami), the Toronto and Montreal regions of Canada, and the west coast of California (San Francisco to San Diego).

makers, are beginning to question the prevailing pattern of sprawling urban development. However, powerful social, economic, and policy forces have been behind the development of the current decentralized urban centers typical of North America. If this pattern is to change, these underlying factors must be understood.

Lifestyle Factors

One of the factors that has supported urban sprawl is the relative wealth of the population. This wealth is reflected in material possessions, two of which are automobiles and homes. In the United States, there are 75 motor vehicles for every 100 people. Since over 20 percent of the population is too young to drive, this means that there is essentially one motor vehicle for every licensed driver in the United States. In addition, about two-thirds of the population live in family-owned homes. With this level of wealth, people make choices about where and how they want to live. Many are attracted to a lifestyle that includes low-density residential settings, with easy access to open space, isolated from the problems of the city. They are also willing to drive considerable distances to live in these settings. Thus, a decentralized housing pattern is possible because of the high rate of automobile ownership which allows for ease of movement.

Economic Factors

Several economic forces operate to encourage sprawl development. First of all, it is less expensive to build on agricultural and other non-urban land than it is to build within established cities. The land is less expensive, the regulations and permit requirements are generally less stringent, and there are fewer legal issues to deal with. An analysis of building costs in the San Francisco Bay area of California determined that it costs between 25 percent and 60 percent more to build in the city than it does to build in the suburbs.

Several tax laws also contributed to encouraging home ownership. The interest on home loans is deductible from income taxes, and, in the past, people could avoid paying capital gains taxes on homes they sold if they bought another home of equal or higher value. Since homes usually increased in value, this created a market for increasingly expensive homes.

Planning and Policy Factors

There are many planning and policy issues that have contributed to sprawl development. First of all, until recently little coordinated effort has been given to planning how development should occur in metropolitan areas. There are several reasons for this. First of all, most metropolitan areas include hundreds of different political jurisdictions (the New York City metropolitan area includes about 700 separate government units and involves the states of New York, New Jersey, and Connecticut.) It is very difficult to integrate the activities of these separate jurisdictions in order to achieve coordinated city planning. In addition, it is very difficult for a small local unit of government to see the "big picture," and many are unwilling to give up their autonomy to a regional governmental body.

Local zoning ordinances have often fostered sprawl by prohibiting the mixing of different kinds of land use. Single family housing, multiple family housing, commercial, and light and heavy industry were restricted to specific parts of the community. In addition, many ordinances specify minimum lot sizes and house sizes. This tends to result in a decentralized pattern of development which is supported by the heavy use of automobiles and the roads and parking facilities necessary to support them.

In addition, many government policies actually subsidize the development of decentralized cities. For example, developers and the people who buy the homes and businesses they build are also able to avoid paying for the full cost of extending services to new areas. The local unit of government picks up the cost of these improvements and the cost is divided among all the tax-payers rather than just those who will benefit. The roads that are needed to support new developments are usually paid for with federal and state monies, so again the cost is not borne by those who benefit most but by the taxpayers of the state or nation. Furthermore, federal and state governments have not supported public transport in an equivalent way.

figure 13.5 **Traffic and Suburbia** Traffic congestion is a common experience for people who work in cities but live in the suburbs. The popularity of automobiles and the desire to live in the suburbs are closely tied.

Problems Associated with Unplanned Urban Growth

As the population increased and metropolitan areas grew, several kinds of problems were recognized. They fall into several broad categories.

Transportation Problems

Most cities experience continual problems with transportation. This is primarily because, as cities grew, little thought was given to how people were going to move around and through the city. Furthermore, when housing patterns and commercial sectors changed, transportation mechanisms had to be changed to meet the shifting needs of the public. This often involved the abandoning of old transportation corridors and the establishment of new ones. Paradoxically, the establishment of new transportation corridors stimulates increased growth in the areas served, and the transportation corridors soon become inadequate. The reliance on the automobile as the primary method of transportation has required the constant building of new highways.

The average person in the United States spends nine hours per week traveling in a car. In many metropolitan areas, the amount of time is much greater, and a significant amount of that time involves traffic jams. A study of traffic in the Washington, D.C., area concluded that the average commuter spent 80 hours per year stuck in traffic, in addition to the time normally needed to make the commute. (See figure 13.5.)

Air Pollution

Reliance on the automobile as the primary method of transportation has resulted in significant air pollution problems in many cities. Most of the large industrial sources of air pollution have been contained. However, the individual car with its single occupant going to work, to shop, or to eat a meal is a constant source of air pollution. A simple solution to this problem is a centralized, efficient public transportation system. However, this is difficult to achieve with a highly dispersed population.

Low Energy Efficiency

Energy efficiency is low for several reasons. First of all, automobiles are the least energy-efficient means of transporting people from one place to another. Secondly, the separation of blocks of homes from places of business and shopping requires that additional distances be driven. Third, congested traffic routes result in hours being spent in stop-and-go traffic which wastes much fuel. Finally, single family homes require more energy (than multifamily dwellings) for heating and cooling.

Loss of Sense of Community

Although the loss is difficult to measure, there is general agreement that in dispersed suburban developments, there is a loss of a sense of community. In many places, people stay within their homes and yards and do not routinely walk through the neighborhood. When they leave the confines of their home, they get in a car and go somewhere. This pattern of behavior reduces human interaction, isolates people from their neighbors, and greatly reduces the sense of community.

Death of the Central City

Currently less than 10 percent of people work in the central city. When people leave the city and move to the suburbs, they take their purchasing power and tax payments with them. Therefore, the city has less income to support the services needed by the public. When the quality of services in urban centers drops, the quality of life declines, the flight from the city increases, and a downward spiral of decay begins. An additional problem is the decline of the downtown business district. When shopping malls are built to accommodate the people in the suburbs, the downtown business district declines. Because people no longer need to come to the city to shop, businesses in the city center fail or leave, which deprives the remaining residents of basic services. They must now travel greater distances to satisfy their basic needs.

Higher Infrastructure Costs

Infrastructure includes all the physical, social, and economic elements needed to support the population. Whenever a new housing or commercial development occurs on the outskirts of the city, municipal services must be extended to the area. Sewer and water services, natural gas and electric services, schools and police stations, roads and airports—all are needed to support this new population. Extending services to these new areas is much more costly than supplying services to areas already in the city because most of the basic infrastructure is already present in the city.

figure 13.6 **Loss of Farmland** As cities grow outward, they eventually grow together to form a regional city. The land between them, once used for farming, becomes developed for residential and commercial purposes. Improved transportation routes and joint facilities (such as airports, shopping centers, and community colleges) hasten this loss of farmland.

Loss of Open Space

One of the important features of a pleasing urban landscape is the presence of open space. Open fields, parks, boulevards, and similar land use allows people to visually escape from the congestion of the city. Unplanned urban growth does not take this important factor into account. Consequently, the buildings must be torn down, and disused spaces must be renovated into parks and other open space at great expense to provide green space in the urban landscape.

Loss of Farmland

Most of the land that has recently been urbanized was previously used for high-value crops. Land that is flat, well drained, accessible to transportation, and close to cities is ideal farmland. However, it is also prime development land. Areas that once supported crops now support housing developments, shopping centers, and parking lots. (See figure 13.6.)

Urban development of farmland is proceeding at a rapid pace. Currently, land is being converted to urban uses at a rate of over 400,000 hectares (1 million acres) per year. About one-third of this land is prime agricultural land. One reason for this conversion is the way the land is taxed. Property is often taxed on what *can* be done with it, not necessarily on what *is* being done with it. For example, if land can be used for both farming and residential development, it is taxed as if it were residential. If a farmer sells a portion of the farm to a developer who builds five houses, local taxing authorities would consider these houses to be the "highest and best use" of the land. All of the farmer's land would then probably be reassessed and the taxes increased substantially. Farmers faced with this situation are often forced to sell their land because they are taxed on its commercial value rather than on its value as farmland. This policy encourages development and forces people out of farming. New policies that assess taxes based on current use of the land and not on the land's highest potential use must be explored.

Several states have established programs that provide protection to farmers who do not want to sell their land to developers. The programs may require farmers, in return for lower taxes on the land, to put their land in a conservation easement that prevents the farmer or future owners from using the land for anything other than farming. Although these programs allow farmers who are committed to their way of life to continue, they do not remove the economic enticement from developers from obtaining farmland for future development.

Water Pollution Problems

A large impervious surface area results in high runoff and flash flood potential. A typical shopping mall has a paved parking lot that is four times larger than the space taken up by the building. Paved parking lots also make sure that pollutants will enter the water rather than the soil.

Floodplain Problems

Because most cities were established along water, many cities are located in areas called **floodplains.** Floodplains are the low areas near rivers and, thus, are subject to periodic flooding. Some floodplains may flood annually, while others flood less regularly. They are generally flat, and so are inviting areas for residential development even though they suffer periodic flooding. A better use of these areas is for open space or for recreation, yet developers continue to build houses and light industry there.

Usually, when a floodplain is developed for residential or commercial use, a retaining wall is built to prevent the periodic flooding natural to the area. This increases the cost of the development, increases the cost of insurance protection, and creates high-water problems downstream. Frequently, tax monies are used to repair the damage that results from the unwise use of floodplains. Floodplain development is one of the insidious problems associated with a rapidly expanding population. As long as the population continues to increase, these less desirable areas are likely to be used for housing, whether they are subject to annual flooding or less frequent damage.

Floods are natural phenomena. Contrary to popular impressions, no evidence supports the premise that floods are worse today than they were one hundred or two hundred years ago, except perhaps on small, isolated watersheds. What has increased is the economic loss from the flooding. In 1993, during an extensive flooding event that involved both the Mississippi and Missouri Rivers, the U.S. Army Corps of Engineers estimated over $1.5 billion in damage to residential and commercial property. (See figure 13.7.) This loss reflects the fact that more cities, industries, railroads, and highways have been constructed on floodplains. Sometimes there are no alternatives to floodplain development, but, too often, risks are simply ignored. Because flooding causes loss of life and property, floodplains should no longer be developed for uses other than agriculture and recreation.

figure 13.7 **Flooding in Floodplain** In 1993, a major flood event affected both the Missouri and Mississippi Rivers. Although dikes and levees were in place to protect the flood plain, the rivers reclaimed the floodplain and billions of dollars of damage was done. This photo shows flooding in Jefferson City, Missouri.

Many communities have enacted **floodplain zoning ordinances** to restrict future building in floodplains. Although such ordinances may prevent further economic losses, what happens to individuals who already live in floodplains? Floodplain building ordinances usually allow current residents to remain. However, these residents often find it extremely difficult to obtain property insurance. Relocation, usually at a financial loss, is the only alternative. Such situations are unfortunate; perhaps proper planning in the future will prevent these problems.

Wetlands Misuse

Since access to water was and continues to be important to industrial development, many cities are located in areas with extensive wetlands. **Wetlands** are

ENVIRONMENTAL CLOSE-UP

Wetlands Loss in Louisiana

The state of Louisiana has extensive coastal wetlands along the Gulf of Mexico that constitute about 40 percent of the wetlands in the United States excluding Alaska and Hawaii. About 3500 square kilometers (1300 square miles) have been lost since 1956. This is an area approximately the size of the state of Rhode Island. Recent rates of loss have ranged from 65 to 90 square kilometers (25–35 square miles) per year. There are several reasons for the loss but human activity is a primary cause.

Much of the Louisiana coastline consists of poorly compacted muds that are easily eroded if exposed to wind and wave action. Faults in the rock that lies below the mud are causing the land to tilt downward which allows the ocean to invade. This is the basic cause for about 60 percent of the wetlands loss. In the past the two actions—loss by tilting and addition of sediments from the Mississippi River—were in balance. The loss due to tilting was replaced by additional sediments from the river. However, two human actions have reduced the amount of sediments being added. Dams and reservoirs upstream on the Missouri and Mississippi Rivers have reduced the amount of sediment delivered to the region, making replacement of lost sediments impossible. In addition, levees on the Mississippi River have prevented the flooding that was a normal part of the behavior of the lower Mississippi River for centuries. The sediments, therefore, are carried out to the Gulf of Mexico and do not contribute to the building of wetlands near the shore. Without the constant addition of sediments, the muds settle and compact, which, along with the subsiding of the land, results in erosion and wetlands loss.

Shipping channels and canals associated with oil and gas production are responsible for about 30 percent of the current wetlands loss. There are nine major shipping channels and 13,000 kilometers (8,000 miles) of canals. Typically, production companies cut canals through the marshes to give barges and other equipment easy access to the drilling platforms and production facilities. When oil or gas is found, it must be transported to land through pipelines. Laying pipelines often requires cutting additional canals through the wetlands. These canals break the wall of vegetation that protects the soft muds, and allow salt water to penetrate into the marshes, killing the vegetation. Both of these activities open up the wetlands to wave action and accelerate the rate of erosion.

The nutria, a rodent introduced from South America, is also contributing to the problem. These animals reproduce rapidly and form such large populations that they eat all the vegetation in a local area. This exposes the muds to erosion and contributes to the loss of wetlands. Trapping these animals for their fur was at one time profitable for the local people and helped to control the animals. However, a shift in public opinion has reduced the acceptability of fur for clothing and depressed the fur market, making the trapping of nutria uneconomical.

A study led by the Louisiana Department of Natural Resources was released in 1998. It called for billions of dollars of federal investment to protect the wetlands. A logical source of these funds would be royalties paid to the federal government by offshore oil production companies. Major features of the plan include changes to current shipping practices, construction of new shipping facilities downstream from New Orleans, and controlled opening of levees along the lower Mississippi to restore some of the flooding that was at one time a normal part of the ecosystem. There are powerful forces in conflict over this plan. Shipping interests feel changes in shipping will cost them money. People who live in low-lying areas currently protected by levees may need to relocate even if controlled flooding is initiated. Changing the water quality by adding freshwater to a brackish water system will change the kinds of organisms present and hurt some traditional fishing activities. However, there is a plan and Congress will need to decide which elements of the plan to authorize and the amount of money to allocate for their implementation.

areas that periodically are covered with water. They include swamps, tidal marshes, coastal areas, and estuaries. Some wetlands, such as estuaries and marshes, are permanently wet, while others, such as many swamps, have standing water during only part of the year. Many wetlands may have standing water for only a few weeks a year, often in the spring of the year when the snow melts. Because wetlands breed mosquitoes and are sometimes barriers to the free movement of people, they have often been considered useless or harmful. Most of them have been drained, filled, or used as dumps. Many modern cities have completely covered over extensive wetland areas and may even have small streams running under streets, completely enclosed in concrete.

Each kind of wetland has unique qualities and serves as a home to many kinds of plants and animals. Wetlands are frequently critical to the reproduction of many kinds of animals. Many fish use estuaries and marshes for spawning. They also provide nesting sites for many kinds of birds and serve as critical habitats for many other species. Waterfowl hunters and commercial and sport fisheries depend upon these habitats to produce and protect the young of the species they harvest. Because most wetlands receive constant inputs of nutrients from the water that drains from the surrounding land, they are highly productive and excellent places for aquatic species to grow rapidly. Human impact on wetlands has severely degraded or eliminated these spawning and nursery habitats.

Besides providing a necessary habitat for fish and other organisms, wetlands provide natural filters for sediments and runoff. This filtration process allows time for water to be biologically cleaned before it enters larger bodies of water, such as lakes and oceans, and reduces the sediment load carried by runoff. Wetlands also protect shorelines from erosion. When destroyed, the natural erosion protection provided by wetlands must be replaced by costly artificial measures, such as breakwalls.

Other Land-Use Considerations

The geologic status of an area must also be considered in land-use decisions. Building cities on the sides of volcanos or on major earthquake-prone faults has led to much loss of life and property. Siting homes and villages on unstable hillsides or in areas subject to periodic fires is also unwise. Yet, every year more houses slide down California hillsides and wildfires consume homes throughout the dry West.

Another problem in some locations is lack of water. Southern California and metropolitan areas in Arizona must import water to sustain their communities. Wise planning would limit growth to whatever could be sustained by available resources. The risk of serious water shortages in Beijing, China, has forced the government to consider a massive 1200-kilometer (745 miles) diversion of water from the Yangtze River to the city. As the population of these cities continues to grow, the strain on regional water resources will increase.

Water-starved cities often cause land-use dilemmas far from the city boundaries. Supplying water and power to cities often involves the construction of dams that flood valleys that may have significant agricultural, scenic, or cultural value. See chapter 16 on water use for a more extensive development of the topic.

Land-Use Planning Principles

Land-use planning is a process of evaluating the needs and wants of the population, the characteristics and values of the land, and various alternative solutions to the use of a particular land surface before changes are made. Planning land use brings with it the desires of many competing interests. The economic and personal needs of the population are a central driving force that requires land-use decisions to be made. However, the unique qualities of particular portions of the land surface prevent some uses, poorly accommodate others, but are highly suitable for others. For example, the floodplain beside a river is unsuitable for building permanent structures, can easily accommodate recreational uses such as parks, but may be most useful as a nature preserve. Agricultural land near cities can be easily converted to housing but may be more valuable for growing fruits and vegetables that are needed by the people of the city. This is particularly true when agricultural land is in short supply near urban centers. The difficulty in making land-use decisions is going through the process of evaluating each piece of land and deciding which of several competing uses to assign it. When land-use decisions are made, the decision process usually involves the public, private landowners, developers, government, and special interest groups. Each interest has special wants and will argue that its desires are most important. Since this is the case, what kinds of principles and processes should people consider when they make land-use decisions irrespective of their personal wants and interests?

A basic rule should be to make as few changes as possible, but, when changes are suggested or required, there are several things that should be considered.

1. *Evaluate and record any unique geologic, geographic, or biologic features of the land.*

   Some land has unique features that should be preserved because of their special value to society. The Grand Canyon, Yellowstone National Park, and many wilderness areas have been set aside to preserve unique physical structures, scenic characteristics, special ecosystems, or unusual organisms. On a more local level, a stream may provide fishing opportunities near a city, or land may have excellent agricultural potential that should take precedence over other uses. New York City purchased and protects a watershed that provides water for the city at a much lower cost than treating water in the Hudson River.

2. *Preserve unique cultural or historical features.*

   Some portions of the landscape, areas within cities, and structures have important cultural, historic, or religious importance that should not be compromised by land-use decisions. In many cities, historic buildings have been preserved and particular sections set aside as historic districts. Sacred sites, many battlefields, and places of unique historic importance are usually protected from development.

3. *Conserve open space and environmental features.*

   It must be recognized that open space and natural areas are not unused, low-value areas. Many studies of human behavior have shown that when people are given choices, they will choose settings that provide a view of nature or that allow one to see into the distance. Some have argued that this is a deep-seated biological need, while others suggest that it is a culturally derived trait, but urban planners know that access to open space and natural areas is an important consideration when determining how to use land. Therefore, it makes sense to protect open space within and near centers of population.

4. *Recognize and calculate the cost of additional changes that will be required to accommodate altered land use.*

   Whenever a land use is altered, additional modifications will be required to accommodate the change in land use. For example, when a new housing development is constructed, schools and other municipal services will be required, roads will need to be improved, and the former use of the land is lost. Frequently, the cost of these changes is not borne by the developer or the homeowner, but becomes the responsibility of the entire community; everyone pays for the additional cost through tax increases. In many areas, basic services, such as provision of water, are severe problems. It makes no sense to build new housing when there is not enough water to support the existing population.

5. *Plan for mixed housing and commercial uses of land in proximity to one another.*

   One of the major problems associated with development in North America is segregation of different kinds of housing from one another and from shopping and other service necessities. Mixing various kinds of uses together (single family housing, apartments, shopping and other service areas, and offices) allows easier connection between uses without reliance on the automobile. Walking and biking become possible when these different uses are within a short distance of one another.

6. *Plan for a variety of transportation options.*

   Plan for transportation options other than the automobile. Currently, most urban and rural areas do not accommodate bicycles. Although they can legally be ridden on streets, it is generally unsafe because of the high speeds of vehicular traffic and poor road surface conditions near the edge of the pavement. Special bike lanes are available in some areas but this is rare in much of North America. Walking is a healthy, pleasant way to get from one task to another. However, crossing wide, busy streets is difficult, many areas lack sidewalks, and related service areas are often far apart. This tends to discourage this mode of transport. Clustering housing and service areas allows for easier planning of bus and rail routes to allow people to get from one place to another without relying on automobiles.

7. *Set limits and require managed growth with compact development patterns.*

   Much of unplanned growth occurs because there is no plan or the land-use plan is not enforced. One very effective tool which promotes efficient use of the land is to establish an urban growth limit for a municipality. An **urban growth limit** establishes a boundary within which development can occur. Development outside the boundary is prohibited. One of the most important outcomes of setting urban growth boundaries is that a great deal of planning must precede the establishment of the limit. This lets all in the community know what is going on and can allow development to occur in logical stages that do not stress the community's ability to supply services. This mechanism also stimulates higher density uses of the urban land.

8. *Encourage development within areas that already have a supportive infrastructure so that duplication of resources is not needed.*

   Because all development of land for human activity requires that services be provided, it makes sense that housing and commercial development occur where the infrastructure is already present. This includes electric, phone, sewer, water, and transportation systems. It includes service industries, such as shopping, banking, restaurants, hotels, and entertainment. It includes schools, hospitals, and police protection. If development occurs far from these services, it is very costly to extend or duplicate those services in the new location. Furthermore, all large cities and most smaller ones have vacant lots or abandoned buildings that have outlived their usefulness or are vacant because of changing business or housing needs. This land is already urban, is close to municipal services, and the buildings can be readily renovated or demolished and replaced. Many will argue that this is too expensive, but, if there are no options, these spaces will be used and can be important in revitalizing inner city spaces.

Mechanisms for Implementing Land-Use Plans

Land-use planning is the construction of an orderly list of priorities for the use of available land. Developing a plan involves gathering data on current use and geological, biological, and sociological information. From these data, projections are made about what human needs will be. All of the data collected are integrated with the projections, and each parcel of land is evaluated and assigned a best use under the circumstances. There are basically three components that contribute to the successful implementation of a land-use development plan: land-use decisions can be assigned to a regional governmental body, the land or its development rights can be purchased, and laws or ordinances can be used to regulate land use.

Establishing State or Regional Planning Agencies

National and regional planning is often more effective than local land-use planning because political boundaries seldom reflect the geological and biological database used in planning. Larger units contain more diverse collections of landscape resources and can afford to hire professional planners. A regional approach is also likely to prevent duplication of facilities and lead to greater efficiency. For example, airport locations should be based on a regional plan that incorporates all local jurisdictions. Three cities only 30 kilometers (20 miles) apart should not build three separate airports when one regional airport could serve their needs better and at a lower cost to the taxpayers.

Although regional planning is increasing in North America, the majority of regional governmental bodies are presently voluntary and lack any power to implement programs. Their only role is to advise the member governments. Unfortunately, members of local governments still seem unwilling to give up power. They may view policy from a narrow perspective and put their own interests above the goals of the region. An elected, multipurpose, regional government, on the other hand, would be ideal for implementing land-use policy. Such governments exist in only a handful of places and show few signs of spreading.

One way to encourage regional planning is to develop policies at the state, provincial, or national level. The first state to develop a comprehensive statewide land-use program was Hawaii. During the early 1960s, much of Hawaii's natural beauty was being destroyed to build houses and apartments for the increasing population. The same land that attracted tourists was being destroyed to provide hotels and supermarkets for them. Local governments had failed to establish and enforce land-use controls. Consequently, in 1961, the Hawaii State Land-Use Commission was founded. This commission designated all land as urban, agricultural, or conservational. Each parcel of land could be used only for its designated purpose. Other uses were allowed only by special permit. To date, the record for Hawaii's action shows that it has been successful in controlling urban growth and preserving the islands' natural beauty, even though the population continues to grow. (See figure 13.8.)

figure 13.8 **Hawaii Land-Use Plan** Hawaii was the first state to develop a comprehensive land-use plan. This development in an agricultural area was stopped when the plan was implemented in the 1960s.

Several states and provinces are attempting to follow Hawaii's lead in state land-use regulation. Some have passed legislation dealing with special types of land use. Examples include wetland preservation, floodplain protection, and scenic and historic site preservation. Although direct state involvement in land-use regulation is relatively new, it is expected to grow. Only large, well-financed levels of government can afford to pay for the growing cost of adequate land-use planning. State, provincial, and regional governments are also more likely to have the power to counter the political and economic influences of land developers, lobbyists, and other special-interest groups when conflicts over specific land-use policies arise.

National governments also have a role to play. Since national governments own and administer the use of large amounts of land, national policy will determine how those lands are used. The designation of lands as wilderness, forests, rangelands, or parks at the federal level often involves a balancing of national priorities with local desires. As with local land-use issues, the conflicts over the use of federal lands are often economic and involve compromise.

Purchasing Land or Use Rights

Probably the simplest way to protect desirable lands is to purchase them. When privately owned land is desired for special purposes, it must be purchased from the owner and the owner has a right to expect to get a fair price for the property. When it is determined that land has a high public value, then either the land or the rights to use it must be purchased. Many environmental organizations purchase lands of special historic, scenic, or environmental value.

In many cases, the owners may not be willing to sell the land but are willing to limit the uses to which the land can be put in the future. Therefore, land-owners may sell the right to develop the land, or may agree to place restrictions on the uses any future owners might consider.

Regulating Use

Many communities are not in a financial position to purchase lands; therefore, they attempt to regulate land use by zoning laws.

figure 13.9 **Zoning** Most communities have a zoning authority that designates areas for particular use. This sign indicates that decisions have been made about the "best" use for the land.

Zoning is a common type of land-use regulation which restricts the kinds of uses to which land in a specific region can be put. When land is zoned, it is designated for specific potential uses. Common designations are agricultural, commercial, residential, recreational, and industrial. (See figure 13.9.)

Most local zoning boards are elected or appointed and often lack specific training in land-use planning. As a result, zoning regulations are frequently made by people who see only the short-term gain and not the possible long-term loss. Often the land is simply zoned so that its current use is sanctioned and is rezoned when another use appears to have a higher short-term value to the

ENVIRONMENTAL CLOSE-UP

Land-Use Planning and Aesthetic Pollution

Unpleasant odors, disagreeable tastes, annoying sounds, and offensive sights, can be aggravating. Yet, it is difficult to get complete agreement on what is acceptable and when some aesthetic boundary has been crossed that is unacceptable. Furthermore, many useful activities generate stimuli that are offensive while the activity itself may be essential or at least very useful. Many of these do not harm us physically but may be harmful from an aesthetic point of view.

Odors are caused by various airborne chemicals. People who live near dairy farms, livestock-raising operations, paper mills, chemical plants, steel mills, and other industries may be offended by the odors originating from these sources, but the products of these activities are needed. Many of these industries discharge wastewater that contains materials that decompose or evaporate and cause odor pollution. People who are constantly exposed to an odor are usually not as offended by it as are people who are newly exposed to it. When an odor is constantly received, the brain ceases to respond to the stimulus. In other words, the person is not aware of the odor.

Chemicals can also affect the taste of things we eat or drink. Some naturally produced chemicals, as well as those discharged into waterways, can affect the taste of food or drinking water. Minute quantities of certain chemicals can affect the taste of our drinking water and our food. Algae in water produce flavors that are offensive to many. Some groundwater sources have sulfur or salts that cause unwanted flavors. In fish, a concentration of 100 ppm of phenol can be tasted. Although this small amount of chemical is not harmful biologically, it does make the fish very unappetizing.

Noise is unwanted sound. It is produced as an incidental byproduct of industry, traffic, and other human activity. People vary considerably in their tolerance of unwanted sound. Many in cities adapt to the constant noise of the city, while those who live in more rural areas would find city noises annoying.

Visual pollution is a sight that offends us. This type of pollution is highly subjective and is, therefore, difficult to define or control. To most people, a dilapidated home or building is offensive, especially if located in an area of higher-priced homes. A heavily littered highway or street is aesthetically offensive to most people, and litter along a wilderness trail is even more unacceptable. Some sources of visual pollution are not so clear-cut, however. To many people, roadside billboards are offensive, but they can be helpful to advertisers and to travelers looking for information.

Since aesthetic pollutants—odors, tastes, sounds, and sights—are extremely difficult to define, it is difficult to establish aesthetic pollution standards. One of the simplest ways to eliminate many of these annoyances is to separate the generator of the offensive stimulus from the general public. Proper land-use planning can greatly reduce the amount of annoying aesthetic pollution. If uses that produce annoying stimuli can be clustered together rather than dispersed, the effect on the public is reduced. It does not make sense to allow new home construction in the vicinity of farming operations that will produce odors or industrial operations or airports that will generate noise. Similarly, allowing homes to be built in areas that have poor quality groundwater that will be used for drinking is a problem—unless an additional decision is made to provide a different source of drinking water. The roadside advertising that many find offensive can be regulated by allowing such signs only in specific areas rather than allowing them to be built just anywhere along the roadway. None of these land-use remedies eliminates the aesthetic pollutant, but, by segregating the public from the source of the annoyance, the impact is reduced.

community. Even when well-designed land-use plans exist, they are usually modified to encourage local short-term growth rather than to provide for the long-range needs of the community. The public needs to be alert to variances from established land-use plans, because once the plan is compromised, it becomes easier to accept future deviations that may not be in the best interests of the community. Many times, individuals who make zoning decisions are real-estate agents, developers, or local business people. These individuals wield significant local political power and are not always unbiased in their decisions. Concerned citizens must try to combat special interests by attending zoning commission meetings and by participating in the planning process.

Special Urban Planning Issues

Urban areas present a large number of planning issues. Transportation, open space, and improving the quality of life in the inner city are significant problems.

Urban Transportation Planning

A growing concern of city governments is to develop comprehensive urban transportation plans. While the specifics of such plans might vary from region to region, urban transportation planning usually involves four major goals:

1. Conserve energy and land resources.
2. Provide efficient and inexpensive transportation within the city, with special attention to people who are unable to drive, such as many elderly, young, handicapped and financially disadvantaged persons.
3. Provide suburban people opportunities to commute efficiently.
4. Reduce urban pollution.

Any successful urban transportation plan should integrate all of these goals, but funding and intergovernmental cooperation are needed to achieve this. The problems associated with current urban transportation will certainly not disappear overnight, but comprehensive planning is the first step to solving them.

Since automobiles are heavily used, transportation corridors and parking facilities must be included in any urban transportation plan.

However, many urban planners recognize that the automobile's disadvantages may outweigh its advantages, so some cities, such as Toronto, London, San Francisco, and New York, have attempted to dissuade automobile use by developing mass transit systems and by allowing automobile parking costs to increase substantially.

The major urban mass transit systems are railroads, subways, trolleys, and buses. In many parts of the world, mass transportation is extremely efficient and effective. However, in the United States where the automobile is the primary method of transportation, mass transportation systems are often underfunded and difficult to establish because mass transit is:

1. Economically feasible only along heavily populated routes
2. Less convenient than the automobile
3. Extremely expensive to build and operate
4. Often crowded and uncomfortable

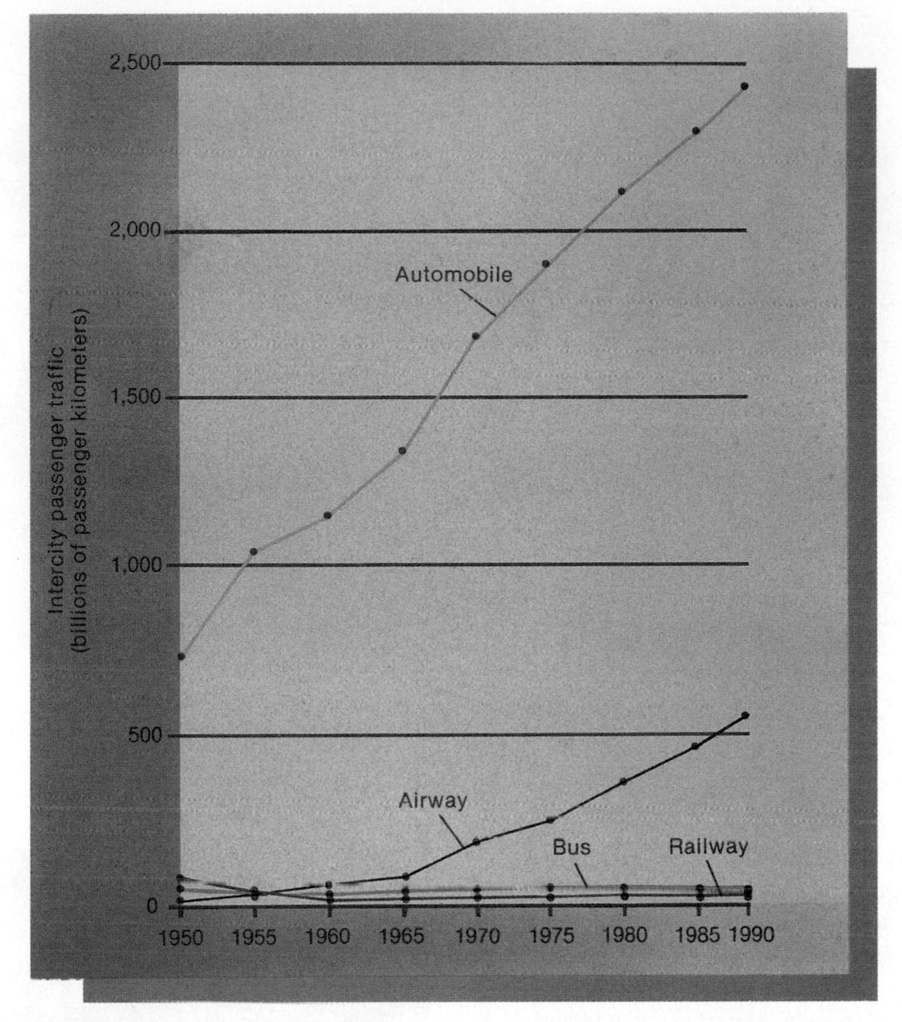

figure 13.10 **Decline of Mass Transportation** Automobile use has increased consistently since 1950, while rail and bus transport has remained low.

Although mass transit meets a substantial portion of the urban transportation needs in some parts of the world, such as in Europe and Russia, its use in North America has declined since the 1940s. (See figure 13.10.) A variety of forces has caused this decline. As people became more affluent, they could afford to own automobiles, which are a convenient, individualized method of transportation. Governments in North America encourage automobile use by financing highways and expressways, by maintaining a cheap energy policy, and by withdrawing support for most forms of mass transportation. Thus, they encourage automobile transportation with hidden subsidies (highway construction and cheap gasoline) but maintain that rail and bus transportation should not be subsidized. North Americans will seek alternatives to private automobile use only when the cost of fuel, or the cost of parking, or the inconvenience of driving becomes too high.

Urban Recreation Planning

Nearly three-fourths of the population of North America live in urban areas. These urban dwellers value open space because it breaks up the sights and sounds of the city and provides a place for recreation. Inadequate land-use planning in the past has rapidly converted urban open spaces to other uses. Until recently, creating a

New York Central Park

Hyde Park, London

figure 13.11 **Urban Open Space** These photos show open space in two urban areas, New York's Central Park and Hyde Park in London. If the land had not been set aside, it would have been developed.

new park within a city was considered an uneconomical use of the land, but people are now beginning to realize the need for parks and open spaces.

Some cities recognized the need for open space a long time ago and allocated land for parks. London, Toronto, and Perth, Australia, have centrally located and well-used parks. New York City set aside approximately 200 hectares (500 acres) for Central Park in the late 1800s. (See figure 13.11.) Boston has developed a park system that provides a variety of urban open spaces. Other cities have not dealt with this need for open space because they have lacked either the foresight or the funding. Recreation is a basic human need. The most primitive tribes and cultures all engaged in games or recreational activities. New forms of recreation are continually being developed. In the congested urban center, cities often must construct special areas where recreation can take place.

figure 13.12 **Urban Recreation** In urban areas, recreation often takes the form of sports programs, playgrounds, and walking. Most cities recognize the need for such activities and develop extensive recreation programs for their citizens.

A major problem with urban recreation is locating recreational facilities near residential areas. Facilities that are not conveniently located may be infrequently used. For example, the hundreds of thousands of square kilometers of national parks in Alaska and the Yukon will be visited every year by a tiny proportion of the population of North America. Large urban centers are discovering that they must provide adequate, low-cost recreational opportunities within their jurisdiction. Some of these opportunities take the form of commercial establishments, such as bowling centers, amusement parks, and theaters. Others must be subsidized by the community. (See figure 13.12.) Playgrounds, organized recreational activities, and open space have usually been combined into an arm of the municipal government known as the parks and recreation department. Cities spend millions of dollars to develop and maintain recreation programs. Often, there is conflict over the allocation of financial and land resources. These are closely tied because open land is scarce in urban areas, and it is expensive. Riverfront property is ideal for park and recreational use, but it is also prime land for industry, commerce, or high-rise residential buildings. Although conflict is inevitable, many metropolitan areas are beginning to see that recreational resources may be as important as economic growth for maintaining a healthy community.

An outgrowth of the trend toward urbanization is the development of **nature centers.** In many urban areas, there is so little natural area left that the people who live there need to be given opportunities to learn about nature. Nature centers are basically teaching institutions that provide a variety of methods for people to learn about and appreciate the natural world. Zoos, botanical gardens, and some urban parks, combined with interpretative centers, also provide recreational experiences. Nature centers are usually located near urban centers, in places where some appreciation of the natural processes and phenomena can be developed. They may be operated by municipal governments or by school systems or other nonprofit organizations.

Table 13.1 Number of People Who Participated in Selected Outdoor Recreational Activities in 1995

Activity	Percent of Population over 7 Years of Age Participating
Exercise walking	30
Swimming	27
Bicycling	24
Fishing	18
Camping	18
Golf	12
Hiking	10
Running	9
Hunting	7
Backpacking	4
Skiing	4

Source: Data from *Statistical Abstract of the United States, 1997.*

Redevelopment of Inner-City Areas

As people moved to the suburbs during the past 50 years, the inner city was abandoned. Many old industrial sites sit vacant. Businesses have moved to the suburban malls. The quality of housing has declined. Services have been reduced. In order to improve the quality of life of the residents of the city, special efforts must be made to revitalize the city. Although activities that will improve the quality of life in the inner city vary from city to city, there are several land-use processes that help. One problem that has plagued industrial cities is vacant industrial and commercial sites. Many of these buildings have remained vacant because the cost of cleanup and renovation is expensive. Such sites have been called **brownfields.** Many of these sites involved environmental contamination and since the EPA required that they be cleaned up to a pristine condition, no one was willing to do so.

A new approach to utilizing these sites is called **brownfields development.** This involves a more realistic approach to dealing with the contamination at these sites. Instead of requiring complete cleanup, the degree of cleanup required is matched to the intended use of the site. Although an old industrial site with specific contamination problems may not be suitable for housing, it may be redeveloped as a new industrial site, since access to the contamination can be controlled. An old industrial site that has soil contamination may be paved to provide parking.

Another important focus in urban redevelopment is the remodeling of abandoned commercial buildings for shopping centers, cultural facilities, or high-density housing. Chattanooga, Tennessee, has received a reputation for revitalizing its inner-city area. The process of revitalization involved extensive planning activities that included the public, public and private funding of redevelopment activities, establishment of an electric bus system to alleviate air pollution, renovation of existing housing, redevelopment of old warehouses into a shopping center, and incorporation of a condemned bridge over the Tennessee River into a portion of a park which is also an important pedestrian connection between a residential area and the downtown business district.

Federal Government Land-Use Issues

Since the federal government manages large amounts of land, the laws and regulations that shape land-use policy are important. For example, the 1960 Multiple Use Sustained Yield Act divided use of national forests into four categories: wildlife habitat preservation, recreation, lumbering, and watershed protection. This act was designed to encourage both economic and recreational use of the forests. However, specific users of this public land are often in conflict, particularly recreational users with timber harvesters.

The 1872 mining law has also been important in federal land management. The law allows anyone to prospect for minerals on public lands and to establish a claim if such minerals are discovered. The miner is then allowed to purchase the rights to extract the mineral for $5.00 per acre. Many feel that the law is obsolete, but it is still in force and public land is still being sold to mining interests at ridiculously low prices. These laws are examples of past policy decisions that are still in effect. Today, one of the major uses of public lands is outdoor recreation.

Many people want to use the natural world for recreational purposes because nature can provide challenges that may be lacking in their day-to-day lives. Whether the challenge is hiking in the wilderness, underwater exploration, climbing mountains, or driving a vehicle through an area that has no roads, these activities offer a sense of adventure. Look at table 13.1. All of these activities use the out-of-doors, but not in the same way. Conflicts develop because some of these activities cannot occur in the same place at the same time. For example, wilderness camping and backpacking often conflict with off-road vehicles.

There is a basic conflict between those who prefer to use motorized vehicles and those who prefer to use muscle power in their recreational pursuits. (See figure 13.13.) This conflict is particularly strong because both groups would like to use the same public land. Both have paid taxes, and both feel that it should be available for them to use as they wish.

Finally, as more rangelands and forests have had vehicular access controlled or eliminated, those who want to use public lands for motorized recreation have become upset.

There are also land-use conflicts between business interests and recreational users of public lands. Federal and state governments give special use permits to certain users of public lands. Many ski resorts in the West make use of public lands. Grazing is also an important use of federal land. Based on "Animal Use Months" established by the Bureau of Land Management or the Forest Service, ranchers are allowed to graze cattle on certain public lands. Technically, failure to comply can mean a loss of grazing rights. However, since the establishment of regulations is highly political, many maintain that the political influence of ranchers allows them to use a public resource without adequately compensating the government. In addition, the regulatory agencies are understaffed and find it difficult to adequately regulate the actions of individual ranchers. As a result, some lands are overgrazed. Many people who want to use the publicly owned rangelands for outdoor recreation resent the control exercised by grazing interests. On the other hand, ranchers resent the intrusion of hikers and campers on land they have traditionally controlled.

An obvious solution to this problem is to allocate land to specific uses and to regulate the use once allocations have been made. Several U.S. governmental agencies, such as the National Park Service, the Bureau of Land Management, the U.S. Forest Service, and the U.S. Fish and Wildlife Service, allocate and regulate the lands they control. However, these agencies have conflicting roles. The U.S. Forest Service has a mandate to manage forested public lands for timber production. This mandate often comes in conflict with recreational uses. Similarly, the Bureau of Land Management has huge tracts of land that can be used for recreation, but it traditionally has been mandated to manage grazing rights.

A particularly sensitive issue is the designation of certain lands as wilderness areas. Obviously, if an area is to be

figure 13.13 **Conflict over Recreational Use of Land** Land may be used for both motorized and nonmotorized activities. The people who participate in these two kinds of recreation are often antagonists over the allocation of land for recreational use.

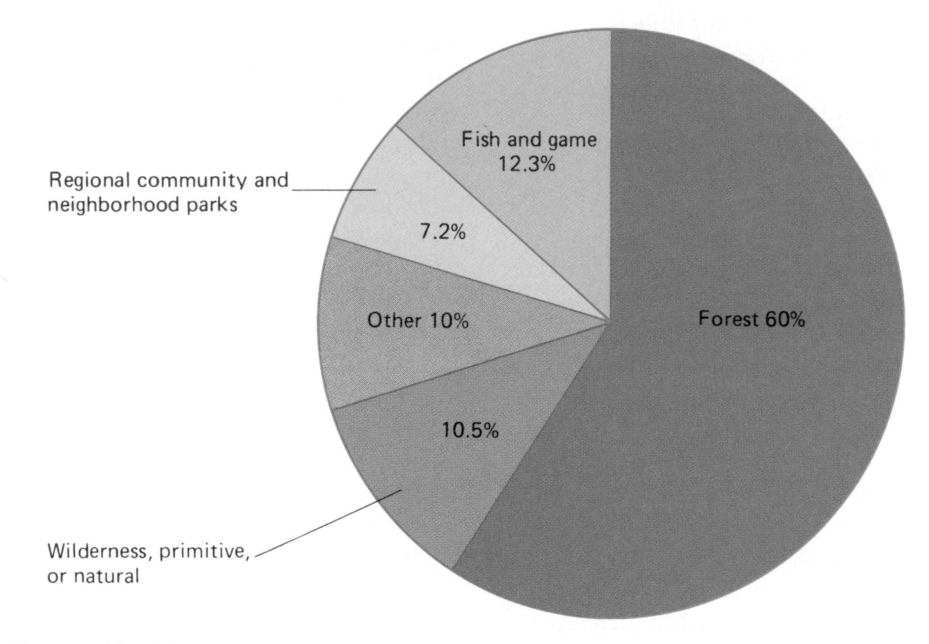

figure 13.14 **U.S. Federal Recreational Lands in 1995** Of the approximately 108 million hectares (267 million acres) of federal recreational lands in the United States, approximately 10 percent is designated as wilderness, primitive, or natural.

Source: Data from the *Statistical Abstract of the United States, 1995.*

wilderness, human activity must be severely restricted. This means that the vast majority of Americans will never see or make use of it. Many people argue that this is unfair because they are paying taxes to provide recreation for a select few. Others argue that if everyone were to use these areas, their charm and unique character would be destroyed and that, therefore, the cost of preserving wilderness is justifiable.

Areas designated as wilderness make up a very small proportion of the total public land available for recreation. (See figure 13.14.) The fact that there are relatively few wilderness areas has resulted in a further problem: The areas are being loved to death. People pressure on this resource has become so great that, in some cases, the wilderness quality is being tarnished. The designation of additional wilderness would relieve some of this pressure.

Issues & Analysis

Decision Making in Land-Use Planning—The Malling of America

The following situation has happened thousands of times during the past 20 years:

A developer has just announced plans to build a large shopping mall on the outskirts of your city in what is now prime farmland. Many jobs will be created by the construction and operation of the proposed mall, which will include stores and three new theaters. Presently, your city has some unemployment, only one theater in the downtown area, and little variety in its retail businesses. On the surface, the proposed mall seems to be only good news. Is this the case? Before answering, look at the entire situation.

- What will happen to the downtown area when the mall is opened?
- If you owned a downtown business, would you favor building the mall?
- What will happen to the taxes on the farms near the new mall?
- If you were a farmer, would you favor the project?
- What effect will paving prime farmland have?
- How will storm water runoff be affected?
- How will future housing development be influenced?
- What other problems can be associated with building the complex?
- If you were the mayor or city manager, would you favor construction of the mall? Why?
- Is the proposed project all goods news after all?

Summary

Historically, waterways served as transportation corridors that allowed for the exploration of new land and for the transport of goods. Therefore, most large urban centers began as small towns located near water. Water served the needs of the towns in many ways, especially as transportation. Several factors resulted in the shift of the population from rural to urban. These included the Industrial Revolution which provided jobs in cities and the addition of foreign immigrants to the cities. As towns became larger, the farmland surrounding them became suburbs surrounding industrial centers. Unregulated industrial development in cities resulted in the degradation of the waterfront and stimulated the development of suburbs around the city as people sought better places to live and had the money to purchase new homes. The rise in automobile ownership further stimulated the movement of people from the cities to the suburbs.

Many problems have resulted from unplanned growth. Current taxation policies encourage residential development of farmland, which results in a loss of valuable agricultural land. Floodplains and wetlands are often mismanaged. Loss of property and life results when people build on floodplains. Wetlands protect our shorelines and provide a natural habitat for fish and wildlife. Transportation problems and lack of open space are also typical in many large metropolitan areas.

Land-use planning involves gathering data, projecting needs, and developing mechanisms for implementing the plan. Good land-use planning should include assessment of the unique geologic, geographic, biological, and historic and cultural features of the land; the costs of providing additional infrastructure; preservation of open space; provision for a variety of transportation options; a mixture of housing and service establishments; redevelopment of disused urban land; and establishment of urban growth limits. Establishing regional planning agencies, purchasing land or its development rights, and zoning are ways to implement land-use planning. The scale of local planning is often not large enough to be effective because problems may not be confined to political boundaries. Regional planning units can afford professional planners and are better able to withstand political and economic pressures. A growing concern of urban governments is to develop comprehensive urban transportation plans that seek to conserve energy and land resources, provide efficient and inexpensive transportation and commuting, and help to reduce urban pollution. Urban areas must also provide recreational opportunities for their residents and seek ways to rebuild decaying inner cities.

Federal governments own and manage large amounts of land, therefore national policy must be developed. This usually involves designating land for particular purposes, such as timber production, grazing land, parks, or wilderness. The recreational use of public land often requires the establishment of rules that prevent conflict between potential users who have different ideas about what appropriate uses should be. Often federal policy is a compromise between competing uses and land is managed for multiple uses.

Interactive Exploration

Check out the website at

http://www.mhhe.com/environmentalscience

and click on the cover of this textbook for interactive versions of the following:

KNOW THE BASICS

brownfields *285*
brownfields development *285*
floodplains *277*
floodplain zoning ordinances *277*
infrastructure *276*
land-use planning *279*
megalopolis *273*
nature centers *285*
ribbon sprawl *273*
tract development *273*
urban growth limit *280*
urban sprawl *273*
wetlands *277*
zoning *281*

On-line Flashcards
Electronic Glossary

IN THE REAL WORLD

What does "open space" mean to you? Think about ways that you envision open space. Do you picture a nature trail through wildlife habitat and stream areas? Do you see an open area that can be used for soccer, picnics, or other community activities? Do you think of wilderness areas with minimal impact? What other scenarios do you envision? See how different interpretations of open space can lead to a Rumble in Gunntown!

Do you like to golf? Do any of your friends or classmates? Is a golf course part of your vision of open space? Take a look at Prairie Dunes Country Club: A Golf Course for Birdies to see if you think the designers of this Kansas golf course have mastered land use for multiple goals.

Another example of an attempt to use land for different goals can be found in Record Conservation Land Purchase Saves Forests in Adirondack Park. Read this story and find out how a diverse group met the challenges they encountered. How does wilderness fit into your vision of open space and land use?

What type of giant is thousands of years old, hundreds of feet in height, and *still* needs protection? Read President Protects Sequoias in New National Monument to see what activities are threatening these noble giants.

Speaking of protection, what protects forests? Take a look at Major Initiative Proposed to Protect National Forests to see what the federal government is doing to protect national forests.

Rules for protecting wetlands have been around for a long time, but wetlands are still being destroyed. Read Stronger Wetland Protection Rules Announced to find out what is being done and why wetlands are so important.

Native arctic Inuit people are able to establish their own land-use practices now that the new territory of Nunavut has been established in Canada. Read through Canada Creates New Territory, Nunavut to see why this is such a historical event for the Inuit people and for Canada.

TEST PREPARATION

Review Questions

1. Why did urban centers develop near waterways? Are they still located near water?
2. Describe the typical changes that have occurred in cities from the time they were first founded until now.
3. Why do people move to the suburbs?
4. Why do some farmers near urban areas sell their land for residential or commercial development? If you were in this position, would you sell?
5. What is a megalopolis?
6. What land uses are suitable on floodplains?
7. What is multiple land use? Can land be used for multiple purposes?
8. Why is it important to provide recreational space in urban planning?
9. How can recreational activities damage the environment? Do you engage in any of those activities?
10. What is the monetary impact of recreational activities?
11. What are some strictly urban-related recreational activities?
12. List some conflicts that arise when an area is designated strictly as wilderness.
13. Describe the steps necessary to develop a land-use plan.
14. What are the advantages of regional or state planning?
15. List three benefits of land-use planning.

Critical Thinking Questions

1. Choose the city where you live. Interview local residents and look at old city maps. What did the city look like 75 years ago? What were the city's boundaries? Where did people do their shopping? How did they get around? How does this compare with the current situation in the city?
2. What historical factors brought members of your family to the city? How does this compare to the factors that are currently contributing to the growth of cities in the developing world?
3. Consider the outer rim of the city closest to you. Which, if any, of the problems associated with unplanned growth are associated with your city? What factors make them a problem? What do you think can be done about them?
4. There has been tremendous development in the arid West of the United States over the past few decades, creating demands for water. How should these demands be met? Should there be limits to this type of development? What kinds of limits, if any?
5. Imagine you are a National Forest Supervisor who is creating a 10-year plan that is in the public comment stage. What interests would be contacting you? What power would each interest have? How would you manage the competing interests of timber, mining, grazing, and recreation or between motorized and non-motorized recreation? What values, beliefs, and perspectives helped you form your recommendations?
6. Imagine that you lived in an area of the country that has the potential to be named a wilderness area. What conflicts do you think would arise from such a declaration? Who might be some of the antagonists? Which perspective do you think is most persuasive? How would you answer the objections of the other perspective?
7. Look at the Issues and Analysis section of this chapter and answer the questions. What values, beliefs, and perspectives lead you, as the imaginary city manager, to reach your decision about the shopping mall?
8. After reading the Environmental Close-up in this chapter concerning wetland loss in Louisiana, what kinds of recommendations would you make to help preserve wetlanda? What do you suppose might happen if nothing is done? What resistance might wetland preservation generate?

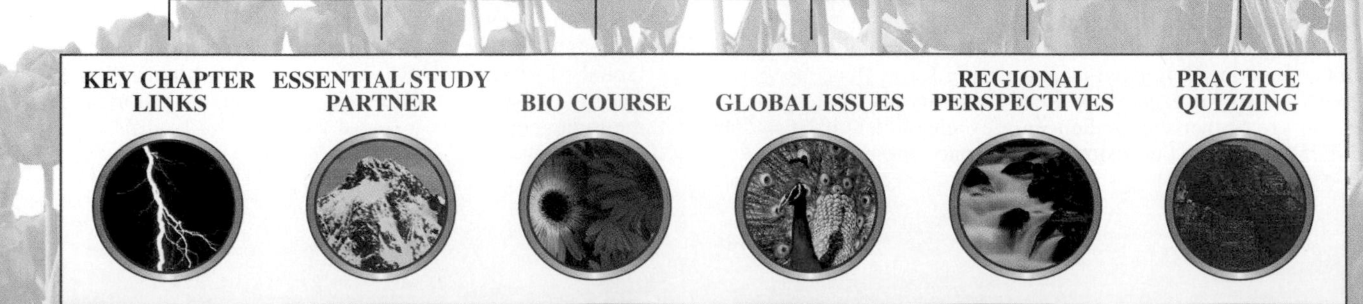

CHAPTER 14

Soil and Its Uses

Objectives

After reading this chapter you should be able to:

- Describe the geologic processes that build and erode the Earth's surface.
- List the physical, chemical, and biological factors involved in soil formation.
- Explain the importance of humus to soil fertility.
- Differentiate between soil texture and soil structure.
- Explain how texture and structure influence soil atmosphere and soil water.
- Explain the role of living organisms in soil formation and fertility.
- Describe the various layers in a soil profile.
- Describe the processes of soil erosion by water and wind.
- Explain how contour farming, strip farming, terracing, waterways, windbreaks, and conservation tillage reduce soil erosion.
- Understand that the misuse of soil reduces soil fertility, pollutes streams, and requires expensive remedial measures.
- Explain how land not suited for cultivation may still be productively used for other purposes.

Chapter Outline

Geologic Processes
Soil and Land
Soil Formation
Soil Properties
Soil Profile
Soil Erosion
Soil Conservation Practices
- Contour Farming
- Strip Farming
- Terracing
- Waterways
- Windbreaks

Conventional Versus Conservation Tillage
Protecting Soil on Nonfarm Land
Global Perspective: *Worldwide Soil Degradation*
Environmental Close-Up: *Land Capability Classes*
Issues & Analysis: *Soil Erosion in Virginia*

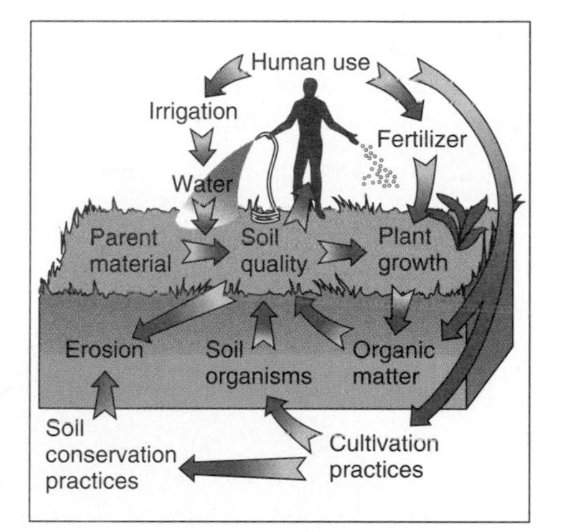

Geologic Processes

We tend to think of the Earth as being stable and unchanging until we recognize such events as earthquakes, volcanic eruptions, floods and windstorms changing the surface of the places we live. There are forces that build new land and opposing forces that tear it down. Much of the building process involves shifting of large portions of the Earth's surface known as plates. The Earth is composed of an outer crust, a plastic mantle, and a central core. The **crust** is an extremely thin, less dense, solid covering over the underlying mantle. The **mantle** is a layer that makes up the majority of the Earth and surrounds a small core made up primarily of iron. The mantle consists of an inner solid portion and an outer portion that is capable of flow. (See figure 14.1.)

Plate tectonics is the concept that the outer surface of the Earth consists of large plates composed of the crust and the outer portion of the mantle and that these plates are slowly moving over the surface of the liquid outer mantle. This combination of crust and outer mantle is known as the **lithosphere.** The heat from the Earth causes slow movements of the outer layer of the mantle similar to what happens when you heat a liquid on the stove, only much slower. The movements of the plates on this plastic outer layer of the mantle are independent of each other. Therefore, some of the plates are pulling apart from one another while others are colliding.

Where the plates are pulling apart from one another, the liquid mantle moves upward to fill the gap and solidifies. Thus, new crust is formed from the liquid mantle. Approximately half of the surface of the Earth has been formed in this way in the past 200 million years. The bottom of the Atlantic and Pacific Oceans and the Rift Valley and Red Sea area of Africa are areas where this is occurring.

If plates are pulling apart on one portion of the Earth, they must be colliding elsewhere. Where plates collide, several things can happen. (See figure 14.2.) Often, one of the plates slides under the other and is melted. Often, when this occurs, some of the liquid mantle makes its way to the surface and volcanoes are formed, which results in the formation of mountains. The west coasts of North and South America have many volcanoes and mountain ranges where the two plates are colliding. The volcanic activity adds new material to the crust. When a collision occurs between two plates under the ocean, the volcanoes may eventually reach the surface and form a chain of volcanic islands such as can be seen in

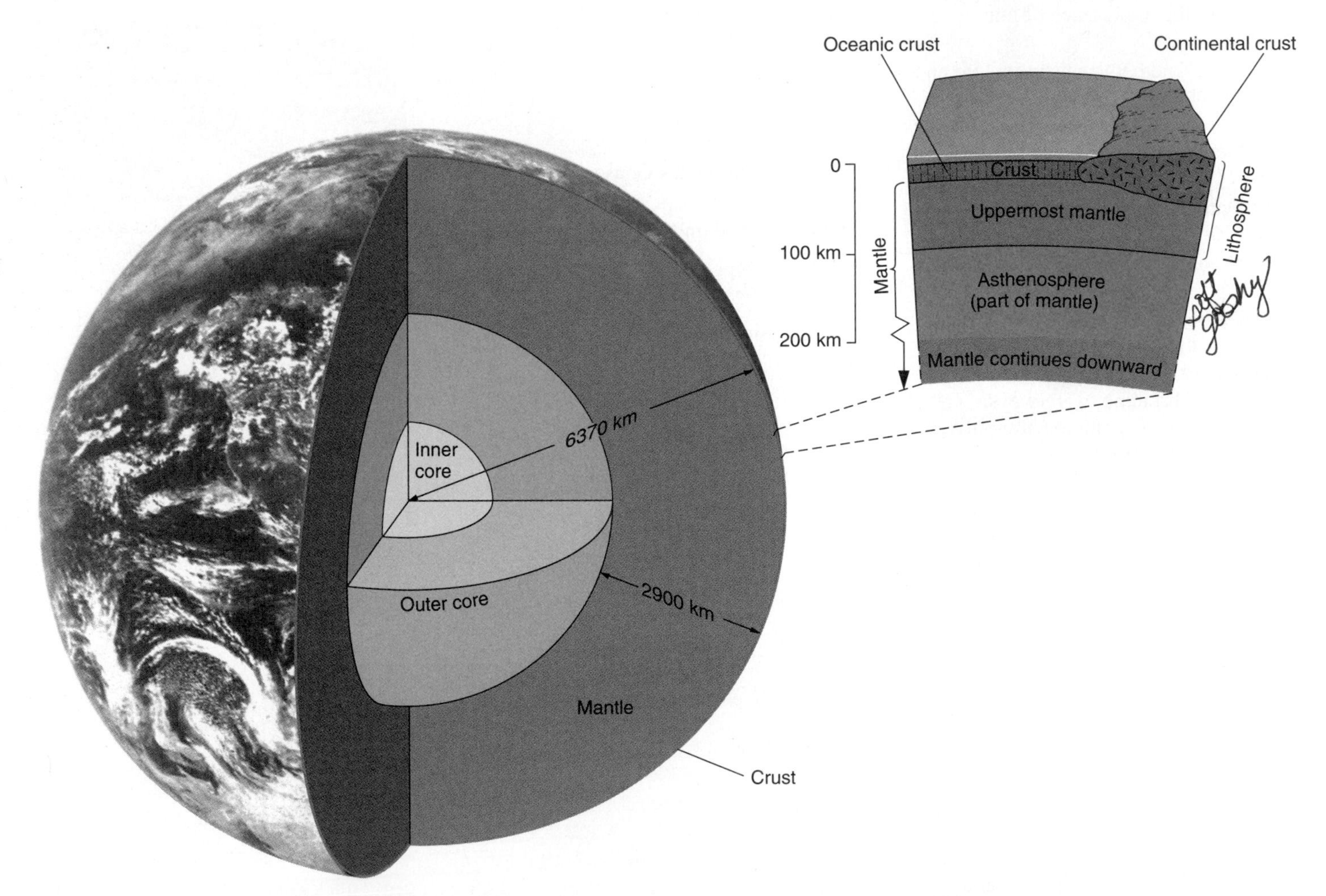

figure 14.1 **Structure of the Earth** The Earth has a solid outer lithosphere which floats on a plastic mantle.

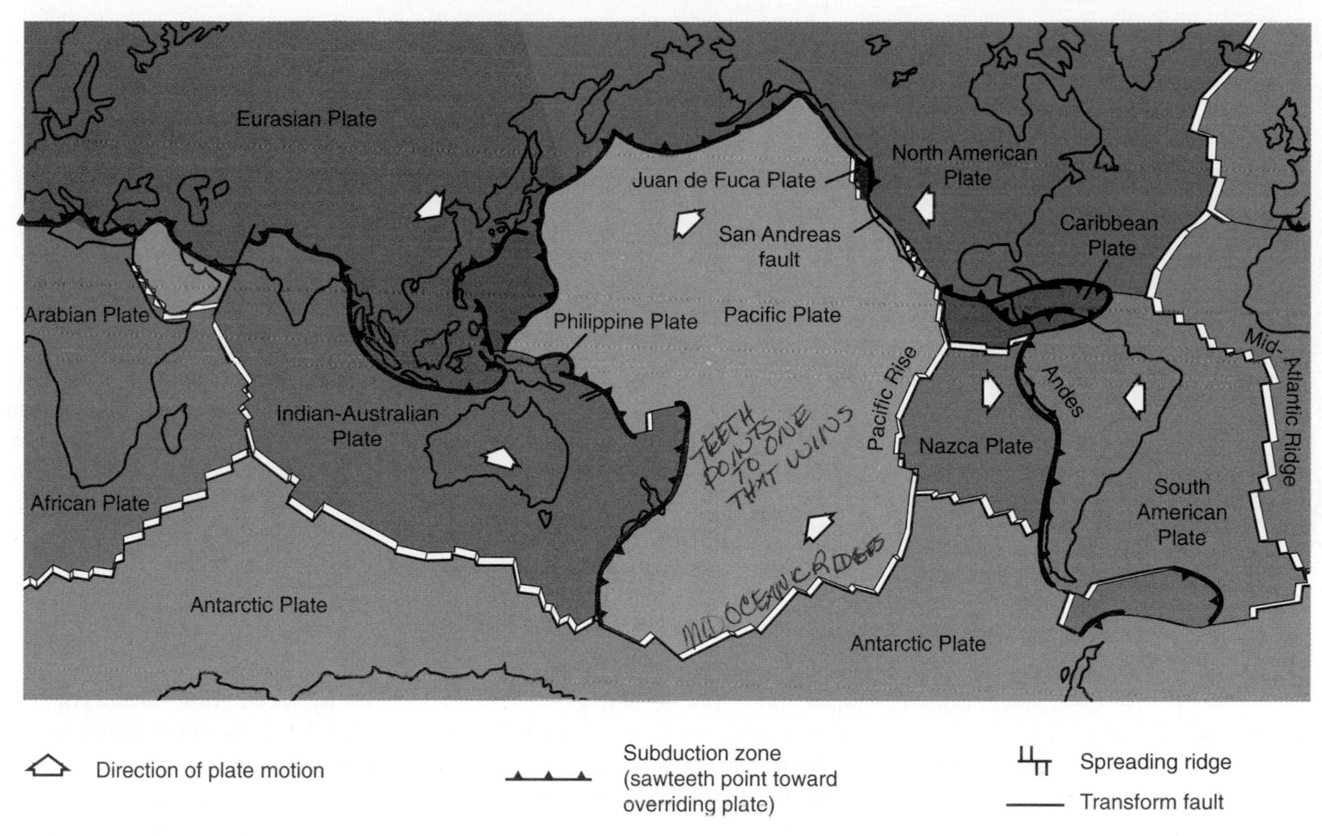

figure 14.2 **Tectonic Plates** The plates which make up the outer surface of the Earth move with respect to one another. They pull apart from one another in some parts of the world and collide with one another in other parts of the world.

the Aleutian Islands and many of the Caribbean Islands. When two continental plates collide, neither plate slides under the other and the crust buckles to form mountains. The Himalayan, Alp, and Appalachian mountain ranges are thought to have formed from the collision of two continental plates. All of these movements of the Earth's surface are associated with earthquakes. The movements of the plates are not slow and steady sliding movements but tend to occur in small jumps. However, what is a small movement between two plates on the Earth is a huge movement for the relatively small structures and buildings produced by humans, these small movements can cause tremendous amounts of damage.

These building processes are counteracted by processes that tend to make the elevated surfaces lower. Gravity provides a force that tends to wear down the high places. Moving water and ice (glaciers) and wind assist in the process; however, their effectiveness is related to the size of the particles. Several kinds of **weathering** processes are important in reducing the size of particles which can then be dislodged by moving water and air. **Mechanical weathering** results from physical forces that reduce the size of rock particles without changing the chemical nature of the rock. Common causes of mechanical weathering are changes in temperature that tend to result in fractures in rock, the freezing of water which expands and tends to split larger pieces of rock into smaller ones, and the actions of plants and animals.

Because rock does not expand evenly, heating a large rock can cause it to fracture, so that pieces of the rock flake off. These pieces can be further reduced in size by other processes, such as the repeated freezing and thawing of water. Water that has seeped into rock cracks and crevices expands as it freezes, causing the cracks to widen. Subsequent thawing allows more water to fill the widened cracks, which are enlarged further by another period of freezing. Alternating freezing and thawing fragments large rock pieces into smaller ones (see figure 14.3). The roots of plants growing in cracks can also exert enough force to break rock.

figure 14.3 **Physical Fragmentation by Freezing and Thawing** The crack in the rock fills with water. As the water freezes and becomes ice, it expands. The pressure of the ice enlarges the crack. The ice melts, and water again fills the crack. The water freezes again and widens the crack. Alternate freezing and thawing splits the rock into smaller fragments.

figure 14.4 **The Painted Desert—An Eroded Landscape** This landscape was created by the action of wind and moving water. The particles removed by these forces were deposited elsewhere and may have become part of the soil in that new location.

The physical breakdown of rock is also caused by forces that move and rub rock particles against each other (abrasion). For example, a glacier causes rock particles to grind against one another, resulting in smaller fragments and smoother surfaces. These particles are deposited by the glacier when the ice melts. In many parts of the world, the parent material from which soil is formed consists of glacial deposits. Wind and moving water also cause small particles to collide, resulting in further weathering. The smoothness of rocks and pebbles in a stream or on the shore is evidence that moving water has caused them to rub together, removing their sharp edges. Similarly, particles carried by wind collide with objects, fragmenting both the objects and the wind-driven particles.

Wind and moving water also remove small particles and deposit them at new locations, exposing new surfaces to the weathering process. For example, the landscape of the Painted Desert in the southwest United States was created by a combination of wind and moving water that removed easily transported particles, while rocks more resistant to weathering remained. (See figure 14.4.)

The activities of organisms can also assist mechanical weathering. The roots of plants can exert considerable force and move particles apart from one another. The burrows of animals expose new surfaces that can be altered by freezing and thawing.

Chemical weathering involves the chemical alteration of the rock in such a manner that it is more likely to fragment or to be dissolved. Some small rock fragments exposed to the atmosphere may be oxidized; that is, they combine with oxygen from the air and chemically change to different compounds. Other kinds of rock may combine with water molecules in a process known as hydrolysis. Often, the oxidized or hydrolyzed molecules are more readily soluble in water and, therefore, may be removed by rain or moving water. Rain is normally slightly acid, and the acid content helps dissolve rocks.

Because of gravity, the prevailing movement of particles is from high elevations to lower ones. This process of loosening and redistributing particles is known as **erosion.** Wind can move sand and dust and can also cause the wearing away of rocky surfaces by sandblasting their surfaces. Glaciers can move large rocks and also cause their surfaces to be rounded by being rubbed against each other and the surface of the Earth. Moving water transports much material in streams and rivers. In addition, wave action along the shores of lakes and the coasts of oceans constantly wears away and transports particles.

Soil and Land

The geologic processes just discussed are involved in the development of both soil and land; however, soil and land are not the same. **Land** is the part of the world not covered by the oceans. **Soil** is a thin covering over the land consisting of a mixture of minerals, organic material, living organisms, air, and water that together support the growth of plant life. The proportions of the soil components vary with different types of soils, but a typical, "good" agricultural soil is about 45 percent mineral, 25 percent air, 25 percent water, and 5 percent organic matter (see figure 14.5). This combination provides good drainage, aeration, and organic matter. Farmers are particularly concerned with soil because the nature of the soil determines the kinds of crops that can be grown and which farming methods must be employed. Urban dwellers should also be concerned about soil because its health determines the quality and quantity of food they will eat. If the soil is so abused that it can no longer grow crops, or if it is allowed to erode, degrading air and water quality, both urban and rural residents suffer. To understand how soil can be protected, we must first understand its properties and how it is formed.

Soil Formation

A combination of physical, chemical, and biological events acting over time is responsible for the formation of soil. Soil building begins with the fragmentation of the **parent material,** which consists of ancient layers of rock or more recent geologic deposits from lava flows or glacial activity. The kind and amount of soil developed depends on the kind of parent material present, the plants and animals present, the climate, the time involved, and the slope of the land. As was discussed earlier, the breakdown of parent material is known as weathering.

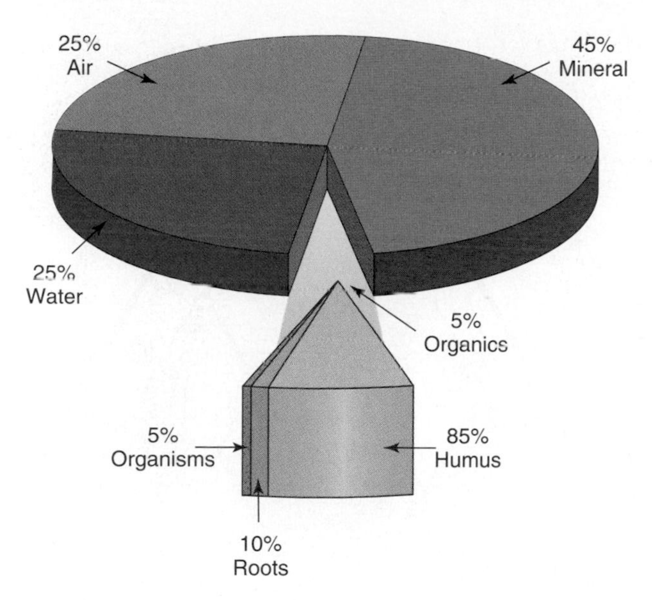

figure 14.5 **The Components of Soil** Although soils vary considerably in composition, they all contain the same basic components: mineral material, air, water, and organic material. The organic material can be further subdivided into humus, roots, and other living organisms. The percentages shown are those that would be present in a good soil.

The climate and chemical nature of the rock material have a great influence on the rate of weathering. Similarly, the size and chemical nature of the particles have a great impact on the nature of the soil that will develop in an area.

The role of organisms in the development of soil is also very important. The first organisms to gain a foothold in this modified parent material also contribute to soil formation. Lichens often form a pioneer community that grows on the surface of rocks and traps small particles. The decomposition of dead lichens and other organic matter releases acids that chemically alter the underlying rock, causing further fragmentation. The release of chemicals from the roots of plants causes further chemical breakdown of rock particles. As other kinds of organisms, like plants and small animals, become established, they contribute, through their death and decay, increasing amounts of organic matter, which are incorporated with the small rock fragments.

The organic material resulting from the decay of plant and animal remains is known as **humus.** It is a very important soil component that accumulates on the surface and ultimately becomes mixed with the top layers of mineral particles. This material contains nutrients that are taken up by plants from the soil. Humus also increases the water-holding capacity and the acidity of the soil so that inorganic nutrients, which are more soluble under acidic conditions, become available to plants. Humus also tends to stick other soil particles together and helps to create a loose, crumbly soil that allows water to soak in and permits air to be incorporated into the soil. Compact soils have few pore spaces, so they are poorly aerated, and water has difficulty penetrating, so it runs off.

Burrowing animals, soil bacteria, fungi, and the roots of plants are also part of the biological process of soil formation. One of the most important burrowing animals is the earthworm. One hectare (2.47 acres) of soil may support a population of 500,000 earthworms that can process as much as 9 tonnes (about 10 U.S. tons) of soil a year. These animals literally eat their way through the soil, resulting in further mixing of organic and inorganic material, which increases the amount of nutrients available for plant use. They often bring nutrients from the deeper layers of the soil up into the area where plant roots are more concentrated, thus improving the soil's fertility. Soil aeration and drainage are also improved by the burrowing of earthworms and other small soil animals, such as nematodes, mites, pill bugs, and tiny insects. They also help to incorporate organic matter into the soil by collecting dead organic material from the surface and transporting it into burrows and tunnels. When the roots of plants die and decay, they release organic matter and nutrients into the soil and provide channels for water and air.

Fungi and bacteria are decomposers and serve as important links in many mineral cycles. (See chapter 5.) They, along with animals, improve the quality of the soil by breaking down organic material to smaller particles and releasing nutrients.

The position on the slope also influences soil development. Soil formation on steep slopes is very slow because materials tend to be moved downslope with wind and water. Conversely, river valleys often have deep soils because they receive materials from elsewhere by these same erosive forces.

Climate and time are also important in the development of soils. In general, extremely dry or cold climates develop soils very slowly, while humid and warm climates develop them more rapidly. Cold and dry climates have slow rates of accumulation of organic matter needed to form soil. Furthermore, chemical weathering proceeds more slowly at lower temperatures and in the absence of water. Under ideal climatic conditions, soft parent material may develop into a centimeter (less than ½ inch) of soil within 15 years. Under poor climatic conditions, a hard parent material may require hundreds of years to develop into that much soil. In any case, soil formation is a slow process.

The amount of rainfall and the amount of organic matter influence the pH of the soil. In regions of high rainfall, basic ions such as calcium, magnesium, and potassium are leached from the soils and more acid materials are left behind. In addition, the decomposition of organic matter tends to increase the soil's acidity. Soil pH is important since it influences the availability of nutrients, which affects the kinds of plants that will grow, which affects the amount of organic matter added to the soil. Since calcium, magnesium, and potassium are important plant nutrients, their loss by leaching reduces the fertility of the soil. Excessively acids soils also cause aluminum ions to become soluble, which in high amounts are toxic to many plants. (See the discussion of acid rain in chapter 17.) Most plants grow well in soils with a pH between 6 and 7, although some plants such as blueberries and potatoes grow well in acid soils. In most agricultural situations, the pH of the soil is usually adjusted by adding chemicals to the soil. Lime can be added to make soils less acid, and acid-forming materials such as sulfates can be added to increase acidity.

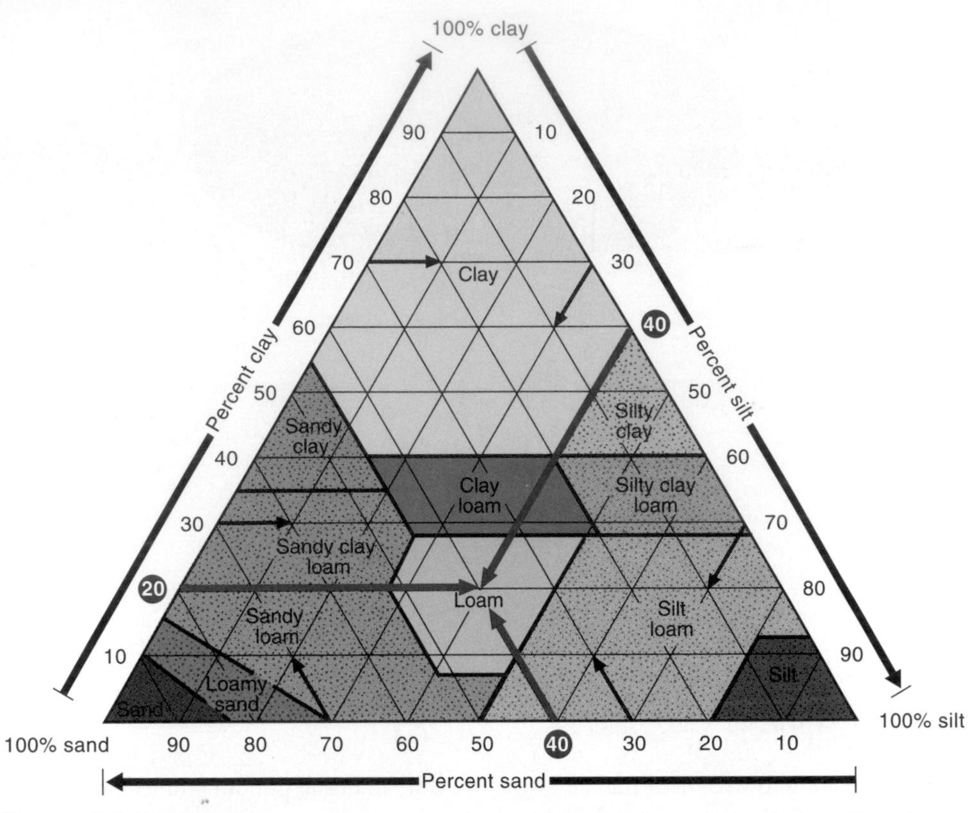

figure 14.6 **Soil Texture** Texture depends upon the percentage of clay, silt, and sand particles in the soil. A loam soil has the best texture for most crops. As shown in the illustration, if a soil were 40 percent sand, 40 percent silt, and 20 percent clay, it would be a loam.

Source: Data from Soil Conservation Service

Soil Properties

Soil properties include soil texture, structure, atmosphere, moisture, biotic content, and chemical composition. **Soil texture** is determined by the size of the mineral particles within the soil. The largest soil particles are gravel, which consists of fragments larger than 2.0 millimeters in diameter. Particles between 0.05 and 2.0 millimeters are classified as sand. Silt particles range from 0.002 to 0.05 millimeters in diameter, and the smallest particles are clay particles, which are less than 0.002 millimeters in diameter.

Large particles, such as sand and gravel, have many tiny spaces between them, which allow both air and water to flow through the soil. Water drains from this kind of soil very rapidly, often carrying valuable nutrients to lower soil layers, where they are beyond the reach of plant roots. Clay particles tend to be flat and are easily packed together to form layers that greatly reduce the movement of water through them. Soils with a lot of clay do not drain well and are poorly aerated. Because water does not flow through clay very well, clay soils tend to stay moist for longer periods of time and do not easily lose minerals to percolating water.

However, rarely does a soil consist of a single size of particle. Various particles are mixed in many different combinations, resulting in many different soil classifications. (See figure 14.6.) An ideal soil for agricultural use is a **loam,** which combines the good aeration and drainage properties of large particles with the nutrient-retention and water-holding ability of clay particles.

Soil structure is different from its texture. **Soil structure** refers to the way various soil particles clump together. The particles in sandy soils do not attach to one another and, therefore, sandy soils have a granular structure. The particles in clay soils tend to stick to one another to form large aggregates. Other soils that have a mixture of particle sizes tend to form smaller aggregates. A good soil is **friable,** which means that it crumbles easily. The soil structure and its moisture content determine how friable a soil is. Sandy soils are very friable, while clay soils are not. If clay soil is worked when it is too wet, it can stick together in massive blocks that will be difficult to break up.

A good soil for agricultural use will crumble and has spaces for air and water. In fact, the air and water content depends upon the presence of these spaces. (See figure 14.7.) In good soil, about one-half to two-thirds of the spaces contain air after the excess water has drained. The air provides a source of oxygen for plant root cells and all the other soil organisms. The relationship between the amount of air and water is not fixed. After a heavy rain, most of the spaces may be filled with water and less oxygen is available to plant roots and other organisms. If some of the excess water does not drain from the soil, the plant roots may die from lack of oxygen. They are literally drowned. On the other hand, if there is not enough soil moisture, the plants wilt from lack of water. Soil moisture and air are also important in determining the numbers and kinds of soil organisms.

Protozoa, nematodes, earthworms, insects, algae, bacteria, and fungi are

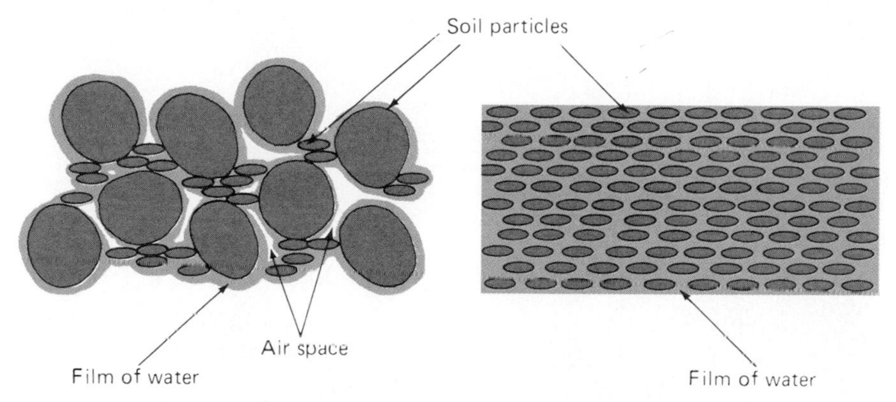

figure 14.7 **Pore Spaces and Particle Size** The soil on the left, which is composed of particles of various sizes, has spaces for both water and air. The particles have water bound to their surfaces (represented by the colored halo around each particle), but some of the spaces are so large that an air space is present. The soil on the right, which is composed of uniformly small particles, has less space for air. Since roots require both air and water, the soil on the left would be better able to support crops than would the soil on the right.

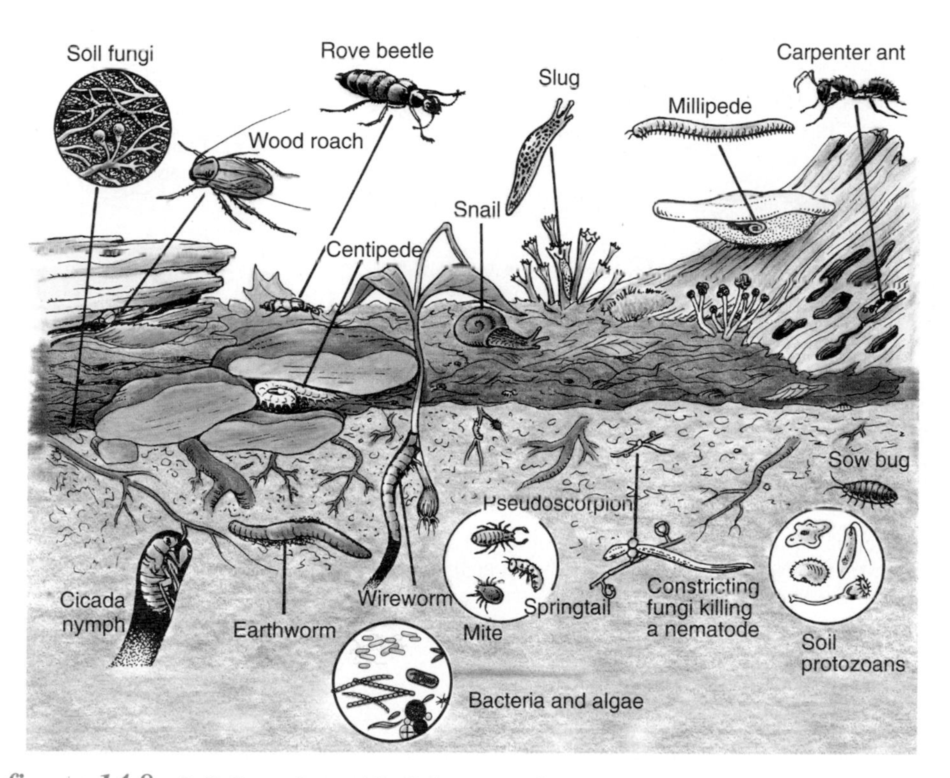

figure 14.8 **Soil Organisms** All of these organisms occupy the soil and contribute to it by rearranging soil particles, participating in chemical transformation, and recycling dead organic matter.

typical inhabitants of soil. (See figure 14.8.) The role of protozoa in the soil is not firmly established, but they seem to act as parasites and predators on other forms of soil organisms and, therefore, help to regulate the populations of those organisms. Nematodes, which are often called wireworms or roundworms, may aid in the breakdown of dead organic matter. Some nematodes are parasitic on the roots of plants. Insects and other soil arthropods contribute to the soil by forming burrows and consuming and fragmenting organic materials, but they are also major crop pests that feed on plant roots. Several kinds of bacteria are able to fix nitrogen from the atmosphere. Algae carry on photosynthesis and are consumed by other soil organisms. Bacteria and fungi are particularly important in the decay and recycling of materials. Their chemical activities change complex organic materials into simpler forms that can be used as nutrients by plants. For example, some of these microorganisms can convert the nitrogen contained in the protein component of organic matter into ammonia or nitrate, which are nitrogen compounds that can be utilized by plants. The amount of nitrogen produced varies with the type of organic matter, type of microorganisms, drainage, and temperature. Finally, it is important to recognize that the soil contains a complicated food chain in which all organisms are subject to being consumed by others. All of these organisms are active within distinct layers of the soil, known as the soil profile.

Soil Profile

The **soil profile** is a series of horizontal layers in the soil that differ in chemical composition, physical properties, particle size, and amount of organic matter. Each recognizable layer is known as a **horizon.** (See figure 14.9.) There are several systems for describing and classifying the horizons in soils. In general, the uppermost layer of the soil contains more nutrients and organic matter than do the deeper layers. The top layer is known as the *A* horizon or topsoil. The *A* horizon consists of small mineral particles mixed with organic matter. Because of the relatively high organic content, it is dark in color. If there is a layer of **litter** (undecomposed or partially decomposed organic matter) on the surface, it is known as the *O* horizon. Forest soils typically have an *O* horizon. Many agricultural soils do not, since the soil is worked to incorporate surface crop residue. As the organic matter decomposes, it becomes incorporated into the *A* horizon. The thickness of the *A* horizon may vary from less than a centimeter (less than ½ inch) on steep mountain slopes to over a

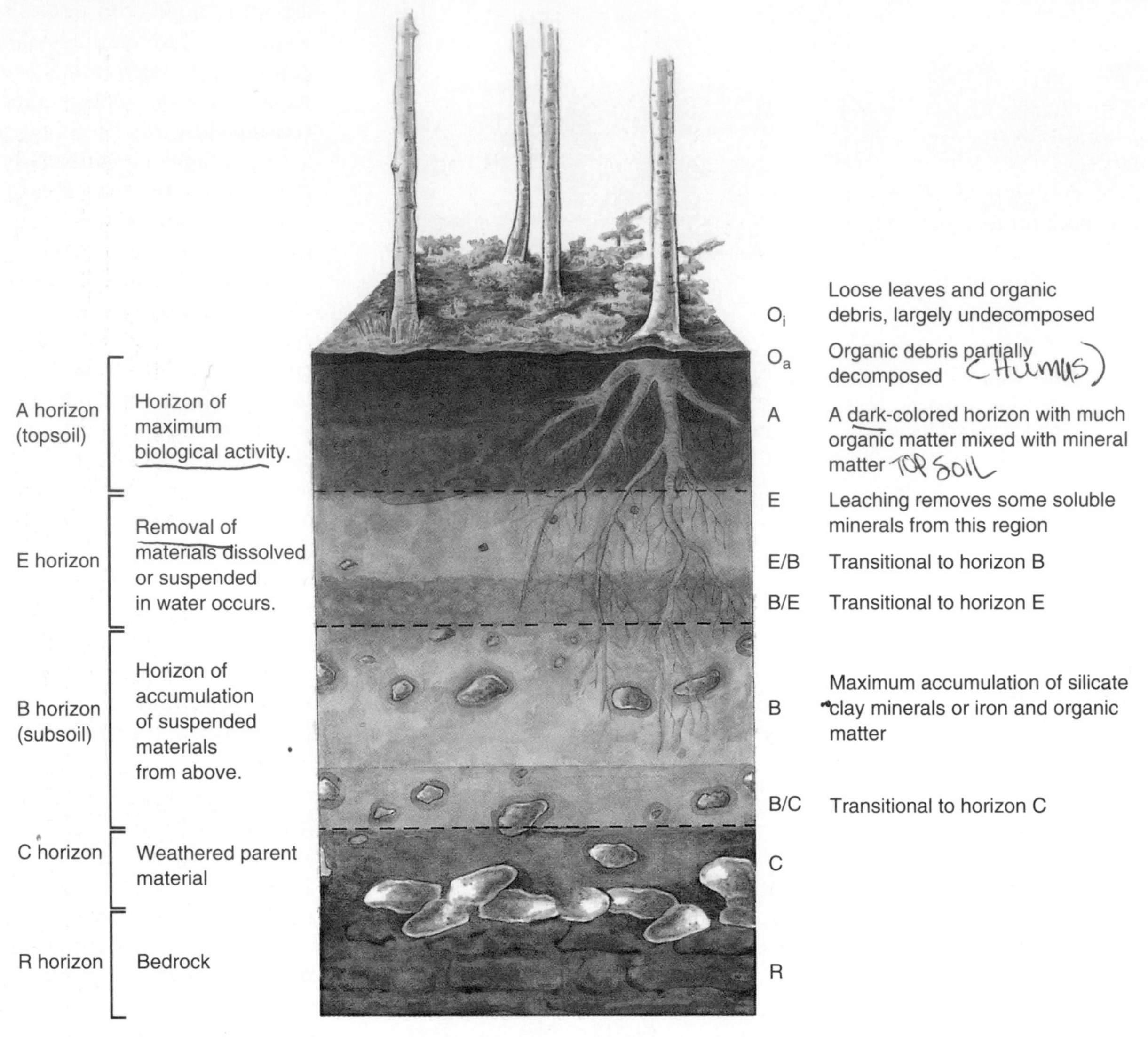

figure 14.9 **Soil Profile** A soil has layers that differ physically, chemically, and biologically. The top layer is known as the *A* horizon and contains most of the organic matter. Organic matter that collects on the surface is known as the *O* horizon. Many soils have a light-colored *E* horizon below the *A* horizon. It is light in color because dark-colored materials are leached from the layer. The *B* horizon accumulates minerals and particles as water carries dissolved minerals downward from the *A* and *E* to the *B* horizon. The *B* horizon is often called the subsoil. Below the *B* horizon is a *C* horizon of weathered parent material.

meter (over 40 inches) in the rich grasslands of central North America. Most of the living organisms and nutrients are found in the *A* horizon. As water moves down through the *A* horizon, it carries dissolved organic matter and minerals to lower layers. This process is known as **leaching.** Because of the leaching away of darker materials such as iron compounds, a lighter-colored layer develops below the *A* horizon that is known as the *E* horizon. Not all soils develop an *E* horizon. This layer usually contains few nutrients because water flowing down through the soil dissolves and transports nutrients to the underlying *B* horizon. The *B* horizon, often called the subsoil, contains less organic material and fewer organisms than the *A* horizon. However, it contains accumulations of nutrients that were leached from higher levels. Often, clay minerals that are leached from the topsoil are deposited in this layer. Because nutrients are deposited in this layer, the *B* horizon in many soils is a valuable source of nutrients for plants, and such subsoils support a well-developed root system. Because the amount of leaching depends on the available rainfall, grasslands soils, which develop under low rainfall, often have a poorly developed *B* horizon, while soils in woodlands that receive higher rainfall usually have a well-developed *B* horizon.

The area below the subsoil is known as the *C* horizon, and it consists of weathered parent material. This parent material contains no organic materials, but it does contribute to some of the soil's properties. The chemical composition of the minerals of the *C* horizon helps to determine the pH of the soil. If the parent material is limestone, the soil will tend to neutralize acids; whereas, if the parent material is granite rock, the soil will not be able to do so. The characteristics of the parent material in the *C* horizon may also influence the soil's rate of water absorption and retention. Ultimately, the *C* horizon rests on bedrock, which is known as the *R* horizon.

Soil profiles and the factors that contribute to soil development are extremely varied. Over 15,000 separate soil types have been classified in North America. However, most of the cultivated land in the world can be classified as either grassland soil or forest soil. (See figure 14.10.)

Because the amount of rainfall in grassland areas is relatively low, it does not penetrate into the soil layers very far. Most of the roots of the grasses and other plants remain near the surface and little leaching of minerals from the topsoil to deeper layers occurs. Since the roots of the plants rot in place when the grasses die, a deep layer of topsoil develops. This lack of leaching also results in a thin layer of subsoil, which is low in mineral and organic content and supports little root growth.

Forest soils develop in areas of more abundant rainfall. Water moves down through the soil so that deeper layers of the soil have a great deal of moisture. The roots of the trees penetrate to this layer and extract the water they need. The leaves and other plant parts that fall to the soil surface form a thin layer of organic matter on the surface. This organic matter decomposes and mixes with the mineral material of the top layers of the soil. The water that moves through the soil tends to carry material from the topsoil to the subsoil where many of the roots of the plants are located. One of the materials that accumulates in the *B* horizon is clay. In some soils, particularly forest soils, clay or other minerals may accumulate and form a relatively impermeable "hardpan" layer that limits the growth of roots and may prevent water from reaching the soil's deeper layers.

Desert soils have very poorly developed horizons. Since there is little rainfall, deserts do not support a large amount of plant growth and much of the soil is exposed. Therefore, there is little organic matter added to the soil and there is little leaching of materials from upper layers to lower layers. Since much of the soil is exposed to wind and water erosion, much of the organic material and smaller particles are carried away by wind or by flash floods when it rains.

In cold, wet climates, typical of the northern parts of Europe, Russia, and Canada, there may be considerable accumulations of organic matter since the rate of decomposition is reduced. The extreme acidity of these soils also reduces the rate of decomposition. Hot, humid climates also tend to have poorly developed soil horizons since the organic matter decays very rapidly, and soluble materials are carried away by the abundant rainfall.

Because tropical rainforests support such a vigorous growth of plants and an incredible variety of plant and animal species, it is often assumed that tropical soils must be very fertile. Consequently, many people have tried to raise crops on tropical soils. It is possible to grow certain kinds of crops that are specially adapted to tropical soils, but raising of most traditional crop species is not successful. In order to understand why this is so, it is important to understand the nature of tropical rainforest soils. Two features of the tropical rainforest climate have a great influence on the nature of the soil. High temperature results in rapid decomposition of organic matter, so that the soils have very little litter and humus. High rainfall tends to leach nutrients from the upper layers of the soil, leaving behind a soil that is rich in iron and aluminum. The high iron content results in a reddish color for most of these soils. Because the nutrients are quickly removed, these soils are very infertile. Furthermore, when the vegetation is removed, the soil is quickly eroded.

In addition to the differences caused by the kind of vegetation and rainfall, topography influences the soil profile. (See figure 14.11.) On a relatively flat area, the topsoil formed by soil-building processes will collect in place and gradually increase in depth. The topsoil formed on rolling hills or steep slopes is often transported down the slope as fast as it is produced. On such slopes, the accumulation of topsoil may not be sufficient to support a cultivated crop. The topsoil removed from these slopes is eventually deposited in the flat floodplains. These regions serve as collection points for topsoil that was produced over extensive areas. As a result, these river-bottom and delta regions have a very deep topsoil layer and are highly productive agricultural land.

Soil Erosion

Erosion is the wearing away and transportation of soil by water, wind, or ice. The Grand Canyon of the Colorado River, the floodplains of the Nile in Egypt, the little gullies on hillsides, and the deltas that develop at the mouths of rivers all attest to the ability of water to move soil. Anyone who has seen muddy water after a rainstorm has observed soil being moved by water. (See figure 14.12.) The force of moving water allows it to carry large amounts of soil. While erosion is a natural process, it is greatly accelerated by agricultural practices that leave the soil exposed. Each year, the Mississippi River transports over 325 million tonnes (360 million U.S. tons) of soil from the central regions of North America to the Gulf of Mexico. This is equal to the removal of a layer of topsoil approximately 1 millimeter (0.04 inches) thick from the entire region. Although the rate of erosion varies from place to place, movement of soil by water occurs in every stream and river in the world. Dry Creek, a small stream in California, has only 500 kilometers (310 miles) of mainstream and tributaries; however, each year, it removes 180,000 tonnes (200,000 U.S. tons) of soil from a 340-square-kilometer (130 square miles) area.

Soil erosion takes place everywhere in the world, but some areas are more exposed than others. Erosion occurs wherever grass, bushes, and trees are disappearing. Deforestation and desertification both leave land open to erosion. In deforested areas, water washes down steep, exposed slopes, taking the soil

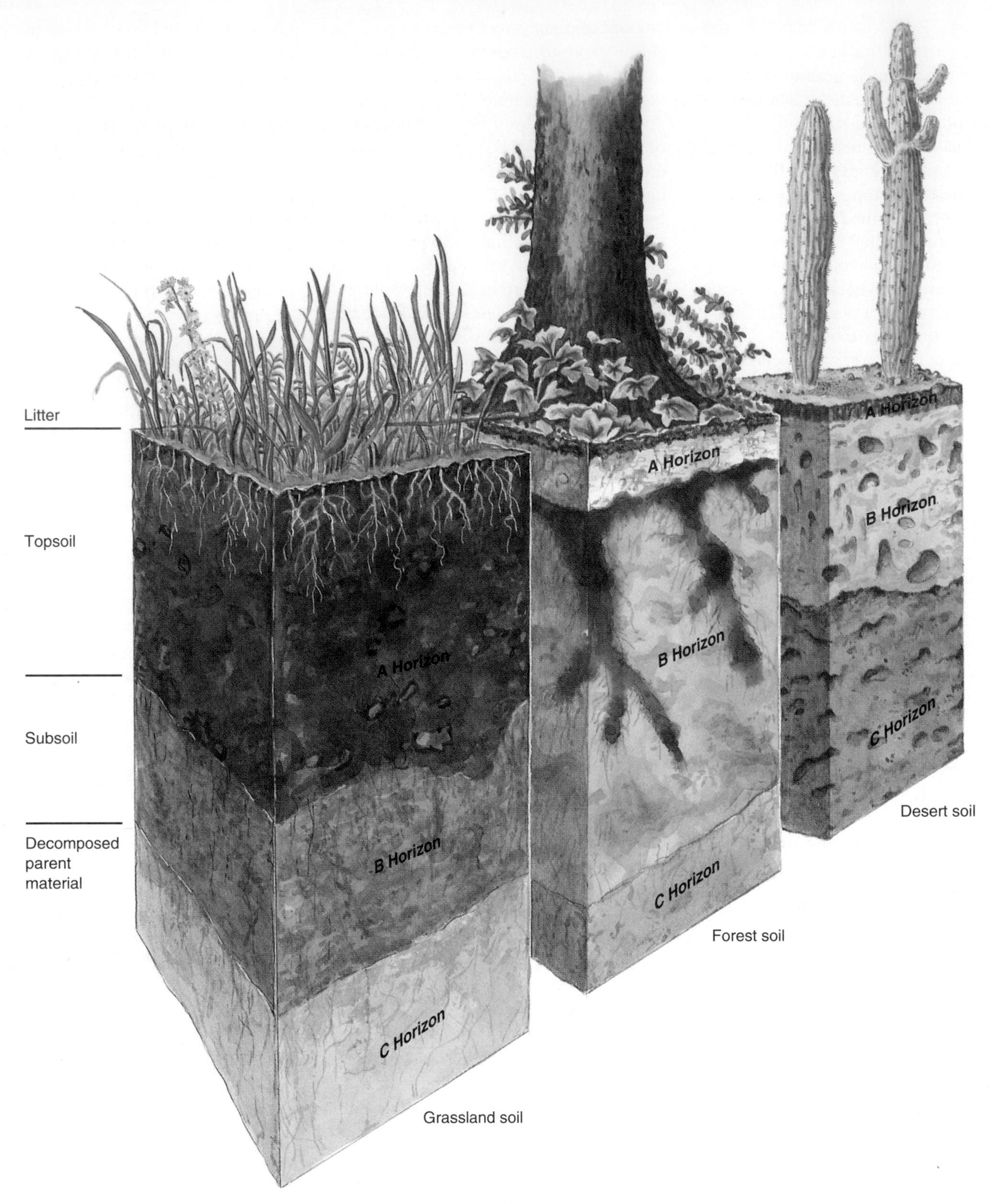

figure 14.10 **Major Soil Types** There are thousands of different soil types, but many of them can be classified into three broad categories. Soils formed in the grasslands have a deep *A* horizon. The shallow *B* horizon does not have sufficient nutrients to support root growth. In forest soils the *A* horizon is thinner, and leaching transfers many nutrients to the *B* horizon. Thus, roots are found in both the *A* and *B* horizons. Desert soils have very thin *A* horizons.

figure 14.11 **The Effect of Slope on a Soil Profile** The topsoil formed on a large area of the hillside is continuously transported down the slope by the flow of water. It accumulates at the bottom of the slope and results in a thicker *A* horizon. The resulting "bottomland" is highly productive because it has a deep, fertile layer of topsoil, while the soil on the slope is less productive.

figure 14.12 **Water Erosion** The force of moving water is able to pick up soil particles and remove them. In cases of prolonged erosion, gullies (such as the one shown here) are likely to form.

with it. In desertified regions, exposed soils, cleared for farming, building, or mining, or overgrazed by livestock, simply blow away. Wind erosion is most extensive in Africa and Asia. Blowing soil not only leaves a degraded area behind but can bury and kill vegetation where it settles. It will also fill drainage and irrigation ditches. When high-tech farm practices are applied to poor lands, soil is washed away and chemical pesticides and fertilizers pollute the runoff. Every year erosion carries away far more topsoil than is created, primarily because of agricultural practices that leave the soil exposed. (See figure 14.13.)

Worldwide, erosion removes about 25.4 billion tonnes (28 billion U.S. tons) of soil each year. In Africa, soil erosion has reached critical levels, with farmers pushing farther onto deforested hillsides. In Ethiopia, for example, soil loss occurs at a rate of between 1.5 billion and 2 billion cubic meters (53–70 billion cubic feet) a year, with some 4 million hectares (about 10 million acres) of highlands considered irreversibly degraded. In Asia, in the eastern hills of Nepal, 38 percent of the land area is fields that have been abandoned because the topsoil has washed away. In the Western Hemisphere, Ecuador is losing soil at 20 times the acceptable rate.

According to the International Fund for Agricultural Development (IFAD), traditional labor-intensive, small-scale soil conservation efforts that combine maintenance of shrubs and trees with crop growing and cattle grazing work best at controlling erosion. In parts of Pakistan, a program begun by IFAD in 1980 to control rainfall runoff, erosion, and damage to rivers from siltation has increased crop yields and livestock productivity by 20 to 30 percent.

Badly eroded soil has lost all of the topsoil and some of the subsoil and is no longer productive farmland. Most current agricultural practices lose soil faster than it is replaced. Farming practices that reduce erosion, such as contour farming and terracing, are discussed later in the chapter.

Wind is also an important mover of soil. Under certain conditions, it can move large amounts. (See figure 14.14.) Wind erosion may not be as evident as water erosion, since it does not leave gullies. Nevertheless, it can be a serious problem. Wind erosion is most common in dry, treeless areas where the soil is exposed. In the Sahel region of Africa, much of the land has been denuded of vegetation because of drought, overgrazing, and improper farming practices. This has resulted in extensive wind erosion of the soil. (See figure 14.15.) In the Great Plains region of North America, there have been four serious periods of wind erosion since European settlement in the 1800s. If this area receives less than 30 centimeters (12 inches) of rain per year, there is not enough moisture to support crops. When this occurs for several years in a row, it is called a drought. Farmers plant crops, hoping for rain. When the rain does not come, they plow their fields again to prepare it for another crop. Thus, the loose, dry soil is left exposed, and wind erosion results. Because of the large amounts of dust in the air during those times, the region is known as the Dust Bowl. During the 1930s, wind destroyed 3.5 million hectares (over 8.5 million acres) of farmland and seriously damaged an additional 30 million hectares (75 million acres) in the Dust Bowl.

Fortunately, many soil conservation practices have been instituted that protect soil However, much soil is being lost to erosion, and more protective measures should be taken.

Soil Conservation Practices

The kinds of agricultural activities that land can be used for are determined by soil structure, texture, drainage, fertility, rockiness, slope of the land, amount and nature of rainfall, and other climatic

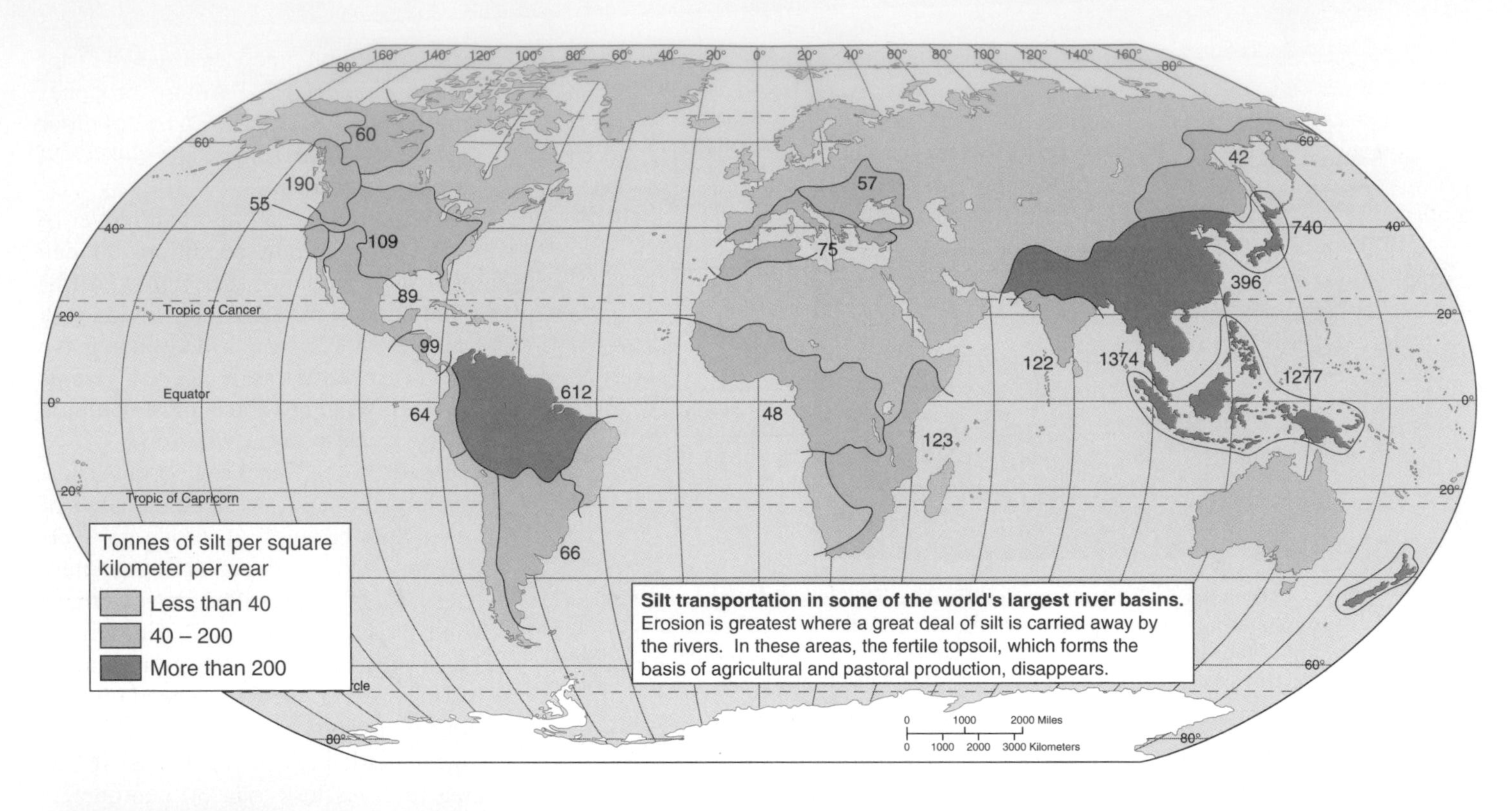

Silt-laden river

figure 14.13 **Worldwide Soil Erosion** Soil erosion is widespread throughout the world. Silty rivers are evidence of poor soil conservation practices upstream.

figure 14.14 **Wind Erosion** The dry, unprotected topsoil from this field is being blown away. The force of the wind is capable of removing all the topsoil and transporting it several thousand kilometers.

conditions. A relatively large proportion—about 20 percent—of U.S. land is suitable for agricultural use. (See figure 14.16.) However, only 2 percent of that land does not require some form of soil conservation practice. This means that nearly all of the soil in the United States must be managed in some way to reduce the effects of soil erosion by wind or water.

Not all parts of the world are as well supplied as the United States with land that has agricultural potential. (See table 14.1.) For example, worldwide, approximately 11 percent of the land surface is suitable for crops, and an additional 24 percent is in permanent pasture. In the United States, about 20 percent is cropland, and 25 percent is in permanent pasture. Contrast this with the continent of Africa, in which only 6 percent is suitable for crops, and 29 percent can be used for pasture. Canada has only 5 percent suitable for crops and 3 percent for pasture. Europe has the highest percentage of cropland with 30 percent, but it has only 17 percent in permanent pasture.

Since there is very little land left that can be converted to agriculture, we must use what we have wisely. There are many techniques that protect soil from erosion while allowing agriculture. Some of the more common methods are discussed here. Whenever soil is lost by water or wind erosion, the topsoil, the most productive layer, is the first to be removed. When the topsoil is lost, the soil's fertility decreases, and larger amounts of expensive fertilizers must be used to restore the fertility that was lost. This raises the cost of the food we buy. In addition, the movement of excessive amounts of soil from farmland into streams has several undesirable effects. First, a dirty stream is less aesthetically pleasing than a clear stream. Second, a stream laden with sediment affects the fish population by reducing visibility, covering spawning sites, and clogging the gills of the fish. Fishing may be poor because of unwise farming practices hundreds of kilometers upstream. Third, the soil carried by a river is eventually deposited somewhere. In many cases, this soil must be removed by dredging to clear shipping channels. We pay for dredging with our tax money, and it is a very expensive operation.

For all of these reasons, proper soil conservation measures should be employed to minimize the loss of topsoil. Figure 14.17 contrasts poor soil conservation practices with proper soil protection. When soil is not protected from the effects of running water, the topsoil is removed and gullies result. This can be prevented by slowing the flow of water over sloping land.

Contour Farming

Contour farming, which is tilling at right angles to the slope of the land, is one of the simplest methods to prevent soil erosion. This practice is useful on gentle slopes and produces a series of small ridges at right angles to the slope. (See figure 14.18.) Each ridge acts as a dam to hold water from running down the incline. This allows more of the water to soak into the soil. Contour farming reduces soil erosion by as much as 50 percent and, in drier regions, increases crop yields by conserving water.

figure 14.15 **Wind Erosion in the Sahel** The semiarid region just south of the Sahara Desert is in an especially vulnerable position. The rainfall is unpredictable, which often leads to crop failure. In addition, population pressure forces people in this region to try to raise crops in marginal areas. This often results in increased wind erosion.

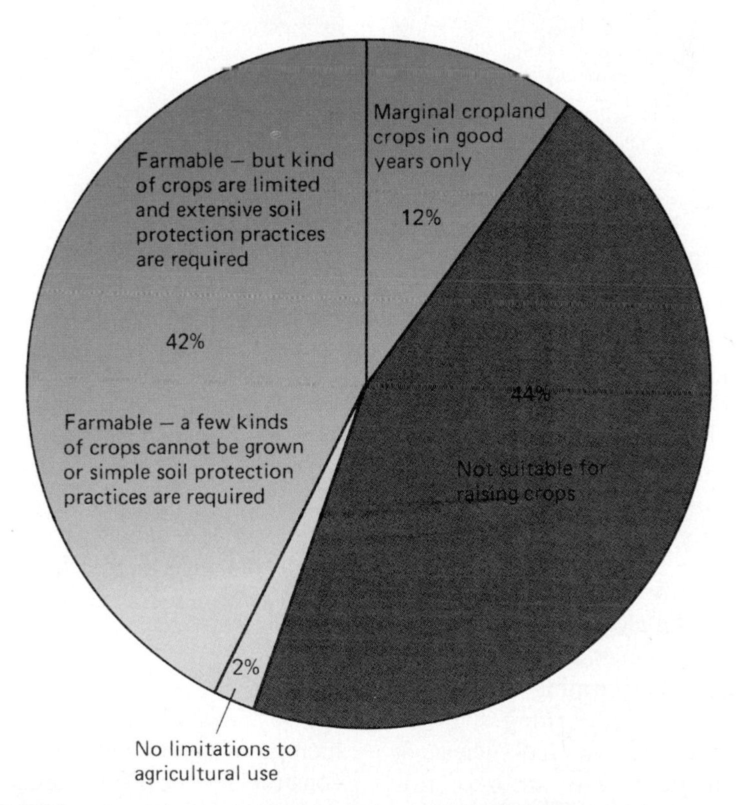

figure 14.16 **U.S. Land Suitable for Agricultural Uses** Only 2 percent of the land in the United States can be cultivated without some soil conservation practices. This 2 percent is primarily flatland, which is not subject to wind erosion. On 42 percent of the remaining land, some special considerations for protecting the soil are required, or the kinds of crops are limited. Twelve percent of the land is marginal cropland that can provide crops only in years when rainfall and other conditions are ideal. Forty-four percent is not suitable for cultivating crops but may be used for other purposes, such as grazing cattle or as forests.

Table 14.1 Percentage of Land Suitable for Agriculture

Country	Percent Cropland	Percent Pasture
World	11.0	26.0
Africa	6.3	28.8
Egypt	2.8	5.0
Ethiopia	12.7	40.7
Kenya	7.9	37.4
South Africa	10.8	66.6
North America	13.0	16.8
Canada	4.9	3.0
United States	19.6	25.0
South America	6.0	28.3
Argentina	9.9	51.9
Venezuela	4.4	20.2
Asia	15.2	25.9
China	10.3	30.6
Japan	12.0	1.7
Europe	29.9	17.1

Source: Data from *World Resources, 1996.*

a.

b.

figure 14.17 **Poor and Proper Soil Conservation Practices** (*a*) This land is no longer productive farmland since erosion has removed the topsoil. (*b*) This rolling farmland shows strip contour farming to minimize soil erosion by running water. It should continue indefinitely to be productive farmland.

Strip Farming

When a slope is too steep or too long, contour farming alone may not prevent soil erosion. However, a combination of contour and strip farming may work. **Strip farming** is alternating strips of closely sown crops like hay, wheat, or other small grains with strips of row crops like corn, soybeans, cotton, or sugar beets. (See figure 14.19.) The closely sown crops retard the flow of water, which reduces soil erosion and allows more water to be absorbed into the ground. The type of soil, steepness, and length of slope dictate the width of the strips and determine whether strip or contour farming is practical.

Terracing

On very steep land, the only practical method of preventing soil erosion is to construct terraces. **Terraces** are level areas constructed at right angles to the slope to retain water and greatly reduce the amount of erosion. (See figure 14.20.) Terracing has been used for centuries in nations with a shortage of level farmland. The type of terracing seen in figure 14.20*a* requires the use of small machines and considerable hand labor and is not suitable for the mechanized farming typical in much of the world. Terracing is an expensive method of controlling erosion since it requires the moving of soil to construct the level areas, protecting the steep areas between terraces, and constant repair and maintenance. Many factors, such as length and steepness of slope, type of soil, and amount of precipitation, determine whether terracing is feasible.

Waterways

Even with such soil conservation practices as contour farming, strip farming, and terracing, farmers must often provide protected channels for the movement of water. **Waterways** are depressions on sloping land where water collects and flows off the land. When not properly maintained, these areas are highly susceptible to erosion. (See figure 14.21.) If a waterway is maintained with a permanent sod covering, the speed of the water is reduced, the roots tend to hold the soil particles in place, and soil erosion is decreased.

Windbreaks

Contour farming, strip farming, terracing, and maintaining waterways are all important ways of reducing water erosion, but wind is also a problem with certain soils, particularly in dry areas of the world. Wind erosion can be reduced if the soil is protected. The best protection is a layer of vegetation over the surface. However, the process of preparing soil for planting and the method of planting often leave the soil exposed to wind. **Windbreaks** are plantings of trees or other plants that protect bare soil from the full force of the wind. Windbreaks reduce the velocity of the wind, thereby decreasing the amount of soil that it can carry away. (See figure 14.22.) In some cases, rows of trees are planted at right angles to the prevailing winds to reduce their force, while in other cases, a kind of strip farming is practiced in which strips of hay or grains are alternated with row crops that leave large amounts of the soil exposed. In some areas of the world, the only way to protect the soil is to not cultivate it at all, but to leave it in a permanent cover of grasses.

Conventional Versus Conservation Tillage

Conventional tillage methods in much of the world require extensive use of farm machinery to prepare the soil for planting and to control weeds. Typically, a field is plowed and then disked or harrowed one to three times before the crop is planted. The plowing, which turns the soil over, has several desirable effects: any weeds or weed seeds are buried, thus reducing the weed problem in the field. Crop residue from previous crops is incorporated into the soil where it will decay faster and contribute to soil structure. Nutrients that had been leached to deeper layers of the soil are brought near the surface. And the dark soil is exposed to the sun so that it warms up faster. This last effect is most critical in areas with short growing seasons. In many areas, fields are plowed in the fall, after the crop has been harvested, and the soil is left exposed all winter.

After plowing, the soil is worked by disks or harrows to break up any clods of earth, kill remaining weeds, and prepare the soil to receive the seeds. After the seeds are planted, there may still be weed problems. Farmers often must cultivate row crops to kill the weeds that begin to grow between the rows. Each trip over the field costs the farmer money, while at the same time increasing the amount of time the soil is exposed to wind or water erosion.

In recent years, several new systems of tillage have developed, as innovations in chemical herbicides and farm equipment have taken place. These tilling practices protect the soil by leaving the crop residue on the soil surface, thus reducing the amount of time it is exposed to erosion forces. **Reduced tillage** is a method that uses less cultivation to control weeds and to prepare the soil to receive seeds but generally leaves 15 to 30 percent of the soil surface covered with crop residue after planting. **Conservation tillage** methods further reduce the amount of disturbance to the soil and leave 30 percent or more of the soil surface covered with crop residue following planting. Selective herbicides are used to kill unwanted vegetation prior to planting the new crop and to control weeds afterward. Several variations of conservation tillage are used:

figure 14.18 **Contour Farming** Tilling at right angles to the slope creates a series of ridges that slows the flow of the water and prevents soil erosion. This soil conservation practice is useful on gentle slopes.

figure 14.19 **Strip Farming** On rolling land, a combination of contour and strip farming prevents excessive soil erosion. The strips are planted at right angles to the slope, with bands of closely sown crops, such as wheat or hay, alternating with bands of row crops, such as corn or soybeans.

1. Mulch tillage involves tilling the entire surface just prior to planting or as planting is occurring.
2. Strip tillage is a method that involves tilling only in the narrow strip that is to receive the seeds. The rest of the

a.

b.

figure 14.20 **Terraces** Since the construction of terraces requires the movement of soil and the protection of the steep slope between levels, terraces are expensive to build. (*a*) The terraces seen here are extremely important for people who live in countries that have little flatland available. They require much energy and hand labor to maintain but make effective agricultural use of the land without serious erosion. (*b*) This modification of the terracing concept allows the use of the large farm machines typical of farming practices in Canada, Europe, and the United States.

a.

b.

figure 14.21 **Protection of Waterways Prevents Erosion** (*a*) An unprotected waterway has been converted into a gully. (*b*) A well-maintained waterway is not cultivated; a strip of grass retards the flow of water and protects the underlying soil from erosion.

soil and the crop residue from the previous crop is left undisturbed.

3. Ridge tillage involves leaving a ridge with the last cultivation of the previous year and planting the crop on the ridge with residue left between the ridges. The crop may be cultivated during the year to reduce weeds.
4. No-till farming involves special planters that place the seeds in slits cut in the soil that still has on its surface the crop residue from the previous crop.

Both reduced tillage and conservation tillage methods reduce the amount of time and fuel needed by the farmer to produce the crop and, therefore, represent an economic savings. By 2000, over half of the cropland in the United States was being farmed using reduced or conservation tillage methods. (See table 14.2.) For many kinds of crops, yields are comparable to that produced by conventional tillage methods.

Other positive effects of reduced tillage, in addition to reducing erosion, are:

1. The amount of winter food and cover available for wildlife increases, which can lead to increased wildlife populations.
2. Since there is less runoff, siltation in streams and rivers is reduced. This results in clearer water for recreation and less dredging to keep waterways open for shipping.
3. Row crops can be planted on hilly land that cannot be converted to such crops under conventional tilling methods. This allows a farmer to convert low-value pasture land

a.

b.

figure 14.22 **Windbreaks** (*a*) In sections of the Great Plains, trees provide protection from wind erosion. The trees along the road protect the land from the prevailing winds. (*b*) In this field, temporary strips of vegetation serve as windbreaks.

Table 14.2 Comparison of Various Tillage Methods

Tillage Method	Fuel Use (liters/hectare)	Time Involved (hours/hectare)
Conventional Plowing	8.08	3.00
Reduced Tillage	5.11	2.20
Mulch Tillage	4.64	2.07
Ridge Tillage	4.12	2.25
No Tillage	2.19	1.21

Source: Data from University of Nebraska, Institute of Agriculture and Natural Resources 1997.

into cropland that gives a greater economic yield.

4. Since fewer trips are made over the field, petroleum is saved, even if the petrochemical feedstocks necessary to produce the herbicides are taken into account.
5. Two crops may be grown on a field in areas that had been restricted to growing one crop per field per year. In some areas, immediately after harvesting wheat, farmers have planted soybeans directly in the wheat stubble.
6. Because conservation tillage reduces the number of trips made over the field by farm machinery, the soil does not become compacted as quickly.

However, there are also some drawbacks to conservation tillage methods:

1. The residue from previous vegetation may delay the warming of the soil, which may, in turn, delay planting some crops for several days.
2. The crop residue reduces evaporation from the soil and the upward movement of water and soil nutrients from deeper layers of the soil, which may retard the growth of plants.
3. The accumulation of plant residue can harbor plant pests and diseases that will require more insecticides and fungicides. This is particularly true if the same crop is planted repeatedly in the same location.

Conservation tillage is not the complete answer to soil erosion problems but may be useful in reducing soil erosion on well-drained soils. It also requires that farmers pay close attention to the condition of the soil and the pests to be dealt with.

In 1972, 12 million hectares (30 million acres) in the United States were under some form of conservation tillage. About 1.3 million hectares (3.2 million acres) were being farmed using no-till methods. By 1992, this had risen to 60 million hectares (150 million acres) of conservation-tilled land, of which 6.2 million hectares (15 million acres) were being farmed using no-till methods. It is estimated that by the year 2010, 95 percent of U.S. cropland will be under some form of reduced-tillage practice.

Protecting Soil on Nonfarm Land

Each piece of land has characteristics, such as soil characteristics, climate, and degree of slope, that influence the way it can be used. When all these factors are taken into consideration, a proper use can be determined for each portion of the planet. Wise planning and careful husbandry of the soil is necessary if the

GLOBAL PERSPECTIVE

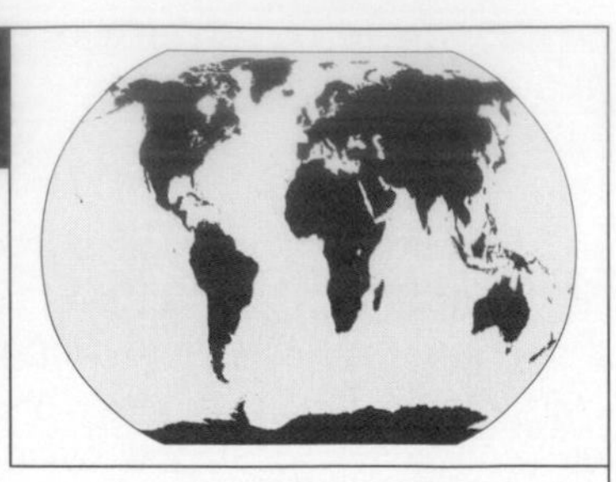

Worldwide Soil Degradation

As humans have used the land and its soil for agriculture, grazing, fuelwood production, and forestry, we have had profound effects on the rates at which soils are formed and lost. Removal of the vegetation has accelerated soil erosion so that most of the world is losing soil faster than it is being formed. The degradation is most severe where the need for food is greatest, since starving people are forced to overexploit their already overused soils. The map and graphs illustrate the magnitude of the soil degradation problem and the relative importance of the causes.

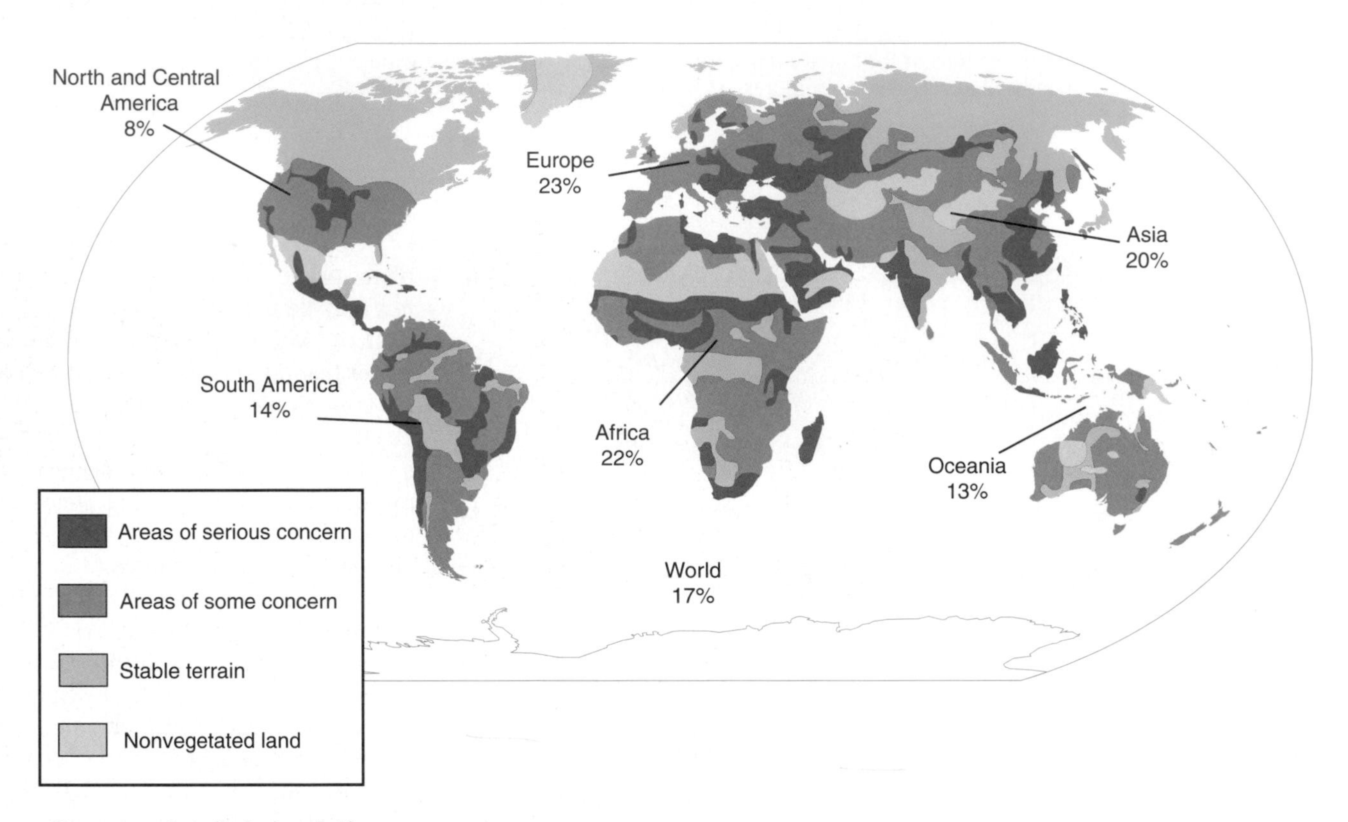

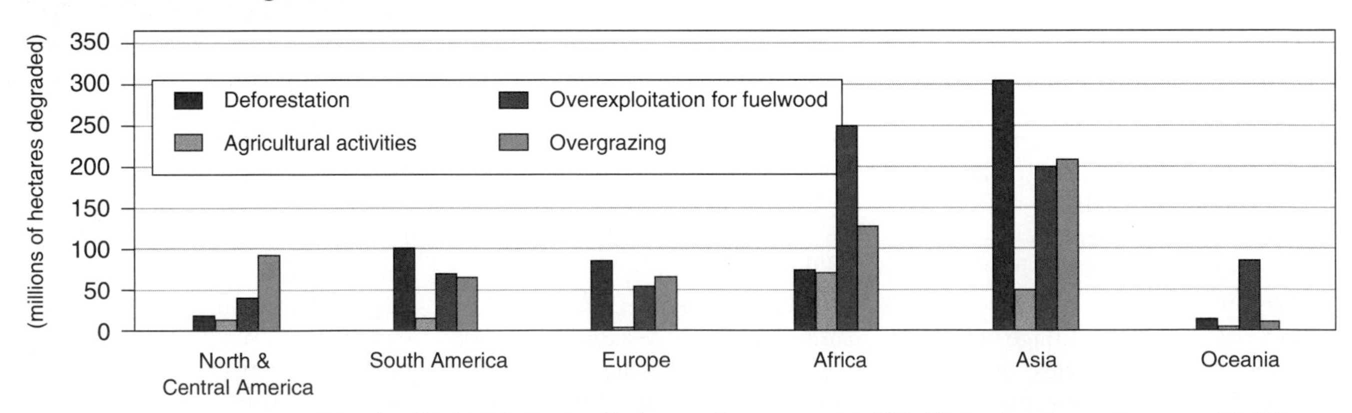

Source: World Resources Institute in collaboration with the United Nations Environment Programme and the United Nations Development Programme, *World Resources 1992–93* (Oxford University Press, New York, 1992), Table 19.4, p. 290.

land and its soil are to provide food and other necessities of life. Not all land is suitable for crops or continuous cultivation. Some has highly erodible soil and must never be plowed for use as cropland, but it can still serve other useful functions. By using appropriate soil conservation practices, much of the land not usable for crops can be used for grazing, wood production, wildlife production, or scenic and recreational purposes. Figure 14.23 shows land that is not suitable for cultivation because it is too arid, but, if it is used properly, it can provide grazing for cattle or sheep.

The land shown in figure 14.24 is not suitable for either crops or grazing because of the steep slope and thin soil. However, it is still a valuable and productive piece of land, since it can be used to furnish lumber, wildlife habitats, and recreational opportunities.

figure 14.23 **Noncrop Use of Land to Raise Food** As long as it is properly protected and managed, this land can produce food through grazing, but it should never be plowed to plant crops because the topsoil is too shallow and the rainfall is too low.

figure 14.24 **Forest and Recreational Use** Although this land is not capable of producing crops or supporting cattle, it furnishes lumber, a habitat for wildlife, and recreational opportunities.

ENVIRONMENTAL CLOSE-UP

Land Capability Classes

Not all land is suitable for raising crops, or urban building. Such factors as the degree of slope, soil characteristics, rockiness, erodibility, and other characteristics determine the best use for a parcel of land. In an attempt to encourage people to use land wisely, the U.S. Soil Conservation Service has established a system to classify land-use possibilities. The table shows the eight classes of land and lists the characteristics and capabilities of each.

Unfortunately, many of our homes and industries are located on type I and II land, which has the least restrictions on agricultural use. This does not make the best use of the land. Zoning laws and land-use management plans should consider the land-use capabilities and institute measures to assure that land will be used to its best potential.

- Can you provide examples from your community where you feel there was not proper land use?
- In your opinion, should there be more or less land-use planning?
- What are some of the consequences of building or farming on land that is not suitable?

	Land Class	Characteristics	Capability	Special Conservation Measures
Land suitable for cultivation	I	Excellent, flat, well-drained land	Cropland	Normal good practices adequate
	II	Good land; has minor limitations, such as slope, sandy soil, or poor drainage	Cropland Pasture	Strip cropping Contour farming
	III	Moderately good land with important limitations of soil, slope, or drainage	Cropland Pasture Watershed	Contour farming Strip cropping Terraces Waterways
	IV	Fair land with severe limitations of soil, slope, or drainage	Pasture Orchards Urban Industry Limited cropland	Crops on a limited basis Contour farming Strip cropping Terraces Waterways
Land not suitable for cultivation	V	Use for grazing and forestry; slightly limited by rockiness, shallow soil, or wetness	Grazing Forestry Watershed Urban Industry	No special precautions if properly grazed or logged; must not be plowed
	VI	Moderate limitations for grazing and forestry because of moderately steep slopes	Grazing Forestry Watershed Urban Industry	Grazing or logging may be limited at times
	VII	Severe limitations for grazing and forestry because of very steep slopes vulnerable to erosion	Grazing Forestry Watershed Recreation Wildlife Urban Industry	Careful management is required when used for grazing or logging
	VIII	Unsuitable for grazing and forestry because of steep slope, shallow soil, lack of water, or too much water	Watershed Recreation Wildlife Urban Industry	Not to be used for grazing or logging; steep slope and lack of soil present problems

ISSUES & ANALYSIS

Soil Erosion in Virginia

In a court case in Virginia, a city attorney filed suit for damages against a land developer. The city contended that the developer was responsible for property damages as a result of erosion from his construction site. During the trial, it was shown that the developer had constructed his subdivision without considering the possible effects of soil erosion. As a result, during a heavy rain, mud from his project had covered the yards, sidewalks, and drives of homes located nearby. As much as 10 meters (30 feet) of soil had eroded from the exposed land of his development and been carried downhill, covering portions of the adjacent community.

Speaking to the developer, the city attorney stated: "After gambling and principally winning over a period of fifteen months, it is apparent that you are attempting to go with your winning streak; and flood, storms, and weather hold no terror for you." The city claimed that the developer had acted in a negligent manner and that he was, therefore, responsible for the damages caused by the deposited mud and should be ordered to pay for the cleanup. The developer claimed no responsibility. His position was that grading large tracts of land was a common construction procedure and that the weather was "an act of God" for which he could not be held accountable.

- If you were a member of the jury hearing the case, would you think the developer acted in a negligent manner by leaving the soil exposed for a long period of time?
- Do you believe that property owners can hold a person responsible for actions taken in an area some distance from their property?
- Do you believe that the erosion and subsequent deposit of mud was "an act of God"?
- Would you find the developer guilty or not guilty? Why?

Summary

The surface of the Earth is in constant flux. The movement of tectonic plates results in the formation of new land as old land is worn down by erosive activity. Soil is an organized mixture of minerals, organic material, living organisms, air, and water. Soil formation begins with the breakdown of the parent material by such physical processes as changes in temperature, freezing and thawing, and movement of particles by glaciers, flowing water, or wind. Oxidation and hydrolysis can chemically alter the parent material. Organisms also affect soil building by burrowing into and mixing the soil, by releasing nutrients, and by their decomposition.

Topsoil contains a mixture of humus and inorganic material, both of which supply soil nutrients. The ability of soil to grow crops is determined by the inorganic matter, organic matter, water, and air spaces in the soil. The mineral portion of the soil consists of various mixtures of sand, silt, and clay particles.

A soil profile typically consists of the *O* horizon of litter; the *A* horizon, which is rich in organic matter; an *E* horizon from which materials have been leached; the *B* horizon, which accumulates materials leached from above; and the *C* horizon, which consists of slightly altered parent material. Forest soils typically have a shallow *A* horizon and an *E* horizon and a deep, nutrient-rich *B* horizon with much root development. Grassland soils usually have a thick *A* horizon containing most of the roots of the grasses. They lack an *E* horizon. There are few nutrients in the thin *B* horizon.

Soil erosion is the removal and transportation of soil by water or wind. Proper use of such conservation practices as contour farming, strip farming, terracing, waterways, windbreaks, and conservation tillage can reduce soil erosion. Misuse reduces the soil's fertility and causes air- and water-quality problems. Land unsuitable for crops may be used for grazing, lumber, wildlife habitats, or recreation.

Interactive Exploration

Check out the website at

http://www.mhhe.com/environmentalscience

and click on the cover of this textbook for interactive versions of the following:

KNOW THE BASICS

chemical weathering *294*
conservation tillage *305*
contour farming *303*
crust *292*
erosion *294*
friable *296*
horizon *297*
humus *295*
land *294*
leaching *298*
lithosphere *292*
litter *297*
loam *296*
mantle *292*
mechanical weathering *293*
parent material *294*
plate tectonics *292*
reduced tillage *305*
soil *294*
soil profile *297*
soil structure *296*
soil texture *296*
strip farming *304*
terraces *304*
waterways *304*
weathering *293*
windbreaks *304*

- **On-line Flashcards**
- **Electronic Glossary**

IN THE REAL WORLD

What role does a prairie wetland have in reducing soil erosion? The story of wetlands in one of the most productive agricultural areas in the world is told in **The Prairie Wetlands of Southwest Minnesota.**

PUT IT IN MOTION

You've heard it before but *see* it this time along with a great explanation of the process. Check out the Plate Tectonics animation to see *and* hear about the process, and for an explanation of why different animals are found only in certain parts of the world.

TEST PREPARATION

Review Questions

1. How are soil and land different?
2. Name the five major components of soil.
3. Describe the process of soil formation.
4. Name five physical and chemical processes that break parent material into smaller pieces.
5. In addition to fertility, what other characteristics determine the usefulness of soil?
6. How does soil particle size affect texture and drainage?
7. Describe a soil profile.
8. Define erosion.
9. Describe three soil conservation practices that help to reduce soil erosion.
10. Besides cropland, what are other possible uses of soil?

Critical Thinking Questions

1. Minimum tillage soil conservation often uses greater amounts of herbicides to control weeds. What do you think about this practice? Why?
2. As populations grow, should we try to bring more land into food production, or should we use technology to aid in producing more food on the land we already have in production of food? What are the tradeoffs?
3. Given what you know about soil formation, how might you explain the presence of a thick *A* horizon in soils in the North American Midwest?
4. Why should non-farmers be interested in soil conservation?
5. Imagine that you are a scientist hired to consult on a project to evaluate land-use practices at the edge of a small city. The area in question has deep ravines and hills. What kinds of agricultural, commercial, and logging practices would you recommend in this area to help preserve the environment?
6. Look at your own community. Can you see examples of improper land use (urban or rural)? What are the consequences of these land-use practices? What recommendations would you make to improve land use?

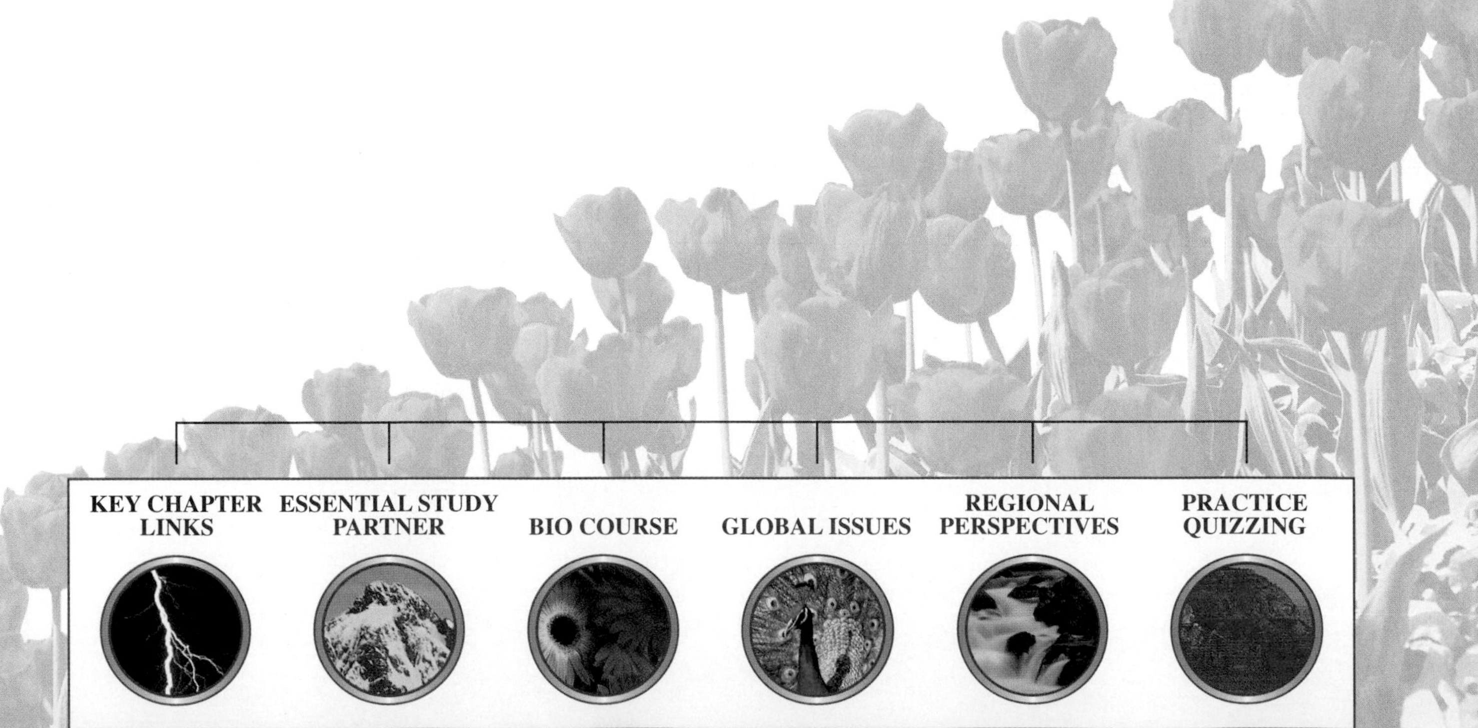

CHAPTER

Agricultural Methods and Pest Management

Objectives

After reading this chapter, you should be able to:

- Explain how the invention of new farm machinery encouraged monoculture farming.
- List the advantages and disadvantages of monoculture farming.
- Explain why chemical fertilizers are used.
- Understand how fertilizers alter soil characteristics.
- Explain why modern agriculture makes extensive use of pesticides.
- Differentiate between persistent pesticides and nonpersistent pesticides.
- Explain how chemicals can be used to delay or accelerate the harvesting of a crop.
- List four problems associated with pesticide use.
- Define biomagnification.
- Define organic farming.
- Explain why integrated pest management depends upon a complete knowledge of the pest's life history.

Chapter Outline

Different Approaches to Agricultural

Fossil Fuel Versus Muscle Power

The Impact of Fertilizer

Agricultural Chemical Use

- Insecticides

Environmental Close-Up: *Regulation of Pesticides*

- Herbicides

Environmental Close-Up: *A New Generation of Insecticides*

- Fungicides and Rodenticides
- Other Agricultural Chemicals

Problems with Pesticide Use

- Persistence
- Bioaccumulation and Biomagnification

Environmental Close-Up: *Politics and the Control of Ethylene Dibromide (EDB)*

- Pesticide Resistance
- Effects on Nontarget Organisms
- Human Health Concerns

Global Perspective: *China's Ravenous Appetite*

Global Perspective: *Contaminated Soils in the Former Soviet Union*

Why Are Pesticides So Widely Used?

Alternatives to Conventional Agriculture

- Techniques for Protecting Soil and Water Resources

Environmental Close-Up: *Food Additives*

- Integrated Pest Management

Issues & Analysis: *Herring Gulls As Indicators of Contamination in the Great Lakes*

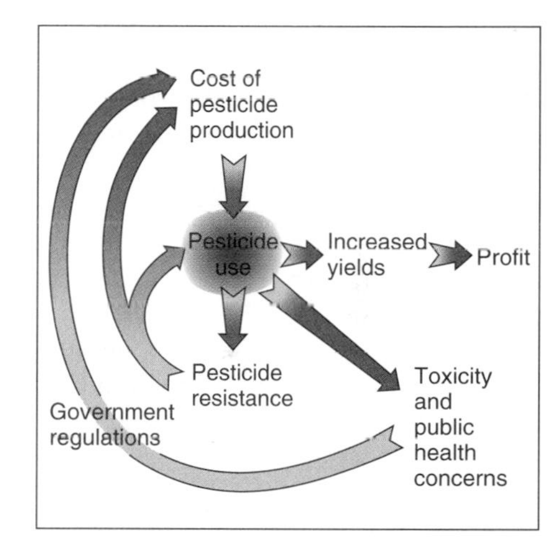

Different Approaches to Agriculture

Around the world, people use a variety of agricultural methods to grow food. In some parts of the world with poor soil and low populations, a form of shifting agriculture known as "slash-and-burn" can be used successfully. This method involves cutting down the trees and burning the trees and other vegetation in a small area of the forest. (See figure 15.1.) The burning releases nutrients that are tied up in the biomass and allows a crop or two to be raised before the soil is exhausted. Once the soil is no longer suitable for raising crops (within two or three years), the site is abandoned. The surrounding forest recolonizes the area, which will return over time to the original forest through the process of succession. This method is particularly useful on thin, nutrient-poor tropical soils and on steep slopes. The small size of the openings in the forest and their temporary existence prevent widespread damage to the soil, and erosion is minimized. While this system of agriculture is successful when human population densities are low, it is not suitable for large, densely populated areas. When populations become too large, the size and number of the garden plots increase and the time between successive uses of the same plot of land decreases. When a large amount of the forest is disturbed and the time between successive uses is decreased, the forest cannot return and repair the damage done by the previous use of the land, and the nature of the forest is changed.

figure 15.1 **Slash-and-Burn Agriculture** In many areas of the world where the soils are poor and human populations are low, crops can be raised by disturbing small parts of the ecosystem followed by several years of recovery. The burning of vegetation releases nutrients that can be used by crops for one or two years before the soil is exhausted. The return of the natural vegetation prevents erosion and repairs the damage done by temporary agricultural use.

The traditional practices of the people who engage in small-scale, shifting agriculture have been developed over hundreds of years and often are more effective for their local conditions than other methods of gardening. Typically, these gardens are planted with a mixture of plants; a system known as **polyculture.** Mixing plants together in a garden often is beneficial, since shade-requiring species may be helped by taller plants, or nitrogen-fixing legumes may provide needed nutrients for species that have a nitrogen requirement. In addition, mixing species may reduce insect pest problems because some plants produce molecules that are natural insect repellents. The small, isolated, temporary nature of the gardens also reduces the likelihood of insect plagues. While today we see this form of agriculture practiced most commonly in tropical areas, it is important to note that many Native American cultures used shifting agriculture and polyculture in temperate areas.

In many areas of the world with better soils, more intense forms of agriculture developed. These usually involved a great deal of manual labor. This style of agriculture is still practiced in much of the world today. Three situations favor this kind of farming: (1) when the growing site does not allow for mechanization, (2) when the kind of crop does not allow it, and (3) when the economic condition of the people does not allow them to purchase the tools and machines used for mechanized agriculture. Crops or terrain that require that fields be small discourage mechanization, since large tractors and other machines cannot be used efficiently on small, oddly shaped fields. Many mountainous areas of the world fit into this category. In addition, some crops require such careful handling in planting, weeding, or harvesting that large amounts of hand labor are required. The planting of paddy rice and the harvesting of many fruits and vegetables are examples.

However, the primary reason for labor-intensive farming is economic. Many densely populated countries have numerous small farms that can be effectively managed with human labor, supplemented by that of draft animals and a few small gasoline-powered engines. (See figure 15.2.) In addition, in the less-developed regions of the world, the cost of labor is low, which encourages the use of hand labor rather than relatively expensive machines to do planting, weeding, and other activities. Mechanization requires large tracts of land that could be accumulated only by the expenditure of large amounts of money or the development of larger cooperative farms from many small units. Even if social and political obstacles to such large landholdings could be overcome, there is still the problem of obtaining the necessary capital to purchase the machines. Large parts of the developing world fit into this category, including much of Africa, many areas in Central and South America, and many areas in Asia. In countries like China and India with a combined population of over two billion people, about 70 percent of the population is rural. Many of these people are engaged in nonmechanized agriculture.

figure 15.2 **Labor-Intensive Agriculture** In many of the less-developed countries of the world the extensive use of hand labor allows for impressive rates of production with a minimal input of fossil fuels and fertilizers. This kind of agriculture is also necessary in areas that have only small patches of land suitable for farming.

figure 15.3 **Monoculture** This wheat field is an example of monoculture, a kind of agriculture that is highly mechanized and requires large fields for the efficient use of machinery. In this kind of agriculture, machines and fossil-fuel energy have replaced the energy of humans and draft animals.

Mechanized agriculture is typical of North America, much of Europe, the republics of the former USSR, parts of South America, and other parts of the world where money and land are available to support this form of agriculture. In large measure, machines and fossil-fuel energy replace the energy formerly supplied by human and animal muscles. Mechanization requires large expanses of fairly level land for the machines to operate effectively. In addition, large tracts of land must be planted in the same crop, a practice known as **monoculture,** for efficient planting, cultivating, and harvesting. (See figure 15.3.) Small sections of land with many kinds of crops require many changes of farm machinery, which takes time. Also, many crops interspersed with one another reduces the efficiency of farming operations because farmers must skip parts of the field, which increases travel time and uses expensive fuel.

Even though mechanized monoculture is an efficient method of producing food, it is not without serious drawbacks. When large tracts of land are prepared for planting, they are often left uncovered by vegetation, and soil erosion increases. Because of problems with erosion, many farmers are now using methods that reduce the time the fields are left bare. (See chapter 14.)

Traditionally, mechanized farming has removed much of the organic matter each year, when the crop was harvested. This tended to reduce the soil organic matter. As agricultural scientists and farmers have recognized the need to improve the organic-matter content of soils, farmers have been leaving increased amounts of organic matter after harvest, or they specifically plant a crop that is later plowed under to increase the soil's organic content.

To ensure that a crop can be planted, tended, and harvested efficiently by machines, farmers rely on hybrid and genet ically modified seed that provides uniform plants with characteristics suitable for mechanized farming. Hybrid seeds can assure that all the plants germinate at the same time, resist the same pests, ripen at the same time, and grow to the same height. These are valuable characteristics to the farmer, but these hybrid plants have little genetic variety. When all the farmers in an area plant the same hybrids, pest control becomes a serious problem. If diseases or pests begin to spread, the magnitude of the problem becomes devastating because all the plants have the same characteristics and, thus, are susceptible to the same diseases. If genetically diverse crops are planted, or crops are rotated from year to year, this problem is not as great.

Because farm equipment is expensive, farmers tend to specialize in a few

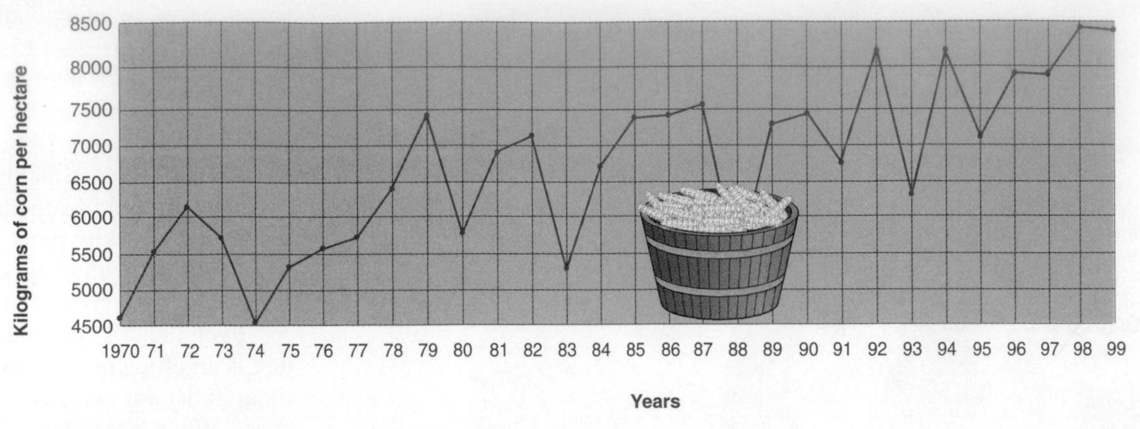

figure 15.4 **Increased Yields Resulting from Modern Technology** The increased yields in the United States and many other parts of the world are the result of a combination of factors, including better genetic qualities in the seed, improved agricultural methods, the application of fertilizers and pesticides, and more efficient machinery.

Source: Data from Department of Agriculture, *Agriculture Statistics*.

crops. This means that the same crop may be planted in the same field several years in a row. This lack of crop rotation may deplete certain essential soil nutrients, thereby requiring special attention to soil chemistry. In addition, planting the same crop repeatedly encourages the growth of insect and fungus pest populations because they have a huge food supply at their disposal. This requires the frequent use of insecticides and fungicides or some other methods of pest control.

Even though there are problems associated with mechanized, monoculture agriculture, it has greatly increased the amount of food available to the world over the past one hundred years. Yields per hectare of land being farmed have increased over much of the world, particularly in the developed world, which includes the United States. (See figure 15.4.) This increase has come about because of improved varieties of crops, irrigation, better farming methods, the use of agricultural chemicals, more efficient machines, and the use of energy-intensive as opposed to labor-intensive technology.

Throughout the 1950s, 1960s, and 1970s, the introduction of new plant varieties and farming methods resulted in increased agricultural production worldwide. This has been called the **Green Revolution.** Both the developed world, which uses highly mechanized farming methods, and the developing world, where labor-intensive farming is typical, have benefited from these advances and food production has increased significantly. However, the population of the world continues to increase and more food is needed.

Fossil Fuel Versus Muscle Power

Mechanized agriculture has substituted the energy stored in petroleum products for the labor of humans. For example, in the United States in 1913, it required 135 hours of labor to produce 2500 kilograms (5,500 pounds) of corn. In 1980, it required only about 15 hours of labor to produce the same amount of corn. The energy supplied by petroleum products replaced the equivalent of 120 hours of labor. Energy is needed for tilling, planting, harvesting, and pumping irrigation water. The manufacture of fertilizer and pesticides also requires the input of large amount of fossil fuels, both as a source of energy for the industrial process and as raw material from which these materials are manufactured. For example, about 5 tonnes (5.5 U.S. tons) of fossil fuel are required to produce about 1 tonne (1.1 U.S. tons) of fertilizer. Since the developed world is dependent on oil to provide energy to manufacture pesticides to support its agriculture, any change in the availability or cost of oil will have a major impact on the world's ability to feed itself.

The Impact of Fertilizer

Various experts estimate that approximately 25 percent of the world's crop yield can be directly attributed to the use of chemical fertilizers. The use of fertilizer has increased significantly over the last few decades and is projected to increase even more (see figure 15.5). However, since fertilizer production relies on energy from fossil fuels, the price and availability of chemical fertilizers is strongly influenced by world energy prices. If the price of oil increases, the price of fertilizer goes up, as does the cost of food. This is felt most acutely in parts of the world where money is in short supply, since the farmers are unable to buy fertilizer, and crop yields fall accordingly.

Fertilizers are valuable because they replace the soil nutrients removed by plants. Some of the chemical building blocks of plants, such as carbon, hydrogen, and oxygen, are easily replaced by carbon dioxide from the air and water from the soil, but others are less easily

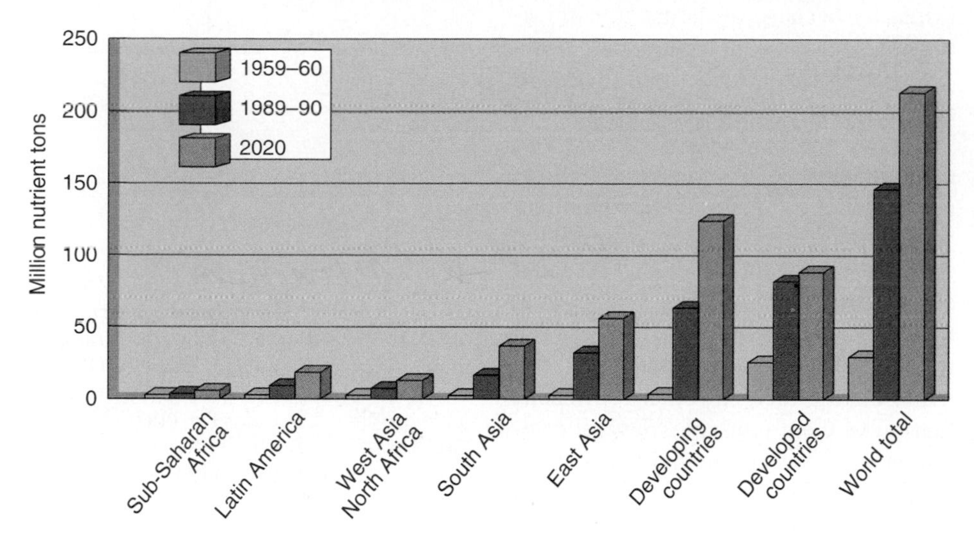

figure 15.5 **Increasing Fertilizer Use** The use of fertilizer is increasing rapidly. Growth in use will be most rapid in the developing countries over the next 20 years.

Source: Balu Bumb and Carlos Baanaunte, *World Trends in Fertilizer Use and Projections to 2020*, 2020 Brief No. 38 (International Food Policy Research Institute, Washington, D.C., 1996), Table 1.

replaced. The three primary soil nutrients often in short supply are nitrogen, phosphorus, and potassium compounds. They are often referred to as **macronutrients** and are the common ingredient of chemical fertilizers. Their replacement is important because, when the crop is harvested, the chemical elements that are a part of the crop are removed from the field. Since many of those elements originated from the soil, they need to be replaced if another crop is to be grown. Certain other elements are necessary in extremely small amounts and are known as **micronutrients.** Examples are boron, zinc, and manganese. As an example of the difference between macronutrients and micronutrients, harvesting a tonne (1.1 U.S. tons) of potatoes removes 10 kilograms (22 pounds) of nitrogen (a macronutrient) but only 13 grams (0.03 pounds) of boron (a micronutrient). When the same crop is grown repeatedly in the same field, certain micronutrients may be depleted, resulting in reduced yields. These necessary elements can be returned to the soil in sufficient amounts by incorporating them into the fertilizer the farmer applies to the field.

Although chemical fertilizers replace inorganic nutrients, they do not replace soil organic matter. Organic material is important because it modifies the structure of the soil, preventing compaction and maintaining pore space, which allows water and air to move to the roots. The decomposition of organic matter produces humus, which helps to maintain proper soil chemistry because it tends to loosely bind many soil nutrients and other molecules and modifies the pH so that nutrients are not released too rapidly. Soil bacteria and other organisms use organic matter as a source of energy. Since these organisms serve as important links in the carbon and nitrogen cycles, the presence of organic matter is important to their function. Thus, total dependency upon chemical fertilizers usually reduces the amount of organic matter and can change the physical, chemical, and biological properties of the soil.

As water moves through the soil, it dissolves soil nutrients (particularly nitrogen compounds) and carries them into streams and lakes, where they may encourage the growth of unwanted plants and algae. This is particularly true when fertilizers are applied at the wrong time of the year, just before a heavy rain, or in such large amounts that the plants cannot efficiently remove them from the soil before they are lost. These ideas are covered in greater detail in chapter 16 on water pollution.

Agricultural Chemical Use

In addition to chemical fertilizers, mechanized monoculture requires large amounts of other agricultural chemicals, such as pesticides, growth regulators, and preservatives. These chemicals have specific scientific names but are usually categorized into broad groups based on their effects. A **pesticide** is any chemical used to kill or control populations of unwanted fungi, animals, or plants, often called **pests.** The term *pest* is not scientific but refers to any organism that is unwanted. Insects that feed on crops are pests, while others, like bees, are beneficial for pollinating plants. Unwanted plants are generally referred to as **weeds.**

Pesticides can be subdivided into several categories based on the kinds of organisms they are used to control. **Insecticides** are used to control insect populations by killing them. Unwanted fungal pests that can weaken plants or destroy fruits are controlled by **fungicides.** Mice and rats are killed by **rodenticides,** and plant pests are controlled by **herbicides.** Since pesticides do not kill just pests but can kill a large variety of living things, including humans, these chemicals might be more appropriately called **biocides.** They kill many kinds of living things. A perfect pesticide is one that kills or inhibits the growth of only the specific pest organism causing a problem. The pest is often referred to as the **target organism.** However, most pesticides are not very specific and kill many **nontarget organisms** as well. For example, most insecticides kill both beneficial and pest species, rodenticides kill other animals as well as rodents, and most herbicides kill a variety of plants, both pests and nonpests.

Many of the older pesticides were very stable and remained active for long periods of time. These are called **persistent pesticides.** Pesticides that break down quickly are called **nonpersistent pesticides.**

Insecticides

If insects are not controlled, they consume a large proportion of the crops produced by farmers. In small garden plots, insects can be controlled by manually removing them and killing them. However, in large fields, this is not practical, so people have sought other ways to control pest insects. In addition, many insects harm humans because they spread diseases, such as sleeping sickness, bubonic plague, and malaria. Mosquitoes are known to carry over 30 diseases harmful to humans. Currently, the World Health Organization estimates that there are between 300 and 500 million new malaria infections per year and that between 1.2 and 2.7 million people die of these infections each year. Most of these deaths are children. Malaria is one of the top five causes of death in children. The discovery of chemicals that could kill insects was celebrated as a major advance in the control of disease and the protection of crops.

Nearly three thousand years ago, the Greek poet Homer mentioned the use of sulfur to control insects. For centuries, it was known that natural plant products could repel or kill insect pests. Plants with insect-repelling abilities were interplanted with crops to help control the pests. Nicotine from tobacco, rotenone from tropical legumes, and pyrethrum from chrysanthemums were extracted and used to control insects. In fact, these compounds are still used today. However, because plant products are difficult to extract and apply and have short-lived effects, other compounds were sought. In 1867, the first synthetic inorganic insecticide, Paris green, was formulated. It was a mixture of acetate and arsenide of copper and was used to control Colorado potato beetles.

The first synthetic organic insecticide to be used was DDT [1,1,1-trichloro-2,2-*bis*-(p-chlorophenyl)ethane]. It was originally thought to be the perfect insecticide. It was long-lasting, relatively harmless to humans, and very deadly to insects. During the first 10 years of its use (1942–1952), DDT is estimated to have saved five million lives, primarily because of its use in controlling disease-carrying mosquitoes. However, after a time, many mosquitoes and other insects became tolerant of DDT. The early success of DDT, and its loss of effectiveness, promoted a search for new synthetic organic compounds that could replace it. Since then, over 60,000 different compounds that have potential as insecticides have been synthesized. However, most of these have never been put into production because of cost, human health effects, or other drawbacks that make them unusable. Several categories of these compounds have been developed. Three that are currently used are chlorinated hydrocarbons, organophosphates, and carbamates.

Chlorinated Hydrocarbons

Chlorinated hydrocarbons are a group of pesticides of complex, stable structure that contain carbon, hydrogen, and chlorine. DDT was the first such pesticide manufactured, but several others have been developed. The chemical structure of DDT is shown in figure 15.6. Other chlorinated hydrocarbons are chlordane, aldrin, heptachlor, dieldrin, and endrin. It is not fully understood how these compounds work, but they are believed to affect the nervous systems of insects, resulting in their death.

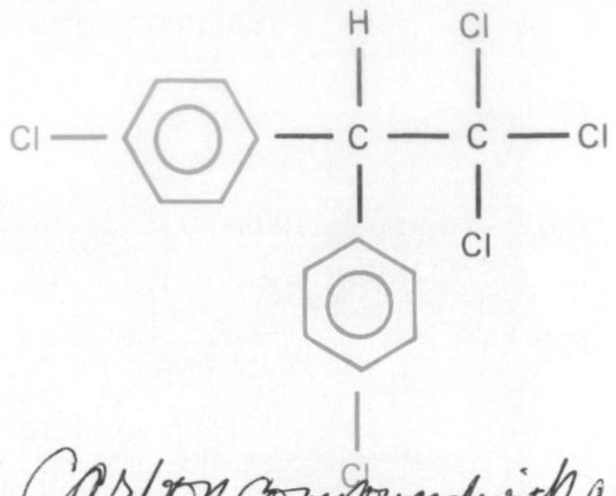

figure 15.6 **The Chemical Structure of DDT** This diagram shows the arrangement of the atoms in a molecule of DDT. The two chlorophenyl portions of the molecule are shown in blue. The other chlorines are shown in red, and the remainder of the ethane molecule appears in black.

One of the major characteristics of these pesticides is that they are very stable chemical compounds. This is both an advantage and a disadvantage. They can be applied once and be effective for a long time. However, since they do not break down easily, they tend to accumulate in the soil and in the bodies of animals in the food chain. Thus, they affect many nontarget organisms, not just the original target insects. In temperate regions of the world, DDT has a half-life (the amount of time required for half of the chemical to decompose) of 10 to 15 years. This means that if 1000 kilograms of DDT were sprayed over an area, 500 kilograms would still be present in the area 10 to 15 years later; 30 years from the date of application, 250 kilograms would still be present. The half-life of DDT varies depending on soil type, temperature, the kinds of soil organisms present, and other factors. In tropical parts of the world, the half-life may be as short as six months. An additional complication is that persistent pesticides may break down into products that are still harmful.

Because of their negative effects, most of the chlorinated hydrocarbons are no longer used in many parts of the world. DDT, aldrin, dieldrin, toxaphene, chlordane, and heptachlor have been banned in the United States and many other developed countries. However, many developing countries still use chlorinated hydrocarbons for insect control to protect crops and public health. Because of their persistence and continued use in many parts of the world, chlorinated hydrocarbons are still present in the food chain although the level of contamination has dropped. These molecules continue to enter parts of the world—where their use as been banned—through the atmosphere and as trace contaminants of imported products.

In 1999, the World Wildlife Fund asked that DDT be banned from use throughout the world because it was showing up worldwide in the bodies of many kinds of animals that were great distances from sources of DDT and was also present in the breast milk of women where DDT is used. This resulted in a great deal of protest from public health

ENVIRONMENTAL CLOSE-UP

Regulation of Pesticides

The Environmental Protection Agency (EPA) is charged by Congress to protect the nation's land, air, and water systems. Under a mandate of national environmental laws focused on air and water quality, solid waste management, and the control of toxic substances, pesticides, noise, and radiation, the agency strives to formulate and implement actions that lead to a compatible balance between human activities and the ability of natural systems to support and nurture life.

To fulfill this mandate, the EPA is charged with the enforcement of various federal laws and acts. One such act governs the registration of pesticides. Section 3 (c) (1) of the Federal Insecticide, Fungicide, and Rodenticide Act (FIFRA) states:

Procedure for Regulation

1. *Statement required. Each applicant for registration of a pesticide shall file with the Administrator a statement which includes:*
   - A. *the name and address of the applicant and of any other person whose name will appear on the labeling;*
   - B. *the name of the pesticide;*
   - C. *a complete copy of the labeling of the pesticide, a statement of all claims to be made for it, and any directions for its use;*
   - D. *except as otherwise provided in subsection (c) (2) (D) of this section, if requested by the Administrator, a full description of the tests made and the results thereof upon which the claims are based, or alternately, a citation to data that appears in the public literature or that previously had been submitted to the Administrator and that the Administrator may consider in accordance with the following provisions;*
   - E. *the complete formula of the pesticide; and*
   - F. *a request that the pesticide be classified for general use, or restricted use, or for both.*

If a corporation wants to manufacture and market a pesticide within the United States, the pesticide must not adversely affect the environment. There is no specific format to test a product's effects on the environment, but some of the tests conducted include the following:

1. **Degradation**—A determination of the physical and chemical methods of decay and an identification of the decay products.
2. **Metabolism**—A determination of the organisms that act on the pesticide and the metabolic products released.
3. **Mobility**—A determination of where the molecules are likely to go and what route they will follow in the ecosystem.
4. **Accumulation**—Do the pesticide molecules accumulate in living tissue? If so, in what form and in what quantities?
5. **Hazard**—Evaluation of the possible hazards to plants, microorganisms, wildlife, and humans, whether target or nontarget organisms.

After receiving the information submitted in the procedure for registration, the Administrator of EPA may register the pesticide for use under the provisions of Section 3 (c) (5) of FIFRA:

Approval of Registration

The Administrator registers a pesticide if he or she determines that, when considered with any restrictions imposed under subsection (d),

1. *its composition is such as to warrant the proposed claims for it;*
2. *its labeling and other material required to be submitted comply with the requirements of this Act;*
3. *it will perform its intended function without unreasonable adverse effects on the environment; and*
4. *when used in accordance with widespread and commonly recognized practice, it will not generally cause unreasonable adverse effects on the environment.*

The Administrator also publishes all applications and supporting data so that various governmental agencies and the public have an opportunity to comment.

The FIFRA is designed to protect the environment from being damaged by the use of pesticides. The law places the burden upon the manufacturer to prove that a pesticide is safe. One source estimates that eight to twelve years and $50 million are needed to bring a major new pesticide from discovery to first registration. These figures do not include the capital cost of constructing a facility to manufacture the pesticide.

---

people concerned about control of malaria, since DDT is still widely used in many parts of the world, where it is sprayed on the walls of houses to kill mosquitoes that spread malaria. The use of DDT is still effective and is much less expensive than materials that would be substituted for it. They fear that the total ban of DDT would result in less effective control of mosquitoes and increased deaths from malaria.

Organophosphates and Carbamates

Because of the problems associated with persistent insecticides, nonpersistent insecticides that decompose to harmless products in a few hours or days were developed. However, like other insecticides, these are not species specific; they kill beneficial insects as well as harmful ones. Although the short half-life prevents the accumulation of toxic material in the environment, it is a disadvantage for farmers, since more frequent applications are required to control pests. This requires more labor and fuel and, therefore, is more expensive.

Both **organophosphates** and **carbamates** work by interfering with the ability of the nervous system to conduct impulses normally. Under normal conditions, a nerve impulse is conducted from one nerve cell to another by means of a chemical known as a neurotransmitter. One of the most common neurotransmitters is acetylcholine. When this chemical is produced at the end of one nerve cell, it causes an impulse to be passed to the next cell, thereby transferring the nerve message. As soon as this transfer is completed, an enzyme known as cholinesterase destroys acetylcholine, so the second nerve cell in the chain is stimulated for only a short time. Organophosphates and carbamates interfere with cholinesterase, preventing it from destroying acetylcholine. This results in nerve cells being continuously stimulated, causing uncontrolled spasms of nervous activity and uncoordination that result in death.

Although these pesticides are less persistent in the environment than are chlorinated hydrocarbons, they are generally much more toxic to humans and other vertebrates because these insecticides affect their nerve cells as well. Persons who use such pesticides must use special equipment and should receive special training because improper use can result in death. Since organophosphates interfere with cholinesterase more strongly than do carbamates, they are considered more dangerous and, for many applications, have been replaced by carbamates.

Common organophosphates are malathion, parathion, and diazinon. Malathion is widely used for such projects as mosquito control, but parathion is a restricted organophosphate because of its high toxicity to humans. Diazinon is widely used in gardens. Sevin, aldicarb, and propoxur are examples of carbamates.

figure 15.7 **The Effect of Herbicides** The grasses in this photograph have been treated with herbicides. The soybeans are unaffected and grow better without competition from grasses.

Herbicides

Herbicides are another major class of chemical control agents. In fact, about 60 percent of the approximately 440 million kilograms (about 1 billion pounds) of pesticides used in U.S. agriculture are herbicides. They are widely used to control unwanted vegetation along power-line rights-of-way, railroad rights-of-way, and highways, as well as on lawns and cropland, where they are commonly referred to as weed killers.

Weeds are plants we do not want to have growing in a particular place. Weed control is extremely important for agriculture since weeds take nutrients and water from the soil, making them unavailable to the crop species. In addition, weeds may shade the crop species and prevent it from getting the sunlight it needs for rapid growth. At harvest time, weeds reduce the efficiency of harvesting machines. Also, weeds generally must be sorted from the crop before it can be sold, which adds to the time and expense of harvesting.

Traditionally, farmers have expended much energy trying to control weeds. Initially, weeds were eliminated with manual labor and the hoe. Tilling the soil also helps to control weeds. Once the crop is planted, row crops like corn or sugar beets may be cultivated to remove weeds from between the rows. All of these activities are expensive in time and fuel. Selective use of herbicides can have a tremendous impact on a farmer's profits.

Many of the recently developed herbicides can be very selective if used appropriately. Some are used to kill weed seeds in the soil before the crop is planted, while others are used after the weeds and the crop begin to grow. In some cases, a mixture of herbicides can be used to control several weed species simultaneously. Figure 15.7 shows the effects of using a herbicide that kills grasses but not other kinds of plants.

Several major types of herbicides are in current use. One type is synthetic plant-growth regulators that mimic natural-growth regulators known as **auxins.** Two of the earliest herbicides were of this type: 2,4-dichlorophenoxyacetic acid (2,4-D) and 2,4,5-trichlorophenoxyacetic acid (2,4,5-T). When applied to broadleaf plants, these chemicals disrupt normal growth, causing the death of the plant. 2,4,-D has been in use for about 50 years and is still one of the most widely used herbicides. Many newer herbicides have other methods of action. Some disrupt photosynthetic activity of plants, causing their death. Others inhibit enzymes, precipitate proteins, stop cell division, or destroy cells directly. Depending on the concentration of the herbicide used, some are toxic to all plants, while others are very selective as

ENVIRONMENTAL CLOSE-UP

A New Generation of Insecticides

The perfect insecticide would harm only target insect species, not be toxic to humans, not be persistent, and would break down into harmless materials. Most currently used insecticides were developed by screening a wide variety of compounds and modifying compounds that showed insect-killing properties. Most are highly toxic to many kinds of organisms and must be used with great care. However, a new generation of insecticides is being developed that starts with knowledge of insect biology and develops a compound that interferes with some essential aspect of the life cycle. Some mimic normal hormones and prevent insects from maturing into adults. Therefore, the insects do not reproduce and the populations are greatly reduced.

Other insecticides attack specific chemical processes in cells and cause death. Nicotine is a normal product of some plants like tobacco, which is toxic to insects. It binds to specific receptors on nerve cells, causing them to fire uncontrollably. However, it is not stable in sunlight. By tinkering with the chemical structure of nicotine, a modified molecule, called imidacloprid, was developed. It is more stable, not toxic to mammals, because they do not have as many of the receptors as insects, and very effective against sucking insects.

Another new insecticide works by turning off the energy-producing processes of mitochondria in cells. All cells of higher organisms have mitochondria and would be injured by this insecticide, but a nontoxic form of the molecule was produced that is converted to a toxic form in the bodies of insects but is not converted in mammals. It is highly toxic to birds and some aquatic organisms, however, which would require strict control of its use.

to which species of plants they affect. One such herbicide is diuron. In proper concentrations and when applied at the appropriate time, it can be used to control annual grasses and broadleaf weeds in over 20 different crops. However, at higher concentrations, it kills all vegetation in an area. Fenuron is a herbicide that kills woody plants. In low concentration, it is used to control woody weed plants in cropland. In high concentrations, it is used on noncroplands, such as power-line rights-of-way. (See figure 15.8.)

figure 15.8 **Herbicide Use to Maintain Rights-of-Way** Power-line rights-of-way are commonly maintained by herbicides that kill the woody vegetation, which might grow so tall that it interferes with the power lines.

Fungicides and Rodenticides

Fungus pests can be divided into two categories. Some are natural decomposers of organic material, but when the organic material being destroyed happens to be a crop or other product useful to humans, the fungus is considered a pest. Other fungi are parasites on crop plants; they weaken or kill the plants, thereby reducing the yield. Fungicides are used as fumigants (gases) to protect agricultural products from spoilage, as sprays and dusts to prevent the spread of diseases among plants, and as seed treatments to protect seeds from rotting in the soil before they have a chance to germinate. Methylmercury is often used on seeds to protect them from spoilage prior to germination. However, since methylmercury is extremely toxic to humans, these seeds should never be used for food. To reduce the chance of a mixup, treated seeds are usually dyed a bright color.

Like fungi, rodents are harmful because they destroy food supplies. In addition, they can carry disease and damage crops in the field. In many parts of the world, such as India, the government pays a bounty to people who kill rats, because this is an inexpensive way to protect the food supply. Several kinds of rodenticides have been developed to control rodents. One of the most widely used is warfarin, a chemical that causes internal bleeding in animals that consume it. It is usually incorporated into a food substance so that rodents eat warfarin along with the bait. Because it is effective in all mammals, including humans, it must be used with care to

figure 15.9 **Chemical Loosening of Cherries to Allow Mechanical Harvesting** By using chemicals and machinery, this farmer can rapidly harvest the cherry crop. This practice reduces the amount of labor required to pick the cherries but requires the application of chemicals to loosen the fruit.

prevent nontarget animals from having access to the chemical. As with many kinds of pesticides, some populations of rodents have become tolerant of warfarin, while others avoid baited areas. In many cases, rodent problems can be minimized by building storage buildings that are rodent proof, rather than relying on rodenticides to control them.

Other Agricultural Chemicals

In addition to herbicides, other agrochemicals are used for special applications. For example, a synthetic auxin sprayed on cotton plants prior to harvest causes the leaves to drop off, which facilitates the harvesting process by reducing clogging of the mechanical cotton picker.

NAA (naphthaleneacetic acid) is used by fruit growers to prevent apples from dropping from the trees and being damaged. This chemical can keep the apples on the trees for up to 10 extra days, which allows for a longer harvest period and fewer lost apples.

Under other conditions, it may be valuable to get fruit to fall more easily. Cherry growers use ethephon to promote loosening of the fruit so that the cherries will fall more easily from the tree when shaken by the mechanical harvester. This method lowers the cost of harvesting the fruit. (See figure 15.9.)

Problems with Pesticide Use

A perfect pesticide would have the following characteristics:

1. It would be inexpensive.
2. It would affect only the target organism.
3. It would have a short half-life.
4. It would break down into harmless materials.

However, the perfect pesticide has not been invented. Many of the more recently developed pesticides have fewer drawbacks than the early pesticides, but none are without problems.

Persistence

Although there has been a trend away from using persistent pesticides in North America and much of the developed world, some are still allowed for special purposes, and they are still in common use in other parts of the world. Because of their stability, these chemicals have become a long-term problem. Persistent insecticides become attached to small soil particles, which are easily moved by wind or water to any part of the world. Persistent pesticides and other pollutants have been discovered in the ice of the poles and are present in detectable amounts in the body tissues of animals, including humans, throughout the world. Thus, chemicals originally sprayed to control mosquito in Africa or to protect a sugarcane field in Brazil may be distributed throughout the world.

Bioaccumulation and Biomagnification

A problem associated with persistent chemicals is that they may accumulate in the bodies of animals. If an animal receives small quantities of persistent pesticides or other persistent pollutants in its food and is unable to eliminate them, the concentration within the animal increases. This process of accumulating higher and higher amounts of material within the body of an animal is called **bioaccumulation.** Many of the persistent pesticides and their breakdown products are fat soluble and build up in the fat of animals. When affected animals are eaten by a carnivore, these toxins are further concentrated in the body of the carnivore, causing disease or death, even though lower-trophic-level organisms are not injured. This phenomenon of acquiring increasing levels of a substance in the bodies of higher-trophic-level organisms is known as **biomagnification.**

The well-documented case of DDT is an example of how biomagnification occurs. DDT is not very soluble in water, but dissolves in oil or fatty compounds. When DDT falls on an insect or is consumed by the insect, the DDT is accumulated in the insect's fatty tissue. Large doses kill insects but small doses do not, and their bodies may contain as much as one part per billion of DDT. This is not very much, but it can have a tremendous effect on the animals that feed on the insects.

If an aquatic habitat is sprayed with a small concentration of DDT, or

ENVIRONMENTAL CLOSE-UP

Politics and the Control of Ethylene Dibromide (EDB)

Since the 1940s, ethylene dibromide (EDB) has been used as a fumigant to protect stored grain from insect pests. More recently, it has been used to fumigate milling machinery and kill nematodes in the soil. It has also been used as a fumigant in fruit shipments to prevent the transportation of insect pests from one part of the world to another.

It was originally thought that EDB dissipated and was not a health hazard. However, residues were found in grain and fruits, and were also reported in drinking water from wells in areas where EDB was used as a soil fumigant. In the 1970s, it was shown that EDB causes cancer in rats and mice at levels of twenty parts per million.

In 1977, the U.S. Environmental Protection Agency first sought to control use of EDB, but manufacturers and users effectively fought its prohibition until September 1984, when the EPA succeeded in banning the use of EDB as a soil fumigant.

When the EPA banned the use of EDB, it was required by law to purchase the remaining stocks of the material from the manufacturers and to dispose of them at its own expense. (This feature of the law was lobbied for very strongly by pesticide manufacturers.) In March 1988, the EPA incinerated the remaining approximately 1.3 million liters (345,000 gallons) of the EDB it had acquired when it banned the substance. The EPA had already paid $2.5 million for these stocks and disposing of them properly cost several million more. The requirement that the EPA acquire and dispose of banned pesticides has had a significant impact on the agency's ability to function.

receives DDT from runoff, small aquatic organisms may accumulate a concentration that is up to 250 times greater than the concentration of DDT in the surrounding water. These organisms are eaten by shrimp, clams, and small fish, which are, in turn, eaten by larger fish. DDT concentrations of large fish can be as much as 2000 times the original concentration sprayed on the area. What was a very small initial concentration has now become so high that it could be fatal to animals at higher trophic levels. This has been of particular concern for birds, since DDT interferes with the production of eggshells, making them much more fragile. This problem is more common in carnivorous birds because they are at the top of the food chain. Although all birds of prey have probably been affected to some degree, those that rely on fish for food seem to have been affected most severely. Eagles, osprey, cormorants, and pelicans are particularly susceptible species. (See figure 15.10.)

Other persistent molecules are known to behave in similar fashion. Mercury, aldrin, chlordane, and other chlorinated hydrocarbons, such as polychlorinated biphenyls (PCBs) used as insulators in electric transformers, are all known to accumulate in ecosystems. PCBs have been strongly implicated in the decline of cormorants in the Great Lakes. As PCB levels have declined, the cormorant population has returned to former levels.

Because of their persistence, their effects on organisms at higher trophic levels, and concerns about long-term human health problems, most chlorinated hydrocarbon pesticides have been banned from use in the United States and some other countries. The use of DDT was prohibited in the United States in the early 1970s. Aldrin, dieldrin, heptachlor and chlordane have also been prohibited from use on crops, although heptachlor and chlordane were still used for termite control until recently. In 1987, Velsicol Chemical Corporation agreed to stop selling chlordane in the United States.

The populations of several species of birds, including the brown pelican, bald eagle, osprey, and cormorant, all of which feed primarily on fish, were severely affected because of biomagnification of persistent organic chemicals. With the control or restriction of most persistent pesticides and several other chemicals, the levels of these chemicals in their body tissues has declined and their populations have rebounded. The bald eagle has been removed from the endangered species list and the pelican has been removed from the endangered species list in part of its range.

Pesticide Resistance

Another problem associated with pesticides is the ability of pest populations (insects, weeds, rodents, fungi) to become resistant to them. Not all organisms within a given species are identical. Each individual has a slightly different genetic composition and slightly different characteristics. If an insecticide is used for the first time on the population of a particular insect pest, it kills all the individuals that are susceptible. Individuals with characteristics that allow them to tolerate the insecticides may live to reproduce.

If only 5 percent of the individuals possess genes that make them resistant to an insecticide, the first application of the insecticide will kill 95 percent of the population and so will be of great benefit in controlling the size of the insect population. However, the surviving individuals that are tolerant of the insecticide will constitute the majority of the breeding populations. Since these individuals possess genetic characteristics for tolerating the insecticide, so will many of their offspring. Therefore, in the next generation, the number of individuals able to tolerate the insecticide will increase, and the second use of the insecticide will not be as effective as the first. Since some species of insect pests

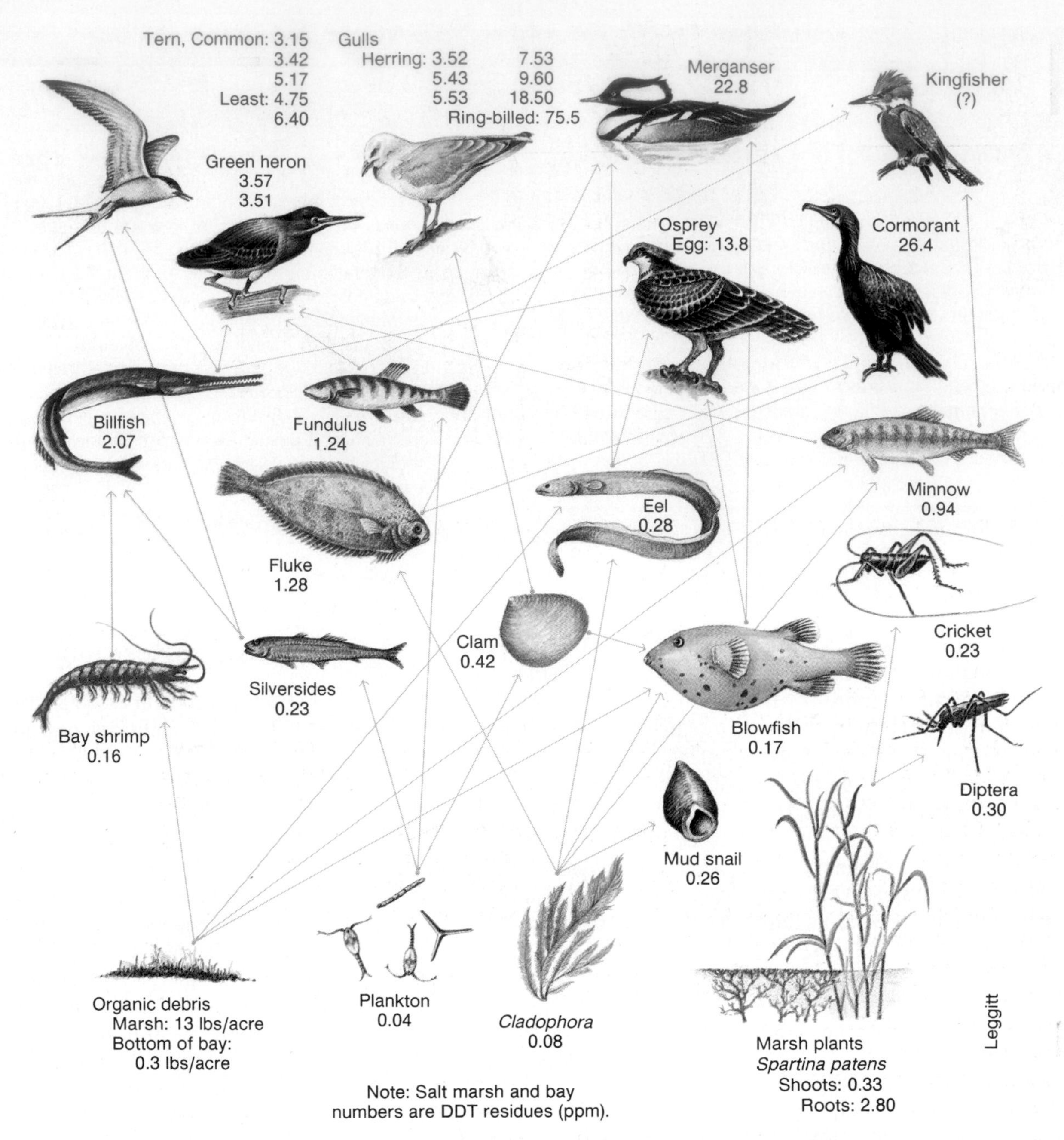

figure 15.10 **Biomagnification** Note how the concentration of DDT increases as it passes through the food chain. At each successive trophic level, the amount of DDT increases because the animals are accumulating the DDT from the bodies of animals they are eating as food.

can produce a new generation each month, this process of selecting individuals capable of tolerating the insecticide can result in resistant populations in which 99 percent of the individuals are able to tolerate the insecticide within five years. As a result, that particular insecticide is no longer as effective in controlling insect pests, and increased dosages or more frequent spraying may be necessary. Figure 15.11 indicates that over 500 species of insects have populations resistant to insecticides.

Cotton growers throughout the world are extremely dependent on the use of insecticides. About 40 percent of the insecticides used in the United States are for controlling pests in cotton. Because of this intensive use, many populations of pest insects have become resistant to the commonly used insecticides. Table 15.1 shows the effect of continual use of insecticides on two pests, the bollworm and the tobacco budworm. The size of the dose necessary to kill the pest increased greatly in a five-year period. In some cases, the

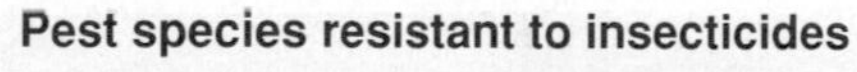

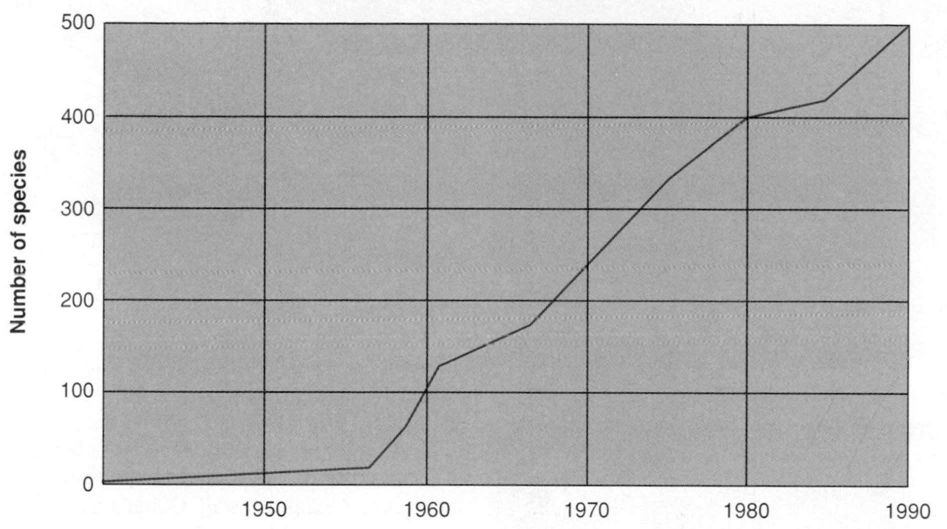

figure 15.11 **Resistance to Insecticides** The continued use of insecticides has constantly selected for genes that give resistance to a particular insecticide. As a result, many species of insects and other arthropods are now resistant to many kinds of insecticides, and the number continues to increase.

Source: Data from George P. Georghiou, University of California at Riverside.

Table 15.1 Average Dose Necessary to Kill Two Cotton Pests

Average Dose Necessary to Kill (milligrams per gram of larva)				
	Bollworm		Tobacco Budworm	
Compound	**1960**	**1965**	**1961**	**1965**
DDT	0.03	1.000+	0.13	16.51
Endrin	0.01	0.13	0.06	12.94
Carbaryl	0.12	0.54	0.30	54.57
Strobane and DDT	0.05	1.04	0.73	11.12
Toxaphene and DDT	0.04	0.46	0.47	3.52

Reprinted with permission from P.L. Adkisson, "Controlling Cotton's Insect Pests: A New System" in *Science*, 216:19–22, April 1982. Copyright © 1982 American Association for the Advancement of Science.

dose increased tenfold and in other a hundredfold or more. By 1965, these insecticides were no longer able to control bollworm or tobacco budworm.

Effects on Nontarget Organisms

Most pesticides are not specific and kill beneficial species as well as pest species. With herbicides, this is usually not a problem because a herbicide is chosen that does not harm the desired crop plant, and generally all other plants are competing pests. However, with insecticides, there are several problems associated with the effects on nontarget organisms. The use of insecticides can harm populations of birds, mammals, and harmless insects. Many insecticides that harm vertebrates are restricted in their use. The insecticide may kill predator and parasitic insects that normally control the pest insects. This allows pest species to increase rapidly following the use of a pesticide because there are no natural checks to their population growth. Additional applications of insecticides are necessary to prevent the pest population from rebounding to levels even higher than the initial one. Once the decision is made to use pesticides, it often becomes an irreversible tactic, because stopping their use would result in rapid increases in pest populations and extensive crop damage.

An associated problem is that the use of insecticides may change the population structures of species present so that a species that was not previously a problem becomes a serious pest. For example, when synthetic organic insecticides came into common use with cotton in the 1940s, the insect parasites and predators were eliminated, and the bollworm and tobacco budworm became major pests. In mid 1990s, a similar situation developed when repeated use of malathion removed predator insects and allowed the populations of beet army worms to become a major pest. The repeated use of insecticides caused a different pest problem to develop.

Human Health Concerns

Short-term and long-term health effects to the persons using the pesticide and the public that consumes the food grown by using pesticides are also concerns. If properly applied, most pesticides can be applied with little danger to the applicator. However, in many cases, people applying pesticides are unaware of how they work and what precautions should be used in their application. In many parts of the world, farmers may not be able to read the caution labels on the packages or do not have access to the protective gear specified for use with the pesticide. Therefore, many incidences of acute poisoning occur each year. In most cases, the symptoms disappear after a period free from exposure. Estimates of the number of poisonings are very difficult to obtain since many go unreported, but, in the United States, pesticide poisonings requiring medical treatment are in the thousands per year. In the developing world, the number of poisonings is in the tens of millions per year. It is clear that some deaths due to misuse of pesticides happen each year.

For most people, however, the most critical health problem related to pesticide use is inadvertent exposure to very

GLOBAL PERSPECTIVE

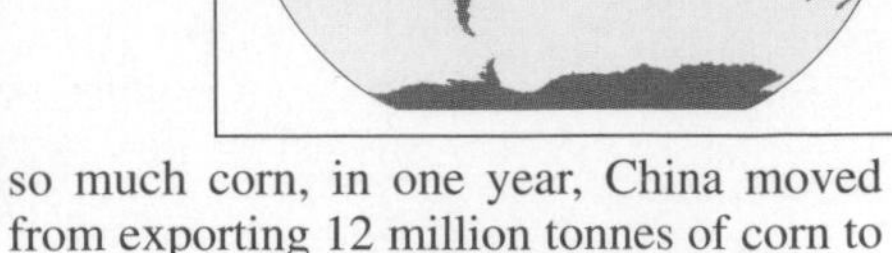

China's Ravenous Appetite

History has shown that when people rise above subsistence level, one of the first things they want is to eat a little better. They tend to leave the basic grains to the animals and eat the animals themselves or animal products in the form of meat, fish, eggs, and butter. They eat more vegetables and use more edible oils. In short, they move a bit higher up the food chain.

A sizable percentage of China's more than 1.2 billion citizens are moving up the food chain. In the early 1980s, the typical urban Chinese diet consisted of rice, porridge, and cabbage. By the middle 1990s, the diet had dramatically changed to include meat, eggs, or fish at least once a day.

Diet is not the only thing changing in China. Even with population control, China's population grows by 11 million people a year, and young people are leaving the countryside and moving to the cities. China's urban population, currently nearly 400 million, should double by 2010. China has 21 percent of the world's population but only 7 percent of the arable land; in other words, it has less arable land than the United States but nearly five times the population. China's farm sector is simply unable to keep up with the surging demand for what people regard as better food.

Total meat consumption in China is growing 10 percent a year; feed for animal consumption is growing 15 percent. Demand for poultry, which requires 2 to 3 kilograms of feed per kilogram of bird, has doubled in five years. China's new diet will make it more dependent on the United States, Canada, and Australia for feed grains. By 1997, China had gone from being a net exporter of grain to importing 16 million tonnes. The switch in corn is even more dramatic. As recently as 1995, China was the second-largest corn exporter in the world. But with the chickens and pigs eating so much corn, in one year, China moved from exporting 12 million tonnes of corn to importing 4 million tonnes.

KFC and McDonald's in Beijing. Demand for poultry and beef is soaring.

China has also become the world's largest importer of fertilizer. In addition, while the government is converting some marginal lands in the north for agriculture, this is more than offset by the loss of fertile, multiple-cropped farmland in the southern coastal provinces. Overall, China's farmland is shrinking by at least 0.5 percent per year.

China also faces a severe water shortage in the arid north. In the Yellow River Valley's large grain belt, irrigation projects tripled crop yields during the 1950s and 1960s, resulting in severe overpumping of groundwater. Today, the water table is falling, aquifers are vanishing, and farmers now have to compete with industry and households for water. The era of the flush toilet has arrived.

China is only a part—but certainly the biggest part—of a larger story. What's true for China is also happening throughout much of Asia. Large, populous countries such as Indonesia, Thailand, and the Philippines are rapidly urbanizing. They are gaining purchasing power and losing farmland, and the people are adding more animal proteins and processed foods to their diets. Given the magnitude of the changes, the world's ability to feed itself in the future is difficult to predict.

- What changes do you anticipate in the food consumption pattern in China in the next decade?
- Is China capable of maintaining a diet similar to that of North Americans? Is such a diet desirable?

small quantities of pesticides. Many pesticides have been proven to cause mutations, produce cancers, or cause abnormal births in experimental animals. There are questions about the effects of chronic, minute exposures to pesticide residues in food or through contamination of the environment.

Although the risks of endangering health by consuming tiny amounts of pesticides are very small compared to other risk factors, such as automobile accidents, smoking, or poor eating habits, many people find pesticides unacceptable and seek to prohibit their sale and use. The U.S. Environmental Protection Agency requires careful studies of the effectiveness and possible side effects of each pesticide licensed for use. It has banned several pesticides from further use because new information suggests that they are not as safe as originally thought. For example, it banned the use of dinoseb, a herbicide, because tests by

GLOBAL PERSPECTIVE

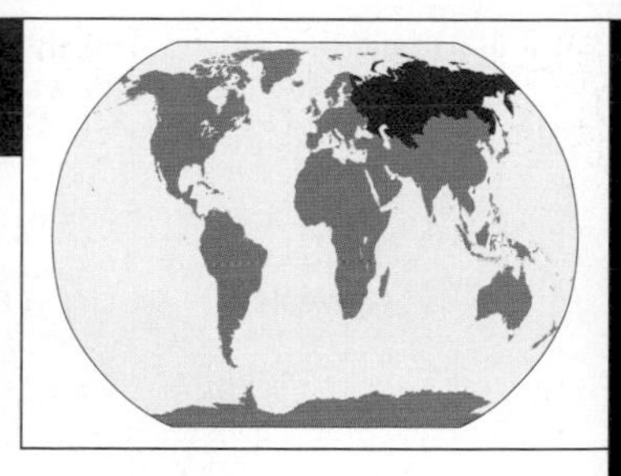

Contaminated Soils in the Former Soviet Union

The newly independent republics of the former Soviet Union have inherited many kinds of environmental problems generated by the failed policies of the Soviet government. One of these policies was a push for increased agricultural production. New land was brought under cultivation, irrigation projects were developed, and extensive use of fertilizer and pesticides stimulated production.

This single-minded approach has resulted in serious environmental damage. In many areas, the soil has been contaminated by excessive fertilizer and pesticide use, poisoning groundwater and tainting food products. It is estimated that 30 percent of the food produced throughout the former Soviet Union is contaminated and should not be eaten. Forty-two percent of all baby food is contaminated by nitrates and pesticides. Latvian ecologists estimate that excessive pesticide use currently results in 14,000 deaths per year and that 700,000 people become ill from pesticides throughout the former Soviet Union annually.

One problem pesticide is DDT. It was officially banned in the Soviet Union in the 1970s, but the drive to improve agricultural production allowed its continued use. Waivers to the ban were allowed, or DDT was simply renamed by government bureaucrats and used under its new label. It is estimated that 10 million hectares (25 million acres) of agricultural land in the former Soviet Union are contaminated with DDT.

The widespread use of agricultural chemicals has also affected the groundwater. Persistent pesticides, nitrates, and heavy metals have entered the groundwater, which many rely on for drinking water. Some areas have cadmium levels three to ten times higher than the amount permitted.

It appears that the fall of Communism in the former Soviet Union did not solve all the problems of the people. These environmental problems are surfacing after years of secrecy and denial under the shortsighted, production-oriented agricultural policy of the failed political system.

the German chemical company Hoechst AG indicated that dinoseb causes birth defects in rabbits. Other studies indicate that dinoseb causes sterility in rats. Similarly, in 1987, Velsicol Chemical Corporation signed an agreement with the U.S. EPA to stop producing and distributing chlordane in the United States. Chlordane had been banned previously for all applications except for termite control. After it was shown that harmful levels of chlordane could exist in treated homes, its use was finally curtailed. Also in 1987, the Dow Chemical Company announced that it would stop producing the controversial herbicide 2,4,5-T. In both of these cases, new pesticides were developed to replace the older types, so that discontinuing them did not result in an economic hardship for farmers or other consumers.

Finally, during the 1990s, the EPA required that many pesticides be reregistered. During this process, manufacturers needed to justify the continued production of the product. In many cases, companies simply did not go through the registration process; therefore, the product will be phased out. In other cases, the variety of uses was modified. For example, chlopyrifos, one of the most widely used insecticides, had its use altered. Through negotiations with the producers, its use in homes and schools will be eliminated while it will still be used in most agricultural settings.

Why Are Pesticides So Widely Used?

Figure 15.12 shows changes in the use of pesticides throughout the world. Use is projected to continue to increase significantly. If pesticides have so many drawbacks, why are they used so extensively? There are three primary reasons. First, the use of pesticides has increased, at least in the short term, the amount of food that can be grown in many parts of the world. In the United States, pests are estimated to consume 33 percent of the crops grown. On a worldwide basis, pests consume approximately 35 percent of crops. This represents an annual loss of $18.2 billion in the United States alone. Farmers, grain-storage operators, and the food industry continually seek to reduce this loss. A retreat from dependence on pesticides would certainly reduce the amount of food produced. Agricultural planners in most countries are not likely to suggest changes in pesticide use that would result in malnutrition and starvation for many of their inhabitants.

The economic value of pesticides is the second reason they are used so extensively. The cost of pesticides is more than offset by increased yields and profits for the farmer. In addition, the production and distribution of pesticides is big business. Companies that have spent millions of dollars developing a pesticide are going to argue very strongly for its continued use. Since farmers and agrochemical interests have a powerful voice in government, they have successfully lobbied for continued use of pesticides.

A third reason for extensive pesticide use is that many health problems are currently impossible to control without insecticides. This is particularly true in areas of the world where insect-borne diseases would cause widespread public health consequences if insecticides were not used.

Alternatives to Conventional Agriculture

Before the invention of synthetic fertilizers, herbicides, fungicides, and other agrochemicals, animal manure and crop rotation provided soil nutrients; a

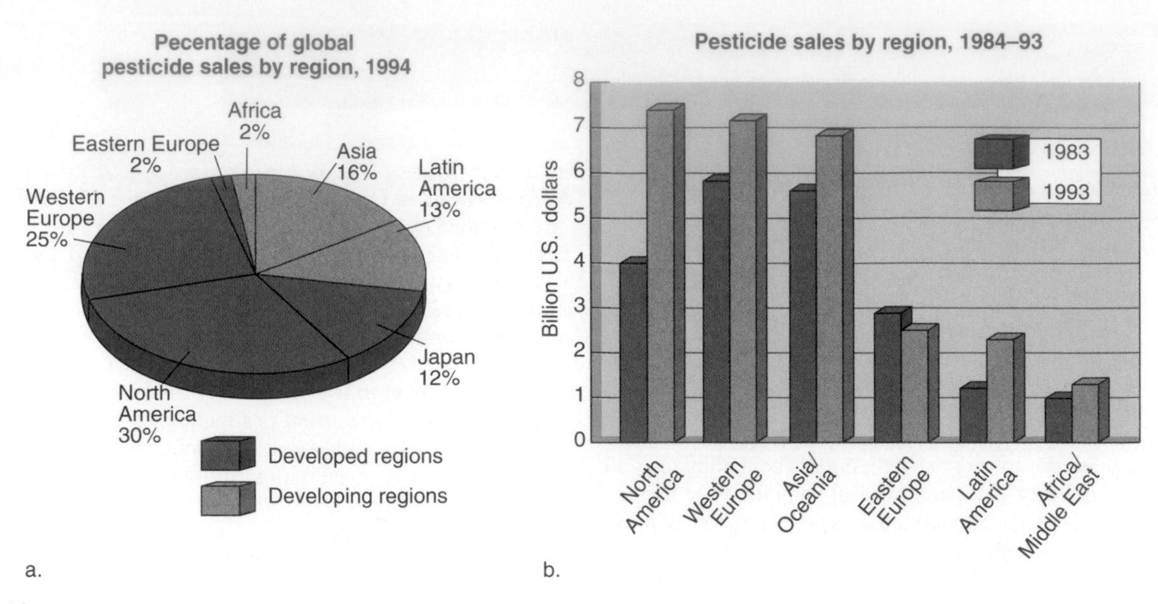

figure 15.12 **Pesticide Sales** The use of pesticides is greatest in the developed world with about 67 percent of the total. However, all parts of the world except for Eastern Europe are experiencing increased use of pesticides.

Source: (*a*) "Upturn in World Agrochemical Sales in 1994," *AGRO: World Crop Protection News,* No. 238 (August 1995), p. 20. (*b*) The Freedonia Group, World Pesticides Report No. 636 (The Freedonia Group, Cleveland, Ohio, 1994) as cited in *AGRO,* No. 225 (1995), p. 16.

mixture of crops prevented regular pest problems; and manual labor killed insects and weeds. With the development of mechanization, larger areas could be farmed, draft animals were no longer needed, and many farmers changed from mixed agriculture, in which animals were an important ingredient, to monoculture. Chemical fertilizers replaced manure as a source of soil nutrients and crop rotation was no longer as important since hay and grain were no longer grown for draft animals and cattle. The larger fields of crops like corn, wheat, and cotton presented opportunities for pest problems to develop and chemical pesticides were used to "solve" this problem.

Today, many people feel current agricultural practices are not sustainable and that we should look for ways to reduce reliance on fertilizer and pesticides, while assuring good yields and controlling the pests that compete with us for the food that we raise. Several alternative approaches have somewhat different but overlapping goals. Some of the terms used to describe these approaches are: *alternative agriculture, sustainable agriculture,* and *organic agriculture.* Alternative agriculture is the broadest term. It includes all nontraditional agricultural methods and encompasses sustainable agriculture, organic agriculture, alternative uses of traditional crops, alternative methods for raising crops, and producing crops for industrial use. The sustainable agriculture movement maintains that current practices are degrading natural resources and seeks methods to produce adequate, safe food in an economically viable manner while enhancing the health of agricultural land and related ecosystems. Organic agriculture advocates avoiding the use of chemical fertilizers and pesticides in the production of food, thus preventing damage to related ecosystems and the food-consuming public. A portion of the proposed definition of organic agriculture by the National Organic Standards Board established by the 1990 Organic Foods Production Act points out a central theme of these three movements.

> Organic agriculture is an ecological production management system that promotes and enhances biodiversity, biological cycles, and soil biological activity. It is based on minimal use of off-farm inputs and management practices that restore, maintain, and enhance ecological harmony.

In order to reduce negative impacts of conventional agriculture while making a profit, farmers must deal with the twin problems of protecting the quality and fertility of the soil while controlling pests of their crops. A wide variety of techniques can be used to reduce the negative impact of conventional agricultural activity while maintaining economic viability for the farmer.

Techniques for Protecting Soil and Water Resources

Conventional farming practices have several negative effects on soil and water. Soil erosion is a problem throughout the world. (See chapter 14.) However, two other problems are also important—compaction of the soil and reduction in soil organic matter. Several changes in agricultural production methods can help to reduce these problems. Reducing the number of times farm equipment travels over the soil will reduce the degree of compaction. Leaving crop residue on the soil and incorporating it into the soil reduces erosion and increases soil organic matter. In addition, the introduction of organic matter into the soil makes compaction less likely.

ENVIRONMENTAL CLOSE-UP

Food Additives

Food additives are chemicals added to food before its sale. They have several purposes:

1. To prolong the storage life of the food.
2. To make the food more attractive by adding color or flavor.
3. To modify nutritive value.

Many kinds of molecules are added to foods to prolong shelf life. Calcium propionate is added to baked goods because they can become contaminated with airborne spores of molds and bacteria, which grow on the food and spoil it. BHT, TBHQ, and other commonly seen alphabetic mixtures have a similar function.

Sometimes additives are used just to make the food appear more attractive. For years, Red Dye II was used to color a variety of foods. Its use was discontinued when it was found to be carcinogenic. Other food colorings are still widely used to increase appeal to the consumer. Many kinds of artificial flavors are added to products as well. Commonly used flavor enhancers are monosodium glutamate, table salt, and citric acid.

Other food additives are used to modify the nutritive value of the food product. Iodine in table salt is a good example. Iodine is a trace element required for proper thyroid functioning. Individuals suffering from a lack of iodine often develop an enlargement of the thyroid gland known as goiter. The addition of iodine to table salt has eliminated goiter in the United States. Most cereals and baked goods and many other products have various vitamins and minerals added to improve their nutritive value. Some additives, like Nutrasweet™, reduce the calories while giving the consumer the sensation of tasting something sweet.

Some additives are an unavoidable residue of some step of the food production industry and could more properly be called contaminants. Pesticide residues are an example. Pesticides are used to grow foods, but they are also used to eliminate pests in the storing, processing, and transportation steps of the food industry.

Diethylstilbestrol (DES) was at one time used in the poultry industry to produce fatter birds. Because there were indications that DES is carcinogenic, the U.S. Food and Drug Administration banned the use of DES in chickens and declared that it was potentially hazardous to humans. Further studies were conducted to determine if DES was safe to use in raising beef. It was used with cattle because an animal gains weight more rapidly with DES in its feed. Because DES was eventually linked to breast cancer in women, it has been banned from all animal feed use in the United States and Europe.

Conventional agriculture has several negative impacts on watersheds: fertilizer runoff stimulates aquatic growth and degrades water resources; pesticides can accumulate in food chains; and groundwater resources can be contaminated by fertilizer, pesticides, or animal waste. Reducing or eliminating these sources of contamination would enhance ecological harmony and reduce a threat to human health. Fertilizer runoff can be lessened by reducing the amount of fertilizer applied and the conditions under which it is applied. Applying fertilizer as plants need it will assure that more of it is taken up by plants and less runs off. Increased organic matter in the soil also tends to reduce runoff. More careful selection, timing, and use of pesticides would decrease the extent to which these materials become environmental contaminants. Precision agriculture is a new technique that addresses many of these concerns. With modern computer technology and geographic information systems, it is now possible, based on the soil and topography, to automatically vary the chemicals applied to the crop at different places within a field. Thus, less fertilizer is used and it is used more effectively.

True organic agriculture which uses neither chemical fertilizers nor pesticides is the most effective in protecting soil and water resources from these forms of pollution but requires several adjustments in the way in which farming is done. Crop rotation is an effective way to enhance soil fertility, reduce erosion, and control pests. The use of nitrogen-fixing legumes, such as clover, alfalfa, beans, or soybeans, in crop rotation increases soil nitrogen, but places other demands on the farmer. For example, it typically requires that cattle be a part of the farmer's operation in order to make use of forage crops and provide organic fertilizer for subsequent crops. Crop rotation also requires a greater investment in farm machinery, since certain crops require specialized equipment. Also, the raising of cattle requires additional expenditures for feed supplements and veterinary care. Critics say that organic farming cannot produce the amount of food required for today's population and that it can be economically successful only in specific cases. Proponents disagree and stress that when the hidden costs of soil erosion and pollution are included, organic agriculture or some modification of it is a viable alternative approach to conventional means of food production. Furthermore, organic farmers are willing to accept lower yields because they do not have to pay for expensive chemical fertilizers and pesticides. In addition, organic farmers often receive premium prices for products that are organically grown. Thus, even with lower yields, they can still make a profit.

Integrated Pest Management

Integrated pest management uses a variety of methods to control pests, rather than relying on pesticides alone. Integrated pest management is a technique that depends on a complete understanding of all ecological aspects of the crop and the particular pests to which it is susceptible to establish pest control strategies. It requires information about the metabolism of the crop plant, the biological interactions between pests and their predators or parasites, the climatic conditions that favor certain pests, and techniques for encouraging beneficial insects. It may involve the selective use of pesticides. Much of the information necessary to make integrated pest management work goes beyond the knowledge of the typical farmer. The metabolic and ecological studies necessary to pinpoint weak points in the life cycles of pests can usually be carried out only at universities or government research institutions. These studies are expensive and must be completed for each kind of pest, since each pest has a unique biology. Once a viable technique has been developed, an educational program is necessary to provide farmers with the information they need in order to use integrated pest management rather than the "spray and save" techniques that they used previously and that pesticide salespersons continually encourage.

figure 15.13 **Beneficial Insect** The ladybird beetle is a predator of many kinds of pest insects, including aphids.

Several methods are employed in integrated pest management. These include sex attractants, male sterilization, the release of natural predators or parasites, the development of resistant crops, the use of natural pesticides, the modification of farming practices, and the selective use of pesticides.

In some species of insects, a chemical called a **pheromone** is released by females to attract males. Males of some species of moths can detect the presence of a female from a distance of up to 3 kilometers (nearly 2 miles). Since many moths are pests, synthetic odors can be used to control them. Spraying an area with the pheromone confuses the males and prevents them from finding females which results in a reduced moth population the following year. In a similar way, a synthetic sex attractant molecule known as Gyplure is used to lure gypsy moths into traps, where they become stuck. Since the females cannot fly and the males are trapped, the reproductive rate drops, and the insect population may be controlled.

Another technique that reduces reproduction is male sterilization. In the southern United States and Central America, the screwworm fly weakens or kills large grazing animals, such as cattle, goats, and deer. The female screwworm fly lays eggs in open wounds on these animals, where the larvae feed. However, it was discovered that the female mates only once in her lifetime. Therefore, the fly population can be controlled by raising and releasing large numbers of sterilized male screwworm flies. Any female that mates with a sterile male fails to produce fertilized eggs and cannot reproduce. In Curaçao, an island 65 kilometers (40 miles) north of Venezuela, a program of introducing sterile male screwworm flies eliminated this disease from the 25,000 goats on the island. In parts of the southwestern United States, the sterile male technique has also been very effective. The screwworm fly has been eliminated from the United States and northern Mexico, and much of Central America may become free of them as well. In 1990, sterile males were released in Libya to begin eliminating screwworm flies that had been introduced with a South American cattle shipment.

During an epidemic of Mediterranean fruitflies in southern California and northern Mexico in the early 1980s, a similar technique was employed. Unfortunately, the X-ray technique used to sterilize the males was ineffective, and most of the flies released were not sterile, which made the problem worse rather than better. Pesticides were eventually used to control the fruitflies. Recent concern about the Mediterranean fruitfly in California has resulted in the controversial aerial spraying of malathion.

The manipulation of predator-prey relationships can also be used to control pest populations. For instance, the ladybird beetle, commonly called a ladybug, is a natural predator of aphids and scale insects. (See figure 15.13.) Artificially increasing the population of ladybird beetles reduces aphid and scale populations. In California during the late 1800s, scale insects on orange trees damaged the trees and reduced crop yields. The introduction of a species of ladybird beetle from Australia quickly brought the pests under control. Years later, when chemical pesticides were first used in the area, so many ladybird beetles were accidentally killed that scale insects once again became a serious problem. When pesticide use was discontinued, ladybird beetle populations rebounded, and the scale insects

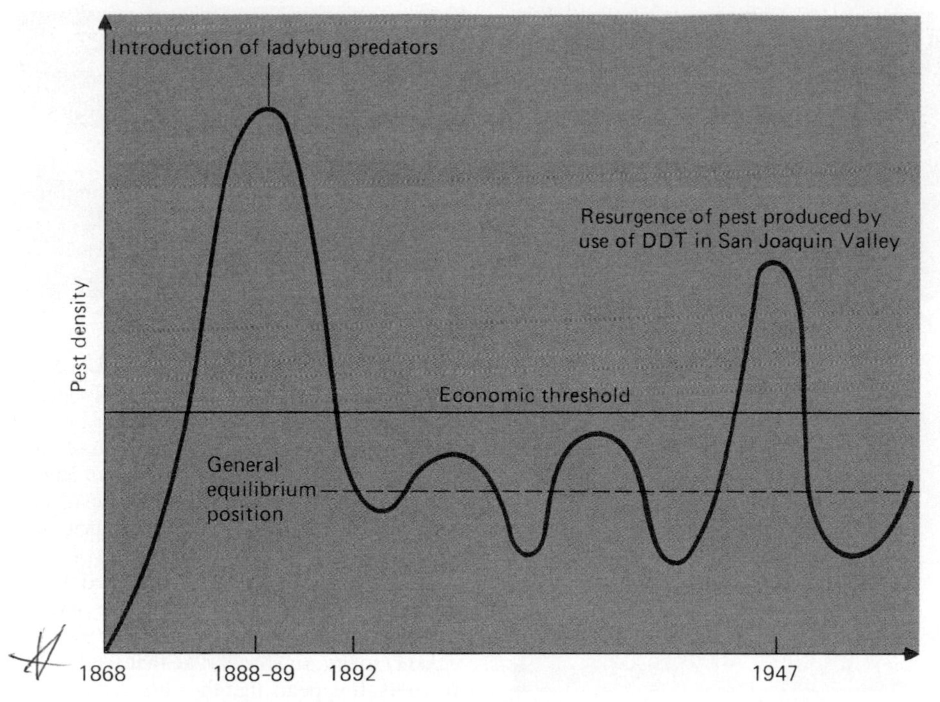

figure 15.14 **Insect Control with Natural Predators** In 1889, the introduction of ladybird beetles (ladybugs) brought the cottony cushion scale under control in the orange groves of the San Joaquin Valley. In the 1940s, DDT reduced the ladybird beetle population and the cottony cushion scale population increased. Stopping the use of DDT allowed the ladybird population to increase, reducing the pest population and allowing the orange growers to make a profit.

were once again brought under control. (See figure 15.14.)

Herbivorous insects can also be used to control weeds. Purple loosestrife (*Lythrum salicaria*) is a wetlands plant accidentally introduced from Europe in the mid-1800s. It takes over sunny wetlands and eliminates native vegetation such as cattails and, therefore, eliminates many species that rely on cattails as food, nesting places, or hiding places (see figure 15.15). The plant is known to exist in most states and provinces in the United States and Canada. In Europe, the plant is not a pest because there are several insect species that attack it in various stages of the plant's life cycle. Since purple loosestrife is a European plant, a search was made to identify candidate species of insects that would control it. The criteria were that the insects must live only on purple loosestrife and not infest other plants and must have the capacity to do major damage to purple loosestrife. Five species of beetles have been identified that will attack the plant in various ways. After extensive studies to determine if introductions of these insects from Europe would be likely to cause other problems, several were selected as candidates to help control purple loosestrife in the United States and Canada. Some combination of these beetles have been released in a large proportion of the states and provinces infested with purple loosestrife. Two species *(Galerucella calmariensis, Galerucella pusilla)* feed on the leaves, shoots, and flowers of the newly growing purple loosestrife; one *(Hylobius transversovittatus)* has larvae that feed on the roots. These species have been introduced and appear to be forming colonies. Two other species *(Nanophyes brevis, Nanophyes marmoratus)* feed on the flowers. Their release was scheduled for the summer of 1998. This multi-pronged attack by several species of beetles is projected to reduce the number of purple loosestrife plants by 90 percent.

The use of specific strains of the bacterium *Bacillus thuringiensis* to control mosquitoes and moths is another example of the use of one organism to control another. A crystalline toxin produced by the bacterium destroys the lining of the gut of the feeding insect, resulting in its death. One strain of *B. thuringiensis* is used to control mosquitoes while another is primarily effective against the caterpillars of leaf-eating moths, including the gypsy moth.

Genetically modified crops are a relatively new innovation. Over the next 20 to 30 years, scientists hope to use biotechnology to produce high-yield plant strains that are more resistant to insects and disease, thrive on less fertilizer, make their own nitrogen fertilizer, do well in slightly salty soils, withstand drought, and use solar energy more efficiently during photosynthesis. Generally, biotechnology involves genetic engineering procedures that insert genes from other species into plants that should allow them to function more efficiently or provide beneficial traits for the farmer. Some crops have had bacterial genes inserted that allow the plant to provide its own pesticide against certain kinds of insect pests. Others have had genes inserted that make the plant resistant to certain herbicides which allow the crop to be sprayed for weed control without affecting the crop plant. To date, however, several factors have limited the success of these undertakings. Most genetically engineered plant varieties have produced yields no higher and often lower than those from traditional strains; the cost of genetically engineered crop strains is too high for most of the world's subsistence farmers; and, as with all new innovations, genetically modified crops have generated controversy. Some countries in Europe have banned the importation of products that are produced from genetically modified plants. This has caused many farmers to abandon use of genetically altered crops because they would not be able to sell them in Europe. There are also ethical issues regarding genetic engineering that farmers, consumers, and scientists are beginning to debate. Will genetically engineered crops have hidden effects

a. Purple loosestrife

b. *Galerucella calmariensis*

c. *Galerucella pusilla*

d. *Hylobius transversovittatus*

figure 15.15 **Biological Control of Purple Loosestrife** Purple loosestrife is a European plant that invades wetlands and prevents the growth and reproduction of native species of plants. Several kinds of European beetles that feed on purple loosestrife have been released as biological control agents. Two species (*b. Galerucella calmariensis, c. Galerucella pusilla*) feed on the leaves, shoots, and flowers of the newly growing purple loosestrife, one (*d. Hylobius transversovittatus*) has larvae that feed on the roots. It appears that they are actually slowing the spread of purple loosestrife.

such as that experienced by the use of pesticides? For example, when monarch butterflies consumed the pollen from corn plants that contained a pesticide gene, many of the monarchs died. Will the special genes inserted into plants to improve them, negatively affect other important plant properties? Will pest species develop resistance to the defenses of genetically altered plants? Will the genes for resistance to herbicides inserted into crop plants be transferred to other plant species by interbreeding and alter weed species so that they become more difficult to control? It was once thought that transfer of genes between species was a rare event. Today, it is clear that it is much more common than formerly thought. Furthermore, it needs to happen only once to produce a strain of weed that could spread rapidly to become a serious problem.

Naturally occurring pesticides found in plants also can be used to control pests. For example, marigolds are planted to reduce the number of soil nematodes, and garlic plants are used to check the spread of Japanese beetles.

Often, modification of farming practices can reduce the impact of pests. In some cases, all crop residues are destroyed to prevent insect pests from finding overwintering sites. For example, shredding and plowing under the stalks of cotton in the fall reduces overwintering sites for boll weevils and reduces their numbers significantly, thereby reducing the need for expensive insecticide applications. Many farmers are also returning to crop rotation, which tends to prevent the buildup of specific pests that typically occurs when the same crop is raised in a field year after year.

Pesticides can also play a part in integrated pest management. Identifying the precise time when the pesticide will have the greatest effect at the lowest possible dose has these advantages: it reduces the amount of pesticide used, and may still allow the parasites and predators of pests to survive. Such precise applications often require the assistance of a trained professional who can correctly identify the pests, measure the size of the population, and time pesticide applications for maximum effect. In several instances, pheromone-baited traps capture insect pests from fields, and an assessment of the number of insects caught can be a guide to when insecticides should be applied.

Integrated pest management will become increasingly popular as the cost of pesticides increases and knowledge about the biology of specific pests becomes available. However, as long as humans raise crops, there will be pests that will outwit the defenses we develop. Integrated pest management is just another approach to a problem that began with the dawn of agriculture.

Herring Gulls as Indicators of Contamination in the Great Lakes

ISSUES & ANALYSIS

Herring gulls nest on islands and other protected sites throughout the Great Lakes region. Since they feed primarily on fish, they are near the top of aquatic food chains and tend to accumulate toxic materials from the food they eat. Eggs taken from nests can be analyzed for a variety of contaminants.

Since the early 1970s, the Canadian Wildlife Service has operated a monitoring program to assess trends in the levels of contaminants in the eggs of herring gulls. In general, the contaminant levels have declined as both the Canadian and U.S. governments have taken action to stop new contaminants from entering the Great Lakes. The figure shows the trends for PCBs and DDE. PCBs are a group of organic compounds that were used as fire retardants, lubricants, insulation fluids in electrical transformers, and in some printing inks. Some forms of PCB are much more toxic than others. Both Canada and the United States have eliminated most uses of PCBs, reducing the levels found in herring gull eggs. DDE is a breakdown product of DDT. DDT use was discontinued in much of the world in the early 1970s. The illustration shows a general decline in the amount of DDE present following the ban on DDT.

- Should the health of a bird species be used to develop policy? Why or why not?
- If PCBs and DDT are no longer being used, why are herring gulls still being contaminated?

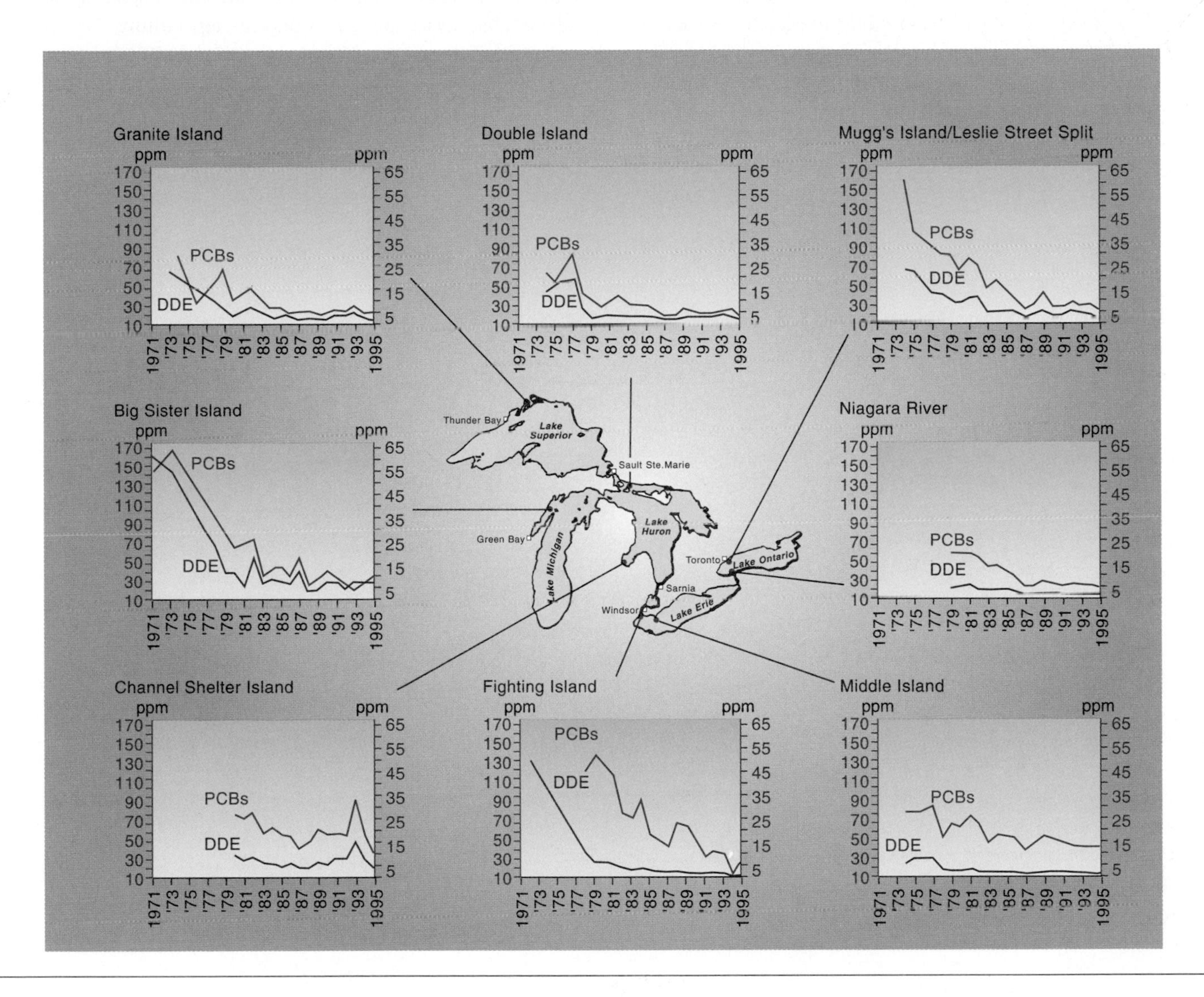

Summary

Although small slash-and-burn garden plots are common in some parts of the world, most of the food in the world is raised on more permanent farms. In countries where population size is high and money is in short supply, much of the farming is labor intensive, making use of human labor for many of the operations necessary to raise crops. However, much of the world's food is grown on large, mechanized farms that use energy rather than human muscle for tilling, planting, and harvesting crops, and for the production and application of fertilizers and pesticides.

Monoculture involves planting large areas of the same crop year after year. This causes problems with plant diseases, pests, and soil depletion. Although chemical fertilizers can replace soil nutrients that are removed when the crop is harvested, they do not replace the organic matter necessary to maintain soil texture, pH, and biotic richness.

Mechanized monoculture is heavily dependent on the control of pests by chemical means. Persistent pesticides are stable and persist in the environment, where they may biomagnify in ecosystems. Consequently, many of the older persistent pesticides have been quickly replaced by nonpersistent pesticides that decompose much more quickly and present less of an environmental hazard. However, most nonpersistent pesticides are more toxic to humans and must be handled with greater care than the older persistent pesticides.

Pesticides can be divided into several categories based on the organism they are used to control. Insecticides are used to control insects, herbicides are used for plants, fungicides for fungi, and rodenticides for rodents. Because of the problems of persistence, biomagnification, resistance of pests to pesticides, and human health concerns, many people are seeking pesticide-free alternatives to raising food. Several different philosophies that seek the same ends—less use of chemicals and better stewardship of soil—are alternative agriculture, sustainable agriculture, and organic agriculture. One ingredient in all of these approaches is the use of integrated pest management, which uses a complete understanding of an organism's ecology to develop pest-control strategies that use no or few pesticides.

Interactive Exploration

Check out the website at

http://www.mhhe.com/environmentalscience

and click on the cover of this textbook for interactive versions of the following:

KNOW THE BASICS

auxins *322*
bioaccumulation *324*
biocides *319*
biomagnification *324*
carbamates *321*
chlorinated hydrocarbons *320*
fungicides *319*
Green Revolution *318*
herbicides *319*
insecticides *319*
integrated pest management *331*
macronutrient *319*
micronutrient *319*
monoculture *317*
nonpersistent pesticides *319*
nontarget organisms *319*
organophosphates *321*
persistent pesticides *319*
pests *319*
pesticides *319*
pheromone *332*
polyculture *316*
rodenticides *319*
target organism *319*
weeds *319*

On-line Flashcards

Electronic Glossary

IN THE REAL WORLD

A decrease in pesticide applications, better productivity with less fertilizer. . . there are many benefits of genetically engineered crops. How could anyone be against this technological advancement? Why are people in such an uproar because genetically altered food is not labeled? Why did Europe ban the use of genetically modified food? Take a look at **Genetically Engineered Crop Tests Evade Scrutiny by Moving to Eastern Europe** and **Transgenic Cotton Tested in India: Farmers Fear Introduction of "Terminator" Crops** to see what we know about the effects of consuming genetically altered food and why farmers in India feel threatened.

Why would the U.S. want to use a persistent herbicide on crops in Colombia? Find out how herbicides are used in the drug war by reading **Colombia to Spray Coca Crops with Stronger Herbicide.**

Persistent chemicals are found not only in Colombia, they are found in Norway, far from widespread use of the specific chemicals. Look over **Norway's Polar Bears Suspected Victims of PCB Contamination** for a story of how the effects of biomagnification and persistence are affecting the reproductive organs of polar bears.

PCB contamination is also an issue in the Great Lakes. Read **PCB Contamination in the Fox River** for the status of a contaminated river that drains into Lake Michigan.

What is the relationship between pesticide use and breast cancer, pesticide use and decreased sperm count? Read the

Endocrine Disrupters on the Gulf Coast case study for information on how the chemicals are affecting human and wildlife health.

Speaking of health, why should we be careful with our use of antibiotics and household antibacterial chemicals? Pesticide resistance is not only for agricultural crops. See the story of how a parasite that causes death to humans and cattle is becoming resistant to drugs in Drug-resistant Strain of Sleeping Sickness Appears in Ethiopia.

PUT IT IN MOTION

Isn't it amazing how an animation can bring some concepts to life and make them much clearer? If you didn't see the Pesticide Resistance animation as an example of natural selection in chapter 5, or if the process is not crystal clear for you yet, check it out now. The graphic representation of the process really helps make it clear and understandable.

Another animation that you may have seen and that deserves a revisit is Biomagnification. This animation clearly explains how toxins accumulate and is a good animation to study for a better understanding of the process of biomagnification.

TEST PREPARATION

Review Questions

1. What is monoculture?
2. List three reasons why fossil fuels are essential for mechanized agriculture.
3. Describe why pesticides are commonly used in mechanized agriculture.
4. Why are fertilizers used? What problems are caused by fertilizer use?
5. How do persistent and nonpersistent pesticides differ?
6. What is biomagnification? What problems does it cause?
7. How do organic farms differ from conventional farms?
8. Name three nonchemical methods of controlling pest populations.
9. What are the advantages and disadvantages of integrated pest management?
10. List three uses of food additives.
11. List three actions farmers could use to reduce the effect of pesticides on the environment.

Critical Thinking Questions

1. If you were a public health official in a developing country, would you authorize the spraying of DDT to control mosquitoes that spread malaria? What would be your reasons?
2. Look at Table 15.1. What caused the changes in the effectiveness of the insecticides? If you were an agricultural extension agent, what alternatives to pesticides might you recommend?
3. Imagine that you are a scientist examining fish in Lake Superior and you find toxaphene in the fish you are studying. Toxaphene was used primarily in cotton farming and has been banned since 1982. How can you explain its presence in these fish?
4. Are the risks of pesticide use worth the benefits? What values, beliefs, and perspectives lead you to this conclusion?
5. Do you think that current agricultural practices are sustainable? Why or why not? What changes in agriculture do you think will need to happen in the next 50 years?
6. Imagine you are an EPA official who is going to make a recommendation about whether an agricultural pesticide can remain on the market or should be banned. What are some of the facts you would need in order to make your recommendation?

   Who are some of the interest groups interested in the outcome of your decision? What arguments might they present regarding their positions? What political pressures might they be able to bring to bear on you?
7. Why are few consumers demanding alternative methods of crop production and why are farmers not using those methods?

KEY CHAPTER LINKS
ESSENTIAL STUDY PARTNER
BIO COURSE
GLOBAL ISSUES
REGIONAL PERSPECTIVES
PRACTICE QUIZZING

CHAPTER

Water Management

Objectives

After reading this chapter, you should be able to:

- Explain how water is cycled through the hydrologic cycle.
- Explain the significance of groundwater, aquifers, and runoff.
- Explain how land use affects infiltration and surface runoff.
- List the various kinds of water use and the problems associated with each.
- List the problems associated with water impoundment.
- List the major sources of water pollution.
- Define biochemical oxygen demand (BOD).
- Differentiate between point and nonpoint sources of pollution.
- Explain how heat can be a form of pollution.
- Differentiate between primary, secondary, and tertiary sewage treatments.
- Describe some of the problems associated with storm-water runoff.
- List sources of groundwater pollution.
- Explain how various federal laws control water use and prevent misuse.
- List the problems associated with water-use planning.
- Explain the rationale behind the federal laws that attempt to preserve certain water areas and habitats.
- List the problems associated with groundwater mining.
- Explain the problem of salinization associated with large-scale irrigation in arid areas.
- List the water-related services provided by local governments.

Chapter Outline

The Water Issue

The Hydrologic Cycle

Human Influences on the Hydrologic Cycle

Kinds of Water Use
- Domestic Use of Water
- Agricultural Use of Water
- Industrial Use of Water
- In-Stream Use of Water

Kinds and Sources of Water Pollution

Environmental Close-Up: *Is It Safe to Drink the Water?*

- Municipal Water Pollution

Global Perspective: *The Cleanup of the Holy Ganges*

- Agricultural Water Pollution
- Industrial Water Pollution
- Thermal Pollution

Global Perspective: *Comparing Water Use and Pollution in Industrialized and Developing Countries*

- Marine Oil Pollution
- Groundwater Pollution

Water-Use Planning Issues
- Water Diversion
- Wastewater Treatment

Environmental Close-Up: *Restoring the Everglades*

- Salinization
- Groundwater Mining
- Preserving Scenic Water Areas and Wildlife Habitats

Global Perspective: *Death of a Sea*

Global Perspective: *ECOPARQUE*

Issues & Analysis: *The California Water Plan*

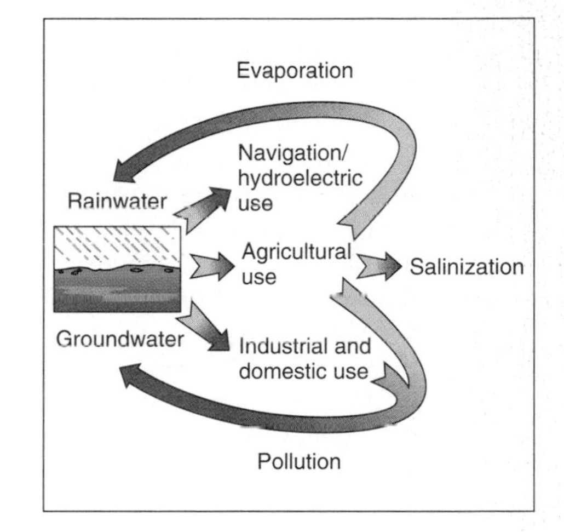

The Water Issue

Water in its liquid form is the material that makes life possible on Earth. All living organisms are composed of cells that contain at least 60 percent water. Furthermore, their metabolic activities take place in a water solution. Organisms can exist only where there is access to adequate supplies of water. Water is also unique because it has remarkable physical properties. Water molecules are polar, that is, one part of the molecule is slightly positive and the other is slightly negative. Because of this, the water molecules tend to stick together and they also have a great ability to separate other molecules from each other. Water's ability to act as a solvent and its capacity to store heat are a direct consequence of its polar nature and these abilities make water extremely valuable for human societal and industrial activities. Water dissolves and carries substances ranging from nutrients to industrial and domestic wastes. A glance at any urban sewer will quickly point out the importance of water in dissolving and transporting wastes. Because water heats and cools more slowly than most other substances, it is used in large quantities for cooling in electric power generation plants and in other industrial processes. Water's ability to retain heat also modifies local climatic conditions in areas near large bodies of water. These areas do not have the wide temperature changes characteristic of other areas.

For most human uses, as well as some commercial and industrial ones, the quality of the water is as important as its quantity. Water must be substantially free of dissolved salts, plant and animal waste, and bacterial contamination to be suitable for human consumption. The oceans which cover approximately 70 percent of the Earth's surface contain over 97 percent of its water. However, saltwater cannot be consumed by humans or used for many industrial processes. Freshwater is free of the salt found in ocean waters. Of the freshwater found on Earth, only a tiny fraction is available for use (See figure 16.1.) Unpolluted freshwater that is suitable for drinking is known as **potable water.** Early human migration routes and settlement sites were influenced by the availability of drinking water. At one time, clean freshwater supplies were considered inexhaustible. Today, despite advances in drilling, irrigation, and purification, the location, quality, quantity, ownership, and control of potable waters remain important human concerns.

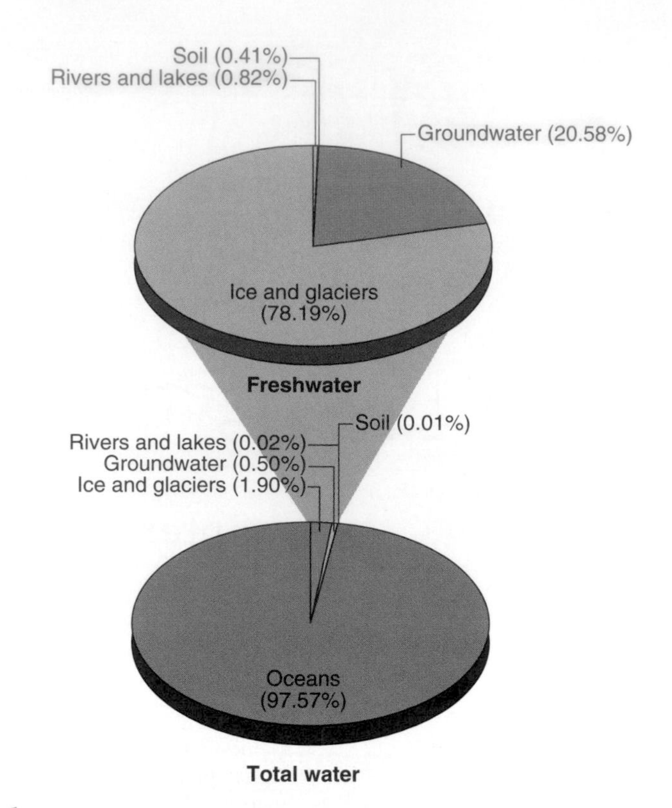

figure 16.1 **Freshwater Resources** Although water covers about 70 percent of the Earth's surface, over 97 percent is saltwater. Of the less than 3 percent that is freshwater, only a tiny fraction is available for human use.

Only recently have we begun to understand that we will probably exhaust our usable water supplies in some areas of the world because of increases in human populations and limitations on the supply available. Some areas of the world have abundant freshwater resources while others have few. In addition, there is an increased demand for freshwater for industrial, agricultural, and personal needs.

Shortages of potable freshwater throughout the world can also be directly attributed to human abuse in the form of pollution. Water pollution has negatively affected water supplies throughout the world. In many parts of the developing world, people do not have access to safe drinking water. The World Health Organization estimates that about 1.4 billion of the world's 6 billion people do not have access to safe drinking water. Even in the economically advanced regions of the world, water quality is a major issue.

Unfortunately, the outlook for the world's freshwater supply is not very promising. According to studies by the United Nations and the International Joint Commission, many sections of the world are currently experiencing a shortage of freshwater and the problem will only intensify. This is particularly true in rapidly developing countries like China and India which have about a third of the world's population. Furthermore, changes in the amount of rain from year to year results in periodic droughts for some areas and devastating floods for others. (See figure 16.2.) However, rainfall is needed to regenerate freshwater and, therefore, is an important link in the cycling of water.

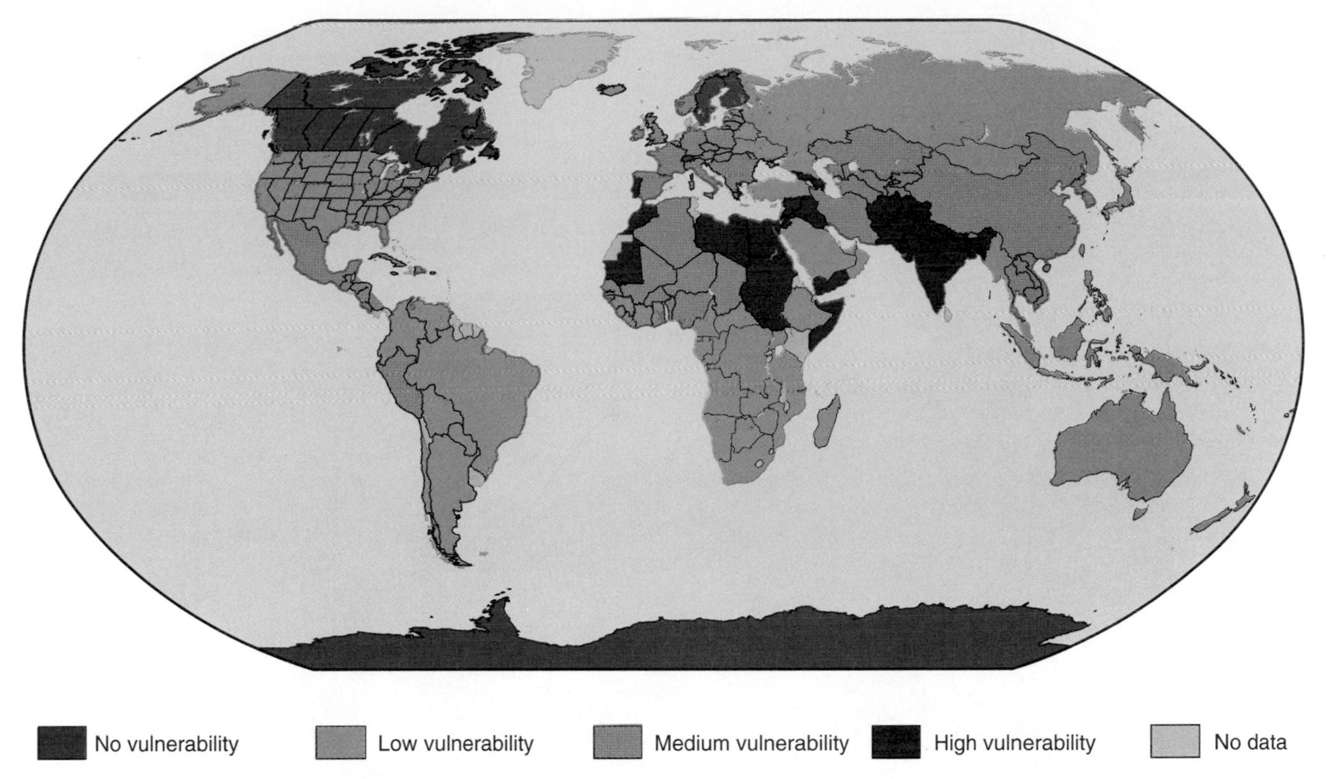

figure 16.2 **Areas of the World Experiencing Water Shortage** Many areas of the world are experiencing water shortages. This map indicates that north and central Africa and parts of Asia are particularly vulnerable.

Source: Data from World Resources Institute.

The Hydrologic Cycle

All water is locked into a constant recycling process called the **hydrologic cycle.** (See figure 16.3.) Two important processes involved in the cycle are the evaporation and condensation of water. Evaporation involves adding energy to molecules of a liquid so that it becomes a gas in which the molecules are farther apart. Condensation is the reverse process in which molecules of a gas give up energy, get closer together, and become a liquid. Solar energy provides the energy that causes water to evaporate from the ocean surface, the soil, bodies of freshwater, and from the surfaces of plants. The water evaporated from plants comes from two different sources. Some is water that has fallen on plants as rain, dew, or snow. In addition, plants take up water from the soil and transport it to the leaves where it evaporates. This process is known as **evapotranspiration**. The water vapor in the air moves across the surface of the earth as the atmosphere circulates. As warm, moist air cools, water droplets form and fall to the land as precipitation. Although some precipitation may simply stay on the surface until it evaporates, most will either sink into the soil or flow downhill and enter streams and rivers which eventually return the water to the ocean. Surface water that moves across the surface of the land and enters streams and rivers is known as **runoff.** Water that enters the soil and is not picked up by plant roots moves slowly downward through the spaces in the soil and subsurface material until it reaches an impervious layer of rock. The water that fills the spaces in the substrate is called **groundwater.** It may be stored for long periods in underground reservoirs.

The porous layer which becomes saturated with water is called an **aquifer.** There are two basic kinds of aquifers: unconfined and confined. An **unconfined aquifer** usually occurs near the land's surface where water enters the aquifer from the land above it. The top of the layer saturated with water is called the **water table.** The lower boundary of the aquifer is an impervious layer of clay or rock that does not allow water to pass through it. Unconfined aquifers are replenished (recharged) primarily by rain that falls on the ground directly above the aquifer and infiltrates the layers below. The water in such aquifers is at atmospheric pressure and flows in the direction of the water table's slope which may or may not be similar to the surface of the land above it. Above the water table and below the land surface is a layer known as the **vadose zone** (also known as the unsaturated zone or zone of aeration) that is not saturated with water. (See figure 16.4.)

A **confined aquifer** is bounded on both the top and bottom by layers that are impervious to water and is saturated with water under greater-than-atmospheric pressure. An impervious confining layer is called an **aquiclude.** If water can pass in and out of the confining layer, the layer is called an **aquitard.** A confined aquifer is primarily replenished by rain and surface water from a recharge zone

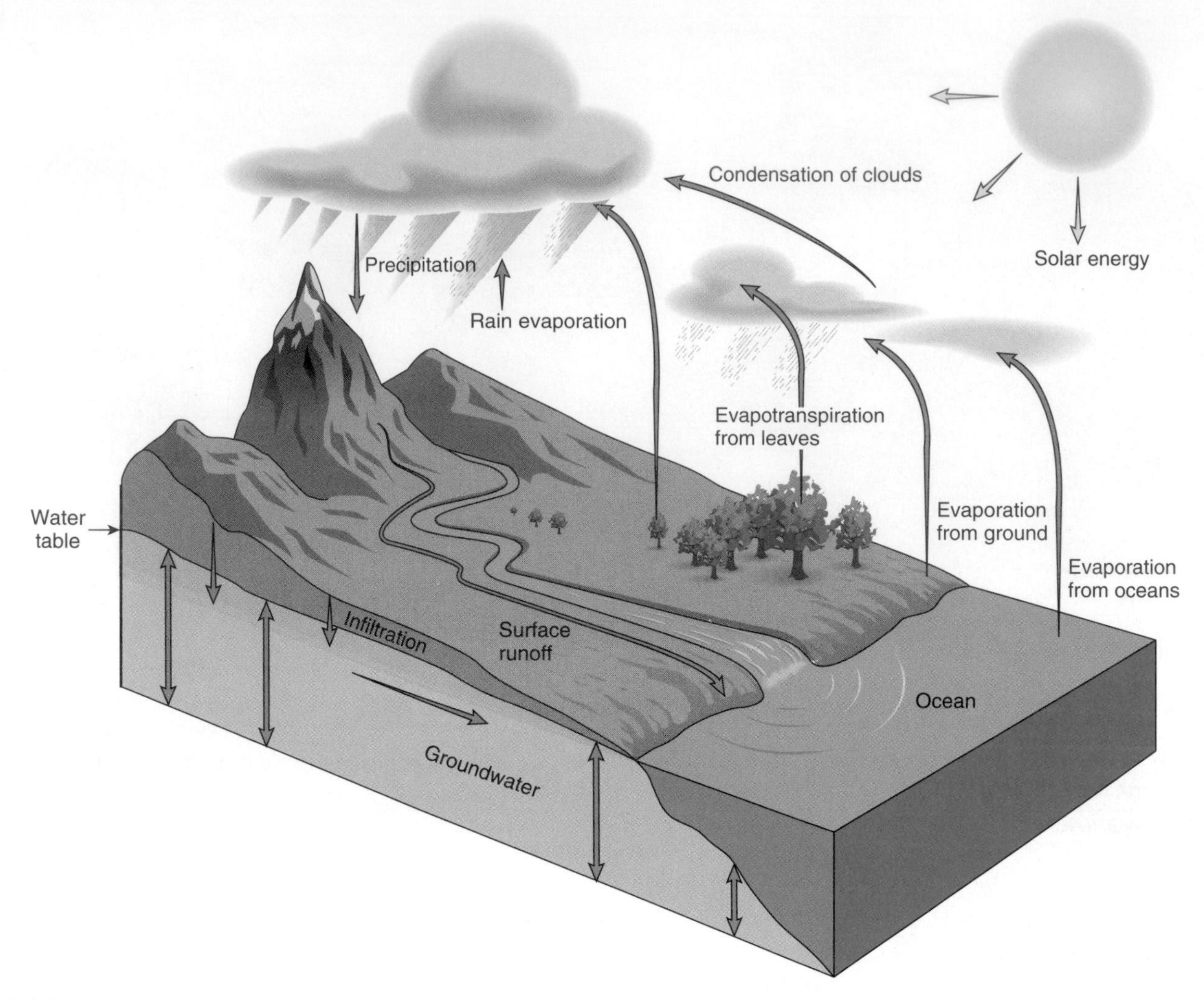

figure 16.3 **The Hydrologic Cycle** The cycling of water through the environment follows a simple pattern. Moisture in the atmosphere condenses into droplets that fall to the earth as rain or snow, supplying all living things with its life-sustaining properties. Water, flowing over the earth as surface water or through the soil as groundwater, returns to the oceans, where it evaporates back into the atmosphere to begin the cycle again.

(the area where water is added to the aquifer) that may be many kilometers from where the aquifer is tapped for use. If the recharge area is at a higher elevation than the place where an aquifer is tapped, water will flow up the pipe until it reaches the same elevation as the recharge area. Such wells are called **artesian wells.** If the recharge zone is above the elevation of the top of the well pipe, it is called a flowing artesian well because water will flow from the pipe. The nature of the substrate in the aquifer influences the amount of water the aquifer can hold and the rate at which water moves through it. **Porosity** is a measure of the size and number of the spaces in the substrate. The greater the porosity, the more water it can contain. The rate at which water moves through an aquifer is determined by the size of the pores, the degree to which they are connected, and any cracks or channels present in the substrate. The rate at which the water moves through the aquifer determines how rapidly water can be pumped from a well per minute.

Human Influences on the Hydrologic Cycle

The way in which land is used has significant impact on evaporation, runoff, and infiltration. When water is used for cooling in power plants or to irrigate crops, the rate of evaporation is increased. Water impounded in reservoirs also evaporates rapidly. This rapid evaporation can affect local atmospheric conditions. Runoff and the rate of infiltration also are greatly influenced by human activity. Removing the vegetation by logging or agriculture increases runoff and decreases infiltration. Because there is more runoff, there is more erosion of soil. Urban complexes with a high percentage of impervious, paved surfaces have increased runoff and reduced infiltration. A major concern in urban areas is providing ways to carry storm water away rapidly. This involves designing and constructing surface

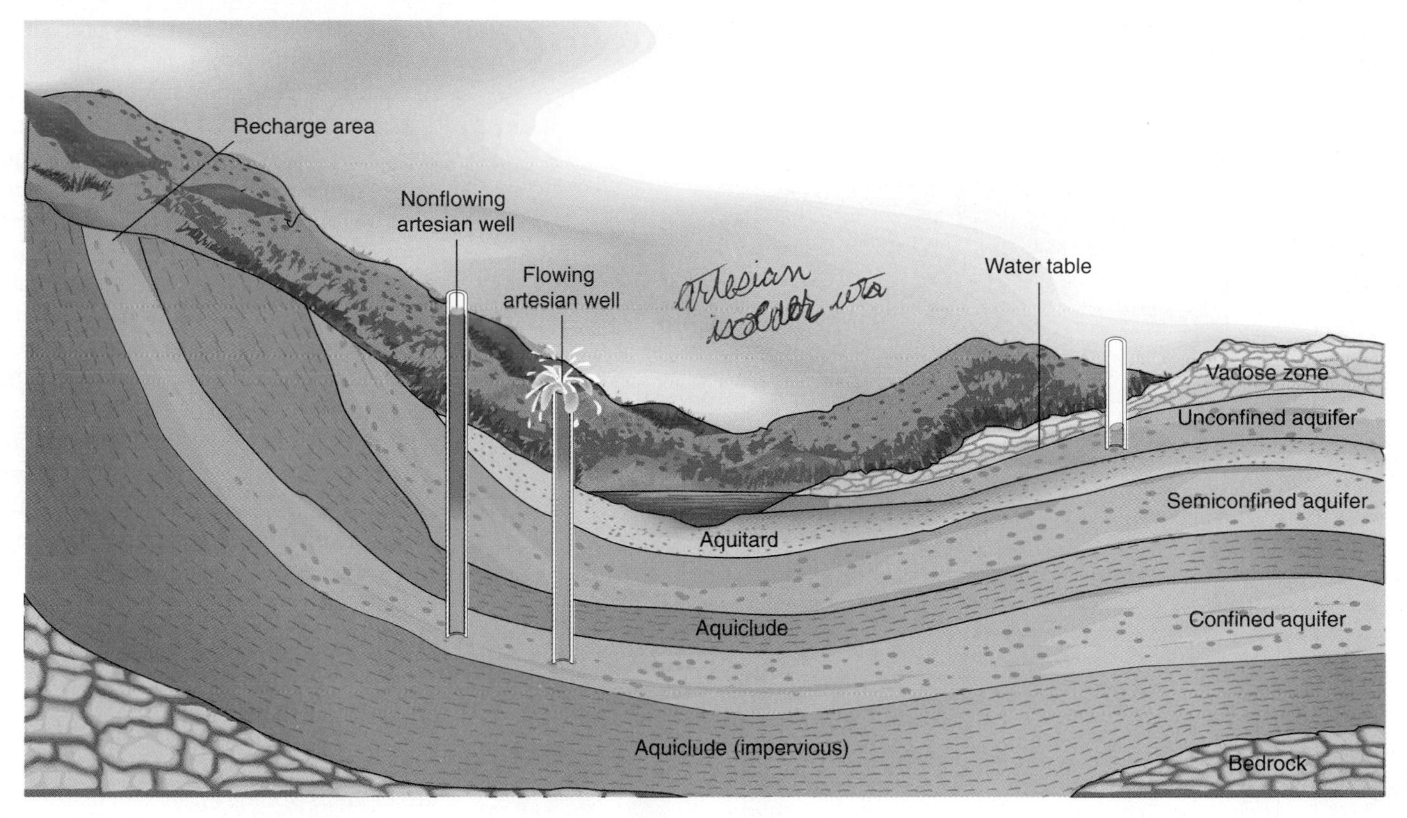

figure 16.4 **Aquifers and Groundwater** Groundwater is found in the pores in layers of sediment or rock. The various layers of sediment and rock determine the nature of the aquifer and how it can be used.

waterways and storm sewers. Often cities have significant flooding problems, when heavy rains overtax the ability of their storm-water management systems to remove excess water.

Cities also have problems supplying water for industrial, agricultural, and municipal use. In many cases, cities rely on surface water sources for their drinking water. The source may be a lake or river, or an impoundment that stores water. In addition, aquifers are extremely important in supplying water. Many large urban areas in the western part of the United States depend upon underground water for their water supply. This groundwater supply can be tapped as long as it is not used faster than it can be replaced. Determining how much groundwater or surface water can be used and what the uses should be is a major concern and is especially important in water-poor areas of the world.

There are several ways to monitor water use from surface and groundwater sources. Water withdrawals are measurements of the amount of water taken from a source. This water may be used temporarily and then returned to its source and used again. For example, when a factory withdraws water from a river for cooling purposes, it returns most of the water to the river; thus, the water can be used later. Water that is incorporated into a product or lost to the atmosphere through evaporation or evapotranspiration cannot be reused in the same geographic area and is said to be consumed. Much of the water used for irrigation is lost to evaporation and evapotranspiration, or is removed with the crop when it is harvested. Therefore, much of the water withdrawn for irrigation is consumed.

Kinds of Water Use

Water use varies considerably around the world depending on availability of water and degree of industrialization. However, use can be classified into four broad categories: (1) domestic use, (2) agricultural use, (3) industrial use, and (4) in-stream use. It is also important to remember that some uses of water are consumptive while others are nonconsumptive

Domestic Use of Water

Over 90 percent of the water used for domestic purposes in North America is supplied by municipal water systems which typically include complex, costly storage, purification, and distribution facilities. However, many rural residents can obtain safe water from untreated private wells. Nearly 37 percent of municipal water supplies come from wells.

Regardless of the water source (surface or groundwater), water supplied to cities in the developed world is treated to assure its safety. Treatment of raw water prior to distribution usually involves some combination of the following processes. The raw water is filtered through sand or other substrates to remove particles. Chemicals may be added to the water that will cause some dissolved materials to be removed. Then,

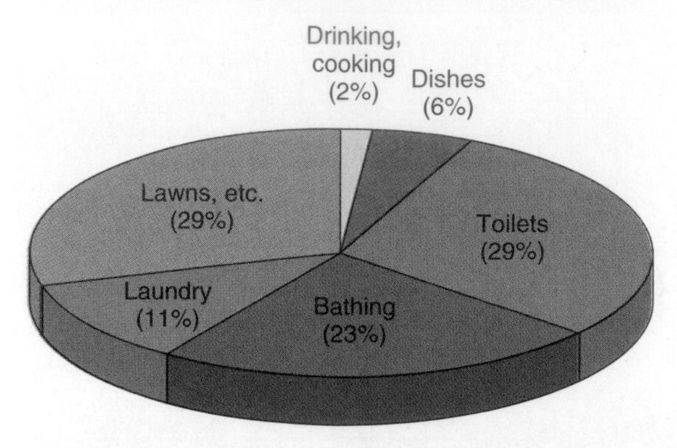

figure 16.5 **Urban Domestic Water Uses** Over 150 billion liters (40 billion gallons) of water are used each day for urban domestic purposes in North America.

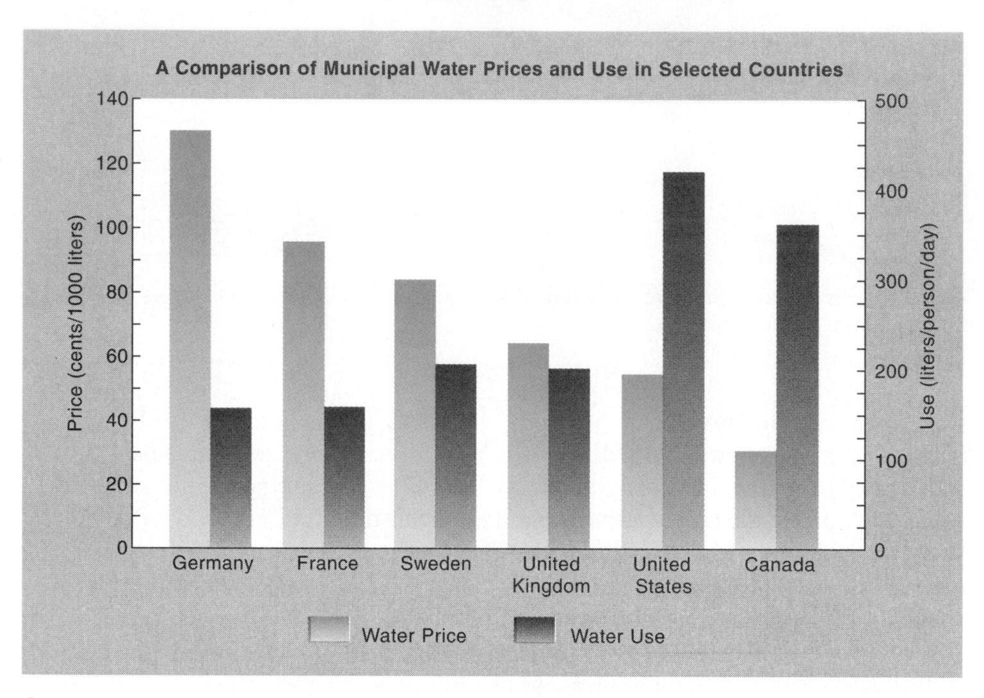

figure 16.6 **Water Use Decreases As Water Price Increases** There is a general correlation between the amount of water that is used and its price.

before the water is released for public use, it is disinfected to remove any organisms that might still be present. In cases where no freshwater is available, expensive desalinization of saltwater may be the only option available.

Domestic activities in highly developed nations require a great deal of water. This domestic use includes drinking, air conditioning, bathing, washing clothes, washing dishes, flushing toilets, and watering lawns and gardens. On average, each person in a North American home uses about 400 liters (about 100 gallons) of water each day. Most of this **domestic water** (about 70 percent) is used as a solvent to carry away wastes (nonconsumptive use), with only a small amount used for drinking and about 30 percent for lawns and gardens (consumptive use). Yet all water that enters the house has been purified and treated to make it safe for drinking. (See figure 16.5.) Natural processes cannot cope with the highly concentrated wastes typical of a large urban area. The unsightly and smelly wastewater also presents a potential health problem, so cities and towns must treat it before returning it to a local water source. Until recently, the cost of water in almost every community has been so low that there was very little incentive to conserve, but shortages of water and increasing purification costs have raised the price of domestic water in many parts of the world and it is becoming evident that increased costs do tend to reduce use. (See figure 16.6.)

Although domestic use of water is a relatively small component of the total water-use picture (see figure 16.7), urban growth has created problems in the development, transportation, and maintenance of quality water supplies. In regions experiencing rapid population growth, such as Asia, domestic use is expected to increase sharply. In North America, rapidly growing cities in the West are experiencing water shortages. (See figure 16.8.) Demand for water in urban areas sometimes exceeds the immediate supply, particularly when the supply is local surface water. This is especially true during the summer, when water demand is high and precipitation is often low. Many communities have begun public education campaigns designed to help reduce water usage. (See figure 16.9.)

In addition to encouraging the public to conserve water, municipalities need to pay attention to losses that occur within the distribution system. Leaking water pipes and mains account for significant losses of water. Even in the developed world, losses may be as high as 20 percent. Poorer countries typically exceed this and some may lose over 50 percent of the water to leaks. Another major cause of water loss has been public attitudes. As long as water is considered a limitless, inexpensive resource, little effort will be made to conserve it. As the cost of water rises and attitudes toward water change, so will usage and efforts to conserve.

Agricultural Use of Water

In North America, groundwater accounts for about 37 percent of the water used in agriculture and surface water about 63 percent. **Irrigation** is the major consumptive use of water in most parts of the world and accounts for about 80 percent of all the water consumed in North America. About 500,000 million liters/day (134,000 million gallons/day or 150 million acre feet/year) are used in irrigation in the United States. The amount of water used for irrigation and livestock continues to increase throughout the world. Future agricultural demand for water will depend on the cost of water for irrigation; the demand for agricultural products, food, and fiber; governmental policies; the development of new technology; and competition for water from a growing human population.

Since irrigation is typical in arid and semiarid areas, local water supplies are often lacking and it is often necessary to transport water great distances to water crops. This is particularly true in the western United States, where about 14 million hectares (35 million acres) of land are irrigated.

There are four commonly used methods for irrigation. Surface or flood irrigation involves supplying the water to crops by having the water flow over the field or in furrows. This requires extensive canals and is not suitable for all kinds of crops. Spray irrigation involves the use of pumps to spray water on the crop. Trickle irrigation uses a series of pipes with strategically placed openings so that water is delivered directly to the roots of the plants. Subirrigation involves supplying water to plants through underground pipes. Often this method is used where soils require draining at certain times of the year. The underground pipes can be used to drain excess water at one time of the year and supply water at others. Each of these methods has its drawbacks and advantages as well as conditions under which it works well.

The construction and maintenance of irrigation structures, such as dams, canals, pipes, and pumps, is expensive. Costs for irrigation water have traditionally been low since many of the dams and canals were constructed with federal assistance and farmers have often used water wastefully. As there has been

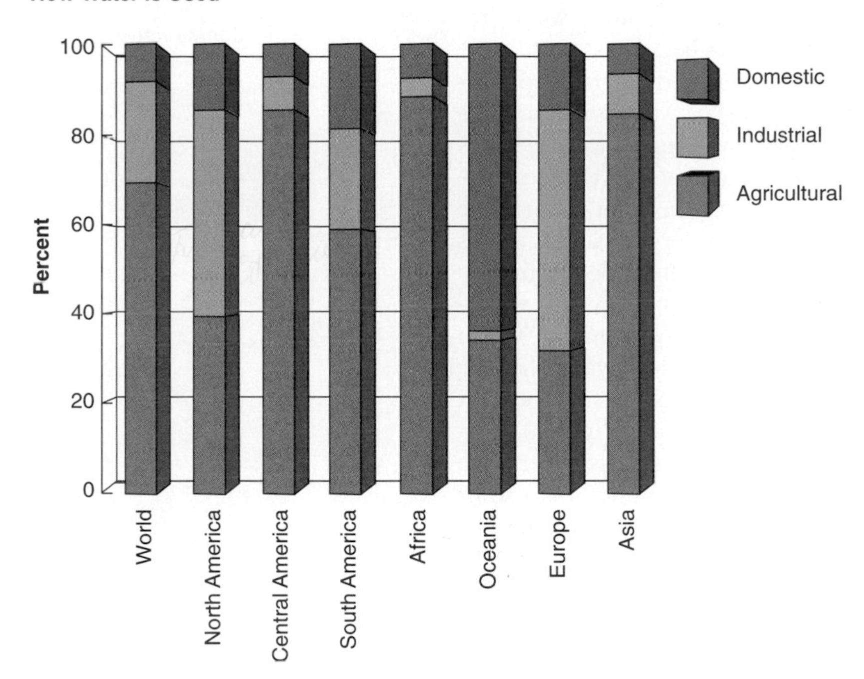

figure 16.7 **World Uses of Water** Domestic, industrial, and agricultural uses dominate the allocation of water resources. However, there is considerable variety in different parts of the world in how these resources are used.

Source: World Resources 1998–99.

figure 16.8 **Urban Expansion** Maintaining a suitable supply of water for growing metropolitan areas can pose major problems, especially in arid areas.

greater competition between urban areas and agriculture for scarce water resources, there has been pressure to raise the cost of water used for irrigation. Increasing the cost of water will stimulate farmers to conserve, just as it does homeowners. Another way to reduce the demand for irrigation water is to reduce the quantity of water-demanding crops grown in dry areas, or change from high water-demanding to lower water-demanding crops. For example, wheat or soybeans require less water than do potatoes or sugar beets. It is also becoming increasingly important to modify irrigation practices to use less water. (See figure 16.10.) For example, the use of trickle irrigation and some variations of spray irrigation use water more efficiently than the more traditional flood irrigation methods.

Many forms of irrigation require a great deal of energy. This is particularly true when pumps are used to deliver the water to the crop. It is estimated that 40 percent of the energy devoted to agriculture in Nebraska is used for irrigation. Increasing energy costs may force some farmers to reduce or discontinue irrigation. In addition, much of western Nebraska relies on groundwater for irrigation, and the water table is dropping rapidly. If a water shortage develops, land values will decline. Land use and water use are interrelated and cannot be viewed independently.

Water savings guide

Conservative use will save water		Normal use will waste water
Wet down, soap-up, rinse off 15 liters (4 gal)	**Shower**	Regular shower 95 liters (25 gal)
May we suggest a shower?	**Tub bath**	Full tub 135 liters (36 gal)
Minimize flushing Each use consumes 20-25 liters (5-7 gal) New toilets use 6 liters (1.6 gal)	**Toilet**	Frequent flushing is very wasteful
Fill basin 4 liters (1 gal)	**Washing hands**	Tap running 8 liters (2 gal)
Fill basin 4 liters (1 gal)	**Shaving**	Tap running 75 liters (20 gal)
Wet brush, rinse briefly 2 liters (1/2 gal)	**Brushing teeth**	Tap running 38 liters (10 gal)
Take only as much as you require	**Ice**	Unused ice goes down drain
Please report immediately	**Leaks**	A small drip wastes 95 liters (25 gal) per week
Turn off light, TV, heaters, and air conditioning when not in room	**Energy**	Wasting energy also wastes water

Thank you for using this column..and <u>not</u> this one

figure 16.9 **Ways to Conserve Water** Minor changes in the way people use water could significantly reduce domestic water use. Note: 1 gallon equals approximately 3.785 liters.

Industrial Use of Water

Industrial water use accounts for nearly half of total water withdrawals in the United States, about 70 percent in Canada, and about 23 percent worldwide. Since most industrial processes involve heat exchanges, 90 percent of the water used by industry is for cooling and is returned to the source, so only a small amount is actually consumed. Industrial use accounts for less than 20 percent of the water consumed in the United States. For example, electric-power generating plants use water to cool steam so that it changes back into water. Many industries, especially power plants, actually can use saltwater for cooling purposes. About 30 percent of the cooling water used by power plants is saltwater. If the water heated in an industrial process is dumped directly into a watercourse, it significantly changes the water temperature. This affects the aquatic ecosystem by increasing the metabolism of the organisms and reducing the water's ability to hold dissolved oxygen.

Industry also uses water to dissipate and transport waste materials. In fact, many streams are now overused for this purpose, especially in urban centers. The use of watercourses for waste dispersal degrades the quality of the water and may reduce its usefulness for other purposes. This is especially true if the industrial wastes are toxic.

Historically, industrial waste and heat were major causes of pollution. However, most industrialized nations have passed laws that severely restrict industrial discharges of wastes or heated water into watercourses. In the United States, the federal role in maintaining water quality began in 1948 with the passage of the Federal Water Pollution Control Act. This act provided federal funds and technical assistance to strengthen local, state, and interstate water-quality programs. It was amended in 1977 and renamed the Clean Water Act. The 1977 Clean Water Act prohibits discharges into surface water from specific sources unless a permit had been obtained from the EPA. Enforcement of this act has been extremely effective in improving surface-water quality. However, many countries in the developing world have done little to control industrial pollution and water quality is significantly reduced by careless use. See

a.

b.

c.

figure 16.10 **Types of Irrigation** Many arid areas require irrigation to be farmed economically. (a) Surface or flood irrigation uses irrigation canals and ditches to deliver water to the crops. The land is graded so that water flows from the source into the fields. Water is siphoned from a canal into ditches between rows of crops. (b) Spray irrigation uses a pump to spray water into the air above the plants. This is an example of central-pivot spray irrigation, in which a long pipe on wheels slowly rotates about a central point. (c) Trickle irrigation conserves water by delivering water directly to the roots of the plants but requires an extensive network of pipes.

Global Perspective: Comparing Water Use and Pollution in Industrialized and Developing Countries p. 354.

In-Stream Use of Water

In-stream water use does not remove water but makes use of it in its channels and basins. Therefore all in-stream uses are nonconsumptive. Major in-stream uses of water are for hydroelectric power, recreation, and navigation. Although in-stream uses do not remove water, they may require modification of the direction, time, or volume of flow and can negatively affect the watercourse.

Electricity from hydroelectric power plants is an important energy resource. Presently, hydroelectric power plants produce about 13 percent of the total electricity generated in the United States. (See figure 16.11.) They do not consume water and do not add waste products to it. However, the dams needed for the plants have definite disadvantages, including the high cost of construction and the resulting destruction of the natural habitat in streams and surrounding lands. The sudden discharge of impounded water from a dam can seriously alter the downstream environment. If the discharge is from the top of the reservoir, the stream temperature rapidly increases. Discharging the colder water at the bottom of the reservoir causes a sudden decrease in the stream's water temperature. Either of these changes is harmful to aquatic life. The impoundment of water also reduces the natural scouring action of a flowing stream. If water is allowed to flow freely, the silt accumulated in the river is carried downstream during times of high water. This maintains the river channel and carries nutrient materials to the river's mouth. But if a dam is constructed, the silt deposits behind the dam, eventually filling the reservoir. Many other dams were constructed to control floodwaters. While dams reduce flooding, they do not eliminate it. In fact, the building of a dam often encourages people to develop the floodplain. As a result, when flooding occurs, the loss of property and lives may be greater.

Because dams create lakes that have a large surface area, evaporation is increased. In arid regions, the amount of water lost can be serious. This is particularly evident in hot climates. Furthermore, flow is often intermittent below the dam, which alters the water's oxygen content and interrupts fish migration. The populations of algae and other small organisms are also altered. Because of all these impacts, dam construction requires careful planning.

figure 16.11 **Dams Interrupt the Flow of Water** The flow of water in most large rivers is controlled by dams. Most of these dams provide electricity. In addition, they prevent flooding and provide recreational areas. However, dams destroy the natural river system.

figure 16.12 **Recreational Use of Water** Marinas provide recreation; however, wetlands destruction and large dredging operations may be necessary to build them.

Dam construction creates new recreational opportunities because reservoirs provide sites for boating, camping, and related recreation. (See figure 16.12.) However, these opportunities come at the expense of a previously free-flowing river. Some recreational pursuits, such as river fishing, are lost. Sailing, waterskiing, swimming, fishing, and camping all require water of reasonably good quality. Water is used for recreation in its natural setting and often is not physically affected. Even so, it is necessary to plan for recreational use, because overuse or inconsiderate use can degrade water quality. For example, waves generated by powerboats can accelerate shoreline erosion and cause siltation. When large numbers of powerboats are used on water, they contribute significantly to water pollution since their exhaust contains unburned hydrocarbons. Newer boat motors are designed to produce less pollution.

Most major rivers and large lakes are used for navigation. North America currently has more than 40,000 kilometers (25,000 miles) of commercially navigable waterways. These waterways must have sufficient water depth to ensure passage of ships and barges. Canals, locks, and dams are used to assure that adequate depths are provided. Often, dredging is necessary to maintain the proper channel depth. Dredging can resuspend in the water contaminated sediments that had been covered over. An additional problem is determining where to deposit the contaminated sediments when they are removed from the bottom. In addition, the flow within the hydrologic system is changed, which, in turn, affects the water's value for other uses.

Most large urban areas rely on water to transport resources. During recent years, the inland waterway system has carried about 10 percent of goods such as grain, coal, ore, and oil. In North America, expenditures for the improvement of the inland waterway system have totaled billions of dollars.

In the past, almost any navigation project was quickly approved and funded, regardless of the impact on other uses. Today, however, such decisions are not made until the impacts on other uses are carefully analyzed.

Table 16.1 Sources and Impacts of Selected Pollutants

Pollutant	Source	Effects on Humans	Effects on Aquatic Ecosystem
Acids	Atmospheric deposition; mine drainage; decomposing organic matter	Reduced availability of fish and shellfish Increased heavy metals in fish	Death of sensitive aquatic organisms; increased release of trace metals from soils, rock, and metal surfaces, such as water pipes
Chlorides	Runoff from roads treated for removal of ice or snow; irrigation runoff; brine produced in oil extraction; mining	Reduced availability of drinking water supplies; reduced availability of shellfish	At high levels, toxic to freshwater organisms
Disease-causing organisms	Dumping of raw and partially treated sewage; runoff of animal wastes from feed lots	Increased costs of water treatment; death and disease; reduced availability and contamination of fish, shellfish, and associated species	Reduced survival and reproduction of aquatic organisms due to disease
Elevated temperatures	Heat trapped by cities that is transferred to water; unshaded streams; solar heating of reservoirs; warm-water discharges from power plants and industrial facilities	Reduced availability of fish	Elimination of cold-water species of fish and shellfish; less oxygen; heat-stressed animals susceptible to disease Inappropriate spawning behavior
Heavy metals	Atmospheric deposition; road runoff; discharges from sewage treatment plants and industrial sources; creation of reservoirs; acidic mine effluents	Increased costs of water treatment; disease and death; reduced availability and healthfulness of fish and shellfish Biomagnification	Lower fish population due to failed reproduction; death of invertebrates leading to reduced prey for fish Biomagnification
Nutrient enrichment	Runoff from agricultural fields, pastures, and livestock feedlots; landscaped urban areas; dumping of raw and treated sewage and industrial discharges; phosphate detergents	Increased water treatment costs; reduced availability of fish, shellfish, and associated species; color and odor associated with algal growth; impairment of recreational uses	Algal blooms occur. Death of algae results in low oxygen levels and reduced diversity and growth of large plants. Reduced diversity of animals; fish kills
Organic molecules	Runoff from agricultural fields and pastures; landscaped urban areas; logged areas; discharges from chemical manufacturing and other industrial processes; combined sewers	Increased costs of water treatment; reduced availability of fish, shellfish, and associated species; odors	Reduced oxygen; fish kills; reduced numbers and diversity of aquatic life
Sediment	Runoff from agricultural land and livestock feed lots; logged hillsides; degraded stream banks; road construction; and other improper land use	Increased water treatment costs; reduced availability of fish, shellfish, and associated species; filling in of lakes, streams, and artificial reservoirs and harbors requiring dredging	Covering of spawning sites for fish; reduced numbers of insect species; reduced plant growth and diversity; reduced prey for predators; clogging of gills and filters
Toxic chemicals	Urban and agricultural runoff; municipal and industrial discharges; leachate from landfills and mines; atmospheric deposits	Increased costs of water treatment; increased risk of certain cancers; reduced availability and healthfulness of fish and shellfish	Reduced growth and survivability of fish eggs and young; fish diseases; death of carnivores due to biomagnification in the food chain

Source: Data, in part, from World Resources 1994–95.

Kinds and Sources of Water Pollution

Water pollution occurs when something enters water that changes the natural ecosystem and/or interferes with water use by segments of society. In an industrialized society, maintaining completely unpolluted water in all drains, streams, rivers, and lakes is probably impossible. (See table 16.1.) But we can evaluate the water quality of a body of water and take steps to preserve or improve its quality by eliminating sources of pollution. Some pollutants seriously affect the quality and possible uses of water. In general, water pollutants can be divided into several broad categories.

Toxic chemicals or acids may kill all organisms present and make the water unfit for human use. If these chemicals are persistent, they may bioaccumulate in individual organisms and biomagnify in food chains.

Dissolved organic matter is a significant water pollution problem because it decays in the water. As the decomposer microorganisms naturally present in water break down the organic matter, they use up available dissolved oxygen (D.O.) from the water. If too much dissolved oxygen is removed, aquatic organisms die. The amount of oxygen required to decay a certain amount of organic matter is called the **biochemical oxygen demand (BOD).** (See figure 16.13.) Measuring the BOD of a body of water is one way to determine how polluted it is. If too much organic matter is added to the water, all of the available oxygen will be used up. Then, anaerobic (not requiring oxygen) bacteria begin to break down wastes. Anaerobic respiration produces

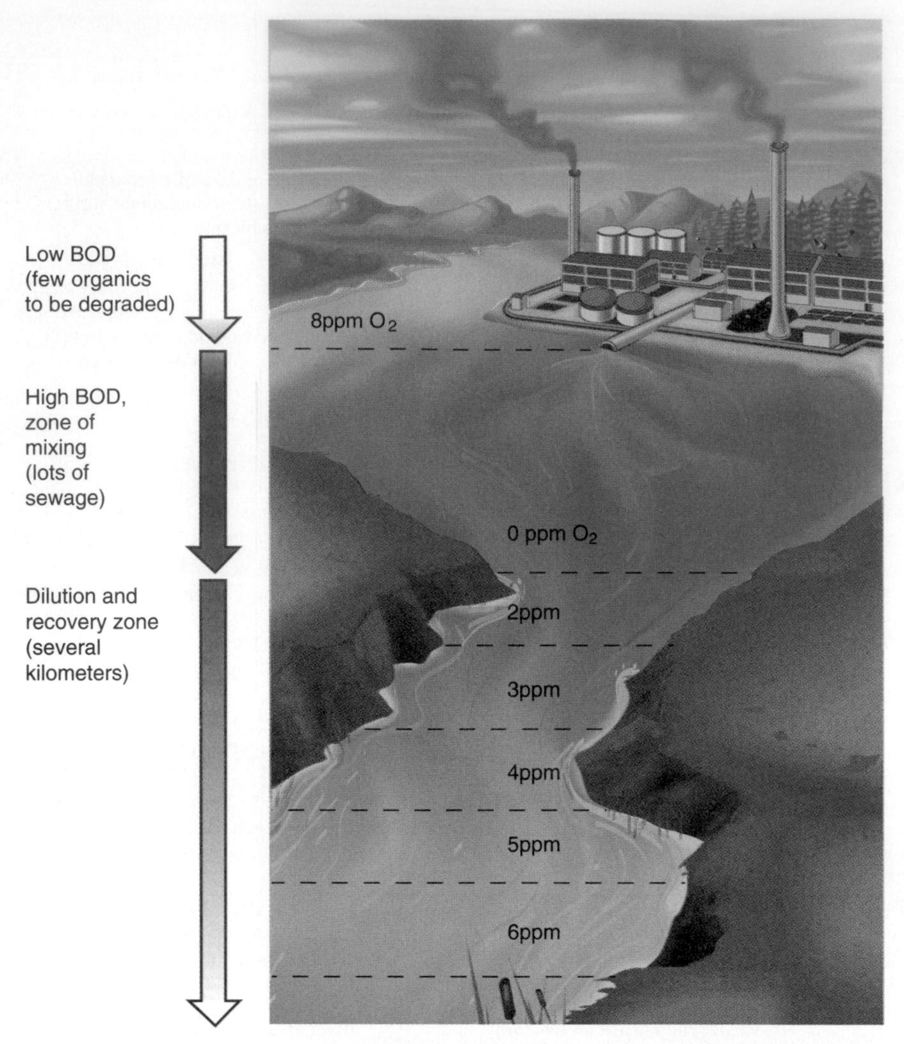

figure 16.13 **Effect of Organic Wastes on Dissolved Oxygen** Sewage contains a high concentration of organic materials. When these are degraded by organisms, oxygen is removed from the water. This is called the biochemical oxygen demand (BOD). There is an inverse relationship between the amount of organic matter and oxygen in the water. The greater the BOD, the more difficult it is for aquatic animals to survive and the less desirable the water is for human use. The more the organic pollution, the greater the BOD.

chemicals that have a foul odor and an unpleasant taste and that generally interfere with the well-being of humans.

Disease-causing organisms are a very important pollution problem in most of the world. Untreated or inadequately treated human or domesticated animal waste is most often the source of these organisms. In the developed world, sewage-treatment and drinking-water treatment plants greatly reduce this public health problem.

Nutrients are also a pollution problem. Additional nutrients in the form of nitrogen and phosphorus compounds increase the rate of growth of aquatic plants and algae. However, phosphates and nitrates are generally present in very limited amounts in unpolluted freshwater and, therefore, are a limiting factor on the growth of aquatic plants and algae. (A **limiting factor** is a necessary material that is in short supply, and because of the lack of it, an organism cannot reach its full potential growth. See chapter 5.) Thus, when phosphates or nitrates are added to the surface water, they acted as a fertilizer and promoted the growth of undesirable algae populations. The excessive growth of algae and aquatic plants due to added nutrients is called **eutrophication.** Algae and larger aquatic plants may interfere with the use of the water by fouling boat propellers, clogging water-intake pipes, changing the taste and odor of water, and causing the buildup of organic matter on the bottom. As this organic matter decays, oxygen levels decrease, and fish and other aquatic species die.

Physical particles also can negatively affect water quality. Particles alter the clarity of the water, can cover spawning sites, act as abrasives that injure organisms, and carry toxic materials attached to them.

Ideally, we should think in terms of eliminating all pollution, but any human use is going to have some minor negative impact. Even activity such as swimming and boating add particles and chemicals to the water. So, determining acceptable water quality involves economic considerations. The cost of removing the last few parts per million of some materials from the water may not significantly improve water quality and may not be economically justifiable. This is certainly true of organic matter, which is biodegradable. However, radioactive wastes and toxins that may accumulate in living tissue are a different matter. Vigorous attempts to remove these materials are often justified because of the materials' potential harm to humans and other organisms.

Sources of pollution are classified as either point sources or nonpoint sources. When a source of pollution can be readily identified because it has a definite source and place where it enters the water, it is said to come from a **point source.** Municipal and industrial discharge pipes are good examples of point sources. Diffuse pollutants, such as from agricultural land and urban paved surfaces, acid rain and runoff, are said to come from **nonpoint sources** and are much more difficult to identify and control. Initial attempts to control water pollution were focused on point sources of pollution, since these were readily identifiable and economic pressure and adverse publicity could be brought to bear on companies that continued to pollute from point sources. In North America, most point sources of water pollution have been identified and are

ENVIRONMENTAL CLOSE-UP

Is It Safe to Drink the Water?

Roughly 1000 contaminants have been detected in the public water supply in the United States, and virtually every major water source is vulnerable to pollution. About 60 percent of the U.S. population relies on surface water from rivers, lakes, and reservoirs that may contain industrial and agricultural wastes and pesticides washed off fields by rain. The other 40 percent uses groundwater that may be tainted by chemicals slowly seeping in from toxic-waste dumps, agricultural activities, and leaking sewage and septic systems. In some areas where groundwater supplies are being gradually depleted, the chemical pollutants are becoming more concentrated.

Most pollutants are probably not concentrated enough to pose significant health hazards; however, there are exceptions. The most widespread danger in water is lead, which can cause high blood pressure and an array of other health problems. Lead is especially hazardous to children, since it impairs the development of brain cells. The U.S. Environmental Protection Agency estimates that at least 42 million Americans are exposed to unacceptably high levels of lead, and the U.S. Public Health Service estimates that perhaps 9 million children are at least slightly affected by it.

The contamination comes from old lead pipes and solder that have been used in plumbing for years. These materials are gradually being replaced in homes and water systems. Individuals may want to have their water tested for lead by an official lab. If the level is too high, they can investigate ways to deal with the problem or switch to bottled water for drinking and cooking. Even then, caution is called for—some bottled waters contain many of the same contaminants that tap water does.

The other four types of contamination in the U.S. water supply, along with their source and risk, are shown in the table. Regardless of the problems, however, the water supply in the United States is among the cleanest in the world.

Common Contaminants of Drinking Water

Substance	Source	Health Effects
Persistent chlorinated organic compounds	Used as solvents in industry; past use as pesticides	Various; including reproductive problems and cancer
Trihalomethanes	Produced by chemical reactions when water is disinfected by chlorination	Liver and kidney damage and possible cancer
Nitrates	Primarily from fertilizer and effluent from concentrated livestock raising	Can react to reduce oxygen uptake by blood; particularly a problem with children
Lead	Old piping and solder in public water distribution systems, homes, and other buildings	Nerve damage, learning difficulties in children, birth defects, possible cancer
Pathogenic bacteria, protozoa, and viruses	Leaking septic tanks and sewers, contamination of water supply from birds and mammals, and inadequate disinfection	Acute gastrointestinal illness and other serious health problems

Source: Data, in part, from the U.S. Environmental Protection Agency.

regulated. In the United States, the EPA is responsible for identifying point sources of pollution, negotiating the permissible levels of pollution allowed from each source, and enforcing the terms of the permits.

Currently, nonpoint sources of water pollution are being addressed, but this is much more difficult to do since regulating many small, individual human acts is necessary. In 1998, President Clinton announced the formation of the Clean Water Action Plan. This plan requires the cooperation of several different branches of government and focuses on watersheds and nonpoint sources of pollution, particularly runoff from urban areas and agricultural land. In addition, plans have been developed to address several long-term water-quality issues, such as acid runoff from mines, contaminated sediments, the presence of toxic chemicals in fish, and restoring degraded wetlands.

Municipal Water Pollution

Municipalities are faced with the double-edged problem of providing suitable drinking water for the population and disposing of wastes. These wastes consist of storm-water runoff, wastes from industry, and wastes from homes and commercial establishments. Wastes from homes consist primarily of organic matter from garbage, food preparation, cleaning of clothes and dishes, and human wastes. Human wastes are mostly undigested food material and a concentrated population of bacteria, such as *Escherichia coli* and *Streptococcus faecalis.* These particular bacteria normally grow in the large intestine (colon) of humans and are present in high numbers in the feces of humans; therefore, they are commonly called **fecal coliform bacteria.** Fecal coliform bacteria are also present in the

GLOBAL PERSPECTIVE

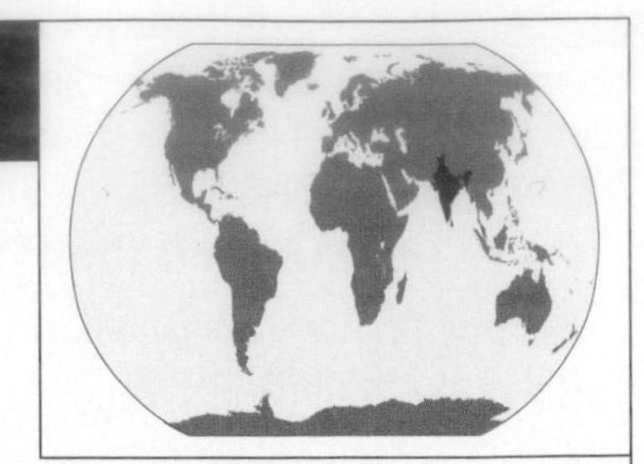

The Cleanup of the Holy Ganges

Every day, thousands of Hindus flock to the banks of the Holy Ganges River in India. There, they drink and bathe in what they believe to be holy water, as partially cremated corpses float past them and nearby drains emit millions of liters of raw sewage. Clean water is one of India's most scarce resources, but, as in many developing nations, the money and technology needed to properly treat sewage are not available in most cities and villages. With the country's population of slightly more than 1 billion people expected to double in 39 years, officials are concerned that Indians will have no choice but to continue dumping raw waste into local waterways, contributing to epidemics of diarrhea and other diseases that kill thousands of people annually.

Keeping the Ganges clean is made especially difficult because faith in the river's incorruptible purity has generated complacency and ambivalence about its pollution among many of the 300 million people who live in the Ganges Basin. More than 1600 million liters (425 million gallons) of untreated municipal sewage, industrial waste, agricultural runoff, and other pollutants are discharged into the river every day. At the same time, officials estimate that more than 1 million people a day bathe or take a "holy dip" in the Ganges, and thousands drink straight from its banks.

The Hindu belief in cremation has led to several environmental problems. A wood cremation takes more than 50 pounds of wood, costing two week's wages for the typical Indian. There are complaints that wood cremations are helping to devour India's forests. Given the high cost of wood, many bodies are not completely cremated, and the partially burned corpses are disposed of in the Ganges.

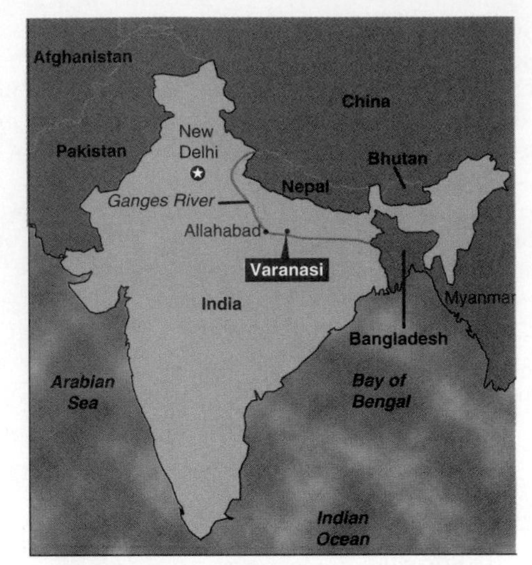

One solution to this problem has been the introduction of 25,000 specially raised snapping turtles that are attracted to the rotten smell of corpses. While this solution may seem extreme, there is another that is more acceptable in the long run. In 1992, as part of the Ganges Action Plan, an electric crematorium was built at the city of Varanasi that charges less than $2 per body and does a thorough job of turning the bodies to ash. There is some concern however, that many Hindus will not want to abandon the traditional ritual of a wood cremation for the more efficient and less costly electric cremation.

Although this entire scenario may seem somewhat unusual to North Americans, it is important to keep in mind how culture and religion affect our environment. To many Indians, the cultural and religious practices of North Americans are equally puzzling.

wastes of other warm-blooded animals, such as birds and mammals. Low numbers of these bacteria in water are not harmful to healthy people. However, because they can be easily identified, their presence in the water is used to indicate the amount of pollution from the fecal wastes of humans and other warm-blooded animals. The numbers of these types of bacteria present in water are directly related to the amount of fecal waste entering the water. When human wastes are disposed of in water systems, other potentially harmful bacteria from humans may be present in amounts too small to detect by sampling. Even in small numbers, these harmful bacteria may cause disease epidemics. The greater the amount of wastes deposited in the water, the more likely it is that there will be populations of disease-causing bacteria. Therefore, fecal coliform bacteria are used as an indicator that other more harmful organisms may be present.

Wastewater from cleaning dishes and clothing contains some organic material along with the soap or detergent, which helps to separate the contaminant from the dishes or clothes. Soaps and detergents are useful because one end of the molecule dissolves in dirt or grease and the other end dissolves in water. When the soap or detergent molecules are rinsed away by the water, the dirt or grease goes with them.

At one time, many detergents contained phosphates as a part of their chemical make-up, which contributed to eutrophication. However, because of the environmental effects of phosphate on aquatic environments, since 1994 most major detergent manufacturers in North America and other developed countries have eliminated phosphates from most of their formulations. Today, the majority of phosphate entering water in North America is from human waste and runoff from farm fields and livestock operations.

Agricultural Water Pollution

Agricultural activities are the primary cause of water pollution problems. Excessive use of fertilizer results in eutrophication in many aquatic habitats, because precipitation carries dissolved nutrients (nitrogen and phosphorus compounds) into streams and lakes. In addition, groundwater may become contaminated with fertilizer and pesticides. The exposure of land to erosion results in increased amounts of sediment being added to water courses. Runoff from animal feedlots carries nutrients, organic matter, and bacteria. Water used to flush irrigated land to get rid of excess salt in the soil carries a heavy load of salt that degrades the water body. And the use of agricultural chemicals results in contamination of sediments and aquatic organisms. One of the largest water pollution problems is agricultural runoff from large expanses of open fields. See chapter 14 for a general discussion of methods for reducing runoff and soil erosion.

Farmers can reduce runoff in several ways. One is to leave a zone of undisturbed, permanently vegetated land, called a conservation buffer, near drains or stream banks. This retards surface runoff because soil covered with vegetation tends to slow the movement of water and allows the silt to be deposited on the surface of the land rather than in the streams. This can be costly because farmers may need to remove valuable cropland from cultivation. One goal of the Clean Water Action Plan is to establish 3.2 million kilometers (2 million miles) of conservation buffer strips. By 2000, the plan was about halfway to its goal. Another way to retard runoff is keep the soil covered with a crop as long as possible. Careful control of the amount and the timing of fertilizer application can also reduce the amount of nutrients lost to streams. This makes good economic sense because any fertilizer that runs off or leaches out of the soil is unavailable to crop plants and results in less productivity.

Industrial Water Pollution

Factories and industrial complexes frequently dispose of some or all of their wastes into municipal sewage systems. Depending on the type of industry involved, these wastes contain organic materials, petroleum products, metals, acids, toxic materials, organisms, nutrients, or particulates. Organic materials and oil add to the BOD of the water. The metals, acids, and specific toxic materials need special treatment, depending on their nature and concentration. In these cases, a municipal wastewater treatment plant will require that the industry pretreat the waste before sending it to the wastewater treatment plant. If this is not done, the municipal sewage treatment plants must be designed with their industrial customers in mind. In most cases, cities prefer that industries take care of their own wastes. This allows industries to segregate and control toxic wastes and design wastewater facilities that meet their specific needs.

Since industries are point sources of pollution they have been relatively easy to identify as pollution sources, have been vigorously regulated, and have responded to mandates that they clean up their effluent. Most companies, when they remodel their facilities, include wastewater treatment as a necessary part of an industrial complex. However, some older facilities continue to pollute. These companies discharge acids, particulates, heated water, and noxious gases into the water. While industrial water pollution in the industrialized world has been significantly regulated, in much of the developing world this is not the case and many lakes, streams, and harbors are severely polluted with heavy metals and other toxic materials, organic matter, and human and animal waste.

A special source of industrial water pollution is mining. By its very nature, mining disturbs the surface of the earth and increases the chances that sediment and other materials will pollute surface waters. Hydraulic mining is practiced in some countries and involves spraying hillsides with high pressure water jets to dislodge valuable ores. Often, chemicals are used to separate the valuable metals from the ores, and the waste from these processes is released into streams as well. Water that drains from current or abandoned coal mines is often very acid. Pyrite is a mineral associated with many coal deposits. It contains sulfur, and, when exposed to weathering, the sulfur reacts with oxygen and sulfuric acid results. In addition, fine coal-dust particles are suspended in the water, which makes the water chemically and physically less valuable as a habitat. Dissolved ions of iron, sulfur, zinc, and copper also are present in mine drainage. Control involves containing mine drainage and treating it before it is released to surface water. Old abandoned mines present a particular problem because no one can be held responsible for cleaning up the drainage.

Thermal Pollution

Amendments to the U.S. Federal Water Pollution Control Act of 1972 (PL 92-500) have mandated changes in how industry treats water. Industries are no longer allowed to use water and return it to its source in poor condition. One of the standards regulates the temperature of the water that is returned to its source. Because many industries use water for cooling, thermal pollution can be a problem. **Thermal pollution** occurs when an industry removes water from a source, uses the water for cooling purposes, and then returns the heated water to its source.

Power plants heat water to convert it into steam, which drives the turbines that generate electricity. For steam turbines to function efficiently, the steam must be condensed into water after it leaves the turbine. This condensation is usually accomplished by taking water from a lake, stream, or ocean to absorb the heat. This heated water is then discharged. The least expensive and easiest method of discharging heated water is to return the water to the aquatic environment, but this can create problems for the inhabitants of the area. Although an increase in temperature of only a few degrees may not seem significant, some aquatic ecosystems are very sensitive to minor temperature changes. Many fish are triggered to spawn by increases in temperature, while others may be inhibited from spawning if the temperature rises. For example, lake trout will not spawn in water above 10°C (50°F). If a lake has a temperature of 8°C (46°F), the lake trout will reproduce, but an increase of 3°C (5°F) would prevent

GLOBAL PERSPECTIVE

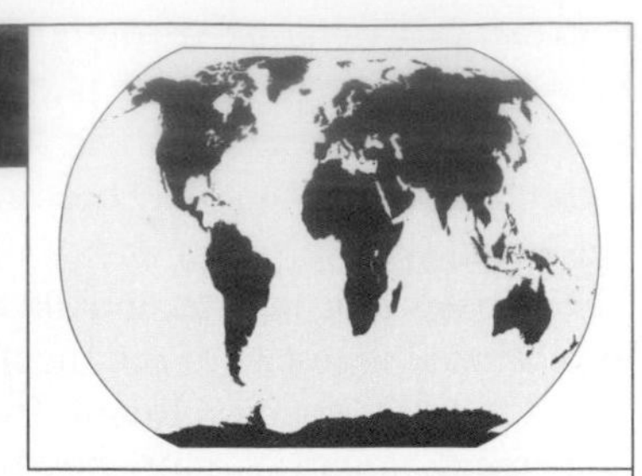

Comparing Water Use and Pollution in Industrialized and Developing Countries

Characteristic	Industrialized Countries	Developing Countries
Water Use		
Domestic water use per capita	1. Heavy per capita use 2. Highest usage in Australia, New Zealand, U.S., and Canada 3. Usage is stabilizing.	1. Small per capita use 2. Water usage increases as living standards go up.
Where water is used	1. Irrigation and industry total about 85 percent. 2. Domestic about 15 percent	1. Irrigation is over 80 percent, particularly high in Asia and Africa. 2. Industry and domestic uses each less than 10 percent.
Access to safe drinking water and wastewater treatment	1. Safe drinking water generally available 2. Wastewater treatment generally available 3. Only small population increases expected	1. Large numbers of people lack safe drinking water. 2. Effective wastewater treatment generally *not* available 3. Rapidly growing urban population will create greater need for safe drinking water and wastewater treatment.
Pollution Control		
Domestic wastewater treatment	1. Most countries treat domestic waste. 2. Central and Eastern European countries lack effective sewage treatment.	1. Almost all sewage is discharged without effective treatment.
Industrial wastes	1. Industrial discharges strictly regulated in most countries 2. Some accidental discharges 3. Discharges from disused industrial sites are a problem. 4. Industrial air pollution has caused acidification of lakes in parts of North America and Europe. 5. Eastern and Central Europe have serious industrial waste problems from historically unregulated industry. Black Sea and Baltic Sea heavily polluted.	1. Largely untreated 2. Little attention to regulating industrial waste 3. Acidification of lakes becoming important in China and tropical Africa
Land-use runoff	1. Fertilizers and pesticides are a continuing problem. 2. Soil conservation practices used, but agricultural runoff is still a significant problem. 3. Runoff from urban areas causes some particles, oil, and other chemicals to enter water.	1. Heavy fertilizer and pesticide use causes serious water-quality problems. 2. Deforestation and poor farming practices cause soil erosion and degrade water quality. 3. Runoff from urban areas causes serious pollution problems; trash, human waste, animal waste, chemical waste all present.

spawning and result in this species' eventual elimination from that lake. Another problem associated with elevated water temperature is that it results in a decrease in the amount of oxygen dissolved in the water.

Ocean estuaries are very fragile. The discharge of heated water into an estuary may alter the type of plants present. As a result, animals with specific food habits may be eliminated because the warm water supports different food organisms. The entire food web in the estuary may be altered by only slight temperature increases.

Cooling water used by industry does not have to be released into aquatic ecosystems. Today in the industrialized world, most cooling water is not released in such a way that aquatic ecosystems are endangered. There are three other methods of discharging the heat. One method is to construct a large shallow pond. Hot water is pumped into one end of the pond, and cooler water is removed from the other end. The heat is dissipated from the pond into the atmosphere and substrate.

A second method is to use a cooling tower. In a cooling tower, the heated water is sprayed into the air and cooled by evaporation. The disadvantage of cooling towers and shallow ponds is that large amounts of water are lost by evaporation. The release of this water into the air can also produce localized fogs.

The third method of cooling, the dry tower, does not release water into the atmosphere. In this method, the heated water is pumped through tubes, and the heat is released into the air. This is the same principle used in an automobile radiator. The dry tower is the most expensive to construct and operate.

Marine Oil Pollution

Marine oil pollution has many sources. One source is accidents, such as oil-drilling blowouts or oil-tanker accidents. The *Exxon Valdez,* which ran aground in Prince William Sound, Alaska, in 1989, released over 42 million liters (11 million gallons) of oil and affected nearly 1500 kilometers (930 miles) of Alaskan coastline. The event had a great effect on the algae and on animal populations of the sound, and the economic impact on the local economy was severe. However, by 1994, there was little visible sign of the oil that had covered the coastline in 1989. A U.S. National Oceanic and Atmospheric Administration (NOAA) study estimates that 50 percent of the oil biodegraded on beaches or in the water; 20 percent evaporated; 14 percent was recovered; 12 percent is at the bottom of the sea, mostly in the Gulf of Alaska; 3 percent lies on shorelines; and less than 1 percent still drifts in the water column. Furthermore, it appears that the animal populations have recovered or are well on their way to doing so. This points out the tremendous resilience of natural ecosystems to recover from disastrous events.

Although accidents such as the *Exxon Valdez* are spectacular events, much more oil is released as a result of small, regular releases from other, less-visible sources. Nearly two-thirds of all human-caused marine oil pollution comes from three sources: (1) runoff from streets, (2) improper disposal of lubricating oil from machines or automobile crankcases, and (3) intentional oil discharges that occur during the loading and unloading of tankers. With regard to the latter, pollution occurs when the tanks are cleaned or oil-contaminated ballast water is released. Oil tankers use seawater as ballast to stabilize the craft after they have discharged their oil. This oil-contaminated water is then discharged back into the ocean when the tanker is refilled. In addition to human-caused oil pollution, there are many places where oil naturally seeps from underlying oil deposits into water.

As the number of offshore oil wells and the number and size of oil tankers have grown, the potential for increased oil pollution has also grown. Many methods for controlling marine oil pollution have been tried. Some of the more promising methods are recycling and reprocessing used oil and grease from automobile service stations and from industries, and enforcing stricter regulations on the offshore drilling, refining, and shipping of oil. As a result of oil spills from shipping tankers, an international agreement was reached in 1992 which required that all new oil tankers be constructed with two hulls—one inside the other. Such double-hulled vessels would be much less likely to rupture and spill their contents. Today, approximately 15 percent of oil tankers are double hulled.

Groundwater Pollution

A wide variety of activities, some once thought harmless, have been identified as potential sources of groundwater contamination. In fact, possible sources of human-induced groundwater contamination span every facet of social, agricultural, and industrial activities. (See figure 16.14.) Once groundwater pollution has occurred, it is extremely difficult to remedy. Pumping groundwater and treating it is very slow and costly. A much better way to deal with the issue of groundwater pollution is to work very hard to prevent the pollution from occurring in the first place.

Major sources of groundwater contamination include:

1. **Agricultural products**

   Pesticides contribute to unsafe levels of organic contaminants in groundwater. Seventy-three different pesticides have been detected in the groundwater in Canada and the United States. Accidental spills or leaks of pesticides pollute groundwater sources with 10 to 20 additional pesticides. Other agricultural practices contributing to groundwater pollution include animal-feeding operations, fertilizer applications, and irrigation practices.

2. **Underground storage tanks**

   For many years in North America, a large number of underground storage tanks containing gasoline and other hazardous substances have leaked. Four liters (1 gallon) of gasoline can contaminate the water supply of a community of 50,000 people. A major program of replacing leaking underground storage tanks has recently been completed in the United States. However, the effects of past leaks and abandoned tanks will continue to be a problem for many years.

3. **Landfills**

   Even though recently constructed landfills have special liners and water collection systems, approximately 90 percent of the landfills in North America have no liners to stop leaks to underlying groundwater, and 96 percent have no system to collect the leachate that seeps from the landfill. Sixty percent of landfills place no restrictions on the waste accepted, and many landfills are not inspected even once a year.

4. **Septic tanks**

   Poorly designed and inadequately maintained septic systems have contaminated groundwater with nitrates, bacteria, and toxic cleaning agents. Over 20 million septic tanks are in use, and up to a third have been found to be operating improperly.

5. **Surface impoundments**

   Over 225,000 pits, ponds, and lagoons are used in North America to store or treat wastes. Seventy-one percent are unlined, and only 1 percent use a plastic or other synthetic, nonsoil liner. Ninety-nine percent of these impoundments have no leak-detection systems. Seventy-three percent have no restriction on the

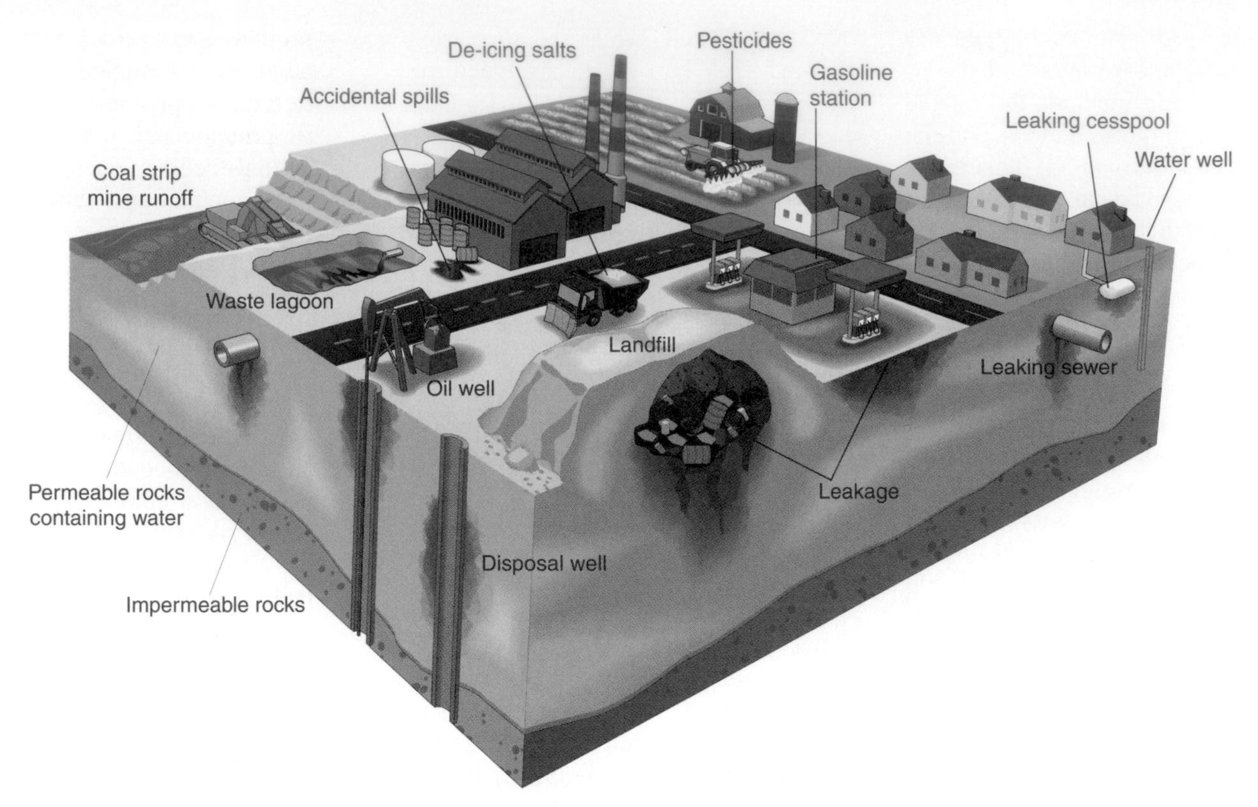

figure 16.14 **Sources of Groundwater Contamination** A wide variety of activities have been identified as sources of groundwater contamination.

waste placed in the impoundment. Sixty percent are not even inspected annually. Many of these ponds are located near groundwater supplies.

Other sources of groundwater contamination include mining wastes, salting for snow control, land application of treated wastewater, open dumps, cemeteries, radioactive disposal sites, urban runoff, construction excavation, fallout from the atmosphere, and animal feedlots.

Water-Use Planning Issues

In the past, wastes were discharged into waterways with little regard for the costs imposed on other users by the resulting decrease in water quality. Furthermore, as the population has grown and the need for irrigation and domestic water has intensified, in many parts of the world there has not been enough water to satisfy everyone's needs. With today's increasing demands for high-quality water, unrestrained waste disposal and unlimited withdrawal of water could lead to serious conflicts about water uses, causing social, economic, and environmental losses at both local and international levels. (See table 16.2.)

Metropolitan areas must deal with a variety of issues and maintain an extensive infrastructure in order to provide three basic water services:

1. Water supply for human and industrial needs
2. Wastewater collection and treatment
3. Storm-water collection and management

Water sources must be identified and preserved for use. Some cities obtain all their municipal water from groundwater and must have a thorough understanding of the size and characteristics of the aquifer they use. Some cities, such as New York City, obtain potable water by preserving a watershed that supplies the water needed by the population. Other cities have abundant water in the large rivers that flow by them but must deal with pollution problems caused by upstream users. In many places where water is in short supply, there is a conflict between municipal and industrial-agricultural needs for water.

Water for human and industrial use must be properly treated and purified. It is then pumped through a series of pipes to consumers. After the water is used, it flows through a network of sewers to a wastewater treatment plant before it is released. Maintaining the infrastructure of pipes, pumps, and treatment plants is expensive.

Metropolitan areas must also deal with great volumes of excess water during storms. This water is known as **storm-water runoff.** Because urban areas are paved and little rainwater can be absorbed into the ground, managing storm water is a significant problem. Cities often have severe local flooding because the water is channeling along streets to storm sewers. If these sewers

Table 16.2 International Water Disputes

River	Countries in Dispute	Issues
Nile	Egypt, Ethiopia, Sudan	Siltation, flooding, water flow/diversion
Euphrates & Tigris	Iraq, Syria, Turkey	Reduced water flow, salinization
Jordan, Yarmuk Litani, & West Bank aquifer	Israel, Jordan, Syria, Lebanon	Water flow/diversion
Indus & Sutlej	India, Pakistan	Irrigation
Ganges-Brahmaputra	Bangladesh, India	Siltation, flooding, water flow
Salween	Myanmar, China	Siltation, flooding
Mekong	Cambodia, Laos, Thailand, Vietnam	Water flow, flooding
Paraná	Argentina, Brazil	Dam, land inundation
Lauca	Bolivia, Chile	Dam, salinization
Rio Grande & Colorado	Mexico, United States	Salinization, water flow, agrochemical pollution
Rhine	France, Netherlands, Switzerland, Germany	Industrial pollution
Maas & Schelde	Belgium, Netherlands	Salinization, industrial pollution
Elbe	Czech Republic, Germany	Industrial pollution
Szamos	Hungary, Romania	Industrial pollution

Source: From Michael Renner, "National Security: The Economic and Environmental Dimensions" in *Worldwatch Paper 89,* 1989. Reprinted by permission of Worldwatch Institute.

are overloaded or blocked with debris, the water cannot escape and flooding occurs.

The Water Quality Act of 1987 requires that municipalities obtain permits for discharges of storm-water runoff so that non-point sources of pollution are controlled. In the past, many cities had a single system to handle both sewage and storm-water runoff. During heavy precipitation or spring thaws, the runoff from streets could be so large that the wastewater treatment plant could not handle the volume. The wastewater was then diverted directly into the receiving body of water without being treated. Because of these new requirements, some cities have created areas in which to store this excess water until it can be treated. This is expensive and, therefore, is done only if federal or state funding is available. Many cities have also gone through the expensive process of separating their storm sewers from their sanitary sewers.

Providing water services is expensive. We must understand that water supplies are limited. We must also understand that water's ability to dilute and degrade pollutants is limited and that proper land-use planning is essential if metropolitan areas are to provide services and limit pollution.

In pursuing these objectives, city planners encounter many obstacles. Large metropolitan areas often have hundreds of local jurisdictions (governmental and bureaucratic areas) that divide responsibility for management of basic water services. The Chicago metropolitan area is a good example. This area is composed of six counties and approximately two thousand local units of government. It has 349 separate water-supply systems and 135 separate wastewater disposal systems. Efforts to implement a water-management plan, when so many layers of government are involved, are complicated and frustrating.

To meet future needs, urban, agricultural, and national interests will need to deal with a number of issues, such as the following:

- Increased demand for water will generate pressure to divert water to highly populated areas or areas capable of irrigated agriculture.
- Increased demand for water will force increased treatment of wastewater and reuse of existing water supplies.
- In many areas where water is used for irrigation, evaporation of water from the soil over many years results in a buildup of salt in the soil. When the water used to flush the salt from the soil is returned to a stream, the quality of the water is lowered.
- In some areas, wells provide water for all categories of use. If the groundwater is pumped out faster than it is replaced, the water table is lowered.
- In coastal areas, seawater may intrude into the aquifers and ruin the water supply.
- The demand for water-based recreation is increasing dramatically and requires high-quality water, especially for water recreation involving total body contact, such as swimming.

Water Diversion

Water diversion is the physical process of transferring water from one area to another. The aqueducts of ancient Rome are early examples of water diversion. Thousands of diversion projects have been constructed since then. New York City, for example, diverts water from Pennsylvania [250 kilometers (155 miles) away], and Los Angeles obtains part of its water supply from the Colorado River. (See Issues and Analysis: The California Water Plan p. 367.)

While diversion is often seen as a necessity in many parts of the world, it often generates controversy. An example of this is the Garrison Diversion Unit in North Dakota, which was originally envisioned as a way to divert water from the Missouri River system for irrigation. (See figure 16.15.) The initial plan to irrigate portions of the Great Plains was developed during the Dust Bowl era of the 1930s. The Federal Flood Control Act of 1944 authorized the construction of the Garrison Diversion Unit. However, one of the original intents of the plan—to divert water from the Missouri River to the Red River—has not been met. Since the Red River flows north into Canada, the project requires international cooperation. The Canadian government has concerns about the effects on water quality of releasing additional water into the Red River. There are also concerns about environmental consequences. Portions of wildlife refuges, native grasslands, and waterfowl breeding marshes would be damaged or destroyed. Some states also have expressed concern about diverting water from the Missouri River, since any water diverted is not available for those downstream on the Missouri River.

While the plan has been modified several times, two sections of canal and a pumping station at Lake Sakakawea have been completed, but not used. Proponents continue to push for legislation to complete the connection between the two canals that would allow water to be diverted to the Red River for municipal use. The original intent of diverting water for irrigation has been eliminated from the most recent proposal.

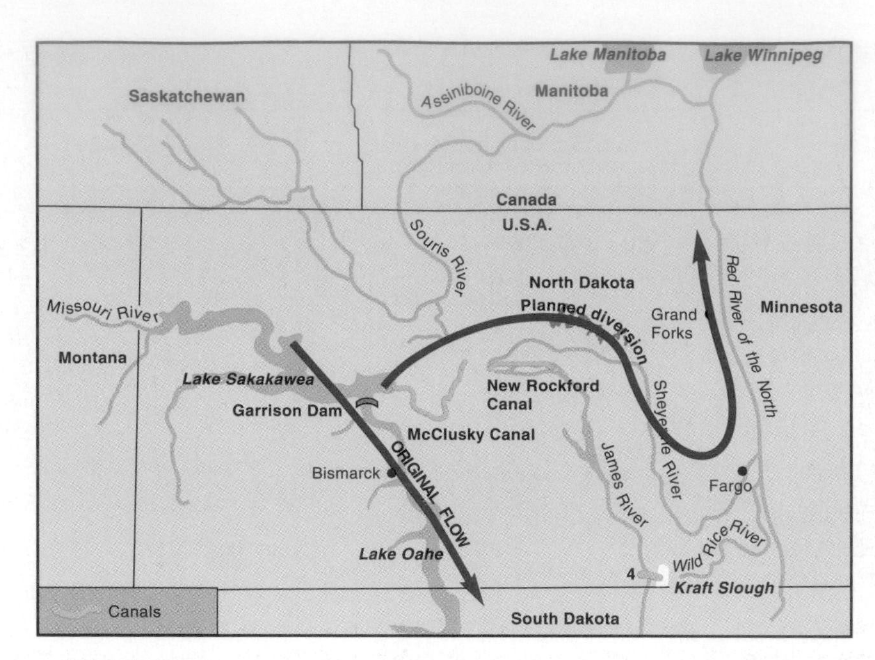

figure 16.15 **The Garrison Diversion Unit** The original intent of this plan was to divert water from the Missouri River to the McClusky Canal and the Sheyenne River and eventually into Lake Winnepeg by way of the Red River. In the process, additional land could be irrigated, and growing populations in the Red River Valley could be served with adequate water. Opposition has stalled the project, and only portions of the canal system and a pumping station have been completed.

Wastewater Treatment

Because water must be cleaned before it is released, most companies and municipalities in the developed world maintain wastewater treatment facilities. The percentage of sewage that is treated, however, varies greatly throughout the world. (See table 16.3.) Treatment of sewage is usually classified as primary, secondary, or tertiary. **Primary sewage treatment** is primarily a physical process that removes larger particles by filtering water through large screens and then allowing smaller particles to settle in ponds or lagoons. Water is removed from the top of the settling stage and either released to the environment or to a subsequent stage of treatment. If the water is released to the environment, it does not have any sand or grit; but it still carries a heavy load of organic matter, dissolved salts, bacteria, and other microorganisms. The microorganisms use the organic material for food, and, as long as there is sufficient oxygen, they will continue to grow and reproduce. If the receiving body of water is large enough and the organisms have enough time, the organic matter will be degraded. In crowded areas, where several municipalities take water and return it to a lake or stream, within a few kilometers of each other, primary water treatment is not adequate and major portions of the receiving body of water are affected.

Table 16.3 Percent of Sewage Treated in Selected Areas

Area	Percent
Europe	72
Mediterranean Sea	30
Caribbean Basin	Less than 10
Southeast Pacific	Almost zero
South Asia	Almost zero
South Pacific	Almost zero
West and Central Africa	Almost zero

Source: Data from World Resources Institute 1994–95.

Secondary sewage treatment is a biological process that usually follows primary treatment. It involves holding the wastewater until the organic material has been degraded by the bacteria and other microorganisms. Secondary treatment facilities are designed to promote the growth of microorganisms. To encourage this action, the wastewater is mixed with large quantities of highly oxygenated water, or the water is aerated directly, as in a trickling filter system or an activated sludge system. In a **trickling filter system,** the wastewater is sprayed over the surface of rock or other substrate to increase the amount of dissolved oxygen. The rock also provides a place for a film of bacteria and

a.

c.

b.

figure 16.16 **Primary and Secondary Wastewater Treatment** Primary treatment is physical; it includes filtrating and settling of wastes. Photograph (*a*) is a settling tank in which particles settle to the bottom. Secondary treatment is mostly biological and includes the concentration of dissolved organics by microorganisms. Two major types of secondary treatment are trickling filter and activated sludge methods. Photograph (*b*) shows a trickling filter system and photograph (*c*) shows an activated sludge system.

other microbes to attach so they are exposed simultaneously to the organic material in the water and to oxygen. These microorganisms feed on the dissolved organic matter and small suspended particles, which then become incorporated into their bodies as part of their cell structure. The bodies of the microorganisms are larger than the dissolved and suspended organic matter, so this process concentrates the organic wastes into particles that are large enough to settle out. This mixture of organisms and other particular matter is called **sewage sludge.** The sludge that settles consists of living and dead microorganisms and their waste products.

In **activated-sludge sewage treatment** plants, the wastewater is held in tanks and has air continuously bubbled through it. The sludge eventually is moved to settling tanks where the water and sludge can be separated. To make sure that the incoming wastewater has appropriate kinds and amounts of decay organisms, some of the sludge is returned to aeration tanks, where it is mixed with incoming wastewater. This kind of process uses less land than a trickling filter. (See figure 16.16.) Both processes produce a sludge that settles out of the water.

The sludge that remains is concentrated and often dewatered (dried) before disposal. Sludge disposal is a major problem in large population centers. In the San Francisco Bay area, 2500 tonnes (2750 U.S. tons) of sludge are produced each day. Most of this is carried to landfills and lagoons, and some is composted and returned to the land as fertilizer. Some municipalities incinerate their sludge. In other areas, if the sewage sludge is free of heavy metals and other contaminants that might affect plant growth or the quality and safety of food products, it is applied directly to agricultural land as a fertilizer and soil conditioner.

In North America and much of the developed world, wastewater receives both primary and secondary sewage treatment. While the water has been cleansed of its particles and dissolved organic matter, it still has microorganisms that might be harmful. Therefore the water discharged from these sewage treatment plants must be disinfected. The least costly method of disinfection is chlorination. However, many people feel that the use of chlorine should be discontinued, since chlorination may be responsible for the creation of chlorinated organic compounds that are harmful. Therefore, other methods of killing

ENVIRONMENTAL CLOSE-UP

Restoring the Everglades

Everglades National Park is a unique, subtropical, freshwater wetland visited by about a million people per year. This unique ecosystem exists because of an unusual set of conditions. South Florida is a nearly flat landscape with a slight decrease in elevation from the north to the south. In addition, the substrate is a porous limestone that allows water to flow through rather easily. Originally, water drained in a broad sheet from Lake Okeechobee southward to Florida Bay. This constant flow of water sustained a vast, grassy wetland interspersed with patches of trees.

After Everglades Park was established in 1947, about 800,000 hectares (2 million acres) of wetlands to the north were converted to farms and urban development. South Florida boomed. Some 4.5 million people currently live in the horseshoe crescent around the Everglades region, and new residents arrive each day. Conversion of land to agriculture and urban development required changes in the natural flow of water. Dams, drainage canals, and water diversion supported and protected the human uses of the area but cut off the essential, natural flow of freshwater to the Everglades. Changes in the normal pattern of water flow to the Everglades resulted in periods of drought and a general reduction in the size of this unique wetland region. Populations of wading birds in the southern Everglades fell drastically as their former breeding and nesting areas dried up. Other wildlife, such as alligators, Florida panthers, snail kites, and wood storks, also were negatively affected because drought reduces suitable habitat during parts of the year and the animals are forced to congregate around the remaining sources of water. The quality of the water is also important. The original wetland ecosystem was a nutrient-poor system. The introduction of nutrients into the water from agricultural activities encouraged the growth of exotic plants that replaced natural vegetation.

As people recognized that the key element necessary to preserve the Everglades was a constant, reliable source of clean freshwater, several steps were taken to modify water use to preserve the Everglades. The Kissimmee River, which flows into Lake Okeechobee, was channelized into an arrow-straight river in the late 1960s. This project destroyed the marshes and allowed nutrients from dairy farms and other agricultural activities to pollute the lake and Everglades Park, to which the water from Lake Okeechobee eventually flowed. To help alleviate this problem, in 1990, the Corps of Engineers began to return the Kissimmee to its natural state, with twisting oxbow curves and extensive wetlands. This allows the plants in the natural wetlands to remove much of the nutrient load before they enter the lake. To reduce the likelihood of further development near the park, in 1989, the U.S. Congress approved the purchase of 43,000 hectares (100,000 acres) for addition to the east section of the park. The state of Florida obtained an additional 60,000 hectares (150,000 acres) as an additional buffer zone for the park.

For several years, the U.S. Army Corps of Engineers and the South Florida Water Management District cooperated in the development of a comprehensive restoration plan that was finished in 2000. The development of the plan involved the input of scientists, politicians, various business interests, and environmentalists. Key components of the plan are:

1. developing facilities to store surface water and pump water into aquifers so that it can be released when needed,
2. developing wetlands to treat municipal and agricultural runoff so that nutrient loads are reduced,
3. using clean wastewater to recharge aquifers and supply water to wetlands in the Miami area,
4. reducing the amount of water lost through levees and redirecting water to the Everglades, and
5. removing barriers to the natural flow of water through the Everglades.

The plan received strong support from Congress in the fall of 2000 when $1.4 billion was allocated to begin implementing the plan. It will require many more billions of dollars and up to 30 years to accomplish all aspects of the plan, but, if the plan continues to be implemented over the next few decades, the Everglades ecosystem will be restored to a more stable condition and will have a more hopeful future.

microorganisms are being explored. The chemical, ozone, also kills microorganisms and has been substituted for chlorine by some facilities. Ultraviolet light and ultrasonic energy can also be used. However, chlorine is inexpensive and very effective, so it continues to be the primary method used.

A growing number of larger sewage treatment plants use additional processes called tertiary sewage treatment. **Tertiary sewage treatment** involves a variety of techniques to remove dissolved pollutants left after primary and secondary treatments. (See table 16.4.) The tertiary treatment of municipal wastewater is often used to remove phosphorus and nitrogen that could increase aquatic plant growth. Some municipalities are using natural or constructed wetlands to serve as tertiary sewage treatment systems. In other cases, the effluent from the treatment facility is used to irrigate golf courses, roadside vegetation, or cropland. The vegetation removes excess nutrients and prevents them from entering streams and lakes where they would present a pollution problem. Tertiary treatment of industrial and other specialized wastewater streams is very costly because it requires specific chemical treatment of the water to eliminate specific problem materials. Many industries maintain their own wastewater facilities and design specific tertiary treatment processes to match the specific nature of their waste products.

As water has become scarce in many parts of the world, people have looked at wastewater as a source of water for other purposes. Water from

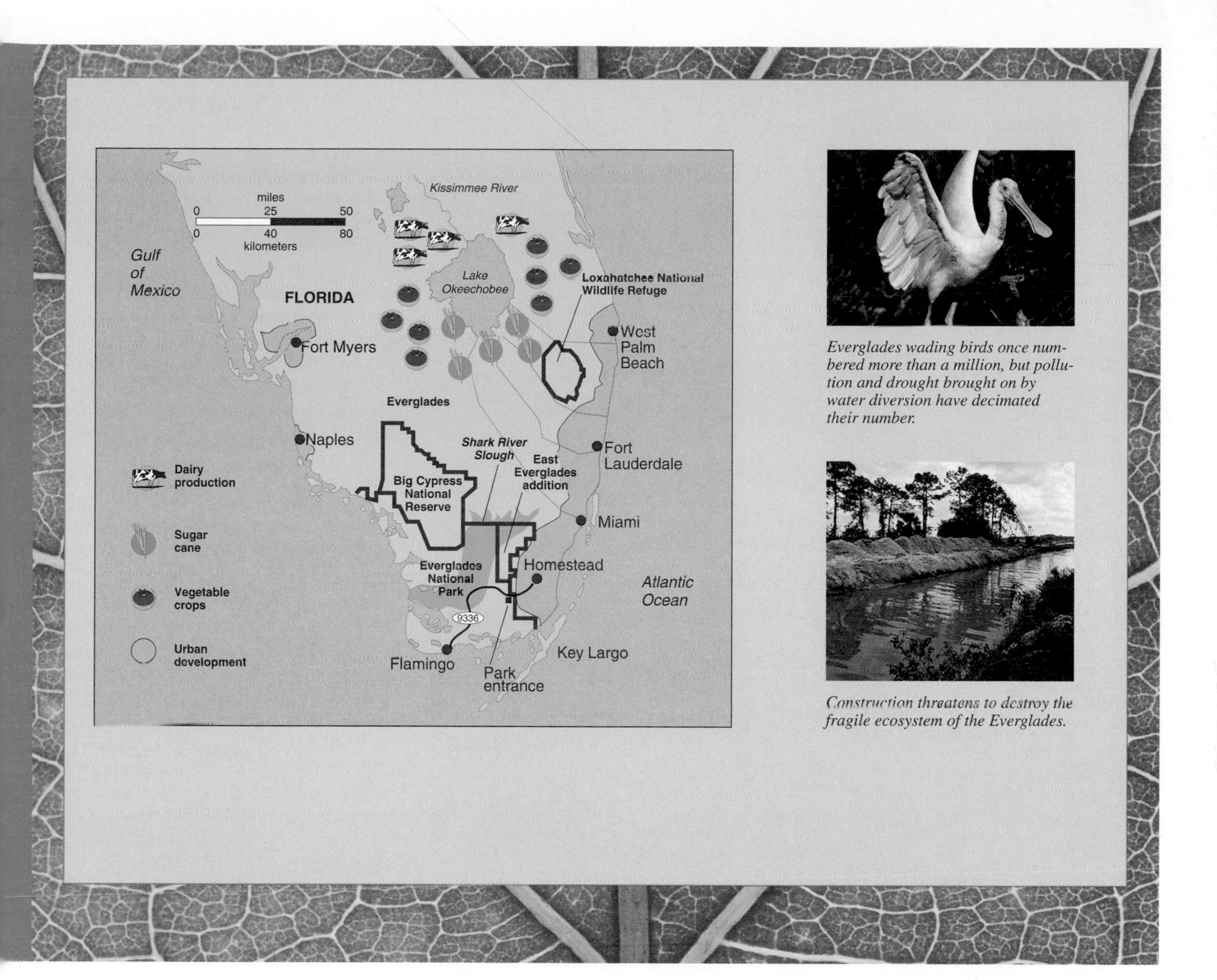

Everglades wading birds once numbered more than a million, but pollution and drought brought on by water diversion have decimated their number.

Construction threatens to destroy the fragile ecosystem of the Everglades.

wastewater plants can be used for irrigation water, industrial purposes, cooling, and many other activities. Ultimately, it is possible to have a closed loop system for domestic water in which the output of the wastewater plant becomes the input for the drinking water supply.

Salinization

Another water-use problem results from **salinization,** an increase in salinity caused by increasing salt concentrations in soil. This is primarily a problem in areas where irrigation is practiced. When water evaporates from soil or plants extract the water they need, the salts present in all natural waters become concentrated. Since irrigation is most common in hot, dry areas that have high rates of evaporation, there is generally an increase in the concentration of salts in the soil and in the water that runs off the land. (See figure 16.17.) Every river increases in salinity as it flows to the ocean. The salinity of the Colorado River water increases 20 times as it passes through irrigated cropland between Grand Lake in north central Colorado and the Imperial Dam in southwest Arizona. The problem of salinity will continue to increase as irrigation increases.

Groundwater Mining

Groundwater mining means that water is removed from an aquifer faster than it is replaced. When this practice continues for a long time, the water table eventually declines. Groundwater mining is common in areas of the western United

Table 16.4 Tertiary Treatment Methods

Kind of Tertiary Treatment	Problem Chemicals	Methods
Biological	Phosphorus and nitrogen compounds	1. Large ponds are used to allow aquatic plants to assimilate the nitrogen and phosphorus compounds from the water before the water is released. 2. Columns containing denitrifying bacteria are used to convert nitrogen compounds into atmospheric nitrogen.
Chemical	Phosphates and industrial pollutants	1. Water can be filtered through calcium carbonate. The phosphate substitutes for the carbonate ion and the calcium phosphate can be removed. 2. Specific industrial pollutants, which are nonbiodegradable, may be removed by a variety of specific chemical processes.
Physical	Primarily industrial pollutants	1. Distillation 2. Water can be passed between electrically charged plates to remove ions. 3. High-pressure filtration through small-pored filters 4. Ion-exchange columns

figure 16.17 **Salinization** As water evaporates from the surface of the soil, the salts it was carrying are left behind. In some areas of the world, this has permanently damaged cropland. Other areas flush the soil to rid it of salt and maintain its fertility.

States and throughout the world. In North America, it is a particular problem due to growing cities and increasing irrigation. In aquifers with little or no recharge, virtually any withdrawal constitutes mining, and sustained withdrawals will eventually exhaust the supply. This problem is particularly serious in communities that depend heavily upon groundwater for their domestic needs.

Groundwater mining can also lead to problems of settling or subsidence of the ground surface. Removal of the water allows the ground to compact, and large depressions may result. For example, in the San Joaquin Valley of California, groundwater has been withdrawn for irrigation and cultivation since the 1850s and groundwater levels have fallen over 100 meters (300 feet). More than 1000 hectares (approximately 2,500 acres) of ground have subsided, some as much as 6 meters (20 feet). Currently, the ground surface in that area is sinking 30 centimeters (12 inches) per year. London, Mexico City, Venice, Houston, and Las Vegas are some other cities that have experienced subsidence as a result of groundwater withdrawal. (See table 16.5.) As people recognize the severity of the problem, public officials are beginning to develop water conservation plans for their cities. Albuquerque, New Mexico, which relies on groundwater for its water supply, has an extensive public education program to encourage people to reduce their water consumption. Since grass demands water, people are encouraged to use desert plants for landscaping or collect rainwater to water their lawns. Finding and correcting leaks, reducing the amount of water used in bathing, and recycling water from swimming pools are other conservation strategies.

Groundwater mining poses a special problem in coastal areas. As the fresh groundwater is pumped from wells along the coast, the saline groundwater moves inland, replacing fresh groundwater with unusable saltwater. (See figure 16.18.) Saltwater intrusion is a serious problem in heavily populated coastal areas throughout the world.

Preserving Scenic Water Areas and Wildlife Habitats

Some bodies of water have unique scenic value. To protect these resources, the way in which the land adjacent to the water is used must be consistent with preserving these scenic areas.

The U.S. Federal Wild and Scenic Rivers Act of 1968 established a system to protect wild and scenic rivers from development. All federal agencies must consider the wild, scenic, or recreational values of certain rivers in planning for the use and development of the rivers and adjacent land. The process of designating a river or part of a river as wild or scenic is complicated. It often encounters local opposition from businesses dependent on growth. Following reviews by state and federal agencies, rivers may be designated as wild and

GLOBAL PERSPECTIVE

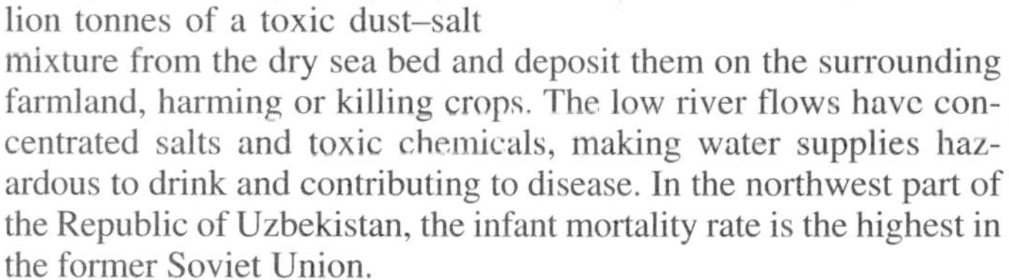

Death of a Sea

The Aral Sea lies on the border between Uzbekistan and Kazakhstan in the former Soviet Union. It was once larger than any of the Great Lakes except Superior, and now it is disappearing. Since the 1920s, Soviet agricultural planners have used up the Aral Sea, diverting its waters for irrigation. The two rivers feeding the Aral were drawn off to irrigate millions of hectares of cotton. An irrigation canal, the world's longest, stretches over 1300 kilometers (800 miles), into Turkmenistan paralleling the boundaries of Afghanistan and Iran. The cotton production plan worked, and, by 1937, the Soviet Union was a net exporter of cotton. The success of the cotton program, however, spelled the end for the Aral Sea.

For a long time, the ecological impact on the sea and surrounding area was largely hidden from public view. Since the 1960s, however, the Aral has lost 75% of its volume and over 50 percent of its surface area, or over 22,000 square kilometers (8,500 square miles) of what are now largely dry, salt-encrusted wastelands. The once-thriving fishing industry that depended on the water is all but gone. Some 20 of the 24 fish species there have disappeared. The fish catch, which totaled 44,000 tonnes (48,000 U.S. tons) a year in the 1950s and supported some 60,000 jobs, has dropped to zero. Abandoned fishing villages dot the sea's former coastline.

Another apparent consequence of the dried-up sea is a host of human illnesses. A high rate of throat cancer is attributed to dust from the drying sea. Each year, winds pick up a million tonnes of a toxic dust–salt mixture from the dry sea bed and deposit them on the surrounding farmland, harming or killing crops. The low river flows have concentrated salts and toxic chemicals, making water supplies hazardous to drink and contributing to disease. In the northwest part of the Republic of Uzbekistan, the infant mortality rate is the highest in the former Soviet Union.

The former fishing center of the sea was a town named Muynak. The town is now landlocked more than 30 kilometers (20 miles) from the water. Less than 25 years ago, Muynak was a seaport. The population of Muynak is down from 40,000 citizens in 1970 to 12,000 today. In 1990, the mayor and last harbormaster of Muynak commented:

> The water continued to go away while the salinity increased. The weather changed for the worse, with the summers getting hotter and the winters colder. The people feel salt on their lips and in their eyes all the time. It's getting harder to open your eyes here.

In 1992 the Central Asian Republics of Uzbekistan, Kazakhstan, Turkmenistan, Tajikistan, and Kaygyzstan signed an international agreement to save the Aral Sea. Additional water is now flowing to the sea but it will take years of cooperative efforts to return the sea to its original size.

Death of a sea

scenic by action of either Congress or the Secretary of the Interior. Sections of over 150 streams comprising about 12,000 kilometers (7,700 miles) in the United States have been designated as wild or scenic.

Many unique and scenic shorelands have also been protected from future development. Until recently, estuaries and shorelands have been subjected to significant physical modifications, such as dredging and filling, which may improve conditions for navigation and construction but destroy fish and wildlife habitats. Recent actions throughout North America have attempted to restrict the development of shorelands. Development has been restricted in some particularly scenic areas, such as Cape Cod National Seashore in Massachusetts and the Bay of Fundy in the Atlantic provinces of Canada.

Historically, poorly drained areas were considered worthless. Subsequently,

Table 16.5 Groundwater Depletion in Major Regions of the World, Circa 1995

Region/Aquifer	Estimates of Depletion
California	Groundwater overdraft exceeds 1.7 billion cubic meters (60 billion cubic feet) per year. The majority of the depletion occurs in the Central Valley, which is referred to as the vegetable basket of the United States.
Southwestern United States	In parts of Arizona, water tables have dropped more than 120 meters (400 feet). Projections for parts of New Mexico indicate that water tables will drop an additional 22 meters (70 feet) by 2020.
High Plains aquifer system, United States	This aquifer underlies nearly 20 percent of all the irrigated land in the United States. To date, the net depletion of the aquifer is in excess of 350 billion cubic meters (12 trillion cubic feet), or roughly 15 times the average annual flow of the Colorado River. Most of the depletion has been in the Texas high plains, which have witnessed a 26 percent decline in irrigated land from 1979 to 1989. Current depletion is estimated to be in excess of 13 billion cubic meters (450 billion cubic feet) a year.
Mexico City and Valley of Mexico	Use exceeds natural recharge by 60 to 85 percent, causing land subsidence and falling water tables.
African Sahara	North Africa has vast non-recharging aquifers where current depletion exceeds 12 billion cubic meters (425 billion cubic feet) a year.
India	Water tables are declining throughout much of the most productive agricultural land in India. In parts of the country, groundwater levels have declined 90 percent during the past two decades.
North China	The water table underneath portions of Beijing has dropped 40 meters (130 feet) during the past 40 years. A large portion of northern China has significant groundwater overdraft.
Arabian peninsula	Groundwater use is nearly three times greater than recharge. At projected depletion rates exploitable groundwater reserves could be exhausted within the next 50 years. Saudi Arabia depends on non-renewable groundwater for roughly 75 percent of its waters. This includes irrigation of 2 million to 4 million tonnes (2.2–4.4 million U.S. tons) of wheat per year.

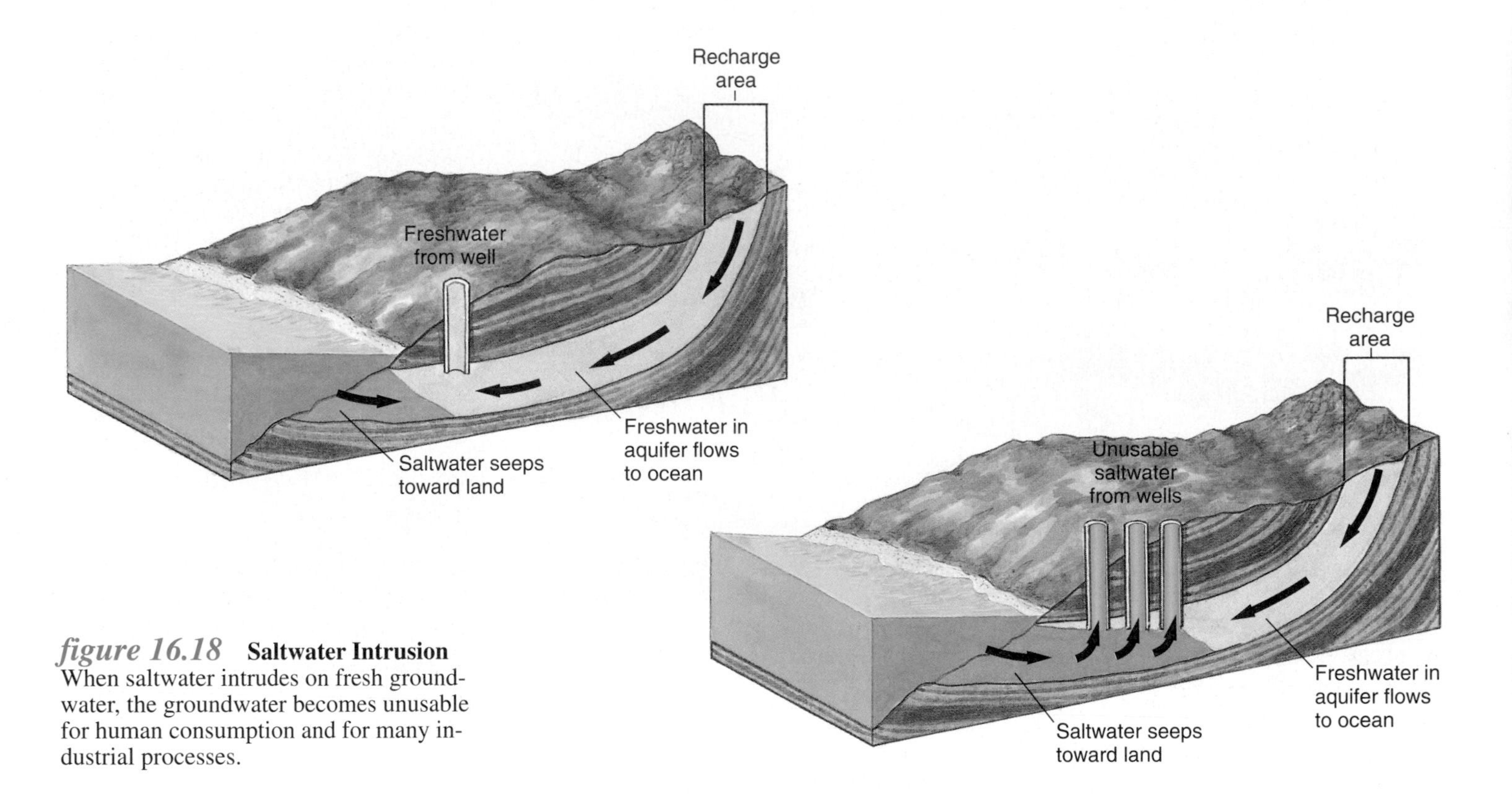

figure 16.18 **Saltwater Intrusion**
When saltwater intrudes on fresh groundwater, the groundwater becomes unusable for human consumption and for many industrial processes.

many of these wetlands were filled or drained and used for building sites. The natural and economic importance of wetlands has been recognized only recently. In addition to providing spawning and breeding habitats for many species of wildlife, wetlands act as natural filtration systems by trapping nutrients and pollutants and preventing them from entering adjoining lakes, streams, or estuaries. Wetlands also slow down floodwaters and permit nutrient-rich particles to settle out. In addition, wetlands can act as reservoirs and release water slowly into lakes, streams, or aquifers, thereby preventing floods. (See figure 16.19.) Coastal estuarine zones and adjoining sand dunes also provide significant natural flood control. Sand dunes act as barriers and absorb damaging waves caused by severe storms.

figure 16.19 **The Value of Wetlands** Wetlands are areas covered by water enough of the year that they support aquatic plant and animal life. Wetlands can be either fresh or saltwater and may be isolated potholes or extensive areas along rivers, lakes, and oceans. We once thought of wetlands as only a breeding site for mosquitoes. Today, we are beginning to appreciate their true value.

Summary

Water is a renewable resource that circulates continually between the atmosphere and the Earth's surface. The energy for the hydrologic cycle is provided by the sun. Water loss from plants is called evapotranspiration. Water that infiltrates the soil and is stored underground in the tiny spaces between rock particles is called groundwater, as opposed to surface water that enters a river system as runoff. There are two basic kinds of aquifers. Unconfined aquifers have an impervious layer at the bottom and receive water which infiltrates from above. The top of the layer of water is called the water table. A confined aquifer is sandwiched between two impervious layers and is often under pressure. The recharge area may be a great distance from where the aquifer is tapped for use. The way in which land is used has a significant impact on rates of evaporation, runoff, and infiltration.

The four human uses of water are domestic, agricultural, in-stream, and industrial. Water use is measured by either the amount withdrawn or the amount consumed. Domestic water is in short supply in many metropolitan areas. Most domestic water is used for waste disposal and washing, with only a small amount used for drinking. The largest consumptive use of water is for agricultural irrigation. Major in-stream uses of water are for hydroelectric power, recreation, and navigation. Most industrial uses of water are for cooling and for dissipating and transporting waste materials.

Major sources of water pollution are municipal sewage, industrial wastes, and agricultural runoff. Nutrients, such as nitrates and phosphates from wastewater treatment plants and agricultural runoff, enrich water and stimulate algae and aquatic plant growth. Organic matter in water requires oxygen for its decomposition and therefore has a large biochemical oxygen demand (BOD). Oxygen depletion can result in fish death and changes in the normal algal community, which leads to visual and odor problems.

GLOBAL PERSPECTIVE

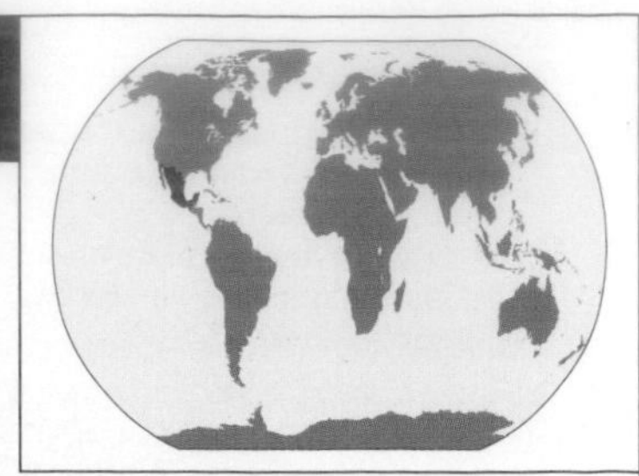

ECOPARQUE

The city of Tijuana, Mexico, is home to approximately 1 million people. Rapid population growth has strained local water resources in this arid region. Both the city and the adjacent U.S.–Mexico border region also have severe water-pollution problems. One source of pollution is the direct discharge of residential wastewater into the Tijuana River and the nearby coastal waters. Because of the water scarcity and the pollution, scientists in Tijuana began thinking of alternative technologies for treating wastewater stating that "water is too valuable to be called waste." One answer is an alternative treatment facility called ECOPARQUE.

ECOPARQUE began in 1986 as a study of decentralized wastewater treatment by El Colegio de la Frontera Norte (COLEF), a college in Tijuana. By 1996, ECOPARQUE had developed into a successful model of sustainable and integrated urban wastewater treatment. The central feature of ECOPARQUE is the basic treatment unit for recapturing wastewater. The unit includes a stainless steel fine screen, a plastic biofilter, and a passive sediment deposit clarifier. These three elements operate without electrical energy or any mechanical parts. For this reason, the process falls under the category of appropriate technology for construction and use in developing countries.

ECOPARQUE and its treated wastewater are being used for a variety of sustainable purposes, including:

- Urban reforestation on hillsides and canyons in Tijuana.
- A nursery that produces a variety of plants, including vegetables and fruits.
- Food production in an urban center that includes an expanding fruit harvest, soon to be joined by beekeeping and an agricultural production unit. There are also plans for raising some species of fish as a food source.
- A meteorological station that will provide local authorities with useful urban planning information.
- An experimental solar field for use at ECOPARQUE and the surrounding neighborhoods.
- A site for environmental and technological education for schoolchildren and scientists.

Given the range of activities at ECOPARQUE, it is easy to see why the scientists who planned the facility believed that water is too valuable to be called waste. While it may not be feasible to build facilities similar to ECOPARQUE in every city in the developing world, the design plans for ECOPARQUE are being made available to other countries. There is hope that many other cities will turn what is a water pollution and health problem into a multifaceted and sustainable environmental and community success story.

Point sources of pollution are easy to identify and resolve. Nonpoint sources of pollution, such as agricultural runoff and mine drainage, are more difficult to detect and control than those from municipalities or industries.

Thermal pollution occurs when an industry returns heated water to its source. Temperature changes in water can alter the kinds and numbers of plants and animals that live in it. The methods of controlling thermal pollution include cooling ponds, cooling towers, and dry cooling towers.

Wastewater treatment consists of primary treatment, a physical settling process; secondary treatment, biological degradation of the wastes; and tertiary treatment, chemical treatment to remove specific components. Two major types of secondary wastewater treatments are the trickling filter and the activated sludge sewage methods.

Groundwater pollution comes from a variety of sources, including agriculture, landfills, and septic tanks. Marine oil pollution results from oil drilling and oil-tanker accidents, runoff from streets, improper disposal of lubricating oil from machines and car crankcases, and intentional discharges from oil tankers during loading or unloading.

Reduced water quality can seriously threaten land use and in-place water use. In the United States and other nations, legislation helps to preserve certain scenic water areas and wildlife habitats. Shorelands and wetlands provide valuable services as buffers, filters, reservoirs, and wildlife areas. Water management concerns of growing importance are groundwater mining, increasing salinity, water diversion, and managing urban water use. Urban areas face several problems, such as providing water suitable for human use, collecting and treating wastewater, and handling storm-water runoff in an environmentally sound manner. Water planning involves many governmental layers, which makes effective planning difficult.

ISSUES & ANALYSIS

The California Water Plan

The management of freshwater is often a controversial subject, involving social, ecological, and economic aspects. A good example is the California Water Plan.

In the early 1900s, it became clear that the growth of Los Angeles, which was then a small coastal town, would be encouraged by irrigating the surrounding land. Los Angeles looked to the Owens Valley, 40 kilometers (25 miles) north, for a source of water. The Los Angeles Aqueduct connecting these two areas was completed in 1913.

After the Owens Valley project, California developed a statewide water program known as the California Water Plan. This plan is necessary because most of the state's population and irrigated land are found in the central and southern regions, but most of the water is in the north. The plan details the construction of aqueducts, canals, dams, reservoirs, and power stations to transport water from the north to the south. In addition to supplying water for southern regions, the aqueducts provide irrigation for the San Joaquin Valley. Eventually, the new land made available for agriculture amounted to about 400,000 hectares (1 million acres).

The California Water Plan has been one of the most controversial programs ever undertaken in California. Its adoption instigated a sectional feud between the moist "north" and the dry "south." Southern California was accused of trying to steal northern water, but because the large population in the southern part of the state carried the vote, the plan was adopted. Environmentalists still claim that the project has irreparably scarred the countryside and upset natural balances of streams, estuaries, vegetation, and wildlife. They argue that providing water to southern California promotes population growth, which leads to further urbanization and land development.

Many questions and controversies center on whether the water is really needed. Ninety percent of the water used in southern California is for irrigation; there is evidently abundant water for domestic and industrial use. A few of the crops raised in the San Joaquin Valley account for a large percentage of the irrigation water. Most notable is rice. Rice is not a native crop and demands intensive irrigation. The cost is borne by all the rate payers, most of whom are urban. California is now one of the nation's most productive agricultural areas.

Allocation of water resources is a matter of economics as well as technology. The California water project has been criticized for using public funds to increase the value of privately held farmland. Furthermore, technological advances in desalination plants may give a new dimension (unforeseen when the water plan was devised) to the problem of water resources. Although more aqueducts, canals, and pumping plants are planned, whether they will be completed is uncertain.

Southern California has been in a drought emergency for several years. Water rates have increased, certain municipal uses have been curtailed, and irrigation has been reduced. Increasing population creates a demand for water that is now causing people to recognize that they will need to choose between using water for municipal purposes or irrigation. There is not enough water to increase both.

- What are the major advantages of having a water plan?
- What problems develop when water is transported to arid regions?
- Should water from northern California be sent to southern California?
- Should crops like rice be raised in California?
- Aside from lack of rainfall, why do you think southern California is in a water crisis?

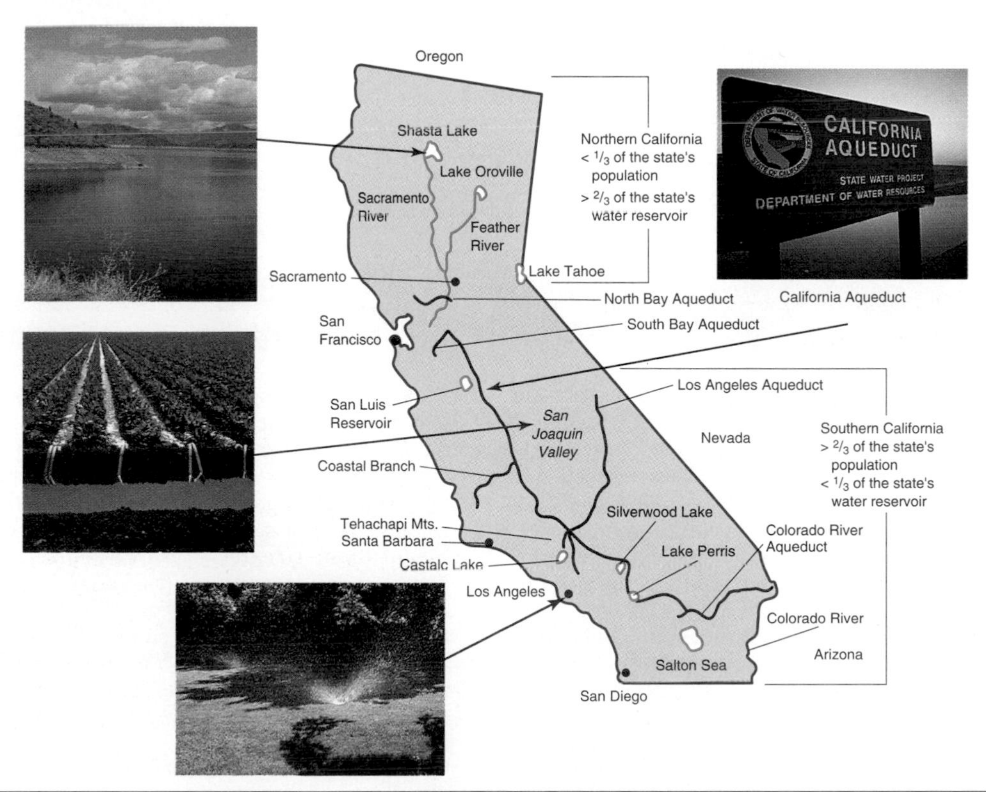

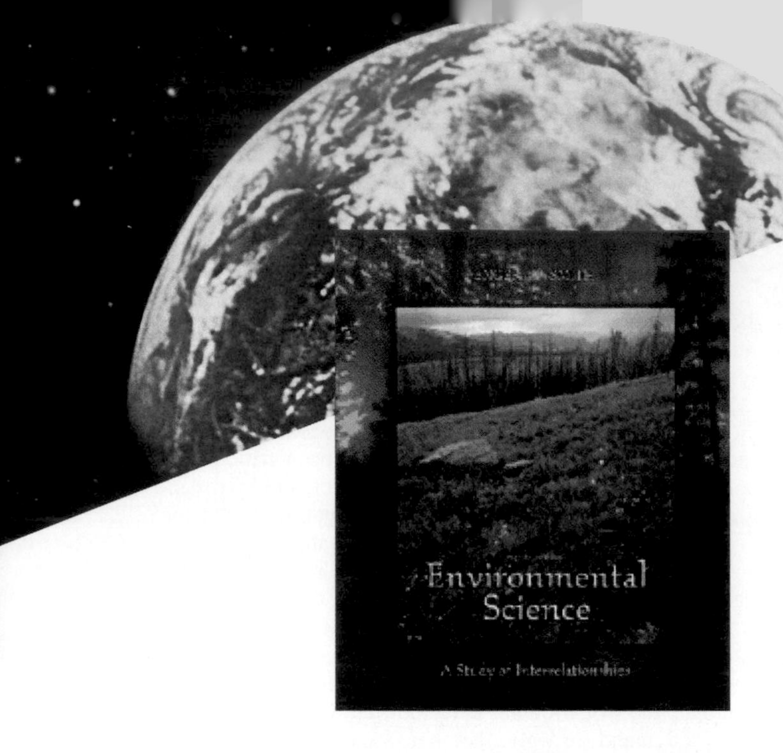

Interactive Exploration

Check out the website at

http://www.mhhe.com/environmentalscience

and click on the cover of this textbook for interactive versions of the following:

KNOW THE BASICS

activated-sludge sewage treatment *359*
aquiclude *341*
aquifer *341*
aquitard *341*
artesian well *342*
biochemical oxygen demand (BOD) *349*
confined aquifer *341*
domestic water *344*
eutrophication *350*
evapotranspiration *341*
fecal coliform bacteria *351*
groundwater *341*
groundwater mining *361*
hydrologic cycle *341*
industrial water use *346*
in-stream water use *347*
irrigation *345*
limiting factor *350*
nonpoint source *350*
point source *350*
porosity *342*
potable waters *340*
primary sewage treatment *358*
runoff *341*
salinization *361*
secondary sewage treatment *358*
sewage sludge *359*
storm-water runoff *356*
tertiary sewage treatment *360*
thermal pollution *353*
trickling filter system *358*
unconfined aquifer *341*
vadose zone *341*
water diversion *357*
water table *341*

On-line Flashcards

Electronic Glossary

IN THE REAL WORLD

Supposedly, the importance of wetlands is understood, but the problems of wetland destruction continue. Check out Stronger Wetland Protection Rules Announced and The Prairie Wetlands of Southwest Minnesota for recent updates on wetlands protection.

Do you know where your drinking water comes from? Is your local source ground or surface water? A detailed explanation of the source of Memphis, Tennessee's water supply is provided in Drinking Water from Wells, Not from the River.

The importance of protecting groundwater from contamination is highlighted in the Protecting Groundwater Resources case study.

Water diversions, the mining of groundwater, and politically charged situations. . . realities when water is a scarce resource. Check out Historic Agreement on Water Allotments in Southern California and Drought Focuses on Long-standing Water Disputes in the Middle East to learn more about conflicts that result from the lack of water. Although water is not scarce in the Great Lakes region, see why Low Water Levels in the Great Lakes are causing concerns in this area.

Restoration projects, protection, and funding for cleanup activities are just a few examples. Check out Everglades Restoration: Greatest Restoration Yet, or Just More of the Same? for an example of a restoration project.

A variety of ways to clean up and improve water quality can be found in the following stories: International Accord to Clean Up the Rhine River, and Food Web Control of Primary Production in Lakes.

Another way to improve water quality is to support local conservation measures. Look over American Heritage River System Created for details on how this program aims to help coordinate efforts to improve water quality.

PUT IT IN MOTION

What is the link between fertilizer runoff and the green soupy lake in an adjacent area? Now is your chance to study it in more detail. Check out the Deoxygenation of Lakes animation for a better understanding of the process.

Normal rainfall is neutral. . . right? What causes acid rain? When is the effect of acid rain most apparent? Study the Acid Rain animation for answers to these questions and to see how acid rain affects aquatic ecosystems such as lakes.

TEST PREPARATION

Review Questions

1. Describe the hydrologic cycle.
2. Distinguish between withdrawal and consumption of water.
3. What are the similarities between domestic and industrial water use? How are they different from in-stream use?
4. How is land use related to water quality and quantity? Can you provide local examples?
5. What is biochemical oxygen demand? How is it related to water quality?
6. How can the addition of nutrients such as nitrates and phosphates result in a reduction of the amount of dissolved oxygen in the water?
7. Differentiate between point and nonpoint sources of water pollution.
8. How are most industrial wastes disposed of? How has this changed over the past 25 years?
9. What is thermal pollution? How can it be controlled?
10. Describe primary, secondary, and tertiary sewage treatment.
11. What are the types of wastes associated with agriculture?
12. Why is storm-water management more of a problem in an urban area than in a rural area?
13. Define groundwater mining.
14. How does irrigation increase salinity?
15. What are the three major water services provided by metropolitan areas?

Critical Thinking Questions

1. Leakage from freshwater distribution systems accounts for significant losses. Is water so valuable that governments should require systems that minimize leakage in order to preserve the resource? Under what conditions would you change your evaluation?
2. Do non-farmers have an interest in how water is used for irrigation? Under what conditions should the general public be involved in making these decisions along with the farmers who are directly involved?
3. Should the United States allow Mexico to have water from the Rio Grande and the Colorado Rivers, both of which originate in the United States and flow to Mexico?
4. Do you believe that large scale hydroelectric power plants should be promoted as a renewable alternative to power plants that burn fossil fuels? What criteria do you use for this decision?
5. What are the costs and what are the benefits of the proposed Garrison Diversion Unit? What do you think should happen with this project?
6. How might you be able to help save freshwater in your daily life? Would the savings be worth the costs?
7. Look at the hydrologic cycle in Figure 16.3. If global warming increases the world-wide temperature, how should increased temperature directly affect the hydrologic cycle?
8. After reading the Issues and Analysis concerning the California Water Plan, do you believe water should be diverted from northern California to southern California?

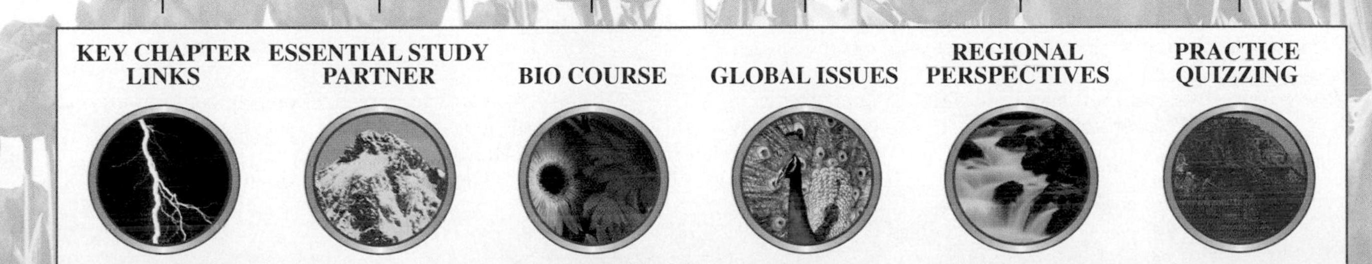

PART FIVE

Pollution and Policy

Addressing environmental problems and concerns is not an easy task. Perceptions vary as to severity of the problems as well as to the best ways of addressing them. There are great differences of opinion about what environmental quality should be or what the current quality of the environment is. The amount of pollution can often be quantified, and pollution may be more or less serious depending on local conditions. Although pollution, by definition, is harmful, some amounts of pollution can be tolerated, depending on the situation. Environmental concerns are both social and scientific by nature. Part Five looks at the interrelationship between environmental concerns and the means by which these concerns are addressed and remediated. This involves a wide range of scientific disciplines as well as social science disciplines, such as law, political science, economics, planning, and international relations.

Chapters 17, 18, and **19** discuss air pollution, solid waste disposal, and hazardous and toxic wastes, respectively. These chapters also describe approaches used to deal with these problems. Throughout this section, economic and political realities are discussed as a part of the pollution equation.

Chapter 20 discusses how environmental decisions are made. It is a complex process, integrating public input, economic demands, political posturing, and legal requirements. As complex as the process is, the decisions we have made have significantly improved the quality of the environment in many parts of the world. There remain, however, very real challenges in other parts of the world. Addressing these challenges will become increasingly important as we progress through the first decade of the new millennium.

CHAPTER 17

Air Quality Issues

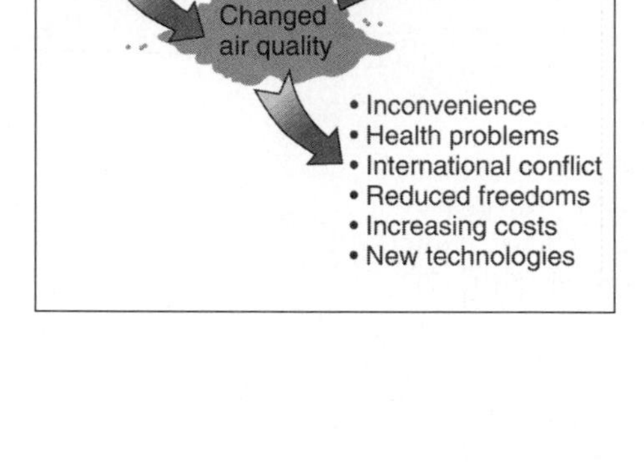

Objectives

After reading this chapter, you should be able to:

- Explain why air can accept and disperse significant amounts of pollutants.
- List the major sources and effects of the five primary pollutants.
- Describe how photochemical smog is formed and how it affects humans.
- Explain how PCV valves, APC valves, catalytic converters, scrubbers, precipitators, filters, and changes in fuel types reduce air pollution.
- Explain how acid rain is formed.
- Understand that humans can alter the atmosphere in such a way that the climate may change.
- Describe the link between chlorofluorocarbon use and ozone depletion.
- Recognize that enclosed areas can trap air pollutants that are normally diluted in the atmosphere.

Chapter Outline

The Atmosphere

Primary Air Pollutants
- Carbon Monoxide (CO)
- Hydrocarbons (HC)

Global Perspective: *Air Pollution in Mexico City*

- Particulates
- Sulfur Dioxide (SO_2)
- Oxides of Nitrogen (NO and NO_2)

Photochemical Smog

Other Significant Air Pollutants

Control of Air Pollution
- Clean Air Act

Acid Deposition

Environmental Close-Up: *Secondhand Smoke*

Global Warming and Climate Change
- Worsening Health Effects
- Rising Sea Level
- Disruption of the Water Cycle
- Changing Forests and Natural Areas
- Challenges to Agriculture and the Food Supply

Addressing Climate Change

Ozone Depletion

Environmental Close-Up: *Radon*

Indoor Air Pollution

Environmental Close-Up: *Noise Pollution*

Issues & Answers: *International Air Pollution*

The Atmosphere

The atmosphere, or air, is normally composed of 79 percent nitrogen, 20 percent oxygen, and a 1 percent mixture of carbon dioxide, water vapor, and small quantities of several other gases. Most of the atmosphere is held close to the earth by the pull of gravitational force, so it gets thinner with increasing distance from the earth. (See figure 17.1.)

Even though gravitational force keeps the air near the earth, the air is not static. As it absorbs heat from the earth, it expands and rises. When its heat content is radiated into space, the air cools, becomes more dense, and flows toward the earth. As the air circulates due to heating and cooling, it also moves horizontally over the surface of the earth because the earth rotates on its axis. The combination of all air movements creates the wind patterns characteristic of different regions of the world. (See figure 17.2.)

As we discussed in chapter 11, pollution is something produced by humans that interferes with our well-being. Because we cause pollution, we may be able to do something to prevent it. There are several natural sources that degrade the quality of the air, such as gases and particles from volcanoes, dust from natural sources, or gases from decomposition of dead plants and animals. However, since these activities are not controlled by humans, they do not fit our definition of pollution. Automobile emissions, chemical odors, and factory smoke are considered air pollution, however, and we will focus on these and similar examples.

The problem of air pollution is directly related to the number of people living in an area and the kinds of activities in which they are involved. When a population is small and its energy use is low, the impact of people is minimal. Their pollution is diluted and the overall negative effect is slight. However, our urbanized, industrialized civilization has a growing population and a history of increasing use of fossil fuels and technological aids. We release large quantities of polluting by-products into our environment as we manufacture products demanded by the population.

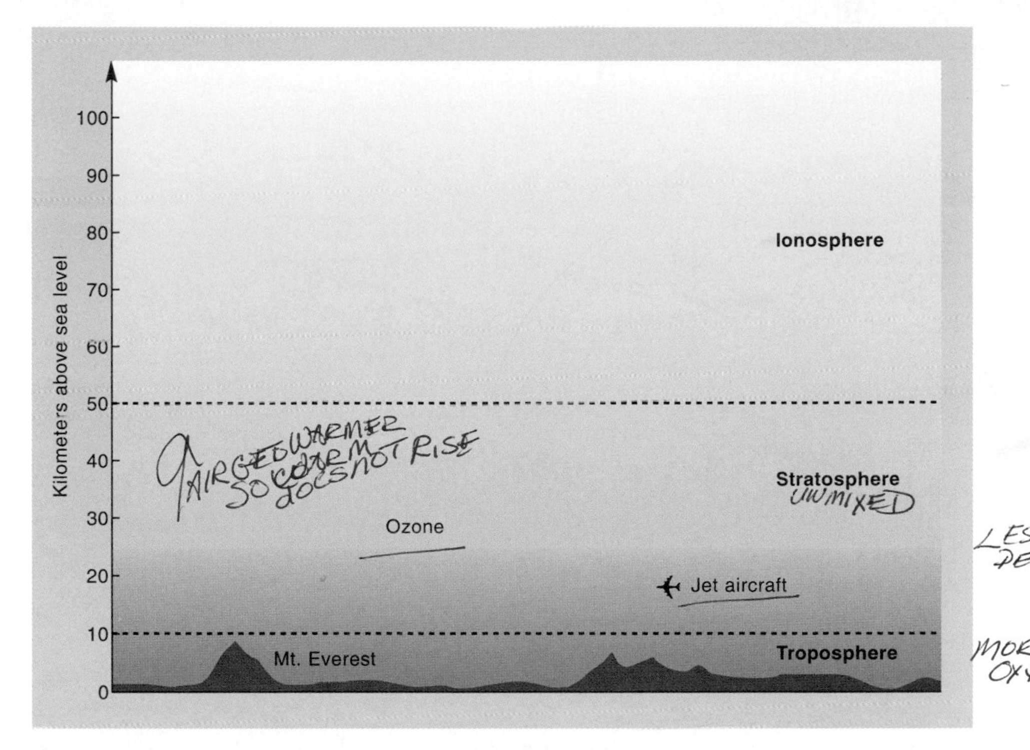

figure 17.1 **The Atmosphere** The atmosphere is divided into the troposphere, the relatively dense layer of gases close to the surface of the earth; the stratosphere, more distant with similar gases but less dense; and the ionosphere, composed of ionized gases.

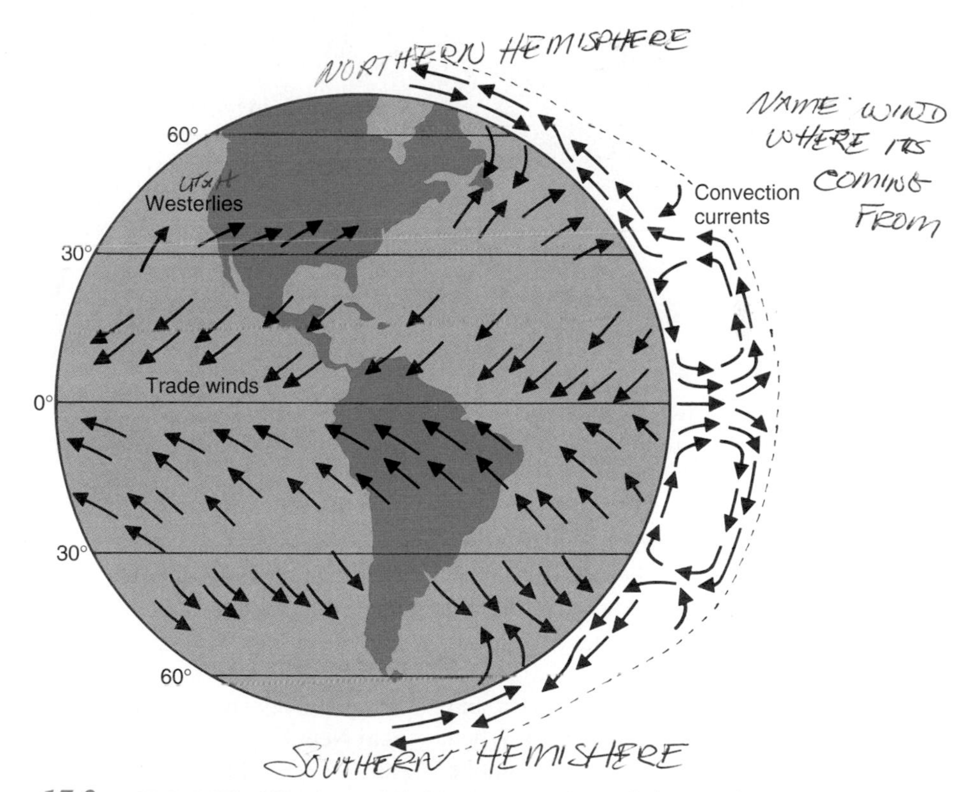

figure 17.2 **Global Wind Patterns** Wind is the movement of air caused by the rotation of the earth and atmospheric pressure changes brought about by temperature differences. Both of these contribute to the patterns of world air movement. In North America, most of the winds are westerlies (from the west to the east).

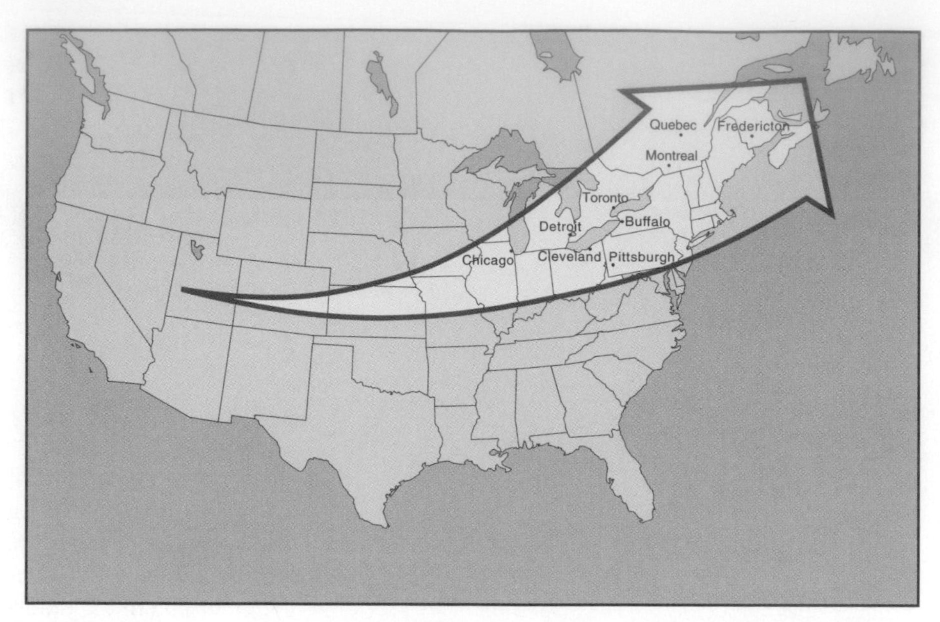

figure 17.3 **Accumulation of Pollutants** As an air mass moves across the continent from west to east, each population center adds its pollutants to the total load in the atmosphere.

figure 17.4 **Air Pollution** Donora, Pennsylvania, was the scene of a serious air-pollution incident. The pollutants from industry were trapped by an inversion, and almost half of the population was affected.

Gases or small particles released into the atmosphere are likely to be mixed, diluted, and circulated, but they are likely to stay near the earth due to gravity. When we put a material into the air, we do not get rid of it; we just dilute it and move it out of the immediate area. When people lived in small groups, the smoke from their fires was diluted. It was produced in such low concentrations that it did not interfere with neighboring groups downwind. In industrialized urban areas, pollutants cannot always be diluted before the air reaches another city. The polluted air from Chicago is further polluted by Gary, Indiana, supplemented by the wastes of Detroit and Cleveland, and finally moves over southeastern Canada and New England to the ocean. (See figure 17.3.) While not every population center adds the same kind or amount of waste, each adds to the total load carried.

Air pollution is not just an aesthetic problem. It also causes health problems. Thousands of deaths have been directly related to poor air quality in cities. A well-documented case of pollution that was harmful to human health occurred in Donora, Pennsylvania. (See figure 17.4.) The city of Donora is located in a valley. In October 1948, the pollutants from a zinc plant and steel mills became trapped in the valley, and a dense smog formed. Within five days, seventeen people died and 5910 persons became ill. The polluted atmosphere affected nearly 50 percent of the city's 12,300 inhabitants. The technology that provided jobs for the people was also killing them.

Many of the megacities of the developing world have extremely poor air quality. Beijing, Seoul, Mexico City, and Cairo exceed World Health Organization guidelines for air quality for at least two pollutants. The causes of this air pollution are open fires, large numbers of poorly maintained motor vehicles, and poorly regulated industrial plants. The World Health Organization estimates that particles in Mexico City contribute to 6400 deaths each year. Not only does poor air quality in such cities increase the death rate, but the general health of the populace is lowered. Chronic coughing and susceptibility to infections are common in these cities. Deaths from air pollution occur primarily among the elderly, the infirm, and the very young. Bronchial inflammations, allergic reactions, and irritation of the mucous membranes of the eyes and nose all indicate that air pollution must be reduced.

Primary Air Pollutants

Around the world, five major types of materials are released directly into the atmosphere in their unmodified forms and in sufficient quantities to pose a health risk. They are carbon monoxide, hydrocarbons, particulates, sulfur dioxide, and nitrogen compounds. This group of pollutants is known as **primary air pollutants.** (See table 17.1.) These materials may interact with one another in the presence of sunlight to form new **secondary air pollutants,**

Table 17.1 Sources of Primary Air Pollutants

Pollutant	Sources
Carbon monoxide	Incomplete burning of fossil fuels Tobacco smoke
Hydrocarbons	Incomplete burning of fossil fuels Tobacco burning Chemicals
Particulates	Burning fossil fuels Farming operations Construction operations Industrial wastes Building demolition
Sulfur dioxide	Burning fossil fuels Smelting ore
Nitrogen compounds	Burning fossil fuels

figure 17.5 **Carbon Monoxide** The major source of carbon monoxide, hydrocarbons, and nitrogen oxides is the internal combustion engine, which is used to provide most of our transportation. The more concentrated the number of automobiles, the more concentrated the pollutants. Carbon monoxide concentrations of a hundred parts per million are not unusual in rush-hour traffic in large metropolitan areas. These concentrations are high enough to cause fatigue, dizziness, and headaches.

such as ozone and other very reactive materials. Secondary air pollutants also form from reactions with natural chemicals in the atmosphere.

Carbon Monoxide (CO)

Carbon monoxide (CO) is produced when organic materials, such as gasoline, coal, wood, and trash, are incompletely burned. The single largest source of carbon monoxide is the automobile. (See figure 17.5.) Although increased fuel efficiency and the use of catalytic converters have reduced carbon monoxide emissions per kilometer driven, carbon monoxide remains a problem because the number of automobiles and the number of kilometers driven have increased. In many parts of the world, automobiles are poorly maintained and may have inoperable pollution control equipment, resulting in even greater amounts of carbon monoxide.

The next largest source of carbon monoxide is smoking tobacco. Currently in the United States and some other countries, there is a great deal of pressure to restrict areas where smoking is permitted to minimize exposure to second-hand cigarette smoke. Restaurants designate nonsmoking sections (some even advertise themselves as smoke-free), public buildings have designated nonsmoking areas, and corporations and colleges are designating their buildings as smoke-free. Smoking is decreasing in the industrialized world today, but in the developing nations, smoking retains its image of glamour and sophistication as a result of extensive marketing campaigns by cigarette companies.

Several hours of exposure to air containing 0.001 percent of carbon monoxide can cause death. Because carbon monoxide remains attached to hemoglobin for a long time, even small amounts tend to accumulate and reduce the blood's oxygen-carrying capacity. The amount of carbon monoxide produced in heavy traffic can cause headaches, drowsiness, and blurred vision. A heavy smoker in congested traffic is doubly exposed and may experience severely impaired reaction time compared to nonsmoking drivers.

Fortunately, carbon monoxide is not a persistent pollutant. Natural processes convert carbon monoxide to other compounds that are not harmful. Therefore, the air can be cleared of its carbon monoxide if no new carbon monoxide is introduced into the atmosphere.

Hydrocarbons (HC)

In addition to carbon monoxide, automobiles emit a variety of **hydrocarbons (HC).** Hydrocarbons are a group of organic compounds consisting of carbon and hydrogen atoms. They are either evaporated from fuel supplies or are remnants of fuel that did not burn completely. The internal combustion engine

GLOBAL PERSPECTIVE

Air Pollution in Mexico City

Mexico City has been labeled the city with the worst air pollution ever recorded. The air over Mexico City exceeded ozone limits set by the World Health Organization on more than 300 days in one year. With about 20 million people, Mexico City is one of the largest cities in the world and grows larger each day. Open sewers and garbage dumps contribute dust and bacteria to the atmosphere. The city has about 35,000 factories and 3.6 million vehicles. Most of these vehicles are older models that are poorly tuned and, therefore, pollute the air with a mixture of hydrocarbons, carbon monoxide, and nitrogen oxides. A large proportion of the air pollution is the result of automobiles. A major source of hydrocarbons is from leakage of LP gas. Fixing the leaks would significantly reduce the hydrocarbon level in the atmosphere. The high altitude [(over 2000 meters) (6500 feet)] results in even greater air pollution from automobiles because automobile engines do not burn fuel efficiently at such high altitudes. Mexico City's location in a valley also allows for conditions suitable for thermal inversions during the winter.

Many foreign companies and governments give special "hazard pay" for working in Mexico City because of the air pollution. Pediatricians estimate that 85 percent of childhood illnesses are related to air pollution and say that the only way to improve the health of many of the children is to get them out of the city. The government has responded with a comprehensive program to clean up the air in the city. Twenty-five million trees have been planted to help clean the air, and 4400 hectares (11,000 acres) of land have been purchased to provide green space for the city.

Public information campaigns encourage people to keep their automobiles tuned to reduce air pollution, lead-free gasoline has been made available and is being used, and catalytic converters are now required on all automobiles manufactured after 1991. Taxis manufactured before 1985 have been banned, and by the year 2002, all pre-1991 taxis must be replaced.

Personal automobiles and taxis are prohibited from being driven on the streets one day a week. This encourages people to carpool or use public transportation. The government is planning to improve public transportation to make it more attractive for people to switch from private automobiles to public transport. Both actions should reduce the number of automobiles releasing pollutants into the air. A polluting, government-owned oil refinery was shut down, and power plants and many industries have switched from oil to natural gas, which pollutes less.

These actions have had significant effects on the quality of the air. Although ozone in the air continues to be a major problem, as it is in metropolitan areas around the world, lead levels, carbon monoxide, and sulfur dioxide have been reduced. In order to further improve air quality, increasingly strong restrictions on polluting industries and further restrictions on the use of private automobiles will be necessary. The Mexican government is considering such actions.

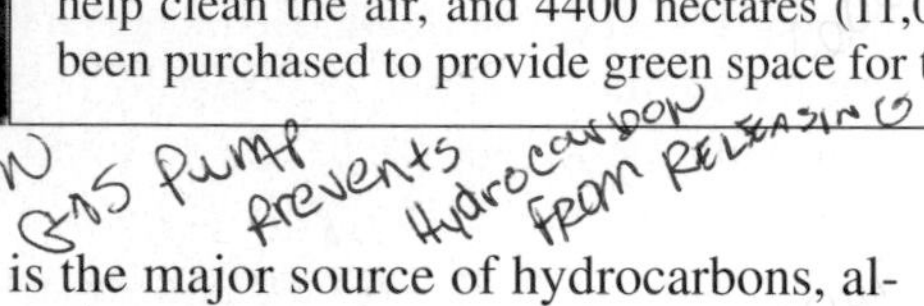

is the major source of hydrocarbons, although refineries and other industries add hydrocarbons to the total atmospheric burden.

Many modifications to automobile engines have reduced the loss of hydrocarbons to the atmosphere. Recycling some gases through the engine, using higher oxygen concentrations in the fuel-air mixture, and using valves to prevent the escape of gases are three of these modifications. In addition, catalytic converters burn exhaust gases more completely so that fewer hydrocarbons leave the tail pipe.

Particulates

Particulates, minute (10 microns and smaller) pieces of solid materials dispersed into the atmosphere, constitute the third largest category of air pollutants. Smoke particles from fires, bits of asbestos from brake linings and insulation, dust particles, and ash from industrial plants contribute to the particulate load. Particulates cause problems ranging from the annoyance of soot settling on a backyard picnic table to the **carcinogenic** (cancer-causing) effects of asbestos. Particulates frequently get attention because they are so readily detected by the public. Heavy black smoke from a factory can be seen without expensive monitoring equipment and generally causes an outcry, whereas the production of colorless gases like carbon monoxide and sulfur dioxide goes unnoticed.

Particles can accumulate in the lungs and interfere with the ability of the lungs to exchange gases. However, this lung damage usually happens to people who are repeatedly exposed to large amounts of particulate matter on the job. Miners and others who work in dusty conditions are most likely to be affected. For most of the population, particulates affect health by acting as centers for the deposition of moisture and gases from the atmosphere. As we breathe air containing particulates, we come in contact with concentrations of other potentially more harmful materials that have accumulated on the particulates. Sulfuric, nitric, and carbonic acids, which irritate the lining of our respiratory system, frequently form on particulates.

Sulfur Dioxide (SO_2)

Sulfur dioxide (SO_2) is a compound of sulfur and oxygen that is produced when sulfur-containing fossil fuels are burned. Coal and oil were produced from organisms that had sulfur in their living structure. When the coal or oil was formed, some of the sulfur was incorporated into the fossil fuel. The sulfur is released as sulfur dioxide when the fuel is burned. Sulfur dioxide has a sharp odor and irritates respiratory tissue. It also reacts with water, oxygen, and other materials in the air to form sulfur-containing acids. The acids can become attached to particles, which, when inhaled, are very corrosive to lung tissue. In 1306, Edward I of England banned the burning of "sea coles," coal found on the seashore, in the city of London. This coal was high in sulfur content and was, in part, responsible for the city's noxious odors. Edward's ban might very well be the earliest environmental legislation concerning air quality.

London, England, was also the site of one of the earliest killer fogs. In 1952, the city was covered with a dense fog for several days. During this time, air over the city failed to mix with layers of air in the upper atmosphere due to temperature conditions. The factories continued to release smoke and dust into this stagnant layer of air, and it became so full of the fog, smoke, and dust that people got lost in familiar surroundings. This combination of smoke and fog has become known as smog. Many London residents developed respiratory discomfort, headaches, and nausea. Four thousand people died in a few weeks. Their deaths have been associated with the high levels of sulfur compounds in the smog. Thousands of others suffered from severe bronchial irritation, sore throats, and chest pains. The 1948 Donora, Pennsylvania, incident already mentioned also involved symptoms related to the particles and sulfur dioxide in the air.

Oxides of Nitrogen (NO and NO_2)

Oxides of nitrogen (NO and NO_2) are the fifth category of primary air pollutants. Several compounds contain nitrogen and oxygen in different combinations; nitrogen oxide (NO) and nitrogen dioxide (NO_2) are the most common. When combustion takes place in air, nitrogen and oxygen molecules from the air may react with each other, and oxides of nitrogen result.

$$N_2 + O_2 \rightarrow 2NO \text{ (nitrogen oxide)}$$

$$2NO + O_2 \rightarrow 2NO_2 \text{ (nitrogen dioxide)}$$

A mixture of nitrogen oxide and nitrogen dioxide is called NO_X. The nitrogen dioxide in the mixture reacts with other compounds to produce photochemical smog, discussed in the next section.

The primary source of nitrogen oxides is the automobile engine. Catalytic converters reduce the amount of nitrogen oxides released from the internal combustion engine, but increased automobile traffic has resulted in significant levels of NO_X in many metropolitan areas. Nitrogen oxides are noteworthy because they are involved in the production of secondary air pollutants.

figure 17.6 **Photochemical Smog** The interaction among hydrocarbons, oxides of nitrogen, and sunlight produces new compounds that are irritants to humans. The visual impact of smog is shown in these photographs taken in Los Angeles, Calif.

Photochemical Smog

Secondary air pollutants are compounds that result from the interaction of various primary air pollutants. **Photochemical smog** is a mixture of pollutants resulting from the interaction of nitrogen oxide with ultraviolet light. (See figure 17.6.) The two most destructive components of photochemical smog are ozone (O_3) and peroxyacetyl nitrates. Both of these materials are excellent oxidizing agents, which means that they react readily with many other compounds, including those found in living things, causing destructive changes. Ozone is particularly harmful because it destroys chlorophyll in plants and injures lung tissue in humans and other animals. Peroxyacetyl nitrates, in addition to being oxidizing agents, are eye irritants. Both ozone and peroxyacetyl nitrates are secondary air pollutants

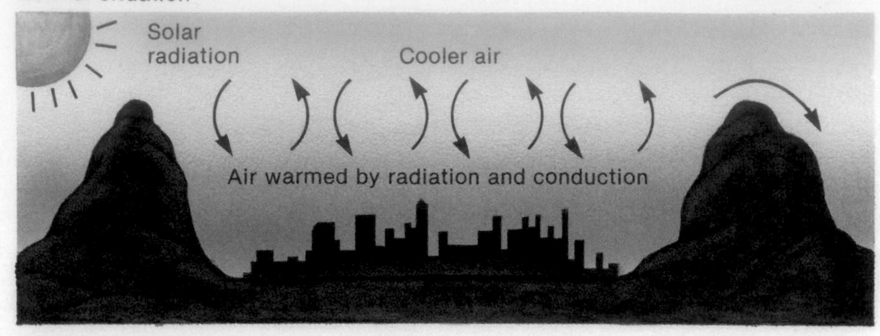

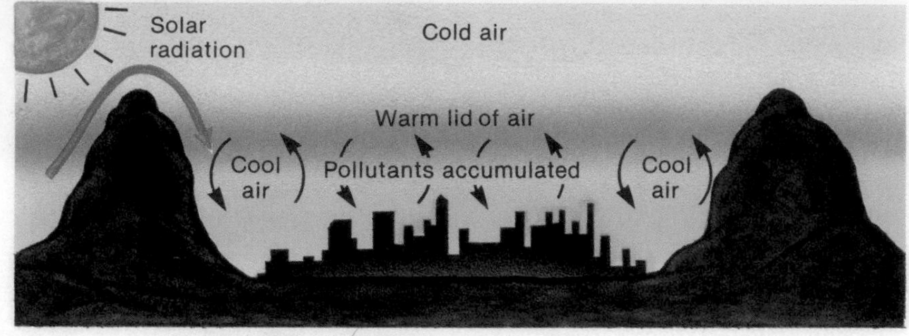

figure 17.7 **Thermal Inversion** Under normal conditions, the air at the earth's surface is heated by the sun and rises to mix with the cooler air above it. When a thermal inversion occurs, a layer of warm air is formed above the cooler air at the surface. The cooler air is then unable to mix with the warm air above and cannot escape because of surrounding mountains. The cool air is trapped, sometimes for several days, and accumulates pollutants. If the thermal inversion continues, the levels of pollution can become dangerously high.

that result from chemical reactions assisted by light.

A typical photochemical smog incident involves an interesting series of events.

1. Morning rush-hour traffic produces large amounts of nitrogen oxide (NO):

$$N_2 + O_2 \rightarrow 2NO$$

2. The nitrogen oxide reacts with molecular oxygen (O_2) from the atmosphere to form nitrogen dioxide (NO_2). It is the nitrogen dioxide in the atmosphere that gives photochemical smog its reddish-brown haze.

$$2NO + O_2 \rightarrow 2NO_2$$

3. Later in the morning, nitrogen dioxide reacts with ultraviolet light to form atomic oxygen (O):

$$NO_2 \xrightarrow[\text{light}]{\text{ultraviolet}} NO + O$$

4. Abundant molecular oxygen in the atmosphere reacts with atomic oxygen to form ozone:

$$O_2 + O \rightarrow O_3$$

5. Hydrocarbons in the atmosphere react with ozone to produce peroxyacetyl nitrates. Various pollution-control devices now reduce the amount of hydrocarbons escaping to the atmosphere, thereby decreasing the amount of peroxyacetyl nitrates formed.
6. As the ozone and peroxyacetyl nitrates react with living things, they cause damage and are converted to less reactive molecules, and the smog eventually clears.

Due to their climate and the geographic features of their locations, such large metropolitan areas as Los Angeles, Salt Lake City, Phoenix, and Denver have more trouble with photochemical smog than do metropolitan areas on the east coast of the United States. Each of these cities is ringed by mountains. The prevailing winds are from the west. As cool air flows into these valleys, it pushes the warm air upward. This warm air becomes sandwiched between two layers of cold air and acts like a lid on the valley, a condition known as a **thermal inversion.** The air is trapped in the valley. (See figure 17.7.) The lid of warm air cannot rise farther because it is covered by a layer of cooler air pushing down on it. It cannot move out of the area because of the ring of mountains. Without normal air circulation, smog accumulates. Harmful chemicals continue to increase in concentration until a major weather change causes the lid of warm air to move up and over the mountains. Then the underlying cool air can begin to circulate, and the polluted air is diluted.

Smog problems could be substantially decreased by reducing the use of internal combustion engines (perhaps eliminating them completely) or by moving population centers away from the valleys that produce thermal inversions. Both of these solutions would require expenditures of billions of dollars and major changes in lifestyle; therefore, people will probably continue to live with the problem.

Other Significant Air Pollutants

In recent years, three other air pollutants have been recognized as significantly affecting the health and welfare of people: radon (see Environmental Close-Up, p. 394), lead, and toxic chemicals. The primary sources of lead are gasoline and paint. For many years, lead was added to gasoline to help engines run more effectively. Recognition that lead emissions were hazardous resulted in the lead additives being removed from gasoline in North America and Europe. This has resulted in a decline in the amount of lead in the atmosphere. (See figure 17.8.) However, many other countries in the world still use leaded gasoline.

Another major source of lead is paints. Many older homes have paints that contain lead, since various lead compounds are colorful pigments. Dust from flaking paint or remodeling or demolition is released into the atmosphere. Although the amount of lead may be small, its presence in the home can result in significant exposure to inhabitants, particularly young children who chew on painted surfaces and often eat paint chips. Batteries are also a major source of lead.

Air toxics are harmful chemicals that are released into the atmosphere on purpose or are released accidentally as a result of leaks or poorly designed manufacturing processes. Materials such as pesticides are purposely released to kill insects or other pests. However, the majority of air toxics are released as a result of manufacturing processes. Although air toxics are important to the entire public, they are most critical for people who are exposed on the job since those people are likely to be exposed often and to higher concentrations. There are literally hundreds of different air toxics.

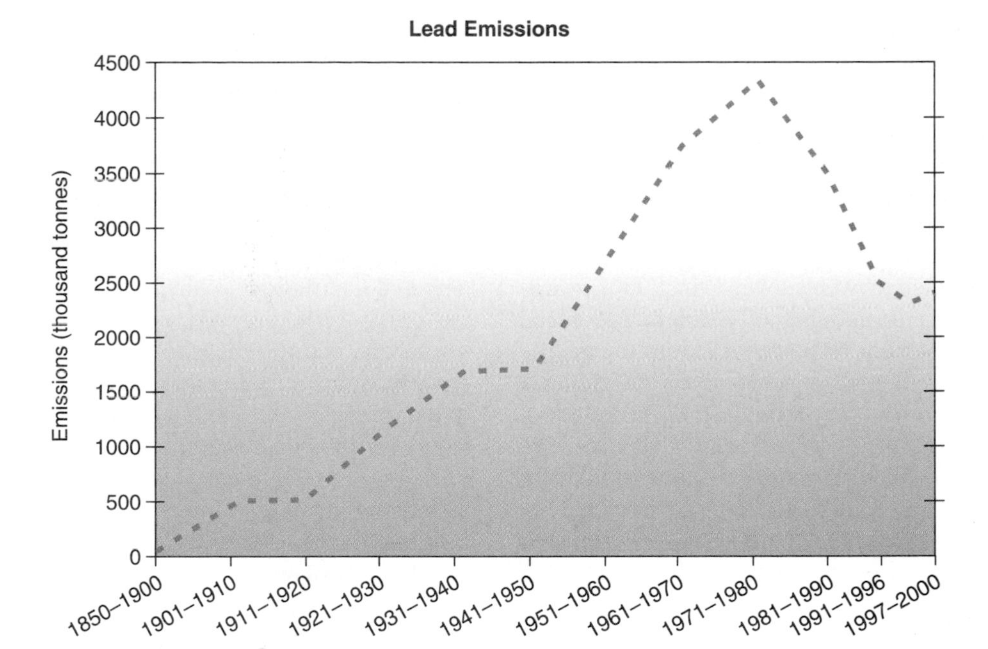

figure 17.8 **Lead Emissions** The amount of lead released into the atmosphere has declined significantly since the early 1980s when the lead additive was removed from gasolines in North America and much of Europe.

Source: United States Environmental Protective Agency Office of Air and Radiation.

Table 17.2 Top Ten Cities for Smog Improvement in United States from 1987 to 2000

	Percent Reduction	Smog days 1987–1989	Smog days 1994–2000
1. Seattle	100.0	14.0	0.0
2. Buffalo	100.0	7.7	0.0
3. Tucson	100.0	4.0	0.0
4. Tacoma	95.5	7.3	0.3
5. Albuquerque	93.2	14.7	1.0
6. Denver	92.5	22.3	2.1
7. Miami	92.3	4.3	0.3
8. Boston	91.7	8.0	0.7
9. Raleigh-Durham	90.9	3.7	0.5
10. Charlotte	88.2	11.3	1.6

Source: Data from the Environmental Protection Agency.

Control of Air Pollution

All of the air pollutants we have examined thus far are produced by humans. That means their release into the atmosphere can be controlled. Methods of controlling air pollution depend upon the type of pollutant and the willingness or ability of industries, governments, and individuals to make changes. It is important to note that positive change is possible. (See table 17.2.)

Eliminating photochemical smog completely would require large-scale changes in driving habits and other aspects of our lifestyle and culture. That is not likely to happen, but the problem can be lessened by reducing the particular primary air pollutants that contribute to photochemical smog: NO_X and hydrocarbons. United States government regulations have pressured the automobile industry to reduce emissions. The positive crankcase ventilation valve (PCV) and gas caps with air-pollution control valves (APC) reduce hydrocarbon loss. Catalytic converters reduce carbon monoxide, oxides of nitrogen, and hydrocarbons in emissions and necessitate the use of lead-free fuel. This lead-free fuel requirement, in turn, significantly reduces the amount of lead (and other metal additives) in the atmosphere.

Particulates are produced primarily in automobiles or by industries burning fuels. In industry, particulate release can be controlled with scrubbers, precipitators, and filters. These devices are effective, but expensive. They can be retrofitted to the smokestack, or they can be designed into the combustion system. They filter the products of combustion and remove some, while allowing others to be released into the atmosphere. Electric charge differentials may also be used to remove the unwanted materials from the emission.

Another major source of particulates are dusts produced by a variety of industrial activities. Transfer of materials such as grain or coal from one container to another generates dust. Mining and other earth-moving activities, including farming operations, cause dust and climatic factors. Improper land use can also be a major source of airborne particles. The clouds of particles from the dust bowl of the 1930s throughout parts of North America were severe. In 2000, it was documented that industrial pollution and dust from Asia travels across the Pacific and degrades air quality over North America. In some instances, levels of airborne particles that originated in China and central Asia had spread as far as Texas and briefly spiked some U.S. pollution levels. Over the past several years, satellites have spotted signs that large amounts of dust and industrial pollution have been rapidly moving across the Pacific toward North America. In 1997, researchers at the Cheeka Peak Observatory in the Olympic Peninsula in Washington state measured carbon monoxide levels that were 10 percent higher than average and fine particulate levels that were 50 percent higher than average. In 2000, large volumes of dust from the Gobi Desert in Mongolia and other Asian deserts traveled in a cloud across the Pacific, reaching as far east as Texas. In Seattle, Vancouver, and other West Coast cities the sky turned a visible milky white from the dust particles.

Industry and automobiles are not the only producers of particulate pollutants. Many people in the world use wood as their primary source of fuel for cooking and heat. In developed countries like the United States and Canada, some people use fireplaces and wood-burning stoves as a primary source of heat, but most use them for supplemental heat or for aesthetic purposes. However, the use of large numbers of wood-burning stoves and fireplaces can generate a significant air-pollution problem, called a brown cloud. Many municipalities, such as Boise, Idaho; Salt Lake City, Utah; and Denver, Colorado (and some entire states), use fines to enforce a ban on wood-burning during severe air-pollution episodes. Many other communities issue pollution alerts and request that people do not use wood-burners. Many of these communities have regulations about the number and efficiency of wood-burning stoves and fireplaces. Some communities, such as Castle Rock, Colorado, prohibit the construction of houses with fireplaces or wood-burning stoves. Many people readily switch to gas fireplaces or high-efficiency wood-burners when they understand that their enjoyment of a rustic stove may be leading to a degraded environment, but most need to be forced to comply by official regulations and the threat of fines. Obviously, accumulations of many small sources of air pollution may cause as big a problem as one large emission source and are frequently more difficult to control.

To control sulfur dioxide, which is produced primarily by electric power generating plants, several possibilities are available. One alternative is to change from high-sulfur fuel to low-sulfur fuel. Switching from a high-sulfur coal to a low-sulfur coal reduces the amount of sulfur released into the atmosphere by 66 percent. Switching to oil, natural gas, or nuclear fuels would reduce sulfur dioxide emissions even more. However, these are not long-term solutions because low-sulfur fuels are in short supply, and nuclear power plants pose a different set of pollution problems. (See chapter 11.)

A second alternative is to remove the sulfur from the fuel before the fuel is used. Chemical or physical treatment of coal before it is burned can remove nearly 40 percent of the sulfur. This is technically possible, but it increases the cost of electricity to the rate payer.

Scrubbing the gases emitted from a smokestack is a third alternative. The technology is available, but, of course, these control devices are costly to install, maintain, and operate. As with auto emissions, governments have required the installation of these devices, but, when industries install them, the cost of construction and operation is passed on to the consumer. The cost of installing scrubbers on a typical power plant is about $200 million. In the United States, an estimated 20 million or more tonnes (25 U.S. tons) of sulfur oxides are released into the atmosphere each year. Installing scrubbers on just 50 of the largest coal-burning plants would reduce this amount by over one-third.

In the past, a common solution was to build taller smokestacks. Tall stacks release their gases above the inversion layers and, therefore, add the sulfur dioxide to the upper atmosphere, where it is diluted before it comes in contact with the population downwind. This works as long as the stack is tall enough and as long as not so much pollution is added to the air that it cannot be diluted to an acceptable level. Taller smokestacks do not reduce pollutants; they simply make pollutants disperse downwind. However, sulfur dioxide reacts with oxygen and dissolves in water in the atmosphere to form sulfuric acid. This acid is washed from the air when it rains or snows. It damages plants and animals and increases corrosion of building materials and metal surfaces.

Clean Air Act

In the United States, the Clean Air Act has been the primary means of controlling air pollution. Initially enacted in 1967 as the Air Quality Act and extensively amended in 1970, 1977, and 1990, the Clean Air Act promulgates a series of detailed control requirements that the federal government implements and the states administer.

The Clean Air Act regulatory programs fall into four categories:

1. All new and existing sources of air pollution are subject to ambient air quality regulation.
2. New sources are subject to more stringent control technology and permitting requirements.
3. Control of specific pollution problems including hazardous air emissions and visibility impairment.
4. A comprehensive operating permit program added with the 1990 amendments.

National ambient air quality standards (NAAQS) are established for six pollutants (criteria pollutants) which present pervasive pollution problems.

1. Sulfur dioxide (SO_2)
2. Nitrogen oxides (NO_X)

3. Particulate matter
4. Carbon monoxide (CO)
5. Ozone
6. Lead

The states have primary responsibility for assuring that air quality is maintained at a level consistent with the NAAQS. State Implementation Plans (SIPs) must include enforceable emission limits, an enforcement program, protection against interstate air pollution, monitoring and emission data requirements, and pre-construction review and notification requirements.

Under the 1990 amendments, 189 substances are regulated, including hazardous organic chemicals and metals. Any source emitting 9.1 tonnes (10 U.S. tons) per year of any listed substance, or 22.7 tonnes (25 U.S. tons) combined, is considered a major source. With regard to acid rain, the 1990 amendments added new regulatory programs to control sulfur dioxide and nitrogen oxides emissions. The 1990 amendments established a program for the phaseout of ozone-depleting substances (CFCs, halons, carbon tetrachloride, and methyl chloroform).

In 2000, the EPA released a study of how effective the Clean Air Act has been. While stating that there is still considerable progress to be made, the report did highlight the successes of the Act. Since passage of the Clean Air Act, the EPA report stated that air pollution has been cut by a third and acid rain by 25 percent. Perhaps most impressive of all: emissions of the six worst air pollutants dropped 33 percent from 1970 to 2000 despite a 31 percent increase in U.S. population, a 114 percent rise in productivity, and a 127 percent jump in the number of kilometers driven by Americans. The EPA estimates that the Clean Air Act's human health, welfare and environmental benefits have outweighed its costs by 40 to 1. But problems remain. The EPA has targeted two major sources of continuing air pollution: the aging coal-fired power plants in the Midwest and motor vehicles such as sport utility vehicles, diesel trucks, and buses. A report issued by the U.S. Public Interest Research Group in 2000 said that even with the Clean Air Act, smog levels exceeded federal standards in 43 states in 1999. Smog helps cause more than 6 million asthma attacks and sends some 160,000 people to hospital emergency rooms each year.

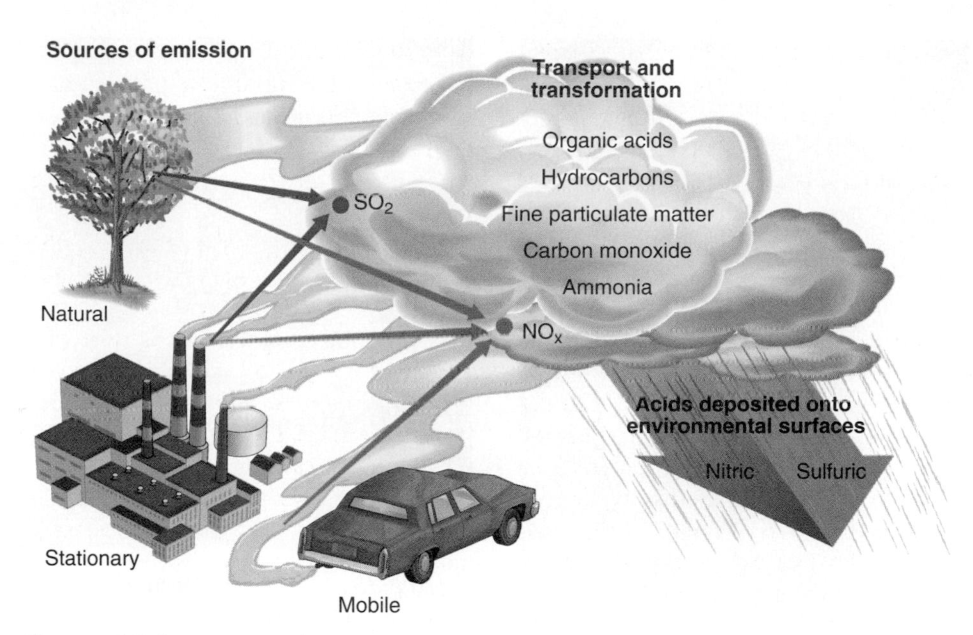

figure 17.9 **Acid Deposition** Molecules from natural sources, power plants, and internal combustion engines react to produce the chemicals that are the source of acid deposition.

Acid Deposition

aka acid Rain

Acid deposition is the accumulation of potential acid-forming particles on a surface. Acids result from natural causes, such as vegetation, volcanos, and lightning; and from human activities, such as coal burning and use of the internal combustion engine. (See figure 17.9.) These combustion processes produce sulfur dioxide (SO_2) and oxides of nitrogen (NO_X). Oxidizing agents, such as ozone, hydroxyl ions, or hydrogen peroxide, along with water, are necessary to convert the sulfur dioxide or nitrogen oxides to sulfuric or nitric acid. Various reactive hydrocarbons (HC) encourage the production of oxidizing agents.

The acid-forming reactants are classified as wet or dry. Wet reactions occur in the atmosphere and come to earth as some form of precipitation: acid rain, acid snow, or acid dew. Dry deposition occurs with the settling of compounds related to the acid on a surface. An acid does not actually form until these materials mix with water. Even though the acids are formed and deposited in several different ways, all of these processes usually are referred to as **acid rain.**

Acid rain is a worldwide problem. Reports of high acid-rain damage have come from Canada, England, Germany, France, Scandinavia, and the United States. Rain is normally slightly acidic, with a pH between 5.6 and 5.7 due to atmospheric carbon dioxide that dissolves to produce carbonic acid. But acid rains sometimes have a concentration of acid a thousand times higher than normal. In 1969, New Hampshire had a rain with a pH of 2.1. In 1974, Scotland had a rain with a pH of 2.4. The average rain in much of the northeastern part of the United States and parts of Ontario has a pH between 4.0 and 4.5.

Acid rain can cause damage in several ways. Buildings and monuments are often made from materials that contain limestone (calcium carbonate, $CaCO_3$), because limestone is relatively soft and easy to work. Sulfuric acid (H_2SO_4), a major component of acid rain, converts limestone to gypsum ($CaSO_4$), which is more soluble and is eroded over many years of contact with acid rain. (See figure 17.10.) Metal surfaces can also be attacked by acid rain.

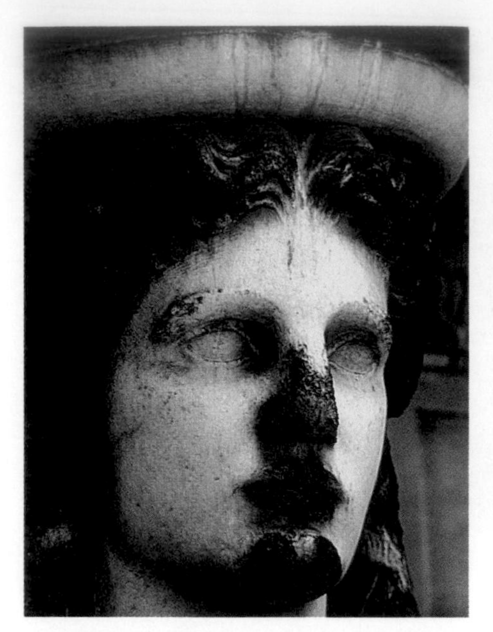

figure 17.10 **Damage Due to Acid Deposition** Sulfuric acid (H_2SO_4), which is a major component of acid deposition, reacts with limestone ($CaCO_3$) to form gypsum ($CaSO_4$). Since gypsum is water soluble, it washes away with rain. The damage to this monument is the result of such acid reacting with the stone.

figure 17.11 **Forest Decline** Many forests at high elevations in northeastern North America have shown significant decline, and dead trees are common.

The effects of acid rain on ecosystems are often more difficult to quantify. Intense sulfur dioxide pollution around smelters is known to cause the death of many kinds of trees and other vegetation. But this is an extreme case and may not be directly comparable to less intense acid rain. However, in many parts of the world, acid rain is suspected of causing the death of many forests and reducing the vigor and rate of growth of others. (See figure 17.11.) In Central Europe, many forests have declined significantly, resulting in the death of about 6 million hectares of trees. Northeastern North America has been affected with significant tree death and reduction in vigor, particularly at higher elevations. Some areas have had 50 percent mortality of red spruce trees.

A clear link between the decline of the forests and acid rain is difficult to establish, but several hypotheses have been formulated. Molecules like sulfur dioxide and ozone are known air pollutants and cause direct damage to plants. The sulfur dioxide also contributes to acid formation. As soil becomes acidic, aluminum is released from binding sites and may interfere with the plant roots' ability to absorb nutrients. A recent long-term study in New Hampshire strongly suggests that the many years of acid precipitation have reduced the amount of calcium in the soil, which is needed by plants for growth. Because there are no easy ways to replace the calcium even if acid rain were to stop, it would still take many years for the forests to return to health. Reduction in the pH of the soil may also change the kind of bacteria in the soil and reduce the availability of nutrients for plants. While none of these factors alone would necessarily result in plant death, each could add to the stresses on the plant and may allow other factors, such as insect infestations, extreme weather conditions (particularly at high elevations), or drought, to further weaken trees and ultimately cause their death.

The effects of acid rain on aquatic ecosystems are much more clear-cut. In several experiments, lakes were purposely converted to acid lakes and the changes in the ecosystems recorded. The experiments showed that as lakes become more acidic, there is a progressive loss of many kinds of organisms. The food web becomes less complicated, many organisms fail to reproduce, and many others die. Most healthy lakes have a pH above 6. At a pH of 5.5, many desirable species of fish have been eliminated; at a pH of 5, only a few starving fish may be found, and none are reproducing. Lakes with a pH of 4.5 are nearly sterile.

There are several reasons for these changes. Many of the early reproductive stages of insects and fish are more sensitive to acid conditions than are the adults. In addition, the young often live in shallow water, which is most affected by a flood of acid into lakes and rivers during the spring snowmelt. The snow and its acids have accumulated over the winter, and the snowmelt releases large amounts of acid all at once. Crayfish and other crustaceans need calcium to form their external skeleton. As the pH of the water decreases, the crayfish are unable to form new skeletons and so they die. Reduced calcium availability also results in some fish with malformed skeletons. (See figure 17.12.) As mentioned earlier, increased acidity also results in the release of aluminum, which impairs the function of a fish's gills.

The extent to which lakes have been acidified is great. About 14,000 lakes in Canada and 11,000 in the United States have been seriously altered by becoming acidic. Many lakes in Scandinavia are similarly affected. The extent to which acid deposition affects an ecosystem depends on the nature of the bedrock in the area and the ecosystem's proximity to acid-forming pollution sources. (See figure 17.13.) Parent material derived from igneous rock is not capable of buffering

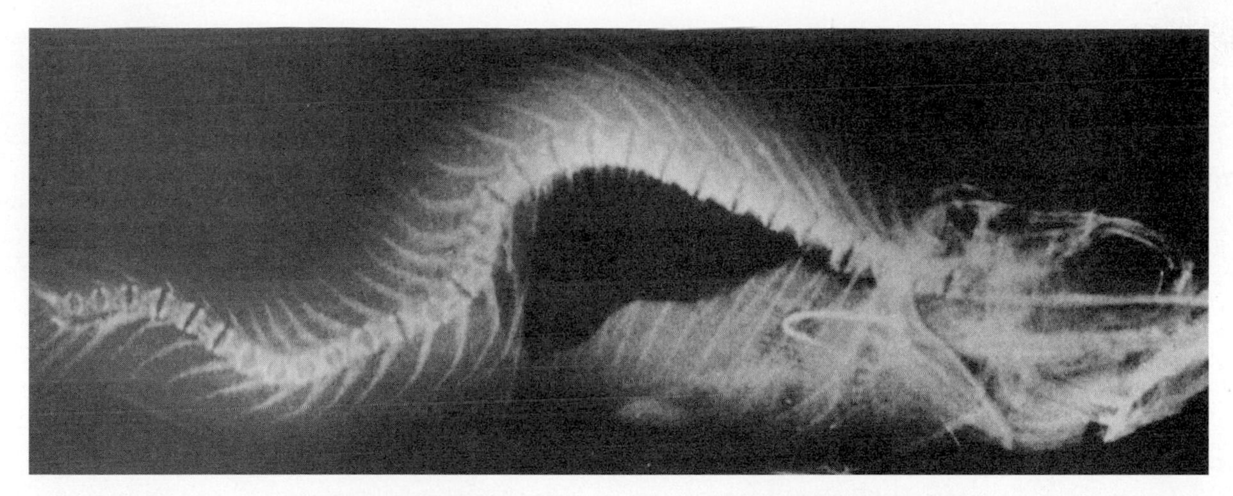

figure 17.12 **Effects of Acid Deposition on Organisms** The low pH of the water in which this fish lived caused the abnormal bone development that ultimately resulted in the death of the fish.

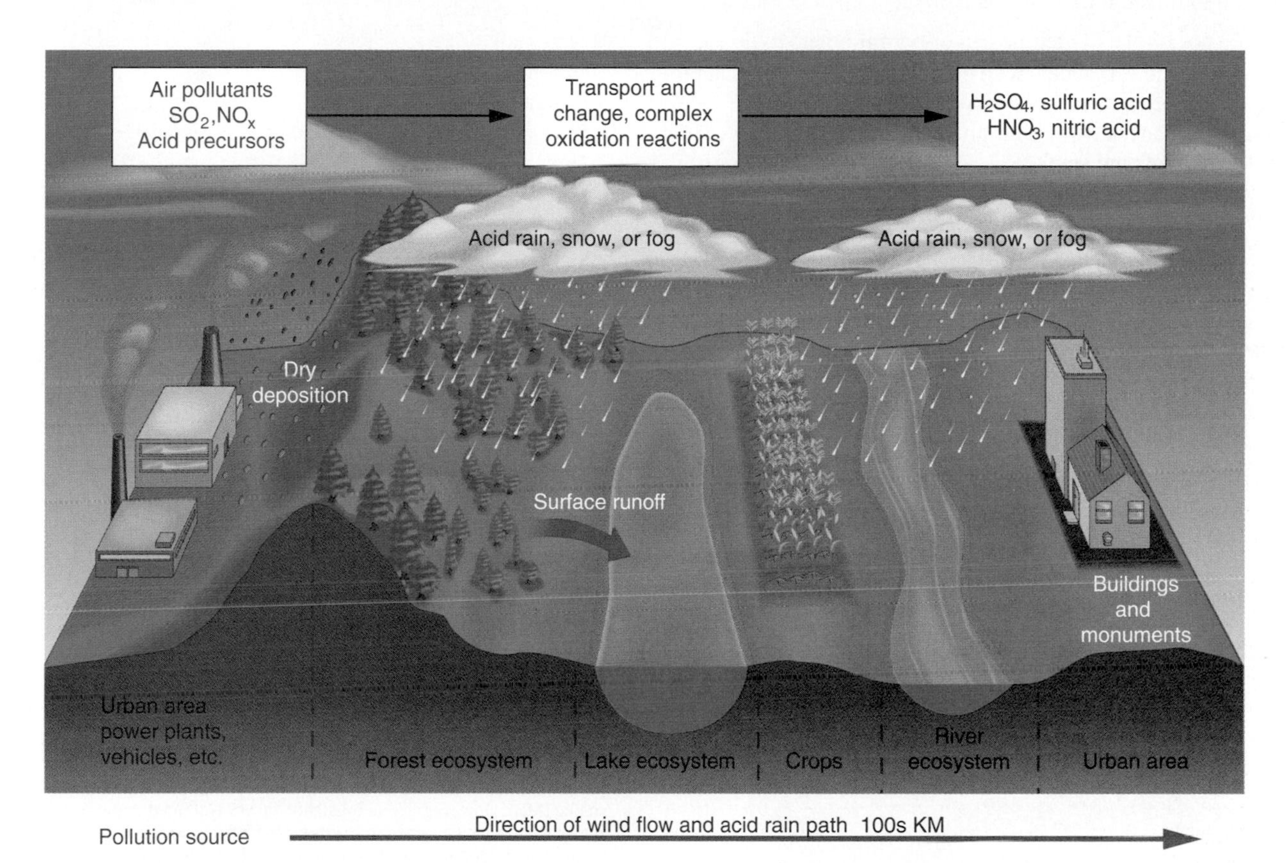

figure 17.13 **Factors That Contribute to Acid Rain Damage** In an aquatic ecosystem, the following factors increase the risk of damage from acid deposition: (1) a lake is located downwind from a major source of air pollution; (2) the area around the lake is hard, insoluble bedrock covered with a layer of thin, infertile soil; (3) the soil has a low buffering capacity; and (4) there is also a low lake surface to watershed area ratio.

Source: Data from U.S. Environmental Protection Agency's "Acid Rain."

the effects of acid deposition, while soils derived from sedimentary rocks such as limestone release bases that neutralize the effects of acids. Because of this, eastern Canada and the U.S. Northeast are particularly susceptible to acid rain. These areas have high amounts of granite rock and are downwind from the major air-pollution sources of North America. Scandinavian countries have a similar geology and receive pollution from industrial areas in the United Kingdom and Europe. Thousands of kilometers of streams and up to 200,000 lakes in eastern Canada and the northeastern United States are estimated to be in danger of becoming acidified because of their location and geology.

ENVIRONMENTAL CLOSE-UP

Secondhand Smoke

Smoking of tobacco has long been associated with a variety of respiratory diseases, including lung cancer, emphysema, and heart disease. Because of these strong links, tobacco products carry warning labels. Many nonsmoking people, however, are exposed to environmental tobacco smoke (secondhand smoke) because they live and work in spaces where people smoke. The U.S. Environmental Protection Agency estimates that approximately 3000 nonsmokers die each year of lung cancer as a result of breathing air that contains secondhand smoke. Young children who are exposed to secondhand smoke are much more likely to have respiratory infections.

In July 1993, the U.S. Environmental Protection Agency recommended several actions to prevent people from being exposed to secondhand indoor smoke. They include recommendations that:

- people not smoke in their homes or permit others to do so;
- all organizations that deal with children have policies that protect children from secondhand smoke;
- every company have a policy that protects employees from secondhand smoke;
- smoking areas in restaurants and bars be placed so that the smoke will have little chance of coming into contact with nonsmokers.

California has some of the most restrictive smoking laws in the United States. In 1998, California passed a no-smoking law that effectively banned all smoking in indoor public places, including restaurants and bars. The law was designed for employees who did not want to inhale secondhand smoke. While there are no no-smoking police going from building to building if an employee complains, the owner of the establishment could pay a fine of up to $7,000.

Below is a chart of various compounds and their toxicity found in secondhand smoke. Note that secondhand smoke is also referred to as environmental tobacco smoke.

Compound	Type of toxicity
Vapor phase:	
Carbon monoxide	T
Carbonyl sulfide	T
Benzene	c
Formaldehyde	c
3-Vinylpyridine	SC
Hydrogen cyanide	T
Hydrazine	c
Nitrogen oxides	T
N-nitrosodimethylamine	C
N-nitrosopyrrolidine	C
Particulate phase:	
Tar	C
Nicotine	T
Phenol	TP
Catechol	CoC
o-Toluidine	C
2-Naphthylamine	C
4-Aminobiphenyl	C
Benz(a)anthracene	C
Benzo(a)pyrene	C
Quinoline	C
N′-nitrosonornicotine	C
NNK	C
N-nitrosodiethanolamine	C
Cadmium	C
Nickel	C
Polonium-210	C

Abbreviations: C, carcinogenic; CoC, cocarcinogenic; SC, suspected carcinogen; T, toxic; TP, tumor promoter. Environmental tobacco smoke (ETS) is diluted in the air before it is inhaled and thus is less concentrated than mainstream smoke (MS). However, active inhalation of MS is limited to the time it takes to smoke each cigarette, whereas exposure to ETS is constant over the period spent in the ETS-polluted environment. This fact is reflected in measurements of nicotine uptake by smokers and ETS-exposed nonsmokers.

Sources: Data from Department of Health and Human Services, 1989; State of California, Report on Secondhand Smoke 1999.

Global Warming and Climate Change

Burning coal, oil, and natural gas to heat our homes, power our cars, and illuminate our cities produces carbon dioxide and other gases as by-products. Deforestation and clearing of land for agriculture also release significant quantities of such gases. Over the last century, we have been emitting gases such as carbon dioxide and methane to the atmosphere faster than natural processes can remove them. During this time, atmospheric levels of these gases have climbed steadily and are projected to continue their steep ascent as global economies grow. (See figure 17.14.)

Records of past climate going back as far as 160,000 years indicate a close correlation between the concentration of greenhouse gases in the atmosphere and global temperatures. Computer simulations of the climate indicate that global temperatures will rise as atmospheric concentrations of carbon dioxide increase.

During the 1980s, scientists, governments, and the public became concerned about the possibility that the world may be getting warmer. The United Nations Environment Programme established an Intergovernmental Panel on Climate Change (IPCC) to study the issue and make recommendations. Its First Assessment was published in 1990. In 1996, the IPCC published its Second Assessment and concluded that climate change is occurring and that it is highly probable that human activity is an important cause of the change. The IPCC has reached several important conclusions:

1. The average temperature of the Earth has increased 0.3–0.6°C (0.5-1.0°F), (1999 was the warmest year on record), and sea level has risen 10–25 cm (4-10 inches) in the last 100 years. (See figure 17.15.)
2. There is a strong correlation between the increase in temperature and the amount of greenhouse gases present in the atmosphere.
3. Human activity greatly increases the amounts of these greenhouse gases.

Total World Emissions of Carbon Dioxide

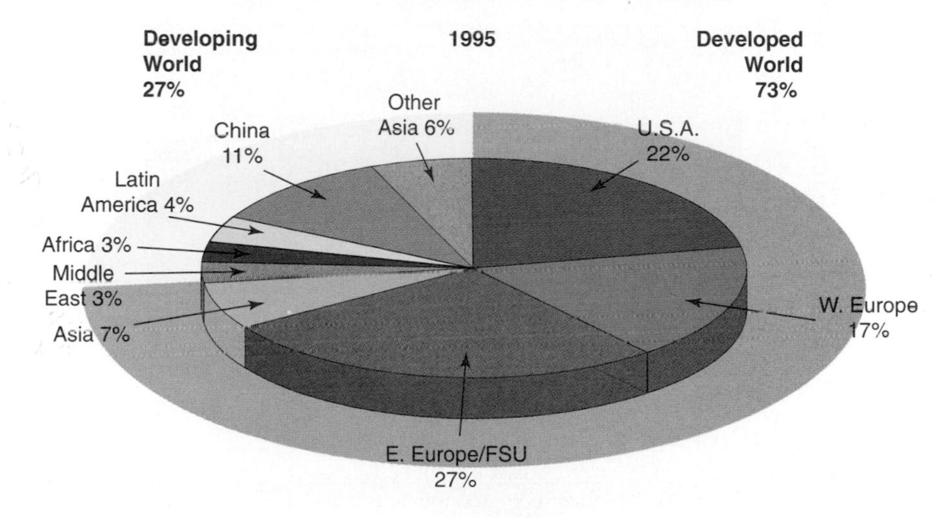

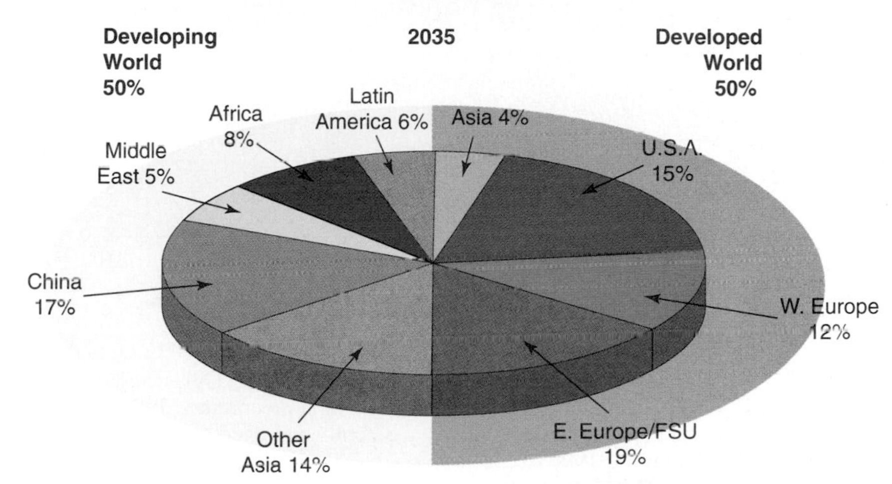

figure 17.14 In 1995, the industrialized nations of the world contributed nearly three-quarters of the global emissions of carbon dioxide, with the U.S. being the largest single emitter. By 2035, developing nations will catch up and contribute half of the global emissions, with China becoming the largest single emitting country. Rapid population growth, industrialization, and increasing consumption per person in the developing world will contribute to this shift.

Source: Data from "Climate Change–State of Knowledge," October 1997, Office of Science and Technology Policy, Washington, D.C. and State of the World 2000.

It is important to recognize that although a small increase in the average temperature of the earth may seem trivial, this increase could set in motion changes that could significantly alter the climate of major regions of the world. (See figure 17.16.)

The effects of global warming and a changing climate will not be felt equally across the planet. Regional climate changes will likely be very different from changes in the global average. Differences from region to region could be in both the magnitude and rate of climate change. Furthermore, not all things, whether they be natural ecosystems or human settlements, are equally sensitive to changes in climate. And, finally, nations and indeed regions within nations vary in their ability to cope and adapt to global warming and a changing climate.

Some nations will likely experience more adverse effects than other nations. Some nations may benefit more than others. Poorer nations are generally more vulnerable to the consequences of global warming. These nations tend to be more

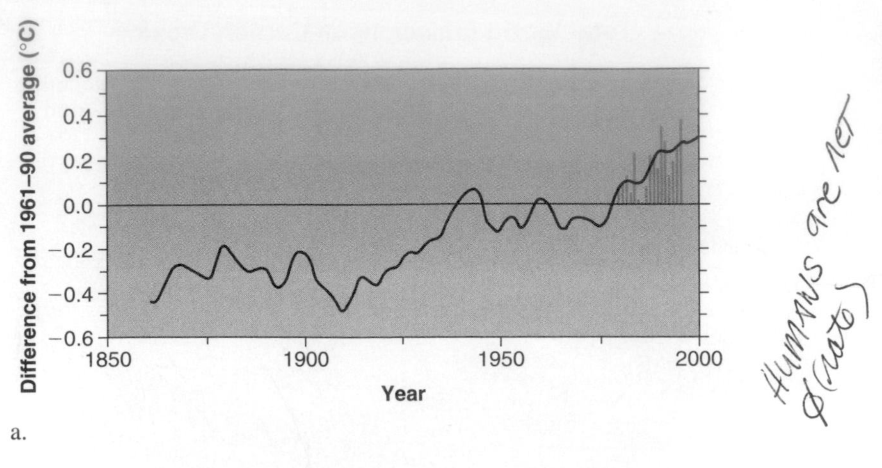

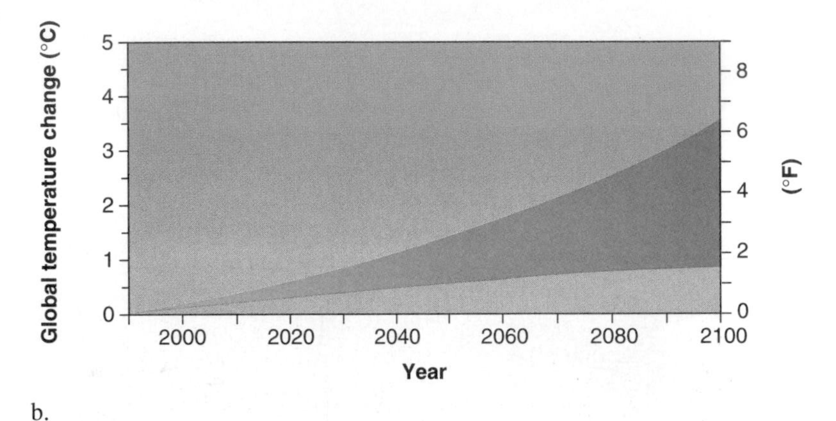

figure 17.15 **Changes in Average Global Temperature** (*a*) Despite considerable variation, there has been a general trend toward increasing temperatures. 1999 was the warmest year on record. (*b*) The possible range of globally averaged surface temperature increase is shown for the period 1990 to 2100.

Source: (*a*) Data from "U.K. Meteorological Office/University of East Anglia" published in Science, January 12, 1996. (*b*) Source: Data from "Global Temperature Change and Range of Globally Averaged Sea Level" in *Climate Changes—State of Knowledge* from Office of Science and Technology Policy, 1997, Office of the President, Washington, D.C., and State of the World 2000.

dependent on climate-sensitive sectors, such as subsistence agriculture, and lack the resources to buffer themselves against the changes that global warming may bring. The Intergovernmental Panel on Climate Change (IPCC) has identified Africa as "the continent most vulnerable to the impacts of projected changes because widespread poverty limits adaptation capabilities."

While the specific consequences resulting from global warming will continue to be debated, areas of concern include worsening human health effects, rising sea levels, disruption of the water cycle, changing forests and natural areas, and challenges to agriculture and the food supply.

What actually causes global warming? An explanation is relatively straightforward. Several gases in the atmosphere are transparent to light but absorb infrared radiation. These gases allow sunlight to penetrate the atmosphere and be absorbed by the earth's surface. This sunlight energy is reradiated as infrared radiation (heat), which is absorbed by the gases. Because the effect is similar to what happens in a greenhouse (the glass allows light to enter but retards the loss of heat), these gases are called greenhouse gases and the warming thought to occur from their increase is called the **greenhouse effect.** (See figure 17.17.) The most important greenhouse gases are carbon dioxide (CO_2), chlorofluorocarbons (primarily CCl_3F and CCl_2F_2), methane (CH_4), and nitrous oxide (N_2O). Table 17.3 lists the relative contribution of each of these gases to the potential for global warming.

Carbon dioxide (CO_2) is the most abundant of the greenhouse gases. It occurs as a natural consequence of respiration. However, much larger quantities are put into the atmosphere as a waste product of energy production. Coal, oil, natural gas, and biomass are all burned to provide heat and electricity for industrial processes, home heating, and cooking. These sources are increasing the amount of carbon dioxide in the atmosphere. Measurement of carbon dioxide levels at the Mauna Loa Observatory in Hawaii show that the carbon dioxide level has increased from about 315 ppm (parts per million) in 1958 to about 362 ppm in 1998. (See figure 17.18 and 17.19.)

A major step toward slowing global warming would be to increase the efficiency of energy utilization. This would also be of value in conserving the shrinking supplies of energy resources. It makes sense to increase energy efficiency, thus reducing carbon dioxide production, even if global warming is not a concern. One way to stimulate a move toward greater efficiency would be the imposition of a carbon tax. A carbon tax would increase the cost of fuels by taxing the amount of carbon put into the atmosphere by their use. This would increase the demand for fuel efficiency because the cost of fuel would rise. It would also stimulate the development of alternative fuels with a lower carbon content and generate funds for research in many aspects of fuel efficiency and alternative fuel technologies.

Another approach to the problem is to increase the amount of carbon dioxide removed from the atmosphere. If enough biomass is present, the excess carbon dioxide can be utilized by vegetation during photosynthesis, thereby reducing the impact of carbon dioxide released by fossil-fuel burning. Australia, the United States, and several other countries have announced plans to plant billions of trees to help remove carbon dioxide from the atmosphere.

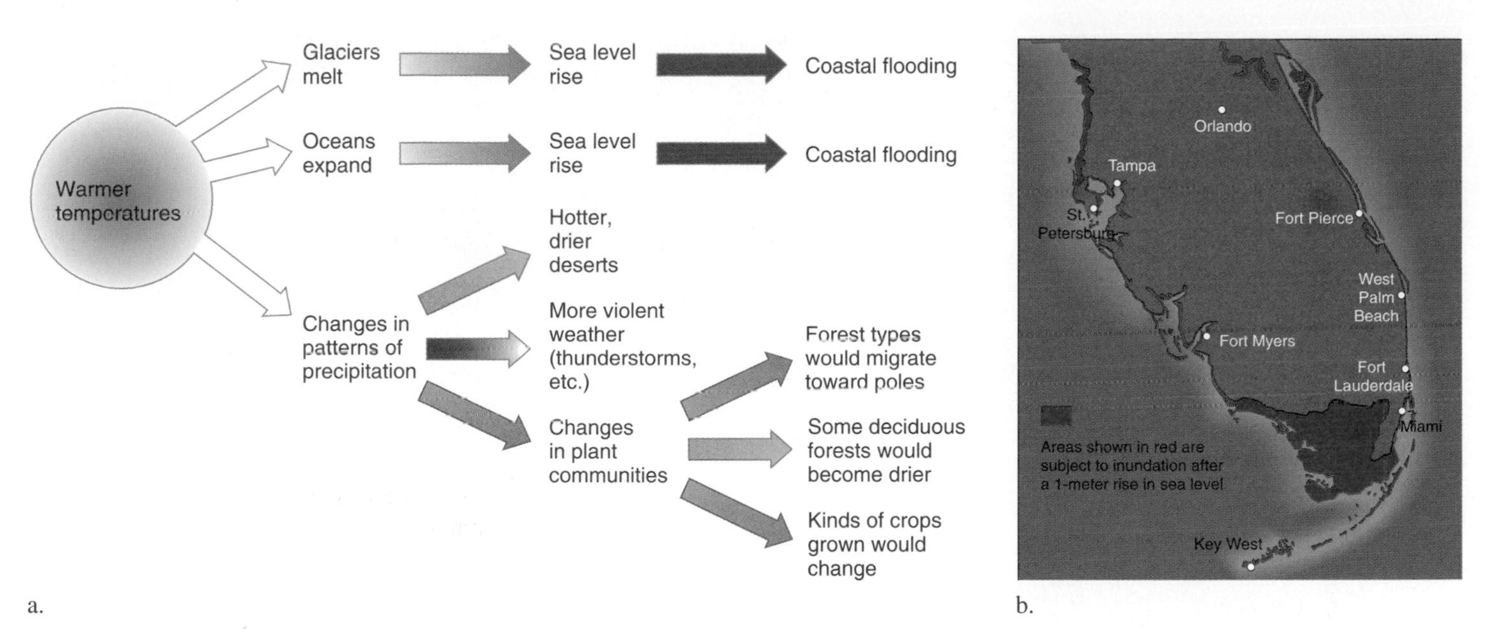

figure 17.16 **Effects of Global Warming** (*a*) Global warming would have several effects on the climate of the world. The climate changes would have important impacts on human and other living things. (*b*) Sea level rise could inundate many low-lying coastal areas in Florida, and will increase the vulnerability of all such areas to storm surges.

Source: (*b*) Elevations from USGS digital data. Prepared by the U.S. Geological Survey, 2000.

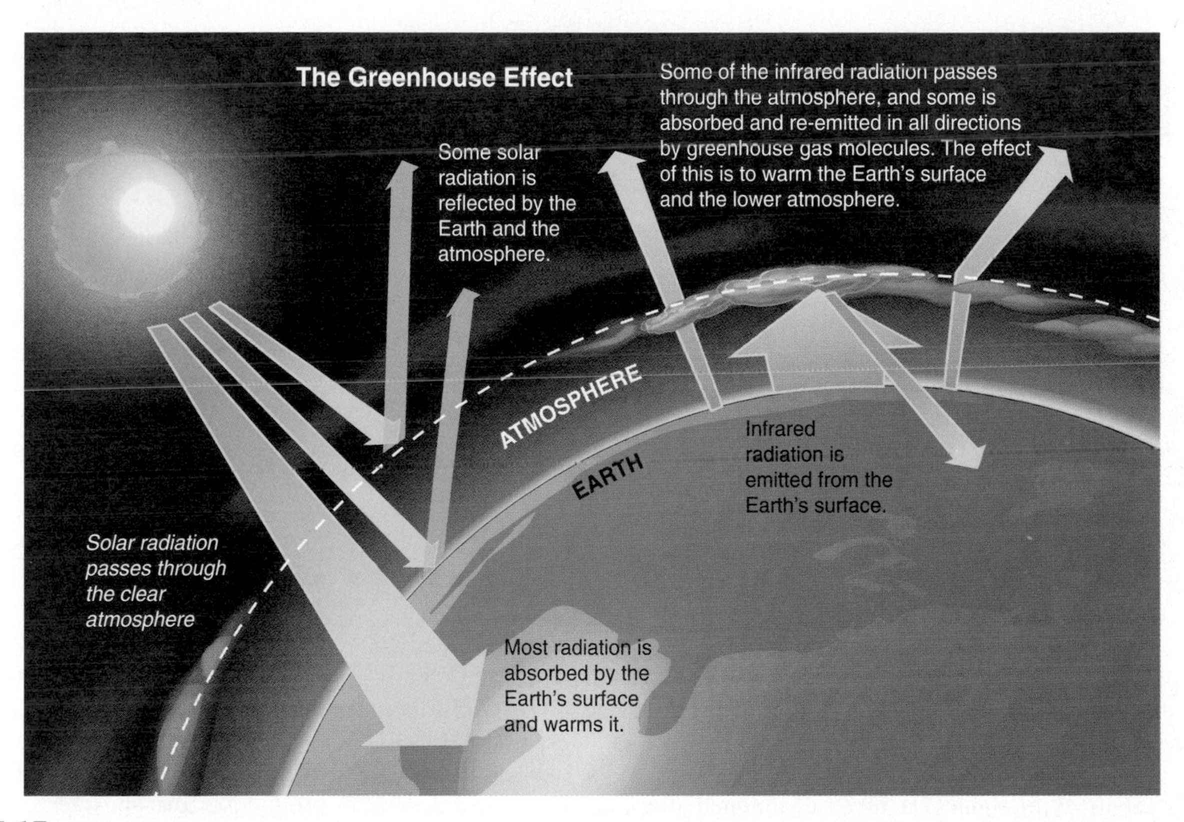

figure 17.17 The greenhouse effect naturally warms the Earth's surface. Without it, Earth would be 60°F cooler than it is today—uninhabitable for life as we know it.

Source: Data from "Climate Change—State of Knowledge," October 1997, Office of Science and Technology Policy, Washington, D.C.

Table 17.3 Major Greenhouse Gases and Their Characteristics

Gas	Atmospheric Concentration (ppm)	Annual Increase (percent)	Life Span (years)	Contribution to Global Warming (percent)	Principal Sources
Carbon Dioxide	355	.4	50–200	55	Coal, Oil, Natural gas, Deforestation
(fossil fuels)				(43)	
(biological)				(12)	
Chlorofluorocarbons	.00085	2.2	50–102	24	Foams, Aerosols, Refrigerants, Solvents
Methane	1.714	.8	12–17	15	Wetlands, Rice, Fossil Fuels, Livestock
Nitrous Oxide	.31	.25	120	6	Fossil Fuels, Fertilizers, Deforestation

Source: World Meteorlogical Organization.

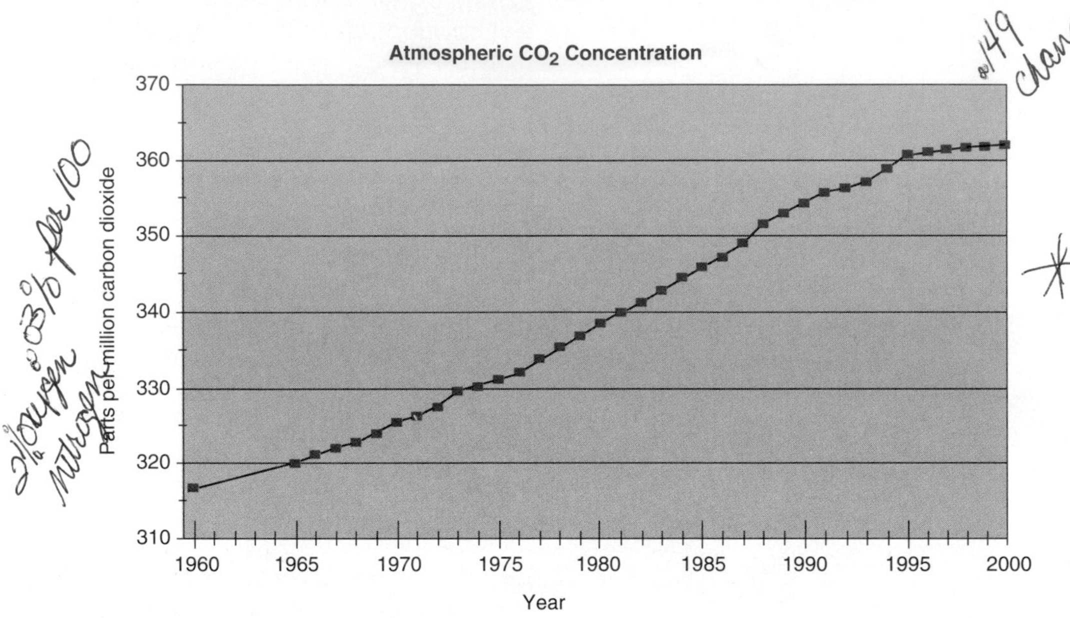

figure 17.18 **Change in Atmospheric Carbon Dioxide** Since the establishment of a carbon dioxide monitoring station at Mauna Loa Observatory in Hawaii, a steady increase in carbon dioxide levels has been observed. Since 1960, the concentration of carbon dioxide in the atmosphere has increased by nearly 14 percent.

Source: Data from Carbon Dioxide Information Analysis Center, 2000, Oak Ridge, Tenn.

Many critics argue that this approach will provide only a short-term benefit since, eventually, the trees will mature and die, and their decay will release carbon dioxide into the atmosphere at some later time.

An associated concern is the destruction of vast areas of rainforest in tropical regions of the world. These ecosystems are extremely efficient at removing carbon dioxide and storing the carbon atoms in the structure of the plant. The burning of tropical rainforests to provide farm or grazing land not only adds carbon dioxide to the atmosphere, but also reduces the ability to remove carbon dioxide from the atmosphere, since the grasslands or farms created do not remove carbon dioxide as efficiently as do the original rainforests. Furthermore, the grazing lands and farms in such regions of the world are often abandoned after a few years and do not return to their original forest condition.

Chlorofluorocarbons are entirely the result of human activity. They were widely used as refrigerant gases in refrigerators and air conditioners, as cleaning solvents, as propellants in aerosol containers, and as expanders in foam products. Although they are present in the atmosphere in minute quantities, they are extremely efficient as greenhouse gases (about 15,000 times more efficient at retarding heat loss than is carbon dioxide).

Since the 1970s, when chlorofluorocarbons were linked to the depletion of the ozone layer in the upper atmosphere, their use as propellants in aerosol cans has been banned in the United States, Canada, Norway, and Sweden, and the European Economic Community agreed to reduce use of chlorofluorocarbons in aerosol cans. However, worldwide, chlorofluorocarbons are still widely used as aerosol propellants. In foam and solvents, care can be taken to recover the chlorofluorocarbons for reuse rather than allow them to escape into the atmosphere. Alternative refrigerants are available, and care could be taken to recycle refrigerant gases rather than just venting them into the air. In 1991, DuPont announced the development of new refrigerants that would not harm the ozone layer. These have been installed in refrigerators and air conditioners in many nations, including the United States. We will need to continue to exploit all these options if we want to reduce chlorofluorocarbon production on an international scale. Until substitutes become available, the mandatory recycling of chlorofluorocarbons will reduce the rate at which new gases are added to the atmosphere.

In 1987, several industrialized countries, including Canada, the United States, the United Kingdom, Sweden,

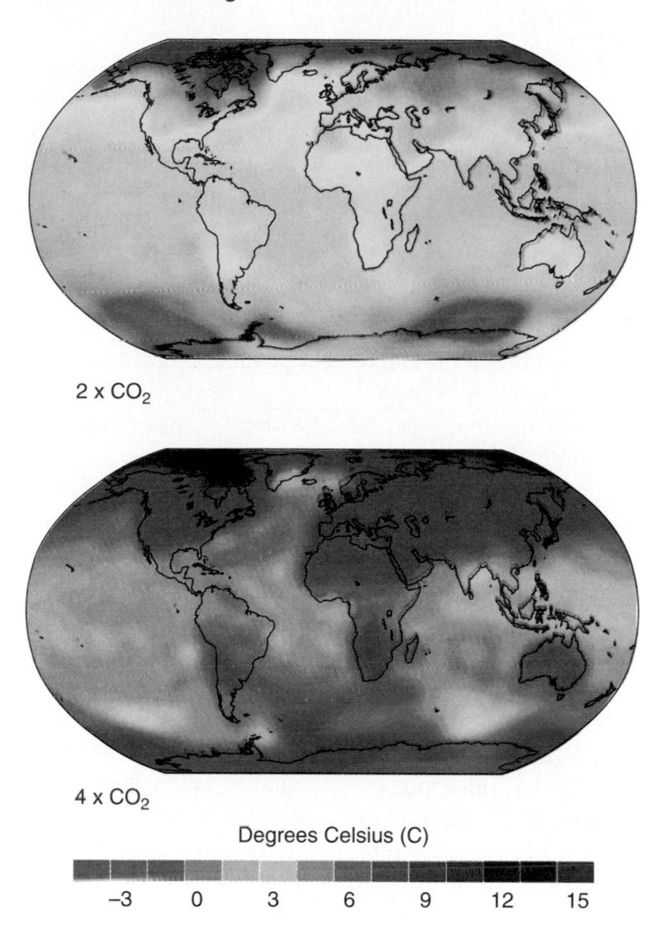

figure 17.19 **Surface Air Warming** Without reductions in CO_2 emissions, atmospheric CO_2 could rise to four times preindustrial levels.

Source: Geophysical Fluid Dynamics Laboratory, (NOAA).

Norway, Netherlands, the Soviet Union, and West Germany, agreed to freeze production of chlorofluorocarbons at present levels and reduce production by 50 percent by the year 2000. This document, known as the Montreal Protocol, was ratified by the U.S. Senate in 1988. As a result of the 1987 Montreal Protocol, chlorofluorcarbon emissions dropped 87 percent from their peak in 1988. Several companies in the United States and Japan (e.g., General Motors, AT&T, Nissan) have announced plans to either phase out chlorofluorocarbon use or reduce use significantly. In 1990, in London, international agreements were reached to further reduce the use of chlorofluorocarbons. A major barrier to these negotiations was the reluctance of the developed countries of the world to establish a fund to help less-developed countries implement technologies that would allow them to obtain refrigeration and air conditioning without the use of chlorofluorocarbons. In 1996, the United States stopped producing chlorofluorocarbons. As a result of these international efforts and rapid changes in technology, the use of chlorofluorocarbons has dropped rapidly, and concentrations of chlorofluorocarbons in the atmosphere have stabilized and are expected to decline in the future.

Methane enters the atmosphere primarily from biological sources. Several kinds of bacteria that are particularly abundant in wetlands and rice fields release methane into the atmosphere. Methane-releasing bacteria are also found in large numbers in the guts of termites and various kinds of ruminant animals such as cattle. Some methane enters the atmosphere from fossil-fuel sources. Control of methane sources is unlikely since the primary sources involve agricultural practices that would be very difficult to change. For example, nations would have to convert rice paddies to other forms of agriculture and drastically reduce the number of animals used for meat production. Neither is likely to occur, since food production in most parts of the world needs to be increased, not decreased.

Nitrous oxide, a minor component of the greenhouse gas picture, enters the atmosphere primarily from fossil fuels and fertilizers. It could be reduced by more careful use of nitrogen-containing fertilizers.

Worsening Health Effects

Climate change will impact human health in a variety of ways. Warmer temperatures increase the risk of mortality from heat stress. For example, in July 1995, more than 700 deaths in Chicago were attributed to a heat wave with temperatures exceeding 32°C (90°F) day and night. Today, such events occur about once every 150 years. Carbon dioxide (CO_2) concentrations of 550 ppm (parts per million) (double the pre-industrial level) could make such events six times more frequent. The potential increases in the heat index, a calculation combining temperature and humidity, illustrate the magnitude of this threat. Washington, D.C., currently has an average July heat index of 30°C (86°F), but if CO_2 levels reach 550 ppm, this could increase to 35°C (95°F), and if concentrations quadrupled to 1100 ppm, it could increase to 43°C (110°F). Climate change will also exacerbate air-quality problems, such as smog, and increase levels of airborne pollen and spores that aggravate respiratory disease, asthma, and allergic disorders. Because children and the elderly are the most vulnerable populations, they are likely to suffer disproportionately with both warmer temperatures and poorer air quality.

Throughout the world, the prevalence of particular diseases and other threats to human health depend largely on local climate. Extreme temperatures can directly cause the loss of life. Moreover, several serious diseases appear only in warm areas. Finally, warm temperatures can increase air and water pollution, which, in turn, harm human health.

The most direct effect of climate change would be the impacts of hotter temperatures themselves. Extremely hot temperatures increase the number of people who die (of various causes) on a given day. For example, people with heart problems are vulnerable because the cardiovascular system must work harder to keep the body cool during hot weather. Heat exhaustion and some respiratory problems increase.

Higher air temperatures also increase the concentration of ozone at ground level. The natural layer of ozone in the upper atmosphere blocks harmful ultraviolet radiation from reaching the Earth's surface, but in the lower atmosphere, ozone is a harmful pollutant. Ozone damages lung tissue and causes particular problems for people with asthma and other lung diseases. Even modest exposure to ozone can cause healthy individuals to experience chest pains, nausea, and pulmonary congestion.

Global warming may also increase the risk of some infectious diseases, particularly those diseases that appear only in warm areas. Diseases that are spread by mosquitoes and other insects could become more prevalent if warmer temperatures enabled those insects to become established farther north. Such "vector-borne" diseases include malaria, dengue fever, yellow fever, and encephalitis. Some scientists believe that algal blooms could occur more frequently as temperatures warm, particularly in areas with polluted waters, in which case diseases (such as cholera) that tend to accompany algal blooms could become more frequent.

Rising Sea Level

Rising sea level erodes beaches and coastal wetland, inundates low-lying areas, and increases the vulnerability of coastal areas to flooding from storm surges and intense rainfall. By 2100, sea level is expected to rise by 15 to 90 cm (6 to 35 inches). (See figure 17.20.) A 50 cm (20 inch) sea level rise will result in substantial loss of coastal land in North America, especially along the southern Atlantic and Gulf coasts, which are subsiding and are particularly vulnerable. The land area of some island nations and countries such as Bangladesh would change dramatically as flooding occurred. The oceans will continue to expand for several centuries after temperatures stabilize. Because of this, the sea level rise associated with CO_2 levels of 550 ppm (double pre-industrial levels) could eventually exceed 100 cm (39.4 inches). A CO_2 level of 1100 ppm could produce a sea level rise of 200 cm (78.7 inches) or even more, depending on the extent to which the Greenland and Antarctic ice sheets melt.

A 50 cm (20 inches) sea level rise would double the global population at risk from storm surges, from roughly 45 million at present to over 90 million, and this figure does not account for any increases in coastal populations. A 100 cm (39.4 inches) rise would triple the number.

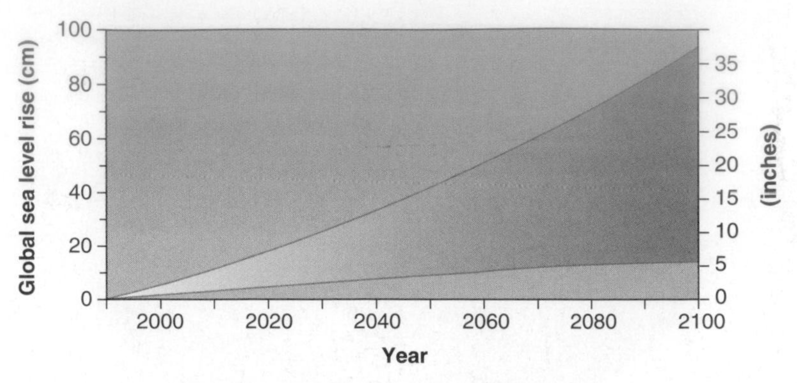

figure 17.20 **Possible Sea Level Rise** The possible range of globally averaged sea level rise is shown for the period 1990 to 2100.

Source: Data from "Climate Change—State of Knowledge," October 1997, Office of Science and Technology Policy, Washington, D.C. and State of the World, 2000.

Disruption of the Water Cycle

Among the most fundamental effects of climate change are intensification and disruption of the water cycle. Droughts and floods and water quality and quantity are the areas of particular concern.

Changing climate is expected to increase both evaporation and precipitation in most areas of the world. In those areas where evaporation increases more than precipitation, soil will become drier, lake levels will drop, and rivers will carry less water.

Lower river flows and lower lake levels could impair navigation, hydroelectric power generation, and water quality, and reduce the supplies of water available for agricultural, residential, and industrial uses. Some areas may experience increased flooding during winter and spring, as well as lower supplies during summer. In California's Central Valley, for example, melting snow provides much of the summer water supply; warmer temperatures would cause the snow to melt earlier and thus reduce summer supplies even if rainfall increased during the spring. More generally, the tendency for rainfall to be more concentrated in large storms as temperatures rise would tend to increase river flooding, without increasing the amount of water available.

Navigation—Climate change could impair navigation by changing average water levels in rivers and lakes, increasing the frequency of both floods (during which navigation is hazardous) and droughts (during which passage is difficult), and necessitating changes in navigational infrastructure. On the other hand, warmer temperatures could extend the ice-free season in many parts of the world.

Hydropower—Changes in the flows of rivers would have a direct impact on the amount of hydropower gen-

Current and Projected Ranges of Sugar Maple

Present range

Overlap

Predicted range

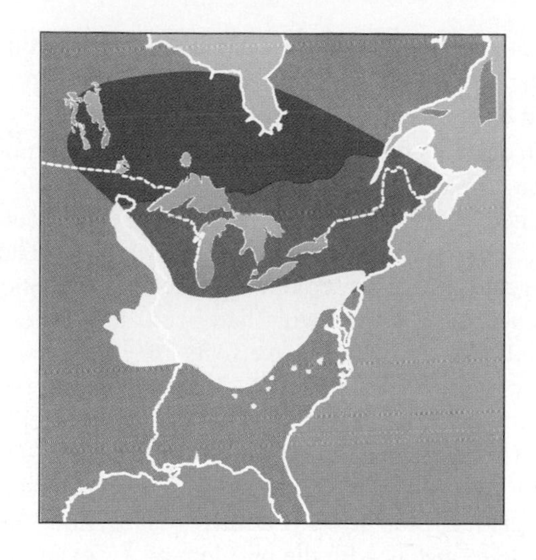

Prediction based on increased temperature

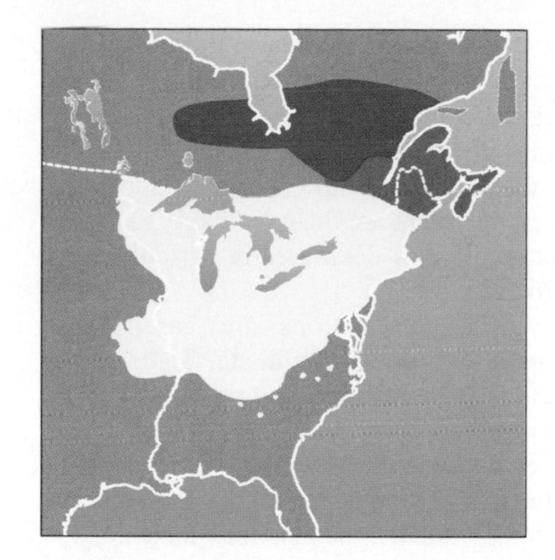

Prediction based on increased temperature and moisture reduction

figure 17.21 **Climatic Shifts** Climatic shifts will force some species to migrate northwards or to higher elevations in order to stay in the appropriate climatic zone. The climatic zone for sugar maple, for example, could shift northwards into Canada. This would compromise the maple syrup industry and the fall foliage colors, both of which make New England famous.

Source: Data from the United States Environmental Protection Agency and United Nations.

erated, because hydropower production decreases with lower flows. Because of the ambiguous projections of changes in future river flows, studies of the impacts of climate change show ambiguous effects on hydropower production.

Water Supply and Demand—In some parts of the world, the most widely discussed potential impact of climate change is the impact on water supply and demand. The potential changes in water supplies would result directly from the changes in runoff and the levels of rivers, lakes, and aquifers.

Environmental Quality and Recreation—Decreased river flows and higher temperatures could harm the water quality of rivers, bays, and lakes. In areas where river flows decrease, pollution concentrations will rise because there will be less water to dilute the pollutants. Increased frequency of severe rainstorms could increase the amount of chemicals that run off from farms, lawns, and streets into rivers, lakes, and bays.

Flood Control—Although the impacts of sea level rise and associated coastal flooding have been more widely discussed, global climate change could also change the frequency and severity of inland flooding, particularly along rivers.

Regarding the issues of water quality and quantity, the areas of greatest vulnerability are those where quality and quantity of water are already problems, such as the arid and semiarid regions of the world. Water scarcity in the Middle East and Africa is likely to be aggravated by climate change, which could increase international tension among countries that depend on water supplies originating outside their borders.

Changing Forests and Natural Areas

Climate change could dramatically alter the geographic distributions of vegetation types. The composition of one-third of the Earth's forests could undergo major changes as a result of climate changes associated with a CO_2 level of 700 ppm. Over the next 100 years, the ideal range for some North American forest species could shift by as much as 500 kilometers (300 miles) to the north, far faster than the forests can migrate naturally. Economically important species, such as the sugar maple, could be lost from the northeastern United States by the end of the next century. (See figure 17.21.)

Such changes could have profound effects on national parks and refuges around the world, leading to reductions in biological diversity and in the benefits provided by ecosystems, such as clean water and recreation. Wetlands are particularly at risk. The wetlands of the prairie pothole region, which support half the waterfowl population of North America, could diminish in area and change dramatically in character in response to climate change.

Challenges to Agriculture and the Food Supply

Climate strongly affects crop yields. A CO_2 concentration of 550 ppm is likely to increase crop yields in some areas by as much as 30 percent to 40 percent, but it will decrease yields in other places by similar amounts, even for the same crop. A warmer climate would reduce flexibility in crop distribution and increase irrigation demands. Expansions of the ranges of pests could also increase vulnerability and result in greater use of pesticides. Despite these effects, total global food production is not expected to be altered substantially by climate change, but there are likely negative regional impacts. Agricultural systems in the developed countries are highly adaptable and can probably cope with the expected range of climate changes without dramatic reductions in yields. It is the poorest countries, already subject to hunger, that are the most likely to suffer significant decreases in agricultural productivity.

These predictions are based on computer models of climate. Some scientists have criticized the predictions as being inaccurate and constructed from sketchy data. However, as more accurate information is gathered and inserted into the models, the general conclusions remain the same. It is difficult for the general public to comprehend these changes or see evidence of them since each of us experiences only our own local weather and climate. Also, there will always be short-term variations in weather patterns. The models, however, are attempting to predict the long-term trends. The consequences of a global warming would be so great that many are suggesting we alter our lifestyles regardless of whether the phenomenon is true, just to be on the safe side.

Some of the most dramatic projections regarding global warming are on natural systems. Here, the rate of change—as well as the absolute magnitude of change—is crucial. Ecosystems long taken for granted in particular regions are predicted to undergo enormous change. For example, as previously mentioned sugar maples—long a fixture in the northeastern United States—and the Everglades system in Florida, both disappear in modeling forecasts. Wetlands and coral reefs may also undergo radical decline due to climate change.

The result is yet another source of accelerated loss of species and additional challenge to efforts to promote biological diversity. The more gradually climate changes, the easier it could be for species to adapt. As a result, ecologists have proposed that emission constraints be calculated so as to limit the rate of warming to no more than a tenth of a degree Celsius (0.18 degree fahrenheit) per decade.

The greatest risks may be those yet to be discovered. Just as the Antarctic ozone hole was an unanticipated surprise, scientists have hypothesized many troubling possibilities, including more frequent or severe hurricanes and a shift in ocean currents responsible for moderating the climate of northern Europe. As we learned in the context of ozone depletion, what we don't know can hurt us—the rate of ozone depletion proved to be much greater than what had been predicted. The most serious effects of climate change too may lie outside current calculation. The evolution of a common international scientific understanding of such complex issues has been a critical step toward addressing the problem, just as with the problem of ozone depletion.

Addressing Climate Change

As experience protecting the ozone layer demonstrates, technological change can be very difficult to initiate but extremely rapid once broad consensus is reached. Once a new direction has been firmly agreed upon, the economic incentives are to move as quickly as possible. Few firms seek to be the last left producing a banned substance, or failing to comply with a new regulation.

The incentive for change is reinforced by the globalization of the economy, which encourages manufacture for the widest possible market acceptance. Thus, many developing nations, given the latitude to continue using CFCs beyond the time afforded their more developed counterparts, quickly discovered that this grace period was a mixed blessing—because it excluded them from the largest export markets. Multinational firms reinforced the trend by supplier specifications that required "CFC free." Financial assistance through the Ozone Multilateral Fund further encouraged a more rapid transition.

A similar process may be more difficult to achieve with respect to climate change because carbon is typically embedded in product manufacture and not distinguishable in products and services. However, the fundamental benefit of cleaner and more energy-efficient production potentially holds similar allure. The policy challenge is to find different paths to the same end, channeling the enormous creativity and resources of the private sector to the search for carbon alternatives.

The initial challenge of reducing greenhouse gas emissions is as much political as economic, as much about how we organize our economies as about how much we consume. The IPCC and numerous national reports have shown that there is no shortage of cost-effective technologies for achieving sizable reductions in greenhouse gas emissions. For example, the IPCC estimates that the market potential for greenhouse gas emission reductions in the building sector—the reduction that can be achieved economically with current technologies and no new policies or measures—is 10 to 15 percent in 2010, 15 to 20 percent in 2020, and 20 to 50 percent in 2040, relative to baseline scenarios.

These opportunities exist today in virtually all nations, including developing economies and nations that are already relatively energy efficient. For example, China's energy intensity (the ratio of commercial energy consumption per unit of GNP) has fallen 50 percent since 1980, an unprecedented rate over such a period. (China remains four times as energy intensive as the United States, however, making it among the world's least energy-efficient economies.) An important reason for China's progress is the recognition that its continued economic success requires the gradual elimination of fossil-fuel subsidies, a philosophy increasingly accepted by many developing nations.

Reductions in greenhouse gas emissions are likely to bring considerable ancillary benefits, potentially offsetting much of the costs. For example, the work bank has recently estimated that in 1999, air pollution in China, primarily attributable to poorly controlled burning of coal, caused 200,000 premature deaths, 1.8 million cases of chronic bronchitis, 1.7 billion restricted activity days, and more than 5 billion cases of respiratory illness.

Improved energy efficiency also benefits development through reduced

Table 17.4 Technologies to Address Climate Change

1. Fuel cell-powered cars and buses that substantially reduce local pollution, greenhouse gas emissions, and oil consumption.
2. Microturbines (building-sized power plants) with lower costs and equivalent efficiency to current scale coal generation.
3. Oxy-fuel firing (combustion using oxygen rather than air) for manufacture of glass, steel, aluminum, and metal casting, reducing energy use up to 45 percent and dramatically reducing local emissions.
4. "Zero energy" buildings that minimize energy requirements and produce more on-site energy (e.g., through photovoltaic rooftiles) than is purchased.
5. High strength, lightweight materials that allow reduced costs and improved efficiency in all transportation modes.
6. Coal gasification combined with CO_2 capture and sequestration to achieve high efficiency, clean burning of coal with near zero carbon emissions.
7. Wind power systems combined with compressed air energy storage at a total cost competitive with coal generation.
8. Generation of electricity directly from sunlight through photovoltaics integrated with roofing materials at a cost competitive with central station power.

Source: Report of the Energy Research and Development Panel of the President's Committee of Advisors on Science and Technology, Federal Energy Research and Development for the Challenges of the Twenty-First Century.

capital needs for hard currency for power plants and related energy infrastructure, now estimated at about $100 billion annually in developing nations.

Many candidates for this longer-term technological transition have been identified and some have already begun to penetrate the market. For example, wind energy systems are improving rapidly, and, in ideal conditions, already compare favorably with conventional coal-burning power plants. Direct conversion of sunlight to electricity is now possible with photovoltaic (PV) and solar thermal technologies. While relatively expensive today, they are already competitive in areas remote from electric utility grids. The costs of these technologies are likely to decline significantly over time because of their small scale, which allows economies of scale and learning by doing. (See table 17.4.)

The U.S. Department of Energy has concluded that, relying primarily on already proven technology, the United States could reduce its carbon emissions by almost 400 million tonnes (440 million U.S. tons) in 2010, or enough to stabilize U.S. emissions in that year at 1990 levels with savings from reduced energy costs roughly equal to the added cost of investment.

Resources and policies to increase investment in renewables and other longer-term technologies will be needed. Good examples from the industrialized nations include a policy in the United Kingdom that reserves a small part of electricity demand for competitive acquisition of designated renewable energy technologies, wind power purchase programs in Denmark and Germany, the "10,000 rooftops" PV program in Japan, and the evolution of "green marketing" campaigns in the United States and Europe to capture consumer willingness to pay modest premiums for electricity from clean energy technologies.

Ozone Depletion

In the 1970s, various sectors of the scientific community became concerned about the possible reduction in the ozone layer in the upper atmosphere surrounding the Earth. **Ozone** is a molecule of three atoms of oxygen (O_3). In 1985, it was discovered that a significant thinning of the ozone layer over the Antarctic occurred during the Southern Hemisphere spring. Some regions of the ozone layer showed 95 percent depletion. Ozone depletion also has been found to be occurring farther north. Measurements in arctic regions suggest a thinning of the ozone layer there also. These findings have caused several countries to become involved in efforts to protect the ozone layer.

The presence of ozone in the outer layers of the atmosphere, approximately 12–25 kilometers (7.5-15.5 miles) from the Earth's surface, shields the Earth from the harmful effects of ultraviolet light radiation. Ozone (O_3) absorbs ultraviolet light and is split into an oxygen molecule and an oxygen atom:

$$O_3 \xrightarrow{\text{Ultraviolet light}} O_2 + O$$

Oxygen molecules are also split by ultraviolet light to form oxygen atoms:

$$O_2 \xrightarrow{\text{Ultraviolet light}} 2O$$

Recombination of oxygen atoms and oxygen molecules allows ozone to be formed again and to be available to absorb more ultraviolet light.

$$O_2 + O \rightarrow O_3$$

This series of reactions results in the absorption of 99 percent of the ultraviolet light energy coming from the sun and prevents it from reaching the Earth's surface. Less ozone in the upper atmosphere would result in more ultraviolet light reaching the Earth's surface, causing increased skin cancers and cataracts in humans and increased mutations in all living things.

Chlorofluorocarbons are strongly implicated in the ozone reduction in the upper atmosphere. Chlorine reacts with ozone in the following way to reduce the quantity of ozone present:

$$Cl + O_3 \rightarrow ClO + O_2$$

$$ClO + O \rightarrow Cl + O_2$$

These reactions both destroy ozone and reduce the likelihood that it will be formed because atomic oxygen (O) is removed as well. It is also important to note that it can take 10 to 20 years for chlorofluorocarbon molecules to get into the stratosphere and then they can react with the ozone for up to 120 years. As a result, the World Meteorological Association predicts ozone depletion will worsen well into the next century before improvements are seen.

The use of chlorofluorocarbons in air conditioners, in refrigerators, and as

ENVIRONMENTAL CLOSE-UP

Radon

In 1985, the clothing worn by an engineer at the Limerick Nuclear Generating Station in Pottstown, Pennsylvania, registered a high radiation level. Initially, the generating station was believed to be the source of the radiation. However, subsequent studies indicated that radon 222 from the engineer's home was the source. Following this incident, there has been an increased interest in radon and its effects.

The source of radon is uranium-238, a naturally occurring element that makes up about three parts per million of the Earth's crust. Uranium-238 goes through fourteen steps of decay before it becomes stable, nonradioactive lead 206. **Radon,** an inert radioactive gas having a half-life of 3.8 days, is one of the products formed during this process.

Since radon is an inert gas, it does not enter into any chemical reactions within the body, but it can be inhaled. Once in the lung, it will undergo radioactive decay, producing other kinds of atoms called "daughters" of radon. These decay products (daughters) of radon—plutonium 218, which has a three-minute half-life; lead 214, which has a twenty-seven minute half-life; bismuth 214, which has a twenty-minute half-life; and polonium 214, which has a millisecond half-life—are solid materials that remain in the lungs and are chemically active.

Increased incidence of lung cancer is the only known health effect associated with radon decay products. It is estimated that the decay products of radon are responsible for about 15,000 lung cancer deaths annually in the United States. This is about 10 percent of lung cancer deaths.

As the radon gas is formed in the rocks, it usually diffuses up through the rocks and soil and escapes harmlessly into the atmosphere. It can also diffuse into groundwater. Radon usually enters a home through an open space in the foundation. A crack in the basement floor or the foundation, the gap around a water or sewer pipe, or a crawl space allows the radon to enter the home. It may also enter in the water supply from wells.

Only 10 percent of the homes in the United States have a potential radon problem. However, the increased publicity about radon has many people worried. In addition, the Environmental Protection Agency and the U.S. Surgeon General recommend that all Americans (other than those living in apartment buildings above the second floor) test their home for radon. If the tests indicate that the level of radon is at or above 4 picocuries/liter, EPA recommends that the homeowner take action to lower the level. This is usually not expensive and consists of blocking the places where radon is entering or venting sources of radon to the outside. People who are concerned about radon should contact their state's public health department or environmental protection agency.

Radon risk evaluation chart

pCi/L	WL	Estimated lung-cancer deaths due to radon exposure (out of 1,000)	Comparable exposure levels	Comparable risk
200	1.0	440–770	One thousand times average outdoor level	More than sixty times nonsmoker risk
				Four-pack-a-day smoker
100	0.5	270–630	One hundred times average indoor level	
				Two thousand chest X rays per year
40	0.2	120–380		
				Two-pack-a-day smoker
20	0.1	60–120	One hundred times average outdoor level	
				One-pack-a-day smoker
10	0.05	30–120	Ten times average indoor level	
				Five times nonsmoker risk
4	0.02	13–50	Level at which EPA suggests remedial action	
				Two hundred chest X rays per year
2	0.01	7–30	Ten times average outdoor level	
				Nonsmoker risk of dying from lung cancer
1	0.005	3–13	Average indoor level	
				Twenty chest X rays per year
0.2	0.001	1–3	Average outdoor level	

Note: Measurement results are reported in one of two ways: (1) pCi/L (picocuries per liter – Measurement of *radon gas*, or (2) WL (working levels) – Measurement of *radon decay products.*

Source: Data from Office of Air and Radiation Programs, U.S. Environmental Protection Agency.

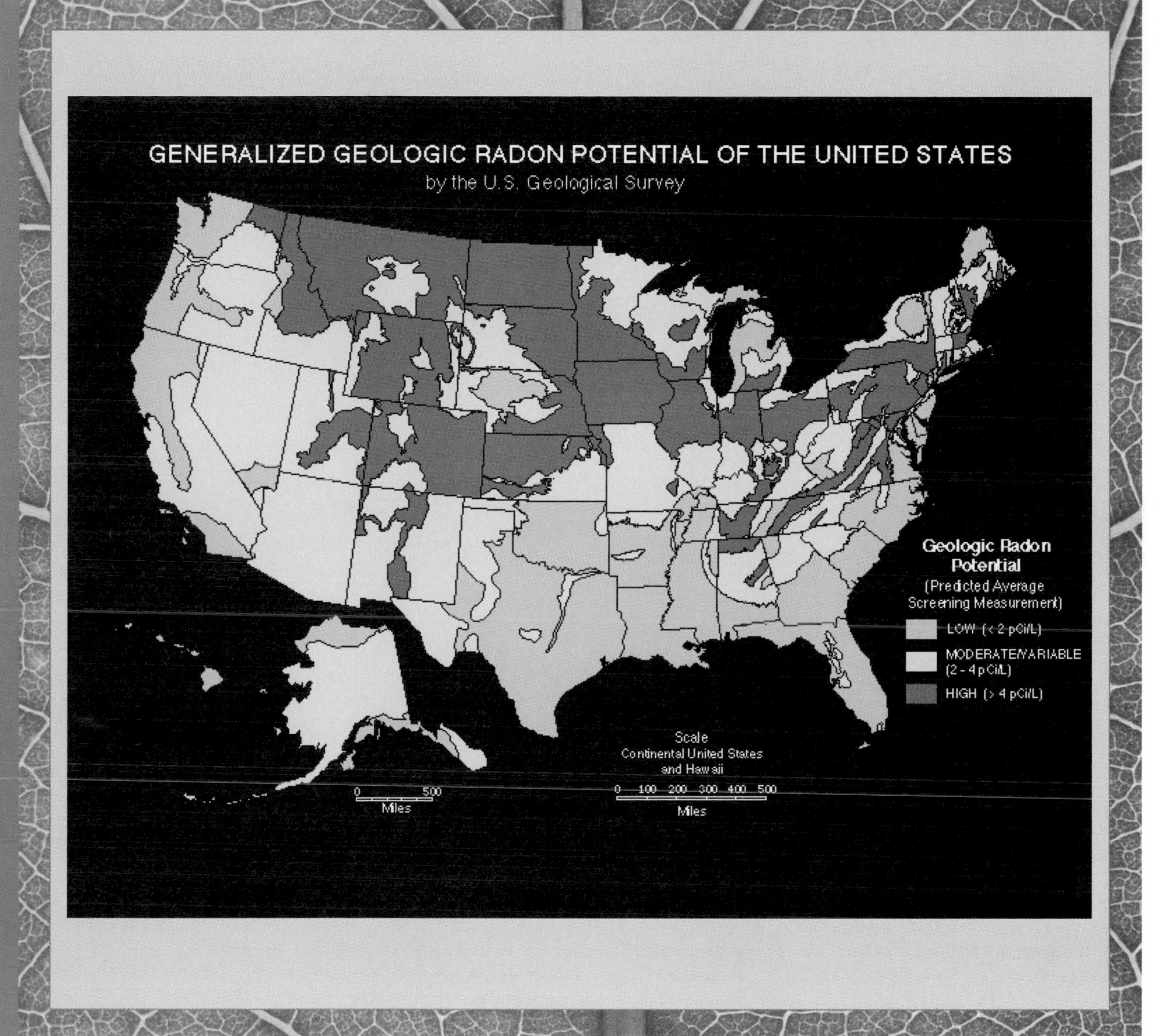
GENERALIZED GEOLOGIC RADON POTENTIAL OF THE UNITED STATES
by the U.S. Geological Survey
Geologic Radon Potential
(Predicted Average Screening Measurement)
LOW (< 2 pCi/L)
MODERATE/VARIABLE (2 - 4 pCi/L)
HIGH (> 4 pCi/L)
Scale
Continental United States and Hawaii
0 100 200 300 400 500
Miles
0 500
Miles

Table 17.5 Indoor Air Pollution Summary

Pollutant	Description	Sources	Effects*
Asbestos	Light, fibrous mineral; fireproof and insulative	Ceilings and floor tiles, insulation, and spackling compounds	Easily inhaled to cause lung damage Cancer
Carbon monoxide (CO)	Odorless, colorless gas	Combustion sources such as charcoal, kerosene heaters, attached garages with poor ventilation, tobacco smoke	Reduces ability of blood to carry oxygen; impairs vision and alertness Dizziness, headaches, fatigue, death by suffocation
Formaldehyde	Pungent gas; preservative and disinfectant	Foam insulation, resin in particleboards, plywood paneling, fiberboard, some carpets, drapery, and upholstery fabrics	Headaches, dizziness, nausea, lethargy, rashes, upper respiratory irritations
Lead (Pb)	Metallic element	House paints manufactured prior to 1976, old plumbing lines and solder, leaded crystal and pottery, old toys	Learning and behavior problems in children, high blood pressure, joint pain, concentration problems, reproductive problems
Biological pollutants and microorganisms	Pollen, dust mites, and pet dander Bacteria, molds, fungi, and viruses	Improperly maintained heating and cooling systems Washrooms, humidifiers, and dehumidifiers Pets and plants Wall-to-wall carpeting	Allergic diseases and skin irritations, influenza and Legionnaires' disease, acute cases of asthma
Nitrogen dioxide (NO_2)	Brownish gas	Gas appliances, fireplaces, wood and coal stoves	Eye and respiratory tract irritations Lowers resistance to respiratory infections Chronic bronchitis
Radon (Rn)	Naturally occurring, radioactive gas	Rocks and soils that contain the decaying radioactive elements of uranium and radium Enters through cracks in foundation and basements	Lung cancer
Tobacco smoke	Mixture of several substances including known human carcinogens	Cigarettes, pipes and cigars, secondhand smoke	Eye, nose and throat irritations, headaches and nausea, coughing and chest discomfort, respiratory infections, lung cancer
Volatile organic compounds (VOCs)	Substances produced by synthetic chemical industry and naturally; vaporize at ordinary temperatures	Some furniture, paint, adhesives, solvents, upholstery and drapery fabrics, construction materials, cleaning compounds, deodorizers, felt-tipped markers, and correction fluid, dry-cleaned clothes, pesticides, wool preservatives, tobacco smoke, etc.	Eye and respiratory irritations, kidney and liver damage in animals Long-term effects are still being studied.

*Effects from prolonged exposure or high concentrations.

Source: Chart compiled by Earth Force, Inc. (www.earthforce.org) with data from the Environmental Protection Agency and Wisconsin Department of Natural Resources.

propellants has resulted in the release of large amounts into the atmosphere. Because chlorofluorocarbons are implicated in both global warming and ozone depletion, considerable international attention is focused on controlling their manufacture and release. (See previous section on global warming and climate changes.)

Indoor Air Pollution

A growing body of scientific evidence indicates that the air within homes and other buildings can be more seriously polluted than outdoor air in even the largest and most industrialized cities. Many indoor air pollutants and pollutant sources are thought to have an adverse effect on human health. These pollutants include asbestos; formaldehyde, which is associated with many consumer products, including certain wood products and aerosols; airborne pesticide residues; chloroform; perchloroethylene (associated particularly with dry cleaning); paradichlorobenzene (from mothballs and air fresheners); and many disease-causing or allergy-producing organisms. (See table 17.5.) Smoking is the most important air pollutant source in the United States in terms of human health. The Surgeon General estimates that 350,000 people in this country die each year from emphysema, heart attacks, strokes, lung cancer, or other diseases caused by tobacco smoking. Banning smoking probably would save more lives than would any other pollution-control measure.

A recent contributing factor to the concern about indoor air pollution is the weatherizing of buildings to reduce heat loss and save on fuel costs. In most older homes, there is a complete ex-

ENVIRONMENTAL CLOSE-UP

Noise Pollution

Noise is referred to as unwanted sound. However, noise can be more than just an unpleasant sensation. Research has shown that exposure to noise can cause physical, as well as mental, harm to the body. The loudness of the noise is measured by decibels (db). Decibel scales are logarithmic, rather than linear. Thus, the change from 40 db (a library) to 80 db (a dishwasher or garbage disposal) represents a ten-thousandfold increase in sound loudness.

The frequency or pitch of a sound is also a factor in determining its degree of harm. High-pitched sounds are the most annoying. The most common sound pressure scale for high-pitched sounds is the A scale, whose units are written "dbA." Hearing loss begins with prolonged exposure (eight hours or more per day) to 80 or 90 dbA levels of sound pressure. Sound pressure becomes painful at around 140 dbA and can kill at 180 dbA. (See the table.)

In addition to hearing loss, noise pollution is linked to a variety of other ailments, ranging from nervous tension headaches to neuroses. Research has also shown that noise may cause blood vessels to constrict (which reduces the blood flow to key body parts), disturbs unborn children, and sometimes causes seizures in epileptics. The U.S. EPA has estimated that noise causes about 40 million U.S. citizens to suffer hearing damage or other mental or physical effects. Up to 64 million people are estimated to live in homes affected by aircraft, traffic, or construction noise.

The Noise Control Act of 1972 was the first major attempt made in the United States to protect the public health and welfare from detrimental noise. This act also attempted to coordinate federal research and activities in noise control, to set federal noise emission standards for commercial products, and to provide information to the public. Subsequent to the passage of the Noise Control Act, many local communities in the United States enacted their own noise ordinances. While such efforts are a step in the right direction, the United States is still controlling noise less than are many European countries. Several European countries have developed quiet construction equipment in conjunction with strongly enforced noise ordinances. The Germans and Swiss have established maximum day and night noise levels for certain areas. Regarding noise-pollution abatement, North America has much to learn from European countries.

Intensity of Noise

Source of Sound	Intensity in decibels
Jet aircraft at takeoff	145
Pain occurs	140
Hydraulic press	130
Jet airplane (160 meters overhead)	120
Unmuffled motorcycle	110
Subway train	100
Farm tractor	98
Gasoline lawn mower	96
Food blender	93
Heavy truck (15 meters away)	90
Heavy city traffic	90
Vacuum cleaner	85
Hearing loss after long exposure	85
Garbage disposal unit	80
Dishwasher	65
Window air conditioner	60
Normal speech	60

change of air every hour. This means that fresh air leaks in around doors and windows and through cracks and holes in the building. In a weatherized home, a complete air exchange may occur only once every five hours. Such a home is more energy efficient, but it also tends to trap air pollutants.

Even though we spend on average almost 90 percent of our time indoors, the movements to reduce indoor air pollution lag behind regulations governing outdoor air pollution. In the United States, the Environmental Protection Agency is conducting research to identify and rank the human health risks that result from exposure to individual indoor pollutants or mixtures of multiple indoor pollutants.

ISSUES & ANALYSIS

International Air Pollution

Acid rain originates as a form of air pollution, but it may damage the environment in the form of water pollution. Also, air pollution that originates in one country may result in water pollution in another country. A particular nation may have stringent environmental controls within its own boundaries but have environmental problems because of the actions of a neighboring country.

Recently, the phenomenon of acid rain and forest fires have underscored the need for international cooperation in dealing with various environmental concerns. For example, an estimated 56 percent of the acid rain falling in Sweden originates outside of that country. The main sources of the pollutants are Germany and the United Kingdom—because the west coast of Sweden receives winds from these countries. When these industrialized countries release more sulfur dioxide into the atmosphere, the result is more acid rain in Sweden. Since the 1930s, lakes in western Sweden have become more acidic by a value of two pH units. Ten thousand lakes have a pH below 6.0, and 5000 lakes have a pH below 5.0. A pH below 5.5 is too acidic for many species of fish.

In the 1930s, the United Kingdom initiated a program to reduce air pollution. This program has been successful in that the air in the United Kingdom has become cleaner. However, since one of the country's solutions was to build taller smokestacks, more of the pollutants from the United Kingdom are transported to Sweden.

- Should the United Kingdom be permitted to disperse air pollutants in a manner that damages the Swedish environment?
- Should there be a series of international agreements to control and regulate the movement of airborne pollutants across international boundaries?
- In addition to air pollutants, are there other forms of pollutants that can naturally be transported across international boundaries?

Summary

The atmosphere has a tremendous ability to accept and disperse pollutants. Carbon monoxide, hydrocarbons, particulates, sulfur dioxide, and nitrogen compounds are the primary air pollutants. They can cause a variety of health problems. Lead and air toxics have also been identified as significant air pollutants.

Photochemical smog is a secondary pollutant, formed when hydrocarbons and oxides of nitrogen are trapped by thermal inversions and react with each other in the presence of sunlight to form peroxyacetyl nitrates and ozone. Elimination of photochemical smog requires changes in technology, such as more fuel-efficient automobiles, special devices to prevent the loss of hydrocarbons, and catalytic converters to more completely burn hydrocarbons.

Acid rain is caused by emissions of sulfur dioxide and oxides of nitrogen in the upper atmosphere, which form acids that are washed from the air when it rains or snows. Direct effects of acid rain on terrestrial ecosystems are difficult to prove, but changes in many forested areas are suspected of being partly the result of additional stresses caused by acid rain. Recent evidence suggests that loss of calcium from the soil may be a major problem associated with acid rain. The effect of acid rain on aquatic ecosystems is easy to quantify. As waters become more acidic, the complexity of the ecosystem decreases, and many species fail to reproduce. The control of acid rain requires the use of scrubbers, precipitators, and filters—or the removal of sulfur from fuels.

Currently, many are concerned about the damaging effects of greenhouse gases: carbon dioxide, methane, and chlorofluorocarbons. These gases are likely to be causing an increase in the average temperature of the earth and, consequently, are leading to major changes in the climate. Human and ecological systems are already vulnerable to a range of environmental pressures, including climate extremes and variability. Global warming is likely to amplify the effects of other pressures and to disrupt our lives in numerous ways. Significant impacts on our health, the vitality of forests and other natural areas, the distribution of freshwater supplies, and the productivity of agriculture are among the probable consequences of climate change. Chlorofluorocarbons are also thought to lead to the destruction of ozone in the upper atmosphere, which results in increased amounts of ultraviolet light reaching the earth. Concern about the effects of chlorofluorocarbons has led to international efforts that have resulted in significant reductions in the amount of these substances reaching the atmosphere. Many commonly used materials release gases into closed spaces (indoor air pollution) where they cause health problems. The most important of these are associated with smoking.

Interactive Exploration

Check out the website at

http://www.mhhe.com/environmentalscience

and click on the cover of this textbook for interactive versions of the following:

KNOW THE BASICS

acid deposition *381*
acid rain *381*
carbon dioxide (CO_2) *386*
carbon monoxide (CO) *375*
carcinogenic *376*
chlorofluorocarbons *388*
greenhouse effect *386*
hydrocarbons (HC) *375*
nitrous oxide *389*
oxides of nitrogen (NO and NO_2) *377*
ozone *393*
particulates *376*
photochemical smog *377*
primary air pollutants *374*
radon *394*
secondary air pollutants *374*
sulfur dioxide (SO_2) *376*
thermal inversion *378*

- On-Line Flashcards
- Electronic Glossary

IN THE REAL WORLD

Is it global warming or a natural cycle? Does it seem that almost everything is attributed to global warming? Check out these articles and see if you think the event is linked to global warming. Why would hurricane damage point to global warming? Take a look at **Hurricane Mitch Brings Unusually Severe and Unusually Late Damage to Honduras** to see what climatologists are saying about the 1998 hurricane season.

Look over **Low Water Levels in the Great Lakes** to find out if lake levels vary much throughout history.

Read the varying opinions of scientists as they react to ice losses in **Portion of Antarctic Ice Sheet Weakening** and **More Large Icebergs Calve from Antarctica's Ice Shelves.**

Why is the ozone layer still being depleted when we aren't using as many chlorofluorocarbons as we did in the past? Read **Antarctic Ozone Hole Continues to Grow** for an explanation of how ozone is broken down and why the hole keeps on expanding. Aerosols in hair spray and refrigerants in air conditioners taught us a valuable lesson.

Have you ever been surprised by how much something that seems so harmless can be so harmful to the environment? Take a look at **Personal Watercraft Pollute Air and Water: Restrictions Proposed** and decide if you think this popular sport is the way to go.

PUT IT IN MOTION

The greenhouse effect is caused by human air pollution such as increased carbon dioxide. . . right? Make sure you have the facts straight with global warming and the greenhouse effect by studying the Global Warming animation.

How do chlorofluorocarbons break down the ozone layer? You've read about it, now see it in action! View Ozone Layer Depletion for an excellent animation that shows how ozone forms and how a CFC molecule affects ozone molecules.

What causes acid rain and why isn't it a problem near all big cities? Check out the Acid Rain animation to see what pollutants lead to the formation of acid rain and how air circulation affects the distribution of these pollutants.

Check out Global Air Circulation for an excellent explanation of world air movement patterns.

Another pattern of air movement that affects our weather in North America is the El Niño-Southern Oscillation.

Look at the animation to see how water and air movement in the Pacific Ocean affects air circulation in many other parts of the world.

TEST PREPARATION

Review Questions

1. List the five primary air pollutants commonly released into the atmosphere and their sources.
2. Define *secondary air pollutants,* and give an example.
3. List three health effects of air pollution.
4. Why is air pollution such a large problem in urban areas?
5. What is photochemical smog? What causes it?
6. Describe three actions that can be taken to control air pollution.
7. What causes acid rain? List three probable detrimental consequences of acid rain.
8. Why is carbon dioxide (a nontoxic normal component of the atmosphere) called a "greenhouse gas"?
9. What would the consequences be if the ozone layer surrounding the earth was destroyed?
10. How does energy conservation influence air quality?

Critical Thinking Questions

1. What could you do to limit the air pollution you create?
2. Do you agree with a ban on smoking, like in California, that includes all indoor public places, even privately owned restaurants and bars? Why or why not?
3. Some developing countries argue that they should be exempt from limits on the production of greenhouse gases and that developed countries should bear the brunt of the changes that appear to be necessary to curb global climate change. What values, beliefs, and perspectives underlie this belief? What do you think about this argument?
4. As a nation, the United States provides many subsidies to make energy cheap, figuring that development depends on cheap energy. If these subsidies were withdrawn, or taxes on energy were added, what effect would this have on your own energy consumption? Would you be willing to support high gasoline prices, in the $3-4/gallon range as in many European countries, if it would cut greenhouse gas emissions?
5. Why do you think air pollution is so much worse in developing countries than in developed countries? What should developed countries do about this, if anything?
6. What common indoor air pollutants are you exposed to? What can you do to limit this exposure?
7. What kinds of noise pollution do you encounter? How important is noise pollution to you? What can you do to reduce noise pollution?
8. Is it possible to have zero emissions of pollutants? What level of risk are you willing to live with?

CHAPTER 18
Solid Waste Management and Disposal

Objectives

After reading this chapter, you should be able to:

- Explain why solid waste has become a problem throughout the world.
- Understand that the management of municipal solid waste is directly affected by economics, changes in technology, and citizen awareness and involvement.
- Recognize that the management of municipal solid waste in the future will require an integrated approach.
- Describe the various methods of waste disposal and the problems associated with each method.
- Understand the difficulties in developing new municipal landfills.
- Define the problems associated with incineration as a method of waste disposal.
- Describe some methods of source reduction.
- Describe composting and how it fits into solid waste disposal.
- List some benefits and drawbacks of recycling.

Chapter Outline

Introduction

The Disposable Decades

The Nature of the Problem

Methods of Waste Disposal
  Landfilling

Environmental Close-Up: *Resins Used in Consumer Packaging*

  Incineration
  Composting
  Source Reduction
  Recycling

Environmental Close-Up: *What You Can Do to Reduce Waste and Save Money*

Environmental Close-Up: *Recycling Is Big Business*

Environmental Close-Up: *Recyclables Market Basket*

Issues & Analysis: *Corporate Response to Environmental Concerns*

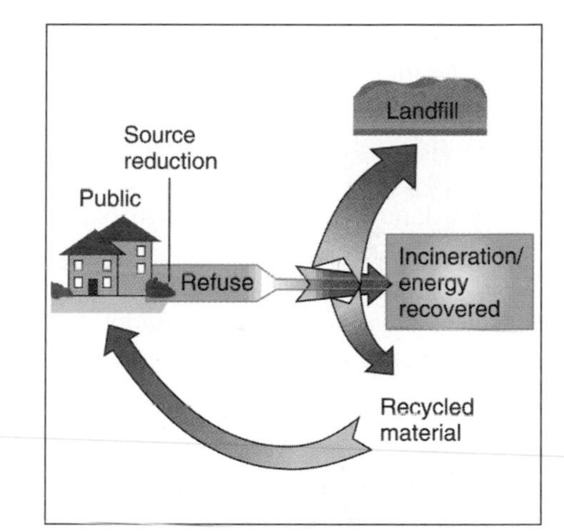

Introduction

In March 1987, a barge laden with 2,900 tonnes (3,200 U.S. tons) of garbage set out in search of a dump. The refuse had been turned away from a landfill in Islip, New York. The barge traveled 10,000 kilometers (6,000 miles) and stopped at several foreign ports, but found no one willing to accept its noxious load. The three-month odyssey took the barge to Mexico, Belize, and the Bahamas before it returned still fully loaded to New York. The futile voyage made headlines, giving many North Americans their first inkling of an impending crisis.

Lack of space for dumping solid waste has become a problem for many large metropolitan areas throughout the world. Communities are concerned about the increasing costs of waste disposal, possible hazards to groundwater, and maintaining air quality. In many areas, there is not suitable land available for new landfills. Problems with solid waste have increased dramatically over the past several decades because of population increases and an attitude that convenience is a very important part of the North American lifestyle.

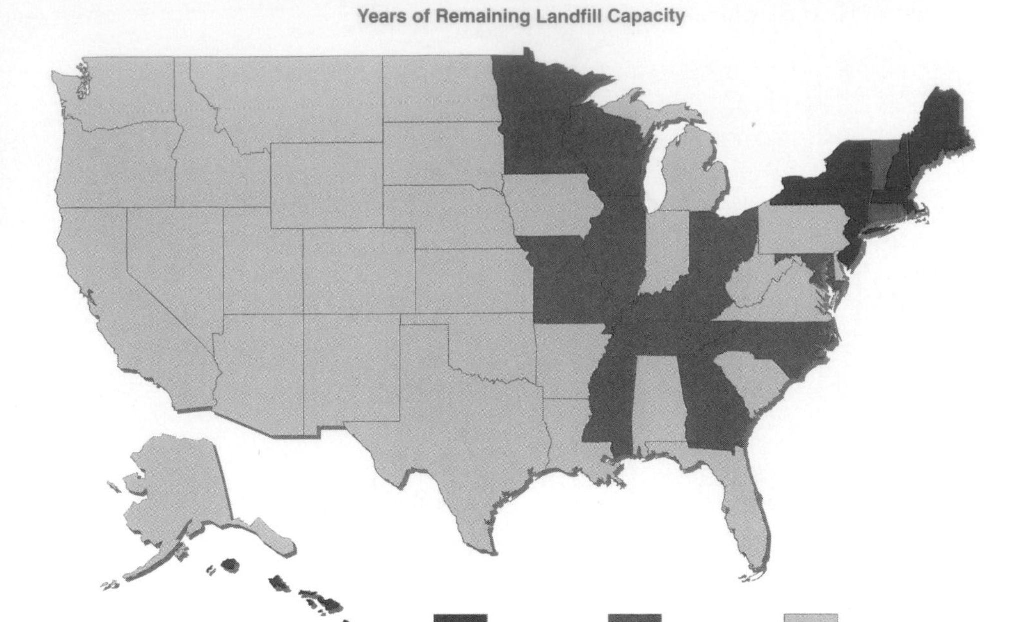

figure 18.1 **Disposable Lifestyles** Landfills are the primary method of dealing with waste in the United States but the disposable lifestyle of the past has created a shortage of space in landfills for many states in the Northeast and central portions of the United States. The development of new landfills is expensive, and in many areas there is not available land appropriate for development.

Source: (*b*) *Directory and Atlas of Solid Waste Disposal Facilities,* 1999.

The Disposable Decades

In 1955, *Life* magazine pictured a happy family in an article entitled "Throwaway Living." A disposable lifestyle was marketed as the wave of the future and as a way to cut down on household chores. "Use it once and throw it away" became a very popular advertising slogan in the 1950s.

The solid waste problems facing us today have their roots in the economic boom that followed World War II. Marketing experts set to work trying new tactics to get consumers to buy and toss or, as economists would say, to "stimulate consumption." In the mid-1950s, marketing consultant Victor Lebow wrote an emotional plea for "forced consumption" in the New York *Journal of Retailing:*

> Our enormously productive economy demands that we make consumption a way of life, that we convert the buying and use of goods into rituals, that we seek our spiritual satisfaction in consumption. . . . We need things consumed, burned up, worn out, replaced, and discarded at an ever-growing rate.

Consumers were quick to adapt to the lifestyle that Lebow envisioned. In fact, it was not long until a disposable, throwaway lifestyle was seen as a consumer's right. The idea was to sell "convenience" to the prosperous postwar consumers. What was initially a convenience was soon to become a "necessity;" at least, that was what the advertisements were telling people.

A good example of this way of thinking is the TV dinner, which was first marketed in 1953. The food-packaging industry was evolving into a major service, closely allied with marketing and advertising. From the first TV dinner, we progressed rapidly to the advent of microwave meals. One such meal consisted of 340 grams (¾ pound) of edible material and six separate layers of packaging, five of them plastic.

After several decades of throwaway living, the disposable lifestyle is changing. Many cities and counties face a shortage of space in old landfills. Solid waste has become a major problem and, in a growing list of communities around the world, it has reached crisis proportions. Many states in the heavily populated eastern United States are currently running out of places to put their garbage. (See figure 18.1.)

The Nature of the Problem

As we begin the twenty-first century, the world must take steps to solve the ever-increasing burden of garbage—or as the professionals say, **municipal solid waste (MSW).** We are all part of the problem, but we can also be part of the solution.

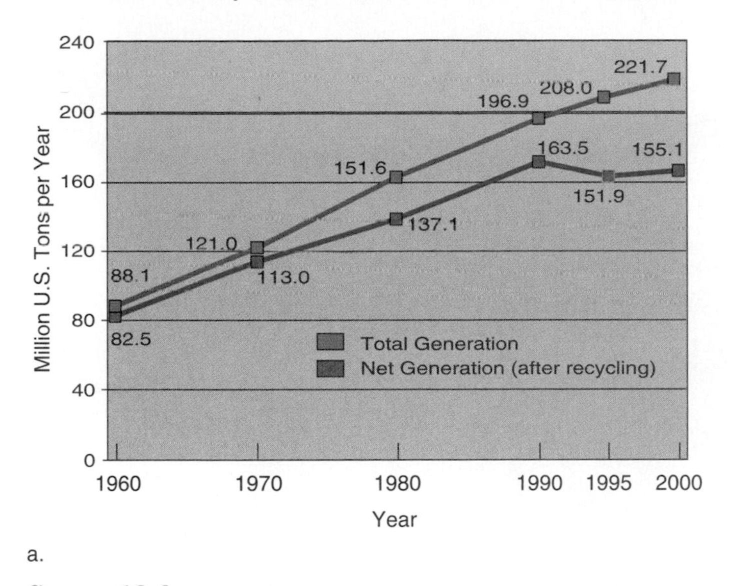

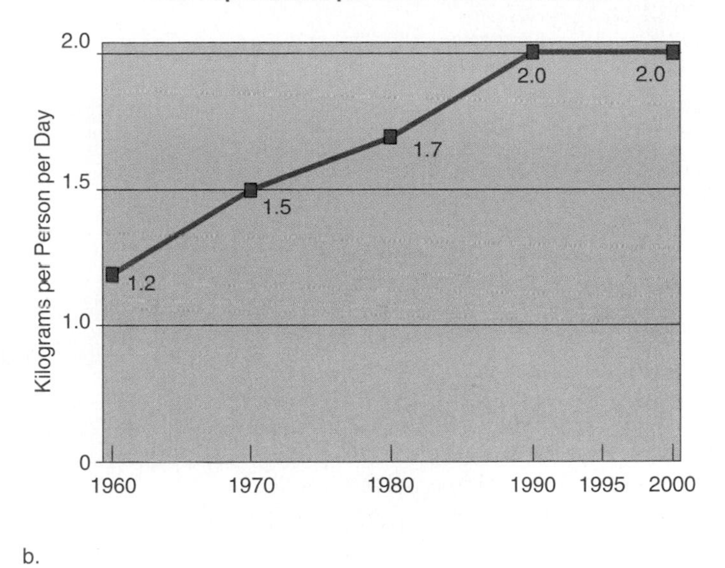

figure 18.2 **Municipal Solid Waste Generation Rates** The generation of municipal solid waste in the United States has increased steadily. However, because of increased recycling rates, the net production rates (after recyclables have been removed) has actually fallen since 1990 (*a*) and the per capita rate has stabilized (*b*).

Source: *Characterization of MSW in the U.S.;* 1999 Update. U.S. EPA, Washington, D.C.

The United States produces about 200 million tonnes (220 million U.S. tons) of municipal solid waste each year. This equates to about 2 kilograms (4.4 pounds) of trash per person per day or 0.73 tonnes (0.8 U.S. tons) per person per year. The amount of municipal solid waste has more than doubled since 1960 and the per capita rate has increased by nearly 70 percent in that same time, although per capita rates began to stabilize about 1990. When recycling is included, the net waste produced has actually fallen since 1990. (See figure 18.2.)

Nations with high standards of living and productivity tend to have more municipal solid waste per person than less-developed countries. (See figure 18.3.) The United States and Canada, therefore, are world leaders in waste production. For example, Toronto, Canada, is running out of places to put its municipal solid waste. Even with a very ambitious plan to reduce waste production by 50 percent, the metropolitan area will run out of space by 2006. The outlying districts are not willing to be the site of a new landfill for Toronto and the metropolitan area is looking for disposal sites north of the city as well as in the United States. Many states that could serve as disposal sites are beginning to pass laws that regulate the importation of waste from other countries and metropolitan Toronto continues to struggle with its waste problem.

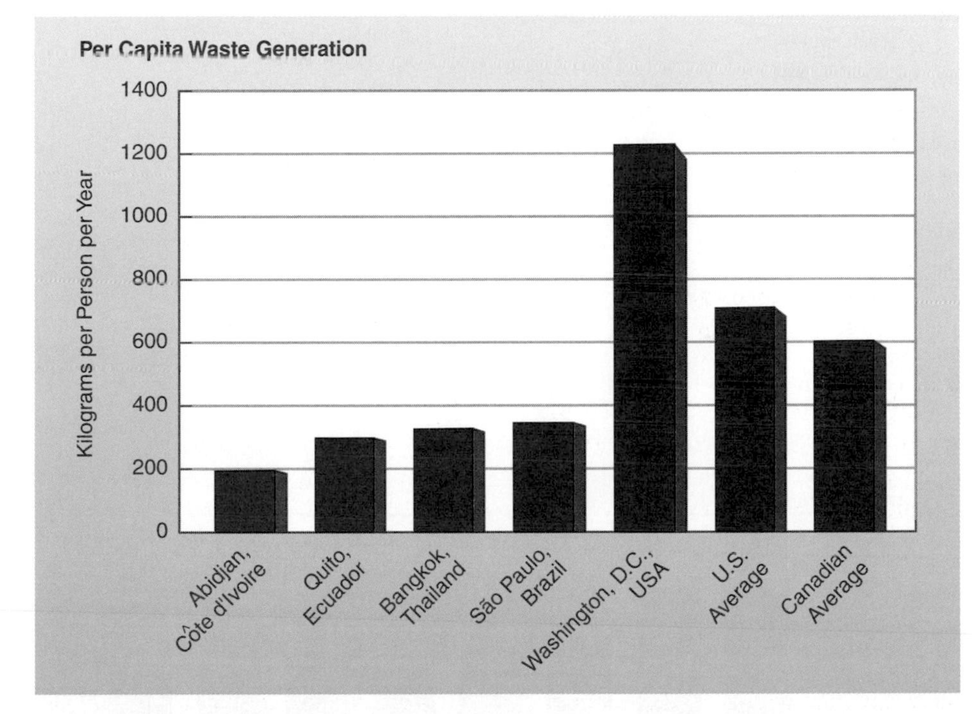

figure 18.3 **Waste Generation and Lifestyle** The waste generation rates of people are directly related to their economic condition. People in richer countries produce more garbage than those in poorer countries.

Source: Data from World Resources 1996–2000 and U.S. EPA.

Archeologists rely on the waste of past societies to tell them about the nature of the culture and lifestyle of ancient civilizations. In the same way

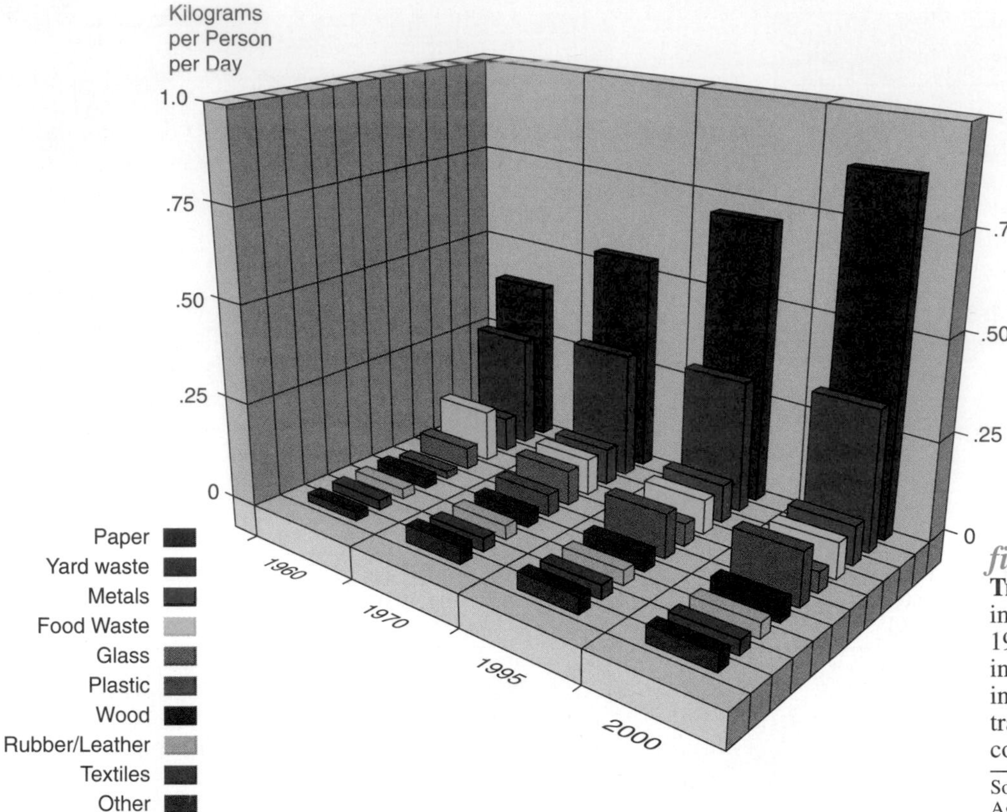

figure 18.4 **The Changing Nature of Trash** While food waste was the third most important component of solid waste in 1960, it slipped to fifth as metal and plastic increased. Changes in lifestyle and packaging have led to a change in the nature of trash. Most of what is currently disposed of could be recycled.

Source: Data from the U.S. Environmental Protection Agency.

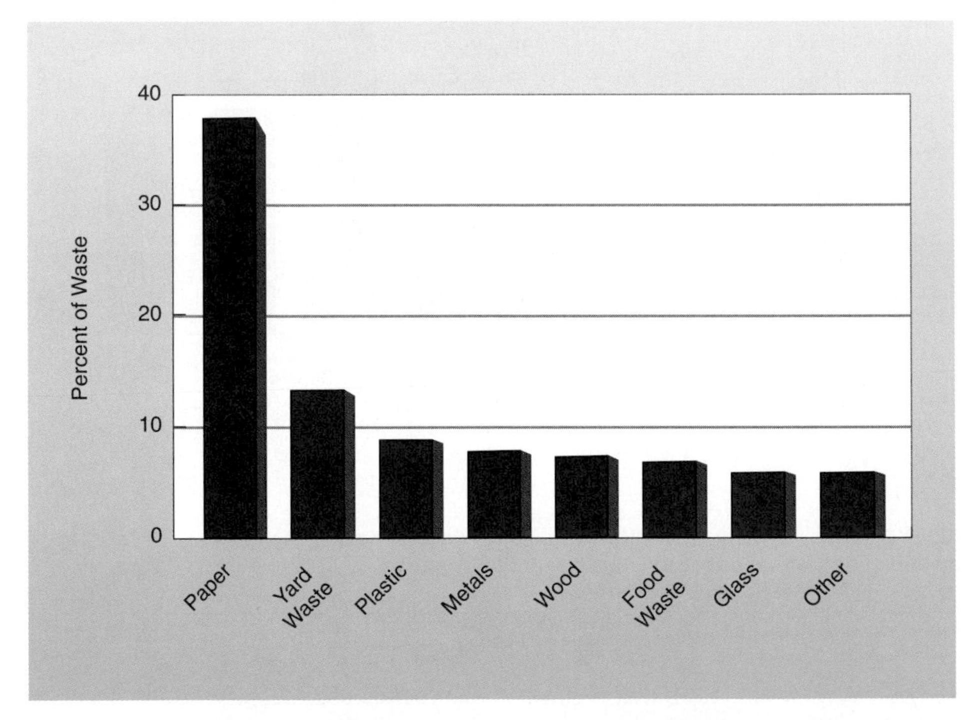

figure 18.5 **Composition of Trash in the United States (2000)** Paper and yard waste are the most common materials disposed of, accounting for 53.5 percent of the waste stream.

Source: Data from the U.S. Environmental Protection Agency.

today, our municipal solid waste is a reflection of our society. Figure 18.4 shows how the composition of our trash has changed since 1960. Notice particularly the increase in the amount of paper and plastic and the effect that recycling has had on the amount of glass in the trash. In the United States, the two most common items in the waste stream are paper products and yard waste, and other significant segments are wood, metal, glass, plastics, and food waste. (See figure 18.5.) An analysis of the composition of our waste will present us with possible approaches to reducing the amount of waste we generate.

Methods of Waste Disposal

Until recently, the disposal of municipal solid waste did not attract much public attention. From prehistory through the present day, the favored means of disposal was simply to dump solid wastes outside of the city or village limits or in the "back 40." Frequently, these dumps were in a wetlands adjacent to a river or lake. To minimize the volume of the waste, the dump was often burned. Unfortunately, this method is still being used in remote or sparsely populated areas in the world. (See figure 18.6.)

As better waste-disposal technologies were developed and as values changed, more emphasis was placed on the environment and quality of life. Simply dumping and burning our wastes is no longer an acceptable practice from an environmental or health perspective. While the technology of waste disposal has evolved during the past several decades, our options are still limited. Realistically, there are no ways of dealing with waste that have not been known for many thousands of years. Essentially, five techniques are used: (1) landfills, (2) incineration, (3) source reduction, (4) composting, and (5) recycling.

Landfills have historically been the primary method of waste disposal because this method is the cheapest and most convenient, and because the threat of groundwater contamination was not initially recognized. As we have recognized some of the problems associated with poorly designed landfills, efforts to reduce the amount of material placed in landfills have been substantial. Although the amount of waste has increased, composting and recycling have removed significant amounts of materials from the waste stream and the amount of material entering landfills has declined. (See figure 18.7.) However, the landfill of today is far different from a simple hole in the ground into which garbage is dumped.

figure 18.6 **Burning Landfills** In the past, it was common practice to burn the waste in landfills to reduce the volume. Waste is still being burned in sparsely populated areas in North America and other parts of the world.

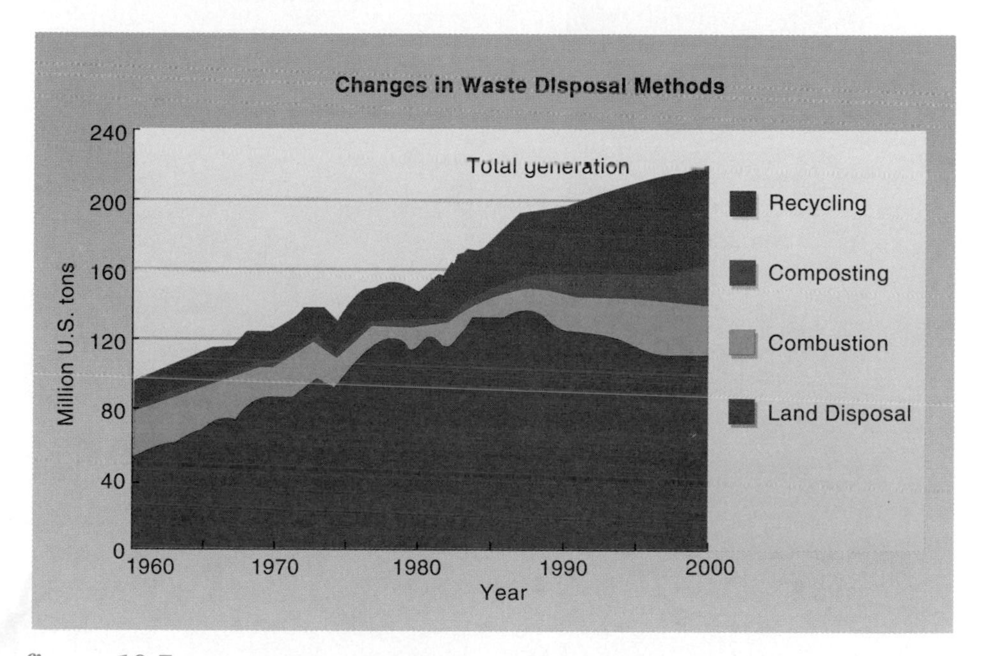

figure 18.7 **Changes in Waste Disposal Methods** The landfill is still the primary method of waste disposal. Historically, landfills have been the cheapest means of disposal. This may not be the case in the future. Notice that recycling and composting have grown over the past decade while the amount going to landfills has declined somewhat.

Source: Data from the U.S. Environmental Protection Agency.

Landfilling

A modern **municipal landfill** is typically a depression in an impermeable clay layer that is lined with an impermeable membrane. Each day's deposit of fresh garbage is covered with a layer of soil. Selection of modern landfill sites must be based on an understanding of groundwater geology, soil type, and sensitivity to local citizens' concerns. Once the site is selected, extensive construction activities are necessary to prepare it for use. New landfills have complex bottom layers to trap contaminant-laden water, called leachate, leaking through the buried trash. In addition, monitoring systems are necessary to detect methane gas production and groundwater contamination. In some cases, methane produced by rotting garbage is collected and used to generate electricity. The water that leaches through the site must be collected and treated. As a result, new landfills are becoming increasingly more complex and expensive. They currently cost up to

ENVIRONMENTAL CLOSE-UP

Resins Used in Consumer Packaging

Thermoplastics, which account for 87 percent of plastics sold, are the most recyclable form of plastics because they can be remelted and reprocessed, usually with only minor changes in their properties. Thermoplastic resins are commonly used in consumer packaging applications:

1. Polyethylene terephthalate (PET) is used extensively in rigid containers, particularly beverage bottles for carbonated beverages and medicine containers.
2. Polyethylene is the most widely used resin. High-density polyethylene (HDPE) is used for rigid containers, such as milk and water jugs, household-product containers, and motor oil bottles.
3. Polyvinyl chloride (PVC) is a tough plastic often used in construction and plumbing. It is also used in some food, shampoo, oil, and household-product containers.
4. Low-density polyethylene (LDPE) is often used in films and bags.
5. Polypropylene (PP) is used in a variety of areas, from yogurt containers to battery cases to disposable diaper linings. It is frequently interchanged for polyethylene or polystyrene.
6. Polystyrene (PS) is best known as a foam in the form of cups, trays, and food containers. In its rigid form, it is used in cutlery.
7. Other. These usually contain layers of different kinds of resins and are most commonly used for squeezable bottles (for example, ketchup).

Currently, HDPE and PET are the two most commonly recycled resins. Efforts to recycle PS are also being explored, especially within the fast-food industry. The percentages listed below are based on each type in relation to all plastics.

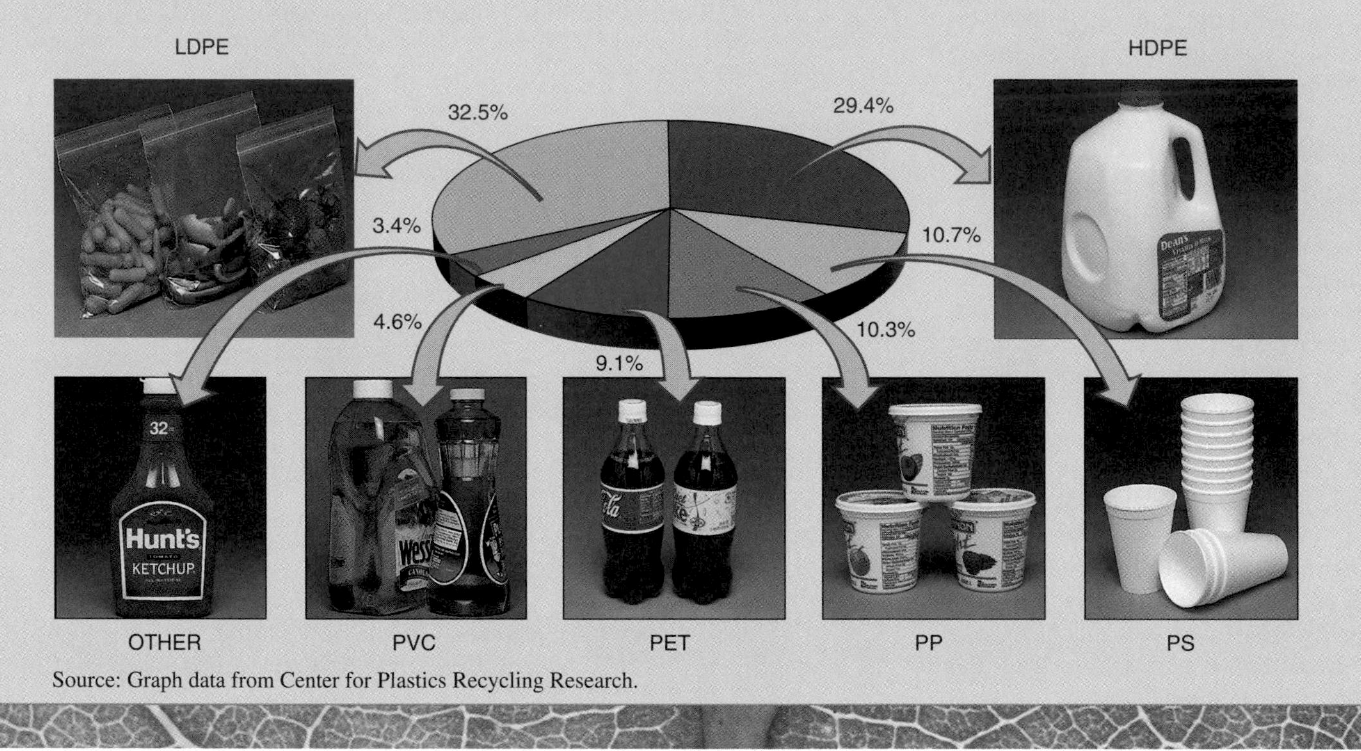

Source: Graph data from Center for Plastics Recycling Research.

$1 million per hectare ($400,000 per acre) to prepare. (See figure 18.8.)

Today, about 57 percent of United States and about 80 percent of Canadian municipal solid waste goes into landfills, but this method is failing to handle the volume. For example, New York City's Fresh Kills Landfill on Staten Island receives 12,600 tonnes (14,000 U.S. tons) of trash each day. Other trash is shipped to Pennsylvania and other states. However, the Fresh Kills Landfill is expected to close at the end of 2002 so alternative arrangements for solid waste disposal are needed.

The problem New York City faces with the closing of its Fresh Kills Landfill is similar to what many communities will be facing in the near future. The number of landfills is declining. In 1988, there were about 8000 landfills, and in 2000, there were under 2500 active sanitary landfills. The capacity, however, has remained relatively constant. New landfills are much larger than old ones. The number of landfills has decreased for two reasons. Many small, poorly run landfills have been closed because they were not meeting regulations. Others have closed because they reached their capacity. (See figure 18.9.)

A prolonged public debate over how to replace lost landfill capacity is developing where population density is high and available land is scarce. Siting new landfills in locations like Toronto,

New York, and Los Angeles is extremely difficult because of (1) the difficulty in finding a geologically suitable site and (2) local opposition, which is commonly referred to as the NIMBY, or "not-in-my-backyard," syndrome. Resistance by the public comes from concern over groundwater contamination, rodents and other vectors of disease, odors, and truck traffic. Public officials look for alternatives to landfills to avoid problems with the public over landfill siting. Although siting a new landfill may be necessary, politicians are often unwilling to take strong positions that might alienate their constituents.

Japan and many Western European countries have already moved away from landfills as the primary method of waste disposal because of land scarcities and related environmental concerns. Switzerland and Japan dispose less than 15 percent of their waste in landfills, compared to 57 percent in the United States. Instead, recycling and incineration are the primary methods. (See figure 18.10.) In addition, the energy produced by incineration can be used for electric generation or heating.

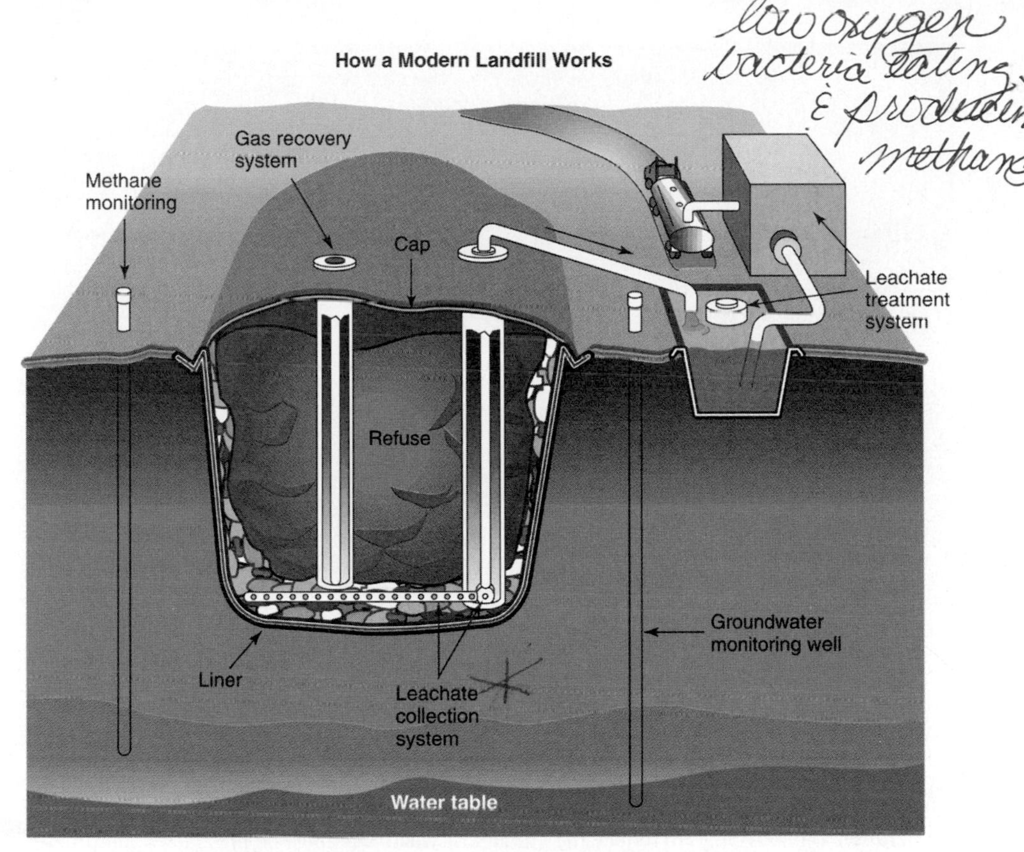

figure 18.8 **A Modern Well-Designed Landfill** A modern sanitary landfill is far different from a simple hole in the ground filled with trash. A modern landfill is a self contained unit that is separated from the soil by impermeable membranes and sealed when filled. Methane gas and groundwater are continuously monitored to assure that wastes are not escaping to the air or groundwater.

Source: National Solid Waste Management Association.

Incineration

Incineration of refuse was quite common in North America and western Europe prior to 1940. However, many incinerators were eliminated because of aesthetic concerns, such as foul odors, noxious gases, and gritty smoke, rather than for reasons of public health. Today, about 16 percent of the municipal solid waste in the United States is incinerated; Canada incinerates about 8 percent. Most incinerators are not used just to burn trash. The heat derived from the burning is converted into steam and electricity. (See figure 18.11.) In 2000, 110 combustors with energy recovery existed in the United States, with the capacity to burn up to 100,000 tonnes of MSW per day.

Most incineration facilities burn unprocessed municipal solid waste. This is not as efficient as some other technologies. About one-fourth of the incinerators use refuse-derived fuel—collected refuse that has been processed into pellets prior to combustion.

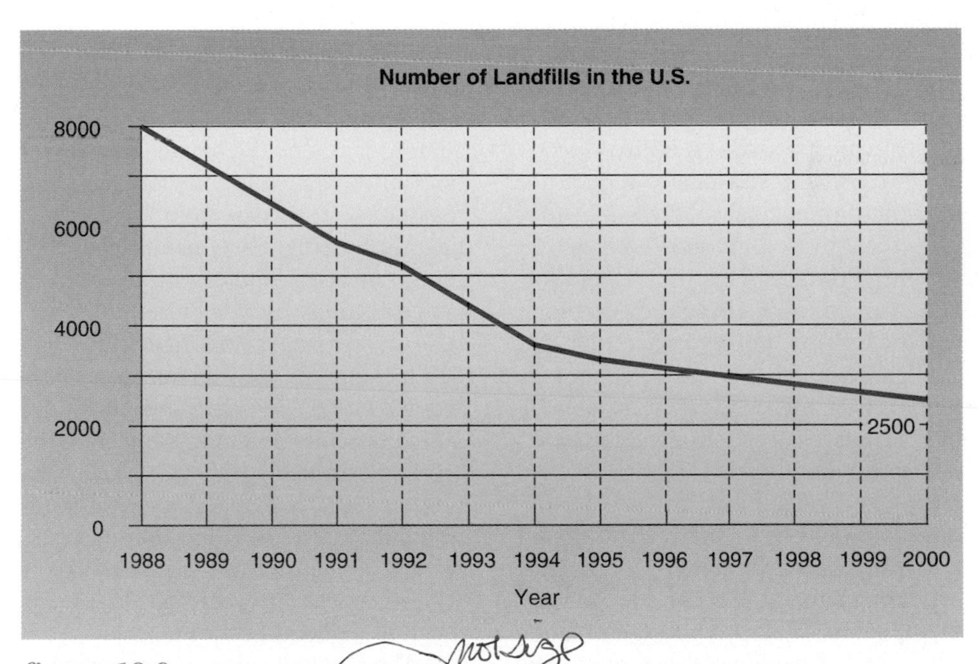

figure 18.9 **Reducing the Number of Landfills** The number of landfills in the United States is declining because they are filling up or because their design and operation do not meet environmental standards.

Source: Data from the U.S. Environmental Protection Agency.

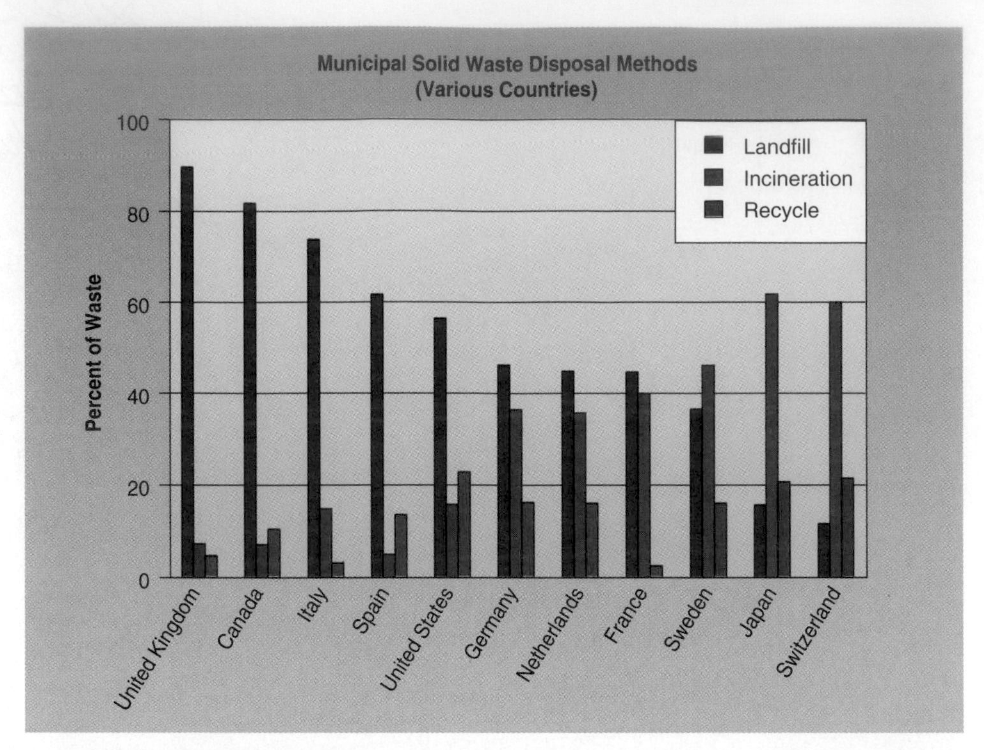

figure 18.10 **Disposal Methods Used in Various Countries** Many countries have difficulty finding adequate space for landfills. Therefore, they rely on other technologies, such as incineration and recycling, to reduce the amount of waste that must be placed in a landfill.

Source: Data from the U.S. Environmental Protection Agency.

The newest means of incineration, a European concept, is called **mass burn.** In the mass-burn technique, municipal solid waste is fed into a furnace, where it falls onto moving grates and is burned at temperatures up to 1300°C (2400°F). The burning waste heats water, and the steam drives a turbine to generate electricity, which is sold to a utility.

Incinerators drastically reduce the amount of municipal solid waste—up to 90 percent by volume and 75 percent by weight. Primary risks of incineration, however, involve air-quality problems and the toxicity and disposal of the ash.

Though mass-burn technology works efficiently in Europe, the technology is not easily transferrable. North American municipal solid waste contains more plastic and toxic materials than European waste, thus creating air-pollution and ash-toxicity concerns. Modern incinerators have electrostatic precipitators, dual scrubbers, and fabric filters called baghouses; however, they still release small amounts of pollutants into the atmosphere, including certain metals, acid gases, and also classes of chemicals known as dioxins and furans, which have been implicated in birth defects and several kinds of cancer. The long-term risks from the emissions are still a subject of debate.

Ash from incineration is also a major obstacle to the construction of waste-to-energy facilities. Small concentrations of heavy metals are present both in the air emissions (fly ash) and residue (bottom ash) from these facilities. Because the ash contains lead, cadmium, mercury, and arsenic in varying concentrations from such items as batteries, lighting fixtures, and pigments, this ash may need to be treated as hazardous waste. The toxic substances are more concentrated in the ash than in the original garbage and can seep into groundwater from poorly sealed landfills. Many cities have had difficulty disposing of incinerator ash, and there is still considerable debate about what is the best method of disposal.

The cost and siting of new incinerators are also major concerns facing many communities. Incinerator construction is often a municipality's single largest bond issue. Incinerator construction costs in North America in 2000 ranged from $45 million to $350 million, and the costs are not likely to decline.

Incineration is also more costly than landfills in most situations. Figure 18.12 shows the cost of landfills in comparison with solid waste incinerators. As long as landfills are available, they will have a cost advantage. When cities are unable to dispose of their trash locally in a landfill and must begin to transport the trash to distant sites, incinerators become more cost effective. The U.S. Environmental Protection Agency (EPA) has not looked favorably on the construction of new waste-to-energy facilities and has encouraged recycling and source reduction as more effective ways to reduce the solid waste problem. Critics have argued that cities and towns have impeded waste reduction and recycling efforts by putting a priority on incinerators and committing resources to them. Proponents of incineration have been known to oppose source reduction. They argue that incinerators need large amounts of municipal solid waste to operate and that reducing the amount of waste generated makes incineration impractical. Many communities that have opposed incineration say that they support a vigorous waste-reduction and recycling effort.

Composting

Composting is the process of harnessing the natural process of decomposition to transform organic materials—anything from manure and corncobs to grass and soiled paper—into compost, a humus-like material with many environmental benefits. In natural surroundings, leaves and branches that fall to the bottom of the forest form a rich, moist layer of mulch that protects the roots of plants and proves a home for nature's most fundamental recyclers: worms, insects, and a host of microorganisms and bacteria.

By properly managing air and moisture, the composting process can transform large quantities of organic material into compost over a relatively short period of time. A good small-scale example is a backyard compost pile. Green materials (grass, kitchen vegetable

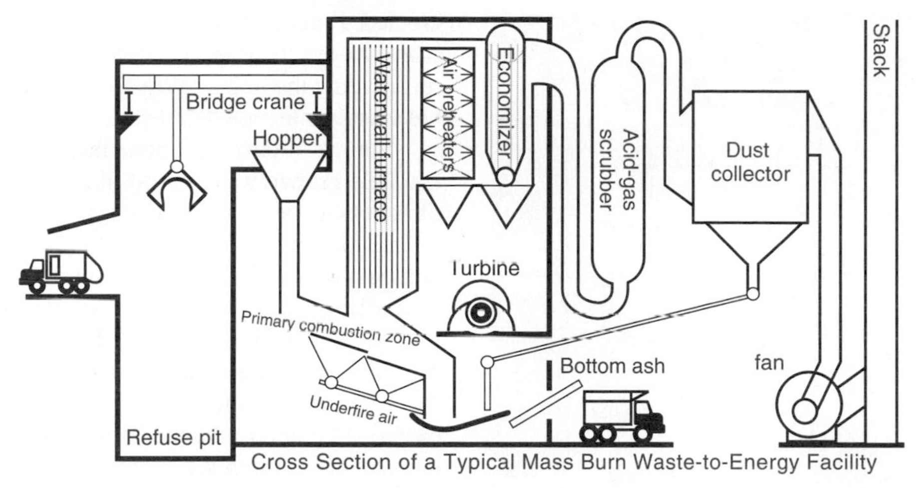

a.

b.

figure 18.11 **A Mass Burn Incineration System** (*a*) The diagram shows how a typical mass burn, waste-to-energy incinerator operates. (*b*) Incineration of municipal solid waste reduces its weight and volume significantly. However, there are concerns about air-quality problems and the toxicity and disposal of the ash.

Source: (*a*) Data from the U.S. Environmental Protection Agency.

scraps, and flower clippings) mixed with brown materials (twigs, dry leaves, and soiled paper towels) at a ratio of 1:3 provides a balance of nitrogen and carbon that helps microbes efficiently decompose these materials.

Large-Scale Composting

Large-scale municipal composting uses the same principles of organic decomposition to process large volumes of organic materials. Composting facilities of various sizes and technological sophistication accept materials such as yard trimmings, food scraps, biosolids, wood shavings, unrecyclable paper, and other organic materials to form the proper balance of nitrogen and carbon. These materials undergo processing—shredding, turning, and mixing—and, depending on the materials, can be turned into compost in a period ranging from 8 to 24 weeks. More than 3000 composting facilities are in use in the United States. In 1999, 47 percent of yard trimmings were composted in the U.S. through municipal programs. (See figure 18.13.) Most municipal programs entail one of the three processes:

Windrow

Compostables are formed into long piles or rows. The piles are periodically "turned" or agitated to promote aeration and homogenization. Sometimes this turning of the rows involved using specialized machinery.

Static Aerated Pile

Compostables are formed into large piles and insulated with a layer of mature compost or other material. Forced aeration can be applied, but no mechanical turning or agitation is done.

Enclosed Vessel

Compostable material is fed into a drum, silo, or other structure where the environmental conditions are closely controlled. This process may also include aeration and mechanical agitation.

In addition to keeping wastes from entering a landfill, composting also has physical, chemical, and biological benefits. The addition of compost to soil will reduce bulk density, improve workability and porosity, and increase its gas and water permeability, thus reducing erosion. Nitrogen, potassium, iron, phosphorus, sulfur, and calcium are all common in compost humus and are beneficial to plant growth. Microorganisms, essential in productive soils, play an important role in organic matter decomposition, which, in turn, leads to humus formation and nutrient availability.

Source Reduction

The most fundamental way to reduce waste is to prevent it from ever becoming waste in the first place. Waste prevention, also known as **source reduction,** is the practice of designing, manufacturing, purchasing, or using materials (such as products and packaging) in ways that reduce the amount or toxicity of trash

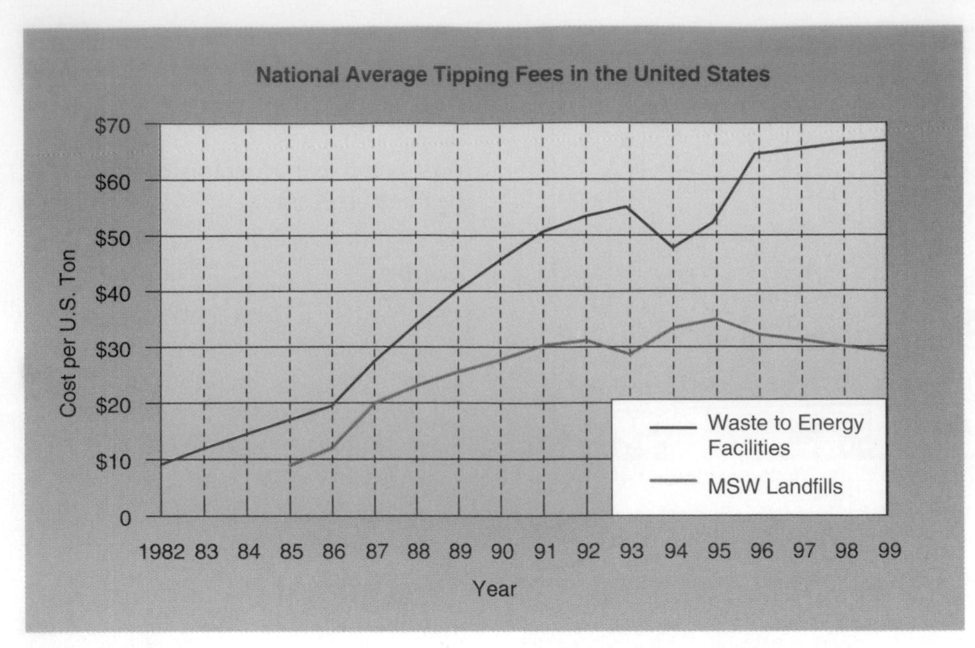

figure 18.12 **Relative Costs of Landfills and Incinerators** In general, the cost of disposing waste by incineration is greater than that of a landfill. However, in areas where landfills are not available and transportation costs for land disposal are high, incineration is a logical alternative.

Source: Data from the U.S. Environmental Protection Agency.

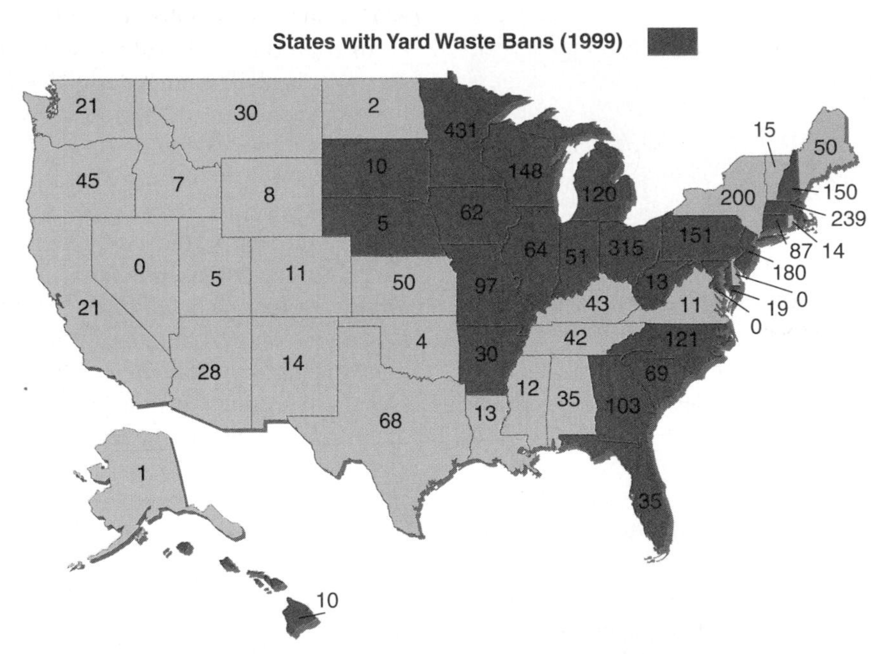

figure 18.13 **States with Yard Waste Bans** Since yard waste is such an important segment of the solid waste stream, many states have passed laws that prohibit the deposition of yard waste in landfills. This will extend the useful life of the landfill. In order to accommodate their citizens, many communities still collect yard waste but have instituted composting programs to deal with the yard waste they collect. The numbers on the map indicate the number of composting programs in each state.

Source: Data from the U.S. Environmental Protection Agency.

created. Reusing items is another way to stop waste at the source because it delays or prevents the entry of those items into the waste collection and disposal system.

Source reduction, including reuse, can help reduce waste disposal and handling costs, because it avoids the costs associated with recycling, municipal composting, landfilling, and combustion. Soft drink bottles are an example of source reduction. Since 1977, the weight of 2-liter plastic soft drink bottles has been reduced from 68 grams (2.4 ounces) each to 51 grams (1.8 ounces). That translates to 114 million kilograms (250 million pounds) of plastic per year that has been kept out of the waste stream. Source reduction also conserves resources and reduces pollution, including greenhouse gases that contribute to global warming.

Source reduction and reuse have many benefits, including saving natural resources, reducing the toxicity of wastes, and reducing costs. Waste is not just created when consumers throw items away. Throughout the life cycle of a product—from extraction of raw materials, to transportation, to processing and manufacturing facilities, to manufacture and use—waste is generated. Reusing items or making them with less material decreases waste significantly. Ultimately, fewer materials will need to be recycled or sent to landfills or waste combustion facilities.

Selecting nonhazardous or less hazardous items is another important component of source reduction. Using less hazardous alternatives for certain items (e.g., cleaning products and pesticides), sharing products that contain hazardous chemicals instead of throwing out leftovers, reading label directions carefully, and using the smallest amount necessary are ways to reduce waste toxicity.

The benefits of preventing waste go beyond reducing reliance on other forms of waste disposal. Preventing waste also can mean economic saving for communities, businesses, schools, and individual consumers.

- **Communities.** In the U.S., over 4000 communities have instituted "pay-as-you-throw" programs where citizens pay for each can or bag of trash they set out for disposal rather than through the tax base or a flat fee. When these households reduce waste at the source, they dispose of less trash and pay lower trash bills.

For example, in 1994 the city of Gainesville, Florida, entered into a contract with two companies for the collection of residential solid waste and recyclable materials (glass, plastic, paper, metal cans). The new contract for solid waste service included a variable rate for residential collections: residents pay $13.50, $15.90, or $19.75 per month according to whether they place 35, 64, or 96 gallons of solid waste at the curb for collection.

Recycling service is unlimited. While residents have had curbside collection of recyclables since 1989, the implementation of this program added brown paper bags, corrugated cardboard, and phone books to the list of items recycled.

The results of the first five years of the program were very positive. The amount of solid waste collected decreased 20 percent, and the recyclables recovered increased 25 percent. This resulted in annual savings of $200,000 to the residents of Gainesville.

- **Businesses.** Industry also has an economic incentive to practice source reduction. When businesses manufacture their products with less packaging, they are buying fewer raw materials. A decrease in manufacturing costs can mean a larger profit margin, with savings that could be passed on to the consumer.
- **Consumers.** Consumers also can share in the economic benefits of source reduction. Buying products in bulk, with less packaging, or that are reusable, frequently means a cost savings.

Recycling

Recycling is one of the best environmental success stories of the late twentieth century. (See figure 18.14.) In the United States, recycling, including composting, diverted 54 million tonnes (60 million U.S. tons) of material away from landfills and incinerators in 1999, up from 34 million tons in 1990—a 68 percent increase in just nine years. In 1990, a thousand U.S. cities had curbside recycling programs. By 1999, the number had grown to nearly 9000 cities. By 1999, mandatory recycling laws for all materials had been passed in fifteen states. In Canada, Toronto, Mississauga, and the Province of Ontario have comprehensive recycling programs.

Container Laws

By 2000, recycling kept 28 percent of municipal waste out of landfills and incinerators in the U.S. This is triple the rate of 1980. Some states and local governments are enacting laws to force businesses to recycle more. The California legislature is considering a law that would require plastic bottles and containers sold in the state to be made with recycled materials. Some local communities—from Portland, Oregon, to Chatham, New Jersey—also have achieved recycling rates of more than 50 percent, in part by expanding the types of trash they will recycle beyond newspapers, aluminum cans, and some types of plastic bottles.

In 1999, Los Angeles reached a record 46 percent recycling rate by expanding its curbside pickup program to include junk mail, cereal boxes, and yard compost. The city also made recycling easy for residents by providing one large recycling bin on wheels.

In October 1972, Oregon became the first state to enact a "Bottle Bill." This statute required a deposit of two to five cents on all beverage containers that could be reused. It banned the sale of one-time-use beverage bottles and cans. The purpose of the legislation was to reduce the amount of litter. Beverage containers were estimated to make up about 62 percent of the state's litter. The bill succeeded in this respect: within two years after it went into effect, beverage-container litter decreased by about 49 percent.

Those opposed to the bottle bill cited a loss of jobs that resulted from the enactment of the bill. In two small can-manufacturing plants, 142 persons did lose their jobs, but hundreds of new employees were needed to handle the returnable bottles. Other arguments against bottle bills focused on the "major" inconvenience to consumers who would need to return the bottles and cans to the store and on retailers' storage problems. Neither of these problems has proven to be serious. Nine other states—Vermont, Maine, Connecticut, New York, Iowa, Rhode Island, Michigan, Delaware, and California—have enacted legislation requiring deposits on bottles, specifically beverage containers. Many states, such as New Jersey, Rhode Island, and California, are turning more toward mandating recycling laws. In 1991, Maine expanded its bottle law to all nondairy beverage containers, including all noncarbonated juice containers holding a gallon or less. Deposits are five cents except on liquor and wine bottles, which carry a fifteen-cent deposit. Excluded are containers for milk and other dairy products, cough syrup, baby formula, soap, and vinegar.

Many argue that a national bottle bill is long overdue. A national bottle bill would reduce litter, save energy and money, and create jobs. It would also help to conserve natural resources. But the lobbying efforts of the soft drink and brewing industries is very strong and the U.S. Congress currently has failed to pass a national container law.

Benefits of Recycling

Some benefits of recycling are readily recognizable, such as conservation of resources and pollution reduction. These perceived benefits provide the primary motivation for participation in recycling programs. The following examples illustrate the resource-conservation and pollution-reducing benefits of recycling:

One Sunday edition of the *New York Times* consumes 62,000 trees. Currently, about 40 percent of all paper that enters the waste stream in North America is recycled.

The United States imports nearly all of its aluminum and recycles over 60 percent of its aluminum beverage cans.

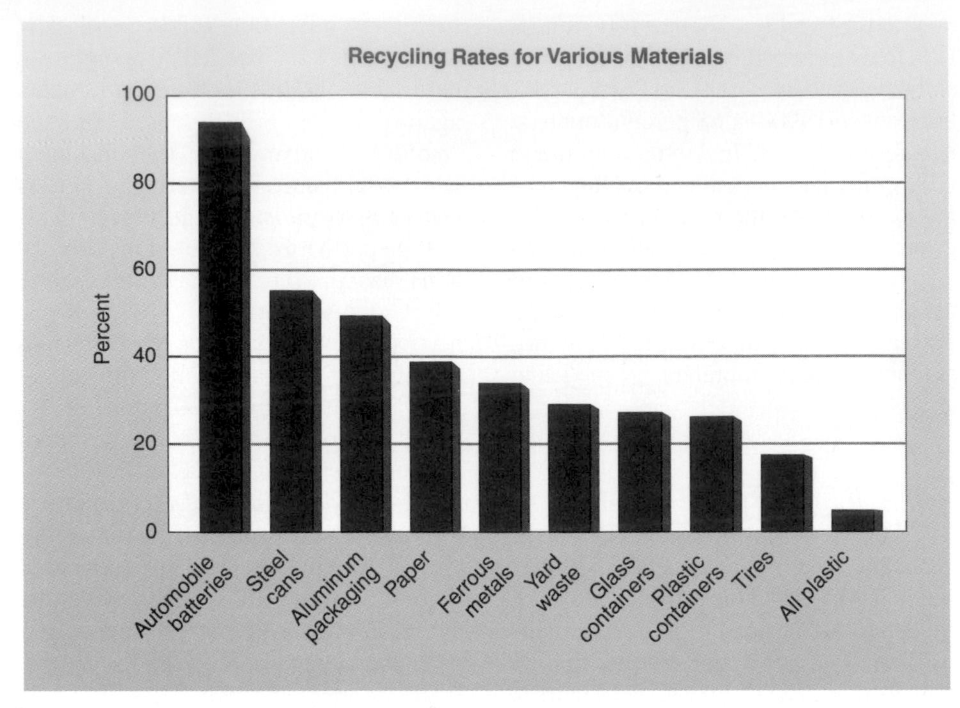

a.

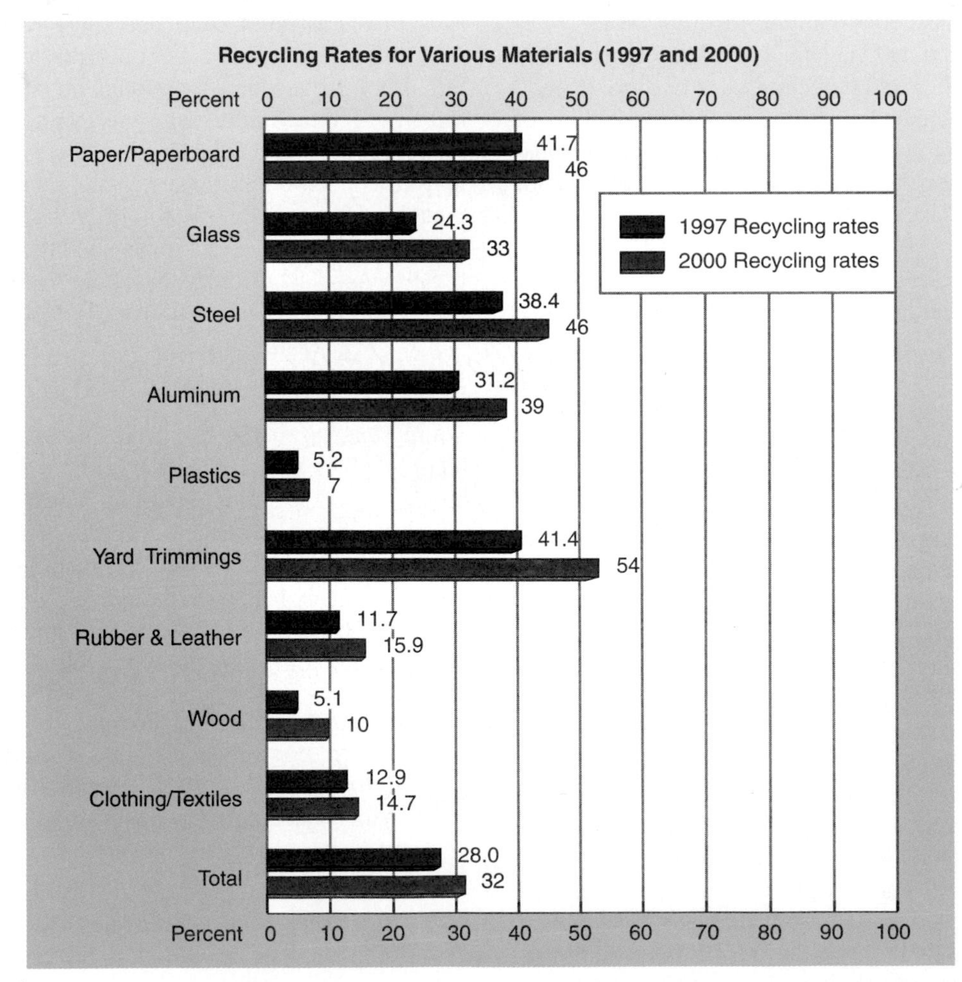

b.

There are no sources of tin within the United States; however, about 2 kilograms (4.4 pounds) of tin can be reclaimed from each 1000 kilograms of metal food cans.

Crushed glass (cullet) reduces the energy required to manufacture new glass by 50 percent. Cullet lowers the temperature requirements of the glassmaking process, thus conserving energy and reducing air pollution. In 2000, 33 percent of the glass in the solid waste stream was recycled.

While recycling is a viable alternative to landfilling or the incineration of municipal solid waste, recycling does present several problems.

Recycling Concerns

Problems associated with recycling tend to be either technical or economic. Technical questions are of particular concern when recycling plastics. (See figure 18.15.) While the plastics used in packaging are recyclable, the technology to do so differs from plastic to plastic. There are many different types of plastic polymers. Since each type has its own chemical makeup, different plastics cannot be recycled together. In other words, a milk container is likely to be high-density polyethylene (HDPE), while an egg container is polystyrene (PS), and a soft-drink bottle is polyethylene terephthalate (PET).

Plastic recycling is still a relatively new field. Industry is researching new technologies that promise to increase the quality of plastics recycled and that will

figure 18.14 **Recycling Rates for Various Materials** Recycling rates for materials that have high value such as automobile batteries are extremely high. Other materials are more difficult to market. But recycling rates today are much higher than in the past as technology and markets have found uses for materials that once were considered valueless.

Source: (a) Data from the U.S. Environmental Protection Agency, *Characterization of MSW in the U.S.*, 2000. (b) *Characterization of Municipal Solid Waste in the United States:* Update. The U.S. Environmental Protection Agency, Washington, D.C.

ENVIRONMENTAL CLOSE-UP

What You Can Do to Reduce Waste and Save Money

You can make a difference. While this statement is sometimes overused, it does speak the truth when it comes to your ability to lessen the stream of solid waste being generated every day. Here are a few ideas that are easy to follow, will save you money, and will help reduce waste.

- Buy things that last, keep them as long as possible, and have them repaired, if possible.
- Buy things that are reusable or recyclable, and be sure to reuse and recycle them.
- Buy beverages in refillable glass containers instead of cans or throwaway bottles.
- Use plastic or metal lunch boxes and metal or plastic garbage containers without throwaway plastic liners.
- Use rechargeable batteries.
- Skip the bag when you buy only a quart of milk, a loaf of bread, or anything you can carry with your hands.
- Buy recycled goods, especially those made by primary recycling, and then recycle them.
- Recycle all newspaper, glass, and aluminum, and any other items accepted for recycling in your community.
- Reduce the amount of junk mail you get. This can be accomplished by writing to Mail Preference Service, Direct Marketing Association Inc., 1120 Avenue of the Americas, New York, NY 10036-6700, or by calling (212) 768-7277. Ask that your name not be sold to large mailing-list companies. Of the junk mail you do receive, recycle as much of the paper as possible.
- Push for mandatory trash separation and recycling programs in your community and schools.
- Choose items that have the least packaging or, better yet, no packaging ("nude products").
- Compost your yard and food wastes, and pressure local officials to set up a community composting program.

ENVIRONMENTAL CLOSE-UP

Recycling Is Big Business

In 1994, Weyerhaeuser Company, one of the world's largest forest products producers, began building a new paper mill in Cedar Rapids, Iowa, an area of North America not known for its abundant forests. The Cedar River Paper Company is a joint venture between Weyerhaeuser and Midwest Recycling Company and is the largest paper recycling plant in the United States. Even with a lack of local trees, the new mill will not be hurting for raw material to make paper and corrugated cardboard.

The mill is supplied with old paper, including scrap boxes from K-Mart. K-Mart signed an agreement with Weyerhaeuser in 1992 to have Weyerhaeuser purchase its cardboard. The agreement provides Weyerhaeuser with a supply of paper for its recycled paper plants and solves a waste disposal problem for K-Mart. The fact that a supply of recycled paper can determine where new paper-making factories will be built indicates just how important the recycling business is becoming. It is projected that by the turn of the century, over a third of the raw material that Weyerhaeuser uses for paper making will come from recycled materials. Weyerhaeuser is the third largest paper recycler and collector in the United States. It has 37 wastepaper processing plants in both Canada and the United States. According to a company spokesperson, the recycling business is becoming as significant as the millions of hectares of forests the company owns.

allow mixing of different plastics. Until such technology is developed, separation of different plastics before recycling will be necessary.

The economics of recycling are also a primary area of concern. The stepped-up commitment to recycling in many developed nations has produced a glut of certain materials on the market. Markets for collected materials fill up just like landfills. Unless the demand for recycled products keeps pace with the growing supply, recycling programs will face an uncertain future. The prices for selected recycled materials are listed in Figure 18.16. Prices for materials can vary widely from year to year depending on demand.

Markets for materials collected in recycling programs grew dramatically during the 1990s. The establishment of a recyclables exchange on the Chicago Board of Trade allows for consumers and producers of recyclable materials to participate in an efficient market so that it is less likely that recyclable materials will be left unclaimed. In 1995, the American Forest and Paper Association stated that it was planning to spend $10 billion by 2002 to retool and build new mills to produce recycled paper.

The long-term success of recycling programs is also tied to other economic incentives, such as taxing issues and the development of and demand for products manufactured from recycled material. Government tax policy needs to be readjusted to encourage recycling efforts. Currently in the United States, it is still cheaper to transport virgin material, such as fresh-cut pulp wood, than to transport collected paper for recycling. Such taxing policy severely inhibits the cost-effectiveness of paper recycling. In addition, on an individual level, we can have an impact by purchasing products made from recycled materials. The demand for recycled products must grow if recycling is to succeed on a large scale. (See Environmental Close-Up: Recyclables Market Basket).

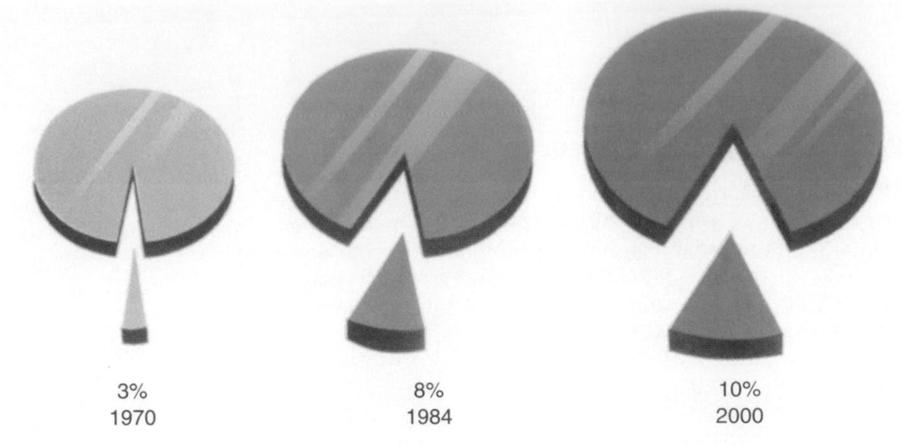

figure 18.15 **Increasing Amounts of Plastics in Trash** Plastics are a growing component of municipal solid waste in North America. Increased recycling of plastics could reverse this trend.

Source: Data from Franklin Associates, Ltd.

Recycling Composite Prices

		1999	2000
Metal			
Ferrous ($/ton)	Used steel cans	71	69
Non-ferrous (¢/lb.)	Aluminum cans	50	99
	Auto batteries	6	6
Plastic (¢/lb.)			
	Green PET	7	11
	Clear PET	6	13
	Mixed HDPE	9	14
Paper ($/ton)			
	Corregated	92	62
	Newspaper	31	53
	High grade office	93	158
	Computer laser	163	193
Glass ($/ton)			
	Clear	39	39
	Green	14	14
	Brown	24	24

figure 18.16 **Recycling Composite Prices** Prices for materials can vary widely from year to year depending on demand.

ENVIRONMENTAL CLOSE-UP

Recyclables Market Basket

Annually in the United States, 17 percent of scrap used tires are recovered. Scrap tires are difficult to dispose of in landfills and waste incinerators. An estimated 2 billion to 4 billion are currently stockpiled. These stockpiles can provide convenient habitats for rodents, serve as breeding grounds for mosquitoes, and pose fire hazards. Of the scrap tires that are used, most are burned for energy. Scrap tires also are used for rubberized asphalt paving, molded rubber products, and athletic surfaces.

About 96 percent of automotive batteries are recovered each year in North America. Although these lead-acid batteries constitute a small portion of the waste stream, they contain metals that may be a concern when disposed of in landfills and incinerators. All three components of automotive batteries are recyclable: the lead, the acid, and the plastic casing.

Seventy percent of all used oil is recovered in North America. Only 10 percent of the amount generated by people who change their own motor oil is returned to collection programs. If disposed of improperly, such as being poured down sewage drains, used oil can contaminate soil, groundwater, and surface water. In some communities, used motor oil is collected at service stations, corporate or municipal collection sites, or at the curbside.

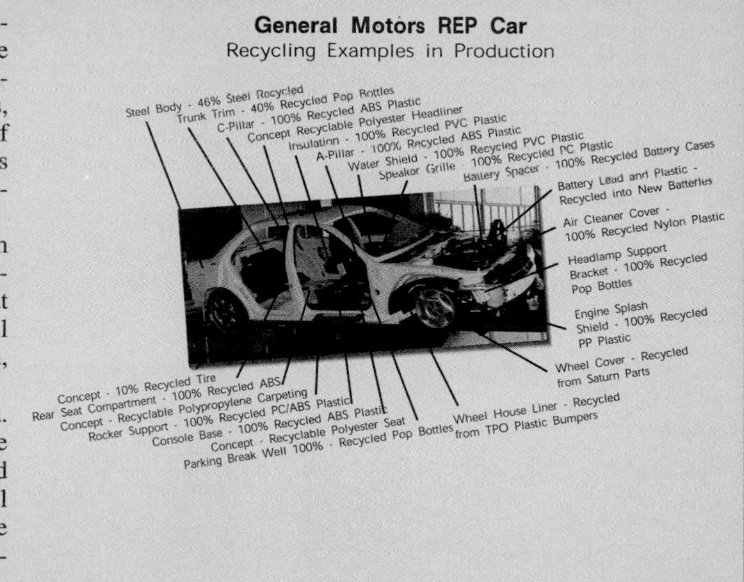

Summary

Beginning with the post–World War II era, increased consumption of consumer goods became a way of life. Products were designed to be used once and then thrown away. By the 1980s, a disposal lifestyle began to cause problems. There simply were no places to dispose of waste. Barges filled with municipal solid waste from the metropolitan areas along the eastern United States were traveling the world trying to dispose of their unwanted cargo.

Municipal solid waste is managed by landfills, incineration, composting, waste reduction, and recycling. Landfills are the primary means of disposal; however, a contemporary landfill is significantly more complex and expensive than the simple holes in the ground of the past. The availability of suitable landfill land is also a problem in large metropolitan areas.

About 16 percent of the municipal solid waste in the United States is incinerated. While incineration does reduce the volume of municipal solid waste, the problems of ash disposal and air quality continue to be major concerns. There are several forms of composting that can keep organic wastes from entering a landfill.

The most fundamental way to reduce waste is to prevent it from ever becoming waste in the first place. Using less material in packaging, producing consumer products in concentrated form, and composting yard waste are all examples of source reduction. On an individual level, we can all attempt to reduce the amount of waste we generate.

About 28 percent of the waste generated in North America is handled through recycling. Recycling initiatives have grown rapidly in North America during the past several years. As a result, the markets for some recycled materials have become very volatile. Recycling of municipal solid waste will be successful only if markets exist for the recycled materials. Another problem in recycling is the current inability to mix various plastics. The plastics industry is working on the development of a more universal plastic.

Future management of municipal solid waste will be an integrated approach involving landfills, incineration, composting, source reduction, and recycling. The degree to which any option will be used will depend on economics, changes in technology, and citizen awareness and involvement.

ISSUES & ANALYSIS

Corporate Response to Environmental Concerns

In the 1990s, McDonald's Corporation announced it would switch from polystyrene to paper for packaging its food products. In making this announcement, McDonald's stated that it was responding to consumer pressure to become more environmentally conscientious. Is using paper to wrap fast-food better for the environment than using polystyrene? The answer is not a simple yes or no.

Advocates for the switch say that polystyrene takes up space in landfills and does not decompose. They also argue that the burning of polystyrene foam in incinerators might release harmful air pollutants.

Opponents of the switch argue that polystyrene can be recycled into useful products, such as insulation board or playground equipment. They further argue that using paper means cutting forests and that, since the paper used to wrap the food is coated with wax, it cannot be recycled.

Many grocery chains now offer several choices of carry-out bags. Some encourage reuse of bags by deducting a small amount from the customer's bill. Others provide a choice of paper or plastic.

- What are the pros and cons of paper and plastic?
- Which alternative do you prefer? Why?
- Is there an alternative to both?

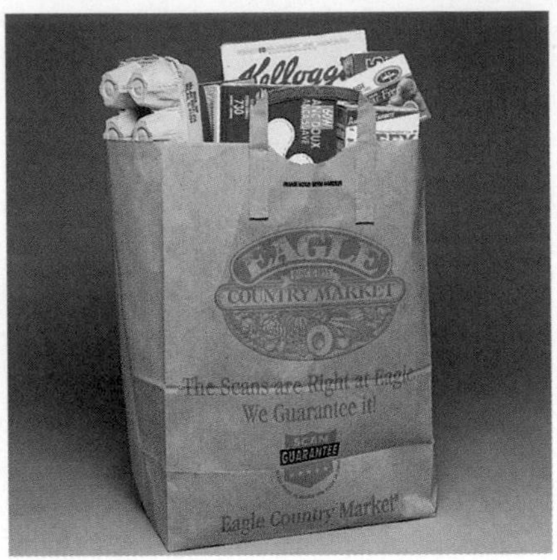

Which is better?

Which do you use?

Is there an alternative?

Interactive Exploration

Check out the website at

http://www.mhhe.com/environmentalscience

and click on the cover of this textbook for interactive versions of the following:

KNOW THE BASICS

incineration *407*
mass burn *408*
municipal landfill *405*
municipal solid waste (MSW) *402*
recycling *411*
source reduction *409*

- On-Line Flashcards
- Electronic Glossary

IN THE REAL WORLD

What do you do with the oilrigs that are no longer in use? One way to decrease the amount of material headed for a landfill or for recycling is to use these huge structures as artificial reefs. Take a look at the Oilrigs as Artificial Reefs case study for an alternative disposal method.

TEST PREPARATION

Review Questions

1. How is lifestyle related to our growing municipal solid waste problem?
2. What four methods are incorporated under integrated waste management?
3. Describe some of the problems associated with modern landfills.
4. What are four concerns associated with incineration?
5. Describe examples of source reduction.
6. Describe the importance of recycling household solid wastes.
7. Name several strategies that would help to encourage the growth of recycling.
8. Describe the various types of composting and the role of composting in solid waste management.

Critical Thinking Questions

1. Why do you suppose consumers were so quick to adapt the "use it once, then throw it away" lifestyle after World War II? What values, beliefs, and perceptions does this reveal?
2. How can you help solve the solid waste problem?
3. Given that you have only so much time, should you spend your time acting locally, as a recycling coordinator for example, or advocating for larger political and economic changes at the national level, changes that would solve the waste problems? Why? Or should you do nothing? Why?
4. How does your school or city deal with solid waste? Can solid waste production be limited at your institution or city? How? What barriers exist that might make it difficult to limit solid waste production?
5. It is possible to have a high standard of living, as in North America and Western Europe, and not produce large amounts of solid waste. How?
6. Often environmental costs are hidden from view, and the "correct" response to an environmental problem is not readily apparent. Read the Issues and Analysis section of this chapter. Which alternative do you prefer? Are there other alternatives?
7. Incineration of solid waste is controversial. Do you support solid waste incineration in general? Would you support an incineration facility in your neighborhood?

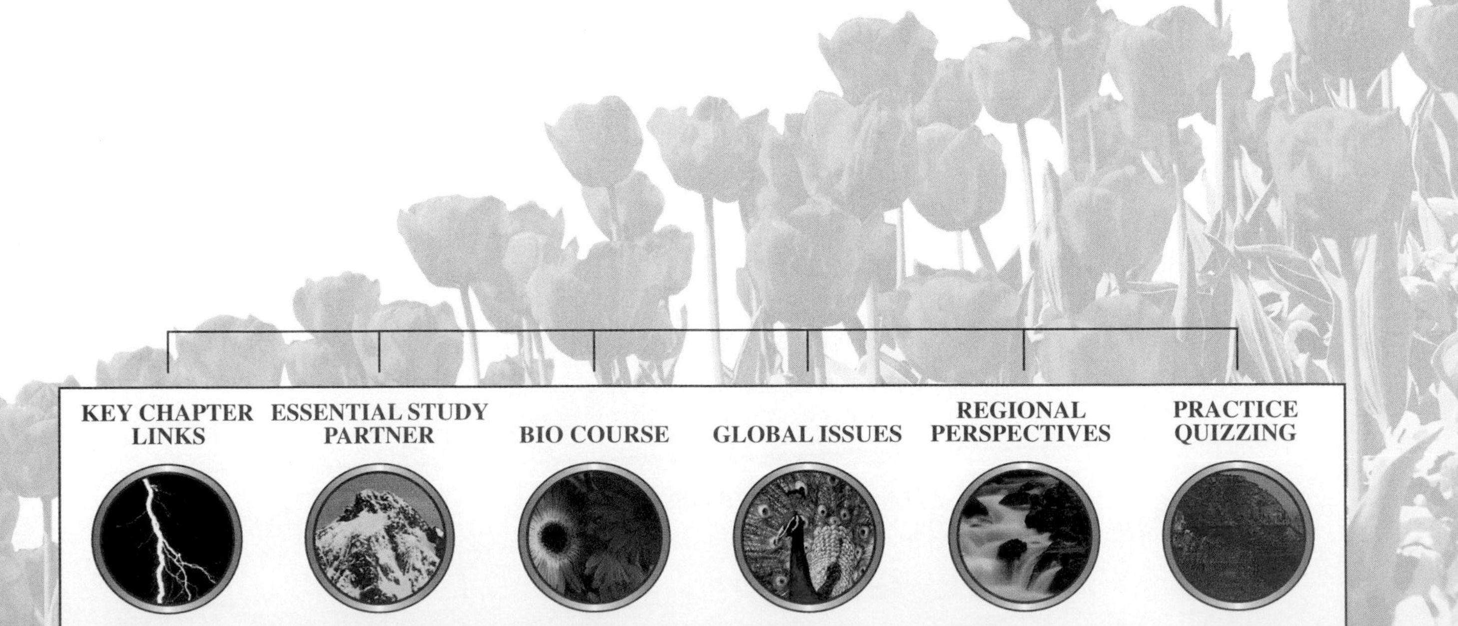

CHAPTER

Regulating Hazardous Materials

Objectives

After reading this chapter, you should be able to:

- Distinguish between hazardous substances and hazardous wastes.
- Distinguish between hazardous and toxic substances.
- Explain the complexity in regulating hazardous materials.
- Describe the four characteristics by which hazardous materials are identified.
- Describe the environmental problems of hazardous and toxic materials.
- Understand the difference between persistent and nonpersistent pollutants.
- Describe the health risks associated with hazardous wastes.
- Explain the problems associated with hazardous-waste dump sites and how such sites developed.
- Describe how hazardous wastes are managed, and list five technologies used in their disposal.
- Describe the importance of source reduction with regard to hazardous wastes.

Chapter Outline

Hazardous and Toxic Materials in Our Environment

Hazardous and Toxic Substances—Some Definitions

Defining Hazardous Waste

Issues Involved in Setting Regulations
- Identification of Hazardous and Toxic Materials

Environmental Close-Up: *Exposure to Toxins*
- Setting Exposure Limits
- Acute and Chronic Toxicity
- Synergism
- Persistent and Nonpersistent Pollutants

Environmental Problems Caused by Hazardous Wastes

Health Risks Associated with Hazardous Wastes

Hazardous-Waste Dumps—A Legacy of Abuse

Global Perspective: *Lead and Mercury Poisoning*

Environmental Close-up: *Computers—A Hazardous Waste*
- Toxic Chemical Releases

Managing Hazardous Waste
- Pollution Prevention
- Waste Minimization
- Recycling of Wastes
- Treatment of Wastes
- Land Disposal

Hazardous-Waste Management Choices
- International Trade in Hazardous Wastes

Global Perspective: *Hazardous Wastes and Toxic Materials in China*
- Hazardous-Waste Program Evaluation

Issues & Analysis: *Love Canal*

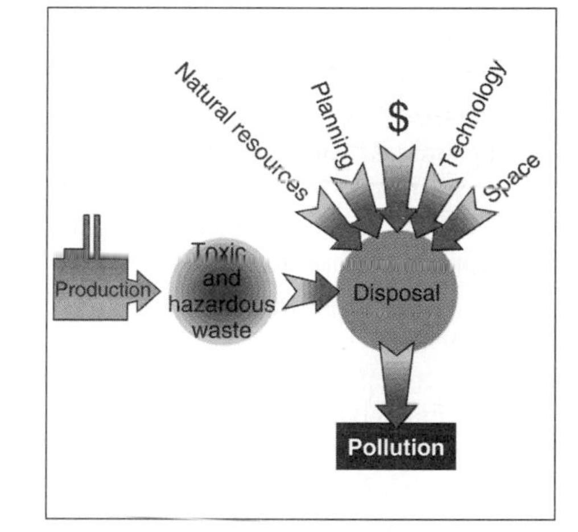

Hazardous and Toxic Materials in Our Environment

Our modern technological society makes use of a large number of substances that are hazardous or toxic. The benefits gained from using these materials must be weighed against the risks associated with their use.

- Pesticides thought to degrade in soils turn up in rural drinking-water wells.
- Underground plumes of toxic chemicals emanating from abandoned waste sites contaminate city water supplies.
- A gas leak at a chemical production plant in Bhopal, India, killed more than 2000 people.
- Pesticides spilled into the Rhine River from a warehouse near Basel, Switzerland, destroyed a half million fish, disrupted water supplies, and caused considerable ecological damage.
- The collapse of an oil storage tank in Pennsylvania spilled over 2 million liters (½ million gallons) of oil into the Monongahela River and threatened the water supply of millions of residents.

Toxic and hazardous products and by-products are becoming a major issue of our time. At sites around the world, accidental or purposeful releases of hazardous and toxic chemicals are contaminating the land, air, and water. The potential health effects of these chemicals range from minor, short-term discomforts, such as headaches and nausea, to serious health problems, such as cancers and birth defects (that may not manifest themselves for years), to major accidents that cause immediate injury or death. Today, names like Love Canal, New York, and Times Beach, Missouri, in the United States; Lekkerkerk in the Netherlands; Vác, Hungary; and Minamata Bay, Japan, are synonymous with the problems associated with the release of hazardous and toxic wastes into the environment.

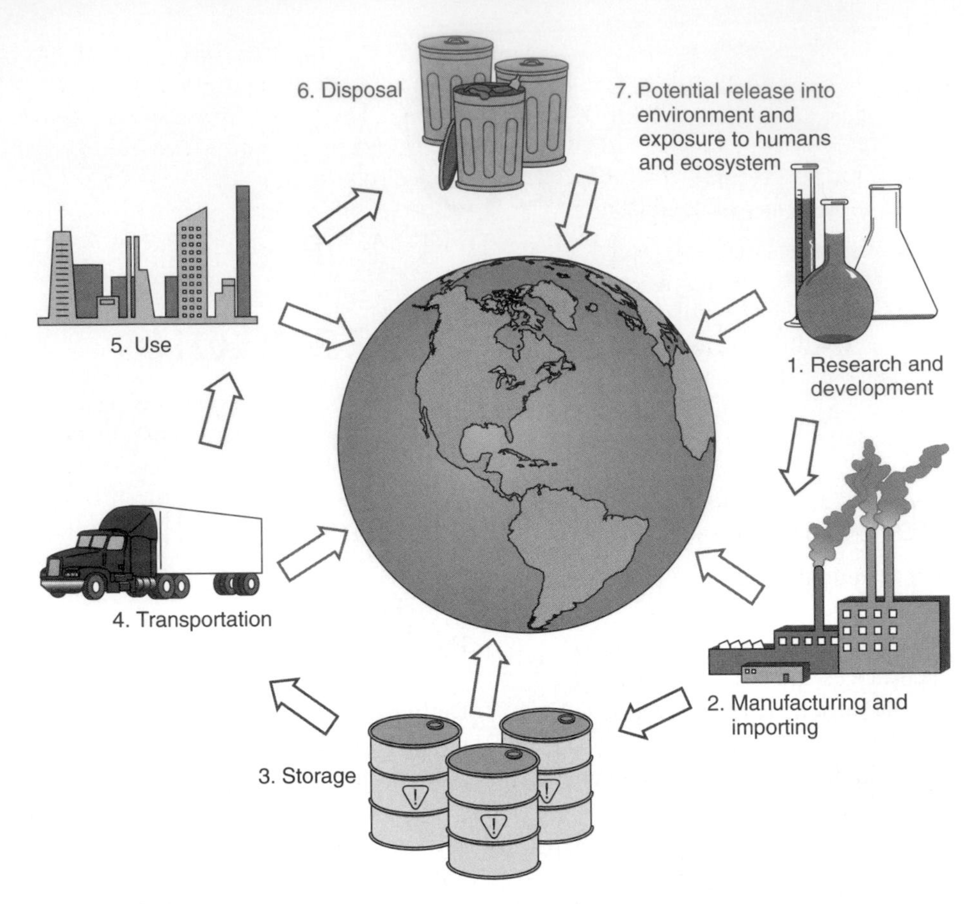

figure 19.1 **The Life Cycle of Toxic Substances** Controlling the problems of hazardous substances is complicated because of the many steps involved in a substance's life cycle.

Increasingly, governments and international agencies are attempting to control the growing problem of hazardous substances in our environment. Controlling the release of these substances is difficult since there are so many places in their cycles of use at which they may be released. (See figure 19.1.)

Hazardous and Toxic Substances—Some Definitions

To begin, it is important to clarify various uses of the words *hazardous* and *toxic* as well as to distinguish between things that are wastes and those that are not. **Hazardous substances** are those that can cause harm to humans or the environment. The U.S. Environmental Protection Agency (EPA) defines hazardous materials as having one or more of the following characteristics:

1. **Ignitability**—Describes materials that pose a fire hazard during routine management. Fires not only present immediate dangers of heat and smoke but also can spread harmful particles over wide areas. Common examples are gasoline, paint thinner, and alcohol.
2. **Corrosiveness**—Describes materials requiring special containers because of their ability to corrode standard materials, or requiring segregation from other materials because of their ability to dissolve toxic contaminants. Common examples are strong acids and bases.
3. **Reactivity** (or explosiveness)—Describes materials that, during routine management, tend to react spontaneously, to react vigorously with air or water, to be unstable to shock or heat, to generate toxic

gases, or to explode. Common examples are gunpowder, which will burn or explode; the metal sodium, which reacts violently with water; and nitroglycerine, which explodes under a variety of conditions.

4. **Toxicity**—Describes materials that, when improperly managed, may release toxicants (poisons) in sufficient quantities to pose a substantial hazard to human health or the environment. Almost everything that is hazardous is toxic in high enough quantities. For example, tiny amounts of carbon dioxide in the air are not toxic, but high levels are.

Some hazardous materials fall into several of these categories. Gasoline, for example, is ignitable, can explode, and is toxic. It is even corrosive to certain kinds of materials. Other hazardous materials meet only one of the criteria. Polychlorinated biphenyls (PCBs) are toxic but will not burn, explode, or corrode other materials. While the terms *toxic* and *hazardous* are often used interchangeably, there is a difference. **Toxic** commonly refers to a narrow group of substances that are poisonous and cause death or serious injury to humans and other organisms by interfering with normal body physiology. **Hazardous,** the broader term, refers to all dangerous materials, including toxic ones, that present an immediate or long-term human health risk or environmental risk.

Another important distinction is the difference between hazardous substances and hazardous wastes. Although the health and safety considerations regarding hazardous substances and hazardous wastes are similar, the legal and regulatory implications are quite different. Hazardous substances are materials that are used in business and industry for the production of goods and services. Typically, hazardous substances are consumed or modified in industrial processes. **Hazardous wastes** are by-products of industrial, business, or household activities for which there is no immediate use. These materials must be disposed of in an appropriate manner, and there are stringent regulations pertaining to their production, storage, and disposal.

There are numerous types of hazardous waste, ranging from materials contaminated with dioxins and heavy metals (such as mercury, cadmium, and lead) to organic wastes. These wastes can also take many forms, from barrels of liquid waste to sludge, old computer parts, used batteries, and incinerator ash. In industrialized countries, industry and mining are the main sources of hazardous wastes, though small-scale industry, hospitals, military establishments, transport services, and small workshops contribute to the generation of large quantities of such wastes in both the industrialized and developing worlds.

Improper handling and disposal of hazardous wastes can affect human health and the environment through the leakage of toxins into groundwater, soil, waterways, and the atmosphere. The environmental and health effects can be immediate (such as an illness caused by exposure to toxins at a particular site) or long-term (such as when contaminated waste leaches into groundwater or soil and then works its way into the food chain). The damage caused by hazardous wastes also takes an economic toll, and cleaning up contaminated sites can be costly for local authorities, particularly in poor communities. Without adequate safeguards, recycling and recovery operations can result in greater health dangers due to the higher level of worker exposure and handling.

While exact figures regarding the amounts of hazardous waste generated internationally are quite difficult to obtain, some information does exist. United Nations Environment Program estimates total annual international generation of hazardous wastes to be between 300 million tonnes (330 million U.S. tons) and 500 million tonnes (550 million U.S. tons) with Organization for Economic Cooperation and Development (OECD) countries accounting for 80 to 90 percent of this quantity. Data from the EPA indicate that the United States generated about 243 tonnes (270 U.S. tons) of hazardous waste in 1999, out of a total of 9 billion tonnes (10 billion U.S. tons) of solid waste. However, in some of the rapidly developing countries of Southeast Asia, the technical and regulatory structures required for proper hazardous-waste management have not kept pace with industrialization. Thailand, for example, generated approximately 2 million tonnes (2.2 million U.S. tons) of hazardous waste in 1990, a figure that is expected to quadruple by 2002.

Defining Hazardous Waste

The definition of hazardous waste varies from one country to another. One of the most widely used definitions, however, is contained in the U.S. **Resource Conservation and Recovery Act** of 1976 (RCRA). The RCRA considers wastes toxic and/or hazardous if they:

> cause or significantly contribute to an increase in mortality or an increase in serious irreversible, or incapacitating reversible, illness; or pose a substantial present or potential hazard to human health or the environment when improperly treated, stored, transported, disposed of, or otherwise managed.

This definition gives one an appreciation for the complexity of hazardous-waste regulation.

Working from the RCRA definition, the EPA compiled a list of hazardous wastes. Listing is the most common method for defining hazardous waste in European countries and in some state laws. The EPA has also required that a hazardous waste be identified by testing it to determine if it possesses any one of the four characteristics discussed earlier: ignitability, reactivity, corrosiveness, and toxicity. If it does, it is subject to regulation under RCRA.

Issues Involved in Setting Regulations

Whether a hazardous substance is a raw material, an ingredient in a product, or a waste, there are problems associated with determining regulations that

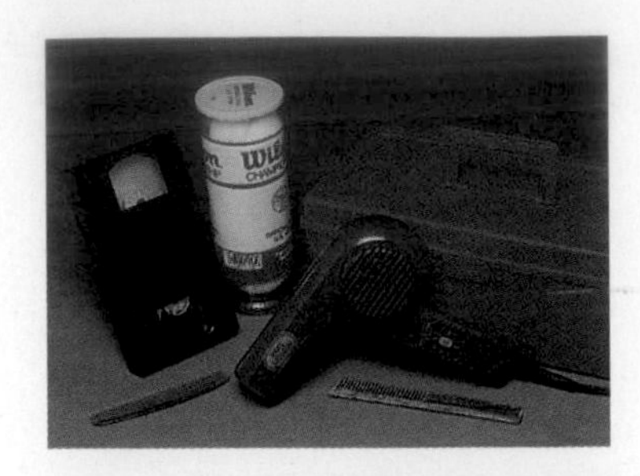

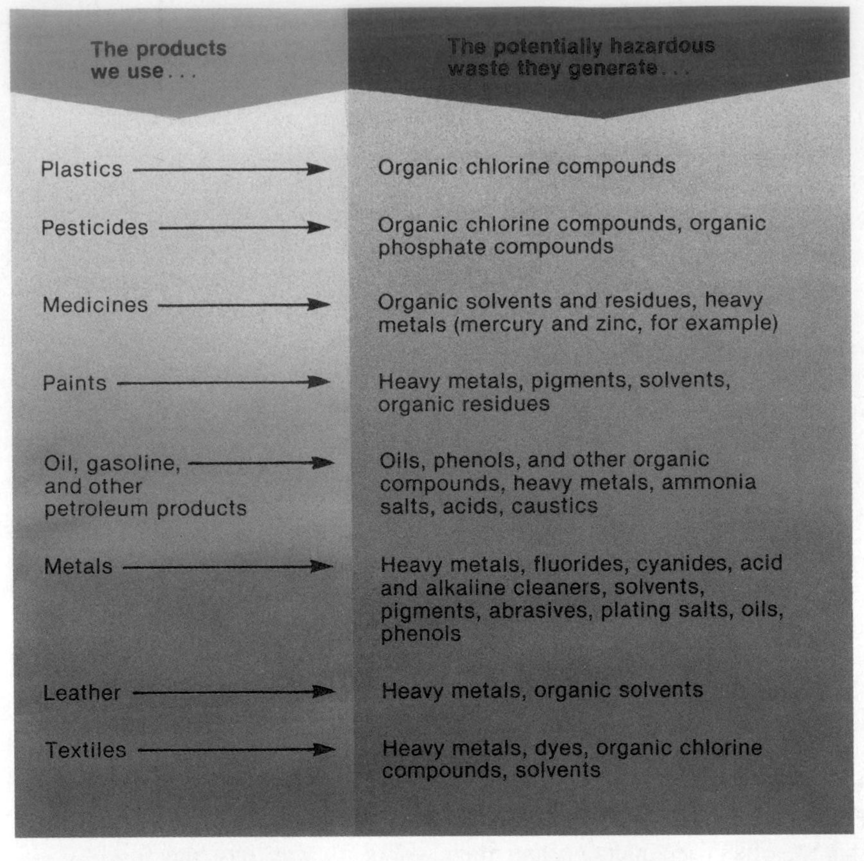

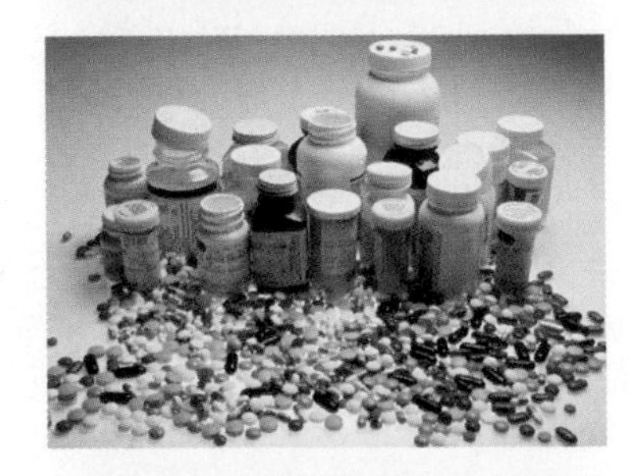

figure 19.2 **Common Materials Can Produce Hazardous Wastes** Many commonly used materials can release toxic or hazardous wastes if not properly disposed of.

pertain to it. In the industrialized countries of Europe and North America, chemical and petrochemical industries produce nearly 70 percent of all hazardous wastes; in developing countries, the figure is 50 to 66 percent. These industries produce many useful materials that are converted into the everyday products we use. Most toxic and hazardous wastes come from chemical and related industries that produce plastics, soaps, synthetic rubber, fertilizers, medicines, paints, pesticides, herbicides, and cosmetics. (See figure 19.2.)

Identification of Hazardous and Toxic Materials

In attempting to regulate the use of toxic and hazardous substances and the generation of toxic and hazardous wastes, most countries simply draw up a list of specific substances which have been scientifically linked to adverse human health or environmental effects. However, since many potentially harmful chemical compounds have yet to be tested adequately, most lists include only the known offenders. Historically, we have often identified toxic materials only after their effects have shown up in humans or other animals. Asbestos was identified as a cause of lung cancer in humans who were exposed on the job, and DDT was identified as toxic to birds when robins began to die and eagles and

ENVIRONMENTAL CLOSE-UP

Exposure to Toxins

We are all exposed to materials that are potentially harmful. The question is, at what levels is such exposure harmful or toxic? One measure of toxicity is **LD_{50}**, the dosage of a substance that will kill (lethal dose) 50 percent of a test population. Toxicity is measured in units of poisonous substance per kilogram of body weight. For example, the deadly chemical that causes botulism, a form of food poisoning, has an LD_{50} in adult human males of 0.0014 milligrams per kilogram. This means that if each of 100 human adult males weighing 100 kilograms consumed a dose of only 0.14 milligrams—about the equivalent of a few grains of table salt—approximately 50 of them will die.

Lethal doses are not the only danger from toxic substances. During the past decade, concern has been growing over minimum harmful dosages, or threshold dosages, of poisons, as well as their sublethal effects.

The length of exposure further complicates the determination of toxicity values. Acute exposure refers to a single exposure lasting from a few seconds to a few days. Chronic exposure refers to continuous or repeated exposure for several days, months, or even years. Acute exposure usually is the result of a sudden accident, such as the tragedy at Bhopal, India, mentioned at the beginning of the chapter. Acute exposures often make disaster headlines in the press, but chronic exposure to sublethal quantities of toxic materials presents a much greater hazard to public health. For example, millions of urban residents are continually exposed to low levels of a wide variety of pollutants. Many deaths attributed to heart failure or such diseases as emphysema may actually be brought on by a lifetime of exposure to sublethal amounts of pollutants in the air.

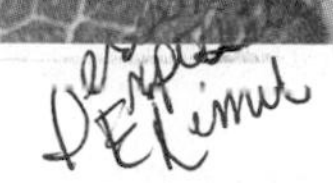

other fish-eating carnivores failed to reproduce. Once these substances are identified as toxic, their use is regulated. Countries contemplating regulation of hazardous and toxic materials and wastes must consider not only how toxic each one is but also how flammable, corrosive, and explosive it is, and whether it will produce mutations or cause cancer.

Setting Exposure Limits

Even after a material is identified as hazardous or toxic, there are problems in determining appropriate exposure limits. Nearly all substances are toxic in sufficiently high doses. The question is, When does a chemical cross over from safe to toxic? There is no easy way to establish acceptable levels. For any new compounds that are to be brought on the market, extensive toxicology studies must be done to establish their ability to do harm. Usually these are tests on animals. (See Environmental Close-Up: Exposure to Toxins.) Typically, the regulatory agency will determine the level of exposure at which none of the test animals is affected (**threshold level**) and then set the human exposure level lower to allow for a safety margin. This safety margin is important because it is known that threshold levels vary significantly among species, as well as among members of the same species. Even when concentrations are set, they may vary considerably from country to country. For example, in the Netherlands, 50 milligrams of cyanide per kilogram of waste is considered hazardous; in neighboring Belgium, the toxicity standard is fixed at 250 milligrams per kilogram.

Acute and Chronic Toxicity

Regulatory agencies must look at both the effects of one massive dose of a substance (**acute toxicity**) and the effects of exposure to small doses over long periods (**chronic toxicity**). Acute toxicity is readily apparent because organisms respond to the toxin shortly after being exposed. Chronic toxicity is much more difficult to determine because the effects may not be seen for years. Furthermore, an acute exposure may make an organism ill but not kill it, while chronic exposure to a toxic material may cause death. A good example of this effect is alcohol toxicity. Consuming extremely high amounts of alcohol can result in death (acute toxicity and death). Consuming moderate amounts may result in illness (acute toxicity and full recovery). Consuming moderate amounts over a number of years may result in liver damage and death (chronic toxicity and death).

Another example of chronic toxicity involves lead. Lead has been used in paints, in gasoline, and in pottery glazes for many years, but researchers discovered that it has harmful effects. The chronic effects on the nervous system are most noticeable in children, particularly when children eat paint chips.

Synergism

Another problem in regulating hazardous materials is assessing the effects of mixtures of chemicals. Most toxicological studies focus on a single compound, even though industry workers may be exposed to a variety of chemicals; and in waste dumps, the compounds are usually found in mixtures. Although the materials may be relatively harmless as separate compounds, once mixed, they may become highly toxic and cause more serious problems than do individual pollutants. This is referred to as **synergism.** For example, all uranium miners are exposed to radioactive gases, but those who smoke tobacco and thus are exposed to the toxins in tobacco smoke have unusually high incidences of lung cancer. Apparently, the radioactive gases found in uranium mines interact synergistically with the carcinogens found in tobacco smoke.

Persistent and Nonpersistent Pollutants

The regulation of hazardous and toxic materials is also influenced by the degree of persistence of the pollutant. **Persistent pollutants** are those that remain in the environment for many years in an unchanged condition. Most of the persistent pollutants are human-made materials. An estimated 30,000 synthetic chemicals are used in the United States. They are mixed in an endless variety of combinations to produce all types of products used in every aspect of daily life. They are part of our food, transportation, clothing, building materials, home appliances, medicine, recreational equipment, and many other items. Our way of life is heavily dependent upon synthetic materials.

An example of a persistent pollutant is DDT. It was used as an effective pesticide worldwide and is still used in some countries because it is so inexpensive and is very effective in killing pests. However, once released into the environment, it accumulates in the food chain and causes death when its concentration is high enough. (See chapter 15 for a discussion of DDT as a pesticide.)

Another widely used group of synthetic compounds of environmental concern are polychlorinated biphenyls (PCBs). PCBs are highly stable compounds that resist changes from heat, acids, bases, and oxidation. These characteristics make PCBs desirable for industrial use but also make them persistent pollutants when released into the environment. At one time, these materials were commonly used in transformers and electrical capacitors. Other uses included inks, plastics, tapes, paints, glues, waxes, and polishes. PCBs are harmful to fish and other aquatic forms of life because they interfere with reproduction. In humans, PCBs produce liver ailments and skin lesions. In high concentration, they can damage the nervous system, and they are suspected carcinogens. In 1970, PCB production was limited to those cases where satisfactory substitutes were not available.

In addition to synthetic compounds, our society uses heavy metals for many purposes. Mercury, beryllium, arsenic, lead, and cadmium are examples of heavy metals that are toxic. When released into the environment, they enter the food chain and become concentrated. In humans, these metals can produce kidney and liver disorders, weaken the bone structure, damage the central nervous system, cause blindness, and lead to death. Because these materials are persistent, they can accumulate in the environment even though only small amounts might be released each year. When industries use these materials in a concentrated form, it presents a hazard not found naturally.

A **nonpersistent pollutant** does not remain in the environment for very long. Most nonpersistent pollutants are biodegradable. Others decompose as a result of inorganic chemical reactions. Still others quickly disperse to concentrations that are too low to cause harm. A biodegradable material is chemically changed by living organisms and often serves as a source of food and energy for decomposer organisms, such as bacteria and fungi. Phenol and many other kinds of toxic organic materials can be destroyed by decomposer organisms.

Other toxic materials, such as many of the insecticides, are destroyed by sunlight or reaction with oxygen or water in the atmosphere. These include the "soft biocides." For example, organophosphates are a type of pesticide that usually decomposes within several weeks. As a result, organophosphates do not accumulate in food chains because they are pollutants for only a short period of time.

Other toxic and hazardous materials such as carbon monoxide, ammonia, or hydrocarbons can be dispersed harmlessly into the atmosphere (as long as their concentration is not too great) where they eventually react with oxygen.

Because persistent materials can continue to do harm for a long time (chronic toxicity), they are particularly important to regulate. Nonpersistent materials need to be kept below threshold levels to protect the public from acute toxicity. They are not likely to present a danger of chronic toxicity since they either disperse or decompose.

Environmental Problems Caused by Hazardous Wastes

Hazardous wastes contaminate the environment in several ways. Many hazardous materials are released directly to the environment. Many molecules that evaporate readily are vented directly to the atmosphere or escape from faulty piping and valves. These materials are often not even thought of as being hazardous waste. Once hazardous wastes are produced, they must be stored. Improper storage or even poor bookkeeping may inadvertently result in a release. Uncontrolled or improper incineration of hazardous wastes, whether on land or at sea, can contaminate the atmosphere and the surrounding environment. The discharge of hazardous substances into the sea or into lakes and rivers often kills fish and other aquatic life. Further, disposal on land in dumps that are later abandoned, or in improperly controlled landfills, can pollute both the soil and the groundwater as materials leach below the site.

Because most hazardous wastes are disposed of on or in land, the most serious environmental effect is contaminated groundwater. In the United States

Table 19.1 Top Fifteen Hazardous Substances 2000

Substance	Source	Toxic effects
Lead	Lead-based paint Lead additives in gasoline	Neurological damage. Affects brain development in children. Large doses affect brain and kidneys in adults and children.
Arsenic	From elevated levels in soil or water	Multiple organ systems affected. Heart and blood vessel abnormalities, liver and kidney damage, impaired nervous system function.
Metallic mercury	Air or water at contaminated sites	Permanent damage to brain, kidneys, developing fetus.
Vinyl chloride	Plastics manufacturing Air or water at contaminated sites	Acute effects: dizziness, headache, unconsciousness, death. Chronic effects: liver, lung, and circulatory damage.
Benzene	Industrial exposure Glues, cleaning products, gasoline	Acute effects: drowsiness, headache, death at high levels. Chronic effects: damages blood-forming tissues and immune system; also carcinogenic.
Polychlorinated biphenyls (PCBs)	Eating contaminated fish Industrial exposure	Probable carcinogens. Acne and skin lesions.
Cadmium	Released during combustion Living near a smelter or power plant Picked up in food	Probable carcinogen, kidney damage, lung damage, high blood pressure.
Benzo[a]pyrene	Product of combustion of gasoline or other fuels In smoke and soot	Probable carcinogen, possible birth defects.
Chloroform	Contaminated air and water Many kinds of industrial settings	Affects central nervous system, liver, and kidneys; probable carcinogen.
Benzo[b]fluoranthene	Product of combustion of gasoline and other fuels Inhaled in smoke	Probable carcinogen.
DDT	From food with low levels of contamination Still used as pesticide in parts of world	Probable carcinogen; possible long-term effect on liver; possible reproductive problems.
Aroclor 1260 (a mixture of PCBs)	From food and air	Probable carcinogens. Acne and skin lesions.
Trichloroethylene	Used as a degreaser, evaporates into air	Dizziness, numbness, unconsciousness, death.
Aroclor 1254 (a mixture of PCBs)	From food and air	Probable carcinogens. Acne and skin lesions.
Chromium (+6)	From food, water, and air Originates from combustion source	Ulcers of the skin, irritation of nose and gastrointestinal tract, also affects kidney and liver.

Source: Data from Agency for Toxic Substances and Disease Registry.

alone, an estimated 100,000 active industrial landfill sites may be possible sources of groundwater contamination, along with 200 special facilities for disposal of both liquid and solid hazardous wastes, and some 180,000 surface impoundments (ponds) for all types of waste. (See chapter 15.) Nearly 2 percent of North America's underground aquifers could be contaminated with such chemicals as chlorinated solvents, pesticides, trace metals, and PCBs.  Once groundwater is polluted with hazardous wastes, the cost of reversing the damage is prohibitive. In fact, if an aquifer is contaminated with organic chemicals, restoring the water to its original state is seldom physically or economically feasible.

Health Risks Associated with Hazardous Wastes

Because most hazardous wastes are chemical wastes, controlling chemicals and their waste products is a major issue in most developed countries. Every year, roughly 1000 new chemicals join the nearly 70,000 in daily use. Many of these hazardous chemicals are toxic, but they pose little threat to human health unless they are used or disposed of improperly. For example, many insecticides are extremely toxic to humans. However, if they are stored, used, and disposed of properly, they do not constitute a human health hazard. Unfortunately, at the center of the hazardous-waste problem is the fact that the products and by-products of industry are often handled and disposed of improperly. Table 19.1 is a list of 15 top toxic materials as identified by the Agency for Toxic Substances and Disease Registry.

Establishing the medical consequences of exposure to toxic wastes is extremely complicated. The problem of linking a particular chemical or other hazardous waste to specific injuries or diseases is further compounded by the lack of toxicity data on most hazardous substances.

Although assessing environmental contamination from toxic wastes and determining health effects is extremely difficult, what little is known is cause for concern. Most older hazardous-waste dump sites, for example, contain dangerous and toxic chemicals along with heavy-metal residues and other hazardous substances.

Hazardous-Waste Dumps—A Legacy of Abuse

In the United States prior to the passage of the Resource Conservation and Recovery Act (RCRA) in 1976, hazardous waste was essentially unregulated. Similar conditions existed throughout most of the industrial nations of the world. The solution to hazardous-waste disposal was simply to bury or dump the wastes without any concern for potential environmental or health risks. Such uncontrolled sites included open dumps, landfills, bulk storage containers, and surface impoundments. These sites were typically located convenient to the industry and were often in environmentally sensitive areas, such as floodplains or wetlands. Rain and melting snow soaked through the sites, carrying chemicals that contaminated underground waters. When these groundwaters reached streams and lakes, they were contaminated as well. When the sites became full or were abandoned, they were frequently left uncovered, thus increasing the likelihood of water pollution from leaching or flooding, and increasing the chances of people having direct contact with the wastes. At some sites, specifically the uncovered ones, the air was also contaminated as toxic vapors rose from evaporating liquid wastes or from uncontrolled chemical reactions. (See figure 19.3.) In North America alone, the number of abandoned or uncontrolled sites is over 25,000, and the list grows yearly. The costs involved in cleaning up the sites are high. Nearly all industrialized countries are faced with costly, even massive, cleanup bills.

As is to be expected from the amount of toxic wastes generated each year, the United States has the highest number of hazardous-waste dumps needing immediate attention. Europeans are also paying a heavy price for their negligence. Every country in Europe (except Sweden and Norway) is plagued by an abundance of toxic-waste sites—both old and new—needing urgent attention. Holland is a good example. Authorities estimate that up to 8 million metric tons of hazardous chemical wastes may be buried in Holland. Estimates for cleaning up those wastes run as high as $7 billion. In the republics of the former Soviet Union and Eastern Europe, many hazardous waste sites have been identified recently, and there is no money to pay for cleanup.

figure 19.3 **Toxic Chemical Storage** This is a site where toxic wastes were improperly stored. Local governments often assume the problems and costs of correcting bankrupt companies' errors.

In the United States, the federal government has become the principal participant in the cleanup of hazardous-waste sites. The program that deals with the cleanup has popularly become known as **Superfund.** Superfund was established when Congress responded to public pressure to clean up hazardous-waste dumps and protect the public against the dangers of such wastes. The **Comprehensive Environmental Response, Compensation, and Liability Act (CERCLA)** (Superfund) was enacted in 1980. CERCLA had several key objectives:

1. To develop a comprehensive program to set priorities for cleaning up the worst existing hazardous-waste sites.
2. To make responsible parties pay for those cleanups whenever possible.
3. To set up a $1.6 billion Hazardous Waste Trust Fund—popularly known as Superfund—to support the identification and cleanup of abandoned hazardous-waste sites.
4. To advance scientific and technological capabilities in all aspects of hazardous-waste management, treatment, and disposal.

GLOBAL PERSPECTIVE

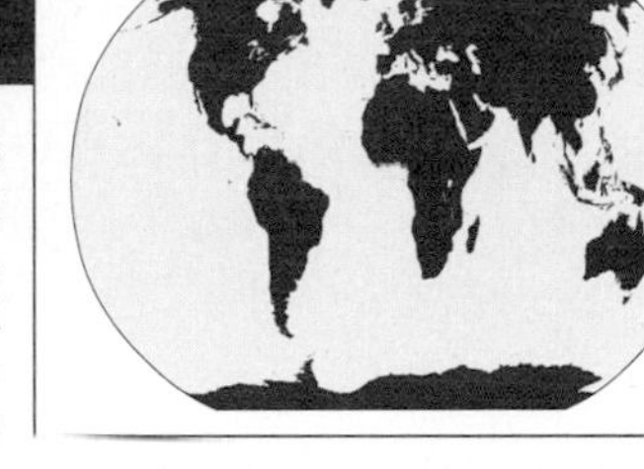

Lead and Mercury Poisoning

Lead and mercury are naturally present in the environment and probably have been a source of pollution for centuries. For example, the lead drinking and eating vessels used by the wealthy Romans may have caused the death of many of them. Another example, believe it or not, comes from *Alice in Wonderland,* published in 1865. One of the characters was the Mad Hatter. At that period in history, mercury was widely used in the treatment of beaver skins for making hats. As a result of exposure to mercury, hat makers often suffered from a variety of mental problems; hence, the phrase "mad as a hatter."

In 1953, a number of physical and mental disorders in the Minamata Bay region of Japan were diagnosed as being caused by mercury: 52 people developed symptoms of mercury poisoning, 17 died, and 23 became permanently disabled. In 1970, an outbreak of mercury poisoning in North America was traced to mercury in the meat of swordfish and tuna. In both incidents, the toxic material was not metallic mercury but a mercurous compound, methylmercury. Metallic mercury is converted to methylmercury by bacteria in the water. Methylmercury enters the food chain and may become concentrated as the result of biological amplification. Sufficient amounts in humans can cause brain damage, kidney damage, or birth defects. Today, regulations reduce the release of mercury into the environment and set allowable levels in foods. The problem persists, however, because it is impossible to eliminate the large amounts of mercury already present in the environment, and it is difficult to prevent the release of mercury in all cases. For example, burning coal releases 3200 tonnes (3500 U.S. tons) of mercury into the Earth's atmosphere each year, and mercury is still "lost" when it is used for various industrial purposes.

Like mercury, lead is a heavy metal and has been a pollutant for centuries. Studies of the Greenland Ice Cap indicate a 1500 percent increase in the lead content today as compared to 800 B.C. These studies reveal that the first large increase occurred during the Industrial Revolution and the second great increase occurred after the invention of the automobile. Oil companies added lead to gasoline to improve performance, and burning gasoline is a major source of lead pollution. There has been a reduction in airborne lead as a result of North America and Europe reducing lead content in gasoline. Another source of lead pollution is older paints. Prior to 1940, indoor and outdoor paints often contained lead.

There is still disagreement over what levels of lead and mercury can cause human health problems. Although ingested lead from paint can cause death or disability, such a strong correlation cannot be made for atmospheric lead.

There are concerns about fish contamination and public health as a result of methylmercury.

A **National Priority List** of hazardous-waste dump sites requiring urgent attention was drawn up for Superfund action. The U.S. Office of Technology Assessment (OTA) estimates that 10,000 sites may eventually be placed on the National Priority List requiring Superfund cleanup. The OTA believes that cleaning up these hazardous dumps may take 50 years and cost up to $100 billion because of the complex mixture of contaminants in most sites. (See table 19.2.) The Government Accounting Office (GAO) believes that the National Priority List could reach more than 4000 sites with cleanup costs of around $40 billion. The EPA has the shortest list of sites (2500) that it says should be placed on the National Priority List at a cost of nearly $30 billion. Table 19.3 lists Superfund sites considered priority one.

Table 19.2 Common Contaminants Found at Superfund Sites (in Order of Occurrence)

Chemical	
Lead	Ethylbenzene
Cadmium	Benzo[a]anthracene
Toluene	Bromodichloromethane
Mercury	Polychlorinated biphenyls
Benzene	Toxaphene
Trichloroethylene	

Source: Data from USEPA, Office of Emergency and Remedial Resource, 1999.

By the late 1990s, the Superfund program was still controversial. Millions of dollars have been spent by both the federal government and industry, but

Table 19.3 Proposed National List of Superfund Cleanup Sites in the United States (Priority One)

State/City/County	Site name
Alabama	
Limestone and Morgan	Triana, Tennessee River
Arkansas	
Jacksonville	Vertac, Inc.
California	
Glen Avon Heights	Stringfellow
Delaware	
New Castle	Army Creek
New Castle County	Tybouts Corner
Florida	
Jacksonville	Pickettville Road Landfill
Plant City	Schuylkill Metals
Indiana	
Gary	Midco I
Iowa	
Charles City	Labounty Site
Kansas	
Cherokee County	Tar Creek, Cherokee County
Maine	
Gray	McKin Company
Massachusetts	
Acton	W.R. Grace
Ashland	Nyanza Chemical
East Woburn	Wells G&H
Holbrook	Baird & McGuire
Woburn	Industri-Plex
Michigan	
Swartz Creek	Berlin & Farro
Utica	Liquid Disposal Inc.
Minnesota	
Brainerd Baxter	Burlington Northern
Fridley	FMC
New Brighton/Arden	New Brighton
St. Louis Park	Reilly Tar
Montana	
Anaconda	Anaconda-Anaconda
Silver Bow/Deer Lodge	Silver Bow Creek
New Hampshire	
Epping	Kes-Epping
Nashua	Sylvester, Nashua
Somersworth	Somersworth Landfill
New Jersey	
Bridgeport	Bridgeport Rent & Oil
Fairfield	Caldwell Trucking
Freehold	Lone Pine Landfill
Gloucester Township	Gems Landfill
Mantua	Helen Kramer Landfill
Marlboro Township	Burnt Fly Bog
Old Bridge Township	CPS/Madison Industries
Pittman	Lipari Landfill
Pleasantville	Price Landfill
New York	
Oswego	Pollution Abatement Services
Oyster Bay	Old Bethpage Landfill
Wellsville	Sinclair Refinery
Ohio	
Arcanum	Arcanum Iron & Metal
Oklahoma	
Ottawa County	Tar Creek
Pennsylvania	
Bruin Boro	Bruin Lagoon
Grove City	Osborne
McAdoo	McAdoo
Rhode Island	
Coventry	Picillo Coventry
South Dakota	
Whitewood	Whitewood Creek
Texas	
Crosby	French, Ltd.
Crosby	Sikes Disposal Pits
Houston	Crystal Chemical
La Marque	Motco

Source: Data from U.S. Environmental Protection Agency.

most of the money has involved litigation. The money has paid lawyers but has not paid for cleanup. One of the primary reasons for this is the way CERCLA was written. It provided that anyone who contributed to a specific hazardous-waste site could be required to pay for the cleanup of the entire site regardless of the degree to which they contributed to the problem. Since many industries that contributed to the problem had gone out of business or could not be identified, those who could be identified were asked to pay for the cleanup. Most businesses found it cost-effective to hire lawyers to fight their inclusion in a cleanup effort rather than to pay for the cleanup. Consequently, cleanup has been slow.

In 1997, the U.S. Congress attempted to reauthorize CERCLA. Unable to reach an agreement, CERCLA has been authorized on a year-to-year basis referred to as a "continuing resolution."

ENVIRONMENTAL CLOSE-UP

Computers—A Hazardous Waste

They once cost thousands of dollars and were considered a major investment meant to last. Now computes sell for hundreds of dollars and are obsolete 18 months after they are out of the bubble wrap. Rapid innovation in computer hardware is dramatically cutting the cost and the useful life of modern computers, creating a solid waste problem in the process.

Computers are more than just your household waste. They contain large amounts of substances such as lead, cadmium, mercury, and chromium that can leach into soil and contaminate groundwater or, if incinerated, can be released into the air. The average PC contains five to eight pounds of lead (to protect the user from radiation) in the cathode ray tube screen alone. Circuit boards typically contain several pounds of cadmium, mercury, and chromium. In 1998, more than 20 million computers became obsolete in the United States, but only 11 percent were recycled. It will no doubt get worse as computers get increasingly cheaper, faster, and more disposable. According to the National Safety Council, by 2005, 350 million machines in the U.S. will have reached obsolescence, with at least 55 million of them expected to end up in landfills. Currently, about 75 percent of all computers ever bought in the U.S. are idling in attics, basements, and office closets, thus creating a major backlog. The good news is that most computers are highly reusable, and up to 97 percent of the parts can be recycled, either as upgraded components for use in other computers, or melted down as scrap.

In 2000, Massachusetts became the first state in the U.S. to initiate a ban on the disposal of household computer screens, TV sets, and other glass picture tubes in landfills and incinerators. The state set up six collection centers to handle the items, and cities and towns must now transport the items to those centers. From there, they will either be refurbished or sent on for recycling. Perhaps, in the not-too-distant future, your used computer will be treated like your old car battery or used car oil—when your computer dies or you upgrade, you will simply take it to your local facility for proper disposal.

Toxic Chemical Releases

In 1987, as the result of EPA requirements, industries in the United States had to report toxic chemicals released into the environment. Any industrial plant that released 23,000 kilograms (50,000 pounds) or more of toxic pollutants was required to file a report. Industrial plants that released under 23,000 kilograms (50,000 pounds) were not required to file; thus, the data are incomplete. In 1999, 22,600 reports were filed, covering 332 toxic chemicals.

About 2 billion kilograms (4.4 billion pounds) of toxic chemicals were reported (see figure 19.4) released into the environment by industry in 1998, compared with nearly 2.3 billion (5 billion pounds) in 1987. Another 800 million kilograms (1800 million pounds)—not counted in the overall figure—were sent to municipal-waste treatment centers or private treatment and storage facilities.

According to the 1999 data, among the chemicals routinely emitted from industrial sources were 77 carcinogens. The most widely released cancer-causing chemical—52 million kilograms (115 million pounds)—was dichloromethane, a chemical often used as an industrial solvent and paint stripper. Industry released a total of 128 million kilograms (280 million pounds) of carcinogens, including such chemicals as arsenic, benzene, and vinyl chloride. Chemical and allied industries accounted for about 60 percent of the total releases.

Managing Hazardous Waste

In the past, the management of hazardous waste was always added on to the end of the industrial process. The effluents from pipes or smokestacks were treated to reduce their toxicity or concentration. For 1999, the EPA reported that the United States produced about 225 million tonnes (250 million U.S. tons) of hazardous waste. This was 52 million tonnes (57 million U.S. tons) less than in 1991. In recent years, it has become obvious that a better way to deal with the problem of hazardous waste is to not produce it in the first place. To this end, the EPA and regulatory agencies in other countries have emphasized pollution prevention and waste minimization. Strong regulatory control requires that industries report the hazardous wastes they produce and that the wastes be stored, transported, and disposed of properly.

The EPA now fosters a **pollution-prevention hierarchy** that emphasizes reducing the amount of hazardous waste produced. This involves the following strategy:

- First—reduce the amount of pollution at the source.
- Second—recycle wastes wherever possible.
- Third—treat wastes to reduce their hazard or volume.
- Fourth—dispose of wastes on land or incinerate them as a last resort. (See figure 19.5.)

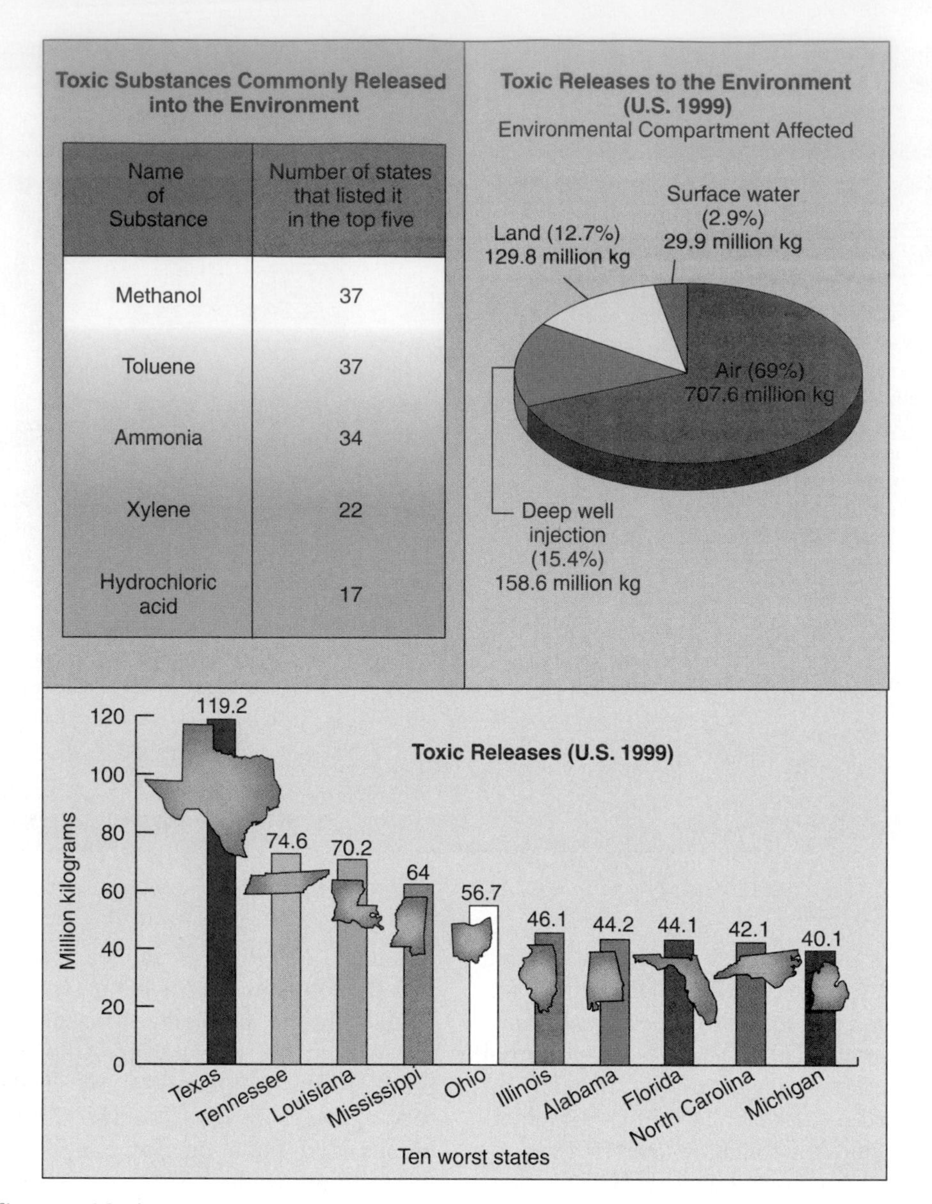

Toxic Substances Commonly Released into the Environment

Name of Substance	Number of states that listed it in the top five
Methanol	37
Toluene	37
Ammonia	34
Xylene	22
Hydrochloric acid	17

figure 19.4 **Toxic Releases** Toxic substances commonly released to the environment in the U.S. and the amount for selected states.

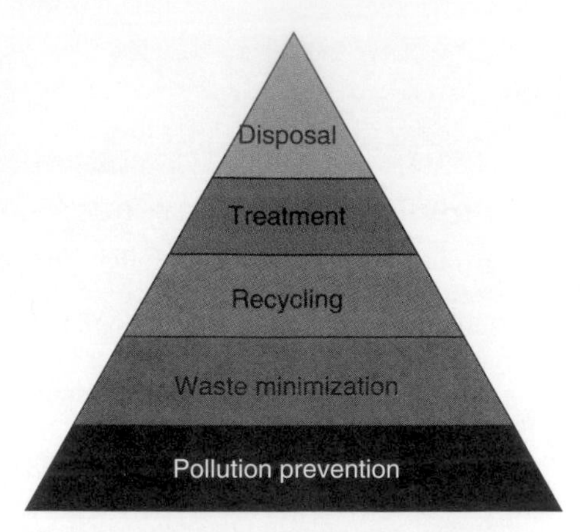

figure 19.5 **Pollution-Prevention Hierarchy** The simplest way to deal with hazardous wastes is to not produce them in the first place. The pollution-prevention hierarchy stresses reductions in the amount of hazardous waste produced by employing several different strategies.

Pollution Prevention

Pollution prevention encourages changes in the operations of business and industry that prevent hazardous wastes from being produced in the first place. Many of these actions are simple to perform and cost little. Primary among them are activities that result in fewer accidental spills, leaks from pipes and valves, loss from broken containers, and similar mishaps. These reductions often can be achieved through better housekeeping and awareness training for employees at little cost. Many industries actually save money because they need to buy less raw material because less is being lost.

Pollution prevention can be applied in unusual ways. In 2000, the United States Army announced that they would begin issuing an environmentally friendly "green bullet" that contains no lead. The bullet contains a nonpolluting tungsten core instead of lead, which contaminates the soil and air around firing ranges. The U.S. military uses between 300 million and 400 million rounds of small-caliber ammunition each year. The military plans to phase out all use of lead by 2003. Lead contamination has closed hundreds of outdoor firing ranges on military bases across the U.S. In 1998, when lead concentrated in firing berms was found to be leaching into Cape Cod's water supply, the Environmental Protection Agency ordered the Massachusetts Military Reservation to stop livefire training.

Waste Minimization

Waste minimization involves changes that industries could make in the way they manufacture products, changes that would reduce the waste produced. For example, it may be possible to change a process so that a solvent that is a hazardous material is replaced with water, which is not a hazardous material. This is an example of source reduction: any change or strategy that reduces the amount of waste produced.

Another strategy is to use the waste produced in a process in another aspect of the process, thus reducing the amount of waste produced. For example, water used to clean equipment might be included as a part of the product rather than being discarded as a contaminated waste.

Another technique that can be used to reduce the amount of waste produced is to clean solvents used in processes. Using a still to purify solvents results in a lower total volume of hazardous waste being produced because the same solvent can be used over and over again.

The simple process of allowing water to evaporate from waste can reduce the total amount of waste produced. Obviously, the hazardous components of the waste are concentrated by this process.

Recycling of Wastes

Often it is possible to use a waste for another purpose and thus eliminate it as a waste. Many kinds of solvents can be burned as a fuel in other kinds of operations. For example, waste oils can be used as fuels for power plants, and other kinds of solvents can be burned as fuel in cement kilns. Care needs to be taken that the contaminants in the oils or solvents are not released into the environment during the burning process, but the burning of these wastes destroys them and serves a useful purpose at the same time.

Similarly, many kinds of acids and bases are produced as a result of industrial activity. Often these can be used by other industries that have a need for them. Ash or other solid wastes can often be incorporated into concrete or other building materials and therefore do not require disposal. Thus, the total amount of waste is reduced.

The exportation of hazardous wastes for recycling from industrialized nations to developing nations raises other questions. Those who support the export of hazardous substances for recycling argue that this practice offers two major benefits: reducing the quantity of such substances that get into the environment through final disposal and slowing down the depletion of natural resources. This argument is undoubtedly correct, provided the receiving country has the proper recycling facilities and adequate environmental standards. Environmentally beneficial trade in hazardous wastes ordinarily requires that there be an established market for these wastes and that the trade be economically viable.

Critics of the export of hazardous wastes for recycling argue that the conditions required for it to be beneficial are unlikely to be fulfilled in practice, especially in countries without the necessary infrastructure and technical capacity. They also point out that the factors that influence the market for recyclable wastes, such as the cost of the available recovery options, are not necessarily conducive to sound waste management. Finally, they object to the suggestion that the export of hazardous wastes for recycling be subject to less stringent rules than the export of such wastes for disposal. It is pointed out that this will discourage waste reduction in the countries of origin and possibly lead to fake recycling schemes, such as the use of the recycling label for disposal operations that would otherwise be prohibited.

Treatment of Wastes

Wastes can often be treated in such a way that their amount is reduced or their hazardous nature is modified. Dangerous acids and bases can be reacted with one another to produce materials that are not hazardous.

Hazardous wastes that are biodegradable can be subjected to the actions of microorganisms that destroy the hazardous chemicals. Many kinds of organic molecules can be handled in this way.

Incineration (thermal treatment) can be used to treat a variety of kinds of wastes, although many people are skeptical of this technique because they feel that toxic materials may be escaping from the smokestacks. Wastes are heated in a flame-powered incinerator. Under controlled conditions, incineration can destroy 99.999 percent of organic wastes, and hazardous waste incinerators must burn 99.9999 percent of certain hazardous materials. Incineration accounts for the disposal of only about 2 percent of the hazardous wastes in North America. In Europe, the amount of hazardous wastes destroyed by incineration is higher, but still amounts to less than 50 percent. The relatively high costs of incineration (compared with landfills) and concerns for the safety of surrounding areas in case of accidents have kept incineration from becoming a major method of treatment or disposal.

Air stripping is sometimes used to remove volatile chemicals from water. Volatile chemicals, which have a tendency to vaporize easily, can be forced out of liquid when air passes through it. Steam stripping works on the same principle, except that it uses heated air to raise the temperature of the liquid and force out volatile chemicals that ordinary air would not. The volatile compounds can be captured and reused or disposed of.

Carbon absorption tanks contain specifically activated particles of carbon to treat hazardous chemicals in gaseous and liquid waste. The carbon chemically combines with the waste or catches hazardous particles just as a fine wire mesh catches grains of sand. Contaminated carbon must then be disposed of or cleaned and reused.

Precipitation involves adding special materials to a liquid waste. These bind to hazardous chemicals and cause them to precipitate out of the liquid and form large particles called floc. Floc that settles can be separated as sludge; floc that remains suspended can be filtered and the concentrated waste can be sent to a hazardous-waste landfill.

Land Disposal

When all other options have been exhausted, any remaining hazardous wastes are typically disposed of on land. (See table 19.4.) For over 80 percent of their hazardous wastes, North America, Europe, and Japan still rely principally on six methods of disposal:

1. Deep-well injection into porous geological formations or salt caverns.
2. Discharge of treated and untreated liquids into municipal sewers, rivers, and streams.
3. Placement of liquid wastes or sludges in surface pits, ponds, or lagoons.
4. Storage of solid wastes in specially lined dumps covered by soil.
5. Storage of liquid and solid wastes in underground caverns and abandoned salt mines.
6. Sending wastes to sanitary landfills not designated for toxic or hazardous wastes.

Table 19.4 Hazardous-Waste Management Methods, United States

Management method	Share of total waste managed (percent)
Land disposal[1]	67
Discharge to sewers, rivers, streams	22
Distillation for recovery of solvents	4
Burning in industrial boilers	4
Chemical treatment by oxidation	1
Land treatment of biodegradable waste	1
Incineration	1
Recovery of metals through ion exchange	less than 1
Total	**100**

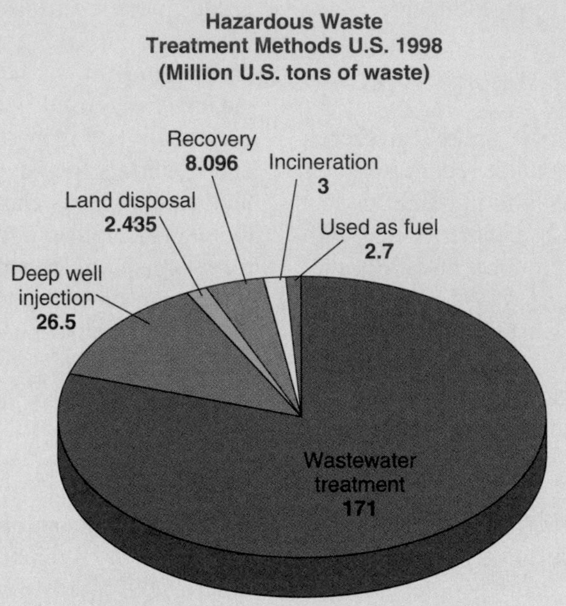

Hazardous Waste Treatment Methods U.S. 1998 (Million U.S. tons of waste)

Recovery **8.096**
Incineration **3**
Land disposal **2.435**
Used as fuel **2.7**
Deep well injection **26.5**
Wastewater treatment **171**

[1]Includes injection wells (25 percent of total), surface impoundments (19 percent), hazardous-waste landfills (13 percent), and sanitary landfills (10 percent).

Source: Data from U.S. Environmental Protection Agency, Office of Solid Waste.

There are techniques that reduce the chance that hazardous materials will escape from these locations and become a problem for the public. Immobilizing a waste puts it into a solid form that is easier to handle and less likely to enter the surrounding environment. Waste immobilization is useful for dealing with wastes, such as certain metals, that cannot be destroyed. Two popular methods of immobilizing waste are fixation and solidification. Engineers and scientists mix materials such as fly ash or cement with hazardous wastes. This either "fixes" hazardous particles, in the sense of immobilizing them or making them chemically inert, or "solidifies" them into a solid mass. Solidified waste is sometimes made into solid blocks that can be stored more easily than can a liquid.

Hazardous-Waste Management Choices

Today, the two most common methods for disposing of hazardous wastes are land disposal and incineration. The choice between these two methods involves both economic decisions and acceptance by the public. In North America, there is abundant land available for land disposal, making it the most economical and most widely used method. In Europe and Japan, where land is in short supply and is expensive, incineration is more economical and is a major method for dealing with hazardous waste. Because of concerns about the emissions from incinerators, significant amounts of hazardous waste are incinerated at sea on specially designed ships.

Laws and regulations dealing with hazardous-waste disposal are driving industrial behavior toward pollution prevention and waste minimization. In addition, because the costs of safe disposal are mounting, waste-handling firms—both private and public—are looking for better and cheaper ways to treat and dispose of hazardous wastes. Strong, enforceable laws have eliminated the economic incentives to pollute. Unfortunately, as strong laws have been enacted, some small companies that were unable or, more likely, unwilling to properly dispose of their wastes have turned to illegal nighttime dumping.

The environmental costs of not managing hazardous wastes, as witnessed in virtually every industrialized country, are astronomical. And because major generators of hazardous wastes remain liable for past mistakes, economic and regulatory incentives for complying with hazardous-waste regulations should continue to encourage responsible management.

International Trade in Hazardous Wastes

The growth in uncontrolled transboundary movements of hazardous wastes has been one of the most contentious environmental issues to appear on the international political agenda. There is particular concern about rich, industrialized countries exporting such wastes to poorer, developing countries lacking the administrative and technological resources to safely dispose of or recycle the waste. For example, in 1999, between 3000 and 4000 tonnes (3300 and 4400 U.S. tons) mercury-contaminated concrete waste packed in plastic bags was found in an open dump in a small town in Cambodia. The waste, labeled as "construction

GLOBAL PERSPECTIVE

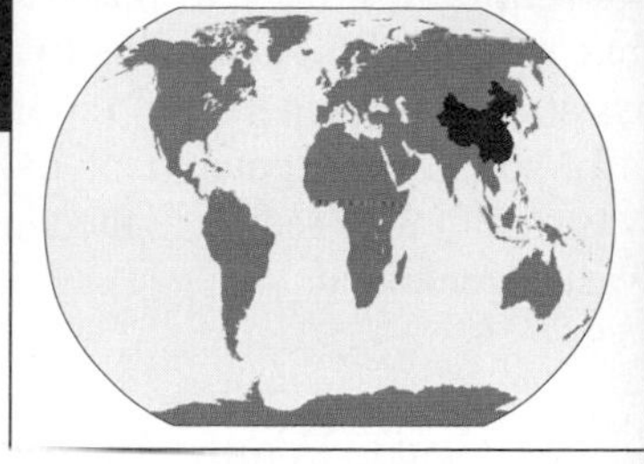

Hazardous Wastes and Toxic Materials in China

Large areas of land are currently utilized in China for the uncontrolled disposal of industrial wastes. Approximately 600 million tonnes (6600 U.S. tons) industrialized waste, of which 50 to 70 percent are hazardous, are generated in China annually. Estimates indicate that in recent years, a total of 5.9 billion tonnes (6500 U.S. tons) industrial waste, occupying 540 million cubic meters (19,000 million cubic feet), have been improperly stored or discarded. The majority of the waste is simply piled on unprotected areas, which causes leaching to surface and groundwater bodies. As a result, environmental accidents are prevalent. For instance, a chromium-residue disposal site in Jinzhou caused groundwater pollution in a 12.5-square-kilometer area (4.8 square miles); as a result, water from 1800 wells in nine villages is no longer potable.

The production and use of chemicals is developing rapidly in China. More than 30,000 classes of chemicals are now produced, of which many are toxic. During production, transportation, storage, and use, many releases and spills occur. For example, in 1995, the toxic chemical storage in Shenzhen exploded, causing significant damage to life, property, and the environment.

Management of hazardous and toxic materials in China is still in its early stages, therefore treatment and disposal technologies are primitive and equipment is poor. None of China's current hazardous-waste disposal sites meet environmental standards. In part, this is due to a lack of funds and management capabilities. A demonstration project, however, is currently under way in China. It is hoped that this will assist them in developing a management system for hazardous and toxic materials.

The "Law of Pollution Prevention and Control of Solid Wastes" has been made a priority item of legislation by the National People's Congress. Research on hazardous-waste management and disposal have been classified as key items in national scientific and technological development plans. Experiments with solid waste declaration and registration, and waste exchanges, are being performed, and standard protocols for chemical testing, toxicity evaluation of synthetic chemicals, laboratory analysis, and risk assessment have begun to be formulated. The long- and short-term objectives of the new project include the following:

- Formulate and strengthen China's hazardous- and toxic-materials control laws and regulations, and establish criteria for sound environmental management of wastes.
- Establish technical support system for hazardous-waste management.
- Design demonstration projects for hazardous-waste treatment and disposal.
- Formulate laws and regulations, criteria, and policies for toxic-chemicals management.
- Construct in Beijing a hazardous-waste incineration plant with an annual capacity of 3000 tonnes (3300 U.S. tons).

This project will introduce hazardous- and toxic-materials control laws and regulations, antipollution criteria, and mitigation measures. The capabilities for the management and control of hazardous materials will be developed and regulatory enforcement will be initiated. It is hoped that the establishment of declaration, registration, licensing, and wastes exchange for hazardous-waste procedures, will result in a reduction in the volume of waste generated in China while providing an incentive for recovery and reuse.

The hazardous-waste disposal demonstration project will provide a model, and increased capabilities and experience can be utilized by China in establishing regional central-disposal facilities for hazardous wastes. Dissemination of the experience gained in the demonstration project should produce a significant improvement in waste-disposal management strategies, resulting in greater awareness, regulation, and environmental protection.

waste" on import documents, came from a Taiwanese petrochemical company. In this case, the waste was tracked down and returned to its point of origin. Unfortunately, most such cases are unreported or detected. International awareness of the problems associated with the trade in hazardous wastes has increased noticeably owing to several factors: the growing amounts of such wastes being generated; closure of old waste disposal facilities and political opposition to the development of new ones; and the dramatically higher costs associated with the disposal of hazardous wastes in industrialized countries (and thus the potential to earn profits by exporting such wastes to developing countries with low disposal costs). The debate over controlling transboundary hazardous-waste movements culminated in 1989 with the creation of the Basel Convention.

The Basel Convention was negotiated under the auspices of the United Nations Environmental Program between 1987 and 1989. The objectives of the convention are to minimize the generation of hazardous wastes and to control and reduce their transboundary movements to protect human health and the environment. To achieve these objectives, the convention prohibits exports of hazardous waste to Antarctica, to countries that have banned such imports as a national policy, and to non-parties (unless those transactions are subject to an agreement that is as stringent as the Basel Convention). Though not part of the original agreement, there is now a broad ban on the export of hazardous wastes from the Northern to the Southern Hemisphere. The waste transfers that are permitted under the Basel regime are subject to the mechanism of prior notification and consent, which requires parties to not export hazardous wastes unless a "competent authority" in the importing country has been properly informed and has consented to the trade.

While the Basel regime may not be perceived as being as successful or significant as some other multilateral environmental agreements, it remains an important part of the international community's attempt to protect the global environment and human health from hazardous materials. Now that the convention has been in place for more than a decade, it is beginning to focus on assisting parties with the environmentally sound management of hazardous wastes and with reducing the amount of wastes generated.

Hazardous-Waste Program Evolution

Fundamentally, the goal of a hazardous-waste management program is to change the behavior of those who generate hazardous wastes so that they routinely store, transport, treat, and dispose of them in an environmentally safe manner. The focus on hazardous-waste management (HWM) typically comes in the second phase of countries' environmental programs, after efforts to address more immediate threats to public health, such as safe drinking water. In the early years of many countries' HWM programs, uncontrolled disposal of hazardous waste is the norm. Few, if any, proper treatment and disposal facilities exist. Information on who is generating waste, what types are being generated, and where it is being disposed of is meager or nonexistent.

The transition from an unregulated environment to a regulated one is complex, but HWM programs typically evolve through the following major stages: identifying the problem and enacting legislation, designating a lead agency, promulgating rules and regulations, developing treatment and disposal capacity, and creating a compliance and enforcement program. Each of these stages takes a number of years, and, at each stage, there are many difficult issues to be resolved. Denmark, Germany, and the United States began this process by passing their first major hazardous waste laws between 1972 and 1976; Canada did so in 1980. In the decade that followed, all four of these countries developed hazardous-waste regulations and requirements, so that by the end of the 1980s, their regulatory systems were largely operational. Laws and policies developed during the 1990s have focused mainly on waste minimization and recycling as well as on harmonization with international standards and the cleanup of contaminated sites.

The more recent evolution of regulatory programs in Hong Kong, Indonesia, Malaysia, and Thailand has followed a similar pattern. By the early 1980s, all of these countries had enacted some form of environmental legislation providing at least limited authority to regulate hazardous waste. However, hazardous-waste management received little attention until the late 1980s, after periods of rapid economic growth and the expansion of the countries' manufacturing sectors. From 1989 to 1998, all of these countries passed major new legislation that addressed hazardous waste or developed regulations outlining comprehensive programs: Malaysia did so in 1989, Hong Kong in 1991, Thailand in 1992, and Indonesia in 1995–98. All four countries now have at least one modern hazardous waste treatment, storage, and disposal facility.

While all countries pass through the same stages of program development, no two countries follow precisely the same path. Differences in geography, demographics, industrial profile, politics, and culture lead countries to make different choices at each stage.

Summary

Public awareness of the problems of hazardous substances and hazardous wastes is relatively recent. The industrialized countries of Europe and North America began major regulation of hazardous materials only during the past 20 years, and most developing countries exercise little or no control over such substances. As a result, many countries are living with serious problems from prior uncontrolled dumping practices, while current systems for management of hazardous and toxic waste remain incomplete and incapable of even identifying all hazardous waste.

A number of fundamental problems are involved in hazardous-waste management. First, there is no agreement as to what constitutes a hazardous waste. Moreover, little is known about the amounts of hazardous wastes generated throughout the world. The issue is further complicated by our limited understanding of the health effects of most hazardous wastes and the fact that large numbers of potentially hazardous chemicals are being developed faster than their health risks can be evaluated.

Hazardous-waste management must move beyond burying and burning. Industries need to be encouraged to generate less hazardous waste in their manufacturing processes. Although toxic wastes cannot be entirely eliminated, technologies are available for minimizing, recycling, and treating wastes. It is possible to enjoy the benefits of modern technology while avoiding the consequences of a poisoned environment. The final outcome rests with governmental and agency policy makers, as well as with an educated public.

ISSUES & ANALYSIS

Love Canal

Love Canal, near Niagara Falls, New York, was constructed as a waterway in the nineteenth century. It was subsequently abandoned and remained unused for many years. In the 1930s, it became an industrial dump. The Hooker Chemical Company purchased the area in 1947 and also used it as a burial site for 20,000 tonnes (22,000 U.S. tons) of chemicals.

Hooker, later to become part of Occidental Petroleum, then sold the property to the local government for one dollar. A housing development and an elementary school were constructed on the site. Soon after the houses were constructed, people began to complain about chemicals seeping into their basements. In 1978, 80 different chemicals were found in this seepage. Approximately one dozen probable carcinogens were identified among these chemicals.

That same year, as a result of these findings, $27 million in government funds were appropriated to purchase homes and permanently relocate 237 families. The funds also provided for the construction of a series of ditches to contain the chemicals and a clay cap to prevent the fumes from entering the atmosphere. But the problems continued.

The remaining 710 families in the Love Canal area were not satisfied with the government's approach to the problem. They cited the fact that women in the area had a 50-percent-higher rate of miscarriages. Of 17 reported pregnancies in the area during 1979, two children were born normal, nine had defects, two were still born, and four pregnancies ended in miscarriage. In addition to the abnormalities in birth, there are other biological problems in the Love Canal area.

Neurologists determined that the speed of the nerve impulses in 37 residents who were examined was slower than normal. They stated that chemical exposure could have caused this damage. In 1980, the EPA released the findings of a study that found that 11 out of 36 residents tested in the Love Canal area had broken chromosomes, which are linked to cancer and birth defects. As a result, the federal government released $5 million to temporarily relocate Love Canal residents to motels or other quarters.

By 1990, a $150-million cleanup effort had sealed off the leaky dump, demolished 238 homes nearest the chemical graveyard, and scoured toxins from neighborhood storm sewers and streams (the two major sources of danger to area homes).

In early 1991, some families began to move back into the area. The Love Canal Area Revitalization Agency is planning to sell 236 homes in Love Canal (renamed Black Creek Village). One of the incentives is that the houses can be purchased at a low cost. The families are confident that the houses are safe, but many environmental groups oppose them. In March of 1998, the last of 2300 families received compensation for medical claims. Individual compensation ranged from $400,000 to as little as $83.

- Who is responsible for providing treatment for the physical and mental problems experienced by the residents in this community?
- There is a question no one ever seems to ask: Why were permits ever awarded to construct a thousand-unit housing development on top of a site known to contain 20,000 tonnes (22,000 U.S. tons) toxic wastes?

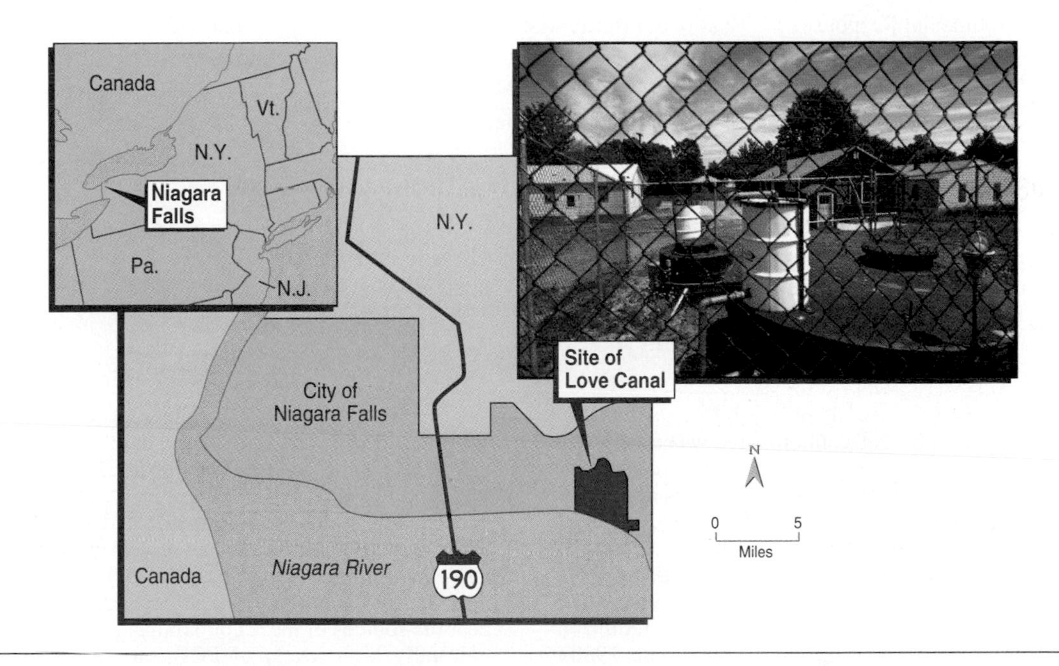

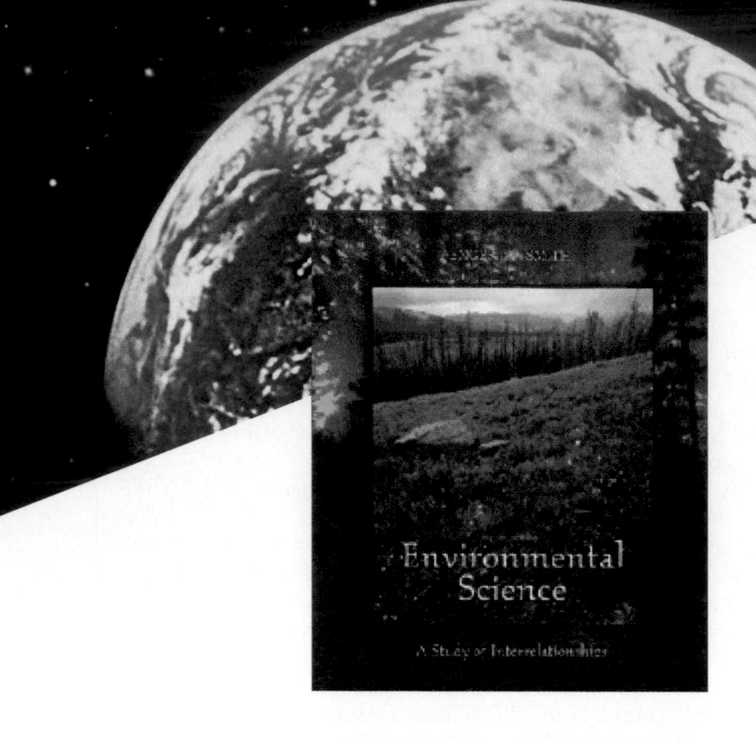

Interactive Exploration

Check out the website at

http://www.mhhe.com/environmentalscience

and click on the cover of this textbook for interactive versions of the following:

KNOW THE BASICS

acute toxicity *423*
chronic toxicity *423*
Comprehensive Environmental Response, Compensation, and Liability Act (CERCLA) *426*
corrosiveness *420*
hazardous *421*
hazardous substances *420*
hazardous wastes *421*
ignitability *420*
incineration *431*
LD_{50} *423*
National Priority List *427*
nonpersistent pollutant *424*
persistent pollutant *424*
pollution-prevention hierarchy *429*
reactivity *420*
Resource Conservation and Recovery Act (RCRA) *421*
Superfund *426*
synergism *424*
threshold level *423*
toxic *421*
toxicity *421*
waste minimization *430*

On-line Flashcards

Electronic Glossary

IN THE REAL WORLD

What should be done with TNT-contaminated wastewater that is left when an army ammunition plant closes? Grow weeds in the water? Check out the relationship between explosives and plants such as American pondweed in Phytoremediation/wetlands Treatment at the Iowa Army Ammunition Plant.

What does a government do with tons of anthrax bacteria, plague, typhus, smallpox, and other disease-causing organisms? The Soviet Union *thought* these toxins were destroyed after the biological weapons program was discontinued in the late 1980s. Now it appears that some of the toxins weren't actually destroyed. Find out what is happening in Biological Weapons Waste Site Threatens to Spread Diseases in Aral Sea Region.

What if you or your family lived near chemical plants that were targeted during the NATO bombing campaign in Yugoslavia? When feeding yourself and your family, environmental cleanup is not always a choice. See what local residents must face in Environmental Costs of Bombing in Yugoslavia.

Do you live near a contaminated site? What are your risks? Even if you do not live near industrial sites, the toxic buildup from industrial effluents can affect you. PCB contamination is adversely affecting Norway's polar bears even though they are not located near the sources of these contaminants. For more information about unusually high levels of PCBs in Arctic humans and animals, read Norway's Polar Bears Suspected Victims of PCB Contamination.

What if you live close to the original site of PCB contamination? Read PCB Contamination in the Fox River for the history and latest news on plans for cleanup.

Wildlife in the Lake Bogoria National Reserve in Kenya is adversely affected by toxic chemical releases from nearby sewage and industrial sources. Check out Flamingos Die in African Rift Valley Lakes for the details of how the buildup of contaminants is killing hundreds of birds.

Sometimes the release of toxic chemicals is not an accident. Read British Petroleum/Amoco Admits to Dumping Toxic Waste on Alaska's North Slope to see how this company managed their toxic wastes.

TEST PREPARATION

Review Questions

1. Explain the problems associated with hazardous-waste dump sites and how such sites developed.
2. Distinguish between acute and chronic toxicity.
3. Give two reasons why regulating hazardous wastes is difficult.
4. In what ways do hazardous wastes contaminate the environment?
5. Describe how hazardous wastes contaminate groundwater.
6. Why is there often a problem in linking a particular chemical or hazardous waste to a particular human health problem?
7. Describe what is meant by the U.S. National Priority List.
8. Describe five technologies for managing hazardous wastes.
9. What is meant by pollution prevention and waste minimization?
10. Describe the pollution-prevention hierarchy.
11. What are RCRA and CERCLA? Why is each important for managing hazardous wastes?

Critical Thinking Questions

1. Scientists at the EPA have to make decisions about thresholds in order to identify which materials are toxic materials. What thresholds would you establish for various toxic materials? What is your reasoning for establishing the limits you do?
2. Go to the EPA's web site (www.epa.gov/enviro/html/ef_overview.html) and identify the major releasers of toxic materials in your area. Were there any surprises? Are there other releasers of toxic materials that might not be required to list their releases?
3. According to the textbook, in North America alone there are over 25,000 abandoned and uncontrolled hazardous-waste dumps. Many were abandoned before the RCRA law of 1976 was passed. Who should be responsible for cleaning up these dumps?
4. Look at the textbook's discussion called "Hazardous-Waste Dumps—A Legacy of Abuse." Do the authors present the information from a particular point of view? What other points of view might there be to this issue? What information do you think these other viewpoints would provide?
5. Many economically deprived areas, Native American reservations, and developing countries that need an influx of cash, have agreed, over significant local opposition, to site hazardous-waste facilities in their areas. What do you think about this practice? Should "outsiders" have a say in what happens within these sovereign territories?
6. After reading about the problem with hazardous wastes and toxic materials in China, do you think that the United States, or any other country, should have the right to intervene if another country is creating significant environmental damage? Why?

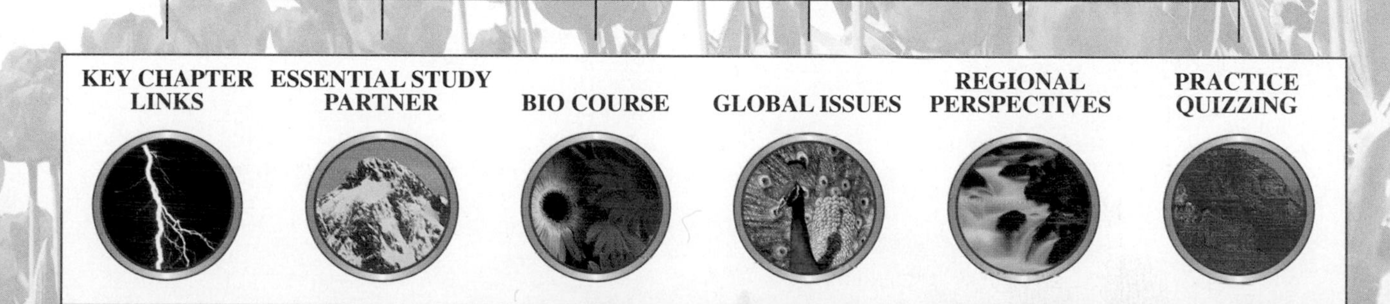

CHAPTER

Environmental Policy and Decision Making

Government
Bureaucracy
Public involvement
Priorities

Objectives

After reading this chapter, you should be able to:

- Explain how the executive, judicial, and legislative branches of the U.S. government interact in forming policy.
- Understand how environmental laws are enforced in the United States.
- Describe the forces that led to changes in environmental policy in the United States during the past three decades.
- Understand the history of the major U.S. environmental legislation.
- Understand why some individuals in the United States are concerned about environmental regulations.
- Understand what is meant by "green" politics.
- Describe the reasons why environmentalism is a growing factor in international relations.
- Understand the factors that could result in "eco-conflicts."
- Understand why it is not possible to separate politics and the environment.
- Explain how citizen pressure can influence governmental environmental policies.

Chapter Outline

New Challenges for a New Century
Learning from the Past
Thinking about the Future
Defining the Future
The Development of Environmental Policy in the United States
  Environmental Backlash—The Wise Use Movement
  The Changing Nature of Environmental Policy
Environmental Policy and Regulation
The Greening of Geopolitics
Environmental Close-Up: *Changing the Nature of Environmental Regulation—The Safe Drinking Water Act*
Global Perspective: *Eco-Terrorism*
International Environmental Policy
Global Perspective: *Environmental Policy and the European Union*
Global Perspective: *Overview of an International Organization—The International Whaling Commission (IWC)*
Global Perspective: *Eco-Labels*
  New International Instruments
It All Comes Back to You

New Challenges for a New Century

We live in remarkable times. This is an era of rapid and often bewildering alterations in the forces and conditions that shape human life. This is evident both in the altered nature of geopolitics in the post–Cold War era and in the growing understanding of the relationship between human beings and the natural world. This relationship varies, however, between the developing and developed countries of the world. One of the major challenges of the foreseeable future will be centered around worldwide environmental impact as developing nations evolve economically.

The end of the Cold War has been accompanied by the swift advance of democracy in places where it was previously unknown and an even more rapid spread of market-based economics. The authority of central governments is eroding, and power has begun to shift to local governments and private institutions. In some countries, freedom and opportunity are flourishing, while in others these changes have unleashed the violence of old conflicts and new ambitions.

Internationally, trade, investment, information, and even people flow across borders largely outside of governmental control. Domestically, deregulation and the shift of responsibilities from federal to state and local governments are changing the relationships among levels of government and between government and the private sector.

Communications technology has enhanced people's ability to receive information and influence events that affect them. This has sparked explosive growth in the number of organizations, associations, and networks formed by citizens, businesses, and communities seeking a greater voice for their interests. As a result, society outside of government—civil society—is demanding a greater role in governmental decisions, while at the same time impatiently seeking solutions outside government's power to decide.

figure 20.1 Struggling to survive in places that can no longer sustain them, growing populations overfish, overharvest, and overgraze.

But technological innovation is changing much more than communication. It is changing the ways in which we live, work, produce, and consume. Knowledge has become the economy's most important and dynamic resource. It has rapidly improved efficiency as those who create and sell goods and services substitute information and innovation for raw materials. During the past 20 years, the amount of energy and natural resources the U.S. economy uses to produce each constant dollar of output has steadily declined, as have many forms of pollution. When U.S. laws first required industry to control pollution, the response was to install cleanup equipment. The shift to a knowledge-driven economy has emphasized the positive connection among efficiency, profits, and environmental protection and helped launch a trend in profitable pollution prevention. More and more people today now understand that pollution is waste, waste is inefficient, and inefficiency is expensive.

Even as their access to information and to means of communication has increased, citizens of the more developed nations are becoming cynical about, and frustrated with, traditional political arrangements that no longer seem responsive to their needs. The confidence of many citizens in the large institutions that affect their lives—such as business; government; the media; and environmental, labor, and civic organizations—is eroding. Individual citizens have lost faith in their ability to influence events and have surrendered to apathy, or worse, to anger.

Since the end of World War II, the world's economic output has increased substantially, allowing widespread improvements in health, education, and opportunity, but also creating growing disparities between rich and poor. Even in the highly developed nations, the gap between rich and poor is widening.

Tomorrow's world will be shaped by the aspirations of a much larger global population. The number of people living on Earth has doubled in the last 50 years. Growing populations demand more food, goods, services, and space. Where there is scarcity, population increase aggravates it. Where there is conflict, rising demand for land and natural resources exacerbates it. Struggling to survive in places that can no longer sustain them, growing populations overfish, overharvest, and overgraze. (See figure 20.1.)

As we begin the new century, it is important that we recognize that economic, environmental, and social goals are integrally linked and that we develop policies that reflect that interrelationship. Thinking narrowly about jobs, energy, transportation, housing, or ecosystems—as if they were not connected—creates new problems even as it attempts to solve old ones. Asking the wrong questions is a sure way to get misleading answers that result in short-term remedies for symptoms, instead of cures for long-term problems.

All of this will require new modes of decision making, ranging from the local to the international level. While trend is not always destiny, the trend

figure 20.2 Scenes such as these were very common in North America only a short time ago. Fortunately, for the most part, such photos are today only historic in nature. Positive change is possible. As is mentioned in the text, trend is not destiny.

that has been evolving over the past several years has been toward more collaborative forms of decision making. Perhaps such collaborative structures will involve more people and a broader range of interests in shaping and making public policy. It is hoped that this will improve decisions, mitigate conflict, and begin to counteract the corrosive trends of cynicism and civic disengagement that seem to be growing.

More collaborative approaches to making decisions can be arduous and time consuming, and all of the players must change their customary roles. For government, this means using its power to convene and facilitate, shifting gradually from prescribing behavior to supporting responsibility by setting goals, creating incentives, monitoring performance, and providing information.

For their part, businesses need to build the practice and skills of dialogue with communities and citizens, participating in community decision making and opening their own values, strategies, and performance to their community and the society.

Advocates, too, must accept the burdens and constraints of rational dialogue built on trust, and communities must create open and inclusive debate about their future.

Does all of this sound too idealistic? Perhaps it is; however, without a vision for the future, where would we be? As was stated previously, trend is not destiny. In other words, we are capable of change regardless of the status quo. This is perhaps nowhere more important than in the world of environmental decision making. (See figure 20.2.)

Learning from the Past

For the past quarter century, the basic pattern of environmental protection in economically developed nations has been to react to specific crises. Institutions have been established, laws passed, and regulations written in response to problems that already were posing substantial ecological and public health risks and costs, or that already were causing deep-seated public concern.

The United States is no exception. The U.S. Environmental Protection Agency (EPA) has focused its attention almost exclusively on present and past problems. The political will to establish the agency grew out of a series of highly publicized, serious environmental problems, like the fire on the Cuyahoga River in Ohio, smog in Los Angeles, and the near extinction of the bald eagle. During the 1970s and 1980s, Congress enacted a series of laws intended to solve these problems, and the EPA, which was created in 1970, was given the responsibility for enforcing most environmental laws.

Despite success in correcting a number of existing environmental problems, there has been a continuing pattern of not responding to environmental problems until they pose immediate and unambiguous risks. Such policies, however, will not adequately protect the environment in the future. People are recognizing that the agencies and organizations whose activities affect the environment must begin to anticipate future environmental problems, and then take steps to avoid them. One of the most important lessons learned during the past quarter century of environmental history is that the failure to think

figure 20.3 The plight of these fishermen, resulting from the collapse of the cod fishery in the northeast United States and Canada and the salmon fishery in the northwest United States and Canada, is an environmental debt inherited from past generations of abuse and misuse.

about the future environmental consequences of prospective social, economic, and technological changes may impose substantial and avoidable economic and environmental costs on future generations.

Thinking about the Future

Thinking about the future is more important today than ever before, because the accelerating rate of change is shrinking the distance between the present and the future. Technological capabilities that seemed beyond the horizon just a few years ago are now outdated. Scientific developments and the flow of information are accelerating. For example, who would have envisioned cellular phones, voice-activated computers, or a widely used Internet only 10 years ago? Similarly, the environmental effects of changes in global economic activity are being felt more rapidly by both nations and individuals. Examples include the collapse of the cod-fishing industry in the Maritimes of Canada and the northeastern United States; the threatened salmon-fishing industry in the Pacific northwest of the United States and British Columbia, Canada; the severe air pollution problems in Mexico; and the shortage of safe drinking water supplies in Russia.

Initiating thought and analysis well in advance of anticipated change can shorten the time needed to respond to such change or allow us to avoid the problem entirely. Because some damage is irreversible, response time is critical.

Thinking about the future is valuable also because the cost of avoiding a problem is often far less than the cost of solving it later. The U.S. experience with hazardous-waste disposal provides a compelling example. Some private companies and federal facilities undoubtedly saved money in the short term by disposing of hazardous wastes inadequately, but those savings were dwarfed by the cost of cleaning up hazardous-waste sites years later. In that case, foresight could have saved private industry, insurance companies, and the federal government (i.e., taxpayers) billions of dollars, while reducing exposure to pollutants and public anxieties in the affected communities.

Thinking about the future has another value, one that goes beyond the immediate costs and benefits of environmental protection. Environmental foresight can preserve the environment for future generations. When one generation's behavior necessitates environmental remediation in the future, an environmental debt is bequeathed to future generations just as surely as unbalanced government budgets bequeath a burden of financial debt. (See figure 20.3.)

By anticipating environmental problems, and by taking steps now to prevent them, the present generation can minimize the environmental and financial debts that its children will incur.

Today, we face new classes of environmental problems that are more diffuse than those of the past and thus demand different approaches. Since the first Earth Day in 1970, the vast majority of the significant "point" sources of air and water pollution—large industrial facilities and municipal sewage systems—that once spewed untreated wastes into the air, rivers, and lakes have been controlled. The most important remaining sources of pollution are diffuse and widespread: sediment, pesticides, and fertilizers that run off farmland; oil and toxic heavy metals that wash off city streets and highways; and air pollutants from automobiles, outdoor grills, and woodstoves. Pollution from these sources cannot always be controlled with sewage treatment plants or the same regulatory techniques used to check emissions from large industries. To make matters more complicated, we now have environmental problems on a global scale—biodiversity loss, ozone depletion, and climate change. These problems will require cooperative international responses. We are also recognizing that controlling pollutants alone, no matter how successful, will not achieve an environmentally sustainable economy, since many global concerns are related to the size of the human population and the unequal distribution of resources.

Defining the Future

We are progressing from an environmental paradigm based on cleanup and control to one including assessment, anticipation, and avoidance. Expenditures to develop technologies that prevent environmental harm are beginning to pay off. Agricultural practices are becoming less wasteful and more sustainable, manufacturing processes are becoming more efficient in the use of resources, and consumer products are being designed with the environment in mind. The infrastructures that supply energy, transportation services, and water supplies are becoming more resource efficient and environmentally benign. Remediation efforts are cleaning up a large portion of existing hazardous-waste sites. Our ability to respond to emerging problems is being aided by more advanced monitoring systems and data analysis tools that continually assess the state of the local, regional, and global environment. Finally, we are developing effective ways of restoring or recreating severely damaged ecosystems to preserve the long-term health and productivity of our natural resource base.

This trend will continue if we are willing to develop strong environmental policies, develop new strategies, and closely coordinate the actions of the public and private sectors toward a set of shared, long-term goals. Such a strategy should include three aims. The first is to articulate a vision of the future and the role environmental technology will play in shaping that future. This vision must be built on an understanding of the strengths and weaknesses of past policies and actions. The second is to define the roles of the many individuals and organizations that are needed to implement the goals. The third is to chart a course by offering suggestions for strategic goals for all the partners in this endeavor.

In the long run, environmental quality is not determined solely by the actions of government, regulated industries, or nongovernment organizations. It is largely a function of the decisions and behavior of individuals, families, businesses, and communities everywhere. Consequently, the extent of environmental awareness and the strength of environmental institutions will be two critical factors driving changes in environmental quality in the future.

A concerned, educated public, acting through responsive local, national, and international institutions, will serve as effective agents for avoiding future environmental problems, no matter what they are. Environmental institutions, strengthened by informed public support, will play a critical role in devising and implementing effective national and international responses to emerging issues.

The Development of Environmental Policy in the United States

Public **policy** is the general principal by which government branches—the **legislative, executive,** and **judicial**—are guided in their management of public affairs. The legislature (Congress) is directed to declare and shape national policy by passing legislation, which is the same as enacting law. The executive (President) is directed to enforce the law while the judiciary (the court system) interprets the law when a dispute arises. (See figure 20.4.)

When Congress considers certain conduct to be against public policy and against the public good, it passes legislation in the form of acts or statutes. Congress specifically regulates, controls, or prohibits activity in conflict with public policy and attempts to encourage desirable behavior.

Through legislation, Congress regulates behavior, selects agencies to implement new programs, and sets general procedural guidelines. When Congress passes environmental legislation, it also declares and shapes the national environmental policy, thus fulfilling its policymaking function. (See figure 20.5.)

Over 80 years ago, President Teddy Roosevelt declared that nothing short of defending your country in wartime "compares in leaving the land even better land for our descendants than it is for us." The environmental issues that Roosevelt strongly believed in, however, did not become major political issues until the early 1970s.

While the publication of Rachel Carson's *Silent Spring* in 1962 is considered to be the beginning of the modern environmental movement, the first Earth Day on April 22, 1970, was perhaps the single event that put the movement into high gear. In 1970, as a result of mounting public concern over environmental deterioration—cities clouded by smog, rivers on fire, waterways choked by raw sewage—many nations, including the United States, began to address the most obvious, most acute environmental problems.

Public opinion polls indicate that a permanent change in national priorities followed Earth Day 1970. When polled in May 1971, 25 percent of the U.S. public declared protecting the environment to be an important goal—a 2500 percent increase over 1969.

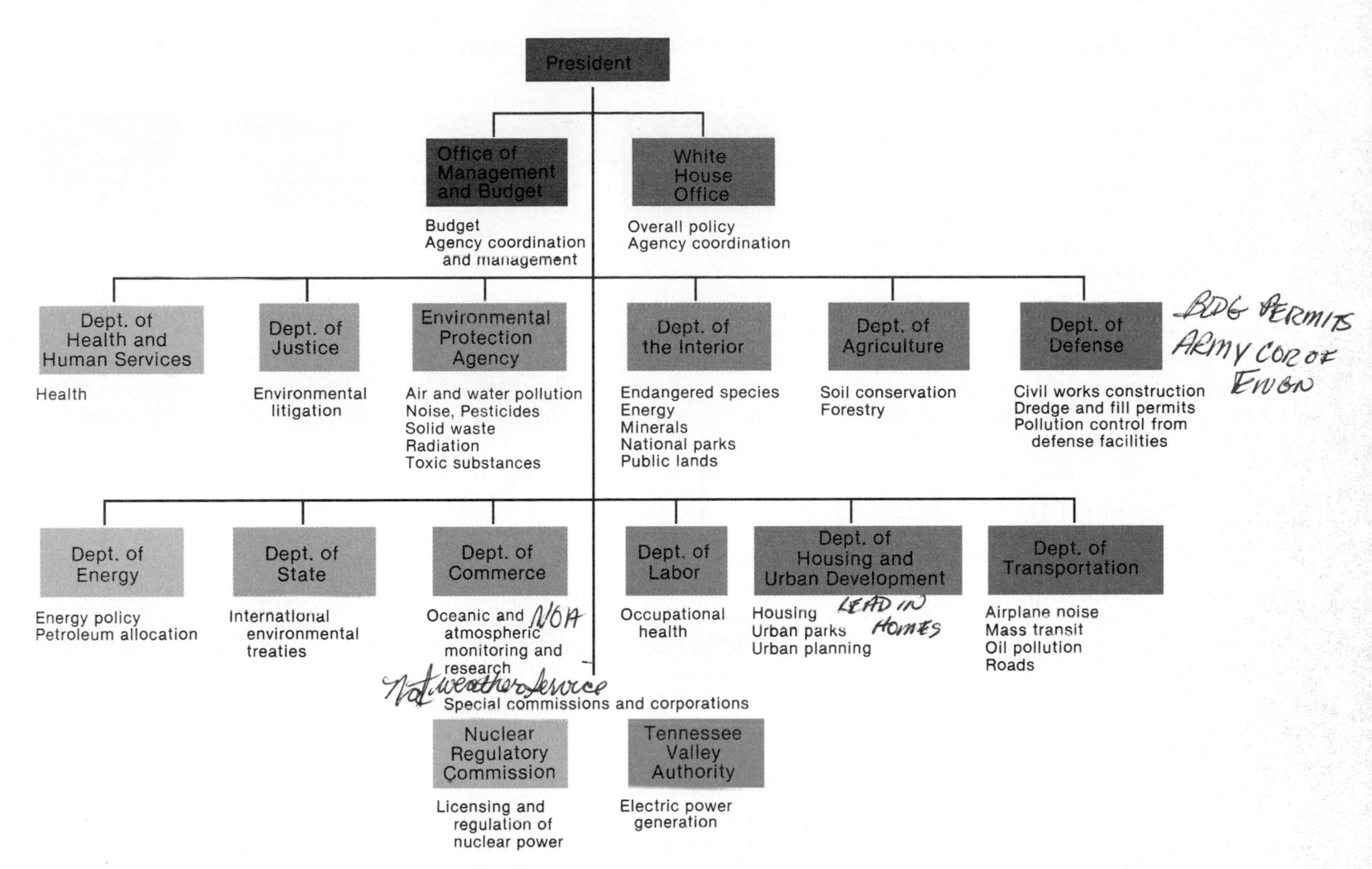

figure 20.4 **Major Agencies of the Executive Branch** Major agencies of the executive branch are shown with their environmental responsibility.

During the 1970s, many important pieces of environmental legislation were enacted in the United States. (See figure 20.5.) Many of the identified environmental problems were so immediate, so obvious, that it was relatively easy to see what had to be done and to summon the political will to do it. (See table 20.1.)

Just as it was beginning to gain momentum, however, the environmental movement began to decline. When the energy crisis threatened to stall the North American economy in the early 1970s, environmental concerns quickly faded. By 1974, President Gerald Ford had proposed an acceleration of his administration's leasing program for offshore gas and oil drilling. A turnaround in environmental policy was even more pronounced in the 1980s during the Reagan administration. Former Vice President Walter Mondale was fond of noting that President Reagan "would rather take a polluter to lunch than to court." During the mid-1980s, the environment was not a priority in the Reagan administration.

Despite the wavering of political will toward environmental concerns in the period from 1970 to 1990, there were some very tangible accomplishments. Among the most visible and quantifiable is the expansion of protected areas. During this period, federal parklands in the United States—excluding Alaska—increased 800,000 hectares, (2 million acres) to 10.5 million. In Alaska, 18.3 million additional hectares (45 million acres) were protected, bringing the state's total to over 232 million hectares (573 million acres). Also, the extent of the waterways included in the National Wild and Scenic Rivers System increased by more than 12 times, to some 15,000 kilometers (9,300 miles).

By the late 1980s, however, a new environmental awareness and concern began to surface as a major political issue. This was in part due to a number of highly visible environmental problems that appeared nightly on the evening news. Images of toxic waste (including hospital waste, such as used syringes) washing up on the nation's beaches, and the pristine waters of Alaska covered in oil from the *Exxon Valdez* spill, made an impact on the public. Once again, the public reacted by organizing and putting pressure on the political system, and, as in 1970, the politicians began to respond. For the first time in the history of the United States, the environment became a key issue in a presidential campaign. In 1988, the environmental records of the two major candidates were hotly debated. Environmentalism was evolving as a major public issue. By the 1992 U.S. presidential election, the environment was established as a major campaign issue, a trend that continues today.

In many respects, the environmental movement of the 1970s and 1980s came of age in the 1990s. The linking of politics and science, and emotionalism and logic, in a new environmental movement, represented a significant integration of human thinking. It has been said that politics have always forged science. Prioritization of issues and political will determine where money will be spent. By the mid-1990s, it appeared that the United States political will to address environmental concerns was on the rise. This was, however, a developing backlash to environmental policies in the United States.

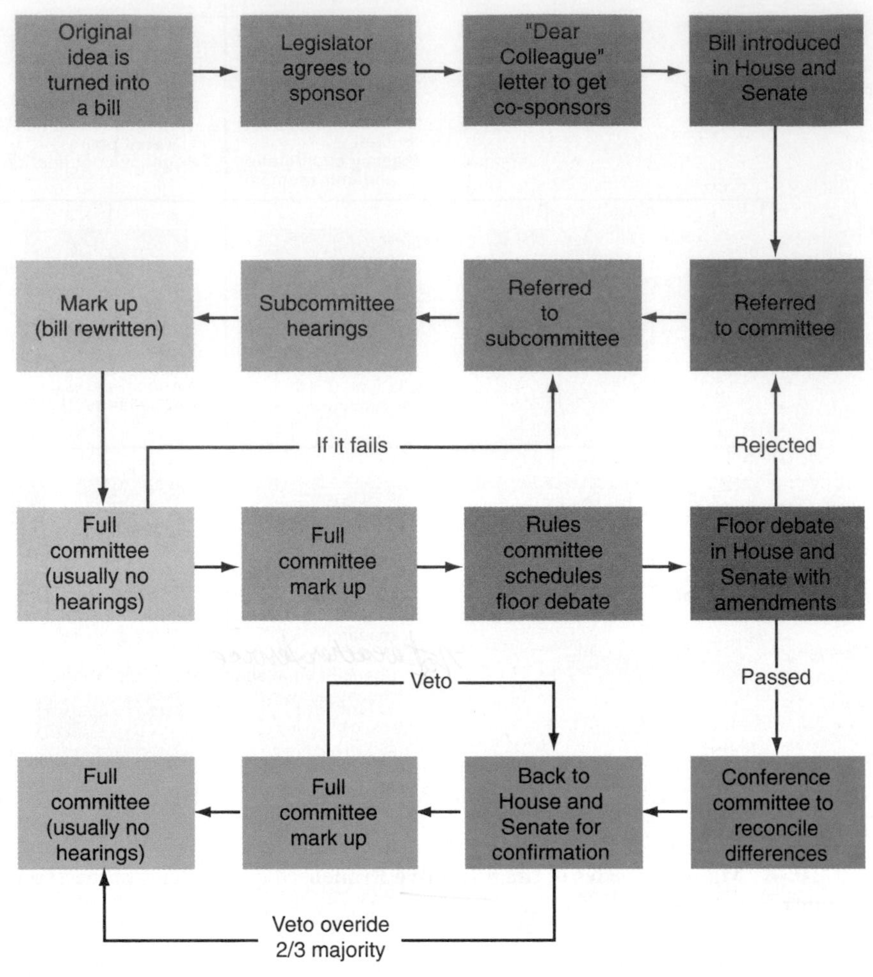

figure 20.5 **Passage of a Law** This figure illustrates the path of a bill from organization to becoming a law. As we can see, the process is not a quick one.

Environmental Backlash—The Wise Use Movement

During the late 1980s, there began to develop a backlash or anti-environmental attitude among sectors of the U.S. populace. A loose-knit organization of several hundred of these groups became known as the Wise Use Movement. The majority of funding for this coalition comes from various interests, including the timber, oil, and coal industries, real estate developers, and ranchers.

In his 1988 book, *The Wise Use Agenda,* Ron Arnold, an early leader of the movement, laid out some of its goals, including:

- Elimination of the National Park Service—to be replaced with privately operated parks.
- Removal of government restrictions on development in wetlands.
- The opening of all national parks, wilderness areas and wildlife refuges to off-road vehicles, commercial development, mining and drilling for oil.
- Cutting of all the remaining old-growth forests in national forests and replacing them with tree plantations.
- Recognize private property rights to water, grazing permits, and mining claims on public lands.

In addition to the above, the Wise Use Movement has been in the forefront of advocating private property rights. They argue that regulations protecting environmentally sensitive areas on private property are unconstitutional "takings." They cite the Fifth Amendment to the U.S. Constitution, which states in part: "nor shall private property be taken for public use, without just compensation." That clause is the basis for the concept of eminent domain, which allows government entities to take land for public projects by paying property owners the land's fair market value.

Wise Use pamphlets argue that extinction is a natural process and that some species were not meant to survive. It has been argued that the movement's signature public relations tactic is to frame complex environmental and economic issues in simple, scapegoating terms that benefit its corporate backers. In the Pacific Northwest, for example, the movement has continually dwelled on a supposed "battle" for survival between spotted owls and the families of the workers involved in harvesting the old-growth timber. "Jobs versus owls" is a good sound bite but it is not the issue. The real issue is much more complicated.

As would be expected, the environmental community has come out strongly against the Wise Use Movement. Environmentalists have renamed it the "earth- and people-abuse movement" and stated that the basic principles of Wise Users are: "If it grows, cut it; if it's swampy, fill it; if it moves, kill it."

Table 20.1 Major U.S. Environmental and Resource Conservation Legislation

Wildlife conservation
- Anadromous Fish Conservation Act of 1965
- Fur Seal Act of 1966
- National Wildlife Refuge System Act of 1966, 1976, 1978
- Species Conservation Act of 1966, 1969
- Marine Mammal Protection Act of 1972
- Marine Protection, Research, and Sanctuaries Act of 1972
- Endangered Species Act of 1973, 1982, 1985, 1988, 1995
- Fishery Conservation and Management Act of 1976, 1978, 1982, 1996
- Whale Conservation and Protection Study Act of 1976
- Fish and Wildlife Improvement Act of 1978
- Fish and Wildlife Conservation Act of 1980 (Nongame Act)

Land use and conservation
- Taylor Grazing Act of 1934
- Wilderness Act of 1964
- Multiple Use Sustained Yield Act of 1968
- Wild and Scenic Rivers Act of 1968
- National Trails System Act of 1968
- National Coastal Zone Management Act of 1972, 1980
- Forest Reserves Management Act of 1974, 1976
- Forest and Rangeland Renewable Resources Act of 1974, 1978
- Federal Land Policy and Management Act of 1976
- National Forest Management Act of 1976
- Soil and Water Conservation Act of 1977
- Surface Mining Control and Reclamation Act of 1977
- Antarctic Conservation Act of 1978
- Endangered American Wilderness Act of 1978
- Alaskan National Interests Lands Conservation Act of 1980
- Coastal Barrier Resources Act of 1982
- Food Security Act of 1985
- Coastal Development Act of 1990

General
- National Environmental Policy Act of 1969 (NEPA)
- International Environmental Protection Act of 1983
- Environmental Education Act of 1990

Energy
- National Energy Act of 1978, 1980

Water quality
- Water Quality Act of 1965
- Water Resources Planning Act of 1965
- Federal Water Pollution Control Acts of 1965, 1972
- Ocean Dumping Act of 1972
- Safe Drinking Water Act of 1974, 1984, 1996
- Clean Water Act of 1977, 1987
- Great Lakes Critical Programs Act of 1990
- Oil Spill Prevention and Liability Act of 1990

Air quality
- Clean Air Act of 1963, 1965, 1970, 1977, 1990

Noise control
- Noise Control Act of 1965
- Quiet Communities Act of 1978

Resources and solid waste management
- Solid Waste Disposal Act of 1965
- Resources Recovery Act of 1970
- Resource Conservation and Recovery Act of 1976
- Waste Reduction Act of 1990

Toxic substances
- Toxic Substances Control Act of 1976
- Resource Conservation and Recovery Act of 1976
- Comprehensive Environmental Response, Compensation, and Liability (Superfund) Act of 1980, 1986, 1990
- Nuclear Waste Policy Act of 1982

Pesticides
- Federal Insecticide, Fungicide, and Rodenticide Control Act of 1972, 1988

The Changing Nature of Environmental Policy

A 2000 survey revealed that most Americans would prefer a federal government more actively involved in environmental protection. Ninety-two percent desired stronger active participation by leaders in government and business in environmental concerns. Forty-one percent believed the general public should bear the primary responsibility for a clean environment, followed by industry at 34 percent and government at 22 percent. The poll also showed that 55 percent of Americans consider the environment to be a serious problem facing the country. Only 26 percent said that a great deal of progress had been made in the past few decades in dealing with environmental problems. It should be noted, however, that these polls, like most polls, often reflect attitudes and not necessarily actions. For example, saying that you support stronger environmental laws does not always translate into individual actions such as purchasing environmentally "friendly" products, active recycling, or support of higher taxes for more governmental regulation of the environment. Saying and doing are not always the same.

While the public was supportive of environmental protection, the 1990s also witnessed a decline in membership of the major environmental organizations. By 1998, membership declines in many of the large environmental organizations was forcing the layoff of staff and the closure of offices. A major part of the problem with the larger environmental groups may be that they have grown into large bureaucracies and, in the process, have lost the trust of many grassroots environmentally concerned individuals. For example, the National Wildlife Federation has an annual budget in excess of $100 million. This includes profits on merchandise of nearly $25 million and nearly $12 million spent on magazines. By any definition, the National Wildlife Federation is a large bureaucracy.

While many of the larger, established environmental organizations are still declining somewhat, many newer, smaller, local and grassroots organizations are being created or are expanding, often in response to environmental threats in their own communities. Estimates are that some 7000 local environmental organizations are active in the United States. On college campuses, interest in the environment is growing. Environmental studies programs are expanding at both the undergraduate and graduate level. Students are not just concerned with local recycling programs or nuclear power plants but are focusing on the broader issues to be faced in the new century. The primary issue is the achievement of long-term sustainability and the fundamental changes in society that this will entail.

Environmental Policy and Regulation

Environmental laws are not a recent phenomenon. As early as 1306, London adopted an ordinance limiting the burning of coal because of the degradation of local air quality. Such laws became more common as industrialization created many sources of air and water pollution throughout the world.

Environmental law in the United States is governed by administrative law. Administrative law is a relatively new concept, having been developed only during the twentieth century. In 1946, Congress passed the Federal Administrative Procedure Act (APA). This act designated general procedures to be used by federal agencies when they exercised their rule-making, adjudicatory, and enforcement powers. This is a rapidly expanding area of law and defines how governmental organizations such as agencies, boards, and commissions develop and implement the regulatory programs they are legislatively authorized to create. Some of the many U.S. federal agencies that impact environmental issues include the Environmental Protection Agency (EPA), the Council on Environmental Quality (CEQ), the National Forest Service, and the Bureau of Land Management.

Administrative law applies to government agencies and to those that are affected by agency actions. In the United States, many federal environmental programs are administered by the states under the authority of federal and related state laws. States often differ from both the federal government and each other in the way they interpret, implement, and enforce federal laws. In addition, each state has its own administrative guidelines that govern and define how state agencies act. All actions of federal agencies must comply with the 1946 Administrative Procedure Act.

The National Environmental Policy Act of 1969 was enacted in 1969 and signed into law by President Nixon on New Year's Day, 1970. It is a short, general statute designed to institutionalize within the federal government a concern for the "quality of the environment." NEPA helps encourage environmental awareness among all federal agencies, not just those that prior to NEPA had to consider environmental factors in their planning and decision making. Until 1970, most federal agencies acted within their delegated authority without considering the environmental impacts of their actions. However, in the 1960s, Congress seriously began to study pollution problems. Because Congress has found that the federal government is both a major cause of environmental degradation and a major source of regulatory activity, all actions of the federal government now fall under NEPA.

NEPA forces federal agencies to consider the environmental consequences of their actions before implementing a proposal or recommendation. NEPA has two purposes: first, to advise the president on the state of the nation's environment, and second, to create an advisory council called the Council on Environmental Quality (CEQ). The CEQ outlines NEPA compliance guidelines. The CEQ also provides the president with consistent expert advice on national environmental policies and problems.

Between 1970 and 1977, the CEQ served only as an advisory council to the president. However, in 1977, through an executive order, President Carter granted the CEQ authority to issue binding regulations. These regulations set out details for matters that are broadly addressed by NEPA. The CEQ applies to all federal agencies except Congress, the judiciary, and the president.

NEPA has been interpreted narrowly by the federal courts. As a result, many states have passed much stronger state environmental protection acts (SEPAs) as well. Today, NEPA analysis is undertaken as part of almost every recommendation or proposal for federal action. This includes not only actions by agencies of the federal government, but also actions of states, local municipalities, and private corporations. NEPA is Congress's mission statement that mandates the means by which the federal government, through the guidance of the CEQ, will achieve its national environmental policy.

In addition to the passage of NEPA, the 1970s saw a series of new environmental laws passed, including the Resource Conservation and Recovery Act (RCRA), the Comprehensive Environmental Response, Compensation and Liability Act (CERCLA) and the Clean Air Act (CAA). All of these acts are broadly worded to identify existing problems that Congress believes can be corrected to protect human health, welfare, and the environment.

Protecting human health, welfare, and the environment is the national policy that Congress has chosen to encourage. For example, under NEPA, the national policy is to "promote efforts which will prevent or eliminate damage to the environment and biosphere and stimulate the health and welfare of man [human]." Similarly, under the Clean Air Act, the policy is to "protect and enhance the quality of the Nation's air resources so as to promote the public health and welfare and the productive capacity of its population."

Each of the above statutes declares national policy on environmental issues and addresses distinct problems. These statutes also authorize the use of some or all of the administrative functions discussed above, such as rule making, adjudication, administrative, civil and criminal enforcement, citizen suits, and judicial review. (See figure 20.6.)

Congress established the Environmental Protection Agency in 1970 as the primary agency to implement the statutes. Administrative functions empower EPA, the states, and private citizens to take the responsibility to enforce the various authorized programs. These administrative functions not only shape environmental law, but also control the daily operations of both the regulated industry and agencies authorized to protect the environment.

To date, much environmental law has reflected the perception that environmental problems are localized in time, space, and media (i.e., air, water, soil). For example, many hazardous-waste sites in the United States have been "cleaned" by simply shipping the contaminated dirt someplace else, which not only does not solve the problem, but creates the danger of incidents during removal and transportation. Environmental regulation has focused on specific phenomena and adopted the so-called "command and control" approach, in which restrictive and highly specific legislation and regulation are implemented by centralized authorities and used to achieve narrowly defined ends.

Such regulations generally have very rigid standards, often mandate the use of specific emission-control technologies, and generally define compliance in terms of "end-of-pipe" requirements. (See figure 20.7.) Examples in the United States include the Clean Water Act (which applied only to surface waters), the Clean Air Act (urban air quality), and the Comprehensive Environmental Response, Compensation, and Liability Act (Superfund), which applied to specific landfill sites.

If properly implemented, command and control methods can be effective in addressing specific environmental problems. For example, rivers such as the Potomac and Hudson in the United States are much cleaner as a result of the Clean Water Act. (See figure 20.8.) Moreover, where applied against particular substances, such as the ban on tetraethyl lead in gasoline in the United States, the command-and-control approach has clearly worked well.

The Greening of Geopolitics

Environmental or "green" politics have emerged from minority status and become a political movement in many nations. Issues such as transboundary water supply and pollution, acid precipitation, and global climate change, have served to bolster the emergence of green politics. Even in Russia and eastern Europe, the public has demanded more

EPA Enforcement Options

1. **Warning Letter**
   - Describes alleged violation
   - Action that must be taken to remedy
     - Deadline for compliance or enforcement action
     - Allow for an opportunity to discuss the situation
2. **Administrative Order**
   - Requires certain activity be ceased, or
   - Requires compliance by a certain date, or
   - Requires certain tests be performed
   - Provides Rights
     - Respond by citing defenses or objections
     - Right to confer with EPA
3. **Permit Action**
   - Revoke or modify existing permit
   - Seek to add conditions to permits that are being negotiated
4. **Civil Enforcement**
   - Injunctive
   - Monetary relief varies by statute per violation, per day
5. **Criminal Proceedings**
   - Fines
   - Incarceration

figure 20.6 **Enforcement Options of the U.S. Environmental Protection Agency** The enforcement options of the U.S. Environmental Protection Agency range from a warning letter to a jail sentence.

Source: U.S. Environmental Protection Agency, Office of Enforcement, Washington, D.C.

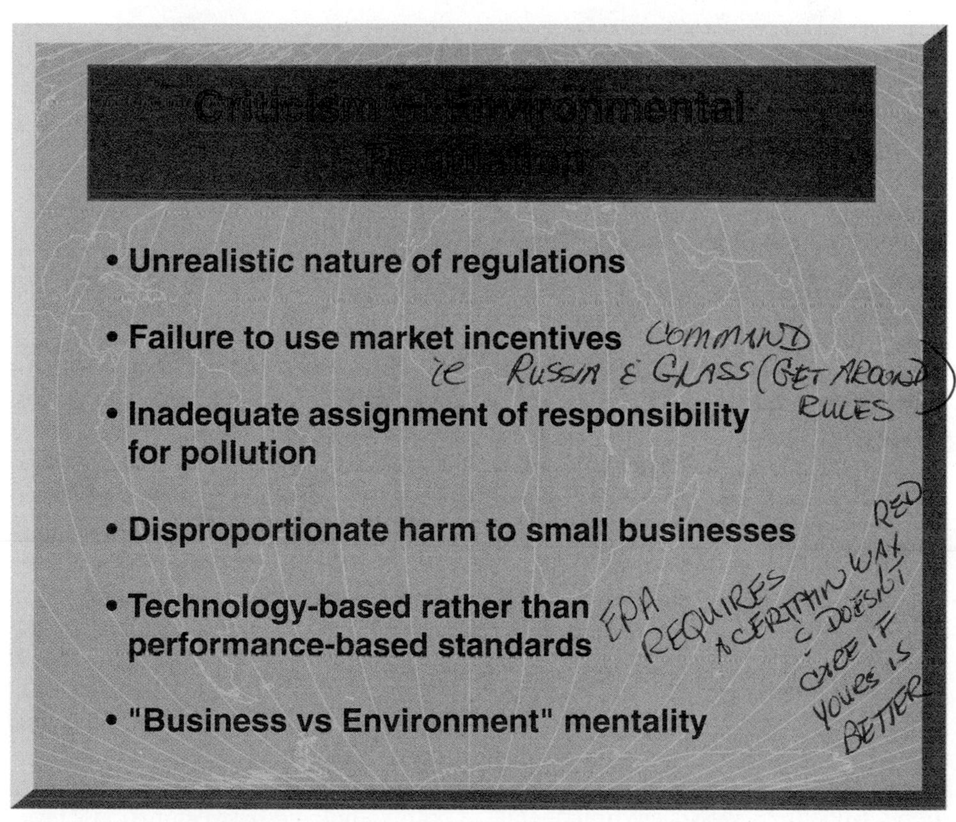

figure 20.7 **Criticisms of Environmental Regulation** While some of the above criticisms are being addressed, others still need to be studied and discussed.

Source: U.S. Environmental Protection Agency, Office of Enforcement, Washington, D.C.

figure 20.8 The environmental quality of rivers has dramatically improved over the past 20 years as a direct result of the Clean Water Act. The upper photo shows a visibly impaired river in New England in the 1970s. The bottom photo shows an environmentally healthy Hudson River in the late 1990s.

figure 20.9 If China's current "modernization" continues, the boom will be fueled by coal to the possible detriment of the planet as a whole.

environmental protection, and leaders seem to be listening.

Concern about the environment is not limited to developed nations. A 1989 treaty signed in Switzerland limits what poorer nations call toxic terrorism—use of their lands by richer countries as dumping grounds for industrial waste. In 1990, more than 100 developing nations called for a "productive dialogue with the developed world" on "protection of the environment." As was covered in chapter 2, the Earth Summit in Rio brought together nearly 180 governments to address world environmental concerns, and the 1997 conference on global warming in Kyoto brought together some 120 nations.

Environmental concern is also a growing factor in international relations. Many world leaders see the concern for environment, health, and natural resources as entering the policy mainstream. A sense of urgency and common cause about the environment is leading to cooperation in some areas. Ecological degradation in any nation is now understood almost inevitably to impinge on the quality of life in others. Drought in Africa and deforestation in Haiti have resulted in large numbers of refugees, whose migrations generate tensions both within and between nations. From the Nile to the Rio Grande, conflicts flare over water rights. The growing megacities of the developing world are areas of potential civil unrest. Sheer numbers of people overwhelm social services and natural resources. The government of the Maldives has pleaded with the industrialized nations to reduce their production of greenhouse gases, fearing that the polar ice caps may melt and inundate the island nation.

Economic progress in developing nations could also bring the possibility for environmental peril and international tension. China, which accounts for 21 percent of the world's population, has the world's third largest recoverable coal reserves. If China's current "modernization" campaign succeeds, the boom will be fueled by coal, to the possible detriment of the planet as a whole. (See figure 20.9.)

Some experts estimate that the developing world, which today produces one-fourth of all greenhouse gas emissions, could be responsible for nearly two-thirds by the middle of this next century. Developing nations have repeatedly indicated that they are not prepared to slow down their own already weak economic growth to help compensate for decades of environmental problems caused largely by the industrialized world.

ENVIRONMENTAL CLOSE-UP

Changing the Nature of Environmental Regulation—The Safe Drinking Water Act

According to the American Water Works Association (AWWA), the 1996 amendments to the Safe Drinking Water Act (SDWA) amount to a radical rewrite of the 1986 amendments. The amendments were written with little substantive input from the regulated community. The 1996 amendments, however, were developed with significant contributions from water suppliers and state and local officials. The new amendments embody a partnership approach that includes major new infusions of federal funds to help water utilities, especially the thousands of smaller systems, comply with the law.

What all this means is that the amended SDWA is more focused on what it was initially intended to achieve. For example, rather than require the EPA to set standards for 25 new contaminants every three years regardless of need or actual threat to public health (in cases, it has been argued, this requirement impeded public health protection by diverting resources from the most critical needs), the new law focuses the EPA's efforts on regulating contaminants known to pose health risks and requires cost-benefit analyses and risk assessments to take place before new standards are set. The act also allows the EPA to adopt interim regulations for contaminants on an emergency basis if they pose urgent health threats. The act provides millions of dollars in new funding for crucial research; links source water to drinking water (and sets out a state-administered system designed to protect drinking water from becoming contaminated in the first place); authorizes a new system of federal grants to states, which will enable water utilities to borrow funds to upgrade their systems; and requires water systems to provide customers with annual updates delineating the sources and quality of the water they provide. It has been stated that the 1996 amendments are good for the public and drinking water industry alike. The law has more flexibility, more state responsibility, more cooperative approaches, and is achieved through partnerships.

All too often we take clean, safe drinking water for granted.

Some developing countries may resist environmental action because they see a chance to improve their bargaining leverage with foreign aid donors and international bankers. Where before the poor nations never had a strategic advantage, they now may have an ecological edge. Ecologically, there could be more parity than there ever was economically or militarily.

A former U.S. ambassador to the United Nations stated that just as the Cold War between the East and West seems to be winding down, "eco-conflicts" between the industrialized North and the developing South may pose a comparable challenge to world peace. National security may no longer be about fighting forces and weaponry alone. It also relates increasingly to watersheds, croplands, forests, climate, and other factors rarely considered by military experts and political leaders, but that, when taken together, deserve to be viewed as equally crucial to a nation's security as are military factors. It is interesting to note that the North Atlantic Treaty Organization (NATO) has developed an office for Scientific and Environmental Affairs, and the U.S. Department of Defense has created an Office of Environmental Security. Environmental offices within traditional military organizations would have been viewed as very unusual until only recently.

In 2000, a dike holding million of liters of cyanide-laced wastewater gave way at a gold-extraction operation in northwestern Romania, sending a deadly waterborne plume across the Hungarian border and down the nation's second largest river. Cyanide separates gold from ore, and mining operations often store cyanide-laced sludge in diked-off lagoons. After devastating the upper Tisza River, the 50-kilometer (30 mile) long pulse of cyanide and heavy metals spilled into the Danube River in northern Yugoslavia. The diluted plume finally filtered into the Danube delta at the Black Sea, more than 1000 kilometers (620 miles) and three weeks after the spill. Scientists across Europe warned that the accident could leave the upper Tisza and the nearby Somes River a poisonous legacy for several years if heavy metals are left to linger in the river sediments.

The Hungarian Academy of Sciences has pointed out that 90 percent of the water entering the tributaries of Hungary's rivers flows through Romania, Ukraine, Slovakia, and Austria. Numerous mines, chemical plants, oil refineries, and other sources of pollutants line those tributaries. The Hungarian government has stated that the management of environmental security cannot be stopped at the borders. The government has filed lawsuits seeking monetary

figure 20.10 **Options and Tradeoffs for the New Century** Unless we become more creative in areas such as transportation and land use, it will be necessary to develop in the United States alone, in the next 50 years, an amount of farmland and scenic countryside as was developed over the past 200 years.

damage against the operators of the Baia-Mare gold extraction lagoon, and it has threatened to sue the Romanian government to help recover the cleanup costs. The accident stoked bilateral tensions: Romanian officials accused the Hungarian side of exaggerating the extent of the damage, while Hungarians asserted that the Romanians are downplaying the spill. The issue of environmental security continues to expand in the dialogue between the two nations.

The increased attention to the environment as a foreign policy and national security issue is only the beginning of what will be necessary to avert problems in the future. The most formidable obstacle may be the entrenched economic and political interests of the world's most advanced nations. If the United States, for example, asks others not to cut their forests, then it will have to be more judicious about cutting its own. If North Americans wish to stem the supply of hardwood from a fragile jungle or furs from endangered species, then they will have to stem demand for fancy furniture and fur coats. If they wish to preserve wilderness from the intrusions of the oil industry, then they will have to find alternative sources of energy and use all fuels more efficiently.

What may be needed is self-discipline on the part of the world's haves and increased assistance to the have-nots. In the world today, a billion people live in a degree of poverty that forces them to deplete the environment without regard to its future. Their governments often are too crippled by international debt to afford the short-term costs of environmental safeguards.

William Ruckelshaus, a former administrator of the U.S. Environmental Protection Agency, believes a historical watershed may be at hand. If the industrialized and developing countries did everything they should, he says, the resulting change would represent a "modification of society comparable in scale to the agricultural revolution of the Neolithic age and to the Industrial Revolution of the past two centuries."

Let's put all of this into another perspective that may ring closer to home. By even conservative estimates, the population of the U.S. is expected to double in the next 50 years. It could be argued then that with twice as many people projected for 2050, it will be necessary to double the total U.S. infrastructure in the next 50 years. Potentially then, this translates to the following: twice as many cars, trucks, planes, airports, parking lots, streets, and freeways; twice as many houses and apartment buildings; twice as many landfills, wastewater treatment plants, hazardous-waste treatment facilities, and chemicals (pesticides and herbicides) for agriculture; in short, twice as much of everything. Under this scenario, it will be necessary to take over and develop in the next 50 years an amount of farmland and scenic countryside equal to the total already developed in the past 200 years. (See figure 20.10.) Megacities such as New York and Los Angeles will double in size and many new megacities will develop. In short, this vision is not very appealing. What happens to wilderness areas, remote and quiet places, habitat for songbirds, waterfowl and other wild creatures? What happens to our quality of life?

Another way of looking at the future would be along the lines of what

GLOBAL PERSPECTIVE

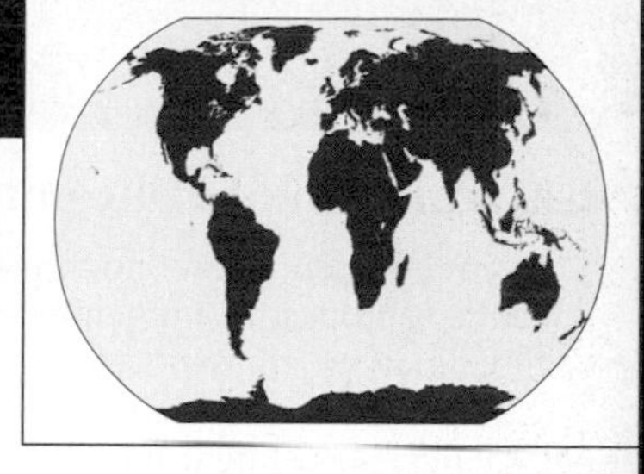

Eco-Terrorism

The effects of war on the natural environment are frequently catastrophic. By its nature, war is destructive, and the environment is an innocent victim. The 1990–91 war in the Persian Gulf, however, witnessed a new role for the environment in war: its degradation was used as a weapon.

The phrase "eco-terrorism" was applied to Iraqi leader Saddam Hussein's use of oil as a weapon in the war. Iraq dumped an estimated 1.1 billion liters of crude oil into the Persian Gulf from Kuwait's Sea Island terminal before the United States made bombing raids on the pumps feeding the facility. The oil spill that resulted from the dumping was the world's largest—over 30 times the size of the 1989 *Exxon Valdez* disaster in Alaska.

Saddam Hussein may have engineered the spill to block allied plans for an amphibious invasion of Kuwait, but he was also probably trying to shut down seaside desalination plants that provide much of the freshwater for Saudi Arabia's Eastern Province. Whatever the military or political effect, the environmental effect was much greater.

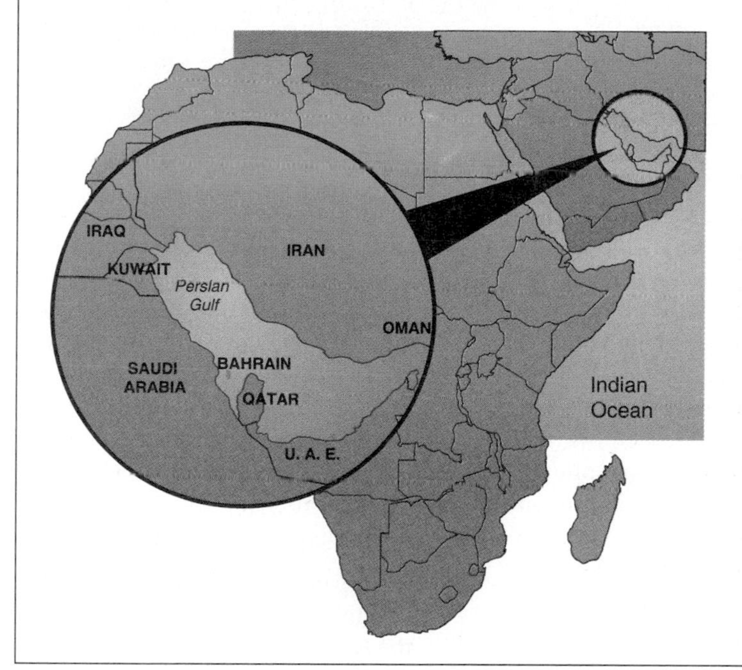

The Persian Gulf is a sensitive ecosystem. The water is very salty, and temperatures vary widely from one season to another. As a result, indigenous animals and plants are finely attuned to specialized conditions. The gulf is also rather isolated, with only one narrow outlet—the Strait of Hormuz, just 55 kilometers across. The Gulf takes up to five years to flush out.

The Persian Gulf waters, shores, and islands are dotted with coral reefs, mangrove swamps, and beds of sea grass, alive with many unique species of birds, fish, and marine mammals. Many of the region's species, such as the bottlenose dolphin, dugong, green turtle, and caspian tern, were already classified as threatened.

Another act of eco-terrorism that took place during the Iraq-Kuwait war was the intentional burning of hundreds of oil wells in Kuwait. The air pollution resulting from the burning oil covered a large area. It took several months to extinguish the fires following the war. Using environmental damage as a weapon of war is a frightening concept. The environmental consequences of such acts of eco-terrorism will not be fully understood for years. Though the war in the Persian Gulf lasted less than one year, the environmental effects on some species could be irreversible.

Source: *International Organizations: A Dictionary and Directory* by Giuseppe Schiavone (St. James Press, 1983).

was previously mentioned about the analogy to the agricultural and industrial revolutions. This translates into a future of profound change. A future in which virtually everything we do will change. This future of profound change will also be one of phenomenal opportunity and excitement. It is a future in which we will farm, build, and transport in entirely new ways. This vision of the future could well become known as the environmental revolution.

International Environmental Policy

"If there must be a war, let it be against environment contamination, nuclear contamination, chemical contamination; against the bankruptcy of soil and water systems; against the driving of people away from the lands as environmental refugees. If there must be war, let it be against those who assault people and other forms of life by profiteering at the expense of nature's capacity to support life. If there must be war, let the weapons be your healing hands, the hands of the world's youth in defense of the environment."

Mustafa Tolba
Former Secretary General
United Nations Environment
Programme

GLOBAL PERSPECTIVE

Environmental Policy and the European Union

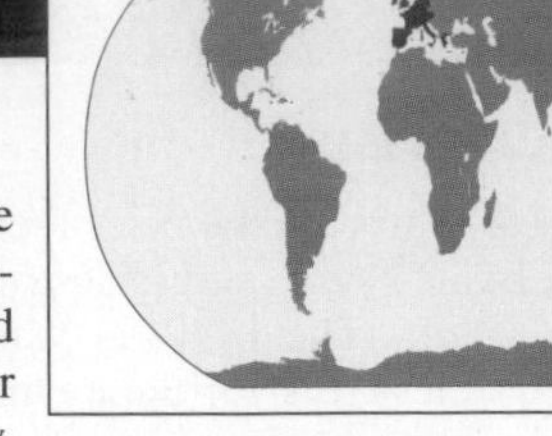

"The environment knows no frontiers" was the slogan of the 1970s, when the European Community—better known today as the European Union—began to develop its first environmental legislation.

The early laws focused on the testing and labeling of dangerous chemicals, testing drinking water, and controlling air pollutants such as SO_2, NO_x, and particulates from power plants and automobiles. Many of the directives from the 1970s and 1980s were linked to Europe's desire to improve the living and working conditions of its citizens.

In 1987, the Single European Act gave this growing body of environmental legislation a formal legal basis, and set three objectives: protection of the environment, human health, and prudent and rational use of natural resources.

The treaty reflected what many governments had already understood: that countries are part of an interconnected and interdependent world of people who are bound together by the air they breathe, the food they eat, the products they use, the wastes they throw away, and the energy they consume.

Similarly, a factory in one European nation may import supplies and raw material from several other neighboring nations; it may consume energy produced from imported gas, produce wastes that affect the air and water quality across the border or downstream, and export products whose wastes become the risk and responsibility of governments and peoples several hundred or many thousand kilometers away.

The 1992 Maastricht Treaty formally established the concept of sustainable development in European Union law. Then, in 1997, the Amsterdam Treaty made sustainable development one of the overriding objectives of the European Union. The Treaty considerably strengthened the commitment to the principle that the European Union's future development must be based on the principle of sustainable development and a high level of protection of the environment. The environment must be integrated into the definition and implementation of all of the Union's other economic and social policies, including trade, industry, energy, agriculture, transport, and tourism.

Under certain conditions, the member states may maintain or introduce environmental standards and requirements which are stricter than those of the Union. The Commission verifies the compatibility of these stricter national programs and the potential effects on other member states.

International coordination and political resolve are necessary if the goals of preserving and protecting the global environment are to be realized. A major step in that direction was the 1972 United Nations Conference in Stockholm, Sweden. This was the first international conference specifically dealing with global environmental concerns. The UN Environment Programme, a separate department of the United Nations that deals with environmental issues, developed out of that conference. The United Nations Conference on the Law of the Sea produced a comprehensive convention that addressed many of the issues concerning jurisdiction over ocean waters and use of ocean resources. This treaty is viewed by many as a model for international environmental protection. Positive results are already evident from application of the agreements on pollution control, marine mammal protection, navigation safety, and other aspects of the marine environment. The issue of deep ocean mining, primarily of manganese nodules, has to date kept many industrial nations, such as Germany and the United States, from ratifying the treaty.

There have been several other successful international conventions and treaties. The Antarctic Treaty of 1961 reserves the Antarctic continent for peaceful scientific research and bans all military activities in the region. The 1979 Convention on Long-Range Transboundary Air Pollution was the first multilateral agreement on air pollution and the first environmental accord involving all the nations of Eastern and Western Europe and North America. The 1987 Montreal Protocol on Substances that Deplete the Stratospheric Ozone Layer addresses the ozone protective shield problem.

Agreed to in 1987, the Montreal Protocol's objective was to phase out the manufacture and use of chemicals depleting the Earth's protective ozone layer (see chapter 17). Over a decade since its adoption, 160 countries are now parties, representing over 95 percent of the Earth's population. Production and consumption of chlorofluorocarbons (CFCs), carbon tetrachloride, halons, and methyl chloroform have been phased out in developed countries, with reduction schedules set for their use in developing countries. In 1999, developing countries ended the production and consumption of CFCs, with phaseout scheduled for 2010.

By the late 1990s, the concentration of some CFCs in the atmosphere had started to decline, and predictions are that the ozone layer could recover by the middle of this century. Among the many reasons for these achievements, three merit attention:

- *Global agreement on the nature and seriousness of the threat.* Even the strongest skeptics could not deny the Antarctic ozone hole, which was first brought to international attention by British scientists in 1985. It was understood for the first time that emissions of ozone-depleting substances were in reality putting our lives and the lives of future generations at risk. Decisive action was required.
- *A cooperative approach, especially between developed and developing countries.* In the developed countries, it was recognized that their industries had contributed significantly to this global problem and that they

GLOBAL PERSPECTIVE

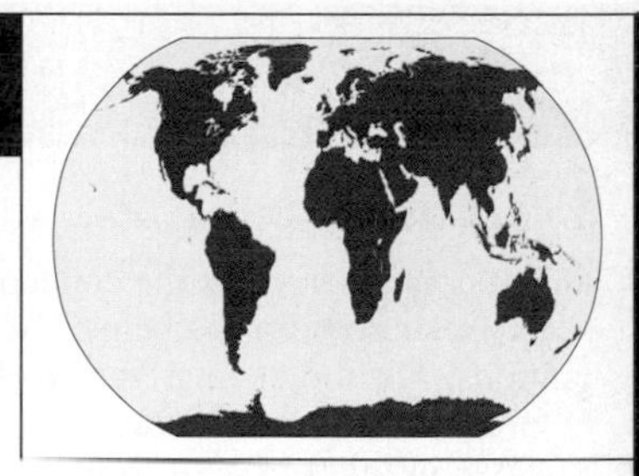

Overview of an International Organization—The International Whaling Commission (IWC)

The first records of whaling can be traced back hundreds of years to the twelfth century in the Bay of Biscay of Spain. During whaling's heydays of the 1930s, 85 percent of the world's catch of whales was coming from the Antarctic, with 95 percent of these catches dominated by Norway and the United Kingdom.

With the possible exhaustion of whale stock, the problem of whales as a common property resource came to the forefront. When a resource is common property, no single user has a right to the resource, nor can they prevent others from sharing in its exploitation. These doctrines of freedom of the seas and common property resources date back to Roman times and were constantly used by states as an argument to justify the overfishing of whale stocks in previous centuries. Seeing a need to keep statistical data on the populations of different whale species, Norway, at the request of the International Council for the Exploration of the Seas (ICES), set up the Bureau of International Whaling Statistics in 1930 to deal with whaling problems. This eventually led to the drafting and signing of the Convention for the Regulation of Whaling, in 1931. The convention "applied to all waters . . . but was only applicable to baleen whales," and provided exemptions for aboriginal subsistence whaling. This convention had little effect on the amount of whaling, because important nations such as Germany and Japan failed to adhere to the rules stated. A new draft of the convention was brought into effect in 1937 and included such countries as Norway, the United Kingdom, and Germany. In 1946, the International Whaling Conference (IWC) was convened by the United States to address the post–World War II whaling issues.

The measures to be taken by this commission were to completely protect certain whale species, specify certain whale sanctuaries, set limits on the number of whales taken, decide when open and closed hunting seasons would begin, and specify the size and age of those whales that could be taken. It is important to point out that the majority of the nations who were original members of the IWC were whaling states and had no real interest in protecting the whale in order to preserve biologic diversity. The formation and actions of the IWC in its early years were purely economically motivated.

The ineffectiveness of the IWC to control the slaughter of whale populations continued into the 1970s until environmentally oriented nongovernmental organizations began to champion the struggle of the whale. Soon millions of people in the world were chanting "save the whale" and began demanding that the IWC pass a moratorium to end whaling. But a powerful veto coalition made up of Japan, Norway, the Soviet Union, Iceland, Chile, and Peru was able to control the vote in the IWC and defeat any measures that were harmful to the whaling industry as a whole.

Seeing that they could not influence the voting of the IWC alone, the United States began to form a new strategy. In order to ensure a moratorium on whaling in the IWC, nonwhaling states from the developing nations were brought in to outnumber the whaling states. The state leading this group of new members was the small island nation of Seychelles. Despite economic threats from Japan, Seychelles held strong and was able to push for a moratorium that was eventually passed by the IWC in 1982. Although many believed that this was the final victory in the struggle to end whaling, the war waged on. Nations such as Norway and Iceland threatened to leave the IWC and eventually did so. Other nations found loopholes in the moratorium and continued to kill large numbers of whales. Today, the future of both the moratorium and the IWC are unclear. Japan rallies developing nations onto its side with promises of economic aid. Other nations such as Norway are refusing to abide by the ruling handed down by the IWC and state that they will resume whaling. With a decline in the United States' ability to influence the IWC, perhaps the only hope for the whale in the future is the changing attitude of the Japanese public toward the killing of whales.

- What are your thoughts on the taking of whales?
- Do individual nations have a right to take whales?
- How would you define "common property resource"?

had to take the lead in stopping emissions and finding alternatives. It was also recognized that solving the ozone problem required a global solution, with all countries committed to eliminating ozone-depleting substances. Innovative partnerships, including early controls for developed countries and a grace period, funding, and technology transfer for developing countries, were set in place.

- *Policy based on expert and impartial advice.* The parties to the Montreal Protocol were able to receive impartial advice from their science, technical, and economic committees. These drew together experts from around the world to evaluate the need for further action and propose options that are technically and economically feasible. Put these three factors together—acknowledgment of the threat, agreement to cooperate, and commitment to take effective action based on expert advice—and you have a potentially

GLOBAL PERSPECTIVE

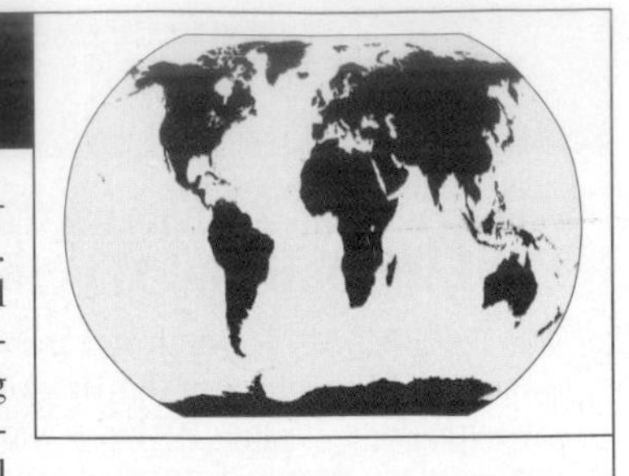

Eco-Labels

Green consumerism is the concept of rational consumption of our scarce resources for the benefit of the environment and future generations. The old saying "the world is enough for everyone's need but not for everyone's greed" calls for a change in our behavior and lifestyle in favor of a sustainable future. Eco-labels have been introduced in a number of countries to help consumers choose products with a proven environmental edge, determined by the product's choice of raw materials, production process, product life cycle, and associated disposal problems. Eco-labels provide evidence that products have met the safety, quality, and environmental protection requirements of the authority that issues the label.

The first eco-label was introduced in Germany in 1978. This distinctive Blue Angel label was followed by other environmental certification, including the White Swan of Northern European countries, Environmental Choice of Canada, Green Label of Singapore, and Environmental Label of China. By 1995, over a dozen countries had adopted this system of providing consumers with information enabling them to become responsible green consumers. The eco-label of the European Union is significant because it is the world's first regional scheme to apply the same minimum standards across national markets.

1992
European Union - 'Eco-label'

1992
Singapore - 'Green Label'

1993
China - 'Environmental Label'

1978
Germany - 'Blue Angel'

1988
Canada - 'Environmental Choice'

1989
Nordic Council - 'White Swan'

Source: From ECCO, *Bulletin of the Environmental Campaign Committee.*

strong recipe to solve global issues, such as climate change and biodiversity loss.

In 1997, the General Assembly of the United Nations held a special session and adopted a comprehensive document entitled Programme for the Further Implementation of Agenda 21 prepared by the Commission on Sustainable Development. It also adopted the program of work of the Commission for 1999–2003. The Commission on Sustainable Development (CSD) was created in 1992 to ensure effective follow-up of the United Nations Conference on Environment and Development (UNCED) held in Rio de Janeiro, Brazil (see chapter 1).

The Commission on Sustainable Development consistently generates a high level of public interest. Over 50 national leaders attend the CSD each year and more than one thousand nongovernmental organizations (NGOs) are accredited to participate in the commission's work. The commission ensures the high visibility of sustainable development issues within the UN system and helps to improve the UN's coordination of environment and development activities. The CSD also encourages governments and international organizations to host workshops and conferences on different environmental and cross-sectoral issues. The results of these expert-level meetings enhance the work of CSD and help it work better with national governments and various nongovernmental partners in promoting sustainable development worldwide.

There is no international legislature with authority to pass laws; nor are there international agencies with power to regulate resources on a global scale. An international court at The Hague in the Netherlands has no power to enforce its decisions. Nations can simply ignore the court if they wish. However, a growing network of multilateral environmental organizations have developed a greater sense of their roles and a greater incentive to work together. These include not only the United Nations Environment Programme but also the Environment Committee of the Organization for Economic Cooperation and Development (OECD) and the Senior Advisors on Environmental Problems of the Economic Commission for Europe. Such institutions perform unique functions that cannot be carried out by governments acting alone or bilaterally.

Is the goal of an environmentally healthy world realistic? There is growing optimism that the community of nations is slowly maturing with regard to our common environment. We have all suffered a loss of innocence about "earth management." Laissez-faire may be good economics, but it can be a prescription for disaster in ecology.

This environmental "coming of age" is reflected in the broadening of intellectual perspective. Governments used to be preoccupied with domestic environmental affairs. Now, they are beginning to broaden their scope to confront problems that cross international borders, such as transboundary air and water pollution, and threats of a planetary nature, such as stratospheric ozone depletion and climatic warming. It is becoming increasingly evident that only decisive mutual action can secure the kind of world we seek.

New International Instruments

Over the past few decades, the global community has responded to emerging environmental problems with unprecedented international agreements—notably the Montreal Protocol, the Framework Convention on Climate Change, the Convention on Biological Diversity, and the Convention to Combat Desertification. There are many others.

In the course of crafting these global agreements, many important lessons have been learned. As the experience with the Montreal Protocol shows, the scientific community can play a crucial role in two ways: first, confirming the links between human activities and global environmental problems; and second, showing what could happen to human health and the global environment if nothing is done.

When the evidence is in hand, an international consensus to act can emerge quickly. The same process is underway in the current international debate on climate change, but the process has been more difficult because the linkages between human activities and global environmental impacts are more complex and still not completely understood. Nevertheless, a global consensus for action is emerging. Another important lesson learned has to do with the structure of international agreements, and the elements that can contribute to an effective structure. In the case of the Montreal Protocol, the agreement was not punitive and favored incentives and results-oriented approaches. All nations participated in the agreement, but at different levels of responsibility in recognition of their differing conditions. The Kyoto Protocol to the Climate Change Convention has benefited from the experience with the Montreal Protocol and included many of the same elements.

A third vital lesson is that, to the extent possible, every interested party must have an opportunity to participate as full partners in the process and to voice their concerns. It is particularly helpful for environmental advocacy groups and the business community to be part of this process. International agreements need to provide incentives to foster public-private partnerships, and to provide a role for business leaders to seek innovative technical solutions.

A fourth lesson concerns the role of governments in implementing these conventions. Government actions need to be consistent and predictable; provide sufficient lead times; favor government-led incentives over direct industry subsidies; and use flexible, market-based solutions where they are appropriate.

Finally, a fifth lesson is that agreements mark the beginning of a process, not the end. Scientists and nongovernmental organizations must continue to further global understanding of environmental problems and communicate what they have learned to the public and policy makers. Policy makers, in turn, must be flexible and respond to changing circumstances with new or modified policy solutions.

It All Comes Back to You

No set of policies, no system of incentives, no amount of information can substitute for individual responsibility when it comes to ensuring that our grandchildren will enjoy a quality of life that comes from a quality environment. Information can provide a basis for action. Vision and ideas can influence perceptions and inspire change. New ways to make decisions can empower those who seek a role in shaping the future. However, all of this will be meaningless unless individuals acting as citizens, consumers, investors, managers, workers, and professionals decide that it is important to them to make choices on the basis of a broader, longer view of their self-interest; to get involved in turning those choices into action; and, most importantly, to be held accountable for their actions.

The combination of political will, technological innovation, and a very large investment of resources and human ingenuity in pursuit of environmental goals has produced enormous benefits over the past two decades. This is an achievement to celebrate, but in a world that steadily uses more materials to make more goods for more people, we must recognize that we will have to achieve more in the future for the sake of the future. We must move toward a world in which zero waste will become the ideal for society that zero defects has become for manufacturing. Finally, we must recognize that the pursuit of one set of goals affects others and that we must pursue policies that integrate economic, environmental, and social goals.

Summary

Politics and the environment cannot be separated. In the United States, the government is structured into three separate branches, each of which impacts environmental policy.

The increase in environmental regulations in the U.S. over the past 30 years has caused concerns in some sectors of society.

The late 1980s and early 1990s witnessed a new international concern about the environment, both in the developed and developing nations of the world. Environmentalism is also seen as a growing factor in international relations. This concern is leading to international cooperation where only tension had existed before.

While there exists no world political body that can enforce international environmental protection, the list of multilateral environmental organizations is growing.

It remains too early to tell what the ultimate outcome will be, but progress is being made in protecting our common resources for future generations. Several international conventions and treaties have been successful. In the final analysis, however, each of us has to adjust our lifestyle to clean up our own small part of the world.

Interactive Exploration

Check out the website at

http://www.mhhe.com/environmentalscience

and click on the cover of this textbook for interactive versions of the following:

KNOW THE BASICS

executive branch *442*
judicial branch *442*
legislative branch *442*
policy *442*

On-line Flashcards
Electronic Glossary

IN THE REAL WORLD

Would you participate in an environmental protest? Would you be willing to go to jail for the cause? Are protests even effective anymore? Protestors in Seattle gathered to bring attention to World Trade Organization decisions that may threaten environmental security in some nations. Check out **Environmental Concerns in the "Battle for Seattle," At World Trade Organization Talks** to see how effective this protest was in bringing attention to the decision-making processes of a powerful international organization.

Is environmental security an issue that will receive more attention in the future? For further information about the cyanide spill that is causing arguments about environmental security between Hungary and Romania, take a look at **Europe's Worst Fish Kill Follows Romanian Mine Spill.**

The Swedish parliament is leading the way in making sure that environmental issues are an important part of national decision making. Check out **Swedish Parliament Plans Budget for Environmental Indicators** to see how Sweden is planning to include environmental indicators in their national budget.

TEST PREPARATION

Review Questions

1. What are the major responsibilities of each of the three branches of the U.S. government?
2. What are some of the enforcement options in U.S. environmental policy?
3. What role does administrative law play in U.S. environmental policy?
4. What are some of the criticisms of U.S. environmental policy?
5. In the past 10 years, how has public opinion in the United States changed concerning the protection of the environment?
6. Why is environmentalism a growing factor in international relations?
7. Give some examples of international environmental conventions and treaties.

Critical Thinking Questions

1. Does Chapter 20 have an overall point of view? If you were going to present the problems of environmental policy making and enforcement to others, what framework would you use?
2. The authors of this text say that "we are progressing from an environmental paradigm based on cleanup and control to one including assessment, anticipation, and avoidance." Do you agree with this assessment? Are there environmental problems that are harder to be proactive about than others?
3. Does a command-and-control approach to environmental problems, an approach that emphasizes regulation and remediation, make sense with global environmental problems such as global climite change, habitat destruction, and ozone depletion?
4. What values, perspectives, and beliefs does the Wise Use Movement exhibit in their response to environmental legislation? How are they similar to, and/or different from your own?
5. How is it best, as a global society with many political demarcations, to preserve the resources that are held in common? What special problems does this kind of preservation entail?
6. Do you agree with William Ruckelshaus that current environmental problems require a change on the part of industrialized and developing countries that would be "a modification in society comparable in scale to the agricultural revolution . . . and Industrial Revolution"? What kinds of changes might that mean in your life? Would these be positive, or negative, changes?
7. New treaties regarding free trade might enable some nations to argue that other nations' environmental legislation is too restrictive, thereby imposing a barrier to trade that is subject to sanction. What special problems and possibilities might the new global economy provide environmental preservation? What do you think about that?

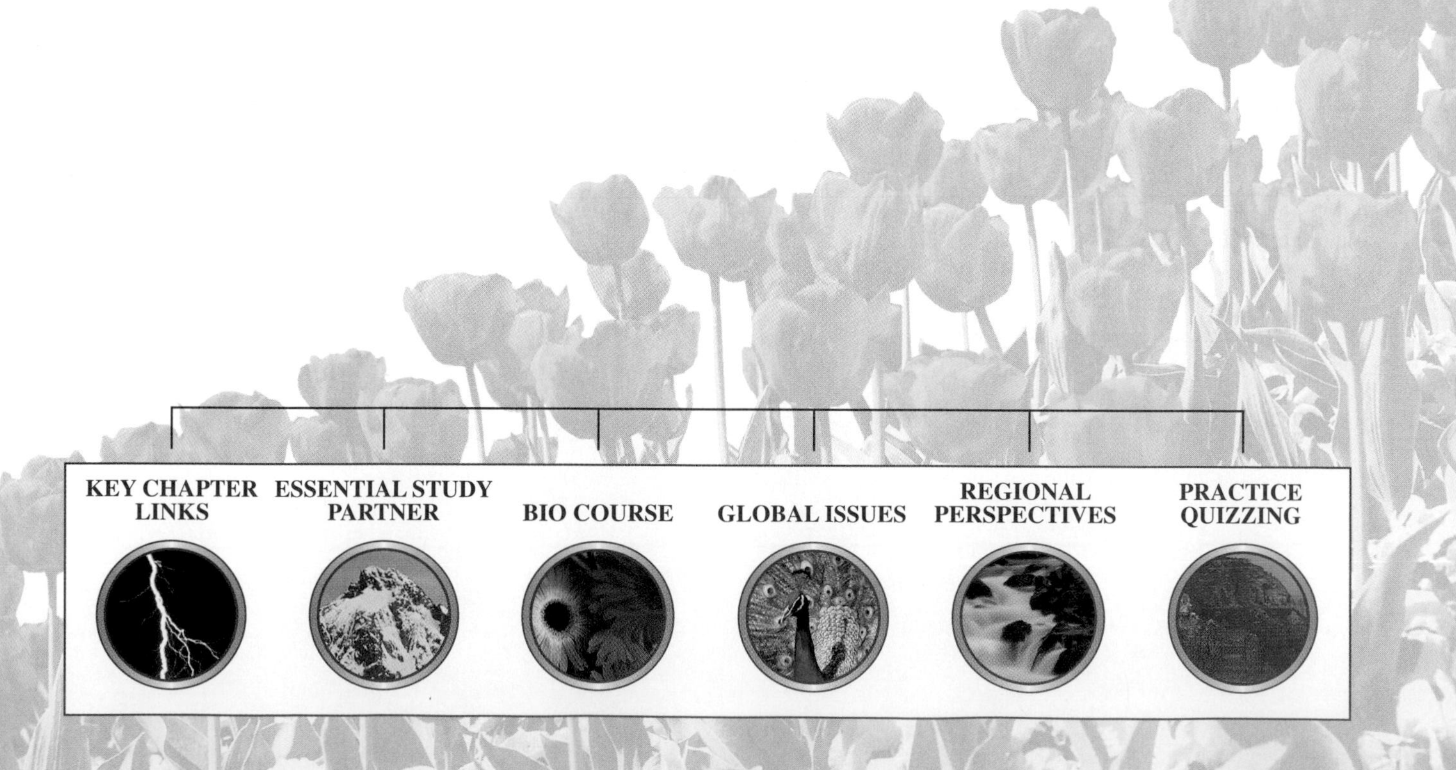

Appendix 1

CRITICAL THINKING

We live in an age of information. Computers, e-mail, the Internet, CD-ROMs, instant news, and fax machines bring us information more quickly than ever before. A simple search of the Internet will provide huge amounts of information. Some of the information has been subjected to scrutiny and is quite valid, some is well-informed opinion, some is naive misinformation, and some is even designed to mislead. How do we critically evaluate the information we get?

Critical thinking involves a set of skills that help us to evaluate information, arguments, and opinions in a systematic and thoughtful way. Critical thinking also can help us better understand our own opinions as well as the points of view of others. It can help us evaluate the quality of evidence, recognize bias, characterize the assumptions behind arguments, identify the implications of decisions, and avoid jumping to conclusions.

Characteristics of Critical Thinking

Critical thinking involves skills that allow us to sort information in a meaningful way and discard invalid or useless information while recognizing that which is valuable. Some key components of critical thinking are:

Recognize the importance of context. All information is based on certain assumptions. It is important to recognize what those assumptions are. Critical thinking involves looking closely at an argument or opinion by identifying the historical, social, political, economic, and scientific context in which the argument is being made. It is also important to understand the kinds of bias contained in the argument and the level of knowledge the presenter has.

Consider alternative views. A critical thinker must be able to understand and evaluate different points of view. Often these points of view may be quite varied. It is important to keep an open mind and to look at all the information objectively and try to see the value in alternative points of view. Often people miss obvious solutions to problems because they focus on a certain avenue of thinking and unconsciously dismiss valid alternative solutions.

Expect and accept mistakes. Good critical thinking is exploratory and speculative, tempered by honesty and a recognition that we may be wrong. It takes courage to develop an argument, engage in debate with others, and admit that your thinking contains errors or illogical components. By the same token, be willing to point out what you perceive to be shortcomings in the arguments of others. It is always best to do this with good grace and good humor.

Have clear goals. When analyzing an argument or information, keep your goals clearly in mind. It is often easy to get sidetracked. A clear goal will allow you to quickly sort information into that which is pertinent and that which may be interesting but not germane to the particular issue you are exploring.

Evaluate the validity of evidence. Information comes in many forms and has differing degrees of validity. When evaluating information, it is important to understand that not all the information from a source may be of equal quality. Often content about a topic is a mix of solid information interspersed with less certain speculations or assumptions. Apply a strong critical attitude to each separate piece of information. Often what appears to be a minor, insignificant error or misunderstanding can cause an entire argument to unravel.

Critical thinking requires practice. As with most skills, you become better if you practice. At the end of each chapter in the text, there are a series of questions that allow you to practice critical thinking skills. Some of these questions are straightforward and simply ask you to recall information from the chapter. Others ask you to apply the information from the chapter to other similar contexts. Still others ask you to develop arguments that require you to superimpose the knowledge you have gained from the chapter on quite different social, economic, or political contexts from your own.

Practice, Practice, Practice.

Appendix 2

METRIC UNIT CONVERSION TABLES

The Metric System

Standard metric units		Abbreviations
Standard unit of mass	Gram	g
Standard unit of length	Meter	m
Standard unit of volume	Liter	L

Common prefixes		Examples
Kilo	1,000	A kilogram is 1,000 grams.
Centi	0.01	A centimeter is 0.01 meter.
Milli	0.001	A milliliter is 0.001 liter.
Micro (μ)	One-millionth	A micrometer is 0.000001 (one-millionth) of a meter.
Nano (n)	One-billionth	A nanogram is 10^{-9} (one-billionth) of a gram.
Pico (p)	One-trillionth	A picogram is 10^{-12} (one-trillionth) of a gram.

Units of Length

Unit	Abbreviation	Equivalent
Meter	m	Approximately 39 in
Centimeter	cm	10^{-2} m
Millimeter	mm	10^{-3} m
Micrometer	μm	10^{-6} m
Nanometer	nm	10^{-9} m
Angstrom	Å	10^{-10} m

Length conversions

1 in = 2.5 cm	1 mm = 0.039 in
1 ft = 30 cm	1 cm = 0.39 in
1 yd = 0.9 m	1 m = 39 in
1 mi = 1.6 km	1 m = 1.094 yd
	1 km = 0.6 mi

To convert	Multiply by	To obtain
Inches	2.54	Centimeters
Feet	30	Centimeters
Centimeters	0.39	Inches
Millimeters	0.039	Inches

Units of Volume

Unit	Abbreviation	Equivalent
Liter	L*	Approximately 1.06 qt
Milliliter	mL	10^{-3} (1 mL = 1 cm^3 = 1 cc)
Microliter	μL	10^{-6} L

Volume conversions

1 tsp = 5 mL	1 mL = 0.03 fl oz
1 tbsp = 15 mL	1 L = 2.1 pt
1 fl oz = 30 mL	1 L = 1.06 qt
1 cup = 0.24 L	1 L = 0.26 gal
1 pt = 0.47 L	
1 qt = 0.95 L	
1 gal = 3.8 L	

To convert	Multiply by	To obtain
Fluid ounces	30	Milliliters
Quarts	0.95	Liters
Milliliters	0.03	Fluid ounces
Liters	1.06	Quarts

*Note: Many people use an upper case "L" for the symbol for liter to avoid confusion with the number one (1). Similarly milliliter is written mL.

Units of Weight

Unit	Abbreviation	Equivalent
Kilogram	kg	10^3 g (approximately 2.2 lb)
Gram	g	Approximately 0.035 oz
Milligram	mg	10^{-3} g
Microgram	μg	10^{-6} g
Nanogram	ng	10^{-9} g
Picogram	pg	10^{-12} g

Weight conversions

1 oz = 28.3 g	1 g = 0.035 oz
1 lb = 453.6 g	1 kg = 2.2 lb
1 lb = 0.45 kg	

To convert	Multiply by	To obtain
Ounces	28.3	Grams
Pounds	453.6	Grams
Pounds	0.45	Kilograms
Grams	0.035	Ounces
Kilograms	2.2	Pounds

Temperature conversions

$$°C = \frac{(°F - 32) \times 5}{9}$$

$$°F = \frac{°C \times 9}{5} + 32$$

Some equivalents

0°C = 32°F
37°C = 98.6°F
100°C = 212°F

Appendix 3

THE PERIODIC TABLE OF THE ELEMENTS

Traditionally, elements are represented in a shorthand form by letters. For example, the formula for water, H_2O, shows that a molecule of water consists of two atoms of hydrogen and one atom of oxygen. These chemical symbols for each of the atoms can be found on any periodic table of the elements. Using the periodic table, we can determine the number and position of the various parts of atoms. Notice that atoms number 3, 11, 19, and so on are in column one. The atoms in this column act in a similar way since they all have one electron in their outermost layer. In the next column, Be, Mg, Ca, and so on act alike because these metals all have two electrons in their outermost electron layer. Similarly, atoms number 9, 17, 35, and so on all have seven electrons in their outer layer. Knowing how fluorine, chlorine, and bromine act, you can probably predict how iodine will act under similar conditions. At the far right in the last column, argon, neon, and so on all act alike. They all have eight electrons in their outer electron layer. Atoms with eight electrons in their outer electron layer seldom form bonds with other atoms.

Periodic Table of the Elements

Key: 1 (Atomic Number) / Hydrogen (Name) / H (Symbol) / 1.0079 (Atomic Weight)

Representative Elements (*s* Series): IA, IIA — Transition Metals (*d* Series of Transition Elements): IIIB–IIB — Representative Elements (*p* Series): IIIA–VIIIA

Period	IA	IIA	IIIB	IVB	VB	VIB	VIIB	VIIIB			IB	IIB	IIIA	IVA	VA	VIA	VIIA	VIIIA
1	1 Hydrogen **H** 1.0079																	2 Helium **He** 4.0026
2	3 Lithium **Li** 6.941	4 Beryllium **Be** 9.0122											5 Boron **B** 10.811	6 Carbon **C** 12.0112	7 Nitrogen **N** 14.0067	8 Oxygen **O** 15.9994	9 Fluorine **F** 18.9984	10 Neon **Ne** 20.179
3	11 Sodium **Na** 22.989	12 Magnesium **Mg** 24.305											13 Aluminum **Al** 26.9815	14 Silicon **Si** 28.086	15 Phosphorus **P** 30.9738	16 Sulfur **S** 32.064	17 Chlorine **Cl** 35.453	18 Argon **Ar** 39.948
4	19 Potassium **K** 39.098	20 Calcium **Ca** 40.08	21 Scandium **Sc** 44.956	22 Titanium **Ti** 47.90	23 Vanadium **V** 50.942	24 Chromium **Cr** 51.996	25 Manganese **Mn** 54.938	26 Iron **Fe** 55.847	27 Cobalt **Co** 58.933	28 Nickel **Ni** 58.71	29 Copper **Cu** 63.546	30 Zinc **Zn** 65.38	31 Gallium **Ga** 69.723	32 Germanium **Ge** 72.59	33 Arsenic **As** 74.992	34 Selenium **Se** 78.96	35 Bromine **Br** 79.904	36 Krypton **Kr** 83.80
5	37 Rubidium **Rb** 85.468	38 Strontium **Sr** 87.62	39 Yttrium **Y** 88.905	40 Zirconium **Zr** 91.22	41 Niobium **Nb** 92.906	42 Molybdenum **Mo** 95.94	43 Technetium **Tc** (99)	44 Ruthenium **Ru** 101.07	45 Rhodium **Rh** 102.905	46 Palladium **Pd** 106.4	47 Silver **Ag** 107.868	48 Cadmium **Cd** 112.40	49 Indium **In** 114.82	50 Tin **Sn** 118.69	51 Antimony **Sb** 121.75	52 Tellurium **Te** 127.60	53 Iodine **I** 126.904	54 Xenon **Xe** 131.30
6	55 Cesium **Cs** 132.905	56 Barium **Ba** 137.34	*57 Lanthanum **La** 138.91	72 Hafnium **Hf** 178.49	73 Tantalum **Ta** 180.948	74 Tungsten **W** 183.85	75 Rhenium **Re** 186.2	76 Osmium **Os** 190.2	77 Iridium **Ir** 192.2	78 Platinum **Pt** 195.09	79 Gold **Au** 196.967	80 Mercury **Hg** 200.59	81 Thallium **Ti** 204.37	82 Lead **Pb** 207.19	83 Bismuth **Bi** 208.980	84 Polonium **Po** (209)	85 Astatine **At** (210)	86 Radon **Rn** (222)
7	87 Francium **Fr** (223)	88 Radium **Ra** (226)	**89 Actinium **Ac** (227)	104 Rutherfordium **Rf** (261)	105 Hahnium **Ha** (262)	106 Seaborgium **Sg** (263)	107 Neilsbohrium **Ns** (261)	108 Hassium **Hs** (265)	109 Meitnerium **Mt** (266)									

Inner Transition Elements (*f* Series)

*Lanthanides	58 Cerium **Ce** 140.12	59 Praseodymium **Pr** 140.907	60 Neodymium **Nd** 144.24	61 Promethium **Pm** 144.913	62 Samarium **Sm** 150.35	63 Europium **Eu** 151.96	64 Gadolinium **Gd** 157.25	65 Terbium **Tb** 158.925	66 Dysprosium **Dy** 162.50	67 Holmium **Ho** 164.930	68 Erbium **Er** 167.26	69 Thulium **Tm** 168.934	70 Ytterbium **Yb** 173.04	71 Lutetium **Lu** 174.97
** Actinides	90 Thorium **Th** 232.038	91 Protactinium **Pa** (231)	92 Uranium **U** 238.03	93 Neptunium **Np** (237)	94 Plutonium **Pu** 244.064	95 Americium **Am** (243)	96 Curium **Cm** (247)	97 Berkelium **Bk** (247)	98 Californium **Cf** 242.058	99 Einsteinium **Es** (254)	100 Fermium **Fm** 257.095	101 Mendelevium **Md** 258.10	102 Nobelium **No** 259.10	103 Lawrencium **Lr** 260.105

Appendix 4

WHAT *YOU* CAN DO TO MAKE THE WORLD A BETTER PLACE IN WHICH TO LIVE

It is easy to complain; in fact, we all tend to do so almost on a daily basis. Complaining, however, never corrects a problem or helps to resolve a dispute. There are actions you can take that *will* help improve the environment. The following list offers a few suggestions, and others will probably come to mind once you have thought about it. Try to expand the list with others in your class, and give your suggestions to your instructor.

1. **Continue your education.** Learning should not stop when you leave school. Become informed about issues; then you can begin to bring about change.
2. **Do not feel responsible for every problem in the world.** You cannot do everything. Concentrate on issues that you feel strongly about and that you can do something about. Focus your energy.
3. **Think about the consequences of your profession and your lifestyle.** If they are damaging to other people or to the environment, adjust your behavior accordingly. Try to persuade friends, family, and coworkers to do the same. Make environmental awareness a family affair.
4. **Work with others.** Attend meetings of your local government and ask officials about their plans to prevent pollution. Often, officials are very responsive to visits of this kind. Being part of a group of people with similar interests gives you support and increases your effectiveness. If you cannot find an appropriate group, start one of your own.
5. **Become active in your community.** Organize a community conference to discuss positive approaches to pollution prevention. Invite public officials, industry and labor representatives, other interested groups, and individual citizens. Get all the facts, and then try to get the appropriate action programs initiated.
6. **Learn about the ecology of your bioregion.** Develop a sense of place that puts you in contact with your local physical environment. Learn about the unique environmental features of your area. What are the most urgent environmental problems?
7. **Vote.** You cannot improve your world by not voting. If you do not like the choices available, work to get individuals on the ballot who represent your interests.
8. **Think globally and act locally.** You need to be aware of global conditions, but you should also work to improve your own particular place.
9. **Do not be discouraged.** It is important to face facts honestly and to be realistic about the state of the world, but it does not help to wallow in despair. Do not dwell on negatives. Do what you can to improve the world, and take pleasure and pride in the small victories and elements of success.
10. **Try to leave things better than you found them.** Pick up a piece of litter on a beach or on your street, plant a tree, recycle your papers . . . the list goes on. Don't wait for the next person to begin—you can make a difference!

Appendix 5

HOW TO WRITE TO YOUR ELECTED OFFICIALS

1. Address your letter properly.
   a. Your representative:
   The Honorable ___________
   House Office Building
   Washington, DC 20515
   Dear Representative__________,
   b. Your senators:
   The Honorable ___________
   Senate Office Building
   Washington, DC 20510
   Dear Senator___________,
   c. The president:
   The President
   The White House
   1600 Pennsylvania Ave. NW
   Washington, DC 20500
   Dear Mr. President,
2. Tell who you are and why you are interested in this subject. Be sure to give your return address.
3. Always be courteous and reasonable. You can disagree with a particular position, but be respectful in doing so. You will gain little by being hostile, or abrasive.
4. Be brief. Keep letters to one page or less. Cover only one subject, and come to the point quickly. Trying to cover several issues confuses the subject and dilutes your impact.
5. Write in your own words. It is more important to be authentic than polished. Don't use form letters or stock phrases provided by others. Try to show how the issue affects the legislator's own district and constituents.
6. If you are writing about a specific bill, identify it by number (for instance, H.R. 321 or S. 123). You can get a free copy of any bill or committee report by writing to the House Document Room, U.S. House of Representatives, Washington, DC 20515 or the Senate Document Room, U.S. Senate, Washington, DC 20510. Copies of bills are also available on-line from The Congressional Record.
7. Ask your legislator to vote a specific way, support a specific amendment, or take a specific action. Otherwise you will get a form response that says: "Thank you for your concern. Of course I support clean air, pure water . . ."
8. If you have expert knowledge or specific relevant experience, share it. But do not try to intimidate, threaten, or dazzle your representative. Legislators see through artifice and posturing; they are professionals in this field!
9. If possible, include some reference to the legislator's past action on this or related issues. Show that you are aware of his or her past record and are following the issue closely.
10. Follow up with a short note of thanks after a vote on an issue that you support. Show your appreciation by making campaign contributions or working for candidates who support issues important to you.
11. Try to meet your senators and representatives when they come home to campaign, or visit their office in Washington if you are able. If they know who you are personally, they are more likely to listen when you call or write.

Glossary

A

abiotic factors Nonliving factors that influence the life and activities of an organism.

abyssal ecosystem The collection of organisms and the conditions that exist in the deep portions of the ocean.

acid Any substance that, when dissolved in water, releases hydrogen ions.

acid deposition The accumulation of potential acid-forming particles on a surface.

acid mine drainage A kind of pollution, associated with coal mines, in which bacteria convert the sulfur in coal into compounds that form sulfuric acid.

acid rain (acid precipitation) The deposition of wet acidic solutions or dry acidic particles from air.

activated sludge sewage treatment Method of treating sewage in which some of the sludge is returned to aeration tanks, where it is mixed with incoming wastewater to encourage degradation of the wastes in the sewage.

activation energy The initial energy input required to start a reaction.

active solar system A system that traps sunlight energy as heat energy and uses mechanical means to move it to another location.

acute toxicity A serious effect, such as a burn, illness, or death, that occurs shortly after exposure to a hazardous substance.

age distribution The comparative percentages of different age groups within a population.

agricultural products Any output from farming: milk, grain, meat, etc.

agricultural runoff Surface water that carries soil particles, nutrients, such as phosphate, nitrates, and other agricultural chemicals, as it runs off agricultural land to lakes and streams.

air stripping The process of pumping air through water to remove volatile materials dissolved in the water.

alpha radiation A type of radiation consisting of a particle with two neutrons and two protons.

alpine tundra The biome that exists above the tree line in mountainous regions.

anthropocentric Human-centered, a theory of moral responsibility that views the environment as a resource for humankind.

aquiclude An impervious confining layer of an aquifer.

aquifer A porous layer of earth material that becomes saturated with water.

aquitard A partially permeable layer in an aquifer.

artesian well The result of a pressurized aquifer being penetrated by a pipe or conduit, within which water rises without being pumped.

atom The basic subunit of elements, composed of protons, neutrons, and electrons.

auxin A plant hormone that stimulates growth.

B

base Any substance that, when dissolved in water, removes hydrogen ions from solution; forms a salt when combined with an acid.

benthic Describes organisms that live on the bottom of marine and freshwater ecosystems.

benthic ecosystems A type of marine or freshwater ecosystem consisting of organisms that live on the bottom.

beta radiation A type of radiation consisting of electrons released from the nuclei of many fissionable atoms.

bioaccumulation The buildup of a material in the body of an organism.

biocentric Life-centered, a theory of moral responsibility that states that all forms of life have an inherent right to exist.

biochemical oxygen demand (BOD) The amount of oxygen required by microbes to degrade organic molecules in aquatic ecosystems.

biocide A kind of chemical that kills many different types of living things.

biodegradable Able to be broken down by natural biological processes.

biodiversity A measure of the variety of kinds of organisms present in an ecosystem.

biomagnification The increases in the amount of a material in the bodies of organisms at successively higher trophic levels.

biomass Any accumulation of organic material produced by living things.

biome A kind of plant and animal community that covers large geographic areas. Climate is a major determiner of the biome found in a particular area.

biotic factors Living portions of the environment.

biotic potential The inherent reproductive capacity.

birthrate The number of individuals born per thousand individuals in the population per year.

black lung disease A respiratory condition resulting from the accumulation of large amounts of fine coal dust particles in miners' lungs.

boiling-water reactor (BWR) A type of light water reactor in which steam is formed directly in the reactor, which is used to generate electricity.

boreal forest A broad band of mixed coniferous and deciduous trees that stretches across northern North America (and also Europe and Asia); its northernmost edge is integrated with the arctic tundra.

brownfields Buildings and land that have been abandoned because they are contaminated and the cost of cleaning up the site is high.

brownfields development The concept that abandoned contaminated sites can be cleaned up sufficiently to allow some specified uses without totally removing all of the contaminants

C

carbamate A class of soft pesticides that work by interfering with normal nerve impulses.

carbon absorption The use of carbon particles to treat chemicals by having the chemicals attach to the carbon particles.

carbon cycle The cyclic flow of carbon from the atmosphere to living organisms and back to the atmospheric reservoir.

carbon dioxide (CO_2) A normal component of the Earth's atmosphere that in elevated concentrations may interfere with the Earth's heat budget.

carbon monoxide (CO) A primary air pollutant produced when organic materials, such as gasoline, coal, wood, and trash, are incompletely burned.

carcinogen A substance that causes cancer.

carcinogenic The ability of a substance to cause cancer.

carnivores Animals that eat other animals.

carrying capacity The optimum number of individuals of a species that can be supported in an area over an extended period of time.

catalyst A substance that alters the rate of a reaction but is not itself changed.

chemical bond The physical attraction between atoms that results from the interaction of their electrons.

chemical weathering Processes that involve the chemical alteration of rock in such a manner that it is more likely to fragment or to be dissolved.

chlorinated hydrocarbon A class of pesticide consisting of carbon, hydrogen, and chlorine, which are very stable.

chlorofluorocarbons (CFC) Stable compounds containing carbon, hydrogen, chlorine and fluorine. They were formerly used as refrigerants, propellants in aerosol containers, and expanders in foam products. They are linked to the depletion of the ozone layer.

chronic toxicity A serious effect, such as an illness or death, that occurs after prolonged exposure to small doses of a toxic substance.

clear-cutting A forest harvesting method in which all the trees in a large area are cut and removed.

climax community Last stage of succession; a relatively stable, long-lasting, complex, and interrelated community of plants, animals, fungi, and bacteria.

coevolution Two or more species of organisms reciprocally influencing the evolutionary direction of the other.

combustion The process of releasing chemical bond energy from fuel.

commensalism The relationship between organisms in which one organism benefits while the other is not affected.

community Interacting groups of different species.

competition An interaction between two organisms in which both require the same limited resource, which results in harm to both.

composting A waste disposal system whereby organic matter is allowed to decay to a usable product.

compound A kind of matter composed of two or more different kinds of atoms bonded together.

Comprehensive Environmental Response, Compensation, and Liability Act (CERCLA) The 1980 U.S. law that addressed the issue of cleanup of hazardous-waste sites.

confined aquifier An aquifier that is bounded on the top and bottom by impermeable confining layers.

conservation To use in the best possible way so that the greatest long-term benefit is realized by society.

conservation ethic An environmental ethic that stresses a balance between total development and absolute preservation.

consumers Organisms that use other organisms as food.

contour farming A method of tilling and planting at right angles to the slope, which reduces soil erosion by runoff.

controlled experiment An experiment in which two groups are compared. One, the control, is used as a basis of comparison and the other, the experimental, has one factor different from the control.

coral reef ecosystem A tropical, shallow-water, marine ecosystem dominated by coral organisms that produce external skeletons.

corporation A business structure that has a particular legal status.

corrosiveness Ability of a chemical to degrade standard materials.

cost-benefit analysis A method used to determine the feasibility of pursuing a particular project by balancing estimated costs against expected benefits.

cover A term used to refer to any set of physical features that conceals or protects animals from the elements or their enemies.

crust The thin, outer, solid surface of the Earth.

D

death phase The portion of the population growth curve of some organisms that shows the population declining.

death rate The number of deaths per thousand individuals in the population per year.

decommissioning Decontaminating and disassembling a nuclear power plant and safely disposing of the radioactive materials.

decomposers Small organisms, like bacteria and fungi, that cause the decay of dead organic matter and recycle nutrients.

demand Amount of a product that consumers are willing and able to buy at various prices.

demographic transition The hypothesis that economies proceed through a series of stages, beginning with growing populations high birth and death rates and low economic development and ending with stable populations with low birth and death rates and high economic development.

demography The study of human populations, their characteristics, and their changes.

denitrifying bacteria Bacteria that convert nitrogen compounds into nitrogen gas.

density-dependent limiting factors Those limiting factors that become more severe as the size of the population increases.

density-independent limiting factors Those limiting factors that are not affected by population size.

desert A biome that receives less than 25 centimeters (10 inches) of precipitation per year.

desertification The conversion of arid and semiarid lands into deserts by inappropriate farming practices or overgrazing.

detritus Tiny particles of organic material that result from fecal waste material or the decomposition of plants and animals.

development ethic Philosophy that states that the human race should be the master of nature and that the Earth and its resources exist for human benefit and pleasure.

dispersal Migration of organisms from a concentrated population into areas with lower population densities.

domestic water Water used for domestic activities, such as drinking, air conditioning, bathing, washing clothes, washing dishes, flushing toilets, and watering lawns and gardens.

E

ecocentrism An approach to environmental responsibility that maintains that the environment deserves direct moral consideration rather than consideration derived merely from human interests.

ecology A branch of science that deals with the interrelationship between organisms and their environment.

economic costs Those monetary costs that are necessary to exploit a natural resource.

economic growth The perceived increase in monetary growth within a society.

ecosystem A group of interacting species along with their physical environment.

ectoparasite A parasite that is adapted to live on the outside of its host.

electron The lightweight, negatively charged particle that moves around at some distance from the nucleus of an atom.

element A form of matter consisting of a specific kind of atom.

emergent plants Aquatic vegetation that is rooted on the bottom but has leaves that float on the surface or protrude above the water.

emigration Movement out of an area that was once one's place of residence.

endangered species Those species that are present in such small numbers that they are in immediate jeopardy of becoming extinct.

endoparasite A parasite that is adapted to live within a host.

energy The ability to do work.

energy cost The amount of energy required to exploit a resource.

environment Everything that affects an organism during its lifetime.

environmental costs Damage done to the environment as a resource is exploited.

environmental justice Fair application of laws designed to protect the health of human beings and ecosystems; that no groups suffer unequal environmental harm.

Environmental Protection Agency (EPA) U.S. government organization responsible for the establishment and enforcement of regulations concerning the environment.

environmental resistance The combination of all environmental influences that tend to keep populations stable.

environmental science An interdisciplinary area of study that includes both applied and theoretical aspects of human impact on the world.

enzyme Protein molecules that speed up the rate of specific chemical reactions.

erosion The processes that loosen and move particles from one place to another.

estuaries Marine ecosystems that consist of shallow, partially enclosed areas where freshwater enters the ocean.

ethics A discipline that seeks to define what is fundamentally right and wrong.

euphotic zone The upper layer in the ocean where the sun's rays penetrate.

eutrophication The enrichment of water (either natural or cultural) with nutrients.

eutrophic lake A usually shallow, warm-water lake that is nutrient rich.

evapotranspiration The process of plants transporting water from the roots to the leaves where it evaporates.

evolution A change in the structure, behavior, or physiology of a population of organisms as a result of some organisms with favorable characteristics having greater reproductive success than those organisms with less favorable characteristics.

executive branch The office of the President of the United States.

experiment An artificial situation designed to test the validity of a hypothesis.

exponential growth phase The period during population growth when the population increases at an ever-increasing rate.

external costs Expenses, monetary or otherwise, borne by someone other than the individuals or groups who use a resource.

extinction The death of a species; the elimination of all the individuals of a particular kind.

F

fecal coliform bacteria Bacteria found in the intestines of humans and other animals, often used as an indicator of water pollution.

first law of thermodynamics A statement about energy that says that under normal physical conditions, energy is neither created nor destroyed.

fissionable The property of the nucleus of some atoms that allows them to split into smaller particles.

fixation A form of waste immobilization in which materials, such as fly ash or cement, are mixed with hazardous waste to prevent the waste from dispersing.

floodplain Lowland area on either side of a river that is periodically covered by water.

floodplain zoning ordinances Designations that restrict future building in floodplains.

food chain The series of organisms involved in the passage of energy from one trophic level to the next.

food web Intersecting and overlapping food chains.

fossil fuels The organic remains of plants, animals, and microorganisms that lived millions of years ago that are preserved as natural gas, oil, and coal.

free-living nitrogen-fixing bacteria Bacteria that live in the soil and can convert nitrogen gas (N_2) in the atmosphere into forms that plants can use.

freshwater ecosystem Aquatic ecosystems that have low amounts of dissolved salts.

friable A soil characteristic that describes how well a soil crumbles.

fungicide A pesticide designed to kill or control fungi.

G

gamma radiation A type of electromagnetic radiation that comes from disintegrating atomic nuclei.

gas-cooled reactor (GCR) A type of nuclear reactor that uses graphite as a moderator and carbon dioxide or helium as a coolant.

geothermal energy The heat energy from the Earth's molten core.

grasslands Areas receiving between 25 and 75 centimeters (10–30 inches) of precipitation per year. Grasses are the dominant vegetation, and trees are rare.

greenhouse effect The property of carbon dioxide (CO_2) that allows light energy to pass through the atmosphere but prevents heat from leaving; similar to the action of glass in a greenhouse.

Green Revolution The introduction of new plant varieties and farming practices that increased agricultural production worldwide during the 1950s, 1960s, and 1970s.

gross national product (GNP) An index that measures the total goods and services generated annually within a country.

groundwater Water that infiltrates the soil and is stored in the spaces between particles in the earth.

groundwater mining Removal of water from an aquifer faster than it is replaced.

H

habitat The specific kind of place where a particular kind of organism lives.

habitat management The process of changing the natural community to encourage the increase in populations of certain desirable species.

hard pesticide A pesticide that persists for long periods of time; a persistent pesticide.

hazardous All dangerous materials, including toxic ones, that present an immediate or long-term human health risk or environmental risk.

hazardous substances Substances that can cause harm to humans or the environment.

hazardous-waste dump A disposal site for hazardous waste in a dump, landfill, or surface impoundment without any concern for potential environmental or health risks.

hazardous wastes Substances that could endanger life if released into the environment.

heavy-water reactor (HWR) A type of nuclear reactor that uses the hydrogen isotope deuterium in the molecular structure of the coolant water.

herbicide A pesticide designed to kill or control plants.

herbivores Primary consumers; animals that eat plants.

horizon A horizontal layer in the soil. The top layer (A horizon) has organic matter. The lower layer (B horizon) receives nutrients by leaching. The C horizon is partially weathered parent material.

host The organism a parasite uses for its source of food.

humus Partially decomposed organic matter typically found in the top layer of the soil.

hydrocarbons (HC) Group of organic compounds consisting of carbon and hydrogen atoms that are evaporated from fuel supplies or are remnants of the fuel that did not burn completely, and that act as a primary air pollutant.

hydrologic cycle Constant movement of water from surface water to air and back to surface water as a result of evaporation and condensation.

hydroxyl ion A negatively charged particle consisting of a hydrogen and an oxygen atom, commonly released from materials that are bases.

hypothesis A logical statement that explains an event or answers a question that can be tested.

I

ignitability Characteristic of materials that results in their ability to combust.

immigration Movement into an area where one has not previously resided.

incineration Method of disposing of solid waste by burning.

industrial ecology A concept that stresses cycling resources rather than extracting and eventually discarding them.

Industrial Revolution A period of history during which machinery replaced human labor.

industrial uses Uses of water for cooling and for dissipating and transporting waste materials.

insecticide A pesticide designed to kill or control insects.

in-stream uses Use of a stream's water flow for such purposes as hydroelectric power, recreation, and navigation.

integrated pest management A method of pest management in which many aspects of the pest's biology are exploited to control its numbers.

interspecific competition Competition between members of different species for a limited resource.

intraspecific competition Competition among members of the same species for a limited resource.

ion An atom or group of atoms that has an electric charge because it has either gained or lost electrons.

irrigation Adding water to an agricultural field to allow certain crops to grow where the lack of water would normally prevent their cultivation.

isotope Atoms of the same element that have different numbers of neutrons.

J

judicial branch That portion of the U.S. government that includes the court system.

K

keystone species One that has a critical role to play in the maintenance of specific ecosystems.

kinetic energy Energy of moving objects.

kinetic molecular theory The widely accepted theory that all matter is made of small particles that are in constant movement.

K-strategists Large organisms that have relatively long lives, produce few offspring, provide care for their offspring, and typically have populations that stabilize at the carrying capacity.

L

lag phase The initial stage of population growth during which growth occurs very slowly.

land The surface of the Earth not covered by water.

landfill A method of disposing of solid wastes that involves burying the wastes in specially constructed sites.

land-use planning The process of evaluating the needs and wants of the population, the characteristics and values of the land, and various alternative solutions before changes in land use are made.

law A fundamental rule of nature that is central to the understanding of many related aspects of nature and is considered true by nearly everyone although it is still possible that it could be disproved. Example; law of gravity.

law of conservation of matter A fundamental rule of nature that states that matter can neither be created nor destroyed.

LD_{50} A measure of toxicity; the dosage of a substance that will kill (lethal dose) 50 percent of a test population.

leaching The movement of minerals from the top layers of the soil to the B horizon by the downward movement of soil water.

legislative branch That portion of the U.S. government that is responsible for the development of laws.

light-water reactor A nuclear reactor that uses ordinary water as a coolant.

limiting factor The one primary condition of the environment that determines population size of an organism.

limnetic zone Region that does not have rooted vegetation in a freshwater ecosystem.

liquefied natural gas Natural gas that has been converted to a liquid by cooling to −162°C (−260°F).

liquid metal fast-breeder reactor (LMFBR) Nuclear fission reactor using liquid sodium as the moderator and heat transfer medium; produces radioactive plutonium-235, which can be used as a nuclear fuel.

lithosphere A combination of the crust and outer layer of the mantle that forms the plates that move over the Earth's surface.

litter A layer of undecomposed or partially decomposed organic matter on the soil surface.

littoral zone Region with rooted vegetation in a freshwater ecosystem.

loam A soil type with good drainage and good texture that is ideal for growing crops.

M

macronutrient A nutrient, such as nitrogen, phosphorus, and potassium, that is required in relatively large amounts by plants.

management ethic An environmental ethic that stresses a balance between total development and absolute preservation.

mangrove swamp ecosystems Marine shoreline ecosystems dominated by trees that can tolerate high salt concentrations.

mantle The layer of the Earth between the crust and the core.

marine ecosystems Aquatic ecosystems that have high salt content.

marsh Area of grasses and reeds that is either permanently flooded or flooded for a major part of the year.

mass burn A method of incineration of solid waste in which material is fed into a furnace on movable metal grates.

matter Substance with measurable mass and volume.

mechanical weathering Physical forces that reduce the size of rock particles without changing the chemical nature of the rock.

megalopolis A large, regional urban center.

micronutrient A nutrient needed in extremely small amounts for proper plant growth; examples are boron, zinc, and magnesium.

migratory birds Birds that fly considerable distances between their summer breeding areas and their wintering areas.

mixture A kind of matter consisting of two or more kinds of matter intermingled with no specific ratio of the kinds of matter.

moderator Material that absorbs the energy from neutrons released by fission.

molecule Two or more atoms chemically bonded to form a stable unit.

monoculture A system of agriculture in which large tracts of land are planted with the same crop.

morals Predominant feeling of a culture about ethical issues.

mortality The number of deaths per year.

multiple land use Land uses that do not have to be exclusionary, so that two or more uses of land may occur at the same time.

municipal landfill An area used for the containment of solid wastes.

municipal solid waste All the waste produced by the residents of a community.

mutualism The association between organisms in which both benefit.

N

natality The number of individuals added to the population through reproduction.

National Priority List A listing of hazardous-waste dump sites requiring urgent attention as identified by Superfund legislation.

natural resources Those structures and processes that can be used by humans for their own purposes but cannot be created by them.

natural selection A process that determines which individuals within a species will reproduce more effectively and therefore results in changes in the characteristics within a species.

nature centers Teaching institutions that provide a variety of methods for people to learn about and appreciate the natural world.

negligible risk A point at which there is no significant health or environmental risk.

neutralization Reacting acids with bases to produce relatively safe end products.

neutron Neutrally charged particle located in the nucleus of an atom.

niche The total role an organism plays in its ecosystem.

nitrifying bacteria Bacteria that are able to convert ammonia to nitrite which can be converted to nitrate.

nitrogen cycle The series of stages in the flow of nitrogen in ecosystems.

nitrogen-fixing bacteria Bacteria that are able to convert the nitrogen gas (N_2) in the atmosphere into forms that plants can use.

nitrous oxide N_2O, one of the oxides of nitrogen.

nonpersistent pesticide A pesticide that degrades in a short period of time.

nonpersistent pollutants Those pollutants that do not remain in the environment for long periods.

nonpoint source Diffuse pollutants, such as agricultural runoff, road salt, and acid rain, that are not from a single, confined source.

nonrenewable energy sources Those energy sources that are not replaced by natural processes within a reasonable length of time.

nonrenewable resources Those resources that are not replaced by natural processes, or those whose rate of replacement is so slow as to be noneffective.

nontarget organism An organism whose elimination is not the purpose of pesticide application.

northern coniferous forest See boreal forest.

nuclear breeder reactor Nuclear fission reactor designed to produce radioactive fuel from nonradioactive uranium and at the same time release energy to use in the generation of electricity.

nuclear chain reaction A continuous process in which a splitting nucleus releases neutrons that strike and split the nuclei of other atoms, releasing nuclear energy.

nuclear fission The decomposition of an atom's nucleus with the release of particles and energy.

nuclear fusion The union of smaller nuclei to form a heavier nucleus accompanied by the release of energy.

nuclear reactor A device that permits a controlled nuclear fission chain reaction.

nucleus The central region of an atom that contains protons and neutrons.

O

observation Ability to detect events by the senses or machines that extend the senses.

oligotrophic lakes Deep, cold, nutrient-poor lakes that are low in productivity.

omnivores Animals that eat both plants and other animals.

organophosphate A class of soft pesticides that work by interfering with normal nerve impulses.

outdoor recreation Recreation that uses the natural out-of-doors for leisure-time activities.

overburden The layer of soil and rock that covers deposits of desirable minerals.

oxides of nitrogen (NO, N_2O, and NO_2) Primary air pollutants consisting of a variety of different compounds containing nitrogen and oxygen.

ozone (O_3) A molecule consisting of three atoms of oxygen, which absorb much of the sun's ultraviolet energy before it reaches the Earth's surface.

P

parasite An organism adapted to survival by using another living organism (host) for nourishment.

parasitism A relationship between organisms in which one, known as the parasite, lives in or on the host and derives benefit from the relationship while the host is harmed.

parent material Material that is weathered to become the mineral part of the soil.

particulates Small pieces of solid materials, such as smoke particles from fires, bits of asbestos from brake linings and insulation, dust particles, or ash from industrial plants, that are dispersed into the atmosphere.

passive solar system A design that allows for the entrapment and transfer of heat from the sun to a building without the use of moving parts or machinery.

patchwork clear-cutting A forest harvest method in which patches of trees are clear-cut among patches of timber that are left untouched.

peat The first stage in the conversion of organic material into coal.

pelagic Those organisms that swim in open water.

pelagic ecosystem A portion of a marine or freshwater ecosystem that occurs in open water away from the shore.

periphyton Attached organisms in freshwater streams and rivers, including algae, animals, and fungi.

permafrost Permanently frozen ground.

persistent pesticide A pesticide that remains unchanged for a long period of time; a hard pesticide.

persistent pollutant A pollutant that remains in the environment for many years in an unchanged condition.

pest An unwanted plant or animal that interferes with domesticated plants and animals or human activity.

pesticide A chemical used to eliminate pests; a general term used to describe a variety of different kinds of pest killers, such as insecticides, fungicides, rodenticides, and herbicides.

pH The negative logarithm of the hydrogen ion concentration; a measure of the number of hydrogen ions present.

pheromone A chemical produced by one animal that changes the behavior of another.

photochemical smog A yellowish-brown haze that is the result of the interaction of hydrocarbons, oxides of nitrogen, and sunlight.

photosynthesis The process by which plants manufacture food. Light energy is used to convert carbon dioxide and water to sugar and oxygen.

photovoltaic cell A means of directly converting light energy into electricity.

phytoplankton Free-floating, microscopic, chlorophyll-containing organisms.

pioneer community The early stages of succession that begin the soil-building process.

plankton Tiny aquatic organisms that are moved by tides and currents.

plate tectonics The concept that the outer surface of the Earth consists of large plates that are slowly moving over the surface of a plastic layer.

plutonium-239 (Pu-239) A radioactive isotope produced in a breeder reactor and used as a nuclear fuel.

point source Pollution that can be traced to a single source.

policy Planned course of action on a question or a topic.

pollution Any addition of matter or energy that degrades the environment for humans and other organism.

pollution costs The private or public expenditures undertaken to avoid pollution damage once pollution has occurred and the increased health costs and loss of the use of public resources because of pollution.

pollution prevention To prevent either entirely or partially the pollution that would otherwise result from some production or consumption activity.

pollution-prevention hierarchy Regulatory controls that emphasize reducing the amount of hazardous waste produced.

polyculture A system of agriculture that mixes different plant species in the same plots of land.

population density A measure of how close organisms are to one another, generally expressed as the number of organisms per unit area.

porosity A measure of the size and number of spaces in an aquifer.

postwar baby boom A large increase in the birthrate immediately following World War II.

potable waters Unpolluted freshwater supplies suitable for drinking.

potential energy The energy of position.

prairies Grasslands.

precipitation Removal of materials by mixing with chemicals that cause the materials to settle out of the mixture.

predator An animal that kills and eats another organism.

preservation To keep from harm or damage; to maintain in its original condition.

preservation ethic Philosophy that considers nature to be so special that it should remain intact.

pressurized-water reactor (PWR) A type of light-water reactor in which the water in the reactor is kept at high pressure and steam is formed in a secondary loop.

prey An organism that is killed and eaten by a predator.

primary air pollutants Types of unmodified materials that, when released into the environment in sufficient quantities, are considered hazardous.

primary consumer An animal that eats plants (producers) directly.

primary sewage treatment Process that removes larger particles by settling or filtering raw sewage through large screens.

primary succession Succession that begins with bare mineral surfaces or water.

probability A mathematical statement about how likely it is that something will happen.

producer An organism that can manufacture food from inorganic compounds and light energy.

profitability The extent to which economic benefits exceed the economic costs of doing business.

proton The positively charged particle located in the nucleus of an atom.

public resources Those parts of the environment that are owned by everyone.

R

radiation Energy that travels through space in the form of waves or particles.

radioactive Describes unstable nuclei that release particles and energy as they disintegrate.

radioactive half-life The time it takes for half of the radioactive material to spontaneously decompose.

radon Radioactive gas emitted from certain kinds of rock; can accumulate in very tightly sealed buildings.

range of tolerance The ability organisms have to succeed under a variety of environmental conditions. The breadth of this tolerance is an important ecological characteristic of a species.

reactivity The property of materials that indicates the degree to which a material is likely to react vigorously to water or air, or to become unstable or explode.

recycling The process of reclaiming a resource and reusing it for another or the same structure or purpose.

reforestation The process of replanting areas after the original trees are removed.

rem A measure of the biological damage to tissue caused by certain amounts of radiation.

renewable energy sources Those energy sources that can be regenerated by natural processes.

renewable resources Those resources that can be formed or regenerated by natural processes.

repeatability An important criterion in the scientific method, it requires that independent investigators be able to repeat an experiment and get the same results.

replacement fertility The number of children per woman needed just to replace the parents.

reserves The known deposits from which materials can be extracted profitably with existing technology under present economic conditions.

Resource Conservation and Recovery Act (RCRA) The 1976 U.S. law that specifically addressed the issue of hazardous waste.

resource exploitation The use of natural resources by society.

resources Naturally occurring substances that can be utilized by people but may not be economic.

respiration The process that organisms use to release chemical bond energy from food.

ribbon sprawl Development along transportation routes that usually consists of commercial and industrial building.

risk assessment The use of facts and assumptions to estimate the probability of harm to human health or the environment that may result from exposures to specific pollutants, toxic agents, or management decisions.

risk management Decision-making process that uses input such as risk assessment, technological feasibility, economic impacts, public concerns, and legal requirements.

rodenticide A pesticide designed to kill rodents.

r-strategist Typically, a small organism that has a short life span, produces a large number of offspring, and does not reach a carrying capacity.

runoff The water that moves across the surface of the land and enters a river system.

S

salinization An increase in the amount of salt in soil due to the evaporation of irrigation water.

savanna Tropical biome having seasonal rainfall of 50 to 150 centimeters (20–60 inches) per year. The dominant plants are grasses, with some scattered fire- and drought-resistant trees.

science A method for gathering and organizing information that involves observation, asking questions about observations hypothesis formation, testing hypothesis, critically evaluating the results, and publishing information so that others can evaluate the process and the conclusions.

scientific method A way of gathering and evaluating information. It involves observation, hypothesis formation, hypothesis testing, critical evaluation of results, and the publishing of findings.

secondary air pollutants Pollutants produced by the interaction of primary air pollutants in the presence of an appropriate energy source.

secondary consumers Animals that eat animals that have eaten plants.

secondary recovery Techniques used to obtain the maximum amount of oil or natural gas from a well.

secondary sewage treatment Process that involves holding the wastewater until the organic material has been degraded by bacteria and other microorganisms.

secondary succession Succession that begins with the destruction or disturbance of an existing ecosystem.

second law of thermodynamics A statement about energy conversion that says that, whenever energy is converted from one form to another, some of the useful energy is lost.

selective harvesting A forest harvesting method in which individual high-value trees are removed from the forest, leaving the majority of the forest undisturbed.

septic tank Underground holding tank into which sewage is pumped and where biological degradation of organic material takes place; used in places where sewers are not available.

seral stage A stage in the successional process.

sere A stage in succession.

sewage sludge A mixture of organic material, organisms, and water in which the organisms consume the organic matter.

sex ratio Comparison between the number of males and females in a population.

soft pesticide A nonpersistent pesticide that breaks down into harmless products in a few hours or days.

soil A mixture of mineral material, organic matter, air, water, and living organisms; capable of supporting plant growth.

soil profile The series of layers (horizons) seen as one digs down into the soil.

soil structure Refers to the way that soil particles clump together. Sand has little structure because the particles do not stick to one another.

soil texture Refers to the size of the particles that make up the soil. Sandy soil has large particles, and clay soil has small particles.

solidification The conversion of liquid wastes to a solid form to allow for more safe storage or transport.

solid waste Unusable or unwanted solid products that result from human activity.

source reduction Reducing the amount of solid waste generated by using less, or converting from heavy packaging materials to lightweight ones.

speciation The process of developing a new species.

species A group of organisms that can interbreed and produce offspring capable of reproduction.

stable equilibrium phase The phase in a population growth curve in which the death rate and birthrate become equal.

standard of living The necessities and luxuries essential to a level of existence that is customary within a society.

steam stripping The use of heated air to drive volatile compounds from liquids.

steppe A grassland.

storm-water runoff Storm water that runs off of streets and buildings and is often added directly to the sewer system and sent to the municipal wastewater treatment facility.

strip farming The planting of crops in strips that alternate with other crops. The primary purpose is to reduce erosion.

submerged plants Aquatic vegetation that is rooted on the bottom and has leaves that stay submerged below the surface of the water.

subsidy A gift given to private enterprise by government when the enterprise is in temporary economic difficulty but is viewed as being important to the public.

succession Regular and predictable changes in the structure of a community, ultimately leading to a climax community.

successional stage A stage in succession.

sulfur dioxide (SO_2) A compound containing sulfur and oxygen produced when sulfur-containing fossil fuels are burned. When released into the atmosphere, it is a primary air pollutant.

Superfund The common name given to the U.S. 1980 Comprehensive Environmental Response, Compensation, and Liability Act, which was designed to address hazardous-waste sites.

supply Amount of a good or service available to be purchased.

supply/demand curve The relationship between available supply of a commodity or service and its demand. The supply and demand change as the price changes.

surface impoundment Pond created to hold liquid materials. Some may hold only water while others may be used to contain polluted water or liquid contaminants.

surface mining (strip mining) A type of mining in which the overburden is removed to procure the underlying deposit.

sustainable development Using renewable resources in harmony with ecological systems to produce a rise in real income per person and an improved standard of living for everyone.

swamp Area of trees that is either permanently flooded or flooded for a major part of the year.

symbiosis A close, long-lasting physical relationship between members of two different species.

symbiotic nitrogen-fixing bacteria Bacteria that grow within a plant's root system and that can convert nitrogen gas (N_2) from the atmosphere to nitrogen compounds that the plant can use.

synergism The interaction of materials or energy that increases the potential for harm.

T

taiga Biome having short, cool summers and long winters with abundant snowfall. The trees are adapted to winter conditions.

target organism The organism a pesticide is designed to eliminate.

technological advances Increasing use of machines to replace human labor.

temperate deciduous forest Biome that has a winter-summer change of seasons and that typically receives 75 to 150 centimeters (30 to 60 inches) or more of relatively evenly distributed precipitation throughout the year.

terrace A level area constructed on steep slopes to allow agriculture without extensive erosion.

tertiary sewage treatment Process that involves a variety of different techniques designed to remove dissolved pollutants left after primary and secondary treatments.

theory A unifying principle that binds together large areas of scientific knowledge.

theory of evolution The widely accepted concept that populations of living things can change genetically over time and that this change can lead to a population that is very well adapted to its environment.

thermal inversion The condition in which warm air in a valley is sandwiched between two layers of cold air and acts like a lid on the valley.

thermal pollution Waste heat that industries release into the environment.

thermal treatment A form of hazardous-waste destruction involving heating waste.

threatened species Those species that could become extinct if a critical factor in their environment were changed.

threshold level The minimum amount of something required to cause measurable effects.

total fertility rate The number of children born per woman per lifetime.

toxic A narrow group of substances that are poisonous and cause death or serious injury to humans and other organisms by interfering with normal body physiology.

toxicity A measure of how toxic a material is.

toxic waste Substances that are poisonous and cause death or serious injury to humans and animals when released into the environment.

tract development The construction of similar residential units over large areas.

transuranic waste Nuclear wastes of the U.S. weapons program that consist primarily of isotopes of plutonium.

trickling filter system A secondary sewage treatment technique that allows polluted water to flow over surfaces harboring microorganisms.

trophic level A stage in the energy flow through ecosystems.

tropical rainforest A biome with warm, relatively constant temperatures where there is no frost. These areas receive more than 200 centimeters (80 inches) of rain per year in rains that fall nearly every day.

tundra A biome that lacks trees and has permanently frozen soil.

U

unconfined aquifer An aquifer that usually occurs near the land's surface, receives water by percolation from above, and may be called a water table aquifer.

underground mining A type of mining in which the deposited material is removed without disturbing the overburden.

underground storage tank Tank located below ground level for the storage of materials, such as oil, gasoline, or other chemicals.

uranium-235 (U-235) A naturally occurring radioactive isotope of uranium used as fuel in nuclear reactors.

urban growth limit A boundary established by municipal government that encourages development within the boundary and prohibits it outside the boundary.

urban sprawl A pattern of unplanned, low-density housing and commercial development outside of cities that usually takes place on previously undeveloped land.

V

vadose zone A zone above the water table and below the land surface that is not saturated with water.

vector An organism that carries a disease from one host to another.

W

waste destruction Destruction of a portion of hazardous waste with harmful residues still left behind.

waste immobilization Putting hazardous wastes into a solid form that is easier to handle and less likely to enter the surrounding environment.

waste minimization A process that involves changes that industries could make in the way they manufacture products that would reduce the waste produced.

waste separation Either separating one hazardous waste from another, or separating hazardous waste from nonhazardous material that it has contaminated.

water diversion The physical process of transferring water from one area to another.

water table The top of the layer of water in an aquifer.

waterways Low areas that water normally flows through.

weathering The physical and chemical breakdown of materials; involved in the breakdown of parent material in soil formation.

weed An unwanted plant.

wetlands Areas that include swamps, tidal marshes, coastal wetlands, and estuaries.

wilderness Designation of land use for the exclusive protection of the area's natural wildlife; thus, no human development is allowed.

windbreak The planting of trees or strips of grasses at right angles to the prevailing wind to reduce erosion of soil by wind.

Z

zero population growth The stabilized growth stage of human population during which births equal deaths and equilibrium is reached.

zoning Type of land-use regulation in which land is designated for specific potential uses, such as agricultural, commercial, residential, recreational, and industrial.

zooplankton Weakly swimming microscopic animals.

Credits:

Photographs

Half Title/Title Pages

Copyright Corbis (bee and thistle)

Dedication Pages

Copyright Corbis (seedling)

Ad Pages

Copyright Corbis (waterfall)
Copyright Digital Vision (bird on cattail)
Copyright Corbis (mountain peak)
Copyright Digital Vision (giraffe)

Front and Endmatter Page Heads

Copyright Corbis (mountain peak at sunset)
Copyright Digital Vision (deer in marsh)
Copyright Digital Vision (butterfly)
Copyright Corbis (starfish)

Part Opening Icons used on BTOC, TOC, and POs

Part 1 Copyright Corbis (stream)
Part 2 Copyright Digital Vision (lightning)
Part 3 Copyright Digital Vision (peacock)
Part 4 Copyright Artville (canyon river)
Part 5 Copyright Corbis (leaves)

Environmental Close-up Boxes

Copyright Corbis (leaf background)

Global Perspective Boxes

Copyright Digital Vision (earth shot)

Interactive EOC Section

Copyright Digital Vision (earth shot)
Copyright Corbis (background mountain range)
Copyright Corbis (background tulips)
Copyright Corbis (mountain peak button)
Copyright Digital Vision (lightning button)
Copyright Corbis (daisy button)
Copyright Digital Vision (peacock button)
Copyright Corbis (stream button)
Copyright Corbis (canyon button)

Point of View Readings

Copyright Digital Vision (antique map background)

Chapter 1

1.3: © Kaku Kurita/Liaison Agency; 1.5 (left): © Steve McCutcheon/Visuals Unlimited; 1.5 (middle): © Vol. 1/Photo Disc; 1.5 (right): © Vol. 86/CORBIS; 1.6 (left): © Vol. 102/CORBIS; 1.6 (middle): © Red Diamond Stock Photos/Bob Coyle, photographer; 1.6 (right): © Vol. 39/Photo Disc; 1.7 (top left): © Bayard H. Brattstrom/Visuals Unlimited; 1.7 (top right): © Vol. 98/CORBIS; 1.8 (top left): © Vol. 14/CORBIS; 1.7 (bottom left): © Vol. 15/CORBIS; 1.8 (top right): © Vol. 72/CORBIS; 1.7 (bottom right): © J. Eastcott and Y. Momatiuk/The Image Works; 1.8 (bottom left): © William J. Weber/Visuals Unlimited; 1.9 (left): © Vol. 25/PhotoDisc; 1.9 (middle): © Matt Bradley/Tom Stack & Associates; 1.9 (right): © Vol. 16/PhotoDisc; 1.10 (left): © Vol. 98/CORBIS; 1.10 (top right): © Vol. 16/PhotoDisc; 1.10 (bottom right): © Martin G. Miller/Visuals Unlimited

Chapter 2

2.1: © Vol. 34/PhotoDisc; 2.2 (top left): © Vol. 80/CORBIS; 2.2 (top right): © Vol. 74/CORBIS; 2.2 (bottom left): © Vol. 80/CORBIS; 2.2 (bottom right): © Vol. 10/PhotoDisc; p. 23 (Emerson): © Granger Collection; p. 23 (Thoreau): © Bettmann/CORBIS; p.23 (Muir): © Bettmann/CORBIS; p.23 (Leopold): © AP/Wide World Photos; p.23 (Carson): © AP/Wide World Photos; 2.3 (top left): © Toni Michaels; 2.4a: © Greg Vaughn/Tom Stack & Associates; 2.3 (top right): © Vol. 154/CORBIS; 2.3 (bottom): © CORBIS website; 2.4b: © Natalie Fobes/Stone Images; 2.5 (top left): © Nigel J.H. Smith/Animals Animals/Earth Scenes; 2.5 (top right): © Vol. 74/CORBIS; 2.5 (middle left): © Vol. 5/PhotoDisc; 2.5 (middle right): © Vol. 102/CORBIS; 2.5 (bottom left): © George Gainsburgh Bernard Photo Productions/Animals Animals/Earth Scenes; 2.5 (bottom right): © Vol. 102/CORBIS

Chapter 3

3.4 (top): © William E. Ferguson, photographer; 3.4 (top left): © Vol. 14/CORBIS; 3.4 (middle left): © McGraw-Hill Companies, Inc./Bob Coyle, photographer; 3.7 (top left): © Cedric Max Dunham/Photo Researchers, Inc.; 3.4 (bottom left): © Vol. 38/CORBIS; 3.4 (bottom right): © Scott Blackman/Tom Stack & Associates; 3.7 (top middle): © Vol. 10/PhotoDisc; 3.4 (middle right): © Vol. 38/CORBIS; 3.7 (top right): © Jan Halaska/The Image Works; 3.7 (bottom left): © Didier Givois/Vandystadt/Photo Researchers, Inc.; 3.7 (bottom middle): © Vol. 10/PhotoDisc; 3.7 (bottom right): © George E. Jones III/Photo Researchers, Inc.; 3.8 (right): © AP/Wide World Photos; 3.9 (right): © Vol. 31/PhotoDisc; p. 59 (left): © C. Allan Morgan/Peter Arnold Inc.; p. 59 (right): © John R. MacGregor/Peter Arnold, Inc.

Chapter 4

p. 68 (left and right): © McGraw-Hill Companies, Inc./Bob Coyle, photographer; 4.8: © Carl Purcell/Photo Researchers, Inc.

Chapter 5

5.3: © Joanne Lotter/Tom Stack & Associates; 5.5: © Vol. 26/PhotoDisc; 5.6: © Stephen Krasemann/Photo Researchers, Inc.; 5.7: © Fritz Polking/Peter Arnold, Inc.; 5.8 (top): © SPL/Photo Researchers, Inc.; 5.8 (bottom): © Manfred Kage/Peter Arnold, Inc.; 5.9: © Noble Proctor/Photo Researchers, Inc.; 5.10: © J. Burgess/SPL/Photo Researchers, Inc.; 5.11: © K. Maslowski/Visuals Unlimited; p. 95: © Marc A. Blovin, National Biological Survey/U.S. Dept. of Interior Fish and Wildlife Service; 5.16 (beans): © Alexander Lowry/Photo Researchers, Inc.; 5.16 (fawn): © Vol. 6/PhotoDisc; 5.16 (fox): © Paul Souders/Stone Images; 5.16 (duck): © Knolan Benfield/Visuals Unlimited; 5.16 (decomposers): © R. Kessel and C. Shih/Visuals Unlimited; 5.16 (nitrate No2): © David Phillips/Visuals Unlimited; 5.16 (nitrate No3): Fred Hossler/Visuals Unlimited

Chapter 6

6.1: © William E. Ferguson, photographer; 6.4: © Larry Mellichamp/Visuals Unlimited; 6.10b: © Leonard Lee Rue, Jr./Photo Researchers, Inc.; p. 114 (left and middle): © Harold Hungerford/University of Southern Illinois-Carbondale; p. 114 (right): © Carl Bollwinkel/University of Northern Iowa; 6.11b: © William E. Ferguson, photo by Stephanie Ferguson; 6.12b: © Vol. 6/CORBIS; p. 116: © Michael J. Balick/Peter Arnold, Inc.; 6.13b: © Eldon Enger; 6.14b: © Vol. 44/PhotoDisc; p. 119: © Jerry Franklin; 6.15b: © Vol. 36/PhotoDisc; 6.16b: © John Shaw/Tom Stack & Associates; 6.19: © Tammy Peluso/Tom Stack & Associates; 6.20: © William E. Ferguson, photographer;

p. 127: © Peter K. Ziminski/Visuals Unlimited

Chapter 7

7.4 (left): © Vol. 102/CORBIS; 7.4 (middle): © Vol. 86/CORBIS; 7.4 (right): © Vol. 19/PhotoDisc; p. 137 (kudza vine): © David M. Dennis/Tom Stack & Associates; p. 137 (dandelions): © Vol. 26/PhotoDisc; p. 137 (mussels): © David M. Dennis/Tom Stack & Associates; p. 137 (starlings): © Maslowski/Visuals Unlimited; p. 137 (moths): © William S. Omerod/Visuals Unlimited; p. 139: © Vol. 44/PhotoDisc

Chapter 8

p. 157: © Sue Cunningham/Stone Images

Chapter 9

9.1: © Irven Devore/Anthro-Photo; 9.2: © Nimatellah/Art Resource, NY; 9.3: © The Field Museum CSGEO 75400C; 9.5 (left): © Vol. 26/PhotoDisc; 9.5 (right): © Van Bucher/Photo Researchers, Inc.; 9.6 (left): © Steve McCurry/Magnum Photos; 9.6 (right): Copyright © Sun Ovens International, Inc.; 9.7: © Margot Granitsas/The Image Works; p. 177 (top and bottom left): Courtesy of General Motors Company; p. 177 (bottom right): Courtesy of General Motors

Chapter 10

Company; 10.8: © Michaud/Photo Researchers, Inc.; 10.10a: © Matt Meadows/Peter Arnold, Inc.; 10.10b: © David J. Cross/Peter Arnold, Inc.; 10.11: © Vol. 39/PhotoDisc; 10.13: © McGraw-Hill Companies, Inc./Bob Coyle, photographer; 10.14: © Natalie Fobes/Stone Images; 10.16: © John Running/Stock Boston; 10.17: © Michelango Durazzo/Magnum Photos; 10.18: © Paolo Koch/Photo Researchers, Inc.; 10.19: © Stephen Krasemann/Photo Researchers, Inc.; 10.21 and 10.22: © Tom McHugh/Photo Researchers, Inc.; 10.23: © Hank Morgan/Photo Researchers, Inc.; 10.25: © Susan Meiselas/Magnum Photos; 10.26: © Hank Morgan/Photo Researchers, Inc.; 10.27 (left): © McGraw-Hill Companies, Inc./Bob Coyle, photographer; 10.27 (middle): © Toni Michaels; 10.27 (right): Courtesy of Andersen Windows, Inc.; p. 212: © B.&C. Alexander/Photo Researchers, Inc.

Chapter 11

11.1 (top): U.S. Air Force; 11.1 (middle): © Tom McHugh/Photo Researchers, Inc.; 11.1 (bottom): © SIU/Visuals Unlimited; 11.12: © TASS/Sovfot/Eastfoto; 11.13: © Alain Morran/Liaison Agency; 11.14: © Alex Tsiaras/Photo Researchers, Inc.; 11.16: © Arthus-Bertrand/Peter Arnold, Inc.

Chapter 12

12.1 (left): © G. Prance/Visuals Unlimited; 12.1 (middle): © McGraw-Hill Higher Education, photographer, Barry W. Barker; 12.1 (right): © Spencer Grant/Photo Edit; 12.2 (Fish kill): © Vol. 44/PhotoDisc; 12.2 (Smoke): © Vol. 44/PhotoDisc; 12.2 (Feed lot): © J.C. Allen & Son; 12.2 (Strip): © Bob Daemmrich/The Image Works; 12.2 (Smog): © Vol. 25/PhotoDisc; 12.2 (Traffic): © Vol. 2/PhotoDisc;12.2 (Nuclear): © Vol. 44/PhotoDisc; 12.2 (Litter): © Vol. 31/PhotoDisc;12.3: © Grant Heilman Photography; 12.4: © Byron Augustin/Tom Stack & Associates; 12.5: © John Cancalosi/Peter Arnold, Inc.; 12.6: © Vol. 31/PhotoDisc; 12.7: © Julia Sims/Peter Arnold, Inc.; p. 247: © Kevin Schafer/Tom Stack & Associates; 12.8 (left): © Vol. 19/PhotoDisc; 12.8 (right): © Les Christman/Visuals Unlimited; 12.9: © J. Eastcott & Y. Momatiuk/The Image Works; p. 252: © Vol. 6/PhotoDisc; 12.13 (top left): © Tom Stack/Tom Stack & Associates; 12.13 (top right): © Jack Fields/Photo Researchers, Inc.; 12.13 (bottom left): © Patrice/Tom Stack & Associates; 12.13 (bottom right): Courtesy of Dept. of Fisheries, Olympia, WA; 12.18: © Dennis Paulson/Visuals Unlimited; 12.19 (Cinnamon teal): © Gary Milburn/Tom Stack & Associates; 12.19 (Black duck): © Richard H. Smith/Photo Researchers, Inc.12.19 (Mallards): © PhotoDisc website; 12.19 (Pintail drake): © Ken Highfill/Photo Researchers; Table 12.4 (left and right): © Vol. 6/CORBIS ; p. 263: © W. Perry Conway/Tom Stack & Associates; 12.21 (top left): © Tom McHugh/Photo Researchers, Inc.; 12.21 (top right): © Steven C. Kauffman/Peter Arnold, Inc.; 12.21 (middle left): © Edward Ross; 12.21 (middle right): © Miguel Castro/Photo Researchers, Inc.; 12.21 (bottom left): © Stan Wayman/Photo Researchers, Inc.; 12.21 (bottom right): © Jeff Le Pore/Photo Researchers, Inc.; p. 266: U.S. Fish and Wildlife Service, photo by John and Karen Hollingsworth

Chapter 13

13.2 (top): © Red Diamond Stock Photos/Bob Coyle, photographer; 13.2 (bottom): © Breck Kent/Animals Animals Earth Scenes; 13.3a: © Vol. 74/CORBIS; 13.3b: © Mark Phillips/Photo Researchers, Inc.; 13.3c: © Vol. 2/PhotoDisc 13.4: © W.T. Sullivan III/SPL/Photo Researchers, Inc.; 13.5: © Bob Daemmrich/The Image Works; 13.6: © James Shaffer; 13.7: Courtesy of Missouri Department of Transportation, photographer Mike Wright; 13.8: USDA; 13.9: © McGraw-Hill Companies, Inc.; p. 282: © Robert W. Ginn/Unicorn Stock Photos; 13.11 (left): © Thomas Hollyman/Photo Researchers, Inc.; 13.11 (right): © Bruce Berg/Visuals Unlimited; 13.12 (left): Courtesy of Roger Loewenberg; 13.12 (right): © Vanessa Vick/Photo Researchers, Inc.; 13.13 (left): © Vol. 2/PhotoDisc; 13.13 (right): © Robert Winslow/Tom Stack & Associates; p. 287: © Eric Hartman/Magnum Photos;

Chapter 14

14.4: © Spencer Swanger/Tom Stack & Associates; 14.11: © John Cunningham/Visuals Unlimited; 14.12: © Larry Miller/Photo Researchers, Inc.; 14.13: © P. Newman/Visuals Unlimited; 14.14: © Grant Heilman Photography; 14.15: © Steve McCurry/Magnum Photos; 14.17a: © Mark Boulton/Photo Researchers, Inc.; 14.17b: © Grant Heilman Photography; 14.18: © Grant Heilman Photography; 14.19: © Joe Munroe/Photo Researchers, Inc.; 14.20a: © Vol. 19/PhotoDisc; 14.20b: Courtesy of USDA Soil Conservation Service; 14.21a: © Grant Heilman Photography; 14.21b: © Larry Lefever/Grant Heilman Photography; 14.22a: © Grant Heilman Photography; 14.22b: © Link/Visuals Unlimited; 14.23: © John Griffin/The Image Works; 14.24: © Francis de Richemond/The Image Works; p. 310: © Earl Roberge/Photo Researchers, Inc.

Chapter 15

15.1: © Jacques Jangoux/Peter Arnold, Inc.; 15.2: © Vol. 111/CORBIS; 15.3: © Earl Roberge/Photo Researchers, Inc.; 15.7: © Larry Lefever/Grant Heilman Photography; 15.8: © Toni Michaels; 15.9: Courtesy of J.E. Nugent, NW Michigan Horticulture Research Foundation; p. 328 (top): © S. Witter/Visuals Unlimited; p. 328 (middle): © Erica Lansner/Stone Images; p. 328 (bottom): © Dave Bartruff/CORBIS; 15.13: © Jeremy Burgess/SPL/Photo Researchers, Inc.; 15.15a-d: © Dr. Bernd Blossey, Dept. of Natural Resources, Cornell University

Chapter 16

16.8: © Pat & Tom Leeson/Photo Researchers, Inc.; 16.10a: © Vol. 19/PhotoDisc; 16.10b: © Vol. 102/CORBIS; 16.10c: © John Colwell/Grant Heilman Photography; 16.11: © Vol. 39/PhotoDisc; 16.12: © Francois Gohier/Photo Researchers, Inc.; p. 352: © Steve Elmore/Tom Stack & Associates; 16.16a: © Ray Pfortner/Peter Arnold, Inc.; 16.16b: © Bob Daemmrich/Stock Boston; 16.16c: © Ray Pfortner/Peter Arnold, Inc.; p. 361 (top): © Vol. 6/PhotoDisc; p. 361 (bottom): © Wendell Mentzen/Index Stock Photography; 16.17: © Mark Gibson; p. 363: © TASS/Sovfoto/Eastfoto; 16.19: © Nada Pecnik/Visuals Unlimited; p. 366: © Brad Smith; p. 367 (top left): © Toni Michaels; p. 367 (middle left): © Gene

Marshall/Tom Stack & Associates; p. 367 (bottom): © William E. Ferguson; p. 367 (right): © Ken W. Davis/Tom Stack & Associates

Chapter 17

17.4: © John Cunningham/Visuals Unlimited; 17.5: © Vol. 31/PhotoDisc; 17.6 (top): © Peggy/Yoram Kahana/Peter Arnold, Inc.; 17.6 (bottom): © Peggy/Yoram Kahana/Peter Arnold, Inc.; 17.10: © Don & Pat Valenti/Tom Stack & Associates; 17.11: © John Shaw/Tom Stack & Associates; 17.12: © CP Picture Archive; p. 384: © Billy E. Barnes/Stock Boston; p. 397: © Vol. 31/PhotoDisc

Chapter 18

18.6: © Rapho Agency/Photo Researchers, Inc.; p. 406 (all): © McGraw-Hill Companies, Inc./Bob Coyle, photographer; 18.11b: © David M. Dennis/Tom Stack & Associates; p. 413: © David Putnan, courtesy of Weyerhaeuser; p. 415: Courtesy of General Motors; p. 416 (all): © McGraw-Hill Companies, Inc./Bob Coyle, photographer

Chapter 19

19.2 (1): © McGraw-Hill Companies, Inc./Bob Coyle, photographer; 19.2 (2): © John Colwell/Grant Heilmann; 19.2 (3): © Vol. 18/PhotoDisc; 19.2 (4): © McGraw-Hill Companies, Inc./Bob Coyle, photographer; 19.2 (5): © McGraw-Hill Companies, Inc./Bob Coyle, photographer; 19.2 (6): © Benelux Press BV/Photo Researchers, Inc.; 19.2 (7): © McGraw-Hill Companies, Inc./Bob Coyle, photographer; p. 423: © Kerry T. Givens/Tom Stack & Associates; 19.3: © Gary Milburn/Tom Stack & Associates; p. 427: © Lowell Georgia/Photo Researchers, Inc.; Table 19.3: © William Campbell/Peter Arnold, Inc.; p. 429: © McGraw-Hill Higher Education, photographer Barry Barker; p. 435: © Ken Sherman/Bruce Coleman/PNI

Chapter 20

20.1 (left): © Michael Dwer/Stock Boston; 20.1 (right): © Ashvin Mehta/Dihodia Picture Agency/The Image Works; 20.2 (left): © Red Diamond Photography/Bob Coyle, photographer; 20.2 (right): © Laurence Loway/Stock Boston; 20.3: © John Neubauer/Rainbow/PNI; 20.8 (top): Courtesy of New York State Department of Environmental Conservation; 20.8 (bottom): © Paul David Mozell/Stock Boston; 20.9: © Forrest Anderson/Liaison Agency; p. 449 (left): © C. Osbourne/Photo Researchers, Inc.; p. 449 (right): © Mary Kate Denny/Photo Edit; 20.10 (left): © Vol. 19/PhotoDisc; 20.10 (right): © A. Ramey/Stock Boston; p. 451: © Peter Turnley/CORBIS; p. 453: © Greenpeace/Morgan

Line Art

Chapter 2

EC Box (p. 25): © General Motors Corporation; Side Bar (p. 29): Reprinted with permission of The Detroit News.

Chapter 3

3.2: Reprinted with permission from "Great Lakes Fish Consumption Study" 1989, Great Lakes National Resource Center, Ann Arbor, MI.

Chapter 10

10.5: Adapted with permission from Arthur N. Strahler, *Planet Earth.* Copyright © 1972 by Arthur N. Strahler.; 10.21a: From *Solar Energy: A Biased Guide, International Library of Ecology Series.* Copyright © 1977 Domus Books. Reprinted by permission.

Chapter 14

14.1: From Charles C. Plummer, David McGeary, and Diane Carlson, *Physical Geology,* 8th edition. Copyright © 1999 McGraw-Hill Company, Inc. All Rights Reserved. Reprinted by permission.; 14.2: From Carla W. Montgomery, *Fundamentals of Geology,* 5th edition. Copyright © 1997 McGraw-Hill Company, Inc. All Rights Reserved. Reprinted by permission.; 14.8: 9.7, p. 136 from *Ecology and Field Biology,* 5th Edition, by Robert Leo Smith. Copyright © 1996 by Addison-Wesley Educational Publishers, Inc. Reprinted by permission of Addison Wesley Longman Publishers, Inc.

Chapter 15

IA graphs (p. 335): From Government of Canada, 1991, *The State of Canada's Environment, 1991.* Reproduced with permission of the Minister of Public Works and Government Services Canada, 1998.

Chapter 16

16.9: Reprinted by permission of the San Francisco Convention & Visitors Bureau.; EC box (p. 360): Copyright, April 2, 1990, U.S. News & World Report.

Chapter 17

17.9: From *Environment,* volume 25, issue 4, pp. 6–9, 1983. Reprinted with permission of the Helen Dwight Reid Educational Foundation. Published by Heldref Publications, 1319 18th St., N.W., Washington, D.C. 20036-1802. Copyright © 1983.

Index

A

Abiotic factors, 81–82
Abyssal ecosystems, 124
Acid deposition
  air pollution, 6, 194, 381–383
  water pollution, 349
Acid mine drainage, 194
Acid rain (acid precipitation)
  Clean Air Act and, 381
  effects, 381–383
  international air quality and, 6, 398
Acids, 70–71
Acquired immunodeficiency syndrome (AIDS), 161
Activated-sludge sewage treatment, 359
Activation energy, 71
Active solar systems, 203–204, 205
Acute toxicity, 423
Adaptability, sustainability and, 53
Additives, food, 331
Administrative law, 446
Administrative Procedure Act, 446
Advertising, green, 51
Aerosol propellants, 388
Aesthetic pollution, 282
Africa
  food production, 155
  global warming and, 386
  groundwater depletion, 364
  HIV/AIDS epidemic, 161
African elephants, 139
Age distribution, 132–133
  population growth and, 149
  U.S., 157, 158
Agenda 21, 31
Agricultural regions
  food production, population growth and, 154–156
  global, 302, 304
  middle North American, 11, 12
  urban sprawl and, 276
  U.S., 302, 303
Agricultural runoff, 353
Agriculture, 315–335. *See also* Soil conservation practices
  air pollution and, 11
  alternative methods, 329–335
  Chinese, 328
  conventional methods, 116, 316–318
  erosion effects, 301, 308
  fertilizer and. *See* Fertilizers
  food additives, 331
  fossil fuel *versus* muscle power, 318
  global warming and, 389, 391–392
  methane production, 389
  natural ecosystems and, 243–244
  pest problems, 316, 317, 318. *See also* Pesticides
  tillage methods, 303–307
Agriculture–*Cont.*
  water pollution, 352–353, 355
  water use, 344–346, 347
AIDS (acquired immunodeficiency syndrome), 161
Air. *See* Atmosphere
Air pollution, 373–398. *See also specific pollutant*
  acid deposition, 6, 194, 381–383, 398
  agricultural origins, 11
  air toxics causing, 378, 379
  atmosphere and, 373–374
  automobiles causing, 176
  Clean Air Act, 380–381
  control, 379–381
  fuelwood causing, 209
  global warming and climate change effects, 385–392
  incineration and, 408
  indoor, 394–397
  international, 5–6, 398
  international conference, 452–454
  land-use planning and, 282
  in Mexico City, 376
  noise causing, 397
  photochemical smog, 377–378, 379
  primary pollutants, 374–377
  radon causing, 394–395
  secondary pollutants, 374–375, 377–378
  secondhand smoke, 384
  technological responses, 392–393
  urban sprawl and, 275
Air quality
  global warming and, 385–392
  improvement, cost-benefits analysis, 49
  international disputes, 5–6, 398
  ozone depletion, 393, 396
  pollutants affecting. *See* Air pollution
  U.S. legislation, 6, 380–381, 445
Air stripping, 431
Air toxics, 379
Air travel trends, 283
Alaska, 121, 212
Alewives, 253
Alpha radiation, 216
Alpine tundra, 120
Alternative agriculture, 330
Amsterdam Treaty, 452
Anaerobic digestion, 206
Animals
  extinction, 260–264
  as power source, 169
  reserves for, 47
  soil formation and, 294, 295
Antarctica, 33, 452
Antarctic Treaty, 452
Anthropocentric environmentalism, 21
Aphids, 332
Aquaculture, 255
Aquatic ecosystems, 120, 250
  acid rain and, 382, 383
  aquaculture, 255
  eutrophication in, 350
  freshwater, 125–126, 253–255
  marine, 121–124, 250–253
  persistent pesticides and, 324–325, 326
  primary succession, 108
  secondary succession, 109
  thermal pollution and, 353–355
Aquatic organisms, 97, 101
Aquiclude, 341
Aquifers, 341
  groundwater in, 343
  groundwater mining, 361–362, 364
  types, 341–342
Aquitard, 341
Arabian peninsula, groundwater, 364
Arab-Israeli War, 178
Aral Sea, 363
Arctic National Wildlife Refuge, 212
Argentina, 153
Aroclor, 425
Arsenic, 425
Artesian wells, 342
Asbestos, 396
Ash toxicity, 408
Aswan Dam, 200
Atmosphere, 373
  greenhouse gases, 386–389
  ozone depletion, 393, 396
  pollution. *See* Air pollution
Atoms. *See also specific atom*
  nutrient cycles, 95–101
  structure, 69–70
AT&T, 54
Australian Stock Exchange, 47
Automobiles
  air pollutants from, 176, 375–376, 377, 378, 379, 381
  alternative fuels, 176–177, 181
  battery recovery, 415
  economic and energy impact, 172
  energy consumption, 174, 175
  historical use, 171
  hybrid electric models, 176
  lead emissions, 378, 379, 427
  off-road, land-use conflicts, 286
  use trends, 283
Auxins, 322–323, 324

B

Bacillus thuringiensis, 333
Bacteria
  carrying capacity, 135–136
  denitrifying, 98
  drinking water contaminants, 351
  fecal coliform bacteria, 351–352
  methane-releasing, 389
  nitrifying, 98
  nitrogen-fixing, 98
Bacteria–*Cont.*
  sewage treatment and, 358–360
  soil and, 295, 297
Bangladesh, 272
Basel Convention, 433–434
Bases (compound), 70–71
Batteries, recovery, 415
Beavers, 82–83, 109, 110
Beetles, pest control, 332–333
Benthic ecosystems, 121–124
Benthic organisms, 121
Benzene, 425
Benzo[a]pyrene, 425
Benzo[b]fluoranthene, 425
Beta radiation, 216
Beverage containers, 411
Bioaccumulation, of persistent pesticides, 324–325
Biocentric environmentalism, 21
Biochemical oxygen demand (BOD), 126
  water pollution and, 349, 350, 353
Biocides, 319
Biodegradable, 239
Biodegradable pollutants, 424, 431
Biodiversity, 243
  global warming and, 392
  loss, 55, 243–244, 261, 262. *See also* Extinction
  protective measures, 262–265
  tropical rainforest resources, 117
Biological energy sources, 169–170
Biological pest control, 332–334
Biological pollutants, indoor, 396
Biological wastewater treatment, 358–360, 362
Biology
  population growth and, 148–149
  radiation effects on, 227–228
Biomagnification, of persistent pesticides, 324–325, 326
Biomass, 94, 205
  carbon dioxide production, 386
  conversion, 205–207
Biomass conversion, 205–207
Biomes, 110–111
  desert, 111–113
  elevation effects, 111, 112
  grassland, 113–114, 115
  savanna, 114–116
  taiga, northern coniferous or boreal forest, 119–120
  temperate deciduous forest, 117–120
  tropical rainforest, 116–117, 118
  tundra, 120, 121
Biophysical world, economics and, 55–56
Biotechnology. *See* Genetic engineering
Biotic factors, 81, 82
Biotic potential, 134
Birds, persistent pesticides and, 325, 326

- Birth control
  - for African elephants, 139
  - Malthus theory and, 148
  - methods, 150
  - use, 151–152
- Birthrates, 131, 148–152
- Bison, 252
- Blackfooted ferret, 265
- Black lung disease, 193
- BOD. *See* Biochemical oxygen demand
- Bog, floating, 108
- Boiling-water reactors (BWRs), 218
- Bollworm, 326–327
- Bombay, India, 272
- Boreal forest, 111, 119–120
- Bottle bills, 411
- Boundary Waters Treaty, 6
- Brazil, 200
- Breast feeding, 150
- Breeder reactors, 221–222
- Brownfields, 285
- Brownfields development, 285
- Burma, 115
- Bus transport trends, 283
- Butterfly, mission blue, 265
- BWRs (boiling-water reactors), 218

C

- Cadmium, 425, 429
- Cairo, Egypt, 113
- California
  - groundwater depletion, 362, 364
  - smoking laws, 384
  - water plan, 367
- California condor, 263
- California Water Plan, 367
- Canada
  - acid rain effects, 382–383
  - bison population, 252
  - cod industry, 8
  - harp seal industry, 8
  - international air quality and, 6
  - population trends, 152, 160
  - waste generation, 403
  - wilderness areas, 8–9, 11
- Canada lynx, 136, 138
- Cancer risk
  - Aral Sea and, 363
  - carcinogens, 41, 376, 429
  - computation, 41
  - drinking water and, 351
  - fish consumption and, 40
  - radiation exposure and, 227, 228
  - radon exposure and, 394
- Canopy studies, 119
- Capability classes, land, 310–311
- Captive breeding, of California condor, 263
- Carbamates, 321–322
- Carbon absorption, 431
- Carbon cycle, 95, 97–98
- Carbon dioxide levels
  - Chinese emissions, 182
  - coal consumption and, 194
  - crop yields and, 391
  - global emissions, 385
  - as greenhouse gas, 385, 386, 388
  - health and, 389
  - indoor effects, 396
  - reduction, 392–393
  - sea levels and, 390
- Carboniferous period, 170
- Carbon monoxide, 375
- Carbon tax, 386
- Carcinogenic effects, 376
- Carcinogens, 41. *See also* Cancer risk
  - industrial releases, 429, 430
  - Love Canal, 435
- Carnivores, 92
- Carrying capacity, 135–136, 141
- Carson, Rachel, 23
- Catalyst, 71
- Cedar River Paper Company, 413
- Cell mitochondria, 323
- CEQ (Council on Environmental Quality), 446
- CERCLA (Comprehensive Environmental Response, Compensation, and Liability Act), 426–429, 446, 447
- CERES Principles, 26–27
- CFCs. *See* Chlorofluorocarbons
- Chemical bonds, 71
- Chemical reactions, 71, 72
- Chemicals. *See also* Hazardous substances; Pesticides; Toxic substances
  - air pollutants, 378
  - Chinese production and use, 433
  - groundwater contamination, 425
  - household, 68
  - Love Canal, effects, 435
  - soil formation and, 295
  - wastewater treatment, 359–360, 362
  - water pollutants, 349
- Chemical weathering, 294
- Chernobyl Nuclear Power Station 4, 225–227, 231
- Chicago, Illinois, 118
- China
  - air pollution, 392
  - diet, population, and agriculture, 328
  - energy consumption and production, 182, 392, 448
  - groundwater depletion, 364
  - industrial waste disposal, 433
  - population control, 151–152
  - Three Gorges Dam, 200, 201
- Chlordane, 329
- Chlorides, as pollutants, 349
- Chlorinated hydrocarbons, 320–321
- Chlorofluorocarbons (CFCs)
  - economic issues, 392
  - as greenhouse gases, 386, 388–389
  - ozone depletion and, 393, 396
  - reduction, 452
- Chloroform, 425
- Chromium, 425, 429
- Chronic toxicity, 423
- CITES (Convention on International Trade in Endangered Animal and Plant Species), 30, 139
- Cities. *See* Urban areas
- Clean Air Act, 6, 380–381, 446, 447
- Clean Water Act, 346, 447, 448
- Clean Water Action Plan, 351, 353
- Clear-cutting, 245–246
- Climagraphs
  - Cairo, Egypt, 113
  - Chicago, Illinois, 118
  - Fairbanks, Alaska, 121
  - Moscow, Russia, 120
  - Rangoon, Burma, 115
- Climagraphs–*Cont.*
  - Singapore, 118
  - Tehran, Iran, 115
- Climate. *See also* Climagraphs; Global warming
  - computer modeling, 392
  - elevation, vegetation and, 111, 112
  - global conferences on, 6–7, 8
  - soil formation and, 294–295
- Climax community, 106, 109–110. *See also* Biomes
- Coal
  - air pollution and, 194, 376–377, 380
  - carbon dioxide production, 385, 386
  - Chinese consumption, 182
  - formation, 190–191
  - global energy consumption, 179
  - historical use, 170
  - reserves, 191
  - U.S. production, 171
  - use issues, 192–194
- Coastal areas
  - global warming and, 387, 390
  - groundwater mining and, 362, 364
  - marine oil pollution, 355
  - protection from development, 363–364
  - wetlands, value, 364–365
- Cod industry, 8
- Coevolution, 86–87
- Cogeneration, of biomass, 207
- Combustion, 75, 207
- Commensalism, 89, 90, 91
- Commercial sites
  - energy use, 173–174
  - redevelopment, 285
- Commission on Sustainable Development (CSD), 454
- Common property, 53–55, 58, 453
- Community(ies), 91. *See also* Ecosystems; Global community; Urban areas
  - climax, 109–110
  - pioneer, 106
  - urban sprawl and, 275
- Competition, 88, 91
- Competitive exclusion principle, 88
- Composting, 408–409, 410
- Compound, 70
- Comprehensive Environmental Response, Compensation, and Liability Act (CERCLA), 426–429, 446, 447
- Compressed natural gas, 181
- Computers, 392, 429
- Condor, California, 263
- Confined aquifer, 341–342
- Congress, U.S. *See* Legislation, U.S.
- Coniferous forest, northern, 111, 119–120
- Conservation
  - animal reserves, 47
  - energy, 210–211
  - extended product responsibility and, 54
  - habitat, 10, 85, 264
  - international organizations, 264
  - land-use planning and, 279
  - recycling and, 411–412
  - soil. *See* Soil conservation practices
- Conservation–*Cont.*
  - U.S. legislation, 445. *See also* Legislation, U.S.
  - water, 346, 362
- Conservation ethic, 24
- Conservation of matter, law of, 68
- Conservation tillage, 305–307
- Consumers
  - eco-labels, 454
  - organisms as, 92
- Contaminants. *See* Pollutants
- Contour farming, 303, 305
- Contraception. *See* Birth control
- Controlled experiment, 67
- Conventional tillage, 305–307
- Conventions
  - Basel Convention, 433–434
  - Convention for the Regulation of Whaling (1931), 453
  - Convention on International Trade in Endangered Animal and Plant Species (CITES), 30, 139
  - Convention on Long-Range Transboundary Air Pollution (1979), 452
  - UN Framework Convention on Climate Change, 6–7, 8
- Conversion systems, 75–76, 407
- Conversion tables, 460–461
- Cooking, energy consumption, 174
- Cooling tower, 354–355
- Cooling water, 354–355
- Coral reef ecosystems, 122, 124
- Corporations, 26. *See also* Industry; *specific company*
- Corrosiveness, 420
- Costa Rica, 57, 131
- Cost-benefit analysis, 49–50
- Cotton, 326–327
- Cottontail rabbits, 133
- Council on Environmental Quality (CEQ), 446
- Cover, 255–256
- Cowbirds, 258
- Critical thinking, 459
- Crop rotation, 317–318, 331, 334
- Crop yield, 317–318, 389, 391–392
- Crude oil, 191, 195, 196. *See also* Oil
- Crust, 292
- CSD (Commission on Sustainable Development), 454
- Cyanide contamination, 449–450

D

- Dams
  - hydroelectric sites, 200, 201, 264
  - salmon and, 255
  - stream water and, 347–348
- Dandelion, 83, 84
- DDE, Great Lakes contamination, 335
- DDT, 320–321, 425
  - bioaccumulation and biomagnification, 324–325, 326
  - chemical structure, 320
  - dosage, for cotton pests, 327
  - Great Lakes contamination, 335
  - source and health effects, 424, 425
  - Soviet soil contamination, 329
- Death phase, 136

Death rates, 131
population growth and, 148–149
radon-related, 394
Deaths
air pollution causing, 374, 375
radon causing, 394
risks table, 42
Debt burden, 56–57, 58
Debt-for-nature exchange, 57
Deciduous forests, temperate, 111, 117–119
Decision making, 39, 55–56. *See also* Land-use planning; Water-use planning
Decomposers, 92, 295
Deforestation, 52
erosion and, 308
tropical, 116–117, 246–247
Delhi, India, 272
Demand (economic), 43–55
Demography, 148
Denitrifying bacteria, 98
Density-dependent limiting factors, 138
Density-independent limiting factors, 138
Department of Energy (DOE)
contaminated properties, 224, 225, 230
radioactive disposal site and, 231–232
Deposit-refund programs, 46
DES (diethylstilbestrol), 331
Desertification
fuelwood and, 208, 209
global risk and degree, 249
of rangelands, 248–249
Deserts, 111–113
global distribution, 111
soils, 299, 300
Detritus, 94
Developed nations
carbon dioxide emissions, 385
demographic transition, 156
food production, 154–155
geopolitics, 447–450
hazardous waste sites, 426–429
water use and pollution control, 354
Developing nations
agricultural techniques, 316, 317
carbon dioxide emissions, 385
demographic transition, 156
economies, 56–58
energy efficiency and, 392–393
extinction and, 262–263
food production and hunger, 154–155
geopolitics, 447–450
global warming and, 385–386, 448
population. *See* Human population growth
rural-to-urban migration, 270–272
water use and pollution control, 354
women in, 149–151, 154
Development
Earth Summit, 31
energy efficiency and, 392–393
planning. *See* Land-use planning
rivers and, 362–363
sustainable, 6, 50–53, 452, 454
Development ethic, 22, 24
Dhaka, Bangladesh, 272
Diet. *See* Food consumption
Diethylstilbestrol (DES), 331
Dinoseb, 328–329
Direct combustion, of biomass, 207
Disease. *See* Illness; *specific disease*
Dispersal, population, 133
Disposable lifestyle, 402, 403
DOE. *See* Department of Energy
Domestic water, 343–344, 345, 346
Donora, Pennsylvania, 374
Drilling, 194–195, 196. *See also* Mining
Drinking water. *See* Potable water
Drugs, tropical forests and, 57, 117
Dry regions, 11–12, 13
Dry tower, 355
Dumps, 426–429, 431
Dust, 379–380

E

Earth, geologic processes, 292–294
Earth Day (1970), 442
Earth Sanctuaries, 47
Earth Summit, 6, 31, 452
Earthworms, 295
Ecocentrism, 22
Eco-conflicts, 449–450
Eco-labels, 454
Ecology, 81. *See also* Ecosystem
environment, 81–82
evolutionary patterns, 86–87
habitat and niche, 82–84
human impact on, 91
industrial, 28
limiting factors, 82
natural selection, 85–86
species definition, 84
Economic costs
of forest utilization, 244
of mineral exploitation, 242–243
of resource exploitation, 241
of wildlife ecosystem management, 255
Economic growth
energy consumption and, 171–172
environmental ethics and, 26
sustainable development and, 51–53
Economic issues, 43–59
biomass conversion and, 206
biophysical world and, 55–56
common ownership, 53–55, 58, 453
concepts, 43–45
corporate attitudes and ethics, 25, 26–28
cost-benefit analysis, 49–50
of developing nations, 56–58
extended product responsibility, 48–49
external costs, 53
geopolitics and, 448–450
human environmental load, 58
market-based instruments, 45–48
recycling, 414
sustainable development and, 50–53
Ecoparque, 366
Ecosystem approach, 7
Ecosystems, 91–101. *See also specific ecosystem*
acid rain effects, 382
agriculture effects, 243–244
biomes, 110–120
climax concept, 109–110
defined, 7, 91
energy flow, 93–94
food chains and webs, 94–95, 96
global warming and, 391, 392
keystone species, 92–93
nutrient cycles, 95–101
organism roles in, 92
organizational levels, 81
services value, defining, 56
succession, 106–109
Eco-terrorism, 451
Ecotourism, 57
Ectoparasites, 89
EDB (ethylene dibromide), 325
Education
population growth and, 150–151, 154
standard of living and, 153
Egypt, Cáiro, 113
Electricity
consumption, 175–176
geothermal sources, 201–202, 208
hybrid electric vehicles, 176
hydroelectric sources, 197–200
solar sources, 203, 204, 205, 206
tidal sources, 200–201
wind sources, 202–203
Electrons, 69–70
Elements, 69, 462
Elephants, 139
Elevation
of ambient temperature, pollution and, 349
climate, vegetation and, 111, 112
Elk, 101
Emergent plants, 125
Emerson, Ralph Waldo, 23
Emigration, of populations, 133
Emission fees and permits, 46
Endangered species, 265
international trade in, 30
protective measures, 262–264
U.S. legislation, 85, 101, 264
Endangered Species Act, 85, 101, 264
Endoparasites, 89
Energy, 72–76
activation, 71
availability, as limiting factor, 135
conservation, 210–211
conversion systems, 75, 76, 407
human population growth and, 141
in photosynthesis, 72
states of matter and, 73, 74
thermodynamic laws, 73–74, 75
types, 72–73
U.S. legislation, 26–27, 445
Energy consumption, 169
automobile and, 172, 176–177
Chinese, 182
economic growth and, 171–172
electrical, 175–176
gasoline prices and, 172–173, 175
global trends, 176–177, 179–180, 183
historical background, 169–171
Energy consumption–*Cont.*
industrial, 174
OPEC and, 178
predicting needs for, 188
regional trends, 173, 176–177, 179–180, 183
residential and commercial, 173–174
transportation, 174–175
Energy costs
of forest utilization, 244
of irrigation, 346
of mineral exploitation, 242–243
of resource exploitation, 241
Energy efficiency
greenhouse gas reduction and, 392–393
urban sprawl and, 275
Energy flow
environmental implications, 74–76
thermodynamic laws, 73–74, 75
through ecosystems, 93–94
Energy sources, 187–212. *See also specific fuel/source*
biological, 169–170
carbon dioxide production, 385, 386
conservation, 210–211
fossil fuels. *See* Fossil fuels
renewable, 197–210
resources and reserves, 188–190
Energy waste
global warming and, 386
sites, DOE responsibility for, 224, 225
England, air pollution, 376–377, 398
Entropy, 73
Environment, 5, 81–82
economics and, 43. *See also* Economic *entries*
ecosystem approach, 7
interrelationships and, 5–7
regional concerns, 7–15
scientific thought and, 67–69
Environmental costs
of forest utilization, 244–245
of freshwater fisheries utilization, 253–255
of marine fisheries utilization, 250–253
of mineral utilization, 242–243
of rangeland utilization, 247–248
of resource exploitation, 241
of waste management failure, 432
Environmental ethics, 20–34
corporate attitudes, 25, 26–28
environmental attitudes, 22, 24
environmental justice, 28–29
global responsibility, 30–31
individual responsibility, 29–30, 34
naturalist philosophers, 23
philosophical questions, 24
societal attitudes, 24, 26
theories of, 21–22
Environmental justice, 28–29
Environmental movement, 442–444
Environmental policy, 438–455. *See also* Political issues
future considerations, 441–442
gasoline pricing, 172–173
geopolitics, 447–451

Environmental policy–*Cont.*
individual action, 413, 455, 463–464
international, 433–434, 451–455
land-use issues, 280–281, 285–287
new challenges, 439–440
past patterns, learning from, 440–441
science *versus,* 6
urban sprawl and, 274–275
U.S. *See* Legislation, U.S.
Environmental Protection Agency (EPA)
air pollution issues, 381
creation, 440, 446
enforcement options, 447
ethylene dibromide and, 325
hazardous material definition, 420–421
hazardous waste management, 429, 430
pesticide regulation, 321, 328–329
Safe Drinking Water Act and, 449
secondhand smoke policy, 384
Environmental resistance, 135
Environmental science, 3, 5
Environmental smoke, 384
Enzymes, 72
EPA. *See* Environmental Protection Agency
EPR (extended product responsibility), 48–49, 54
Erosion
defined, 294, 299
global, 301, 302
monoculture and, 317
soil, 299, 301, 308, 312. *See also* Soil conservation practices
Escherichia coli, 351–352
Estuaries, 124
development and, 363–364
thermal pollution, 354
Ethanol production, 207
Ethephon, 324
Ethics, 21. *See also* Environmental ethics
extinction and, 262
genetically modified plants, 333–334
Ethylene dibromide (EDB), 325
EU (European Union), 452
Euphotic zone, 121
Europe
acid rain, 382–383, 398
cyanide contamination, 449–450
energy consumption, 177, 183
environmental policy, 452
hazardous waste sites, 426
mass burn technology, 408, 409
whaling industry, 453
European Union (EU), 452
Eutrophication, 350
Eutrophic lakes, 125
Evaporation
dams and, 347
global warming and, 390
salinization and, 361, 362
Evapotranspiration, 341
Everglades National Park, 360–361
Evolution, 68, 85–87
Executive branch of government, 442, 443
Exotic species, 255, 257
Experiments, 67
Exponential growth phase, 134
Exports
food, 155
hazardous wastes, 431, 432–434
Extended producer responsibility, 48
Extended product responsibility (EPR), 48–49, 54
External costs, 53
Extinction, 86, 260
causes, 260–261, 262
considerations, 261–262
preventive measures, 262–265
probability, 260
Extractive reserves
federal policy, 285
fossil fuel, 183, 188–190, 191, 192, 193, 197, 212
Mendes, Chico and, 28
Exxon Valdez, 355

F

Fairbanks, Alaska, 121
Family size, 149–150, 151
Farming, salmon, 255. *See also* Agriculture
Farmland, 276, 287
Fecal coliform bacteria, 351–352
Federal Water Pollution Control Act, 346, 353
Federal Wild and Scenic Rivers Act, 362–363
Fermentation, 207
Ferret, blackfooted, 265
Fertility, 149–150
Fertilizers
impact and use trends, 318–319
nutrient cycles and, 98–101, 319–320
Soviet soil contamination, 329
water pollutants, 352–353
Fire, forest management via, 266
First law of thermodynamics, 73
Fish
acid rain effects, 382, 383
Aral Sea, 363
cancer risk and, 40
freshwater ecosystems, 125–126, 250, 253–255
Great Lakes. *See* Great Lakes
limiting factors, 82, 83
marine ecosystems, 121–124, 250–253
mercury poisoning and, 427
persistent pesticides and, 325, 326
Pfiesteria piscicida and, 255
water temperature and, 353–354
Fisheries
cod, 8
extinction and, 262
global catch rates, 251
international disputes, 252
major harvesters, 253
Native Americans and, 257
utilization costs, 250–255
whaling industry, 54–55, 453
Fish farming, 255
Fish hatcheries, 7
Fission, nuclear, 217
Fissionable, 217
Fleas, 89
Floating bog, 108
Flooding
global warming and, 390, 391
urban sprawl and, 277
Flood irrigation, 345
Floodplains, 277
Floodplain zoning ordinances, 277
Florence, Italy, 173
Florida, 360–361, 411
Food additives, 331
Food chain, 13, 94, 95
Food supply
Chinese, 328
developing nations and, 154–155
global warming and, 391–392
population growth and, 154–156
production on nonfarm land, 309
Food web, 94–95, 96
ocean estuary, 354
Forest ecosystems, 116–120, 244–247. *See also specific type of forest*
acid rain and, 382–383
biodiversity loss and extinction, 261
canopy studies, 119
changes, 244
climax communities, 109–110
deforestation. *See* Deforestation
desertification, 208, 209
environmental issues, 10, 12–13, 14
global convention on, 31
global distribution, 111
global warming and, 391
logging. *See* Logging
management via fire, 266
plantation forestry, 246
preservation, 57, 127
recreational use, 309
salmon and, 7
slash-and-burn agriculture, 316
soils, 299, 300
U.S. legislation, 285
utilization costs, 244–245
Forest soils, 299, 300
Formaldehyde, 396
Fossil fuels
agricultural use, 318
air pollution from, 194, 195–196, 197, 375–377
formation, 190–192, 193
global warming and, 385, 386
Industrial Revolution and, 170–171
reserves, 183, 188–190, 191, 192, 193, 197, 212
use issues, 192–197, 199, 200
Free-living nitrogen-fixing bacteria, 98
Freezing, rock fragmentation and, 293
Freshwater, 340, 367
Freshwater ecosystems, 125–126, 250, 253–255
Friable, 296
Fruitflies, 332
Fuels. *See also* Air pollution; *specific fuel*
alternative vehicular, 176–177, 181
Fuels–*Cont.*
fossil. *See* Fossil fuels
taxes on, 172–173
Fuelwood, 207–209
air pollutant, 209, 380
erosion and, 308
historical use, 169–170
U.S. energy production via, 171
Fungi, 295, 297
Fungicides, 319, 323–324
Fusion, nuclear, 222–223

G

Galápagos tortoise, 265
Galerucella sp., 333, 334
Game species, 258
Gamma radiation, 216
Ganges River, 352
Garbage. *See* Solid waste disposal; Waste products
Garrison Diversion Unit, 357–358
Gas-cooled reactors (GCRs), 219
Gases
energy as, 73
global warming and, 385–389. *See also* Global warming
industrial emissions, 380
in secondhand smoke, 384
Gasification, 207
Gasoline, lead emissions, 378, 379, 427
Gasoline prices
fluctuations, 178, 190
government policy and, 172–173
OPEC and, 178, 190
variability, 175
GCRs (gas-cooled reactors), 219
General Motors, 25, 176
Genes
chemical exposure and, 435
pesticide resistance and, 325–326, 327
Genetic engineering
modified seed, monoculture and, 317
pest/disease resistant crops, 333–334
Geology
basic processes, 292–294
land-use planning and, 279
Geopolitics, 447–451. *See also* Global community
Georgia-Pacific Corporation, 48
Geothermal energy, 201–202, 208
Giant panda, 265
Global community
air pollution, 398. *See also* Global warming
disputes, 6, 252, 357, 398
eco-terrorism, 451
energy consumption, 176–177, 179–180, 183
environmental complexities, 5–7
environmental conferences, 6–7, 31, 34, 452
environmental ethics, 30–31
environmental policy, 433–434, 447–455
hazardous waste trade, 432–434
HIV/AIDS epidemic, 161
ozone recovery efforts, 452–454

Global community–*Cont.*
population growth. *See* Human population growth
urbanization, 157
Global warming, 385–393
addressing, 392–393
agricultural yields and, 391–392
causes, 385, 386–389
computer modeling, 392
ecosystems and, 391
health effects, 389–390
regional effects, 385–386, 387
science *versus* policy, 6
sea level and, 387, 390
water cycle and, 390–391
Glossary, 465–472
GNP (gross national product), 153
Gothenburg, Sweden, 173
Grasslands, 113–114
bison and, 252
climagraph, 115
climax communities, 109–110
geographic distribution, 111
soils, 299, 300
succession, 114
Grassland soils, 299, 300
Great Lakes
native fish species, 254
pollutants, 95, 335
regional concerns, 13–14, 15
species invasion, 137, 253–255
Great Plains, erosion, 301
Green advertising, 51
Green bullet, 430
Greenhouse effect, 386, 387
Greenhouse gases, 386–389
developing nations and, 448
Kyoto Protocol and, 6–7, 8, 34
Montreal Protocol, 389, 452, 455
reduction, 392–393
Greenland, 8
Green politics, 447–451
Green Revolution, 318
Gross national product (GNP), 153
Groundwater, 341
in aquifers, 343
depletion, 361–362, 364
hazardous waste contamination, 424–425
pollutants, 355–356
Soviet, contamination, 329
Groundwater mining, 361–362, 364
Gulf of Mexico, 101
Gymnogyps californianus, 263
Gypsum, 381, 382

H

Habitat
alteration, extinction and, 261, 262
analysis and management, 255–256
conservation, 10, 85, 264
destruction, 55, 244–245
niche and, 82–83, 84
protection, 258–259
Habitat Conservation Plan, 10
Habitat destruction, 55, 244–245
Habitat management, 256
Half-life, 216
Harp seals, 8
Harvesting. *See* Fisheries; Hunting; Logging
Hawaii, 281
Hazardous, 421
Hazardous substances. *See also* Hazardous waste; Toxic substances
benefit-risk associations, 420
definitions, 420–421
regulatory issues, 421–424
source reduction, 410
top fifteen, 425
Hazardous waste
Chinese levels, 433
computer components, 429
definitions, 421
environmental impact, 421, 424–425, 432
health risks, 421, 425–426
international trade in, 432–434
Love Canal, 435
management, 421, 429–434
radioactive, 224, 225
statistical data, 421
types, 421
U.S. legislation, 426–429, 432
Hazardous waste disposal
political issues, 231–232, 432–433
radioactive materials, 230–233
sites, 426–429
HC. *See* Hydrocarbons
Headwaters Forest, 10
Health. *See also* Cancer risk; Illness
global warming and, 389–390
hazardous waste and, 425–426, 435
perceived risks, 43
pesticides and, 321–322
smog effects, 377, 381
standard of living and, 153
Hearing loss, 397
Heat
energy conversion and, 72–76, 407
sensible *versus* latent, 73
solar source, 203–204
thermal pollution, 228, 229, 353–355
wood source, 208
Heat mining, 208
Heat stress, 389
Heavy metal pollutants
in computers, 429
persistent, 424
in water, 349
Heavy-water reactors (HWRs), 218
Herbicides, 319, 322–323
Herbivores, 91, 92, 97–98
Herring gulls, 335
HEVs (hybrid electric vehicles), 176
High-level radioactive waste, 230–232
HIV (human immunodeficiency virus), 161
Holy Ganges River, 352
Hong Kong, 173
Hooker Chemical Company, 435
Horizons, soil, 297–299, 300, 301
Host, 88
Household chemicals, 68
Housing
indoor pollutants, 394–397
land-use planning and, 280
urban sprawl and, 273
Human immunodeficiency virus (HIV), 161
Human impact, 238–266
on aquatic ecosystems, 250–255
changing role of, 239
on ecology, 91
on forest ecosystems, 244–247, 266
on hydrologic cycle, 342–343
on natural resources, 240–243
on nutrient cycles, 100–101
on rangeland ecosystems, 247–249
on soil degradation, 308
on species extinction and biodiversity, 262–265
on terrestrial ecosystems, 243–244
on wilderness areas, 249–250
on wildlife ecosystems, 255–262
Human population growth, 55, 138, 140–142
AIDS and, 161
current trends, 147–148
doubling time, 140
factors influencing, 148–152
food supply and, 154–156
future trends, 158–160
historic growth curve, 138, 140
implications, 147–148
Malthus theory, 148
North American trends, 150
pollution and, 239
poverty and, 153–154
selected nations, 149
size limitation, 141–142
social factors, 141
standard of living and, 152–153
U.S. trends, 131, 153, 157–158, 159, 160, 450
Human populations. *See also* Human population growth; Population characteristics
age distribution, 132
AIDS infected, 161
demographic transition, 156
ecological impact, 91
economic output and fossil-fuel consumption, 188
nutrient cycles and, 100
pollution and, 147, 154–156
species extinction and, 260–261
urban. *See* Urban areas
Humus, 295
Hungary, 131, 449–450
Hunger, food supply and, 154–156
Hunter-gatherer society, 169
Hunting, 256–257
HWRs (heavy-water reactors), 218
Hybrid electric vehicles (HEVs), 176
Hybrid seed, 317
Hydrocarbons (HC)
loss reduction, 379
primary air pollutants, 375–376
secondary air pollutants, 378
Hydroelectric power, 7, 197–200. *See also* Dams
global consumption, 198
global warming and, 390–391
site development, 198
stream water and, 347
Hydrogen, vehicular, 181
Hydrologic cycle, 341–342
global warming and, 390–391
human impact on, 342–343
Hydroxide ion, 70
Hylobius transversovitatus, 333, 334
Hypothesis, 67

I

Ice sheets, 390
IEEP (International Environmental Education Programme), 7
IFAD (International Fund for Agricultural Development), 301
Ignitability, 420
Illegal immigrants, 160
Illness
air pollution and, 374, 375, 376, 377, 381, 396
black lung disease, 193
cancer, Aral Sea and, 363
genetic engineering and, 333–334
global warming and, 389–390
hazardous substances causing, 425
Love Canal residents and, 435
noise pollution and, 397
pesticides and, 327–329
tobacco-related, 384
water pollution and, 349, 350, 351–352
Immigration
illegal, 160
population growth and, 152
of populations, 133
U.S. policy, 157–158
Imports
energy, 182
food, 155
Incineration
hazardous/toxic waste, 431, 432
solid waste, 407–408, 409, 410
India
groundwater depletion, 364
Holy Ganges River cleanup, 352
population control, 152
population trends, 272
Individual action, 29–30, 34, 413, 455, 463–464
Indonesia, 272
Indoor air pollution, 394–397
Industrial ecology, 28
Industrial energy use, 174
Industrialized nations. *See* Developed nations
Industrial pollution
hazardous sites, 426–429, 433
reduction, 379–380
toxic chemical releases, 429, 430
water contaminants, 353
water transport, 346
Industrial regions, 13–15
Industrial Revolution, 170–171
Industrial solvents, 430–431
Industrial water use, 346–347
Industry. *See also specific company*
demographic transition and, 156
environmental ethics, 25, 26–28
hazardous waste management, 429–432
pollution. *See* Industrial pollution
site redevelopment, 285
source reduction, 411
water use, 346–347
Wise Use Movement and, 444
Information programs, 46, 376

Infrastructure, urban sprawl and, 276
Inner-city areas, 275–276, 285
Inorganic chemicals, household, 68
Inorganic matter, 71
Insecticides, 319
bioaccumulation and biomagnification, 324–325
chlorinated hydrocarbons, 320–321
new generation, 323
nontarget organisms and, 327
organophosphates and carbamates, 321–322
resistance to, 320, 325–327
Insects
chemical control of. *See* Insecticides
integrated management, 333–334
natural selection, 86
new species, 119
Institutional commitment, 53
In-stream water use, 347–348
Integrated pest management, 331–334
Interdependence, sustainability and, 53
Intergovernmental Panel on Climate Change (IPCC), 385, 386, 392
Internal combustion engine, 375–376, 377, 378
International Environmental Education Programme (IEEP), 7
International Fund for Agricultural Development (IFAD), 301
International Joint Commission, 6
International relations, 447–455. *See also* Global community
International Union for the Conservation of Nature (IUCN), 264
International Whaling Commission (IWC), 453
Interspecific competition, 88
Intraspecific competition, 88
Invading species, 137
IPCC (Intergovernmental Panel on Climate Change), 385, 386, 392
Iran, 115
Irrigation, 345–346, 347
Aral Sea and, 363
salinization and, 361, 362
water management issues, 357
Isle Royale, Michigan, 143
Isotopes, 70, 216
Itaipu hydroelectric plant, 200
Italy, 173
IUCN (World Conservation Union), 264
Ivory sales, 139
IWC (International Whaling Commission), 453

J

Jakarta, Indonesia, 272
Judicial branch of government, 442

K

Karachi, Pakistan, 272
Kenya, 153
Keystone species, 92–93
Kinetic energy, 72, 74
Kinetic molecular theory, 69
Kirtland's warbler, 258
K-Mart, 413
K-strategists, 138
Kuwait, 178, 189, 451
Kyoto Conference on Climate Change, 6–7
Kyoto Protocol, 7, 8, 34, 455

L

Labor-intensive agriculture, 316, 317
Ladybird beetles, 332–333
Lagos, Nigeria, 272
Lag phase, 134
Lake Okeechobee, 360–361
Lakes
acid rain and, 382
dams and, 347–348
ecosystem, 125–126
global warming and, 390, 391
Lamphrey, 253, 254
Land
capability classes, 310–311
groundwater mining and, 362
nonfarm, protection, 307, 309
versus soil, 294
waste disposal, 431–432
Landfills
cost, 410
groundwater contamination, 424–425
hazardous waste disposal, 431
hazardous waste sites, 426–429
solid waste disposal, 405–407, 408
U.S. capacity, 402
water pollution, 355
Land use
federal issues, 285–287
historical forces, 270–272
Louisiana wetlands loss, 278
regional concerns, 7–13
rural to urban shift, 157, 270–271, 328
sprawl factors, 273–275
subsidies, 47
unplanned urban growth, problems with, 275–279
urban issues, 272
urban to suburb shift, 272–273, 274
U.S. legislation, 285, 445
waterways and, 270
Land-use planning, 269–287
aesthetic pollution and, 282
decision making in, 287
implementation, 280–282
mall development, 287
need for, 270
principles, 279–280
urban issues, 274–275, 282–285
U.S. legislation, 285
Land use subsidies, 47
Latent heat, 73
Law, 68
Law of conservation of matter, 68
Laws of thermodynamics, 73–74, 75
LD_{50} (lethal dose), 423
Leaching, 298
Lead
air pollutant, 378–379, 427
chronic effects, 423, 425, 426
in computers, 429
historical increases, 427
indoor effects, 396
military use, 430
source, 425
water pollutant, 351
Legislation, EU, 452
Legislation, U.S., 445
air quality, 6, 380–381
endangered species, 85, 101, 264
energy resources, 26–27
EPA and, 440
hazardous waste, 426–429, 432
land-use, 285
legislative process, 442–445
noise pollution, 397
recycling, 411
regulation via, 446–447, 448
smoking restrictions, 384
water quality, 346, 351, 353, 357
Legislative branch of government, 442
writing to, 464
Leopold, Aldo, 22, 23
Less developed nations. *See* Developing nations
Lethal dose (LD_{50}), 423
Life cycle
product, 48
toxic substances, 420
Lifestyle
energy consumption and, 172
environmental impact, 32
urban sprawl and, 274
waste products and, 402, 403–404
Lighting, 76, 210, 211
Light-water reactors (LWRs), 218
Limestone, 381, 382
Limiting factors, 82, 83, 135
eutrophication and, 350
population growth and, 138, 140–141
Limnetic zone, 125
Liquefied natural gas, 196
Liquid energy, 73
Liquid metal fast-breeder reactors (LMFBRs), 222
Lithosphere, 292
Litigation
owl protection, 247
Romanian cyanide release, 449–450
Virginia soil erosion, 312
Litter, soil, 297, 298, 300
Littoral zone, 125
Livestock prey, 101
LMFBRs (liquid metal fast-breeder reactors), 222
Loam, 296
Logging
costs, 244–245
methods, 245–246
Pacific Northwest, 127, 247
regional concerns, 10, 12–13
tropical rainforest, 116
Log phase, 134
London, England, 377
Loosestrife, purple, 333, 334
Los Alamos National Laboratory, 208
Louisiana wetlands, 278
Love Canal, 435
Low-level radioactive waste, 231, 232–233
Loxodonta africana, 139
LWRs (light-water reactors), 218
Lynx, Canada, 136, 183
Lythrum salicaria, 333, 334

M

Maastricht Treaty, 452
McDonald's Corporation, 416
Macronutrients, 319
Malaria, 321
Mall development, 287
Malnutrition, 154
Malthus, Thomas, 148
Management ethic, 24
Mangrove swamp ecosystems, 124
Manila, Philippines, 272
Mantle, 292
Manufacturing costs, 243
Marine ecosystems, 121–124, 250–253
Marine oil, 355
Market-based instruments, 45–48
Marshes, 126
Mass burn technology, 408, 409
Mass transit systems, 283
Matter
conservation, law of, 68
inorganic and organic, 71
states, 73, 74
structure, 69
types and characteristics, 70
Mechanical weathering, 293–294
Medicine, tropical forests and, 57, 117
Megalopolis, 273
Mendes, Chico, 28
Mercury
in computers, 429
in concrete waste, 432–433
Mercury poisoning, 425, 427
Metals
in computers, 429
in ores, 242
pollutants, 349, 424
Methane gas
as greenhouse gas, 385, 386, 388, 389
production, 206, 207
Methanol/flex-fuel, vehicular, 181
Methylmercury, 427
Metric conversion tables, 460–461
Mexico
Ecoparque, 366
groundwater depletion, 364
international air quality and, 6
Mexico City air pollution, 376
population trends, 160
Micronutrients, 319
Microorganisms
hazardous waste treatment, 431
indoor pollutants, 396
wastewater treatment, 358–360, 362
water pollutants, 349–350, 351–352
Midwest Recycling Company, 413
Migration. *See also* Immigration
rural-to-urban, 157, 270–272, 328
suburban, 272–275
Migratory birds, 259–260

Mineral resources
federal extraction policy, 285
human impact on, 241–243
recycling and, 243
utilization costs and steps, 242–243
Mine tailings, 228, 229
Mining
acid mine drainage, 194
coal, 192, 193
environmental impact, 242–243
groundwater, 361–362, 364
heat, 201–202, 208
in nuclear fuel cycle, 223, 224
tropical rainforest, 117
U.S. legislation, 285
water pollution from, 353
Minnesota Mining and Manufacturing Company, 54
Minorities, environmental justice and, 29
Mission blue butterfly, 265
Mississippi River, 101, 277, 278
Missouri River, 277, 278, 357–358
Mitochondria, 323
Mixtures, 70
Moderator, 217
Molecular theory, kinetic, 69
Molecules, 70
air pollutants, 377, 378
water pollutants, 349
Monoculture, 317–318
Montreal Protocol, 389, 452, 455
Moose prey, 143
Morals, 21. *See also* Environmental ethics
Mortality. *See also* Death rates
defined, 131
global warming and, 389, 390
Moscow, Russia, 120
Moss, 82, 83
Mountains
air pollution and, 378
global distribution, 111
Muir, John, 23
Multiple Use Sustained Yield Act, 285
Municipal landfill, 405–407, 408
cost, 410
water pollution, 355
Municipal solid waste, 351–352, 402
composition, 404
disposal. *See* Solid waste disposal
generation, 402, 403
Municipal water systems, 343–344, 345, 346
pollution, 351–352
Mutualism, 90, 91
Mycorrhizae, 90

N

NAAQS (national ambient air quality standards), 380–381
Nanophyes sp., 333
Naphthaleneacetic acid, 324
Natality, 131
National ambient air quality standards (NAAQS), 380–381
National Environmental Policy Act, 446
National Parks. *See* Parklands
National Priority List, 427–428
National security, 449
National Wildlife Federation, 445
Native Americans, 252, 257
Natural ecosystems. *See* Ecosystems
Natural gas
carbon dioxide production, 385, 386
formation, 191–192, 193
global consumption, 177, 179
historical use, 171
reserves, 183, 193, 197
U.S. production, 171
use issues, 196–197
Natural predators, insect control, 332–333
Natural resources, 188, 240–243. *See also* Resource exploitation; *specific resource*
Natural selection, 85–87
Nature, views of, 20–21, 22
Nature centers, urban, 285
Navigation, global warming and, 390
Negligible risk, 42
Nervous system, 321–322
Nest parasitism, 90
Neutrons, 69–70
Nevada, 232
New Mexico, 362, 364
Niche, ecological, 82–83, 84
Nicotine, 320, 323
Nigeria, 272
Nitrates, 351
Nitrifying bacteria, 98
Nitrogen cycle, 98–99
Nitrogen dioxide
acid deposition and, 381
Clean Air Act and, 380, 381
indoor effects, 396
primary air pollutant, 377
reduction, 379
secondary air pollutant, 378
Nitrogen-fixing bacteria, 98
Nitrogen oxide
acid deposition and, 381
Clean Air Act requirements and, 380, 381
primary air pollutant, 377
reduction as pollutant, 379
secondary air pollutant, 378
Nitrous oxide (N_2O), 386, 388, 389
N_2O (nitrous oxide), 386, 388, 389
Noise Control Act, 397
Noise pollution, 397
land-use planning and, 282
U.S. legislation, 397, 445
Nomadic herding, 248
Nonpersistent pesticides, 319
Nonpersistent pollutants, 424
Nonpoint source pollution, 350, 351
Nonrenewable resources, 241
energy, 188. *See also* Fossil fuels; *specific fuel*
human impact on, 240–241
Nontarget organisms, 319, 327
North America
agricultural regions, 11
dry regions, 11–12
energy consumption, 177, 183
forested regions, 12–13, 14
grassland succession, 114
industrial regions, 13–15
North America–*Cont.*
migratory waterfowl routes, 259
nuclear power plants, 220
population trends, 160
regional cities, 273, 274
rural-to-urban shift, 157, 270–272
waterways, importance, 270, 271
wilderness regions, 8–9, 11
Northeastern United States, 13–14, 15
Northern coniferous forest, 111, 119–120
Northern spotted owl, 247
Nuclear breeder reactors, 222
Nuclear chain reaction, 217, 218
Nuclear fission, 217
Nuclear fusion, 222–223
Nuclear power, 215–233
accidents, 225–227
fuel cycle, 223, 224
fusion technology, 222–223
global consumption, 179
historical development, 217
nature, 216–217
opposition to, 227
radiation exposure, 226, 227–228, 229
Soviet legacy, 231
thermal pollution, 228, 229
weapons production and, 223–224, 225
Nuclear reactors
breeder reactors, 221–222
at Chernobyl, 225–227
decommissioning costs, 229–230
distribution, 220, 221
plant construction programs, 219–220
plant life extension, 220–221
at Three Mile Island, 225
types and functions, 217–219, 220, 221
waste disposal, 230–233
Nuclear weapons, 223–224, 225
Nucleus, 69
Numbat, 245
Nutria, wetlands loss and, 278
Nutrient cycles, 95
carbon, 95, 97–98
fertilizer and, 98–101, 319–320
human impact on, 100–101
nitrogen, 98–99
phosphorus, 99–100
Nutrients, as water pollutants, 349, 350

O

O_3. *See* Ozone
Observation, 67
Oceans
aquaculture, 255
common property resource problems, 54–55, 453
marine oil pollution, 355
resource exploitation, 252
UN conference, 452
OECD (Organization for Economic Cooperation and Development), 177
Offshore drilling, 194, 195
Oil
agricultural use, 318
air pollution and, 376–377
Oil
Arctic National Wildlife Refuge and, 212
carbon dioxide production, 385, 386
formation, 191–192
global consumption, 177, 179
historical use, 171, 189
marine, water pollution, 355
North American recovery, 415
OPEC supply, 178
reserves, 183, 190, 192
U.S. legislation, 26–27
U.S. production, 171
use issues, 194–196, 197
Oil industry
CERES Principles and, 26–27
offshore drilling, 194, 195
OPEC and, 178, 189–190
price fluctuations, 189–190
Oil Protection Act, 26–27
Oil spills, 195–196, 197, 451
Old-growth forests, 127, 247
Oligotrophic lakes, 125
OPEC (Organization of the Petroleum Exporting Countries), 178, 189–190
Open space
land-use planning and, 279
urban planning, 284–285
urban sprawl and, 276
Ores, metal content, 242
Organic agriculture, 330, 331
Organic matter, 71
composting, 408–409, 410
Organic molecules
waste treatment, 431
water pollutants, 349
Organism interactions, 87–90
categorization difficulties, 90
competition, 88
disease-causing, 349, 350, 351–352
human population growth and, 141
as limiting factor, 135, 136
nonpersistent pollutants and, 424
predation, 87
symbiosis, 88
Organization for Economic Cooperation and Development (OECD), 177
Organization of the Petroleum Exporting Countries (OPEC), 178, 189–190
Organophosphates, 321–322
Outdoor recreation, 285. *See also* Recreational resources
Overburden, 192
Overgrazing, 308
Overnutrition, 155
Owl, northern spotted, 247
Ownership, common, 53–55, 58, 453
Oxygen
atomic structure, 70
in lake/pond ecosystem, 125–126
Ozone (O_3)
air pollution and, 377–378, 390
depletion, 393, 396
Montreal Protocol and, 452–454

P

Pacific Lumber Company, 10
Pacific Northwest, 10, 127, 247

Packaging
  McDonald's Corporation policy, 416
  resin content, 406
  waste generation, 402, 404
Paint, lead in, 379, 427
Pakistan, 272
Panda, 265
Paper mills, 413
Paraguay, 200
Parasites, 88, 91
Parasitism, 88–89
Parent material, 294, 300
Parklands
  Everglades National Park, 360–361
  human impact on, 250
  urban planning, 284–285
  Wise Use Movement and, 444
  Yellowstone National Park, 9, 101
Particles
  in primary sewage treatment, 358, 359
  water pollution and, 350
Particulates, 376
  control of, 379–380
  in secondhand smoke, 384
Partnership for a New Generation of Vehicles (PNGV), 176
Passive solar systems, 203, 204
Patch-work clear-cutting, 246
Pay-as-you-throw programs, 410–411
PCBs. *See* Polychlorinated biphenyls
Pelagic ecosystems, 121, 122
Pelagic organisms, 121
Pelletising, of biomass, 206–207
Perceived *versus* true risks, 42–43
Performance bond programs, 46
Periodic table of the elements, 462
Periphyton, 126
Permafrost, 120
Peroxyacetylnitrates, 377–378
Persian Gulf War, 178, 189, 451
Persistent chlorinated organic compounds, 351
Persistent pesticides, 319, 324–325, 326
Persistent pollutants, 424
Persistent trillium, 265
Pesticides, 319–329, 330
  bioaccumulation and biomagnification, 324–325
  fungicides and rodenticides, 319, 323–324
  global sales, 329, 330
  herbicides, 319, 322–323
  human health and, 327–329
  insecticides, 320–322
  integrated pest management and, 334
  nontarget organisms and, 327
  persistence, 324
  reasons for use, 329, 330
  regulation, 321
  resistance to, 86, 320, 325–327
  Soviet contamination, 329
  U.S. legislation, 445
  water pollution, 355
Pests, 319
  extinction, 261
  integrated management, 331–334
Petroleum products
  agricultural use, 318
  global demand, 180
  global resources, 180, 183
Pfiesteria piscicida, 255
pH, 70–71
  acid rain effects, 381, 382
  soil and, 295–296, 298–299, 382
Pheromones, 332
Philippines, 272
Phosphates, 352
  organophosphates, 321–322
Phosphorus cycle, 99–100
Photochemical smog, 377–378, 379
Photosynthesis, 72
Photovoltaic cells, 204, 206
Physical wastewater treatment, 358, 359, 362
Phytoplankton, 121
Pioneer community, 106
Pipelines, 171
Place, 55–56
Plankton, 121
Planning. *See* Land-use planning; Water-use planning
Plantation forestry, 246
Plants. *See also* Agriculture; Food production
  acid rain and, 382
  climate and elevation effects, 111, 112
  extinction, 261, 262, 264
  genetic engineering, 317, 333–334
  lake and pond, 125
  pest control via, 320, 334
  soil formation and, 294, 295
Plastics
  recycling issues, 412–413, 414
  resins in, 406
  source reduction, 410
Plate tectonics, 292–293
Plutonium-239, 222, 224
PNGV (Partnership for a New Generation of Vehicles), 176
Poaching, 261
Point source pollution, 350–351, 353
Poisoning, pesticide, 327–329
Polar ice cap, 111
Policy. *See* Environmental policy
Political issues. *See also* Environmental policy
  Arctic National Wildlife Refuge, 212
  ethylene dibromide and, 325
  green, 447–451
  Native American fishing rights, 257
  population growth and, 151–152
  waste disposal and, 231–232, 432 433
Pollutants. *See also specific pollutant*
  Great Lakes, 95, 335
  hazardous waste sites, 426–429
  persistent and nonpersistent, 424
  risk values, computation, 41
  Soviet soil, 329
  Superfund sites, 426–428
  wastewater, treatment, 358–360, 362
Pollution. *See also specific type of pollution*
  control costs, 53
  economic solutions, 53
  energy conversion and, 76
  food demands and, 154–156
  forms, 240
  fossil fuel sources, 194, 195–196, 197, 375–377
  historical basis, 239–240
  human populations and, 147, 154–156, 239–240
  land-use planning and, 282
  nuclear sources, 224, 230–233
  prevention, 53, 54, 429–430
  recycling and, 411–412
  renewable energy sources, 199–200, 202, 206, 209, 210
Pollution costs, 53
Pollution-prevention costs, 53
Pollution-prevention hierarchy, 429, 430
Polychlorinated biphenyls (PCBs), 325
  Great Lakes, 335
  sources and health effects, 424, 425
Polyculture, 316
Polystyrene, 406, 416
Ponds
  aquaculture, 255
  cooling water, 354
  ecosystem, 125–126
  primary succession, 108
  secondary succession, 109, 110
Population characteristics, 131. *See also* Human populations; Wildlife ecosystems
  carrying capacity, 135–136
  density and spatial distribution, 133
  insecticides and, 327
  natality and mortality, 131
  sex ratio and age distribution, 132–133
Population density, 133, 147
Population growth
  factors influencing, 133
  growth curve, 134–135, 136
  human. *See* Human population growth
  invading species, 137
  reproductive strategies and, 138
Population growth curve, 134–135, 136
Population management. *See* Wildlife ecosystems
Porosity, 342
Postwar baby boom, 157, 158
Potable water, 340
  contaminants, 351
  Safe Drinking Water Act, 449
Potential energy, 72, 74
Poverty
  environmental impact, 450
  population growth and, 153–154
  southern U.S., 15
Power plants. *See also* Hydroelectric power
  hydroelectric, 198, 347
  nuclear. *See* Nuclear reactors
  sulfur dioxide emission control, 380
Prairies. *See* Grasslands
Precept (automobile), 176
Precipitation. *See also* Climagraphs
  desert, 111–112, 113
  forest, 116, 117, 118, 119, 120
  global warming and, 390
  grassland, 113, 115
  savanna, 114, 115
  vegetation and, 112
  waste treatment, 431
Predation, 87
Predator-prey populations, 87
  bison-Native Americans, 252
  Canada lynx-varying hare, 136
  Isle Royale, 143
  pest management and, 332–333
  wolf-multiple species, 101, 143
  Yellowstone National Park, 101
Predators, 87
  humans, 91
  management, 258–259
  natural, insect control, 332–333
  wolves, 101, 143
Preservation ethic, 24
Preservation/protection
  Arctic, 212
  coastal areas, 363–364
  EPA and. *See* Environmental Protection Agency
  forest ecosystems, 57, 127
  habitat, 258–259
  land-use planning and, 279
  nonfarm land, 307, 309
  preservation ethic, 24
  soil, 330–331
  water resources, 331, 356, 362–365
Pressurized-water reactors (PWRs)
  decommissioning contaminants, 230
  description, 218, 219
Prey, 87. *See also* Predator-prey populations
Primary air pollutants, 374–377
Primary consumers, 92
Primary sewage treatment, 358, 359
Primary succession, 106
  aquatic, 108
  terrestrial, 106–107
Private property rights, 444
Probability, 39
Processing, of crude oil, 195, 196
Producers, 92
Profitability, corporate, 26, 27
Propane, vehicular, 181
Property
  common, 53–55, 58, 453
  private, rights, 444
Protection. *See* Preservation/protection
Protons, 69–70
Pu-239, 222, 224
Public land-use conflicts, 285–287
Public policy, 442. *See also* Environmental policy; Legislation, U.S.
Public transportation
  energy consumption, 174–175
  land-use planning and, 280
  in Mexico City, 376
  urban planning, 283
  urban sprawl and, 275
Purple loosestrife, 333, 334

PWRs. *See* Pressurized-water reactors
Pyrolysis, 207

Q

Quail, 256
Questioning, 67

R

Rabbits, cottontail, 133
Radiation, 216
Radiation exposure, 226, 227–228, 229
Radioactive, 216
Radioactive half-life, 216
Radioactive materials
  disposal, 224, 230–233
  transportation, 223, 228
Radon, 394, 396
  air pollutant, 378
  geologic potential, U.S., 395
Rail transport trends, 283
Rainfall. *See* Precipitation
Rainforests, temperate, 127. *See also* Tropical rainforests
Ranching, tropical rainforest, 116–117
Rangeland ecosystems, 247–249, 286
Range of tolerance, 82
Rangoon, Burma, 115
Raw materials
  human population growth and, 141
  as limiting factor, 135
  recyclables as, 413
Reactivity, 420–421
Reclamation, surface mine, 192, 194
Recreational resources
  forest-based, 309
  global warming and, 391
  land-use conflicts, 286–287
  urban planning, 283–285
  U.S. legislation, 285
  water-based, 348, 357
Recycling, 410–415
  benefits, 411–412
  as big business, 413
  of computers, 429
  concerns, 412, 414
  as energy source, 209–210
  Georgia-Pacific Corporation, 48
  green advertising and, 51
  market basket, 415
  mineral, 243
  rates, 412
  symbols, 51
  U.S. legislation, 411
  waste, 209–210, 431
Red-cockaded woodpeckers, 246
Reduced-emission vehicles, 181
Reforestation, 246
Refrigerants, 388–389
Refund programs, 46
Refuse. *See* Solid waste disposal; Waste products
Regional cities, 273, 274
Remote areas. *See* Wilderness areas
Rems (roentgen equivalent man), 227–228
Renewability, sustainability and, 52
Renewable energy sources, 188, 197
  biomass conversion, 205–207
  fuelwood. *See* Fuelwood
  geothermal power, 201–202, 208
  global consumption, 179
  hydroelectric power, 197–200
  solar energy, 203–205, 206
  solid waste, 209–210
  tidal power, 200–201
  wind power, 202–203
Renewable resources, 241
Repeatability, 68
Replacement fertility, 149
Reproduction. *See also* Birth control
  acid rain and, 382
  biological factors, 148–149
  California condor, 263
  natural selection and, 85–86
  political factors, 151–152
  population strategies and fluctuations, 138
  reproductive potential, 256, 258
  social factors, 149–151
Reserves
  animal, 47
  extractive. *See* Extractive reserves
  versus resources, 188
Reservoirs, 199–200, 347–348
Residential energy use, 173–174
Resins, packaging, 406
Resistance
  environmental, 135
  pesticide, 86, 320, 325–327
Resource Conservation and Recovery Act, 421, 446
Resource exploitation
  humans and, 147–148, 240–243
  ocean-based, 252
  societal environmental ethics and, 26
  sustainable development and, 52
  tropical rainforest-based, 117
  utilization costs, 241, 242–243
Resources, 43, 188. *See also* Resource exploitation; *specific resource or class of resource*
Resource subsidies, 47
Resource waste
  economic solutions, 53
  U.S. legislation, 445
Respiration, 72, 73
Respiratory system
  air pollutants and, 375, 376, 377, 381, 396
  radon effects, 394
  tobacco smoke and, 384
Reuse
  of computers, 429
  source reduction and, 410
Ribbon sprawl, 273
Rio Declaration on Environment and Development, 31
Risk assessment, 39–41
Risk management, 41–42
Risks, 39–43
  assessment, 39–41
  management, 41–42
  true and perceived, 42–43
Rivers. *See also specific river*
  cyanide contamination, 449–450
  dams and, 347–348
  development and, 362–363
  ecosystem, 126
Rivers–*Cont.*
  global warming and, 390–391
  water diversion, 357–358
Rodenticides, 319, 323–324
Rodents, 323
Roentgen equivalent man (rems), 227–228
Romania, 449–450
Rome, Italy, 173
R-strategists, 138
Rubber industry, 28
Runoff, 341
  agricultural, 353
  marine oil, 355
  reduction, 353
  storm-water, 356–357
Rural areas, migration from, 157, 270–272, 328
Russia, 120, 363. *See also* Soviet Union, former

S

Safe Drinking Water Act, 449
Salinization, 361, 362, 363
Salmon
  dams and, 255
  farming, 255
  in Great Lakes, 253, 255
  human impact on, 7
  political friction over, 6
Salt. *See* Salinization
Saltwater intrusion, 362, 364
Sandstone, 191
Savannas, 111, 114–116
Scale insects, 332–333
Scandinavia, 382–383, 398
Scavengers, humans as, 91
Scenic water areas, 362–365
Science, 6, 67
Scientific method, 67–69
Scrap tires, 415
Screwworm flies, 332
Scrubbers, 380
Sea lamphrey, 253, 254
Sea level, global warming and, 387, 390
Secondary air pollutants, 374–375, 377–378
Secondary consumers, 92
Secondary recovery of oil, 195
Secondary sewage treatment, 358, 359
Secondary succession, 106, 109, 110
Secondhand smoke, 384
Second law of thermodynamics, 73–74, 75
Sediment, as pollutant, 349
Seeds, 317
Selective harvesting, 246
Sensible heat, 73
Septic tanks, 355
Seral stage, 107
Sere, 107
Sewage sludge, 359
Sewage systems
  industrial waste and, 353
  wastewater treatment, 358–360, 362
Sex attractants, 332
Sex ratio, 132
Shipping industry, 278, 355
Shorelines, 122–123
Shrimp industry, 59
Singapore, 118, 173
Single European Act, 452
Size limitations, human populations, 141
Slash-and-burn agriculture, 116, 316
Slope, 295, 301
Smog, 377–378, 381
Smokestacks, 380
Smoking
  secondhand smoke, 384
  tobacco effects, 375, 384, 396
Social factors, population growth and, 141, 149–151
Soft drink containers, 410, 411
Soil, 291–312. *See also* Soil conservation practices
  acid rain and, 382
  compaction, 330
  erosion, 299, 301, 308, 312
  fertilizer and, 319–320
  formation, 294–296
  global degradation, 308
  versus land, 294
  monoculture and, 317
  on nonfarm land, protection, 307, 309
  properties, 296–297
  protection, 330–331
  soil profile, 297–299, 300, 301
  Soviet, contamination, 329
  tillage methods, 305–307
Soil conservation practices, 301–303, 304
  contour farming, 303, 305
  conventional *versus* conservation tillage, 305–307
  strip farming, 304, 305
  terracing, 304, 306
  waterways, 304, 306
  windbreaks, 304–305, 307
Soil profile, 297–299, 300, 301
Soil structure, 296
Soil texture, 296
Solar energy, 203–205, 206, 393
Solar radiation, greenhouse effect, 386, 387
Solid energy, 73
Solid waste disposal, 401–416. *See also* Waste products; Wastewater
  composting, 408–409, 410
  incineration, 407–408, 409, 410
  landfilling, 405–407, 408
  methods, changes in, 404–405
  problems, 402–404
  radioactive material, 224, 230–233
  recycling, 411–415
  source reduction, 409–411
  transport, 346
  U.S. legislation, 445
Solvents, industrial, 430–431
Source reduction, 409–411
Southern United States, 14–15
Soviet Union, former
  Aral Sea decline, 363
  Chernobyl accident, 225–227, 231
  nuclear legacy, 231
  soil contamination, 329
Space (physical)
  in economics *versus* ecology, 55–56
  soil and, 296, 297

Speciation, 86
Species, 84
community and ecosystem interactions, 91–101
evolutionary patterns, 86–87
extinction, 86, 260–265
global warming and, 392
invading/introduced, 137, 253–255, 257–258
natural selection, 85–86
organism interactions, 87–90
Sport fishers, 253, 257
Sport hunters, 256–257
Spray irrigation, 345
Stable equilibrium phase, 135
Standard of living, 152–153
Steam engine, 170
Steam stripping, 431
Steppes. *See* Grasslands
Sterilization, 332
Storm-water runoff, 356–357
Streams, 126, 347–348
Streptococcus faecalis, 351–352
Strip farming, 304, 305
Strip mining. *See* Surface mining
Strix occidentalis caurina, 247
Subirrigation, 345
Submerged plants, 125
Subsidy(ies), 46–48
Subsoil, 298, 300
Substitution, sustainability and, 52
Suburbs, 172, 272–275
Succession, 106
grassland, 114
primary, 106–108
secondary, 109, 110
Successional stage, 107
Sulfur dioxide, 376–377
acid rain effects, 381–382
control, 380, 381
Sulfuric acid
in acid rain, 381, 382
mining contamination, 194
Supply (economic), 43–45
Supply/demand curve, 44–45
Surface impoundments, 355–356
Surface irrigation, 345
Surface mining, 242–243
coal, 192, 193
reclamation, 192, 194
Sustainability management framework, 25
Sustainable agriculture, 330
Sustainable development, 6, 50–53
Amsterdam Treaty, 452
UN commission, 454
Swamps, 126
coal formation and, 190–191
mangrove ecosystem, 124
Sweden, 173, 398
Symbiosis, 88
Symbiotic nitrogen-fixing bacteria, 98
Symbiotic relationships
commensalism, 89–90
mutualism, 90
parasitism, 88–89
Synergism, of hazardous/toxic substances, 423–424

T

Taiga, 111, 119–120
Tapeworms, 89
Target organisms, 319
Taxes, 46, 172–173
Technology, pollution and, 239–240, 392–393
Tehran, Iran, 115
Tellico Dam project, 200, 264
Temperate forests
changes, 244
deciduous, 111, 117–119
rainforest, 127
Temperature. *See also* Climagraphs; Global warming
ambient, pollution and, 349
aquatic ecosystems and, 353–355
conversion tables, 461
coral reefs and, 122
global changes, 385, 386, 387
lakes and, 125
terrestrial ecosystems and, 111, 112, 116
Terracing, 304, 306
Terrestrial ecosystems. *See also* Biomes
primary succession, 106–107
secondary succession, 109
temperature and, 111, 112, 116
utilization and modification, 243–244
Terrorism, 448, 451
Tertiary sewage treatment, 360, 362
Thawing, rock fragmentation and, 293
Theory, 68–69. *See also specific theory*
Thermal inversion, 378
Thermal pollution, 228, 229, 353–355
Thermodynamic laws, 73–74, 75
Thermoplastics, 406
Third World debt, 58
Thoreau, Henry David, 23
Thought, critical, 459
Threatened species, 247, 262
Three Gorges Dam, 200, 201
3M Company, 54
Three Mile Island nuclear plant, 225
Threshold level, 423
Tidal power, 200–201
Tillage methods, 303–307
Time
demographic transition and, 156
in economics *versus* ecology, 55
soil formation and, 295
Tire disposal, 415
Tobacco, smoking, 375, 384, 396
Tobacco budworm, 326–327
Tokyo, Japan, 173
Tolerance, range of, 82
Topography, 299
Topsoil
erosion, 299, 301, 302
profile, 297–298, 300
Tortoise, 265
Total fertility rate, 149
Toxic, 421
Toxicity, 421, 423
Toxic substances. *See also* Hazardous substances
air pollutants, 378, 379, 384, 408
benefit-risk associations, 420
chemical releases, 429, 430
Chinese levels, 433
definitions, 420, 421
exposure, 423
Toxic substances–*Cont.*
life cycle, 420
Love Canal, 435
regulatory issues, 421–424
in secondhand smoke, 384
U.S. legislation, 445
water pollutants, 13, 349
Toxic terrorism, 448
Tract development, 273
Tradable emission permits, 46
Trade, endangered species, 30
Traffic, 173, 275. *See also* Public transportation
Tragedy of the commons, 58
Transportation
of coal, 193–194
energy consumption, 174–175
of hazardous wastes, 431, 432–434
of minerals, 243
of natural gas, 196–197
public. *See* Public transportation
of radioactive material, 223, 228
Transuranic wastes, 230
Trash. *See* Solid waste disposal; Waste products
Trash power, 209–210
Trawls, 252–253
Trees, carbon dioxide and, 386, 388. *See also* Logging
Trichloroethylene, 425
Trickle irrigation, 345
Trickling filter system, 358
Trihalomethanes, 351
Trillium, persistent, 265
Trophic levels, 93, 154, 155
Tropical forests, 244, 246–247
Tropical rainforests, 116–117
biodiversity loss and extinction in, 261
carbon dioxide reduction, 388
economic exploitation, 116–117
geographic distribution, 111
Singapore climagraph, 118
Tropical soils, 299
True *versus* perceived risks, 42–43
Tundra, 111, 120, 121
Turtle excluder devices, 59

U

UNCED (United Nations Conference on Environment and Development), 6, 31, 452
Unconfined aquifer, 341
Underground mining, 192, 193
Underground storage tanks, 355
United Kingdom, 376–377, 398
United Nations Commission on Sustainable Development (CSD), 454
United Nations Conference on Environment and Development (UNCED), 6, 31, 452
United Nations Conference on the Law of the Sea, 452
United Nations Environment Programme, 385, 433, 451, 452
United Nations Framework Convention on Climate Change, 6–7, 8
United States
acid rain effects, 382–383
agencies, 443. *See also specific agency*
agricultural lands, 302, 303
carbon emissions, reduction, 393
energy consumption, 177, 180, 181
energy production, 171
environmental activism, 445–446
groundwater depletion, 362, 364
hazardous waste sites, 426–429
immigration policy, 157–158
international air quality and, 6
land-use policy and conflicts, 285–287
legislation. *See* Legislation, U.S.
legislators, writing to, 464
population trends, 131, 153, 157–158, 159, 160, 450
radioactive waste disposal sites, 230, 232–233
radon potential, 395
standard of living, 153
waste generation, 402, 403–404
Wise Use Movement, 444
United States Army, 430
Unplanned urban growth. *See* Urban sprawl
Uranium
mine tailings, 228, 229
in nuclear fuel cycle, 223, 224
nuclear power requirements, 221
Pu-239 and, 222
as radon source, 394
U-235, 217, 223, 224
Urban areas
air pollution, 374, 378, 380
Chinese, 328
global urbanization, 157
planning issues, 282–285
population trends, 157, 272
rural-to-urban shift, 157, 270–272, 328
sprawl, 273–279
urban-to-suburb shift, 272–273, 274
waste disposal. *See* Solid waste disposal
water sources and use, 343–344, 345, 346
water-use planning, 356–357, 358–361
waterways and, 270, 271
Urban growth limit, 280
Urban planning, 282–285
Urban sprawl
contributing factors, 273–275
development and types, 273
problems associated with, 275–279

V

Vadose zone, 341
Valdez Principles, 26–27
Value
economic, analysis, 50
of ecosystem services, 56
Values, 52
Varying hare, 136, 138
Vegetation. *See* Plants
Vehicles. *See* Automobiles
Vinyl chloride, 425

Virginia, 312
Visual pollution, 282
Volatile organic compounds (VOCs), 396

W

Warbler, Kirtland's, 258
Waste immobilization, 432
Waste minimization, 430–431
Waste products
  disposal. *See* Hazardous waste disposal; Solid waste disposal
  as energy source, 209–210
  hazardous. *See* Hazardous waste
  human population growth and, 141
  industrial ecology and, 28
  municipal, 351–352, 402–403
  radioactive, 224, 225
  reduction, 409–411, 413
  resource, economic solutions, 53
  U.S. legislation, 445
Waste-to-energy facilities, 408, 409
Wastewater
  domestic, 344, 352
  Ecoparque and, 366
  treated, use, 360–361
  treatment, 356, 358–361, 362
  waste minimization and, 430
Water, 339–367
  agricultural use, 331, 344–346, 347
  California Water Plan, 367
  characteristics, 340
  conservation, 346, 362
  developed *versus* less-developed nations and, 354
  domestic use, 343–344, 345, 346
  global use, 345
  global warming and, 390–391
  hydrologic cycle, 341–343, 390–391
  industrial use, 346–347
Water–*Cont.*
  in-stream use, 347–348
  international disputes, 357
  issues surrounding, 340
  planning use of. *See* Water-use planning
  pollution. *See* Water pollution
  potable, 340, 351, 449
  regional concerns, 12, 13–15
  rock weathering and, 293
Water diversion, 357–358
Water erosion, 299, 301, 302
Waterfowl, migratory, 259–260
Water molecule, 71
Water pollution, 340. *See also specific pollutant*
  agricultural, 352–353, 355
  in developed *versus* less-developed nations, 354
  groundwater, 355–356
  industrial, 353
  land-use planning and, 282
  marine oil, 355
  municipal, 351–352
  Soviet, 329
  thermal, 353–355
  types and sources, 349–351
  urban sprawl and, 277
Water quality, 340
  fisheries and, 253
  global warming and, 391
  U.S. legislation, 346, 351, 353, 357, 445
Water Quality Act, 357
Water shortages, 340–341
  Chinese levels, 328
  urban sprawl and, 279
Water table, 341
Water treatment
  air stripping, 431
  raw water, 343–344
  wastewater, 356, 358–361, 362
Water-use planning, 356–364
  groundwater mining, 361–362, 364
Water-use planning–*Cont.*
  preservation issues, 356, 362–365
  salinization, 361
  wastewater treatment, 356, 358–361, 362
  water diversion, 357–358
Waterways
  importance, 270, 271
  recreational use, 348, 357
  resource transport via, 348
  scenic, 362–365
  soil conservation, 304, 306
  waste disposal into, 431
  waste transport via, 346
Wealth inequality, 58
Weapons, nuclear, 223–224, 225
Weathering, 293–294, 295
Weed control, 322–323, 333
Weeds, 319
Well water, 342, 343, 357
Western North America, 11–12, 13, 14
Wetlands
  global warming and, 391
  Louisiana loss, 278
  urban sprawl and, 277–278
  value, 364–365
Weyerhaeuser Company, 413
Whaling, 54–55, 453
Whooping crane, 265
Wilderness areas, 8, 250
  Arctic National Wildlife Refuge, 212
  fires, 266
  human impact on, 249–250
  land-use conflicts, 286
  North American, 8–9, 11, 12, 13
  protected, growth, 250
Wildlife ecosystems, 255
  Arctic refuge, 212
  habitat analysis and management, 255–256
  migratory waterfowl management, 259–260
Wildlife ecosystems–*Cont.*
  population assessment and management, 256–258
  predator and competitor control, 258–259
  U.S. legislation, 445
Wind
  global patterns, 373
  as pollution vector, 374, 380. *See also* Air pollution
  as power source. *See* Wind power
Windbreaks, 304–305, 307
Wind erosion, 301, 302, 303
Wind power, 202–203
  green marketing, 393
  rock weathering and, 293
Windrows, 409
Wise Use Movement, 444
Wolves
  on Isle Royale, 143
  population control, 258
  in Yellowstone National Park, 101
Women, 149–151, 154
Woodpeckers, red-cockaded, 246
Wood products, recycling, 48. *See also* Fuelwood
World Bank, 57–58
World Conservation Union (IUCN), 264

Xerox, 54

Yard waste, 409, 410
Yellowstone National Park, 9, 101

Zebra mussels, 137, 255
Zero population growth, 149
Zoning ordinances, 274, 277, 281–282
Zooplankton, 12

Southern Living®

2015 Annual Recipes

Oxmoor House®

Beer-Braised Pot Roast (page 29)

clockwise from top left:

• Meatball Sliders with Tomato Sauce (page 28)
• Slow-Cooker Chicken Cacciatore with Spaghetti (page 30)
• Italian Turkey Meatloaves (page 35)
• Chicken Paillard with Citrus Salad and Couscous (page 33)

Triple-Chocolate
Buttermilk Pound Cake
(page 44)

Mustard Greens with Yogurt-Parmesan Dressing and Bacon Croutons (page 48)

Citrus-Braised Chicken Thighs (page 64)

clockwise from top left:

- Chicken, Farro, and Vegetable Salad with Lemon Vinaigrette (page 64)
- The Ultimate Fried Egg Sandwich with BBQ Bacon (page 61)
- Slow-Cooker Chicken Lettuce Cups (page 92)
- Bucatini, Ham, and Asparagus (page 89)

clockwise from top left:

• Coconut-and-Pecan Strawberry Shortcakes (page 86)
• Coconut Cream Cake (page 93)
• Strawberry-Blueberry Relish (page 107)
• Raspberry "Rhubars" (page 111)

Pork Tenderloin Sliders (page 101)

Skillet Pork Chops with Sautéed Peaches (page 142)

Peach Divinity Icebox Pie
(page 121)

Shrimp Boil with Green Olives and Lemon (page 124)

clockwise from top left:

• Pasta with Heirloom Tomatoes, Goat Cheese and Basil (page 159)
• Burgers with Green Tomato Mayonnaise (page 159)
• Broiled Pork Chops with Basil Butter and Summer Squash (page 194)
• Pork Chop Sandwiches with Gravy and Grits (page 194)

clockwise from top left:

- Ultimate Cheese Pizza (page 207)
- Buttermilk-Lady Pea Soup with Bacon (page 217)
- Zucchini Stuffed with Lady Peas (page 217)
- Lady Pea-and-Corn Patties (page 217)

Biscuit Cinnamon Sweet Rolls
(page 224)

Spicy Pumpkin Soup with Avocado Cream (page 238)

Our Year at Southern Living®

Dear Friends,

Our Test Kitchen staff gathers around a big table every day to taste dozens of dishes with a simple goal in mind: to find recipes you'll be proud to serve those you love.

We ask tons of questions during the process. Will grandmother approve? Will it impress company? Is it practical for a busy weeknight? I've got a 9-month-old daughter, Vivian, and the way she reacts to what I feed her is my daily reminder why those questions matter.

Weeknight cooking is no easy task, and sometimes I simply hand Vivian a pouch of premade baby food. She happily slurps it down. When I cook her dinner, however, she not only takes down her whole supper (I'm lucky to have a good eater), but she does what we call the Yummy Dance, a combo of pursed lips, wide eyes, and fluttering feet that's her way of saying "more, please!" It happens only when we serve her something fresh or homemade. Her Yummy Dance makes me proud; it's how I know she can taste the difference.

There's nothing like food, family, and fellowship to remind you why life is good.

That difference is the comfort and joy that comes with a homemade meal. It's pride that comes with nailing grandma's prized recipe. It's an appreciation for food that was grown close to where we live. It's what we all know, but can't necessarily explain, that makes the way we eat distinctly Southern. In our Test Kitchen, we ask so many questions and test so many dishes to make sure our recipes deliver that difference to you.

I hope this year's collection inspires you to set the table, cook, and eat with those you love, especially when, as Vivian taught me, you're having a tough day. There's nothing like food, family, and fellowship to remind you why life is good.

Thank you for your loyal support of *Southern Living,* in particular as we head into 2016, our 50th Anniversary year. We've got a lot of exciting things planned and can't wait to celebrate with you.

Very fondly,

Whitney

Whitney Wright
Deputy Food Director and General Manager

Contents

17 Our Year at *Southern Living*
20 Top-Rated Recipes

23 January
24 **The South is Your Oyster**
28 **Slow-Cooked Perfection**
32 THE *SL* TEST KITCHEN ACADEMY: **Slow-Cooker Success**
33 QUICK-FIX SUPPERS: **5 Hearty Main-Dish Salads**
35 THE NEW SUNDAY SUPPER: **Mighty Good Meatloaf**
37 *COOKING LIGHT:* **A Delicious New Diet**
38 COMMUNITY COOKBOOK

39 February
40 **The Next Generation of Soul Food**
43 **Buttermilk Revival**
47 SOUTHERN GREENS: **Eat Your Greens**
50 QUICK-FIX SUPPERS: **Ten-Dollar Dinners**
52 THE *SL* TEST KITCHEN ACADEMY: **Spectacular Fresh Butter & Buttermilk**
53 *COOKING LIGHT:* **A Light Pork Supper**
54 *SOUTHERN LIVING* RECIPE REVIVAL: **Blue Cheese-Pecan Grapes**
55 THE SLOW DOWN: **Sauté, Simmer, Savor**
56 SAVE ROOM: **Dark Magic**
58 COMMUNITY COOKBOOK

59 March
60 **The Saturday Morning Chef**
63 WHAT TO COOK NOW: **Thigh and Mighty**
66 *COOKING LIGHT:* **A Fresh Fiesta**
67 THE *SL* TEST KITCHEN ACADEMY: **Candied Carrot Curls**
68 SAVE ROOM: **Crazy for Carrot Cake**
69 THE SLOW DOWN: **Try a New Stew**
70 COMMUNITY COOKBOOK

71 April
72 **Fresh from Garden to Plate**
78 **The Pitmaster's Guide to Brisket**
84 HOSPITALITY: **Easiest Easter Ever**
87 THE *SL* TEST KITCHEN ACADEMY: **Cooking Perfect Pasta**
88 QUICK-FIX SUPPERS: **30-Minute Spring Pastas**
90 **The Pursuit of Perfect Biscuits**
91 *COOKING LIGHT:* **Crab Takes the Cake**
92 THE SLOW DOWN: **Saucy Chicken Lettuce Cups**
93 SAVE ROOM: **Our Best Coconut Cake**
94 COMMUNITY COOKBOOK

99 May
100 **Host Derby Day**
103 **An Editor's Escape**
104 QUICK-FIX SUPPERS: **Bake, Grill, Sear, or Roast: It's Fish Season**
106 THE *SL* TEST KITCHEN ACADEMY: **Buying Fresh Seafood**
107 WHAT TO COOK NOW: **Chowchows to Relish**
109 THE SLOW DOWN: **Slow-Cooker Ribs**
110 *COOKING LIGHT:* **A Healthy Slice**
111 SAVE ROOM: **Raise the Bar**
112 COMMUNITY COOKBOOK

117 June
118 **The Pies of Summer**
123 **Taste of the Gulf**
126 **Food Awards 2015**
132 COOKING: **The *SL* Guide to Grilled Chicken**
141 THE *SL* TEST KITCHEN ACADEMY: **Quick Fruit Pickles**
142 *COOKING LIGHT:* **Dinner in a Breeze**
143 THE SLOW DOWN: **Green Beans Like You've Never Had**
144 COMMUNITY COOKBOOK

145 July
146 **The Ultimate Lazy Lunch**
150 **Romancing the Stone**
156 **Loving Summer Tomatoes**
160 QUICK-FIX SUPPERS: **Chop, Chop!**
195 THE NEW SUNDAY SUPPER: **Steak Nights**
197 SOUTHERN CLASSIC: **Summer Succotash**
198 THE *SL* TEST KITCHEN ACADEMY: **Summer Tomatoes**
199 THE SLOW DOWN: **Taco Night Reimagined**
200 *COOKING LIGHT:* **Light and Luscious Summer Pasta**
201 SAVE ROOM: **Easy Ice Cream**
202 COMMUNITY COOKBOOK

203 August

- 204 HOSPITALITY: **Backyard Pizzeria**
- 208 THE *SL* TEST KITCHEN ACADEMY: **Grilled Pizza**
- 209 SOUTHERN CLASSIC: **Classic Sweet Tea**
- 210 THE SLOW DOWN: **Fresh, Versatile Tomato Sauce**
- 212 QUICK-FIX SUPPERS: **Fast, Fresh, and Filling**
- 216 THE NEW SUNDAY SUPPER: **One Pot, Three Ways**
- 218 SAVE ROOM: **Cool Down with Easy Granitas**
- 220 COMMUNITY COOKBOOK

221 September

- 222 **Baked with Love**
- 225 QUICK-FIX SUPPERS: **Dinner in 20 Minutes Flat**
- 228 THE *SL* TEST KITCHEN ACADEMY: **Five Time Saving Tools**
- 229 *COOKING LIGHT:* **Better Than Takeout**
- 230 SOUTHERN CLASSIC: **Classic Pimiento Cheese**
- 231 SAVE ROOM: **The No-Peel Apple Cake**
- 232 **The Hunt Breakfast**
- 234 COMMUNITY COOKBOOK

235 October

- 236 **The Great Pumpkin Cookbook**
- 240 SOUTHERNERS: **Cooking for My Family**
- 241 WHAT TO COOK NOW: **Beautiful Beef Dinners**
- 244 **Make It a Movie Night**
- 246 THE *SL* TEST KITCHEN ACADEMY: **The *SL* Beef Primer**
- 247 THE SLOW DOWN: **Bolognese Made Easy**
- 248 COMMUNITY COOKBOOK

249 November

- 250 **Make-Ahead Thanksgiving**
- 259 ENTERTAINING WITH JULIA: **Host a Gumbo Party**
- 261 **Pecan Delights**
- 264 HOSPITALITY: **Soulful Soups and Stews**
- 266 WHAT TO COOK NOW: **The Best Pancakes Ever**
- 267 THE *SL* TEST KITCHEN ACADEMY: **Our Secrets to the Best Pancakes**
- 268 COMMUNITY COOKBOOK

269 December

- 270 HOLIDAY 2015: **Appetizers & Cocktails**
- 276 HOLIDAY 2015: **Mains & Sides**
- 284 HOLIDAY 2015: **Desserts & Cookies**
- 295 **The Baker's Brunch**
- 299 ENTERTAINING WITH JULIA: **Ham for the Holidays**
- 301 THE *SL* TEST KITCHEN ACADEMY: **Smart Cookies**
- 302 COMMUNITY COOKBOOK

303 Bonus Favorites

- 304 **Quick-and-Easy**
- 310 **Holiday Tea**
- 313 **Baked Greats**
- 316 **Get the Party Started**
- 318 **Soups and Stews**
- 320 **Holiday Sideboard**
- 324 **Main Attractions**
- 326 **Save Room**
- 328 **A Big Batch of Cookie Favorites**
- 331 **Good Luck New Year**
- 333 **Southern Baker**

341 Metric Equivalents

343 Indexes

- 343 **Menu Index**
- 344 **Recipe Title Index**
- 348 **Month-by-Month Index**
- 352 **General Recipe Index**

367 Favorite Recipes Journal

Top-Rated Recipes

We cook, we taste, we refine, we rate, and at the end of each year our Test Kitchen shares the highest-rated recipes from each issue exclusively with *Southern Living Annual Recipes* readers.

JANUARY

- Oyster-Bacon Pot Pie (page 25) A gussied-up riff on chowder, golden puff pastry crowns a creamy, briny filling. Instead of individual portions, make this recipe in a lightly greased 11- x 7-inch baking dish.
- Meatball Sliders with Tomato Sauce (page 28) Two recipes in one: delicious, easy meatballs and a slow-simmered tangy tomato sauce
- Beer-Braised Pot Roast (page 29) A coffee rub, and then stout beer and beef stock simmer yield a rich, delicious gravy.
- Italian Turkey Meatloaves (page 35) Mushrooms keep this ground turkey loaf moist while pesto amps up the flavor.

FEBRUARY

- Peanut Chicken Stew (page 41) With West African roots, this nut soup is a hearty stand-alone meal and perfect fortification for a cold winter day.
- Triple Chocolate Buttermilk Pound Cake (page 44) Our February cover cake is worth every bite and calorie too!
- The Ultimate Classic Collards (page 47) Simmered greens aren't complete without the smoky flavor a ham hock lends the pot liquor. A splash of vinegar brightens the flavors for serving.
- The Ultimate Chocolate Pie (page 56) A chocolate wafer crumb crust and toasted pecan topping gild this three-layer chocolate pie.
- Chocolate Whipped Cream (page 56) Chocolate syrup whisked into heavy cream before whipping is the perfect topping for desserts and the third layer in our Ultimate Chocolate Pie.
- Chocolate-Mayonnaise Cake (page 57) Delivering a moist crumb and mellowing overt sweetness, mayonnaise is the surprise ingredient in this deliciously rich cake.
- Chocolate-Cream Cheese Frosting (page 57) Cream cheese, butter, and powdered sugar are the stars of cream cheese frosting, but a generous dose of rich cocoa takes it over the top.

MARCH

- Shrimp 'n' Grits Pie (page 61) Overnight slow-cooker grits provide the base crust for this easy weeknight spin on the Lowcountry favorite.
- Grits Cakes with Poached Eggs and Country Gravy (page 61) Serve this as a hearty weekday breakfast of champions or a decidedly Southern weekend brunch main course.
- Pecan-Rosemary Bacon (page 62) Woodsy rosemary and Southern pecans add surprising flavor and nutty crunch to bacon. Try this uncommon bacon on grilled burgers too.
- BBQ Bacon (page 62) Barbecue rub spices melt into the fat while the bacon cooks, yielding a lacquered effect with loads of flavor.
- Sesame Chicken Thigh Paillard with Peanut Sauce (page 63) The French method of pounding meat thinly means it cooks in no time, making this tender chicken dish perfect for busy nights.
- Peanut Sauce (page 63) This no-cook flavorful sauce goes with Sesame Chicken Thigh Paillard but also is great as a dipping sauce or tossed with cold noodles.
- The Ultimate Carrot Cake (page 68) Moist, rich, and crowned with candied carrot curls, this one's for keeps.
- Brown Sugar-Cream Cheese Frosting (page 68) Brown sugar adds depth and makes this a new frosting favorite.

APRIL

- Garden Potato Salad (page 74) Fresh herbs brighten a salad of just-dug new potatoes.
- Kale-and-Blueberry Slaw with Buttermilk-Tarragon Dressing (page 75) Early season flavors collide in this delicious, colorful slaw that's sure to be the star of the picnic table.
- Smoked Beef Brisket Tostadas (page 81) Cookbook author Chris Prieto of North Carolina's PRIME Barbecue shows us a new way with beef brisket that's a keeper.
- Roasted Leg of Lamb and Lemon-Herb Salt (page 84) Salt, herbs, and citrus rind get whirled in a blender for a flavorful crust on this holiday-worthy roast.
- Roasted Beets, Carrots, and Sweet Onions (page 85) Roasting brings out the inherent sweetness of baby spring vegetables in a rainbow of hues.
- Cheese Grits Soufflé with Mushroom Gravy (page 85) Grits, Gruyère, and sherry-laced mushrooms take cooked grits from humble to heavenly.
- Coconut-and-Pecan Strawberry Shortcakes (page 86) Coconut and pecans added to shortcake dough provide a flavorful new spin on this classic springtime dessert.
- Shrimp and Peas with Farfalle (page 89) Lemony breadcrumbs add a delightful crunchy topping to this fresh spring pasta dish.
- Bucatini, Ham, and Asparagus (page 89) Hollow, spaghetti-like bucatini provides the perfect, slurpable base for this light pasta dish.

- Coconut Cream Cake (page 93) Gorgeous airy topping and rich and fragrant coconut make this four-layer cake one of the best versions of this classic we've tasted.
- Fluffy Coconut Whipped Cream Frosting (page 93) A touch of coconut and vanilla extracts flavor this cloud-like topping for our Coconut Cream Cake.

MAY

- Chilled Sweet Pea Soup with Mint and Cream (page 100) We love this bright, minty soup because it adds a pop of spring color to the buffet and may be made up to two days ahead of time.
- Italian-Style Salsa Verde (page 101) Stirred into an egg scramble or drizzled atop a burger, this all-purpose condiment is vibrant, packed with herbs, and gives dishes garden-fresh flavor.
- Snapper Baked in Parchment with Spring Vegetables (page 104) This French method for steaming delicate fish allows the juices to cook with the vegetables, concentrating flavors for a standout dish.
- Pan-Seared Grouper with Balsamic Brown Butter Sauce (page 105) Pan-searing gives this restaurant-quality dish a flavorful brown crust.
- Grilled Triggerfish (page 105) Don't be gun-shy: Our tips for flipping fish on the grill will have you using this method again and again.
- Sweet Corn Relish (page 107) The perfect little condiment to jazz up tacos, spoon over grilled pork or chicken, or scoop up with a chip.
- Napa Cabbage-and-Sweet Pepper Slaw (page 108) On its own or mounded atop a pulled pork sandwich, this crisp, refreshing slaw is a versatile new favorite.
- Easy BBQ Ribs (page 109) Tangy, tender, juicy, and slow-cooked to perfection, this recipe for Easy BBQ Ribs is our newest and best.

JUNE

- Lemon-Buttermilk Icebox Pie (page 120) This bright and refreshing lemon pie is elevated by an optional lemon-berry topping.
- Peach Divinity Icebox Pie (page 121) A gingersnap crust marries beautifully with the season's first crop of juicy, ripe peaches.
- Shrimp Boil with Green Olives and Lemon (page 124) The simmering liquid is loaded with flavor, which infuses the shrimp as it cooks.
- Roasted Potato-and-Okra Salad (page 124) This fresh yet hearty side dish features roasted potatoes instead of the traditional boiled ones, which add depth of flavor.
- Wonder Wings (page 133) A game-day favorite packed with Nashville Hot heat keeps you coming back for more.
- Buttery Nashville Hot Sauce (page 133) The origin of this iconic sauce for chicken is Prince's Hot Chicken on Nashville's north side. Now you can make it wherever you live.
- Crispy Salt-and-Pepper Grilled Chicken Thighs (page 135) Sometimes the most basic recipes are best, and this three-ingredient one is a prime example.
- Pickled Strawberries (page 141) Add interest to the pickle plate with these colorful strawberries pickled with rosemary. The all-purpose brine is ideal for pickling peaches and melons too.

JULY

- Buttermilk-Plum Ice Cream (page 155) This recipe for homemade ice cream is well worth the effort. Plums give this creamy, fruity ice cream a unique flavor twist.
- Green Tomato Soup with Lump Crabmeat (page 157) This vibrant, peppery soup and easy crab salad may be made up to a day ahead and stashed in the fridge until mealtime.
- Pasta with Heirloom Tomatoes, Goat Cheese, and Basil (page 159) Fresh garden-ripe tomatoes and Italian basil are a classic summer pairing that's perfect on pasta.
- Bourbon-Butter-Salted Pecan Ice Cream (page 201) Try a small twist on classic Butter-Pecan Ice Cream by spiking the mix with a bit of bourbon.
- Coffee Liqueur Cookies-and-Cream Ice Cream (page 201) If you like Cookies-and-Cream Ice Cream, this is a winner. Coffee-flavored liqueur is added to crushed, cream-filled chocolate sandwich cookies for a delightful twist on the classic flavor.

AUGUST

- Summer Corn-and-Golden Potato Chowder (page 213) Simmering scraped corn cobs directly in the soup concentrates the corn flavor, making the most of the season's sweet bounty.
- Buttermilk Lady Pea Soup with Bacon (page 217) Sweet summer peas and smoky bacon collide in a creamy soup. Even served warm, buttermilk has a cooling effect on the palate, just right for summer meals.

SEPTEMBER

- Pumpkin-Honey-Beer Bread Pudding with Apple Brandy-Caramel Sauce (page 224) This flavorful dessert combines two autumn favorites: pumpkin and apples.
- Pumpkin-Honey-Beer Bread (page 224) Canned pumpkin keeps this delicious beer bread extra moist. It is fantastic on its own or in our Pumpkin-Honey-Beer Bread Pudding with Apple Brandy-Caramel Sauce.
- Caramel Apple Cake (page 231) This cake celebrates the autumn apple harvest with a delicious yet versatile caramel sauce drizzle, which is also great on pancakes, French toast, baked apples, pound cake, or apple pie, and takes ice cream to another level.

OCTOBER

- Spicy Pumpkin Soup with Avocado Cream (page 238) Sweet, spicy, and smoky, this soup gets a hit of richness from a creamy buttermilk and avocado drizzle.
- Beer-Battered Pumpkin with Dipping Sauce (page 238) Pumpkin and fresh sage leaves become otherworldly when batter-fried and dunked in a creamy sauce.
- Pumpkin-and-Turnip Green Lasagna (page 238) We like no-boil lasagna noodles because they soak up the deliciousness from the sauce.
- Creole Seafood Jambalaya (page 240) A family favorite of New Orleans chef John Besh

NOVEMBER

- Parsnip-Potato Soup (page 252) The distinctive sweetness of parsnips adds flavor and interest to a classic.
- Easy Butter Rolls (page 252) Buttery and cloudlike, these simple dinner-time rolls are destined to be a favorite.
- Holiday Spice Cake (page 253) Make one to enjoy and another to give this Thanksgiving.
- Roasted Vegetable Salad with Apple Cider Vinaigrette (page 255) Rustic and simple to prepare, roasting vegetables concentrates their innate sweetness.
- Pumpkin-Espresso Tiramisú (page 256) Coffee and pumpkin may sound like an odd pairing, but this elegant dessert will make you a fan.
- Chocolate-Pecan Mousse Tart (page 262) A picture-perfect dessert for the holiday table.
- Pecan Pound Cake (page 263) Go nutty for this rich, dense pound cake accented with Southern pecans.
- Shrimp-and-New Potato Chowder (page 264) Clams and oysters usually get top chowder billing, but we think the shrimp makes this soup.
- The Original "Pam-Cakes" (page 266) Pancake perfection brought to you by our Test Kitchen pro Pam Lolley.

DECEMBER

- Creamy Shrimp Dip with Crispy Wonton Chips (page 272) Meet your new favorite chip-and-dip.
- Grapefruit-Beet-Goat Cheese Flatbread (page 273) Sweet, cool-season favorites join flavorful forces in an inspired topping for flatbread.
- Roasted Fennel-and-Prosciutto Flatbread (page 273) Fennel adds sweet texture and a distinctive anise flavor to this company-worthy appetizer or light main course.
- Grilled Pork Loin Steaks with Cherry-Plum Sauce (page 278) Pork is an economical and juicy change of pace from usual bacon-wrapped filets of beef.
- Spice-Rubbed Tenderloin with Mustard-Cream Sauce (page 279) This rapid-cooking cut makes entertaining or weeknights easy on the cook.
- Gnocchi Mac and Cheese (page 282) Trade Italian pasta for Italian dumpling and you've got an impressive new spin on a Southern cheesy classic.
- Christmas Salad (page 283) The perfect start to the holiday meal.
- Brussels Sprouts with Parmesan Cream Sauce (page 283) A creamy cheese sauce makes the humble Brussels sprout a side-dish star.
- Spice Cake with Cranberry Filling (page 286) Our cover-worthy cake is company-worthy too. Make it the pièce de résistance of your holiday sideboard.
- Praline Cream-Beignet Tower (page 286) Over the top but worth the effort, this is a dessert to impress.
- Coconut-Macadamia Nut Pound Cake (page 289) This inspired spin on pound cake is all about tropical flavors.
- Layered Eggnog Cream with Puff Pastry (page 290) A dessert that mixes two holiday favorites in one decadent dessert.
- Smoky Sausage-and-Grits Casserole (page 295) This rib-sticking brunch or dinner side-dish comfort is a cinch to make.
- Christmas Morning Cinnamon Rolls (page 296) The aroma of these gooey breakfast flavors is sure to get lazy-bones to rise and dine.
- Cream Cheese Pastries (page 298) Tender and sweet, these treats are great for giving.

January

24 **The South is Your Oyster** This is the season to collect these Southern favorites

28 **Slow-Cooked Perfection** These hearty one-pot recipes are the perfect escape on a chilly evening

32 THE *SL* TEST KITCHEN ACADEMY **Slow-Cooker Success** Discover some of our best tips for cooking with this handy appliance

33 QUICK-FIX SUPPERS **5 Hearty Main-Dish Salads** Try these great ways to power up your basic bowl of greens and veggies

35 THE NEW SUNDAY SUPPER **Mighty Good Meatloaf** Pair leftovers with veggies for two tasty new dinners

37 *COOKING LIGHT* **A Delicious New Diet** *Cooking Light* magazine lightens a hearty soup

38 COMMUNITY COOKBOOK Virginia Willis' newest book *Lighten Up, Y'all!* delivers delicious Southern classics with a lightened Southern twist

The South is Your Oyster

THE BEST OYSTERS IN THE COUNTRY GROW PLUMP AND SWEET ON SOUTHERN TIDES; WHETHER EATEN BY THE BUSHEL, PECK, OR DOZEN, THIS IS THE SEASON TO CELEBRATE THEM

FROM PEACHES TO PEANUTS, iconic ingredients are the seasonal currency of the edible South, and right now, oysters are the culinary coin of the realm. At home, we celebrate the high season differently from one region to the next—with backyard roasts in the Lowcountry; with elegant stews and fat, flame-licked grilled oysters along the Gulf; and with casseroles, well, everywhere. On the road, we elbow up to some of the finest raw bars in the world for the ritual of a glistening dozen, or three, on the half shell.

Don't take any of this culture for granted. Only the hardiest watermen still gather by hand, tong, or dredge. And the ravages of storms, spills, and overharvesting continue to make headlines this season, particularly along the Gulf. Meanwhile, dozens of new oyster farms, aka "gardens," have taken root in Southern waters, producing year-round for the half shell market. Farming emphasizes quality over quantity and the hyper-local notion—down to the GPS coordinates of a saltwater plot—that temperature, salinity, and minerals give an oyster its distinct flavor and character. Aficionados of the generic Gulf oyster may scoff at a $2 branded bivalve, but when it comes to the health of Southern waters, variety matters. However you take yours—wild or farmed, raw or cooked, Gulf or East Coast—the hour of the pearl is now.

CLASSIC OYSTER STEW

There are countless versions of this simple, elegant stew. To achieve the perfect texture of just-cooked oysters, poach them in the milk until their edges begin to curl, set aside, and return them to the stew just before serving.

- **1 pt. shucked fresh oysters, undrained**
- **2 cups milk**
- **Kosher salt**
- **Freshly ground black pepper**
- **1/4 cup butter**
- **1 shallot, minced**
- **1 small garlic clove, minced**
- **2 Tbsp. all-purpose flour**
- **1 cup half-and-half**
- **2 Tbsp. sherry**
- **1/2 tsp. Worcestershire sauce**
- **1/8 tsp. celery salt**
- **Fresh lemon juice**
- **Dash of hot sauce (such as Tabasco)**
- **Oyster crackers, saltine crackers, or buttered toast**

1. Drain oysters, reserving oyster liquor (about 1 cup). Heat milk and oyster liquor in a small saucepan over medium heat, whisking occasionally to prevent scorching, 3 to 4 minutes or until mixture just begins to steam. Add oysters, and season with desired amount of salt and pepper. Cook 4 to 5 minutes or until the edges of the oysters just begin to curl. Remove pan from heat. Using a slotted spoon, transfer oysters to a plate to prevent them from overcooking.

2. Melt butter in a large saucepan over medium heat. Add shallot and garlic, and cook, stirring often, 4 minutes or until tender. Sprinkle flour over shallot mixture, and cook, whisking constantly, 1 to 2 minutes or until completely incorporated and bubbly. Gradually whisk in half-and-half and next 3 ingredients. Bring to a boil, whisking constantly. Gradually stir in reserved milk mixture and oysters. Reduce heat to medium-low, and cook, stirring occasionally, just until warmed through. Season to taste with salt, pepper, lemon juice, and hot sauce. Serve with crackers.

MAKES 4 to 6 servings. **HANDS-ON** 35 min., **TOTAL** 35 min.

OYSTER-BACON POT PIE

On the fence about oysters? Consider this Chesapeake Bay-inspired number your gateway dish. This gussied-up riff features a golden puff pastry crown over a creamy, briny filling. You can also make this recipe in a lightly greased 11- x 7-inch baking dish. Seal puff pastry sheet over filling, brush with egg wash, and bake as directed.

- 1 **qt. shucked fresh oysters, undrained**
- 4 **thick bacon slices, diced**
- 3 **Tbsp. butter**
- 8 **oz. fresh button mushrooms, thinly sliced**
- 6 **green onions, sliced**
- 1 **celery rib, chopped**
- 1 **jalapeño pepper, seeded and minced**
- 1 **garlic clove, minced**
- 2 **Tbsp. fresh lemon juice**
- 1/4 **cup dry white wine**
- 2/3 **cup all-purpose flour**
- 3/4 **cup heavy cream**
- 1/4 **tsp. table salt**
- 1/4 **tsp. ground red pepper**
- 1/4 **tsp. freshly grated nutmeg**
- 1 **tsp. Old Bay seasoning**
- 1/2 **(17.3-oz.) package frozen puff pastry sheets, thawed**
- 1 **large egg**

Oyster-Bacon Pot Pie

1. Place an oven rack in lower third of oven, and preheat oven to 400°. Drain oysters, reserving 1 1/2 cups oyster liquor. Cook bacon in a Dutch oven over medium heat, stirring occasionally, 8 minutes or until crisp. Drain bacon on paper towels; reserve 3 Tbsp. drippings in Dutch oven.
2. Add butter and next 4 ingredients to Dutch oven; sauté 5 minutes. Add garlic and lemon juice; cook 1 minute. Add wine, and cook 2 minutes. Sprinkle with flour; cook, stirring constantly, 1 minute. Stir in cream, next 4 ingredients, and reserved oyster liquor; bring to a boil. Boil, whisking constantly, 2 minutes.
3. Remove from heat; stir in oysters and bacon. Spoon mixture into 6 lightly greased 12-oz. ramekins. Cut pastry sheets into circles slightly larger than ramekins, and place 1 on top of filling in each ramekin. Whisk together egg and 1 Tbsp. water; brush mixture over pastry.
4. Bake at 400° on lower oven rack 30 to 35 minutes or until browned and bubbly. Let stand 15 minutes before serving.

MAKES 6 servings. **HANDS-ON** 50 min.; **TOTAL** 1 hour, 50 min.

AN EASY LOWCOUNTRY OYSTER ROAST

This simple method translates the regional ceremony of roasting South Carolina cluster oysters on a large metal slab over an open fire to the backyard grill. Just cover them with wet burlap or a wet, clean towel with no detergent smell. To serve, place multiple oyster knives and gloves on a table and encourage folks to shuck their own. Plan on at least 1 dozen oysters per person, and grill them in batches. Be sure to scrub and rinse oysters well before roasting; discard any with broken shells.

- 2 **dozen fresh oysters in the shell**
- **Burlap or towel soaked in water**
- **Cocktail sauce**
- **Saltine crackers**

Preheat grill to 400° to 450° (high) heat. Arrange oysters in a single layer on grill; cover with wet burlap or towel. Cook, covered with grill lid, 10 to 12 minutes or until oysters open. Using tongs, carefully transfer roasted oysters to a platter. Serve warm with cocktail sauce and saltine crackers.

MAKES 2 dozen. **HANDS-ON** 15 min., **TOTAL** 25 min.

LEARN TO READ AN OYSTER

Crassostrea virginica, aka the Atlantic or Virginia oyster, is a shape-shifter that takes on chameleon-like colors and tastes of the water in which it beds. It can grow wild to the size of a hand. Farmed, it can take the gentle curve of a comma or sharpen into a thin blade.

WILD

Wild oysters have a thicker, coarser, sometimes barnacled shell with a shallower cup. Because they grow on oyster reefs on the seafloor, wilds often taste earthier, richer, and more robust.

FARMED

Farmed oysters form a smoother, deeper cup and thinner, decorative shell as they tumble in protective cages. Since they feed higher in the water column, they often taste cleaner with a distinct mineral flavor.

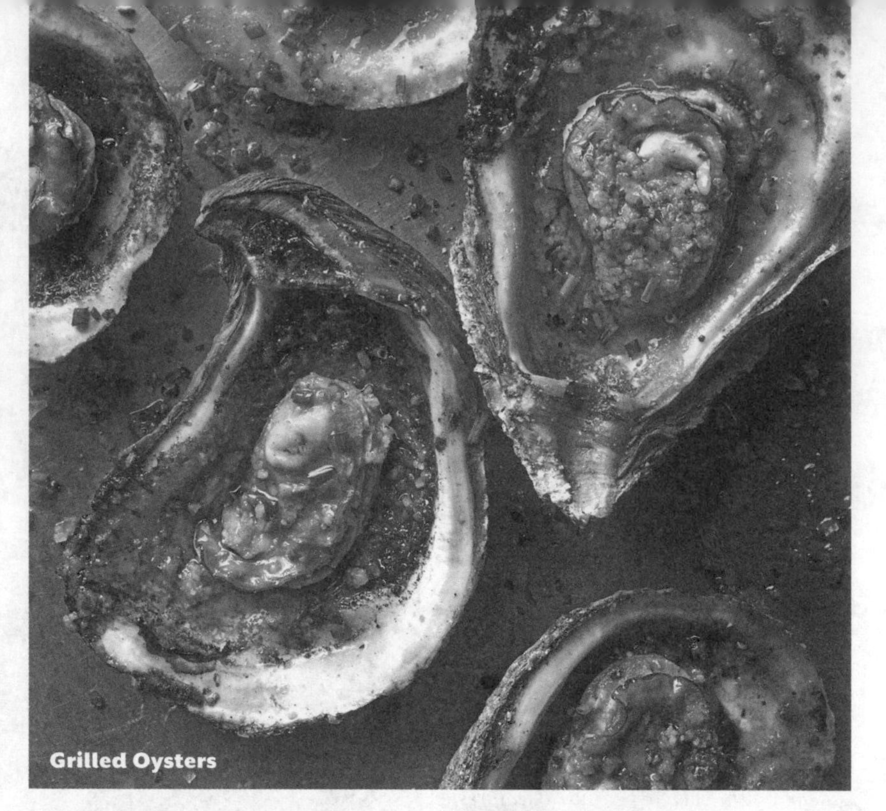

Grilled Oysters

GRILLED OYSTERS

The secret to this dish, a chargrilled homage to Gulf oyster houses, is a knockout garlic-herb butter.

- **2 cups butter, softened**
- **1/2 cup finely grated Parmesan cheese**
- **1/4 cup finely chopped parsley**
- **2 garlic cloves, minced**
- **1 Tbsp. Worcestershire sauce**
- **1 tsp. paprika**
- **1/2 tsp. ground red pepper**
- **1/2 tsp. hot sauce**
- **2 dozen large fresh oysters on the half shell**

1. Preheat grill to 450°. Pulse first 8 ingredients in a food processor until well combined.
2. Arrange oysters in a single layer on grill. Spoon 2 tsp. butter mixture into each oyster; grill, uncovered, 7 minutes or until edges curl.

MAKES 4 to 6 servings. **HANDS-ON** 25 min., **TOTAL** 25 min.

Broiled Oysters

Preheat broiler with oven rack 3 inches from heat. Prepare recipe as directed, placing oysters in a single layer in a jelly-roll pan. Broil 4 minutes or until edges curl and butter drips over the shell.

HANDS-ON 20 min., **TOTAL** 20 min.

GREEN APPLE MIGNONETTE

Pair this bright condiment, inspired by the new wave of fancy raw bars who make tart sauces with everything from cucumbers to rhubarb, with any fresh oyster on the half shell.

- **1/3 cup finely chopped Granny Smith apple (about 1/4 large apple)**
- **2 Tbsp. minced shallot**
- **1/2 cup rice vinegar**
- **2 tsp. minced fresh cilantro**
- **1 1/2 tsp. fresh lime juice**
- **1 tsp. sugar**
- **Kosher salt**
- **Freshly ground black pepper**

Stir together first 6 ingredients, and add salt and pepper to taste. Let stand 20 minutes before serving.

MAKES about 1/2 cup. **HANDS-ON** 10 min., **TOTAL** 30 min.

THE 10 BEST OYSTER HOUSES IN THE SOUTH

In truth, the best oyster bar is whichever one you're closest to, but these merit the journey

ACME OYSTER HOUSE

NEW ORLEANS, LA

For many, the French Quarter spot and its worn marble counters is the final word in raw bars. It's also host of the World Oyster Eating Championship. (Record is 47 dozen in 8 minutes.) *acmeoyster.com*

BOWENS ISLAND RESTAURANT

CHARLESTON, SC

At the end of a furrowed dirt road, join the Lowcountry culinary ritual of prying open clusters of oysters, plucked from the pluff mud and roasted under burlap. *843/795-2757*

DRAGO'S SEAFOOD RESTAURANT

METAIRIE, LA

Order the charbroiled: Shucked oysters bathed in garlicky butter hunker atop a gas grill as flames leap high and char the shell. *dragosrestaurant.com*

FELIX'S RESTAURANT & OYSTER BAR

NEW ORLEANS, LA

With its stand-up oyster bar, Felix's is the less raucous, equally delicious answer to Acme, just across the street. *felixs.com*

GILHOOLEY'S OYSTER BAR

DICKINSON, TX

Exemplar of Gulf Coast barbecued oysters: bivalves smoked in a wood-fired pit, topped with garlic butter and Parmesan. Note: no children allowed. *281/339-3813*

INDIAN PASS RAW BAR

PORT ST. JOE, FL

Located in an old commissary, the divey Indian Pass Raw Bar trades in live music, a constant crowd, and oysters straight from the nearby waters of Apalachicola Bay. *indianpassrawbar.com*

KIMBALL HOUSE

DECATUR, GA

A leading supporter of the farmed oyster movement, Kimball House offers, give or take, 22 varieties, each described with heady tasting notes. *kimball-house.com*

MERROIR

TOPPING, VA

At the Rappahannock Oyster Co. tasting room, slurp down a dozen just feet from the water where they were harvested. Or try Angels on Horseback, baked oysters with thick, crisp slices of Edwards ham. *rroysters.com*

THE ORDINARY

CHARLESTON, SC

Chef Mike Lata serves staggering towers of clams and oysters (including local Caper's Blades, in this stylish redo of a former bank. *eattheordinary.com*

SALTINE OYSTER BAR

JACKSON, MS

Newly opened in August, Saltine touts raw oysters from all of the coasts (Gulf, East, West), plus clever twists like Nashville-style hot fried oysters. *saltinerestaurant.com*

Slow-Cooked Perfection

SIMMER POT ROAST AND VEGGIES IN YOUR SLOW COOKER FOR A COMPANY-WORTHY MEAL, OR KEEP FAMILY DINNER COMFORTING AND CASUAL WITH A BIG BOWL OF CHILI

MEATBALL SLIDERS WITH TOMATO SAUCE

This fun dish is two recipes in one: delicious, easy meatballs and a slow-simmered tangy tomato sauce. If you have any extra sauce, toss it with your favorite green vegetable, spoon it over a bowl of creamy grits, or pair it with pork chops or other meats.

2 Tbsp. extra virgin olive oil
2 Tbsp. tomato paste
1 Tbsp. red wine vinegar
3 large fresh basil sprigs
2 garlic cloves, minced
2 (28-oz.) cans whole tomatoes, crushed
5 tsp. kosher salt, divided
1 tsp. freshly ground black pepper, divided
1 lb. ground chuck
1 lb. ground pork
1/2 cup panko (Japanese breadcrumbs)
1/4 cup freshly grated Parmigiano-Reggiano cheese
1/4 cup ricotta cheese
1/4 cup loosely packed chopped fresh flat-leaf parsley
1 large egg
1 large egg yolk
24 slider buns or dinner rolls, split and lightly toasted
12 (1-oz.) provolone cheese slices, halved
Garnish: fresh basil leaves

1. Stir together first 6 ingredients in a 7-qt. slow cooker; stir in 2 tsp. salt and 1/2 tsp. pepper. Cover and cook on LOW 8 hours.

2. Meanwhile, place ground chuck and ground pork in a large bowl. Add next 6 ingredients and remaining 1 Tbsp. salt and 1/2 tsp. pepper; mix gently just until blended. Gently shape mixture into 24 meatballs. (Do not pack.) Place meatballs in a single layer on a baking sheet; cover well, and chill until ready to use.

3. Let meatballs stand at room temperature 20 minutes. Carefully submerge meatballs in tomato mixture in slow cooker. Cover and cook on HIGH 1 hour or until meatballs are done, stirring after 30 minutes.

4. Preheat broiler with oven rack 7 inches from heat. Arrange bottom halves of 12 buns in a single layer on each of 2 baking sheets. Place 1 meatball on each bun; top with sauce and 1 halved provolone cheese slice. Repeat with remaining buns, meatballs, and cheese.

5. Broil sliders, 1 baking sheet at a time, 2 to 3 minutes or until cheese melts. Top with top halves of buns.

MAKES 2 dozen. **HANDS-ON** 35 min.; **TOTAL** 9 hours, 35 min.

UNCLE JACK'S MAC-AND-CHEESE

Classic cafeteria-style goodness, this recipe yields enough to serve as a side for a crowd, or it could be dinner for a family of four. Recipe by Pam Rawlinson, North Augusta, South Carolina.

- **1 (16-oz.) package elbow macaroni**
- **1 1/2 cups heavy cream**
- **1 (12-oz.) can evaporated milk**
- **4 large eggs, lightly beaten**
- **1/2 cup butter, melted**
- **1 1/2 tsp. table salt**
- **1/2 tsp. freshly ground black pepper**
- **4 cups (16 oz.) shredded extra-sharp Cheddar cheese, divided**
- **Vegetable cooking spray**

1. Cook macaroni according to package directions. Stir together cream, next 5 ingredients, cooked macaroni, and 2 1/2 cups cheese in a large bowl.

2. Pour macaroni mixture into a lightly greased (with cooking spray) 6-qt. slow cooker; sprinkle remaining 1 1/2 cups cheese over macaroni mixture.

3. Cover and cook on HIGH 3 hours; reduce slow cooker to LOW, and cook 1 hour.

MAKES 8 to 10 servings. **HANDS-ON** 25 min.; **TOTAL** 4 hours, 25 min.

BEER-BRAISED POT ROAST

This is no ordinary pot roast. First, it's rubbed with coffee. Then it simmers in dark stout beer and beef stock, yielding a deeply delicious gravy. Small carrots with tops and pearl onions elevate it further. Just be sure to pile the veggies on top of the beef after all other ingredients are in your cooker, so they'll cook perfectly and keep a vibrant color.

- **1 lb. small carrots with tops, peeled**
- **1 (3- to 4-lb.) boneless chuck roast, trimmed**
- **2 tsp. kosher salt**
- **1 tsp. freshly ground black pepper**
- **2 Tbsp. instant dark roast coffee**
- **2 Tbsp. olive oil**
- **3 Tbsp. tomato paste**
- **4 garlic cloves, chopped**
- **2 (12-oz.) bottles stout beer**
- **2 Tbsp. fresh thyme leaves**
- **2 (1-oz.) containers home-style concentrated beef stock (from a 4.66-oz. package)**
- **2 lb. pearl onions**
- **1 Tbsp. balsamic vinegar**
- **2 Tbsp. cornstarch**
- **Hot cooked grits**
- **Garnish: fresh thyme sprigs**

1. Cut tops from carrots, leaving 1 inch of greenery on each. Sprinkle roast with salt and pepper. Rub coffee over roast, and let stand at room temperature 10 minutes. Cook roast in hot oil in a large skillet over medium-high heat 3 to 5 minutes on each side or until browned, reserving drippings in skillet. Place roast in a 6-qt. slow cooker.

2. Add tomato paste and garlic to hot drippings, and sauté 1 minute. Slowly add beer, whisking constantly. Stir in thyme and concentrated beef stock; bring to a boil. Boil, stirring occasionally, 8 minutes or until mixture reduces to about 3 cups.

3. Pour beer mixture over roast. Top roast with onions and carrots. Cover and cook on LOW 8 to 10 hours or until roast is fork-tender. Transfer roast to a serving platter, and shred into large chunks, discarding any large pieces of fat. Spoon vegetables onto platter around roast.

4. Skim fat from cooking liquid; transfer cooking liquid to a large saucepan. Whisk in vinegar. Whisk together cornstarch and 2 Tbsp. water in a small bowl until smooth; add to mixture in pan, stirring until blended. Bring mixture to a boil, and boil, whisking often, 1 minute or until sauce reaches desired thickness. Serve sauce with roast, vegetables, and hot cooked grits.

Note: We tested with Guinness Extra Stout beer and Knorr Homestyle Concentrated Beef Stock.

MAKES 6 to 8 servings. **HANDS-ON** 35 min.; **TOTAL** 8 hours, 45 min.

SLOW-COOKER CHICKEN CACCIATORE WITH SPAGHETTI

Spend 20 minutes in the kitchen prepping this gorgeous, hearty sauce. Once the sauce simmers in the cooker, all that's left to do is cook a pound of pasta, and serve.

- 6 garlic cloves, minced
- 2 green bell peppers, chopped
- 2 red bell peppers, chopped
- 1 yellow onion, chopped
- 1 (8-oz.) package sliced cremini mushrooms
- 1 Tbsp. kosher salt
- 1 tsp. dried crushed red pepper
- 1 tsp. freshly ground black pepper
- 3 Tbsp. tomato paste
- 1/2 cup white wine
- 1 (28-oz.) can fire-roasted diced tomatoes, drained
- 2 skinned, bone-in chicken breasts (about 1 1/2 lb.)
- 2 skinned, bone-in chicken leg quarters (about 1 1/2 lb.)
- 1 (16-oz.) box spaghetti
- 1 1/2 cups pitted kalamata olives, halved
- 1/4 cup freshly grated Parmesan cheese
- 2 Tbsp. butter
- Garnishes: fresh basil, fresh parsley, shaved fresh Parmesan cheese

1. Place first 5 ingredients in a 6-qt. slow cooker; stir in salt, crushed red pepper, and black pepper. Whisk together tomato paste and wine, and add to slow cooker. Add drained tomatoes and chicken. Cover and cook on LOW 8 hours.
2. Uncover and carefully remove chicken from slow cooker, using tongs. Increase slow cooker temperature to HIGH. Cover and cook tomato mixture 30 more minutes or until sauce thickens to desired consistency.
3. Meanwhile, cook pasta according to package directions. Remove chicken meat from bones; discard bones. Shred meat. Stir olives and next 2 ingredients into sauce. Serve immediately over spaghetti.

MAKES 6 to 8 servings. **HANDS-ON** 20 min.; **TOTAL** 9 hours, 10 min.

RAISING THE BAR ON CHILI

Take your pick from our three outrageously good chili recipes, or try them all! Serve it buffet-style, and kick it up a notch by setting out assorted vibrant, crunchy, creamy, fiery, and tangy toppings.

BEEF-AND-BEAN CHILI

This classic red chili has chunks of stew meat and ground beef along with tomatoes, red beans, and spices. Fire-roasted tomatoes add a note of smokiness to the mix. If you can't find fire-roasted, use plain tomatoes and add a pinch of smoked paprika.

- 3/4 lb. ground chuck
- 1 Tbsp. olive oil
- 1 lb. chuck-eye steak, cut into bite-size pieces
- 1 1/2 cups chopped onion
- 1/4 cup crushed garlic cloves (about 10 large)
- 1 1/2 Tbsp. kosher salt
- 1 Tbsp. ground cumin
- 1 tsp. chili powder
- 1/2 tsp. ground red pepper
- 1 lb. dried red kidney beans
- 1 (28-oz.) can crushed fire-roasted tomatoes
- 1 (8-oz.) can tomato sauce
- 3 cups unsalted chicken cooking stock
- Toppings: shredded sharp Cheddar cheese, fresh jalapeño pepper slices, radish slices, chopped avocado

1. Brown beef in hot oil in a large skillet over medium-high heat, stirring often, 6 minutes or until meat crumbles and is no longer pink. Transfer browned beef to a 6-qt. slow cooker. Add chuck-eye cubes to skillet, and sauté over medium-high heat 6 to 8 minutes or until browned on all sides. Add chuck-eye cubes to slow cooker.
2. Add onion and next 8 ingredients to slow cooker; stir in stock. Cover and cook on HIGH 8 hours or until beans are tender. Serve with desired toppings.

MAKES 4 qt. **HANDS-ON** 30 min.; **TOTAL** 8 hours, 30 min.

PORK-AND-BLACK BEAN CHILI

Fresh tomatillos look like small green tomatoes wrapped in thin papery skin. Remove skin, and rinse before chopping. If you can't find fresh, look for canned on the Latin aisle at the grocery.

- **3 Tbsp. all-purpose flour**
- **2 tsp. ground black pepper**
- **1 1/2 tsp. ground cumin**
- **4 1/2 tsp. kosher salt, divided**
- **3 lb. boneless country-style pork ribs, cut into 1-inch pieces**
- **1/4 cup vegetable oil, divided**
- **9 medium tomatillos, diced**
- **3 cups chopped poblano peppers**
- **2 cups diced white onion**
- **3/4 cup reduced-sodium chicken broth**
- **1 1/2 Tbsp. dried oregano**
- **1 1/2 Tbsp. chili powder**
- **3 garlic cloves, minced**
- **2 (15-oz.) cans black beans, drained and rinsed**
- **2 Tbsp. plain cornmeal**
- **Chipotle Cream**
- **Garnishes: corn chips, lime wedges, fresh cilantro**

1. Stir together first 3 ingredients and 2 1/2 tsp. salt in a large bowl; toss with pork. Sauté half of pork in 2 Tbsp. hot oil in a large skillet over medium-high heat 10 minutes or until browned. Transfer to a 7-qt. slow cooker. Repeat procedure with remaining oil and pork.
2. Stir tomatillos, next 7 ingredients, and remaining 2 tsp. salt into pork mixture. Cover and cook on LOW 8 hours.
3. Uncover and stir in cornmeal. Cover and cook on LOW 1 hour. Serve with Chipotle Cream.

MAKES 6 to 8 servings. **HANDS-ON** 20 min.; **TOTAL** 9 hours, 20 min.

Chipotle Cream

Stir together 1 cup **sour cream,** 3 finely chopped canned **chipotle peppers** (in adobo sauce), and 1 Tbsp. **lime juice.** Add **salt** to taste.

MAKES about 1 cup. **HANDS-ON** 5 min., **TOTAL** 5 min.

SMOKY TURKEY-AND-SWEET POTATO CHILI

Chipotle chiles add a kick, but sweet potatoes balance the heat. Adjust the spice in this homey chili by increasing or decreasing the amount of chipotle to suit your taste. Buy the large dried lima beans if you can find them. They really look amazing and also give the chili a decidedly Southern accent.

- **1 1/4 lb. ground turkey**
- **2 Tbsp. olive oil**
- **1 Tbsp. tomato paste**
- **1 cup Mexican beer**
- **1 cup dried pinto beans**
- **1 1/2 Tbsp. chopped canned chipotle peppers in adobo sauce**
- **1 Tbsp. kosher salt**
- **2 tsp. ground cumin**
- **1 tsp. smoked paprika**
- **1/2 tsp. freshly ground black pepper**
- **2 medium-size green bell peppers, coarsely chopped**
- **1 (8-oz.) package dried lima beans**
- **1 large onion, coarsely chopped**
- **5 cups unsalted chicken cooking stock**
- **2 1/2 cups 1/2-inch peeled sweet potato cubes**
- **Garnishes: green onions, cilantro, sweet mini pepper slices**

1. Season turkey with kosher salt and freshly ground black pepper. Sauté turkey in hot oil in a large skillet over medium-high heat 4 minutes or until browned. Transfer mixture to a 6-qt. slow cooker.
2. Add tomato paste to skillet, and cook, stirring often, 30 seconds. Add beer, and bring to a boil, stirring to loosen browned bits from bottom of skillet. Boil 2 to 3 minutes or until reduced by half; stir into turkey mixture. Add beans and next 8 ingredients; stir in stock. Cover and cook on HIGH 7 hours. Stir in sweet potatoes; cover and cook on HIGH 1 hour or until potatoes are tender.

Note: We tested with Modelo Especial beer and Swanson Unsalted Chicken Cooking Stock.

MAKES about 4 1/2 qt. **HANDS-ON** 30 min.; **TOTAL** 8 hours, 30 min.

Smoky Turkey-and-Sweet Potato Chili

THE *SL* TEST KITCHEN ACADEMY

ROBBY MELVIN, DIRECTOR OF THE SOUTH'S MOST TRUSTED KITCHEN, SHARES NEW RULES FOR

SLOW-COOKER SUCCESS

The slow cooker is pure magic, but it plays by a different set of rules. When I started working on recipes for our new slow-cooker column, I learned by trial and error. The result: our new *Southern Living* rule book that'll help you cook like a pro.

1 The slow cooker does most of the work, but take time to brown the meat. This extra little step makes a huge difference. If you're looking to add another layer of flavor, brown before you cook. If it's appearance and texture you're after, broil afterward.

2 Pick the right cuts. Short ribs, pork shoulder, chuck roast, and other inexpensive, tough cuts become succulent and tender with the low, moist heat. Leaner, small cuts like boneless chicken breast and pork tenderloin become tough and dry when cooked for long periods of time.

3 Trim visible fat from meats before they go into the cooker. This keeps grease to a minimum in the cooking liquid and yields silky gravies and sauces.

4 Since there's no evaporation in the slow cooker, start with less liquid than you might think—less than you would add to a stew or braise—to yield the desired balance of flavors.

5 When including wine or alcohol in a recipe, add only a little. Too much taints the dish with an alcohol burn and a sharp acidic note.

6 Powerful flavors, such as fresh rosemary, become dominant, so add them judiciously before cooking or simply to garnish.

7 Use fresh herbs and spices whenever possible. Dried and ground versions often don't blend well in a long cooking process.

8 Low and slow isn't the only option. Kick the cooker up to high to cook dried beans.

9 Add dairy products like cream, milk, and yogurt during the last 10 minutes of cooking (to avoid curdling).

10 Although cooking with the lid sealed tight is ideal for slow-cooking, sometimes you want to remove the lid for the last 30 minutes or so to allow for some evaporation.

5 Hearty Main-Dish Salads

POWER UP YOUR BOWL OF BASIC GREENS AND VEGGIES BY ADDING NUTS, GRAINS, MEATS, FISH, OR FOWL

SHRIMP AND PESTO-RICE SALAD

Serve this brand-new twist on the old rice salad warm, or make it ahead, chill, and serve cold.

- **1 lb. peeled, large raw shrimp, deveined**
- **1/2 tsp. ground black pepper**
- **1/2 cup olive oil, divided**
- **2 1/4 tsp. kosher salt, divided**
- **2 tsp. loosely packed lime zest, divided**
- **1/2 cup loosely packed fresh mint leaves**
- **1 jalapeño pepper, seeded and diced**
- **2 large garlic cloves, smashed**
- **1/4 cup fresh lime juice**
- **2 (5-oz.) packages fresh baby spinach**
- **4 cups warm cooked long-grain rice**
- **1/2 cup smoked almonds, chopped**

1. Preheat oven to 425°. Toss together shrimp, black pepper, 1/4 cup oil, 1 tsp. salt, and 1 tsp. lime zest. Spread shrimp in a single layer on a parchment paper-lined baking sheet; bake 5 minutes or until shrimp are pink and done.

2. Meanwhile, process mint, next 3 ingredients, 1 (5-oz.) package spinach, and remaining 1 1/4 tsp. salt and 1 tsp. lime zest in a food processor until finely chopped. With processor running, pour remaining 1/4 cup oil through chute in a slow, steady stream, processing until combined.

3. Toss remaining 1 (5-oz.) package spinach with warm rice. (Spinach will wilt slightly.) Stir pesto into rice mixture; sprinkle with almonds. Top with shrimp.

MAKES 6 servings. **HANDS-ON** 20 min., **TOTAL** 25 min.

CHICKEN PAILLARD WITH CITRUS SALAD AND COUSCOUS

Israeli couscous is also sometimes labeled pearl couscous.

VINAIGRETTE

- **1/4 cup fresh orange juice**
- **2 Tbsp. fresh lime juice**
- **2 Tbsp. fresh lemon juice**
- **2 Tbsp. rice wine vinegar**
- **2 Tbsp. honey**
- **1/2 tsp. kosher salt**
- **1/3 cup olive oil**
- **1 Tbsp. chopped fresh flat-leaf parsley**
- **1 tsp. fresh thyme leaves**

CHICKEN

- **4 (6- to 8-oz.) skinned and boned chicken breasts**

COUSCOUS

- **1/2 Tbsp. butter**
- **1 1/2 tsp. olive oil**
- **1 cup uncooked Israeli couscous**
- **1 1/2 cups chicken broth**
- **1 tsp. kosher salt, divided**
- **1 tsp. black pepper, divided**

SALAD

- **8 oz. haricots verts (French green beans), trimmed**
- **1 cup arugula**
- **1/2 cup thinly sliced celery**
- **1/2 cup celery leaves**
- **1 navel orange, peeled and sectioned**
- **1 grapefruit, peeled and sectioned**

1. Prepare Vinaigrette: Whisk together first 6 ingredients in a medium bowl. Add olive oil in a slow, steady stream, whisking constantly until smooth and well blended. Stir in parsley and thyme leaves. Reserve 1/3 cup plus 2 Tbsp. vinaigrette for use in couscous and salad.

2. Prepare Chicken: Place chicken between 2 sheets of plastic wrap, and flatten to about 1/2-inch thickness, using a rolling pin or flat side of a meat mallet. Place chicken in a zip-top plastic bag, and pour remaining vinaigrette over chicken, turning to coat. Seal and marinate in refrigerator for 30 minutes. Turn bag over after 15 minutes.

3. Meanwhile, prepare Couscous: Melt butter with 1 1/2 tsp. oil in a medium saucepan over medium-high heat. Add couscous; cook, stirring constantly, 3 minutes or until toasted. Stir in broth. Bring mixture to a boil; cover, reduce heat to low, and cook, stirring occasionally, 15 minutes or until liquid is absorbed. Uncover and fluff. Stir in 1/4 tsp. each salt and pepper and reserved 2 Tbsp. vinaigrette.

4. Preheat oven to 400°. Remove chicken from bag; discard marinade. Place chicken on a wire rack in an aluminum foil-lined jelly-roll pan. Sprinkle with 3/4 tsp. each salt and pepper.

5. Bake at 400° for 12 to 15 minutes or until a meat thermometer inserted in thickest portion of chicken registers 165°. Let chicken stand at room temperature 5 minutes.

6. Prepare Salad: Cook beans in boiling salted water 3 minutes or until crisp-tender; drain. Halve beans lengthwise; toss with arugula, next 4 ingredients, and remaining 1/3 cup vinaigrette.

7. Divide couscous among 4 plates; top each with a chicken breast, and serve with salad.

Note: We tested with Peloponnese All Natural Grande Pearl Shaped Couscous.

MAKES 4 servings. **HANDS-ON** 30 min.; **TOTAL** 1 hour, 25 min.

PEAS AND KALE SALAD WITH BACON VINAIGRETTE

Ensure that 2015 is a lucky and prosperous year when you serve this hearty Southern salad for your New Year's Day festivities.

- **10 slices thick applewood-smoked bacon**
- **1 (1-lb.) package frozen black-eyed peas**
- **3 cups reduced-sodium chicken broth**
- **1/2 tsp. freshly ground black pepper**
- **2 garlic cloves, smashed**
- **1 medium-size fresh thyme sprig**
- **1 1/2 tsp. kosher salt, divided**
- **1 cup thinly sliced yellow onion**
- **1/3 cup apple cider vinegar**
- **2 tsp. light brown sugar**
- **2 tsp. Dijon mustard**
- **1/4 cup olive oil**
- **1 cup loosely packed fresh flat-leaf parsley leaves**
- **1 (5-oz.) package baby kale**
- **1 (8-oz.) package sweet mini peppers, thinly sliced**
- **Chopped fresh flat-leaf parsley**

1. Preheat oven to 425°. Arrange bacon in a single layer on a wire rack in a jelly-roll pan; bake 20 minutes. Turn bacon; bake 5 more minutes or until crisp. Cool 5 minutes; chop. Reserve 3 Tbsp. plus 2 tsp. drippings.
2. Stir together peas, next 4 ingredients, 1/2 tsp. salt, and 2 tsp. bacon drippings in a medium saucepan. Bring to a boil over medium-high heat. Reduce heat to medium, and simmer, stirring occasionally, 20 minutes or until peas are tender. Drain peas, and discard garlic and thyme.
3. Sauté onion in remaining 3 Tbsp. hot drippings in a large skillet over medium heat 1 minute. Stir in vinegar, next 2 ingredients, and remaining 1 tsp. salt, and cook, stirring constantly, 30 seconds or until smooth and slightly thickened. Gradually add oil, stirring constantly. Stir in pea mixture, and cook 1 minute or until thoroughly heated.
4. Toss together parsley and kale in a large bowl. Top with peppers and warm pea mixture. Sprinkle with chopped bacon and parsley, and serve immediately.

MAKES 4 to 6 servings. **HANDS-ON** 25 min.; **TOTAL** 1 hour, 15 min.

CHARRED STEAK SALAD WITH SPICY DRESSING

- **1 cup mayonnaise**
- **3 Tbsp. Asian chili-garlic sauce**
- **2 Tbsp. rice vinegar**
- **2 Tbsp. fresh lime juice**
- **2 tsp. kosher salt, divided**
- **1 tsp. freshly ground black pepper, divided**
- **1 (2-lb.) flank steak**
- **1 Tbsp. olive oil**
- **2 fresh rosemary sprigs**
- **1 lb. broccoli florets**
- **1 lb. cauliflower florets**
- **1/4 cup extra virgin olive oil**
- **3 cups loosely packed gourmet lettuce (such as arugula, baby kale, or baby spinach)**
- **French fried onions**

1. Preheat grill to 350° to 400° (medium-high) heat. Whisk together first 4 ingredients in a medium bowl.
2. Sprinkle 1 tsp. salt and 1/2 tsp. pepper over steak. Grill steak, covered with grill lid, 8 to 10 minutes on each side or to desired degree of doneness. Brush steak with olive oil, using rosemary sprigs as a basting brush. Let stand 10 minutes.
3. Meanwhile, preheat broiler with oven rack 7 inches from heat. Toss together broccoli, next 2 ingredients, and remaining 1 tsp. salt and 1/2 tsp. pepper in a large bowl. Spread broccoli mixture in a single layer in a jelly-roll pan, and broil 3 to 5 minutes or until charred and tender. Transfer to large bowl.
4. Toss together broccoli mixture and lettuce. Cut steak diagonally across the grain into thin strips. Arrange steak on a serving platter; top with broccoli mixture. Sprinkle with fried onions, and serve with dressing.

MAKES 4 to 6 servings. **HANDS-ON** 35 min., **TOTAL** 35 min.

ROMAINE SALAD WITH COUNTRY HAM AND EGGS

- **4 oz. country ham, diced**
- **2 1/2 cups 1-inch cubed bread**
- **1/4 tsp. ground black pepper**
- **2 Tbsp. olive oil**
- **1/3 cup white wine vinegar**
- **2 Tbsp. stone-ground mustard**
- **1 Tbsp. anchovy paste**
- **1 Tbsp. honey**
- **1/4 tsp. kosher salt**
- **2/3 cup olive oil**
- **12 oz. romaine lettuce hearts**
- **1/2 small red onion, sliced**
- **4 oz. sliced mushrooms**
- **1/2 tsp. white vinegar**
- **4 large eggs**
- **Shaved Parmesan cheese**

1. Sauté ham, bread, and pepper in hot oil in a large skillet over medium heat 1 minute or until oil is absorbed. Reduce heat to low, and cook, stirring occasionally, 7 minutes or until bread toasts. Remove from heat.
2. Process white wine vinegar and next 4 ingredients in a blender until combined. With blender running, slowly add oil, processing until smooth. Slice lettuce lengthwise. Arrange lettuce, onion, and mushrooms on 4 serving plates.
3. Pour water to depth of 3 inches into a large saucepan. Bring to a boil; reduce heat to a simmer. Add white vinegar. Break eggs, and slip into water, 1 at a time. Simmer 3 minutes or to desired degree of doneness. Remove eggs with a slotted spoon. Top each salad with egg, crouton mixture, and desired amount of dressing and Parmesan cheese.

MAKES 4 servings. **HANDS-ON** 30 min., **TOTAL** 30 min.

Mighty Good Meatloaf

PAIR LEFTOVERS WITH VIBRANT VEGGIES FOR FRESH AND HEARTY NEW DINNERS THAT ARE A SNAP TO PREPARE

SUNDAY STRATEGIST

Bake two meatloaves on Sunday; serve one, and use the other for two weeknight meals.

MONDAY
Turkey-Stuffed Peppers

TUESDAY
Rotini with Crumbled Turkey and Tomato Sauce

Italian Turkey Meatloaves

SUNDAY

ITALIAN TURKEY MEATLOAVES

Round out the meal with Brussels sprouts and your favorite rice pilaf.

- 1 lb. sliced button mushrooms
- 2 cups finely chopped onion
- 1 cup grated Parmesan cheese
- 3/4 cup Italian breadcrumbs
- 1 Tbsp. minced fresh garlic
- 2 large eggs, lightly beaten
- 1 (7-oz.) container refrigerated basil pesto
- 2 Tbsp. tomato paste
- 1 1/2 tsp. table salt
- 1 tsp. freshly ground black pepper
- 2 (8-oz.) cans tomato sauce
- 2 (20-oz.) packages lean ground turkey
- Vegetable cooking spray

1. Preheat oven to 400°. Process mushrooms in a food processor until finely ground. Stir together onion, next 5 ingredients, and mushrooms in a large bowl. Combine tomato paste, salt, pepper, and 1 (8-oz.) can tomato sauce; stir into mushroom mixture.

2. Add turkey to mushroom mixture, and combine, using hands, until well blended. Line 2 (9- x 5-inch) loaf pans with heavy-duty aluminum foil; coat lightly with cooking spray. Divide turkey mixture between prepared pans. Place pans on a foil-lined baking sheet.

3. Bake at 400° for 30 minutes. Spread remaining 8-oz. can tomato sauce over meatloaves. Bake 30 more minutes or until a meat thermometer inserted in center of meatloaf registers 165°.

4. Let meatloaves stand 10 minutes. Slice and serve 1 loaf. Cool second loaf completely; cover and chill for later use.

MAKES 2 loaves. (Each loaf makes 6 servings.) **HANDS-ON** 25 min.; **TOTAL** 1 hour, 35 min.

SAUTÉED SPROUTS

Halve and slice 1 lb. trimmed **Brussels sprouts.** Sauté sprouts and 1 tsp. minced **garlic** in 1 Tbsp. hot **olive oil** in a large skillet over medium-high heat 2 minutes. Toss with 1/2 tsp. **table salt,** 1 tsp. **fresh lemon juice,** and 1/2 tsp. **black pepper.**

MAKES 4 to 6 servings. **HANDS-ON** 15 min., **TOTAL** 15 min.

MONDAY NIGHT

TURKEY-STUFFED PEPPERS

- 6 bell peppers, tops removed
- 1/2 cup uncooked plain couscous
- Reduced-sodium chicken broth
- 1/2 Italian Turkey Meatloaf
- 1 tsp. extra virgin olive oil
- 1/2 cup toasted pine nuts
- 1/2 cup chopped fresh flat-leaf parsley
- 1 1/2 Tbsp. fresh lemon juice
- 1/2 cup (2 oz.) shredded mozzarella cheese
- 1/3 cup grated Parmesan cheese

1. Preheat oven to 400°. Discard pepper seeds and membranes. Stand peppers on end on a baking sheet. Bake 15 minutes or until tender.
2. Meanwhile, cook couscous according to package directions, using chicken broth instead of water. Crumble meatloaf; cook in hot oil in a nonstick skillet over medium-high heat, stirring gently, 3 minutes. Toss nuts, next 2 ingredients, and meatloaf mixture with couscous. Add salt and pepper to taste. Divide couscous mixture among peppers (about 1/2 cup each).
3. Bake at 400° for 15 minutes or until peppers are very tender. Increase oven temperature to broil, with oven rack 7 inches from heat. Combine cheeses, and sprinkle over peppers. Broil peppers 2 to 4 minutes or until cheese melts and begins to brown.

MAKES 6 servings. **HANDS-ON** 20 min.; **TOTAL** 45 min., not including meatloaf

TUESDAY NIGHT

ROTINI WITH CRUMBLED TURKEY AND TOMATO SAUCE

READY IN 45 MINUTES!

- 2 pt. grape tomatoes
- 1 Tbsp. olive oil
- 1 Tbsp. chopped fresh garlic
- 1/2 Italian Turkey Meatloaf
- 1/2 cup reduced-sodium chicken broth
- 12 oz. rotini or other short pasta
- 2 Tbsp. butter
- 3/4 cup shaved Parmesan cheese
- 1/2 cup fresh basil leaves, chopped

1. Sauté tomatoes in 1 Tbsp. hot oil in a large skillet over medium-high heat 10 minutes or until blistered and slightly wilted. Stir in garlic, and cook, stirring constantly, 30 seconds. Crumble meatloaf; add meatloaf and broth to skillet. Cook, stirring often, 3 minutes or until thoroughly heated. Remove from heat, and add salt and pepper to taste.
2. Cook pasta according to package directions, reserving 1 1/2 cups pasta water. Add pasta water to turkey mixture, and return skillet to heat. Cook mixture 2 to 3 minutes or until liquid is reduced to about 1 cup. Stir in butter. Add pasta and 1/2 cup cheese to turkey mixture; cook, stirring often, 1 to 2 minutes or until thoroughly heated. Sprinkle with basil and remaining 1/4 cup cheese. Serve immediately.

MAKES 6 servings. **HANDS-ON** 30 min.; **TOTAL** 45 min., not including meatloaf

A Delicious New Diet

COOKING LIGHT MAGAZINE SERVES UP YOUR FAVORITE FOODS—ONLY HEALTHIER

JUST IN TIME for New Year's resolutions, our friends at *Cooking Light* are rolling out the best-looking, best-tasting diet we've ever seen. The new Cooking Light Diet isn't about sacrificing flavor or abandoning your kitchen. This is a diet for cooks. The meals are good enough—and satisfying enough—to serve your whole family, all while meeting your weight-loss goals. Try this light, healthy potato soup from the plan and you'll see what we mean.

LOADED POTATO SOUP

- 4 (6-oz.) red potatoes
- 1/2 cup chopped onion
- 2 tsp. olive oil
- 1 1/4 cups nonfat, reduced-sodium chicken broth
- 3 Tbsp. all-purpose flour
- 2 cups 1% low-fat milk
- 1/4 cup reduced-fat sour cream
- 1/2 tsp. table salt
- 1/4 tsp. freshly ground black pepper
- 3 bacon slices, halved
- 1/3 cup (1 1/2 oz.) shredded Cheddar cheese
- 4 tsp. thinly sliced green onions

1. Pierce potatoes with a fork. Microwave at HIGH 13 minutes or until tender. Cut in half; cool slightly.
2. Sauté onion in hot oil in a medium saucepan over medium-high heat 3 minutes. Add broth. Combine flour and 1/2 cup milk; stir into broth. Stir in remaining 1 1/2 cups milk. Bring to a boil, stirring often. Boil 1 minute. Remove from heat; stir in sour cream, salt, and pepper.
3. Place bacon in a single layer on a paper towel-lined microwave-safe plate. Cover with a paper towel; microwave at HIGH 4 minutes. Crumble bacon.
4. Peel potatoes; discard potato skins. Coarsely mash potatoes into soup. Top with cheese, green onions, and bacon.

MAKES 4 servings (serving size: about 1 1/4 cups) **HANDS-ON** 20 min., **TOTAL** 20 min.

NUTRITIONAL INFORMATION
(per serving)

CALORIES: 325; FAT: 11.1g (SATURATED FAT: 5.2g); PROTEIN: 13.2g; FIBER: 3g; CARBOHYDRATES: 43.8g; SODIUM: 670mg

Community Cookbook

BUILD A BETTER LIBRARY, ONE GREAT BOOK AT A TIME

If you think Southern classics such as pimiento cheese and smothered chicken don't fit into a sensible diet, Virginia Willis' lightened-up versions will show you how they can. With *Lighten Up, Y'all,* Willis reminds us why she's one of the most creative cooks in the South.

HOPPIN' JOHN AND LIMPIN' SUSAN

Legend has it that Limpin' Susan was the wife of Hoppin' John.

- **1 bacon slice, cut into 1/2-inch pieces**
- **1 sweet onion, chopped**
- **2 cups fresh or frozen black-eyed peas, thawed**
- **2 cups reduced-sodium chicken broth**
- **1 dried chipotle pepper or ancho chile, halved and seeded**
- **1 cup uncooked long-grain white rice**
- **8 oz. fresh or frozen okra, thawed and sliced**
- **2 green onions, chopped**
- **Hot sauce**

1. Cook bacon in a medium saucepan over medium-high heat 3 minutes or until fat renders. Drain bacon on paper towels; reserve 1 tsp. bacon drippings in pan. Return bacon to pan. Add onion, and cook 4 minutes. Add peas, broth, dried pepper, and 2 cups water; season with kosher salt and freshly ground black pepper. Bring to a boil over high heat. Reduce heat to low; simmer, uncovered, 20 minutes or until peas are just tender (almost al dente) and about 2 cups of liquid remain. Add salt and pepper to taste.
2. Stir in rice and okra. Cover and simmer over low heat about 20 minutes or until rice is tender. Do not remove the lid during this part of cooking.
3. Remove pan from heat, and let stand, covered, 10 minutes. Uncover and discard chile. Fluff with a fork; add green onions and salt and pepper to taste. Serve immediately with hot sauce.

MAKES 8 servings. **HANDS-ON** 30 min.; **TOTAL** 1 hour, 45 min.

CLAIRE'S CREAM CHEESE SWIRL BROWNIES

If you don't have pastry flour, just process whole wheat flour for 1 minute in a food processor.

- **Vegetable cooking spray**
- **4 oz. reduced-fat cream cheese**
- **1 cup plus 2 Tbsp. sugar, divided**
- **2 1/2 tsp. pure vanilla extract, divided**
- **1 large egg yolk, at room temperature**
- **3/4 cup whole wheat pastry flour**
- **1/2 cup unsweetened cocoa**
- **3/4 tsp. baking powder**
- **1/2 tsp. fine sea salt**
- **1/4 cup canola oil**
- **6 oz. best-quality semisweet chocolate, finely chopped**
- **1/2 cup low-fat buttermilk**
- **1/2 cup unsweetened applesauce**
- **1 large egg white, at room temperature**
- **1 large egg, at room temperature**

1. Preheat oven to 325°. Spray an 8-inch square pan with cooking spray.
2. Beat cream cheese, 2 Tbsp. sugar, and 1/2 tsp. vanilla with an electric mixer until creamy and smooth. Add yolk to cream cheese mixture. Stir to combine.
3. Whisk together flour, cocoa, baking powder, and salt in a small bowl.
4. Heat oil and chocolate in a medium saucepan over medium heat, whisking until chocolate is melted. Whisk in remaining 1 cup sugar, and stir until melted. Add buttermilk, applesauce, and remaining 2 tsp. vanilla. Remove from heat. Add egg white and whole egg, whisking constantly until incorporated to prevent eggs from curdling. Add flour mixture, stirring until just combined. Transfer brownie batter to prepared pan.
5. Using a tablespoon, drop 9 dollops of cream cheese mixture on brownie batter. Draw the tip of a sharp knife or skewer through the 2 batters in a crisscross fashion to create a swirled effect.
6. Bake at 325° for 40 minutes or until top is just firm to the touch, rotating halfway through baking. Cool completely in pan on a wire rack (about 1 hour).
7. Coat a knife with cooking spray, and cut into 16 squares. Refrigerate in an airtight container up to 3 days.

MAKES 16 servings. **HANDS-ON** 40 min.; **TOTAL** 2 hours, 20 min.

February

40 **The Next Generation of Soul Food** Caroline Randall Williams shares some of her favorite recipes from her cookbook, *Soul Food Love*

43 **Buttermilk Revival** From sweet to savory, use the South's most beloved ingredient to create these dishes

47 SOUTHERN GREENS **Eat Your Greens** Try a new crop of recipes using flavorful Southern fronds

50 QUICK-FIX SUPPERS **Ten-Dollar Dinners** Cook smarter on a shoestring budget

52 THE *SL* TEST KITCHEN ACADEMY **Spectacular Fresh Butter & Buttermilk** These creamy homemade specialties put store-bought versions to shame

53 *COOKING LIGHT* **A Light Pork Supper** Soaking a lean pork loin in a flavorful brine keeps it moist and delicious

54 *SOUTHERN LIVING* RECIPE REVIVAL **Blue Cheese-Pecan Grapes** This February 1995 recipe gets a fresh twist

55 THE SLOW DOWN **Sauté, Simmer, Savor** Warm up your winter with a tasty stew perfect for potlucks and Sunday suppers

56 SAVE ROOM **Dark Magic** Decadent chocolate desserts let you put a sweet spell on your Valentine

58 COMMUNITY COOKBOOK Cheryl and Griff Day share some treats from their cookbook *Back in the Day Bakery: Made with Love*

The Next Generation of Soul Food

MOVING TO MISSISSIPPI GAVE ONE SOUTHERN DAUGHTER INSIGHT INTO HER FAMILY'S CULINARY LEGACY AND PROMPTED HER TO WRITE HER OWN FUTURE

I'M A SOUTHERN GIRL. Georgia and Alabama are in my blood. Nashville is my home. Mississippi is my right now. And us Southern girls, we're supposed to be able to cook. And by cook, I don't mean follow a tidy little recipe or warm something up; I mean feed your family and friends in mind and body with things that are delicious and not give two blinks about how healthy they are. Except that when you love your family, you want them to live—and live well. I don't have enough fingers and toes to count the number of people in my family who have gotten fat and sick from what modern memory remembers as "soul food."

I was raised to go to well-stocked specialty grocery stores, farmers' markets, even the farms themselves. My world turned upside down in 2010, when I moved to the Mississippi Delta to teach school for two years. In the Delta, almost without exception, the best groceries—the only groceries—come from Walmart. They have the widest selection, with the freshest vegetables, and the consistency isn't matched anywhere else in the region. With their produce aisle as my only grocery store, I learned how to eat healthfully and soulfully, day in and day out, on a teacher's budget.

"What you got to eat today, Miss Williams?" That's what my students, who ate mostly take-out, started to ask me. "You fix that chicken? Bring me some." It cost me three dollars (give or take). It smelled like home cooking. It looked like real food.

I remember a student asking me one day, "Miss Williams, why you always try to eat so healthy?" The question startled me. It was such a good question. I couldn't answer my student quickly. I was thinking.

Growing up, most of my family was large. My mother, who called me Baby Girl, thought I was perfect just the way I was and let me eat whatever I wanted. I watched her become heavier and heavier without any real concern—she was just following the model of the many women who came before her. It wasn't until my mom reached her largest, and I watched how hard it was for her to try to do something about it, that I really began to worry.

I didn't tell my student all of that. "I'm trying to take care of my body," I told her.

One day my dear friend Ruthie Collins, the parent-teacher liaison at the school where I taught, shared a conversation she had with her mother-in-law that changed how I looked at my family's foodways. Ruthie used to bake a pound cake once a week, but her mother-in-law criticized her for it. Her mother-in-law said that when she was young, they had a pound cake at holidays, yes, but not every week! That was too much! It made the cake less special. That's when I had my little revelation:

ABOUT THE AUTHOR

Caroline Randall Williams is an award-winning poet and young adult novelist currently pursuing her MFA at the University of Mississippi. Her first cookbook, *Soul Food Love*, which she co-authored with her mother, Alice Randall, is due out this month.

Recipes adapted from Soul Food Love. *Copyright © 2015 by Alice Randall and Caroline Randall Williams. Published by Clarkson Potter/Publishers, an imprint of Random House LLC, in 2015*

We've begun to mistake celebration food for everyday food.

When I think about what the future of food looks like, I find myself thinking that it looks like the past of Ruthie's mother-in-law more than it looks like the past of my grandmothers. Ruthie's mother-in-law knew that excess every day would spoil the real pleasures of a meaningful feast. And now I know it too.

The foods we now think of as "soul food" are not the ones our families were eating day in and day out; they are the celebration foods that have claimed our attention over time. All that extra sugar, the flour, the cream—those things were luxuries. The food at the soul of our community, the food that kept us on our feet and marching forward, was clean and delicious.

My future children are going to eat differently than I did as a kid. I ate out; my kids will eat in. I thought cooking was for special occasions; my kids will know cooking is for every day. I thought "soul food" was a guilty pleasure. My kids will know "soul food" is a healthy truth.

But I'm not a mama yet.

For now, standing on the shoulders of these brilliant, big, black women, I go on ahead and feed my friends from my small kitchen. I feed them from my history, from our history, our past, our present, and from the fresh start of what I hope our future looks like. And that, as we like to say in my family, is how you entertain like Mama and stay healthy like Baby Girl.

PEANUT CHICKEN STEW

A lot of times, when people say a soup is thick enough to be a meal, they're playing you. But this soup sustains. This recipe takes its cues from West Africa, particularly Senegal and The Gambia, known for peanut, or groundnut, stews.

- **3 cups chopped cooked chicken**
- **1½ cups creamy peanut butter**
- **1 (28-oz.) can diced tomatoes, drained**
- **1 Tbsp. curry powder**
- **1 tsp. ground red pepper**
- **4 to 6 cups Sweet Potato Broth (recipe, page 42)**
- **½ cup chopped roasted unsalted peanuts**
- **Garnish: chopped cilantro**

Stir together first 5 ingredients and 4 cups Sweet Potato Broth in a medium stockpot. Bring to a simmer over medium heat, stirring occasionally, and simmer, stirring occasionally, 20 minutes or until thickened. Stir in up to 2 cups broth, ½ cup at a time, until desired consistency is reached. Add salt to taste. Sprinkle with peanuts, and serve immediately.

MAKES 6 servings. **HANDS-ON** 15 min.; **TOTAL** 1 hour, 45 min., including broth

SALMON CROQUETTES WITH DILL SAUCE

Back in the day, salmon croquettes usually meant rich bindings and fillers (eggs, flour, cracker crumbs) to hold them together. And they were typically fried in an inch of bacon grease. In my house, the binder is egg only, and the patties are pan-seared in a little olive oil.

- **1½ cups fat-free plain Greek yogurt**
- **¼ cup Dijon mustard**
- **2 Tbsp. chopped fresh dill**
- **1 tsp. lemon zest**
- **1 Tbsp. fresh lemon juice**
- **Pinch of ground red pepper**
- **2 (14.75-oz.) cans salmon, packed in water**
- **2 cups finely chopped celery**
- **4 large eggs, beaten**
- **1 cup finely chopped onion**
- **1 to 2 tsp. table salt**
- **1 to 2 tsp. freshly ground black pepper**
- **2 to 3 Tbsp. olive oil**
- **Garnish: blistered green onions**

1. Whisk together yogurt and next 5 ingredients in a small bowl.

2. Drain salmon; remove and discard skin and bones. Flake salmon, and place in a medium bowl. Stir celery and next 4 ingredients into salmon; shape mixture into 6 (4-inch) patties (about ¾ cup per patty).

3. Cook patties, in batches, in 2 Tbsp. hot oil in a large skillet over medium-high heat 5 minutes on each side or until golden. (Add 1 Tbsp. oil, if needed, for second batch.) Serve with yogurt mixture.

MAKES 6 servings. **HANDS-ON** 50 min., **TOTAL** 50 min.

FIERY GREEN BEANS

Fire!! was an influential literary journal that put out only one issue. According to Langston Hughes, the point was to "burn up a lot of the old, dead conventional Negro-white ideas of the past." In the spirit of Fire!!, we challenge the notion of what a soul food green bean recipe can be. This recipe—one we got from a young black chef at a hip, more than a little bougie, burger joint in Nashville—is a luminous example of the new old school.

- **2 Tbsp. fresh lemon juice**
- **1 Tbsp. chopped fresh flat-leaf parsley**
- **1 tsp. chopped fresh cilantro**
- **1 tsp. dried crushed red pepper**
- **1 green onion (white and light green parts only), thinly sliced**
- **1 lb. fresh green beans or haricots verts (French green beans), trimmed**
- **2 Tbsp. olive oil**
- **Kosher salt**
- **Freshly ground black pepper**

1. Preheat oven to 375°. Stir together first 5 ingredients. Arrange green beans in a single layer on an aluminum foil-lined baking sheet. Drizzle with olive oil; toss. Bake 10 minutes. Increase oven temperature to broil, and broil 4 to 6 minutes or until browned in spots and crisp-tender.
2. Remove from oven. Drizzle beans with lemon juice mixture; toss. Season to taste with kosher salt and freshly ground black pepper. Serve hot, or cool completely, chill, and serve cold. These just get better as they sit.

MAKES 4 servings. **HANDS-ON** 10 min., **TOTAL** 25 min.

AFRICAN CHICKPEA SOUP

My Uncle Paul and Aunt Sonia Bontemps, who helped found an African-American genealogical society, took Grandma on a trip to revisit the scenes of her youth and explore our family history. One myth, which can't be proved, is that the Bontemps family descended from Madagascar. The possibility encouraged us in the exploration of Madagascan foodstuffs and foodways, culminating in this hearty and healthy recipe.

- **2 garlic cloves, chopped**
- **1 Tbsp. olive oil**
- **1 tsp. dried crushed red pepper**
- **1 tsp. ground coriander**
- **½ tsp. ground red pepper**
- **¼ tsp. ground cardamom**
- **Pinch of ground turmeric**
- **4 cups Sweet Potato Broth**
- **1 cup unsweetened coconut milk**
- **1 bunch fresh mustard greens, chopped**
- **2 (15-oz.) cans chickpeas, drained and rinsed**
- **Garnish: dried crushed red pepper**

1. Sauté garlic in hot oil in a large saucepan over medium heat 1 minute; add 1 tsp. dried crushed red pepper and next 4 ingredients. Cook 1 to 2 more minutes or until fragrant. (Toasting the spices opens up their flavor.) Stir in Sweet Potato Broth, coconut milk, and greens. Bring to a gentle boil; add chickpeas. Reduce heat to low, and simmer about 1½ hours or until greens are soft. Season to taste with salt and pepper.

MAKES 8 servings. **HANDS-ON** 20 min.; **TOTAL** 3 hours, including broth

Sweet Potato Broth

Heat 2 Tbsp. olive oil in a large stockpot over medium heat. Add 1 medium **onion,** sliced; 3 **celery ribs,** chopped; and 1 **carrot,** chopped. Cook, stirring often, 8 to 10 minutes or until vegetables are tender. Add 1 peeled and quartered large **sweet potato,** 5 whole **cloves,** desired amount of **kosher salt** and **freshly ground black pepper,** and 6 cups water. Increase heat to high, and bring to a boil. Reduce heat to medium-low, and simmer 30 to 35 minutes or until sweet potato is tender. Discard cloves. Let mixture stand 15 minutes. Process, in batches, in a food processor or blender until smooth. Season with salt and pepper. Use immediately, or cool completely, and refrigerate in an airtight container up to 5 days.

Note: You can freeze cooked and cooled mixture up to 2 months.

MAKES 6 cups. **HANDS-ON** 25 min.; **TOTAL** 1 hour, 10 min.

HONEY PEANUT BRITTLE

Many enslaved African-Americans came from beekeeping countries; others interacted with certain Native American groups who bartered with beeswax and honey. This candy deliciously celebrates those all-but-forgotten intertwinings in early American society. It also celebrates George Washington Carver, who advocated for everything peanut.

- **4 cups dry-roasted peanuts**
- **1 cup sugar**
- **1 cup honey**
- **½ tsp. fresh lemon juice**
- **1 Tbsp. baking soda**
- **1 Tbsp. unsalted butter**
- **Pinch of table salt**

1. Bring first 4 ingredients to a boil in a deep pot over high heat, stirring constantly until sugar dissolves. Boil mixture until a candy thermometer registers 300°. Remove from heat. (If you don't have a candy thermometer, drop a bit of the mixture into a glass of ice water to test the temperature. If it hardens, you're all set.)
2. Carefully stir in baking soda and remaining ingredients. (Mixture will fluff up.) Spread mixture onto a well-buttered baking sheet, and cool completely (about 1 hour). Break into pieces, and store cooled brittle in an airtight container in a cool, dry place up to 1 week.

MAKES about 8 oz. **HANDS-ON** 30 min.; **TOTAL** 1 hour, 30 min.

Buttermilk Revival

WHY THERE'S NEVER BEEN A BETTER TIME TO COOK WITH THE SOUTH'S MOST BELOVED INGREDIENT

WHETHER POURED FROM A MEASURING CUP OR SIPPED FROM A GLASS, old-fashioned buttermilk is irreplaceable in the South. A byproduct of churned butter—the leavings in the bottom of the churn—it is light, yet substantial, with golden flecks of butter floating through it like gold dust in a miner's pan. Buttermilk anchors the core of the Southern recipe canon: It tenderizes fried chicken, makes biscuits light as air, and delivers pillow-soft pound cakes.

Yet the origins are modest. No one knows which smart cook first embraced buttermilk, but it was likely happenstance. Starting way back, many families kept a cow or had access to fresh milk ("sweet milk"). But they lacked refrigeration. Leaving milk at room temperature overnight before churning starts the growth of the natural, harmless, active cultures similar to the good-for-us stuff found in yogurt, sour cream, and other fermented dairy foods. And those cultures, aside from making the butter taste better, work culinary wonders. Resourceful cooks learned to make the most of it.

But this tangy elixir isn't confined just to cooking. Connoisseurs sip and chug the stuff. A glass at bedtime has been known to soothe the tummies of hearty eaters and late-night revelers. And while many Southern traditions celebrate buttermilk, few garner more devotion and sentiment than the curious practice of combining it in a glass with crumbles of leftover cornbread. It is hard to find a native of the Mountain South who doesn't crave this concoction as a meal, snack, or dessert. It's equally hard to find folks from elsewhere who understand the appeal.

Real liquid buttermilk should always be our first choice in cooking. Some well-intentioned but misguided recipes encourage us to replace it with reconstituted powder or milk curdled with lemon juice. Other than being acidic, curdled milk bears no resemblance to buttermilk, so it can never deliver the goods. Plus, we live in a region where buttermilk is easy to find. And our growing interests in fermented foods, grass-fed cows, and local products are giving farmsteads a boost. Gleaming bottles of locally made buttermilk are returning to our neighborhood markets.

For that matter, it's simpler than you might think to churn your own in a food processor.

Buttermilk remains our peerless dairy queen. It's simple, yet special. It's familiar, yet there's nothing else like it. As inimitable artisan dairy farmer Earl Cruze reminds us, buttermilk might not solve all of the world's problems, but it'll sure help.

TRIPLE-CHOCOLATE BUTTERMILK POUND CAKE

CAKE

- 2 cups all-purpose flour
- 3/4 cup unsweetened cocoa
- 1/2 tsp. baking powder
- 1 tsp. table salt
- 1 1/2 cups butter, softened at room temperature
- 3 cups granulated sugar
- 5 large eggs, at room temperature
- 1 1/4 cups buttermilk
- 2 tsp. instant espresso
- 2 tsp. vanilla extract
- 1 cup 60% cacao bittersweet chocolate morsels
- Shortening

CHOCOLATE GLAZE

- 3/4 cup semisweet chocolate morsels
- 3 Tbsp. butter
- 1 Tbsp. light corn syrup
- 1/2 tsp. vanilla extract

BUTTERMILK GLAZE

- 1 cup powdered sugar
- 1 to 2 Tbsp. buttermilk
- 1/4 tsp. vanilla extract

1. Prepare Cake: Preheat oven to 325°. Whisk together flour and next 3 ingredients. Beat 1 1/2 cups butter in a medium bowl at medium-high speed with an electric mixer until smooth. Gradually add granulated sugar, beating until light and fluffy. Add eggs, 1 at a time, beating just until yolk disappears. Combine 1 1/4 cups buttermilk and next 2 ingredients. Add flour mixture to egg mixture alternately with buttermilk mixture, beginning and ending with flour mixture. Beat at low speed after each addition. Fold in bittersweet chocolate morsels. Pour batter into a well-greased (with shortening) and floured 12-cup Bundt pan. Sharply tap pan on counter to remove air bubbles.

2. Bake at 325° for 1 hour and 15 minutes to 1 hour and 25 minutes or until a wooden pick inserted in center comes out clean. Cool in pan on a wire rack 20 minutes. Remove from pan; cool completely on rack.

3. Prepare Chocolate Glaze: Combine semisweet chocolate morsels, 3 Tbsp. butter, and 1 Tbsp. corn syrup in a microwave-safe glass bowl. Microwave at MEDIUM (50% power) 1 to 1 1/2 minutes or until morsels begin to melt, stirring after 1 minute. Stir until smooth. Stir in 1/2 tsp. vanilla.

4. Prepare Buttermilk Glaze: Whisk together powdered sugar, 1 Tbsp. buttermilk, and 1/4 tsp. vanilla in a small bowl until smooth. Add up to 1 Tbsp. buttermilk, if desired. Drizzle warm glazes over cooled cake.

MAKES 10 to 12 servings. **HANDS-ON** 45 min.; **TOTAL** 4 hours, 25 min.

Triple-Chocolate Buttermilk Pound Cake

Mini Triple-Chocolate Buttermilk Pound Cakes

Prepare recipe as directed through Step 1, pouring batter into 2 lightly greased (with vegetable cooking spray) 12-cup Bundt brownie pans, filling each about three-fourths full. Bake at 325° for 26 to 30 minutes. Cool in pans on wire racks 10 minutes; remove from pans, and cool completely. Proceed with recipe as directed in Steps 3 and 4.

MAKES 2 dozen. **HANDS-ON** 45 min.; **TOTAL** 1 hour, 25 min.

MANGO-BUTTERMILK SHAKES

Process 4 cups frozen diced **mango,** 1 cup **buttermilk,** 1 Tbsp. **sugar,** and 1/2 tsp. **vanilla extract** in a blender until smooth. Add up to 2/3 cup more buttermilk, 1 Tbsp. at a time, processing to desired consistency. Stir in 1/2 tsp. **ground cardamom,** if desired. Serve immediately.

MAKES 3 cups. **HANDS-ON** 5 min., **TOTAL** 5 min.

Fluffy Buttermilk Waffles

FLUFFY BUTTERMILK WAFFLES

- 2 cups cake flour
- 3 tsp. baking powder
- 1 tsp. baking soda
- ½ tsp. table salt
- 4 large eggs, separated
- 2 cups buttermilk
- ½ tsp. cream of tartar
- 4 Tbsp. butter, melted
- 1 tsp. vanilla extract
- Syrup

1. Whisk together first 4 ingredients in a large bowl. Whisk together egg yolks and buttermilk in a small bowl. Beat egg whites and cream of tartar at high speed with an electric mixer until stiff peaks form.

2. Using a fork, stir yolk mixture into flour mixture just until dry ingredients are moistened. Stir in butter and vanilla. (Batter will be lumpy.) Fold in egg white mixture just until incorporated. (Do not overmix.)

3. Pour about ½ cup batter for each waffle into a preheated, oiled Belgian-style waffle iron; cook according to manufacturer's instructions until waffles are golden brown and crisp. Keep warm in a single layer on a baking sheet in a 200° oven up to 30 minutes. Serve with syrup.

Note: Batter will also make 24 pancakes.

MAKES about 12 waffles. **HANDS-ON** 25 min., **TOTAL** 45 min.

CHOCOLATE-BUTTERMILK PUDDING

PUDDING

- ¾ cup granulated sugar
- 3 Tbsp. unsweetened cocoa
- 2 Tbsp. cornstarch
- ¼ tsp. table salt
- 1 cup buttermilk
- 1¼ cups heavy cream
- 2 large egg yolks
- 1 large egg
- 1 (4-oz.) semisweet chocolate baking bar, finely chopped
- 1 Tbsp. butter
- 1 tsp. vanilla extract

BUTTERMILK CREAM

- ¾ cup heavy cream
- ¼ cup powdered sugar
- ¼ cup buttermilk

1. Prepare Pudding: Whisk together first 4 ingredients in a large saucepan. Slowly whisk in 1 cup buttermilk to make a smooth paste. Whisk together 1¼ cups cream, egg yolks, and egg in a 2-cup glass measuring cup. Slowly add cream mixture to buttermilk mixture, whisking constantly until well blended.

2. Cook pudding over medium heat, whisking constantly, about 10 minutes or until mixture begins to boil and thicken. Remove from heat.

3. Add chopped semisweet chocolate; whisk until smooth. Add butter and vanilla, and whisk until butter melts and pudding is smooth.

4. Transfer to a bowl, and place plastic wrap directly on warm pudding (to prevent a film from forming). Chill 4 to 24 hours.

5. Prepare Buttermilk Cream: Beat ¾ cup cream, ¼ cup powdered sugar, and ¼ cup buttermilk at high speed with an electric mixer until soft peaks form. (Do not overbeat.) Divide pudding among 6 (6- to 8-oz.) bowls; dollop with buttermilk cream.

MAKES 6 servings. **HANDS-ON** 40 min.; **TOTAL** 4 hours, 40 min.

HERBED BUTTERMILK RANCH DRESSING

- 1 cup buttermilk
- ¼ cup mayonnaise
- ¼ cup sour cream
- 1 small garlic clove, pressed
- 2 Tbsp. minced shallot
- 1 Tbsp. finely chopped fresh flat-leaf parsley
- 2 tsp. finely chopped fresh chives
- 1 tsp. kosher salt
- 1 tsp. whole grain Dijon mustard
- ½ tsp. freshly ground black pepper
- ¼ tsp. hot sauce

Combine all ingredients in a 1-qt. glass jar with a tight-fitting lid. Cover and shake vigorously to blend. Chill 30 minutes. Refrigerate in covered jar up to 1 week. Shake well before serving.

MAKES about 1 1/2 cups. **HANDS-ON** 15 min., **TOTAL** 45 min.

Lemony Ranch

Prepare Herbed Buttermilk Ranch Dressing as directed, adding ¼ cup fresh lemon juice, 3 Tbsp. extra virgin olive oil, and 2 tsp. honey to jar before shaking.

MAKES about 2 cups. **HANDS-ON** 20 min., **TOTAL** 50 min.

Blue Cheese Ranch

Prepare Herbed Buttermilk Ranch Dressing as directed, adding 2 oz. mashed blue cheese, 2 oz. crumbled blue cheese, 2 tsp. chopped chives, 1/2 tsp. ground black pepper, and 1/2 tsp. paprika to jar before shaking.

MAKES about 2 cups. **HANDS-ON** 20 min., **TOTAL** 50 min.

Avocado Ranch

Prepare Herbed Buttermilk Ranch Dressing as directed, adding 1 ripe mashed avocado, 2 Tbsp. minced jalapeño pepper, 2 Tbsp. hot sauce, 4 tsp. fresh lime juice, and 1/2 tsp. kosher salt to jar before shaking.

MAKES about 2 1/2 cups. **HANDS-ON** 20 min., **TOTAL** 50 min.

BUTTERMILK-VEGETABLE CURRY

- 2 Tbsp. butter
- 1 Tbsp. vegetable oil
- 1 cup thinly sliced shallots (3 to 4 medium)
- 1 serrano pepper, seeded and minced
- 2 Tbsp. minced fresh ginger
- 1 Tbsp. red curry powder
- 1 tsp. ground coriander
- 1 tsp. ground cumin
- 1 tsp. kosher salt
- ¼ tsp. freshly ground black pepper
- ¼ tsp. ground red pepper
- 2 Tbsp. tomato paste
- 3 cups ½-inch cubed new potatoes
- 2 cups reduced-sodium chicken broth
- 2 cups cauliflower florets
- 2 Tbsp. cornstarch
- 2 Tbsp. cold water
- 2 cups buttermilk
- 1 (5-oz.) package fresh baby spinach
- Hot cooked long-grain rice
- Garnish: serrano pepper slices

1. Melt butter with oil in a large saucepan over high heat; add shallots and next 2 ingredients; sauté 2 minutes. Add curry powder and next 5 ingredients; cook, stirring constantly, 1 minute. Stir in tomato paste; cook, stirring constantly, 30 seconds. Stir in potatoes. Reduce heat to medium. Cover and cook, stirring occasionally, 5 minutes. Add broth and cauliflower. Increase heat to high; bring to a boil. Cover partially, reduce heat to medium-low, and simmer 6 minutes or until potatoes are tender. Reduce heat to low.

2. Meanwhile, whisk together cornstarch and 2 Tbsp. cold water in a small bowl until smooth. Whisk into buttermilk. Slowly stir buttermilk mixture into vegetable mixture, and cook, stirring occasionally, 2 to 3 minutes or until mixture begins to simmer.

3. Stir in spinach in batches, and cook 2 to 3 minutes or until wilted. Remove from heat, and serve immediately over rice.

MAKES 6 to 8 servings. **HANDS-ON** 55 min., **TOTAL** 55 min.

Buttermilk-Vegetable Curry

Eat Your Greens

A NEW CROP OF RECIPES STARRING THE FLAVORFUL SOUTHERN FRONDS THAT HAVE GRACED OUR STOCKPOTS FOR GENERATIONS

MANY SOUTHERNERS argue that low-and-slow is the only way to cook cool-weather greens like turnip and collard. Although we agree a mound of velvety, savory greens that simmered all day is food that feeds the soul, there are more ways than one to enjoy our region's leafy veggies. Birmingham gardener Mary Beth Shaddix grew a mess of Southern varieties for our Test Kitchen that we used to develop modern recipes we think you'll love. And while discovering new twists and time-saving techniques, we perfected a recipe every Southern cook should know: The Ultimate Classic Collards.

POTLIKKER

This robust broth, found at the bottom of a pot of slow-cooked greens, has inspired song, is fodder for debates (ranging from the correct spelling to a spirited discussion on dunking vs. crumbling your cornbread into it), and has been used as a medicinal elixir. Southern food writer John T. Edge calls it "a backbone dish of the American South." I call it liquid gold and spoon it over pork, chicken, and even fish. I sop it up with cornbread (guess that puts me in the dunking camp) and save any precious extra in the fridge to fortify soup.
—Robby Melvin, Test Kitchen Director

FETTUCCINE WITH SMOKY TURNIP GREENS, LEMON, AND GOAT CHEESE

Don't discard the cooking water when you drain the pasta. Reserve 1 cup to ladle over the cooked greens, and you'll instantly create a creamy sauce.

- **1 (8-oz.) package fettuccine**
- **8 oz. country ham, cut into 1-inch strips**
- **3 Tbsp. olive oil**
- **3 shallots, finely chopped**
- **4 garlic cloves, minced**
- **1 lb. turnip greens, stems removed and chopped**
- **1 1/4 tsp. dried crushed red pepper, divided**
- **Kosher salt**
- **Freshly ground black pepper**
- **1 cup freshly grated Parmigiano-Reggiano cheese**
- **3 Tbsp. butter**
- **3 Tbsp. fresh lemon juice**
- **4 oz. crumbled goat cheese**
- **1 Tbsp. lemon zest**

1. Cook pasta according to package directions, reserving 1 cup pasta water.
2. Sauté ham in hot oil in a Dutch oven over medium heat 8 minutes or until crisp. Add shallots; sauté 2 minutes. Add garlic, and sauté 1 minute.
3. Stir in greens in batches, and cook, stirring constantly, 5 minutes. Add 1/4 tsp. red pepper, and season with salt and black pepper. Stir in cheese, next 2 ingredients, and 1/2 cup reserved pasta water; cook, stirring constantly, until butter melts and mixture thickens. Stir in up to 1/2 cup pasta water, to reach desired consistency.
4. Divide pasta into individual bowls; top with greens mixture, goat cheese, lemon zest, and remaining 1 tsp. red pepper.

MAKES 6 to 8 servings. **HANDS-ON** 40 min., **TOTAL** 55 min.

THE ULTIMATE CLASSIC COLLARDS

Tangy vinegar brightens the earthy flavors, and a touch of honey rounds out the smoke from the ham hock.

- **3 (1-lb.) packages fresh collard greens**
- **12 smoked bacon slices, chopped**
- **2 medium-size yellow onions, chopped**
- **3 garlic cloves, minced**
- **3 cups reduced-sodium chicken broth**
- **1/4 cup apple cider vinegar**
- **2 Tbsp. honey**
- **1 (12- to 16-oz.) smoked ham hock**
- **Kosher salt**
- **Freshly ground black pepper**

Remove and chop collard stems. Chop collard leaves. Cook bacon in a large Dutch oven over medium heat, stirring occasionally, 12 to 15 minutes or until almost crisp. Add onion, and sauté 8 minutes or until onion is tender. Add garlic, and sauté 1 minute. Stir in chicken broth and next 2 ingredients; add ham hock. Increase heat to high, and bring to a boil. Add collards in batches. Reduce heat to medium-low; cover and cook 2 hours or to desired tenderness. Remove meat from ham hock; chop meat, and discard bone. Stir chopped meat into collards. Season with kosher salt and freshly ground pepper.

MAKES 6 to 8 servings. **HANDS-ON** 50 min.; **TOTAL** 2 hours, 50 min.

ANY GREENS SAUTÉ

Greens can be cooked in minutes over high heat, making it possible to serve collards on a busy weeknight. Try this recipe with whatever green you've got on hand, or mix a few varieties together.

- **1 shallot, minced**
- **2 garlic cloves, minced**
- **2 Tbsp. olive oil**
- **1 lb. collard, turnip, or mustard greens, washed, trimmed, and cut into thin strips**
- **1 1/2 tsp. apple cider vinegar**
- **1/4 tsp. dried crushed red pepper**
- **Kosher salt**
- **Freshly ground black pepper**

Cook shallot and garlic in hot oil in a Dutch oven over medium heat, stirring often, 1 minute. Stir in greens in batches, and cook, stirring often, 7 to 8 minutes or until crisp-tender. Stir in vinegar and red pepper. Season with kosher salt and black pepper. Serve immediately.

MAKES 4 servings. **HANDS-ON** 25 min., **TOTAL** 25 min.

TWICE-BAKED GREEN POTATOES

In Step 1, the potatoes can also be cooked in the microwave, but we prefer the crisp texture of the skins after baking in the oven.

- **4 (12-oz.) russet potatoes**
- **1 small shallot, minced**
- **1 garlic clove, minced**
- **2 Tbsp. olive oil**
- **1/2 lb. greens, washed, trimmed, and chopped**
- **2 tsp. apple cider vinegar**
- **1/8 tsp. ground red pepper**
- **1/2 cup sour cream**
- **1/2 cup butter, melted**
- **1 1/2 tsp. kosher salt**
- **1/2 tsp. ground black pepper**
- **1/2 tsp. hot sauce**
- **1 cup shredded sharp Cheddar cheese, divided**
- **Vegetable cooking spray**

1. Preheat oven to 400°. Pierce potatoes with a fork; bake directly on oven rack 1 hour or until tender. Cool 10 minutes.
2. Sauté shallot and garlic in hot oil in a Dutch oven over medium high heat 1 minute. Stir in greens and next 2 ingredients. Add salt and black pepper to taste. Cook 10 minutes or until tender.
3. Cut potatoes in half lengthwise; carefully scoop pulp into a large bowl, leaving shells intact. Mash together potato pulp, sour cream, next 4 ingredients, and 1/2 cup cheese. Add greens mixture. Spoon into potato shells, and place on a lightly greased (with cooking spray) baking sheet.
4. Bake at 400° for 15 minutes. Top with remaining cheese, and bake 3 to 5 minutes or until cheese melts and potatoes are thoroughly heated.

MAKES 8 servings. **HANDS-ON** 30 min., **TOTAL** 2 hours

MUSTARD GREENS WITH YOGURT-PARMESAN DRESSING AND BACON CROUTONS

To tenderize mustard greens and mellow their fiery flavor, dress the raw greens about 15 minutes before serving. If you prefer them spicy, drizzle the dressing at the table, as we suggest in the recipe.

- **3 cups 1-inch French bread baguette cubes**
- **5 bacon slices**
- **1 Tbsp. minced shallot**
- **2 Tbsp. white wine vinegar**
- **1 tsp. lemon zest**
- **2 Tbsp. fresh lemon juice**
- **1 cup Greek yogurt**
- **2 Tbsp. extra virgin olive oil**
- **1/2 cup (2 oz.) finely grated Parmigiano-Reggiano cheese**
- **1 bunch mustard greens, washed, trimmed, and torn**
- **1/2 tsp. kosher salt**
- **1/4 tsp. freshly ground black pepper**

1. Preheat oven to 350°. Arrange bread cubes in a single layer in a jelly-roll pan. Place bacon slices over bread cubes, so that most of bread is covered. Bake 25 to 30 minutes or until bacon and bread are crisp.
2. Meanwhile, stir together shallot and next 3 ingredients in a medium bowl. Let stand 10 minutes. Whisk in yogurt, oil, and grated cheese.
3. Crumble bacon; toss bacon and croutons with greens. Arrange salad on a platter. Drizzle 3 to 4 Tbsp. dressing over salad. Sprinkle with salt and pepper. Serve immediately with remaining dressing.

MAKES 4 to 6 servings. **HANDS-ON** 30 min., **TOTAL** 55 min.

COLLARD-AND-OLIVE PESTO

Make this pesto and freeze portions in an ice cube tray. Pop out the "cubes," and keep frozen in a zip-top plastic bag for up to one month.

Process 1 cup pitted **green olives** and 4 **garlic cloves** in a food processor until finely chopped. Add 1 lb. washed, trimmed, and chopped **collard greens,** 1 tsp. **lemon zest,** 2 Tbsp. fresh **lemon juice,** 1 Tbsp. **white wine vinegar,** 2 tsp. **kosher salt,** and 1 tsp. freshly ground black **pepper;** pulse 6 to 8 times or until collards are finely chopped. With processor running, pour 1 cup **extra virgin olive oil** through food chute in a slow, steady stream. Add 1 cup grated **Parmesan cheese,** and pulse until smooth. Keep refrigerated in an airtight container up to 3 days.

MAKES 3 cups. **HANDS-ON** 20 min., **TOTAL** 20 min.

3 WAYS WITH COLLARD-AND-OLIVE PESTO

PESTO MASHERS Stir 3 to 4 Tbsp. pesto into hot cooked mashed potatoes.

PESTO EGG SALAD Add 1 to 2 Tbsp. pesto to 3 cups egg salad.

PESTO MAYO Stir 1 Tbsp. pesto into 1/3 cup mayonnaise. Use as a dipping sauce or sandwich spread.

GLORIOUS GREENS

Six hearty, healthy, grown-in-the-South varieties we're cooking right now

1 Swiss Chard: Not only is chard beautiful to behold with its vibrant-colored stems, but it's also one of the world's most nutritious greens. Enjoy it raw, wilted, or cooked.

2 Sweet Potato: Look for vitamin-rich sweet potato greens at farmers' markets. Once common in African soups and stews, these slightly sweet greens are hip again.

3 Mustard: Decidedly peppery, mustard greens are more tender than collards, making them a good candidate for raw salads. Cooking mellows their bite.

4 Turnip: Tender baby greens have a mild bite, but older, large leaves are sometimes tough and bitter. Simmer the big greens to tame and soften them.

5 Collard: Soulful and synonymous with Southern cooking, collard greens are a strongly flavored, sturdy-leafed member of the cabbage family.

6 Cress: Varieties abound—broadleaf, curly, garden, and watercress—all with a tangy, pepper-meets-lemon flavor. Cress conjures images of tea sandwiches and salads, but you can also cook it.

Ten-Dollar Dinners

COOK SMARTER WITH RECIPES YOU CAN WHIP UP ON A SHOESTRING BUDGET

GARLICKY BEEF-AND-BEAN STIR-FRY

Cut costs at the supermarket by rethinking how you buy meats. Here, we use a less expensive cut of beef and stretch it by adding colorful fresh veggies to the entrée. Thinly slice the meat and stir-fry it quickly to keep it tender.

- **4 Tbsp. sugar**
- **6 Tbsp. soy sauce**
- **3 Tbsp. fresh lime juice**
- **1 tsp. dried crushed red pepper**
- **8 tsp. minced garlic**
- **4 Tbsp. peanut oil, divided**
- **1 (10-oz.) sirloin steak, thinly sliced across the grain**
- **1 lb. fresh green beans, cut into 2-inch pieces**
- **2 red bell peppers, cut into 1/4- to 1/2-inch-wide strips**
- **2 tsp. cornstarch**
- **3 cups hot cooked rice**

1. Combine first 5 ingredients. Gradually whisk in 3 Tbsp. oil; transfer to a large zip-top plastic bag. Add steak; seal. Let stand at room temperature 15 minutes.
2. Pour steak and marinade into a bowl. Transfer steak to a wok, reserving marinade. Stir-fry steak in 1 Tbsp. oil over medium-high heat 1 1/2 minutes or until browned. Remove steak. Add beans and bell peppers to wok; stir-fry 3 minutes.
3. Whisk cornstarch into reserved marinade. Stir cornstarch mixture into vegetable mixture. Stir-fry 30 seconds or until sauce thickens. Stir steak into vegetable mixture, and stir-fry 30 seconds. Remove from heat, and serve over rice.

MAKES 4 servings. **HANDS-ON** 20 min., **TOTAL** 30 min.

Garlicky Beef-and-Bean Stir-Fry

CHICKEN CUTLETS WITH HERBED MUSHROOM SAUCE

Cleanup is easy with this one-pan dinner (you'll make the sauce in the same pan you used to cook the cutlets). Serve the dish over mashed potatoes or with a loaf of crusty bread to sop up the earthy sauce. To cut costs, purchase fresh herbs labeled "poultry blend." You'll get an assortment of herbs for the price of one package.

- **2 (8-oz.) skinned and boned chicken breasts**
- **1 tsp. kosher salt**
- **1/2 tsp. freshly ground black pepper**
- **1/4 cup all-purpose flour**
- **2 Tbsp. chopped fresh thyme**
- **2 tsp. chopped fresh rosemary**
- **1/4 cup canola oil, divided**
- **1 (8-oz.) package fresh button mushrooms, quartered**
- **2 Tbsp. red wine vinegar**
- **1 1/2 cups reduced-sodium chicken broth**
- **2 fresh sage sprigs**
- **1 tsp. Dijon mustard**
- **3 Tbsp. butter**
- **Garnish: fresh thyme sprigs**

1. Halve each chicken breast lengthwise to form 4 cutlets. Place each cutlet between 2 sheets of heavy-duty plastic wrap, and flatten to 1/4-inch thickness, using a rolling pin or flat side of a meat mallet. Discard plastic.

Sprinkle salt and pepper over both sides of chicken.
2. Place flour in a shallow dish, and stir in chopped thyme and rosemary. Dredge chicken in flour mixture; shake off excess. Cook 2 cutlets in 4 1/2 tsp. hot oil in a large skillet over medium-high heat 2 to 3 minutes on each side or until done. Remove cutlets from pan, and repeat procedure with 4 1/2 tsp. oil and remaining 2 dredged cutlets.
3. Add remaining 1 Tbsp. oil to drippings in skillet; add mushrooms, and sauté over medium-high heat 4 to 6 minutes or until browned and all moisture evaporates. Stir in vinegar, and cook, stirring occasionally, 1 to 3 minutes or until liquid evaporates. Stir in broth and sage; bring to a boil. Boil until liquid is reduced to about 1/4 cup. Remove from heat; stir in mustard. Whisk in butter, 1 Tbsp. at a time. Spoon sauce over cutlets.

MAKES 4 servings. **HANDS-ON** 45 min., **TOTAL** 45 min.

HONEY-GLAZED PORK TENDERLOIN WITH HOMEMADE APPLESAUCE

- 5 **apples (such as Gala or Fuji), peeled and chopped**
- 3 **Tbsp. butter, melted**
- 1 **Tbsp. lemon zest**
- 2 **Tbsp. fresh lemon juice**
- 1 **Tbsp. sugar**
- 1 1/4 **tsp. kosher salt, divided**
- 2 **(12-oz.) pork tenderloins, trimmed**
- 3 **Tbsp. olive oil**
- 1 1/4 **tsp. ground cumin**
- 1 **tsp. black pepper**
- 1/2 **tsp. apple pie spice**
- 2 **Tbsp. honey**
- 1 **Tbsp. butter, melted**

1. Preheat oven to 425°. Sauté apples in 3 Tbsp. melted butter in a large saucepan over medium-high heat 5 minutes. Stir in zest, juice, sugar, 1/4 tsp. salt, and 1 cup water; bring mixture to a boil. Reduce heat to medium, and simmer, stirring occasionally, 40 minutes or until chopped apples are tender.
2. Meanwhile, brush pork with oil. Stir together cumin, next 2 ingredients, and remaining salt; rub over pork. Place pork in a roasting pan.
3. Bake at 425° for 12 minutes. Whisk together honey and 1 Tbsp. melted butter; brush half of mixture over pork, and bake 1 minute. Turn pork, and brush with remaining honey mixture. Bake 2 more minutes or until a meat thermometer inserted in thickest portion registers 145°. Let stand 5 minutes. Mash apple mixture to desired consistency, and serve with pork.

MAKES 4 servings. **HANDS-ON** 40 min.; **TOTAL** 1 hour, 20 min.

BRUSSELS SPROUTS AND SAUTÉED CABBAGE

Sauté 6 cups thinly sliced **cabbage** in 2 Tbsp. **canola oil** in a skillet over medium-high heat 4 to 6 minutes or until wilted. Remove pan from heat; stir in 4 oz. sliced **Brussels sprouts,** and season with **salt** and **pepper.**

MAKES 4 servings. **HANDS-ON** 15 min., **TOTAL** 15 min.

CHEESY BACON-AND-TWO-ONION TART

- 5 **smoked bacon slices**
- 1 **red onion, sliced**
- 3/4 **cup heavy cream**
- 1/4 **tsp. kosher salt**
- 1/4 **tsp. ground black pepper**
- 3 **large eggs, lightly beaten**
- 1/2 **(17.3-oz.) package frozen puff pastry sheets, thawed**
- 1 1/2 **cups (6 oz.) freshly shredded Swiss cheese, divided**
- 1 **cup coarsely chopped green onions**
- 1/3 **cup grated Parmesan cheese**
- 1 **large egg yolk**
- 2 **Tbsp. milk**

1. Preheat broiler with oven rack 7 inches from heat. Cook bacon in a skillet over medium heat until crisp; crumble. Reserve 1 Tbsp. drippings. Place red onion on a baking sheet; brush with drippings. Broil 4 minutes per side; reduce oven temperature to 400°. Whisk together cream, salt, pepper, and eggs.
2. Unfold pastry; roll into a 12-inch square on a floured surface, and fit into a greased 9-inch tart or quiche pan. Sprinkle 1 cup Swiss cheese and half of crumbled bacon onto crust; top with green onions and red onions.
3. Pour cream mixture over onions. Fold excess dough toward center. Sprinkle center with Parmesan and remaining bacon and Swiss cheese. Combine yolk and milk; brush over dough. Bake at 400° for 40 minutes or until golden. Let stand 10 minutes. Serve with Pea-and-Green Onion Soup.

MAKES 4 servings. **HANDS-ON** 50 min.; **TOTAL** 1 hour, 35 min.

PEA-AND-GREEN ONION SOUP

Sauté 2 chopped **green onions** in 1 Tbsp. **bacon drippings** in a saucepan over medium heat 1 minute. Stir in 1 1/2 cups frozen **English peas** and water to cover. Bring to a simmer; cook 25 minutes. Process until smooth. Stir in 3 Tbsp. **cream;** add **salt** and **pepper** to taste. Add water, if desired.

MAKES 4 servings. **HANDS-ON** 15 min., **TOTAL** 30 min.

RESIDENT BAKER PAM LOLLEY SHOWS YOU HOW TO MAKE

SPECTACULAR FRESH BUTTER & BUTTERMILK

We all love butter. But have you ever tasted *homemade* butter? It is creamy and sweet with a hint of tang that puts grocery store sticks to shame. This mostly hands-free technique yields the same results as an old-fashioned butter churn (minus a workout) and yields amazing butter plus its built-in bonus: fresh buttermilk.

1 CULTURE THE CREAM Heat 4 cups **heavy cream** in a saucepan over low heat until a thermometer registers 70°. Pour cream into a large bowl; stir in 1 cup store-bought **buttermilk.** Cover tightly with plastic wrap, and let stand at room temperature, free from drafts, 24 hours. (Place the bowl in an oven that's not in use, and close the door.) Mixture will thicken and resemble pancake batter. You'll get the best results if you use cream that is not ultra-pasteurized.

2 STRAIN AND DRAIN Process thickened mixture in a food processor 5 to 8 minutes or until small bits of butter begin to form. Pour through a fine wire-mesh strainer into a large bowl. Let strainer stand over bowl 10 minutes, occasionally pressing butter with a rubber spatula to drain buttermilk thoroughly. Use buttermilk immediately, or pour into a glass jar, cover tightly, and chill up to 2 weeks. Run cold water over butter in strainer; squeeze to form a ball.

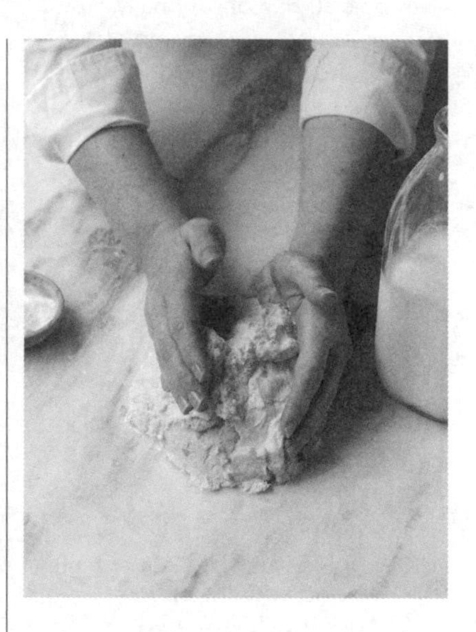

3 RINSE, SHAPE, CHILL Knead butter in **ice water** until water becomes cloudy. Drain; repeat process until water remains clear. Knead in 1/2 tsp. **kosher salt.** Shape into a disk, wrap, and chill up to 3 weeks.

MAKES 1 pt. buttermilk and 8 to 12 oz. butter.
HANDS-ON 50 min., **TOTAL** 25 hours

A Light Pork Supper

LEAN PORK LOIN IS A HEALTHY CUT, AND SOAKING IT IN A FLAVORFUL BRINE KEEPS IT MOIST

ROASTED CIDER-BRINED PORK LOIN WITH GREEN TOMATO CHUTNEY

Cook frozen lady peas or limas to round out the menu.

- **1 cup granulated sugar**
- **1/2 cup kosher salt**
- **2 qt. apple cider**
- **1 (2 1/2-lb.) pork loin roast, trimmed**
- **1/2 cup packed brown sugar**
- **1/3 cup diced red onion**
- **1/3 cup apple cider vinegar**
- **1/3 cup golden raisins**
- **1 Tbsp. minced garlic**
- **3/4 tsp. minced peeled fresh ginger**
- **1/4 tsp. ground red pepper**
- **1/4 tsp. ground cinnamon**
- **1/8 tsp. kosher salt**
- **1 lb. green tomatoes, diced**
- **Vegetable cooking spray**

1. Bring first 2 ingredients and 1 qt. water to a boil in a large stockpot; stir until sugar and salt dissolve. Remove from heat. Stir in cider and 1 qt. water; cool completely. Add pork; cover and chill 12 hours, turning occasionally.
2. Bring brown sugar and next 9 ingredients to a boil in a saucepan over medium heat; reduce heat to low. Simmer, stirring occasionally, 1 hour and 45 minutes.
3. Preheat oven to 375°. Remove pork from brine; pat dry. Transfer to a broiler pan coated with cooking spray. Bake 55 minutes or until a meat thermometer registers 145°. Let stand 15 minutes; slice. Serve with tomato chutney.

MAKES 10 servings (serving size: 3 oz. cooked pork and about 3 Tbsp. chutney).
HANDS-ON 35 min., **TOTAL** 15 hours

NUTRITIONAL INFORMATION
(per serving)

CALORIES: 237; FAT: 5.9g (SATURATED FAT: 1.7g); PROTEIN: 22.1g; CARBOHYDRATES: 23.8g; SODIUM: 540mg

BLUE CHEESE-PECAN GRAPES

(FROM FEBRUARY 1995)

Fresh From the Vine

THE ORIGINAL: A classic example of the Southern remedy for ho-hum appetizers: Add cream cheese. But we didn't stop there. The grapes were rolled in toasted pecans to add a tasty crunch.
THE REVIVAL: An haute hors d'oeuvre you can whip up in mere minutes. The sophisticated sweet-tart and pungent flavors, all found in the original, are simply magnified and presented in a fresh twist.

BLUE CHEESE CROSTINI WITH BALSAMIC-ROASTED GRAPES

Keep this recipe in your back pocket to use as your go-to for last-minute company. You can also serve the grapes with chicken or pork in lieu of a sauce. Or toss them in salads.

- 2 cups halved seedless red grapes
- 2 Tbsp. balsamic vinegar
- 1 1/2 Tbsp. minced shallot
- 2 tsp. olive oil
- 1/2 tsp. light brown sugar
- 1/4 tsp. kosher salt
- 1 (12-oz.) French bread baguette, cut into 15 to 20 (1/2-inch-thick) slices
- 3 Tbsp. butter, softened
- 2 oz. crumbled blue cheese, softened
- 1/3 cup chopped toasted pecans
- Garnish: chopped fresh thyme

1. Preheat oven to 425°. Toss together first 6 ingredients. Arrange grapes in a single layer in a small roasting pan; bake 15 to 20 minutes or until grapes wilt and liquid forms a thin syrup. Remove from oven.

2. Increase oven temperature to broil with oven rack 7 inches from heat. Arrange bread slices in a single layer on a baking sheet. Stir together butter and cheese, and spread evenly over bread slices. Broil 2 to 3 minutes or until browned and bubbly. Spoon grape mixture over toasted bread, and sprinkle with pecans.

MAKES 8 to 10 appetizer servings.
HANDS-ON 20 min., **TOTAL** 35 min.

Sauté, Simmer, Savor

FROM POTLUCKS TO SUNDAY SUPPERS, THIS RECIPE WILL WARM UP YOUR WINTER

MEXICAN STEW

Classic posole, *a delicious slow-simmered Mexican pork stew often reserved for special occasions, is typically a time-consuming labor of love.*

- **2 lb. boneless pork shoulder roast, chopped**
- **2 Tbsp. olive oil**
- **2 cups chopped white onion**
- **1 cup chopped carrot**
- **1 Tbsp. ground cumin**
- **2 tsp. kosher salt**
- **1/2 tsp. ground red pepper**
- **1 tsp. freshly ground black pepper**
- **5 garlic cloves, coarsely chopped**
- **3 (15.5-oz.) cans hominy, drained and rinsed**
- **1 (28-oz.) can crushed tomatoes**
- **4 cups unsalted chicken cooking stock, divided**
- **2 dried ancho chiles**
- **Lime wedges**
- **Sliced radishes**
- **Garnish: fresh cilantro**

1. Sauté pork in hot oil in a large skillet over medium-high heat 6 minutes. Place pork in a 7-qt. slow cooker. Stir onion, next 8 ingredients, and 2 cups stock into slow cooker with pork.

2. Microwave remaining 2 cups stock in a microwave-safe bowl on HIGH 3 minutes or until boiling. Add dried chiles, and let stand 10 minutes. Remove and discard tops of chiles. Process broth and chiles in a blender until smooth. Stir chile mixture into pork mixture. Cover and cook on LOW 7 to 8 hours or until pork is very tender. Serve with lime wedges and radishes.

Note: We tested with Swanson Unsalted Chicken Cooking Stock.

MAKES 3 1/2 qt. **HANDS-ON** 30 min.; **TOTAL** 7 hours, 40 min.

Dark Magic

PUT A SWEET SPELL ON YOUR VALENTINE WITH DECADENT CHOCOLATE DESSERTS

THE ULTIMATE CHOCOLATE PIE

For the crust, crumble three-fourths of 1 (9-oz.) box of chocolate wafer cookies to yield 2 cups.

CRUST

- 2 cups chocolate wafer cookie crumbs
- 1/2 cup finely chopped toasted pecans
- 1/4 cup sugar
- 1/2 cup butter, melted
- Vegetable cooking spray

FILLING

- 3/4 cup sugar
- 1/4 cup cornstarch
- 1/4 cup unsweetened cocoa
- 1/8 tsp. table salt
- 2 cups half-and-half
- 4 egg yolks
- 1 (4-oz.) semisweet chocolate baking bar, finely chopped
- 1/2 (4-oz.) 60% cacao bittersweet chocolate baking bar, finely chopped
- 2 Tbsp. butter
- 1 tsp. vanilla extract

MOUSSE

- 3/4 cup milk chocolate morsels
- 1 cup plus 3 Tbsp. heavy cream

TOPPING

- Chocolate Whipped Cream

1. Prepare Crust: Preheat oven to 350°. Pulse first 3 ingredients in a food processor 4 to 5 times. Transfer crumb mixture to a medium bowl; stir in 1/2 cup melted butter. Press on bottom, up sides, and onto rim of a lightly greased (with cooking spray) 9-inch deep-dish pie plate. Bake 10 minutes. Cool on a wire rack.

2. Prepare Filling: Whisk together 3/4 cup sugar and next 3 ingredients in a large saucepan. Whisk together half-and-half and egg yolks in a large bowl. Gradually whisk egg mixture into sugar mixture. Cook over medium heat, whisking constantly, 6 to 8 minutes or just until mixture begins to boil. Cook, whisking constantly, 1 more minute; remove from heat. Whisk in semisweet chocolate and next 3 ingredients. Place plastic wrap directly on warm filling. Let stand 30 minutes. Spread filling in cooled crust; place plastic wrap directly on filling, and chill 30 minutes.

3. Prepare Mousse: Microwave milk chocolate morsels and 3 Tbsp. heavy cream in a medium bowl at MEDIUM (50% power) for 1 to 1 1/2 minutes or until melted, stirring at 30-second intervals. Let stand 30 minutes, stirring occasionally. Beat 1 cup heavy cream at medium-high speed with an electric mixer until soft peaks form. Gently fold half of whipped cream into milk chocolate mixture until blended and smooth; fold in remaining whipped cream. Spread mousse over filling. Cover and chill 8 to 24 hours or until set. Top with Chocolate Whipped Cream just before serving.

Note: We tested with Ghirardelli Semi-Sweet Chocolate Baking Bar and Ghirardelli 60% Cacao Bittersweet Chocolate Baking Bar.

MAKES 8 to 10 servings.
HANDS-ON 1 hour, 10 min.; **TOTAL** 11 hours, including Chocolate Whipped Cream

Chocolate Whipped Cream

Whisk together 3 cups **heavy cream** and 2 Tbsp. **chocolate syrup** in a large bowl. Beat cream mixture and 1 tsp. **vanilla extract** at medium-high speed with an electric mixer until foamy; gradually add 1/4 cup **sugar,** beating until soft peaks form.

MAKES about 6 cups. **HANDS-ON** 10 min., **TOTAL** 10 min.

CHOCOLATE-MAYONNAISE CAKE

If you've never known the glory of a mayonnaise cake, you're in for a treat! The staple ingredient yields an ultra-moist crumb and keeps the sugar from taking over your taste buds.

- **2 cups all-purpose flour**
- **2/3 cup unsweetened cocoa**
- **1 tsp. baking soda**
- **1 tsp. ground cinnamon**
- **1/4 tsp. table salt**
- **1/4 tsp. baking powder**
- **Shortening**
- **3 large eggs**
- **1 2/3 cups firmly packed light brown sugar**
- **2 tsp. vanilla extract**
- **1 cup mayonnaise**
- **1 1/3 cups hot water**
- **Chocolate-Cream Cheese Frosting**

1. Preheat oven to 350°. Whisk together flour and next 5 ingredients in a medium bowl. Grease (with shortening) and flour a 13- x 9-inch pan.
2. Beat eggs, sugar, and vanilla at medium-high speed with a heavy-duty electric stand mixer about 3 minutes or until mixture is very light brown and ribbons form when beater is lifted. Add mayonnaise, and beat at low speed until combined.
3. Add flour mixture to egg mixture alternately with hot water, beginning and ending with flour mixture. Beat at low speed just until blended after each addition. Pour batter into prepared pan.
4. Bake at 350° for 30 to 35 minutes or until a wooden pick inserted in center comes out clean. Cool completely on a wire rack (about 1 hour). Spread Chocolate-Cream Cheese Frosting on cake.

MAKES 12 to 15 servings. **HANDS-ON** 20 min.; **TOTAL** 2 hours, including frosting

Chocolate-Cream Cheese Frosting

- **1 (8-oz.) package cream cheese, softened**
- **1/2 cup butter, softened**
- **2 tsp. vanilla extract**
- **1 (32-oz.) package powdered sugar**
- **1/2 cup unsweetened cocoa**
- **5 to 6 Tbsp. heavy cream**

Beat first 3 ingredients at medium speed with an electric mixer until creamy. Whisk together powdered sugar and cocoa in a medium bowl; gradually add to butter mixture alternately with 5 Tbsp. cream. Beat at low speed until blended after each addition. (If needed, add up to 1 Tbsp. cream, 1 tsp. at a time, to reach desired consistency.) Increase speed to medium, and beat 1 to 2 minutes or until light and fluffy.

MAKES about 5 cups. **HANDS-ON** 10 min., **TOTAL** 10 min.

NUTTY CHOCOLATE THUMBPRINTS

COOKIES

- **1 2/3 cups all-purpose flour**
- **2/3 cup unsweetened cocoa**
- **1/2 tsp. baking powder**
- **1/2 tsp. table salt**
- **1/2 cup butter, softened**
- **1 cup firmly packed light brown sugar**
- **3/4 cup powdered sugar**
- **3/4 cup creamy peanut butter**
- **2 large eggs**
- **1 tsp. vanilla extract**
- **Parchment paper**

FILLING

- **1/4 cup creamy peanut butter**
- **2 Tbsp. butter, softened**
- **1/2 (4-oz.) 60% cacao bittersweet chocolate baking bar, chopped**
- **1 1/2 cups powdered sugar**
- **2 to 3 Tbsp. milk, at room temperature**

1. Preheat oven to 350°. Whisk together first 4 ingredients in a medium bowl. Beat 1/2 cup butter at medium-high speed with an electric mixer until fluffy. Add brown sugar and 3/4 cup powdered sugar; beat until well blended. Beat in 3/4 cup peanut butter, scraping down sides as needed. Add eggs, 1 at a time, beating until blended after each addition. Beat in vanilla. Reduce speed to medium-low, and gradually add flour mixture, beating just until blended.
2. Shape dough into 30 balls. Place 12 balls 2 inches apart on a parchment paper-lined baking sheet. Press thumb into each ball, forming an indentation. Bake at 350° for 12 minutes or until set; cool 5 minutes. Transfer cookies to a wire rack, and cool 15 minutes. Repeat with remaining dough.
3. Beat 1/4 cup peanut butter and 2 Tbsp. butter at medium speed until smooth. Microwave chopped chocolate in a microwave-safe bowl at HIGH 1 to 2 minutes or until smooth, stirring every 30 seconds. Add melted chocolate to peanut butter mixture, and beat at medium speed just until blended. Gradually add 1 1/2 cups powdered sugar to peanut butter mixture alternately with 2 Tbsp. milk, beginning with sugar. Beat at low speed just until blended after each addition. Beat in up to 1 Tbsp. milk, 1 tsp. at a time, to reach desired consistency. Spoon filling into a zip-top plastic bag; snip 1 corner of bag to make a small hole, and pipe filling into indentations in cookies.

MAKES 30 cookies. **HANDS-ON** 40 min.; **TOTAL** 1 hour, 30 min.

Community Cookbook

BUILD A BETTER LIBRARY, ONE GREAT BOOK AT A TIME

From the moment their bakery opened, Cheryl and Griff Day always dreamed it would become a community gathering spot. Twelve years, countless homespun sweets and rustic breads, and one award-winning cookbook later, their community extends well beyond their charming Savannah digs. If you're hungry for another heaping helping of their baking magic, check out *Back in the Day Bakery: Made with Love.*

PB&J SAMMIES

- **1 1/2 cups all-purpose flour**
- **1 tsp. baking soda**
- **1/4 tsp. fine sea salt**
- **1/2 cup unsalted butter, softened**
- **1/2 cup granulated sugar**
- **1/2 cup firmly packed light brown sugar**
- **1/2 cup creamy peanut butter**
- **1 large egg**
- **1/2 tsp. vanilla extract**
- **3/4 cup of your favorite jam**

1. Position racks in middle and lower third of oven; preheat to 375°. Whisk together first 3 ingredients in a large bowl.
2. Beat butter and next 3 ingredients at medium speed with a heavy-duty electric stand mixer 3 to 5 minutes or until light and fluffy. Add egg and vanilla; beat 1 to 2 minutes or until egg is incorporated. Add flour mixture in 3 batches; beat at low speed until just combined.
3. Using a 1-inch cookie scoop, drop dough by level scoopfuls 2 inches apart onto 2 parchment paper-lined baking sheets. (Use heaping tablespoonfuls, if you don't have a 1-inch scoop.) Flatten tops of each dough portion slightly, using a fork to create a crisscross pattern on each cookie.
4. Bake cookies at 375° for 5 minutes, placing 1 baking sheet on middle oven rack and 1 sheet on lower oven rack. Rotate pans. Bake 5 to 6 more minutes or until cookies are golden brown. Cool 5 minutes on pans. Transfer cookies to wire racks; cool completely (about 20 minutes).
5. Spoon about 1 Tbsp. jam onto bottom side of 1 cooled cookie; top with another cookie to make a sandwich. Repeat with remaining jam and cookies.

MAKES about 1 dozen sandwich cookies. **HANDS-ON** 30 min., **TOTAL** 1 hour

COTTON CANDY MERINGUES

- **4 large egg whites**
- **Parchment paper**
- **1/4 tsp. cream of tartar**
- **1 cup sugar**
- **2 tsp. cornstarch**
- **1 tsp. white wine vinegar**
- **1 1/2 tsp. vanilla extract**
- **Pastel-colored liquid gel food coloring (optional)**

1. Place egg whites in a bowl, and let stand until they reach room temperature (about 20 to 30 minutes). Position oven rack in lower third of oven. Preheat oven to 250°.
2. Line a baking sheet with parchment paper. Using a 3-inch round cutter as a guide, draw 12 circles at least 1 inch apart on the parchment paper. Turn paper over.
3. Beat egg whites and cream of tartar at high speed with a heavy-duty electric stand mixer, using whisk attachment, until soft peaks form. Whisk together sugar and cornstarch, and gradually add to egg white mixture, beating at low speed until well blended. Increase speed to high, and beat until stiff peaks form.
4. Fold in vinegar and vanilla. If desired, add a few drops of food coloring, and gently fold into egg white mixture until well blended. (You can also divide the meringue and add different colors of gel to each, if desired.)
5. Gently spoon mounds of meringue, about 2 inches high, into each circle on parchment paper-lined baking sheet. Smooth sides with a butter knife.
6. Bake at 250° for 1 hour and 15 minutes to 1 hour and 30 minutes or until set. Turn oven off, and let meringues stand in oven for 1 1/2 hours. Store meringues in an airtight container at room temperature up to 5 days.

MAKES 1 dozen. **HANDS-ON** 30 min.; **TOTAL** 3 hours, 15 min.

March

60 **The Saturday Morning Chef** Southern breakfast starts with three ingredients: bacon, eggs, and grits. Master these basics and then become a weekend-kitchen rock star

63 WHAT TO COOK NOW **Thigh and Mighty** Consider this adaptable, budget-friendly, and delectable cut of meat for your next gathering

66 *COOKING LIGHT* **A Fresh Fiesta** Shake up taco night with this Asian-inspired recipe

67 THE *SL* TEST KITCHEN ACADEMY **Candied Carrot Curls** Embellish your next batch of pastries with these beautiful and tasty garnishes

68 SAVE ROOM **Crazy for Carrot Cake** Transform a classic dessert

69 THE SLOW DOWN **Try a New Stew** Smoked sausage, beans, and Parmesan cheese lend great flavor to this new dinnertime staple

70 COMMUNITY COOKBOOK Try these baked goods from country music star Martina McBride

The Saturday Morning Chef

EVERY GREAT SOUTHERN BREAKFAST IS BUILT FROM THREE ESSENTIALS: BACON, EGGS, AND GRITS. MASTER THE BASICS. THEN SPIN THEM INTO A SPECTACULAR MEAL WITH OUR GUSSIED-UP TWISTS. YOU'LL BE THE CULINARY STAR OF THE WEEKEND KITCHEN.

MASTER THE BASICS

Nail the essentials with our Test Kitchen's foolproof recipes and tried-and-true techniques. After you've perfected these building blocks, try our easy gourmet variations on the following pages.

1 BACON

OVEN-ROASTED BACON

Preheat oven to 350°. Place 12 thick **bacon** slices in a single layer on a wire rack coated with **vegetable cooking spray;** place rack in an aluminum foil-lined jelly-roll pan. Bake 40 to 50 minutes or until browned and crisp; cool 5 minutes.

MAKES 6 servings. **HANDS-ON** 5 min., **TOTAL** 50 min.

2 EGGS

POACHED EGGS

Pour water to depth of 3 inches into a large saucepan. Bring to a boil; reduce heat, and simmer. Add 2 Tbsp. **white vinegar.** Break 6 large **eggs,** 1 at a time, into a ramekin, and slip each egg into water, as close to surface as possible. Simmer 3 to 5 minutes or to desired degree of doneness.

MAKES 6 servings. **HANDS-ON** 5 min., **TOTAL** 15 min.

FRIED EGGS

Preheat broiler with oven rack 6 inches from heat. Heat 1 Tbsp. **olive oil** in a large ovensafe nonstick skillet over medium heat. Gently break 4 large **eggs** into skillet; sprinkle with **salt** and **pepper.** Cook 2 minutes. Place skillet in oven, and broil 1 minute.

MAKES 4 servings. **HANDS-ON** 10 min., **TOTAL** 10 min.

3 GRITS

STOVE-TOP GRITS

Bring 2 tsp. **kosher salt** and 1 qt. **water** to a boil in a heavy saucepan over high heat. Whisk in 1 cup **stone-ground grits,** and cook, whisking constantly, 45 seconds. Return to a boil; cover and reduce heat to medium-low. Cook 20 to 25 minutes or until tender. Stir in 2 1/2 Tbsp. **butter.** Serve immediately.

MAKES 6 servings. **HANDS-ON** 5 min., **TOTAL** 30 min.

OVERNIGHT SLOW-COOKER GRITS

Stir together 2 cups **stone-ground grits,** 1/2 cup melted **butter,** and 1 Tbsp. **kosher salt** in a 5-qt. slow cooker; stir in 6 cups **water.** Cover and cook on LOW 8 hours. Stir just before serving.

MAKES about 8 cups. **HANDS-ON** 5 min.; **TOTAL** 8 hours, 5 min.

Note: For quicker slow-cooker grits, cook on HIGH 4 hours.

THE ULTIMATE FRIED EGG SANDWICH WITH BBQ BACON

- 8 (1/2-inch-thick) country-style bread slices
- 2 1/2 Tbsp. butter, melted
- 4 (3/4-oz.) processed American cheese slices
- 2 cups chopped butter lettuces
- 8 BBQ Bacon slices (recipe, page 62)
- 1/4 cup mayonnaise
- 4 Fried Eggs (recipe, page 60)

1. Preheat broiler with oven rack 5 inches from heat. Brush bread with butter. Broil on an aluminum foil-lined baking sheet 1 to 2 minutes on each side or until lightly toasted.
2. Top each of 4 bread slices with 1 cheese slice, 1/2 cup lettuce, and 2 bacon slices. Spread mayonnaise over remaining bread slices.
3. Carefully place 1 Fried Egg on each sandwich; top with remaining bread slices.

MAKES 4 servings. **HANDS-ON** 20 min., **TOTAL** 20 min., not including bacon or eggs

WAFFLED BACON & CHEDDAR GRITS

- 2 cups Overnight Slow-Cooker Grits, cooled completely (recipe, page 60)
- 1 cup (4 oz.) shredded sharp Cheddar cheese
- 4 Oven-Roasted Bacon slices, crumbled (recipe, page 60)
- Melted butter

Preheat waffle iron to medium-high heat. Stir together first 3 ingredients. Brush hot waffle iron with melted butter. Place about 3/4 cup grits mixture in prepared waffle iron; close and cook 4 minutes or until grits are golden and crisp. Repeat with remaining grits mixture.

MAKES 4 waffles. **HANDS-ON** 20 min.; **TOTAL** 20 min., not including grits or bacon

Waffled Gruyère & Ham Grits

Prepare recipe as directed, omitting Cheddar cheese and bacon. Stir in 1 cup shredded **Gruyère cheese** and 1/2 cup **chopped ham** before cooking.

Waffled Jalapeño & Monterey Jack Cheese Grits

Prepare recipe as directed, omitting Cheddar cheese and bacon. Stir in 1 cup shredded **Monterey Jack cheese** and 1 **jalapeño pepper,** seeded and chopped, before cooking.

SHRIMP 'N'GRITS PIE

Use a whisk to beat the eggs vigorously and your grits pie will have a light and airy texture.

- 4 cups Overnight Slow-Cooker Grits (recipe, page 60)
- 1 cup jarred marinara sauce
- 1/2 cup grated Parmesan cheese
- 3 large eggs
- Vegetable cooking spray
- 2 Tbsp. butter
- 1 Tbsp. bacon drippings
- 1 lb. medium-size raw shrimp, peeled and deveined
- 1/3 cup crumbled feta cheese
- 3 thick bacon slices, cooked and crumbled
- 3 Tbsp. torn fresh basil leaves

1. Preheat oven to 425°. Stir together first 4 ingredients. Spread mixture in a 10-inch pie plate coated with cooking spray, and bake 1 hour or until browned. Let stand 20 minutes.
2. Melt butter with bacon drippings in a large skillet over medium-high heat; add shrimp, and sauté 3 minutes or until pink. Transfer shrimp to a medium bowl. Toss together shrimp and feta; add salt and pepper to taste. Top pie with shrimp mixture, bacon, and basil.

MAKES 6 servings. **HANDS-ON** 30 min.; **TOTAL** 1 hour, 50 min., not including grits

GRITS CAKES WITH POACHED EGGS AND COUNTRY GRAVY

Trim away uneven edges for a clean look. Use a paring knife, round cookie cutter, or pizza wheel.

GRITS CAKES

- 1 1/2 cups Overnight Slow-Cooker Grits (recipe, page 60)
- Vegetable cooking spray
- 2 Tbsp. bacon drippings

GRAVY

- 1/4 cup plus 1 Tbsp. butter
- 1/3 cup all-purpose flour
- 3 3/4 cups milk
- 1 tsp. kosher salt
- 1/2 tsp. freshly ground black pepper
- 1/2 tsp. hot sauce

ADDITIONAL INGREDIENTS

- 6 Poached Eggs (recipe, page 60)
- Sliced chives

1. Prepare Grits Cakes: Spoon 1/4 cup grits into each cup of a lightly greased (with cooking spray) 6-cup muffin pan, smoothing tops. Cover and chill 4 to 12 hours. Remove cakes from pan, and sprinkle with desired amount of salt and pepper. Cook cakes in hot drippings in a large skillet over medium-high heat 3 minutes on each side or until browned. Keep warm in a 200° oven.
2. Prepare Gravy: Melt 1/4 cup butter in a large skillet over medium-low heat. Whisk in flour. Cook, whisking constantly, 1 minute. Gradually whisk in milk; reduce heat to medium, and cook, whisking constantly, 10 minutes or until thickened. Stir in 1 tsp. salt, 1/2 tsp. pepper, 1/2 tsp. hot sauce, and remaining 1 Tbsp. butter.
3. Assemble: Place each warm grits cake on a plate; top with 2 Tbsp. gravy and 1 Poached Egg. Sprinkle with chives, salt, and pepper.

MAKES 6 servings. **HANDS-ON** 30 min.; **TOTAL** 4 hours, 20 min., not including grits or eggs

SOUTHWEST CORNMEAL BACON

Preheat oven to 350°. Stir together 1/3 cup plain yellow **cornmeal,** 1/4 cup firmly packed **light brown sugar,** 1/4 tsp. **ground cumin,** and 1/4 tsp. **ground red pepper.** Dredge 12 thick **bacon** slices in cornmeal mixture, shaking off excess. Arrange bacon in a single layer on a wire rack coated with **vegetable cooking spray;** place in an aluminum foil-lined jelly-roll pan. Bake 40 to 50 minutes or until crisp.

MAKES 6 servings. **HANDS-ON** 10 min., **TOTAL** 55 min.

PECAN-ROSEMARY BACON

Preheat oven to 350°. Stir together 1 cup finely chopped **pecans,** 6 Tbsp. **light brown sugar,** 1 1/4 tsp. chopped fresh **rosemary,** and 3/4 tsp. freshly ground **black pepper.** Dredge 12 thick **bacon** slices in pecan mixture, pressing to adhere. Arrange bacon in a single layer on a wire rack coated with **vegetable cooking spray;** place in an aluminum foil-lined jelly-roll pan. Bake 40 to 50 minutes or until crisp.

MAKES 6 servings. **HANDS-ON** 15 min., **TOTAL** 1 hour

BBQ BACON

Preheat oven to 350°. Stir together 5 Tbsp. **light brown sugar,** 1 Tbsp. **chili powder,** 1 tsp. **ground cumin,** and 1/4 tsp. **ground red pepper.** Dredge 12 thick **bacon** slices in sugar mixture, pressing to adhere. Arrange bacon in a single layer on a wire rack coated with **vegetable cooking spray;** place in an aluminum foil-lined jelly-roll pan. Bake 40 to 50 minutes or until crisp.

MAKES 6 servings. **HANDS-ON** 10 min., **TOTAL** 55 min.

Thigh and Mighty

THE HUMBLE CHICKEN THIGH HAS A LOT GOING FOR IT—LOW PRICE, VERSATILITY, AND GREAT FLAVOR, MAKING IT OUR FAVORITE CUT TO COOK

1

TECHNIQUE TAKEAWAY:
PAILLARD

SESAME-CHICKEN THIGH PAILLARD WITH PEANUT SAUCE

Paillard is a French cooking term used to describe a thinly pounded piece of meat that cooks quickly. This recipe gives you a flavorful weeknight meal that's on the table in less than 45 minutes.

- 6 **skinned and boned chicken thighs**
- 1 1/2 **tsp. kosher salt, divided**
- 1/2 **tsp. black pepper, divided**
- 1/4 **cup all-purpose flour**
- 2 **large eggs, lightly beaten**
- 2 **cups panko (Japanese breadcrumbs)**
- 1/4 **cup sesame seeds**
- 1/4 **cup canola oil**
- **Peanut Sauce**

1. Place each chicken thigh between 2 sheets of plastic wrap, and flatten to 1/4-inch thickness, using a rolling pin or flat side of a meat mallet. Sprinkle thighs with 1 tsp. salt and 1/4 tsp. pepper.
2. Place flour in a shallow dish. Place eggs in a second dish. Stir together panko, sesame seeds, and remaining 1/2 tsp. salt and 1/4 tsp. pepper in a third dish. Dredge chicken in flour; shake off excess. Dip in eggs, and dredge in panko mixture, pressing to adhere.
3. Cook 3 chicken thighs in 1 Tbsp. hot oil in a large nonstick skillet over medium heat 3 minutes. Add 1 Tbsp. oil, turn chicken, and cook 3 minutes or until done. Keep warm in a 200° oven. Repeat procedure with remaining oil and chicken. Serve with Peanut Sauce.

MAKES 4 to 6 servings. **HANDS-ON** 35 min.; **TOTAL** 40 min., including sauce

Peanut Sauce

Process 1/2 cup **creamy peanut butter,** 1/3 cup **soy sauce,** 1/4 cup **cilantro leaves,** 3 Tbsp. **rice wine vinegar,** 3 Tbsp. **lime juice,** 3 Tbsp. **honey,** 3 Tbsp. **sesame oil,** and 1 tsp. **Dijon mustard** in a blender or food processor until smooth.

MAKES 1 1/3 cups. **HANDS-ON** 5 min., **TOTAL** 5 min.

2

TECHNIQUE TAKEAWAY:
BRAISING

CITRUS-BRAISED CHICKEN THIGHS

Braising gives you tender, tasty results and a pan sauce to boot.

- **Kosher salt**
- **Black pepper**
- **8 bone-in, skin-on chicken thighs, trimmed**
- **2 Tbsp. olive oil**
- **2 carrots, sliced**
- **1 yellow onion, sliced**
- **3 garlic cloves, minced**
- **1 cup fresh orange juice**
- **2 Tbsp. fresh lemon juice**
- **1/2 tsp. ground cumin**
- **1/2 cup green olives, pitted**
- **1 Tbsp. chopped fresh flat-leaf parsley**

1. Sprinkle salt and pepper over chicken. Cook 4 chicken thighs in 1 Tbsp. hot oil in a large Dutch oven over medium-high heat 6 minutes on each side. Remove chicken; wipe Dutch oven clean. Repeat with remaining 1 Tbsp. oil and 4 chicken thighs. Reserve 1 Tbsp. drippings in Dutch oven.

2. Reduce heat to medium; add carrots, and cook, stirring occasionally, 2 minutes. Add onions, and cook, stirring occasionally, 5 to 7 minutes or until tender. Add garlic, and cook, stirring occasionally, 1 minute. Stir in orange juice, lemon juice, and cumin. Increase heat to high, and bring to a boil.

3. Add chicken and olives. Reduce heat to medium-low; cover and simmer 35 to 40 minutes or until meat pulls away from bone. Just before serving, stir in parsley, and add salt and pepper to taste.

MAKES 4 to 6 servings. **HANDS-ON** 55 min.; **TOTAL** 1 hour, 30 min.

Citrus-Braised Chicken Thighs

3

TECHNIQUE TAKEAWAY:
GRILLING

CHICKEN, FARRO, AND VEGETABLE SALAD WITH LEMON VINAIGRETTE

Grilling the chicken and vegetables gives this salad a smoky, sweet flavor. If you prefer, substitute sliced celery for fennel and pearl barley or wheat berries for farro.

- **4 skinned and boned chicken thighs, trimmed**
- **Lemon Vinaigrette, divided**
- **1 cup uncooked farro**
- **2 1/4 tsp. kosher salt, divided**
- **2 garlic cloves, peeled**
- **3/4 tsp. black pepper, divided**
- **1/2 red onion, cut into wedges**
- **1 fennel bulb, thinly sliced**
- **10 sweet mini peppers, halved and seeded**
- **1 1/2 tsp. olive oil**
- **1/2 cup loosely packed fresh flat-leaf parsley leaves**
- **1/3 cup torn fresh basil leaves**
- **1 Tbsp. fresh thyme leaves**

1. Place chicken and 1/4 cup Lemon Vinaigrette in a 1-gal. zip-top plastic freezer bag. Seal and turn to coat. Chill 30 minutes. (Reserve and chill remaining vinaigrette.)

2. Meanwhile, cook farro according to package directions, adding 1 tsp. salt and 2 garlic cloves before bringing to a boil. Drain and rinse; discard garlic, and transfer farro to a large bowl.

3. Preheat grill to 350° to 400° (medium-high) heat. Remove chicken from marinade, discarding marinade in bag. Sprinkle chicken with 1/4 tsp. each salt and black pepper. Grill chicken, covered with grill lid, 4 to 5 minutes on each side or until done. Transfer to a platter; cover with foil.

4. Toss together onion, next 3 ingredients, and remaining 1 tsp. salt and 1/2 tsp. black pepper. Place vegetables in a large grill basket, and grill, covered with grill lid, 10 to 12 minutes or until vegetables start to char and soften, stirring and turning every 2 minutes. Transfer vegetables to bowl, and cover with foil.
5. Coarsely chop chicken; toss with farro, vegetables, parsley, basil, thyme, and 1/4 cup chilled reserved Lemon Vinaigrette. Season with salt and pepper, and serve with remaining vinaigrette.

MAKES 6 to 8 servings. **HANDS-ON** 50 min.; **TOTAL** 1 hour, 30 min., including vinaigrette

Lemon Vinaigrette

- 2 **tsp. lemon zest**
- 2 **Tbsp. fresh lemon juice**
- 2 **Tbsp. white wine vinegar**
- 1 **Tbsp. Dijon mustard**
- 1 **Tbsp. honey**
- 1 **small garlic clove, pressed**
- 1/2 **tsp. kosher salt**
- 1/4 **tsp. ground black pepper**
- 1/2 **cup olive oil**

Whisk together first 8 ingredients; add oil in a slow, steady stream, whisking constantly until smooth.

MAKES about 3/4 cup. **HANDS-ON** 10 min., **TOTAL** 10 min.

4

TECHNIQUE TAKEAWAY:
MARINATING

GARLIC-YOGURT-MARINATED CHICKEN THIGHS

Tangy yogurt tenderizes the meat and balances the heat from the jalapeño peppers. Grill indoors on a grill pan, or rev up the outdoor grill. Steamed sugar snap peas round out the meal.

- 8 **garlic cloves, smashed**
- 2 **jalapeño peppers, tops removed**
- 1 **large shallot, quartered**
- 1 **Tbsp. kosher salt**
- 1 **tsp. ground cumin**
- 6 **Tbsp. olive oil, divided**
- 1 **lime**
- 1 **Tbsp. fresh lime juice**
- 3/4 **cup plain 2% reduced-fat Greek yogurt**
- 6 **Tbsp. honey**
- 8 **skinned and boned chicken thighs**

1. Process first 5 ingredients and 2 Tbsp. oil in a food processor until finely ground. Cut a thin slice from each end of lime. Peel lime; cut away bitter white pith. Add peeled lime and 1 Tbsp. juice to processor; process until smooth.
2. Whisk together yogurt, honey, 2 Tbsp. oil, and garlic mixture in an 11- x 7-inch baking dish. Add chicken to marinade, turning to coat. Cover and chill 8 hours.
3. Preheat oven to 400°. Heat remaining 2 Tbsp. oil in an ovenproof grill pan over medium-high heat. Remove chicken from marinade, discarding marinade. Cook chicken in hot oil 4 minutes; turn chicken. Transfer pan to oven, and bake 10 minutes or until done.

MAKES 4 servings. **HANDS-ON** 25 min.; **TOTAL** 8 hours, 35 min.

5

TECHNIQUE TAKEAWAY:
ROASTING

ROASTED CHICKEN THIGHS WITH HERB BUTTER

Roasting chicken is a great weeknight technique because it's hands-free and hard to mess up. Let the cooked chicken stand for about 10 minutes before serving so the meat has time to relax and soak up any cooking juices.

- 1/2 **cup butter, softened**
- 2 **garlic cloves, minced**
- 2 **Tbsp. chopped fresh oregano**
- 2 **Tbsp. chopped fresh flat-leaf parsley**
- 2 **Tbsp. sliced fresh chives**
- 1 1/2 **tsp. kosher salt**
- 1 **tsp. loosely packed lemon zest**
- 1/2 **tsp. freshly ground black pepper**
- 8 **bone-in, skin-on chicken thighs**
- 1 **Tbsp. olive oil**
- **Goat Cheese Mashed Potatoes**

1. Preheat oven to 425°. Stir together first 8 ingredients in a small bowl until well combined.
2. Loosen skin from each chicken thigh without totally detaching skin; spread butter mixture under skin. (Discard any remaining butter mixture.) Replace skin; brush thighs with oil, and sprinkle with desired amount of salt and pepper. Place thighs on a lightly greased (with cooking spray) wire rack in an aluminum foil-lined jelly-roll pan.
3. Bake at 425° for 50 minutes or until a meat thermometer inserted into thickest portion of a chicken thigh registers 165°. Let stand 10 minutes. Serve with Goat Cheese Mashed Potatoes.

MAKES 4 to 6 servings. **HANDS-ON** 35 min.; **TOTAL** 2 hours, 15 min., including potatoes

Goat Cheese Mashed Potatoes

- 4 **russet potatoes (about 2 lb.), peeled and cut into 2-inch cubes**
- 2 **oz. goat cheese**
- 3 **Tbsp. unsalted butter**
- 1/2 **tsp. kosher salt**
- 1/4 **tsp. ground black pepper**
- 2/3 **cup milk**
- **Garnishes: chopped fresh chives, crumbled goat cheese**

Bring potatoes with water to cover by 2 inches to a boil in a medium saucepan over high heat; reduce heat to medium. Simmer 25 minutes or until tender. Drain and return to pan. Add goat cheese and next 3 ingredients; mash with a potato masher until cheese and butter melt. Gradually add milk, mashing to desired consistency.

MAKES 4 to 6 servings. **HANDS-ON** 25 min., **TOTAL** 50 min.

A Fresh Fiesta

BAJA AND BEIJING COLLIDE IN A HEALTHY NEW WAY WITH TACO NIGHT

SESAME CHICKEN TACOS

- **6 skinned and boned chicken thighs, cut into small pieces**
- **3 Tbsp. low-sodium soy sauce, divided**
- **1/4 tsp. kosher salt**
- **1/4 cup plus 1 1/2 tsp. cornstarch, divided**
- **2 Tbsp. canola oil**
- **1 1/2 Tbsp. honey**
- **1 Tbsp. dark sesame oil**
- **2 tsp. rice vinegar**
- **1 tsp. sambal oelek (chile paste)**
- **1 large garlic clove, minced**
- **3 Tbsp. coarsely chopped dry-roasted peanuts**
- **3/4 cup celery slices**
- **8 (6-inch) fajita-size corn tortillas, warmed**
- **1/3 cup sliced green onions**
- **1/2 red bell pepper, sliced**

1. Place chicken and 1 Tbsp. soy sauce in a large zip-top plastic bag; seal bag. Let stand at room temperature 30 minutes. Remove chicken; discard marinade. Sprinkle chicken with salt. Place 1/4 cup cornstarch in a shallow dish. Toss chicken with cornstarch.
2. Sauté half of coated chicken in 1 Tbsp. hot oil in a large skillet over medium-high heat 6 minutes or until done. Remove chicken; drain on paper towels. Repeat procedure with remaining 1 Tbsp. oil and coated chicken.
3. Whisk together honey, next 3 ingredients, and remaining 2 Tbsp. soy sauce and 1 1/2 tsp. cornstarch in a small microwave-safe bowl. Microwave at HIGH 1 1/2 minutes or until thickened, stirring twice. Stir in garlic. Toss together honey mixture, chicken, peanuts, and celery; divide among tortillas. Top with green onions and bell pepper slices.

MAKES 4 servings (serving size: 2 tacos).
HANDS-ON 30 min., **TOTAL** 1 hour

NUTRITIONAL INFORMATION
(per serving)

CALORIES: 418; FAT: 19.1g (SATURATED FAT: 2.5g); PROTEIN: 25.2g; CARBOHYDRATES: 39.3g; SODIUM: 531mg

PAM LOLLEY, MASTER OF EDIBLE FLOURISHES, SHARES A RECIPE FOR

CANDIED CARROT CURLS

Channel your inner pastry chef and elevate your cake with these vibrant embellishments that are a cinch to make. Best of all, they are as tasty as they are beautiful, adding sweetness, texture, and whimsy. Make up to five days ahead, and layer between sheets of wax paper in an airtight container. Store at room temperature.

1 PREPARE CARROTS Preheat oven to 225°. Line a baking sheet with parchment paper, and lightly grease with vegetable cooking spray. Remove 15 to 20 long strips from 1 or 2 large peeled carrots, using a vegetable peeler. (Strips will get wider as you get close to the core of the carrot.)

2 COOK AND DRAIN Bring 1 cup water and 1 cup sugar to boil in a large heavy-duty saucepan over medium-high heat. Add carrot strips, and reduce heat to medium-low. Simmer carrot strips 15 minutes. Drain in a wire-mesh strainer, and cool 5 minutes.

3 BAKE THE STRIPS Spread cooked carrot strips 1 inch apart in a single layer on prepared baking sheet. Bake at 225° for 30 minutes. As the carrot strips bake, they will begin to look translucent. Remove from oven. (Strips will be warm but cool enough to handle.)

4 CREATE THE CURLS Working quickly, wrap each carrot strip around the handle of a wooden spoon, forming curls. Gently slide off spoon. Sprinkle with sugar, if desired. Let curls sit at room temperature until completely dry (about 30 minutes).

MAKES 15 to 20 curls. **HANDS-ON** 15 min.; **TOTAL** 1 hour, 35 min.

Crazy for Carrot Cake

PERFECTLY SPICED CAKE, BROWN-SUGAR KISSED FROSTING, AND WHIMSICAL CURLS TAKE THE HUMBLE CLASSIC TO NEW HEIGHTS

THE ULTIMATE CARROT CAKE

These layers are tender, so remove from pans carefully!

- 1 1/2 cups chopped pecans
- 1 Tbsp. butter, melted
- 1/8 tsp. kosher salt
- 2 1/2 cups all-purpose flour
- 2 tsp. baking soda
- 1 1/2 tsp. ground cinnamon
- 1/2 tsp. ground nutmeg
- 1/2 tsp. table salt
- 1/2 cup butter, softened
- 1 cup granulated sugar
- 1 cup firmly packed light brown sugar
- 1/2 cup canola oil
- 3 large eggs
- 3/4 cup buttermilk
- 2 tsp. vanilla extract
- 3 cups grated carrots
- 1 cup peeled and grated Granny Smith apple
- 1 cup sweetened flaked coconut
- Brown Sugar-Cream Cheese Frosting
- Candied Carrot Curls (recipe, page 67)

1. Preheat oven to 350°. Toss together first 3 ingredients; spread in a single layer in a foil-lined pan. Bake 10 minutes or until toasted, stirring once.
2. Stir together flour and next 4 ingredients. Beat butter and both sugars at medium speed with an electric mixer until blended. Add oil; beat until blended. Add eggs, 1 at a time, beating just until blended.
3. Add flour mixture to butter mixture alternately with buttermilk, beginning and ending with flour mixture. Stir in vanilla; fold in carrots, next 2 ingredients, and 1 cup toasted pecans. Spoon batter into 3 greased (with shortening) and floured 9-inch round cake pans.
4. Bake at 350° for 23 to 28 minutes or until a wooden pick inserted in center comes out clean. Cool in pans on wire racks 15 minutes. Remove from pans, and cool completely. Spread Brown Sugar-Cream Cheese Frosting between layers and on top and sides of cake. Top with Candied Carrot Curls and remaining toasted pecans.

MAKES 12 servings. **HANDS-ON** 1 hour; **TOTAL** 4 hours, 35 min., including frosting and carrot curls

Brown Sugar-Cream Cheese Frosting

Beat 2 (8-oz.) packages **cream cheese,** softened; 1/2 cup **butter,** softened; and 1/4 cup firmly packed **light brown sugar** at medium speed with an electric mixer until creamy. Add 2 tsp. **vanilla extract,** and beat until blended. Gradually add 7 cups **powdered sugar,** beating at low speed until blended. Increase speed to high; beat 1 minute or until smooth.

MAKES about 5 cups. **HANDS-ON** 10 min., **TOTAL** 10 min.

Try a New Stew

YOU'RE THREE STEPS AWAY FROM A KID-FRIENDLY, COMPANY-WORTHY MEAL

SAUSAGE-AND-BEAN STEW

This dish is so easy: Just brown your favorite smoked sausage, toss the ingredients together, and let the slow cooker do the rest. All you need to round out the meal is a simple green salad and loaf of crusty bread.

- **1 lb. smoked sausage**
- **2 Tbsp. olive oil**
- **1 (48-oz.) container reduced-sodium chicken broth**
- **3 cups dried great Northern beans**
- **2 cups chopped yellow onion**
- **3/4 cup chopped carrot**
- **1/2 cup chopped celery**
- **10 garlic cloves, sliced**
- **1 Tbsp. chopped fresh sage**
- **2 tsp. kosher salt**
- **1 tsp. freshly ground black pepper**
- **1 Parmesan cheese rind**
- **1 fresh rosemary sprig**
- **Garnishes: chopped fresh parsley, Parmesan cheese**

1. Cut sausage into 2-inch pieces, and halve sausage pieces lengthwise, cutting to but not through other side. Cook half of sausage in 1 Tbsp. hot oil in a large skillet over medium-high heat 2 minutes on each side or until browned. Transfer sausage and drippings to a 7-qt. slow cooker. Repeat with remaining sausage and oil.
2. Add 1 cup broth to skillet, and bring to a boil over medium-high heat, stirring to loosen browned bits from bottom of skillet. Boil 1 minute or until reduced by half. Stir beans, next 8 ingredients, boiled broth mixture, 3 cups water, and remaining broth into slow cooker. Cover and cook on HIGH 7 hours or until beans are tender. Discard cheese rind.
3. Stir mixture with rosemary sprig 30 seconds to 1 minute; discard rosemary. Remove 1 cup beans, and process in a blender 1 to 2 minutes or until smooth. Stir pureed bean mixture into slow cooker. (Repeat procedure with 1 more cup beans, if a thicker consistency is desired.) Serve immediately.

MAKES about 4 qt. **HANDS-ON** 35 min.; **TOTAL** 7 hours, 35 min.

Community Cookbook

BUILD A BETTER LIBRARY, ONE GREAT BOOK AT A TIME

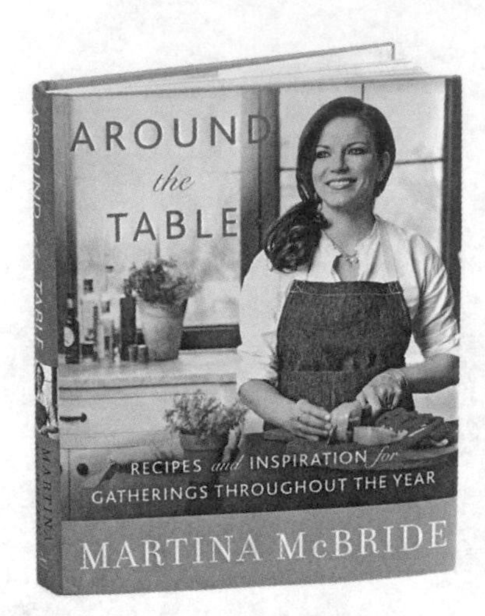

When the *SL* Test Kitchen started to test recipes from Martina McBride's new cookbook, we quickly found out this country star doesn't just write Grammy-winning songs—she also creates cookies that are big hits! *Around the Table* is filled with her most beloved dishes, family stories, and easy party ideas from a woman who knows how to keep a crowd entertained.

TRIPLE CHOCOLATE CRANBERRY OATMEAL COOKIES

The prepared dough keeps well, covered, in the refrigerator. Let it sit at room temperature for about 20 minutes, and then scoop and bake.

- **1 cup all-purpose flour**
- **1/2 tsp. baking soda**
- **1/2 tsp. ground cinnamon**
- **1/4 tsp. kosher salt**
- **10 Tbsp. unsalted butter, softened**
- **1/2 cup granulated sugar**
- **1/2 cup packed light brown sugar**
- **1 large egg**
- **1 tsp. vanilla extract**
- **1 Tbsp. orange zest**
- **1 cup old-fashioned oats**
- **1/2 cup semisweet chocolate chunks**
- **1 cup milk chocolate chips**
- **1 cup white chocolate chips**
- **1/2 cup dried cranberries**

Preheat the oven to 350°. Line 2 (17- x 13-inch) rimmed baking sheets with parchment paper. Whisk together the flour, baking soda, cinnamon, and salt in a medium bowl, and set aside. In a large bowl, beat the butter, granulated sugar, and brown sugar with an electric mixer until smooth and fluffy. Add egg, vanilla, and orange zest, and beat until blended. Add the flour mixture and oats, and stir until blended. Stir in chocolate chunks, 1/2 cup of the milk chocolate chips, 1/2 cup of the white chocolate chips, and the cranberries. Drop batter by rounded Tbsp. onto prepared baking sheets, at least 2 inches apart. Bake for 12 to 14 minutes or until the edges are lightly browned. Cool on baking sheets for 5 minutes, then transfer cookies to a rack, and cool completely. In separate small bowls, microwave remaining 1/2 cup milk chocolate chips and 1/2 cup white chocolate chips at HIGH power for 1 minute, stirring after 30 seconds. Using a small spoon, drizzle the melted chocolate over cookies. Let stand until the chocolate sets, about 1 hour.

MAKES about 2 1/2 dozen. **HANDS-ON** 30 min.; **TOTAL** 2 hours, 5 min.

CHERRY-ROSEMARY MUFFINS

These sweet and savory muffins have a unique and delicious flavor combination.

- **4 cups all-purpose flour**
- **1 tsp. kosher salt**
- **1 1/2 Tbsp. baking powder**
- **1 1/2 cups sugar**
- **2 medium oranges**
- **2 large eggs, beaten**
- **1 1/2 cups milk**
- **1/2 cup butter, melted**
- **1 1/2 cups dried cherries, lightly chopped**
- **2 Tbsp. chopped fresh rosemary**

Preheat oven to 375°. Lightly grease or line 2 (12-cup) muffin tins. In a large bowl, whisk together the flour, salt, baking powder, and sugar. Grate 2 Tbsp. of zest from the oranges, and set zest aside. Then squeeze the oranges to get 1/2 cup juice. In a medium bowl, combine the eggs, orange juice, milk, and butter. Stir the egg mixture into the flour mixture until just combined. Fold in orange zest, cherries, and rosemary. Spoon batter into muffin tins, filling three-fourths of each cup. Bake for 18 to 20 minutes or until a wooden pick inserted into the center comes out clean.

MAKES 2 dozen. **HANDS-ON** 15 min., **TOTAL** 35 min.

April

72 **Fresh from Garden to Plate** A Maryland couple shows how a kitchen garden keeps summer bounty on the menu all season long, for healthy meals that are as local as it gets

78 **The Pitmaster's Guide to Brisket** Master of smoke, Chris Prieto, shares insider tips for making perfect brisket, from achieving smoky, flavorful bark to moist and tender meat

84 HOSPITALITY **Easiest Easter Ever** Showstopping recipes any cook can master

87 THE *SL* TEST KITCHEN ACADEMY **Cooking Perfect Pasta** Tips from the South's most trusted kitchen for perfectly cooked pasta guaranteed to make an Italian grandma proud

88 QUICK-FIX SUPPERS **30-Minute Spring Pastas** For simple, surefire suppers

90 **The Pursuit of Perfect Biscuits** Our Editor in Chief gets schooling in the kitchen

91 *COOKING LIGHT* **Crab Takes the Cake** Crunchy, tender crab cakes are simple to make

92 THE SLOW DOWN **Saucy Chicken Lettuce Cups** East meets South in this hands-off recipe

93 SAVE ROOM **Our Best Coconut Cake** Classic flavor piled high in a layer cake to die for

94 COMMUNITY COOKBOOK **Southern Made Fresh** Cheesemaker and author Tasia Malakasis puts a fresh spin on Southern classics with a nod to her Greek heritage

Fresh from Garden to Plate

HOW ONE MARYLAND COUPLE BUILT A GARDEN THAT FEEDS THEIR FAMILY ALL SUMMER LONG

Homeowners and gardeners Chase (right) and Brien Poffenberger

FEW KITCHEN GARDENS ARE MORE ROMANTIC than those grown in the Colonial Revival style. Brien Poffenberger saw this firsthand while working summers as a college student at Thomas Jefferson's Monticello. He absorbed the classic aesthetic and developed a taste for garden-fresh food.

In 1993, while still in his twenties, Brien bought a 1760s stone structure in Sharpsburg, Maryland, that would become his home. With a preservationist's eye, he rejoiced in the neglect that had left plaster, flooring, and other original details intact. He explains, "It is an adage that poverty is a friend of preservation. I was lucky. Almost nothing had been done to it."

In 1998, Chase Poffenberger became his bride and partner in a painstaking renovation, making the house into a comfortable, 21st-century home while preserving its charms. The first of their two daughters was born in 2000. By 2008, the work on the house was complete, and Brien turned his attention to the backyard.

Back to His Roots

Brien is a local boy. His ancestors arrived in the 1730s and settled in western Maryland less than a mile from his current home. Until recently, the family made their living on the land; Brien is only two generations removed from farming.

"What really would have been in the backyard in the 18th century would have been hog and sheep pens and a horse stable," Brien explains. "What I have built is our idealized version, but I try to be true to the kinds of vegetables and plants that would have grown here originally."

Chase recalls, "When we got married, there was one raised bed. Brien had just started growing herbs. It has really developed as a passion for him. It's his stress relief and his spiritual connection with the earth. There's nothing more satisfying, relaxing, and inspiring than watching things grow."

Writing a New Chapter

The verdant patch behind the Poffenberger home is only 40 feet wide, but it's 120 feet deep. When Brien designed it, he visualized it as three 40-foot squares. The one closest to the porch is the dooryard, where vegetables and herbs thrive. A chicken coop is built into the corner to resemble an outhouse in Colonial Williamsburg. The hens, an old American breed called Dominique, spend nights in the safe enclosure and then scratch about the garden during the day. It didn't take long for Brien to learn how destructive chickens can be, so he decided to put their natural talents to work.

"I was digging in our two raised beds, and so were my chickens. I thought, 'Wait, I can put this to use.' I made two 6-foot wire coops with no bottoms. I move them up and down the garden and put the chickens inside. I rotate them, and they do 90% of the work I used to do," Brien says.

A little white fence creates a division between the dooryard garden and the 80-foot-deep section that ends with two little outbuildings connected by an arbor and bench. This is mainly lawn, but it's surrounded by a narrow bed where almost anything will grow. There's always something new in the garden. "One year it was artichokes; another, it was cotton. Most recently, it was ancient einkorn wheat," Brien says.

As Local As It Gets

When it comes to eating local produce, it's impossible to do better than your backyard. Although Chase and Brien both garden and cook, the practical division of labor has made her the primary cook. "It takes a little bit of preparation, because I have a 90-minute commute each way to work," Chase says. "But I still cook four or five nights a week. It's relaxing to me. When the food is fresh, it's easy to get dinner on the table."

SPINACH-AND-THREE-HERB PESTO

This versatile pesto is great on grilled bread and hard-cooked eggs. Or toss a spoonful with your favorite pasta.

- **1 1/3 cups grated Parmesan cheese**
- **1 cup firmly packed fresh baby spinach**
- **2/3 cup olive oil**
- **1/2 cup firmly packed fresh basil leaves**
- **1/4 cup firmly packed fresh flat-leaf parsley**
- **1/4 cup toasted chopped pecans**
- **3 Tbsp. cold water**
- **1 Tbsp. fresh lemon juice**
- **1 Tbsp. fresh tarragon leaves**
- **2 garlic cloves, chopped**
- **3/4 tsp. kosher salt**

Process all ingredients in a food processor until smooth, stopping to scrape down sides as needed.

MAKES about 1 1/2 cups. **HANDS-ON** 10 min., **TOTAL** 10 min.

CREAMY BASIL-BLACK PEPPER CUCUMBERS

- **2 1/2 lb. cucumbers, peeled and cut into spears**
- **1 1/2 tsp. kosher salt**
- **1/2 cup Greek yogurt**
- **3 Tbsp. extra virgin olive oil**
- **1 tsp. lime zest**
- **2 Tbsp. fresh lime juice**
- **1 tsp. freshly ground black pepper**
- **1/2 cup firmly packed fresh basil leaves, chopped**
- **Garnish: lime peel strips**

Toss together cucumbers and salt in a large bowl, and let stand 5 minutes. Whisk together yogurt and next 4 ingredients; gently stir into cucumber mixture. Cover and chill 1 to 24 hours. Add basil, and toss to combine. Let stand 10 minutes before serving. Season with salt.

MAKES 8 servings. **HANDS-ON** 15 min.; **TOTAL** 1 hour, 30 min.

GARDEN POTATO SALAD

New potatoes are baby spring potatoes with beautifully thin skins. We love how red-skinned potatoes contrast with vibrant green peas and herbs, but you can use any kind of potato you like.

- **3 lb. new potatoes, halved**
- **1 1/2 tsp. kosher salt, divided**
- **4 oz. fresh snow peas or sugar snap peas**
- **3 Tbsp. coarse-grained Dijon mustard**
- **3 Tbsp. fresh lemon juice**
- **1 tsp. sugar**
- **1/4 tsp. freshly ground black pepper**
- **2/3 cup olive oil**
- **1 cup loosely packed fresh herbs (such as basil, chives, mint, and dill), coarsely chopped**

1. Bring potatoes, water to cover, and 1 tsp. salt to a boil in a large Dutch oven over medium-high heat. Reduce heat to medium-low, and cook 10 to 15 minutes or until tender; drain. Cool 30 minutes.

2. Cook snow peas in 2 cups boiling water in a medium saucepan over medium-high heat 1 minute or until crisp-tender; drain, pressing between paper towels. Cut peas into 1/2-inch pieces. Cover with plastic wrap, and chill until ready to use.

3. Whisk together mustard, next 3 ingredients, and remaining 1/2 tsp. salt in a medium bowl; gradually add olive oil in a slow, steady stream, whisking until smooth.

4. Gently toss together potatoes and ½ cup dressing in a large bowl, and let stand 30 minutes. Just before serving, gently stir in peas, herbs, and remaining dressing. Add salt and pepper to taste.

MAKES 8 servings. **HANDS-ON** 15 min.; **TOTAL** 1 hour, 25 min.

MINTY LEMONADE

Transform lemonade in seconds with a handful of fresh mint leaves. Don't chop them—keep leaves whole to avoid murky lemonade (and bits of green in your teeth). To maximize the mint flavor, cup the leaves in the palm of one hand, and clap your hands together a few times to release the herb's aroma.

KALE-AND-BLUEBERRY SLAW WITH BUTTERMILK DRESSING

We like fresh tarragon in this creamy, zippy dressing, but other herbs work just as well.

- 6 **Tbsp. apple cider vinegar**
- 3 **Tbsp. grated onion**
- 1/2 **tsp. Worcestershire sauce**
- 1/4 **tsp. hot sauce (such as Tabasco)**
- 1 **garlic clove, minced**
- 1/2 **Granny Smith apple, grated**
- 1 **cup buttermilk**
- 6 **Tbsp. mayonnaise**
- 6 **Tbsp. sour cream**
- 3 **Tbsp. finely chopped fresh tarragon**
- 1/2 **to 1 tsp. kosher salt**
- 1/4 **to 1/2 tsp. freshly ground black pepper**
- 1/4 **to 1/2 tsp. sugar**
- 6 **radishes, thinly sliced**
- 4 **medium carrots, cut into thin strips**
- 1 **bunch kale, trimmed and thinly sliced**
- 1/2 **small head red cabbage, shredded**
- 1 **cup fresh blueberries**
- 1 **cup fresh raspberries**

1. Stir together first 6 ingredients in a jar with a tight-fitting lid; let stand 5 minutes. Add buttermilk and next 3 ingredients. Cover jar with lid; shake vigorously until blended and smooth. Add salt, pepper, and sugar to taste.
2. Toss together radishes, next 5 ingredients, and ½ cup dressing in a large bowl; let stand 30 minutes. Season with salt and pepper. Serve with remaining dressing.

MAKES 6 servings. **HANDS-ON** 30 min.; **TOTAL** 1 hour, 5 min.

Kale-and-Blueberry Slaw with Buttermilk Dressing

CUCUMBERS WITH GINGER, RICE VINEGAR, AND MINT

- 2 1/2 **lb. cucumbers, chopped**
- 1 1/2 **tsp. kosher salt**
- 1/2 **cup rice vinegar**
- 1 1/2 **tsp. fresh ginger, minced**
- 2 **garlic cloves, minced**
- 1/4 **tsp. dried crushed red pepper**
- 1/4 **cup loosely packed fresh mint leaves, chopped**
- **Freshly ground black pepper**

Toss together cucumbers and salt in a large bowl, and let stand 5 minutes. Whisk together vinegar and next 3 ingredients. Pour over cucumber mixture; cover and chill 1 to 24 hours. Add mint, and toss to combine. Let stand 10 minutes before serving. Season with salt and pepper.

MAKES 8 servings. **HANDS-ON** 15 min.; **TOTAL** 1 hour, 20 min.

WHAT'S GROWING IN THE GARDEN

Chase and Brien Poffenberger enjoy a continuous supply of homegrown produce from spring through fall. Here are our top picks for reliable and easy-to-grow fruits, veggies, and herbs—paired with fresh recipes from the *Southern Living* Test Kitchen, on pages 73-75.

BASIL

▸ **WHEN TO PLANT:** After your last frost in spring

▸ **HOW LONG UNTIL HARVEST:** Start to pick individual leaves as soon as the plant grows.

▸ **WHAT IT NEEDS:** This herb loves the heat. Give it sun, moisture, and rich soil.

BLUEBERRIES

▸ **WHEN TO PLANT:** Fall, winter, or spring

▸ **HOW LONG UNTIL HARVEST:** Plants fruit lightly when young and then heavier as they grow. The main harvest is in June.

▸ **WHAT THEY NEED:** Blueberries require quite acid (pH 4-5-5.5), moist, well-drained soil that contains lots of organic matter. Plant two or more different selections for cross-pollination and heavier crops.

CARROTS

▸ **WHEN TO PLANT:** Early spring or fall. Plant more every two weeks for a longer harvest.

▸ **HOW LONG UNTIL HARVEST:** Sweet, crunchy roots will be ready to pull in four to six weeks.

▸ **WHAT THEY NEED:** Loose, deep soil. If you have heavy clay, choose shorter-growing, rounded kinds.

CUCUMBERS

▸ **WHEN TO PLANT:** After your last frost in spring. Use either seeds or transplants.

▸ **HOW LONG UNTIL HARVEST:** Seven to eight weeks from seed. Transplants are ready sooner.

▸ **WHAT THEY NEED:** Full sun and good air circulation. Vining types require a trellis to climb.

EGGS

▸ **RECOMMENDED BREED:** Brien likes Dominique chickens, an heirloom breed known for having a mild temperament and pretty brown eggs.

▸ **HOW LONG THEY LAY EGGS:** March through November for five to seven years

▸ **WHAT CHICKENS NEED:** A secure coop to protect them from predators. They should be allowed to forage by day and then be sheltered in their coop at night.

HONEY

▸ **WHEN TO START:** Spring, after doing research

▸ **HOW LONG UNTIL HARVEST:** Don't expect to collect very much the first season as the hive grows large enough to support itself and have extra honey. The second season should be productive.

▸ **WHAT BEES NEED:** Start with a proper hive setup and access to water; the bees will do the rest. Be sure to consult with an experienced beekeeper before you begin.

KALE

▸ **WHEN TO PLANT:** Set out transplants in early spring or fall. Kale is a cool-weather green that will not survive when the weather gets warm.

▸ **HOW LONG UNTIL HARVEST:** Pick the outer leaves after a few weeks so the inner ones will continue to develop. For optimal flavor, harvest after a frost.

▸ **WHAT IT NEEDS:** Full sun is best.

NEW POTATOES

▸ **WHEN TO PLANT:** Early spring or fall. Use certified seed potatoes, which are actually small potatoes that you cut into pieces, each one having an eye that will sprout.

▸ **HOW LONG UNTIL HARVEST:** Eight weeks or so after planting

▸ **WHAT THEY NEED:** Full sun and well-drained, moist soil finished with a thick layer of mulch

RADISHES

▸ **WHEN TO PLANT:** Early spring and fall from seed. Plant more every two weeks for nonstop veggies.

▸ **HOW LONG UNTIL HARVEST:** Spring radishes are ready to pull and eat in about three weeks. Summer radishes take about five weeks.

▸ **WHAT THEY NEED:** Sow seeds in a shallow furrow. Pull extra seedlings so the growing plants have about 2 inches between them.

SNOW PEAS

▸ **WHEN TO PLANT:** Early spring or fall from seeds you've soaked in water for several hours

▸ **HOW LONG UNTIL HARVEST:** The tender pods are ready to be picked about two months after planting.

▸ **WHAT THEY NEED:** Plant in a sunny spot with a trellis for the vines to climb.

SPINACH

▸ **WHEN TO PLANT:** Early spring and again in fall. Use either seeds or transplants.

▸ **HOW LONG UNTIL HARVEST:** Seedlings are ready in about five to six weeks. Transplants can be harvested sooner. Snip outer leaves so the inner leaves keep growing.

▸ **WHAT IT NEEDS:** Full sun and moist, well-drained soil. Amend soil with compost before planting.

STRAWBERRIES

▸ **WHEN TO PLANT:** It depends on where you live. In the Lower and Coastal South, late fall is best. In the Upper and Middle South, plant in early spring.

▸ **HOW LONG UNTIL HARVEST:** June-bearing types, the most dependable for the South, flower in spring and fruit in late spring and early summer.

▸ **WHAT THEY NEED:** Full sun and fertile, well-drained soil. Mulch between plants to keep soil moist and berries clean.

BASIL
EGGS
RADISHES
BLUEBERRIES
HONEY
SNOW PEAS
CARROTS
KALE
SPINACH
CUCUMBERS
NEW POTATOES
STRAWBERRIES

The Pitmaster's Guide to Brisket

OUR STEP-BY-STEP PRIMER FOR HOME COOKS

A LOT OF SOUTHERNERS PASSIONATELY INSIST THAT "BARBECUE" IS A NOUN, NOT A VERB, and that it should be used only to refer to pork. Until just recently, you might come across slow-cooked chicken in barbecue restaurants in the Carolinas, Georgia, and Tennessee, but you were about as likely to find barbecued alligator as you were beef. Brisket was an obscure regional specialty found only in exotic places like Texas and Kansas City.

Today, though, brisket is the star of the barbecue scene. At Texas' top joints, diners wait in line for hours to sample the tender smoked beef, and the brisket often sells out by early afternoon. No line is longer than the one at Franklin Barbecue in Austin, where pitmaster Aaron Franklin's famous brisket earned a visit from the President of the United States. Franklin even landed a starring role in a nationally aired credit card commercial, appearing alongside famed Japanese chef Nobu Matsuhisa.

Brisket is now coming into vogue in parts of the South that were previously die-hard pork territory. It's piled atop stainless steel trays at Heirloom Market BBQ in Atlanta. South Carolina pitmasters like Aaron Siegel from Charleston's Home Team BBQ make pilgrimages to the Lone Star State to tour the brisket masters' pits at places such as Kreuz Market in Lockhart, bringing back techniques like wrapping the meat in butcher paper midway through the cook. One noted Austin pitmaster, John Lewis of La Barbecue, is even opening a brisket joint in the heart of Charleston. And Houston native Christopher Prieto (above) has brought his expertise to Prime Barbecue in North Carolina, where he also cooks and teaches classes.

Pitmaster Christopher Prieto

Once you've tasted good brisket, you immediately understand its appeal. When cooked properly and sliced against the grain, it has a texture that's tender but still hefty, brimming with juice and capped with a salty, smoky bark. It's a remarkable example of how low-and-slow cooking can transform tough, undesirable cuts of meat into something tempting and delicious.

Brisket takes naturally to a dry rub, which lets cooks unleash their creativity and cloak it in mystery too—arcane blends of a dozen or more spices, secret formulas inscribed on grease-stained slips of paper and hidden from prying eyes. (I must note that many celebrated Texas cooks, like Franklin and Lewis, use nothing more than salt and pepper, but where's the fun in that?)

More than anything, it's the sheer sensual abandon that draws barbecue lovers to brisket. Those folks down in Austin aren't queuing up for hours for modest forkfuls of chopped pork with coleslaw on the side. They're ravenous for butcher paper-lined platters mounded with thick slices of red meat—crisp bark outside, gleaming with fat, suitable for eating with fingers. It's a scandalous, over-the-top indulgence, and boy, do we love it. —Robert Moss

BRISKET STEP BY STEP

Pitmaster Christopher Prieto considers beef brisket the king of barbecue. This meat comes from the chest of the cow and consists of two different muscles known as the point and flat. The point (the fatty end) has a rib-eye texture. The flat, which runs along the whole brisket, is lean like a sirloin. When you cook a giant hunk of protein, the difference between a doorstop and tender, juicy meat all comes down to technique and time. Here is Christopher's tried-and-true method to give you instant meat cred (full recipe, page 80).

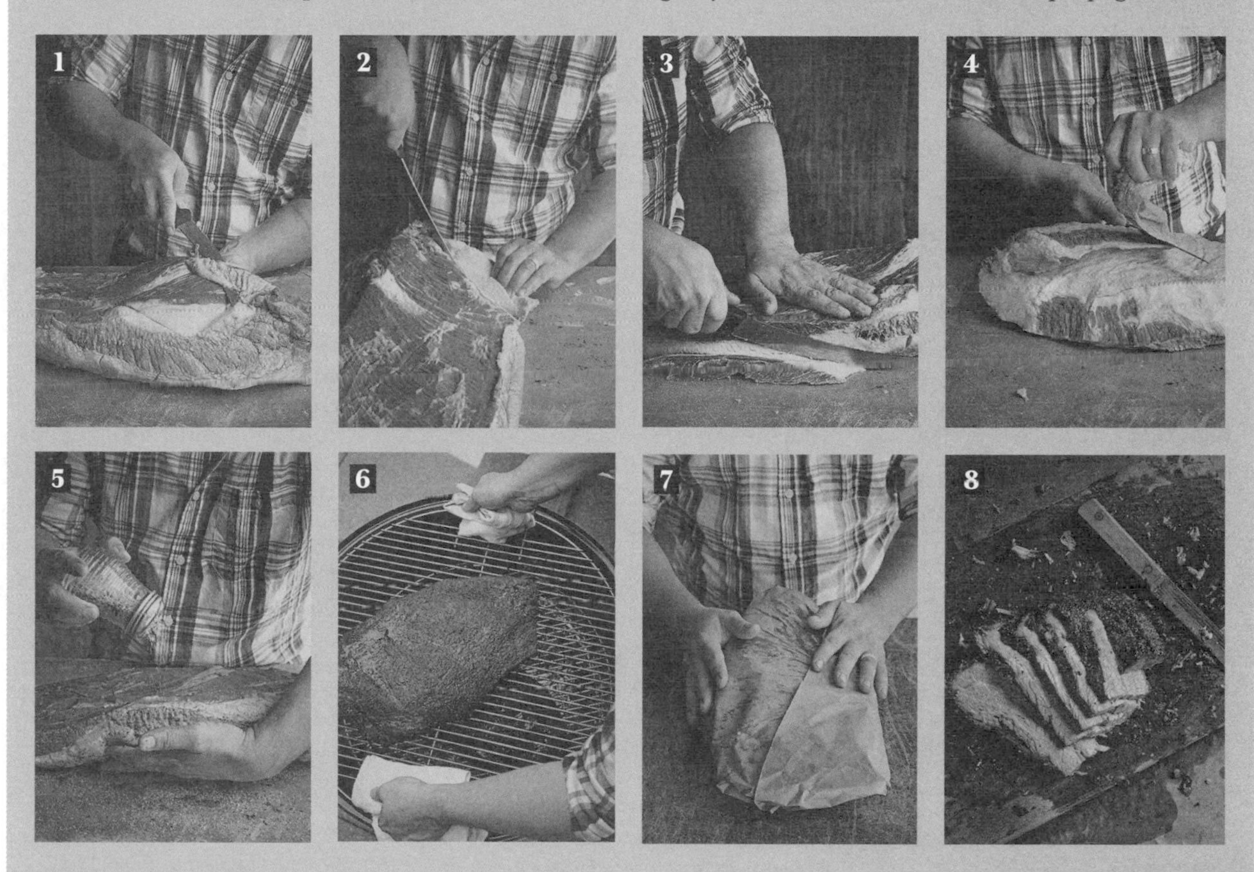

1. TRIM FAT. Remove any large or hard patches of fat from the flat (the meat that runs along the entire brisket) so the rub will adhere to the meat and create a nice bark during cooking.
2. REMOVE HARD FAT. Next, remove the hard fat heel that sits between the point and the flat. This fat should be removed because it will not render during cooking.
3. TRIM AND SHAPE. Continue trimming excess fat from the brisket, as well as any fat that appears discolored. Trim from both sides of the brisket, and square off the edges. This will help ensure an even cook.
4. CHECK THE FAT CAP, AND FINISH TRIMMING. Check the side with the fat cap (the thick layer of white fat, common in pastured animals, that covers one side of the meat) for any hard or large patches of fat. Trim the fat cap to about ¼ inch of fat throughout. The finished brisket should be nice and square, with even fat and marbling. (You can also ask your butcher to trim the brisket for you.)
5. SEASON THE MEAT. Brush or rub the entire brisket with Worcestershire sauce so the seasoning will stick. Once the brisket is moist, apply a heavy coating of rub, and let the brisket sit in the refrigerator for a minimum of 4 hours before placing it on the smoker.
6. SMOKE IT. CHECK THE BARK. Let the brisket sit at room temp 1 hour, then place it, fat side down, on the top food grate, and close the smoker. Smoke 5 hours or until thermometer inserted into the center of the brisket, where the point and flat meet, registers 165°. At this point, the brisket should have a rich, dark mahogany color and will no longer take on any smoke while cooking.
7. REMOVE AND WRAP. Remove the brisket from the smoker, and wrap it tightly in wax-free butcher paper. Return it to the smoker, and cook 3 to 5 more hours, checking temperature hourly until thermometer inserted into the center of the brisket registers 200°.
8. REST AND SLICE. Remove the brisket from the smoker, and allow it to rest (wrapped) for 2 hours, preferably in an empty cooler or insulated food storage container. Unwrap and place brisket on a cutting board, reserving drippings in butcher paper (to mix into your barbecue sauce). Cut the meat across the grain into ¼-inch-thick slices.

5 SECRETS TO BETTER BRISKET

START WITH A FRESH CUT
▪ A frozen brisket will make for tough results every time.

USE A BULLET SMOKER
▪ This basic, old-school smoker is ideal for this recipe. Kamado-style smokers have less distance between the heat source and the meat, so the brisket tends to cook too quickly.

USE CHARCOAL BRIQUETTES
▪ Charcoal briquettes are essential. Christopher prefers high-heat briquettes, such as the Kingsford Competition Briquets.

BUILD YOUR FIRE WITH THE MINION METHOD
▪ Place chimney starter in the center of the charcoal grate, and fill with briquettes. Add briquettes to the grate around the chimney. Start the fire in the chimney. When coals in chimney are white hot, pour them into the center of the grate. You'll get a slow burn outward that's perfect for smoking.

DON'T OVERSMOKE
▪ Wait to put your meat on the smoker until after the intense early smoke clears. By waiting, you will avoid oversmoking and camouflaging the other flavors.

TRUE SMOKED BEEF BRISKET

- **1 (12- to 14-lb.) beef brisket, trimmed**
- **½ cup Worcestershire sauce**
- **1 cup Beef Rub**
- **3-4 pecan, hickory, or oak wood chunks**
- **Butcher paper**
- **El Sancho Barbecue Sauce**

1. Brush or rub brisket with Worcestershire sauce. Sprinkle Beef Rub over brisket. Chill 1 to 4 hours.
2. Let brisket stand 1 hour. Meanwhile, prepare charcoal fire in smoker according to manufacturer's instructions. Place wood chunks on coals. Maintain internal temperature at 250° to 260° for 15 to 20 minutes.
3. Place brisket, fat side down, on top food grate. Smoke brisket, covered with smoker lid, 5 hours or until a thermometer inserted in center of brisket (where the point and flat meet) registers 165°. Remove brisket, and wrap tightly in butcher paper. Return brisket to smoker, and smoke 3 to 5 more hours or until thermometer inserted in brisket registers 200°. (Check temperature each hour.)
4. Remove brisket from smoker; open butcher paper, and let stand 2 to 4 minutes or until no longer steaming. Let brisket stand, loosely covered with butcher paper, 2 hours.
5. Meanwhile, prepare El Sancho Barbecue Sauce. Cut brisket across the grain into ¼-inch-thick slices. Serve with El Sancho Barbecue Sauce.

MAKES 10 to 15 servings. **HANDS-ON** 20 min.; **TOTAL** 12 hours, 40 min., including rub and sauce

Beef Rub

- **1 cup kosher salt**
- **1 cup coarsely ground black pepper**
- **3 Tbsp. granulated garlic**
- **2 tsp. ground red pepper**

Stir together all ingredients. Store at room temperature in an airtight container up to 1 month.

MAKES 2¾ cups. **HANDS-ON** 5 min., **TOTAL** 5 min.

El Sancho Barbecue Sauce

Turn up the Texas on this sweet and spicy barbecue sauce by stirring in ¼ cup of reserved brisket drippings just before serving.

- **1 cup apple cider vinegar**
- **½ cup ketchup**
- **⅓ cup tomato paste**
- **¼ cup yellow mustard**
- **¼ cup Worcestershire sauce**
- **1 Tbsp. hot sauce (such as Tabasco)**
- **1 Tbsp. onion powder**
- **1 Tbsp. granulated garlic**
- **1 Tbsp. kosher salt**
- **2 tsp. hickory liquid smoke**
- **1 tsp. coarsely ground black pepper**
- **1 cup sugar**
- **½ cup honey**

Stir together first 11 ingredients in a medium saucepan. Stir in sugar and honey, and bring to a boil over medium heat. Reduce heat to medium-low; simmer, stirring occasionally, 30 minutes. Use immediately, or cool completely, and refrigerate in an airtight container up to 1 week.

MAKES 2½ cups. **HANDS-ON** 15 min., **TOTAL** 45 min.

SMOKED BEEF BRISKET TOSTADAS

Top these tostadas with fresh avocado, lime juice, crumbled queso fresco, diced jalapeños, chopped tomatoes, and shredded lettuce.

- **½ cup salsa**
- **1 Tbsp. chopped fresh cilantro**
- **1 (16-oz.) can fat-free refried beans**
- **6 corn tostada shells**
- **1 lb. chopped True Smoked Beef Brisket (without sauce), warmed**
- **¼ cup chopped red onion**

1. Preheat oven to 400°. Stir together salsa and cilantro. Spread beans on tostada shells. Place shells in a jelly-roll pan; top with brisket, onion, and salsa mixture.
2. Bake at 400° for 8 to 10 minutes or until thoroughly heated. Serve immediately.

MAKES 4 to 6 servings. **HANDS-ON** 10 min.; **TOTAL** 20 min., not including brisket

COWBOY BRISKET SANDWICH

- **8 thick bread slices**
- **4 (¾-oz.) pepper Jack cheese slices**
- **2 cups chopped True Smoked Beef Brisket**
- **Pickled jalapeño pepper slices**

Preheat oven to 400°. Arrange bread slices in a single layer on a baking sheet. Top 4 bread slices with pepper Jack cheese slices. Bake 5 minutes or until bread is toasted and cheese is melted. Top cheese-covered bread slices with brisket (about ½ cup per slice), jalapeño pepper slices, and remaining 4 bread slices. Serve immediately.

MAKES 4 servings. **HANDS-ON** 5 min.; **TOTAL** 10 min., not including brisket

Smoked Beef Brisket Tostadas

WATERMELON, ARUGULA & PECAN SALAD

Arugula is a peppery, tender green that pairs perfectly with juicy summer watermelon, sharp Gorgonzola cheese, and a spicy-sweet vinaigrette. Baby spinach and mâche are great alternatives to arugula.

- **½ baby watermelon, cut into thin wedge slices**
- **½ (6-oz.) package arugula**
- **Pepper Jelly Vinaigrette**
- **1 cup crumbled Gorgonzola cheese**
- **¾ cup chopped pecans, toasted**

Combine watermelon and arugula in a large bowl; add vinaigrette, tossing gently to coat. Transfer watermelon mixture to a serving platter, and sprinkle evenly with cheese and pecans.

MAKES 6 to 8 servings. **TOTAL** 30 min., including Pepper Jelly Vinaigrette

Pepper Jelly Vinaigrette

Pepper jelly is a perfect condiment served with cream cheese or roasted meat. Here it is the secret ingredient in this tangy-hot dressing. Red and green pepper jellies are equally delicious.

- **¼ cup rice wine vinegar**
- **¼ cup pepper jelly**
- **1 Tbsp. fresh lime juice**
- **1 Tbsp. grated onion**
- **1 tsp. table salt**
- **¼ tsp. freshly ground black pepper**
- **¼ cup vegetable oil**

Whisk together first 6 ingredients. Gradually add oil in a slow, steady stream, whisking until blended. Store in an airtight container in refrigerator for up to 1 week.

MAKES ¾ cup. **TOTAL** 5 min.

Grilled Peach & Avocado Salad

GRILLED PEACH & AVOCADO SALAD

This recipe calls for firm avocados so that they can stand up to the grill. When choosing avocados at the store, look for those that are almost ripe. Check under the "button" on the stem end. If it's green, it's the one you want; if it's brown, it's overripe.

- **4 large peaches, peeled, divided**
- **7 Tbsp. canola oil, divided**
- **2 Tbsp. Champagne vinegar or white wine vinegar**
- **½ tsp. honey**
- **¼ tsp. kosher salt**
- **⅛ tsp. freshly ground black pepper**
- **2 firm avocados, peeled and quartered**
- **6 cups loosely packed arugula**
- **½ cup freshly grated Manchego or Parmesan cheese**

1. Preheat grill to 350° to 400° (medium-high) heat.
2. Chop 1 peach, and place in a blender; process peach, 6 Tbsp. canola oil, vinegar, and honey until smooth. Add kosher salt and freshly ground pepper.
3. Halve remaining 3 peaches. Gently toss peaches and avocados in remaining 1 Tbsp. canola oil; add salt and pepper to taste.
4. Grill peach halves and avocado quarters, covered with grill lid, 2 minutes on each side or until charred. Slice and serve over arugula. Top with peach vinaigrette and cheese.

MAKES 6 servings. **TOTAL** 20 min.

MEMPHIS SLAW

The cabbage is always coarsely chopped in Memphis slaw, and the flavors are big and bold—it has to stand up to the standout pulled pork and ribs it's typically served with.

- **1 Tbsp. firmly packed light brown sugar**
- **2 tsp. kosher salt**
- **1 tsp. paprika**
- **½ tsp. dry mustard**
- **½ tsp. dried oregano**
- **½ tsp. freshly ground black pepper**
- **¼ tsp. granulated garlic**
- **¼ tsp. ground coriander**
- **¼ tsp. onion powder**
- **½ cup mayonnaise**
- **¼ cup apple cider vinegar**
- **½ head cabbage (about 1 lb.)**
- **1 cup diced green bell pepper**
- **1 cup diced red onion**

1. Whisk together brown sugar and next 8 ingredients in a bowl. Whisk in mayonnaise and vinegar until sugar dissolves.
2. Cut cabbage into thick slices; cut slices crosswise. Fold cabbage, bell pepper, and red onion into mayonnaise mixture until coated.
3. Let stand 1 hour before serving, tossing occasionally.

MAKES 8 cups. **TOTAL** 1 hour, 15 min.

BARBECUE DEVILED EGGS

Take this traditional picnic favorite to a whole new level with smoked pork butt.

- **1 dozen hard-cooked eggs, peeled**
- **¼ cup mayonnaise**
- **⅓ cup finely chopped smoked pork butt**
- **1 Tbsp. Dijon mustard**
- **¼ tsp. table salt**
- **½ tsp. freshly ground black pepper**
- **⅛ tsp. hot sauce**
- **Paprika (optional)**

1. Slice eggs in half lengthwise, and carefully remove yolks, keeping egg whites intact.
2. Mash together yolks and mayonnaise, and stir in pork and next 4 ingredients; blend well.
3. Spoon yolk mixture evenly into egg white halves, and let chill for 1 hour before serving. Garnish with paprika, if desired.

MAKES 2 dozen. **TOTAL** 1 hour, 30 min.

SWEET POTATO CHIPS

A candy thermometer is an essential tool for this recipe to maintain the oil at the correct temperature for frying. Cooking the potatoes in small batches also helps to fry them evenly.

- **2 sweet potatoes, peeled (about 2 lb.)**
- **Peanut oil**

1. Cut sweet potatoes into 1/16-inch-thick slices, using a mandoline.
2. Pour peanut oil to depth of 3 inches into a Dutch oven; heat over medium-high heat to 300°.
3. Fry potato slices, in small batches, stirring often, 4 to 4½ minutes or until crisp.
4. Drain on a wire rack over paper towels.
5. Immediately sprinkle with desired amount of kosher salt. Cool completely, and store in an airtight container at room temperature up to 2 days.

MAKES 6 to 8 servings. **TOTAL** 40 min.

CLASSIC BAKED MACARONI & CHEESE

Everyone will love this extra-cheesy mac and cheese. You can substitute cavatappi or corkscrew pasta for the elbow macaroni if you want to switch up the classic look.

- **1 (8-oz.) package elbow macaroni**
- **2 Tbsp. butter**
- **2 Tbsp. all-purpose flour**
- **2 cups milk**
- **½ tsp. table salt**
- **½ tsp. freshly ground black pepper**
- **¼ tsp. ground red pepper**
- **1 (8-oz.) block sharp Cheddar cheese, shredded and divided**

1. Preheat oven to 400°. Prepare pasta according to package directions. Keep warm.
2. Melt butter in a large saucepan or Dutch oven over medium-low heat; whisk in flour until smooth. Cook, whisking constantly, 2 minutes. Gradually whisk in milk, and cook, whisking constantly, 5 minutes or until thickened. Remove from heat. Stir in salt, black and red pepper, 1 cup shredded cheese, and cooked pasta.
3. Spoon pasta mixture into a lightly greased 2-qt. baking dish; top with remaining 1 cup cheese.
4. Bake at 400° for 20 minutes or until bubbly. Let stand 10 minutes before serving.

MAKES 8 servings. **TOTAL** 1 hour, 15 min.

Classic Baked Macaroni & Cheese

Easiest Easter Ever

FRESH, CELEBRATORY RECIPES THAT ANY COOK CAN CONQUER

ROASTED LEG OF LAMB WITH LEMON-HERB SALT

Leg of lamb is an easy yet holiday-worthy roast. Buzz together the flavorful salt seasoning in a food processor, and rub it onto the meat at least 12 hours before roasting. Trust an instant-read meat thermometer to ensure the lamb is the perfect rosy pink in the center.

- **⅓ cup loosely packed fresh rosemary leaves**
- **⅓ cup loosely packed fresh oregano leaves**
- **3 tsp. lemon zest**
- **¼ cup kosher salt**
- **¼ tsp. freshly ground black pepper**
- **1 (5- to 7-lb.) bone-in leg of lamb, trimmed**
- **2 Tbsp. olive oil**

1. Process first 5 ingredients in a food processor 2 to 3 minutes or until blended. Rub mixture over lamb, and place on a rack in a roasting pan. Chill 12 to 24 hours.

2. Let lamb stand at room temperature 30 minutes. Meanwhile, preheat oven to 450°. Brush lamb with olive oil.

3. Bake at 450° for 45 minutes; reduce oven temperature to 350°, and bake 1 more hour or until a meat thermometer inserted into thickest portion registers 145°. Let lamb stand 20 minutes before serving.

MAKES 8 servings. **HANDS-ON** 15 min.; **TOTAL** 14 hours, 50 min.

ROASTED BEETS, CARROTS, AND SWEET ONIONS

Springtime is the right time for finding sweet baby beets and carrots in the market. Look for small, firm beets and slender carrots with their bright, feathery greens still attached—a sign of freshness. To make this dish even prettier, mix golden or red-and-white striped beets (sometimes called Chioggia, Candy Stripe, or Bull's-eye) with crimson beets. Look for bunches of carrots in mixed colors as well.

- **1 lb. cipollini onions, peeled**
- **3 garlic cloves**
- **2 bay leaves**
- **3 Tbsp. olive oil, divided**
- **2 Tbsp. white balsamic vinegar, divided**
- **3 tsp. kosher salt, divided**
- **1½ tsp. cracked black pepper, divided**
- **8 baby beets with tops, peeled and halved**
- **1 Tbsp. honey**
- **2 fresh rosemary sprigs**
- **1 orange, halved**
- **1 lb. small carrots with tops**
- **4 fresh thyme sprigs**

1. Preheat oven to 500°. Toss together first 3 ingredients, 1 Tbsp. olive oil, 1 Tbsp. vinegar, 1 tsp. salt, and ½ tsp. black pepper in a medium bowl.
2. Toss together beets, next 2 ingredients, 1 Tbsp. olive oil, 1 tsp. salt, ½ tsp. black pepper, and remaining 1 Tbsp. vinegar in a large bowl. Place beets in a single layer in center of a large piece of heavy-duty aluminum foil. Bring up sides of foil over beets; double fold top and side edges to seal, making a packet. Place packet in 1 end of a jelly-roll pan.
3. Bake beets at 500° for 15 minutes.
4. Meanwhile, cut 1 orange half into 6 wedges. Squeeze juice from remaining orange half to equal 1½ Tbsp. Cut tops from carrots, leaving 1 inch of greenery on each. Toss together carrots, orange juice, orange wedges, thyme sprigs, and remaining olive oil, salt, and pepper.
5. Remove pan from oven, and place onion mixture and carrot mixture in pan next to foil packet, spreading in a single layer. Return to oven, and bake beets, onions, and carrots 25 minutes. Cool 10 minutes; combine vegetables, and serve immediately.

MAKES 8 servings. **HANDS-ON** 25 min.; **TOTAL** 1 hour, 15 min.

SHAVED VEGETABLE SALAD WITH TOASTED ALMONDS

All it takes is a sharp vegetable peeler to transform a few favorite veggies into an inventive salad. To shave the asparagus into ribbons, place each spear flat on your work surface so you can pull the peeler evenly down its length.

- **2 tsp. lemon zest**
- **2 Tbsp. fresh lemon juice**
- **2 Tbsp. white wine vinegar**
- **1 Tbsp. minced shallot**
- **½ tsp. kosher salt**
- **¼ tsp. black pepper**
- **½ cup olive oil**
- **1 lb. large fresh asparagus**
- **2 medium-size yellow squash**
- **½ cup thinly sliced radishes**
- **½ cup almond slivers, toasted**
- **½ cup crumbled feta cheese**

1. Whisk together first 6 ingredients in small bowl. Add oil in a slow, steady stream, whisking constantly, until smooth.
2. Snap off and discard tough ends of asparagus. Cut squash and asparagus into thin, ribbon-like strips, using a vegetable peeler; place in a large bowl. Add radishes, vinaigrette, ¼ cup almonds, and ¼ cup feta; toss to combine. Season with salt and pepper; top with remaining ¼ cup each almonds and feta. Serve immediately.

MAKES 8 servings. **HANDS-ON** 30 min., **TOTAL** 30 min.

CHEESE GRITS SOUFFLÉ WITH MUSHROOM GRAVY

Warm, sherry-laced mushroom gravy gives this simple grits soufflé an extra layer of depth. Gruyère cheese melts perfectly into the hot grits and adds subtle nutty flavor.

- **2 Tbsp. butter**
- **2 tsp. minced garlic**
- **1 tsp. table salt**
- **½ tsp. freshly ground black pepper**
- **1½ cups uncooked stone-ground grits**
- **2 cups (8 oz.) shredded Gruyère cheese**
- **6 large eggs, separated**
- **¼ tsp. cream of tartar**
- **Vegetable cooking spray**
- **Mushroom Gravy (page 86)**

1. Preheat oven to 400°. Melt butter in a large saucepan over medium heat; add garlic, and sauté 1 minute. Stir in salt, pepper, and 6 cups water. Increase heat to medium-high, and bring to a boil. Gradually whisk in grits. Reduce heat to low, and simmer, whisking often, 15 minutes or until thickened. Remove from heat; add cheese, and whisk until melted.
2. Whisk egg yolks until thick and pale. Gradually whisk about 1 cup of hot grits mixture into yolks; whisk yolk mixture into remaining hot grits mixture.
3. Beat egg whites and cream of tartar in a large bowl at high speed with an electric mixer until stiff peaks form. Fold one-fourth of egg whites into grits mixture. Fold grits mixture into remaining egg whites, and pour into a greased (with cooking spray) and floured 3-qt. baking dish.
4. Bake at 400° for 40 to 45 minutes or until puffed and golden brown. Serve immediately with Mushroom Gravy.

MAKES 6 to 8 servings. **HANDS-ON** 40 min.; **TOTAL** 1 hour, 55 min.

Mushroom Gravy

- ½ cup chopped shallots
- 1 tsp. minced garlic
- 1 Tbsp. olive oil
- 1 (8-oz.) package sliced baby portobello mushrooms
- ½ tsp. kosher salt
- ¼ cup dry sherry
- 3 cups vegetable broth
- 2 Tbsp. all-purpose flour
- 2 Tbsp. butter
- 1 tsp. fresh thyme leaves
- ¼ tsp. freshly ground black pepper

1. Sauté shallots and garlic in hot oil in a medium nonstick skillet over medium-high heat 1 minute. Add mushrooms and ¼ tsp. salt; sauté 5 to 6 minutes or until tender. Add sherry; cook, stirring occasionally, 30 to 45 seconds or until liquid is reduced to about 2 Tbsp. Stir in broth and remaining ¼ tsp. salt. Bring mixture to a boil, and boil about 10 minutes or until mixture is reduced to about 1½ cups.
2. Whisk together flour and 2 Tbsp. water in a small bowl until smooth. Whisk flour mixture into mushroom mixture. Reduce heat to medium, and simmer, whisking occasionally, 2 minutes or until slightly thickened. Remove from heat; stir in butter, 1 Tbsp. at a time, until butter melts. Stir in thyme and pepper. Serve immediately, or cool completely, and refrigerate up to 2 days. Heat in a saucepan over low heat.

MAKES 2½ cups. **HANDS-ON** 35 min., **TOTAL** 35 min.

COCONUT-AND-PECAN STRAWBERRY SHORTCAKES

For a creative, updated take on shortcake, we added coconut and pecans to the dough.

SHORTCAKES

- 4 cups all-purpose flour
- ½ cup sugar
- 5 tsp. baking powder
- 1½ tsp. table salt
- ¾ cup butter, cubed
- 1½ cups sweetened flaked coconut, toasted
- 4 large eggs
- 2 tsp. vanilla extract
- ¼ tsp. coconut extract
- 1 cup plus 1 Tbsp. heavy cream, divided
- 1 cup finely chopped toasted pecans
- Parchment paper
- 8 pecan halves
- Sweetened flaked coconut

STRAWBERRIES

- 1 qt. strawberries, halved
- 1 Tbsp. fresh lime juice
- 3 Tbsp. sugar

ADDITIONAL INGREDIENT

- Sweet Mascarpone Cream

1. Prepare Shortcakes: Preheat oven to 400°. Whisk together flour and next 3 ingredients in a large bowl. Cut butter into flour mixture with a pastry blender until mixture resembles small peas.
2. Process toasted coconut in a food processor until finely chopped. Whisk together eggs, extracts, and 1 cup cream. Add egg mixture, chopped pecans, and coconut to flour mixture, and stir with a wooden spoon just until blended. Knead dough in bowl about 10 times, using lightly floured hands. (Dough will be soft and moist.)
3. Turn dough out on a floured surface, and gently pat to a 1-inch thickness. Cut into 8 (2½- to 3-inch) squares, reshaping dough once. Place shortcakes on a parchment paper-lined baking sheet, and chill 15 minutes. Press 1 pecan half into each shortcake; brush with remaining 1 Tbsp. cream. Sprinkle with sweetened flaked coconut.
4. Bake at 400° for 18 minutes or until golden brown. Cool on pan on a wire rack 20 minutes.
5. Meanwhile, prepare Strawberries: Stir together strawberries, lime juice, and 3 Tbsp. sugar.
6. Assemble: Split shortcakes in half horizontally. Place each shortcake bottom on a dessert plate, and top with about ⅓ cup strawberry mixture, desired amount of Sweet Mascarpone Cream, and shortcake tops.

MAKES 8 servings. **HANDS-ON** 40 min.; **TOTAL** 1 hour, 35 min., including cream

Sweet Mascarpone Cream

- 1 (8-oz.) container mascarpone cheese, softened
- ½ cup powdered sugar
- 1 tsp. vanilla extract
- ½ tsp. coconut extract
- 1 cup heavy cream

1. Gently stir together mascarpone cheese, powdered sugar, vanilla extract, and coconut extract in a large bowl.
2. Beat cream in a medium bowl at medium speed with an electric mixer until stiff peaks form. Gently fold whipped cream into mascarpone mixture. Use immediately.

MAKES 4 cups. **HANDS-ON** 10 min., **TOTAL** 10 min.

SPARKLING ELDERFLOWER LEMONADE

Elderflower liqueur is sweet, delicate, and floral, like nectar from a spring bouquet.

Microwave ½ cup **sugar** and 1 cup water in a microwave-safe bowl at HIGH 2 minutes. Stir until sugar dissolves. Cool completely (15 minutes). Stir together sugar syrup, ½ cup **elderflower liqueur** (such as St-Germain), ⅓ cup **fresh lemon juice,** and 1½ cups ice cubes. Let stand 20 minutes. Stir in 3 cups chilled **sparkling wine,** and serve immediately.

MAKES about 5 cups. **HANDS-ON** 10 min., **TOTAL** 45 min.

THE *SL* TEST KITCHEN ACADEMY

ROBBY MELVIN, DIRECTOR OF THE SOUTH'S MOST TRUSTED KITCHEN, SHARES TIPS FOR

COOKING PERFECT PASTA

Cooking pasta seems simple, but working in restaurants taught me that it's not the foolproof boil and strain that I thought. Follow these tips to get trattoria-style results at home.

BITE-SIZE
Pick a short shape like penne or rigatoni when you have other bite-size ingredients, such as chopped chicken or vegetables.

SHELLS
These sturdy little cups trap creamy sauces and bits of big flavor like chopped bacon or fresh herbs.

FESTIVE
The twists and turns of shapes like bow ties (farfalle) make dishes feel more fun. Great for kids or entertaining.

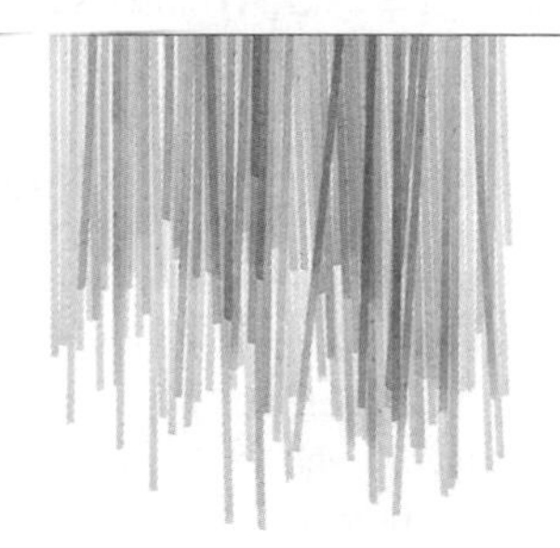

THIN STRANDS
Delicate strands like angel hair are best for thin sauces, such as olive oil or melted butter.

RIDGED
Ridges on pasta are for grabbing smooth sauces, such as Alfredo or classic marinara.

THICK STRANDS
Use noodles like fettuccine or linguine to stand up to hearty sauces like Bolognese.

1 BOIL BIG Pasta will cook evenly if it tumbles freely in boiling water. Bring your water to a rip-roaring boil before adding the pasta. We recommend 4 qt. of water for up to 16 oz. of pasta.

2 SEASON THE WATER The pro's simple secret to restaurant-quality pasta? Salty cooking water! When we say salty, we mean it; the water should taste like the sea. Our guideline is ¼ cup kosher salt for 4 qt. water.

3 STIR OFTEN Perfect pasta is not a hands-free endeavor. Stir the pot every minute to ensure even cooking and to prevent gummy, stuck-together noodles.

30-Minute Spring Pastas

PICK YOUR PASTA, ADD A FEW SIMPLE INGREDIENTS, AND ENJOY FOUR SUREFIRE SUPPERS

CHICKEN-AND-ARTICHOKE PENNE

A handheld Microplane grater makes it easy to add fresh Parmesan flavor.

- **¼ cup plus 1¼ tsp. kosher salt, divided**
- **½ lb. penne pasta**
- **1 lb. chicken breast tenders**
- **1¼ tsp. black pepper, divided**
- **3 Tbsp. olive oil**
- **1 pt. cherry tomatoes**
- **1 (9-oz.) package frozen artichoke hearts, thawed**
- **1 (6-oz.) package fresh baby spinach**
- **1 cup freshly grated Parmesan cheese**

1. Bring 4 qt. water to a boil in a stockpot over high heat. Add ¼ cup salt; return to a boil. Add pasta; boil, stirring occasionally, 11 minutes or until al dente. Drain, reserving 1 cup pasta water.
2. Sprinkle chicken with 1 tsp. each salt and pepper. Cook chicken in hot oil in a large skillet over medium-high heat 4 to 5 minutes on each side or until golden. Remove chicken from skillet. Add tomatoes to skillet; cook, stirring often, 4 minutes. Stir in artichokes, ¼ tsp. each salt and pepper, and ½ cup reserved pasta water. Reduce heat to medium, and cook, stirring often, 3 minutes or until tomatoes burst.
3. Stir pasta, spinach, and ¾ cup cheese into tomato mixture. Add up to ½ cup remaining pasta water, stirring constantly, until thin sauce forms. Chop chicken; toss with pasta mixture. Top with remaining cheese.

MAKES 6 to 8 servings. **HANDS-ON** 30 min., **TOTAL** 30 min.

SHRIMP AND PEAS WITH FARFALLE

Lemony breadcrumbs add a delightful crunch to this dish.

- **¼ cup kosher salt**
- **½ lb. farfalle pasta**
- **1 lb. fresh English peas, shelled, or 1 1/2 cups frozen peas, thawed**
- **6 Tbsp. butter, divided**
- **½ cup panko (Japanese breadcrumbs)**
- **2 tsp. lemon zest, divided**
- **1 lb. peeled, medium-size raw shrimp, deveined**
- **1 Tbsp. fresh lemon juice**
- **2 Tbsp. finely chopped fresh flat-leaf parsley**
- **¼ tsp. black pepper**

1. Bring 4 qt. water to a boil in a stockpot over high heat. Add ¼ cup salt; return to a boil. Add pasta; boil, stirring occasionally, 11 minutes. Add peas, and cook 3 minutes or until peas are tender. Drain, reserving 1 cup pasta water.
2. Melt 2 Tbsp. butter in a medium skillet over medium heat; add breadcrumbs, and cook, stirring often, 4 minutes or until brown. Add 1 tsp. zest, and cook, stirring constantly, 2 minutes. Remove from heat.
3. Sauté shrimp in remaining butter in stockpot over medium-high heat 4 minutes or just until shrimp turn pink. Stir in lemon juice and ½ cup reserved pasta water. Stir in pasta mixture. Stir in up to ½ cup pasta water until desired consistency is reached. Transfer mixture to a platter. Top with panko, parsley, black pepper, and remaining lemon zest.

MAKES 6 servings. **HANDS-ON** 30 min., **TOTAL** 30 min.

BUCATINI, HAM, AND ASPARAGUS

Bucatini is long, thin pasta that's hollow in the center, like little garden hoses. Spaghetti and linguine are good substitutes.

- **½ lb. fresh asparagus**
- **¼ cup kosher salt**
- **½ lb. bucatini pasta**
- **¼ cup butter**
- **2 Tbsp. olive oil**
- **1 (8-oz.) package cremini mushrooms, sliced**
- **1 cup diced cooked ham**
- **1 (4-oz.) goat cheese log, crumbled and divided**
- **½ tsp. black pepper, divided**
- **2 Tbsp. chopped fresh chives**

1. Cut asparagus in half lengthwise, discarding tough ends. Bring 4 qt. water to a boil in a stockpot over high heat. Add ¼ cup salt; return to a boil. Add pasta; boil, stirring occasionally, 11 minutes. Stir in asparagus, and cook 2 minutes or until crisp-tender. Drain, reserving 1 cup pasta water.
2. Melt butter with olive oil in large skillet over medium-high heat. Add mushrooms, and sauté 6 minutes. Add ham, and cook, stirring occasionally, 2 minutes or until browned. Stir in ½ cup reserved pasta water, half of goat cheese, and ¼ tsp. pepper.
3. Stir pasta mixture into mushroom mixture. Stir in up to ½ cup additional pasta water until desired consistency is reached. Remove from heat. Top with remaining goat cheese and pepper; sprinkle with chives.

MAKES 6 servings. **HANDS-ON** 30 min., **TOTAL** 30 min.

PASTA SHELLS WITH SPRING VEGETABLES

Pea tendrils, the tender tips of pea vines that taste just like the peas, make an unexpected and charming garnish.

- **½ lb. fresh asparagus**
- **1 (8-oz.) package sugar snap peas**
- **1 cup ricotta cheese**
- **2 tsp. lemon zest, divided**
- **4 tsp. fresh lemon juice**
- **¼ cup plus ¼ tsp. kosher salt, divided**
- **½ lb. large pasta shells**
- **1 Tbsp. olive oil**
- **4 radishes, thinly sliced**
- **1 Tbsp. chopped fresh mint**
- **Garnish: pea tendrils**

1. Cut asparagus into 1-inch pieces, discarding tough ends. Cut snap peas in half.
2. Stir together ricotta, 1 tsp. zest, lemon juice, and ¼ tsp. salt in a medium bowl.
3. Bring 4 qt. water to a boil in a stockpot over high heat. Add ¼ cup salt; return to a boil. Add pasta; boil, stirring occasionally, 11 minutes or until al dente. Add asparagus and sugar snap peas, and cook 2 to 3 minutes or until tender. Drain pasta mixture, reserving ¼ cup pasta water. Return pasta to pot. Stir in olive oil, radishes, and reserved pasta water.
4. Spread half of ricotta mixture in serving dish; top with pasta mixture. Dollop with remaining ricotta mixture. Sprinkle with mint and remaining lemon zest. Add pepper to taste. Serve immediately.

MAKES 6 servings. **HANDS-ON** 30 min., **TOTAL** 30 min.

The Pursuit of Perfect Biscuits

WHY EVERY *SOUTHERN LIVING* RECIPE IS A LABOR OF LOVE

IT WAS A COLD SATURDAY MORNING in January, and the biscuit mix seemed to be calling to me from a shelf in the pantry. A Christmas gift, it came in an elegant little cloth bag with a card that promised "perfect buttermilk biscuits." I proudly told the kids that Daddy was making homemade biscuits, rolled up my sleeves, and got to work. Half an hour later, I opened the oven and pulled out...hockey pucks. They were hard, crumbly, and not even cooked in the center. "Maybe you ought to stick to pancakes," said my son, Austin.

Dejected, I knew right away what my mistake had been. I hadn't used the *Southern Living* recipe. A few months earlier, *SL* Test Kitchen Director Robby Melvin had led an effort to produce the ultimate no-fail buttermilk biscuit recipe. I went right to the source and asked for a tutorial. (This job has its perks.) I wanted to be the heroic dad who could make great biscuits, and failing twice was not an option.

We met in the kitchen early one morning, and the lesson took about an hour. Robby walked me through the essential steps, explaining not only his technique but also how he got there. He was careful, exacting, and patient, and I could see this was a labor of love. He and the Test Kitchen staff had tried dozens of versions, experimenting with ingredients, temperatures, and cooking times. They discovered that grating a stick of frozen—not just cold—butter distributed it throughout the flour perfectly. They recommend stirring the buttermilk into the flour exactly 15 times. ("You don't want to overwork the dough," explains Robby.) And the lightest, fluffiest biscuits they baked were made from White Lily Self-Rising Flour. That's not a commercial, just a fact.

Type "biscuit recipes" into Google and you will get about 9,550,000 results. "Buttermilk biscuits" gets you about 1,550,000. That's a lot of biscuit recipes, and the words "ultimate," "best," "authentic," and "perfect" are scattered all over them. Now, I'm not saying that *Southern Living* has the only buttermilk biscuit recipe you should ever try. But I bet you can't find a better one on Google. And I can tell you that when I made this recipe for my kids on a recent Saturday morning, they couldn't believe the light, flaky, buttery beauties that the Hockey Puck Guy had set before them. In an instant I went from goat to hero.

I hope all of you get the same results when the biscuit urge strikes—or when you're cooking any recipe in this magazine. Whether it's a Saturday morning with the family or an Easter brunch, we know how important it is to get these things right.

SID EVANS, EDITOR IN CHIEF
SID@SOUTHERNLIVING.COM

OUR FAVORITE BUTTERMILK BISCUIT

- **1/2 cup butter (1 stick), frozen**
- **2 1/2 cups self-rising flour**
- **1 cup chilled buttermilk**
- **Parchment paper**
- **2 Tbsp. butter, melted**

1. Preheat oven to 475°. Grate frozen butter using large holes of a box grater. Toss together grated butter and flour in a medium bowl. Chill 10 minutes.

2. Make a well in center of mixture. Add buttermilk, and stir 15 times. Dough will be sticky.

3. Turn dough out onto a lightly floured surface. Lightly sprinkle flour over top of dough. Using a lightly floured rolling pin, roll dough into a 3/4-inch-thick rectangle (about 9 x 5 inches). Fold dough in half so short ends meet. Repeat rolling and folding process 4 more times.

4. Roll dough to 1/2-inch thickness. Cut with a 2 1/2-inch floured round cutter, reshaping scraps and flouring as needed.

5. Place dough rounds on a parchment paper-lined jelly-roll pan. Bake at 475° for 15 minutes or until lightly browned. Brush with melted butter.

MAKES 12 to 14 biscuits.
HANDS-ON 25 min., **TOTAL** 50 min.

Crab Takes the Cake

CRUNCHY ON THE OUTSIDE, TENDER ON THE INSIDE, FEW HEALTHY INDULGENCES ARE THIS FAST AND FABULOUS

LIGHT CRAB CAKES

Fresh lump crabmeat is key to this simple recipe. It's pricey, but a little goes a long way.

- **8 oz. lump crabmeat**
- **2 Tbsp. finely chopped green onions**
- **1 Tbsp. finely chopped fresh flat-leaf parsley**
- **2 Tbsp. canola mayonnaise**
- **1 tsp. fresh lemon juice**
- **1 tsp. Dijon mustard**
- **½ tsp. Old Bay seasoning**
- **½ tsp. Worcestershire sauce**
- **⅛ tsp. kosher salt**
- **⅛ tsp. ground red pepper**
- **1 large egg, lightly beaten**
- **⅔ cup panko (Japanese breadcrumbs), divided**
- **1 Tbsp. olive oil**
- **1 lemon, quartered**

1. Pick crabmeat, removing any bits of shell. Stir together green onions, next 9 ingredients, and ⅓ cup panko in a large bowl. Add crabmeat, and stir gently to combine.
2. Using wet hands, shape mixture into 4 balls. Dredge balls in remaining ⅓ cup panko. Gently flatten each ball to form a 4-inch patty.
3. Cook patties in hot oil in a large skillet over medium-high heat 3 minutes on each side or until golden. Serve with lemon wedges.

MAKES 4 servings (serving size: 1 crab cake).
HANDS-ON 25 min., **TOTAL** 25 min.

NUTRITIONAL INFORMATION
(per serving)

CALORIES: 181; FAT: 7.8g (SATURATED FAT: 0.9g); PROTEIN: 16.3g; CARBOHYDRATES: 8.8g; SODIUM: 482mg

Saucy Chicken Lettuce Cups

EAST MEETS SOUTH IN THIS RESTAURANT FAVORITE THAT'S EASY TO MAKE AT HOME

SLOW-COOKER CHICKEN LETTUCE CUPS

- **1 yellow onion, cut into wedges**
- **1 Granny Smith apple, cut into wedges**
- **6 garlic cloves, smashed**
- **4 uncooked thick bacon slices, chopped**
- **1 (4-lb.) whole chicken**
- **⅓ cup plus 1 ½ tsp. apple cider vinegar, divided**
- **Kosher salt**
- **Black pepper**
- **⅔ cup mayonnaise**
- **⅛ tsp. ground red pepper**
- **12 Bibb lettuce or radicchio leaves**
- **Toppings: green onion slices, sweet mini pepper slices, spicy pickles**
- **Garnish: dried crushed red pepper**

1. Place first 4 ingredients in a 7-qt. slow cooker. Add chicken, ⅓ cup vinegar, and ½ cup water. Sprinkle chicken with salt and black pepper. Cover and cook on LOW 5 to 6 hours or until meat easily pulls from bone. Remove meat from bones; discard skin and bones. Shred meat. Pour cooking liquid through a fine wire-mesh strainer into a measuring cup; discard solids. Let cooking liquid stand 10 minutes. Skim fat, and reserve ⅓ cup cooking liquid. Toss remaining cooking liquid with chicken.
2. Whisk together mayonnaise, red pepper, reserved ⅓ cup cooking liquid, and remaining 1 ½ tsp. vinegar; season with salt and pepper. Spoon chicken into lettuce leaves; drizzle with sauce. Add toppings. Serve immediately.

MAKES 4 to 6 servings. **HANDS-ON** 20 min. **TOTAL** 5 hours, 30 min.

Our Best Coconut Cake

THE NOSTALGIA OF COCONUT CREAM PIE MEETS THE FANFARE OF A LAYER CAKE IN ONE DELICIOUS DESSERT

COCONUT CREAM CAKE

Look for packaged coconut shavings in your grocer's natural foods section.

FILLING

- ½ cup sugar
- ¼ cup cornstarch
- ⅛ tsp. table salt
- 4 large egg yolks
- 1 cup half-and-half
- 1 cup coconut milk
- 1 cup sweetened flaked coconut
- 3 Tbsp. butter
- 1 tsp. vanilla extract
- ¼ tsp. coconut extract

CAKE LAYERS

- 1 cup butter, softened
- 2 cups sugar
- 4 large eggs
- 3 cups cake flour
- 1 Tbsp. baking powder
- ½ tsp. table salt
- ½ cup milk
- ½ cup coconut milk
- 1 tsp. vanilla extract
- ½ tsp. coconut extract
- Shortening

ADDITIONAL INGREDIENTS

- 1 cup heavy cream
- Fluffy Coconut Frosting

GARNISH

- Toasted coconut shavings

1. Prepare Filling: Whisk together first 3 ingredients in a heavy saucepan. Whisk egg yolks and next 2 ingredients in a glass bowl. Gradually whisk egg mixture into sugar mixture. Cook over medium heat, whisking constantly, 6 to 7 minutes or until mixture just starts to bubble. Cook, whisking constantly, 1 more minute; remove from heat.
2. Whisk flaked coconut and next 3 ingredients into sugar-egg mixture; transfer to a medium bowl. Place plastic wrap directly on warm filling to prevent film from forming. Let stand 30 minutes; chill 4 to 24 hours.
3. Meanwhile, prepare Cake Layers: Preheat oven to 350°. Beat 1 cup butter at medium speed with a heavy-duty electric stand mixer until creamy. Gradually add 2 cups sugar, beating until light and fluffy. Add 4 eggs, 1 at a time, beating just until blended after each addition.
4. Stir together cake flour and next 2 ingredients in a bowl. Stir together milk and ½ cup coconut milk in a measuring cup. Add flour mixture to butter mixture alternately with milk mixture, beginning and ending with flour mixture. Beat at low speed just until blended after each addition. Stir in 1 tsp. vanilla and ½ tsp. coconut extract. Spoon batter into 2 greased (with shortening) and floured 9-inch cake pans.
5. Bake at 350° for 25 to 30 minutes or until a wooden pick inserted in center comes out clean. Cool in pans on wire racks 10 minutes. Remove from pans to wire racks, and cool completely (about 1 hour).
6. Beat 1 cup heavy cream at high speed with an electric mixer until stiff peaks form. Gently fold whipped cream into chilled Filling.
7. Assemble cake: Split each layer in half horizontally with a serrated knife to make 4 layers. Place 1 layer, cut side up, on a serving platter; spread with 1⅓ cups Filling. Repeat with remaining Cake Layers and Filling, ending with a Cake Layer, cut side down. Spread top and sides of cake with Fluffy Coconut Frosting.

MAKES 10 to 12 servings. **HANDS-ON** 50 min.; **TOTAL** 6 hours, 25 min., including frosting

Fluffy Coconut Frosting

Beat 3 cups **heavy cream,** 1 tsp. **vanilla extract,** and ½ tsp. **coconut extract** at medium-high speed with an electric mixer 1 to 2 minutes or until foamy. Gradually add ⅓ cup **sugar,** beating 2 minutes or until stiff peaks form. (Do not overbeat.)

MAKES about 6 cups. **HANDS-ON** 10 min., **TOTAL** 10 min.

Community Cookbook

BUILD A BETTER LIBRARY, ONE GREAT BOOK AT A TIME

Cheesemaker Tasia Malakasis turned the tiny town of Elkmont, Alabama, into the goat cheese capital of the South. In her second cookbook, she shares simple and fresh Southern recipes inspired by her Greek heritage and her award-winning Belle Chèvre creamery.

STUFFED PEPPERS WITH CHÈVRE, PECANS, AND GOLDEN RAISINS

"Both sides of my family—the Greeks and the Southerners—love stuffed peppers."

- **12 sweet mini peppers**
- **¼ tsp. table salt**
- **¼ tsp. freshly ground black pepper**
- **4 oz. goat cheese, softened**
- **¼ cup pecan halves, toasted and chopped**
- **¼ cup thinly sliced fresh basil, divided**
- **¼ cup golden raisins, divided**
- **1 Tbsp. olive oil**
- **Garnish: small fresh basil leaves**

1. Preheat broiler. Broil peppers on an aluminum foil-lined baking sheet 5 inches from heat, turning occasionally, 8 minutes or until peppers look blackened.
2. Place peppers in a large zip-top plastic freezer bag; seal and let stand 10 minutes to loosen skins. Reduce oven temperature to 450°.
3. Peel peppers. Make a slit in one side of each pepper; sprinkle with salt and black pepper.
4. Stir together cheese, pecans, 2 Tbsp. basil, and 2 Tbsp. raisins in a small bowl. Spoon cheese mixture evenly into peppers, pressing gently to close slits.
5. Return peppers to baking sheet, and drizzle with oil. Bake at 450° for 10 minutes or until cheese is bubbly.
6. Arrange peppers on a serving plate; sprinkle with remaining 2 Tbsp. basil and 2 Tbsp. raisins.

MAKES 1 dozen. **HANDS-ON** 21 min., **TOTAL** 49 min.

HOMEMADE STRAWBERRY MILK

"This is a great drink for the kids, and it's always a huge hit. Try the strawberry syrup over ice cream too!"

- **1 cup fresh strawberries, chopped**
- **½ cup sugar**
- **1½ cups milk**
- **Garnish: fresh strawberries**

1. Freeze 2 small freezer-safe glasses 10 minutes.
2. Meanwhile, bring strawberries, sugar, and 1 cup water to a boil in a small saucepan; boil 10 minutes or until slightly thickened and reduced. Pour strawberry mixture through a fine wire-mesh strainer into a small bowl, reserving strawberries for another use. Cover syrup, and chill 2 hours.
3. Pour ¾ cup milk and 3 Tbsp. strawberry syrup into each frozen glass, stirring to blend. Refrigerate remaining strawberry syrup in an airtight container.

MAKES 2 servings. **HANDS-ON** 18 min.; **TOTAL** 2 hours, 36 min.

PEPPERED WHITE BEAN MASH

"This is an excellent side dish, but I also love to smear it on bread with a drizzle of fruity olive oil."

- **2 Tbsp. olive oil, divided**
- **2 garlic cloves, minced**
- **1 Tbsp. chopped fresh thyme**
- **1 Tbsp. fresh lemon juice (about 1 small lemon)**
- **2 (19-oz.) cans cannellini beans, drained and rinsed**
- **1 tsp. table salt**
- **1 tsp. freshly ground black pepper**

1. Heat 1 Tbsp. oil in a large skillet over medium heat. Add garlic, and sauté 1 minute.
2. Add thyme, lemon juice, beans, and 1 cup water. Cook, stirring often, over medium heat 4 minutes or until thickened. Stir in salt and pepper, and place bean mixture into a large bowl. Mash with a potato masher, leaving some beans whole. Drizzle mash with remaining 1 Tbsp. oil.

MAKES 4 servings. **HANDS-ON** 5 min., **TOTAL** 11 min.

SPICY ANDOUILLE SPANISH TORTILLA

"A Spanish tortilla has nothing at all to do with flour or masa tortillas we acquaint with tacos. In fact, it's most closely related to a frittata, but loaded with potatoes. Add some cheese, cherry tomatoes, and a Vidalia onion, and you have a complete one-dish meal."

- **½ lb. andouille sausage***
- **2 Tbsp. olive oil**
- **1 small sweet onion, vertically cut into thin slices**
- **2 garlic cloves, minced**
- **2 lb. Yukon gold potatoes, peeled and thinly sliced**
- **1 tsp. table salt, divided**
- **8 large eggs**
- **¼ tsp. freshly ground black pepper**
- **¾ cup grape tomatoes, halved**
- **4 oz. Manchego cheese, shredded**
- **Garnish: chopped green onions**

1. Preheat oven to 350°. Cut sausage lengthwise in half; cut crosswise into ¼-inch-thick half-moon-shaped slices. Cook sausage in a 12-inch ovenproof skillet over medium-high heat, stirring occasionally, 5 minutes or until browned. Remove from skillet; drain on paper towels.
2. Heat oil in skillet over medium-high heat. Add onion, and sauté 5 minutes or until tender. Add garlic, and sauté 30 seconds. Stir in potato, ⅔ cup water, and ½ tsp. salt. Reduce heat to medium. Cover and cook, stirring occasionally, 10 minutes or until potatoes are tender and water has evaporated.
3. Whisk together eggs, pepper, and remaining ½ tsp. salt. Stir in sausage. Pour egg mixture over potato mixture in pan. Reduce heat to medium-low, and cook 4 minutes or until mixture begins to set. Sprinkle with tomato and cheese.
4. Bake, uncovered, at 350° for 5 minutes or until top is set. Turn oven to broil; broil 2 minutes or until cheese is golden and bubbly.
5. Slide tortilla onto a serving plate. Cut into 6 wedges.

Spicy Andouille Spanish Tortilla

*Spicy smoked sausage may be substituted.

MAKES 6 servings. **HANDS-ON** 50 min., **TOTAL** 50 min.

POOL PARTY WATERMELON PUNCH

"What says summer more than a big ruby watermelon? Usually, I'm a little overly ambitious when buying one at the store, and then I'm left trying to find ways to use it all up at home. This recipe is great for this dilemma. Just cut it up into pieces, and let the blender do all the work for you."

- **1 (6-lb.) seedless watermelon**
- **1 Tbsp. sugar**
- **1½ cups chilled Lillet Blanc**
- **¼ cup fresh lime juice (about 2 large limes)**
- **1 (750-milliliter) bottle Prosecco, chilled**

1. Remove rind from watermelon; cut flesh into large chunks. Process half of watermelon chunks in a blender 1 minute or until liquefied. Pour through a fine wire-mesh strainer into a large bowl, using the back of a spoon to squeeze out juice; discard pulp. Repeat procedure with remaining half of watermelon chunks.
2. Measure 3 cups watermelon juice, and pour into a 3-qt. pitcher. (Reserve remaining watermelon juice for another use.) Add sugar, stirring until dissolved. Cover and chill 2 hours.
3. Stir in Lillet Blanc and lime juice. Gently stir in Prosecco, and serve immediately.

MAKES 10 servings. **HANDS-ON** 10 min.; **TOTAL** 2 hours, 10 min.

SUMMER CORN SOUP WITH CORN SALSA

"I love that this creamy soup is deceivingly thick, even though it has no cream in it at all. Use any extra soup stock, which is made from the corn cobs, in other vegetable-based soups."

- **5 ears fresh corn**
- **1½ tsp. table salt, divided**
- **2 Tbsp. extra virgin olive oil, divided**
- **2 Tbsp. chopped red onion**
- **2 Tbsp. chopped fresh cilantro**
- **2 Tbsp. fresh lime juice (about 2 limes)**
- **½ tsp. sugar**
- **½ small jalapeño pepper, seeded and diced**
- **1½ cups chopped white onion**
- **2 garlic cloves, minced**
- **Garnish: chopped fresh cilantro, or chopped fresh chives**

1. Cut corn kernels from cobs to measure 5 cups, and place in a bowl, reserving cobs. Place corn cobs, 6 cups water, and ½ tsp. salt in a Dutch oven. Bring to a boil; reduce heat, partially cover, and simmer 30 minutes.

2. Meanwhile, heat 1 tsp. oil in a small saucepan over medium-high heat. Add 1 cup corn kernels; sauté 5 minutes or until tender and beginning to brown. Transfer to a small bowl, and add red onion, next 4 ingredients, 2 tsp. oil, and ½ tsp. salt, stirring well. Cover corn salsa, and let stand until ready to serve.

3. Remove and discard cobs from corn stock. Pour stock through a fine wire-mesh strainer into a bowl to measure 3 cups.

4. Heat remaining 1 Tbsp. oil over medium heat in a large saucepan. Add white onion; sauté 5 minutes. Add garlic; sauté 30 seconds. Add remaining 4 cups corn kernels and remaining ½ tsp. salt. Cook, stirring often, 4-5 minutes or just until corn is tender. Add stock; bring to a simmer; cover and simmer 15 minutes.

5. Place half of soup in a blender. Remove center piece of blender lid (to allow steam to escape); secure lid on blender. Place a clean towel over opening in lid (to avoid splatters). Blend until smooth; pour into a large bowl. Repeat procedure with remaining half of soup.

6. Return soup to pan, and cook over medium-high heat 2 minutes or until thoroughly heated. Ladle soup into bowls. Top with corn salsa.

MAKES 5 cups soup, 1 cup salsa. **HANDS-ON** 1 hour, 12 min.; **TOTAL** 1 hour, 12 min.

PEANUTTY BRAISED CHICKEN

"Peanut butter gives this dish exotic flavor and creates a velvety sauce when combined with tomatoes."

- **1 cup creamy peanut butter**
- **1 ½ cups chicken broth, warmed**
- **2 lb. skinned and boned chicken breasts, cut into ½-inch cubes**
- **1 ¼ tsp. kosher salt, divided**
- **2 Tbsp. vegetable oil**
- **1 medium onion, chopped**
- **1 red bell pepper, chopped**
- **3 garlic cloves, finely chopped**
- **1 (14.5-oz.) can diced tomatoes, undrained**
- **Hot cooked rice**
- **Toppings: sliced green onions, fresh cilantro leaves, chopped peanuts**

1. Place peanut butter in a medium bowl; gradually whisk in chicken broth until blended.
2. Pat chicken dry with paper towels; sprinkle with 1 tsp. salt, and toss well. Heat oil in a 4- to 5-qt. heavy oven-proof Dutch oven over medium-high heat. Brown chicken, in batches, in hot oil 3-5 minutes. Transfer to a bowl.
3. Add onion and bell pepper to Dutch oven; sauté 4 minutes or until onion begins to brown. Add garlic; sauté 1 minute. Stir in chicken along with any accumulated juices, peanut butter mixture, tomatoes, and remaining ¼ tsp. salt. Bring to a simmer; reduce heat to low, and simmer 10 minutes or until desired degree of doneness.
4. Serve chicken mixture over rice. Sprinkle with toppings.

MAKES 6 servings. **HANDS-ON** 34 min., **TOTAL** 34 min.

BLT BUTTERMILK BLUE SLAW

"My son eats only one type of lettuce. Even though I grow and bring home an abundance of beautiful tender varieties, he opts, always, for iceberg. This recipe is a great way to incorporate it into a tangy slaw. Add a sliced hard-cooked egg to make it a meal!"

- **¼ cup buttermilk**
- **3 Tbsp. mayonnaise**
- **½ tsp. table salt**
- **¼ tsp. freshly ground black pepper**
- **½ cup crumbled blue cheese**
- **½ cup thinly sliced red onion**
- **3 Tbsp. coarsely chopped fresh parsley**
- **1 lb. multicolored baby heirloom tomatoes, cut into thin wedges**
- **1 head iceberg lettuce (about 1 ½ lb.), quartered and shredded**
- **5 applewood-smoked bacon slices, cooked and crumbled**

1. Stir together first 4 ingredients in a bowl. Gently stir in cheese.
2. Combine onion and next 3 ingredients in a large bowl. Drizzle dressing over lettuce mixture; sprinkle with bacon, and toss gently. Serve immediately.

MAKES 6 servings. **HANDS-ON** 15 min., **TOTAL** 15 min.

BELLE CHÈVRE RASPBERRY TIRAMISÙ

"Although tiramisù is traditionally made with mascarpone cheese, I prefer to use a honey-flavored goat cheese for the lightness and bright flavor it imparts."

- **1 cup hot espresso or hot strong brewed coffee**
- **2 Tbsp. sugar**
- **1 (6-oz.) container Belle Chèvre's honey-flavored goat cheese**
- **1 Tbsp. honey**
- **¼ tsp. vanilla extract**
- **1 cup heavy cream**
- **1 (3.5-oz.) package crisp ladyfingers, broken in half crosswise**
- **2 oz. semisweet chocolate, finely chopped**
- **1 ½ cups fresh raspberries**
- **Unsweetened cocoa**
- **Garnish: raspberries**

1. Combine coffee and sugar in a 1-cup glass measuring cup, stirring until sugar dissolves.
2. Stir together cheese, honey, and vanilla in a small bowl until blended. Beat heavy cream until soft peaks form; gently fold in cheese mixture.
3. Dip half of ladyfinger halves in espresso mixture; place 4 ladyfinger halves in bottoms of 5 (10-oz.) ramekins. Sprinkle evenly with half of chocolate; top with half of raspberries. Spoon half of cheese mixture over raspberries, spreading evenly.
4. Dip remaining half of ladyfinger halves in espresso mixture; repeat layers, beginning with dipped ladyfinger halves and ending with cream mixture. Cover each ramekin with plastic wrap, and chill for at least 4 hours.
5. Unwrap ramekins; dust desserts lightly with cocoa. Serve cold.

MAKES 5 servings. **HANDS-ON** 15 min.; **TOTAL** 4 hours, 15 min.

BAKED FRENCH TOAST WITH BANANA BRÛLÉE

"This is the perfect dish to serve a crowd when you don't want to be chained to the kitchen but still want to impress. Prepare this the night before, and bake it in the morning while you relax over a cup of coffee."

- **1 cup buttermilk**
- **1 cup milk**
- **⅓ cup granulated sugar**
- **1 Tbsp. Grand Marnier (optional)**
- **½ tsp. ground cinnamon**
- **⅛ tsp. freshly grated nutmeg**
- **⅛ tsp. table salt**
- **6 large eggs**
- **1 (12-oz.) French bread loaf, cut into 16 (1 ½-inch-thick) slices**
- **Butter**
- **2 ripe bananas, sliced**
- **⅓ cup firmly packed brown sugar**
- **Maple syrup**

1. Whisk together first 8 ingredients in a medium bowl. Dip bread slices on both sides into egg mixture. Place bread slices in a single layer in a buttered 13- x 9-inch baking dish. Pour remaining egg mixture over bread slices. Cover and chill overnight.
2. Preheat oven to 350°. Remove baking dish from refrigerator, and let stand while oven preheats. Uncover dish, and bake at 350° for 30 minutes or until set and bottom is lightly browned.
3. Remove baking dish from oven, and top French toast with banana slices. Sprinkle brown sugar over banana slices. Caramelize brown sugar using a kitchen torch, holding torch 1-2 inches from bananas and moving torch back and forth. Serve with maple syrup.

MAKES 8 servings. **HANDS-ON** 12 min.; **TOTAL** 12 hours, 42 min.

100 **Host Derby Day** Don your finest hat and dish up Derby-inspired classics that impress

103 **An Editor's Escape** A bourbon-spiked Old Fashioned makes porch sitting an occasion

104 QUICK-FIX SUPPERS **Bake, Grill, Sear, or Roast: It's Fish Season** Master the fine art of cooking your favorite catch

106 THE *SL* TEST KITCHEN ACADEMY **Buying Fresh Seafood** Get to know your fishmonger and get schooled in what's in season and abundant to make the tastiest picks for your plate

107 WHAT TO COOK NOW **Chowchows to Relish** Fresh-picked and quick-pickled, these flavor-packed accents take dishes from ordinary to extraordinary

109 THE SLOW DOWN **Slow-Cooker Ribs** A hands-off method yields barbecue perfection

110 *COOKING LIGHT* **A Healthy Slice** Put pizza back on the weeknight table with our healthy toppings and time-saving shortcuts

111 SAVE ROOM **Raise the Bar** Love our make-ahead treat that celebrates spring

112 COMMUNITY COOKBOOK Only delicious dishes come out of *A Southern Gentleman's Kitchen* when author Matt Moore is doing the cooking

Host Derby Day

CELEBRATE THE SOUTH'S SWANKIEST HORSE RACE WITH EASY AND ELEGANT FOOD

THE *SL* MINT JULEP

Our go-to recipe for the quintessential Derby-day drink

Place 3 mint leaves and 1 Tbsp. Mint Simple Syrup in a chilled 8-oz. glass. Muddle leaves to release flavors. Pack glass with crushed ice; pour 3 Tbsp. bourbon over ice, and top with 2 Tbsp. club soda.

THE BLUSH LILY

Our version of Churchill Downs' Oaks Lily

Fill a cocktail shaker with crushed ice; add 6 Tbsp. cranberry juice, 3 Tbsp. vodka, 2 Tbsp. fresh lime juice, and a splash of Triple Sec. Cover with lid, and shake vigorously until chilled. Strain into a highball glass filled with crushed ice.

MINT SIMPLE SYRUP

Simple syrup is a wonder because the sugar is fully dissolved and a sludge of sugar crystals won't appear at the bottom of your drink.

Bring 1 cup sugar and 1 cup water to a boil in a medium saucepan over high heat. Boil, stirring often, 5 minutes or until sugar dissolves. Remove from heat; add 10 to 12 fresh mint sprigs, and cool completely (about 10 minutes). Pour into a glass jar with a tight-fitting lid; cover and chill 24 hours. Discard mint. Keep for up to 2 weeks. Makes 2 cups.

JALAPEÑO CHEESE STRAWS

We know: You've got a cheese straw recipe you can make with your eyes closed. But give this subtly spicy recipe a try.

1 (8-oz.) block extra-sharp Cheddar cheese, grated
¼ cup grated Parmesan cheese
½ cup butter, softened
3 Tbsp. diced pimiento, drained
1 jalapeño pepper, seeded and minced
2 tsp. half-and-half
1 tsp. kosher salt
⅛ to ¼ tsp. ground red pepper (optional)
1½ cups all-purpose flour
Parchment paper

1. Preheat oven to 350°. Beat Cheddar cheese, next 6 ingredients, and, if desired, red pepper at medium speed with a heavy-duty electric stand mixer until blended. Gradually add flour, beating at low speed just until combined.
2. Turn dough out onto a well-floured surface. Divide dough in half; flatten each half into a square. Roll each square to ⅛-inch thickness (about 12 x 12 inches). Cut into ½- x 12-inch strips, using a fluted pastry wheel.
3. Holding 1 dough strip at each end, twist strip until tightly curled. Pinch ends to seal. Repeat with remaining strips, and place ½ inch apart on 2 parchment paper-lined baking sheets.
4. Bake at 350° for 10 minutes, placing 1 sheet on middle oven rack and 1 sheet on lower oven rack. Rotate pans front to back, and top rack to bottom rack. Bake 10 to 12 more minutes or until edges begin to brown. Cool on baking sheets on wire racks 5 minutes. Remove to wire racks, and cool completely.

MAKES about 4 dozen. **HANDS-ON** 30 min.; **TOTAL:** 1 hour, 15 min.

CHILLED SWEET PEA SOUP WITH MINT AND CREAM

We love this bright, minty soup because it adds a pop of spring color to the buffet. Make it up to 48 hours ahead of time.

2 Tbsp. butter
3 medium leeks (white and light green parts only), rinsed, drained, and chopped
1 (32-oz.) container reduced-sodium chicken broth
1 (16-oz.) package frozen sweet peas
¼ cup chopped fresh mint leaves
2 tsp. kosher salt
½ tsp. freshly ground black pepper
1 cup sour cream
2 Tbsp. fresh lemon juice
Garnish: sliced chives

1. Melt butter in a large saucepan over medium-low heat. Add leeks, and cook, stirring occasionally, 6 to 8 minutes or until tender. Stir in chicken broth, and increase heat to high. Bring to a boil. Add peas, and cook, stirring occasionally, 3 minutes or until peas are tender. Remove from heat, and stir in mint, salt, and pepper.
2. Process pea mixture, in batches, in a blender or food processor until smooth. Transfer mixture to a bowl, and whisk in ½ cup sour cream. Season with salt and pepper, and pour into 2-oz. glasses. Chill 30 minutes to 1 hour. Whisk together lemon juice and remaining sour cream, and dollop on each serving.

MAKES 1½ qt. **HANDS-ON** 30 min., **TOTAL** 1 hour

PORK TENDERLOIN SLIDERS

One carefree way to serve these cuties is to slice the tenderloins on a large cutting board, set out fresh rolls and homemade condiments, and let guests build their own sandwiches.

- **2 pork tenderloins (about 2 ½ lb.), trimmed**
- **3 Tbsp. olive oil, divided**
- **2 tsp. kosher salt**
- **1 tsp. freshly ground black pepper**
- **¼ cup firmly packed dark brown sugar**
- **2 Tbsp. Dijon mustard**
- **3 Tbsp. fresh thyme leaves**
- **2 Tbsp. chopped fresh rosemary**
- **20 slider buns or dinner rolls, split**
- **Italian-Style Salsa Verde (recipe at right), Blackberry-Honey Mustard Sauce, or Bacon-and-Sweet Onion Jam (recipes, page 102)**

1. Preheat oven to 400°. Rub pork tenderloins with 1 Tbsp. oil, and sprinkle with salt and pepper. Stir together sugar and next 3 ingredients; rub over pork.

2. Cook pork in remaining 2 Tbsp. hot oil in a skillet over medium-high heat 5 minutes, browning on all sides. Place tenderloins on a wire rack in a jelly-roll pan.

3. Bake at 400° for 20 minutes or until a meat thermometer inserted in thickest portion registers 155°. Remove from oven, and let stand 10 minutes. Slice and serve on slider buns with sauces, or wrap tenderloin whole and refrigerate up to 3 days.

MAKES 20 sliders. **HANDS-ON** 25 min.; **TOTAL** 55 min., not including sauces

Pork Tenderloin Sliders

ITALIAN-STYLE SALSA VERDE

This all-purpose condiment is vibrant and packed with herbs and makes everything taste garden fresh. It brightens up our sliders (above) and can also be stirred into scrambled eggs or drizzled over steak.

- **1 small jalapeño pepper**
- **2 medium-size banana peppers**
- **½ cup extra virgin olive oil**
- **⅓ cup finely chopped fresh flat-leaf parsley**
- **4 ½ tsp. chopped fresh chives**
- **1 Tbsp. minced fresh oregano**
- **2 garlic cloves, minced**
- **1 tsp. kosher salt**

1. Preheat broiler with oven rack 6 inches from heat. Broil jalapeño 3 to 4 minutes on each side or until blackened. Place blackened jalapeño in a small bowl, cover with plastic wrap, and let stand 10 minutes. Meanwhile, broil banana peppers 1 to 2 minutes on each side or just until blistered and slightly softened. Cool completely (about 10 minutes), and chop. Peel and finely chop jalapeño, discarding seeds.

2. Stir together oil, next 5 ingredients, and chopped peppers in a small bowl. Cover and let stand 30 minutes. Serve at room temperature, or cover and refrigerate up to 1 week.

MAKES about 1 cup. **HANDS-ON** 30 min.; **TOTAL** 1 hour, 5 min.

BLACKBERRY-HONEY MUSTARD SAUCE

You won't find another condiment as pretty as this tangy and sweet blackberry mustard.

- **⅓ cup sugar**
- **1 (6-oz.) container fresh blackberries**
- **¼ cup honey**
- **1 Tbsp. dry mustard**
- **3 Tbsp. Dijon mustard**
- **2 Tbsp. fresh lemon juice**
- **1 tsp. kosher salt**
- **⅓ cup extra virgin olive oil**

1. Bring first 2 ingredients to a boil in a small saucepan over medium-high heat, stirring occasionally and mashing berries with the back of a wooden spoon. Reduce heat to medium, and simmer, stirring often and mashing berries, 2 to 3 minutes or until slightly thickened. Remove from heat, and pour mixture through a fine wire-mesh strainer into a blender, pressing with spoon to release juices; discard solids.
2. Add honey and next 4 ingredients to blender; process on low 20 seconds. Increase blender speed to high, and process 30 seconds. With blender running, add oil in a slow, steady stream, processing until smooth. Transfer mixture to a small bowl; cover and chill 1 to 12 hours.

MAKES about 1¼ cups. **HANDS-ON** 20 min.; **TOTAL** 1 hour, 20 min.

BACON-AND-SWEET ONION JAM

- **4 uncooked thick applewood-smoked bacon slices, chopped**
- **1 Tbsp. butter**
- **2 medium-size sweet onions, chopped**
- **4 large shallots, chopped**
- **½ cup balsamic vinegar**
- **3 Tbsp. light brown sugar**
- **2½ tsp. kosher salt**
- **2 Tbsp. chopped fresh chives**
- **2 tsp. chopped fresh thyme**

1. Cook bacon in a medium skillet over medium-low heat, stirring occasionally, 8 to 10 minutes or until crisp. Remove bacon, and drain on paper towels, reserving drippings in skillet.
2. Add butter to drippings, and stir until butter melts. Increase heat to medium; add onions, and sauté 10 to 12 minutes or until tender. Add shallots, vinegar, sugar, and kosher salt, and cook, stirring constantly, 1 to 2 minutes or until sugar dissolves.
3. Reduce heat to low, and cook, stirring occasionally, 20 to 25 minutes or until onions are very tender and brown. Remove from heat; stir in chives, thyme, and bacon. Cool completely (about 20 minutes), or refrigerate in an airtight container up to 3 days.

MAKES 1½ cups. **HANDS-ON** 35 min.; **TOTAL** 1 hour, 15 min.

HONEY CUSTARD WITH BERRIES

If you don't have white balsamic vinegar, substitute Champagne vinegar or apple cider vinegar.

- **1 cup heavy cream**
- **4 large egg yolks**
- **¼ cup honey**
- **¼ cup white wine (such as Sauvignon Blanc)**
- **1 Tbsp. white balsamic vinegar**
- **6 to 8 cups mixed fresh berries (such as strawberries, blueberries, blackberries, and raspberries)**
- **Garnish: fresh mint**

1. Beat cream in a medium bowl at medium-high speed with an electric mixer until soft peaks form. Cover and chill until ready to use.
2. Whisk together egg yolks, honey, wine, and balsamic vinegar in top of a double boiler. Bring water in bottom pan to a light boil. Cook egg yolk mixture, whisking constantly, 8 to 10 minutes or until thick and foamy.
3. Remove from heat. Fill a large bowl with ice. Place top of double boiler containing egg yolk mixture in ice, and whisk 5 minutes or until completely cool. Remove from ice bath, and fold in whipped cream. If desired, cover and chill up to 24 hours. Spoon berries into bowls, and serve with custard.

MAKES 8 to 10 servings. **HANDS-ON** 30 min., **TOTAL** 30 min.

An Editor's Escape

A FEW KEY ELEMENTS TURNED A SMALL, UNDERUSED PORCH INTO THE MOST INVITING SPOT IN THE HOUSE

IT'S AN ODD LITTLE PORCH, stuck on the side of the house like an afterthought. At only 130 square feet, it's too small for a dining table or a sofa, but it's the ideal place to have a cocktail, read a book, or watch the kids play in the yard. When my wife, Susan, and I bought our Birmingham house, the porch was enclosed in glass, but we knew it had to be screened. We wanted a place to catch a breeze, to escape the madness of our busy lives. The challenge was making this tiny, awkward space comfortable.

We started with the idea of four chairs and a table. It was a ridiculously simple concept, so the elements had to be unique. We asked local furniture designer Michael Morrow to come up with something that had an organic feel but with a modern twist. His solution: a coffee table made from four slices of a maple tree, a whimsical piece that's also sturdy. To me it makes the whole room.

Susan wanted the seating to be cozy but stylish, and she loved the leather-wrapped rattan fretwork of these chairs, which adds visual flair. Then we just needed a rug to cover the unattractive brick floor (which would have cost a fortune to replace or refinish), a simple bar, and a couple of fiddleleaf fig trees to liven up the room. Now this little oasis is one of our favorite spots in the house.

MY GO-TO OLD FASHIONED

I'm a bourbon fan, and there's something that just feels right about drinking an old fashioned on a screened porch. This recipe, from Julian Van Winkle of Old Rip Van Winkle Distillery, ran in our May 2012 issue.

Place 1 to 2 **brown sugar cubes** on a cocktail napkin. Sprinkle 2 to 3 drops of **orange bitters** and 2 to 3 drops of **Angostura bitters** over sugar cubes. (Napkin will soak up excess bitters.) Transfer cubes to a 10-oz. old fashioned glass. Add 1 fresh **orange slice** and a few drops of **bourbon** to glass. Mash sugar cubes and orange slice, using a muddler, until sugar is almost dissolved. (Avoid mashing the rind; doing so will release a bitter flavor.) Add 1½ to 2 oz. bourbon, and fill glass with ice cubes. Stir until well chilled. Add a touch more bourbon, if desired. —Sid Evans

Bake, Grill, Sear, or Roast: It's Fish Season

TRIED-AND-TRUE TECHNIQUES FOR MASTERING YOUR FAVORITE FISH AT HOME

PARCHMENT IS BEST FOR: Fish fillets that are about 1 inch thick, such as **snapper, flounder, sole, striped bass,** and **salmon. EXPERT ADVICE:** Packets must be tightly sealed so they don't come undone while baking. Make small, snug, overlapping folds to seal each bundle, and then twist the tail ends tightly closed.

GRILLING IS BEST FOR: Thin fillets or steaks that are ½ to 1 inch thick, such as **triggerfish, tilapia, mahi mahi, tuna, swordfish,** and **trout. EXPERT ADVICE:** First, make sure your grill is clean and hot. Flip once, and don't rush. The fish is ready to turn when it releases easily from the grate without tugging or tearing.

SEARING IS BEST FOR: Nearly any type of fish, but is easiest with firm fillets that are at least 1½ inches thick, such as **grouper, halibut, sea bass,** and **striped bass. EXPERT ADVICE:** Make sure the pan is hot before adding fish. Press very lightly with a spatula while cooking for even searing.

ROASTING IS BEST FOR: Shrimp or large fillets of firm fish, such as **snapper, cod, sea bass,** and **salmon.** This is the best technique for whole, dressed fish. **EXPERT ADVICE:** Don't crowd the pan. Spread the shrimp or pieces of fish in a single layer to avoid soggy or tough results.

TECHNIQUE NO. 1

BAKING IN PARCHMENT PAPER

This method makes French-style cooking easy. Each parcel of fish and veggies steams with its own delicious juices (a technique called en papillote*). The package concentrates the flavors, and when opened at the table, it releases a dramatic cloud of aromatic steam.*

SNAPPER BAKED IN PARCHMENT WITH SPRING VEGETABLES

- ½ **lb. fresh asparagus**
- 8 **sweet mini peppers**
- 1 **small sweet onion, thinly sliced**
- 4 **pickled okra pods, halved lengthwise**
- 8 **oz. small Yukon gold potatoes, sliced**
- 1 **Tbsp. chopped fresh flat-leaf parsley**
- ½ **tsp. chopped fresh dill**
- 2 **Tbsp. plus 1½ tsp. olive oil**
- 1½ **tsp. kosher salt**
- ½ **tsp. ground black pepper**
- 4 **(17-inch) parchment paper or aluminum foil squares**
- 4 **(6- to 8-oz.) fresh snapper fillets, skin on**
- 1 **lemon, quartered**
- 4 **Tbsp. butter**

1. Preheat oven to 400°. Snap off tough ends of asparagus, and discard. Toss together asparagus, next 6 ingredients, 1 Tbsp. olive oil, and ½ tsp. each salt and black pepper. Divide mixture among parchment paper squares; top each with 1 snapper fillet, skin side up. Sprinkle with remaining

salt, and drizzle with remaining oil. Squeeze juice from lemon over fillets; and top each with 1 Tbsp. butter. Bring parchment paper sides up over mixture; fold top, and twist ends to seal. Place packets on a baking sheet.
2. Bake at 400° for 15 to 20 minutes or until a thermometer registers 140° to 145° when inserted through paper into fish. Place packets on plates, and cut open. Serve immediately.

MAKES 4 servings. **HANDS-ON** 20 min., **TOTAL** 35 min.

TECHNIQUE NO. 2

GRILLING

Grilling adds smoky flavor fast. Follow our Expert Advice (on facing page) to flip fillets effortlessly.

GRILLED TRIGGERFISH

- **4 (6-oz.) triggerfish fillets**
- **Vegetable cooking spray**
- **2 Tbsp. extra virgin olive oil**
- **½ tsp. table salt**
- **¼ tsp. freshly ground black pepper**
- **Strawberry-Blueberry Relish (recipe, page 107)**

1. Pat fillets dry with paper towels, and let stand at room temperature 10 minutes. Meanwhile, coat cold cooking grate of grill with cooking spray, and preheat grill to 400° (medium-high) heat.
2. Brush both sides of fish with oil; sprinkle with salt and pepper. Place fish on grate, and grill, covered with grill lid, 4 minutes or until grill marks appear and fish no longer sticks to grate.
3. Carefully turn fish over, using a metal spatula, and grill, without grill lid, 2 minutes or just until fish separates into moist chunks when gently pressed. Serve with Strawberry-Blueberry Relish.

MAKES 4 servings. **HANDS-ON** 10 min., **TOTAL** 20 min.

TECHNIQUE NO. 3

PAN-SEARING

This is the secret to restaurant-quality fish. The magic happens on the bottom of the pan where the fish forms an even, crisp crust. The sauce seals the deal.

PAN-SEARED GROUPER WITH BALSAMIC BROWN BUTTER SAUCE

- **4 (4- to 6-oz.) fresh grouper fillets**
- **1 tsp. kosher salt**
- **¼ tsp. freshly ground black pepper**
- **2 Tbsp. olive oil**
- **4 Tbsp. butter**
- **1 Tbsp. balsamic vinegar**
- **1 tsp. minced shallot**
- **1 tsp. fresh lemon juice**

1. Preheat oven to 425°. Pat fish dry with paper towels, and let stand at room temperature 10 minutes. Sprinkle fillets with salt and pepper.
2. Heat oil in a large ovenproof skillet over medium-high heat. Carefully place fillets, top side down, in hot oil. Cook 3 to 4 minutes or until the edges are lightly browned. Transfer skillet to oven.
3. Bake at 425° for 4 to 5 minutes or until fish is opaque. Remove skillet from oven, and place fish, seared side up, on a platter.
4. Wipe skillet clean. Cook butter in skillet over medium heat 2 to 2½ minutes or until butter begins to turn golden brown. Pour butter into a small bowl. Whisk in vinegar, shallot, and lemon juice. Season with salt and pepper.

MAKES 4 servings. **HANDS-ON** 20 min., **TOTAL** 30 min.

TECHNIQUE NO. 4

ROASTING

This recipe requires almost no effort for maximum reward. Just 10 minutes in the oven plus our special sauce equals tender, juicy shrimp with a kick.

ROASTED GULF SHRIMP WITH ROMESCO SAUCE

- **1½ lb. large, peeled, and deveined raw shrimp**
- **1 Tbsp. olive oil**
- **1 tsp. kosher salt**
- **½ tsp. freshly ground black pepper**
- **Parchment paper**
- **1 (12-oz.) jar roasted red peppers, drained**
- **¼ cup slivered almonds, toasted**
- **2 garlic cloves**
- **2 Tbsp. extra virgin olive oil**
- **1½ Tbsp. red wine vinegar**
- **1 Tbsp. fresh lemon juice**
- **Garnishes: chopped fresh flat-leaf parsley, sliced chives**

1. Preheat oven to 425°. Toss together shrimp and next 3 ingredients in a large bowl. Place shrimp in a single layer in a parchment paper-lined jelly-roll pan.
2. Process roasted peppers and next 5 ingredients in a blender until smooth. Drizzle ½ cup sauce over shrimp.
3. Roast at 425° for 10 minutes or until shrimp turn pink. Serve immediately with remaining sauce.

MAKES 4 to 6 servings. **HANDS-ON** 20 min., **TOTAL** 30 min.

THE *SL* TEST KITCHEN ACADEMY

ROBBY MELVIN, DIRECTOR OF THE SOUTH'S MOST TRUSTED KITCHEN, SHARES TIPS FOR

BUYING FRESH SEAFOOD

Talk to the fish guy at the seafood counter, and ask what's in season and what's local. When fish is abundant, the quality goes up and the price goes down.

1 FIND A STORE where fish experts can answer questions and guide your selection. Busy markets get frequent shipments and are likely to have the freshest product.

2 TRUST YOUR NOSE Fresh fish and shellfish smell like the briny sea or a clean creek. A fishy smell is a red flag.

3 PAY ATTENTION to how the fish is displayed. Seafood should be surrounded by plenty of clean, crushed ice. Fish packaged under plastic wrap should have little or no accumulated liquid.

- Fillets should have taut, shiny flesh that springs back when gently pressed. There should be no discoloration, dryness, or mushiness around the edges.
- Shrimp should look translucent with no black spots or dark edges. Shells keep raw shrimp firm and moist, so for best flavor and texture, buy shrimp with the shells on, and peel at home. It's worth the work!

4 KEEP IT COOL Fresh seafood needs to be properly chilled at all times. Ask for a separate small bag of crushed ice to place on top of the wrapped fish to keep it cold in the cart and on the way home. Store fish in the coldest part of the refrigerator (in the back), and cook it within two days of purchase.

5 BE FLEXIBLE Let freshness guide your purchase. More than one type of fish will work in almost any recipe. Share your recipe with your fish guy and he should be able to offer substitution options.

Chowchows to Relish

CHOWCHOW (NOUN): A TANGY CONCOCTION OF FRESH-PICKED AND QUICK-PICKLED PRODUCE BELOVED BY SOUTHERN COOKS

STRAWBERRY-BLUEBERRY RELISH

Sweet berries and jalapeño heat make this a flavorful accompaniment for your favorite grilled fish.

- **½ cup white wine vinegar**
- **½ cup firmly packed light brown sugar**
- **2 Tbsp. minced fresh ginger**
- **1 tsp. lime zest**
- **½ tsp. kosher salt**
- **1 jalapeño pepper, seeded and minced**
- **2 Tbsp. fresh lime juice**
- **2 cups chopped fresh strawberries**
- **1 cup fresh blueberries**
- **1 cup diced cucumber**
- **3 Tbsp. minced red onion**
- **2 Tbsp. chopped fresh cilantro**

1. Bring first 5 ingredients to a boil in a small saucepan over medium-high heat; reduce heat to low, and simmer, stirring occasionally, 5 minutes. Add jalapeño, and simmer, stirring occasionally, 5 minutes. Remove from heat, and let stand 30 minutes. Stir in lime juice.

2. Stir together strawberries and next 4 ingredients in a medium bowl. Add vinegar mixture, and stir to coat. Serve immediately, or refrigerate in an airtight container up to 2 days.

MAKES about 3 cups. **HANDS-ON** 30 min., **TOTAL** 1 hour

Strawberry-Blueberry Relish

SWEET CORN RELISH

Just what you need to jazz up grilled chicken, spoon over a pork chop, or stuff into a cheesy quesadilla.

- **1 cup apple cider vinegar**
- **¼ cup sugar**
- **1 tsp. kosher salt**
- **1 tsp. prepared horseradish**
- **2 cups fresh corn kernels (about 4 ears)**
- **2 medium zucchini, diced**
- **½ cup diced plum tomato**
- **½ cup thinly sliced green onions**
- **1 Tbsp. chopped fresh flat-leaf parsley**

1. Stir together vinegar, sugar, and salt in a small saucepan. Bring to a boil over medium-high heat; reduce heat to medium, and simmer 10 minutes or until reduced to about ½ cup. Remove from heat, and let stand 15 minutes. Stir in horseradish.

2. Stir together corn and next 4 ingredients in a medium bowl. Add vinegar mixture, and toss to coat. Serve immediately, or cover and chill up to 3 days.

MAKES about 3 cups. **HANDS-ON** 20 min., **TOTAL** 45 min.

GREEN TOMATO RELISH

This is a classic tomato relish recipe, but with a tangy twist from green tomatoes.

- 1½ lb. green tomatoes, quartered
- 1 red bell pepper
- 1 poblano pepper
- 1 large white onion, sliced
- 1 cup matchstick carrots
- 2 cups distilled white vinegar (5% acidity), divided
- 1 cup sugar
- 1 Tbsp. kosher salt
- 1 Tbsp. mustard seeds
- ¼ tsp. ground turmeric
- ⅛ tsp. ground cloves
- ⅛ tsp. ground allspice

1. Remove and discard seeds and ribs from tomatoes and peppers. Cut into thin slices.
2. Bring onion, carrots, tomatoes, peppers, ½ cup water, and 1 cup vinegar to a boil in a large stockpot over high heat; reduce heat to medium-low, and simmer, stirring occasionally, 30 minutes. Drain.
3. Return vegetables to stockpot, and stir in sugar, next 5 ingredients, and remaining 1 cup vinegar. Bring to a boil over medium-high heat; reduce heat to medium-low, and simmer 5 minutes. Cool completely. Transfer to a jar; screw on lid, and chill 30 minutes. Store in refrigerator up to 1 week.

MAKES about 4 cups. **HANDS-ON** 40 min.; **TOTAL** 2 hours, 10 min.

THREE-PEPPER CHOWCHOW

Slightly sweet, this pretty relish will transform tacos or a bowl of rice and beans.

- 1 small yellow onion, diced
- ½ cup distilled white vinegar (5% acidity)
- ½ cup fresh lemon juice
- ¼ cup sugar
- 1 tsp. kosher salt
- ¼ tsp. dried crushed red pepper
- 1 yellow bell pepper, diced
- 1 red bell pepper, diced
- 4 banana peppers, diced
- 1 Tbsp. minced fresh thyme

Bring first 6 ingredients and ½ cup water to a boil in a medium saucepan over medium-high heat. Reduce heat to medium, and simmer, stirring occasionally, 5 to 10 minutes or until reduced to about ¾ cup. Stir in peppers, and cook, stirring occasionally, 2 minutes. Remove from heat, and let stand 30 minutes. Stir in thyme. Serve immediately, or refrigerate in an airtight container up to 3 months.

MAKES about 3 cups. **HANDS-ON** 15 min., **TOTAL** 50 min.

NAPA CABBAGE-AND-SWEET PEPPER SLAW

Put this on your pulled pork sandwich and you'll never go back to regular coleslaw again.

- 1 cup rice vinegar
- ¼ cup honey
- 2¼ tsp. kosher salt
- 4 cups chopped napa cabbage
- 1 cup seeded and thinly sliced sweet mini peppers
- ¼ large red onion, thinly sliced
- ¼ cup firmly packed fresh flat-leaf parsley leaves
- 2 Tbsp. chopped fresh mint

1. Stir together first 3 ingredients and 1 cup water in a small saucepan. Bring to a boil over high heat. Reduce heat to medium, and simmer 5 minutes. Remove from heat.
2. Toss together cabbage and next 4 ingredients in a medium bowl. Transfer to a jar; add vinegar mixture, and screw on lid. Chill 8 to 12 hours, turning occasionally. Store in refrigerator up to 3 days.

MAKES about 4½ cups. **HANDS-ON** 25 min.; **TOTAL** 8 hours, 25 min.

Slow-Cooker Ribs

OUR "LOW AND SLOW" METHOD IS LIKE MAGIC

EASY BBQ RIBS

- **1 slab baby back pork ribs (3 ½ to 4 lb.)**
- **Rib Rub, divided**
- **½ cup pineapple juice**
- **2 tsp. rice vinegar**
- **1 tsp. minced garlic**
- **Fancy Barbecue Sauce**

1. Rinse slab, and pat dry with paper towels. Remove thin membrane from back of ribs by slicing into it with a knife and then pulling it off. (This will make ribs more tender.) Reserve ½ Tbsp. Rib Rub; sprinkle both sides of ribs with remaining rub. Wrap ribs in plastic wrap; chill 8 to 12 hours.
2. Stir together pineapple juice, next 2 ingredients, and reserved ½ Tbsp. Rib Rub in a 6-qt. slow cooker. Cut slab in half, and stack halves in slow cooker. Cover and cook on HIGH 1 hour. Reduce heat to LOW, and cook 4 to 5 hours or until ribs are tender.
3. Preheat broiler with oven rack 10 inches from heat. Remove ribs from slow cooker, and place in a single layer in a heavy-duty aluminum foil-lined jelly-roll pan. Brush both sides of ribs with Fancy Barbecue Sauce. Broil ribs 2 to 3 minutes on each side or until caramelized. Serve with remaining barbecue sauce.
Note: Ribs may be cooked entirely on LOW for 6 to 7 hours or until tender.

MAKES 4 to 6 servings. **HANDS-ON** 30 min.; **TOTAL** 13 hours, 45 min., including rub and sauce

Rib Rub

Stir together 2 Tbsp. **dark brown sugar,** 1 ½ Tbsp. **smoked paprika,** 1 ½ tsp. **ground cumin,** 1 ½ tsp. **kosher salt,** ½ tsp. **garlic salt,** ½ tsp. **onion salt,** ½ tsp. **dry mustard,** ¼ tsp. **freshly ground black pepper,** ¼ tsp. **ground red pepper,** and ¼ tsp. **ground ginger.**

MAKES about ¼ cup. **HANDS-ON** 5 min., **TOTAL** 5 min.

Fancy Barbecue Sauce

Stir together 1 cup **tomato-based bottled barbecue sauce,** ¼ cup **honey,** 1 ½ tsp. **apple cider vinegar,** ¼ tsp. **dry mustard,** and ¼ tsp. **dried crushed red pepper** in a small saucepan. Bring to a boil over medium heat, stirring often; boil 1 minute. Remove from heat. Cover to keep warm.
Note: We tested with KC Masterpiece Original Barbecue Sauce.

MAKES about 1 ¼ cups. **HANDS-ON** 10 min., **TOTAL** 10 min.

A Healthy Slice

TAKE ADVANTAGE OF SMART SUPERMARKET SHORTCUTS TO MAKE A HOMEMADE PIZZA THAT'S HEALTHIER (AND EVEN FASTER) THAN DELIVERY

QUICK BBQ CHICKEN PIZZAS

Think pizza is off-limits while watching your waistline? Think again. Try this easy recipe that's ready in 30 minutes. Find more family-friendly, healthy dishes in the Cooking Light Diet, a great new meal planner from Cooking Light *magazine. Get custom meal plans based on what you like to eat. The meal planner counts the calories so you don't have to!*

- **Vegetable cooking spray**
- **¾ cup chopped red onion**
- **1½ cups shredded skinned and boned deli-roasted chicken breast**
- **¼ cup unsalted chicken cooking stock**
- **¼ cup reduced-sodium marinara sauce**
- **¼ cup bottled barbecue sauce**
- **1 (9-oz.) package 3 (7-inch) prebaked thin-and-crispy pizza crusts**
- **2 oz. fresh mozzarella cheese, sliced**
- **⅓ cup chopped green onions**
- **½ tsp. freshly ground black pepper**

1. Preheat oven to 450°. Heat a saucepan over medium heat. Coat pan with cooking spray. Add onion; sauté 2 minutes or until tender. Stir in chicken and next 3 ingredients; cook 3 minutes or until thoroughly heated. Remove pan from heat.

2. Spread about ½ cup chicken mixture over each crust; top with mozzarella cheese. Place pizzas directly on oven rack (for a crisp crust in less time). Bake at 450° for 9 minutes or until cheese melts. Top with green onions and pepper. Cut each pizza into 4 wedges.

MAKES 4 servings (serving size: 3 wedges).
HANDS-ON 20 min., **TOTAL** 30 min.

NUTRITIONAL INFORMATION
(per serving)

CALORIES: 337; FAT: 11g;
PROTEIN: 21.3g;
CARBOHYDRATES: 37.4g;
SODIUM: 538mg

Raise the Bar

TANGY RHUBARB AND SWEET RASPBERRIES MINGLE IN A PORTABLE, MAKE-AHEAD DESSERT

RASPBERRY "RHUBARS"

CRUST

- 2 cups all-purpose flour
- ½ cup powdered sugar
- ½ tsp. baking soda
- ¼ tsp. table salt
- ¾ cup cold butter, cubed
- ¾ cup toasted slivered almonds, coarsely chopped
- Vegetable cooking spray

RHUBARB-RASPBERRY FILLING

- ¾ cup granulated sugar
- ¼ cup cornstarch
- ¼ tsp. table salt
- 1 lb. rhubarb or 1 (1-lb.) package frozen rhubarb, thawed, cut into ½-inch-thick slices
- 2 (6-oz.) containers fresh raspberries
- 1 tsp. vanilla extract

CREAM CHEESE BATTER

- 2 (8-oz.) packages cream cheese, softened
- ¾ cup granulated sugar
- 2 large eggs
- 1 tsp. lemon zest
- 1 Tbsp. fresh lemon juice
- Garnish: powdered sugar

1. Prepare Crust: Preheat oven to 350°. Stir together flour and next 3 ingredients in a large bowl; cut butter into flour mixture with a pastry blender until crumbly. Stir in almonds. Press mixture onto bottom of a lightly greased (with cooking spray) 13- x 9-inch pan. Bake 15 to 20 minutes or until lightly browned. Cool on a wire rack until ready to use.

2. Prepare Filling: Stir together ¾ cup granulated sugar and next 2 ingredients in a large heavy saucepan; stir in rhubarb and 1 container raspberries. Let stand 15 minutes, stirring occasionally. Bring mixture to a boil over medium heat, stirring constantly. Reduce heat to low, and simmer 3 minutes, stirring constantly. Remove from heat, and stir in vanilla and remaining raspberries.

3. Prepare Cream Cheese Batter: Beat cream cheese and ¾ cup granulated sugar with an electric mixer until smooth. Add eggs, 1 at a time, and beat just until blended after each addition. Add lemon zest and juice, beating well. Spread Rhubarb-Raspberry Filling over Crust. Gently spread Cream Cheese Batter over filling.

4. Bake at 350° for 25 to 30 minutes or until set. Cool on a wire rack 1 hour. Cover and chill 4 to 8 hours. Cut into bars.

MAKES about 2 dozen 2- x 2-inch bars.
HANDS-ON 40 min.; **TOTAL** 6 hours, 35 min.

Community Cookbook

BUILD A BETTER LIBRARY, ONE GREAT BOOK AT A TIME

Nashville cook Matt Moore wants all Southern gentlemen to feel comfortable whipping up a great meal. But you don't need a Y chromosome to find inspiration in Moore's collection of 130 recipes with classic and New South flavors. All Southern cooks—men and women alike—will want to keep this book close at hand.

PULLED PORK BARBECUE NACHOS

- 2 Tbsp. unsalted butter
- 2 Tbsp. all-purpose flour
- 2 cups milk
- ½ tsp. kosher salt
- 3 cups (12 oz.) shredded Monterey Jack cheese
- 1 lb. smoked pulled pork
- 1 cup bottled vinegar-based barbecue sauce
- 1 (15-oz.) can black beans, drained and rinsed
- 1 (12-oz.) package blue corn tortilla chips
- 3 plum tomatoes, seeded and diced
- ¼ cup finely chopped fresh cilantro
- ½ cup sour cream
- ½ cup guacamole
- ¼ cup pickled sliced jalapeño peppers

1. Melt butter in a medium saucepan over medium heat; whisk in flour until smooth. Cook, whisking constantly, 1 minute. Gradually whisk in milk; cook, whisking constantly, until mixture is thickened and bubbly. (Be careful not to scorch milk.) Stir in salt. Remove from heat, and stir in cheese until blended and smooth; keep warm.
2. Combine pork and barbecue sauce in a medium saucepan. Cook over medium heat, stirring occasionally, 2 to 3 minutes or until thoroughly heated; keep warm. Cook beans in a small saucepan over medium heat, stirring occasionally, 2 to 3 minutes or until thoroughly heated; keep warm.
3. Arrange half of chips in a single layer on a large serving tray. Layer with half each of beans, pork mixture, tomato, and cilantro. Drizzle with 1 cup cheese sauce. Repeat layers with remaining chips, beans, pork mixture, tomato, and cilantro. Drizzle with 1 cup cheese sauce. Top with sour cream, guacamole, and jalapeño peppers. Serve with remaining cheese sauce.

MAKES 8 to 10 servings. **HANDS-ON** 19 min.; **TOTAL** 19 min.

TUESDAY NIGHT HAMBURGER STEAK

- 1 lb. ground chuck
- 1½ tsp. kosher salt, divided
- 1 tsp. freshly ground black pepper, divided
- ½ tsp. garlic powder
- 1 Tbsp. extra virgin olive oil
- 1 medium Vidalia or sweet onion, coarsely chopped
- 1 (8-oz.) package fresh mushrooms, quartered
- 2 Tbsp. all-purpose flour
- 1¼ cups beef stock
- Dash of Worcestershire sauce
- Garnish: finely chopped fresh parsley

1. Shape meat with hands into 4 (3-inch) patties (each about ¾ inch thick). Season both sides of patties with 1 tsp. salt, ½ tsp. pepper, and garlic powder.
2. Place a 12-inch cast-iron skillet over medium heat 1 minute or until hot. Add patties, and cook 4 minutes on each side. Transfer patties to a plate, reserving drippings in skillet.
3. Add oil to reserved drippings in skillet, and cook over medium heat until hot. Add onion, mushrooms, and remaining ½ tsp. salt and ½ tsp. pepper; sauté 8 minutes or until onion is slightly browned and tender.
4. Sprinkle flour over onion mixture. Cook, stirring constantly with a wooden spoon, 1 minute or until flour is thoroughly blended and smooth. Slowly stir in beef stock, stirring to loosen browned bits from bottom of skillet. Stir in Worcestershire sauce, and bring mixture to a light boil, stirring constantly.
5. Reduce heat to medium-low, add patties to skillet, and simmer 10 minutes or until patties are thoroughly cooked and gravy thickens. Place patties on individual serving plates, and top each with gravy.

MAKES 4 servings. **HANDS-ON** 46 min., **TOTAL** 46 min.

GAME DAY VENISON CHILI

Surrounded by plenty of cold beer, college football, and friends—I can honestly say that I'm in paradise when I'm eating a bowl of warm venison chili. If you don't have a freezer full of venison, don't fret! You can simply sub lean ground beef, turkey, or even bison, which can be found at major supermarkets. I like to make up a big pot of this chili to share; refrigerate any leftovers in an airtight container, and heat them up for a tasty lunch.

- 2 **Tbsp. extra virgin olive oil**
- 1 **large Vidalia or sweet onion, finely diced (about 2 cups)**
- 1 **green bell pepper, finely diced (about 1 ½ cups)**
- 4 **garlic cloves, minced**
- 2 **jalapeño peppers, seeded and finely diced**
- 2 **lb. ground venison**
- 3 **Tbsp. chili powder**
- 2 **Tbsp. plus 1 ½ tsp. ground cumin**
- 1 **Tbsp. kosher salt**
- 1 **tsp. freshly ground black pepper**
- 1 **cup amber beer**
- 1 **(28-oz.) can tomato sauce**
- 1 **(28-oz.) can petite diced tomatoes**
- 1 **(14-oz.) can dark red kidney beans, drained and rinsed**
- 1 **(14-oz.) can black beans, drained and rinsed**
- **Toppings: shredded sharp Cheddar cheese, sour cream, finely chopped green onions**

1. Heat a Dutch oven over medium heat 1 minute or until hot; add oil. Add onion and bell pepper; sauté 8 minutes. Add garlic and jalapeño; sauté 2 minutes or just until fragrant.
2. Add venison and next 4 ingredients, and cook, stirring often, 6 minutes or until meat crumbles and is no longer pink. Add beer, and cook, stirring with a wooden spoon to loosen any browned bits from bottom of Dutch oven, 2 minutes.
3. Stir in tomato sauce and next 3 ingredients; reduce heat to medium-low, and simmer, partially covered and stirring occasionally, 30 minutes. Remove from heat, and serve with desired toppings.

Note: We tested with Dos Equis Amber.

MAKES 8 to 10 servings. **HANDS-ON** 30 min., **TOTAL** 1 hour

QUICK COOK SHRIMP AND CHORIZO PAELLA

Brimming with fresh seafood, paella is a real crowd-pleaser. I like to build on great Spanish flavors with local Southern ingredients like Gulf shrimp, Vidalia onion, and green peas.

- 1 **lb. unpeeled, large raw Gulf shrimp**
- ½ **lb. dried chorizo sausage, cut into ½-inch slices**
- 1 **medium Vidalia or sweet onion, finely diced**
- 1 **(5-oz.) package yellow rice mix**
- 1 **(14.5-oz.) can petite diced tomatoes, undrained**
- 2 ¼ **cups reduced-sodium chicken broth**
- 1 ½ **cups fresh or frozen green peas**
- **Garnish: chopped fresh parsley**

Quick Cook Shrimp and Chorizo Paella

1. Peel shrimp, leaving tails on; devein, if desired.
2. Place a large skillet over medium-high heat 1-2 minutes or until hot. Add chorizo, and cook, stirring constantly, 3 minutes or until browned. Transfer chorizo to a paper towel-lined plate, reserving drippings in skillet.
3. Add onion to reserved hot drippings in skillet, and sauté 5 minutes or until translucent and tender. Stir in rice mix, and cook, stirring constantly, 1-2 minutes or until rice is well coated.
4. Add tomatoes, and cook, stirring with a wooden spoon to loosen any browned bits from bottom of skillet, 1 minute. Add chicken broth, and bring to a boil over high heat. Cover, reduce heat to low, and simmer, stirring once, 15 minutes.
5. Fold chorizo, shrimp, and peas into rice mixture; cover and cook 5 minutes or until rice is al dente and shrimp turn bright pink and firm. Serve immediately.

MAKES 6 servings. **HANDS-ON** 42 min., **TOTAL** 42 min.

GRILLED CORN AND TEQUILA-LIME BUTTER

I serve fresh, grilled corn at all of my summertime cookouts. This quick dish simply goes on the grill—easy as that. I like to keep the husks on so that the corn is nice and juicy and you get smoky, grilled flavors in each bite. Grilling with the husks on also makes for a rustic, authentic presentation.

CORN

- **8 ears fresh corn with husks**
- **1 tsp. kosher salt**

TEQUILA-LIME BUTTER

- **½ cup unsalted butter, softened**
- **1 tsp. seeded, finely diced jalapeño pepper**
- **1 tsp. lime zest**
- **1½ tsp. fresh lime juice**
- **1 tsp. tequila**
- **½ tsp. kosher salt**

1. Prepare Corn: Preheat grill to 300° to 350° (medium heat). Pull husks down, keeping husks attached and intact; remove and discard silks. Pull husks back up to cover corn. Place corn in salted water to cover in a large bowl or stockpot, and let stand 10 minutes.

2. Meanwhile, prepare Tequila-Lime Butter: Combine butter and next 5 ingredients in a small bowl, stirring with a wooden spoon until blended. Spoon butter mixture onto plastic wrap; roll tightly, forming a log. Chill until ready to use.

3. Remove corn from water. Place corn in husks on cooking grate, and grill, covered with grill lid, 15 to 20 minutes or until husks are blackened, turning every 5 minutes. Place corn on a serving plate, and fold back or remove husks.

4. Cut butter mixture into 8 pieces. Spread 1 piece of butter mixture on each cob until melted, turning to coat each cob completely. Serve immediately.

MAKES 8 servings. **HANDS-ON** 27 min., **TOTAL** 37 min.

NASHVILLE FARMERS' MARKET SUCCOTASH

One of the best ways to celebrate the bounty of fresh ingredients is with a traditional succotash. Primarily corn and lima beans (butter beans), succotash is one of those dishes that can take on many forms based on what's fresh.

- **4 ears fresh corn, husks removed**
- **2 Tbsp. unsalted butter**
- **2 Tbsp. extra virgin olive oil**
- **1 Vidalia or sweet onion, finely chopped**
- **1 red bell pepper, finely chopped**
- **1 Tbsp. kosher salt**
- **2½ tsp. freshly ground black pepper**
- **3 garlic cloves, minced**
- **2 cups fresh lima beans***
- **3 cups chopped tomatoes**
- **½ lb. fresh okra, cut into 1-inch pieces**
- **1 cup dry white wine****

1. Hold each cob upright on a cutting board, and carefully cut downward, cutting corn kernels from cobs into a large bowl to equal 2¾ cups. Discard cobs.
2. Place a 12-inch skillet over medium-high heat 1 minute or until hot; melt butter with olive oil in hot skillet. Add onion and next 3 ingredients, and sauté 8 minutes or until onion is tender and translucent. Add garlic, and sauté 1 minute or until fragrant.
3. Stir in corn kernels, lima beans, and next 2 ingredients. Add wine, and cook, stirring to loosen any browned bits from bottom of skillet, 1 minute. Reduce heat to medium-low, and simmer, partially covered and stirring occasionally, 25 minutes or until lima beans and okra are tender. Serve immediately.

*Frozen lima beans, thawed, may be substituted.
**Reduced-sodium chicken broth may be substituted.

MAKES 6 servings. **HANDS-ON** 48 min., **TOTAL** 48 min.

GOO GOO CLUSTER BROWNIES À LA MODE

In any event, I like to make up a big batch of brownies, fortified with this delicious treat, and serve them warm with a hearty helping of vanilla ice cream on the side.

- **Vegetable cooking spray**
- **3 Original Goo Goo Clusters**
- **1 cup sugar**
- **½ cup all-purpose flour**
- **⅓ cup unsweetened cocoa**
- **½ tsp. baking powder**
- **¼ tsp. kosher salt, divided**
- **½ cup vegetable oil**
- **2 large eggs**
- **Vanilla ice cream**

1. Preheat oven to 350°. Line bottom and sides of an 8-inch square pan with aluminum foil, allowing 2 inches to extend over sides; lightly grease foil with cooking spray. Coarsely chop Goo Goo Clusters into bite-size pieces to equal 1¼ cups.
2. Whisk together sugar, next 3 ingredients, and ⅛ tsp. salt in a large bowl until blended. Make a well in center of mixture; add oil and eggs to well, and stir until batter is blended and smooth. Fold three-fourths of chopped candy into batter. Pour into prepared pan.
3. Bake at 350°, on center oven rack, for 30 minutes or until a wooden pick inserted in center comes out clean. Remove from oven, and immediately sprinkle with remaining ⅛ tsp. salt and remaining chopped candy. Cool completely on a wire rack (about 1 hour).
4. Lift brownies from pan, using foil sides as handles. Gently remove foil, and cut into 9 squares. To serve, reheat brownies, and serve warm with ice cream.

MAKES 9 servings. **HANDS-ON** 10 min.; **TOTAL** 1 hour, 40 min.

BACON OLD-FASHIONED

When it comes to bacon, we Southerners are always getting creative. Like, say, adding bacon flavor to bourbon. That's what gives this traditional cocktail a makeover. Here's the deal: Cook 4 to 5 strips of your favorite hickory-smoked bacon in a skillet. Eat that delicious bacon—mmm. Pour 1 (750-milliliter) bottle of your favorite bourbon and the bacon drippings from skillet into a 4-cup glass measuring cup; let stand at room temperature for 24 hours. Pop the measuring cup into the freezer for a couple of hours. Pull it out, and skim off the bacon fat (it will congeal and separate from the alcohol). Pour the ice-cold bourbon through a cheesecloth-lined strainer into a glass jar to remove any particles. Your bourbon is now flavored with the smoky bacon fat for use in all of your favorite cocktails.

- **1 sugar cube or ½ tsp. granulated sugar**
- **2-3 dashes Angostura bitters**
- **¼ cup bacon-infused bourbon**
- **Garnishes: bacon strips, orange peel strip, or maraschino cherry**

Muddle sugar, bitters, and 1 tsp. water in an old-fashioned glass to release flavors and blend ingredients. Swirl sugar mixture to coat inside of glass. Add 1 large ice cube, and pour bourbon over ice. Serve immediately.

MAKES 1 serving. **HANDS-ON** 5 min., **TOTAL** 5 min.

PAN-SEARED SKIRT STEAK AND CHIMICHURRI

Traditionally used as fajita meat, skirt steak is gaining popularity these days as a stand-alone meat. It's my preferred cut of beef because it offers great flavor, tenderness, and affordability. I like to pan-sear this all-in-one cut over really high heat in my cast-iron skillet to create a nice char on the outside of the meat (open a window or turn on the fan in the kitchen). Instead of wasting time on a marinade, I prefer to reverse the process by slicing the meat thin and pouring all of the chimichurri sauce over the meat prior to serving. Even better, this chimichurri sauce is super versatile. It also goes well on pork, chicken, and firm cuts of fish, such as shark and swordfish.

- **2 lb. skirt steak**
- **¾ cup plus 2 Tbsp. extra virgin olive oil, divided**
- **2 ½ tsp. kosher salt, divided**
- **2 tsp. freshly ground black pepper, divided**
- **¼ cup finely chopped red onion**
- **2 large garlic cloves, minced**
- **¼ cup red wine vinegar**
- **½ cup chopped fresh parsley**

1. Place a 12-inch cast-iron skillet over medium-high heat 2-3 minutes or until hot. Coat steak with 2 Tbsp. oil, and season both sides of steak with 1 ½ tsp. salt and 1 tsp. pepper. Cook steak in hot skillet 3 to 4 minutes on each side or until a meat thermometer inserted into thickest portion registers 130° (medium-rare). Transfer steak to a cutting board, and cover loosely with aluminum foil; let stand 5 minutes.
2. Meanwhile, make chimichurri by stirring together onion, garlic, remaining 1 tsp. salt, and remaining 1 tsp. pepper in a small bowl. Stir in vinegar; let stand 5 minutes. Stir in parsley. Add remaining ¾ cup oil in a slow, steady stream, whisking constantly until smooth.
3. Cut steak diagonally across grain into ¼-inch-thick slices, and place on a serving platter. Pour chimichurri over sliced steak, and serve immediately.

MAKES 6 servings. **HANDS-ON** 21 min., **TOTAL** 26 min.

June

118 **The Pies of Summer** Chill out and satisfy a sweet tooth with swoon-worthy icebox pies

123 **Taste of the Gulf** Alabama's "one-pot" is another regional spin on the Southern seafood boil that deserves a spot on your summertime table...with all the fixin's, of course

126 **Food Awards 2015** We've tasted and tested over 300 of the South's newest artisan food products to bring you two dozen worth adding to your shopping list

132 COOKING **The *SL* Guide to Grilled Chicken** From wing to thigh, and all parts in between, our Test Kitchen shares surprising new twists for the South's favorite bird

141 THE *SL* TEST KITCHEN ACADEMY **Quick Fruit Pickles** No-cook refrigerator pickles are a busy cook's trick for preserving summer's fleeting fruits

142 *COOKING LIGHT* **Dinner in a Breeze** One-dish pork in 20 minutes is just peachy

143 THE SLOW DOWN **Green Beans Like You've Never Had** Gussy up green beans and tomatoes with red curry paste and peanuts for a side that's anything but ordinary

144 COMMUNITY COOKBOOK New ways with Mason jars make infusing liquids kitchen alchemy

The Pies Of Summer

ICEBOX DESSERTS ARE THE HEROES OF THE SOUTHERN SUMMER. NASHVILLE PASTRY CHEF LISA DONOVAN SHARES HER EASY-AS-PIE RECIPES FOR THESE COOL AND CREAMY TREATS

IT WAS ALWAYS my job to put the graham crackers in a big plastic bag and bang them with a wooden spoon until they were smashed enough to add the salt, sugar, and melted butter, and stir. Eventually, my mom got a small food processor, which left me feeling both very well-to-do and a little disappointed that my job had been replaced by a motor.

Icebox pie—or refrigerator casserole as I call it—has come back in vogue with a bit of Southern belle sashay. Technically, there is nothing to it: a crumb crust and a cream filling. Even as a pastry chef who requires from-scratch recipes for everything, with refrigerator casseroles, I only labor to turn off the part of my brain that's inclined to try to reinvent the wheel. From the satiny sheen of the filling to the mounds of whipped cream or fresh fruit piled on top, there is much beauty in a refrigerator casserole, but none more than its sheer simplicity.

My mom's Four-Layer Surprise—a stratified combo of pecan crust, cream cheese with powdered sugar, chocolate pudding, and whipped cream—was my favorite thing to eat growing up. I had a thing for those types of one-pan comforts. I still do. It's the most honest and warmest kind of love letter that I know. When I drive from Nashville through the entirety of Alabama to see my mom and dad, I know exactly what will be waiting for me in her kitchen.

The refrigerator casserole didn't originate in the South. It has centuries-old roots in European desserts, such as the French charlotte, English trifle, or Welsh tiffin. So, while Southerners can't really stake a claim on icebox desserts, we can sure be proud of the fact that we do them better than anyone else. After all, as Mississippi native and chef Martha Foose will tell you, "our love of sweetened condensed milk is strong." And cool, easy cookery with minimal oven use is a veritable requirement for the months of sultry weather we endure.

If love is a casserole, which I believe it is, then an icebox dessert is the first kiss—simple, sweet, and perfect on warm summer nights.

CRUMB CRUST

- **1½ cups crushed cookies or crackers, such as:**
  - **vanilla wafers**
  - **graham crackers**
  - **gingersnaps**
  - **saltine crackers**
  - **round buttery crackers**
- **¼ cup sugar**
- **1 tsp. kosher salt (omit when using buttery crackers or saltines)**
- **6 Tbsp. butter, melted**
- **Vegetable cooking spray**

Process crushed cookies or crackers, sugar, and (if used) salt in a food processor until finely crushed and well combined. Add melted butter, and process until thoroughly combined. Press on bottom, up sides, and onto lip of a lightly greased (with cooking spray) 9-inch regular pie plate or 9-inch deep-dish pie plate. Freeze 30 minutes to 1 hour or while preparing fillings.

Note: For baked piecrusts, preheat oven to 325°. Bake crust 8 to 10 minutes or until lightly browned.

MAKES 1 (9-inch) regular piecrust or 1 (9-inch) deep-dish piecrust.
HANDS-ON 15 min., **TOTAL** 15 min.

CRUSHED
VANILLA
WAFERS
CRUSHED
GRAHAM
CRACKERS
CRUSHED
GINGERSNAPS
CRUSHED
SALTINE
CRACKERS
CRUSHED
ROUND BUTTERY
CRACKERS

LEMON-BUTTERMILK ICEBOX PIE

- 1 (14-oz.) can sweetened condensed milk
- 1 Tbsp. loosely packed lemon zest
- ½ cup fresh lemon juice
- 3 large egg yolks
- ¼ cup buttermilk
- Graham Cracker Crust, baked (recipe, page 118)
- Vegetable cooking spray

1. Preheat oven to 325°. Whisk together first 3 ingredients in a bowl.
2. Beat egg yolks with a handheld mixer in a medium bowl at high speed 4 to 5 minutes or until yolks become pale and ribbons form on surface of mixture when beater is lifted. Gradually whisk in sweetened condensed milk mixture, and whisk until thoroughly combined; whisk in buttermilk. Pour mixture into prepared crust.
3. Bake at 325° for 20 to 25 minutes or until set around edges. (Pie will be slightly jiggly.) Cool on a wire rack 1 hour. Cover pie with lightly greased (with cooking spray) plastic wrap, and freeze 4 to 6 hours.

MAKES 8 servings. **HANDS-ON** 20 min.; **TOTAL** 6 hours, 15 min., including crust

KEY LIME-BUTTERMILK ICEBOX PIE WITH BAKED BUTTERY CRACKER CRUST

Prepare Lemon-Buttermilk Icebox Pie as directed, substituting baked Buttery Cracker Crust for Graham Cracker Crust and **Key lime juice** for lemon juice. Top with **Sweetened Whipped Cream.**

HANDS-ON 20 min.; **TOTAL** 6 hours, 20 min., including crust and whipped cream

Sweetened Whipped Cream

Beat 2 cups **heavy cream** and 1 tsp. **vanilla extract** at medium-high speed with an electric mixer until foamy; gradually add 1/4 cup **powdered sugar,** beating until soft peaks form.

MAKES about 3 1/2 cups. **HANDS-ON** 5 min., **TOTAL** 5 min.

STRAWBERRY-LEMON-BUTTERMILK ICEBOX PIE WITH BAKED GINGERSNAP CRUST

Prepare Lemon-Buttermilk Icebox Pie as directed, substituting baked Gingersnap Crust for Graham Cracker Crust and whisking 1/2 cup **strawberry jam** into condensed milk mixture. Microwave 3 Tbsp. strawberry jam in a medium-size microwave-safe bowl at HIGH 20 seconds. Stir 16 oz. sliced **strawberries** into jam. Top pie with strawberry mixture just before serving.

HANDS-ON 25 min.; **TOTAL** 6 hours, 25 min., including crust

Lemon-Blueberry Topping

(You will need 2 cups fresh **blueberries.**) Bring 1 cup blueberries, 1/3 cup **sugar,** 3 Tbsp. **water,** 2 Tbsp. **fresh lemon juice,** and 1 Tbsp. **lemon zest** to a boil in a small saucepan over medium-high heat. Reduce heat to low, and simmer, stirring occasionally, 8 to 10 minutes or until mixture has thickened and berries begin to break down. Remove from heat, and stir in 1 cup blueberries. Cool completely (about 1 hour); cover and chill until ready to use.

MAKES about 1 1/2 cups. **HANDS-ON** 20 min.; **TOTAL** 1 hour, 20 min.

MANGO-LEMON-BUTTERMILK ICEBOX PIE WITH BAKED SALTINE CRACKER CRUST

Bring 1/2 cup **water** to a boil over medium-high heat in a saucepan. Stir in 1 cup chopped fresh or thawed frozen **mango.** Remove from heat, and stir in 1 Tbsp. **fresh lemon juice.** Process mixture in a blender until smooth. Prepare Lemon-Buttermilk Icebox Pie as directed, substituting baked Saltine Cracker Crust for Graham Cracker Crust and 3/4 cup mango puree and 1/4 cup lemon juice for 1/2 cup lemon juice. Discard remaining puree. Whisk 2 Tbsp. **mango schnapps** into condensed milk mixture along with puree and lemon juice. Top with **Sweetened Whipped Cream** (recipe, page 120) and chopped fresh mango just before serving.

HANDS-ON 30 min.; **TOTAL** 6 hours, 25 min., including crust and whipped cream

PEACH DIVINITY ICEBOX PIE

- **3 large eggs**
- **3/4 cup sugar**
- **4 tsp. unflavored gelatin**
- **1/4 cup boiling water**
- **1 cup peach preserves**
- **Pinch of kosher salt**
- **1 cup heavy cream**
- **Gingersnap Crust, baked (recipe, page 118)**
- **2 medium peaches, thinly sliced**
- **Sweetened Whipped Cream (recipe, page 120)**
- **Garnish: fresh mint leaves, peach slices**

1. Pour water to a depth of 1 1/2 inches into a 3 1/2-qt. saucepan; bring to a boil over medium-high heat. Reduce heat to medium, and simmer. Whisk together eggs and sugar in a 2 1/2-qt. glass bowl. Place bowl over simmering water, and cook, whisking constantly, 5 to 6 minutes or until mixture becomes slightly thick and sugar dissolves. Remove from heat.
2. Place gelatin in a small bowl, and gradually add 1/4 cup boiling water, whisking constantly until gelatin is completely dissolved.
3. Beat egg mixture at high speed with a handheld electric mixer 8 to 10 minutes or until ribbons form on surface of mixture when beater is lifted. Add gelatin mixture, and beat 1 minute. Fold in peach preserves and salt.
4. Beat heavy cream in a medium bowl at medium-high speed 2 to 3 minutes or until soft peaks form. Fold one-third of whipped cream into egg-and-preserves mixture; gently fold in remaining whipped cream.
5. Spoon filling into baked crust; arrange peach slices over filling. Cover with plastic wrap, and freeze 2 hours. Dollop with Sweetened Whipped Cream just before serving.

Tip: When beating the egg mixture, stop the mixer occasionally, and lift the beater. If drips form ribbons that hold shape for 1 to 2 seconds on surface of mixture, you're ready to add the gelatin.

MAKES 8 servings. **HANDS-ON** 40 min.; **TOTAL** 3 hours, 10 min., including crust and whipped cream

CHOCOLATE-BOURBON-BUTTERSCOTCH ICEBOX CAKE

- ½ cup cornstarch
- 3 cups milk
- 4 large eggs
- ½ cup butter
- 2 cups firmly packed dark brown sugar
- 2 cups heavy cream
- 1 Tbsp. vanilla extract
- Pinch of sea salt
- ¼ cup bourbon
- 64 chocolate wafers (about 1½ [9-oz.] packages)
- Sweetened Whipped Cream (recipe, page 120)
- Garnish: crushed chocolate wafers

1. Place cornstarch in a medium bowl. Gradually whisk in milk and eggs.
2. Melt butter in a Dutch oven over medium heat, and stir in brown sugar. Cook, stirring constantly, about 10 minutes or until mixture bubbles and becomes shiny.
3. Gradually whisk cream into butter mixture. (Mixture will bubble and clump but will get smooth as it cooks.) Cook, whisking constantly, 4 to 5 minutes or until smooth.
4. Gradually whisk about one-fourth of hot butter mixture into egg mixture, 1/2 cup at a time. Gradually add egg mixture to remaining hot butter mixture, whisking constantly.
5. Increase heat to medium-high. Add vanilla and salt, and cook, whisking constantly, about 5 minutes or until mixture thickens and just begins to bubble. Remove from heat, and stir in bourbon. Place plastic wrap directly onto warm filling (to prevent a film from forming). Cool completely (about 2 hours).
6. Place 16 cookies in a single layer on bottom of a 9-inch square pan. Spread 1 1/2 cups pudding over cookies. Repeat layers 3 times, using remaining pudding on top layer. Cover and chill 8 to 24 hours. Just before serving, top with Sweetened Whipped Cream.

Note: We tested with Nabisco Famous Chocolate Wafers.

MAKES 10 to 12 servings. **HANDS-ON** 45 min.; **TOTAL** 10 hours, 50 min., including whipped cream

Taste of the Gulf

DOWN IN FAIRHOPE, ALABAMA, IT'S ALL ABOUT THE MAGIC OF A ONE-POT SHRIMP BOIL AND A DOZEN FRIENDS

MY FRIENDS MENA AND HENRY MORGAN down in Fairhope, Alabama, call it a "one-pot," but that may be a Gulf Coast thing. Similar dishes are called a "Lowcountry boil," a "Beaufort boil," or "Frogmore stew." There are probably as many variations on the name as there are on the recipe, but for the Morgans, what matters is super-fresh shrimp, a simple setup, and a festive atmosphere. The party happens on their dock overlooking Mobile Bay, and they try to do it when there's a nice breeze to keep the mosquitoes away. A pretty sunset is a bonus, as are the schooners that sometimes cross the Bay.

Though Henry is known to occasionally wear women's sunglasses (to put his guests at ease), he is serious about the food. For him, a one-pot has to have the freshest shrimp you can find, Conecuh Sausage, and green olives, which give it a little tang to balance out the starches. Mena makes sure everyone feels relaxed and welcome, sometimes inviting over friends and neighbors (or even the occasional magazine editor) at the last minute. Over the years they've perfected the one-pot formula, but it's always a little imperfect. That's their secret, and that's the spirit to embrace when throwing a one-pot of your own. —Sid Evans

SHRIMP BOIL WITH GREEN OLIVES AND LEMON

The simmering liquid is packed with flavor, which transfers nicely to the shrimp.

- **2 medium-size yellow onions, chopped**
- **3 celery ribs, chopped**
- **4 garlic cloves**
- **4 bay leaves**
- **8 black peppercorns**
- **2 dried chile peppers**
- **3 lemons, sliced**
- **6 fresh flat-leaf parsley sprigs**
- **6 fresh thyme sprigs**
- **¼ cup kosher salt**
- **3 lb. medium-size raw shrimp**
- **1½ cups pitted Spanish olives**
- **Shrimp Spice Mix**
- **Classic Cocktail Sauce**
- **White Cocktail Sauce**
- **Lemon wedges**

1. Bring first 10 ingredients and 4 qt. water to a boil in a large Dutch oven over medium-high heat. Reduce heat to medium-low; cover and simmer 20 minutes.
2. Add shrimp to Dutch oven. Increase heat to high, and return to a simmer. Add Spanish olives. Remove from heat, and cover with a tight-fitting lid. Let stand 5 minutes.
3. Drain shrimp and olives, discarding remaining solids. Place shrimp-and-olive mixture in a jelly-roll pan, and sprinkle with Shrimp Spice Mix. Serve immediately with cocktail sauces and lemon wedges, or chill 15 to 20 minutes or until cold.

MAKES 6 to 8 servings. **HANDS-ON** 15 min.; **TOTAL** 55 min., including spice mix

Shrimp Spice Mix

Stir together 1 1/2 Tbsp. **paprika,** 2 tsp. **kosher salt,** 1 tsp. **lemon pepper,** 1/2 tsp. **dried oregano,** and 1/4 tsp. **ground red pepper** in a small bowl.

MAKES about 1/2 cup. **HANDS-ON** 5 min., **TOTAL** 5 min.

Shrimp Boil with Green Olives and Lemon

Classic Cocktail Sauce

- **1 cup ketchup**
- **½ cup prepared horseradish**
- **1 Tbsp. fresh lemon juice**
- **1 Tbsp. Worcestershire sauce**
- **1 tsp. hot sauce**
- **Kosher salt and freshly ground black pepper**

Stir together first 5 ingredients in a small bowl until well blended. Season with salt and pepper. Serve immediately, or cover and chill up to 2 days.

MAKES 1 1/2 cups. **HANDS-ON** 5 min., **TOTAL** 5 min.

White Cocktail Sauce

- **½ cup mayonnaise**
- **½ cup sour cream**
- **½ cup prepared horseradish**
- **1 Tbsp. fresh lime juice**
- **1 tsp. thinly sliced chives**
- **1 tsp. Worcestershire sauce**
- **½ tsp. hot sauce**
- **Kosher salt and freshly ground black pepper**

Stir together first 7 ingredients in a small bowl until well blended. Season with salt and pepper. Serve immediately, or cover and chill up to 2 days.

MAKES 1 1/2 cups. **HANDS-ON** 5 min., **TOTAL** 5 min.

ROASTED POTATO-AND-OKRA SALAD

This fresh yet hearty side dish features roasted potatoes instead of the boiled ones traditionally served as part of the Lowcountry-style spread.

- **1 lb. baby red potatoes**
- **1 lb. small Yukon gold potatoes**
- **1 lb. fresh whole okra**
- **2 Tbsp. olive oil, divided**
- **3 tsp. kosher salt, divided**
- **1 tsp. freshly ground black pepper, divided**
- **Lemon-Thyme Vinaigrette**
- **3 green onions, thinly sliced**
- **Garnish: fresh thyme sprigs**

1. Preheat oven to 450°. Bring potatoes and water to cover to a boil in a large saucepan over medium-high heat. Reduce heat to medium-low, and simmer 15 to 20 minutes or until fork-tender. Drain and cool completely (about 20 minutes).

2. Toss together okra, 1 Tbsp. olive oil, 1 tsp. salt, and 1/2 tsp. pepper, and place in a single layer in a jelly-roll pan. Cut potatoes in half, and toss with remaining olive oil, salt, and pepper. Place potatoes in a single layer in another jelly-roll pan.

3. Bake okra and potatoes at 450° for 20 to 25 minutes or until tender and golden brown. Remove from oven, and let stand 5 minutes.

4. Cut okra pods in half lengthwise. Toss together potatoes, okra, vinaigrette, and green onions in a medium bowl. Serve immediately.

MAKES 6 to 8 servings. **HANDS-ON** 20 min.; **TOTAL** 1 hour, 40 min., including vinaigrette

Lemon-Thyme Vinaigrette

- **1 small shallot, minced**
- **3 Tbsp. white wine vinegar**
- **1 tsp. lemon zest**
- **2 Tbsp. fresh lemon juice**
- **1 tsp. Dijon mustard**
- **¼ tsp. fresh thyme leaves**
- **¼ cup extra virgin olive oil**

Stir together shallot and white wine vinegar in a small bowl. Let stand 10 minutes. Stir in zest and next 3 ingredients. Add oil in a slow, steady stream, whisking until smooth.

MAKES about 1/2 cup. **HANDS-ON** 10 min., **TOTAL** 20 min.

CRISPY ANDOUILLE HUSH PUPPIES

This crunchy, savory bite is a delicious dockside starter.

- **Vegetable oil**
- **1½ cups self-rising white cornmeal mix**
- **1 cup diced andouille sausage**
- **¾ cup self-rising flour**
- **¾ cup finely chopped sweet onion**
- **1 large egg, lightly beaten**
- **⅔ cup lager beer**
- **⅓ cup buttermilk**

1. Pour oil to a depth of 3 inches into a Dutch oven; heat to 375°. Stir together cornmeal and next 3 ingredients in a large bowl. Add egg, beer, and buttermilk; stir just until moistened. Let stand 10 minutes.

2. Using a 1-inch cookie scoop, drop batter into hot oil, and fry, in batches, 2 to 3 minutes on each side or until golden brown. Drain on a wire rack over paper towels. Keep warm in a 200° oven.

MAKES 8 to 10 servings. **HANDS-ON** 20 min., **TOTAL** 35 min.

BOOZY PEACH SHORTCAKES WITH SWEET CREAM

- **6 fresh peaches, sliced**
- **1/3 cup peach schnapps**
- **2 Tbsp. fresh lime juice**
- **1 tsp. vanilla extract**
- **1 Tbsp. granulated sugar**
- **¼ cup loosely packed fresh mint leaves, torn**
- **Pinch of kosher salt**
- **1 cup heavy cream**
- **1 Tbsp. powdered sugar**
- **6 refrigerated jumbo biscuits (from a 16.3-oz. can)**
- **1 large egg, lightly beaten**
- **3 tsp. Demerara sugar**

Boozy Peach Shortcakes with Sweet Cream

1. Stir together first 7 ingredients in a medium bowl. Cover and let stand 1 hour, stirring occasionally.

2. Beat heavy cream at medium-high speed with an electric mixer until foamy; gradually add powdered sugar, beating until soft peaks form. Cover and chill until ready to use.

3. Place biscuits on a baking sheet, and brush with egg. Sprinkle Demerara sugar over biscuits, and bake according to package directions.

4. Split biscuits, and place bottom halves on individual plates. Spoon whipped cream over each biscuit, and top with peach slices and remaining biscuit half. Serve immediately.

Note: We tested with Pillsbury Grands Homestyle Buttermilk refrigerated biscuits.

MAKES 6 servings. **HANDS-ON** 20 min.; **TOTAL** 1 hour, 35 min.

FOOD
AWARDS
2015

THE SOUTHERN LIVING TEST KITCHEN TASTE-TESTED MORE THAN 300 OF THE NEWEST ARTISAN PRODUCTS TO FIND THE BEST SOUTHERN-MADE FOODS. THESE ARE OUR 24 FAVORITES

We kicked off our three-day taste test by sampling nearly 60 jars of jams and jellies and ended with a side-by-side comparison of 46 different kinds of coffee. In between the sugar headaches and caffeine rushes, not only did we find the perfect pickle and the most refreshing summer beer, but also we discovered the stories of the people who make them. There's Texan Sam Addison of **Pogue Mahone Pickles,** who embarked on an Indiana Jones-like quest to create the ideal pickle while in culinary school, and **Fullsteam Brewery's** Sean Lilly Wilson, who brews beer with locally grown—even crowd-sourced—ingredients like basil and persimmons. We savored **Cloister Honey** harvested by beekeeping couple Joanne Young and Randall York from their urban hives (one of which is located on the rooftop of Charlotte's Ritz-Carlton hotel) and swooned over Elizabeth Heiskell's **Debutante Farmer Bloody Mary mix,** inspired by the heirloom tomatoes she grows on her 25-acre farm outside Oxford, Mississippi. They all remind us that the best Southern recipes mix innovation with tradition. To read more of the stories behind our winning artisans and their products, visit *southernliving.com/foodawards.*

❶
Texas Sweet Heat Pickles
Pogue Mahone Pickles
MADE IN **Austin, TX**

Cool, crisp, and zingy all at once. The sweetness of these bread-and-butter-style pickles—our favorite of all the food we tasted—tempers the piquant punch of jalapeño, mustard seeds, and black peppercorns.

❷
Summer Basil Farmhouse Ale
Fullsteam Brewery
MADE IN **Durham, NC**

The beer equivalent of lemonade. The intensity of the North Carolina-grown basil varies depending on when the basil is harvested during summer.

❸
Pork Clouds
Bacon's Heir
MADE IN **Atlanta, GA**

A gourmet take on gas station rinds, kettle-cooked in olive oil for a light and airy texture with a satisfying crunch. In flavors like Malabar Black Pepper, Rosemary-and-Sea Salt, and Cinnamon Ceylon, it's an extraordinary high-low snack.

---

❶
Tupelo Honey Shortbread Bites
Willa's Shortbread
MADE IN **Madison, TN**

Tiny cookies sweetened with buttery, floral Tupelo honey.

❷
Original Salted Caramels
Shotwell Candy Co.
MADE IN **Memphis, TN**

Semi-hard caramels, infused with Kentucky bourbon and vanilla for a rich, deep flavor.

❸
Buttermilk Ice Cream
Southern Craft Creamery
MADE IN **Marianna, FL**

Supremely creamy with a subtle lemon-like tang. Milk from the creamery's own family-run dairy provides the velvety base.

❹
Wildflower Honeycomb Chocolate Bar
Chocolat by Adam Turoni
MADE IN **Savannah, GA**

Gilded dark chocolate bars that ooze with raw wildflower honeycomb at first bite. Nearby business Savannah

❶

Jalapeño, Serrano, and Habanero Sauces

Yellowbird Sauce
MADE IN **Austin, TX**

A brighter, less astringent hot sauce that uses citrus juice as the backdrop for fiery peppers in its three different varieties. This bottle has become our main squeeze.

❷

Habanero-Infused White-Vinegar

Frankie V's Kitchen
MADE IN **Dallas, TX**

A high-octane pepper sauce made by steeping organic habaneros in organic white vinegar. It delivers blazing acidity.

❸

Seasoning Salts

Beautiful Briny Sea
MADE IN **Atlanta, GA**

Hand-harvested sea salt blended with herbs and spices. Granulated honey makes Friends Forever savory-sweet; French Picnic adds Dijon mustard and herbes de Provence with a subtle hint of lavender.

❶
Ghost Pepper Whipped Honey
Cloister Honey
MADE IN **Charlotte, NC**

Thick, spreadable honey with a creamy sweetness that soothes the slow-burning bite from notoriously hot ghost peppers. A dollop heats up a ho-hum cheese board—or try it as a glaze on smoked wings.

❷
Strawberry Balsamic Preserves
Blackberry Farm
MADE IN **Walland, TN**

A scarlet jam made with strawberries grown in the gardens of the farmstead-style resort. The tart twist of balsamic vinegar keeps the chunky preserves vibrant, not cloyingly sweet.

❸
Mission Almond Butter
Big Spoon Roasters
MADE IN **Durham, NC**

A three-ingredient nut butter that blends wildflower honey and sea salt with heirloom Mission almonds, which are creamier and richer than regular almonds.

❶
Twenty Strong
King Bean Coffee Roasters
MADE IN **Charleston, SC**

A dark, chocolaty roast with a fruity hint of berries. These beans, the company's 20th anniversary blend, produce a black-coffee lover's dream: a full-bodied brew, with low acidity and plenty of pep.

❷
Front Porch Special
Piper and Leaf Artisan Tea Co.
MADE IN **Huntsville, AL**

A black tea blend dotted with cornflowers. It's spiked with delicate notes of cooling spearmint, jasmine, and a tinge of bittersweet orange from bergamot peel.

❸
Goat Cheese Cheesecake
Belle Chèvre
MADE IN **Elkmont, AL**

A light and airy cheesecake, delicately scented with vanilla and lemon zest. Alabama-made goat cheese, the signature ingredient, lends a cloud-like texture and some tangy tartness.

❶
Bloody Mary Mix
Debutante Farmer
MADE IN **Oxford, MS**

Bursting with the vine-fresh verve of summer tomatoes and fresh lime juice (for zip without the sinus-clearing horseradish). This bottle is our new brunch must-have.

❷
No. 6 Oaxacan Old Fashioned
Bittermilk
MADE IN **Charleston, SC**

A subtly spicy blend of Mexican chiles, cocoa nibs, spices, and sweet raisins. Mix with mezcal or rum.

❸
Pineapple Gum Syrup
Liber & Co.
MADE IN **Austin, TX**

A fruit-forward mixer, made with fresh pineapple juice and cane sugar, that tastes like a tropical resort bar in a bottle.

❹
Bristow Gin
MADE IN **Jackson, MS**

A botanical—not piney—gin with hints of orange peel and minty hyssop. It pairs perfectly with Jack Rudy's Elderflower Tonic.

❺
Belle Meade Bourbon Sherry Cask Finish
MADE IN **Nashville, TN**

A special batch of nine-year-old bourbon, matured in sherry casks from Spain for a darker color and sweeter, nutty taste.

❻
Richland Rum
MADE IN **Richland, GA**

A butterscotch-like rum made with sugarcane grown on the owners' estate. It goes down so smooth we love to drink it straight up.

❼
Elderflower Tonic
Jack Rudy Cocktail Co.
MADE IN **Charleston, SC**

A floral twist on traditional tonic, with subtle hints of pear and lychee. We recommend trying it with gin, bourbon, rum, or even tequila.

❽
Cucumber & Lavender Bitters
El Guapo Bitters
MADE IN **New Orleans, LA**

Nonalcoholic bitters that impart herbal freshness. Just a few drops elevate a gin and tonic.

THE *SL* GUIDE TO GRILLED CHICKEN

EIGHT WAYS TO TAKE YOUR GRILLED CHICKEN GAME FROM GOOD TO GREAT

No. 1

DISCOVER THE WONDER WING

Our easy, indirect-to-direct-heat grilling technique turns out tender wings with crisp skin. Pair them with any of our three sauces for the greatest finger-lickin' appetizer of the summer.

WONDER WINGS

- **3 lb. chicken wings**
- **2 tsp. vegetable oil**
- **1 tsp. kosher salt**
- **½ tsp. freshly ground black pepper**
- **Alabama White Sauce, Buttery Nashville Hot Sauce, or Vietnamese Peanut Sauce**

1. Light 1 side of grill, heating to 350° to 400° (medium-high) heat; leave other side unlit. Dry each wing well with paper towels. Toss together wings and oil in a large bowl. Sprinkle with salt and pepper, and toss to coat.
2. Place chicken over unlit side of grill, and grill, covered with grill lid, 15 minutes on each side. Transfer chicken to lit side of grill, and grill, without grill lid, 10 to 12 minutes or until skin is crispy and lightly charred, turning every 2 to 3 minutes. Toss wings immediately with desired sauce. Let stand, tossing occasionally, 5 minutes before serving.

MAKES 6 to 8 servings. **HANDS-ON** 25 min.; **TOTAL** 1 hour, not including sauces

Alabama White Sauce

Whisk together 1/3 cup **mayonnaise,** 3 Tbsp. chopped fresh **chives,** 1 Tbsp. prepared **horseradish,** 4 tsp. **apple cider vinegar,** 2 tsp. **Creole mustard,** 1 tsp. coarsely ground **pepper,** 1/4 tsp. **sugar,** and 1 finely grated **garlic clove.**

MAKES about 2/3 cup

Buttery Nashville Hot Sauce

Cook 1/4 cup melted **butter,** 3 to 4 tsp. **ground red pepper,** 2 tsp. **dark brown sugar,** 3/4 tsp. **kosher salt,** 1/2 tsp. **smoked paprika,** and 1/2 tsp. **garlic powder** in a small saucepan over medium heat, stirring constantly, 1 minute or until fragrant. Remove from heat, and stir in 1 Tbsp. **apple cider vinegar.**

MAKES about 1/3 cup

Vietnamese Peanut Sauce

Sauté 3 finely chopped large **garlic** cloves in 1 Tbsp. **vegetable oil** in a small saucepan over medium heat 1 to 2 minutes or until golden. Stir in 1/3 cup **fish sauce,** 1/3 cup firmly packed **light brown sugar,** and 2 to 3 tsp. **Asian chili-garlic sauce.** Bring to a simmer over medium heat, and simmer, stirring occasionally, 4 to 5 minutes or until thickened and reduced to about 1/2 cup. Stir in 3 Tbsp. finely chopped **toasted peanuts.** Sprinkle coated wings with 1/4 cup torn **cilantro** and **mint leaves.**

MAKES about 1/2 cup

No. 2

RETHINK THE KABOB

Mixing vegetables and meat on the same skewer is a no-go; you'll end up overcooking something. Instead, grill vegetable-only skewers to go with the meat-only kabobs. Choose thighs, a great cut of chicken for marinating because when they are deboned, the butcher leaves lots of nooks and crannies that trap the marinade. Finally, use a wide, flat skewer to keep the food from twirling around, making cooking and eating easier.

PERFECT CHICKEN KABOBS

- **3 lb. skinned and boned chicken thighs**
- **Sticky Sesame Sorghum, Chardonnay-Herb, or Buttermilk Tandoori Marinade (recipes, page 134)**
- **8 to 10 flat wooden skewers**
- **2 Tbsp. vegetable oil**
- **Kosher salt and freshly ground black pepper**

1. Trim excess fat from thighs. Cut each thigh lengthwise into 1 1/2-inch-wide strips. Reserve and chill 1/4 cup of desired marinade. Place chicken and remaining marinade in a large zip-top plastic freezer bag. Seal and turn bag to coat. Chill, turning bag occasionally, 3 to 8 hours.
2. Meanwhile, soak wooden skewers in water 30 minutes. Preheat grill to 350° to 400° (medium-high) heat. Remove chicken from marinade, discarding marinade. Pat chicken dry with paper towels. Thread 3 chicken strips in a loose accordion style onto each skewer. Brush with vegetable oil, and sprinkle with salt and pepper.
3. Grill kabobs, covered with grill lid and turning occasionally, 11 to 14 minutes or until chicken is done. During last 5 minutes of grilling, brush kabobs with reserved 1/4 cup marinade.

MAKES 6 to 8 servings. **HANDS-ON** 40 min.; **TOTAL** 3 hours, 35 min., including marinade

Buttermilk Tandoori Marinade

Whisk together 1 cup **buttermilk;** 3 **garlic** cloves, finely grated; 1 Tbsp. **paprika,** 2 tsp. finely grated fresh **ginger;** 1 1/2 tsp. **garam masala;** 1 1/2 tsp. **turmeric;** 1 tsp. **kosher salt;** and 1/2 tsp. freshly ground black **pepper.**

MAKES about 1 1/4 cups

Sticky Sesame-Sorghum Marinade

Whisk together 1/3 cup **reduced-sodium soy sauce;** 1/4 cup **sorghum syrup;** 2 Tbsp. **seasoned rice wine vinegar;** 2 Tbsp. **toasted sesame oil;** 2 tsp. **hot mustard;** and 2 **garlic** cloves, finely grated.

Note: We tested with Hengstenberg Hot Mustard.

MAKES about 3/4 cup

Chardonnay-Herb Marinade

Whisk together 1/2 cup **Chardonnay;** 1/4 cup **olive oil;** 1 **shallot,** minced; 2 **garlic** cloves, finely grated; 3 Tbsp. chopped fresh flat-leaf **parsley;** 2 Tbsp. chopped fresh **basil;** 1 Tbsp. chopped fresh **thyme;** 1 Tbsp. **Dijon mustard;** 2 tsp. **sugar;** 1 tsp. **kosher salt;** and 1/2 tsp. freshly ground black **pepper.**

MAKES 1 1/2 cups

No. 3

RELY ON THE THIGH

Inexpensive, flavorful, and nearly impossible to overcook, chicken thighs are the MVP of poultry cuts. Sprinkle salt on the thighs ahead of time, and chill in the fridge to pull out excess water—an easy way to guarantee tender meat and crisp skin.

CRISPY SALT-AND-PEPPER GRILLED CHICKEN THIGHS

- **12 bone-in, skin-on chicken thighs (about 5 lb.), trimmed**
- **Parchment paper**
- **2 tsp. kosher salt**
- **1 tsp. freshly ground black pepper**

1. Pat chicken dry with paper towels, and place thighs in a single layer in a parchment paper-lined jelly-roll pan. Sprinkle skin with salt, and chill, uncovered, 8 to 12 hours.
2. Light 1 side of grill, heating to 350° to 400° (medium-high) heat; leave other side unlit. Sprinkle thighs with pepper.
3. Arrange chicken, skin side up, on unlit side of grill, and grill, covered with grill lid, 1 hour and 15 minutes or until skin is golden and meat thermometer inserted into thickest portion registers 165°, turning pieces as needed for even cooking. Transfer chicken, skin side down, to lit side of grill. Grill, without grill lid, 4 to 5 minutes or until skin is crispy and grill marks appear. Let stand 5 minutes before serving.

MAKES 6 to 8 servings. **HANDS-ON** 15 min.; **TOTAL** 9 hours, 40 min.

Crunchy Jerk Tacos with Watermelon-Mango Salsa

Stir together 1 cup finely chopped **watermelon;** 1 cup finely chopped **mango;** 1/4 cup finely chopped **red onion;** 1/4 cup chopped fresh **cilantro;** 1 **jalapeño pepper,** minced; and 2 Tbsp. **fresh lime juice** in a medium bowl. Stir in **kosher salt** and freshly ground **black pepper** to taste, and let stand 10 minutes. Sauté 2 cups **shredded cooked chicken thighs** (including skin) and 1 1/2 tsp. **jerk seasoning** in 1 tsp. hot **canola oil** in a medium skillet over medium heat 3 to 4 minutes or until warm. Pour canola oil to a depth of 3 inches into a Dutch oven; heat over medium-high heat to 350°. Fry 4 (6-inch) **tortillas,** in batches, 2 minutes on each side or until lightly browned and crispy. Gently fold warm fried tortillas over a rolling pin to form a taco shape. Drain on paper towels. Spoon chicken onto fried tortillas; top with salsa and desired amount of crumbled **queso fresco.**

MAKES 4 servings. **HANDS-ON** 25 min.; **TOTAL** 35 min., not including chicken

Ginger Ale Glaze

Stir 2 cups **ginger ale** or ginger beer, 2 Tbsp. **light brown sugar,** 2 tsp. **grated fresh ginger,** and 1/8 tsp. **ground red pepper** together in a small saucepan. Bring to a boil over high heat; reduce heat to medium, and cook, stirring occasionally, 15 to 20 minutes or until reduced to about 1/4 cup.

MAKES about ¼ cup. **HANDS-ON** 10 min., **TOTAL** 25 min.

No. 4 RETIRE THE BEER CAN...

...and opt for a can of ginger ale instead. The spicy soft drink puts a tangy twist on your whole bird. Look for canned ginger beer (Gosling's makes a good one) to add even more spicy, fresh ginger flavor.

GINGER "BEER CAN" CHICKEN

- 1 **(13- x 9-inch) disposable aluminum baking pan**
- 2 **Tbsp. kosher salt**
- 2 **tsp. paprika**
- 2 **tsp. sugar**
- 1 **tsp. black pepper**
- 3/4 **tsp. ground ginger**
- ½ **tsp. ground red pepper**
- 2 **(3 ½- to 4-lb.) whole chickens**
- 2 **(12-oz.) cans ginger ale**
- **Ginger Ale Glaze**

1. Light 1 side of grill, heating to 400° to 475° (medium-high to high) heat; leave other side unlit, placing disposable baking pan under grill grate. Stir together salt and next 5 ingredients; sprinkle mixture inside cavity and on outside of each chicken.
2. Reserve just 3/4 cup ginger ale from each can to put toward the 2 cups needed in glaze. Place chickens upright onto each can, fitting can into cavity. Pull legs forward to form a tripod.
3. Pour water to a depth of 1 inch into disposable pan. Place chickens upright on unlit side of grill above pan. Grill, covered with grill lid and rotating chickens occasionally, 1 hour and 20 minutes to 1 hour and 30 minutes or until golden and a meat thermometer inserted in thickest portion registers 165°. Let stand 10 minutes. Remove chickens from cans, and carve. Serve with Ginger Ale Glaze.

MAKES 8 to 10 servings. **HANDS-ON** 15 min.; **TOTAL** 2 hours, 10 min., including glaze

No. **5**

SAY GOODBYE TO DRY

Because boneless, skinless chicken breast is lean, it easily goes dry and rubbery when cooked. Use our Brown Sugar Brine and we promise you'll never serve a dry chicken breast again.

BROWN SUGAR-BRINED CHICKEN

- ½ **cup apple cider vinegar**
- ¼ **cup kosher salt**
- ¼ **cup firmly packed dark brown sugar**
- 1 **tsp. dried crushed red pepper**
- 6 **large fresh thyme sprigs**
- 4 **(4-inch) rosemary sprigs**
- 4 **garlic cloves, crushed**
- 8 **to 12 (6- to 8-oz.) skinned and boned chicken breasts**
- **Freshly ground black pepper**

1. Stir together first 7 ingredients and 2 qt. water in a large stockpot. Heat over medium-high heat, stirring occasionally, until salt and sugar dissolve. Remove from heat, and cool 1 hour. Place chicken breasts in brine; cover and chill 1 to 3 hours.
2. Preheat grill to 350° to 400° (medium-high) heat. Drain chicken, discarding brine. Pat chicken dry with paper towels, and sprinkle with desired amount of black pepper. Grill chicken, covered with grill lid, 5 to 7 minutes on each side or until done. Remove from grill. Let stand 5 minutes before serving.

Note: This brine will season up to 12 chicken breasts.

MAKES 8 to 12 servings. **HANDS-ON** 30 min.; **TOTAL** 2 hours, 35 min.

Tomato-Peach Salad With Chicken

Whisk together ¼ cup **red pepper jelly,** ¼ cup **olive oil,** 2 tsp. **lime zest,** 3 Tbsp. **fresh lime juice,** and 3 Tbsp. chopped fresh **basil;** season with **kosher salt** and freshly ground **black pepper.** Fold in 1½ cups diced fresh **peaches** and 1½ lb. assorted small and medium-size **heirloom tomatoes,** cut into wedges. Spoon over **torn butter lettuce,** and top with slices of **grilled chicken breast.**

MAKES 6 servings. **HANDS-ON** 20 min.; **TOTAL** 20 min., not including chicken

No. 6

SIMPLIFY SMOKING

You don't need to be a pitmaster to smoke chicken. Try our no-fancy-equipment-needed technique.

SMOKED MOJO LEG QUARTERS

- ⅓ cup olive oil
- 1 Tbsp. loosely packed orange zest
- 1 Tbsp. loosely packed lime zest
- 1/3 cup fresh orange juice
- 1/4 cup fresh lime juice
- 1 bunch fresh cilantro, stems removed
- 6 large garlic cloves
- 2 red jalapeño peppers (seeded, if desired)
- ¼ cup fresh oregano leaves
- 1¼ tsp. ground cumin
- 1 tsp. kosher salt
- ½ tsp. black pepper
- 6 chicken leg quarters (about 5½ lb.)
- 6 cups applewood chips

1. Pulse first 12 ingredients in a food processor 20 seconds or until finely chopped. Reserve and chill 1/2 cup of mixture for later use. Place chicken in a large bowl; add remaining mixture, and toss gently to coat. Cover and chill 4 to 8 hours.
2. Place 3 cups dry applewood chips in center of 1 (18-inch) square piece of heavy-duty aluminum foil; repeat with remaining chips and another piece of foil. Wrap to form 2 envelopes. Pierce several holes in each packet.
3. Light 1 side of grill, heating to 350° to 400° (medium-high) heat; leave other side unlit. Place 1 applewood smoke packet, with holes facing up, on lit side.
4. Pat excess moisture from chicken with paper towels, leaving marinade on skin. Sprinkle chicken with desired amount of salt and pepper. Carefully move smoke packet to unlit side of grill. Place chicken over lit side, and grill, without grill lid, 4 to 6 minutes on each side or just until grill marks appear. Transfer chicken, skin side up, back to unlit side of grill; carefully move smoke packet back to lit side for the remainder of cooking. Grill, covered with grill lid and turning pieces occasionally, 1 hour and 30 minutes or until skin is golden brown and meat thermometer inserted into thickest portion registers 165°, replacing smoke packet halfway through. Let stand 10 minutes; serve with reserved sauce.

MAKES 6 servings. **HANDS-ON** 35 min., **TOTAL** 6 hours

SMOKING 101

STEP 1 Pick your wood chip. Here, our favorites for chicken.

APPLE: Mild fruit flavor; gives the skin a pretty, golden hue

PECAN: Hickory's milder, more sophisticated cousin

MAPLE: A robust smoke flavor with hits of sweetness

STEP 2 Make an aluminum foil "smoke envelope." Place 3 cups dry wood chips in the center of an 18-inch square of heavy-duty foil. Wrap to form an envelope, and pierce 3 to 5 holes in the packet.

STEP 3 Smolder the packet over direct heat to get it smoking, and follow the steps in our recipe (at left).

STEP 4 Let the packet cool completely; discard.

No. 7

MAKE A BETTER (CHICKEN) BURGER

Our BBQ sandwich-meets-burger is a slam dunk for grilling season.

CAROLINA CHICKEN BURGERS WITH CREAMY ANCHO SLAW

- **2 lb. lean ground chicken breast**
- **⅓ cup mayonnaise**
- **1 Tbsp. coarse-grained Dijon mustard**
- **1 tsp. kosher salt**
- **½ tsp. freshly ground black pepper**
- **6 sesame seed hamburger buns**
- **Carolina Burger Sauce**
- **Creamy Ancho Slaw**
- **Dill pickle slices**

1. Preheat grill to 350° to 400° (medium-high) heat. Gently combine first 5 ingredients in a large bowl. Shape mixture into 6 patties.

2. Grill patties, covered with grill lid, 5 to 6 minutes on each side or until a meat thermometer inserted into thickest portion registers 165°. Grill buns 1 minute on each side or until toasted. Spread Carolina Burger Sauce on bottom half of each bun. Top with patties, Creamy Ancho Slaw, pickles, and top halves of buns.

MAKES 6 servings. **HANDS-ON** 20 min., **TOTAL** 1 hour, 5 min., including sauce and slaw

Carolina Burger Sauce

Stir together 3/4 cup **ketchup;** 1/3 cup **apple cider vinegar;** 1/4 cup coarse-grained **Dijon mustard;** 1 large **shallot,** minced; 1 Tbsp. **dark brown sugar;** and 1/4 tsp. **dried crushed red pepper** in a medium saucepan. Bring to a boil over medium heat; reduce heat to low, and simmer, stirring occasionally, 8 to 10 minutes or until thickened. Season with **kosher salt** and **freshly ground black pepper.** Refrigerate in an airtight container up to 2 weeks.

MAKES about 1 cup. **HANDS-ON** 20 min., **TOTAL** 20 min.

Creamy Ancho Slaw

Whisk together 1/2 cup **mayonnaise,** 2 Tbsp. **apple cider vinegar,** 1 1/4 tsp. **ancho chile powder,** and 1 tsp. **sugar** in a large bowl. Add 1/2 small head **savoy cabbage,** shredded or finely chopped; 1 large **celery** rib, chopped; and 1/2 cup grated **carrot.** Toss to coat, and add **kosher salt** and **freshly ground black pepper** to taste. Let stand 15 minutes before serving, tossing occasionally.

MAKES 6 servings. **HANDS-ON** 10 min., **TOTAL** 25 min.

No. 8

MASTER THE BASICS

Memorize these simple tips to guarantee grilling success.

1. Start Clean

A well-kept grill ensures great flavor. (You wouldn't eat that charred gunk stuck to the grates, would you?) Here's how to get rid of that gunk. Each time you preheat your grill, cover the grates with a sheet of heavy-duty aluminum foil. After 10 minutes, use tongs to crumple the foil into a ball, and vigorously wipe away all the old char. Turn the ball over, add about 2 Tbsp. of vegetable oil, and "brush" the grates with the oil. This helps keep food from sticking as you grill.

2. Understand Indirect vs. Direct

What is direct heat?
Cooking directly over flame or coals. It's cooking over high heat, usually for a short period of time (under 1 hour).

What is indirect heat?
Cooking off to the side of your flame or coals. Think of it like roasting in an oven—the heat is all around the meat, not just under it. You'll need a grill lid to make it work. And you'll cook on indirect heat for a longer period of time (1 hour or more). Knowing how and when to use direct and indirect heat is key to keeping chicken tender while crisping or charring the skin.

3. Use a Chimney Starter

These funny-looking cylinders are inexpensive, reliable, and the best way to get charcoal hot without lighter fluid (yuck).

Here's what you do:

1. Fill the space under the chimney's wire rack with crumpled newspaper.
2. Fill the space on top of the wire rack with charcoal.
3. Use a match or two to light the newspaper, and set the starter (with coals on top) on the grill.
4. Wait and watch while thermodynamics channels the heat evenly through the coals.
5. When the coals are lightly covered with gray ash (after 20 to 30 minutes), slide on a heavy-duty oven mitt and dump the hot coals into the bottom of your grill. (If you need indirect heat, place the coals on just one side.)
6. Pop on the grill grate and you're good to grill.

BE A GEAR MINIMALIST

You don't need tons of tools. We recommend:

TWO THERMOMETERS

One to temp your grill (the Tel-Tru BQ300 is accurate and durable; *teltru.com*) and another to temp your meat (the ThermoWorks ThermoPop; *thermoworks.com*).

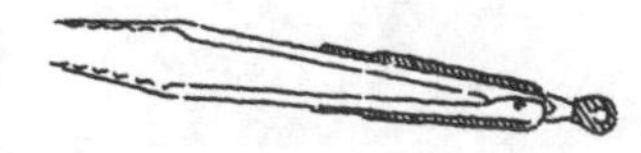

LONG TONGS

Ditch the meat fork and spatula. If food is ready to be flipped, it'll lift off grates easily, and tongs won't pierce the meat and let out juices.

ROBBY MELVIN, DIRECTOR OF THE SOUTH'S MOST TRUSTED KITCHEN, SHARES TIPS FOR

QUICK FRUIT PICKLES

Summertime means plenty of fresh fruit picked at its peak. Alas, perfection is fleeting. But you can preserve the harvest—quickly, easily, and a jar at a time—by making these fruit pickles. Spoon these gems onto salads or add them to a cheese plate. Serve as a relish or sandwich side. Add a splash of the delicious leftover fruity brine to vinaigrettes, iced tea, or cocktails.

FRUIT PICKLE BRINE

This all-purpose brine is ideal for pickling peaches, strawberries, and melons. The fruit is ready to serve after 1 hour in the brine, but tastes best after 8 to 12 hours.

- ½ **cup white balsamic vinegar**
- ¼ **cup sugar**
- 2 **tsp. kosher salt**

Bring all ingredients and 1 cup water to a boil in a medium saucepan over high heat. Remove from heat. Cool completely (about 1 hour), stirring occasionally.

MAKES about 1 3/4 cups. **HANDS-ON** 5 min.; **TOTAL** 1 hour, 10 min.

Pickled Strawberries

Place 1 lb. fresh **strawberries** and 2 fresh **rosemary** sprigs in a 24-oz. jar with a tight-fitting lid. Pour **Fruit Pickle Brine** over strawberries; seal and chill 1 to 12 hours. Store in refrigerator up to 1 week.

MAKES 1 (24-oz.) jar. **HANDS-ON** 5 min.; **TOTAL** 2 hours, 15 min., including brine

Pickled Melon

Cut 1 (1-lb.) **honeydew melon** into chunks; place melon and 2 fresh **thyme** sprigs in a 24-oz. jar with a tight-fitting lid. Pour **Fruit Pickle Brine** over melon; seal and chill 1 to 12 hours. Store in refrigerator up to 1 week.

MAKES 1 (24-oz.) jar. **HANDS-ON** 5 min.; **TOTAL** 2 hours, 15 min., including brine

Pickled Peaches

Place about 1 lb. fresh **peaches,** cut into slices, and 2 fresh **mint** sprigs in a 24-oz. jar with a tight-fitting lid. Pour **Fruit Pickle Brine** over peaches; seal and chill 1 to 12 hours. Store in refrigerator up to 1 week.

MAKES 1 (24-oz.) jar. **HANDS-ON** 5 min.; **TOTAL** 2 hours, 15 min., including brine

Dinner in a Breeze

Try this one-dish, 20-minute recipe featuring summer's tastiest fruit

SKILLET PORK CHOPS WITH SAUTÉED PEACHES

Serve this light, seasonal dish with couscous, rice, or your favorite salad.

- **2 tsp. olive oil**
- **4 (4-oz.) center-cut boneless pork loin chops, trimmed**
- **½ tsp. kosher salt**
- **½ tsp. black pepper**
- **2 Tbsp. thinly sliced shallots**
- **2 tsp. chopped fresh thyme**
- **2 peaches, each cut into 8 wedges**
- **½ cup dry white wine**
- **½ cup reduced-sodium fat-free chicken broth**
- **2 tsp. honey**
- **2 tsp. butter**

Heat a skillet over medium-high heat; add oil, and swirl to coat. Sprinkle chops with salt and pepper. Cook chops 3 minutes on each side or until done. Remove chops from skillet; cover with foil to keep warm. Cook shallots, thyme, and peaches 2 minutes. Add wine, stirring to loosen browned bits from skillet; bring to a boil. Boil 2 minutes or until reduced to about 1/3 cup. Stir in broth and honey; bring to a boil. Boil 2 minutes. Remove from heat; stir in butter. Spoon sauce over chops.

MAKES 4 servings (serving size: 1 chop, 4 peach wedges, and about 1 1/2 Tbsp. broth mixture). **HANDS-ON** 15 min., **TOTAL** 20 min.

NUTRITIONAL INFORMATION
(per serving)
CALORIES: 235; FAT: 8.6G (SATURATED FAT: 2.8G); PROTEIN: 26.2G; CARBOHYDRATES: 13.6G; FIBER: 1.1G; SODIUM: 433MG

Green Beans Like You've Never Had

Easy-to-find ingredients will have you falling for this twist on a classic Southern combo, and it's as simple as plugging in your slow cooker

STEWED GREEN BEANS AND TOMATOES

Our update to this favorite Southern side will remind you of great Asian takeout. Don't be intimidated by red curry paste; it's a subtly spicy blend of fresh ginger, lemongrass, garlic, and chiles. Look for it at Walmart or online at amazon.com.

- **2 cups diced fresh tomatoes (about 3 medium)**
- **1 cup coconut milk**
- **2 Tbsp. red curry paste**
- **1 Tbsp. finely chopped fresh ginger**
- **1½ tsp. kosher salt**
- **2 garlic cloves, finely chopped**
- **1¼ lb. fresh green beans, trimmed**
- **1 Tbsp. fresh lime juice**
- **Hot cooked long-grain rice**
- **⅓ cup roasted salted peanuts, chopped**
- **2 Tbsp. chopped fresh cilantro (optional)**

1. Whisk together first 6 ingredients in a 6-qt. slow cooker. Add green beans, and cook on LOW 7 hours, stirring once after 3 1/2 hours.

2. Stir in lime juice. Serve over rice, and top with chopped peanuts and, if desired, fresh cilantro.

MAKES 4 servings. **HANDS-ON** 20 min.; **TOTAL** 7 hours, 20 min.

Community Cookbook

BUILD A BETTER LIBRARY, ONE GREAT BOOK AT A TIME

First, product designers Eric Prum and Josh Williams took the Mason jar, a fixture in their Virginia childhoods, to the next level with The Mason Shaker, a lid that turns a jar into a cocktail shaker. Now they have a new recipe book and new tool, The Mason Tap, a lid with a spout. Inspired by adding fresh peaches to a jar of whiskey, they share ways to infuse oils, spirits, and water with fresh flavor.

GARLIC WITH CONFIT OIL, THYME, AND BLACK PEPPER

"Infused with deliciously fragrant roasted garlic, this olive oil infusion often serves as the backbone flavoring for much of what we end up cooking up in the kitchen. Endlessly versatile in its use, from dipping crusty bread before the meal begins to finishing a juicy grilled steak, this is an infusion we try to keep on hand at all times."

1 head of garlic
Sea salt
Ground black pepper
1 cup olive oil
5 whole black peppercorns
4 sprigs of fresh thyme, divided

1. Preheat the oven to 300°. Wash the head of garlic, and cut the top off horizontally. Season the exposed edges with salt and pepper, and place facedown in a deep baking dish.
2. Add the olive oil, peppercorns, and 3 sprigs fresh thyme to the baking dish; cover with aluminum foil, and roast for 1 hour.
3. Remove from the oven, and let cool for 30 minutes, covered.
4. Uncover and remove the garlic, placing on a towel to drain. Then strain the oil through cheesecloth, and add to an 8-oz. Mason jar, adding remaining 1 sprig of fresh thyme.
5. Squeeze the cloves of garlic from the head, and add them to the Mason jar. The infusion and roasted garlic cloves will keep in the refrigerator for up to 1 month.

MAKES 8 oz. **HANDS-ON** 30 min., **TOTAL** 2 hours

Authors' note: Don't let the roasted garlic cloves in the above infusion go to waste. You can use them in addition to the infused oil, and they come with a caramelized, umami-packed richness you won't want to miss.

CUCUMBER-MINT WATER

"As much as we love the hot days of summer, it's sometimes hard to escape that sticky feeling. Enter this cooling infusion. Lime, cucumber, and mint work together to create a refreshing sip that we keep in the fridge all summer long."

12 slices of cucumber
4 slices of lime
4 large sprigs of mint
28 oz. of water

1. Combine the cucumber, lime, and mint in a 32-oz. Mason jar. Muddle until lightly crushed.
2. Add the water; seal and shake for 30 seconds to combine.
3. Serve over ice. The infusion will keep in the refrigerator for up to 24 hours.

MAKES 32 oz. **HANDS-ON** 5 min., **TOTAL** 5 min.

Authors' note: For all our water infusions, we suggest using filtered tap water or bottled spring water for the freshest-tasting results. Don't have either? Tap water can work in a pinch.

July

146 **The Ultimate Lazy Lunch** Summer ease, a Gulf breeze, and Julia Reed's Grand Aïoli equal perfect beachside entertaining

150 **Romancing the Stone** Sweet desserts that celebrate summer's bountiful stone fruits

156 **Loving Summer Tomatoes** Sara Foster weaves garden tomatoes into salads, soups, condiments, and even cookies!

160 QUICK-FIX SUPPERS **Chop, Chop!** Inspired seasonal recipes for a favorite cut of pork

195 THE NEW SUNDAY SUPPER **Steak Nights** Summertime and grilling is easy, so fire up the grill once for three easy meals

197 SOUTHERN CLASSIC **Summer Succotash** The South's most-loved veggie medley, perfected

198 THE *SL* TEST KITCHEN ACADEMY **Summer Tomatoes** Tips for buying and storing and warning: Never refrigerate!

199 THE SLOW DOWN **Taco Night Reimagined** Take the slow road to a street food favorite

200 *COOKING LIGHT* **Light and Luscious Summer Pasta** The buttermilk secret

201 SAVE ROOM **Easy Ice Cream** Churn out four fantastic flavors effortlessly

202 COMMUNITY COOKBOOK A chef takes farmstand produce to new heights in ways the whole family will love

The Ultimate Lazy Lunch

WHEN BEACH SEASON ARRIVES, JULIA REED LIKES TO KEEP THINGS SIMPLE, FAST, AND A LITTLE BIT FRENCH

WHEN I WAS a kid, we spent a few weeks each summer at the Frangista Beach on Florida's Gulf Coast near Destin. Each "suite" in the one-story cinder block affair featured a kitchen with a noisy fridge and a Formica-topped "dining table." Easy summer lunches meant my mother's tuna fish salad sandwiches. It was heaven.

The Frangista was torn down years ago, and we have since migrated down the beach to Seaside. Our current base of operations has a big front porch, and the dining table is a French farm table that stretches out to seat at least 10. The only thing that hasn't changed is that the lunches, and the living, remain easy. The good news is that "easy" can also mean elegant. It just takes a bit of planning and some inspiration. Mine comes from another hot-weather culture—the south of France.

First of all, that particular South is home to some excellent rosé, one of the world's great (and inexpensive) summer wines. It is also a source of delicious no-fuss hors d'oeuvres: olives (Niçoise, picholine, Lucques) and almonds toasted with sea salt. Most importantly, their straightforward, ingredient-centered menu staples are eminently adaptable to our own food culture.

For years, my go-to main course was a classic Niçoise salad—but with fresh Gulf tuna subbed for the French oil-cured variety. I still serve it, but lately I've become obsessed with an equally simple platter of goodies—a grand aïoli. I've always been a huge fan of aïoli itself—what's not to love about mayonnaise with olive oil and garlic? But a "grand aïoli" refers to the complete dish, an assortment of simply poached seafood and vegetables served with the sauce.

Gorgeous to look at and a cinch to make ahead, a grand aïoli is a fixture at French village fetes, where it is always accompanied by the aforementioned rosé. It's also the ultimate lazy-day menu. Guests help themselves from the platter and dip into bowls passed at the table.

Giant clamshells are my favorite summer wine coolers.

Don't have the real deal? Affordable resin versions are easy to find and far lighter to carry. (Try *zgallerie.com*.) Chilled wine cups from Mississippi's McCarty's Pottery (*mccartyspottery.com*) keep wine cold like an ice bucket.

The routine becomes talk, dip, and drink—at our house, the beach usually becomes a distant memory.

In France, the components include poached salt cod and snails along with potatoes, hard-boiled eggs, green beans, and cauliflower and are so sacrosanct that the locals get riled up over the slightest alteration. But because I am on the Redneck Riviera rather than the Côte d'Azur, I can safely play around with whatever's close by. Snails are replaced with fresh Gulf shrimp, salt cod with hours-old grouper. (Season filets with salt and white pepper, and place in simmering, salted water or fish stock for about five minutes.) New potatoes and hard-boiled eggs are still musts, but I often replace the green beans with okra and add whatever else my buddy Paul (from Paul's Pick of the Crop) has on hand from the Seaside Farmers Market. The only other necessary addition is plenty of sliced French bread. Brushed with olive oil and toasted, it gets slathered with aïoli too. Because you can never have enough of the latter, I always offer a variation or two.

For a prelunch cocktail, I take my cues from another south-of-France favorite, a vin d'orange (wine fortified with vodka and infused with citrus). The French concoction has to sit around for weeks, so I cut to the chase with orange-infused rum made into a classic gimlet. For dessert, I serve local ice cream, and if I'm very ambitious, I'll make chocolate meringue cookies with the egg whites left over from making the aïoli. Bon appétit!

Grand Aïoli

GRAND AÏOLI

- **3 garlic cloves, peeled and grated on the finest holes of your grater**
- **1/4 tsp. kosher salt**
- **2 large pasteurized egg yolks, at room temperature**
- **2 tsp. warm water**
- **1/2 cup pure olive oil**
- **1/2 cup extra virgin olive oil**
- **Fresh lemon juice**
- **Freshly ground black pepper**
- **Poached shrimp, hard-cooked eggs, boiled new potatoes, green beans, okra, and beets**

Drape a kitchen towel over a small saucepan, and set a small metal bowl over it. Mash garlic and salt together in bowl; whisk in yolks and 2 tsp. warm water. Add oil, 1 tsp. at a time, whisking constantly. Add extra virgin olive oil in a slow, steady stream, whisking constantly. (If mixture gets too stiff, add more water.) Stir in lemon juice and pepper to taste. Serve with shrimp, eggs, and vegetables.

MAKES 1 cup. **HANDS-ON** 15 min., **TOTAL** 15 min.

Lime-Sriracha Aïoli:

Omit lemon juice and black pepper. Prepare recipe as directed, stirring in 2 tsp. each fresh lime juice and Asian hot chili sauce (such as Sriracha) after whisking in extra virgin olive oil.

Saffron Aïoli:

Omit lemon juice and black pepper. Stir together 2 tsp. warm water and 1/2 tsp. crumbled saffron threads. Prepare recipe as directed, substituting saffron mixture for plain warm water. Stir in a pinch of dried crushed red pepper before serving.

To me, flowers look out of place at the beach.

Get creative with other natural beauties instead. I used my mother's coral collection (below), but you could make a runner with overlapping palmetto leaves or use sand dollars as place cards. And don't be afraid to mix old and new china and glassware.

ORANGE-INFUSED RUM GIMLETS

Slice 2 or 3 navel oranges, and place in a 2-qt. pitcher. Add 1 (750-milliliter) bottle white rum; cover and let stand 8 to 12 hours. Stir in 3/4 cup fresh lime juice, 1/2 cup orange liqueur, and 1/2 cup simple syrup*, and fill pitcher with ice. Serve immediately.

MAKES 2 qt. **HANDS-ON** 10 min.; **TOTAL** 8 hours, 10 min.

*To make simple syrup, bring 1/2 cup sugar and 1/2 cup water to a boil in a small saucepan over medium heat, stirring occasionally. Boil 1 minute or until sugar dissolves. Cool 30 minutes.

For dessert, stick silver spoons straight into pints of small-batch ice cream. (I love Southern Craft Creamery from Seaside's Modica Market.) Let guests fill their own bowls.

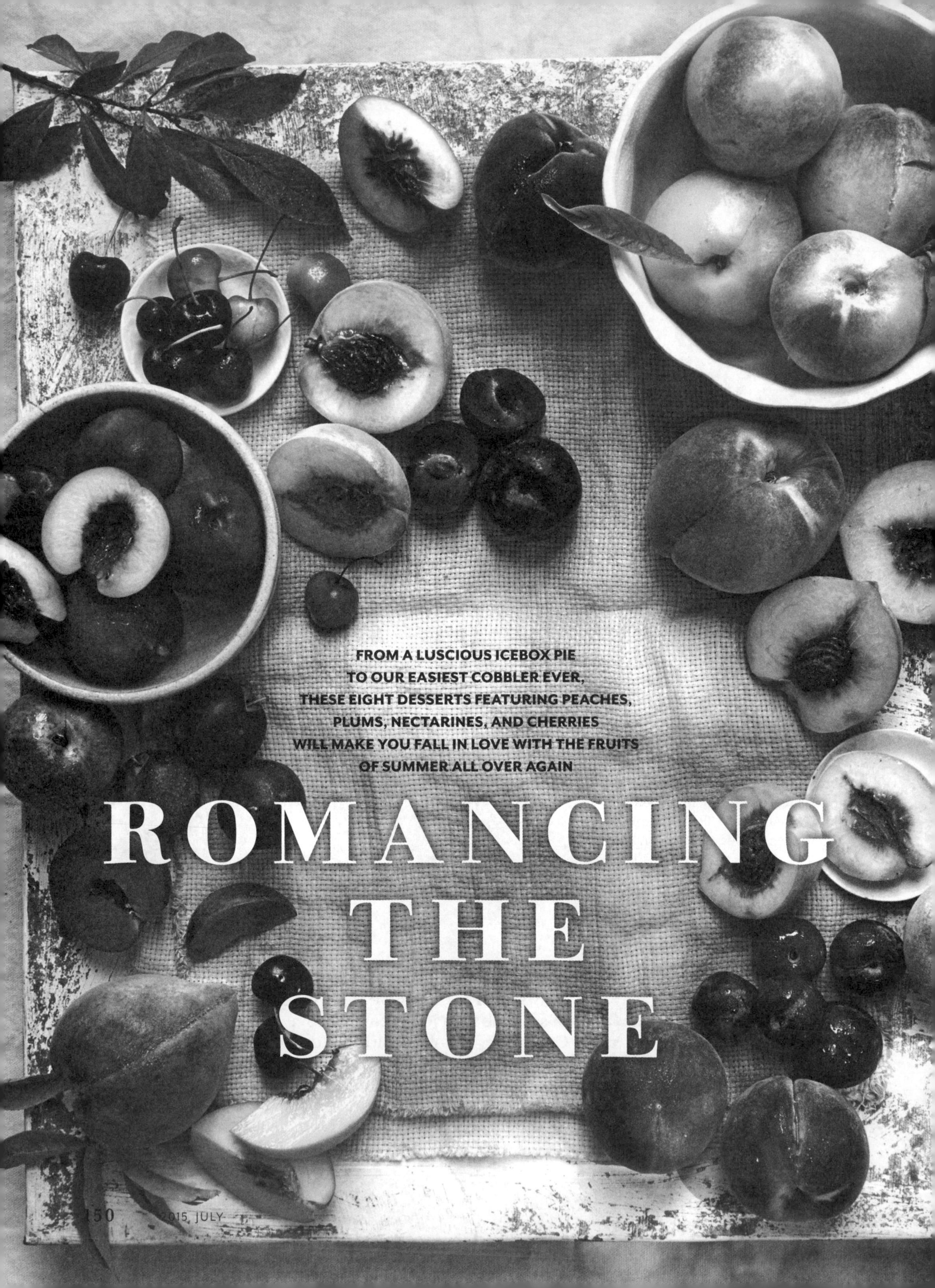

FROM A LUSCIOUS ICEBOX PIE
TO OUR EASIEST COBBLER EVER,
THESE EIGHT DESSERTS FEATURING PEACHES,
PLUMS, NECTARINES, AND CHERRIES
WILL MAKE YOU FALL IN LOVE WITH THE FRUITS
OF SUMMER ALL OVER AGAIN

ROMANCING THE STONE

Ginger-Peach Shortbread Cobbler, page 153

Mixed Stone Fruit Pie

AMERICA'S LOVE AFFAIR

with stone fruit is iconic—a cherry pie in the windowsill shouts U.S.A. as loudly as the red bandana tucked in Bruce Springsteen's jeans. But nowhere does that romance burn more intensely than in the South. We haul baskets of peaches at local U-Pick farms—our shirts damp with sweat from the summer heat. We fire up stove burners for bubbling, jewel-colored jams and jellies, and open the oven door to syrupy cobblers and pies. We hold cherry pit spitting contests on the porch. We know exactly how it feels when the sun bakes the sticky trails of juice from a just-bitten plum on our arms. There are few fruits that inspire such longing for Southerners; nor are there many others as delicious eaten fresh off the tree as they are underneath a lattice crust. "The perfect peach is like the perfect tomato," says chef Hugh Acheson. "Rare are the foods that are so exquisite on their own, like stone fruits, that nothing else needs to be done to them." That's why our newest desserts show off what we love about stone fruits: their natural beauty, juicy sweetness, and blushing colors.

Orange-and-Basil Macerated Cherries

Lime-and-Mint Macerated Cherries

GINGER-PEACH SHORT-BREAD COBBLER

- **1 cup butter, softened**
- **1/2 cup loosely packed light brown sugar**
- **1/8 tsp. kosher salt**
- **2 1/4 cups plus 3 Tbsp. all-purpose flour, divided**
- **1 (1/2-inch) piece fresh ginger, peeled**
- **3/4 cup turbinado sugar, divided**
- **7 to 9 medium-size peaches (such as 'Elberta'), halved**
- **2 Tbsp. fresh lemon juice**
- **Butter**
- **1 large egg white, lightly beaten**
- **2 tsp. granulated sugar**

1. Preheat oven to 400°. Beat first 3 ingredients and 2 1/4 cups flour at medium speed with a heavy-duty electric stand mixer just until combined. Place dough on a lightly floured surface; roll to 1/4-inch thickness. Cut about 14 rounds with a 2 1/2-inch fluted round cutter. Place rounds in a single layer on a baking sheet; cover and chill until ready to use.
2. Pulse ginger and 1/4 cup turbinado sugar in a food processor 5 to 6 times or until well combined. Stir together ginger mixture and remaining 1/2 cup turbinado sugar in a large bowl. Add peaches, lemon juice, and remaining 3 Tbsp. flour; toss to coat. Place peach halves, slightly overlapping, in a greased (with butter) 10-inch cast-iron skillet, and pour any remaining juice over peaches.
3. Bake at 400° for 15 minutes. Remove from oven, and nestle shortbread dough rounds among peaches. Brush rounds with egg white, and sprinkle with granulated sugar. Bake at 400° for 17 to 20 minutes or until golden brown.

Note: We tested with Sugar in the Raw Natural Cane Turbinado Sugar.

MAKES 6 to 8 servings. **HANDS-ON** 40 min.; **TOTAL** 1 hour, 10 min.

MIXED STONE FRUIT PIE

- **1 cup graham cracker crumbs**
- **1 cup sliced almonds**
- **6 Tbsp. butter, melted**
- **1/4 cup sugar**
- **1 cup fruit juice (such as cranberry or grape)**
- **3 tsp. cornstarch**
- **1/4 cup honey**
- **1/8 tsp. table salt**
- **2 cups heavy cream**
- **2 (8-oz.) packages cream cheese, softened**
- **1 (14-oz.) can sweetened condensed milk**
- **1 Tbsp. loosely packed lemon zest**
- **2 Tbsp. fresh lemon juice**
- **1/2 tsp. vanilla extract**
- **1/4 tsp. table salt**
- **1/4 tsp. almond extract**
- **4 cups mixed stone fruit slices**

1. Preheat oven to 325°. Pulse first 4 ingredients in a food processor 15 times or until almonds are finely ground. Press on bottom and up sides of a 10-inch pie plate. Bake at 325° for 10 to 12 minutes or until golden brown. Cool completely (about 2 hours).
2. Meanwhile, whisk together fruit juice and cornstarch in a small bowl. Stir together honey and salt in a small saucepan, and cook over medium-low heat 3 to 4 minutes or until mixture begins to bubble. Whisk in fruit juice mixture. Increase heat to medium, and bring to a boil, whisking often. Remove from heat, and cool completely (about 10 minutes). Cover and chill until ready to use.
3. Beat heavy cream in a medium bowl at medium speed with an electric mixer until stiff peaks form.
4. Beat cream cheese at medium speed with a heavy-duty electric stand mixer until smooth; gradually add condensed milk, beating until blended after each addition. Add next 5 ingredients, and beat at medium-high speed until smooth and creamy. Fold in whipped cream. Pour mixture into cooled pie crust, and chill 12 to 24 hours.
5. Place fruit slices in a large bowl; add half of chilled juice mixture, tossing to coat. (If mixture is too thick, place in a microwave-safe bowl, and microwave at HIGH 30 seconds.) Spoon fruit onto chilled pie, and serve immediately with remaining juice mixture. Or chill remaining juice mixture, and reserve for another use, such as a topping for ice cream.

MAKES 8 to 10 servings. **HANDS-ON** 45 min.; **TOTAL** 5 hours, 35 min.

ORANGE-AND-BASIL MACERATED CHERRIES

- **1 lb. fresh cherries, pitted**
- **1 Tbsp. firmly packed orange zest**
- **2 Tbsp. fresh orange juice**
- **2 Tbsp. dark brown sugar**
- **1 Tbsp. chopped fresh basil**
- **1/8 tsp. kosher salt**
- **Ricotta cheese**

1. Cut half of cherries in half crosswise. Place whole cherries and cherry halves in a large bowl; add zest and next 4 ingredients, tossing to coat. Cover and chill 1 hour.
2. Remove cherries from refrigerator, and let stand 10 minutes. Spoon over ricotta cheese, and serve immediately.

MAKES 6 to 8 servings. **HANDS-ON** 20 min.; **TOTAL** 1 hour, 30 min.

LIME-AND-MINT MACERATED CHERRIES

- **1 lb. fresh cherries, pitted**
- **1 Tbsp. firmly packed lime zest**
- **2 Tbsp. fresh lime juice**
- **2 Tbsp. sugar**
- **1 Tbsp. fresh mint leaves**
- **1/8 tsp. kosher salt**
- **Ricotta cheese**

1. Cut half of cherries in half crosswise. Place whole cherries and cherry halves in a large bowl; add zest and next 4 ingredients, tossing to coat. Cover and chill 1 hour.
2. Remove cherries from refrigerator, and let stand 10 minutes. Spoon over ricotta cheese, and serve immediately.

MAKES 6 to 8 servings. **HANDS-ON** 20 min.; **TOTAL** 1 hour, 30 min.

Nectarine Tarts with Honey

Buttermilk-Plum Ice Cream

Buttermilk-Glazed Cherry Sheet Cake

Nectarine-Chile Ice Pops

NECTARINE TARTS WITH HONEY

- 1 (17.3-oz.) package frozen puff pastry sheets, thawed
- Parchment paper
- 2 lb. nectarines (such as 'Red-gold'), cut into 1/4-inch slices
- 6 Tbsp. sugar
- 1/4 cup honey

Preheat oven to 425°. Cut each pastry sheet into 6 (3-inch) squares, and place in a single layer on 2 parchment paper-lined baking sheets. Arrange nectarine slices on each pastry square; sprinkle with sugar. Bake at 425° for 15 to 20 minutes or until pastry is golden brown, rotating pans top to bottom halfway through. Cool on a wire rack 5 minutes. Drizzle with honey just before serving.

MAKES 12 servings. **HANDS-ON** 15 min., **TOTAL** 35 min.

BUTTERMILK-PLUM ICE CREAM

- 2 cups half-and-half
- 1 cup buttermilk
- 1 cup heavy cream
- 6 large egg yolks
- 1 1/3 cups plus 3 Tbsp. sugar, divided
- 2 cups peeled, chopped black plums (such as 'Methley')
- 1 Tbsp. light corn syrup
- 1/8 tsp. table salt
- 2 tsp. fresh lemon juice

1. Whisk together first 4 ingredients and 1 1/3 cups sugar in a medium saucepan. Cook over medium-low heat, whisking often, 10 to 12 minutes or until mixture just begins to boil. Pour mixture into a medium-size metal bowl. Fill a large bowl halfway with ice water. Place bowl containing custard mixture in ice water, and cool completely (about 30 minutes), whisking occasionally. Cover and chill custard mixture 2 hours.
2. Meanwhile, stir together plums, corn syrup, salt, and remaining 3 Tbsp. sugar in a small saucepan, and cook over medium heat, stirring occasionally, 12 minutes or until plums are very tender. Remove from heat; let stand 15 minutes. Stir in lemon juice. Cover and chill plum mixture 2 hours.
3. Pour custard mixture into freezer container of a 2-qt. electric ice-cream maker, and freeze according to manufacturer's instructions. (Instructions and times may vary.) Transfer ice cream to a freezer-safe bowl or pan, and freeze 1 hour. Meanwhile, freeze plum mixture 30 minutes. Dollop frozen plum mixture over frozen ice cream, and gently swirl.

Note: We tested with Cuisinart Pure Indulgence Frozen Yogurt-Sorbet & Ice Cream Maker.

MAKES about 2 qt. **HANDS-ON** 35 min., **TOTAL** 5 hours

BUTTERMILK-GLAZED CHERRY SHEET CAKE

- 1 cup butter, softened
- 2 cups plus 1 tsp. granulated sugar, divided
- 4 large eggs
- 1 Tbsp. fresh lemon juice
- 1 tsp. almond extract
- 2 tsp. baking powder
- 1/2 tsp. table salt
- 3 1/2 cups plus 2 Tbsp. all-purpose flour, divided
- 1 cup milk
- Butter
- 3 cups fresh cherries, pitted and halved
- 1/2 cup plus 2 Tbsp. powdered sugar
- 3 Tbsp. buttermilk

1. Preheat oven to 325°. Beat butter at medium speed with an electric mixer until fluffy. Gradually add 2 cups granulated sugar, and beat until well blended. Add eggs, 1 at a time, beating well after each addition and stopping to scrape sides of bowl as needed. Add lemon juice and almond extract, and beat until blended. Increase speed to medium-high, and beat 1 to 2 minutes or until pale yellow.
2. Stir together baking powder, salt, and 3 1/2 cups flour. Add to butter mixture alternately with milk, beginning and ending with flour mixture. Beat at low speed just until blended after each addition. Line bottom and sides of 13- x 9-inch pan with aluminum foil, allowing 2 to 3 inches of foil to extend over sides. Grease (with butter) and flour foil. Pour batter into prepared pan. Toss together cherries and remaining 2 Tbsp. flour and 1 tsp. granulated sugar, and scatter over batter.
3. Bake at 325° for 1 hour to 1 hour and 10 minutes or until set and browned on top. Cool on a wire rack 1 hour.
4. Whisk together powdered sugar and buttermilk, and drizzle over cake. Let cake stand until glaze sets (about 5 minutes). Lift cake from pan, using foil sides as handles. Cut into squares.

MAKES 12 to 16 servings. **HANDS-ON** 30 min.; **TOTAL** 2 hours, 35 min.

NECTARINE-CHILE ICE POPS

- 2/3 cup sugar
- 1 Tbsp. light corn syrup
- 1/4 tsp. table salt
- 2 (1/4-inch-thick) jalapeño pepper slices (with seeds)
- 4 large nectarines, peeled and quartered
- 1 1/2 tsp. firmly packed lime zest
- 1/4 cup fresh lime juice
- 10 food-safe wooden craft sticks

1. To make a sugar syrup, bring first 3 ingredients and 3 Tbsp. water to a boil in a small saucepan over medium heat. Boil, stirring constantly, 1 to 2 minutes or just until sugar dissolves. Add pepper slices, and cook, stirring occasionally, 1 minute. Remove from heat; let stand 15 minutes. Discard jalapeño pepper slices.
2. Process sugar syrup, nectarines, and next 2 ingredients in a food processor until smooth, stopping to scrape down sides as needed. Pour mixture into 10 (2-oz.) plastic pop molds. Top with lids of pop molds, and freeze 1 hour. Insert craft sticks, leaving 1 1/2 to 2 inches sticking out, and freeze 7 more hours or until sticks are solidly anchored and pops are completely frozen.

MAKES 10 frozen pops. **HANDS-ON** 20 min.; **TOTAL** 8 hours, 35 min.

Loving Summer Tomatoes

FIVE FRESH, NO-FUSS RECIPES FROM **SARA FOSTER** FOR EASY ENTERTAINING

Sara Foster, cookbook author and founder of Foster's Market in Durham, North Carolina, is an expert at creating inventive recipes that bring fresh appreciation to classic farmstead ingredients. Come summertime, that means tomatoes—from familiar standards to snazzy heirlooms—fresh off the vine and warm from the sun. When nearby farmers' markets are overflowing with ripe tomatoes, Sara snaps up plenty of her favorite varieties, such as 'German Johnson,' 'Pineapple,' 'Green Zebra,' 'Sun Gold,' and 'Zebra Cherry.' Whether cooking in her cozy home or in the bustling kitchen at Foster's, Sara's seasonal Southern recipes make summertime entertaining easy and enjoyable.

Pasta with Heirloom Tomatoes, Goat Cheese, and Basil, **page 159**

Tomato, Watermelon, and Feta Skewers with Mint and Lime

TOMATO, WATERMELON, AND FETA SKEWERS WITH MINT AND LIME

No time for skewers? Gently toss the ingredients in a salad bowl, or dice the tomato, watermelon, and cheese into small cubes to serve as a fresh relish for grilled meats.

- **2 large heirloom tomatoes, cored and cut into 1-inch pieces**
- **3 cups 1-inch watermelon cubes (about 1/4 of a 3-lb. watermelon)**
- **8 oz. feta cheese, cubed**
- **2 1/2 Tbsp. fresh lime juice**
- **2 Tbsp. chopped fresh mint**
- **1 Tbsp. extra virgin olive oil**
- **1 tsp. kosher salt**
- **1/2 tsp. freshly ground black pepper**
- **36 (3-inch) wooden skewers**

Gently toss together tomatoes and next 7 ingredients in a large bowl. Cover and chill 30 minutes to 1 hour. Thread 1 tomato piece, 1 watermelon cube, and 1 feta cube onto a skewer, and place in a serving bowl. Repeat with remaining skewers. Drizzle with remaining marinade, and serve immediately.

MAKES about 3 dozen skewers. **HANDS-ON** 30 min., **TOTAL** 1 hour

GREEN TOMATO SOUP WITH LUMP CRABMEAT

Make this vibrant, peppery soup and easy crab salad up to a day ahead, and then stash them in the fridge to chill.

- **1 poblano pepper**
- **2 Tbsp. butter**
- **2 Tbsp. olive oil**
- **2 medium onions, chopped**
- **4 celery ribs, chopped**
- **2 jalapeño peppers, seeded and chopped**
- **4 garlic cloves, minced**
- **2 1/2 lb. firm green tomatoes, cored and coarsely chopped**
- **6 cups chicken or vegetable broth**
- **2 bay leaves**
- **Kosher salt**
- **Freshly ground black pepper**
- **3 cups loosely packed arugula**
- **14 fresh basil leaves**
- **1/2 bunch fresh cilantro, stems removed**
- **2 1/2 Tbsp. fresh lemon juice**
- **1 to 2 tsp. hot sauce**
- **Lump Crabmeat Salad**

1. Preheat broiler with oven rack 5 inches from heat. Cut poblano pepper in half lengthwise; remove seeds. Broil pepper halves, skin sides up, on an aluminum foil-lined baking sheet 2 to 3 minutes on each side or until blistered. Place pepper halves in a zip-top plastic freezer bag; seal and let stand 10 minutes to loosen skins. Peel pepper halves, and chop.

2. Melt butter with oil in a Dutch oven over medium-high heat. Reduce heat to low; add onions, and cook, stirring often, 15 minutes. Add celery, jalapeño peppers, and chopped poblano pepper; cook, stirring often, 5 minutes. Add garlic; cook, stirring constantly, 2 minutes. Add tomatoes, broth, and bay leaves. Season with salt and pepper. Increase heat to medium-high, and bring to a boil. Reduce heat to low, and simmer, stirring occasionally, 15 to 20 minutes or until tomatoes are tender. Remove from heat, and discard bay leaves. Stir in arugula, basil, and cilantro. Let cool 30 minutes.

3. Process soup, in batches, in a food processor or blender until smooth. Stir in lemon juice and hot sauce; add salt and pepper to taste. Cover and chill 8 to 24 hours. Ladle chilled soup into serving bowls; top each serving with about 2 Tbsp. Lump Crabmeat Salad.

MAKES about 3 qt. **HANDS-ON** 45 min.; **TOTAL** 9 hours, 45 min., including salad

Lump Crabmeat Salad

Stir together 8 oz. **fresh lump crabmeat** (picked through to remove bits of shell), 2 Tbsp. **olive oil,** 1 Tbsp. fresh **lime juice,** 1 Tbsp. each chopped fresh **basil** and **cilantro,** and 1 seeded and sliced **jalapeño** pepper. Season with **sea salt** and freshly ground black **pepper.** Serve immediately, or refrigerate up to 2 days.

MAKES 2 cups. **TOTAL** 5 min.

Green Tomato Soup with Lump Crabmeat

CORNMEAL THUMBPRINT COOKIES WITH TOMATO JAM

This clever take on beloved thumbprint cookies is sweet and savory. Refrigerated dough keeps for days, and the baked cookies freeze well.

- **1/2 cup butter, softened**
- **1/2 cup powdered sugar**
- **1 large egg, separated**
- **1 tsp. vanilla extract**
- **3/4 cup all-purpose flour**
- **3/4 cup plain yellow cornmeal**
- **1 tsp. firmly packed lemon zest**
- **1/2 tsp. kosher salt**
- **1/4 tsp. ground nutmeg**
- **1/2 cup chopped pecans, lightly toasted**
- **3 Tbsp. granulated sugar**
- **Parchment paper**
- **1/2 cup Tomato Jam**

1. Beat butter and powdered sugar at medium speed with a heavy-duty electric stand mixer 2 to 3 minutes or until pale and fluffy. Add egg yolk and vanilla, beating until blended.
2. Stir together flour and next 4 ingredients in a small bowl. Gradually add flour mixture to butter mixture, beating at low speed until blended after each addition.
3. Turn dough out onto a lightly floured surface; knead 3 to 4 times. Shape dough into a 1-inch-thick disk. Wrap tightly in plastic wrap, and chill 2 to 24 hours.
4. Preheat oven to 350°. Stir together pecans and granulated sugar in a small bowl. Shape chilled dough into 1-inch balls. Lightly beat egg white. Dip each ball into egg white; dredge in pecan mixture. Place 1 inch apart on a parchment paper-lined baking sheet. Press thumb into each ball, forming an indentation.
5. Bake at 350° for 10 minutes. Remove from oven, and, using back of a spoon, press indentations again. Spoon about 1/2 tsp. Tomato Jam into each indentation. Bake at 350° for 8 to 10 more minutes or until golden brown. Cool on baking sheets 5 minutes; transfer to a wire rack, and cool completely. Store in an airtight container up to 4 days.

MAKES about 2 dozen. **HANDS-ON** 45 min.; **TOTAL** 4 hours, 25 min., including jam

Tomato Jam

- **1 1/2 lb. assorted tomatoes**
- **1 cup sugar**
- **1/4 cup apple cider vinegar**
- **1 Tbsp. firmly packed orange zest**
- **3 Tbsp. fresh orange juice**
- **1 Tbsp. grated fresh ginger**
- **2 tsp. sea salt**
- **1 tsp. ground coriander**
- **4 whole cloves**

1. Bring 6 qt. water to a boil in a large Dutch oven over high heat. Add tomatoes, and boil 1 minute to loosen skins. Drain and plunge tomatoes into ice water to stop the cooking process. Peel tomatoes over a medium saucepan, letting juices drip into saucepan. Core and chop tomatoes; place in saucepan. Stir in sugar and next 7 ingredients. Bring to a low boil over medium heat; reduce heat to low, and simmer 30 to 40 minutes or until thickened.
2. Remove and discard cloves. Mash tomato mixture to desired consistency, using a potato masher. Remove from heat, and cool 20 minutes. Refrigerate in an airtight container up to 1 month.

MAKES about 2 cups. **HANDS-ON** 25 min.; **TOTAL** 1 hour, 15 min.

Burgers with Green Tomato Mayonnaise

BURGERS WITH GREEN TOMATO MAYONNAISE

This is a fabulous new way to add even more tomato flavor to a great burger. Try the tangy, chunky mayo on fish sandwiches and crab cakes too.

- **2 1/2 lb. ground chuck**
- **1/2 cup Worcestershire sauce**
- **1 1/4 tsp. freshly ground black pepper, divided**
- **1 tsp. kosher salt**
- **8 whole wheat hamburger buns**
- **Tomato slices**
- **Green Tomato Mayonnaise**
- **Fresh baby spinach**

1. Preheat grill to 350° to 400° (medium-high) heat. Gently combine ground chuck, 1/4 cup Worcestershire sauce, and 1/4 tsp. pepper in a large bowl. Using your hands, gently shape meat into 8 (4-inch) patties (about 3/4 inch thick). Transfer patties to a jelly-roll pan, and drizzle with remaining 1/4 cup Worcestershire sauce. Sprinkle with 1/2 tsp. each salt and pepper. Turn patties over, and sprinkle with remaining salt and pepper.
2. Grill patties, covered with grill lid, 6 to 8 minutes on each side. Transfer burgers to a platter, and cover loosely with aluminum foil. Let stand 5 minutes.
3. Meanwhile, grill buns 1 to 2 minutes or until grill marks appear. Serve burgers on buns with tomato slices, Green Tomato Mayonnaise, and baby spinach.

MAKES 8 servings. **HANDS-ON** 20 min.; **TOTAL** 40 min., including mayonnaise

Green Tomato Mayonnaise

Stir together 1/2 cup **mayonnaise;** 1 (7- to 8-oz.) **green tomato,** cored and diced; 4 fresh **basil** leaves, cut into thin strips; 1 Tbsp. chopped fresh **chives;** and 2 tsp. **Dijon mustard.** Serve immediately, or refrigerate in an airtight container up to 5 days.

MAKES about 1 1/2 cups. **HANDS-ON** 15 min., **TOTAL** 15 min.

PASTA WITH HEIRLOOM TOMATOES, GOAT CHEESE, AND BASIL

When preparing this, start with half the vinaigrette. Cooked pasta absorbs dressing as it sits, so add the rest just before serving for a last-minute burst of flavor.

- **12 oz. uncooked cavatappi pasta**
- **3 Tbsp. olive oil, divided**
- **4 oz. country ham, thinly sliced**
- **1 1/2 lb. heirloom tomatoes, chopped**
- **4 cups loosely packed arugula**
- **1 cup crumbled goat cheese**
- **15 fresh basil leaves, cut into thin strips**
- **3 Tbsp. chopped fresh flat-leaf parsley**
- **1/2 tsp. dried crushed red pepper**
- **Herb Vinaigrette**
- **Kosher salt**
- **Freshly ground black pepper**

1. Prepare pasta according to package directions for al dente. Toss together pasta and 1 Tbsp. olive oil in a large bowl.
2. Cook ham in remaining hot olive oil in a skillet over medium-high heat 1 minute on each side or until crispy. Drain on paper towels. Break ham into small pieces.
3. Add tomatoes, next 5 ingredients, and 6 Tbsp. vinaigrette to pasta; toss gently. Add salt and pepper to taste; top with ham. Just before serving, stir in remaining vinaigrette.

MAKES 8 to 10 servings. **HANDS-ON** 25 min.; **TOTAL** 30 min., including vinaigrette

Herb Vinaigrette

Process 2 Tbsp. **red wine vinegar,** 1 Tbsp. **fresh lemon juice,** and 4 small **garlic cloves** in a food processor until smooth. With processor running, pour 1/2 cup **canola oil** through food chute in a slow, steady stream, processing until blended. Add 12 fresh **basil** leaves and 1/2 cup fresh flat-leaf **parsley** leaves, and process until smooth. Add 1/2 cup grated **Parmesan** cheese; pulse to combine. Stir in **kosher salt** and freshly ground black **pepper** to taste. Serve immediately, or refrigerate in an airtight container up to 5 days.

MAKES about 3/4 cup. **HANDS-ON** 10 min., **TOTAL** 10 min.

Chop, Chop!

FRESH SPINS ON SUMMER'S MOST APPROACHABLE MAIN: THE PORK CHOP

CHILE-RUBBED CHOPS WITH SWEET POTATOES AND GRILLED OKRA

We love a rub. It's easier and less messy than a marinade and gives through-and-through flavoring.

- **1 tsp. ancho chile powder**
- **2 tsp. dried oregano**
- **1 tsp. celery seeds**
- **1 tsp. paprika**
- **1 tsp. dried thyme**
- **2 1/2 tsp. kosher salt, divided**
- **4 (1/2-inch-thick) bone-in pork loin chops**
- **2 1/2 Tbsp. olive oil, divided**
- **3 or 4 flat wooden skewers**
- **2 medium-size sweet potatoes**
- **1 lb. fresh whole okra**
- **1/4 tsp. ground black pepper**
- **2 Tbsp. butter**

1. Stir together first 5 ingredients and 1 1/2 tsp. salt. Brush pork chops with 1 Tbsp. oil, and rub mixture over both sides. Chill, uncovered, 8 to 12 hours.
2. Soak skewers in water 20 minutes. Preheat grill to 350° to 400° (medium-high) heat. Pierce potatoes several times with a fork. Cover with damp paper towels, and microwave at HIGH 8 minutes or until tender, turning after 3 minutes. Let stand 10 minutes.
3. Toss together okra, pepper, and remaining olive oil and salt; thread okra onto skewers. Grill pork, covered with grill lid, 5 minutes on each side or until a meat thermometer inserted in thickest portion registers 155°. At the same time, grill okra 5 minutes, turning occasionally.
4. Cut potatoes in half, and top with butter; season with salt and pepper. Serve with pork and okra.

MAKES 4 servings. **HANDS-ON** 25 min.; **TOTAL** 8 hours, 35 min.

Grilled Oysters (page 26)

clockwise from top left:

- Oyster-Bacon Pot Pie (page 25)
- Turkey-Stuffed Peppers (page 36)
- Charred Steak Salad with Spicy Dressing (page 34)
- Peas and Kale Salad with Bacon Vinaigrette (page 34)

Rotini with Crumbled Turkey and Tomato Sauce (page 36)

clockwise from top left:

- Beef-and-Bean Chili and Pork-and-Black Bean Chili (pages 30-31)
- Lemony Ranch, Blue Cheese Ranch, and Avocado Ranch Buttermilk Dressing (page 46)
- Fluffy Buttermilk Waffles (page 45)

Chocolate-Buttermilk Pudding (page 45)

Fettuccine with Smoky Turnip Greens, Lemon and Goat Cheese (page 47)

The Ultimate Classic Collards (page 47)

clockwise from top left:

- Blue Cheese Crostini with Balsamic-Roasted Grapes (page 54)
- Honey-Glazed Pork Tenderloin with Homemade Applesauce (page 51)
- Shrimp 'n' Grits Pie (page 61)
- Grits Cakes with Poached Eggs and Country Gravy (page 61)

The Ultimate Carrot Cake (page 68)

Roasted Chicken Thighs with Herb Butter (page 65)

Sesame-Chicken Thigh Paillard with Peanut Sauce (page 63)

Sausage-and-Bean Stew (page 69)

clockwise from top left:

- Spinach-and-Three-Herb Pesto (page 73), Minty Lemonade (page 74), Cucumbers with Ginger, Rice Vinegar, and Mint, (page 75), Creamy Basil-Black Pepper Cucumbers (page 73), and Kale-and-Blueberry Slaw with Buttermilk Dressing (page 75)
- Shrimp and Peas with Farfalle (page 89)
- Pasta Shells with Spring Vegetables (page 89)

Shaved Vegetable Salad with Toasted Almonds (page 85)

Roasted Leg of Lamb with Lemon-Herb Salt (page 84)

Pan-Seared Skirt Steak and Chimichurri (page 116)

Quick Cook Shrimp and Chorizo Paella (page 113)

Quick BBQ Chicken Pizzas
(page 110)

Snapper Baked in Parchment with Spring Vegetables (page 104)

clockwise from top left:

• Mango-Lemon-Buttermilk Icebox Pie (page 121), Lemon-Buttermilk Icebox Pie with Lemon-Blueberry Topping (pages 120-121), Key Lime-Buttermilk Icebox Pie (page 120), Strawberry-Lemon-Buttermilk Pie (page 120)
• Lemon-Buttermilk Icebox Pie (page 120)
• Chocolate-Bourbon-Butterscotch Icebox Cake (page 122)

Tomato-Peach Salad with Chicken (page 137)

Seared Flank Steak with Lime-Wasabi Sauce (page 196)

Buttermilk-Plum Ice Cream (page 155)

Brisket Tacos with Summer Salsa (page 199)

Green Tomato Soup with Lump Crabmeat (page 157)

clockwise from top left:

- Classic Sweet Tea (page 209)
- Crispy Eggplant, Tomato, and Provolone Stacks with Basil (page 215)
- Summer Corn-and-Golden Potato Chowder (page 213)
- Open-Faced Shrimp-and-Avocado Sandwiches (page 214)

Summer Pasta Salad with
Lime Vinaigrette (page 212)

Slow-Cooker Tomato Sauce over Spaghetti **(page 210)**

Judy's Bloody Mary Mix (page 233)

Caramel Apple Cake
(page 231)

The Ultimate Grilled Cheese and Tomato-and-Red Pepper Soup (page 227)

Pumpkin-Honey-Beer Bread Pudding with Apple Brandy-Caramel Sauce (page 224)

HOMEMADE SHAKE-AND-BAKE PORK CHOPS WITH MUSTARD SAUCE

- 1 1/4 cups panko (Japanese breadcrumbs)
- 3 Tbsp. olive oil
- 1/2 tsp. dried oregano
- 1/2 tsp. dried parsley
- 1/4 tsp. garlic powder
- 1/4 tsp. onion powder
- 2 1/4 tsp. kosher salt, divided
- 1 1/8 tsp. freshly ground black pepper, divided
- 4 (1/2-inch-thick) bone-in pork loin chops
- 2 Tbsp. butter
- 1 large shallot, minced
- 1 cup reduced-sodium chicken broth
- 1/4 cup Dijon mustard
- 2 Tbsp. heavy cream
- 2 tsp. fresh lemon juice
- 1 Tbsp. chopped fresh flat-leaf parsley

1. Preheat oven to 425°. Stir together first 6 ingredients, 2 tsp. salt, and 1 tsp. pepper in a bowl. Transfer mixture to a 1-gal. zip-top plastic freezer bag.
2. Sprinkle pork chops on both sides with remaining 1/4 tsp. salt and 1/8 tsp. pepper. Place 2 chops in breadcrumb mixture, and shake to coat well. Place chops on a wire rack on a baking sheet. Repeat with remaining chops.
3. Bake at 425° for 15 minutes; turn chops over, and bake 10 more minutes or until a meat thermometer inserted in thickest portion registers 155°. Let stand 5 minutes.
4. Melt butter in a medium skillet over medium heat. Add shallot, and sauté 3 minutes or until softened. Increase heat to medium-high; add broth, and bring to a boil. Boil 1 minute. Stir in mustard, cream, and lemon juice, and cook, stirring occasionally, 2 to 3 minutes or until sauce is slightly thickened. Stir in chopped parsley. Serve immediately with pork chops.

Note: We tested with Swanson Natural Goodness 33% Less Sodium Chicken Broth.

MAKES 4 servings. **HANDS-ON** 20 min., **TOTAL** 50 min.

EASY SUMMER GREEN BEANS

Bring 1 qt. water, 1 tsp. **kosher salt,** 1/4 tsp. freshly ground black **pepper,** and 1/4 tsp. **sugar** to a boil in a medium saucepan over medium-high heat. Add 1 lb. **fresh green beans;** cover and reduce heat to medium. Cook 10 minutes or to desired degree of doneness. Drain beans, and drizzle with 1/2 Tbsp. **olive oil** and 1/2 Tbsp. fresh **lemon juice.** Add more olive oil and lemon juice, if desired.

MAKES 4 servings. **HANDS-ON** 25 min., **TOTAL** 25 min.

Broiled Pork Chops with Basil Butter and Summer Squash

Pork Chop Sandwiches with Gravy and Grits

BROILED PORK CHOPS WITH BASIL BUTTER AND SUMMER SQUASH

The basil butter is the star here: It seasons the pork and adds an herby, rich flavor to the roasted squash.

- **1/2 cup butter, softened**
- **1/2 cup chopped fresh basil**
- **1 Tbsp. minced shallot**
- **2 1/2 tsp. kosher salt, divided**
- **1 1/2 tsp. freshly ground black pepper, divided**
- **2 lb. mixed summer squash, chopped**
- **1 small red onion, chopped**
- **3 medium tomatoes, seeded and chopped**
- **4 (3/4-inch-thick) frenched pork loin chops**

1. Stir together first 3 ingredients, 1 tsp. salt, and 1/2 tsp. pepper in a small bowl. Cover and chill basil-butter mixture until ready to use (up to 48 hours).

2. Preheat broiler with oven rack 5 inches from heat. Toss together squash, next 2 ingredients, 1 tsp. salt, and 1/2 tsp. pepper in a large bowl. Spread mixture in a single layer in a jelly-roll pan, and broil 10 to 12 minutes or until tender and lightly browned, stirring every 5 minutes. Transfer to a serving bowl, and stir in 2 Tbsp. basil-butter mixture. Cover with aluminum foil to keep warm.

3. Sprinkle chops with remaining salt and pepper, and place on a wire rack in an aluminum foil-lined jelly-roll pan. Top each chop with a heaping tablespoon of basil-butter mixture. Broil 8 to 10 minutes or until a meat thermometer inserted in thickest portion registers 155°. Let stand 5 minutes before serving. Serve pork chops with squash mixture.

MAKES 4 servings. **HANDS-ON** 35 min., **TOTAL** 40 min.

PORK CHOP SANDWICHES WITH GRAVY AND GRITS

Breakfast for dinner is always a welcome surprise. Use biscuits to make the sandwiches, and serve with fresh fruit to balance out the meal.

- **6 frozen buttermilk biscuits (from a 41.6-oz. package)**
- **6 thin boneless pork loin chops (about 1 3/4 lb.)**
- **1 1/2 tsp. kosher salt, divided**
- **1/2 tsp. ground black pepper, divided**
- **2 Tbsp. olive oil**
- **1/2 cup butter, divided**
- **1/3 cup all-purpose flour**
- **3 3/4 cups milk**
- **1/2 tsp. hot sauce**
- **1 cup uncooked quick-cooking grits**
- **1/2 cup (2 oz.) freshly shredded Cheddar cheese**

1. Cook biscuits according to package directions.

2. Meanwhile, sprinkle pork chops with 1 tsp. salt and 1/4 tsp. pepper. Cook in hot oil in a large skillet over medium-high heat, in batches, 2 to 3 minutes on each side or until browned. Transfer chops to a platter, and cover with aluminum foil to keep warm. Wipe skillet clean.

3. Melt 1/4 cup butter in skillet over medium-low heat. Whisk in flour. Cook, whisking constantly, 1 minute, scraping browned bits from bottom of skillet. Gradually whisk in milk; increase heat to medium, and cook, whisking constantly, 10 minutes or until thickened. Stir in hot sauce and remaining 1/2 tsp. salt and 1/4 tsp. pepper.

4. Cook grits according to package directions; stir in cheese and remaining butter. Season with salt and pepper. Split biscuits; top with pork chops and desired amount of gravy. Serve with grits.

MAKES 4 to 6 servings. **HANDS-ON** 30 min., **TOTAL** 40 min.

MAKE IT LAST
Stretch Sunday's supper into two more meals.
MONDAY
Chopped Salad with Steak
TUESDAY
Southern Cheese Steak Sandwiches
Serve Sunday's flank steak with seared tomatoes and Lime-Wasabi Sauce.
Steak Nights
FIRE UP THE GRILL ONCE FOR THREE DIFFERENT DINNERS, MAKING THE MOST OF FLANK STEAK AND YOUR TIME

SUNDAY

SEARED FLANK STEAK WITH LIME-WASABI SAUCE

Zesty red wine marinade is the key for this super-tender and deeply flavored steak. The longer it marinates, the better.

- 2 (1 1/2- to 2-lb.) flank steaks
- Red Wine Marinade
- 2 Tbsp. olive oil
- 1 1/4 lb. mixed cherry tomatoes
- 4 to 6 cups loosely packed arugula
- Lime-Wasabi Sauce

1. Place steaks and marinade in a 2-gal. zip-top plastic freezer bag; seal bag, and turn to coat. Chill 2 to 12 hours.
2. Preheat grill to 400° to 450° (high) heat. Grill steak, covered with grill lid, 5 to 7 minutes on each side or to desired degree of doneness. Remove from grill, and let stand 10 minutes.
3. Heat oil in a skillet over medium-high heat; add tomatoes, and cook, stirring occasionally, 10 minutes or until seared. Season with salt and pepper.
4. Cut 1 flank steak thinly across the grain; serve with tomatoes, arugula, and sauce. (Reserve remaining steak for salad and sandwiches.)

MAKES 4 to 6 servings. **HANDS-ON** 30 min.; **TOTAL** 2 hours, 45 min.

Red Wine Marinade

Whisk together 1 cup dry **red wine,** 3 Tbsp. **soy sauce,** and 1 Tbsp. minced **garlic** in a bowl; add black **pepper** to taste. Whisk in 1 cup **olive oil.**

MAKES about 2 cups. **HANDS-ON** 5 min., **TOTAL** 5 min.

Lime-Wasabi Sauce

Stir together 2 Tbsp. fresh **lime juice,** 3 Tbsp. **wasabi sauce,** 1 Tbsp. minced fresh **ginger,** 1 Tbsp. **rice vinegar,** 1/4 tsp. **kosher salt,** and 1/8 tsp. black **pepper** in a small bowl. Whisk in 1/2 cup **olive oil.**

MAKES about 1 cup. **HANDS-ON** 10 min., **TOTAL** 10 min.

MONDAY NIGHT

CHOPPED SALAD WITH STEAK

Chill plates one hour before serving to help keep the salad cold after serving.

- 6 cups chopped romaine lettuce
- 1 small red onion, thinly sliced
- 1 red bell pepper, chopped
- 1 yellow bell pepper, chopped
- 1 cucumber, peeled and sliced
- 1/2 cup pitted kalamata olives, chopped
- 1/4 cup loosely packed fresh flat-leaf parsley, chopped
- 3/4 lb. Seared Flank Steak, chopped
- Crumbled blue cheese
- Blue cheese dressing

Toss together first 7 ingredients in a large bowl. Divide salad mixture among plates; top each with steak and desired amount of blue cheese. Serve with blue cheese dressing.

MAKES 4 servings. **HANDS-ON** 20 min.; **TOTAL** 40 min., not including steak

TUESDAY NIGHT

SOUTHERN CHEESE STEAK SANDWICHES

Simple with significant flavor, this Southern spin on the classic cheese steak is deluxe comfort food.

- 1 sweet onion, sliced
- 1 Tbsp. olive oil
- 6 oz. sharp Cheddar cheese, grated
- 1/2 cup mayonnaise
- 1 (4-oz.) jar diced pimiento, drained
- 4 sub rolls, split
- 3/4 lb. Seared Flank Steak, sliced

1. Preheat broiler with oven rack 8 to 9 inches from heat. Sauté onion in hot oil in a saucepan over medium heat 10 minutes or until tender. Pulse cheese, mayonnaise, and pimiento in a food processor until combined. Season with salt and pepper. Spread each roll with about 1/4 cup pimiento cheese mixture.
2. Sauté steak in a small skillet over medium-high heat 2 to 3 minutes or until heated through. Divide steak and onions among rolls, and place on a baking sheet. Broil 2 to 3 minutes or until cheese is melted. Serve immediately with remaining pimiento cheese mixture.

MAKES 4 servings. **HANDS-ON** 25 min.; **TOTAL** 25 min., not including steak

Summer Succotash

OUR NO-FAIL, NO-FRILLS, KEEP-FOREVER RECIPE

CLASSIC SUCCOTASH

- **2 cups fresh lima beans**
- **1/2 small yellow onion**
- **4 fresh thyme sprigs**
- **1 garlic clove**
- **3 uncooked bacon slices**
- **1 medium-size sweet onion, chopped**
- **3 cups fresh corn kernels (about 6 ears)**
- **1 pt. cherry tomatoes, halved**
- **2 Tbsp. unsalted butter**
- **1 Tbsp. red wine vinegar**
- **1 1/2 Tbsp. chopped fresh dill**
- **1 1/2 Tbsp. chopped fresh chives**

1. Bring first 4 ingredients and water to cover to a boil in a saucepan over medium-high heat; reduce heat to medium, and simmer, stirring occasionally, 20 minutes or until beans are tender. Drain beans, reserving 3/4 cup cooking liquid. Discard yellow onion, thyme, and garlic.

2. Cook bacon in a large skillet over medium heat 7 minutes or until crisp, turning once. Remove bacon, reserving 2 Tbsp. drippings in skillet. Drain bacon on paper towels, and crumble.

3. Sauté chopped sweet onion in hot drippings over medium-high heat 5 minutes. Stir in corn, and cook, stirring often, 6 minutes or until corn is tender. Stir in tomatoes, cooked lima beans, and 3/4 cup reserved cooking liquid; cook, stirring occasionally, 5 minutes. Stir in butter and next 3 ingredients. Season with salt and pepper. Sprinkle with crumbled bacon.

MAKES 6 servings. **HANDS-ON** 45 min.; **TOTAL** 1 hour, 10 min.

THE *SL* TEST KITCHEN ACADEMY

OUR TEST KITCHEN DIRECTOR, ROBBY MELVIN, SHARES TIPS FOR CHOOSING AND STORING

SUMMER TOMATOES

Summer's greatest prize is a sandwich made with juicy sliced tomatoes. I keep mine simple and classic with fresh white bread, mayo, kosher salt, and pepper. Knowing what to look for (hint: Use your nose!) and how to store your 'maters will make all the difference in flavor and texture.

1 APPEARANCE Some tomatoes, especially heirloom varieties, are grown for flavor rather than for looks. They might be odd and riddled with cracks, but they are often the most delicious. No matter the color, shape, or size, choose tomatoes with firm skin and no soft, mushy spots.

2 SCENT All tasty tomatoes have one thing in common: They smell like a tomato. A perfectly ripe tomato should smell like a glass of tomato juice, especially around the stem. A tomato without an aroma will lack in flavor.

3 STORAGE All you need to know is this: *Never* store tomatoes in the refrigerator. Cold temps turn them mealy and rock hard. The ideal temperature for keeping tomatoes is 60° to 65°, so place them in the coolest part of the kitchen, stem-side up, out of direct sun. Arrange miniature tomatoes in a shallow dish so air can circulate around them.

Taco Night Reimagined

IT'S TOO HOT TO TURN ON THE OVEN! KICK BACK, RELAX, AND LET YOUR SLOW COOKER DO ALL THE WORK

BRISKET TACOS WITH SUMMER SALSA

- **2 uncooked bacon slices, cut into 1-inch pieces**
- **1 medium-size white onion, chopped (about 1 cup)**
- **2 tsp. kosher salt**
- **1 tsp. freshly ground black pepper**
- **1 (3- to 3 1/2-lb.) beef brisket, trimmed**
- **1 cup reduced-sodium chicken broth**
- **3 canned chipotle peppers in adobo sauce**
- **3 Tbsp. adobo sauce from can**
- **3 garlic cloves, peeled and smashed**
- **1 Tbsp. ground cumin**
- **1 Tbsp. Worcestershire sauce**
- **1 Tbsp. honey**
- **1 tsp. dried oregano**
- **2 Tbsp. apple cider vinegar**
- **10 to 12 (8-inch) flour tortillas, warmed**
- **Crunchy Summer Salsa**

1. Place bacon and onion in a 6- to 8-qt. slow cooker. Stir together salt and pepper; sprinkle over all sides of brisket. Place brisket in slow cooker.

2. Process broth and next 7 ingredients in a blender for 30 seconds or until smooth; pour mixture over brisket. Cover and cook on LOW 7 hours or until brisket is fork-tender. Transfer brisket to a 13- x 9-inch baking dish; cover with aluminum foil to keep warm.

3. Pour sauce through a fine wire-mesh strainer into a medium saucepan, and cook over medium-high heat, stirring occasionally, 15 to 20 minutes or until reduced to 1/3 cup. Stir in vinegar.

4. Coarsely chop brisket; spoon over warm tortillas. Drizzle with sauce, and top with Crunchy Summer Salsa.

MAKES 6 to 8 servings. **HANDS-ON** 20 min.; **TOTAL** 7 hours, 30 min., including salsa

Crunchy Summer Salsa

Stir together 1 cup diced fresh **peaches;** 1/2 cup diced **cucumber,** 1 **jalapeño** pepper, seeded (if desired) and diced; 1 **garlic clove,** minced; 3 Tbsp. chopped fresh **cilantro;** 2 Tbsp. fresh **lime juice;** and 1/2 tsp. **kosher salt** in a small bowl.

MAKES about 1 3/4 cups. **HANDS-ON** 10 min., **TOTAL** 10 min.

Light and Luscious Summer Pasta

THE SECRET TO THIS IRRESISTIBLE SALAD IS OUR KICKED-UP BUTTERMILK DRESSING

ORZO SALAD WITH SPICY BUTTERMILK DRESSING

Creamy avocado and a fiery buttermilk dressing make this good-for-you pasta salad feel like a hearty meal. The recipe is easy to double, and you'll have most of the ingredients in your pantry, so it's the perfect solution for a last-minute party or potluck.

- 1 cup uncooked orzo
- 1 cup frozen whole-kernel corn, thawed and drained
- 12 cherry tomatoes, quartered
- 3 green onions, sliced
- 1 (15-oz.) can black beans, drained and rinsed
- 1/4 cup low-fat buttermilk
- 3 Tbsp. fresh lime juice
- 2 Tbsp. light sour cream
- 2 Tbsp. canola mayonnaise
- 1 tsp. chili powder
- 1/2 tsp. kosher salt
- 1/4 tsp. black pepper
- 1/4 tsp. ground red pepper
- 2 garlic cloves, crushed
- 3 Tbsp. chopped fresh cilantro
- 1 peeled avocado, cut into 8 wedges
- 1 Tbsp. chopped fresh flat-leaf parsley

1. Cook orzo according to package directions, omitting salt and fat. Drain and rinse; drain well. Toss together orzo, corn, and next 3 ingredients in a large bowl.

2. Whisk together buttermilk, next 8 ingredients, and 2 Tbsp. cilantro in a small bowl. Drizzle over orzo mixture; toss. Top with avocado, parsley, and remaining cilantro.

MAKES 4 servings (serving size: 1 3/4 cups orzo mixture, 2 avocado wedges, 3/4 tsp. cilantro, and 3/4 tsp. parsley). **HANDS-ON** 30 min., **TOTAL** 30 min.

NUTRITIONAL INFORMATION
(per serving)
CALORIES: 424; FAT: 15.3g; (SATURATED FAT: 2.3g); PROTEIN: 12.7g; FIBER: 10.1g; CARBOHYDRATES: 63.8g; SODIUM: 607mg

Easy Ice Cream

Use these short and sweet recipes to churn out four fantastic flavors at home

COFFEE LIQUEUR COOKIES-AND-CREAM ICE CREAM

- 1 (14-oz.) can sweetened condensed milk
- 1 (5-oz.) can evaporated milk
- 2 cups whole milk
- 2 Tbsp. sugar
- 1 tsp. vanilla extract
- 1/8 tsp. table salt
- 1 cup crushed cream-filled chocolate sandwich cookies
- 3 Tbsp. coffee-flavored liqueur

1. Whisk together first 6 ingredients; cover and chill 2 hours.
2. Pour milk mixture into freezer container of a 1-qt. electric ice-cream maker, and freeze according to manufacturer's instructions. (Instructions and times will vary.)
3. Remove container with ice cream from ice-cream maker, and freeze 30 minutes.
4. Stir crushed cookies and liqueur into ice-cream mixture. Transfer mixture to an airtight container or a loaf pan covered tightly with aluminum foil; freeze 3 to 4 hours or until firm.

MAKES about 1 qt. **HANDS-ON** 10 min., **TOTAL** 6 hours

BLUEBERRY-LEMON ZEST ICE CREAM

Omit cookies and liqueur. Bring 2 cups coarsely chopped **blueberries,** 2 Tbsp. **sugar,** and 2 Tbsp. water to a boil in a small saucepan over medium heat; reduce heat to low, and simmer 10 minutes, stirring often. Cool 30 minutes; cover and chill 2 to 3 hours. Meanwhile, prepare recipe as directed through Step 3. Stir 2 tsp. **lemon zest** into prepared ice-cream mixture, and swirl in chilled blueberry mixture. Proceed as directed in Step 4. **HANDS-ON** 25 min.; **TOTAL** 6 hours, 45 min.

BOURBON-BUTTER-SALTED PECAN ICE CREAM

Omit cookies and liqueur; increase vanilla to 2 tsp. Prepare recipe as directed through Step 3. Meanwhile, cook 1 cup coarsely chopped **pecans** and 1/2 Tbsp. **butter** in a small skillet over medium heat, stirring constantly, 7 to 8 minutes or until toasted and fragrant. Spread pecans on wax paper, and sprinkle with 1/4 tsp. **kosher salt;** cool completely. Stir pecans and 2 Tbsp. **bourbon** into prepared ice-cream mixture, and proceed as directed in Step 4. **HANDS-ON** 10 min., **TOTAL** 6 hours

STRAWBERRY-BASIL ICE CREAM

Omit vanilla, cookies, and liqueur; reduce whole milk to 1 1/2 cups. Prepare recipe as directed through Step 1. Meanwhile, pulse 1 (16-oz.) container fresh **strawberries,** coarsely chopped; 2 Tbsp. chopped fresh **basil;** and 2 Tbsp. fresh **lemon juice** in a food processor 5 or 6 times or until finely chopped. Stir strawberry mixture into chilled milk mixture. Proceed as directed in Steps 2 through 4.
HANDS-ON 10 min., **TOTAL** 6 hours

Community Cookbook

BUILD A BETTER LIBRARY, ONE GREAT BOOK AT A TIME

If you've ever felt lost at the farmers' market, chef Hugh Acheson has the answer. In his latest book, he shares 200 recipes and meal ideas centered around fruits and vegetables, with a side of fatherly advice for getting kids to eat them too.

LEEK FONDUTA

Fonduta is simply the Italian version of fondue.

- **2 lb. leeks, white and pale green parts, washed and thinly sliced in half-rounds**
- **3 garlic cloves, thinly sliced**
- **2 Tbsp. unsalted butter**
- **1 tsp. sea salt, plus more to taste**
- **1 cup heavy cream**
- **1/2 cup crème fraîche**
- **1 cup finely grated Parmigiano-Reggiano cheese**
- **2 Tbsp. chopped fresh flat-leaf parsley leaves**
- **1 tsp. firmly packed lemon zest**
- **1 tsp. freshly cracked black pepper**
- **1 baguette, cut into 1-inch-thick slices and toasted**

In a large saucepan, combine the leeks, garlic, butter, and the 1 tsp. sea salt. Place the pot over medium heat, and cook until the leeks are tender and translucent (about 20 minutes). Add the cream and crème fraîche, and bring the liquid to a boil. Reduce the heat, and simmer for 5 minutes. Add the Parmigiano-Reggiano, parsley, lemon zest, black pepper, and more sea salt to taste, and stir to incorporate. Transfer the fonduta to a serving bowl, and serve with baguette slices.

MAKES 4 to 6 servings. **HANDS-ON** 50 min.; **TOTAL** 1 hour, 5 min.

CARROT SOUP WITH BROWN BUTTER, PECANS, AND YOGURT

This soup brings out the natural sweetness of the carrots.

- **1 lb. carrots**
- **4 Tbsp. unsalted butter, divided**
- **1 medium yellow onion, minced**
- **2 sprigs fresh thyme**
- **1 red jalapeño, minced**
- **2 Tbsp. ground sesame seeds (not tahini—pulse them in a food processor or spice grinder)**
- **1 qt. chicken stock**
- **Kosher salt**
- **1/2 cup plus 2 Tbsp. plain Greek yogurt**
- **1/4 cup crushed pecans**
- **1 Tbsp. sherry vinegar**
- **2 Tbsp. chopped carrot tops**
- **Maple syrup to taste**

1. Peel the carrots, and cut 1 carrot into very thin rounds; reserve for garnish. Cut the rest of the carrots into 1/2-inch pieces.
2. Melt 2 Tbsp. of the butter in a medium saucepan over medium heat. Add the onion, and cook, stirring occasionally, until soft (about 10 minutes). Add the thyme sprigs, jalapeño, sesame, and 1/2-inch cut carrots. Cook for 10 more minutes, stirring occasionally, and then add the stock and kosher salt to taste. Bring to a boil; reduce the heat, and simmer until the carrots are very tender (about 15 minutes). Remove from the heat, and remove the thyme sprigs.
3. Puree the carrot mixture in a blender. Pour the soup back in saucepan, stir in the 1/2 cup yogurt, adjust seasoning with salt, and place the lid on the pan.
4. Melt the remaining 2 Tbsp. butter in a small sauté pan over medium-high heat, and cook until the solids begin to brown. Add the pecans. Toss and toast for about 1 minute. Remove from the heat, and add the vinegar.
5. Serve the soup in bowls. Dollop each serving with the remaining yogurt and the pecan brown butter. Sprinkle with the carrot coins and carrot tops, and finish with a drizzle of maple syrup.

MAKES 4 to 6 servings. **HANDS-ON** 45 min., **TOTAL** 1 hour

Carrot Soup with Brown Butter, Pecans, and Yogurt

August

204 HOSPITALITY **Backyard Pizzeria** Beat the heat! Pizza-oven perfection right from the grill top

208 THE *SL* TEST KITCHEN ACADEMY **Grilled Pizza** From dough to delicious with our how-tos from crust to toppings

209 SOUTHERN CLASSIC **Classic Sweet Tea** What every Southern cook should know to master the region's favorite refresher

210 THE SLOW DOWN **Fresh, Versatile Tomato Sauce** Go from bumper crop to tasty tomato sauce with minimal effort or attention

212 QUICK-FIX SUPPERS **Fast, Fresh, and Filling** Satisfying summer suppers highlighting the season's best produce

216 THE NEW SUNDAY SUPPER **One Pot, Three Ways** Peas please! A summer ingredient turned into a trio of meals

218 SAVE ROOM **Cool Down with Easy Granitas** Grown-up snow cones for cocktail hour or dessert

220 COMMUNITY COOKBOOK Savor summertime harvest long after summer with this canning guide

Backyard Pizzeria

BRICK-OVEN-STYLE PIZZAS ARE JUST A BLAZING GRILL AWAY

Ultimate Cheese Pizza, **page 207**

IT'S NO COINCIDENCE many of the South's best pizzerias tout "wood-fired" or "brick oven" at the top of their menus. Those traditional, high-heat ovens make a delicious difference in the final pizza, especially the crust. A little secret: You can get the same results at home without building or buying a specialty oven. Simply fire up your grill.

Ultimate Cheese Pizza, page 207

Smoked Sausage, Grilled Corn, and Sweet Onion Pizza, page 207

Smoked Chicken Pizza with White Barbecue Sauce, page 207

GRILLED PINEAPPLE DESSERT PIZZA

Prepare **Homemade Pizza Dough** (recipe, facing page), and grill both sides of rounds according to Our Grilled Pizza Technique (facing page). Cool crusts 10 minutes. Stir together 3 Tbsp. light **brown sugar** and 1/8 tsp. ground **cinnamon.** Cut 1 fresh **pineapple** into 1/2-inch slices, and spread brown sugar mixture on both sides. Grill, covered with grill lid, 1 minute on each side or until grill marks appear. Stir together 1 (8-oz.) package **cream cheese,** softened; 3 Tbsp. light brown sugar; and 1/8 tsp. ground cinnamon. Spread about 2 Tbsp. cream cheese mixture on each grilled pizza crust, and top with grilled pineapple slices. Drizzle with warm jarred **caramel topping,** and sprinkle with toasted **pecans** and fresh **mint.**

MAKES 8 (6- to 8-inch) pizzas.
HANDS-ON 20 min.; **TOTAL** 2 hours, 40 min.

HOMEMADE PIZZA DOUGH

See our tips for grilling pizzas in Test Kitchen Academy on page 208.

- **2 cups warm water (100° to 110°)**
- **1 (1/4-oz.) envelope active dry yeast**
- **2 tsp. sugar**
- **1 Tbsp. kosher salt**
- **6 cups all-purpose flour, divided**
- **1/2 cup olive oil**
- **Plain yellow cornmeal**
- **2 to 3 Tbsp. olive oil**

1. Stir together first 3 ingredients in a small bowl. Let stand 5 minutes.
2. Stir together salt and 5 1/2 cups flour in a large bowl; make a well in center of mixture. Pour yeast mixture and 1/2 cup olive oil into well, and stir to form a soft dough.
3. Turn dough out onto a lightly floured surface, and knead 5 minutes, gradually kneading in remaining 1/2 cup flour. Place in a well-greased bowl, turning to grease top.
4. Cover and let rise in a warm place (80° to 85°), free from drafts, 1 hour and 30 minutes or until doubled in bulk.
5. Punch dough down. Turn dough out onto lightly floured surface; cover and let rest 5 minutes. Shape dough into 8 balls. Roll each dough ball into a 6- to 8-inch circle. Lightly sprinkle 2 large baking sheets with cornmeal. Place 2 dough rounds on each sheet; lightly brush each round with olive oil. Grill as directed using our Grilled Pizza Technique (at right). Repeat with remaining dough balls.

Note: Dough may be wrapped in plastic wrap and refrigerated up to 24 hours after Step 3. Let dough stand 20 minutes before shaping and grilling. Dough also may be frozen after shaping into balls. Wrap each ball individually with plastic wrap, and freeze in a small zip-top plastic freezer bag up to 1 month. Thaw overnight in refrigerator. Allow dough balls to rest, covered with plastic wrap, for 20 minutes. Shape as directed.

MAKES 8 (6- to 8-inch) pizzas.
HANDS-ON 40 min.; **TOTAL** 2 hours, 20 min.

ULTIMATE CHEESE PIZZA

Prepare **Homemade Pizza Dough** as directed. Place rounds directly on grill grate (cornmeal side down), and grill, covered with grill lid, 4 minutes or until golden brown. Carefully flip dough. Spread each pizza crust with 2 Tbsp. **Slow-Cooker Tomato Sauce** (recipe, page 210). Top each with 3 fresh **mozzarella** slices, 1 oz. shredded **Parmesan** cheese, and 1 oz. shredded **Gouda** cheese. Grill, covered with grill lid, 3 to 4 minutes or until cheese is browned and bubbly. Top with fresh **basil.**

SMOKED CHICKEN PIZZA WITH WHITE BARBECUE SAUCE

WHITE BARBECUE SAUCE

- **1 1/2 cups mayonnaise**
- **1/4 cup white wine vinegar**
- **1 Tbsp. freshly ground black pepper**
- **1 Tbsp. spicy brown mustard**
- **2 tsp. prepared horseradish**
- **1 1/2 tsp. sugar**
- **1 tsp. kosher salt**
- **1 tsp. minced garlic**

REMAINING INGREDIENTS

- **Homemade Pizza Dough**
- **1 lb. chopped smoked chicken**
- **2 cups sliced tomatoes**
- **8 oz. shredded Monterey Jack cheese**
- **Garnish: fresh opal basil**

1. Whisk together first 8 ingredients.
2. Preheat grill to 400° (high) heat. Place Homemade Pizza Dough rounds directly on grill grate (cornmeal side down), and grill, covered with grill lid, 4 minutes or until golden brown. Carefully flip dough.
3. Spread each pizza crust with 2 Tbsp. White Barbecue Sauce, and top with smoked chicken, tomatoes, and cheese. Grill, covered with grill lid, 3 to 4 minutes or until cheese is browned and bubbly.

SMOKED SAUSAGE, GRILLED CORN, AND SWEET ONION PIZZA

- **2 cups torn kale leaves**
- **1 tsp. kosher salt**
- **1/4 cup extra virgin olive oil**
- **1/4 cup roasted almonds**
- **2 garlic cloves, chopped**
- **1/2 cup grated Parmesan cheese**
- **Homemade Pizza Dough**
- **1 lb. smoked sausage, sliced**
- **1 cup grilled corn kernels**
- **2 sweet onions, grilled and sliced**
- **16 oz. shredded Havarti cheese**

1. Pulse kale and salt in a food processor 10 to 12 times or until finely chopped.
2. With processor running, pour olive oil through food chute in a slow stream. Add almonds and garlic, and pulse 4 to 5 times. Add cheese, and pulse until well blended.
3. Preheat grill to 400° (high) heat. Place Homemade Pizza Dough rounds directly on grill grate (cornmeal side down), and grill, covered with grill lid, 4 minutes or until golden brown. Carefully flip dough.
4. Spread each pizza crust with 2 Tbsp. kale mixture. Top each with sausage, corn, onion, and cheese. Grill, covered with grill lid, 3 to 4 minutes or until cheese is browned and bubbly.

MASTER OUR GRILLED PIZZA TECHNIQUE

1. Preheat grill to high (at least 400°) heat. Slide pizza dough rounds directly onto grill grate (cornmeal side down) from baking sheet, and grill, covered with grill lid, 3 to 4 minutes or until golden brown. Carefully flip dough.
2. Top cooked side of crust with desired sauce and toppings.
3. Grill, covered with grill lid, 3 to 4 minutes or until cheese is melted and bubbly.

THE *SL* TEST KITCHEN ACADEMY

WHITNEY WRIGHT, A COOK FROM THE SOUTH'S MOST TRUSTED KITCHEN, SHARES TIPS FOR

GRILLED PIZZA

The trick to pizzeria-quality pies at home? The grill! A smoking-hot grill mimics the kind of heat professionals get from those inferno-like pizza ovens. Bonus: You don't have to turn on your oven during summer. I recommend making your own dough (recipe on page 207); it's got better flavor and texture than commercial doughs, which contain preservatives. Don't have time to knead? Call your local pizza place and ask if they'll sell you a ball of their fresh dough.

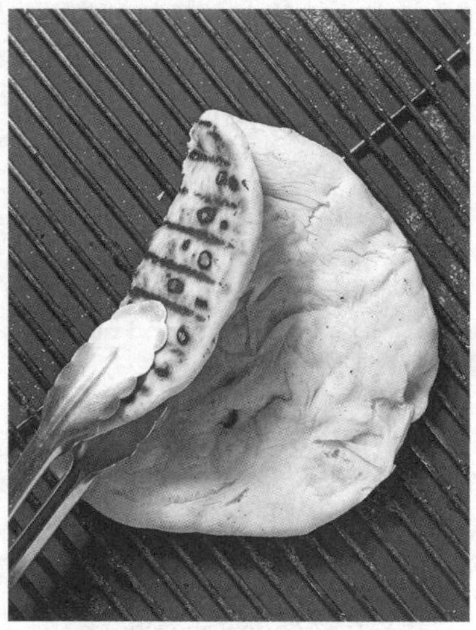

The Dough

- A sprinkle of cornmeal adds texture to the crust and helps prevent it from sticking to the grill grates. Use it generously!
- If you're having a pizza party, divide the dough and shape into individual balls ahead of time. Your guests will have an easier time rolling their dough if it has "relaxed" in the fridge for at least 1 hour.

The Grill

- Make sure the grill is very hot (at least 400°) before you lay on the dough. High heat helps prevent sticking and begins cooking the crust right away, which keeps the dough from falling through the grate.
- Cook for 3 to 4 minutes, using tongs to check the bottom occasionally for char. Flip when the crust has nice grill marks and is fully cooked on one side but not yet crisp.

The Finish

- Immediately after flipping the crusts, add your sauce and toppings; this means you've got to be organized and have everything ready to go. Cover with the grill lid, and cook 3 to 5 minutes or until the crust is crisp and cheese is melted and bubbly. If you want to top off with fresh herbs, sprinkle them on after taking the pizzas off the grill.

Classic Sweet Tea

WHY SOUTHERN COOKS WORTH THEIR SALT SHOULD MASTER THIS RECIPE

No. 1:

There is simply no more iconic Southern beverage (despite strong arguments that could be made for both bourbon and buttermilk).

No. 2:

Our geography demands it. Worldwide, sweet beverages are most popular and refreshing in hot and humid places.

No. 3:

It's a rite of passage. Being offered a glass of sweet tea instead of milk is how Southern children know they're growing up.

Making perfect sweet tea isn't difficult, but it isn't random either. It's the little things that make all the difference.

Tea Tip #1
Begin with fresh water each time. Using water left in a kettle can give the tea a flat taste.

Tea Tip #2
A bit of baking soda keeps the tea clear and guards against excess bitterness sometimes found in tea's natural tannins.

Tea Tip #3
Tea bags should be fresh and aromatic. Stale bags have no flavor. Store tea in an airtight container at room temperature.

Tea Tip #4
Cover brewed tea with plastic wrap before refrigerating to prevent it from absorbing food odors.

CLASSIC SWEET TEA

We love this recipe because it's got strong tea flavor without being bitter, and it's sweet but not cloying.

- **12 regular-size tea bags**
- **1/8 tsp. baking soda**
- **1 qt. distilled or bottled water**
- **1 qt. ice cubes**
- **1 1/4 cups Simple Syrup**

1. Place tea bags and baking soda in a large heatproof glass pitcher.
2. Bring water just to a rolling boil in a saucepan or kettle, and immediately pour over tea bags, making sure bags are submerged. Cover and steep 7 minutes. Remove tea bags without squeezing; discard tea bags.
3. Add ice, and stir until ice melts. Stir in Simple Syrup, and serve over ice.

Note: We tested with Luzianne Tea.

MAKES 2 qt. **HANDS-ON** 5 min.; **TOTAL** 50 min., including Simple Syrup

Simple Syrup

This versatile syrup is the best way to sweeten tea because the sugar is already dissolved.

Bring 1 cup water and 1 cup **sugar** to a boil in a small saucepan over medium heat, stirring occasionally. Boil 1 minute or until sugar dissolves. Remove from heat, and cool 30 minutes. Refrigerate in an airtight container up to 2 weeks.

MAKES 1 1/2 cups. **HANDS-ON** 5 min., **TOTAL** 35 min.

Fresh, Versatile Tomato Sauce

TURN SUMMER TOMATOES INTO A DINNERTIME STAR

SLOW-COOKER TOMATO SAUCE

Our sauce makes a great pizza topper (pizza recipes, page 207) and is the base for three delicious variations served with pasta or rice (facing page).

- **8 lb. fresh tomatoes (such as plum or 'San Marzano')**
- **2 large sweet onions, chopped**
- **6 garlic cloves, chopped**
- **1 (6-oz.) can tomato paste**
- **1/4 cup white wine vinegar**
- **2 Tbsp. sugar**
- **1 1/2 Tbsp. dried Italian seasoning**
- **1 Tbsp. kosher salt**
- **1 tsp. freshly ground black pepper**
- **1/2 cup chopped fresh basil**
- **1 Tbsp. fresh thyme leaves**

1. Cut a small "X" in bottom of each tomato. Bring 8 qt. water to a boil in a large stockpot; add tomatoes, and boil 30 to 60 seconds or until peel begins to separate from tomato flesh. Remove with a slotted spoon. Rinse immediately with cold running water, or plunge into ice water to stop the cooking process; drain. Peel tomatoes, and discard skin. Cut tomatoes in half, and gently squeeze to remove seeds. Coarsely chop tomatoes.

2. Stir together onions, next 7 ingredients, and chopped tomatoes in a 7-qt. slow cooker. Cover and cook on HIGH 4 to 5 hours or until tomatoes have broken down and are very soft. Cool 20 minutes. Stir in basil and thyme.

3. Process mixture with a handheld blender until smooth, or process mixture, in batches, in a food processor or blender. Season with salt and pepper.

Note: We recommend freezing the sauce in 3-cup portions, which is the amount of sauce found in most bottles of commercial sauce. Put the 3 cups of sauce in a 1-qt. freezer-safe jar or freezer bag to allow for expansion as the sauce freezes.

MAKES about 13 cups. **HANDS-ON** 45 min.; **TOTAL** 5 hours, 5 min.

Top spaghetti or your favorite pasta with our better-than-bottled version.

TRY THESE TWISTS!

TURN OUR SLOW-COOKER TOMATO SAUCE INTO THREE SPECTACULAR MEALS

DIABLO FARFALLE

Spicy but not over-the-top, this sauce is family-friendly. Make what Italians call frutti di mare *by adding clams, mussels, and shrimp during the last 5 minutes; cover and cook until clams and mussels open and shrimp is pink.*

- **1 shallot, finely chopped**
- **1 Tbsp. olive oil**
- **2 garlic cloves, minced**
- **1/2 cup red wine**
- **3 cups Slow-Cooker Tomato Sauce (recipe, facing page)**
- **1/2 tsp. dried crushed red pepper**
- **Kosher salt**
- **12 oz. farfalle pasta, cooked according to package directions**
- **1/2 cup chopped fresh basil**
- **Shredded Parmesan cheese**

Sauté shallots in hot oil in a large skillet over medium heat 1 minute. Add garlic, and sauté 1 minute. Add wine, and cook, stirring often, 3 minutes or until reduced to about half. Stir in tomato sauce and red pepper; add salt to taste. Bring to a simmer; cook 5 minutes. Stir in cooked pasta and basil. Serve immediately with Parmesan.

MAKES 4 to 6 servings. **HANDS-ON** 30 min.; **TOTAL** 30 min., not including tomato sauce

CREOLE RICE

Create a dinner featuring the seasonings of classic Creole cooking: onion, bell pepper, and celery. The sauce is perfect for pasta, but we love it on rice. Stir in cooked shrimp and smoked sausage, and you've got jambalaya.

- **1/2 cup chopped onion**
- **1/2 cup chopped bell pepper**
- **1/4 cup chopped celery**
- **2 Tbsp. olive oil**
- **2 garlic cloves, minced**
- **3 cups Slow-Cooker Tomato Sauce (recipe, facing page)**
- **1 1/2 tsp. Creole seasoning**
- **1 tsp. sugar**
- **Hot cooked pasta or long-grain rice**

Sauté first 3 ingredients in hot oil in a large skillet over medium-high heat 5 minutes; stir in garlic, and sauté 1 minute. Stir in tomato sauce and next 2 ingredients. Bring mixture to a boil; reduce heat to low, and simmer, stirring occasionally, 10 minutes. Serve immediately over hot cooked pasta or rice.

MAKES 4 to 6 servings. **HANDS-ON** 25 min.; **TOTAL** 35 min., not including tomato sauce

VODKA-CREAM SAUCE WITH PENNE

This is an old-school favorite with fancy flair. The vodka perks up the tomatoes, and the cream makes the sauce rich.

- **4 garlic cloves, peeled and smashed**
- **2 Tbsp. olive oil**
- **3 cups Slow-Cooker Tomato Sauce (recipe, facing page)**
- **1/4 tsp. dried crushed red pepper**
- **1/4 cup vodka**
- **1/2 cup heavy cream**
- **1/2 cup (2 oz.) shredded Parmesan cheese**
- **12 oz. penne pasta, cooked according to package directions**
- **1/4 cup chopped fresh flat-leaf parsley**

Sauté garlic in hot oil in a large skillet over medium heat 4 to 5 minutes or until golden brown. Add tomato sauce and red pepper, and cook, stirring often, 5 minutes. Stir in vodka, and reduce heat to low; cook, stirring often, 15 minutes. Remove garlic cloves. Add cream and Parmesan cheese; stir until well incorporated and cheese is melted. Stir in cooked pasta, and top with parsley. Sprinkle with shredded Parmesan cheese, if desired. Serve immediately.

MAKES 4 to 6 servings. **HANDS-ON** 40 min.; **TOTAL** 40 min., not including tomato sauce

Vodka-Cream Sauce

Fast, Fresh, and Filling

THESE LIGHT AND BRIGHT SUPPERS MAKE THE MOST OF SUMMER'S BEST PRODUCE

Use firm, thin-skinned veggies that don't need to be peeled.

SUMMER PASTA SALAD WITH LIME VINAIGRETTE

Our colorful pasta salad is so satisfying it qualifies as dinner.

- **8** **oz. farfalle pasta, prepared according to package directions for al dente**
- **1** **(1 1/2- to 2 1/2-lb.) whole deli-roasted chicken, skin removed and meat shredded**
- **1** **medium zucchini (about 5 oz.), thinly sliced**
- **2** **small yellow squash (about 5 oz.), thinly sliced**
- **2** **nectarines, coarsely chopped**
- **1/3** **cup coarsely chopped fresh flat-leaf parsley**
- **Lime Vinaigrette**
- **1/4** **cup toasted sliced almonds**

Rinse prepared pasta with cold water, and drain well. Gently stir together pasta, chicken, and next 5 ingredients in a large bowl. Add salt and pepper to taste. Sprinkle with almonds, and serve immediately.

MAKES 6 servings. **HANDS-ON** 30 min.; **TOTAL** 40 min., including vinaigrette

Lime Vinaigrette

Whisk together 1 Tbsp. **lime zest;** 2 Tbsp. fresh **lime juice;** 2 Tbsp. **white balsamic vinegar;** 1 Tbsp. **Creole mustard;** 1 1/2 tsp. **honey;** 1 small **garlic** clove, pressed; 1/2 tsp. each **kosher salt** and freshly ground black **pepper** in a small bowl. Add 1/2 cup **olive oil** in a slow, steady stream, whisking constantly until smooth.

MAKES about 3/4 cup. **HANDS-ON** 10 min., **TOTAL** 10 min.

SUMMER CORN-AND-GOLDEN POTATO CHOWDER

This brothy, aromatic chowder is light enough for summer. Simmering the scraped corn cobs directly in the soup concentrates the corn flavor, making the most of the season's sweet bounty.

- **4 Tbsp. butter**
- **5 cups fresh corn kernels; reserve 2 cobs**
- **3 fresh thyme sprigs**
- **1 large sweet onion, diced (about 2 cups)**
- **1 1/2 tsp. kosher salt**
- **4 cups low-sodium chicken broth**
- **1 large Yukon gold potato, diced**
- **3/4 cup half-and-half**
- **Garnishes: fresh basil, fresh dill**

1. Melt butter in a small stockpot over medium heat. Add corn kernels, thyme sprigs, diced onion, and salt. Cook, stirring occasionally, 15 to 20 minutes or until corn is tender but not browned.

2. Stir in broth and diced potato. Increase heat to high; add reserved corn cobs, and bring to a boil. Reduce heat to medium, and simmer, stirring occasionally, 8 to 10 minutes or until potatoes are tender. Discard cobs and thyme.

3. Process 1/2 cup of corn mixture in a blender until smooth. Return processed mixture to stockpot, and stir in half-and-half. Serve immediately.

MAKES 4 to 6 servings. **HANDS-ON** 25 min., **TOTAL** 40 min.

OPEN-FACED SHRIMP-AND-AVOCADO SANDWICHES

For a sturdy sandwich base, use a thicker cut of bread.

- 4 thick bacon slices
- 1 lb. peeled, medium-size raw shrimp, deveined
- 1/2 tsp. kosher salt, divided
- 1/4 tsp. freshly ground black pepper
- 1/4 cup fresh lemon juice, divided
- 1 (5-oz.) package arugula
- 1 1/2 cups cherry tomatoes, halved
- 1 cup fresh corn kernels (about 2 ears)
- 1/4 cup thinly sliced red onion
- 1 Tbsp. extra virgin olive oil
- 4 (1-inch-thick) country bread slices, toasted
- 1 garlic clove, crushed
- 2 Tbsp. mayonnaise
- 2 avocados, peeled and sliced
- 2 Tbsp. chopped fresh dill

1. Cook bacon in a medium skillet over medium heat 12 minutes or until crisp; remove bacon, and drain on paper towels, reserving 2 Tbsp. drippings in skillet. Cut bacon into 1-inch pieces.

2. Cook shrimp and 1/2 tsp. salt in hot drippings over medium-high heat 2 minutes on each side or until bright pink. Remove from heat, and stir in pepper and 2 Tbsp. lemon juice.

3. Toss together arugula, tomatoes, corn kernels, onion slices, olive oil, bacon, and remaining lemon juice.

4. Rub 1 side of each bread slice lightly with garlic. Spread about 1/2 Tbsp. mayonnaise on each bread slice; top with avocado slices and shrimp mixture. Sprinkle with dill, and serve with arugula mixture.

MAKES 4 servings. **HANDS-ON** 35 min., **TOTAL** 35 min.

CRISPY EGGPLANT, TOMATO, AND PROVOLONE STACKS WITH BASIL

Add bacon or prosciutto just before baking for a more savory meal.

- **1/2 cup all-purpose flour**
- **1 tsp. kosher salt**
- **1/2 tsp. freshly ground black pepper**
- **2 large eggs, lightly beaten**
- **2 Tbsp. cold water**
- **2 1/2 cups panko (Japanese breadcrumbs)**
- **1/2 cup finely grated Parmesan cheese**
- **2 tsp. dried Italian seasoning**
- **1 tsp. firmly packed lemon zest**
- **1/2 tsp. granulated garlic**
- **2 large eggplants, cut into 20 (1/2-inch-thick) slices**
- **Vegetable oil**
- **Parchment paper**
- **10 provolone cheese slices**
- **10 tomato slices**
- **1 cup loosely packed fresh basil leaves**

1. Preheat oven to 400°. Stir together first 3 ingredients in a medium bowl. Whisk together eggs and cold water in a second bowl. Stir together panko and next 4 ingredients in a third bowl.
2. Dredge each eggplant slice in flour mixture; dip in egg mixture, and dredge in panko mixture.
3. Pour oil to a depth of 3/4 inch in a large skillet. Fry eggplant slices, in 2 batches, in hot oil over medium-high heat 2 to 2 1/2 minutes on each side or until browned. Drain on a wire rack in a jelly-roll pan.
4. Place 10 eggplant slices in a single layer on a parchment paper-lined baking sheet; top each with 1 provolone slice.
5. Bake at 400° for 3 to 5 minutes or until cheese is melted. Top each with 1 tomato slice and 1 basil leaf. Add desired amount of salt and pepper, and top with remaining eggplant slices.

MAKES 4 to 5 servings. **HANDS-ON** 35 min., **TOTAL** 35 min.

MAKE IT LAST

On Sunday, simmer a pot of lady peas; mix some with mushrooms, and stuff into zucchini.

MONDAY
Lady Pea-and-Corn Patties

TUESDAY
Buttermilk-Lady Pea Soup with Bacon

For a quick side, toss together chopped heirloom tomatoes, olive oil, salt, and pepper. Top with fresh basil.

One Pot, Three Ways

TURN A POT OF SAVORY LADY PEAS INTO THREE EASY, SUMMERY MEALS

SUNDAY

LADY PEAS

Bring 9 cups **fresh lady peas;** 1/2 small yellow **onion,** quartered; 1 tsp. **kosher salt;** and 9 cups water to a boil in a stockpot over high heat. Reduce heat to medium, and simmer 45 minutes or until peas are tender, skimming foam often. Drain, reserving 2 cups cooking liquid; discard onion. Refrigerate peas and cooking liquid up to 2 days.

MAKES 9 cups. **HANDS-ON** 10 min., **TOTAL** 55 min.

ZUCCHINI STUFFED WITH LADY PEAS

- 8 medium-size zucchini
- 2 Tbsp. butter
- 1/2 small yellow onion, chopped
- 1 1/2 cups cooked lady peas
- 4 oz. chopped fresh mushrooms
- 1 large tomato, chopped
- 1/2 tsp. minced garlic
- 2 tsp. kosher salt, divided
- 1 tsp. black pepper, divided
- 2 Tbsp. chopped fresh basil
- 1 cup (4 oz.) finely shredded Parmesan cheese, divided
- 1/2 cup panko (Japanese breadcrumbs)
- 1 Tbsp. butter, melted

1. Preheat oven to 375°. Cut zucchini in half lengthwise; scoop pulp into a bowl, leaving 1/4-inch shells intact. Chop pulp. Microwave zucchini shells in a microwave-safe dish covered with plastic wrap at HIGH 4 minutes; transfer to a foil-lined jelly-roll pan.
2. Melt 2 Tbsp. butter in a skillet over medium heat; add chopped zucchini pulp, onion, next 4 ingredients, 1 1/2 tsp. salt, and 1/2 tsp. pepper, and cook 10 minutes. Stir in basil and 3/4 cup cheese.
3. Divide mixture among shells. Stir together breadcrumbs, melted 1 Tbsp. butter, and remaining cheese, salt, and pepper; sprinkle over stuffed shells.
4. Bake at 375° for 20 minutes or until thoroughly heated.

MAKES 4 servings. **HANDS-ON** 55 min.; **TOTAL** 1 hour, 15 min., not including peas

MONDAY NIGHT

LADY PEA-AND-CORN PATTIES

- 1 cup fresh corn kernels
- 1 tsp. olive oil
- 1 1/2 cups cooked lady peas
- 1 to 2 Tbsp. reserved cooking liquid from peas
- 2 green onions, sliced
- 1 Tbsp. finely chopped fresh flat-leaf parsley
- 1 Tbsp. chopped cilantro
- 1/2 tsp. freshly ground black pepper
- 1 1/4 tsp. kosher salt
- 2 large eggs, lightly beaten
- 1 1/2 cups panko (Japanese breadcrumbs)
- 2 Tbsp. butter
- 2 Tbsp. olive oil

Sauté corn in 1 tsp. hot olive oil in a medium skillet over medium-high heat 3 minutes or until tender. Process 1 cup lady peas in a food processor until smooth, adding up to 2 Tbsp. reserved cooking liquid as needed. Stir together green onions, next 5 ingredients, whole peas, pureed peas, corn, and 1/2 cup panko. Gently shape mixture into 8 patties; cover and chill 30 minutes. Dredge patties in remaining panko. Melt 1 Tbsp. butter with 1 Tbsp. olive oil in a large skillet over medium heat; add 4 patties, and cook 3 minutes on each side or until lightly browned. Drain on paper towels. Add remaining oil and butter to skillet, and repeat procedure with remaining patties.

MAKES 4 servings. **HANDS-ON** 45 min.; **TOTAL** 1 hour, 15 min., not including peas

TUESDAY NIGHT

BUTTERMILK-LADY PEA SOUP WITH BACON

Even served warm, buttermilk has a cooling effect on the palate, just right for summer meals.

- 4 uncooked bacon slices, coarsely chopped
- 1 small shallot, minced
- 1/2 tsp. minced garlic
- 5 cups cooked lady peas
- 3/4 to 1 cup reserved cooking liquid from peas
- 3/4 tsp. kosher salt
- 1/2 tsp. freshly ground black pepper
- 2 cups buttermilk

1. Cook bacon in a large saucepan over medium heat 6 to 7 minutes or until crisp; remove bacon, and drain on paper towels, reserving 1 tsp. drippings in saucepan. Finely chop cooked bacon. Sauté shallots and garlic in hot drippings over medium heat 2 minutes or until tender.
2. Process 3 cups peas and 3/4 cup reserved cooking liquid in a food processor until smooth. Add pureed pea mixture, remaining peas, salt, and pepper to shallot mixture. Stir in buttermilk. Cook over medium heat, stirring often, 8 minutes. (Do not boil.) Stir in up to 1/4 cup cooking liquid, 1 Tbsp. at a time, until desired consistency is reached. Spoon soup into 4 bowls, and top with bacon.

MAKES about 5 cups. **HANDS-ON** 25 min.; **TOTAL** 25 min., not including peas

Orange Cream Granita

Chocolate Milk Granita

Raspberry Limeade Granita

Cucumber-Basil Granita

Cool Down with Easy Granitas

THESE FRESH AND FRUITY SHAVED-ICE DESSERTS ARE A BREEZE TO MAKE. THROWING A PARTY? ADD YOUR FAVORITE SPIRIT FOR SLUSHY COCKTAILS

Want a richer taste? Serve with whipped heavy cream.

If your granita freezes solid before you can stir it, let it stand at room temperature for 15 to 20 minutes. Shave with a fork to create crystals. Freeze until ready to serve.

RASPBERRY LIMEADE GRANITA

- **1 cup fresh raspberries**
- **1/2 cup fresh lime juice**
- **1 1/2 cups Simple Syrup (recipe, page 209)**
- **1 cup cold water**
- **Chilled rum (optional)**

Process first 3 ingredients and 1 cup cold water in a blender 30 seconds or until smooth. Pour into a 9- x 5-inch loaf pan. Freeze 4 to 5 hours or until ice crystals form, stirring every hour. If desired, top each serving with a shot of chilled rum.

MAKES 4 to 6 servings. **HANDS-ON** 5 min.; **TOTAL** 4 hours, 40 min., including Simple Syrup

You can make granitas up to a week in advance. Just be sure to cover the pans with plastic wrap.

ORANGE CREAM GRANITA

- **1 1/2 tsp. firmly packed orange zest**
- **2 cups fresh orange juice (including pulp)**
- **1/2 cup Simple Syrup (recipe, page 209)**
- **1/2 cup half-and-half**
- **2 Tbsp. fresh lemon juice**
- **1 tsp. vanilla extract**
- **Whipped cream (optional)**
- **Chilled orange liqueur (such as Grand Marnier), optional**

Whisk together first 6 ingredients in a medium bowl; pour into a 9- x 5-inch loaf pan. Freeze 4 to 5 hours or until ice crystals form, stirring every hour. If desired, top each serving with whipped cream or a shot of chilled orange liqueur.

MAKES 4 to 6 servings. **HANDS-ON** 15 min.; **TOTAL** 4 hours, 50 min., including Simple Syrup

CUCUMBER-BASIL GRANITA

- **4 cups seedless cucumber slices**
- **3 Tbsp. fresh lime juice**
- **1/2 cup Simple Syrup (recipe, page 209)**
- **1 cup fresh basil leaves**
- **Pinch of salt**
- **Chilled herbal gin (such as Hendricks), optional**

Process first 3 ingredients in a blender 20 seconds. Add basil and salt; process 20 seconds. Pour mixture into a 9- x 5-inch loaf pan. Freeze 4 to 5 hours or until ice crystals form, stirring every hour. If desired, top each serving with a shot of chilled gin.

MAKES 4 to 6 servings. **HANDS-ON** 10 min.; **TOTAL** 4 hours, 45 min., including Simple Syrup

CHOCOLATE MILK GRANITA

This one may take a little longer to freeze, but the rich chocolate flavor is worth it.

- **1 cup sugar**
- **2/3 cup unsweetened cocoa**
- **2/3 cup semisweet chocolate morsels**
- **1 cup half-and-half**
- **1 tsp. vanilla extract**
- **Cold brewed coffee (optional)**
- **Irish cream liqueur (such as Baileys), optional**

Stir together sugar, cocoa, and 3 cups water in a saucepan. Bring to a low boil over medium heat, whisking occasionally. Boil, whisking constantly, 1 minute. Remove from heat. Add chocolate morsels, and whisk until melted. Whisk together half-and-half and vanilla; whisk into chocolate mixture. Pour into a 9- x 5-inch loaf pan. Freeze 5 to 6 hours or until ice crystals form, stirring every hour. If desired, drizzle each serving with cold coffee or Irish cream liqueur.

MAKES 4 to 6 servings. **HANDS-ON** 10 min.; **TOTAL** 5 hours, 10 min.

Community Cookbook

BUILD A BETTER LIBRARY, ONE GREAT BOOK AT A TIME

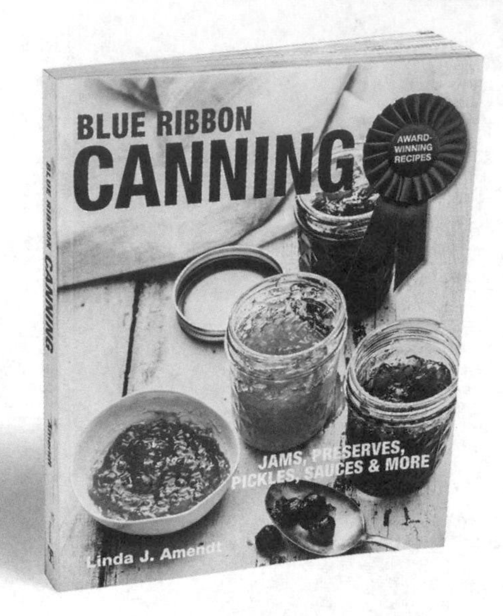

Canning is a long-honored summer pastime in the South. From peaches to cucumbers, we have too much goodness to enjoy in just one season. Linda J. Amendt's collection of state fair award-winning recipes offers classic and modern ways to put up summer's bounty to savor year-round.

BLACKBERRY CHAMBORD JAM

This jam blends blackberries and Chambord, a black raspberry liqueur that brings out the flavor of the fruit.

- **5 1/2 cups seedless blackberry pulp (about 6 [18-oz.] packages blackberries, crushed and pressed through a fine-mesh sieve twice)**
- **1 (1.75-oz.) box powdered pectin**
- **1/2 tsp. unsalted butter (optional)**
- **7 cups granulated sugar**
- **1/4 cup Chambord**

1. In an 8-qt. stainless steel stockpot, combine the blackberry pulp and pectin. Add the butter, if using. Bring the mixture to a full rolling boil over medium-high heat, stirring constantly.
2. Add the sugar, and stir until completely dissolved. Return the mixture to a full rolling boil, stirring constantly. Boil for 1 minute, stirring constantly.
3. Remove the pot from the heat, and skim off any foam. Add the Chambord, and stir until well blended. Let the jam rest for a few minutes, stirring occasionally.
4. Ladle the jam into hot jars, leaving 1/4-inch headspace. Remove any air bubbles by gently tapping the jars. Wipe the jar rims and threads with a clean, damp paper towel. Cap the jars with hot lids, and screw on the bands.
5. Process 4-oz., 8-oz., and pint jars in a water bath canner for 10 minutes. Remove from the water bath canner, and let cool for 12 to 24 hours. Check the seals, and remove the screw bands. Store jars in a cool, dry, dark place for up to 1 year.

Serving Suggestion: Enjoy this jewel-toned jam on toasted bread, muffins, and scones or as a topping for vanilla ice cream. It also makes a wonderful filling for individual fruit tarts and shaped cookies.

MAKES 6 to 7 (8-oz.) jars or 3 pt.

OLD TIMEY PICKLES

Chop up these pickles, and use them in slaw or chicken salad.

- **2 cups apple cider vinegar**
- **1 cup granulated sugar**
- **1 Tbsp. whole cloves**
- **1 Tbsp. whole allspice**
- **1 tsp. canning or pickling salt**
- **2 qt. sliced pickling cucumbers (about 12 to 15 cucumbers)**

1. In an 8-qt. stainless steel stockpot, combine the vinegar, sugar, cloves, allspice, and salt. Bring to a boil over medium heat; then reduce the heat to low, and simmer for 5 minutes.
2. Add the cucumber slices, and bring the mixture to a full boil over medium-high heat. Reduce the heat, and simmer until the cucumbers lighten in color, 3 to 5 minutes. Remove the pot from the heat.
3. Using a slotted spoon, pack the cucumbers into hot jars, leaving ½-inch headspace. Ladle the hot syrup into the jars, covering the pickles and maintaining the ½-inch headspace. Remove any air bubbles. Wipe the jar rims and threads with a clean, damp paper towel. Apply hot lids and screw bands.
4. Process pint jars in a water bath canner for 15 minutes. Remove from the water bath canner, and let cool for 12 to 24 hours. Check the seals, and remove the screw bands. Store jars in a cool, dry, dark place for up to 1 year.

MAKES 3 to 4 pt.

September

222 **Baked with Love** Sweet and savory oven-baked goodness guaranteed to please all

225 QUICK-FIX SUPPERS **Dinner in 20 Minutes Flat** Speedy approaches to favorite Southern dishes your family will love

228 THE *SL* TEST KITCHEN ACADEMY **Five Time-Saving Tools** Tricks of the trade for getting dinner done fast

229 *COOKING LIGHT* **Better Than Takeout** A guilt-free pork-and-veggie stir-fry packed with flavor, not fat

230 SOUTHERN CLASSIC **Classic Pimiento Cheese** Stop looking: We have the only recipe you'll ever need for this beloved Southern spread

231 SAVE ROOM **The No-Peel Apple Cake** Apple skins add interest and color to this pretty fall cake

232 **The Hunt Breakfast** An al fresco morning repast that celebrates history, culture, and the season

234 COMMUNITY COOKBOOK Buttermilk-soaked, deep-fried, unadorned or smothered in gravy, *SL* Contributing Editor Rebecca Lang celebrates the many incarnations of Southern fried chicken in her newest book

Baked with Love

WARM THE HEARTS OF FAMILY AND FRIENDS WITH EASY FALL BAKED GOODS DESIGNED TO PRESENT AS GIFTS

APPLE-CHEDDAR BREAD

Try this bread sandwiched with ham and apple slices, or cut it into cubes and toast it to make croutons for salads and soups.

- **1 3/4 cups all-purpose flour**
- **1 Tbsp. baking powder**
- **1/2 tsp. table salt**
- **1/8 tsp. ground red pepper**
- **4 thick bacon slices, cooked and crumbled**
- **1 cup peeled and diced Granny Smith apple (about 1 small apple)**
- **3/4 cup (3 oz.) shredded extra-sharp Cheddar cheese**
- **1/2 cup toasted chopped pecans**
- **1 tsp. finely chopped fresh rosemary**
- **3 large eggs**
- **1/3 cup milk**
- **1/3 cup canola oil**
- **Shortening**

1. Preheat oven to 350°. Whisk together flour and next 3 ingredients in a large bowl until thoroughly combined. Stir together bacon and next 4 ingredients in a medium bowl. Whisk together eggs, milk, and oil in a small bowl.
2. Add egg mixture to flour mixture, and stir just until dry ingredients are moistened; stir in bacon mixture just until incorporated. Transfer dough to a greased (with shortening) and floured 8 1/2- x 4 1/2-inch loaf pan.
3. Bake at 350° for 55 minutes to 1 hour or until a wooden pick inserted in center comes out clean, shielding bread with aluminum foil after 40 to 45 minutes to prevent excessive browning. Cool bread in pan on a wire rack 10 minutes. Remove from pan, and cool 30 minutes before slicing.

MAKES 1 loaf. **HANDS-ON** 25 min., **TOTAL** 2 hours

SPINACH-FETA SCONES

Day-old scones are firm enough to slice open without crumbling. Tuck in a few paper-thin slices of country ham for an unforgettable ham "biscuit."

- **2 1/2 cups self-rising flour**
- **1 Tbsp. sugar**
- **1/2 cup cold butter, cut into 1/2-inch cubes**
- **1 cup chopped fresh spinach**
- **1 cup crumbled feta cheese**
- **1 1/4 cups heavy cream**
- **Wax paper**
- **Parchment paper**
- **2 Tbsp. heavy cream**

1. Preheat oven to 450°. Stir together first 2 ingredients in a large bowl. Cut butter into flour mixture with a pastry blender until crumbly and mixture resembles small peas. Freeze 5 minutes. Stir in spinach and feta until combined. Add 1 cup cream, stirring just until dry ingredients are moistened. Stir in up to 1/4 cup more cream, 1 Tbsp. at a time, if needed.
2. Turn dough out onto lightly floured wax paper; gently press or pat dough into an 8-inch round. (Mixture will be crumbly.) Cut round into 8 wedges. Place wedges 2 inches apart on a parchment paper-lined baking sheet. Brush tops with 2 Tbsp. cream just until moistened.
3. Bake at 450° for 14 to 16 minutes or until golden.

MAKES 8 scones. **HANDS-ON** 20 min., **TOTAL** 40 min.

BANANA-NUT MUFFINS

This batter can also be baked as 2 (8- x 4-inch) loaves. (You'll need to increase the bake time to about 1 hour.) Spread the Cream Cheese-Honey Filling on the warm bread, or use as a topping for French toast.

- **1 cup butter, softened**
- **2 cups firmly packed light brown sugar**
- **3 large eggs**
- **1/4 cup sour cream**
- **1 tsp. vanilla extract**
- **3 1/4 cups all-purpose flour**
- **1 tsp. ground cinnamon**
- **3/4 tsp. baking powder**
- **3/4 tsp. baking soda**
- **3/4 tsp. table salt**
- **1/4 tsp. ground nutmeg**
- **2 1/2 cups mashed bananas (about 5 medium)**
- **1 cup toasted chopped pecans**
- **Vegetable cooking spray**

1. Preheat oven to 350°. Beat butter at medium speed with a heavy-duty electric stand mixer until creamy; add brown sugar, and beat until light and fluffy. Add eggs, 1 at a time, beating just until blended after each addition. Add sour cream and vanilla, and beat just until blended.
2. Stir together flour and next 5 ingredients. Gradually add flour mixture to butter mixture, beating at low speed just until blended. Stir in bananas and pecans just until blended. Spoon batter into 2 lightly greased (with cooking spray) 12-cup muffin pans, filling three-fourths full.
3. Bake at 350° for 25 to 30 minutes or until a wooden pick inserted in center comes out clean. Cool in pans 10 minutes. Remove from pans, and cool completely on wire racks (about 30 minutes).
4. Make a small hole in top of each muffin, using the handle of a small wooden spoon. Spoon Cream Cheese-Honey Filling into a zip-top plastic freezer bag. Snip 1 corner of bag to make a tiny hole. Pipe a generous amount of filling into each muffin.

MAKES 2 dozen. **HANDS-ON** 30 min.; **TOTAL** 1 hour, 40 min., including filling

Cream Cheese-Honey Filling

Beat 1 (8-oz.) package **cream cheese,** softened; 3 Tbsp. **honey;** and 1/8 tsp. ground **cinnamon** with an electric mixer at medium speed in a small bowl until blended.

MAKES about 1 1/2 cups. **HANDS-ON** 5 min., **TOTAL** 5 min.

PUMPKIN-HONEY-BEER BREAD

- 2 cups sugar
- 1 cup canola oil
- 2/3 cup beer (at room temperature)
- 1/4 cup honey
- 4 large eggs
- 1 (15-oz.) can pumpkin
- 3 1/2 cups all-purpose flour
- 2 tsp. table salt
- 2 tsp. baking soda
- 1 tsp. baking powder
- 1 tsp. pumpkin pie spice
- Shortening

1. Preheat oven to 350°. Beat first 4 ingredients at medium speed with a heavy-duty electric stand mixer until well blended. Add eggs, 1 at a time, beating just until blended after each addition. Add pumpkin, and beat at low speed just until blended.
2. Whisk together flour and next 4 ingredients in a medium bowl until well blended. Add flour mixture to pumpkin mixture, and beat at low speed just until blended. Divide batter between 2 greased (with shortening) and floured 9- x 5-inch loaf pans.
3. Bake at 350° for 55 minutes to 1 hour and 10 minutes or until a wooden pick inserted in center comes out clean, shielding with aluminum foil after 45 to 50 minutes to prevent excessive browning if necessary. Cool bread in pans on a wire rack 10 minutes. Remove from pan, and cool 30 minutes before slicing.

Note: We tested with Back Forty Truck Stop Honey Brown Ale.

MAKES 2 loaves. **HANDS-ON** 15 min.; **TOTAL** 1 hour, 50 min.

MINI PUMPKIN-HONEY-BEER BREAD

Prepare recipe as directed, spooning batter into 6 lightly greased (with cooking spray) 5 3/4- x 3 1/4-inch disposable aluminum loaf pans. Decrease bake time to 35 to 40 minutes or until a wooden pick inserted in center comes out clean.

MAKES 6 loaves. **HANDS-ON** 15 min.; **TOTAL** 1 hour, 30 min.

PUMPKIN-HONEY-BEER BREAD PUDDING WITH APPLE BRANDY-CARAMEL SAUCE

- 1 (9- x 5-inch) Pumpkin-Honey-Beer Bread loaf, cut into 1/2-inch cubes (recipe, at left)
- Vegetable cooking spray
- 4 large eggs
- 1/2 cup granulated sugar
- 1/4 tsp. table salt
- 1/4 tsp. pumpkin pie spice
- 2 cups milk
- 1 1/2 cups heavy cream
- 1 Tbsp. Demerara sugar (optional)
- Apple Brandy-Caramel Sauce

1. Preheat oven to 400°. Spread bread cubes in a single layer in a lightly greased (with cooking spray) jelly-roll pan. Bake 12 to 15 minutes or until lightly toasted. Remove cubes from oven to a wire rack, and cool 15 minutes. Reduce temperature to 350°.
2. Whisk together eggs and next 3 ingredients in a large bowl until well blended. Whisk in milk and heavy cream until well blended.
3. Stir bread cubes into egg mixture until coated. Let stand 20 minutes.
4. Spread bread mixture in a lightly greased (with cooking spray) 11- x 7-inch baking dish; sprinkle with Demerara sugar, if desired.
5. Bake at 350° for 1 hour to 1 hour and 10 minutes or until set in the center, shielding with aluminum foil after 45 minutes to prevent excessive browning if necessary. Cool 15 minutes before serving with Apple Brandy-Caramel Sauce.

MAKES 8 to 10 servings. **HANDS-ON** 15 min.; **TOTAL** 2 hours, 15 min., not including bread or sauce

Apple Brandy-Caramel Sauce

Bring 1/2 cup firmly packed **light brown sugar,** 1/4 cup **butter,** 1/4 cup **heavy cream,** and a pinch of table **salt** to a boil in a small saucepan over medium heat, stirring constantly. Boil, stirring constantly, 1 minute. Remove from heat, and stir in 1 Tbsp. **apple brandy.** Whisk in 1 Tbsp. **powdered sugar;** cool 15 minutes before serving.

MAKES about 2/3 cup. **HANDS-ON** 10 min., **TOTAL** 25 min.

BISCUIT CINNAMON SWEET ROLLS

- 1/4 cup frozen shortening, cut into small pieces
- 1/4 cup cold butter, cut into small cubes
- 2 1/2 cups self-rising flour
- 1 cup buttermilk
- Wax paper
- 6 Tbsp. butter, softened
- 1/4 cup granulated sugar
- 1/4 cup firmly packed light brown sugar
- 1/2 tsp. ground cinnamon
- Vegetable cooking spray
- Creamy Glaze

1. Preheat oven to 450°. Cut shortening and butter into flour with a pastry blender or fork in a medium bowl until crumbly and mixture resembles small peas. Freeze 10 minutes.
2. Make a well in center of flour mixture; add buttermilk, and stir with a fork just until dough comes together.
3. Turn dough out onto a heavily floured surface, and knead 8 to 10 times. Transfer dough to a heavily floured piece of wax paper about 18 inches long. Roll dough into a 14- x 10-inch rectangle.
4. Spread dough with softened butter, leaving a 1/2-inch border. (Make sure butter is very soft, and spread it gently.) Stir together sugars and cinnamon; sprinkle over butter. Lift and tilt wax paper, and roll up dough, jelly-roll fashion, starting at 1 long side and using wax paper as a guide. Cut dough into 14 to 16 (1-inch-thick) slices. Place rolls in a lightly greased (with cooking spray) 9-inch round pan.
5. Bake at 450° for 13 to 15 minutes or until rolls are golden brown. Cool in pan on a wire rack 5 minutes.
6. Meanwhile, prepare Creamy Glaze, and drizzle over rolls.

MAKES 12 rolls. **HANDS-ON** 20 min., **TOTAL** 1 hour

Creamy Glaze

Whisk together 1 cup **powdered sugar,** 1/2 tsp. **vanilla** extract, and 3 Tbsp. **heavy cream.** Whisk in up to 1 Tbsp. cream, 1 tsp. at a time, until smooth and creamy. Use immediately.

MAKES 1/2 cup. **HANDS-ON** 5 min., **TOTAL** 5 min.

Dinner in 20 Minutes Flat

10 SOUTHERN SPINS TO HELP YOU ON THE BUSIEST BACK-TO-SCHOOL EVENINGS

SPEEDY HOMEMADE MAC AND CHEESE

- 1/4 cup plus 1 1/2 tsp. kosher salt, divided
- 1 qt. milk
- 6 Tbsp. butter, cut into pieces
- 6 Tbsp. all-purpose flour
- 1 lb. pasta (such as penne, cavatappi, or rotini)
- 1 (8-oz.) package shredded extra-sharp Cheddar cheese
- 1 (8-oz.) package shredded Monterey Jack cheese
- 1 tsp. hot sauce (such as Tabasco)
- 1/2 tsp. freshly ground black pepper
- 1 1/2 cups panko (Japanese breadcrumbs)
- 2 tsp. olive oil

1. Preheat broiler with oven rack 8 to 9 inches from heat.
2. Bring 1/4 cup salt and 4 qt. water to a boil in a large covered Dutch oven over high heat.
3. Meanwhile, microwave milk in a microwave-safe 1-qt. glass measuring cup covered with plastic wrap at HIGH 3 minutes. While milk is heating, melt butter in a 12-inch cast-iron skillet over medium heat. Reduce heat to medium-low; add flour, and cook, whisking constantly, 2 minutes. Gradually whisk in hot milk. Increase heat to medium-high, and bring to a low boil, whisking often.
4. Add pasta to boiling water, and cook 8 minutes.
5. Meanwhile, continue to cook sauce, whisking often, 6 minutes. Remove from heat; whisk in cheeses, hot sauce, 1 1/2 tsp. salt, and 1/2 tsp. pepper. Cover.
6. Stir together panko and olive oil.
7. Drain pasta, and fold into cheese sauce. Sprinkle with panko mixture.
8. Broil 1 to 2 minutes or until breadcrumbs are golden brown. Serve immediately.

MAKES 6 to 8 servings. **HANDS-ON** 20 min., **TOTAL** 20 min.

Southern Pimiento Mac and Cheese

Substitute **extra-sharp Cheddar cheese** for Monterey Jack cheese. Prepare recipe as directed through Step 5, whisking 2 Tbsp. fresh **lemon juice,** 1 Tbsp. **Worcestershire sauce,** and 1/2 cup grated sweet **onion** into sauce along with cheese. Proceed with recipe as directed, stirring 2 (7-oz.) jars **diced pimiento,** drained, and 2 (2.52-oz.) packages **fully cooked bacon,** diced, into pasta mixture just before topping with panko. Sprinkle with 2 Tbsp. chopped **fresh chives** before serving.

Barbecue Mac and Cheese

Substitute **Gouda cheese** for Monterey Jack cheese and crumbled savory **cornbread** for panko. Prepare recipe as directed through Step 5, stirring 1 lb. **pulled pork barbecue** (without sauce) into pasta mixture after adding cheese. Sprinkle 1/2 cup chopped **green onions** over cornbread mixture before broiling, and drizzle with 1/2 cup bottled **barbecue sauce** after broiling.

Mexican Mac and Cheese

Substitute **pepper Jack cheese** for Monterey Jack cheese and crushed **tortilla chips** for panko. Prepare recipe as directed through Step 4. While pasta cooks, sauté 1 lb. Mexican **chorizo** in 1 Tbsp. hot **olive oil** in a large skillet over medium-high heat 4 to 5 minutes or until crumbled and cooked. Proceed with recipe, folding chorizo and 2 cups **cherry tomatoes,** halved, into cheese sauce along with pasta in Step 7.

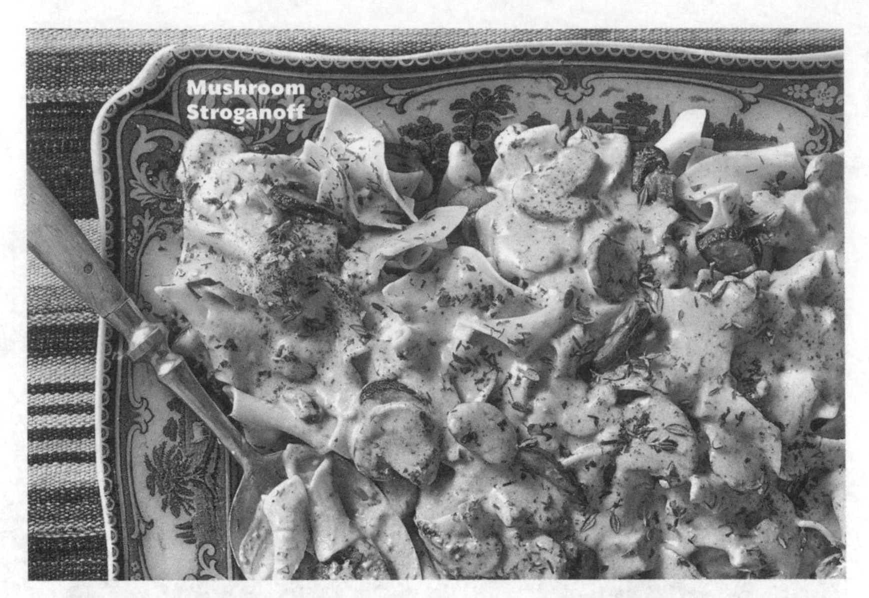
Mushroom Stroganoff

MUSHROOM STROGANOFF

- 1 (8-oz.) package egg noodles
- 1 Tbsp. olive oil
- 2 Tbsp. butter, divided
- 2 (8-oz.) packages sliced cremini mushrooms
- 2 garlic cloves, minced
- 2 fresh thyme sprigs
- 1 1/2 tsp. smoked paprika
- 1 tsp. kosher salt
- 1/2 tsp. freshly ground black pepper
- 2 Tbsp. all-purpose flour
- 1 1/2 cups chicken broth
- 1 cup sour cream
- 1 Tbsp. chopped fresh flat-leaf parsley

1. Cook egg noodles according to package directions. Remove from heat, and cover to keep warm.
2. Heat olive oil and 1 Tbsp. butter in a large skillet over medium-high heat. Add mushrooms, and cook 5 minutes. Stir in garlic and next 4 ingredients, and cook, stirring occasionally, 3 minutes or until mushrooms are slightly browned.
3. Add flour and remaining butter, and cook, stirring constantly, 2 minutes. Whisk in broth, and bring to boil, whisking constantly; boil, whisking constantly, 5 minutes. Remove from heat. Stir in sour cream and parsley, and serve over noodles.

MAKES 4 servings. **HANDS-ON** 20 min., **TOTAL** 20 min.

CHICKEN-AND-BLACK BEAN CHIMICHANGAS

- 1 lb. shredded deli-roasted chicken
- 1 (15-oz.) can black beans, drained and rinsed
- 1 (4-oz.) can mild chopped green chiles
- 1/4 cup salsa verde
- 1/2 tsp. kosher salt
- 1/4 tsp. freshly ground black pepper
- 1/4 cup chopped fresh cilantro
- 4 (10-inch) flour tortillas
- 1 cup (4 oz.) shredded Monterey Jack cheese
- 1/3 cup canola oil
- Toppings: guacamole, sour cream, chopped tomatoes

1. Stir together first 7 ingredients in a large bowl. Divide chicken mixture among tortillas, placing mixture just below center of each tortilla. Sprinkle with cheese. Fold sides of tortilla over filling, and roll up.
2. Fry chimichangas, in 2 batches, in hot oil in a large skillet over medium-high heat 3 to 4 minutes on each side or until browned and crispy. Drain on paper towels. Serve with desired toppings.

MAKES 4 servings. **HANDS-ON** 20 min., **TOTAL** 20 min.

SKILLET STEAK AND WILTED KALE

Sprinkle this restaurant-style steak salad with crumbled blue cheese for even more flavor.

- 1 bunch kale, stems removed
- 2 (3/4-inch-thick) beef strip steaks
- 1 tsp. kosher salt
- 1/2 tsp. ground black pepper
- 1 Tbsp. olive oil
- 2 Tbsp. butter
- 1 large sweet onion, such as Vidalia, thinly sliced
- 1 Tbsp. red wine vinegar
- 1 tsp. honey

1. Preheat oven to 425°. Chop kale. Sprinkle steaks with salt and pepper. Cook steaks in hot oil in a large skillet over medium-high heat 3 minutes on each side. Place on a wire rack in a jelly-roll pan, and bake 6 minutes.
2. Meanwhile, wipe skillet clean, and melt butter over medium-high heat. Add onion, and sauté 5 minutes or until tender.
3. Remove steaks from oven, and loosely cover with foil to keep warm. Add kale to onion mixture, and cook, stirring constantly, 5 minutes or until kale is wilted. Stir in vinegar and honey; season with salt and pepper. Serve with sliced steak.

MAKES 4 servings. **HANDS-ON** 20 min., **TOTAL** 20 min.

Skillet Steak and Wilted Kale

THE ULTIMATE GRILLED CHEESE

Use mayonnaise instead of butter for added flavor and to keep the bread from burning.

- **1/4 cup mayonnaise**
- **8 Italian bread slices**
- **8 American cheese slices**

1. Spread 1 1/2 tsp. mayonnaise on 1 side of each bread slice.
2. Heat a large nonstick skillet over medium heat. Place 2 bread slices, mayonnaise side down, in skillet; top each with 2 cheese slices and 1 bread slice, mayonnaise side up. Cook 3 minutes on each side or until golden brown. Repeat with remaining bread slices and cheese.

MAKES 4 sandwiches, **HANDS-ON** 15 min., **TOTAL** 15 min.

TOMATO-AND-RED PEPPER SOUP

Red pepper adds a tangy zest to this classic crowd-pleaser.

- **1 (28-oz.) can whole tomatoes**
- **1 (12-oz.) jar roasted red peppers, drained**
- **1/4 cup half-and-half**
- **1 1/2 tsp. kosher salt**
- **1 tsp. sugar**
- **1/2 tsp. freshly ground black pepper**
- **2 garlic cloves**

Process all ingredients and 1/4 cup water in a food processor until smooth, stopping to scrape down sides as needed. Transfer mixture to a medium-size saucepan, and cook over medium-high heat, stirring often, 8 minutes or until hot. Serve immediately.

MAKES 4 to 6 servings. **HANDS-ON** 15 min., **TOTAL** 15 min.

Dixie Chicken Salad with Grapes, Honey, Almonds, and Broccoli

DIXIE CHICKEN SALAD WITH GRAPES, HONEY, ALMONDS, AND BROCCOLI

- **2 large eggs, lightly beaten**
- **6 (4-oz.) chicken cutlets, 1/4 to 1/2 inch thick**
- **1 1/2 cups seasoned breadcrumbs**
- **1/2 cup apple cider vinegar**
- **2 Tbsp. honey**
- **1 Tbsp. Dijon mustard**
- **1/2 tsp. kosher salt**
- **1/4 tsp. freshly ground black pepper**
- **1/2 cup olive oil**
- **1 (5-oz.) package spring lettuce mix**
- **3 cups broccoli florets**
- **1/2 cup halved seedless red grapes**
- **1/2 cup halved seedless green grapes**
- **1/2 cup sliced honey-roasted almonds**

1. Preheat oven to 425°. Whisk together eggs and 3 Tbsp. water in a small bowl. Dip chicken in egg mixture, and dredge in breadcrumbs, pressing firmly to adhere. Place on an aluminum foil-lined baking sheet. Bake at 425° for 15 minutes or until chicken is brown and done.
2. Meanwhile, whisk together vinegar, honey, Dijon mustard, salt, pepper, and olive oil. Toss together lettuce, broccoli, red grapes, and green grapes; season with salt and pepper. Top with chicken and sliced almonds; serve with vinaigrette.

MAKES 4 to 6 servings. **HANDS-ON** 20 min., **TOTAL** 20 min.

THE *SL* TEST KITCHEN ACADEMY

ROBBY MELVIN, DIRECTOR OF THE SOUTH'S MOST TRUSTED KITCHEN, SHARES

FIVE TIME-SAVING TOOLS

In the South, fixing homemade meals is how we show we care. But homemade doesn't have to mean long, tedious prep work. Use these five time-saving tools and you'll get dinner on the table faster without sacrificing from-scratch flavor.

1 MICROPLANE ZESTER/GRATER "Grate garlic cloves or fresh ginger for fast flavor without the mincing." $14.95; *surlatable.com*

2 CITRUS JUICER "Place citrus, cut-side down, in the bowl of this juicer, and squeeze for more juice—and no seeds." $14.95; *surlatable.com*

3 EXTRA MEASURING CUPS AND SPOONS "Don't lose your momentum by cleaning spoons and cups between uses. Just grab a fresh set to speed up hands-on time."

4 FOOD CHOPPER/PROCESSOR "This machine not only slices and dices but also shreds cheese and whips up piecrusts." Models and prices vary; *target.com*

5 PRECUT VEGGIES "Pass go and collect 20 extra minutes by picking up precut veggies and onions at the grocery store."

Better Than Takeout

THIS ZESTY STIR-FRY OFFERS RESTAURANT-STYLE FLAVOR FOR A FRACTION OF THE CALORIES

SMOKY PORK STIR-FRY

Smoked paprika and dark sesame oil give this stir-fry rich flavor. Serve over precooked brown rice or soba noodles.

- **2 tsp. canola oil**
- **10 oz. pork tenderloin, trimmed and cut into bite-size pieces**
- **1/2 tsp. smoked paprika**
- **1/4 tsp. kosher salt**
- **2 tsp. dark sesame oil**
- **1 1/2 cups thinly sliced orange bell pepper (1 medium)**
- **1 cup fresh snow peas**
- **1 Tbsp. minced peeled fresh ginger**
- **1 garlic clove, minced**
- **3 Tbsp. rice vinegar**
- **1 Tbsp. reduced-sodium soy sauce**
- **2 tsp. sugar**
- **1 tsp. Asian chili-garlic sauce**
- **3 cups tricolor coleslaw**
- **3 green onions, thinly sliced**

1. Heat a large skillet over high heat. Add canola oil; swirl to coat. Sprinkle pork with paprika and salt. Add pork to skillet; sauté 3 minutes or until browned. Remove pork from pan.

2. Return pan to medium-high heat. Add sesame oil; swirl to coat. Add bell pepper, peas, ginger, and garlic; stir-fry 3 minutes or until vegetables are crisp-tender, stirring often. Whisk together vinegar, soy sauce, sugar, and chili-garlic sauce in a bowl. Add pork and soy sauce mixture to pan; cook 1 minute. Stir in coleslaw; cook 1 minute or until slightly wilted. Remove pan from heat; sprinkle with chopped green onions.

MAKES 4 servings (serving size: about 1 cup).
HANDS-ON 25 min., **TOTAL** 25 min.

NUTRITIONAL INFORMATION
(per serving)
CALORIES: 165; FAT: 6.4g (SATURATED FAT: 1g); PROTEIN: 17g; FIBER: 3g; CARBOHYDRATES: 10g; SODIUM: 323mg

OUR ALL-TIME FAVORITE, BEST-EVER, CAN'T-STOP-EATING-IT RECIPE FOR

Classic Pimiento Cheese

WHETHER SPELLED pimiento or pimento and pronounced *puh-minner, puh-minnah,* or just *minner,* homemade pimiento cheese gives all cooks a shot at greatness. No other iconic Southern food is as easy to perfect at home.

The undisputed core ingredients—cheese, mayonnaise, and pimientos—are easy to find. Beyond those basics, however, everything is up for debate, especially additional ingredients and texture. People are particular about their pimiento cheese.

Publicly, we Southerners support all pimiento cheese. Privately, we prefer the way it's made at our house, or was made by a favorite aunt or the sweet lady who runs that little cinder block store on the way to the beach. That's one of the charms of pimiento cheese: It's feasible. Each of us easily can make a version that lives up to our expectations and recollections.

Southerners kept their love of pimiento cheese hidden in plain sight for decades, but it recently soared in popularity as people around the country tried a bite and jumped on the bandwagon. Pimiento cheese is a powerful Southern culinary ambassador. In the last few years, it has inspired a documentary, recipe contests, an explosion of small-batch purveyors, and at least one masters' thesis (at UNC-Chapel Hill).

There's also the annual hoopla over the legendary pimiento cheese sandwich served at the Masters Tournament in Augusta, Georgia. Few golfers will don the champion's Green Jacket, but for $1.50, even a duffer can peel open a green sandwich wrapper. When that guarded recipe was altered a couple of years ago, you'd have thought someone had mowed down every azalea in Augusta. People don't like it when you mess with their pimiento cheese.

The simple concoction is special, but not reserved for special occasions. It is the food of everyday folks, the stuff of everyday life. It's not possible to make pimiento cheese highfalutin.

Good pimiento cheese comes together in minutes, keeps for days, and stays on our minds for years.

CLASSIC PIMIENTO CHEESE

- **1/3 cup mayonnaise**
- **3 Tbsp. cream cheese, at room temperature**
- **2 tsp. Worcestershire sauce**
- **2 tsp. fresh lemon juice or apple cider vinegar**
- **1 1/2 tsp. dried mustard**
- **1 1/2 tsp. hot sauce**
- **1 tsp. sugar**
- **1/2 tsp. kosher salt**
- **1/4 tsp. freshly ground black pepper**
- **2 Tbsp. finely grated onion**
- **1 (12-oz.) jar diced pimiento**
- **8 oz. coarsely shredded extra-sharp white Cheddar cheese**
- **8 oz. coarsely shredded sharp yellow Cheddar cheese**

Stir together first 9 ingredients in a medium bowl until smooth. Stir in onion. Fold in pimiento and cheeses. Cover and chill 8 to 12 hours. Let stand at room temperature 30 minutes, and stir well before serving.

MAKES 3 1/2 cups. **HANDS-ON** 20 min.; **TOTAL** 8 hours, 50 min.

The No-Peel Apple Cake

THIS STUNNING DESSERT LOOKS ELABORATE BUT REQUIRES JUST 20 MINUTES OF HANDS-ON PREP

CARAMEL APPLE CAKE

CAKE

- 1 1/3 **cups firmly packed light brown sugar**
- 3/4 **cup butter, softened**
- 3 **large eggs**
- 1 **tsp. vanilla extract**
- 2 **cups all-purpose flour**
- 1 **tsp. baking powder**
- 1 **tsp. table salt**
- 1 **tsp. ground cinnamon**
- 1/2 **tsp. baking soda**
- 3/4 **cup buttermilk**
- **Shortening**

APPLES

- 2 **lb. McIntosh apples (about 6 apples, 6 to 7 oz. each)**
- 1/2 **cup firmly packed light brown sugar**
- 1 **tsp. cornstarch**
- 1/4 **tsp. ground cinnamon**
- **Pinch of table salt**
- 2 **Tbsp. butter**

ADDITIONAL INGREDIENT

Apple Brandy-Caramel Sauce (recipe, page 224)

1. Prepare Cake: Preheat oven to 350°. Beat first 2 ingredients at medium speed with a heavy-duty electric stand mixer until light and fluffy. Add eggs, 1 at a time, beating just until blended after each addition; stir in vanilla.

2. Whisk together flour and next 4 ingredients in a medium bowl. Add flour mixture to sugar mixture alternately with buttermilk, beginning and ending with flour mixture. Beat just until blended after each addition. Spread batter in a greased (with shortening) and floured 9- x 2-inch round cake pan.

3. Bake at 350° for 50 minutes or until a wooden pick inserted in center comes out clean, shielding with aluminum foil after 35 to 40 minutes to prevent excessive browning, if necessary. Cool in pan on a wire rack 10 minutes. Remove from pan, and cool completely (about 1 hour).

4. Meanwhile, prepare Apples. Cut apples into 1/2-inch-thick wedges. Toss together apples, 1/2 cup brown sugar, and next 3 ingredients. Melt 2 Tbsp. butter in a large skillet over medium-high heat; add apple mixture, and sauté 5 to 6 minutes or until crisp-tender and golden. Cool completely (about 30 minutes).

5. Arrange sautéed apples over cooled cake, and drizzle with desired amount of warm Apple Brandy-Caramel Sauce; serve with remaining sauce.

MAKES 8 to 10 servings. **HANDS-ON** 20 min.; **TOTAL** 2 hours, 20 min.

The Hunt Breakfast

AT THIS OUTDOOR BRUNCH, **JULIA REED** SKIPS THE EARLY-MORNING WAKE-UP CALL BUT STILL CELEBRATES THE OPENING OF HUNTING SEASON

ONE OF MY ALL-TIME favorite fall rituals is the hunt breakfast, a restorative feast that's likely been around since William the Conqueror carved out the New Forest as the place to pursue the royal deer. Several centuries later, fox hunters got in on the act, heading off to the hounds after a bracing shot of sherry or port and returning to sideboards piled with steaming silver chafing dishes. That tradition made its way across the Atlantic not long after the first settlers did—George Washington often hosted his fox-hunting friends—but these days, my own hunt breakfast is likely to follow a morning in the dove field.

To take advantage of the gorgeous weather, the event is staged outside, a setting that—based on a brief art history survey—gives us Southerners much in common with the French. For example, breakfasts in 18th-century English paintings feature men in red coats dining around oak tables in chilly-looking halls. The French depictions, on the other hand, capture fresh-air gatherings full of well-dressed women, dogs peering out from crisp damask cloths, bottles being poured, and servants carving great haunches of meat. Jean-François de Troy's *Un Déjeuner de Chasse*, which resides in the Louvre, is described by the museum as "an outdoor meal" in which the morning's "action is supplanted by the pleasures of the table, flirtation and amorous plotting."

Obviously, the latter setting is where I'd rather be. While the tradition can include serving a previous hunt's spoils, I hedge my bets with a visit to the butcher. My go-to menu consists of grillades, the Creole dish of beef or pork medallions simmered in mildly spicy red gravy, served with cheese grits and curried fruit. The fruit, a staple in the Mississippi Delta where I grew up, is mentioned in my pal Susan Puckett's book *Eat Drink Delta* as a favorite of the Longreen Foxhounds, a club that hunts near Glendora, Mississippi.

To begin, I pass around trays of ham biscuits and my mother's Bloody Marys; for serving, silver is the order of the day. If my ancestors had managed to win a hunt cup or two, I'd use those for centerpieces. Instead, I haul out Champagne coolers. In 1889's *The Steward's Handbook and Guide to Party Catering,* Jessup Whitehead, the British-expat author, suggests hunt breakfast decor that consists of white tablecloths and "silver antique jardinieres" filled with "light foliage interspersed with yellow and red flowers." As luck would have it, my friends Keith and Jon Meacham, whose Nashville lawn provides the bucolic backdrop for the breakfast here, have a trellis full of yellow 'Teasing Georgia' roses as well as a pair of English springer spaniels, who are not only picturesque but also very useful for retrieving birds.

For our table, Keith and I forwent the damask in favor of a linen print that our dear friend Suzanne Rheinstein designs for Lee Jofa. The fabric, Gore House Green, pairs beautifully with Keith's Spode Woodland plates, which are perfect for a feast celebrating a hunt. It should be noted, however, that the festivities need not be preceded by a quest for game. I've served this exact menu to hungover guests at countless postwedding brunches, and it's also perfect pre-football fare.

GRILLADES

SEASONING MIX

- 1 1/2 tsp. onion powder
- 1 1/2 tsp. garlic powder
- 1 1/2 tsp. ground red pepper
- 1 tsp. kosher salt
- 1 tsp. white pepper
- 1 tsp. paprika
- 1 tsp. freshly ground black pepper
- 1/2 tsp. dry mustard
- 1/2 tsp. dried thyme
- 1/2 tsp. gumbo filé

GRILLADES

- 2 lb. pork (or veal) shoulder steaks, thinly sliced
- 1 cup all-purpose flour, divided
- 7 Tbsp. vegetable oil
- 1 cup chopped yellow onion
- 1 cup chopped celery
- 1 cup chopped green bell pepper
- 1 1/2 tsp. minced garlic
- 4 bay leaves
- 3 cups beef or chicken broth
- 1/2 cup red wine
- 1 1/2 cups canned whole tomatoes, drained and chopped
- 1 Tbsp. Worcestershire sauce
- 1 tsp. dried thyme
- Hot cooked cheese grits

1. Stir together all Seasoning Mix ingredients in a small bowl. Sprinkle about 2 tsp. Seasoning Mix on both sides of pork slices. In a jelly-roll pan, stir together 1/2 cup flour and 1 tsp. Seasoning Mix. Dredge pork in flour mixture, shaking off excess.

2. Heat oil in a large skillet over medium-high heat. Add pork, and fry about 2 minutes on each side or until golden brown. Transfer pork to a plate, reserving drippings in skillet.

Grillades

Judy's Bloody Mary Mix

JUDY'S BLOODY MARY MIX

- 6 cups tomato juice
- 1 1/4 cups lime juice
- 1/2 cup Worcestershire sauce
- 4 dashes of hot sauce (such as Tabasco)
- 1 Tbsp. kosher salt
- 1 Tbsp. prepared horseradish
- Cracked pepper

Stir together first 6 ingredients. Add cracked pepper to taste. Refrigerate in an airtight container up to 3 days.

MAKES about 2 qt. **HANDS-ON** 5 min., **TOTAL** 5 min.

3. Sprinkle remaining 1/2 cup flour over drippings. Cook over high heat, whisking constantly, about 3 minutes or until roux is medium brown. Immediately add onion, celery, bell pepper, and garlic, and stir with a wooden spoon until well blended. Add bay leaves and another 2 tsp. Seasoning Mix. Continue cooking, stirring constantly, about 5 minutes.

4. Bring broth to a boil in a medium saucepan, and add to vegetable mixture, stirring until well incorporated. Add wine, next 3 ingredients, and pork, and bring to a boil over high heat. Reduce heat to low, and cook, stirring occasionally, 40 minutes. Midway through, taste to check seasoning. You'll have some Seasoning Mix left over; feel free to add more. Serve hot with cheese grits.

MAKES 4 to 6 servings. **HANDS-ON** 40 min.; **TOTAL** 1 hour, 20 min.

Community Cookbook

BUILD A BETTER LIBRARY, ONE GREAT BOOK AT A TIME

We Southerners all know and love fried chicken, but there's more than one approach to frying a bird. In her latest cookbook, *SL* Contributing Editor Rebecca Lang explores both classic and surprising ways to enjoy this beloved dish.

BUTTERMILK-SOAKED, BACON-FRIED CHICKEN IN GRAVY

CHICKEN

- **1 1/2 cups buttermilk**
- **2 Tbsp. hot sauce (such as Tabasco)**
- **1 chicken (about 2 1/2 lb.), cut into 4 pieces**
- **2 cups all-purpose flour**
- **1 tsp. salt**
- **1/2 tsp. freshly ground black pepper**
- **12 oz. bacon, chopped into 1/2-inch pieces**
- **Vegetable oil, for frying**

GRAVY

- **1/4 cup all-purpose flour**
- **2 cups heavy cream**
- **1 Tbsp. dry sherry**
- **1/2 tsp. salt**

1. For the marinade, whisk together the buttermilk and hot sauce. Pour into a large zip-top plastic bag, and add the chicken. Seal the bag, rub to coat the meat, and refrigerate for 6 hours.
2. In a bowl, whisk together the flour, salt, and pepper.
3. In a large heavy skillet, cook the bacon over medium heat until crispy. Using a slotted spoon, remove the bacon from the skillet, and drain. Reserve drippings in the skillet.
4. Add enough vegetable oil to the skillet drippings to be 1 inch deep, and heat over medium heat to 325°. Set a wire rack over a rimmed baking sheet.
5. Remove the chicken from the marinade, and dredge in the flour mixture.
6. Carefully place the chicken in the hot oil, and fry, turning often, for 20 to 24 minutes or until cooked through and juices run clear. Maintain a frying temperature of 320°. Drain the chicken on the wire rack.
7. For the gravy, carefully pour off the hot oil, reserving about 1/4 cup in the skillet. Over low heat, whisk in the 1/4 cup flour, 1 Tbsp. at a time, and cook for 1 minute, whisking constantly. Gradually add the heavy cream, and cook for 6 minutes or until creamy. Stir in the sherry and salt, and cook for 1 minute.
8. Serve the gravy over the chicken with reserved bacon.

MAKES 4 servings.

TANGY FRIED CHICKEN WITH DIJON

- **1/2 cup Dijon mustard**
- **2 tsp. herbes de Provence**
- **2 tsp. salt, divided**
- **1 1/2 tsp. freshly ground black pepper, divided**
- **1 chicken (about 3 lb., 12 oz.), cut into 8 pieces**
- **Canola oil, for frying**
- **3 cups all-purpose flour**

1. In a small mixing bowl, whisk together the mustard, herbes de Provence, 1 tsp. of salt, and 1/2 tsp. of pepper.
2. Rub the mustard mixture all over each piece of chicken, and let sit on a rimmed baking sheet at room temperature for 30 minutes.
3. In a large heavy skillet, heat 1 1/2 inches of canola oil over medium heat to 340°. Set a wire rack over a rimmed baking sheet.
4. In a shallow bowl, whisk together flour and remaining salt and pepper. Working with half of the chicken at a time, dredge the pieces in the flour mixture, shaking off the excess.
5. Carefully place the chicken in the hot oil. Fry, turning often, for 18 to 24 minutes or until brown and juices run clear. Maintain a frying temperature of 320°. Drain the chicken on a wire rack.

MAKES 4 to 6 servings.

October

236 **The Great Pumpkin Cookbook** A recipe from the pumpkin patch for appetizers, desserts, and every course in between

240 SOUTHERNERS **Cooking for My Family** At home with New Orleans chef John Besh

241 WHAT TO COOK NOW **Beautiful Beef Dinners** Whether you're serving guests fancy or family style, try these inspired ways with ground beef

244 **Make it a Movie Night** Snacks worthy of a starring role at your next showing

246 THE *SL* TEST KITCHEN ACADEMY **The *SL* Beef Primer** Know your cuts to make dishes that are a cut above

247 THE SLOW DOWN **Bolognese Made Easy** The hands-off route to an Italian favorite

248 COMMUNITY COOKBOOK Emeril Lagasse shares his wisdom and favorite recipes from a lifetime of cooking

The Great Pumpkin Cookbook

SWEET, SPICY, SAVORY—SIX DISHES STARRING FALL'S PRIME PICK

Spicy Pumpkin Soup with Avocado Cream, page 238

PEAR-AND-PUMPKIN TART

- 1 (17.3-oz.) package frozen puff pastry sheets, thawed
- 1/2 (3-lb.) sugar pumpkin, peeled, seeded, and cut into 1/4-inch-thick slices
- 1 firm Bartlett pear, cut into 1/4-inch-thick slices
- 1/2 tsp. kosher salt
- 1/4 tsp. freshly ground black pepper
- 2 tsp. olive oil, divided
- 2 cups loosely packed arugula leaves
- 1/4 cup crumbled blue cheese
- 1/4 cup fresh pomegranate seeds
- 1 tsp. red wine vinegar

1. Preheat oven to 425°. Unfold puff pastry sheets, and place side by side on a baking sheet, overlapping short sides 1/2 inch. Press seam to seal. Score a 1/2-inch border on all sides, using a knife. Do not cut through pastry.

2. Toss together pumpkin slices, next 3 ingredients, and 1 tsp. olive oil in a large bowl. Spread mixture in a single layer on prepared pastry sheets, leaving a 1/2-inch border.

3. Bake at 425° for 20 to 22 minutes or until pastry is golden brown. Cool on a wire rack 10 minutes.

4. Toss together arugula, next 3 ingredients, and 1 tsp. olive oil in a medium bowl. Add salt and pepper to taste. Sprinkle mixture over tart, and cut into desired shapes.

MAKES 6 to 8 appetizer servings. **HANDS-ON** 20 min., **TOTAL** 50 min.

Pear-and-Pumpkin Tart

PUMPKIN-CHOCOLATE BROWNIES

- 1 1/4 cups semisweet chocolate morsels
- 1 cup unsalted butter, cut into pieces
- 3 (1-oz.) unsweetened chocolate baking squares, chopped
- 3 large eggs
- 1 cup plus 2 Tbsp. granulated sugar
- 2 Tbsp. cold brewed coffee
- 1 Tbsp. vanilla extract
- Parchment paper
- 2/3 cup all-purpose flour
- 1 1/2 tsp. baking powder
- 1 tsp. kosher salt, divided
- 1 (15-oz.) can pumpkin
- 3 large eggs
- 1/2 cup heavy cream
- 1/3 cup firmly packed light brown sugar
- 1 1/2 tsp. pumpkin pie spice

1. Preheat oven to 350°. Pour water to a depth of 1 inch in the bottom of a double boiler over medium heat; bring to a boil. Reduce heat, and simmer. Place first 3 ingredients in top of double boiler over simmering water. Cook, stirring occasionally, 5 to 6 minutes or until melted. Remove from heat; cool 10 minutes.

2. Whisk together 3 eggs, granulated sugar, and next 2 ingredients in a large bowl. Gradually whisk warm chocolate mixture into egg mixture; cool 10 minutes.

3. Grease a 13- x 9-inch baking pan with butter. Line bottom and sides of pan with parchment paper, allowing 2 to 3 inches to extend over sides. Grease (with butter) and flour parchment paper.

4. Sift flour, baking powder, and 1/2 tsp. salt in a bowl. Whisk into chocolate mixture. Pour batter into prepared pan, reserving 2/3 cup.

5. Whisk together pumpkin, next 4 ingredients, and remaining 1/2 tsp. salt; pour over brownie batter in pan. Top with reserved brownie batter, and swirl batter gently 3 times in 1 direction and 3 times in the opposite direction with a knife or the end of a wooden spoon.

6. Bake at 350° for 45 to 50 minutes or until a wooden pick inserted in center comes out with a few moist crumbs. Cool completely on a wire rack (about 2 hours). Lift brownies from pan, using parchment paper sides as handles.

Pumpkin-Chocolate Brownies

Gently remove parchment paper, and cut brownies into 24 squares.

Note: To make ahead, refrigerate cooled uncut brownies, uncovered, overnight; cut into squares while cold.

MAKES 2 dozen. **HANDS-ON** 40 min.; **TOTAL** 4 hours, 5 min.

OUR EASIEST PUMPKIN PIE EVER

- 1 (14.1-oz.) package refrigerated piecrusts
- 1 1/2 cups plus 2 Tbsp. buttermilk, divided
- Parchment paper
- 1 (15-oz.) can pumpkin
- 3/4 cup sugar
- 2 tsp. ground cinnamon
- 1/2 tsp. kosher salt
- 1 tsp. vanilla extract
- 2 large eggs
- 1 large egg yolk

1. Preheat oven to 425°. Fit 1 piecrust into a 9-inch metal pie pan according to package directions, pressing excess dough onto rim of pie pan. Cut shapes from remaining piecrust to use around pie edge. (We used a 1/2-inch round cutter.) Brush 1 Tbsp. buttermilk around pie edge; arrange shapes around pie edge, pressing to adhere. Brush shapes with 1 Tbsp. buttermilk. Prick bottom and sides of piecrust 8 to 10 times with a fork. Line piecrust with parchment paper, and fill with pie weights. Bake 15 minutes.

2. Whisk together pumpkin, next 6 ingredients, and remaining 1 1/2 cups buttermilk in a large bowl. Pour mixture into piecrust.

3. Bake at 425° for 10 minutes. Reduce heat to 325°, and bake 35 to 40 more minutes or until edge of filling is slightly puffed and center is slightly jiggly. Cool on a wire rack 1 hour. Store in refrigerator up to 2 days.

MAKES 8 servings. **HANDS-ON** 15 min.; **TOTAL** 2 hours, 15 min.

SPICY PUMPKIN SOUP WITH AVOCADO CREAM

- **1 cup diced yellow onion**
- **3 Tbsp. olive oil, divided**
- **1 1/2 tsp. kosher salt, divided**
- **2 garlic cloves, chopped**
- **1 Tbsp. ground cumin**
- **1 (29-oz.) can pumpkin**
- **6 to 6 1/2 cups reduced-sodium chicken broth**
- **1 canned chipotle pepper in adobo sauce**
- **1 Tbsp. adobo sauce from can**
- **1 medium avocado, peeled and diced**
- **1/2 cup whole buttermilk**
- **2 Tbsp. fresh lime juice**
- **2 Tbsp. extra virgin olive oil**
- **8 oz. smoked sausage, sliced**
- **1 cup black beans, drained and rinsed**
- **1/2 tsp. smoked paprika**

1. Place onions, 2 Tbsp. olive oil, and 1 tsp. salt in a Dutch oven over medium heat; cover and cook 5 to 6 minutes or until translucent. Stir in garlic and cumin; cook 2 minutes. Whisk in pumpkin and 6 cups broth; add chipotle pepper and 1 Tbsp. adobo sauce. Increase heat to medium-high, and simmer, stirring occasionally, 12 minutes.

2. Process soup, in batches, in a food processor or blender 1 minute. Add up to 1/2 cup broth, 2 Tbsp. at a time, to reach desired consistency.

3. Process avocado, next 3 ingredients, and remaining 1/2 tsp. salt in a blender until smooth. Add up to 1/4 cup water, 1 Tbsp. at a time, to reach desired consistency.

4. Cook smoked sausage in remaining 1 Tbsp. olive oil in a large skillet over medium heat, stirring occasionally, 3 minutes. Stir in black beans and paprika, and cook 1 minute. Ladle soup into serving bowls; top with sausage mixture and avocado cream.

MAKES 6 to 8 servings. **HANDS-ON** 55 min., **TOTAL** 55 min.

Our Easiest Pumpkin Pie Ever, page 237

BEER-BATTERED PUMPKIN WITH DIPPING SAUCE

- **Vegetable oil**
- **1 1/3 cups all-purpose flour**
- **1/4 cup cornstarch**
- **1 Tbsp. plus 1/2 tsp. kosher salt, divided**
- **1 (12-oz.) bottle cold light beer (such as Corona Light)**
- **1/2 (3-lb.) sugar pumpkin, peeled, seeded, and cut into 1/2-inch-thick wedges**
- **15 large fresh sage leaves**
- **1 garlic clove**
- **1/2 cup low-fat Greek yogurt**
- **1/4 cup buttermilk**
- **1/4 tsp. hot sauce (such as Tabasco)**
- **1/4 tsp. freshly ground black pepper**

1. Pour oil to a depth of 1 1/2 inches in a Dutch oven; heat over medium-high heat to 350°.

2. Whisk together flour, cornstarch, and 1 Tbsp. salt in a large bowl; whisk in beer. Dip pumpkin wedges in batter, allowing excess batter to drip off. (Pumpkin should be very lightly coated.)

3. Gently lower pumpkin into hot oil, using tongs. Fry pumpkin, in 3 batches, 3 to 4 minutes or until tender inside and light brown outside, turning once. Place fried pumpkin on a wire rack in a jelly-roll pan; season with salt. Repeat procedure with sage leaves, frying 1 minute on each side.

4. Place peeled garlic clove on a cutting board; smash garlic, using flat side of knife, to make a paste. Whisk together garlic, yogurt, next 3 ingredients, and remaining 1/2 tsp. salt. Serve yogurt mixture with fried pumpkin and sage leaves.

MAKES 4 to 6 appetizer servings. **HANDS-ON** 40 min., **TOTAL** 40 min.

PUMPKIN-AND-TURNIP GREEN LASAGNA

- **1 lb. mild Italian sausage, casings removed**
- **2 Tbsp. olive oil, divided**
- **2 garlic cloves, finely chopped**
- **2 1/2 tsp. kosher salt, divided**
- **1 (1-lb.) package fresh turnip greens, chopped**
- **1 1/2 qt. milk**
- **6 Tbsp. butter**
- **6 Tbsp. all-purpose flour**
- **1/2 tsp. dry mustard**
- **2 cups (8 oz.) shredded Parmesan cheese, divided**
- **3/4 tsp. freshly ground black pepper, divided**
- **1/2 tsp. ground nutmeg, divided**
- **1 (29-oz.) can pumpkin**
- **1 lb. no-boil lasagna noodles**
- **Vegetable cooking spray**

1. Preheat oven to 375°. Cook sausage in 1 Tbsp. olive oil in a large skillet over medium-high heat, stirring often, 4 to 5 minutes or until meat crumbles and is no longer pink. Remove sausage to a plate, using a slotted spoon; reserve drippings in skillet. Reduce heat to medium.

2. Stir garlic, 1/2 tsp. salt, half of turnip greens, 1/2 cup water, and remaining 1 Tbsp. oil into hot drippings. Cook 1 minute, stirring to loosen browned bits from bottom of skillet. Stir in remaining turnip greens. Reduce heat to medium-low, and cook 5 to 6 minutes or until greens are tender and water has evaporated. Remove from heat.

3. Microwave milk, in batches, in a microwave-safe measuring cup covered with plastic wrap at HIGH for 2 to 3 minutes or until very warm. Melt butter in a large saucepan over medium heat. Whisk in flour, and cook, whisking constantly, 1 minute. Gradually whisk in warm milk; cook, whisking often, 12 to 14 minutes or until mixture thickens and comes to a low boil. Remove from heat, and whisk in dry mustard, 1 cup Parmesan cheese, 1/4 tsp. pepper, 1/4 tsp. nutmeg, and 1 tsp. salt.

4. Whisk together pumpkin, 3/4 cup Parmesan cheese, remaining 1/4 tsp. ground nutmeg, 1 tsp. salt, and 1/2 tsp. pepper in a large bowl.

5. Place 1 layer of lasagna noodles in a lightly greased (with cooking spray) 13- x 9-inch baking dish, covering bottom completely. (Use pieces of noodles to fill in any gaps.) Spread 1 cup sauce over noodles; top with cooked sausage. Add another layer of noodles and 1 cup sauce; top with half of pumpkin mixture. Add another layer of noodles and 1 cup sauce; top with turnip green mixture. Add another layer of noodles and 1 cup sauce; top with remaining pumpkin mixture. Add another layer of noodles and 1 cup sauce; top with remaining 1/4 cup Parmesan cheese.

6. Bake at 375° for 40 minutes or until top is golden brown. Let stand 15 minutes before serving. Sprinkle with freshly grated Parmesan cheese. Serve with remaining sauce.

MAKES 6 to 8 servings. **HANDS-ON** 1 hour; **TOTAL** 1 hour, 55 min.

ODE TO PUMPKIN PIE'S BEST FRIEND

Once regarded as a miracle that fed arctic explorers and prevented disease, canned foods don't get much love these days. But one famous can deserves praise by pumpkin pie bakers everywhere: Libby's. Although making your own puree may seem like a noble effort, you'll likely end up with a hot mess. That's because pumpkins at the grocery store don't compare to those Libby's has developed since 1929. For its 90 million pies' worth of cans, Libby's uses a proprietary variety of Dickinson pumpkin, which has a thinner skin and a thicker layer of sweet flesh. As for that velvety texture? There's no blender that will produce a puree as smooth and free of fibrous strands. So go ahead and crank that can opener. Some things are best left to the professionals.

—HANNAH HAYES

5 MORE WAYS TO USE CANNED PUMPKIN

1. ROAST IT: Spread puree on a parchment paper-lined baking sheet, and roast at 350° until it begins to brown. It will create a deeper taste.

2. DRINK IT: Pour a teaspoon or two of puree into bourbon-based cocktails.

3. SAUCE IT: Stir puree into a cream-based pasta sauce, and serve with ravioli.

4. SPREAD IT: Add a few tablespoons of puree to cream cheese.

5. STIR IT: Spoonfuls of puree, plus honey, cinnamon, and berries added to oatmeal make a quick, good-for-you breakfast.

Cooking for My Family

CHEF JOHN BESH ON SIMPLE, SOULFUL AT-HOME MEALS

CHEF JOHN BESH is making his mama's seafood gumbo, dropping in smoked pork, sausage, and blue crab. It's a dish he prepares for his annual tailgate party at LSU. Besh owns 12 restaurants and has written 4 books, including *Besh Big Easy: 101 Home Cooked New Orleans Recipes*. In New Orleans, Besh is a hero. After Katrina he made it his mission to preserve the city's culinary history. In his new book, he trades the technique of a professional kitchen for the foods his mother and grandmother cooked.

He talks of his family and the daily meals that have become more precious to him. "It all starts at the family table," he explains.

As a boy, Besh helped care for his father, Ted, a former pilot, paralyzed by a drunk driver. He cooked to be able to make his dad breakfast. "My Dad was my hero," he says. "He came back from what happened to him, started another career, and cared for us. That's who I try to emulate."

Ted Besh died in 2014, but John talks to his mom, Imelda, daily. Losing his father and sister Kathleen (in 2006) changed him and his priorities. With his oldest son at Notre Dame and three boys at home, making time to cook for family took on a whole new meaning.

"We have a strict no-phone rule at the table," he says. "Even when we gather to enjoy simple meals, we're going to look at each other."

Growing up, it was a pregame tradition for his dad to take him to Mandina's Restaurant on Canal before football games at Tulane Stadium to cheer for the Saints. Apart from the LSU party, tailgating at home is much more his speed.

The menu stays fairly consistent. He serves the food family style in his huge open kitchen, encouraging his guests to stir the pot between plays.

Nothing formal here. Just good food and football. "Tailgating is one way we share love with one another, regardless of the team. But I sure hope LSU wins."

CREOLE SEAFOOD JAMBALAYA

"My family has been making a version of this dish for generations."

- **1/2 lb. bacon, diced**
- **1 lb. fresh pork sausage, casings removed**
- **1/2 lb. andouille sausage, diced**
- **3 Tbsp. lard**
- **4 skinned and boned chicken thighs, cut into 1-inch cubes**
- **Kosher salt**
- **Freshly ground black pepper**
- **1 large onion, diced**
- **1 bell pepper, diced**
- **3 celery ribs, diced**
- **3 garlic cloves, minced**
- **2 cups converted white rice**
- **1 tsp. dried thyme**
- **2 bay leaves**
- **1 1/2 Tbsp. pimentón de la Vera or smoked paprika**
- **1 tsp. ground red pepper**
- **1 Tbsp. celery salt**
- **1 cup canned crushed tomatoes**
- **2 cups basic chicken stock**
- **1 1/2 lb. raw Louisiana white shrimp or other wild American shrimp, peeled and deveined**
- **1 bunch green onions, chopped**

1. Heat a large Dutch oven over high heat until hot, and then reduce heat to medium. (This will allow the heat to be uniform all over, preventing those little hot spots that are likely to burn.) Cook bacon, sausages, and lard in the hot pot, stirring slowly with a long wooden spoon, for 10 minutes. Season chicken thighs with kosher salt and black pepper. Add the chicken to pot, and cook, stirring often, 5 minutes or until chicken is brown.

2. Increase heat to medium-high. Add onion to pot, and cook about 15 minutes or until soft. Add bell pepper, celery, and garlic, and cook 5 minutes. Continue stirring occasionally so everything in the pot cooks evenly.

3. Add rice, thyme, bay leaves, pimentón, red pepper, and celery salt to pot, and cook, stirring often, 3 minutes. Increase heat to high, and add tomatoes and chicken stock. Bring to a boil. Reduce heat to medium-low; cover pot, and simmer 15 minutes.

4. After the rice has simmered for 15 minutes, fold in the shrimp and green onions. Turn off the heat, and let everything continue to cook in the hot covered pot 10 more minutes. Remove the lid, fluff the jambalaya, and serve.

MAKES 6 to 8 servings. **HANDS-ON** 1 hour, 20 min.; **TOTAL** 1 hour, 45 min.

Beautiful Beef Dinners

EASY ENOUGH FOR A WEEKNIGHT MEAL ... ELEGANT ENOUGH FOR WEEKEND GUESTS

Shepherd's Pie with Potato Crust, page 242

Tex-Mex Meatballs in Red Chile Sauce, page 242

Spiced Beef Kabobs with Herbed Cucumber-and-Tomato Salad, page 243

Greek Baked Ziti, page 242

SHEPHERD'S PIE WITH POTATO CRUST

Instead of mashed potatoes, use thinly sliced potatoes to form the gorgeous, scalloped crust. We used a mandoline for slicing, but a sharp knife will also work.

FILLING

- **1 1/2 lb. lean ground chuck**
- **2 Tbsp. vegetable oil**
- **1 cup dry red wine**
- **1 Tbsp. tomato paste**
- **3 Tbsp. butter**
- **1 1/2 cups chopped onion**
- **1 1/2 cups chopped carrots**
- **3 garlic cloves, minced**
- **3 Tbsp. all-purpose flour**
- **2 cups beef broth**
- **1 Tbsp. Worcestershire sauce**
- **1 Tbsp. chopped fresh rosemary leaves**
- **1 tsp. chopped fresh thyme leaves**
- **2 1/2 tsp. kosher salt**
- **1 tsp. freshly ground black pepper**
- **1 1/2 cups fresh or frozen English peas**

CRUST

- **2 sweet potatoes (about 1 1/2 lb.), peeled and very thinly sliced**
- **2 large Yukon gold potatoes (about 1 1/2 lb.), peeled and very thinly sliced**
- **Vegetable cooking spray**
- **2 Tbsp. butter, melted**

ADDITIONAL INGREDIENT

- **2 Tbsp. chopped fresh flat-leaf parsley**

1. Prepare Filling: Cook ground chuck in hot oil in a large skillet over medium-high heat, stirring occasionally, 8 minutes or until meat crumbles and is no longer pink. Add wine and tomato paste, and cook, stirring occasionally, 4 to 5 minutes or until wine is completely evaporated. Remove mixture from skillet, and drain. Wipe skillet clean.

2. Melt butter in skillet over medium-high heat. Add onion and carrots, and sauté 4 to 5 minutes or until slightly browned and tender. Add garlic, and cook, stirring constantly, 30 seconds. Stir in flour, and cook, stirring constantly, 1 minute. Stir in broth, and bring to a boil, stirring constantly. Reduce heat to medium; stir in Worcestershire sauce and next 4 ingredients. Simmer 10 minutes or until thickened. Stir in peas and beef mixture. Remove from heat.

3. Prepare Crust: Preheat oven to 375°. Microwave sweet potatoes and 1/4 cup water in a microwave-safe bowl covered with plastic wrap at HIGH 5 minutes. Repeat procedure with Yukon gold potatoes.

4. Arrange about two-thirds of potato slices, edges slightly overlapping, in a lightly greased (with cooking spray) 9-inch deep-dish pie plate, covering bottom and sides. Spoon in filling. Arrange remaining potato slices, edges slightly overlapping, over filling. Brush potatoes with melted butter.

5. Bake at 375° for 40 to 45 minutes or until potatoes are tender. Increase oven temperature to broil, and broil 2 to 3 minutes or until golden brown and crispy. Sprinkle with chopped fresh parsley. Let stand 10 minutes before serving.

MAKES 4 to 6 servings.
HANDS-ON 1 hour, 10 min.; **TOTAL** 2 hours

GREEK BAKED ZITI

Inspired by the classic Greek comfort dish known as pastitsio, *this pasta casserole uses fresh and flavorful ingredients.*

- **12 oz. ziti pasta**
- **1 small yellow onion, chopped**
- **1 Tbsp. olive oil**
- **2 garlic cloves, minced**
- **1 1/2 lb. lean ground beef**
- **2 (15-oz.) cans tomato sauce**
- **1 Tbsp. fresh lemon juice**
- **1 1/2 tsp. dried oregano**
- **1 tsp. sugar**
- **1/2 tsp. ground cinnamon**
- **1 1/2 tsp. kosher salt, divided**
- **3 Tbsp. butter**
- **3 Tbsp. all-purpose flour**
- **3 cups milk**
- **1 cup grated Parmesan cheese**
- **1/2 tsp. freshly ground black pepper**
- **Vegetable cooking spray**
- **1 (8-oz.) package shredded mozzarella cheese**
- **1/3 cup fine, dry breadcrumbs**

1. Preheat oven to 350°. Cook pasta in a Dutch oven according to package directions.

2. Meanwhile, sauté onion in hot oil in a large skillet over medium-high heat 4 to 5 minutes or until tender. Add garlic; sauté 30 seconds. Add beef; cook, stirring occasionally, 5 minutes or until meat crumbles and is longer pink. Drain mixture, and return to skillet.

3. Stir tomato sauce, next 4 ingredients, and 1 tsp. salt into meat mixture. Bring to a simmer over medium-high heat, and cook, stirring occasionally, 2 minutes. Remove from heat.

4. Melt butter in a large saucepan over low heat. Whisk in flour, and cook, whisking constantly, 2 minutes. Gradually whisk in milk. Increase heat to medium, and cook, whisking constantly, 5 to 7 minutes or until thickened and bubbly. Stir in Parmesan cheese, pepper, and remaining 1/2 tsp. salt. Add sauce to pasta, stirring to coat.

5. Transfer pasta mixture to a lightly greased (with cooking spray) 13- x 9-inch baking dish. Top with beef mixture, mozzarella cheese, and breadcrumbs.

6. Bake at 350° for 20 to 25 minutes or until mixture is bubbly and cheese is melted. Let stand 10 minutes before serving.

MAKES 6 to 8 servings. **HANDS-ON** 50 min.; **TOTAL** 1 hour, 20 min.

TEX-MEX MEATBALLS IN RED CHILE SAUCE

Let guests doctor their own plates with crunchy, colorful toppings.

- **1 poblano pepper**
- **1 small white onion, coarsely chopped**
- **2 garlic cloves**
- **1/2 cup firmly packed fresh cilantro leaves with tender stems**
- **1/2 cup finely crushed corn chips**
- **1/4 cup milk**
- **2 large eggs**
- **1 1/2 tsp. kosher salt**
- **1/2 tsp. freshly ground black pepper**
- **2 lb. ground chuck**
- **Vegetable cooking spray**

2 (10-oz.) cans red chile enchilada sauce
2 cups reduced-sodium chicken broth
2 to 2 1/2 Tbsp. sugar, divided
1 cup sour cream
1 tsp. lime zest
2 Tbsp. fresh lime juice
6-inch corn tortillas, warmed
Toppings: Cotija cheese, radishes, toasted shelled pumpkin seeds (pepitas), cilantro sprigs, avocado, red leaf lettuce

1. Preheat broiler with oven rack 5 inches from heat. Broil poblano pepper on an aluminum foil-lined baking sheet 6 to 8 minutes or until blistered, turning occasionally. Place poblano in a zip-top plastic freezer bag; seal and let stand 10 minutes to loosen skin. Peel poblano; remove and discard stem and seeds. Pulse poblano, onion, garlic, and cilantro in a food processor until finely chopped.
2. Stir together corn chips and milk in a large bowl; let stand about 5 minutes or until chips soften. Stir in eggs, salt, pepper, and poblano mixture. Fold in beef. Shape into 35 (1 1/2-inch) meatballs (about 2 tablespoonfuls each). Place 1 1/2 inches apart on a lightly greased (with cooking spray) rack in an aluminum foil-lined jelly-roll pan.
3. Preheat oven to 400°. Bake 10 to 12 minutes or until browned. Transfer meatballs to a large Dutch oven; add enchilada sauce, chicken broth, and 1 Tbsp. sugar. Bring to a boil over high heat. Reduce heat to medium, and simmer 15 to 20 minutes or until meatballs are cooked through and sauce is slightly thickened, turning meatballs halfway through.
4. Meanwhile, whisk together sour cream, zest, and lime juice in a small bowl. Season with salt, and add desired amount of remaining sugar. Serve meatballs with sour cream mixture, tortillas, and toppings.

MAKES 8 to 10 servings. **HANDS-ON** 40 min.; **TOTAL** 1 hour, 20 min.

SPICED BEEF KABOBS WITH HERBED CUCUMBER-AND-TOMATO SALAD

Serve this zesty beef with toasty bread and a tangy herb salad.

8 (10-inch) flat wooden skewers
3 firm white bread slices, torn into small pieces
1 1/2 lb. ground beef
1/2 cup finely chopped yellow onion
1/4 cup chopped fresh flat-leaf parsley leaves
1/4 cup chopped fresh cilantro leaves
1 1/2 tsp. kosher salt
1 tsp. ground cumin
1/2 tsp. ground allspice
1/2 tsp. ground coriander
1/2 tsp. freshly ground black pepper
1/2 tsp. ground red pepper
Vegetable cooking spray
8 pita rounds, warmed
Yogurt-Tahini Sauce
Herbed Cucumber-and-Tomato Salad

1. Soak skewers in water 30 minutes. Meanwhile, place bread and water to cover in a small bowl; let stand 10 minutes. Using hands, squeeze excess liquid from bread; transfer bread to a large bowl. Stir in beef and next 9 ingredients.
2. Coat cold grill grate with cooking spray. Preheat grill to 400° (medium-high) heat. Divide ground beef mixture into 8 portions, and chill 10 minutes. Press and flatten each portion around a skewer, forming a 4- to 5-inch oval.
3. Grill, covered with grill lid, 7 to 8 minutes or until golden brown, turning once. Serve warm with pita rounds, Yogurt-Tahini Sauce, and Herbed Cucumber-and-Tomato Salad.

MAKES 8 servings. **HANDS-ON** 30 min.; **TOTAL** 50 min., including sauce and salad

Yogurt-Tahini Sauce

Whisk together 3/4 cup **Greek yogurt;** 1 **garlic** clove, minced; 2 Tbsp. **tahini;** 1 Tbsp. fresh **lemon juice;** 1/2 tsp. ground **cumin;** and 1/8 tsp. **kosher salt** in a small bowl. Drizzle with **olive oil,** and sprinkle with **black pepper.**

MAKES 1 cup. **HANDS-ON** 5 min., **TOTAL** 5 min.

Herbed Cucumber-and-Tomato Salad

3 Tbsp. extra virgin olive oil
2 Tbsp. red wine vinegar
1 tsp. chopped fresh oregano
1 1/2 lb. Kirby cucumbers, peeled, seeded, and sliced
1 cup grape tomatoes, halved
1/2 cup thinly sliced red onion
2 Tbsp. fresh mint leaves
2 Tbsp. fresh flat-leaf parsley leaves
2 Tbsp. chopped fresh dillweed
Kosher salt and freshly ground black pepper
1/2 cup crumbled feta cheese

Whisk together first 3 ingredients in a small bowl. Toss together cucumbers, next 5 ingredients, and vinegar mixture in a large bowl. Season with salt and pepper. Sprinkle with feta.

MAKES 4 to 6 servings. **HANDS-ON** 15 min., **TOTAL** 15 min.

GROUND BEEF TIPS

STORING: Refrigerate or freeze ground beef as soon as possible after purchase. If it will be used within one or two days, keep meat in its original packaging and store in the refrigerator.

FREEZING: Original packaging won't protect meat from freezer burn. To freeze, slip the package into a zip-top plastic freezer bag or rewrap recipe-size portions of uncooked ground beef in heavy-duty aluminum foil. Be sure to label, including the weight and date. Use within four months.

THAWING: The best way to thaw ground beef is in the refrigerator, which prevents the texture from becoming mealy. Cook thawed meat within two days.

Make it a Movie Night

WHAT COULD BE BETTER THAN A FAVORITE FILM AND BIG BOWLS OF FAMILY-PLEASING SNACKS HANDPICKED FOR MUNCHING? FOR A CHANGE OF PACE FROM THE USUAL BUTTERED POPCORN, TRY OUR SMOKE-MEETS-SPICE VARIATION OR A NUTTY SPIN ON THE CLASSIC CARAMEL CORN. BET THEY'LL BE THE STARS OF THE SHOW!

SMOKY BARBECUE POPCORN

- 6 Tbsp. butter
- 1 tsp. Worcestershire sauce
- 1 Tbsp. sugar
- 2 tsp. smoked paprika
- 2 tsp. chili powder
- 2 tsp. garlic powder
- 2 tsp. kosher salt
- 15 cups popped popcorn (from about 3/4 cup kernels)

Melt butter in a small saucepan over medium heat. Remove from heat, and stir in Worcestershire sauce. Stir together sugar and next 4 ingredients in a small bowl. Slowly drizzle butter mixture over popcorn, gently stirring as you drizzle. Sprinkle with spice mixture, gently stirring to coat.

MAKES 15 cups. **HANDS-ON** 15 min., **TOTAL** 15 min.

CARAMEL-PEANUT-POPCORN SNACK MIX

There's no trick to this treat. All you need is 30 minutes to make this scary-good mix for a Halloween party or your next movie night.

- 15 cups popped popcorn (about 3/4 cup kernels)
- Vegetable cooking spray
- 1 cup plus 2 Tbsp. firmly packed dark brown sugar
- 1/2 cup butter
- 1/2 cup dark corn syrup
- 1/4 tsp. kosher salt
- 1 cup lightly salted dry-roasted peanuts
- Wax paper
- 1 (10.5-oz.) package candy-coated peanut butter pieces (such as Reese's Pieces)

1. Preheat oven to 325°. Spread popcorn in an even layer on a lightly greased (with cooking spray) heavy-duty aluminum foil-lined 18- x 13-inch pan. Stir together brown sugar and next 3 ingredients in a small saucepan over medium-low heat; bring to a simmer, and simmer, stirring constantly, 1 minute. Pour over popcorn, and stir gently to coat.

2. Bake at 325° for 25 minutes, stirring every 5 minutes. Add peanuts during last 5 minutes. Remove from oven, and spread on lightly greased (with cooking spray) wax paper. Cool completely (about 20 minutes). Break apart large pieces, and stir in candy pieces. Store in an airtight container up to 1 week.

Note: Cook kernels without oil in the microwave. Place 1/4 cup kernels in a 2 1/2-qt. microwave-safe bowl; completely cover bowl top with a microwave-safe plate. Cook at HIGH 3 to 4 minutes or until kernels have popped. Repeat with remaining kernels.

MAKES about 17 1/2 cups. **HANDS-ON** 30 min., **TOTAL** 50 min.

Spiced with the right ingredients, ground ginger, allspice, and nutmeg,this classic fall drink is transformed into a festive treat. Optional: Serve with rum.
Bring **1 gal. apple cider, 1/3 cup sorghum syrup, 3/4 cup fresh lemon juice, 4 (3-inch) cinnamon sticks, 1 1/2 tsp. ground allspice, 1 tsp. ground ginger,** and **1/2 tsp. ground nutmeg** to a boil in a Dutch oven over high heat; reduce heat to medium, and simmer, stirring occasionally, 15 minutes. Remove cinnamon sticks. Serve hot.

MAKES 17 cups. **HANDS-ON** 5 min., **TOTAL** 25 min.

THE *SL* TEST KITCHEN ACADEMY

BY ROBBY MELVIN, DIRECTOR OF THE SOUTH'S MOST TRUSTED KITCHEN

Does the meat counter seem like a mystery? What's the difference between top and bottom sirloin anyway? We're here to help. Our handy chart shows where each cut comes from, how to cook it best, and the average price per pound.

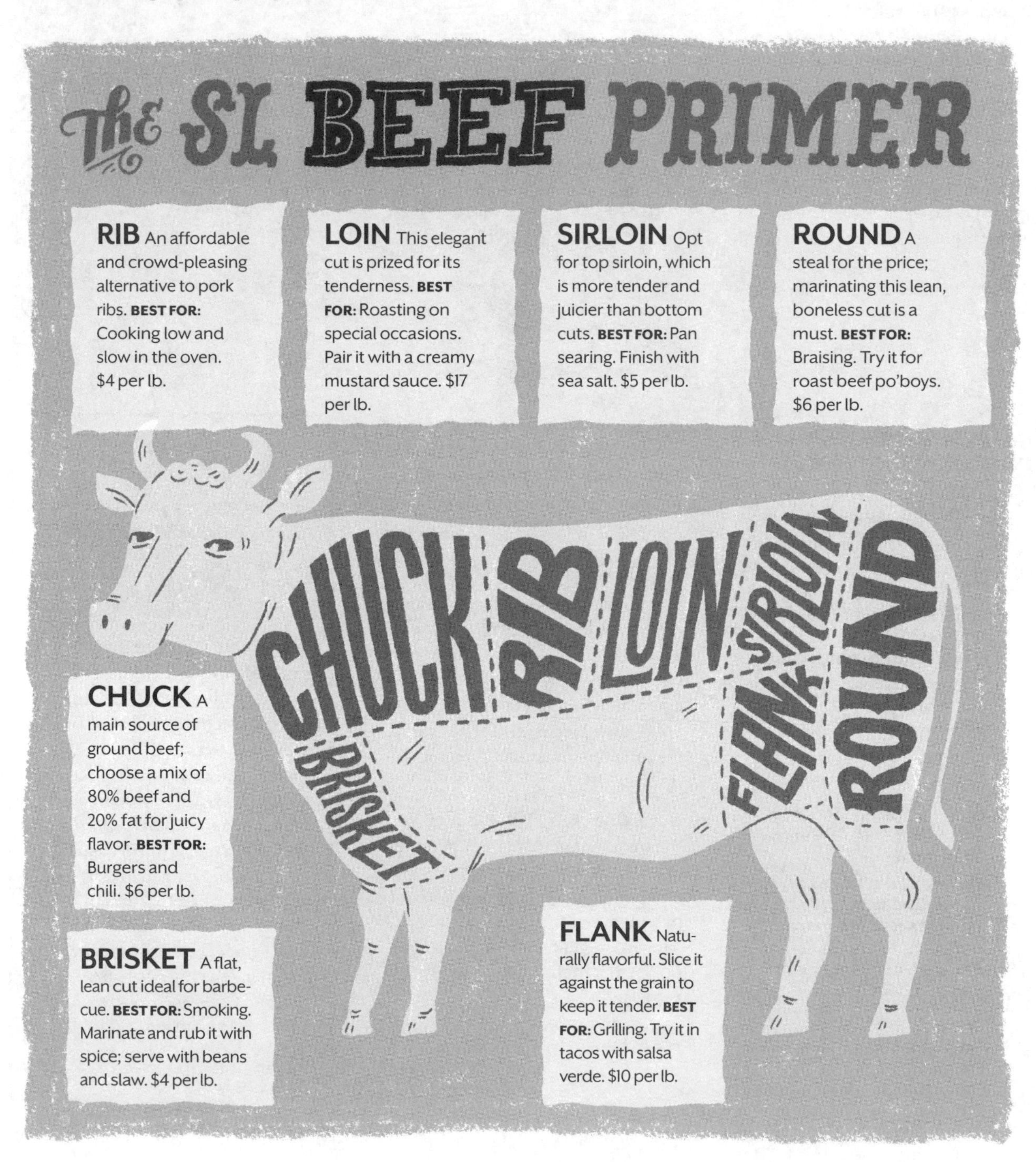

Bologenese Made Easy

STEP UP PASTA NIGHT WITH THIS MEATY SLOW-COOKER SAUCE

SLOW-COOKER BOLOGNESE SAUCE OVER PAPPARDELLE PASTA

Traditionally, Bolognese sauce requires hours of babysitting as it simmers on the stove, but our recipe is virtually hands-off.

- 1 medium-size yellow onion, finely chopped
- 2 celery ribs, finely chopped
- 1 medium carrot, finely chopped
- 1 Tbsp. olive oil
- 2 garlic cloves, minced
- 2 Tbsp. tomato paste
- 2 lb. lean ground beef
- 1 Tbsp. kosher salt
- 2 tsp. sugar
- 2 tsp. dried Italian seasoning
- 1/2 tsp. ground black pepper
- 1/4 tsp. ground nutmeg
- 1 bay leaf
- 1/2 cup red wine
- 2 (28-oz.) cans whole peeled tomatoes
- 2/3 cup heavy cream
- 1 lb. pappardelle pasta or wide fettuccine
- Freshly shaved or grated Parmigiano-Reggiano cheese

1. Sauté first 3 ingredients in hot oil in a large skillet over medium-high heat 8 minutes or until tender. Add garlic and tomato paste; cook, stirring constantly, 30 seconds. Transfer to a 6-qt. slow cooker.
2. Add beef to skillet, and cook, stirring often, 7 minutes or until meat crumbles and is no longer pink; drain. Stir in kosher salt and next 5 ingredients. Stir in wine, and cook, stirring occasionally, 7 minutes or until almost all liquid evaporates. Add to slow cooker, and stir to combine.
3. Drain tomatoes, reserving liquid. Using your hands, crush tomatoes, and break them apart. Add to slow cooker, and stir to combine.
4. Cover and cook on LOW 6 hours. Stir in cream. Check sauce; it should be thick and creamy. (If soupy, uncover and cook 30 more minutes. If dry, stir in reserved canned tomato liquid, 1 Tbsp. at a time, to thin sauce.) Discard bay leaf.
5. Cook pasta according to package directions. Drain well. Toss together pasta and 4 cups sauce. (Reserve remaining sauce for another use.) Spoon into serving dishes, and sprinkle with desired amount of cheese. Serve immediately.

Note: Refrigerate saúce up to 5 days, or freeze for up to 3 months.

MAKES 6 to 8 servings. **HANDS-ON** 40 min.; **TOTAL** 6 hours, 40 min.

Community Cookbook

BUILD A BETTER LIBRARY, ONE GREAT BOOK AT A TIME

Ever since he lied about his age in middle school so he could sign up for a cooking class, Emeril Lagasse has been collecting recipes. All 100 recipes in this book have stories—served with valuable tips, step-by-step tutorials, and a lifetime of lessons in technique.

EASY BARBECUE SHRIMP

- **24 large head-on shrimp (about 2 lb.), peeled, tails left on; shells and heads reserved**
- **1 Tbsp. cracked black peppercorns**
- **2 tsp. Emeril's Creole Seasoning or other Creole seasoning**
- **1 1/2 tsp. chopped fresh rosemary**
- **3 Tbsp. olive oil**
- **1 cup dry white wine**
- **3 cups water**
- **3/4 cup Worcestershire sauce**
- **Juice from 2 lemons (about 1/4 cup)**
- **1/2 cup chopped onion**
- **1 Tbsp. minced garlic**
- **1 Tbsp. hot sauce**
- **1/4 cup (1/2 stick) unsalted butter, cut into 8 pieces**
- **Jalapeño Biscuits**

1. In a medium bowl, toss the shrimp with half of the cracked pepper, 1 tsp. of the Creole seasoning, and the rosemary until evenly coated. Cover and refrigerate until ready to use.
2. Heat 1 Tbsp. of the oil in a 12-inch skillet over high heat. Add the shrimp shells and heads and the remaining 1/2 Tbsp. pepper and 1 tsp. Creole seasoning, and cook, stirring a few times, for 3 minutes. Add the wine, water, Worcestershire, lemon juice, onion, garlic, and hot sauce. Bring to a boil, reduce the heat to a simmer, and let gently bubble for 45 minutes. Strain through a coarse strainer; you should have about 1 cup of barbecue base.
3. Heat a 14-inch skillet over high heat. Add the remaining 2 Tbsp. oil and then the shrimp; cook for 2 minutes, searing on both sides. Pour in the barbecue base, reduce the heat to medium, and simmer until the shrimp are cooked through, about 1 minute. Remove from the heat; whisk in the butter, 1 piece at a time, not adding another until the previous piece is fully incorporated in the sauce.
4. Transfer the shrimp to a serving platter or small individual plates. Spoon the sauce over the shrimp, and serve immediately with the Jalapeño Biscuits.

MAKES 4 first-course servings.
HANDS-ON 45 min.; **TOTAL** 2 hours, 5 min., including biscuits

JALAPEÑO BISCUITS

- **1 cup all-purpose flour**
- **1 tsp. baking powder**
- **1/8 tsp. baking soda**
- **1/4 tsp. kosher salt**
- **1/4 cup cold unsalted butter, cut into small pieces**
- **2 Tbsp. chopped seeded jalapeño pepper**
- **1/4 cup buttermilk, or as needed**

1. Preheat the oven to 375°. Line a baking sheet with parchment paper.
2. Sift the dry ingredients into a small bowl. Work the butter into the flour with a fork until the mixture is crumbly. Stir in the jalapeño. Mix in the buttermilk a little at a time, adding just enough so that it comes together into a smooth ball of dough. Do not overwork the dough.
3. On a lightly floured work surface, roll the dough into a 7-inch round, 1/2 inch thick. Using a 1-inch cookie cutter, press out 12 rounds. Reroll scraps to make additional biscuits. Transfer the rounds to the baking sheet. Bake until the tops are golden and the bottoms browned, about 15 minutes. Serve warm.

MAKES 24 mini biscuits. **HANDS-ON** 20 min., **TOTAL** 35 min.

November

250 **Make-Ahead Thanksgiving** If our Test Kitchen hosted this holiday, this would be the easy-on-the-cook menu served

259 ENTERTAINING WITH JULIA **Host a Gumbo Party** When sweater weather hits, it's time to cook up a roux, make Louisiana's favorite stew, and raise a glass with family and friends

261 **Pecan Delights** From bars and tarts to toffee, pies, and cakes, the pecan proves its worth as a versatile favorite Southern staple

264 HOSPITALITY **Soulful Soups and Stews** Gussied up bowlfuls perfect for entertaining

266 WHAT TO COOK NOW **The Best Pancakes Ever** Foolproof pancake perfection

267 THE *SL* TEST KITCHEN ACADEMY **Our Secrets to the Best Pancakes** Visual cues to take you from batter to golden pancake goodness

268 COMMUNITY COOKBOOK Just in time for fall, over 150 of our favorite recipes for baking from our kitchen to yours

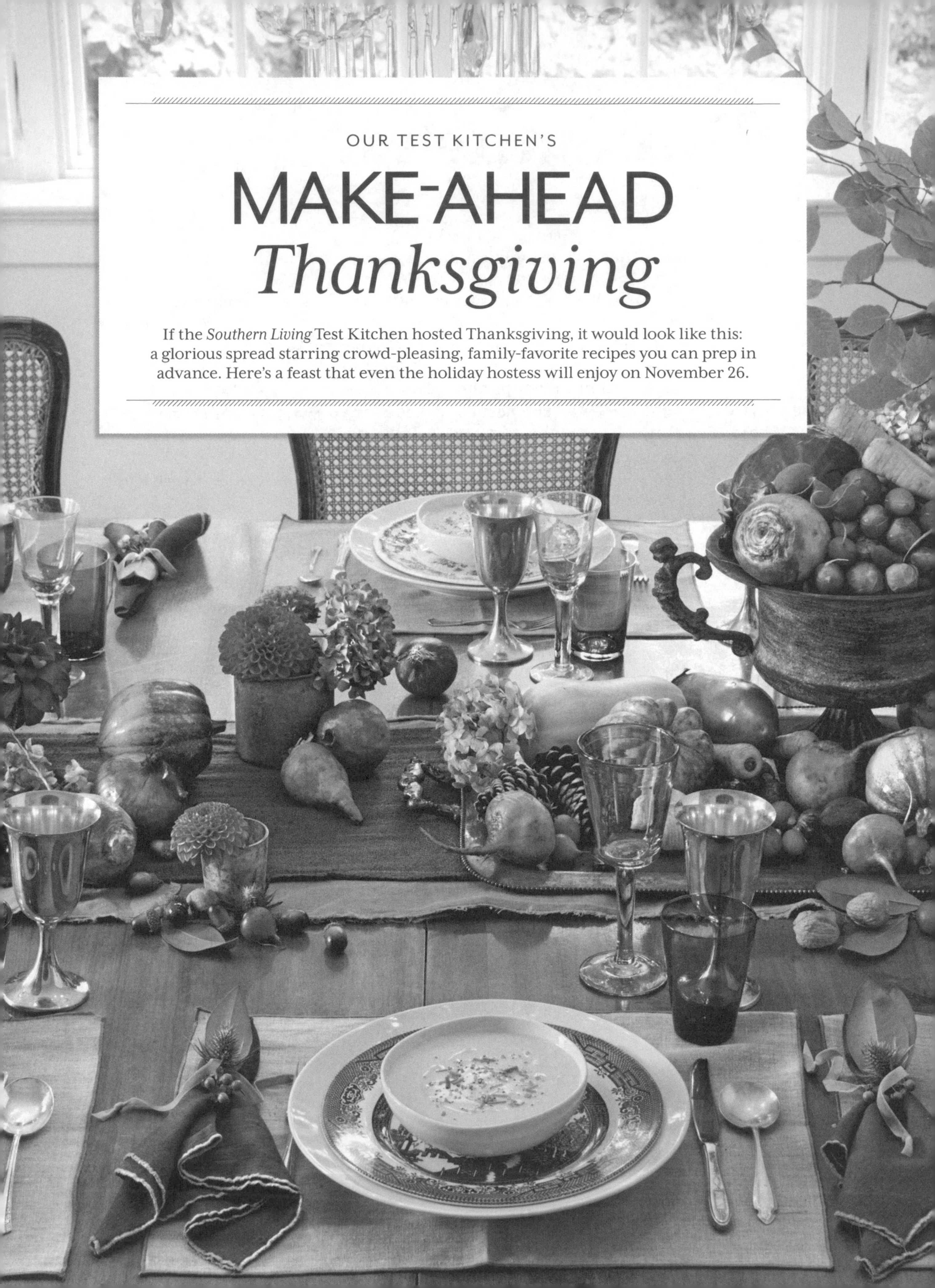

OUR TEST KITCHEN'S

MAKE-AHEAD *Thanksgiving*

If the *Southern Living* Test Kitchen hosted Thanksgiving, it would look like this: a glorious spread starring crowd-pleasing, family-favorite recipes you can prep in advance. Here's a feast that even the holiday hostess will enjoy on November 26.

Oh, the Thanksgiving paradox: a house filled with loved ones who have traveled from near and far to see you, yet your quality time is spent with the stove. You try to seem a calm, cheery hostess, but in truth, you're frazzled. Your mind races with forgotten ingredients, you overmix the mashed potatoes, and the gravy burns. We're here to break the cycle. This year, we've designed every dish on our menu to be made ahead so all you have to do on November 26 is reheat and assemble. As for the grocery list? Don't worry, we've written it all out for you—it's on page 258. A festive, fuss-free Thanksgiving? Now *that's* something to be thankful for.

ONE MONTH AHEAD

SIMPLE CHICKEN STOCK

This easy homemade chicken stock adds loads of fresh flavor to our Make-Ahead Gravy.

- **Carcass and skin from 1 deli-roasted chicken**
- **2 qt. reduced-sodium chicken broth**
- **Optional extras: 2 yellow onions, quartered; 3 celery ribs, chopped; 2 carrots, chopped; 4 to 5 fresh thyme sprigs; 6 to 8 fresh flat-leaf parsley stems**

1. Bring carcass and skin, broth, any or all optional extras, and 4 cups water just to a boil in a stockpot or large Dutch oven over high heat. Reduce heat to low, and simmer 2 to 3 hours or until carcass falls apart and stock tastes like chicken soup. Add water to cover if bones become exposed.
2. Remove and discard large bones. Pour liquid through a fine wire-mesh strainer; discard solids. Cool completely (about 1 hour).
3. Pour into portion-size containers or zip-top plastic freezer bags. Refrigerate up to 1 week or freeze up to 6 months.

Note: We tested with Swanson Natural Goodness 33% Less Sodium Chicken Broth.

MAKES about 6 cups. **HANDS-ON** 5 min.; **TOTAL** 3 hours, 15 min.

MAKE-AHEAD GRAVY

Make Ahead: *Pour cooled gravy into qt.-size zip-top plastic freezer bags; seal and lay bags flat on a baking sheet. Place in freezer. Once frozen, the bags are easily stackable. Thaw three to four days in the fridge before reheating on the stove-top.*

- **1/2 cup butter**
- **1/2 cup chopped yellow onion**
- **1/2 cup all-purpose flour**
- **4 to 5 cups Simple Chicken Stock or reduced-sodium chicken broth, warmed**
- **Kosher salt and freshly ground black pepper**

1. Melt butter in a medium saucepan over medium heat. Add onions, and cook, stirring occasionally, 5 minutes. Sprinkle flour over onions, stirring constantly; cook, stirring constantly, 2 minutes or until flour is golden brown.
2. Gradually whisk in 4 cups stock. Cook, stirring often, 5 minutes or just until mixture comes to a boil and is smooth and thick. (If mixture is too thick, add up to 1 more cup stock, 1/4 cup at a time, until desired consistency is reached.)
3. Season with salt and pepper. Cool completely (about 30 minutes). Cover and store in refrigerator up to 3 days.

Note: To add turkey flavor to gravy, skim fat from drippings in roasting pan, and stir 2 to 3 Tbsp. water into drippings. Bring to a simmer over medium-high heat, stirring and scraping to loosen browned bits from bottom of pan. Pour drippings mixture into warm gravy, and stir gently.

Note: We tested with Swanson Natural Goodness 33% Less Sodium Chicken Broth.

MAKES 5 to 6 cups. **HANDS-ON** 20 min.; **TOTAL** 50 min., not including Simple Chicken Stock

Sides

EASY BUTTER ROLLS

Make Ahead: *After Step 3, shape the dough into 1-inch balls, and place them on a baking sheet. Place baking sheet in freezer, and freeze completely. Transfer frozen balls to a large zip-top plastic freezer bag. The day before baking, place three dough balls in each cup of a lightly greased muffin pan, cover with plastic wrap, and chill until ready to bake as directed.*

- 1 cup milk
- 1 cup butter, divided
- 1 (1/4-oz.) package active dry yeast
- 1/2 cup warm water (100° to 110°)
- 1/2 cup plus 1 tsp. sugar, divided
- 2 large eggs
- 1 tsp. table salt
- 5 cups bread flour
- Vegetable cooking spray

1. Heat milk and 1/2 cup butter in a small saucepan over medium-low heat just until butter melts. Remove from heat, and cool mixture until a thermometer registers 100° to 110° (about 10 minutes).
2. Stir together yeast, 1/2 cup warm water (100° to 110°), and 1 tsp. sugar in a glass measuring cup, and let stand 5 minutes.
3. Beat eggs at medium speed with a heavy-duty electric stand mixer; add remaining 1/2 cup sugar and 1 tsp. salt, beating to combine. Add milk mixture and yeast mixture, beating until combined. Reduce speed to low, and gradually add flour, beating until blended. Place dough in a large lightly greased (with cooking spray) bowl, turning to grease top. Cover with plastic wrap, and chill 8 hours to 5 days.
4. Turn dough out onto a lightly floured surface, and knead 2 or 3 times. Gently shape dough into 72 (1-inch) balls.
5. Place 3 dough balls in each cup of 2 lightly greased (with cooking spray) 12-cup muffin pans. Microwave remaining 1/2 cup butter in a microwave-safe bowl at HIGH 1 minute or until melted. Brush rolls with half of melted butter.
6. Cover muffin pans with plastic wrap, and let rise in a warm place (80° to 85°), free from drafts, 45 minutes to 1 hour or until doubled in bulk.
7. Preheat oven to 375°. Bake rolls for 11 to 13 minutes or until golden brown. Brush with remaining melted butter.

MAKES 2 dozen. **HANDS-ON** 30 min.; **TOTAL** 9 hours, 40 min.

Sorghum Butter

Stir together 1 cup **butter,** softened, and 1/2 cup **sorghum syrup** in a small bowl until blended. Serve immediately, or cover and chill up to 1 month.

MAKES 1 1/2 cups. **HANDS-ON** 5 min., **TOTAL** 5 min.

Herb Butter

Stir together 1 cup **butter,** softened; 2 tsp. chopped **fresh thyme;** 2 tsp. chopped **fresh sage,** and 2 tsp. chopped **fresh flat-leaf parsley** in a small bowl until blended. Serve immediately, or cover and chill up to 1 month.

MAKES about 1 cup. **HANDS-ON** 5 min., **TOTAL** 5 min.

PARSNIP-POTATO SOUP

Make Ahead: *Once it's cooled, freeze this silky soup in zip-top plastic freezer bags using the Make-Ahead Gravy method (page 251). Thaw in refrigerator three to four days before using. Reheat soup over medium-low heat. Add salt and pepper if needed, and garnish just before serving.*

- 3 Tbsp. butter
- 4 fresh thyme sprigs
- Kitchen string
- 5 cups (1 1/2 lb.) peeled and thinly sliced parsnips
- 3 leeks, white and light green parts only, thinly sliced
- 1 russet potato, peeled and chopped
- 1/2 cup thinly sliced celery
- 2 garlic cloves, minced
- 2 1/2 tsp. kosher salt
- 1 1/2 tsp. minced fresh ginger
- 3 cups reduced-sodium chicken broth
- 1/2 cup heavy cream
- Garnishes: sour cream, fresh chives, olive oil

1. Melt butter in a medium Dutch oven over medium heat. Tie thyme sprigs together with kitchen string. Add parsnips, next 6 ingredients, and thyme to Dutch oven, and cook, stirring often, 10 minutes. Stir in chicken broth and 2 cups water, and bring to a boil. Reduce heat to medium-low; cover and simmer 15 minutes.

2. Discard thyme. Process parsnip mixture, in batches, in a blender or food processor until smooth, stopping to scrape down sides as needed. Return mixture to Dutch oven, and stir in heavy cream.

3. Cook over medium heat, stirring often, 5 minutes or until hot. Serve immediately.

Note: We tested with Swanson Natural Goodness 33% Less Sodium Chicken Broth.

MAKES about 2 1/2 qt. **HANDS-ON** 50 min.; **TOTAL** 1 hour, 5 min.

THREE WEEKS AHEAD

HOLIDAY SPICE CAKE

Make Ahead: *Use these cake layers for Cherry-Spice Cake Trifle (page 254). Wrap each cake layer twice with plastic wrap, and then once with aluminum foil. Place in a single layer on a baking sheet, and freeze. Once they are frozen, remove from baking sheet, and stack in freezer.*

- **1 cup butter, softened**
- **1 cup granulated sugar**
- **1 cup firmly packed light brown sugar**
- **4 large eggs**
- **1 ½ cups mashed cooked sweet potatoes**
- **2 tsp. vanilla extract**
- **3 cups all-purpose flour**
- **1 Tbsp. baking powder**
- **1 tsp. ground cinnamon**
- **½ tsp. baking soda**
- **½ tsp. table salt**
- **½ tsp. ground ginger**
- **¼ tsp. ground nutmeg**
- **1 cup buttermilk**
- **Shortening**

1. Preheat oven to 350°. Beat butter at medium speed with a heavy-duty electric stand mixer until creamy. Gradually add granulated sugar and light brown sugar, and beat until light and fluffy. Add eggs, 1 at a time, beating just until blended after each addition. Add sweet potatoes and vanilla, beating just until blended.

2. Stir together flour and next 6 ingredients. Add to butter mixture alternately with buttermilk, beginning and ending with flour mixture. Beat at low speed just until blended after each addition. Spoon batter into 2 greased (with shortening) and floured 9-inch round, 2-inch-deep cake pans.

3. Bake at 350° for 35 to 40 minutes or until a wooden pick inserted in center comes out clean. Cool in pans on a wire rack 10 minutes; remove from pans to wire rack, and cool completely (about 1 hour).

MAKES 2 layers. **HANDS-ON** 30 min.; **TOTAL** 2 hours, 45 min.

GRANDMOTHER CARTER'S CORNBREAD DRESSING

Make Ahead: *Freeze the unbaked dressing mixture in 2 (1-gal.) zip-top plastic freezer bags, making sure to press out all the excess air. Thaw in refrigerator five days before Thanksgiving, and continue recipe with Step 4.*

- **14 cups reduced-sodium chicken broth, divided**
- **2 cups chopped celery**
- **2 cups chopped sweet onion**
- **Cornbread Crumbles (page 254)**
- **1 (14-oz.) package herb-seasoned stuffing mix**
- **2 cups cooked long-grain rice**
- **8 large eggs, lightly beaten**
- **2 Tbsp. chopped fresh sage**
- **2 Tbsp. poultry seasoning**
- **2 tsp. freshly ground black pepper**
- **1 1/2 tsp. table salt**
- **Vegetable cooking spray**

1. Bring 8 cups chicken broth to a boil in a large Dutch oven over medium-high heat; stir in chopped celery and onion. Reduce heat to medium, and simmer 20 to 30 minutes or until celery and onion are very tender. Remove from heat, and stir in 4 cups chicken broth. Cool 30 minutes.

2. Meanwhile, stir together Cornbread Crumbles and next 2 ingredients in a very large bowl.

3. Preheat oven to 350°. Stir broth mixture, eggs, and next 4 ingredients into cornbread mixture until thoroughly combined. If needed, stir in up to 2 more cups chicken broth, 1/4 cup at a time, until mixture is slightly soupy.

4. Spoon mixture into 2 (13- x 9-inch) lightly greased (with cooking spray) baking dishes. Bake at 350° for 50 to 60 minutes or until done and top is light brown.

Note: We tested with Swanson Natural Goodness 33% Less Sodium Chicken Broth.

MAKES 10 to 12 servings per baking dish. **HANDS-ON** 20 min.; **TOTAL** 3 hours, 20 min., including Cornbread Crumbles

Grandmother Carter's Cornbread Dressing

Cornbread Crumbles

- 3 cups self-rising white cornmeal mix
- 1 cup all-purpose flour
- 2 Tbsp. sugar
- 3 cups buttermilk
- 3 large eggs, lightly beaten
- 1/2 cup butter, melted
- Vegetable cooking spray

1. Preheat oven to 425°. Stir together cornmeal mix, flour, and sugar in a large bowl; whisk in buttermilk, eggs, and butter. Pour batter into a lightly greased (with cooking spray) 13- x 9-inch baking pan.
2. Bake at 425° for 30 minutes or until golden brown. Remove from oven, invert onto a wire rack, and cool completely (about 30 minutes). Crumble cornbread.

MAKES 16 cups crumbles. **HANDS-ON** 10 min.; **TOTAL** 1 hour, 10 min.

TWO WEEKS AHEAD

MASHED POTATOES WITH GREENS

Make Ahead: *Prepare this recipe through Step 4, and freeze. Thaw potato mixture in the fridge at least 24 hours before you plan to serve, and let it stand 20 minutes before baking as directed.*

- 1/2 cup butter, softened and divided
- 6 cups shredded kale, chard, cabbage, or other leafy greens
- 1/2 cup thinly sliced green onions, white and light green parts only
- 1/4 cup reduced-sodium chicken broth
- Kosher salt and freshly ground black pepper to taste
- 4 lb. small russet or Yukon gold potatoes, peeled and cut into 2-inch pieces
- 1 Tbsp. kosher salt
- 4 oz. cream cheese, softened
- 1 cup milk

1. Melt 2 Tbsp. butter in a large stockpot over medium-high heat. Add kale and green onions; stir to coat. Add broth; cover and cook, stirring often, 10 minutes or until tender. Add salt and pepper to taste. Transfer to a bowl; cover to keep warm.
2. Bring potatoes, 1 Tbsp. salt, and water to cover to a boil in stockpot over high heat. Reduce heat to medium, and simmer 20 minutes or just until potatoes are tender. Drain potatoes, and let stand 3 minutes or until dry. Return to stockpot. Mash with a potato masher until smooth; stir in cream cheese and 4 Tbsp. butter. Fold in kale mixture.
3. Microwave milk in a microwave-safe measuring cup at HIGH 1 to 2 minutes or until warm. Stir 1/2 cup warm milk into potato mixture. Add up to 1/2 cup more milk, 1 Tbsp. at a time, and stir until mixture thickens. (Mixture will firm up as it chills in Step 4.) Season with salt and pepper.
4. Transfer mixture to a greased (with butter) 2 1/2-qt. gratin dish or baking dish. Dot with remaining 2 Tbsp. butter. Cover dish tightly with plastic wrap, and then with aluminum foil. Chill 8 hours to 5 days, or freeze up to 2 weeks.
5. Preheat oven to 350°. Remove plastic wrap and foil, and bake 30 minutes or until thoroughly heated. Serve warm.

Note: We tested with Swanson Natural Goodness 33% Less Sodium Chicken Broth.

MAKES 8 to 10 servings. **HANDS-ON** 30 min.; **TOTAL** 9 hours, 30 min.

THREE DAYS AHEAD

CHERRY-SPICE CAKE TRIFLE

Make Ahead: *Thaw Holiday Spice Cake layers (page 253), and prepare the recipe through Step 3. Assemble trifle the morning of Thanksgiving, and refrigerate. Remove 1 hour before serving.*

FILLING

- 1 (13-oz.) jar cherry preserves
- 3/4 cup granulated sugar
- 1/4 cup fresh orange juice
- 4 cups fresh or frozen cranberries, divided

CUSTARD

- 1 cup firmly packed light brown sugar
- 5 Tbsp. cornstarch
- 1/4 tsp. table salt
- 3 1/2 cups milk
- 1 1/2 cups heavy cream
- 1/4 cup butter, cut into pieces
- 1 tsp. vanilla bean paste
- 1/4 tsp. ground cinnamon
- 1/8 tsp. ground nutmeg

CAKE

- Holiday Spice Cake (page 253), cut into 1-inch cubes

TOPPING

- 1 tsp. vanilla extract
- 2 cups heavy cream
- 6 Tbsp. powdered sugar

1. Prepare Filling: Bring first 3 ingredients and 3 cups cranberries to a boil in a saucepan over medium-high heat; reduce heat to low, and boil, stirring often, 5 to 6 minutes or until berries begin to pop. Remove from heat, and stir in remaining 1 cup cranberries. Transfer mixture to a bowl; cool completely (about 30 minutes). Cover and chill 8 to 24 hours.
2. Meanwhile, prepare Custard. Whisk together brown sugar, cornstarch, and salt in a large heavy saucepan; whisk in milk and 1 1/2 cups cream. Bring mixture to a boil over medium heat, whisking constantly. Boil 1 minute, whisking constantly. Remove from heat.
3. Whisk in butter and next 3 ingredients. Transfer mixture to a medium bowl, and place plastic wrap directly on mixture (to prevent a film from forming). Cool completely (about 30 minutes). Chill 8 to 24 hours.
4. Assemble trifle. Layer about one-third of cake cubes in a 4-qt. bowl. Top with one-third each cranberry filling and custard. Repeat layers twice.
5. Prepare Topping: Beat vanilla and 2 cups heavy cream at medium-high speed until foamy; gradually add powdered sugar, beating until soft

The Turkey

Dry-Brined-and-Marinated Smoked Turkey, page 256

peaks form. Dollop over trifle. Serve immediately, or chill up to 8 hours.

MAKES 15 to 20 servings. **HANDS-ON** 45 min.; **TOTAL** 9 hours, 15 min.

ROASTED VEGETABLE SALAD WITH APPLE CIDER VINAIGRETTE

Make Ahead: *Store roasted vegetables in a zip-top plastic freezer bag or an airtight container in the fridge. Before serving, return to room temperature and check seasoning.*

- **1 lb. parsnips, peeled and cut lengthwise**
- **1 lb. carrots, peeled and cut lengthwise**
- **1 lb. small golden beets, peeled and coarsely chopped**
- **10 to 12 garlic cloves**
- **1 cup frozen pearl onions, thawed**
- **1 lb. small Brussels sprouts, trimmed and halved**
- **3 fresh rosemary or thyme sprigs**
- **3 small bay leaves**
- **3 Tbsp. butter, melted**
- **1 1/2 Tbsp. olive oil**
- **Kosher salt and freshly ground black pepper**
- **Apple Cider Vinaigrette**
- **1 head radicchio, separated into leaves**

1. Preheat oven to 425°. Divide first 8 ingredients between 2 aluminum foil-lined jelly-roll pans. Drizzle with butter and oil; toss to coat. Spread vegetables in a single layer in each pan, leaving about 1 inch between pieces. Season with salt and pepper.
2. Bake both pans at 425° for 20 minutes, placing 1 pan on middle oven rack and 1 pan on lower oven rack. Rotate pans front to back, and top rack to bottom rack. Bake 20 to 25 more minutes or until vegetables are tender.
3. Gently loosen vegetables, and add salt and pepper to taste. Cool completely (about 20 minutes). Discard herb sprigs and bay leaves. Place vegetables in a zip-top plastic freezer bag, and refrigerate 2 hours to 2 days.
4. To serve, let vegetables stand 20 minutes or until room temperature. Add 1/4 cup Apple Cider Vinaigrette; toss to coat.
5. Arrange radicchio leaves on a serving platter; top with roasted vegetables. Drizzle 1/4 cup vinaigrette over salad. Season with salt and pepper. Serve salad with remaining vinaigrette.

MAKES 8 to 10 servings. **HANDS-ON** 30 min.; **TOTAL** 3 hours, 55 min., including vinaigrette

Apple Cider Vinaigrette

Make Ahead: *Store vinaigrette, covered, in the fridge. Let stand 10 minutes or until room temperature. Shake well, and check seasoning before using.*

- **3/4 cup extra virgin olive oil**
- **1/4 cup apple cider**
- **1/4 cup apple cider vinegar**
- **2 Tbsp. finely chopped shallot**
- **1 Tbsp. whole grain Dijon mustard**
- **1 Tbsp. honey**
- **1 1/2 tsp. kosher salt**
- **1 tsp. fresh thyme leaves**
- **1/2 tsp. freshly ground black pepper**

Combine all ingredients in a glass jar with a tight-fitting lid. Cover with lid, and shake well. Shake jar again just before serving.

MAKES about 1 1/4 cups. **HANDS-ON** 5 min., **TOTAL** 5 min.

Dessert

TWO DAYS AHEAD

PUMPKIN-ESPRESSO TIRAMISÙ

Make Ahead: *Loosely wrap baked and cooled tiramisù in plastic wrap, and refrigerate until ready to serve.*

- 1 cup granulated sugar
- 1 Tbsp. all-purpose flour
- 2 large eggs
- 2 large egg yolks
- 3 cups heavy cream, divided
- 1 tsp. vanilla bean paste
- 2 (8-oz.) packages cream cheese, softened
- 1 cup canned pumpkin
- 1/2 cup brewed espresso or dark roast coffee
- 3 Tbsp. brandy
- 3 (7-oz.) packages crisp ladyfingers, divided
- 3 Tbsp. powdered sugar

Garnish: ground nutmeg

1. Whisk together sugar and flour in a large heavy saucepan. Whisk together eggs, egg yolks, and 2 cups cream in a bowl. Whisk cream mixture into sugar mixture, and cook over medium-low heat, whisking constantly, 15 minutes or until very thick. (Mixture will come to a simmer during the last 2 to 3 minutes.) Remove from heat; whisk in vanilla bean paste, and transfer to a medium bowl. Place plastic wrap directly on warm custard (to prevent a film from forming). Cool completely (about 1 1/2 hours).
2. Whisk cream cheese into cream mixture until smooth; whisk in pumpkin.
3. Stir together espresso and brandy. Brush flat sides of about 24 ladyfingers with espresso mixture. Stand ladyfingers around edge of a 10-inch springform pan, placing rounded sides against pan. Line bottom of pan with additional ladyfingers, cutting if necessary to cover bottom of pan completely. Brush espresso mixture over ladyfingers on bottom of pan.
4. Spread one-third of cream cheese mixture over ladyfingers on bottom. Repeat layers twice with remaining ladyfingers, espresso mixture, and cream cheese mixture, ending with cream cheese mixture. (Reserve any remaining ladyfingers for another use.) Cover and chill 4 hours to 2 days.
5. Beat remaining 1 cup cream at high speed with an electric mixer until foamy; gradually add powdered sugar, beating until soft peaks form. Dollop over cream cheese mixture.

MAKES 10 to 12 servings. **HANDS-ON** 45 min.; **TOTAL** 6 hours, 15 min.

DRY-BRINED-AND-MARINATED SMOKED TURKEY

Make Ahead: *Dry-brine the turkey two days before Thanksgiving, and smoke it the next day. Cut the turkey into pieces (legs, breast, wings, etc.), and place in a zip-top plastic freezer bag. Store in refrigerator. About an hour before dinner, place turkey pieces in a single layer on a baking sheet, and cover with aluminum foil. Warm in 300° oven until heated.*

- 1 (12- to 14-lb.) whole fresh or frozen, thawed, turkey
- Turkey Dry Rub
- 1 cup apple cider
- 2 Tbsp. dark brown sugar
- 1 Tbsp. kosher salt
- 3 cups hickory wood chips

1. Pat turkey dry with paper towels. Reserve 2 Tbsp. Turkey Dry Rub; sprinkle remaining dry rub over entire turkey, rubbing into skin. Chill turkey 10 to 24 hours.
2. Prepare smoker according to manufacturer's instructions, bringing internal temperature to 250° to 260°; maintain temperature 15 to 20 minutes.
3. Stir together apple cider, brown sugar, salt, and reserved 2 Tbsp. Turkey Dry Rub. Attach needle to marinade injector, and fill with cider mixture according to package directions. Inject top of turkey, thighs, and legs at 1-inch intervals with cider mixture. Add wood chips to coals, and smoke turkey, maintaining temperature inside smoker between 250° and 260°, for 5 to 6 hours or until a meat thermometer inserted into thickest portion registers 160°.
4. Remove turkey, cover loosely with aluminum foil, and let stand 10 to 15 minutes before slicing.

MAKES 10 to 12 servings. **HANDS-ON** 30 min.; **TOTAL** 15 hours, 40 min., including dry rub

Turkey Dry Rub

Stir together 2 Tbsp. **kosher salt,** 2 Tbsp. **dark brown sugar,** 2 tsp. **freshly ground black pepper,** 2 tsp. **smoked paprika,** and 1/2 tsp. **garlic powder** together in a small bowl.

MAKES about 1/3 cup. **HANDS-ON** 5 min., **TOTAL** 5 min.

THANKSGIVING DAY

SLOW-COOKER SWEET POTATOES WITH BACON

Make Ahead: *Pop this sweet potato mixture into your slow cooker on Thanksgiving morning. It will be ready to serve as soon as everyone sits down to dinner.*

- **4 lb. slender sweet potatoes, peeled and cut into 1-inch-thick slices**
- **1/2 cup frozen orange juice concentrate, thawed**
- **4 Tbsp. butter, melted**
- **3 Tbsp. light brown sugar**
- **2 tsp. kosher salt**
- **2 tsp. chopped fresh rosemary**
- **2 tsp. cornstarch**
- **1 Tbsp. cold water**
- **1/2 cup loosely packed fresh flat-leaf parsley leaves, finely chopped**
- **1 Tbsp. firmly packed orange zest**
- **2 garlic cloves, minced**
- **3 cooked bacon slices, crumbled**

1. Place sweet potatoes in a 5- to 6-qt. slow cooker. Stir together orange juice concentrate and next 4 ingredients in a small bowl. Pour over sweet potatoes, tossing to coat.
2. Cover and cook on LOW for 5 1/2 to 6 hours or until potatoes are tender.
3. Transfer potatoes to a serving dish, using a slotted spoon. Increase slow cooker to HIGH. Whisk together cornstarch and 1 Tbsp. cold water until smooth. Whisk cornstarch mixture into cooking liquid in slow cooker. Cook, whisking constantly, 3 to 5 minutes or until sauce thickens. Spoon sauce over potatoes.
4. Stir together parsley, orange zest, and garlic. Sprinkle potatoes with parsley mixture and crumbled bacon.

MAKES 8 servings. **HANDS-ON** 30 min., **TOTAL** 6 hours

THE TABLETOP

❶ THE NEW CORNUCOPIA

Two weeks ahead, shop for a variety of fall produce like purple cabbage, turnips, beets, pomegranates, radishes, and parsnips. Cut off any green tops, and then wash and dry produce. Brush lightly with olive oil. Refrigerate until ready to use.

One day before, place a large tray in the center of your table. Add a small footed urn or planter slightly off center for an asymmetrical, more organic design. Place crumpled aluminum foil in urn to add height to your arrangement. Start by filling urn and tray with the biggest items, then arrange smaller pieces around them. Group smaller fruits and veggies together for more impact. Add flowers, if desired, to urn, and use wooden florist picks to secure items as needed. Brush produce with a mix of equal parts water and olive oil.

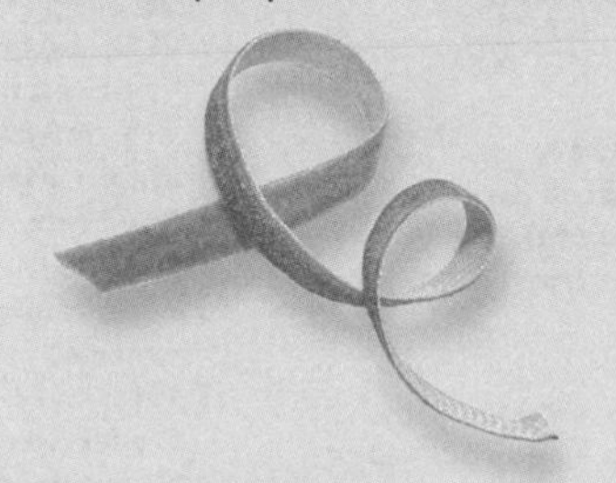

❷ NAPKIN RING

Two to four days before, iron each napkin into a folded square, using starch. Loosely fold into a triangle by holding the corner and letting napkin fall into a natural point. Gently gather napkin and tie with thin velvet ribbon. Tuck a magnolia leaf, thistle blossom, and sprig of hypericum berries into each.

❸ GRATITUDE JAR

Four days before, buy thistle blossoms, a dahlia blossom, hypericum berries, hydrangeas, and a metallic paint pen. Gather magnolia leaves and one or two pinecones. Wrap top of a large widemouthed jar with ribbon. Using paint pen, write "gratitude" on a magnolia leaf, and punch with a hole puncher; thread with ribbon, and secure it to jar. Tuck in pinecones, thistle, blossoms, berries, and other natural items as desired.

Before your Thanksgiving meal, invite guests to write what they're thankful for on magnolia leaves and drop them into the jar. Read them aloud at the table.

The Southern Living *Make-Ahead*

GROCERY GAME PLAN

RIP OUT THIS PAGE, and take it with you on your weekly grocery shopping trip. It provides the exact amounts needed or a standard ingredient size for each recipe. Before you go, read recipe ingredients carefully, and cross off what you already have. Don't forget to stock up on plenty of pantry staples (listed below).

Tip: Even if you don't make our exact menu, try this easy-to-follow format. Our Test Kitchen faithfully organizes our shopping lists by areas of the grocery store: a refrigerated section for proteins, dairy, and frozen goods; a produce area that includes fruits, vegetables, and herbs; and a pantry group for everything else. You can efficiently shop for all of your necessary ingredients at once, rather than scrambling back and forth amid holiday shoppers.

PANTRY STAPLES

- ☐ **Reduced-sodium chicken broth (We prefer Swanson Natural Goodness 33% Less Sodium Chicken Broth.)**
- ☐ **Granulated sugar**
- ☐ **Brown sugar (dark and light)**
- ☐ **Powdered sugar**
- ☐ **All-purpose flour**
- ☐ **Olive oil**
- ☐ **Extra virgin olive oil**
- ☐ **Vegetable cooking spray**
- ☐ **Shortening**
- ☐ **Black pepper, kosher salt, table salt**
- ☐ **Kitchen string**

1 MONTH AHEAD

Refrigerated:

- ☐ **1 deli-roasted chicken**
- ☐ **2 lb. butter**
- ☐ **1/2 cup heavy cream**
- ☐ **1 cup milk**
- ☐ **1/2 dozen large eggs**

Produce:

- ☐ **3 yellow onions**
- ☐ **1 bunch celery**
- ☐ **2 large carrots**
- ☐ **1 bunch fresh flat-leaf parsley**
- ☐ **1 bunch fresh thyme**
- ☐ **1 bunch fresh sage**
- ☐ **1 1/2 lb. parsnips**
- ☐ **1 russet potato**
- ☐ **3 leeks**
- ☐ **1 garlic bulb**
- ☐ **2-inch piece fresh ginger**

Pantry:

- ☐ **1 package active dry yeast**
- ☐ **5 cups bread flour**
- ☐ **1/2 cup sorghum syrup**

3 WEEKS AHEAD

Refrigerated:

- ☐ **1 1/2 dozen large eggs**
- ☐ **1 qt. buttermilk**
- ☐ **1 lb. butter**

Produce:

- ☐ **2 large sweet onions**
- ☐ **1 1/4 lb. sweet potatoes**
- ☐ **1 bunch celery**
- ☐ **1 bunch fresh sage**

Pantry:

- ☐ **1 package herb-seasoned stuffing mix**
- ☐ **1 cup long-grain rice**
- ☐ **1 small jar poultry seasoning**
- ☐ **3 cups self-rising white cornmeal mix**
- ☐ **1 bottle vanilla extract**
- ☐ **1 container baking soda**
- ☐ **1 container baking powder**
- ☐ **1 small jar ground nutmeg**
- ☐ **1 small jar ground cinnamon**
- ☐ **1 small jar ground ginger**

2 WEEKS AHEAD

Refrigerated:

- ☐ **1/2 cup butter**
- ☐ **4 oz. cream cheese**
- ☐ **1 cup milk**

Produce:

- ☐ **6 cups shredded kale, cabbage, or other leafy greens**
- ☐ **1 bunch green onions**
- ☐ **4 lb. small russet or Yukon gold potatoes**

1 WEEK AHEAD

Refrigerated:

- ☐ **3 1/2 cups milk**
- ☐ **6 1/2 cups heavy cream**
- ☐ **1/2 lb. butter**
- ☐ **1/2 dozen large eggs**
- ☐ **2 (8-oz.) packages cream cheese**
- ☐ **Sour cream (for garnish)**
- ☐ **1 cup frozen pearl onions**
- ☐ **1 (12- to 14-lb.) fresh or frozen turkey**
- ☐ **1/2 cup frozen orange juice concentrate**
- ☐ **3 bacon slices**

Produce:

- ☐ **4 cups fresh or frozen cranberries**
- ☐ **2 oranges**
- ☐ **1 lb. parsnips**
- ☐ **1 lb. carrots**
- ☐ **1 lb. small golden beets**
- ☐ **1 lb. Brussels sprouts**
- ☐ **4 lb. slender sweet potatoes**
- ☐ **1 head radicchio**
- ☐ **2 garlic bulbs**
- ☐ **2 shallots**
- ☐ **1 bunch fresh thyme**
- ☐ **1 bunch fresh rosemary**
- ☐ **1 bunch fresh flat-leaf parsley**
- ☐ **1 bunch fresh chives (for garnish)**
- ☐ **1 small jar bay leaves**

Pantry:

- ☐ **3 (7-oz.) packages crisp ladyfingers**
- ☐ **1 (13-oz.) jar cherry preserves**
- ☐ **1 cup canned pumpkin**
- ☐ **1 (12-oz.) package espresso or dark roasted coffee**
- ☐ **1 small container cornstarch**
- ☐ **1 jar vanilla bean paste**
- ☐ **1 small jar smoked paprika**
- ☐ **1 small jar garlic powder**
- ☐ **1 1/4 cups apple cider**
- ☐ **1/4 cup apple cider vinegar**
- ☐ **1 Tbsp. honey**
- ☐ **1 Tbsp. whole grain Dijon mustard**

Miscellaneous:

- ☐ **3 Tbsp. brandy**
- ☐ **3 cups hickory wood chips**

Host a Gumbo Party

JULIA REED SHARES HER PRE-TURKEY DAY TRADITION, ALONG WITH HER OWN SEAFOOD-LADEN RECIPE

IN MUCH OF LOUISIANA where I now live, "gumbo weather" is not a cutesy term. It's an actual season, beginning sometime in November when the temperature finally starts to drop and continuing at least through the Super Bowl, a huge gumbo day. (When I found an enormous pearl from an oyster in a seafood gumbo I'd made for Super Bowl XLIV in 2010, we put it on top of the TV as a talisman. I'm still convinced that's why the Saints won for the first time ever, but I digress.) There are "gumbo weather" cookbooks and blogs; there's even a folk blues band by that name in Texas. But for most of us, the words denote that joyous period when you can open the doors, put a pot on the stove, and maybe even wear a sweater.

The great characteristic of gumbo is that it's an infinitely flexible culinary art form—there are as many versions as there are people who make them. Cajun gumbos feature a roux so dark it all but hides the seemingly mysterious ingredients, giving rise to the ubiquitous newspaper description of Louisiana politics as "a roiling gumbo of corruption." New Orleans chef Donald Link makes a roux using 4 cups of flour and 3 cups of oil as a base for the seafood gumbo in his James Beard Award-winning cookbook, *Real Cajun*. Most Creole versions, by contrast, are far lighter, with a base that includes no more than a couple of tablespoons of oil and flour, and they often feature tomatoes and okra.

The dish, named either for the Bantu word for okra *(ki ngombo)* or the Choctaw word for filé (*kombo*, the powder made from dried sassafras leaves that often serves as a thickener), originated in the 18th century and is said to have been inspired by everything from the French bouillabaisse to West African stews. Whatever, it's now its own delicious thing, rightly described as "an original conception" in the 1901 *Picayune's Creole Cook Book*, which also refers to the "occult science of making a good 'Gombo a la Creole.' " In actuality, it's not that hard. A dark roux requires more patience than skill, but the version I include here (page 260) takes little time. This gumbo also features tomatoes, okra, and the de rigueur trio of chopped onion, bell pepper, and celery.

The gumbo pot is an excellent place to utilize leftover Thanksgiving or Christmas turkey, but during the holidays, I like to give folks a break from all that fowl with a Creole seafood gumbo that utilizes the area's bounty. I throw in andouille, which adds deep flavor to the stock as well as a Cajun touch. The Cajuns often serve their gumbos with a scoop of potato salad, but I serve mine over rice with toasty French bread and a green salad. In my house, the menu has become a pre-Thanksgiving tradition, as well as a super-easy dinner to keep on the stove and serve up as out-of-town visitors trickle in. For dessert, the most fitting choice is a classic New Orleans bread pudding, inspired by the one served at the beloved Bon Ton Café. For it, I replaced their raisins with pecans and added orange zest because citrus is in season. My great pal Peter Patout, whose magical French Quarter courtyard was the setting for our party here, let me use his aunt's famous preserved kumquats to decorate our plates, but you can make your own garnish by simmering sliced kumquats in a simple syrup for 20 minutes. The best news is the pudding takes even less time than the gumbo, and you can make it ahead. (Get my recipe at *southernliving.com/breadpudding;* for gumbo, turn the page.)

Simmering andouille with the stock lends the base a tasty depth; adding the seafood at the last minute means it's perfectly cooked.

JULIA'S GUIDE TO GUMBO

THE GENIUS OF THIS MENU IS THAT MOST OF THE WORK CAN BE DONE AHEAD OF TIME

THE RECIPE

BEST EVER SEAFOOD GUMBO

- 1 lb. fresh lump crabmeat
- 3 lb. medium-size raw shrimp (with heads)
- Pinch of kosher salt
- 6 Tbsp. oil, divided
- 2 lb. frozen sliced okra, thawed*
- 1/2 tsp. kosher salt
- 1 lb. andouille sausage, thinly sliced
- 2 Tbsp. all-purpose flour
- 2 cups finely chopped yellow onion
- 1 cup finely chopped celery
- 1 cup finely chopped green bell pepper
- 4 garlic cloves, minced
- 1 cup chopped green onions, divided
- 1 (6-oz.) can tomato paste
- 3 bay leaves
- 1 tsp. finely chopped fresh thyme
- 2 tsp. kosher salt
- 1 tsp. hot sauce (such as Tabasco)
- 1/4 tsp. ground red pepper
- 1/2 tsp. ground black pepper
- 1 Tbsp. Worcestershire sauce
- 1 (14.5-oz.) can whole tomatoes, drained and coarsely chopped, liquid reserved
- 1 pt. oysters (between 2 dozen and 3 dozen)
- 1/4 cup finely chopped fresh flat-leaf parsley
- Hot cooked long-grain rice

1. Pick crabmeat, removing any bits of shell.
2. Remove shrimp shells and heads. Cover and refrigerate shrimp, and place shells and heads in a large stockpot. Add 8 cups water and a generous pinch of salt, and bring mixture to a boil over high heat. Reduce heat to medium-low, and simmer, partially covered, 1 hour. Cool 30 minutes; pour stock through a fine wire-mesh strainer, and reserve for use in Step 4. Discard shells and heads.
3. Heat 2 Tbsp. oil in a large heavy skillet over medium heat, and add okra. Sprinkle with 1/2 tsp. salt, and sauté 10 minutes. Set aside okra, and wipe out skillet. Heat another tablespoon of oil in skillet over medium-high heat, and sauté andouille about 8 minutes or until brown. Set aside.
4. Heat remaining 3 Tbsp. oil in a large heavy pot over high heat. Add flour, reduce heat to medium, and cook, stirring constantly, about 10 minutes or until roux is a medium-to-dark brown. Add yellow onions, celery, and bell pepper, and sauté 4 to 5 minutes or until vegetables begin to soften. Add garlic and 1/2 cup green onions, and cook 3 more minutes. Stir in tomato paste and next 7 ingredients. Stir in tomatoes and 1/3 cup of their liquid; gradually stir in shrimp stock. Add reserved sausage and okra. Bring to a boil over high heat, and cover. Reduce heat to medium-low, and simmer 30 minutes.
5. Stir in shrimp and remaining green onions, and simmer 3 minutes. Stir in crabmeat, oysters, and parsley, and simmer 1 minute. Remove from heat. Taste and add salt, if desired. Serve over rice.

* Fresh okra may be substituted for frozen.

MAKES about 5 qt. **HANDS-ON** 1 hour, 35 min.; **TOTAL** 2 hours, 25 min.

Drink Pairings

A light red wine would be a fine choice for this meal, but my go-to for all things seafood, including gumbo, is a Muscadet, particularly the one from Domaine de la Pépière that we served (page 259). With the bread pudding, we drank Peter's homemade calamondin (citrus) liqueur, but Cointreau makes a fine substitute. Or bring out your finest cognac.

Pecan Delights

FROM POUND CAKE TO PRALINE CREAM PUFFS, WE'RE GRATEFUL TO HAVE SEVEN NEW TREATS STARRING THE SOUTH'S FAVORITE NUT

PUMPKIN-PECAN STREUSEL PIE

Create the chunky texture by leaving those extra-big blueberry-size pieces in the streusel mix.

- 3 Tbsp. powdered sugar
- 1/2 (14.1-oz.) package refrigerated piecrusts
- 3 Tbsp. finely chopped toasted pecans
- Wax paper
- Vegetable cooking spray
- 1 (15-oz.) can pumpkin
- 1 1/4 cups half-and-half
- 1/2 cup firmly packed light brown sugar
- 3 large eggs, lightly beaten
- 1 Tbsp. all-purpose flour
- 1 tsp. pumpkin pie spice
- 1/4 tsp. table salt
- Pecan Streusel Topping
- Garnish: toasted pecan halves

1. Preheat oven to 350°. Sprinkle work surface with powdered sugar. Unroll piecrust, and place on prepared surface. Sprinkle piecrust with chopped pecans. Place wax paper over piecrust and pecans, and lightly roll pecans into crust. Fit piecrust, pecan side up, in a lightly greased (with cooking spray) 9-inch pie plate; fold edges under, and crimp.

2. Whisk together pumpkin and next 6 ingredients in a large bowl until smooth and well incorporated; pour into prepared piecrust.

3. Bake at 350° for 45 minutes, shielding edges with aluminum foil after 20 minutes, if needed, to prevent excessive browning. Carefully top pie with Pecan Streusel Topping. Reduce oven temperature to 325°, and bake 25 minutes, covering pie with aluminum foil after 5 to 10 minutes to prevent excessive browning. Cool completely on a wire rack (about 2 hours).

MAKES 8 servings. **HANDS-ON** 20 min.; **TOTAL** 3 hours, 40 min., including streusel

Pecan Streusel Topping

Stir together 1 1/2 cups **pecan halves and pieces;** 1 cup firmly packed **light brown sugar;** 1/4 cup **all-purpose flour;** 1/4 cup **butter,** melted; and 1/2 tsp. **pumpkin pie spice** in a medium bowl until well combined.

MAKES about 1 3/4 cups. **HANDS-ON** 10 min., **TOTAL** 10 min.

PECAN PIE BARS

CRUST

- 2 1/2 cups all-purpose flour
- 2/3 cup powdered sugar
- 3/4 cup chopped toasted pecans
- 1/2 tsp. kosher salt
- 1 cup butter, cubed
- Vegetable cooking spray

FILLING

- 3/4 cup cane syrup
- 1/2 cup firmly packed dark brown sugar
- 6 Tbsp. heavy cream
- 6 Tbsp. butter
- 1/4 tsp. kosher salt
- 1 tsp. vanilla extract
- 3 cups coarsely chopped toasted pecans

1. Prepare Crust: Preheat oven to 350°. Pulse flour and next 3 ingredients in a food processor 8 to 10 times or until thoroughly combined and pecans are finely chopped; add butter, and pulse 8 or 9 times or until mixture resembles coarse meal. Line bottom and sides of a 13- x 9-inch pan with aluminum foil, allowing 2 to 3 inches to extend over sides; lightly grease foil with cooking spray. Press mixture onto bottom of prepared pan.
2. Bake at 350° for 20 to 25 minutes or until lightly browned. Cool completely on a wire rack (about 25 minutes).
3. Prepare Filling: Bring syrup and next 4 ingredients to a boil in a large saucepan over medium heat, stirring constantly until butter melts and mixture is smooth; boil, stirring constantly, 1 minute. Remove from heat, and stir in vanilla and 3 cups coarsely chopped toasted pecans. Pour hot filling over cooled crust.
4. Bake at 350° for 14 to 16 minutes or until filling bubbles in center. Cool bars completely in pan on a wire rack (about 1 hour). Chill bars 1 hour before cutting.

MAKES 32 bars. **HANDS-ON** 20 min.; **TOTAL** 3 hours, 25 min.

Pecan Pie Bars

CHOCOLATE-PECAN MOUSSE TART

CRUST

- 1/2 cup butter, softened
- 1/2 (8-oz.) package cream cheese, softened
- 1/4 cup powdered sugar
- 1 1/2 cups all-purpose flour
- 1/4 tsp. table salt
- 1/2 cup finely chopped toasted pecans
- Vegetable cooking spray

MOUSSE

- 1 (12-oz.) package semisweet chocolate morsels
- 2 1/2 cups heavy cream, divided
- 1 Tbsp. bourbon
- 1 tsp. vanilla extract
- 1/2 cup coarsely chopped toasted pecans

1. Prepare Crust: Preheat oven to 400°. Beat first 3 ingredients at medium speed with a heavy-duty electric stand mixer until creamy. Gradually add flour and salt, beating at low speed just until blended after each addition. Add finely chopped pecans, and beat just until blended. Shape dough into a disk, and wrap with plastic wrap. Chill 1 hour.
2. Place dough disk on a lightly floured surface, and roll into a 13- x 9-inch rectangle. Gently place dough in a lightly greased (with cooking spray) 12- x 8-inch rectangular tart pan with removable bottom. Press dough into pan and up sides; trim off excess dough along edges.
3. Bake at 400° for 14 to 17 minutes or until lightly browned. Remove pan to a wire rack, and cool completely (about 30 minutes).
4. Meanwhile, prepare Mousse: Microwave chocolate morsels and 1/2 cup cream in a medium-size microwave-safe glass bowl at HIGH 1 1/2 minutes or until melted, stirring every 30 seconds. Stir in bourbon and vanilla; let stand 5 minutes.
5. Beat remaining 2 cups cream at medium-high speed with an electric mixer 2 to 3 minutes or until medium peaks form; fold cream into chocolate mixture. Spread in cooled tart shell. Top with coarsely chopped pecans. Chill 4 to 24 hours.

Note: To make in a 9-inch round deep-dish tart pan, prepare crust as directed, and roll into a 12-inch circle; increase baking time for crust to 22 to 27 minutes.

MAKES 8 to 10 servings. **HANDS-ON** 30 min.; **TOTAL** 5 hours, 50 min..

PECAN-ESPRESSO TOFFEE

- Parchment paper
- Vegetable cooking spray
- 1 1/4 cups butter
- 1 cup granulated sugar
- 1/3 cup firmly packed light brown sugar
- 1 Tbsp. instant espresso
- 1 Tbsp. dark molasses
- 1/2 tsp. table salt
- 2 cups chopped toasted pecans
- 1 cup bittersweet chocolate morsels
- 1 cup semisweet chocolate morsels

1. Line a 15- x 10-inch jelly-roll pan with parchment paper, and lightly coat with cooking spray.
2. Melt butter in a large heavy saucepan over medium heat; stir in granulated sugar, next 4 ingredients, and 1/3 cup water. Cook, stirring constantly, until a candy thermometer registers 290° (soft crack stage), about 17 to 20 minutes. Remove pan from heat, and stir in pecans. Immediately pour mixture into prepared pan. Spread in an even layer, and sprinkle bittersweet and semisweet chocolate morsels over top. Let stand 5 minutes.

Spread and swirl chocolate using an offset spatula. Chill 1 hour or until firm. Break toffee into pieces. Store in an airtight container in refrigerator up to 7 days. Serve cold or at room temperature.

MAKES 10 to 12 servings. **HANDS-ON** 35 min.; **TOTAL** 1 hour, 40 min.

DECADENT CREAM PUFFS WITH PRALINE SAUCE AND TOASTED PECANS

- **1/2 cup butter**
- **1 cup all-purpose flour**
- **1/8 tsp. table salt**
- **4 large eggs**
- **Vegetable cooking spray**
- **Parchment paper**
- **4 cups butter-pecan ice cream**
- **Praline Sauce**
- **1 cup coarsely chopped toasted pecans**

1. Preheat oven to 400°. Bring 1 cup water to a boil in a large saucepan over medium-high heat. Reduce heat to low, and add butter, stirring until melted.
2. Stir in flour and salt, beating vigorously with a wooden spoon 1 minute or until mixture leaves sides of saucepan. Remove from heat, and cool 5 minutes. Add eggs, 1 at a time, beating with wooden spoon until smooth. Drop by rounded 1/4-cupfuls, 2 inches apart, onto a lightly greased (with cooking spray) parchment paper-lined baking sheet.
3. Bake at 400° for 30 to 35 minutes or until puffed and golden brown. Remove from oven, and, using a wooden pick, poke a small hole into side of each cream puff to allow steam to escape. Cool completely on baking sheet on a wire rack (about 30 minutes).
4. Cut each cream puff in half horizontally; remove and discard any soft dough inside. Spoon 1/2 cup ice cream onto each bottom half; top cream puffs with remaining halves. Spoon 3 Tbsp. Praline Sauce over each puff; sprinkle with toasted pecans. Serve immediately with remaining sauce.

MAKES about 8 servings. **HANDS-ON** 30 min.; **TOTAL** 1 hour, 25 min., including sauce

Praline Sauce

Bring 1 cup firmly packed **light brown sugar,** 1/2 cup **half-and-half,** 1/2 cup **butter,** and a pinch of **salt** to a boil in a small saucepan over medium heat, stirring constantly. Cook, stirring constantly, 1 minute. Remove from heat, and stir in 1 tsp. **vanilla** extract. Let stand 10 minutes. Serve warm.

MAKES about 1 1/2 cups. **HANDS-ON** 10 min., **TOTAL** 20 min.

PECAN POUND CAKE

- **2 cups butter, softened**
- **1 1/4 cups granulated sugar**
- **1 1/4 cups firmly packed light brown sugar**
- **6 large eggs**
- **1 Tbsp. vanilla extract**
- **1 1/2 tsp. baking powder**
- **1/4 tsp. table salt**
- **4 cups all-purpose flour, divided**
- **1 cup milk**
- **4 cups chopped toasted pecans**
- **Shortening**
- **Citrus Glaze**

1. Preheat oven to 325°. Beat butter at medium speed with a heavy-duty electric stand mixer until creamy. Gradually add sugars, beating 3 to 5 minutes or until light and fluffy. Add eggs, 1 at a time, beating just until blended after each addition. Stir in vanilla.
2. Stir together baking powder, salt, and 3 3/4 cups flour in a medium bowl. Add flour mixture to butter mixture alternately with milk, beginning and ending with flour mixture. Beat at low speed just until blended after each addition. Stir together pecans and remaining 1/4 cup flour in a small bowl; add to batter, and stir just until combined. Pour batter into a greased (with shortening) and floured 10-inch tube pan.
3. Bake at 325° for 1 hour and 15 minutes to 1 hour and 30 minutes or until a long wooden pick inserted in center comes out clean, shielding with aluminum foil after 55 minutes to prevent excessive browning. Cool in pan on a wire rack 15 minutes; remove cake from pan to wire rack. Cool 20 minutes.
4. Spoon Citrus Glaze over cake. Cool completely (about 1 hour).

MAKES 10 to 12 servings. **HANDS-ON** 25 min.; **TOTAL** 3 hours, 25 min., including glaze

Citrus Glaze

Whisk together 2 1/2 cups **powdered sugar;** 2 Tbsp. **butter,** melted; 1 tsp. **orange zest;** 1 tsp. **lemon zest;** 2 Tbsp. fresh **orange juice;** and 1 Tbsp. fresh **lemon juice** until smooth. Whisk in up to 1 Tbsp. more lemon juice, 1 tsp. at a time, until desired consistency is reached.

MAKES about 3/4 cup. **HANDS-ON** 10 min., **TOTAL** 10 min.

SPICY-SWEET PECANS

- **3/4 cup granulated sugar**
- **1 Tbsp. light brown sugar**
- **2 tsp. chopped fresh rosemary**
- **1 tsp. kosher salt**
- **1 tsp. ground cinnamon**
- **1/2 tsp. ground ginger**
- **1/4 tsp. ground nutmeg**
- **1/4 tsp. ground red pepper**
- **1 large egg white**
- **4 cups pecan halves**
- **Vegetable cooking spray**
- **Wax paper**

1. Preheat oven to 275°. Stir together first 8 ingredients in a medium bowl.
2. Whisk together egg white and 1 Tbsp. water in a separate medium bowl until foamy. (No liquid should remain.) Add pecans, stirring to coat.
3. Add pecan mixture to sugar mixture, stirring until evenly coated. Spread pecans in a single layer on a lightly greased (with cooking spray) aluminum foil-lined half-sheet pan (about 17 x 12 inches).
4. Bake at 275° for 50 to 55 minutes or until sugar mixture hardens and nuts are toasted, stirring every 15 minutes. Spread immediately in a single layer on wax paper; cool completely (about 30 minutes). Store in an airtight container up to 7 days.

MAKES 4 cups. **HANDS-ON** 10 min.; **TOTAL** 1 hour, 30 min.

Soulful Soups and Stews

FROM SHRIMP CHOWDER TO RICH CHICKEN SOUP, THESE QUICK AND EASY RECIPES ARE ALL DRESSED UP FOR HOLIDAY ENTERTAINING

SOUTHWEST PORK STEW

Make this stew with pulled or shredded pork from your favorite barbecue joint or deli-roasted chicken from the grocery store.

- 1 (16-oz.) jar salsa verde
- 2 cups reduced-sodium chicken broth
- 1 (15-oz.) can black beans, drained and rinsed
- 1 (15-oz.) can fire-roasted diced tomatoes
- 1 lb. shredded smoked pork, without sauce
- 1 tsp. ground cumin
- Kosher salt and freshly ground black pepper

Cook salsa verde in a large saucepan over medium-high heat, stirring occasionally, 2 minutes. Stir in chicken broth and next 4 ingredients, and bring to a boil. Reduce heat to medium, and simmer, stirring occasionally, 10 minutes. Season with salt and pepper, and serve immediately.

Note: We tested with Swanson Natural Goodness 33% Less Sodium Chicken Broth.

MAKES 2 qt. **HANDS-ON** 10 min., **TOTAL** 20 min.

Chicken-and-Prosciutto Tortelloni Soup, page 265

SHRIMP-AND-NEW POTATO CHOWDER

Add shrimp a few minutes before serving so they don't overcook.

- 2 Tbsp. butter
- 3 bunches green onions, sliced
- 1 1/2 lb. new potatoes, diced
- 2 cups reduced-sodium chicken broth
- 1 1/2 cups heavy cream
- 1/2 cup dry white wine
- 1 tsp. kosher salt
- 1/4 tsp. black pepper
- 1/2 lb. medium-size raw shrimp, peeled and deveined
- 2 tsp. hot sauce

1. Melt butter in a medium Dutch oven over medium heat. Add green onions, and cook, stirring often, 1 minute.
2. Add potatoes and next 5 ingredients, and increase heat to high. Bring to a boil. Reduce heat to medium-low, and cook, stirring occasionally, 25 minutes or until potatoes are tender.
3. Stir in shrimp and hot sauce, and cook 3 minutes.

Note: We tested with Swanson Natural Goodness 33% Less Sodium Chicken Broth.

MAKES about 2 qt. **HANDS-ON** 25 min., **TOTAL** 50 min.

KING RANCH CHICKEN SOUP

This restaurant-style Tex-Mex dish is both rich and a cinch to prepare.

- 2 Tbsp. butter
- 1 cup chopped yellow onion
- 1 cup chopped green bell pepper
- 1 garlic clove, minced
- 2 (10-oz.) cans diced tomatoes and green chiles
- 1 (10 3/4-oz.) can cream of mushroom soup
- 1 (10 3/4-oz.) can cream of chicken soup

- 5 cups chicken broth
- 1 (1 1/2- to 2 1/2-lb.) whole deli-roasted chicken, skin removed and meat shredded
- 1 tsp. dried oregano
- 1 tsp. ground cumin
- 1 tsp. chili powder
- 8 (6-inch) fajita-size corn tortillas, cut into 1/2-inch strips and halved
- Kosher salt
- Garnishes: shredded cheese, fresh cilantro

1. Melt butter in a large Dutch oven over medium-high heat. Add onions and peppers, and sauté 6 to 7 minutes or until tender. Add garlic, and sauté 1 minute. Stir in diced tomatoes, cream of mushroom soup, and cream of chicken soup; combine thoroughly. Stir in broth and next 4 ingredients.
2. Increase heat to medium-high, and bring to a boil. Reduce heat to low, and simmer, stirring occasionally, 5 minutes. Stir in tortilla strips, and simmer 2 more minutes. Add salt to taste.

Note: We tested with Swanson 100% Natural Chicken Broth.

MAKES about 4 qt. **HANDS-ON** 35 min., **TOTAL** 35 min.

King Ranch Chicken Soup

CHICKEN-AND-PROSCIUTTO TORTELLONI SOUP

Any flavor tortelloni will work well in this dish.

- 1 (8-oz.) package chopped fresh onions, peppers, and celery
- 1 Tbsp. olive oil
- 1/2 tsp. dried Italian seasoning
- 1 (14.5-oz.) can diced tomatoes with roasted garlic
- 5 cups reduced-sodium chicken broth
- 1/4 tsp. table salt
- 1/2 tsp. ground black pepper
- 1 (9-oz.) package refrigerated chicken-and-prosciutto tortelloni
- 1 (6-oz.) package fresh baby spinach

Sauté onions, peppers, and celery in hot oil in a Dutch oven over medium-high heat 3 minutes; add Italian seasoning, and sauté 1 minute. Stir in tomatoes and next 3 ingredients. Increase heat to high; bring to a boil. Stir in tortelloni, and return to a boil. Reduce heat to low, and simmer 8 minutes or until tortelloni are tender. Remove from heat, and stir in spinach.

Note: We tested with Swanson Natural Goodness 33% Less Sodium Chicken Broth.

MAKES 2 qt. **HANDS-ON** 20 min., **TOTAL** 20 min.

CHEESY GARLIC FRENCH BREAD

Preheat oven to 375°. Stir together 1/4 cup **butter,** melted, and 1 **garlic** clove, pressed. Split 1 (12-oz.) **French bread** loaf; spread with butter mixture. Top with 3 Tbsp. grated **Parmesan cheese.** Place halves together on a baking sheet. Bake at 375° for 8 to 10 minutes or until cheese is melted. Slice and serve warm.

MAKES 6 to 8 servings. **HANDS-ON** 5 min., **TOTAL** 15 min.

White Bean, Sausage, and Turnip Green Stew

WHITE BEAN, SAUSAGE, AND TURNIP GREEN STEW

- 1/4 cup plus 1 1/2 tsp. kosher salt, divided
- 1 (1-lb.) package fresh chopped turnip greens
- 2 cups diced smoked sausage
- 2 Tbsp. olive oil
- 3 medium turnips (about 1 lb.), chopped
- 2 leeks, chopped
- 1 red bell pepper, chopped
- 3 garlic cloves, minced
- 2 (15-oz.) cans white beans, drained and rinsed
- 1 (32-oz.) container reduced-sodium chicken broth
- 1/4 cup grated Parmigiano-Reggiano cheese
- 1 tsp. ground black pepper

1. Bring 2 qt. water and 1/4 cup kosher salt to a boil in a large stockpot over high heat. Add turnip greens, and boil 5 minutes or until tender. Drain.
2. Cook sausage in hot oil in a large Dutch oven over medium heat, stirring often, 5 minutes or until browned. Stir in turnips, and cook 5 minutes. Add leeks and next 3 ingredients, and cook, stirring often, 5 minutes or until leeks are tender.
3. Increase heat to high. Add turnip greens and chicken broth, and bring to a boil. Reduce heat to medium, and simmer, stirring occasionally, 10 minutes. Stir in Parmigiano-Reggiano cheese, black pepper, and remaining 1 1/2 tsp. salt.

Note: We tested with Swanson Natural Goodness 33% Less Sodium Chicken Broth.

MAKES 2 qt. **HANDS-ON** 30 min., **TOTAL** 50 min.

The Best Pancakes Ever

OUR FOOLPROOF RECIPE FOR FLUFFY STACKS IS PURE GOLD

The original "Pam-cakes"

THE ORIGINAL "PAM-CAKES"

Some people toast special occasions with Champagne, but here at Southern Living, *we often celebrate with "Pam-cakes." Developed by our Test Kitchen's Pam Lolley, these flapjacks are deceptively light and unbelievably fluffy.*

- **1 3/4 cups all-purpose flour**
- **2 tsp. sugar**
- **1 1/2 tsp. baking powder**
- **1 tsp. baking soda**
- **1 tsp. table salt**
- **2 cups buttermilk**
- **2 large eggs**
- **1/4 cup butter, melted**

1. Stir together flour and next 4 ingredients in a large bowl. Whisk together buttermilk and eggs; gradually stir into flour mixture. Gently stir in butter. (Batter will be lumpy.) Let stand 5 minutes.
2. Pour about 1/4 cup batter for each pancake onto a hot (350°) buttered griddle.
3. Cook pancakes 3 to 4 minutes or until tops are covered with bubbles and edges look dry and cooked. Turn and cook 3 to 4 minutes or until golden brown. Place pancakes in a single layer on a baking sheet, and keep warm in a 200° oven up to 30 minutes.

MAKES about 20 pancakes.
HANDS-ON 30 min., **TOTAL** 35 min.

APPLE-SPICE PANCAKES

Save leftover Apple Cider Syrup for drizzling over vanilla ice cream.

Cut 3 small **McIntosh apples** (about 5 oz. each) into thin slices, and toss with 1/4 cup **sugar** and 1/4 tsp. **apple pie spice.** Prepare recipe at left as directed through Step 1. Drain apples. Pour about 1/4 cup batter onto a hot (350°) buttered griddle, and gently place 4 to 5 apple slices on each pancake. Proceed with recipe as directed in Step 3. Serve warm with **Apple Cider Syrup.**

MAKES about 20 pancakes. **HANDS-ON** 40 min.; **TOTAL** 1 hour, 5 min., including syrup

Apple Cider Syrup

Bring 1 cup **apple cider** to a boil in a small saucepan over medium heat. Boil, stirring occasionally, 12 to 14 minutes or until reduced to 1/4 cup. Meanwhile, microwave 3/4 cup **apple jelly** in a microwave-safe bowl at HIGH 30 seconds to 1 minute or until melted. Whisk melted apple jelly and 1/8 tsp. **apple pie spice** into cider.

MAKES about 1 cup. **HANDS-ON** 5 min., **TOTAL** 20 min.

CHOCOLATE CHIP-TOASTED PECAN PANCAKES

Add a generous dollop of whipped cream for an unexpected dessert.

Prepare recipe at left as directed in Step 1. Pour about 1/4 cup batter onto a hot (350°) buttered griddle, and sprinkle each pancake with 1 Tbsp. **dark chocolate chips** and 2 tsp. finely chopped **toasted pecans.** Proceed with recipe as directed in Step 3. Top with **powdered sugar** and **shaved chocolate.**

MAKES about 20 pancakes.
HANDS-ON 35 min., **TOTAL** 40 min.

BUBBLES ARE YOUR CUE
Flip pancakes when you see medium-size bubbles around the edges and a few popping up in the middle.

LUMPS ARE GOOD
It sounds strange, but lumps signal tender—not tough—cakes. Let the batter stand 5 minutes before pouring it onto the hot griddle.

OUR SECRETS TO THE BEST
PANCAKES

Pam Lolley, one of our cooks in The South's Most Trusted Kitchen, shares her technique for light and fluffy pancakes. Hosting a crowd? Double our recipe (page 266). Freeze any leftover cakes, and simply pop them in the toaster oven as needed.

FANCY FLIPPING IS A NO-NO
Slide the spatula underneath the entire pancake, and carefully turn it so that it doesn't splatter or smear.

BOTH SIDES SHOULD BE GOLDEN BROWN
After 2 minutes, gently lift the edge with your spatula to check the color.

Community Cookbook

BUILD A BETTER LIBRARY, ONE GREAT BOOK AT A TIME

From frosted Bundts to gooey orange rolls, *The Southern Baker* celebrates our favorite *Southern Living* Test Kitchen recipes handed down through the years, many from our own families. With over 150 recipes—along with our best tips, tricks, and secrets—it's an heirloom-worthy compilation to give to a new generation of Southern bakers.

APPLE-CREAM CHEESE BUNDT CAKE

This delicious apple Bundt cake features a sweet cream cheese filling and homemade praline frosting. For extra crunch, garnish with additional toasted pecans.

CREAM CHEESE FILLING

- 1 (8-oz.) package cream cheese, softened
- 1/4 cup butter, softened
- 1/2 cup granulated sugar
- 1 large egg
- 2 Tbsp. all-purpose flour
- 1 tsp. vanilla extract

APPLE CAKE BATTER

- 3 cups all-purpose flour
- 1 cup granulated sugar
- 1 cup firmly packed light brown sugar
- 2 tsp. ground cinnamon
- 1 tsp. table salt
- 1 tsp. baking soda
- 1 tsp. ground nutmeg
- 1/2 tsp. ground allspice
- 3 large eggs, lightly beaten
- 3/4 cup canola oil
- 3/4 cup applesauce
- 1 tsp. vanilla extract
- 3 cups peeled and finely chopped Gala apples (about 1 1/2 lb.)
- 1 cup toasted finely chopped pecans
- Praline Frosting

1. Prepare Cream Cheese Filling: Beat first 3 ingredients at medium speed with an electric mixer until blended and smooth. Add egg, flour, and vanilla; beat just until blended.
2. Prepare Apple Cake Batter: Preheat oven to 350°. Stir together 3 cups flour and next 7 ingredients in a large bowl; stir in eggs and next 3 ingredients, stirring just until dry ingredients are moistened. Stir in chopped apples and pecans.
3. Spoon two-thirds of apple mixture into a greased and floured 14-cup Bundt pan. Spoon Cream Cheese Filling over apple mixture, leaving a 1-inch border around edges of pan. Swirl filling gently with a knife. Spoon remaining apple mixture over Cream Cheese Filling.
4. Bake at 350° for 1 hour to 1 hour and 15 minutes or until a long wooden pick inserted in center comes out clean. Cool cake in pan on a wire rack 15 minutes; remove from pan to wire rack, and cool completely (about 2 hours).
5. Pour frosting immediately over cooled cake.

MAKES 12 servings. **HANDS-ON** 40 min.; **TOTAL** 4 hours, 10 min.

Praline Frosting

- 1/2 cup firmly packed light brown sugar
- 1/4 cup butter
- 3 Tbsp. milk
- 1 tsp. vanilla extract
- 1 cup powdered sugar

Bring first 3 ingredients to a boil in a 2-qt. saucepan over medium heat, whisking constantly; boil mixture 1 minute, whisking constantly. Remove from heat; stir in vanilla. Gradually whisk in powdered sugar until smooth; stir gently 3 to 5 minutes or until mixture begins to cool and thickens slightly. Use immediately.

MAKES about 1 cup. **HANDS-ON** 10 min., **TOTAL** 10 min.

December

270 HOLIDAY 2015 **Appetizers & Cocktails** Eight appetizers that are as lovely as the ornaments on the tree with cocktail pairings to suit every spirit

276 HOLIDAY 2015 **Mains & Sides** Magnificent holiday main attractions for every type of crowd

284 HOLIDAY 2015 **Desserts & Cookies** Showstopping finales that can be prepped in 30 minutes or less and finished before guests arrive or while you mingle

295 **The Baker's Brunch** Our Test Kitchen breakfast whisperer shares her family's holiday morning favorites honed over three decades of baking

299 ENTERTAINING WITH JULIA **Ham for the Holidays** Why this holiday centerpiece endures on the Southern table

301 THE *SL* TEST KITCHEN ACADEMY **Smart Cookies** Get the 411 for baking up the best batches of cookies around

302 COMMUNITY COOKBOOK For almost 35 years, we have brought you the best holiday recipes, entertaining, and decorating ideas in the pages of *Christmas with Southern Living*

Inside-Out Hot Brown Bites

APPETIZERS & COCKTAILS

*From a beet-studded flatbread to a decadent shrimp dip, our **8 amazing appetizers** might be prettier than the ornaments on your tree. But don't let their impressive looks fool you—these bites are a breeze. Plus, **5 festive cocktails** to please your party crowd*

APPETIZERS: *Inside-Out Hot Brown Bites ✦ Creamy Shrimp Dip with Crispy Wonton Chips ✦ Chicken Liver Mousse Crostini with Pepper Jelly ✦ Smoked Trout Crostini with Radishes and Dill Cream ✦ Roasted Fennel-and-Prosciutto Flatbread ✦ Grapefruit-Beet-Goat Cheese Flatbread ✦ Cheddar-Horseradish-Walnut Cheese Ball ✦ Blue Cheese-Red Currant Cheese Ball ✦ Feta-Olive-Fresh Herb Cheese Ball ✦ Petite Sweet Potato Biscuits with Pulled Pork and Slaw*

COCKTAILS: *Spiced Chocolate Stout Beer Cocktail ✦ Retro Rum Punch ✦ Rosemary Collins ✦ Peanut Whiskey and Cola ✦ Southern Lemon Americano*

8 AMAZING APPETIZERS

INSIDE-OUT HOT BROWN BITES

We've reimagined this Kentucky sandwich combination of bacon, turkey, and cheese sauce into beautiful bite-sized cups.

- **Vegetable cooking spray**
- 1 1/2 **(5-oz.) containers finely shredded Parmesan cheese**
- 1 2/3 **cups milk**
- 1/4 **cup butter**
- 3 **Tbsp. all-purpose flour**
- 1/2 **cup (2 oz.) shredded medium Cheddar cheese**
- 1/8 **tsp. kosher salt**
- 1/8 **tsp. freshly ground black pepper**
- 4 **oz. thinly sliced deli turkey, cut into 2-inch squares**
- 4 **cooked bacon slices, crumbled**
- 1/2 **cup diced fresh tomato**
- **Garnish: fresh flat-leaf parsley leaves**

1. Preheat oven to 350°. Line 2 baking sheets with aluminum foil, and lightly coat with cooking spray. Spoon Parmesan cheese by tablespoonfuls 1/2 inch apart onto prepared baking sheets, forming 12 (2 1/2-inch) rounds on each sheet.
2. Bake 1 sheet at 350° for 7 to 9 minutes or until edges are lightly browned and beginning to set. Working quickly, transfer cheese rounds to a lightly greased (with cooking spray) 24-cup miniature muffin pan, pressing gently into each cup to form shells. Repeat procedure with second baking sheet.
3. Microwave milk in a microwave-safe measuring cup 30 seconds at HIGH or until warm. Melt butter in a small saucepan over medium-high heat. Whisk in flour; cook, whisking constantly, 1 minute. Gradually whisk in warm milk. Bring to a boil, and boil, whisking constantly, 1 to 2 minutes or until thickened. Whisk in shredded Cheddar cheese, kosher salt, and black pepper.
4. Increase oven temperature to 425°. Line each Parmesan shell with 2 turkey pieces, and fill each with 1 tsp. cheese sauce. Bake 5 minutes. Remove from pan to wire rack, and top with crumbled bacon and diced tomato.

MAKES 2 dozen. **HANDS-ON** 45 min., **TOTAL** 45 min.

CREAMY SHRIMP DIP WITH CRISPY WONTON CHIPS

If you don't have time to make fried wontons, use fancy potato chips or crispy tortilla strips.

- 1 **lb. large cooked, peeled, and deveined shrimp, chopped**
- 2 **green onions, finely chopped**
- 1 **shallot, finely chopped**
- 1/2 **cup mayonnaise**
- 1/2 **cup sour cream**
- 2 **Tbsp. fresh tarragon leaves, finely chopped**
- 2 **tsp. rice vinegar**
- 1/2 **tsp. kosher salt**
- 1/4 **tsp. freshly ground black pepper**
- **Crispy Wonton Chips**
- **Garnish: fresh chives**

Chicken Liver Mousse Crostini with Pepper Jelly

Stir together first 9 ingredients in a medium bowl until well blended. Cover and chill 30 minutes to 1 day. Serve with Crispy Wonton Chips.

MAKES 8 to 10 servings. **HANDS-ON** 15 min.; **TOTAL** 55 min., including wontons

Crispy Wonton Chips

Pour **vegetable oil** to a depth of 1 inch into a Dutch oven, and heat over medium-high heat to 350°. Cut each **wonton** wrapper from 1 (6-oz.) package into quarters. Fry, in batches, 15 to 20 seconds on each side or until golden brown and crisp. Drain on a wire rack over paper towels; sprinkle with **kosher salt.**

MAKES 8 to 10 servings. **HANDS-ON** 10 min., **TOTAL** 10 min.

CHICKEN LIVER MOUSSE CROSTINI WITH PEPPER JELLY

For a different presentation, make this mousse in a 1 1/2-qt. ovenproof dish instead of small jars.

- 1/2 **cup butter, divided**
- 1 **cup finely chopped yellow onion**
- 1 **garlic clove, minced**
- 1/2 **cup heavy cream**
- 1 **lb. chicken livers**
- 1/2 **tsp. kosher salt**
- 1/4 **tsp. freshly ground black pepper**
- 2/3 **cup red pepper jelly**
- 1 **(8- to 9-oz.) French bread baguette, sliced and toasted**

1. Preheat oven to 300°. Melt 2 Tbsp. butter in a small saucepan over medium heat. Stir in chopped onion. Cover and cook, stirring occasionally, 6 to 7 minutes or until tender. Add minced garlic, and sauté 10 seconds. Stir in cream, and bring to a simmer. Cover, reduce heat to low, and cook 6 minutes.
2. Add remaining 6 Tbsp. butter, and cook, stirring constantly, about 1 minute or until butter is melted and well blended. Remove from heat.

3. Process chicken livers, kosher salt, black pepper and onion mixture in a blender or food processor 30 to 45 seconds or until smooth, stopping to scrape down sides as needed.
4. Place 7 (8-oz.) widemouthed, ovenproof jars with tight-fitting lids in a 13- x 9-inch baking dish. Divide liver mixture among jars, filling each halfway. Add warm water to baking dish to a depth of 1 1/4 inches. Cover jars and baking dish tightly with heavy-duty aluminum foil.
5. Bake at 300° for 30 minutes or until mousse is set. Remove from oven, transfer jars to a wire rack, and cool completely (about 30 minutes). Screw on lids, and chill 1 hour.
6. Just before serving, spoon 1 1/2 Tbsp. red pepper jelly into each jar, and serve with toasted baguette slices.

Note: We tested with Kerr Wide Mouth Half Pint Jars.

MAKES 12 to 14 servings. **HANDS-ON** 35 min.; **TOTAL** 2 hours, 35 min.

SMOKED TROUT CROSTINI WITH RADISHES AND DILL CREAM

For an even more colorful spread, replace half the trout with smoked salmon.

- **8 thick, firm white bread slices, crusts removed**
- **1 Tbsp. olive oil**
- **1/4 tsp. ground black pepper**
- **1/2 tsp. kosher salt, divided**
- **1 (8-oz.) container sour cream**
- **1/2 tsp. fresh lemon juice**
- **1 tsp. prepared horseradish**
- **1 Tbsp. chopped fresh dillweed**
- **1 1/4 cups thinly sliced radishes**
- **1 (8-oz.) smoked trout fillet, flaked into 1/2-inch pieces**
- **Garnish: fresh dillweed**

1. Preheat oven to 350°. Cut each bread slice into 4 triangles. Brush with olive oil, and sprinkle with pepper and 1/4 tsp. salt. Place in a single layer on a baking sheet; bake 15 to 20 minutes or until golden.
2. Stir together sour cream, next 3 ingredients, and remaining 1/4 tsp. salt. Spoon about 1/2 Tbsp. sour cream mixture onto each bread triangle. Top with radish slices and trout.

MAKES 8 to 10 servings. **HANDS-ON** 20 min., **TOTAL** 35 min.

ROASTED FENNEL-AND-PROSCIUTTO FLATBREAD

Try substituting sweet onion for fennel and bacon for prosciutto.

- **1 lb. bakery pizza dough**
- **2 fennel bulbs**
- **2 1/2 Tbsp. olive oil, divided**
- **1 tsp. finely chopped fresh thyme**
- **1 tsp. finely chopped fresh oregano**
- **2 oz. thinly sliced prosciutto**
- **Vegetable cooking spray**
- **1 1/2 cups (6 oz.) shredded fontina cheese**
- **1/4 tsp. dried crushed red pepper**
- **1 tsp. coarse sea salt**
- **1 Tbsp. bottled balsamic glaze**

1. Preheat oven to 425°. Remove pizza dough from refrigerator, and let stand 30 minutes or until ready to use.
2. Meanwhile, trim and discard root end of fennel bulbs. Trim stalks from bulbs, and chop fronds to equal 2 tsp. Thinly slice fennel bulbs lengthwise, and place on an aluminum foil-lined baking sheet. Drizzle with 2 Tbsp. olive oil. Sprinkle with thyme and oregano. Bake at 425° for 35 minutes or until edges are golden brown.
3. Cook prosciutto in a large nonstick skillet over medium-high heat 1 to 2 minutes on each side or until browned and crisp. Break prosciutto into large pieces.
4. Turn dough out on a lightly floured surface, and roll into a 17- x 13-inch rectangle (about 1/4 inch thick). Place dough rectangle on a lightly greased (with cooking spray) baking sheet. Brush dough with remaining 1/2 Tbsp. olive oil. Bake at 425° for 15 to 20 minutes or until golden brown. Remove crust from oven, and increase oven temperature to broil.
5. Top crust with fontina cheese, fennel slices, and prosciutto. Broil 1 minute. Sprinkle with dried crushed red pepper, reserved 2 tsp. chopped fennel fronds, and coarse sea salt. Drizzle with balsamic glaze.

Roasted Fennel-and-Prosciutto Flatbread

Note: We tested with Acetum Blaze Original Balsamic Glaze.

MAKES 10 servings. **HANDS-ON** 30 min.; **TOTAL** 1 hour, 20 min.

GRAPEFRUIT-BEET-GOAT CHEESE FLATBREAD

Letting bakery pizza dough rest at room temperature for about 30 minutes makes it easier to knead.

- **1 1/2 lb. golden beets**
- **1 lb. bakery pizza dough**
- **Vegetable cooking spray**
- **1/2 Tbsp. olive oil**
- **1/4 tsp. kosher salt**
- **1/4 tsp. freshly ground black pepper**
- **6 oz. goat cheese, crumbled**
- **2 small grapefruit, peeled and sectioned**
- **2 Tbsp. fresh mint leaves, thinly sliced**
- **1/4 cup toasted pine nuts**
- **2 Tbsp. honey**
- **1/4 tsp. coarse sea salt**

1. Preheat oven to 350°. Trim beet stems to 1 inch; gently wash beets, and place in a 13- x 9-inch baking dish. Add 1/3 cup water, and cover with aluminum foil.
2. Bake at 350° for 1 hour or until tender. Uncover and cool completely (about 30 minutes).
3. Meanwhile, remove pizza dough from refrigerator, and let stand 30 minutes.
4. Increase oven temperature to 425°. Turn dough out on a lightly floured surface, and roll into a 17- x 13-inch rectangle (about 1/4 inch thick). Place dough rectangle on a lightly greased (with cooking spray) baking sheet.
5. Brush dough with olive oil; sprinkle with salt and pepper. Bake at 425° for 15 to 20 minutes or until golden brown.
6. Cut cooled beets into thin slices. Top crust with beet slices, goat cheese, grapefruit segments, sliced mint, and toasted pine nuts, leaving a 1/2-inch border. Drizzle flatbread with honey, and sprinkle with sea salt.

MAKES 10 servings. **HANDS-ON** 20 min.; **TOTAL** 2 hours, 5 min.

CHEESE BALLS, THREE WAYS

For another festive presentation, you can shape the cheese mixture into yule logs.

BASE MIXTURE

- **3 (8-oz.) packages cream cheese, softened**
- **1/2 cup butter, softened**
- **1 Tbsp. lemon zest**
- **2 Tbsp. fresh lemon juice**
- **1/2 tsp. Worcestershire sauce**
- **1/4 tsp. hot sauce (such as Tabasco)**
- **1/2 tsp. kosher salt**

Cheddar-Horseradish-Walnut Cheese Ball

- **4 oz. sharp Cheddar cheese, finely shredded**
- **2 Tbsp. prepared horseradish**
- **1/2 cup finely chopped toasted walnuts**
- **Water crackers**

Blue Cheese-Red Currant Cheese Ball

- **4 oz. crumbled blue cheese**
- **2 Tbsp. red currant jam or jelly**
- **1 cup dried currants, coarsely chopped**
- **Thin gingersnaps**

Feta-Olive-Fresh Herb Cheese Ball

- **4 oz. feta cheese, crumbled**
- **1/4 cup green olives, pitted and finely chopped**
- **1/4 cup finely chopped fresh flat-leaf parsley**
- **2 Tbsp. green onions, finely chopped**
- **Cucumber slices**

1. Prepare Base Mixture: Beat cream cheese and next 6 ingredients at medium speed with a heavy-duty electric stand mixer 2 to 3 minutes or until well combined. Divide mixture into 3 portions; place each portion in a medium bowl. Cover and chill until ready to use.
2. Prepare Cheddar-Horseradish-Walnut Cheese Ball: Stir together Cheddar cheese, horseradish, and 1 portion of Base Mixture until combined. Chill 5 minutes. Shape into a ball, and roll in walnuts to coat. Serve with water crackers.
3. Prepare Blue Cheese-Red Currant Cheese Ball: Stir together blue cheese, jam, and 1 portion of Base Mixture until combined. Chill 5 minutes. Shape into a ball, and roll in chopped dried currants to coat. Serve with gingersnaps.
4. Prepare Feta-Olive-Fresh Herb Cheese Ball: Stir together crumbled feta cheese, chopped green olives, and 1 portion of Base Mixture until combined. Chill 5 minutes. Meanwhile, stir together chopped parsley and green onions. Shape feta mixture into a ball, and roll in parsley mixture to coat. Serve with cucumber slices.

MAKES 10 servings per cheese ball. **HANDS-ON** 30 min., **TOTAL** 45 min.

Petite Sweet Potato Biscuits with Pulled Pork and Slaw

PETITE SWEET POTATO BISCUITS WITH PULLED PORK AND SLAW

The sweet potato biscuits freeze beautifully—thaw, bake, and top them just before serving.

- **1 cup finely chopped red cabbage**
- **1/2 cup shredded carrots, chopped**
- **1 tsp. kosher salt**
- **1 Tbsp. mayonnaise**
- **1 Tbsp. red wine vinegar**
- **1/3 cup barbecue sauce, divided**
- **1/4 cup sliced green onions**
- **Kosher salt and freshly ground black pepper**
- **12 Sweet Potato Biscuits**
- **2 Tbsp. butter, melted**
- **1/2 lb. warm barbecue pork (without sauce), chopped**
- **1 Tbsp. chopped fresh chives**

1. Toss together first 3 ingredients in a small bowl. Let stand 30 minutes. Rinse and drain well. Whisk together mayonnaise, vinegar, and 1 Tbsp. barbecue sauce in a medium bowl. Stir in cabbage mixture and green onions. Add salt and pepper to taste.
2. Preheat oven to 450°. Split biscuits, and brush with butter. Place in a single layer on a baking sheet. Bake 5 minutes or until golden.

3. Top biscuit halves evenly with pork, remaining barbecue sauce, cabbage mixture, and chives.

MAKES 2 dozen. **HANDS-ON** 15 min.; **TOTAL** 1 hour, 25 min., including biscuits

Sweet Potato Biscuits

- **1 1/2 cups mashed cooked sweet potatoes**
- **1 cup buttermilk**
- **6 Tbsp. butter, melted**
- **2 Tbsp. sugar**
- **1/8 tsp. baking soda**
- **3 1/3 cups self-rising flour, divided**
- **Vegetable cooking spray**

1. Preheat oven to 400°. Stir together first 3 ingredients in a large bowl. Add sugar, baking soda, and 3 cups flour, stirring just until dry ingredients are moistened.
2. Turn dough out onto a lightly floured surface; knead 8 to 10 times, adding up to 1/3 cup more flour to prevent dough from sticking. Roll dough to 3/4 inch thick; cut with a 2-inch round cutter. Place biscuits on a lightly greased (with cooking spray) baking sheet.
3. Bake at 400° for 15 to 20 minutes or until golden brown.

MAKES about 2 dozen. **HANDS-ON** 20 min., **TOTAL** 35 min.

CROWD-PLEASING COCKTAILS

THESE DRINKS ARE GUARANTEED TO PUT ALL YOUR GUESTS—FROM UNEXPECTED DROP-INS TO PICKY IN-LAWS—IN THE HOLIDAY SPIRIT

1
FOR THE HIPSTER SON
Spiced Chocolate Stout Beer Cocktail

Combine 2 Tbsp. **brandy** and 2 Tbsp. **Spicy Simple Syrup** in a pint glass over ice. Top with **chocolate stout**, and stir. Garnish with a **cinnamon stick.**

MAKES 1 serving. **HANDS-ON** 5 min., **TOTAL** 5 min.

Simple Syrup

Bring 1 cup **sugar** and 1 cup **water** to a boil in a saucepan over medium-high heat. Simmer 5 minutes or until sugar dissolves. Cool completely (about 30 minutes).

MAKES 1 1/2 cups. **HANDS-ON** 5 min., **TOTAL** 35 min.

Spicy Simple Syrup

Stir 1/2 tsp. each **ground red pepper** and **ground cinnamon** into 1 cup hot **Simple Syrup.** Cool completely.

2
FOR THE SPIRITED COUSINS
Retro Rum Punch

Stir together 1 cup **dark spiced rum,** 1 cup **light rum,** 2 cups **pineapple juice,** 2 cups **cranberry juice,** and 2 (12-oz.) bottles **ginger beer** or ale in a punch bowl. Add ice, and garnish with **frozen cranberries** and **sliced pineapple.** Serve liberally.

MAKES 9 cups. **HANDS-ON** 5 min., **TOTAL** 5 min.

3
FOR THE DROP-IN NEIGHBORS
Rosemary Collins

Place 1/4 cup **gin**, 2 Tbsp. **Simple Syrup,** 2 Tbsp. **lemon juice**, and ice in a cocktail shaker. Cover with lid, and shake vigorously until thoroughly chilled. Strain into a highball glass filled with ice cubes; top with **club soda.** Garnish with a **rosemary sprig.**

MAKES 1 serving. **HANDS-ON** 5 min., **TOTAL** 5 min.

4
FOR THE SALTY UNCLE
Peanut Whiskey and Cola

Dip rim of a rocks glass in water; dip in 1 Tbsp. crushed **dry-roasted peanuts** to coat. Fill with ice. Add 1/4 cup **bourbon**, and, if desired, 1/4 tsp. **vanilla extract.** Top with **cola soft drink** (such as Coca-Cola).

MAKES 1 serving. **HANDS-ON** 5 min., **TOTAL** 5 min.

5
FOR THE REFINED MOTHER-IN-LAW
Southern Lemon Americano

Combine 2 Tbsp. **Campari,** 2 Tbsp. **sweet vermouth,** 1 tsp. **Meyer lemon juice**, and ice in a highball glass. Top with **club soda,** and garnish with **Meyer lemon peel.**

MAKES 1 serving. **HANDS-ON** 5 min., **TOTAL** 5 min.

Classic Roasted Duck with Orange-Bourbon-Molasses Glaze

MAINS

SIDES

*Feast your eyes on an impressive lineup featuring **7 marvelous mains**, including new classics like our Creole-spiced beef tenderloin and unexpected additions like shrimp piccata. Then mix and match with any of these **8 spectacular sides**, from a risotto twist on dirty rice to a red-and-green salad*

MAINS: *Classic Roasted Duck with Orange-Bourbon-Molasses Glaze ✦ Veal Chops Milanese with Lemon and Herbs ✦ Grilled Pork Loin Steaks with Cherry-Plum Sauce ✦ Spice-Rubbed Tenderloin with Mustard-Cream Sauce ✦ Grilled Cornish Hens with Herb Brine ✦ Hoppin' John Cakes with Tomato-Jalapeño Gravy ✦ Creamy Shrimp Piccata* ***SIDES:*** *Dirty Rice Risotto ✦ Broccolini with Pecans and Cane Syrup Vinaigrette ✦ Roasted Garlic Duchess Potatoes ✦ Charred Ambrosia with Toasted Coconut and Marshmallow Crème ✦ Gnocchi Mac and Cheese ✦ Fontina-Chive Yorkshire Puddings ✦ Christmas Salad ✦ Brussels Spouts with Parmesan Cream Sauce*

7 MARVELOUS MAINS

CLASSIC ROASTED DUCK WITH ORANGE-BOURBON-MOLASSES GLAZE

Best ducks are dry ducks. Pat them with paper towel before, during, and after refrigerating.

- 2 (6- to 7-lb.) whole ducks
- Kitchen string
- 2 tsp. kosher salt
- 1/2 tsp. freshly ground black pepper
- 1 cup orange marmalade
- 1/4 cup bourbon
- 3 Tbsp. molasses
- 1 Tbsp. fresh lemon juice
- 1/4 tsp. ground ginger
- 1/4 tsp. dried crushed red pepper

1. Remove giblets from ducks, and reserve for another use. Rinse ducks, and pat dry with paper towels. Remove excess fat and skin. Tie legs together with kitchen string, and chill, uncovered, 10 to 24 hours.
2. Preheat oven to 450°. Let ducks stand at room temperature 15 minutes. Prick legs, thighs, and breasts with a fork. Rub ducks with salt and black pepper, and place, breast side up, on a wire rack in an aluminum foil-lined jelly-roll pan. Bake at 450° for 45 minutes.
3. Meanwhile, stir together orange marmalade and next 5 ingredients in a small saucepan, and bring to a boil over high heat. Reduce heat to medium, and cook, stirring often, 10 to 15 minutes or until reduced to about 1 cup.
4. Remove ducks from oven. Reduce oven temperature to 350°. Carefully spoon fat from pan. Brush ducks with orange marmalade glaze. Bake at 350° for 20 to 25 minutes or until a meat thermometer inserted in thickest portion registers 180°. Let stand 15 minutes before serving.

MAKES 8 to 10 servings. **HANDS-ON** 30 min.; **TOTAL** 11 hours, 50 min.

VEAL CHOPS MILANESE WITH LEMON AND HERBS

Shopping tip: Ask your butcher to "french" the chops for you so you take home clean meat that's cartilage and fat free.

- 6 (6- to 7-oz.) frenched veal chops
- 2 1/2 cups panko (Japanese breadcrumbs)
- 1/2 cup grated Parmigiano-Reggiano cheese
- 2 Tbsp. fresh rosemary leaves
- 2 Tbsp. fresh thyme leaves
- 1 tsp. lemon zest
- 2 large eggs
- 2 Tbsp. kosher salt
- 1 1/2 tsp. freshly ground black pepper
- 3/4 cup olive oil
- 6 Tbsp. butter
- Garnishes: fresh flat-leaf parsley, lemon wedges

1. Place each veal chop between 2 sheets of heavy-duty plastic wrap, and flatten to 1/4-inch thickness, using a rolling pin or flat side of a meat mallet.
2. Process panko and next 4 ingredients in a food processor until herbs are finely chopped; place mixture in a shallow dish. Whisk together eggs and 1 Tbsp. water in another shallow dish. Sprinkle each veal chop with 1 tsp. salt and 1/4 tsp. pepper. Dip each chop in egg mixture (do not dip the bone), and dredge in panko mixture, pressing to adhere.
3. Preheat oven to 400°. Cook 1 chop in 2 Tbsp. hot oil and 1 Tbsp. butter in a large skillet over medium heat 2 minutes on each side or until golden brown. Repeat procedure with remaining chops, oil, and butter, wiping skillet clean after each chop. Place chops on wire racks in 2 jelly-roll pans.
4. Bake at 400° for 15 minutes. Serve immediately.

MAKES 6 servings. **HANDS-ON** 45 min., **TOTAL** 1 hour

GRILLED PORK LOIN STEAKS WITH CHERRY-PLUM SAUCE

Make the Cherry-Plum Sauce three to four days ahead, and reheat over low heat. If it's too thick, add water, a little at a time, until it reaches desired consistency.

- 1 (4 1/2- to 5-lb.) pork loin roast, trimmed
- 4 tsp. kosher salt
- 2 tsp. freshly ground black pepper
- 1 1/2 tsp. finely chopped fresh rosemary
- 4 tsp. olive oil, divided
- 8 thick bacon slices
- Cherry-Plum Sauce

1. Cut roast into 8 (2-inch-thick) slices, reserving end pieces for another use. Stir together salt, pepper, and rosemary in a small bowl. Rub each pork slice with 1/2 tsp. olive oil. Sprinkle each pork slice with about 1 tsp. salt mixture, pressing to adhere. Wrap 1 bacon slice around edge of each pork slice, and secure with a wooden pick. Let pork stand at room temperature 30 minutes.
2. Light 1 side of grill, heating to 350° to 400° (medium-high) heat; leave other side unlit.
3. Place pork slices over unlit side, and grill, covered with grill lid, 8 to 9 minutes on each side. Transfer pork to lit side, and grill, covered with grill lid, 4 to 5 minutes on each side or until a meat thermometer inserted into thickest portion registers 145° or to desired degree of doneness. Let stand 5 minutes. Serve with warm Cherry-Plum Sauce.

MAKES 8 servings. **HANDS-ON** 45 min.; **TOTAL** 1 hour, 40 min., including sauce

Cherry-Plum Sauce

- 1 shallot, minced
- 1 1/2 tsp. olive oil
- 1 (5-oz.) package dried tart cherries
- 1/2 cup plum jam
- 1 Tbsp. balsamic vinegar
- 1 Tbsp. Dijon mustard
- 1/2 tsp. kosher salt
- 1/4 tsp. freshly ground black pepper

Sauté shallot in hot oil in a small saucepan over medium heat 2 to 3 minutes or until softened. Stir in cherries, next 5 ingredients, and 3/4 cup water, and bring to a simmer. Reduce heat to medium-low, and cook, stirring often, 5 minutes or until mixture has thickened slightly and cherries are plumped. Serve warm.

MAKES about 1 1/2 cups. **HANDS-ON** 20 min., **TOTAL** 20 min.

SPICE-RUBBED TENDERLOIN WITH MUSTARD-CREAM SAUCE

Beef tenderloin, a luxurious and pricey meat that serves a large crowd, often goes on sale around the Christmas holidays.

- 2 1/2 tsp. kosher salt
- 1 tsp. freshly ground black pepper
- 1 tsp. dried thyme
- 1 tsp. garlic powder
- 1/2 tsp. ground cumin
- 1/2 tsp. paprika
- 1/2 tsp. ground red pepper
- 1 (5- to 6-lb.) beef tenderloin, trimmed
- 1 Tbsp. olive oil
- Vegetable cooking spray
- Mustard-Cream Sauce

1. Preheat oven to 500°. Stir together kosher salt and next 6 ingredients. Rub tenderloin evenly with olive oil. Sprinkle salt mixture over tenderloin, pressing to adhere. Cover and let stand at room temperature 30 minutes. Place tenderloin on a lightly greased (with cooking spray) rack in a roasting pan.
2. Bake at 500° for 15 minutes; reduce oven temperature to 375°, and bake 25 to 30 minutes or until a meat thermometer inserted in thickest portion registers 130° (for medium-rare) or to desired degree of doneness. Remove from oven; let stand 10 minutes before slicing. Serve with Mustard-Cream Sauce.

MAKES 10 to 12 servings. **HANDS-ON** 15 min.; **TOTAL** 1 hour, 50 min., including sauce

Grilled Cornish Hens with Herb Brine

Mustard-Cream Sauce

- 1 shallot, minced (about 3 Tbsp.)
- 1 Tbsp. olive oil
- 1 garlic clove, minced
- 1 cup dry white wine
- 1/4 cup Creole mustard
- 2 tsp. sugar
- 1 (8-oz.) container sour cream
- 1 tsp. kosher salt
- 1/4 tsp. freshly ground black pepper

Sauté shallot in hot oil in a medium skillet over medium heat 2 minutes or until soft; add garlic, and sauté 1 minute. Stir in wine, mustard, and sugar; bring to a boil. Cook, stirring constantly, 3 minutes or until mixture has thickened and is reduced by about half. Remove from heat, and whisk in sour cream, salt, and pepper. Serve immediately.

MAKES about 2 cups. **HANDS-ON** 15 min., **TOTAL** 15 min.

GRILLED CORNISH HENS WITH HERB BRINE

If you don't own kitchen shears, that's ok—a pair of heavy-duty scissors will work just as well.

- 3 Tbsp. kosher salt
- 1 1/2 tsp. freshly ground black pepper
- 1/2 tsp. garlic powder
- 1/2 tsp. onion powder
- 2 Tbsp. fresh thyme leaves
- 2 Tbsp. fresh flat-leaf parsley leaves
- 6 (20-oz.) Cornish hens
- Parchment paper

1. Process first 6 ingredients in a food processor 15 to 20 seconds or until well combined.
2. Pat hens dry with paper towels. Place each hen, breast side down, on a cutting board. Cut hens, using kitchen shears, along both sides of backbone, separating backbone from hen. Discard bone. Open hens as you would a book; turn breast side up, and press firmly against breastbone with the heel of your hand until bone cracks. Tuck wing tips under. Place hen in a parchment paper-lined jelly-roll pan. Repeat procedure for each hen.
3. Sprinkle salt mixture over hens, and chill, uncovered, 10 to 24 hours.
4. Heat grill to 350° to 400° (medium-high) heat. Grill hens, skin side down and covered with grill lid, 8 minutes on each side or until a meat thermometer inserted in thickest portion registers 165°. Remove from grill, and let stand 5 minutes. Serve immediately.

MAKES 6 to 8 servings. **HANDS-ON** 45 min., **TOTAL** 11 hours

HOPPIN' JOHN CAKES WITH TOMATO-JALAPEÑO GRAVY

We suggest cooking rice one to two days ahead of time so it has time to dry out.

- **2 (15-oz.) cans black-eyed peas, drained and rinsed**
- **2 cups cooked white long-grain rice**
- **1/4 cup butter, divided**
- **1 cup diced red bell pepper**
- **1 cup diced celery**
- **1 cup diced yellow onion**
- **3 tsp. kosher salt, divided**
- **3/4 tsp. freshly ground black pepper, divided**
- **4 cups finely chopped fresh turnip greens**
- **2 large eggs**
- **1 3/4 cups panko (Japanese breadcrumbs), divided**
- **2 garlic cloves, minced**
- **1 jalapeño pepper, seeded and finely chopped**
- **1/2 cup dry white wine**
- **1 (24-oz.) can crushed tomatoes**
- **1 Tbsp. Creole seasoning**
- **1/4 cup heavy cream**
- **1/2 cup vegetable oil, divided**

1. Reserve 1/2 cup black-eyed peas. Pulse rice and remaining peas in a food processor 8 times or until coarsely chopped. Transfer to a large bowl; add reserved peas.
2. Heat 2 Tbsp. butter in a large skillet over medium heat. Stir together bell pepper, celery, and onion; add half of mixture to skillet. Stir in 1 tsp. salt and 1/4 tsp. black pepper. Cook, stirring occasionally, 5 minutes or until slightly softened. Stir in turnip greens. Cook, stirring occasionally, 4 minutes or until greens have wilted. Stir into pea mixture.
3. Stir in eggs, 3/4 cup panko, and 1/2 tsp. salt. Using wet hands, shape mixture into 14 balls. Place balls on a baking sheet, 1 inch apart. Flatten each ball to form a 3/4-inch-thick patty. Freeze 15 minutes or until ready to use.
4. Meanwhile, heat remaining 2 Tbsp. butter in a medium saucepan over medium-high heat. Add remaining bell pepper mixture, 1 tsp. salt, and 1/4 tsp. black pepper. Cook, stirring occasionally, 5 to 6 minutes or until softened. Add garlic and jalapeño pepper, and cook, stirring occasionally, 1 minute. Stir in wine; cook, stirring occasionally, 3 minutes or until reduced to 1/4 cup. Stir in tomatoes and Creole seasoning. Reduce heat to low, and simmer 8 minutes. Remove from heat, and whisk in cream. Cover to keep warm.
5. Heat 1/4 cup vegetable oil in a large skillet over medium heat. Stir together remaining 1 cup panko, 1/2 tsp. salt, and 1/4 tsp. black pepper. Dredge cakes in panko mixture. Fry 7 cakes in hot oil 4 minutes on each side or until golden. Drain on paper towels. Place on a wire rack in a jelly-roll pan, and keep warm in a 200° oven. Repeat procedure with remaining oil and cakes. Serve immediately with Tomato-Jalapeño Gravy.

MAKES 7 to 8 servings. **HANDS-ON** 1 hour, 10 min.; **TOTAL** 1 hour, 10 min.

CREAMY SHRIMP PICCATA

Save on prep time by buying shrimp peeled and deveined.

- **8 oz. spaghetti**
- **2 Tbsp. olive oil**
- **1 1/2 lb. peeled and deveined large raw shrimp**
- **3 Tbsp. butter**
- **2 shallots, minced (about 1/3 cup)**
- **3 garlic cloves, minced**
- **1/4 cup dry white wine**
- **1 cup heavy cream**
- **1 1/2 tsp. lemon zest**
- **1 tsp. kosher salt**
- **1/2 tsp. freshly ground black pepper**
- **1/2 cup (2 oz.) freshly shredded Parmigiano-Reggiano cheese**
- **2 Tbsp. fresh lemon juice**
- **1/4 cup capers, drained and rinsed**
- **Herbed Breadcrumbs**
- **Garnishes: smoked paprika, chopped fresh flat-leaf parsley**

1. Cook spaghetti according to package directions. Drain, reserving 1/4 cup hot pasta water.
2. Heat olive oil in a large skillet over medium heat. Cook shrimp, stirring often, 3 to 4 minutes or just until pink. Remove from skillet. Add butter to skillet. Cook shallots in melted butter, stirring often, about 3 minutes or until softened. Add garlic, and cook 30 seconds.
3. Stir in wine, and cook, stirring often, 2 to 3 minutes or until reduced to about 1 Tbsp. Stir in heavy cream and next 3 ingredients, and cook, stirring often, 5 minutes or until slightly thickened. Stir in cheese, lemon juice, spaghetti, and reserved 1/4 cup pasta water. Reduce heat to medium-low, and cook 5 minutes. Stir in shrimp and capers, and cook, stirring constantly, 5 minutes or until heated through. Remove from heat, and sprinkle with Herbed Breadcrumbs.

Creamy Shrimp Piccata

MAKES 6 servings. **HANDS-ON** 40 min.; **TOTAL** 1 hour, 5 min., including breadcrumbs

Herbed Breadcrumbs

- **5 sourdough French bread slices, torn into large pieces**
- **1 Tbsp. olive oil**
- **1/2 Tbsp. butter, melted**
- **1 Tbsp. chopped fresh flat-leaf parsley**
- **1 tsp. chopped fresh thyme**

1. Preheat oven to 400°. Process bread in food processor 20 seconds or until coarsely chopped. Stir together breadcrumbs, oil, and butter. Spread in an even layer on an aluminum foil-lined baking sheet.
2. Bake at 400° for 15 minutes or until lightly browned, stirring every 5 minutes. Stir in parsley and thyme; add salt and pepper to taste. Bake 2 more minutes. Use immediately.

MAKES 1 1/2 cups. **HANDS-ON** 10 min., **TOTAL** 25 min.

8 SPECTACULAR SIDES

DIRTY RICE RISOTTO

Slowly adding hot broth to the mixture helps release the starch from the rice, giving the risotto its characteristic creamy texture.

- **5 1/4 cups chicken broth, divided**
- **8 oz. chicken livers**
- **2 Tbsp. olive oil**
- **1 lb. ground pork**
- **1 medium-size yellow onion, finely chopped (about 1 cup)**
- **3 celery ribs, finely chopped (about 1 cup)**
- **1 green bell pepper, finely chopped**
- **1/4 cup butter, divided**
- **3 garlic cloves, minced**
- **1 cup Arborio rice**
- **1/2 cup dry white wine**
- **2 1/2 tsp. Creole seasoning**
- **2 green onions, thinly sliced**
- **1/4 cup fresh flat-leaf parsley**
- **1 1/2 Tbsp. fresh lemon juice**
- **Garnish: celery slices**

1. Bring 5 cups broth to a simmer in a large saucepan over medium-high heat. Maintain at a low simmer until ready to use.
2. Meanwhile, sauté livers in hot oil in a Dutch oven over medium heat 5 minutes or until cooked. Remove livers, and finely chop. Add ground pork to Dutch oven, and cook, stirring occasionally, 10 minutes or until deep golden brown. Remove from Dutch oven using a slotted spoon, and drain on paper towels; stir into chopped liver mixture.
3. Add onion, celery, bell pepper, and 1 Tbsp. butter to Dutch oven, and sauté 10 minutes or until tender. Stir in garlic, and sauté 1 minute.
4. Add 1 Tbsp. butter to Dutch oven, and stir until melted. Add rice, and cook, stirring constantly, 1 minute or until fragrant. Stir in wine, and cook, stirring often, 2 minutes or until nearly dry. Add 1 cup hot broth, and cook, stirring constantly, until liquid is absorbed. Repeat with remaining hot broth, 1 cup at a time, until liquid is absorbed. (Total cooking time is 20 to 25 minutes.) Remove from heat.
5. Stir in liver-and-pork mixture, Creole seasoning, next 3 ingredients, and remaining 2 Tbsp. butter. Stir in remaining 1/4 cup broth, if desired. Serve immediately.

Note: We tested with Swanson 100% Natural Chicken Broth.

MAKES 6 to 8 servings. **HANDS-ON** 1 hour, 20 min.; **TOTAL** 1 hour, 20 min.

BROCCOLINI WITH PECANS AND CANE SYRUP VINAIGRETTE

Easy, in-a-pinch substitutions: Use molasses for cane syrup and broccoli florets in place of fresh Broccolini.

- **1/4 cup kosher salt**
- **3 bunches fresh Broccolini (about 1/2 lb.), trimmed**
- **Cane Syrup Vinaigrette**
- **1 cup toasted chopped pecans**
- **Garnish: pomegranate seeds**

Bring salt and 8 qt. water to a boil in a stockpot over high heat. Gently stir in Broccolini, and cook 3 to 4 minutes or until tender. Drain and transfer to a large bowl. Toss together Broccolini and 2 Tbsp. Cane Syrup Vinaigrette. Top with toasted pecans, and serve immediately with remaining vinaigrette.

MAKES 6 to 8 servings. **HANDS-ON** 10 min., **TOTAL** 25 min., including vinaigrette

Cane Syrup Vinaigrette

- **2/3 cup olive oil**
- **1/3 cup apple cider vinegar**
- **2 1/2 Tbsp. pure cane syrup**
- **1 Tbsp. Creole mustard**
- **1/4 tsp. kosher salt**
- **1/4 tsp. freshly ground black pepper**
- **1/8 tsp. ground red pepper**

Whisk together all ingredients in a medium bowl.

MAKES about 1 1/4 cups. **HANDS-ON** 5 min., **TOTAL** 5 min.

Broccolini with Pecans and Cane Syrup Vinaigrette

ROASTED GARLIC DUCHESS POTATOES

Want to make the ornate star presentation and don't have a pastry bag? A 1-gal. zip-top plastic freezer bag will do the trick. Simply snip a small hole in one corner of the bag, and pipe.

- **1 garlic bulb**
- **2 Tbsp. olive oil**
- **3 Tbsp. plus 1/4 tsp. kosher salt, divided**
- **3 lb. russet potatoes, peeled and cubed**
- **5 Tbsp. butter**
- **1/4 cup heavy cream**
- **2 large egg yolks**
- **Parchment paper**

1. Preheat oven to 425°. Cut off pointed end of garlic bulb; place bulb on a piece of aluminum foil. Drizzle with oil, and sprinkle with 1/4 tsp. kosher salt. Fold foil to seal. Bake 30 minutes; cool 15 minutes. Squeeze pulp from garlic cloves into a small bowl, and mash with a spoon until smooth.
2. Place potatoes and remaining 3 Tbsp. salt in a large saucepan with cold water to cover by 2 inches. Bring to a boil over medium heat; boil 25 minutes or until potatoes are tender. Drain and return to pan. Reduce heat to medium-low. Cook, gently stirring often, 5 minutes or until dry. Remove from heat.
3. Bring butter and cream to a simmer in a small saucepan over medium heat. (Do not boil.) Remove from heat.
4. Transfer potatoes to a large bowl, and mash with a potato ricer or masher until smooth and free of lumps. Add hot cream mixture, egg yolks, and roasted garlic, and stir until thoroughly combined. Spoon potato mixture into a pastry bag fitted with a star tip. Pipe 3-inch-wide mounds 2 inches apart on 2 parchment paper-lined baking sheets.
5. Bake at 425° for 20 to 25 minutes or until tops are lightly browned. Serve immediately.

MAKES 8 to 10 servings. **HANDS-ON** 30 min., **TOTAL** 2 hours

Roasted Garlic Duchess Potatoes

CHARRED AMBROSIA WITH TOASTED COCONUT AND MARSHMALLOW CRÈME

Because the marshmallow-egg white mixture isn't cooked, it's necessary to use pasteurized egg whites.

- **1/2 cup plus 1 Tbsp. sugar, divided**
- **1 fresh pineapple, peeled, cored, and cut into 3/4-inch-thick rings**
- **2 oranges, peeled and cut into 3/4-inch-thick rounds**
- **2 red grapefruit, peeled and cut into 3/4-inch-thick rounds**
- **3 pasteurized egg whites**
- **Pinch of kosher salt**
- **1 (7-oz.) jar marshmallow crème**
- **1 cup whole red maraschino cherries, drained and rinsed**
- **1 1/4 cups unsweetened flaked coconut, toasted**

1. Preheat broiler with oven rack 8 to 9 inches from heat.
2. Heat a cast-iron skillet over medium-high heat. Sprinkle 5 Tbsp. sugar on both sides of pineapple slices and next 2 ingredients. Place fruit slices in skillet, and cook, in batches, 30 to 45 seconds on each side or until caramelized. Cool completely on a wire rack in a jelly-roll pan (about 10 minutes).
3. Meanwhile, beat egg whites and salt at high speed with a heavy-duty electric stand mixer until mixture is foamy. Gradually add remaining 1/4 cup sugar, beating until stiff peaks form. Beat one-third of marshmallow crème into egg white mixture; repeat 2 times with remaining marshmallow crème, beating until smooth (about 1 minute) after each addition.
4. Coarsely chop fruit. Stir together chopped fruit and cherries in a large bowl. Spoon into individual bowls or glasses. Dollop with marshmallow crème mixture. If desired, brown mixture using a kitchen torch, holding torch 2 inches from marshmallow crème and moving torch back and forth. Top with coconut.

Note: To broil in oven, place half of chopped fruit in an 11- x 7-inch broiler-safe glass baking dish; top with cherries and 1 cup coconut. Add remaining fruit, and spread marshmallow crème mixture over top. Broil 1 minute. Remove from oven, and sprinkle with remaining 1/4 cup coconut.

MAKES 4 to 6 servings. **HANDS-ON** 35 min., **TOTAL** 35 min.

GNOCCHI MAC AND CHEESE

Turn this kid-friendly side into an entrée by serving with a salad and warm, crusty bread.

- **2 (16-oz.) packages potato gnocchi**
- **1/4 cup plus 2 1/2 tsp. kosher salt**
- **3 Tbsp. butter**
- **2 shallots, finely chopped**
- **3 Tbsp. all-purpose flour**
- **2 Tbsp. fresh thyme leaves, finely chopped**
- **2 1/2 cups milk**
- **2 Tbsp. Dijon mustard**
- **1/4 tsp. hot sauce**
- **4 oz. sharp Cheddar cheese, grated**
- **4 oz. extra-sharp white Cheddar cheese, grated**
- **Vegetable cooking spray**

1. Preheat oven to 375°. Bring 3 qt. water to a boil in a large saucepan. Add both packages of gnocchi and 1/4 cup kosher

salt. Cook about 3 minutes or until gnocchi float.
2. Melt butter in a large saucepan over medium heat. Add shallots, and sauté 30 seconds or until fragrant. Add flour and thyme, and cook, stirring constantly, 2 to 3 minutes or until mixture is golden brown and smooth. Gradually whisk in milk, and increase heat to high. Bring to a boil, whisking occasionally. Reduce heat to medium-low, and simmer, whisking constantly, 5 minutes or until slightly thickened and mixture coats a spoon. Stir in mustard and hot sauce. Remove from heat.
3. Add Cheddar cheeses, and stir until melted. Stir in gnocchi and remaining 2 1/2 tsp. salt, and transfer to a lightly greased (with cooking spray) 11- x 7-inch or 2-qt. baking dish.
4. Bake at 375° for 25 minutes or until gnocchi is puffed and sauce is golden and bubbly. Increase oven temperature to broil, and broil 2 minutes or until top is slightly browned. Let stand 5 minutes, and serve immediately.

Note: We tested with Gia Russa Gnocchi with Potato.

MAKES 6 to 8 servings. **HANDS-ON** 30 min., **TOTAL** 1 hour

FONTINA-CHIVE YORKSHIRE PUDDINGS

For the fluffiest, puffiest puddings, make sure the muffin pan is super hot.

- **6 bacon slices, finely chopped**
- **1 1/2 cups all-purpose flour**
- **1 tsp. kosher salt**
- **4 large eggs**
- **1 3/4 cups milk**
- **4 oz. grated fontina cheese**
- **2 Tbsp. thinly sliced fresh chives**
- **Garnish: sliced fresh chives**

1. Preheat oven to 450°. Heat a 12-cup muffin pan in oven 15 minutes.
2. Meanwhile, cook bacon in a medium skillet over medium heat, stirring occasionally, 8 to 9 minutes or until crisp; remove bacon, using a slotted spoon, and drain on paper towels. Reserve drippings.
3. Whisk together flour and salt in a large bowl. Whisk together eggs and milk in a medium bowl. Gently whisk egg mixture into flour mixture until well blended. Stir in cheese, chives, and cooked bacon.

Fontina-Chive Yorkshire Puddings

4. Spoon 1 tsp. bacon drippings into each cup of hot muffin pan; place muffin pan in oven for 2 minutes. Carefully remove pan. Divide batter evenly between cups.
5. Bake at 450° for 25 minutes until puffed and golden brown. (Center will still be wet.) Serve immediately.

MAKES 8 to 10 servings. **HANDS-ON** 25 min., **TOTAL** 55 min.

CHRISTMAS SALAD

A warm vinaigrette will help soften these sturdy greens, but their structure will hold up for up to an hour or two on a holiday buffet.

- **8 oz. thick bacon, coarsely chopped**
- **1 cup sliced red onion**
- **1 1/2 tsp. kosher salt, divided**
- **3/4 tsp. freshly ground black pepper, divided**
- **1/2 cup red wine vinegar**
- **2 Tbsp. honey**
- **1 bunch Lacinato kale, stemmed and coarsely chopped (about 7 cups loosely packed)**
- **1 bunch red Swiss chard, stemmed and coarsely chopped (about 8 cups loosely packed)**
- **2 small beets, thinly sliced**
- **1/2 cup toasted pumpkin seeds**
- **1/2 cup dried cherries**
- **1/2 cup crumbled feta cheese**

1. Cook bacon in a large skillet over medium heat, stirring occasionally, 6 to 8 minutes or until crisp; remove bacon, using a slotted spoon, and drain on paper towels. Reserve 6 Tbsp. dripping in skillet. Add onion, 1/2 tsp. salt, and 1/4 tsp. pepper, and sauté 2 minutes. Remove from heat.
2. Add red wine vinegar to skillet, and stir to loosen browned bits from bottom of skillet. Whisk in honey.
3. Toss together kale, chard, onion mixture, and remaining 1 tsp. salt and 1/2 tsp. black pepper in a large bowl. Transfer to a serving platter. Top with bacon, beet slices, and next 3 ingredients. Serve immediately.

MAKES 8 to 10 servings. **HANDS-ON** 20 min., **TOTAL** 20 min.

BRUSSELS SPROUTS WITH PARMESAN CREAM SAUCE

Buy in bulk and save. Pick up 3- to 5-lb. packages of Brussels sprouts at warehouse clubs like Costco Wholesale or Sam's Club.

- **3 lb. fresh Brussels sprouts**
- **1 1/2 Tbsp. olive oil**
- **2 tsp. kosher salt**
- **1/2 tsp. freshly ground black pepper**
- **2 cups heavy cream**
- **3 fresh rosemary sprigs**
- **4 garlic cloves**
- **1/4 cup freshly grated Parmesan cheese**
- **Toppings: shaved Parmesan cheese, sliced chives, freshly ground black pepper**

1. Preheat oven to 425°. Remove discolored leaves from Brussels sprouts, cut off stem ends, and quarter. Toss with olive oil, salt, and pepper, and place in a single layer on a baking sheet. Bake 25 minutes or until golden and tender, stirring halfway through.
2. Meanwhile, stir together cream, rosemary, and garlic in a medium saucepan. Cook over medium heat, stirring occasionally, 15 to 20 minutes or until reduced to about 1 cup. Discard rosemary and garlic. Stir in Parmesan cheese. Drizzle over Brussels sprouts. Serve immediately with toppings.

MAKES 6 to 8 servings. **HANDS-ON** 25 min., **TOTAL** 35 min.

Spice Cake with Cranberry Filling

DESSERTS & COOKIES

For a showstopping finale, wow your guests with ***8 dazzling desserts,*** *starring this year's Christmas spice layer cake topped with a sweet, snowy forest scene. And fill tins and trays with an assortment of our delectable* ***9 bakery-worthy cookies,*** *all with a prep time of 30 minutes or less*

DESSERTS: Spice Cake with Cranberry Filling ✦ Praline Cream-Beignet Tower ✦ Gingerbread Baked Alaska ✦ Triple Chocolate Brownie-Mousse Stacks ✦ Chocolate-Peppermint Cheesecake ✦ Coconut-Macadamia Nut Pound Cake ✦ Hot Chocolate ✦ Our Best Homemade Marshmallows ✦ Layered Eggnog Cream with Puff Pastry **COOKIES:** Belgian Spice Cookies ✦ Coconut Snowballs ✦ Red Velvet Thumbprints ✦ Pecan-Cranberry Shortbread ✦ Chocolate-Peppermint Crackle Cookies ✦ 5-Ingredient Sugar Cookies ✦ Layered Eggnog Blondies ✦ Chewy Ginger Cookies ✦ Glazed Fruitcake Bars

9 DAZZLING DESSERTS

SPICE CAKE WITH CRANBERRY FILLING

Pipe a ring of frosting around the cake layer, just inside the top edge, to keep filling from oozing over.

CRANBERRY FILLING

- 2 cups fresh or frozen whole cranberries
- 1 cup granulated sugar
- 3 Tbsp. fresh orange juice
- 2 Tbsp. cornstarch
- 1 Tbsp. cold water
- 1/2 cup chopped fresh or frozen cranberries
- 2 Tbsp. butter

CAKE LAYERS

- 1 cup butter, softened
- 2 cups granulated sugar
- 4 large eggs
- 3 cups all-purpose flour
- 2 tsp. baking powder
- 1 tsp. ground cinnamon
- 1/2 tsp. table salt
- 1/2 tsp. baking soda
- 1/2 tsp. ground ginger
- 1/4 tsp. ground nutmeg
- 1 1/2 cups buttermilk
- 2 tsp. vanilla extract
- 4 (8 1/2-inch) round disposable pans
- Shortening

APPLE CIDER FROSTING

- 1 cup apple cider
- 1 cup butter, softened
- 1/4 tsp. table salt
- 1 (32-oz.) package powdered sugar
- 2 tsp. vanilla extract
- 4 to 5 Tbsp. whole milk

1. Prepare Filling: Stir together first 3 ingredients in a small saucepan. Whisk together cornstarch and cold water in a small bowl until smooth. Stir cornstarch mixture into cranberry mixture, and cook over medium-low heat, stirring often, 4 to 5 minutes or until cranberry skins begin to split and mixture comes to a boil. Cook, stirring constantly, 1 more minute or until thickened and translucent. Stir in chopped cranberries and 2 Tbsp. butter, and cook, stirring constantly, 1 minute or until butter is melted. Remove from heat, and cool completely (about 30 minutes). Cover and chill 8 to 24 hours.
2. Prepare Cake Layers: Preheat oven to 350°. Beat 1 cup butter at medium speed with a heavy-duty electric stand mixer until creamy. Gradually add 2 cups granulated sugar, beating until light and fluffy. Add eggs, 1 at a time, beating just until blended after each addition.
3. Whisk together flour and next 6 ingredients until well blended. Add flour mixture to butter mixture alternately with buttermilk, beginning and ending with flour mixture. Beat at low speed just until blended after each addition. Stir in 2 tsp. vanilla. Pour batter into 4 greased (with shortening) and floured 8 1/2-inch round disposable cake pans.
4. Bake at 350° for 18 to 20 minutes or until a wooden pick inserted in center comes out clean. Cool in pans on wire racks 10 minutes; remove from pans. Cool completely on wire racks (about 1 hour).
5. Meanwhile, prepare Frosting: Cook cider in a small saucepan over medium heat, stirring often, 10 to 15 minutes or until reduced to 1/4 cup. Cool completely (about 20 minutes).
6. Beat 1 cup butter and 1/4 tsp. salt at medium speed with a heavy-duty electric stand mixer until creamy. Gradually add powdered sugar alternately with reduced apple cider, 2 tsp. vanilla, and 4 Tbsp. milk, beating well after each addition. If needed, add up to 1 Tbsp. milk, 1 tsp. at a time, and beat until desired consistency is reached.
7. Place 1 cake layer on a cake plate or platter. Spoon 1 cup Apple Cider Frosting into a zip-top plastic freezer bag. Snip 1 corner of bag to make a small hole. Pipe a ring of frosting around cake layer just inside the top edge. Spread cake layer with about 2/3 cup chilled Cranberry Filling, spreading to edge of piped frosting. Repeat with 2 more layers. Top with remaining cake layer. Spread remaining frosting over top and sides of cake.

MAKES 10 to 12 servings. **HANDS-ON** 1 hour, 15 min.; **TOTAL** 11 hours, 15 min.

PRALINE CREAM-BEIGNET TOWER

Make praline filling and dough up to two days ahead of time.

PRALINE CREAM FILLING

- 3/4 cup firmly packed light brown sugar
- 1/4 cup cornstarch
- 1/4 tsp. table salt
- 2 1/2 cups half-and-half
- 4 large egg yolks
- 3 Tbsp. butter
- 1 Tbsp. pecan liqueur

BEIGNETS

- 1 (1/4-oz.) envelope active dry yeast
- 1 1/2 cups warm water (100° to 110°), divided
- 1/2 cup plus 1 tsp. granulated sugar, divided
- 1 cup evaporated milk
- 2 large eggs, lightly beaten
- 1 tsp. table salt
- 1/4 cup shortening
- 6 1/2-7 cups bread flour
- Vegetable cooking spray

ADDITIONAL INGREDIENTS

- Vegetable oil
- 120 heavy-duty 2-inch wooden picks
- 1 (17 7/8- x 4 15/16-inch) plastic foam cone
- Powdered sugar

1. Prepare Filling: Whisk together first 3 ingredients in a large heavy saucepan; whisk together half-and-half and egg yolks in a medium bowl. Gradually whisk half-and-half mixture into brown sugar mixture, and cook over medium heat, whisking constantly, about 7 minutes or until mixture just begins to bubble. Cook, whisking constantly, 1 more minute; remove from heat, and whisk in butter and pecan liqueur. Whisk until butter melts. Spoon mixture into a medium bowl, and place plastic wrap directly onto filling (to prevent a film from forming). Cool 30 minutes. Chill 4 to 24 hours.

2. Meanwhile, prepare Beignet dough: Combine yeast, 1/2 cup warm water (100° to 110°), and 1 tsp. granulated sugar in bowl of a heavy-duty stand mixer; let stand 5 minutes. Add evaporated milk, eggs, salt, and remaining 1/2 cup granulated sugar to yeast mixture, beating at medium speed until combined.

3. Microwave remaining 1 cup warm water in a 2-cup glass measuring cup until hot (about 115°); stir in shortening until melted. Add to yeast mixture. Beat at low speed, gradually adding 4 cups flour, until smooth. Gradually add remaining 2 1/2 to 3 cups flour, beating until a sticky dough forms. Transfer to a lightly greased (with cooking spray) bowl; lightly grease top of dough with cooking spray. Cover and chill 4 to 24 hours.

4. Turn dough out onto a heavily floured surface; roll half of dough to 1/4-inch thickness. (Refrigerate remaining dough until ready to use.) Cut into 2-inch squares with a pizza wheel or a sharp knife. Repeat with remaining half of dough.

5. Pour oil to a depth of 2 to 3 inches into a large Dutch oven; heat to 360°. Fry dough, in batches, 1 to 2 minutes on each side or until golden brown. Drain on a wire rack.

6. Insert a small metal tip (we used a #17) into a large pastry bag. Using a long wooden skewer, poke a hole in the side of each beignet. Fill pastry bag with half of chilled Praline Filling. Pipe a small amount into each beignet. Repeat with remaining filling and beignets.

7. Assemble: Starting at bottom, insert 1 pick into cone. Press 1 beignet onto pick. Repeat with remaining picks and beignets, covering entire cone. Dust entire beignet-covered cone heavily with powdered sugar.

MAKES 12 to 15 servings. **HANDS-ON** 2 hours; **TOTAL** 6 hours, 35 min.

GINGERBREAD BAKED ALASKA

Be sure your plastic wrap extends well beyond your bowl to cover your cake and ice cream mixture completely—this will also give you plenty of plastic wrap to hold on to when removing mixture from the bowl.

CAKE

- **2 cups firmly packed dark brown sugar**
- **1 cup butter, softened**
- **3 large eggs**
- **1 Tbsp. grated fresh ginger**
- **1 tsp. vanilla extract**
- **3 cups all-purpose flour**
- **1 tsp. baking soda**
- **1/2 tsp. baking powder**
- **1/2 tsp. table salt**
- **1/4 tsp. ground nutmeg**
- **1/4 tsp. ground allspice**
- **1 1/2 cups buttermilk**
- **1/4 cup minced crystallized ginger**
- **Shortening**

ICE CREAM

- **Vegetable cooking spray**
- **1/2 gal. vanilla ice cream, softened**
- **1 Tbsp. lemon zest**
- **2 Tbsp. fresh lemon juice**

SWISS MERINGUE

- **5 large egg whites**
- **1 1/4 cups granulated sugar**
- **1 tsp. vanilla extract**

1. Prepare Cake: Preheat oven to 350°. Beat brown sugar and butter at medium speed with a heavy-duty electric stand mixer until light and fluffy. Add eggs, 1 at a time, beating just until blended after each addition. Stir in grated ginger and 1 tsp. vanilla.

2. Whisk together flour and next 5 ingredients in a small bowl. Add flour mixture to brown sugar mixture alternately with buttermilk, beginning and ending with flour mixture. Beat at low speed just until blended after each addition. Stir in crystallized ginger. Spoon batter into 3 greased (with shortening) and floured 8-inch round cake pans.

3. Bake at 350° for 23 to 28 minutes or until a wooden pick inserted in center comes out clean. Cool in pans on wire racks 10 minutes. Remove from pans to wire racks, and cool completely (about 1 hour).

4. Cut a 6-inch circle from 1 cake layer. (We used a 6-inch round cake pan as a guide.) Discard scraps, or reserve for another use.

Gingerbread Baked Alaska

5. Prepare Ice Cream: Lightly grease a 5-qt. metal bowl with cooking spray. Line bowl with plastic wrap, allowing 4 to 5 inches to extend over sides.
6. Stir together softened ice cream, zest, and lemon juice. Spoon 1 1/2 cups ice-cream mixture into prepared bowl. Place 6-inch cake layer on ice-cream mixture, pressing gently. Top with half of remaining ice-cream mixture.
7. Place 1 (8-inch) cake layer on ice-cream mixture, pressing gently. Top with remaining ice-cream mixture and remaining 8-inch cake layer, pressing gently. Fold extended plastic wrap over cake layer to cover completely. Freeze 12 to 24 hours.
8. Uncover cake and ice cream, and invert bowl onto serving platter, keeping plastic wrap around cake and ice cream. Return to freezer until ready to use.
9. Prepare Swiss Meringue: Pour water to a depth of 1 1/2 inches into a small saucepan; bring to a boil over medium-high heat. Reduce heat to medium-low, and maintain at a simmer. Whisk together egg whites and 1 1/4 cups granulated sugar in bowl of a heavy-duty electric stand mixer. Place bowl over simmering water, and cook, whisking constantly, 3 minutes or until sugar dissolves and a candy thermometer registers 140°. Whisk in 1 tsp. vanilla.
10. Beat mixture at medium-high speed with heavy-duty stand mixer, using whisk attachment, 8 to 10 minutes or until stiff peaks form and meringue has cooled completely.
11. Remove cake and ice cream from freezer; remove and discard plastic wrap. Spread meringue over top and sides, completely covering cake and ice cream. Brown meringue using a kitchen torch, holding torch 2 inches from meringue and moving torch back and forth. Serve immediately.

MAKES 8 to 10 servings. **HANDS-ON** 1 hour; **TOTAL** 14 hours, 35 min.

Triple Chocolate Brownie-Mousse Stacks

TRIPLE CHOCOLATE BROWNIE-MOUSSE STACKS

Allow enough time to make fillings and immediately assemble stacks. You can chill the assembled stacks up to 24 hours.

BROWNIES
- **Vegetable cooking spray**
- 3/4 **cup butter**
- 1 **(4-oz.) bittersweet dark chocolate baking bar, chopped**
- 1 1/2 **cups sugar**
- 1 **tsp. vanilla extract**
- 4 **large eggs**
- 1 **cup all-purpose flour**
- 1/4 **tsp. baking powder**
- 1/4 **tsp. table salt**

MILK CHOCOLATE MOUSSE
- 1/2 **(12-oz.) package milk chocolate morsels (1 cup)**
- 1/4 **cup creamy peanut butter**
- 1 **cup heavy cream**

WHITE CHOCOLATE MOUSSE
- 1 **cup white chocolate morsels**
- 1 1/4 **cups heavy cream, divided**

ADDITIONAL INGREDIENTS
- 8 **large paper clips**
- **Heavy-duty aluminum foil**
- **Garnish: shaved chocolate**

1. Prepare Brownies: Preheat oven to 350°. Line bottom and sides of a 13- x 9-inch pan with aluminum foil, allowing 2 to 3 inches to extend over sides; lightly grease foil with cooking spray. Microwave butter and chopped bittersweet chocolate in a large microwave-safe bowl at HIGH 1 1/2 to 2 minutes or just until melted and smooth, stirring every 30 seconds. Whisk in sugar and vanilla. Add eggs, 1 at a time, whisking just until blended after each addition. Stir together flour, baking powder, and salt in a small bowl. Whisk flour mixture into chocolate mixture until blended. Pour mixture into prepared pan.
2. Bake at 350° for 18 to 20 minutes or until a wooden pick inserted in center comes out with a few moist crumbs. Cool completely on a wire rack (about 1 hour). Lift brownies from pan, using foil sides as handles. Cut 8 circles, using a 3-inch round cutter. Reserve scraps for another use.
3. Prepare Milk Chocolate Mousse: Microwave milk chocolate morsels and peanut butter in a small microwave-safe glass bowl at MEDIUM (50% power) 1 1/2 to 2 minutes or until melted and smooth, stirring every 30 seconds. Cool 5 minutes.
4. Beat 1 cup heavy cream at medium speed with an electric mixer until soft peaks form; fold cream into milk chocolate mixture. Chill while making White Chocolate Mousse.
5. Prepare White Chocolate Mousse: Microwave white chocolate morsels and 1/4 cup cream in a small microwave-safe glass bowl at MEDIUM (50% power) 1 1/2 to 2 minutes or until melted and smooth, stirring every 30 seconds. Cool 5 minutes.
6. Beat remaining 1 cup cream at medium speed with an electric mixer until soft peaks form; fold cream into white chocolate mixture. Chill while preparing foil molds in Step 7.
7. Assemble stacks: Wash and dry paper clips. Cut heavy-duty aluminum foil into 8 (6- x 10-inch) pieces. Fold

each piece in half to form a 3- x 10-inch strip. Wrap each strip around a 3-inch diameter can. (This helps to create a smooth curve.) Wrap 1 curved foil strip around each brownie; secure with a large paper clip. Immediately spoon Milk Chocolate Mousse into a zip-top plastic freezer bag. (Do not seal.) Snip 1 corner of bag to make a small hole (about 1/2 inch). Pipe mousse onto brownies, dividing mixture evenly. Use a small spoon to gently level. Repeat procedure with White Chocolate Mousse. Chill 2 hours; remove foil to serve.

MAKES 8 servings. **HANDS-ON** 45 min.; **TOTAL** 4 hours, 30 min.

CHOCOLATE-PEPPERMINT CHEESECAKE

Don't worry if your cheesecake cracks; the light and fluffy layer of whipped cream will cover it.

CRUST

- **2 cups chocolate wafer crumbs (about 35 wafers)**
- **3 Tbsp. granulated sugar**
- **5 Tbsp. butter, melted**
- **Vegetable cooking spray**

FILLING

- **1 cup semisweet chocolate morsels**
- **1/4 cup heavy cream**
- **4 (8-oz.) packages cream cheese, softened at room temperature**
- **1 cup granulated sugar**
- **1 tsp. vanilla extract**
- **1 tsp. peppermint extract**
- **4 large eggs**

ADDITIONAL INGREDIENTS

- **Whipped Peppermint Cream**
- **Garnish: crushed peppermint candies**

1. Prepare Crust: Preheat oven to 350°. Stir together first 3 ingredients in a medium bowl. Press mixture on bottom and 1 inch up sides of a lightly greased (with cooking spray) 9-inch springform pan. Bake 10 minutes. Let stand at room temperature until ready to use.

2. Prepare Filling: Reduce oven temperature to 325°. Microwave chocolate morsels and cream in a small microwave-safe bowl at MEDIUM (50% power) 1 to 1 1/2 minutes or until melted and smooth, stirring at 30-second intervals. Cool 10 minutes.

3. Beat cream cheese and 1 cup sugar at medium-low speed with a heavy-duty electric stand mixer just until smooth. Add chocolate mixture and extracts, and beat at low speed just until blended. Add eggs, 1 at a time, beating at low speed just until yellow disappears after each addition; pour into prepared crust.

4. Bake at 325° for 50 minutes to 1 hour or until center of cheesecake jiggles and is almost set. Remove cheesecake from oven, and gently run a knife around outer edge of cheesecake to loosen from sides of pan. (Do not remove sides.) Cool cheesecake completely on a wire rack (about 2 hours). Cover and chill 8 hours to 2 days.

5. Remove sides of pan, and spread cheesecake with Whipped Peppermint Cream.

MAKES 10 to 12 servings. **HANDS-ON** 30 min.; **TOTAL** 11 hours, 50 min., including topping

Whipped Peppermint Cream

Beat 2 cups **heavy cream** and 3/4 tsp. **peppermint extract** at medium-high speed until foamy; gradually add 3/4 cup **powdered sugar,** beating until soft peaks form.

MAKES about 4 cups. **HANDS-ON** 10 min., **TOTAL** 10 min.

Coconut-Macadamia Nut Pound Cake

COCONUT-MACADAMIA NUT POUND CAKE

Lightly toasting the macadamia nuts brings out their rich, buttery flavor.

- **1 1/2 cups butter, softened**
- **1 (8-oz.) package cream cheese, softened**
- **3 cups sugar**
- **6 large eggs**
- **3 cups all-purpose flour**
- **1/4 cup gold rum**
- **1 tsp. vanilla extract**
- **1/2 tsp. coconut extract**
- **1 cup chopped, roasted, salted macadamia nuts, lightly toasted**
- **Shortening**
- **1/2 cup sweetened shredded coconut**
- **Snowy Rum Glaze**
- **Garnishes: toasted coconut, chopped macadamia nuts**

1. Preheat oven to 325°. Beat butter and cream cheese at medium speed with a heavy-duty electric stand mixer 2 to 3 minutes or until creamy. Gradually add sugar, beating at medium speed 5 to 7 minutes or until light and fluffy. Add eggs, 1 at a time, beating just until yellow disappears.

2. Add flour to butter mixture alternately with rum, beginning and ending with flour. Beat batter at low speed just until blended after each addition. Stir in vanilla and next 3 ingredients. Pour batter into a well greased (with shortening) and floured 15-cup Bundt pan.

3. Bake at 325° for 1 hour and 10 minutes to 1 hour and 20 minutes or until a long wooden pick inserted in center comes out clean. Cool in pan on a wire rack 10 to 15 minutes; remove from pan to wire rack, and cool completely (about 2 hours). Spoon Snowy Rum Glaze over cooled cake.

MAKES 10 to 12 servings. **HANDS-ON** 20 min.; **TOTAL** 3 hours, 35 min., including glaze

Snowy Rum Glaze

Whisk together 2 cups **powdered sugar,** 2 Tbsp. **milk,** 1 Tbsp. **gold rum,** and 1 tsp. **vanilla** extract. If needed, stir in additional 1 Tbsp. milk, 1 tsp. at a time, to reach desired consistency.

MAKES about 1 cup. **HANDS-ON** 5 min., **TOTAL** 5 min.

HOT CHOCOLATE

Like your hot cocoa extra rich? Increase heavy cream to 3 cups and decrease milk to 1 cup.

- **2 cups milk**
- **1/2 cup sugar**
- **1/2 cup unsweetened cocoa**
- **Pinch of table salt**
- **2 cups heavy cream**
- **2 tsp. vanilla extract**
- **Our Best Homemade Marshmallows**

Bring first 4 ingredients to a low boil in a saucepan over medium heat, whisking often. Whisk in cream, and cook, whisking often, 3 to 5 minutes or until bubbles begin to form around edges of pan. Remove from heat; stir in vanilla. Serve immediately with Our Best Homemade Marshmallows.

MAKES 4 1/2 cups. **HANDS-ON** 15 min., **TOTAL** 15 min.

OUR BEST HOMEMADE MARSHMALLOWS

- **Parchment paper**
- **Vegetable cooking spray**
- **Powdered sugar**
- **3 envelopes unflavored gelatin**
- **1 cup cold water, divided**
- **1 1/4 cups granulated sugar**
- **1 1/4 cups light corn syrup**
- **1/4 tsp. kosher salt**
- **2 tsp. vanilla extract**

1. Line bottom of an 11- x 7-inch baking dish with parchment paper; lightly grease with cooking spray. Dust sides and bottom of dish with powdered sugar. Sprinkle gelatin over 1/2 cup cold water in bowl of a heavy-duty electric stand mixer.
2. Stir together granulated sugar, next 2 ingredients and remaining 1/2 cup cold water in a medium saucepan over medium-high heat. Cook, stirring often, until a candy thermometer registers 240° (about 12 to 15 minutes; lower heat as necessary to prevent mixture from boiling over).
3. Gradually add hot sugar mixture to gelatin mixture, beating at medium-low speed and using whisk attachment. Gradually increase speed to high, and beat 7 minutes or until bottom of bowl is slightly warm and mixture is very thick. Stir in vanilla.
4. Working quickly, spread marshmallow mixture into prepared dish; smooth with a lightly greased offset spatula. Dust heavily with powdered sugar.
5. Let marshmallow mixture stand, uncovered, in a cool, dry place 8 to 24 hours. (Mixture should be dry enough to release from baking dish.) Cut mixture into 1 1/2-inch squares.

MAKES 28 marshmallows. **HANDS-ON** 30 min.; **TOTAL** 8 hours, 30 min.

Peppermint Marshmallows

Substitute 1 tsp. **peppermint extract** for vanilla extract. Prepare recipe as directed through Step 3, adding 4 drops **red food coloring gel,** and swirling into marshmallow mixture, using a lightly greased rubber spatula. Spread in prepared dish, and dot with 6 drops red food coloring gel. Swirl, using a wooden skewer. Proceed with recipe as directed in Step 5.

Bourbon-Caramel Marshmallows

Substitute 2 Tbsp. **bourbon** for vanilla extract. Prepare recipe as directed through Step 3, adding 2 Tbsp. warm jarred **caramel topping,** and swirling into marshmallow mixture, using a lightly greased rubber spatula. Spread in prepared dish; drizzle with 2 Tbsp. warm jarred caramel topping. Swirl using a wooden skewer. Proceed with recipe as directed in Step 5.

LAYERED EGGNOG CREAM WITH PUFF PASTRY

- **1 (17.3-oz.) package frozen puff pastry sheets**
- **1 1/2 cups fresh blackberries**
- **1 Tbsp. granulated sugar**
- **1 tsp. fresh lemon juice**
- **1 1/2 Tbsp. bourbon, divided**
- **Parchment paper**
- **2 cups heavy cream**
- **1/2 cup refrigerated eggnog**
- **1/2 cup powdered sugar**
- **1/4 tsp. ground nutmeg**
- **Garnish: fresh mint**

1. Preheat oven to 400°. Let puff pastry sheets stand at room temperature 20 minutes. Meanwhile, stir together blackberries, next 2 ingredients, and 1 Tbsp. bourbon in a small bowl. Cover and chill until ready to use.
2. Unfold pastry sheets onto a floured surface, and cut into 3 strips along fold marks. Cut each strip into 4 (3- x 2-inch) rectangles, discarding scraps. Place rectangles in a single layer on a parchment paper-lined baking sheet. Place parchment paper over pastry rectangles, and top with a second baking sheet to weight the dough.
3. Bake weighted pastry rectangles at 400° for 25 minutes or until golden brown. Remove and discard top layer of parchment paper. Cool completely on wire rack (about 15 minutes). Separate each rectangle into 2 thinner pieces, using a small knife.
4. Beat cream and eggnog at medium-high speed with an electric mixer until foamy; gradually add powdered sugar and nutmeg, beating until soft peaks form. Stir in remaining 1/2 Tbsp. bourbon.
5. Spoon mixture into a zip-top plastic freezer bag. Twist end to loosely close. Snip 1 corner of bag to make a small hole (about 1/2 inch). Place a pastry rectangle on a dessert plate. Pipe a 1/2-inch-thick layer of eggnog mixture on pastry rectangle; top with second pastry rectangle. Pipe another layer of eggnog mixture, and top with third pastry rectangle. Dust with powdered sugar. Repeat with remaining pastry rectangles and eggnog mixture. Serve immediately with blackberry mixture.

MAKES 16 servings. **HANDS-ON** 30 min.; **TOTAL** 1 hour, 5 min.

9 SUPERFAST COOKIES

BELGIAN SPICE COOKIES

To create a lacy design on these cookies (also known as Speculoos), place a stencil or doily over the cooled cookies before dusting with spices and sugar.

- 1 **cup all-purpose flour**
- 1/4 **tsp. baking powder**
- 1/4 **tsp. baking soda**
- 1/4 **tsp. table salt**
- 3 **tsp. pumpkin pie spice, divided**
- 1/2 **cup butter, softened**
- 1/4 **cup turbinado sugar**
- 1/4 **cup firmly packed light brown sugar**
- **Parchment paper**
- 1/2 **cup powdered sugar**

1. Stir together first 4 ingredients and 1 tsp. pumpkin pie spice in a small bowl. Beat butter and next 2 ingredients in a medium bowl at medium speed with an electric mixer 2 to 3 minutes or until combined. Stir in flour mixture just until blended. Shape dough into a 2-inch-thick log; wrap in plastic wrap, and chill 4 hours to 3 days.

Belgian Spice Cookies

2. Preheat oven to 325°. Slice dough log into 1/4-inch-thick rounds. Place cookies 1 inch apart on a parchment paper-lined baking sheet.
3. Bake at 325° for 10 to 14 minutes or until cookies are fragrant and dry. Cool completely on a wire rack (about 30 minutes).
4. Stir together powdered sugar and remaining 2 tsp. pumpkin pie spice. Place stencil on cookie, if desired. Sift powdered sugar mixture over cookie.

MAKES about 2 dozen. **HANDS-ON** 10 min., **TOTAL** 5 hours

COCONUT SNOWBALLS

If the white chocolate is too firm after it is heated, stir in about 1/4 tsp. coconut oil.

- 1/2 **cup unsalted butter, softened**
- 1/2 **cup powdered sugar**
- 1 **tsp. pure coconut extract or vanilla extract**
- 1 **cup all-purpose flour**
- 1/2 **cup unsweetened shredded coconut**
- **Parchment paper**

Coconut Snowballs

- 4 **oz. white chocolate, chopped and melted according to package directions**
- **Garnish: shaved coconut**

1. Preheat oven to 400°. Beat butter and sugar at medium speed with an electric mixer until creamy. Add extract; beat 30 seconds. Gradually add flour, beating at low speed until combined after each addition. Stir in shredded coconut. (If dough is soft, divide into half, and chill 30 minutes to 5 days.)
2. Drop dough by level spoonfuls 2 inches apart onto 2 parchment paper-lined baking sheets, using a 1-inch cookie scoop.
3. Bake, in batches, at 400° for 7 to 9 minutes or until cookies are golden brown on bottom. Cool completely on a wire rack (about 30 minutes). Spread each cooled cookie with about 1/2 tsp. melted chocolate.

MAKES about 2 dozen. **HANDS-ON** 10 min.; **TOTAL** 1 hour, 5 min.

RED VELVET THUMBPRINTS

Transfer frosting mixture to a zip-top bag, and snip the corner to use as a pastry bag.

- 3/4 **cup granulated sugar**
- 1/2 **cup unsalted butter, softened**
- 1 **large egg**
- 1 1/2 **tsp. red liquid food coloring**
- 1 1/4 **cups all-purpose flour**
- 1 **Tbsp. unsweetened cocoa**
- 1/2 **tsp. table salt**
- 1 **tsp. vanilla extract**
- **Parchment paper**
- 2 **oz. cream cheese, softened at room temperature**
- 1/4 **cup white chocolate morsels, melted according to package directions and cooled**
- **Garnish: white chocolate morsels**

1. Preheat oven to 325°. Beat sugar and butter at medium speed with an electric mixer until creamy. Add egg and food coloring; beat 30 seconds.
2. Sift flour with next 2 ingredients in a small bowl. Add flour mixture to butter mixture, beating at low speed until combined. Stir in vanilla.

3. Shape dough into 1-inch balls, and place 1/2 inch apart on parchment paper-lined baking sheets. Press thumb or end of a wooden spoon into each ball, forming an indentation.
4. Bake at 325° for 10 to 15 minutes or until cookies are fragrant and dry. While cookies are still warm, press indentations again. Cool cookies completely on a wire rack (about 30 minutes).
5. Stir together cream cheese and melted white chocolate in a small bowl. Fill centers of cookies with cream cheese mixture.

MAKES about 2 dozen. **HANDS-ON** 12 min.; **TOTAL** 1 hour, 20 min.

PECAN-CRANBERRY SHORTBREAD

Making shortbread with melted butter skips the wait for it to soften.

- 3/4 **cup toasted pecan halves, divided**
- 1 **cup powdered sugar**
- 10 **Tbsp. unsalted butter, melted**
- 1/2 **cup dried sweetened cranberries**
- 1/4 **tsp. table salt**
- 1/2 **tsp. orange zest**
- 1 **cup all-purpose flour**
- **Vegetable cooking spray**
- **Parchment paper**
- 1/4 **cup semisweet chocolate morsels**
- 1/2 **tsp. shortening**

1. Preheat oven to 325°. Process 1/2 cup pecans in a food processor until finely ground. Add powdered sugar and next 4 ingredients; pulse until cranberries are coarsely chopped. Add flour, and pulse just until blended.
2. Spread dough in a lightly greased (with cooking spray) 9-inch tart pan with removable bottom; press remaining pecans into dough.
3. Bake at 325° for 30 to 35 minutes or until edges are golden. Remove sides of pan, and cut shortbread into 8 or 10 wedges; transfer to a parchment paper-lined baking sheet. Bake 10 minutes or until firm. Transfer wedges to a wire rack, and cool completely (about 30 minutes).

Chocolate-Peppermint Crackle Cookies

4. Microwave chocolate morsels and shortening in a microwave-safe bowl at HIGH 1 minute or until melted and smooth, stirring halfway through. Drizzle melted chocolate over cookies; let cookies stand 10 minutes or until chocolate is set.

MAKES 8 to 10 cookies. **HANDS-ON** 10 min.; **TOTAL** 1 hour, 40 min.

CHOCOLATE-PEPPERMINT CRACKLE COOKIES

For the best "crackle" effect, reroll balls in powdered sugar just before baking.

- 2 **oz. unsweetened chocolate baking bars, chopped**
- 2 **Tbsp. butter, cut into pieces**
- 1/4 **tsp. table salt**
- 1 **cup granulated sugar**
- 1/2 **tsp. peppermint extract**
- 2 **large eggs**
- 3/4 **cup all-purpose flour**
- 1/4 **cup unsweetened cocoa**
- 1 **tsp. baking powder**
- 1 **cup powdered sugar**
- **Parchment paper**

1. Microwave first 3 ingredients in a medium-size microwave-safe bowl at HIGH 1 to 2 minutes or until butter and chocolate are melted and smooth, stirring at 30-second intervals. Stir in granulated sugar and next 2 ingredients until blended. Stir together flour and next 2 ingredients in a small bowl. Stir

5-Ingredient Sugar Cookies

flour mixture into chocolate mixture until well blended. Cover bowl with plastic wrap, and chill 1 hour to 5 days.
2. Preheat oven to 350°. Carefully drop dough by level spoonfuls into powdered sugar, using a 1 1/2-inch cookie scoop, and roll to coat. Place coated cookies 1 inch apart on 2 parchment paper-lined baking sheets.
3. Bake, in batches, at 350° for 10 to 12 minutes or until cookies are crackled. Cool on a wire rack 20 minutes.

MAKES about 15 cookies. **HANDS-ON** 15 min., **TOTAL** 2 hours

5-INGREDIENT SUGAR COOKIES

When glazing, use a shallow bowl that will accommodate the entire cookie.

- 1/2 **cup butter, softened**
- 3 **cups powdered sugar, divided**
- 1 **large egg**
- 1 1/2 **tsp. vanilla extract**
- 1 1/2 **cups all-purpose flour**
- **Parchment paper**
- **White nonpareils**

1. Beat butter and 1 cup sugar at medium speed with an electric mixer until creamy. Add egg and vanilla; beat 30 seconds. Add flour, beating at low speed until combined.
2. Place dough on lightly floured parchment paper, and roll to 1/4 inch

Layered Eggnog Blondies

thick. Transfer rolled dough and parchment paper to a baking sheet, and chill 15 minutes.
3. Preheat oven to 375°. Cut dough with lightly floured 2-inch cutters; place 1/2 inch apart on 2 parchment paper-lined baking sheets. Reroll scraps, and repeat process.
4. Bake, in batches, at 375° for 8 to 10 minutes or until cookies are golden brown around edges. Cool on wire rack 30 minutes.
5. Whisk together 2 1/2 Tbsp. water and remaining 2 cups powdered sugar. Dip top of each cookie in glaze, and transfer to a wire rack in a jelly-roll pan. Sprinkle with nonpareils. Let stand 1 hour or until set.

MAKES about 2 1/2 dozen. **HANDS-ON** 20 min., **TOTAL** 2 hours

LAYERED EGGNOG BLONDIES

These rich and creamy treats have a split personality: one part blondie, one part eggnog cheesecake.

- **Vegetable cooking spray**
- **1 1/2 cups crushed vanilla wafers**
- **3 Tbsp. granulated sugar**
- **1/4 tsp. table salt**
- **5 Tbsp. butter, melted**
- **1 (8-oz.) package cream cheese, softened**
- **1/2 cup powdered sugar**
- **1/2 cup refrigerated eggnog**
- **1/4 tsp. ground nutmeg**
- **1/4 tsp. ground cinnamon**
- **1 large egg**
- **3/4 cup heavy cream**
- **1 tsp. vanilla extract**
- **2 Tbsp. powdered sugar**

1. Preheat oven to 350°. Lightly grease (with cooking spray) an aluminum foil-lined 8-inch square pan. Pulse vanilla wafers and next 2 ingredients in a food processor until blended. Add butter, and pulse until blended. Press mixture into bottom of pan.
2. Bake at 350° for 8 to 10 minutes or until lightly browned. Reduce oven temperature to 325°. Beat cream cheese and 1/2 cup powdered sugar at medium speed with an electric mixer. Add eggnog in a steady stream, beating at medium speed. Stir in nutmeg, cinnamon, and egg; pour over vanilla wafer mixture.
3. Bake at 325° for 30 to 35 minutes or until outer 2 inches are set. Cool 1 hour.
4. Beat cream at medium-high speed, using whisk attachment, until foamy. Gradually add vanilla and 2 Tbsp. powdered sugar, beating just until soft peaks form.
5. Remove blondies from pan; discard foil. Cut blondies into squares, and dollop with whipped cream. Refrigerate blondies in an airtight container up to 1 week.

MAKES about 1 1/2 dozen. **HANDS-ON** 8 min.; **TOTAL** 1 hour, 45 min.

CHEWY GINGER COOKIES

Fresh ginger and black pepper bring a kick to this classic cookie. Prep tip: Swirl a liquid measuring cup with a thin coat of vegetable oil before measuring molasses for the most accurate (and least sticky) result.

- **2/3 cup firmly packed dark brown sugar**
- **1/2 cup shortening**
- **1 large egg**
- **1/4 cup unsulphured molasses**
- **2 tsp. finely grated peeled fresh ginger**
- **1 1/2 cups all-purpose flour**
- **1 1/2 tsp. baking soda**
- **1 1/2 tsp. pumpkin pie spice**
- **1/4 tsp. table salt**
- **1/8 tsp. freshly ground black pepper**
- **2/3 cup or 1 (4-oz.) container gold sanding sugar**
- **Parchment paper**

1. Preheat oven to 350°. Beat dark brown sugar and shortening in a medium bowl at medium speed with an electric mixer until creamy. Add egg, molasses, and ginger; beat

30 seconds. Sift flour with next 4 ingredients in a small bowl. Add flour mixture to shortening mixture, and beat at low speed until combined.

2. Place sanding sugar in a small bowl. Drop dough by level spoonfuls into sanding sugar, using a 1 1/2-inch cookie scoop; roll to coat. Place coated cookies 3 inches apart on 2 parchment paper-lined baking sheets.

3. Bake, in batches, at 350° for 10 to 12 minutes or until cookies are fragrant and browned around edges. Cool on pans 5 minutes; transfer to a wire rack, and cool completely (about 30 minutes).

Note: Sanding sugar can be ordered online or purchased wherever specialty cake supplies are sold.

MAKES 1 1/2 dozen. **HANDS-ON** 15 min.; **TOTAL** 1 hour, 25 min.

GLAZED FRUITCAKE BARS

These dense, chewy fruitcake bars are a fresh version of the kind passed around at holiday parties. We love raisins and dried cherries in this recipe, but use your favorite combinations.

- **Parchment paper**
- **Vegetable cooking spray**
- 1/2 **cup all-purpose flour**
- 1/4 **tsp. baking soda**
- 1/4 **tsp. baking powder**
- 1/8 **tsp. ground cloves**
- 1/4 **tsp. table salt**
- 1/2 **cup firmly packed light brown sugar**
- 2 **cups coarsely chopped walnuts**
- 1 1/4 **cups mixed dried fruit**
- 1/4 **cup minced crystallized ginger**
- 1 **large egg, lightly beaten**
- 1 **Tbsp. butter, melted**
- 4 **tsp. bourbon or brandy**
- 1 **cup powdered sugar**
- **Garnish: chopped walnuts**

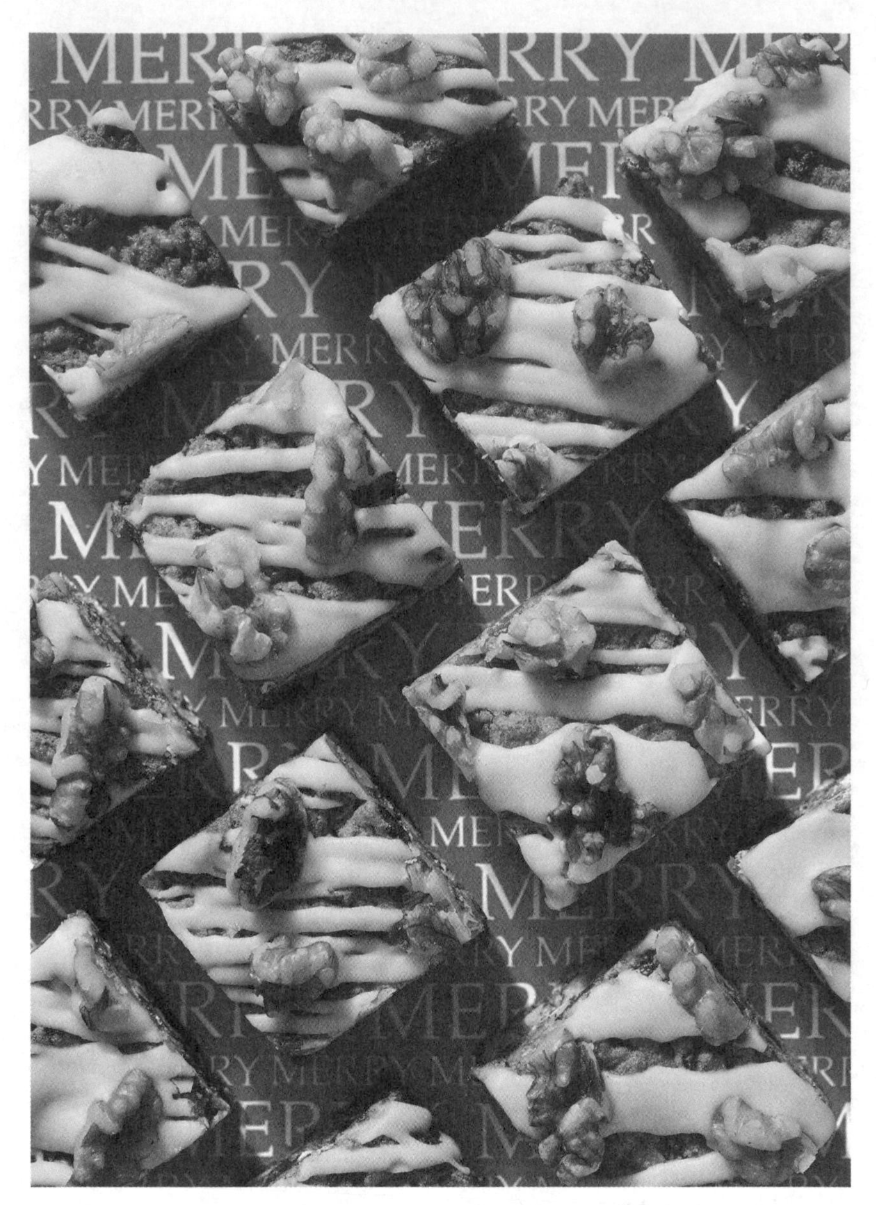

1. Preheat oven to 325°. Line bottom and sides of an 8-inch square pan with parchment paper, allowing 2 inches to extend over sides; lightly grease with cooking spray.

2. Stir together flour and next 4 ingredients in a large bowl; add sugar and next 3 ingredients, tossing to break up any clusters of dried fruit. Add egg, stirring until blended. Press batter into prepared pan.

3. Bake at 325° for 35 to 40 minutes or until set. Cool in pan 20 minutes. Lift fruitcake from pan, using parchment paper sides as handles, and cool completely on a wire rack (about 1 hour).

4. Stir together butter and bourbon in a small bowl. Whisk in powdered sugar; drizzle glaze over cooled fruitcake. Let stand 20 minutes or until glaze is set. Cut into 1 1/2-inch bars.

MAKES about 3 dozen. **HANDS-ON** 15 min.; **TOTAL** 2 hours, 30 min.

The Baker's Brunch

FOR 32 YEARS, PAM LOLLEY—OUR TEST KITCHEN'S BAKER EXTRAORDINAIRE—HAS BEEN PERFECTING HER CHRISTMAS MENU. IT'S NO WONDER HER FAMILY CONSIDERS CINNAMON ROLLS AND GRITS AMONG THEIR FAVORITE GIFTS.

When Test Kitchen Professional Pam Lolley's eight grandchildren—all younger than 8—come over for brunch on Christmas, they bypass the presents under the tree and make a beeline straight for the kitchen. Cinnamon rolls rise in the oven, and their grandmother swats their hands away as they try to sneak a piece of the warm cream cheese pastry on the counter. "If I didn't make those two treats, there would be a riot," Pam says. Pam started her make-ahead Christmas brunch tradition when she was pregnant with her third child and wanted a stress-free way to feed a crowd. Today, she preps all of it in advance, from the glazed ham to the grits casserole—except for the muffins, which she saves as a special activity to share with her grandkids. "Every year, another one is old enough to stand on the stool next to me to help stir the batter," she says. "Cooking with them during the holidays is my Christmas present."

SMOKY SAUSAGE-AND-GRITS CASSEROLE

You can assemble this up to four days ahead, and keep in the fridge; then let it stand at room temp for 30 minutes before baking.

- **1 1/2 lb. smoked sausage, chopped**
- **1/2 tsp. table salt**
- **1 1/2 cups uncooked quick-cooking grits**
- **2 (8-oz.) blocks sharp Cheddar cheese, shredded**
- **1 cup milk**
- **1 1/2 tsp. chopped fresh thyme**
- **1/4 tsp. garlic powder**
- **1/4 tsp. black pepper**
- **4 large eggs, lightly beaten**
- **Vegetable cooking spray**

1. Preheat oven to 350°. Brown sausage in a large skillet over medium-high heat, stirring often, 7 to 9 minutes or until cooked; drain on paper towels.
2. Bring salt and 4 1/2 cups water to a boil in a large Dutch oven over high heat. Whisk in grits, and return to a boil. Cover, reduce heat to medium, and simmer 5 minutes or until thickened, whisking occasionally. Remove from heat; add cheese, stirring until completely melted. Stir in milk and next 4 ingredients. Stir in sausage. Spoon mixture into a lightly greased (with cooking spray) 13- x 9-inch baking dish.
3. Bake at 350° for 50 minutes to 1 hour or until golden and cooked through. Let stand for 5 minutes before serving.

MAKES 12 to 14 servings. **HANDS-ON** 30 min.; **TOTAL** 1 hour, 25 min.

MINI CHOCOLATE CHESS TARTS

If you make tarts often, Pam suggests investing in a tart tamper. It will make quick work of fitting dough rounds into the pan. Another quick-prep tip: Stack piecrusts to cut multiple rounds in one turn.

- **2 (14.1-oz.) packages refrigerated piecrusts**
- **Vegetable cooking spray**
- **1 1/2 cups sugar**
- **3 Tbsp. unsweetened cocoa**
- **1 1/2 Tbsp. cornmeal**
- **1 Tbsp. all-purpose flour**
- **1/8 tsp. table salt**
- **1/2 cup butter, melted**
- **2 tsp. white vinegar**
- **1 tsp. vanilla extract**
- **3 large eggs, lightly beaten**
- **Garnishes: sweetened whipped cream, cocoa**

1. Preheat oven to 350°. Unroll piecrusts on a flat surface. Cut 60 rounds, using a 2 1/2-inch cutter. Press rounds onto bottoms and up sides of a lightly greased (with cooking spray) 24-cup miniature muffin pan. Refrigerate remaining piecrust rounds until ready to use.
2. Whisk together sugar and next 4 ingredients in a medium bowl. Whisk in melted butter, vinegar, and vanilla. Add eggs, whisking until well blended. Spoon about 1 heaping teaspoonful chocolate mixture into each piecrust shell.
3. Bake at 350° for 22 minutes or until set and crust is golden brown. Cool completely on a wire rack (about 30 minutes). Repeat with remaining piecrust rounds and chocolate mixture.

Note: Store in an airtight container in the refrigerator up to 3 days. Top with whipped cream and cocoa just before serving.

MAKES 5 dozen. **HANDS-ON** 30 min.; **TOTAL** 1 hour, 20 min.

5 BRUNCH BEAUTIES

CHRISTMAS MORNING CINNAMON ROLLS

Make sure your butter is very soft when you spread it on the rolled dough so the dough doesn't tear.

- **1 (1/4-oz.) envelope active dry yeast**
- **1/4 cup warm water (105° to 115°)**
- **1 cup plus 1 tsp. granulated sugar, divided**
- **1/2 cup butter, softened**
- **1 tsp. table salt**
- **2 large eggs, lightly beaten**
- **1 cup milk**
- **1 Tbsp. fresh lemon juice**
- **5 cups bread flour, divided**
- **1/4 tsp. ground nutmeg**
- **Vegetable cooking spray**
- **1/2 cup very soft butter**
- **1/2 cup firmly packed light brown sugar**
- **1 Tbsp. ground cinnamon**
- **1 cup toasted chopped pecans**
- **Cream Cheese Icing**

1. Stir together first 2 ingredients and 1 tsp. granulated sugar in a glass measuring cup; let stand 5 minutes.
2. Beat 1/2 cup softened butter at medium speed with a heavy-duty electric stand mixer until creamy. Gradually add 1/2 cup granulated sugar and 1 tsp. salt, beating at medium speed until light and fluffy. Add eggs and next 2 ingredients, beating until blended. Stir in yeast mixture.
3. Stir together 4 1/2 cups flour and 1/4 tsp. nutmeg. Gradually add flour mixture to butter mixture, beating at low speed 1 minute or until well blended.
4. Heavily flour a flat surface; turn dough out, and knead until smooth and elastic (about 5 minutes), adding up to 1/2 cup more bread flour as needed to prevent dough from sticking. Place dough in a lightly greased (with cooking spray) large bowl, turning to grease top. Cover and let rise in a warm place (80° to 85°), free from drafts, 1 1/2 to 2 hours or until doubled in bulk.
5. Punch dough down; turn out onto a lightly floured surface. Roll into a 16- x 12-inch rectangle. Spread with 1/2 cup very soft butter, leaving a 1-inch border at edges. Stir together brown sugar, cinnamon, and remaining 1/2 cup granulated sugar, and sprinkle sugar mixture over butter. Top with pecans.
6. Roll up dough, jelly-roll fashion, starting at 1 long side; cut into 16 slices (about 1 inch thick). Place rolls, cut sides down, in 2 lightly greased (with cooking spray) 9-inch square pans. Cover with plastic wrap.
7. Let rolls rise in a warm place (80° to 85°), free from drafts, 1 hour or until doubled in bulk.
8. Preheat oven to 350°. Bake 20 to 22 minutes or until rolls are golden brown. Cool in pans 5 minutes. Spread Cream Cheese Icing over rolls. Serve warm.

Christmas Morning Cinnamon Rolls

Note: To make ahead, prepare through Step 6; cover with foil. Freeze up to 1 month. Thaw covered rolls in refrigerator overnight. Let rise as directed, allowing extra rising time. Bake and frost as directed.

MAKES 16 rolls. **HANDS-ON** 30 min.; **TOTAL** 3 hours, 40 min., including icing

Cream Cheese Icing

Beat 1 (3-oz.) package **cream cheese,** softened, and 2 Tbsp. **butter,** softened, at medium speed with an electric mixer 3 to 4 minutes or until creamy. Gradually add 2 1/4 cups **powdered sugar,** beating at low speed until blended. Stir in 1 tsp. **vanilla extract** and 1 Tbsp. **milk.** Stir in up to 1 Tbsp. more milk, 1 tsp. at a time, until smooth.

MAKES about 1 1/2 cups. **HANDS-ON** 10 min., **TOTAL** 10 min.

APPLE-PECAN STREUSEL MUFFINS

These muffins are simple enough to make with kids. Reheat by wrapping in foil and popping into a 325° oven for 15 minutes.

- 1 **cup butter, softened**
- 3/4 **cup granulated sugar**
- 3/4 **cup firmly packed light brown sugar**
- 2 **large eggs**
- 2 **cups all-purpose flour**
- 1 **tsp. baking powder**
- 1/2 **tsp. ground cinnamon**
- 1/4 **tsp. baking soda**
- 1/4 **tsp. table salt**
- 1 **(8-oz.) container sour cream**
- 1 **cup chopped toasted pecans**
- 1 **cup peeled and chopped Granny Smith apple (about 1 [8-oz.] apple)**
- 1 **Tbsp. vanilla extract**
- **Vegetable cooking spray**
- **Pecan Streusel**

1. Preheat oven to 350°. Beat butter at medium speed with a heavy-duty electric stand mixer until creamy. Gradually add granulated sugar and brown sugar, beating until light and fluffy. Add eggs, 1 at a time, beating at low speed just until blended after each addition.

Apple-Pecan Streusel Muffins

2. Stir together flour and next 4 ingredients in a small bowl; add to butter mixture alternately with sour cream, beginning and ending with flour mixture. Beat at low speed just until blended after each addition. Stir in pecans, apples, and vanilla.

3. Spoon batter into well-greased (with cooking spray) muffin pans, filling three-fourths full. Sprinkle about 1 Tbsp. Pecan Streusel over each muffin.

4. Bake at 350° for 22 to 25 minutes or until a wooden pick inserted in center comes out clean. Cool in pans on wire racks 5 minutes; remove from pans to wire racks, and cool 15 minutes.

MAKES 18 to 20 muffins. **HANDS-ON** 20 min.; **TOTAL** 1 hour, 10 min., including streusel

Pecan Streusel

Stir together 1/2 cup firmly packed **brown sugar,** 2 Tbsp. **all-purpose flour,** and 1 tsp. **ground cinnamon** in a small bowl. Stir in 1 cup chopped **pecans** and 2 Tbsp. melted **butter** until crumbly.

MAKES 1 1/2 cups. **HANDS-ON** 5 min., **TOTAL** 5 min.

FESTIVE FRESH FRUIT SALAD WITH MINT-LIME SIMPLE SYRUP

Use any left over Simple Syrup in cocktails and to sweeten iced tea. Or spoon it over any fresh fruit.

MINT-LIME SIMPLE SYRUP

- 1 **cup sugar**
- 1 **cup loosely packed fresh mint leaves**
- 1/4 **cup fresh lime juice**

FRUIT SALAD

- 2 **(16-oz.) containers fresh strawberries, halved**
- 1 **(6-oz.) container fresh raspberries**
- 2 **cups seedless green grapes**
- 2 **kiwifruit, peeled and thinly sliced**

1. Prepare Syrup: Stir together first 2 ingredients and 1 cup water in a saucepan over medium-high heat. Bring to a boil, stirring occasionally, and boil 1 minute or until sugar dissolves. Remove from heat. Stir in lime juice, and cool 30 minutes.

2. Pour mixture through a fine wire-mesh strainer into an airtight container. Cover and chill syrup 4 hours.

3. Prepare Salad: Gently toss together strawberries and next 3 ingredients in a large bowl. Add 1/2 cup Mint-Lime Simple Syrup, and gently stir to coat. Serve immediately. Cover and chill any remaining syrup up to 2 weeks.

MAKES 8 to 10 servings. **HANDS-ON** 15 min.; **TOTAL** 4 hours, 45 min.

CREAM CHEESE PASTRIES

Pam found this recipe in an old issue of Southern Living *and was making this family favorite long before joining the staff in 2003.*

- 1 **(8-oz.) container sour cream**
- 1/2 **cup butter, cubed**
- 1 **tsp. table salt**
- 1/2 **cup plus 1 tsp. sugar, divided**
- 2 **(1/4-oz.) envelopes active dry yeast**
- 1/2 **cup warm water (100° to 110°)**
- 2 **large eggs, lightly beaten**
- 4 **cups bread flour**
- **Cream Cheese Filling**
- **Parchment paper**
- **Powdered Sugar Glaze**

1. Heat first 3 ingredients and 1/2 cup sugar in a small saucepan over medium-low heat, stirring occasionally, until butter melts. Cool to 100° to 110°.
2. Combine yeast, warm water, and remaining 1 tsp. sugar in a large bowl; let stand 5 minutes. Stir in sour cream mixture and eggs; gradually add bread flour, beating at low speed with an electric mixer. (Dough will be soft.) Cover and chill 8 to 24 hours.
3. Divide dough into 4 portions. Turn out each portion onto a heavily floured surface, and knead 4 or 5 times.
4. Roll each portion into a 12- x 8-inch rectangle, and spread each rectangle with one-fourth of Cream Cheese Filling, leaving a 1-inch border. Carefully roll up, starting at 1 long side; press seam, and fold ends under to seal. Place 2 loaves, seam side down, on each of 2 parchment paper-lined baking sheets. Make 6 cuts across each loaf; cover with plastic wrap.
5. Let loaves rise in a warm place (80° to 85°), free from drafts, 1 hour or until doubled in bulk.
6. Preheat oven to 375°. Bake 2 loaves at a time 15 minutes or until golden brown. (Refrigerate remaining loaves while first batch is baking.) Drizzle warm loaves with Powdered Sugar Glaze.

Note: To make ahead, prepare recipe as directed through Step 4. Refrigerate up to 24 hours. Let rise as directed in Step 5, allowing a little extra rising time. Proceed as directed in Step 6.

MAKES 4 loaves. **HANDS-ON** 1 hour; **TOTAL** 10 hours, 40 min., including filling and glaze

Cream Cheese Filling

- 2 **(8-oz.) packages cream cheese, softened**
- 3/4 **cup sugar**
- 1 **large egg**
- 2 **tsp. vanilla extract**

Beat all ingredients at medium speed with an electric mixer until smooth.

MAKES about 2 1/2 cups. **HANDS-ON** 5 min., **TOTAL** 5 min.

Powdered Sugar Glaze

Stir together 2 1/2 cups sifted **powdered sugar,** 2 tsp. **vanilla extract,** and 3 Tbsp. **milk** in a medium bowl. Stir in up to 1 Tbsp. more milk, 1 tsp. at a time, until desired consistency is reached.

MAKES 1 cup. **HANDS-ON** 5 min., **TOTAL** 5 min.

Holiday Ham with Apricot Glaze

HOLIDAY HAM WITH APRICOT GLAZE

Refrigerate any leftover ham in individual servings. In the days after Christmas when your house is still full and your energy is waning, this will make it easy for guests to fix a meal on their own.

- 1 **(7- to 8-lb.) fully cooked, bone-in spiral-cut ham half**
- **Vegetable cooking spray**
- 1 **(18-oz.) jar apricot preserves**
- 1/2 **cup bourbon**
- 1/2 **cup whole grain mustard**
- 2 **Tbsp. dark brown sugar**

1. Preheat oven to 325°. Place ham, cut side down, on a lightly greased (with cooking spray) rack in a heavy-duty aluminum foil-lined broiler pan; let stand at room temperature 30 minutes.
2. Stir together preserves and next 3 ingredients in a saucepan. Bring to a simmer over medium-low heat. Simmer, stirring occasionally, 2 minutes. Remove from heat, and reserve 1 1/2 cups. Brush remaining glaze over ham.
3. Bake, uncovered, at 325° on lowest oven rack 2 hours or until a meat thermometer inserted in thickest portion registers 140°, basting every 30 minutes with reserved 1 1/2 cups glaze. Let stand 15 minutes before serving.

MAKES 8 to 10 servings. **HANDS-ON** 10 min.; **TOTAL** 2 hours, 45 min.

Ham for the Holidays

JULIA REED EXPLAINS WHY SHARING A COUNTRY HAM IS THE MOST GENEROUS THING A HOSTESS CAN DO

DOROTHY PARKER FAMOUSLY said, "Eternity is a ham and two people." If the Roundtable wit had wanted to see time fly, she should've visited my childhood home during the holidays, when my Nashville grandparents arrived toting an enormous Tennessee country ham. The long-cured and smoked ham was an anomaly—and a delicacy—in the wilds of Mississippi, and my father watched it like a hawk. Much to his dismay, my mother served it at our annual Christmas night party, sliced paper thin on a silver platter, where the neighbors fell upon it like wolves. Invariably, my father would circle the table, sentry-like, and invariably my mother would yell at him to stop. Finally, we solved the problem with two hams, one for him and one for everybody else.

Tennessee is smack in the middle of what is often referred to as the "ham belt," the temperate zone running from Missouri through Kentucky and Tennessee to Virginia and the Carolinas. I have no idea where my grandmother's hams came from, but her cook, Ernestine Turner, always prepared them flawlessly. These days you can buy already cooked country hams and even presliced country hams, but I prefer to bake it myself, a surprisingly easy enterprise. It's less salty (you can soak it as long as overnight) and the texture seems far superior. Ernestine coated hers with a mix of fine breadcrumbs and brown sugar and I do the same.

My mother's ham was part of a larger Christmas buffet, and to this day she also offers it alongside the turkey at Thanksgiving and Christmas dinner. But I love to serve it as a cocktail or breakfast/brunch snack accompanied with one or two sauces. The Deep South classic Jezebel provides a perfect kick, and I'm especially crazy about chutney mayonnaise. (Mix a jar of Major Grey's chutney with a couple of cups of Hellmann's or, preferably, homemade mayonnaise.) Virginians often serve ham with cracker-like beaten biscuits. I take the easier (and, to me, tastier) route with Marshall's frozen buttermilk variety, and my mother is partial to yeast rolls. Here, I've adapted a grits muffin recipe from the talented Felicia Willett, chef at Felicia Suzanne's in Memphis. Ham is always great with cheese grits, after all, and now it's incorporated into a delicious delivery system.

A SIMPLE PARTY MENU

Milk punch is a hallowed New Orleans tradition where it's often made with bourbon and brandy, but I prefer rum to the latter.

THE HAMS | **THE FOOD**

Julia Porks Out

HERE ARE HER TOP RECOMMENDATIONS

S. Wallace Edwards & Sons

SURRY, VA

The Edwards family has been producing classic Virginia country hams for 75 years. They also offer a delicious Spanish-style ham call Surryano (a play on Serrano), but the uncooked, bone-in ham is my mainstay. The cooking directions on its cloth bag couldn't be easier; *edwardsvaham.com*

Benton's Smoky Mountain Country Ham

MADISONVILLE, TN

Allan Benton, one of the great American cure masters, is also the nicest man in the world and a favorite of top chefs from Thomas Keller to John Besh. He cures his hams almost twice as long as most, but they're worth the wait (as is his extraordinary bacon); *bentonscountryhams2.com*

The Curehouse

LOUISVILLE, KY

Former chef Jay Denham is a relative newcomer to the cured meats game. But his stint in Italy, as well as his internship with Kentucky legend Nancy Newsom, served him well. His Mountain Ham is cured, never smoked, along the lines of prosciutto and Serrano with a silky texture and a rich, nutty flavor; *fossilfarms.com*

MILK PUNCH

- **2 qt. whole milk**
- **1 qt. heavy cream**
- **4 cups bourbon**
- **2 cups dark rum**
- **2 Tbsp. vanilla extract**
- **Simple Syrup (page 275)**
- **Garnish: whole nutmeg, grated**

Whisk together milk, cream, spirits, and vanilla in a large pitcher or bowl until well blended. Stir in Simple Syrup to taste. Chill the punch well (about 1 hour). Pour into highball glasses, and grate nutmeg on top.

Note: Leftover punch freezes well. Even better, if you don't thaw it out completely, you get Milk Punch slushies!

MAKES about 24 servings. **HANDS-ON** 10 min.; **TOTAL** 1 hour, 10 min.

CHEESE-GRIT-AND-CHIVE MUFFINS

- **Butter for greasing muffin pans**
- **1½ cups all-purpose flour**
- **1 tsp. baking powder**
- **½ tsp. baking soda**
- **½ tsp. kosher salt**
- **2 large eggs**
- **¾ cup buttermilk**
- **½ cup butter, melted**
- **1 cup cooked grits**
- **1 cup grated sharp Cheddar cheese**
- **1 Tbsp. chopped fresh chives**
- **Ground red pepper**

1. Preheat oven to 350°. Grease 24 cups of a miniature muffin pan with butter.
2. Whisk together flour, baking powder, baking soda, and salt in a large bowl. Whisk together eggs, buttermilk, and melted butter in another bowl. Fold grits into egg mixture. Add grits mixture to flour mixture, and stir well. Stir in cheese and chives and a healthy pinch of ground red pepper. Taste for seasoning.
3. Spoon batter into muffin cups, filling them about three-fourths full. Bake at 350° for 30 to 35 minutes or until muffins are golden brown and separating from edges of pan.

MAKES about 2 dozen. **HANDS-ON** 15 min., **TOTAL** 45 min.

SMART COOKIES

Tried-and-true tricks for whipping up a perfect dozen—or three

CUT CORNERS (IN A GOOD WAY)

Baking bar cookies or brownies? Line your pan with aluminum foil that hangs over the sides so it's easy to lift them out, start cutting, and clean the pan.

MAKE A LITTLE NOISE

When cookies are done, sharply tap the baking sheet against the counter. It will force the cookies to settle faster, creating a crisp outside and a chewy center.

CRACK THE RIGHT EGGS

You'll want to bake with room temperature eggs so they easily incorporate into your dough. In a rush? Just set them in a bowl of warm tap water for 10 minutes.

AVOID A STICKY SITUATION

When a recipe calls for you to measure something like honey, molasses, or peanut butter, first coat the cup with cooking spray.

REACH FOR A SPOON

When measuring flour, spoon—don't pack—the dry ingredient into the measuring cup. This technique helps yield tender cookies.

Community Cookbook

BUILD A BETTER LIBRARY, ONE GREAT BOOK AT A TIME

The holidays are about gathering friends and family and creating warm winter memories over home-cooked meals. *Christmas with Southern Living 2015* captures that spirit with more than 100 recipes that cover everything from Christmas morning's streusel muffins to New Year's Eve's cocktails and canapés. Grab a copy and cozy up with our Peppermint Swirl Hot Chocolate (on page 51).

SPEAKEASY SPARKLER

This festive Champagne sparkler gets a lift from moonshine and limoncello. You can purchase any number of fine bottles of handcrafted moonshine at your local package store.

- 1 **sugar cube**
- **Dash of lemon bitters**
- 2 1/2 **Tbsp. moonshine**
- 2 **Tbsp. Italian lemon liqueur**
- 1/3 **cup Champagne or dry sparkling wine**

1. Place sugar cube in a Champagne flute. Sprinkle sugar cube with lemon bitters.
2. Combine moonshine and liqueur in a cocktail shaker; add ice cubes. Cover with lid, and shake vigorously until thoroughly chilled (about 30 seconds). Strain into Champagne flute, and top with Champagne.

Note: We tested with Midnight Moon moonshine and limoncello for lemon liqueur.

MAKES 1 serving. **HANDS-ON** 5 min., **TOTAL** 5 min.

GOUDA GRITS

Flavorful Gouda cheese and chicken broth take these easy grits to the next level.

- 4 **cups chicken broth**
- 3 1/2 **cups milk**
- 2 **tsp. table salt**
- 1/2 **tsp. freshly ground black pepper**
- 2 **cups uncooked quick-cooking grits**
- 6 **oz. Gouda cheese, shredded**
- 1/4 **cup butter**

1. Bring broth, milk, salt, and freshly ground black pepper to a boil in a large Dutch oven over medium heat. Gradually whisk in grits.
2. Reduce heat, and simmer, uncovered, 10 minutes or until thick, whisking often.

3. Remove from heat, and stir in cheese and butter, whisking until cheese melts.

MAKES 8 servings. **HANDS-ON** 18 min., **TOTAL** 18 min.

CHEESY PIMIENTO CORNBREAD

Two Southern staples come together in this rich, savory cornbread.

- 1/4 **cup butter**
- 1 1/4 **cups stone-ground yellow cornmeal**
- 3/4 **cup all-purpose flour**
- 1 **tsp. baking soda**
- 1 **tsp. baking powder**
- 1 **tsp. table salt**
- 1/4 **tsp. freshly ground black pepper**
- 4 **oz. extra-sharp Cheddar cheese, shredded**
- 1 1/4 **cups buttermilk**
- 2 **large eggs**
- 1 **(4-oz.) jar diced pimientos, well drained**

1. Preheat oven to 425°. Place butter in a 9-inch cast-iron skillet. Place skillet in oven, and heat at 425° for 4 minutes or until butter melts. Remove skillet from oven.
2. Meanwhile, combine cornmeal and next 6 ingredients in a large bowl. Whisk together buttermilk, eggs, and pimientos; add to dry ingredients, stirring just until moistened. Pour batter over melted butter in hot skillet.
3. Bake at 425° for 25 minutes or until golden brown and a wooden pick inserted in center comes out clean.

MAKES 8 to 10 servings. **HANDS-ON** 5 min., **TOTAL** 30 min.

Bonus Favorites

304 **Quick-and-Easy** No time? Put these recipes in your arsenal for busy weeknights

310 **Holiday Tea** Try these bite-size savories and sweets for afternoon entertaining

313 **Baked Greats** From popovers and cakes to muffins and quick breads, these are recipe box-worthy

316 **Get the Party Started** Raise a glass and toast the season with addictive nibbles and inspired cocktails

318 **Soups and Stews** Cool-season comforts that warm the soul

320 **Holiday Sideboard** We have all the fixin's for a memorable holiday feast

324 **Main Attractions** Elegant centerpiece roasts for the holiday table

326 **Save Room** Grand finales for holiday feasts

328 **A Big Batch of Cookie Favorites** Perfect for holiday giving and anytime devouring

331 **Good Luck New Year** Start off the year the Southern way with Hoppin' John and a menu for entertaining

333 **Southern Baker** Some of our best recipes ever, inspired by Southern ingredients and methods

Quick-and-Easy

WHEN YOU WANT DINNER ON THE TABLE IN NO TIME, THESE RECIPES FIT THE BILL FABULOUSLY

GRILL

KOREAN FLANK STEAK

- 1/4 **cup firmly packed light brown sugar**
- 1/4 **cup soy sauce**
- 2 **tsp. dark sesame oil**
- 1 **tsp. grated fresh ginger**
- 1/4 **tsp. table salt**
- 2 **(1-lb.) flank steaks**
- **Vegetable cooking spray**
- **Garnish: sliced green onions**

1. Combine first 5 ingredients in a shallow dish or large zip-top plastic freezer bag; add steaks. Cover or seal, and chill 8 hours.
2. Preheat grill to 450° (high) heat. Remove steaks from marinade, discarding marinade. Coat steaks with cooking spray. Grill steak, covered with grill lid, 2 minutes on each side or to desired degree of doneness. Remove from grill, and let stand 5 minutes. Cut steak diagonally across the grain into thin slices.

MAKES 6 servings. **HANDS-ON** 3 min., **TOTAL** 8 hours, 12 min.

Korean Flank Steak

PESTO CHICKEN GRILL

- 1 **cup zesty Italian dressing**
- 4 **(8- to 10-oz.) skinned and boned chicken breasts**
- **Vegetable cooking spray**
- 1/2 **tsp. table salt**
- 1/4 **tsp. black pepper**
- 1 **medium-size yellow bell pepper, halved and seeded**
- 1 **medium-size sweet onion, cut into 1/2-inch-thick slices**
- 1/2 **(7-oz.) container refrigerated pesto sauce (about 1/2 cup)**
- **Garnish: sliced fresh basil**

1. Place dressing and chicken in a large zip-top plastic freezer bag. Seal bag, turning to coat; chill at least 8 hours or overnight.
2. Coat a cold cooking grate with cooking spray, and place on grill. Preheat grill to 350° to 400° (medium-high) heat. Remove chicken from marinade, discarding marinade. Sprinkle chicken with salt and pepper.
3. Grill chicken, covered, 10 to 12 minutes on each side or until done. At the same time, grill bell pepper and onion slices 4 minutes on each side or until tender and grill marks appear.
4. Remove vegetables from grill; coarsely chop, and place in a medium bowl. Add pesto, tossing to coat. Remove chicken from grill. Let stand 5 minutes. Spoon vegetable mixture over chicken.

MAKES 4 servings. **HANDS-ON** 30 min.; **TOTAL** 8 hours, 55 min.

GRILLED CHICKEN CAPRESE

Try these easy open-faced sandwiches when you can purchase perfect summer-ripe tomatoes at their peak.

- **Vegetable cooking spray**
- 4 **cups diced tomato**
- 1 **Tbsp. olive oil**
- 1 1/4 **tsp. kosher salt, divided**
- 3/4 **tsp. freshly ground black pepper, divided**
- 2 1/4 **lb. chicken breast cutlets (8 cutlets)**
- 8 **(1-inch-thick) diagonally cut French bread slices**
- 1 **(16-oz.) log presliced fresh mozzarella cheese**
- 1/4 **cup balsamic glaze**
- **Garnish: fresh basil leaves**

1. Coat a cold cooking grate with cooking spray, and place on grill. Preheat grill to 350° to 400° (medium-high) heat.
2. Combine tomato, olive oil, 1/4 tsp. salt, and 1/4 tsp. pepper in a bowl; toss gently.
3. Sprinkle chicken with remaining 1 tsp. salt and remaining 1/2 tsp. pepper. Coat chicken with cooking spray, and place on cooking grate. Grill chicken, covered with grill lid, 2 to 3 minutes on each side or until done. Remove from grill; keep warm.
4. Coat both sides of bread slices with cooking spray. Grill bread slices 30 to 45 seconds; turn slices over. Top each bread slice with 2 cheese slices. Grill 30 seconds or until bread is toasted.
5. Arrange chicken cutlets evenly over cheese; top each sandwich with 1/2 cup tomato mixture. Drizzle each sandwich with 1 1/2 tsp. balsamic glaze.

MAKES 8 servings. **HANDS-ON** 27 min., **TOTAL** 27 min.

GRILLED TERIYAKI SHRIMP KABOBS

- **4 (10-inch) wooden skewers**
- **1 (6-oz.) package uncooked rice vermicelli**
- **1 1/2 cups fresh pineapple chunks**
- **1 1/2 lb. peeled and deveined large raw shrimp with tails**
- **1 (8-oz) package sweet mini peppers**
- **2 Tbsp. olive oil**
- **1/2 tsp. table salt**
- **1/2 tsp. freshly ground black pepper**
- **2/3 cup triple ginger teriyaki sauce**
- **1/4 cup chopped fresh cilantro**

1. Soak skewers in water 30 minutes. Preheat grill to 350° to 400° (medium-high) heat. Cook vermicelli according to package directions. Drain and keep warm.
2. Thread pineapple alternately with shrimp and peppers onto each skewer. Brush skewers with olive oil, and sprinkle with salt and pepper.
3. Reserve 1/3 cup teriyaki sauce for serving. Grill kabobs, covered with grill lid, 3 to 4 minutes on each side or just until shrimp turn pink, basting with remaining 1/3 cup teriyaki sauce during last 2 minutes of cooking. Discard basting sauce. Sprinkle kabobs with cilantro. Serve kabobs over vermicelli with reserved 1/3 cup teriyaki sauce.

MAKES 6 servings. **HANDS-ON** 16 min., **TOTAL** 16 min.

GRILLED SALMON WITH ORANGE-FENNEL SALAD

This healthy meal is full of fresh flavors and interesting texture combinations.

- **Vegetable cooking spray**
- **3 navel oranges**
- **1 fennel bulb**
- **2 Tbsp. olive oil**
- **1 tsp. table salt**
- **1/2 tsp. freshly ground black pepper**
- **4 (6-oz.) salmon fillets (1 1/2 inches thick)**
- **4 tsp. Caribbean seasoning**
- **1/3 cup orange marmalade**
- **Garnish: fresh dill**

1. Coat a cold cooking grate with cooking spray, and place on grill. Preheat grill to 350° to 400° (medium-high) heat.
2. Meanwhile, peel oranges and cut crosswise into 1/4-inch-thick slices. Rinse fennel thoroughly. Trim and discard root end of fennel bulb. Trim stalks from bulb; chop fronds to measure 2 Tbsp. Set aside for later use. Reserve remaining fronds for another use. Cut bulb lengthwise into quarters; cut quarters crosswise into very thin strips.
3. Place orange slices, fennel strips, olive oil, salt, and pepper in a bowl; toss gently to coat. Cover and chill while preparing salmon.
4. Sprinkle both sides of salmon with Caribbean seasoning; coat with cooking spray. Place salmon, skin side up, on cooking grate. Grill, covered with grill lid, 2 to 3 minutes. Turn salmon over; brush with marmalade. Grill, covered, 2 to 3 minutes or to desired degree of doneness.
5. Place 1 cup orange-fennel salad on each of 4 plates. Top each serving with 1 salmon fillet; sprinkle evenly with 2 Tbsp. reserved fennel fronds.

MAKES 4 servings. **HANDS-ON** 17 min., **TOTAL** 17 min.

NO COOK

CITRUS CHICKEN AND BEET SALAD

The convenience of deli-roasted chicken and canned or pickled beets makes this salad a good choice for a weeknight supper.

- **4 oranges, divided**
- **2 Tbsp. sugar**
- **2 Tbsp. white wine vinegar**
- **1/2 tsp. table salt**
- **1/4 tsp. ground black pepper**
- **1/4 cup canola oil**
- **6 cups spring greens mix**
- **2 cups shredded deli-roasted chicken breast**
- **1 (15-oz.) can small whole beets, drained and quartered**
- **1/3 cup honey-roasted sliced almonds**

1. Grate zest from 1 orange to equal 1 tsp. Cut orange in half; squeeze juice from orange into a measuring cup to equal 1/3 cup. Peel remaining 3 oranges, and cut crosswise into 1/4-inch-thick slices.
2. Whisk together orange zest, orange juice, sugar, and next 3 ingredients in a small bowl. Whisk in oil.
3. Toss together spring greens, chicken, beets, and orange slices; arrange on a serving platter. Whisk dressing, and drizzle over salad. Sprinkle with almonds.

Note: We tested with Almond Accents Honey Roasted Flavored Sliced Almonds.

MAKES 4 servings. **HANDS-ON** 23 min., **TOTAL** 23 min.

PROSCIUTTO AND TOMATO SANDWICHES

If blue cheese is too pungent for your taste, use goat cheese or green onion whipped cream cheese in the mayonnaise spread.

- 1/3 **cup mayonnaise**
- 1/4 **cup crumbled blue cheese**
- 8 **(1-oz.) Italian sandwich bread slices**
- 2 **cups loosely packed arugula**
- 1 **(4-oz.) package thinly sliced prosciutto***
- 1 **large heirloom tomato, sliced**
- **Freshly ground black pepper**

1. Stir together mayonnaise and blue cheese in a small bowl. Spread mayonnaise mixture over 1 side of 4 bread slices. Layer 4 bread slices, mayonnaise sides up, evenly with arugula, prosciutto, tomato slices, and pepper. Top with remaining 4 bread slices, mayonnaise sides down.

*Thinly sliced country ham or bacon may be substituted.

MAKES 4 servings. **HANDS-ON** 11 min., **TOTAL** 11 min.

CORNBREAD-VEGETABLE SALAD

This is a great use for leftover cornbread or corn muffins. No leftovers? Pick up cornbread in the bakery section, or make your own semihomemade from a mix.

- 6 **slices precooked bacon, crumbled and divided**
- 3 **cups crumbled cornbread**
- 1 **cup bottled refrigerated buttermilk Ranch dressing**
- 1 **(8-oz.) container prechopped tricolored bell pepper**
- 1 1/2 **cups (6-oz.) shredded sharp Cheddar cheese**

1. Reserve 2 Tbsp. bacon crumbles. Combine cornbread, next 2 ingredients, 1 1/4 cups cheese, and remaining bacon crumbles in a medium bowl; toss to coat. Sprinkle with remaining cheese and bacon.

MAKES 5 servings. **HANDS-ON** 4 min., **TOTAL** 4 min.

PIZZA/PASTA

PHILLY CHEESESTEAK PIZZA

Precooked beef roast makes this pizza a quick and easy fix. You can find it with the lunchmeat in your local grocery store.

- 1 **lb. bakery pizza dough, at room temperature**
- **Parchment paper**
- 1 **(14-oz.) package frozen pepper and onion stir-fry vegetables, thawed**
- 1 **Tbsp. olive oil**
- 1 **(17-oz.) package fully cooked beef roast au jus**
- 8 **oz. sharp provolone cheese, shredded**

1. Preheat oven to 450°. Stretch dough into a 12-inch circle on 17- x 12-inch parchment paper-lined half baking sheet.
2. Sauté vegetables in hot oil in a large nonstick skillet over medium-high heat 5 minutes or until tender.
3. Heat beef according to package directions. Shred beef in juices in container using 2 forks. Remove beef with a slotted spoon. Top pizza dough with beef and sautéed vegetables. Sprinkle with cheese.
4. Bake at 450° for 18 minutes or until crust is golden.

MAKES 4 servings. **HANDS-ON** 13 min., **TOTAL** 31 min.

CHICKEN, ARTICHOKE, AND SUN-DRIED TOMATO PIZZA

Serve this hearty pizza with a simple green salad.

- 1 1/2 **Tbsp. olive oil**
- 1 **lb. bakery pizza dough, at room temperature**
- 1/3 **cup jarred sun-dried tomato pesto sauce***
- 2 **cups shredded deli-roasted chicken**
- 1 **(7.5-oz.) jar marinated quartered artichoke hearts, drained**
- 1 **(8-oz.) package shredded Italian three-cheese blend**
- **Garnish: fresh oregano leaves**

1. Preheat oven to 450°. Brush a baking sheet with olive oil. Stretch pizza dough evenly into a 12-inch circle on baking sheet.
2. Spread pesto sauce over dough, leaving a 1/2-inch border. Top with chicken and artichoke hearts. Sprinkle with cheese.
3. Bake at 450° for 12 minutes or until edges of crust are golden brown and cheese melts.

*Traditional basil pesto sauce may be substituted.

MAKES 4 servings. **HANDS-ON** 8 min., **TOTAL** 20 min.

Chicken, Artichoke, and Sun-Dried Tomato Pizza

GRILLED TOMATO-PEACH PIZZA

Vegetable cooking spray
- 2 tomatoes, sliced
- 1/2 tsp. table salt
- 1 large peach, peeled and sliced
- 1 lb. bakery pizza dough, at room temperature
- 1/2 (16-oz.) package fresh mozzarella, sliced
- 4 to 6 fresh basil leaves

Garnishes: coarsely ground black pepper, olive oil

1. Coat cold cooking grate of grill with cooking spray, and place on grill. Preheat grill to 300° to 350° (medium) heat.
2. Sprinkle tomatoes with salt; let stand 15 minutes. Pat tomatoes dry with paper towels.
3. Grill peach slices, covered with grill lid, 2 to 3 minutes on each side or until grill marks appear.
4. Place dough on a large baking sheet coated with cooking spray; lightly coat dough with cooking spray. Roll dough to 1/4-inch thickness (about 14 inches in diameter). Slide pizza dough from baking sheet onto cooking grate.
5. Grill, covered with grill lid, 2 to 3 minutes or until lightly browned. Turn dough over, and reduce temperature to 250° to 300° (low) heat; top dough with tomatoes, grilled peaches, and mozzarella. Grill, covered with grill lid, 5 minutes or until cheese melts. Arrange basil leaves evenly over pizza. Serve immediately.

MAKES 4 servings. **HANDS-ON** 26 min., **TOTAL** 26 min.

SAGE-BROWN BUTTER SQUASH AGNOLOTTI

- 3 (8-oz.) packages refrigerated roasted butternut squash ravioli or other butternut squash-filled pasta
- 1 lemon
- 1/2 cup butter
- 2 Tbsp. chopped fresh sage leaves
- 1/4 tsp. garlic salt
- 1/4 tsp. freshly ground black pepper
- 1/3 cup toasted chopped walnuts

Garnish: fresh sage leaves

1. Cook pasta according to package directions; drain.
2. Grate zest from lemon to equal 1 tsp. Cut lemon in half; squeeze juice from lemon into a measuring cup to equal 1 1/2 tsp.
3. Cook butter in a small heavy saucepan over medium heat, stirring constantly, 5 minutes or just until butter begins to turn golden brown. Immediately remove pan from heat. (Butter will continue to darken if left in saucepan.) Stir in lemon zest and juice, sage, garlic salt, and pepper. Drizzle butter mixture over pasta, tossing to coat. Sprinkle with toasted walnuts. Serve immediately.

Note: We tested with Monterey Gourmet Foods Butternut Squash Ravioli.

MAKES 6 servings. **HANDS-ON** 10 min., **TOTAL** 10 min.

Eggs in Fire Roasted Tomato Sauce

BREAKFAST

EGGS IN FIRE ROASTED TOMATO SAUCE

Enjoy a very satisfying savory meal with a piece of toasted baguette bread for dipping in the delicious sauce. You can also add a couple tablespoons cooked, crumbled bacon.

- 2 (14.5-oz) cans diced fire roasted tomatoes with garlic
- 2 Tbsp. extra virgin olive oil
- 1/2 tsp. dried oregano
- 1/2 tsp. table salt, divided
- 1/2 tsp. freshly ground black pepper, divided
- 12 large eggs
- 6 Tbsp. freshly grated Parmesan cheese
- 6 French bread baguette slices, toasted

1. Preheat oven to 400°. Process first 3 ingredients, 1/4 tsp. salt, and 1/4 tsp. pepper in a blender until smooth. Pour into microwave-safe bowl; cover and microwave at HIGH 2 minutes, stirring after 1 minute.
2. Place 6 (10-oz.) shallow baking dishes on 2 baking sheets. Divide warm sauce among dishes. Break 2 eggs, and slip into sauce in each dish. Sprinkle evenly with remaining 1/4 tsp. salt and remaining 1/4 tsp. pepper. Sprinkle 1 Tbsp. cheese over eggs in each dish.
3. Bake at 400° for 15 minutes or until egg whites are set. Serve hot with toasted baguette slices for dipping.

MAKES 6 servings. **HANDS-ON** 21 min., **TOTAL** 21 min.

BACON AND CHEDDAR BELGIAN WAFFLES

- **1 (2.52-oz.) package precooked bacon**
- **2 cups all-purpose baking mix**
- **1 cup (4 oz.) shredded sharp Cheddar cheese**
- **1 large egg, lightly beaten**
- **1 1/2 cups milk**
- **1/4 cup butter, melted**

1. Prepare bacon according to package directions; chop.
2. Stir together baking mix, cheese, and bacon. Whisk together egg, milk, and melted butter in small bowl; whisk into dry ingredients.
3. Cook batter in a preheated, oiled Belgian-style waffle iron until golden (about 1 1/4 cups batter each, spreading slightly). Serve with butter and maple syrup, if desired.

Note: If you don't have a Belgian waffle iron, use 1/2 cup batter for each waffle in a traditional waffle iron.

MAKES 6 servings. **HANDS-ON** 18 min., **TOTAL** 18 min.

PEACH PANCAKES

- **2 cups chopped fresh or frozen peaches, divided**
- **1/4 cup firmly packed light brown sugar, divided**
- **1/4 tsp. ground ginger**
- **2 cups just-add-water pancake mix**
- **2 cups peach nectar, divided**

1. Stir together 1 cup peaches, 2 Tbsp. brown sugar, and ginger. Stir together pancake mix, 1 cup peach nectar, and 1/2 cup water. Stir in peach mixture.
2. Combine remaining 1 cup peaches, remaining 2 Tbsp. brown sugar, and remaining 1 cup peach nectar in a small saucepan. Bring to a boil over medium heat; cook, stirring occasionally, 5 minutes or until syrupy.
3. Pour about 1/4 cup batter for each pancake onto a hot, lightly greased griddle or large nonstick skillet. Cook pancakes over medium heat 2 minutes or until tops are covered with bubbles and edges look dry and cooked; turn and cook other side about 30 seconds. Serve with remaining peach syrup.

Note: We tested with Krusteaz Buttermilk Pancake Mix.

MAKES 6 servings. **HANDS-ON** 20 min., **TOTAL** 22 min.

DINNER IN A DISH

BAKED BEEF AND FRIED PEPPER BURRITOS

- **1 (15-oz.) package fully cooked beef roast au jus, drained**
- **1 cup (4 oz.) shredded Mexican four-cheese blend**
- **1/2 cup drained jarred fried peppers**
- **1/2 cup corn and black bean salsa**
- **4 (9-inch) soft taco-size flour tortillas**
- **Vegetable cooking spray**

1. Preheat oven to 400°. Shred beef with 2 forks.
2. Place 1/3 cup beef, 1/4 cup cheese, 2 Tbsp. peppers, and 2 Tbsp. salsa just below center of each tortilla. Fold opposite sides of tortilla over filling, and roll up. Place burritos on a lightly greased baking sheet. Coat tops of burritos with cooking spray. Bake at 400° for 10 minutes or until tops are golden brown.

Note: We tested with Mancini Fried Peppers.

MAKES 4 servings. **HANDS-ON** 9 min., **TOTAL** 19 min.

FRIED APPLE PORK CHOPS

- **4 bone-in center-cut pork loin chops (about 2 3/4 lb.)**
- **1/2 tsp. table salt**
- **1/2 tsp. freshly ground black pepper**
- **2 Tbsp. canola oil**
- **1/2 cup apple cider**
- **1 (15-oz.) can fried apples**
- **2 Tbsp. butter**
- **Garnish: fresh thyme leaves**

1. Sprinkle pork chops on both sides with salt and pepper. Heat oil in a large nonstick skillet over medium-high heat. Add pork chops; cook 4 minutes on each side or until golden brown. Remove pork chops from pan and keep warm.
2. Add apple cider to pan; reduce heat to low. Cook 30 seconds, stirring to loosen browned bits from bottom of skillet. Return pork chops to skillet; add fried apples. Cook 2 minutes or until apples are thoroughly heated, breaking up apples with a spoon during cooking. Stir in butter until melted. Serve pork chops with sauce.

MAKES 4 servings. **HANDS-ON** 13 min., **TOTAL** 13 min.

SIDES

LEMON AND SHALLOT BRUSSELS SPROUTS

- 1 lb. fresh Brussels sprouts
- 2 Tbsp. olive oil
- 2 Tbsp. chopped shallots
- 1/2 tsp. table salt
- 1/4 tsp. freshly ground black pepper
- 1 Tbsp. lemon zest
- 2 Tbsp. fresh lemon juice
- 2 Tbsp. butter

1. Remove discolored leaves from Brussels sprouts. Cut off stem ends, and cut lengthwise into quarters.
2. Heat oil in a large skillet over medium-high heat. Add shallots; sauté 3 minutes or until crisp-tender. Reduce heat to medium; add Brussels sprouts, salt, and pepper. Cook, stirring often, 8 minutes or until brown. Add lemon zest and juice; cook 1 minute. Add butter, tossing until melted. Serve immediately.

MAKES 4 servings. **HANDS-ON** 19 min., **TOTAL** 19 min.

BALSAMIC GREEN BEANS WITH TOMATOES AND FETA

- 1 (12-oz.) package steam-in-bag fresh green beans
- 1 Tbsp. olive oil
- 2 garlic cloves, minced
- 1 cup chopped seeded tomato
- 1/4 tsp. table salt
- 1/4 tsp. freshly ground black pepper
- 2 Tbsp. balsamic vinegar
- 1 (4-oz.) package crumbled feta cheese

1. Prepare green beans according to package directions.
2. Heat oil in a large skillet over medium-high heat. Add garlic; sauté 30 seconds. Add green beans and tomato; cook 1 minute, stirring often, until tomato begins to soften. Add salt, pepper, and vinegar; cook, stirring often, 2 minutes. Remove from heat; sprinkle with cheese.

MAKES 4 servings. **HANDS-ON** 9 min., **TOTAL** 13 min.

DESSERTS

CINNAMON APPLE TARTS

Serve with ice cream and caramel topping, if desired.

- 1 (17.3-oz.) package frozen puff pastry sheet, thawed
- 1 (12-oz.) package frozen harvest apples, thawed
- 2 Tbsp. bottled cinnamon sugar

1. Preheat oven to 425°. Cut each puff pastry sheet into 6 rectangles. Spoon 2 Tbsp. apples into center of each rectangle. Pull corners together over apples, pinching edges to seal. Place on a baking sheet lined with parchment paper. Sprinkle each tart with 1/2 tsp. cinnamon sugar.
2. Bake at 425° for 20 minutes or until golden brown. Serve immediately.

Note: We tested with Stouffer's Harvest Apples.

MAKES 1 dozen. **HANDS-ON** 10 min., **TOTAL** 30 min.

Cinnamon Apple Tarts

BLUEBERRY YOGURT POPS

Using whole buttermilk makes a rich and creamy flavor.

- 3/4 cup buttermilk
- 1/4 cup wild blueberry preserves
- 1 (6-oz.) container wild blueberry Greek yogurt

Stir together all ingredients in a 2-cup glass measuring cup. Pour into 4 (3-oz.) ice pops molds. Freeze 6 hours or until firm.

Note: We tested with Smucker's blueberry preserves and Liberte Wild Blueberry Greek Yogurt.

MAKES 4 servings. **HANDS-ON** 5 min.; **TOTAL** 6 hours, 5 min

MOLTEN HAZELNUT BROWNIES

- 1 cup hazelnut spread
- 2/3 cup all-purpose flour
- 2 large eggs, beaten
- 1/2 cup caramel topping

1. Preheat oven to 350°. Stir together first 3 ingredients in a medium bowl. Divide batter evenly among 4 greased muffin cups. (Muffin cups will be full.)
2. Bake at 350° for 15 minutes or until a wooden pick inserted in center comes out with a few moist crumbs. Cool in pan 10 minutes; carefully remove from muffin cups. Serve warm or cool. Drizzle each brownie with 2 Tbsp. caramel topping just before serving.

Note: We tested with Nutella hazelnut spread.

MAKES 4 servings. **HANDS-ON** 5 min., **TOTAL** 30 min.

Holiday Tea

TRY THESE INSPIRED TWO-BITE NIBBLES THAT ARE PERFECT FOR AFTERNOON ENTERTAINING

HOT APPLE TEA

Sweeten with honey or sugar, if desired.

- **3 Tbsp. fresh lemon juice**
- **1/4 tsp. whole cloves**
- **1/4 tsp. whole allspice**
- **2 (3-inch) cinnamon sticks**
- **1 (12-oz.) can frozen apple juice concentrate**
- **10 English Breakfast tea bags**
- **Garnish: thin Granny Smith apple slices**

1. Place first 5 ingredients and 8 cups water in a Dutch oven. Bring to a boil, stirring until juice melts; remove from heat. Add tea bags. Cover; steep 5 minutes. Remove tea bags.
2. Pour tea through a strainer, discarding solids.

MAKES 8 servings. **HANDS-ON** 5 min., **TOTAL** 10 min.

CHICKEN SALAD TOMATO CUPS

Shredded chicken salad, paired here with fresh herbs, is served up in delicate tomato cups: perfect finger food for an afternoon tea.

- **1/2 cup mayonnaise**
- **1/3 cup finely chopped red onion**
- **1/3 cup finely chopped celery**
- **2 Tbsp. chopped fresh basil**
- **1 Tbsp. chopped fresh chives**
- **1 Tbsp. fresh lemon juice**
- **1/8 tsp. freshly ground pepper**
- **6 oz. deli-roasted chicken, shredded (1 1/2 cups)**
- **18 Campari tomatoes, halved crosswise**
- **Garnish: fresh chives**

1. Stir together first 7 ingredients in a medium bowl; gently stir in chicken.
2. Carefully scoop out and discard tomato pulp, leaving a 1/8-inch shell. Place tomato shells upside down on several layers of paper towels; let drain 5 minutes.
3. Spoon about 1 Tbsp. chicken salad into each tomato shell.

MAKES 9 servings. **HANDS-ON** 32 min., **TOTAL** 37 min.

SMOKED TROUT AND WATERCRESS TEA SANDWICHES

These delicate finger sandwiches are filled with a creamy smoked trout spread accented with lemon zest, capers, and dill. The spicy watercress offsets the salty, smoky trout while the very thin white bread slices keep the focal point on the star attraction.

- **1/4 cup sour cream**
- **1/4 cup mayonnaise**
- **1 Tbsp. finely chopped fresh chives**
- **1 tsp. chopped fresh dill**
- **1 tsp. chopped drained capers**
- **1/2 tsp. lemon zest**
- **1/4 tsp. ground white pepper**
- **5 oz. smoked trout, skin removed and flaked (1 cup)**
- **12 (1/2-oz.) thin white bread slices, crusts removed**
- **1 1/2 cups torn watercress leaves**

1. Stir together first 7 ingredients in a medium bowl. Gently fold in trout.
2. Spread 1/4 cup trout mixture onto 1 side of each of 6 bread slices. Top each with 1/4 cup watercress. Top with remaining bread slices. Cut sandwiches diagonally into quarters.

MAKES 24 appetizer servings. **HANDS-ON** 25 min., **TOTAL** 25 min.

FIG AND BACON PALMIERS

These are the ideal make-ahead recipe for afternoon tea.

- **1 cup chopped small Mission figs**
- **1 cup slivered almonds, toasted**
- **1 cup (4 oz.) crumbled blue cheese**
- **6 oz. cream cheese, softened**
- **1 (17.3-oz.) package frozen puff pastry sheets, thawed**
- **1/4 cup sugar, divided**
- **8 bacon slices, cooked, crumbled, and divided**
- **Parchment paper**
- **1/2 cup fig preserves, finely chopped**

1. Process first 2 ingredients in a food processor until finely chopped. Add cheeses; pulse until blended.
2. Carefully unfold 1 sheet of puff pastry on a work surface sprinkled with 2 Tbsp. sugar, pressing out seams.
3. Spread half of cream cheese mixture over pastry to within 1/2 inch of edges. Press half of bacon into cream cheese mixture. Roll 2 opposite sides, jelly-roll fashion, to meet in center. Brush water between rolled sides, and press lightly to seal. Wrap roll in plastic wrap. Repeat procedure with remaining dough, sugar, cream cheese mixture, and bacon. Chill rolls overnight.
4. Preheat oven to 400°.
5. Unwrap rolls. Cut each roll into 16 (1/2-inch-thick) slices. Place slices 2 inches apart on 2 large baking sheets lined with parchment paper.
6. Bake at 400° for 20 minutes or until golden brown. Immediately brush tops of palmiers with fig preserves. Serve warm.

MAKES 32 servings. **HANDS-ON** 40 min., **TOTAL** 9 hours

Smoked Trout and Watercress Tea Sandwiches, Fig and Bacon Palmiers, and Pumpernickel Tea Sandwiches

PUMPERNICKEL TEA SANDWICHES

Smear a bit of herb-flavored egg salad on these dainty sandwiches and top them with arugula and radishes.

- **6 large hard-cooked eggs, peeled**
- **1/3 cup mayonnaise**
- **1/4 cup minced fresh chives**
- **1/4 cup finely chopped fresh mint**
- **2 tsp. lemon zest**
- **2 tsp. Dijon mustard**
- **1 1/3 cups loosely packed arugula**
- **1 cup very thinly sliced radishes**
- **1 Tbsp. lemon juice**
- **1 Tbsp. olive oil**
- **14 pumpernickel bread slices**

1. Mash eggs with mayonnaise and next 4 ingredients until well blended. Add table salt and black pepper to taste. Cover and chill up to 1 day.
2. Toss together arugula, radishes, lemon juice, and olive oil. Add salt and pepper to taste. Toss to coat.
3. Spread egg salad mixture on 1 side of each bread slice; top 7 slices with arugula mixture. Top with remaining 7 bread slices, egg salad side down. Trim crusts from sandwiches; cut each sandwich into 3 rectangles with a serrated knife.

MAKES 21 appetizer servings.
HANDS-ON 5 min.; **TOTAL** 15 min., plus 1 day for chilling

HAZELNUT CHIP SHORTBREAD

Pulsing mini-morsels in a food processor produces little bits of chocolate that freckle the dough.

- **1/2 cup semisweet chocolate mini-morsels**
- **1 1/2 cups all-purpose flour**
- **1/2 cup powdered sugar**
- **1/2 cup cornstarch**
- **1 cup butter, cut into pieces and softened**
- **1/2 cup hazelnuts, toasted, chopped, and divided**
- **Parchment paper**

1. Preheat oven to 325°. Pulse mini-morsels in a food processor 10 times or until morsels are almost ground. Remove to a bowl. Add flour, powdered sugar, cornstarch, and butter to food processor. Cover; process just until a dough forms and mixture holds its shape. Stir in 1/4 cup hazelnuts and ground mini-morsels. Turn dough out onto a lightly floured surface. Roll dough to about 1/2-inch thickness; cut with a 2-inch scallop-shaped cutter. Press remaining 1/4 cup hazelnuts into top of each cookie. Place 1 inch apart on parchment paper-lined baking sheets.
2. Bake at 325° for 18 to 20 minutes or until golden brown. Cool on baking sheets 5 minutes; transfer to wire racks, and cool completely (about 30 minutes).

Note: To toast and skin hazelnuts, place nuts in a single layer in a shallow pan. Bake at 350° for 5 to 10 minutes or until skins begin to split. Transfer warm nuts to a colander; using a towel, rub briskly to remove skins.

MAKES 26 servings. **HANDS-ON** 15 min.; **TOTAL** 1 hour, 37 min.

BROWNED BUTTER SCONES WITH FAUX CLOTTED CREAM

Browned butter adds depth of flavor to these pure vanilla scones.

- **3/4 cup unsalted butter**
- **3 cups all-purpose flour**
- **2/3 cup sugar**
- **1 Tbsp. baking powder**
- **1/2 tsp. table salt**
- **1 cup whipping cream**
- **1 tsp. vanilla extract**
- **1/2 tsp. almond extract**
- **1 large egg**
- **Parchment paper**
- **2 Tbsp. whipping cream**
- **1 Tbsp. sugar**
- **Faux Clotted Cream**
- **Preserves**

1. Cook butter in a small heavy saucepan over medium heat, stirring constantly, 6 to 8 minutes or until butter begins to turn golden brown. Remove from heat immediately. Pour butter into a small bowl. Cover and freeze until firm (about 1 1/2 hours).
2. Preheat oven to 425°. Stir together flour, 2/3 cup sugar, baking powder, and salt. Cut in frozen browned butter with a pastry blender until mixture is crumbly. Combine 1 cup whipping cream, extracts, and egg, whisking until blended. Add to flour mixture, stirring with a fork until a shaggy dough forms.
3. Scoop dough by 2 heaping Tbsp. onto parchment paper-lined baking sheets. Brush with 2 Tbsp. whipping cream. Sprinkle with 1 Tbsp. sugar.
4. Bake at 425° for 12 to 14 minutes or until browned. Cool on wire racks. Serve with Faux Clotted Cream and preserves.

MAKES 28 servings. **HANDS-ON** 30 min.; **TOTAL** 2 hours, 20 min.

Faux Clotted Cream

- **1/2 (8-oz.) container mascarpone cheese, softened**
- **3/4 cup heavy whipping cream**
- **1 1/2 Tbsp. honey**
- **1/2 tsp. vanilla extract**

Combine all ingredients in a large mixing bowl. Beat at medium speed with an electric mixer until soft peaks form. Store in refrigerator.

MAKES 2 1/4 cups. **HANDS-ON** 5 min., **TOTAL** 5 min.

PETITS FOURS PRESENTS

These tiny morsels made with a dense almond cake and filled with strawberry jam are sure to impress your guests in both taste and presentation. Quick Almond Buttercream can be easily doubled or tripled for decorating layer cakes, cookies, and other baked goods. If you don't want to take the time to make the frosting, simply purchase ready-made frosting at the supermarket.

CAKE

- **1 1/4 cups butter, softened**
- **2 cups sugar**
- **1 (7-oz.) tube almond paste**
- **6 large eggs**
- **3 cups all-purpose flour**
- **1/2 tsp. baking powder**
- **1/2 tsp. table salt**
- **1/2 cup sour cream**
- **1 tsp. vanilla extract**
- **1 cup strawberry jam**

GLAZE

- **10 cups powdered sugar**
- **1/4 cup meringue powder**
- **1 cup plus 2 Tbsp. half-and-half**
- **2 tsp. almond extract**

QUICK ALMOND BUTTERCREAM

- **1/2 cup butter, softened**
- **2 cups powdered sugar**
- **2 Tbsp. half-and-half**
- **1/2 tsp. almond extract**
- **Green food coloring paste**

1. Prepare Cake: Preheat oven to 325°. Beat butter at medium speed with an electric mixer until creamy. Gradually add sugar and almond paste; beat 5 minutes or until light and fluffy. Add eggs, 1 at a time, beating until blended after each addition.
2. Combine flour, baking powder, and salt; gradually add to butter mixture alternately with sour cream, beginning and ending with flour mixture. Beat at low speed just until blended after each addition, stopping to scrape bowl as needed. Stir in vanilla. Pour batter into a greased and floured 13- x 9-inch pan.
3. Bake at 325° for 55 minutes or until a wooden pick inserted in center comes out clean. Cool in pan on a wire rack 20 minutes; remove from pan to wire rack, and cool completely (about 1 hour).
4. Trim crusts from all surfaces, making sure top of cake is flat. Slice cake in half horizontally. Spread cut side of bottom half with strawberry jam; replace top half. Cover; freeze 1 hour or until firm.
5. Cut cake into 40 (1 1/2-inch) squares; brush away loose crumbs. Place squares 2 inches apart on wire racks in jelly-roll pans.
6. Prepare Glaze: Beat 10 cups powdered sugar, meringue powder, half-and-half, and almond extract at medium speed with an electric mixer until smooth.
7. Pour glaze over cake squares, completely covering top and sides. Spoon up all excess glaze; continue pouring glaze until all cakes have been covered. Let stand 1 hour or until glaze dries completely.
8. Prepare Quick Almond Buttercream: Beat butter at medium speed with an electric mixer until creamy; gradually add 2 cups powdered sugar, beating until light and fluffy. Add half-and-half and almond extract; beat 1 minute. Stir in food coloring paste.
9. Trim any excess glaze from bottom of each cake square. Spoon Quick Almond Buttercream into a piping bag fitted with a small round tip, and decorate as desired.

Note: We tested with Wilton Meringue Powder, which can be found at craft and cake-decorating stores.

MAKES 40 servings. **HANDS-ON** 1 hour, 15 min.; **TOTAL** 5 hours, 30 min.

Baked Greats

WOW FAMILY AND FRIENDS WITH BAKE-SHOP-QUALITY PASTRIES AND BREADS THAT ARE A CINCH TO MAKE

BLUE CHEESE-PECAN POPOVERS

Any leftover popovers are great for breakfast the next day. Place them on a baking sheet, and bake at 375° for 5 minutes. Spread with fig jam and enjoy!

- 1/4 **cup pecans, finely chopped**
- 6 **large eggs**
- 2 **cups all-purpose flour**
- 2 **cups milk**
- 1/4 **cup butter, melted and divided**
- 1/2 **tsp. table salt**
- 2 **Tbsp. chopped fresh chives**
- 2 **oz. crumbled Roquefort cheese**

1. Preheat oven to 350°. Bake pecans in a single layer in a shallow pan 8 to 10 minutes or until toasted and fragrant, stirring halfway through.
2. Increase oven temperature to 375°. Process eggs, flour, milk, 2 Tbsp. butter, and salt in a blender until smooth, stopping to scrape down sides as needed.
3. Grease popover pans with remaining 2 Tbsp. butter. Pour batter evenly into prepared pans, filling three-fourths full. Sprinkle with chives, cheese, and pecans.
4. Bake at 375° for 40 to 45 minutes or until browned and puffy. Pierce each popover in several places with a thin wooden skewer. Bake 5 more minutes or until crisp. Serve immediately.

MAKES 1 dozen. **HANDS-ON** 16 min.; **TOTAL** 1 hour, 46 min.

Blue Cheese-Pecan Popovers

APPLE FRITTER PULL-APART BREAD

The sweet inspiration for this recipe is a combination of an apple fritter and monkey bread.

- 1/2 **cup butter**
- 1/4 **cup firmly packed light brown sugar**
- 1 1/4 **cups granulated sugar, divided**
- 1/2 **tsp. vanilla extract**
- 2 **(16.3-oz.) cans refrigerated buttermilk biscuits**
- 1 **Tbsp. ground cinnamon**
- 2 1/2 **cups peeled and chopped Fuji apples (2 large)**
- **Vanilla Glaze**

1. Preheat oven to 350°. Melt butter in a medium saucepan over medium heat. Stir in brown sugar and 1/4 cup granulated sugar; cook, stirring constantly, 3 minutes or until sugar dissolves. Remove from heat; stir in vanilla.
2. Cut biscuits into quarters. Stir together cinnamon and remaining 1 cup granulated sugar in a medium bowl; add half of biscuit pieces, tossing to coat. Arrange coated biscuit pieces in a lightly greased 12-cup Bundt pan; top with chopped apples. Toss remaining half of biscuit pieces in cinnamon-sugar; arrange over apples. Pour butter mixture evenly over biscuits.
3. Place Bundt pan on middle oven rack and a foil-lined baking sheet on lower oven rack. Bake at 350° for 45 minutes or until top is golden brown.
4. Carefully invert bread onto a platter, scraping any syrup in pan over bread. Let cool 10 minutes; drizzle with half of Vanilla Glaze. Let stand 15 more minutes, and drizzle with remaining Vanilla Glaze. Serve warm.

MAKES 12 servings. **HANDS-ON** 20 min.; **TOTAL** 1 hour, 40 min.

Vanilla Glaze

- 1 **cup powdered sugar**
- 1 **Tbsp. milk**
- 1 **tsp. vanilla extract**

Stir together all ingredients in a small bowl.

MAKES 1/2 cup. **HANDS-ON** 3 min., **TOTAL** 3 min.

SOUTHERN ANADAMA BREAD

1 (1/4-oz.) envelope active dry yeast
1/2 cup warm water (100° to 110°)
1 tsp. sugar
1/2 cup boiling water
1/2 cup molasses
3 Tbsp. butter
2 tsp. table salt
1/2 cup regular grits
3 to 3 1/2 cups bread flour
1 large egg white, lightly beaten
Whipped Honey-Rosemary Butter

1. Combine yeast, warm water (100° to 110°), and sugar in a small bowl. Let stand 5 minutes or until foamy.
2. Stir together boiling water and next 3 ingredients in a large bowl until butter melts. Gradually stir in grits; let cool to 100° to 110° (about 4 minutes). Stir in yeast mixture. Stir in 3 cups flour, 1 cup at a time, until dough is smooth, but not sticky. Turn dough out onto a lightly floured surface, and knead until smooth and elastic, adding remaining 1/2 cup flour, if necessary, to prevent sticking.
3. Place dough in a large oiled bowl, turning to coat top. Cover and let rise in a warm place (80° to 85°), free from drafts, 1 hour or until doubled in bulk. Punch dough down; turn out onto a lightly floured surface, and knead several times. Roll dough into a 14- x 7-inch rectangle. Starting at 1 short side, roll up dough jelly-roll fashion, ending seam side down. Fold ends under, and pinch seam to seal. Place dough, seam side down, in a greased 9- x 5-inch loaf pan. Cover and let rise in a warm place (80° to 85°), free from drafts, 50 minutes or until doubled in bulk.
4. Preheat oven to 375°. Brush top of dough with egg white and make several shallow slashes diagonally across top of loaf with a knife. Bake at 375° for 45 minutes or until bread is dark brown and sounds hollow when tapped. Remove bread from pan; let cool completely on a wire rack. Serve with Whipped Honey-Rosemary Butter.

MAKES 16 servings. **HANDS-ON** 26 min.; **TOTAL** 3 hours, 46 min.

Whipped Honey-Rosemary Butter

1/2 cup butter, softened
2 Tbsp. honey
1 Tbsp. chopped fresh rosemary

Beat butter and honey at medium speed with an electric mixer until fluffy. Add rosemary, and beat until blended.

MAKES 1/2 cup. **HANDS-ON** 4 min., **TOTAL** 4 min.

COFFEE CAKE MUFFINS WITH BROWN BUTTER ICING

Get more of the streusel topping in these individual cakes that have streusel inside and on top.

2 1/2 cups all-purpose flour, divided
1/2 cup firmly packed brown sugar
1 tsp. ground cinnamon, divided
1/3 cup butter, softened
1 cup chopped walnuts
1/2 cup granulated sugar
2 1/2 tsp. baking powder
1/4 tsp. table salt
3/4 cup sour cream
1/3 cup butter, melted
1/4 cup milk
1 tsp. vanilla extract
1 large egg
12 paper baking cups
Vegetable cooking spray
Brown Butter Icing

1. Preheat oven to 400°. Combine 1/2 cup flour, brown sugar, and 1/2 tsp. cinnamon in a medium bowl; add softened butter and pinch with fingers until mixture is crumbly. Stir in walnuts. Cover and chill until ready to use.
2. Combine remaining 2 cups flour, granulated sugar, baking powder, salt, and remaining 1/2 tsp. cinnamon in a large bowl. Make a well in center of flour mixture. Whisk together sour cream, melted butter, milk, vanilla, and egg; add to dry mixture, stirring just until moistened.
3. Place paper baking cups in a 12-cup muffin pan, and coat with cooking spray. Spoon half of batter evenly into cups, filling about one-third full. Spoon 1 Tbsp. streusel mixture into each cup. Spoon remaining batter evenly over streusel; sprinkle evenly with remaining streusel mixture, pressing gently, if necessary.
4. Bake at 400° for 18 to 20 minutes or until a wooden pick inserted in center comes out clean. Cool in pan on a wire rack 5 minutes. Remove from pan to wire rack, and cool completely (about 20 minutes).
5. Drizzle with Brown Butter Icing.

MAKES 1 dozen. **HANDS-ON** 18 min.; **TOTAL** 1 hour, 12 min.

Brown Butter Icing

The easy thing about this icing is that the butter is browned in the microwave. Microwaves vary in wattage, so cook the butter the lesser time, and then watch closely for browning.

1/4 cup butter, cut into pieces
1 1/2 cups powdered sugar
1 tsp. vanilla extract
1 to 2 Tbsp. milk

Place butter in a 1-qt. microwave-safe bowl or 4-cup glass measuring cup; cover with a paper plate. Microwave at HIGH 3 minutes to 3 minutes and 30 seconds or just until solids in bottom of bowl turn light brown, swirling butter in bowl after 3 minutes. Let cool 5 minutes. Add powdered sugar, vanilla, and 1 Tbsp. milk. Whisk until smooth. Add additional milk, 1 tsp. at a time, if necessary, until drizzling consistency. Use immediately.

MAKES 2/3 cup. **HANDS-ON** 5 min., **TOTAL** 12 min.

BOURBON-SPIKED MAPLE-BACON MINI CAKES

These mini Bundt cakes with three of our favorite ingredients—bourbon, maple syrup, and bacon—make not only a sweet treat for the family, but they're also perfect for holiday gifts. For a serious flavor boost, splurge on Benton's artisanal bacon and Grade B maple syrup; both can be ordered online from specialty stores.

- **11 bacon slices**
- **1 1/3 cups butter, softened**
- **1 3/4 cups sugar**
- **4 large eggs**
- **2/3 cup sour cream**
- **1/4 cup maple syrup**
- **2 2/3 cups all-purpose flour**
- **2 Tbsp. bourbon or whiskey**
- **1 1/2 tsp. vanilla extract**
- **Maple-Bourbon Glaze**

1. Preheat oven to 325°. Cook bacon in a large skillet over medium-high heat 8 to 10 minutes or until crisp; remove bacon, and drain on paper towels.
2. Beat butter at medium speed with an electric mixer 2 to 3 minutes or until creamy. Gradually add sugar, beating 5 to 7 minutes. Add eggs, 1 at a time, beating just until yellow disappears. Stir together sour cream and maple syrup.
3. Add flour to butter mixture alternately with sour cream mixture, beginning and ending with flour. Beat at low speed just until blended after each addition, stopping to scrape bowl as needed. Stir in bourbon and vanilla.
4. Crumble 7 bacon slices; gently fold into batter. Pour batter into 12 greased and floured miniature Bundt pans.
5. Bake at 325° for 28 to 33 minutes or until a wooden pick inserted in center comes out clean. Cool in pans on wire racks 10 minutes; remove from pans to wire racks, and cool completely (about 1 hour).
6. Drizzle Maple-Bourbon Glaze over cakes. Crumble remaining 4 bacon slices, and sprinkle over glaze.

MAKES 12 servings. **HANDS-ON** 43 min.; **TOTAL** 2 hours, 15 min.

Maple-Bourbon Glaze

The glaze sets quickly so drizzle over the cakes immediately after adding the powdered sugar.

- **2/3 cup firmly packed light brown sugar**
- **1/2 cup butter**
- **1/2 cup bourbon or whiskey**
- **1/2 tsp. maple extract**
- **2 cups powdered sugar**

Combine brown sugar and butter in a small heavy saucepan; cook over medium heat, stirring constantly, 3 minutes or until bubbly. Remove from heat; whisk in bourbon and maple extract. Gradually whisk in powdered sugar until smooth, adding 1 to 2 Tbsp. water if necessary for right consistency.

MAKES 1 3/4 cups. **HANDS-ON** 8 min., **TOTAL** 8 min.

Get the Party Started

WHIP UP THESE TEST KITCHEN HITS FOR COCKTAIL HOUR OR AS GREAT STARTS TO A FORMAL DINNER PARTY

TURKEY-ARTICHOKE CORNUCOPIAS

A cornucopia, or "horn of plenty," is a symbol of abundance and overflowing gifts. These bite-size treats are perfect at every holiday occasion. Leftover chicken salad works equally well in these pastries.

- **2 cups finely chopped roasted turkey**
- **1/2 cup drained and finely chopped marinated artichoke hearts**
- **1/2 cup mayonnaise**
- **1/3 cup finely shredded Parmesan cheese**
- **2 Tbsp. finely chopped green onions**
- **1/4 tsp. table salt**
- **Cream horn molds**
- **Vegetable cooking spray**
- **1/2 (17.3-oz) package frozen puff pastry sheets, thawed**
- **1 large egg**

1. Stir together first 6 ingredients in a medium bowl; cover and chill at least 1 hour.
2. Preheat oven to 375°. Coat cream horn molds with cooking spray.
3. Unfold pastry sheet on a lightly floured surface; gently smooth out lines with a rolling pin. Cut dough crosswise into 18 (1/2-inch) strips using a pizza cutter. (Do not separate strips.) Whisk together egg and 1 Tbsp. water; brush tops of strips with egg mixture.
4. Starting at tip end of cream horn molds, wrap 1 strip of pastry, egg side out, about halfway up each mold, overlapping slightly and forming a cornucopia. (Do not stretch pastry.) Repeat procedure with remaining molds and pastry strips. If working in batches, chill remaining pastry strips until ready to use. Place cream horn molds 2 inches apart on a lightly greased baking sheet, with ends of pastry strips facing down. Brush with egg mixture.
5. Bake at 375° for 18 to 20 minutes or until lightly browned. Transfer to wire rack; cool 5 minutes. Remove molds; let pastries cool completely (about 20 minutes).
6. Spoon turkey salad into a large zip-top plastic bag. Snip 1 corner of bag to make a hole about 3/4 inch in diameter. Gently pipe turkey salad into each pastry horn.

Note: You can order cream horn molds from Amazon.com or King Arthur Flour, or purchase at a local kitchen store.

MAKES 18 servings. **HANDS-ON** 45 min.; **TOTAL** 1 hour, 21 min.

Cranberry-Chutney Cheese Ball

CRANBERRY-CHUTNEY CHEESE BALL

Make ahead of time for fantastic flavor.

- **1 (8-oz.) package cream cheese, softened**
- **1 (8-oz.) block extra-sharp white Cheddar cheese, shredded**
- **1/2 cup mango chutney**
- **1/2 tsp. curry powder**
- **1/8 tsp. ground red pepper**
- **1 cup sweetened dried cranberries**
- **1/3 cup chopped green onions**
- **1 cup chopped pistachios**
- **Flatbread crackers**

1. Beat cream cheese at medium speed with an electric mixer in a large bowl until smooth. Reduce speed to low; add Cheddar cheese, and beat until well blended.
2. Chop any large pieces of chutney. Add chutney, curry powder, and red

pepper to cream cheese mixture, beating at low speed until blended. Stir in cranberries and green onions.
3. Shape mixture into a ball; wrap in plastic wrap, and chill 3 to 24 hours.
4. Roll cheese ball in chopped pistachios. Let stand 10 minutes before serving. Serve with crackers.

Note: We tested with Cracker Barrel Extra-Sharp Cheddar Cheese and Craisins.

MAKES 8 to 10 appetizer servings. **HANDS-ON** 10 min.; **TOTAL** 3 hours, 20 min.

CHEESE BISCUIT BARBECUE BITES

Grits add texture and a twist to these cheesy buttermilk biscuits. Paired with barbecued pork, these sliders are sure to be an instant hit at your holiday party.

- **2 cups self-rising flour**
- **2/3 cup uncooked quick-cooking grits**
- **1/4 tsp. freshly ground black pepper**
- **1/2 cup cold butter, cut into pieces**
- **4 oz. sharp Cheddar cheese, shredded**
- **2 Tbsp. finely chopped green onions**
- **3/4 cup buttermilk**
- **2 Tbsp. butter, melted**
- **3/4 lb. shredded barbecued pork without sauce, warmed**
- **1/2 cup barbecue sauce**

1. Preheat oven to 425°. Stir together first 3 ingredients in a large bowl; cut 1/2 cup butter into flour mixture with a pastry blender or fork until mixture resembles small peas and dough is crumbly. Stir in cheese and green onions. Add buttermilk, stirring just until dry ingredients are moistened.
2. Turn dough out onto a lightly floured surface, and knead lightly 3 or 4 times. Pat or roll dough to 1/2-inch thickness; cut with a 1 1/2-inch round cutter, and place on a lightly greased baking sheet. Brush tops of biscuits with 2 Tbsp. melted butter.
3. Bake at 425° for 10 minutes or until golden brown.
4. Combine pork and barbecue sauce in small bowl. Split biscuits; fill with pork mixture before serving.

MAKES 45 servings. **HANDS-ON** 30 min., **TOTAL** 50 min.

CRANBERRY TEA-INI

To make frozen sugared cranberries, dip frozen berries in water, and roll in granulated sugar; refreeze.

- **1/2 cup fresh or frozen cranberries, thawed**
- **2 Tbsp. superfine sugar**
- **1 cup ice cubes**
- **1/2 cup sweet tea-flavored vodka**
- **1/4 cup cranberry juice**
- **2 Tbsp. dry vermouth**
- **Garnishes: 4 frozen sugared cranberries, 2 lemon zest twists**

Place 1/2 cup cranberries and superfine sugar in a cocktail shaker. Muddle cranberries against sides of shaker to release flavors; add ice, vodka, cranberry juice, and vermouth. Cover with lid, and shake vigorously until thoroughly chilled (about 30 seconds). Strain into 2 chilled martini glasses. Serve immediately.

MAKES 2 servings. **HANDS-ON** 5 min., **TOTAL** 5 min.

RUBY RED NEGRONI

Our twist on the classic Negroni features fresh ruby red grapefruit juice for a festive drink to ring in the holidays.

- **2 cups fresh red grapefruit juice (3 large grapefruit)**
- **1 cup gin**
- **1 cup Campari**
- **1 cup sweet vermouth**
- **Crushed ice**
- **1 1/2 cups club soda**
- **Garnish: fresh orange slices**

1. Stir together first 4 ingredients in a pitcher. Cover and chill.
2. Fill 6 double old-fashioned glasses with crushed ice. Divide juice mixture evenly among glasses. Top each serving with 1/4 cup club soda, and stir gently. Serve immediately.

MAKES 6 servings. **HANDS-ON** 7 min., **TOTAL** 7 min.

Cranberry Tea-Ini and Ruby Red Negroni

Soups and Stews

RESTORATIVE WINTERTIME BOWLFULS NEVER TASTED SO GOOD

BRAISED BEEF-AND-MUSHROOM STEW

For an even heartier meal, serve this stew over mashed potatoes or creamy polenta.

- 3 lb. beef stew meat
- 1/2 tsp. table salt
- 3/4 tsp. freshly ground black pepper, divided
- 1/2 cup all-purpose flour
- 1/4 cup butter, divided
- 1/4 cup olive oil
- 3 garlic cloves, minced
- 3 (8-oz.) packages fresh mushrooms, quartered
- 1 large onion, cut into wedges
- 1 cup dry red wine
- 4 cups beef broth
- 1 Tbsp. tomato paste
- 2 tsp. fresh thyme leaves
- 1 tsp. chopped fresh rosemary
- 4 large carrots, peeled and cut into 1-inch pieces (3 cups)
- 1 Tbsp. all-purpose flour
- 1/4 cup cold water
- Garnish: fresh thyme or rosemary sprigs

1. Sprinkle beef with salt and 1/2 tsp. black pepper. Place 1/2 cup flour in a shallow dish. Dredge beef in flour; shake off excess.

2. Heat 1 Tbsp. butter and 2 Tbsp. oil in a large Dutch oven over medium-high heat until butter is melted. Add half of beef to pan. Cook 8 to 10 minutes or until browned on all sides. Remove beef from pan. Repeat procedure with 1 Tbsp. butter, remaining 2 Tbsp. oil, and remaining beef.

3. Heat remaining 2 Tbsp. butter in pan until butter is melted. Add garlic, mushrooms, and onion. Cook over medium-high heat 22 minutes or until mushrooms are browned, stirring occasionally. Stir in red wine. Bring to a boil; cook, uncovered, 10 minutes or until wine is reduced by half. Return beef to pan; add beef broth and tomato paste. Bring to a boil; cover, reduce heat, and simmer 1 hour or until beef is tender.

4. Add thyme and next 2 ingredients. Cover and cook over medium heat 30 minutes or until carrots are tender.

5. Combine 1 Tbsp. flour and 1/4 cup cold water, stirring until blended and smooth. Gradually stir flour mixture and remaining 1/4 tsp. black pepper into stew. Cook, uncovered, 10 minutes or until thickened, stirring occasionally.

MAKES 10 servings. **HANDS-ON** 30 min.; **TOTAL** 2 hours, 33 min.

PUMPKIN SOUP WITH RED PEPPER RELISH

Top off this pretty soup with toasted pumpkin seeds.

- 3 **Tbsp. butter**
- 5 **carrots, chopped**
- 1 **large sweet onion, chopped**
- 2 **garlic cloves, minced**
- 5 **cups chicken broth**
- 3/4 **tsp. table salt**
- 1/2 **tsp. ground cumin**
- 1/4 **tsp. freshly ground black pepper**
- 2 **(15-oz.) cans pumpkin**
- 1/4 **cup whipping cream**
- **Red Pepper Relish**

1. Melt butter in a large Dutch oven over medium heat. Sauté carrots, onion, and garlic 5 to 6 minutes or until tender. Add chicken broth, salt, cumin, black pepper, and pumpkin. Bring to a boil; reduce heat to medium-low, and simmer 10 minutes, stirring occasionally. Cool 10 minutes.
2. Process pumpkin mixture, in batches, in a blender until smooth, stopping to scrape down sides as needed. Return mixture to Dutch oven; stir in whipping cream. Cook just until thoroughly heated.
3. Ladle soup into bowls, and top with Red Pepper Relish before serving.

MAKES 10 cups. **HANDS-ON** 24 min., **TOTAL** 40 min.

Red Pepper Relish

- 1 **(12-oz.) jar roasted red bell peppers, drained and finely chopped**
- 2 **Tbsp. chopped fresh cilantro**
- 1 **Tbsp. olive oil**
- 1 **tsp. fresh lime juice**
- 1/4 **tsp. table salt**
- 1/4 **tsp. smoked paprika**

Stir together all ingredients.

MAKES 1 cup. **HANDS-ON** 5 min., **TOTAL** 5 min.

HAM-AND-CORN CHOWDER

This is a nice way to use up leftover holiday ham.

- 1 1/2 **cups finely chopped ham**
- 1 **Tbsp. olive oil**
- 2 **(16-oz.) packages frozen baby gold and white corn, thawed and divided**
- 1 **small onion, chopped**
- 1 **small red bell pepper, chopped**
- 2 **celery ribs, chopped**
- 2 **garlic cloves, minced**
- 2 **Tbsp. all-purpose flour**
- 1 **lb. unpeeled red potatoes, cut into 1/2-inch cubes**
- 2 **cups milk, divided**
- 2 **tsp. thyme leaves**
- 1 **(14-oz.) can chicken broth**
- 1 **cup heavy cream**
- 1 **cup (4 oz.) shredded sharp Cheddar cheese**
- 1/2 **tsp. freshly ground black pepper**
- 1/2 **tsp. table salt**

1. Sauté ham in hot oil in a Dutch oven over medium heat until lightly browned. Remove ham from Dutch oven using a slotted spoon; reserve drippings. Sauté 1 package corn and next 4 ingredients in hot drippings 5 minutes or until tender.
2. Sprinkle flour over vegetables; cook, stirring constantly, 1 minute. Add potatoes, 1 cup milk, thyme, and broth. Bring to a boil; reduce heat, and simmer 15 minutes or until potatoes are almost tender, stirring occasionally.
3. Process remaining 1 package corn and remaining 1 cup milk in a blender; stir into potato mixture. Simmer 15 minutes, stirring occasionally. Return ham to soup; stir in cream and remaining ingredients. Cook 2 minutes or until cheese melts.

MAKES 8 servings. **HANDS-ON** 51 min., **TOTAL** 51 min.

Ham-and-Corn Chowder

Holiday Sideboard

A GREAT SIDE DISH SHOULD NEVER BE AN AFTERTHOUGHT. WE'VE GOT OUR TOP PICKS FOR THE SEASON.

WINTER CITRUS-AVOCADO SALAD

Beautiful curly endive (also called frisée) and slightly bitter radicchio are a perfect foil for winter's best juicy citrus and creamy avocado.

- **6 cups curly endive leaves (3 heads)**
- **2 cups radicchio**
- **1/2 cup pistachios**
- **1/3 cup firmly packed fresh parsley leaves**
- **12 (6-inch-long) chives, cut into 1-inch pieces**
- **Rosemary-Citrus Vinaigrette**
- **2 red grapefruits, peeled and sectioned**
- **4 small clementines, peeled and sliced**
- **2 medium avocados, sliced**

Place first 5 ingredients in a serving bowl. Add desired amount of Rosemary-Citrus Vinaigrette; toss to coat. Top with grapefruit sections, clementine slices, and avocado. Serve with any remaining dressing.

MAKES 8 servings. **HANDS-ON** 16 min., **TOTAL** 16 min.

Rosemary-Citrus Vinaigrette

- **1/3 cup fresh red grapefruit juice (about 1/2 grapefruit)**
- **1/4 cup extra virgin olive oil**
- **3 Tbsp. rice vinegar**
- **1 Tbsp. honey**
- **1 tsp. chopped fresh rosemary**
- **1/2 tsp. table salt**
- **1/4 tsp. freshly ground black pepper**

Whisk together all ingredients.

MAKES 3/4 cup. **HANDS-ON** 2 min., **TOTAL** 2 min.

BUTTERNUT-PECAN SAUTÉ

Try sweet potato in this dish as a savory alternative to sweet potato casserole topped with pecans.

- **1 butternut squash (about 2 lb.)**
- **3 Tbsp. butter, divided**
- **1 1/2 cups chopped onion**
- **2 tsp. chopped fresh sage**
- **1/2 tsp. table salt**
- **1/4 tsp. freshly ground black pepper**
- **1 large garlic clove, minced**
- **1/3 cup coarsely chopped pecans**
- **1/2 cup (2 oz.) shredded Parmesan cheese**

1. Microwave squash at HIGH 1 minute. Peel squash; cut squash in half lengthwise. Remove and discard seeds and membranes; cut squash into 3/4-inch cubes to measure 5 cups.
2. Melt 2 Tbsp. butter in a large nonstick skillet. Add squash cubes, onion, and next 4 ingredients; cook over medium heat, stirring often, 16 minutes or until squash is tender.
3. Meanwhile, melt remaining 1 Tbsp. butter in a small nonstick skillet over medium-low heat; add pecans. Cook, stirring often, 5 to 6 minutes or until toasted and fragrant.
4. Sprinkle squash with cheese and pecans; cover and cook 1 minute or until cheese melts.

MAKES 4 to 6 servings. **HANDS-ON** 18 min., **TOTAL** 35 min.

POTATO GRATIN WITH BACON AND COMTÉ

Comté cheese is a wonderful melting cheese, ideal for topping this gratin. You'll notice its outstanding flavor at first bite.

- **3 1/4 cups heavy cream**
- **2 garlic cloves, minced**
- **1/4 tsp. freshly ground nutmeg**
- **2 1/2 lb. Yukon gold potatoes, peeled**
- **1 tsp. table salt**
- **3/4 tsp. freshly ground black pepper**
- **1 1/2 cups shredded Comté cheese, divided**
- **4 cooked hickory-smoked bacon slices, crumbled**

1. Preheat oven to 400°. Cook cream, garlic, and nutmeg in a heavy nonaluminum saucepan over medium heat, stirring often, 5 to 6 minutes or just until it begins to simmer (do not boil); remove from heat.
2. Using a mandolin or sharp knife, cut potatoes into 1/8-inch-thick slices.
3. Arrange one-third potato slices in a thin layer on bottom of a buttered 13- x 9-inch baking dish. Sprinkle with 1/4 tsp. salt, 1/4 tsp. black pepper, 1/2 cup cheese, and one-third crumbled bacon. Repeat layers twice. (Do not sprinkle last 1/2 cup cheese over top layer.)
4. Pour hot cream mixture over potatoes. Sprinkle with remaining 1/4 tsp. salt. Sprinkle top with remaining 1/2 cup cheese.
5. Bake, uncovered, at 400° for 42 to 45 minutes or until golden brown and potatoes are tender.

MAKES 8 to 10 servings. **HANDS-ON** 22 min.; **TOTAL** 1 hour, 10 min.

Cheesy Pimiento Corn Casserole

CHEESY PIMIENTO CORN CASSEROLE

Unlike traditional corn casserole made with cornbread mix, this recipe produces a creamy, cheesy casserole somewhat like a squash casserole.

- 2 **Tbsp. butter**
- 1/2 **cup chopped onion**
- 2 **garlic cloves, minced**
- 1 **(8-oz.) block sharp white Cheddar cheese, shredded and divided**
- 2 **(15.25-oz.) cans whole kernel gold and white super sweet corn, drained**
- 1 **(14 3/4-oz.) can cream-style corn**
- 1 **(8-oz.) container sour cream**
- 1 **(4-oz.) jar diced pimiento, drained**
- 1/2 **cup self-rising yellow corn-meal mix**
- 1/2 **tsp. table salt**
- 1/4 **tsp. freshly ground black pepper**
- 2 **large eggs, lightly beaten**
- 1/2 **cup French fried onions**

1. Preheat oven to 350°. Melt butter in a small skillet over medium heat; add chopped onion and garlic, and sauté 3 minutes or until tender.
2. Stir together onion mixture, 1 1/2 cups cheese, whole kernel corn, and next 7 ingredients in a large bowl. Pour into a lightly greased 11- x 7-inch baking dish. Sprinkle with French fried onions and remaining 1/2 cup cheese.
3. Bake at 350° for 50 minutes or just until set. Let stand 10 minutes before serving.

MAKES 8 to 10 servings. **HANDS-ON** 10 min.; **TOTAL** 1 hour, 10 min.

ROASTED BRUSSELS SPROUTS WITH SAGE PESTO

Nothing brings out the natural sweetness in vegetables quite like roasting. If you've never been a fan of Brussels sprouts (think childhood memories of the frozen variety), this recipe will make you a convert: sweet, crispy Brussels sprouts tossed in a pesto of pungent sage, bright citrus, rich pistachios, and a sprinkling of kosher salt. Brussels sprouts never tasted so good.

- 2 **lb. fresh Brussels sprouts**
- 1 **Tbsp. olive oil**
- 1/2 **cup pistachios**
- 1/3 **cup loosely packed fresh sage leaves**
- 2 **tsp. lemon zest**
- 2 **Tbsp. fresh lemon juice**
- 1 **tsp. kosher salt**
- 1/2 **tsp. freshly ground black pepper**
- 1/3 **cup olive oil**
- **Garnishes: lemon slices, chopped pistachios**

1. Preheat oven to 425°. Remove discolored leaves from Brussels sprouts. Cut off stem ends; cut Brussels sprouts in half. Toss together Brussels sprouts and 1 Tbsp. oil; spread in an 18- x 13-inch half-sheet pan. Bake at 425° for 20 to 25 minutes or until lightly browned and tender, stirring once.
2. Meanwhile, process pistachios and next 5 ingredients in a food processor until smooth, stopping to scrape down sides as needed. With processor running, pour 1/3 cup oil through food chute in a slow, steady stream, processing until smooth.
3. Remove Brussels sprouts from oven; toss with sage pesto in a medium bowl. Serve immediately.

MAKES 6 to 8 servings. **HANDS-ON** 10 min., **TOTAL** 30 min.

HERBED GRITS DRESSING WITH LEEKS AND MUSHROOMS

Croutons made from chilled cubed grits are the base of this dressing.

- **3 1/2 cups chicken broth**
- **1 cup milk**
- **1 1/2 cups uncooked stone-ground grits**
- **1 cup (4 oz.) shredded white Cheddar cheese**
- **3/4 tsp. table salt, divided**
- **1/4 tsp. freshly ground black pepper**
- **2 medium leeks, thinly sliced (2 cups)**
- **2 garlic cloves, minced**
- **2 Tbsp. olive oil**
- **2 (4-oz.) packages fresh gourmet blend mushrooms**
- **1 1/2 Tbsp. chopped fresh thyme**
- **1 Tbsp. chopped fresh parsley**
- **1/2 cup whipping cream**
- **2 large eggs, lightly beaten**

1. Bring broth and milk to a boil in a large saucepan. Gradually stir in grits. Cover, reduce heat, and simmer 10 minutes or until thickened, stirring occasionally. Stir in cheese, 1/4 tsp. salt, and black pepper, stirring until cheese melts. Spoon grits into a greased 13- x 9-inch baking dish. Chill until firm (about 45 minutes).
2. Preheat oven to 450°. Unmold grits onto a large cutting board, sliding a knife or a spatula under grits to loosen from dish. Cut grits into 3/4-inch cubes. Place cubes in a single layer on a large greased baking sheet.
3. Bake at 450° for 20 minutes; turn grits cubes, and bake 10 to 12 more minutes or until crisp and browned. Remove from oven. Reduce oven temperature to 350°.
4. Meanwhile, sauté leeks and garlic in hot oil in a large skillet over medium heat 4 minutes or until almost tender. Add mushrooms and remaining 1/2 tsp. salt; sauté 6 minutes or until browned and liquid has evaporated, stirring occasionally. Toss together leek mixture, grits cubes, thyme, and parsley in a large bowl. Whisk together whipping cream and eggs; pour over dressing, tossing gently to coat. Spoon dressing loosely into a greased 11- x 7-inch baking dish.
5. Bake, uncovered, at 350° for 30 to 35 minutes or until browned.

MAKES 8 servings. **HANDS-ON** 35 min.; **TOTAL** 2 hours, 25 min.

TART CHERRY-CHESTNUT SAUSAGE DRESSING

This flavorful dressing boasts sweet Italian sausage, dried Montmorency cherries, and roasted chestnuts. Enjoy it as a moist, hearty side dish or as a stuffing for pork, turkey, or chicken.

- **1 (12-oz.) French bread loaf, cut into 3/4-inch cubes**
- **12 oz. Italian pork sausage, casings removed**
- **1/3 cup unsalted butter**
- **1 large sweet onion, chopped**
- **2 celery ribs, chopped**
- **1 fennel bulb, chopped**
- **1 cup dried cherries, chopped**
- **1 cup whole roasted and peeled chestnuts, chopped**
- **2 cups chicken broth**
- **1/2 cup dry white wine**
- **2 tsp. crushed fennel seeds**
- **1 tsp. kosher salt**
- **3/4 tsp. freshly ground black pepper**
- **1 large egg, beaten**

1. Preheat oven to 350°. Spread bread cubes in a single layer on a large baking sheet.
2. Bake at 350° for 10 minutes or until toasted. Cool completely (about 10 minutes). Transfer to a very large bowl. Increase oven temperature to 400°.
3. Brown sausage in a large skillet over medium-high heat, stirring often, 5 to 6 minutes or until meat crumbles and is no longer pink; drain. Add mixture to bread in bowl. Melt butter in skillet over medium heat; add onion, celery, and fennel. Sauté mixture 8 minutes or until tender. Remove from heat; add to bread in bowl. Stir in cherries and chestnuts.
4. Combine broth and next 5 ingredients. Add broth mixture to bread, tossing well. Spoon into a lightly greased 13- x 9-inch baking dish.
5. Bake dressing, covered, at 400° for 30 minutes. Uncover and bake 20 more minutes or until top is browned and crusty.

MAKES 12 servings. **HANDS-ON** 24 min.; **TOTAL** 1 hour, 34 min.

Herbed Grits Dressing with Leeks and Mushrooms

PECAN-AND-WILD RICE STUFFING

Prepare this stuffing and use it to fill the cavity of Cornish hens, if desired.

- **2/3 cup coarsely chopped pecans**
- **3 (2.75-oz.) packages quick-cooking wild rice**
- **1/4 cup butter**
- **1/2 cup chopped onion**
- **1/2 cup chopped celery**
- **2/3 cup chopped dried apricots**
- **1/2 cup chicken broth**
- **1 tsp. table salt**
- **1 tsp. orange zest**
- **1/2 tsp. freshly ground black pepper**
- **1/4 cup chopped fresh parsley**

1. Preheat oven to 350°. Bake pecans in a single layer in a shallow pan 5 to 6 minutes or until lightly toasted and fragrant, stirring halfway through. Set aside.
2. Cook rice according to package directions. Cool slightly.
3. Melt butter in a large skillet over medium heat. Add onion and celery, and sauté 5 minutes or until tender. Stir in next 5 ingredients and reserved rice; cook 2 minutes or until liquid is absorbed. Remove from heat, and stir in parsley and reserved pecans.

MAKES 8 to 10 servings. **HANDS-ON** 25 min., **TOTAL** 25 min.

CROWN PORK ROAST WITH FIG-FENNEL STUFFING

Make your holiday extra-special by serving this herb-rubbed crown roast filled with moist Fig-Fennel Stuffing. Have your butcher french the chops and tie the roast in a crown to save time.

- **1 Tbsp. fennel seeds**
- **1 Tbsp. coriander seeds**
- **1 Tbsp. chopped fresh thyme**
- **2 tsp. kosher salt**
- **2 tsp. freshly ground black pepper**
- **2 tsp. lemon zest**
- **5 large garlic cloves, minced**
- **3 Tbsp. olive oil**
- **1 (16-rib) crown pork roast, trimmed and tied (10 to 11 lb.)**
- **Fig-Fennel Stuffing**
- **1 Tbsp. butter**
- **1 Tbsp. chopped shallot**
- **1 Tbsp. all-purpose flour**
- **1/3 cup dry white wine**
- **1/4 cup chicken broth**
- **1/2 cup whipping cream**
- **1/4 tsp. freshly ground black pepper**
- **Garnishes: fresh thyme sprigs, fresh figs, fresh currants**

1. Preheat oven to 350°. Place a small skillet over medium-high heat until hot; add fennel and coriander seeds, and cook, stirring constantly, 1 to 2 minutes or until toasted. Cool 5 minutes; coarsely crush seeds.
2. Combine crushed fennel and coriander seeds, chopped thyme, and next 4 ingredients in a small bowl. Stir in olive oil until blended. Rub spice mixture evenly over all sides of pork roast. Place roast in a lightly greased roasting pan. Spoon 3 1/2 cups Fig-Fennel Stuffing into center of roast. Cover stuffing with a square of aluminum foil. Cap the end of each bone with aluminum foil to prevent tips from burning.
3. Bake at 350° for 2 to 2 1/2 hours or until a meat thermometer inserted 2 inches into meat between ribs registers 145°. Carefully transfer roast and stuffing to a serving platter. Let pork roast stand 15 minutes before carving. Reserve 2/3 cup pan drippings for gravy.
4. Melt butter in a small saucepan over medium heat; add shallot and cook 3 to 4 minutes or until tender. Whisk in flour; cook, stirring constantly, 1 minute. Gradually whisk in wine, broth, and reserved drippings. Bring to a boil, whisking constantly. Whisk in whipping cream and 1/4 tsp. black pepper. Reduce heat to medium-low; simmer, stirring occasionally, 1 to 2 minutes or until thickened.
5. To serve, carve roast between bones using a sharp knife. Serve with pan gravy.

MAKES 12 servings. **HANDS-ON** 35 min.; **TOTAL** 4 hours, 40 min.

Fig-Fennel Stuffing

- **1 (16-oz.) day-old crusty Italian bread loaf, cut into 3/4-inch cubes**
- **3 Tbsp. butter**
- **1 red onion, chopped**
- **1 cup chopped celery**
- **1 small fennel bulb, chopped**
- **1/3 cup Marsala**
- **1 (7-oz.) package dried Mission figs, chopped**
- **3 cups chicken broth**
- **1 Tbsp. chopped fresh thyme**
- **1 tsp. kosher salt**
- **1/2 tsp. freshly ground black pepper**
- **2 large eggs**

1. Preheat oven to 350°. Spread bread cubes in a single layer on 2 large baking sheets. Bake at 350° for 15 minutes or until toasted. Let cool 15 minutes; transfer to a very large bowl. Increase oven temperature to 375°.
2. Melt butter in a large skillet over medium heat; add onion, celery, and fennel. Sauté 6 to 8 minutes or until tender. Add Marsala, and cook 1 to 2 minutes or until liquid evaporates.
3. Add onion mixture and figs to bread in bowl, stirring well. Whisk broth and next 4 ingredients in a small bowl; pour over bread mixture, tossing well. Reserve 3 1/2 cups stuffing for pork roast. Spoon remaining stuffing into a lightly greased 11- x 7-inch baking dish.
4. Bake, covered, at 375° for 30 minutes. Uncover and bake 15 more minutes or until top is browned and crusty.

MAKES 13 cups. **HANDS-ON** 20 min.; **TOTAL** 1 hour, 35 min.

Main Attractions

ELEGANT LOINS, ROASTS, AND RACKS THAT ARE SHOWSTOPPING CENTERPIECES FOR ANY FEAST

Butterflied and Barbecued Turkey

BUTTERFLIED AND BARBECUED TURKEY

A turkey cooked low and slow on direct heat will result in a smoky-flavored, tender, succulent bird without having to use a smoker. Butterflying the turkey allows the skin to brown and the meat to cook quicker.

- 1 **(10- to 12-lb.) whole fresh or frozen turkey, thawed**
- 1 1/2 **Tbsp. olive oil**
- 1 1/2 **tsp. table salt**
- 3/4 **tsp. garlic powder**
- 3/4 **tsp. dried Italian seasoning**
- 3/4 **tsp. paprika**
- 1/2 **tsp. chili powder**
- 1/2 **tsp. ground red pepper**
- 1/2 **tsp. freshly ground black pepper**
- **Sweet-and-Tangy Barbecue Sauce, divided**
- **Vegetable cooking spray**

1. Remove giblets and neck, and rinse turkey with cold water. Drain cavity well; pat dry with paper towels.
2. Cut turkey, using kitchen shears, along both sides of backbone, separating backbone from the turkey. Remove and discard backbone. Press turkey until flattened, and brush entire bird with oil.
3. Combine salt and next 6 ingredients; rub spice mixture evenly on turkey skin and underside. Place turkey in a large shallow dish, large oven bag, or 2-gal. zip-top plastic bag. Cover or seal, and chill in refrigerator 8 to 24 hours.
4. Reserve 1 cup Sweet-and-Tangy Barbecue Sauce to serve with cooked turkey. Light 1 side of grill, heating to 400° to 500° (high) heat; leave other side unlit. Coat unlit side with cooking spray. Shield wing tips and legs with aluminum foil to prevent excessive browning. Place turkey, skin side up, over unlit side, and grill, covered with grill lid, 2 to 3 hours or until a meat thermometer inserted into thickest portion of thigh registers 170°, basting with remaining Sweet-and-Tangy Barbecue Sauce during last 30 minutes of cooking.
5. Remove turkey from grill to a large shallow dish. Cover with foil, and let stand 15 minutes before carving. Serve with reserved Sweet-and-Tangy Barbecue Sauce.

MAKES 10 to 12 servings. **HANDS-ON** 3 hours, 3 min.; **TOTAL** 3 hours, 3 min., plus 8 hours for chilling

Sweet-and-Tangy Barbecue Sauce

- 1 **(14-oz.) can chicken broth**
- 1 3/4 **cups white wine vinegar**
- 1 3/4 **cups ketchup**
- 1/2 **cup finely chopped onion**
- 1/3 **cup orange marmalade**
- 1/4 **cup firmly packed brown sugar**
- 2 1/2 **Tbsp. ancho chile powder**
- 2 **Tbsp. garlic salt**
- 2 **tsp. freshly ground black pepper**

Combine all ingredients in a large saucepan; bring to a boil. Reduce heat, and simmer, stirring often, 45 minutes or until thickened.

MAKES 3 cups. **HANDS-ON** 11 min., **TOTAL** 56 min.

TUSCAN BEEF TENDERLOIN

Be sure to buy a trimmed tenderloin—the butcher will be able to do this for you.

- 2 Tbsp. olive oil
- 1 (6-oz.) bag fresh spinach, chopped
- 1/2 tsp. dried crushed red pepper
- 1 tsp. lemon zest
- 1/2 tsp. table salt
- 2 garlic cloves, minced
- 2 Tbsp. butter
- 1 (8-oz.) package fresh cremini mushrooms, chopped
- 1/3 cup grated Asiago cheese
- 1/3 cup finely chopped roasted red bell pepper
- 1/2 tsp. table salt
- 1/4 tsp. freshly ground black pepper
- 1 (4 1/2- to 5-lb.) trimmed beef tenderloin
- 2 tsp. table salt
- 2 tsp. fennel seeds, crushed
- 2 tsp. freshly ground black pepper
- 6 garlic cloves, pressed

1. Preheat oven to 425°. Heat oil in a large nonstick skillet over medium-high heat. Add half of spinach to hot oil, and cook, stirring constantly, until spinach begins to wilt. Add remaining spinach, and cook 1 minute or until all spinach has wilted. Add crushed red pepper and next 3 ingredients; sauté 1 minute. Transfer to a large bowl.
2. Melt butter in skillet over medium-high heat. Add mushrooms; cook 10 minutes or until well browned, stirring once after 8 minutes. Add to spinach in bowl. Stir in cheese and next 3 ingredients.
3. Cut beef tenderloin lengthwise down center, cutting to within 1/2 inch of other side. (Do not cut all the way through tenderloin.) Spoon spinach mixture down center of tenderloin. Fold tenderloin over spinach mixture, and tie with kitchen string, securing at 2-inch intervals. Place on a lightly greased rack in a roasting pan.
4. Combine 2 tsp. salt and next 3 ingredients; rub over outside of beef.
5. Bake at 425° for 30 minutes or until a meat thermometer inserted into thickest portion of tenderloin registers 130° (rare). Let stand 15 minutes before slicing.

MAKES 10 to 12 servings. **HANDS-ON** 25 min.; **TOTAL** 1 hour, 15 min.

PISTACHIO-CRUSTED RACK OF LAMB

A simple sauce of apricot preserves and whole grain mustard provides the finishing touch to the nutty crunch of the lamb coating.

- 3 garlic cloves
- 1 (3/4-oz.) French bread baguette slice, torn
- 1/2 cup pistachios
- 1/4 cup firmly packed fresh flat-leaf parsley
- 1 Tbsp. lemon zest
- 1/2 tsp. ground coriander
- 2 (8-rib) lamb rib roasts (2 lb. each), trimmed
- 1 tsp. table salt
- 1/2 tsp. freshly ground black pepper
- 2 Tbsp. olive oil
- 3 Tbsp. whole grain mustard, divided
- 3/4 cup apricot preserves

1. Preheat oven to 450°. With food processor running, drop garlic through food chute, processing until minced. Add bread and next 4 ingredients; process until minced.
2. Sprinkle lamb with salt and black pepper. Brown 1 lamb roast 2 minutes on each side in 1 Tbsp. hot oil in a large skillet over medium-high heat. Remove lamb from skillet; repeat procedure with remaining lamb roast and oil.
3. Brush meaty side of each lamb roast with 1 Tbsp. mustard; press breadcrumb mixture into mustard. Place lamb roasts, meaty sides up, on a lightly greased roasting rack.
4. Bake at 450° for 30 to 33 minutes or until a meat thermometer inserted into thickest portion of lamb registers 130°. Remove from oven and loosely cover with aluminum foil. Let stand 10 minutes.
5. Meanwhile, stir together apricot preserves and remaining 1 Tbsp. mustard in a small saucepan over low heat, and cook, stirring often, 1 to 2 minutes or until preserves melt. Cut each roast into 8 chops, and serve with preserves mixture.

MAKES 8 servings. **HANDS-ON** 18 min., **TOTAL** 1 hour

Pistachio-Crusted Rack of Lamb

Save Room

CHOOSE FROM THESE THREE CLASSIC CAKES TO PUNCTUATE YOUR HOLIDAY MEAL

PEPPERMINT RED VELVET BUNDT CAKES

Miniature red velvet cakes with a peppermint-cream cheese swirl are drizzled with snow white glaze and sprinkled with chocolate curls for a showstopping and palate-pleasing holiday finish.

- 1 **cup butter, softened**
- 2 1/2 **cups sugar**
- 6 **large eggs**
- 3 **cups all-purpose flour**
- 1/4 **tsp. baking soda**
- 1 **(8-oz.) container sour cream**
- 2 **tsp. vanilla extract**
- 1 **(8-oz.) package cream cheese, softened**
- 1 **(4-oz.) white chocolate baking bar, melted**
- 1/2 **tsp. peppermint extract**
- 1/4 **cup unsweetened cocoa**
- 3 **Tbsp. red liquid food coloring (1 1/2 oz.)**
- **Powdered Sugar Glaze**
- **Garnish: red-and-white chocolate curls**

1. Preheat oven to 325°. Beat butter at medium speed with an electric mixer until creamy. Gradually add sugar, beating until light and fluffy. Add eggs, 1 at a time, beating just until blended after each addition.
2. Stir together flour and baking soda; add to butter mixture alternately with sour cream, beginning and ending with flour mixture. Beat at low speed until blended after each addition, stopping to scrape bowl as needed. Stir in vanilla.
3. Beat cream cheese, melted white chocolate, and peppermint extract at medium speed with an electric mixer until creamy. Add 1 cup cake batter to cream cheese mixture, beating just until blended.
4. Add cocoa to remaining cake batter; stir well. Stir in red food coloring. Spoon two-thirds red velvet batter evenly into 12 (3/4-cup) greased and floured miniature Bundt pans. Drop cream cheese batter by tablespoonfuls onto red velvet batter in pans. Spoon remaining red velvet batter over cream cheese batter; gently swirl with a knife.
5. Bake at 325° for 25 to 28 minutes or until a wooden pick inserted in center comes out clean. Cool in pans on wire racks 10 minutes; remove from pans to wire racks, and cool completely (about 1 hour). Drizzle cooled cakes with Powdered Sugar Glaze.

MAKES 12 cakes. **HANDS-ON** 35 min.; **TOTAL** 2 hours, 15 min.

Powdered Sugar Glaze

- 3 1/2 **cups powdered sugar**
- 1/3 **cup milk**
- 1/2 **tsp. vanilla extract**

Whisk together all ingredients in a medium bowl until smooth.

MAKES 1 1/2 cups. **HANDS-ON** 2 min.; **TOTAL** 2 min.

MOCHA-HAZELNUT DACQUOISE

Look for hazelnuts without the skin. If you can find them only with skins, rub in a clean kitchen towel after toasting to remove the skins. Gently fold the ground hazelnuts into the egg whites to prevent them from deflating.

- **Parchment paper**
- 1 1/4 **cups hazelnuts**
- 3/4 **cup powdered sugar**
- 1 **Tbsp. cornstarch**
- 6 **large egg whites**
- 1/2 **tsp. cream of tartar**
- 3/4 **cup granulated sugar**
- **Mocha Cream Filling**
- **Chocolate shavings**

1. Line 2 large baking sheets with parchment paper. Draw 2 (8-inch) circles on each sheet of parchment by tracing an 8-inch round cake pan. Turn parchment paper over.
2. Preheat oven to 350°. Bake hazelnuts in a single layer in a shallow pan 8 to 10 minutes or until lightly toasted and fragrant, stirring halfway through. Cool completely (about 20 minutes). Reduce oven temperature to 275°.
3. Coarsely chop 1/4 cup hazelnuts for top of cake. Process remaining 1 cup hazelnuts, powdered sugar, and cornstarch in a food processor until hazelnuts are finely ground.
4. Beat egg whites and cream of tartar at high speed with an electric mixer until foamy. Gradually add granulated sugar, 1 Tbsp. at a time, beating until stiff peaks form and sugar dissolves (about 2 to 4 minutes). Gently fold in ground hazelnut mixture in 4 additions, blending after each.
5. Spoon one-fourth meringue batter into center of each circle on parchment paper-lined baking sheets. Spread evenly to edges of each circle using a small offset spatula.
6. Bake at 275° for 1 hour and 15 minutes or until crisp and lightly golden. Turn off oven; let meringues stand in closed oven 1 hour. Remove from oven; cool completely on baking sheets on wire racks (about 30 minutes). Carefully peel off parchment paper.
7. Place 1 meringue on a serving plate. Top with one-fourth Mocha Cream Filling, spreading almost to edges of meringue. Repeat layers twice. Top with remaining meringue layer. Spread remaining Mocha Cream Filling on top of meringue. Sprinkle with chopped hazelnuts and chocolate shavings. Cover and chill 8 to 24 hours before serving.

MAKES 10 to 12 servings. **HANDS-ON** 33 min.; **TOTAL** 3 hours, 48 min., plus 1 day for chilling

Mocha Cream Filling

- 2 (4-oz.) semisweet chocolate baking bars, chopped
- 3 cups heavy whipping cream, divided
- 1/4 cup powdered sugar
- 2 Tbsp. coffee liqueur
- 1 Tbsp. instant espresso

1. Place chopped chocolate in a large bowl. Bring 3/4 cup cream to a simmer in a small saucepan over medium-high heat. Pour cream over chocolate; let stand 1 minute. Whisk until chocolate is melted and smooth. Cool to room temperature (about 30 minutes).
2. Beat chocolate mixture, sugar, liqueur, espresso, and remaining 2 1/4 cups cream at high speed with an electric mixer until stiff peaks form.

MAKES 5 cups. **HANDS-ON** 10 min., **TOTAL** 40 min.

WHITE GERMAN CHOCOLATE CAKE

This German white chocolate cake is wonderfully dense and moist. It will keep for up to one week, covered, in the refrigerator.

- Parchment paper
- 1 (4-oz.) white chocolate baking bar, chopped
- 1/4 cup milk
- 2 1/2 cups all-purpose flour
- 1 tsp. baking soda
- 1/4 tsp. table salt
- 1 cup butter, softened
- 1 cup granulated sugar
- 4 large eggs, separated
- 2 tsp. vanilla extract
- 1 cup buttermilk
- Coconut-Pecan Frosting
- Garnish: white chocolate curls

1. Preheat oven to 350°. Lightly grease 3 (9-inch) round cake pans; line bottoms with parchment paper, and lightly grease paper.
2. Microwave white chocolate and milk in a medium-size microwave-safe bowl at HIGH for 30 to 45 seconds or until chocolate is melted and smooth, stirring once halfway through. Cool 5 minutes.
3. Combine flour and next 2 ingredients in a medium bowl.
4. Beat butter at medium speed with an electric mixer until creamy; gradually add sugar, beating until light and fluffy. Add egg yolks, 1 at a time, beating just until blended after each addition. Add chocolate mixture and vanilla; beat on low speed until blended. Add flour mixture alternately with buttermilk, beginning and ending with flour mixture. Beat at low speed just until blended after each addition.
5. Beat egg whites at high speed until medium-stiff peaks form; gently fold into batter. Spoon batter into prepared pans.
6. Bake at 350° for 20 to 22 minutes or until a wooden pick inserted in center comes out clean. Cool in pans on wire racks 10 minutes; remove from pans to wire racks. Carefully remove parchment paper, and discard. Cool completely (about 1 hour).
7. Spread Coconut-Pecan Frosting between layers and on top and sides of cake.

MAKES 12 servings. **HANDS-ON** 31 min.; **TOTAL** 2 hours, 50 min.

Coconut-Pecan Frosting

- 2 1/2 cups chopped pecans
- 3 (5-oz.) cans evaporated milk
- 1 1/4 cups granulated sugar
- 1 cup butter
- 2/3 cup firmly packed light brown sugar
- 5 large egg yolks, lightly beaten
- 3 1/2 cups sweetened flaked coconut
- 1 Tbsp. vanilla extract

1. Preheat oven to 350°. Bake pecans at 350° in a single layer in a shallow pan 8 to 10 minutes or until toasted and fragrant, stirring halfway through. Cool completely (about 20 minutes).
2. Meanwhile, cook evaporated milk and next 4 ingredients in a heavy 3-qt. saucepan over medium heat, stirring constantly, 3 to 4 minutes or until butter melts and sugar dissolves. Cook, stirring constantly, 10 to 12 minutes or until mixture is bubbly and reaches a pudding-like thickness.
3. Remove pan from heat; stir in coconut, vanilla, and pecans. Transfer mixture to a bowl. Let stand, stirring occasionally, 1 hour or until cooled and spreading consistency.

MAKES 5 1/2 cups. **HANDS-ON** 20 min.; **TOTAL** 1 hour, 35 min.

A Big Batch of Cookie Favorites

CHRISTMAS AND COOKIES GO HAND IN HAND, AND WE'VE GOT INSPIRED NEW RECIPES TO ENJOY OR GIVE

SPICED SORGHUM SNOWFLAKES

Perfect for gift giving or tucking inside a stocking, these beautiful, crisp cookies are flavored with sweet sorghum syrup and a blend of gingerbread spices.

1/2 cup butter, softened
1/4 cup granulated sugar
1/4 cup firmly packed dark brown sugar
2 tsp. orange zest
1 large egg
3 Tbsp. hot water
1 tsp. baking soda
1/2 cup sorghum syrup
3 cups all-purpose flour
2 tsp. ground ginger
1 tsp. ground cinnamon
1/2 tsp. ground allspice
1/4 tsp. ground nutmeg
1/4 tsp. table salt
Parchment paper
Royal Icing
White sparkling sugar, nonpareils, sugar pearls

1. Beat butter and sugars at medium speed with a heavy-duty electric stand mixer until fluffy. Add orange zest and egg, beating until smooth.
2. Stir together hot water and baking soda in a small bowl until baking soda is dissolved. Stir in sorghum syrup.
3. Stir together flour and next 5 ingredients; add to butter mixture alternately with sorghum mixture, beginning and ending with flour mixture.
4. Divide dough into 2 equal portions; flatten each into a disk. Cover and chill at least 1 hour or until firm.
5. Preheat oven to 325°. Place 1 portion of dough on a lightly floured surface, and roll to 1/4-inch thickness. Cut with assorted sizes of snowflake-shaped cutters. Place cookies 1 inch apart on parchment paper-lined baking sheets. Repeat procedure with remaining dough disk. Once cookies are cut and placed on baking sheets, freezing them about 10 minutes allows them to better hold their shape during baking.
6. Bake at 325° for 13 to 15 minutes or until cookies are puffed and slightly darker around the edges. Cool on pans 1 minute; transfer to wire racks, and cool completely (about 30 minutes).
7. Spoon Royal Icing into a zip-top plastic freezer bag. Snip 1 corner of bag to make a small hole. Pipe icing in decorative designs on each cookie. Sprinkle with white sparkling sugar, and decorate with nonpareils and sugar pearls. Let icing harden at least 1 hour.

Note: We tested with snowflake-shaped cutters ranging in size from 1 3/4 inch to 4 inches.

MAKES about 7 dozen. **HANDS-ON** 3 hours, 25 min.; **TOTAL** 5 hours, 25 min.

Royal Icing

1 (16-oz.) package powdered sugar
3 Tbsp. meringue powder
1/2 cup warm water

Beat all ingredients at low speed with an electric mixer until blended. Beat at high speed 4 minutes or until glossy and stiff peaks form, adding a few drops of additional water, if necessary, for desired consistency.

MAKES 3 cups. **HANDS-ON** 5 min., **TOTAL** 5 min.

MAPLE-PRALINE COOKIES

This cookie is special enough for a cookie swap or gift giving; it's like biting into praline candy on top of a drop cookie.

- 3/4 **cup butter, softened**
- 1/4 **cup shortening**
- 1 **cup firmly packed brown sugar**
- 1 **Tbsp. maple syrup**
- 1 **large egg**
- 2 **cups all-purpose flour**
- 1/2 **tsp. baking soda**
- 1/4 **tsp. table salt**
- **Maple-Praline Frosting**
- **48 pecan halves, lightly toasted**

1. Beat butter and shortening at medium speed with an electric mixer until creamy; gradually add sugar, beating until light and fluffy. Add maple syrup and egg, beating until blended.
2. Stir together flour, baking soda, and salt; gradually add to butter mixture, beating until blended. Cover and chill dough 1 hour.
3. Preheat oven to 350°. Drop dough by rounded tablespoonfuls onto ungreased baking sheets.
4. Bake at 350° for 9 to 10 minutes or until edges are lightly browned. Cool on baking sheets 1 minute. Transfer to wire racks, and cool completely (about 10 minutes).
5. Spread Maple-Praline Frosting over cookies. Top each cookie with 1 pecan half.

MAKES 4 dozen. **HANDS-ON** 30 min.; **TOTAL** 2 hours, 22 min.

Maple-Praline Frosting

- 1 **cup firmly packed light brown sugar**
- 1/3 **cup whipping cream**
- 2 **Tbsp. maple syrup**
- 1 **cup powdered sugar**

Combine brown sugar and whipping cream in a 2-qt. saucepan. Cook over medium heat, stirring constantly, until mixture comes to a boil; boil 4 minutes (do not stir). Remove from heat; stir in maple syrup. Gradually stir in powdered sugar until smooth. Use immediately. If frosting hardens, stir in additional whipping cream, 1/2 tsp. at a time, as needed.

MAKES 1 cup. **HANDS-ON** 12 min., **TOTAL** 12 min.

CHOCOLATE-CAPPUCCINO WHOOPIE PIES

These little handheld treats are the perfect combination of not-too-sweet cookies joined together with a decadent espresso-cream cheese filling.

- 2 **cups all-purpose flour**
- 1/2 **cup unsweetened cocoa**
- 1 **tsp. baking soda**
- 1/2 **tsp. table salt**
- 1/2 **cup butter, softened**
- 1 **cup firmly packed light brown sugar**
- 1 **large egg**
- 1 **tsp. vanilla extract**
- 1 **cup buttermilk**
- **Parchment paper**
- **Espresso Filling**
- 1/2 **cup crushed chocolate-coated coffee beans**

1. Preheat oven to 350°. Whisk together first 4 ingredients in a medium bowl.
2. Beat butter and brown sugar at medium speed with a heavy-duty electric stand mixer 2 minutes or until fluffy. Add egg and vanilla; beat just until blended. Add flour mixture alternately with buttermilk, beginning and ending with flour mixture. Beat at low speed until blended after each addition, scraping bowl as needed.
3. Drop dough by tablespoonfuls 2 inches apart onto parchment paper-lined baking sheets.
4. Bake at 350° for 10 minutes or until cookies are set and tops spring back when touched. Cool on pans 5 minutes; transfer to wire racks, and cool completely (about 10 minutes).
5. Spread about 1 1/2 Tbsp. Espresso Filling on 1 flat side of half of cooled cookies; top with remaining cookies, pressing gently. Roll edges of whoopie pies in crushed coffee beans.

MAKES 25 whoopie pies. **HANDS-ON** 28 min.; **TOTAL** 1 hour, 28 min.

Espresso Filling

- 1 1/2 **(3-oz.) packages cream cheese, softened**
- 6 **Tbsp. butter, softened**
- 1 **Tbsp. instant espresso**
- 3 **Tbsp. milk**
- 1 1/2 **tsp. vanilla extract**
- 3 **cups powdered sugar**

Beat first 5 ingredients at medium speed with an electric mixer until smooth. Gradually add powdered sugar, beating mixture at low speed until blended.

MAKES 2 1/2 cups. **HANDS-ON** 5 min., **TOTAL** 5 min.

SUGAR COOKIE STARS

Use your favorite three-inch holiday cookie cutter if you don't have a star cutter on hand.

- 1 **cup butter, softened**
- 1 **cup granulated sugar**
- 3/4 **tsp. almond extract**
- 1 **large egg**
- 2 1/4 **cups all-purpose flour**
- 1/4 **tsp. table salt**
- **White Frosting**
- 3/4 **cup light blue sparkling sugar**

1. Beat butter and granulated sugar at medium speed with an electric mixer until light and fluffy. Add almond extract and egg, beating until blended. Combine flour and salt. Gradually add to butter mixture, beating at low speed just until blended.
2. Divide dough into 2 equal portions; flatten each into a disk. Cover and chill 20 minutes.
3. Preheat oven to 350°. Place 1 portion of dough on a lightly floured surface, and roll to a 1/8-inch thickness. Cut with a 3-inch star-shaped cookie cutter. Place cookies 2 inches apart on ungreased baking sheets.

Repeat procedure with remaining dough disk.

4. Bake at 350° for 10 to 12 minutes or until edges are lightly browned. Cool on baking sheets 5 minutes; transfer to wire racks, and cool completely (about 10 minutes).

5. Spread White Frosting over cookies. Sprinkle with sparkling sugar.

MAKES about 3 dozen. **HANDS-ON** 1 hour; **TOTAL** 2 hours, 10 min.

White Frosting

- **1 cup butter, softened**
- **3 cups powdered sugar, sifted**
- **3 Tbsp. heavy cream**
- **1/2 tsp. almond extract**

Beat butter and powdered sugar at low speed with an electric mixer until blended. Increase speed to medium, and beat 3 minutes. Add heavy cream and almond extract, beating to desired consistency.

Note: Purchase sparkling sugar at cook stores or crafts stores.

MAKES 3 cups. **HANDS-ON** 5 min., **TOTAL** 5 min.

ALMOND POINSETTIA COOKIES

This recipe bakes up into beautiful flower shapes that are ready for decorating. Shop online for a poinsettia cookie cutter.

- **2 cups all-purpose flour**
- **1/2 cup almond flour**
- **1/4 tsp. table salt**
- **3/4 cup butter, softened**
- **2 oz. cream cheese, softened**
- **2 oz. almond paste**
- **1 cup granulated sugar**
- **1 large egg yolk**
- **1 tsp. almond extract**
- **Royal Red Icing**
- **1/4 cup yellow candy sprinkles**
- **1/4 cup fine red sanding sugar**

1. Combine first 3 ingredients in a small bowl. Beat butter and cream cheese at medium speed with an electric mixer until creamy. Add almond paste and granulated sugar, beating until light and fluffy. Add egg yolk and almond extract, beating just until blended. Gradually add flour mixture to cream cheese mixture, beating to blend after each addition.

2. Divide dough into 2 equal portions; flatten each into a disk. Cover and chill 2 hours.

3. Preheat oven to 350°. Place 1 portion of dough on a lightly floured surface, and roll to a 1/4-inch thickness. Cut with a 3 1/2- to 4-inch poinsettia cookie cutter. Place cookies 1 inch apart on ungreased baking sheets. Repeat procedure with remaining dough disk.

4. Bake at 350° for 13 to 15 minutes or until edges are golden brown. Cool on baking sheets 2 minutes. Transfer to wire racks, and cool completely (about 30 minutes).

5. Spoon Royal Red Icing into a small zip-top plastic freezer bag. Snip 1 corner of bag to make a small hole. Working with 1 cookie at a time, pipe icing to outline cookie. Use icing to fill in cookie. Lightly sprinkle yellow candies in center. Sprinkle petals with red sugar.

MAKES about 2 1/2 dozen. **HANDS-ON** 1 hour; **TOTAL** 4 hours, 25 min.

Royal Red Icing

- **1 (16-oz.) package powdered sugar**
- **3 Tbsp. meringue powder**
- **3/4 tsp. red paste food color**
- **4 to 6 Tbsp. warm water**

Beat first 3 ingredients and 4 Tbsp. water at low speed with an electric mixer until blended. Add up to 2 Tbsp. additional water, 1 tsp. at a time, until desired consistency is reached.

Note: Royal Icing dries rapidly. Work quickly, keeping extra icing tightly covered at all times. Place a damp paper towel directly on surface of icing (to prevent a crust from forming) while icing cookies.

MAKES about 3 cups. **HANDS-ON** 10 min., **TOTAL** 10 min.

CANDY CANE BISCOTTI

Before you drizzle cookies with chocolate, stand them up on a wire rack set over wax paper to catch the drips.

- **3/4 cup granulated sugar**
- **1/2 cup butter, softened**
- **2 large eggs, lightly beaten**
- **2 1/2 cups all-purpose flour**
- **2 tsp. baking powder**
- **1/4 tsp. table salt**
- **1 Tbsp. peppermint schnapps**
- **1 tsp. vanilla extract**
- **3/4 cup crushed soft peppermint sticks, divided**
- **Wax paper**
- **1 (4-oz.) dark chocolate baking bar, chopped**
- **Coarse sugar**

1. Preheat oven to 350°. Beat sugar and butter at medium speed with an electric mixer until creamy. Add eggs, 1 at a time, beating until blended after each addition. Combine flour, baking powder, and salt; gradually add to butter mixture, beating until blended. Stir in peppermint schnapps and vanilla. Stir in 1/2 cup crushed peppermint.

2. Divide dough in half. Shape each portion of dough into a 9- x 2-inch log on a lightly greased baking sheet, using lightly floured hands.

3. Bake at 350° for 28 to 30 minutes or until firm. Transfer to wire racks; cool completely (about 1 hour). Cut each log diagonally into 1/2-inch-thick slices with a serrated knife, using a gentle sawing motion. Place slices, cut sides up, on baking sheets.

4. Bake at 350° for 10 minutes; turn cookies over, and bake 8 more minutes. Transfer to wire racks set over wax paper, and cool completely (about 30 minutes).

5. Microwave chocolate in a small microwave-safe bowl at HIGH 30 to 60 seconds or until melted and smooth, stirring at 30-second intervals. Drizzle chocolate over tops of biscotti; sprinkle with coarse sugar and remaining 1/4 cup crushed peppermint. Let stand until chocolate is set.

MAKES about 2 1/2 dozen. **HANDS-ON** 14 min.; **TOTAL** 2 hours, 30 min.

Good Luck New Year

RING IT IN WITH ALL THE FIXIN'S AND GET YOUR ANNUAL DOSE OF GOOD FORTUNE TOO

EASY HOPPIN' JOHN

Boil-in-bag rice, canned peas, and fully cooked bacon keep your kitchen work to a minimum in this classic dish that's good for lunch or dinner.

- 1 (3.5-oz.) package boil-in-bag whole grain brown or white rice
- 1/2 tsp. table salt
- 4 fully cooked bacon slices
- 3 Tbsp. butter
- 3/4 cup chopped celery
- 3/4 cup finely chopped red bell pepper
- 1/2 cup chopped onion
- 4 garlic cloves, minced
- 1 (15-oz.) can black-eyed peas, drained and rinsed
- 1 tsp. Asian hot chili sauce (such as Sriracha)
- 1/4 tsp. freshly ground black pepper
- Garnish: celery leaves

1. Cook rice according to package directions, adding 1/2 tsp. salt to water; drain well. Heat bacon according to package directions; crumble.
2. Melt butter in a large skillet over medium heat; add celery and next 3 ingredients, and sauté 5 to 7 minutes or until vegetables are tender. Stir in rice, peas, hot chili sauce, and black pepper; sauté 1 to 2 minutes or until thoroughly heated. Sprinkle with crumbled bacon.

MAKES 4 to 6 servings. **HANDS-ON** 14 min., **TOTAL** 14 min.

PORK CHOPS WITH BOURBON-ROSEMARY-MUSTARD SAUCE

Finishing the pork chops in the oven ensures juicy chops.

- 6 (1-inch-thick) bone-in center-cut pork chops
- 1/2 tsp. table salt
- 1 tsp. freshly ground black pepper, divided
- 2 Tbsp. butter
- 1 Tbsp. olive oil
- 2 garlic cloves, minced
- 1/2 cup chicken broth
- 1/4 cup bourbon or whiskey
- 1/4 cup country-style Dijon mustard
- 1/4 cup heavy cream
- 1 Tbsp. chopped fresh rosemary

1. Preheat oven to 400°. Sprinkle pork chops with salt and 1/2 tsp. black pepper.
2. Melt 1 Tbsp. butter with 1 1/2 tsp. oil in a large ovenproof skillet. Brown half of pork chops 2 to 3 minutes on each side. Remove to a plate. Repeat procedure with remaining butter, oil, and pork chops. Arrange all pork chops in skillet, overlapping slightly; bake at 400° for 10 minutes or until a meat thermometer inserted in thickest portion of chops registers 155°. Remove pork chops from skillet to a platter, reserving drippings in skillet. Cover chops, and let stand 10 minutes or until meat thermometer inserted in thickest portion of chops registers 160°.
3. Add garlic to drippings in skillet; sauté over medium heat 1 minute. Add broth, bourbon, mustard, cream, and remaining 1/2 tsp. black pepper. Cook, stirring occasionally, 4 minutes or until sauce is reduced to 2/3 cup. Stir in rosemary; cook 1 minute. Serve sauce over pork chops.

MAKES 6 servings. **HANDS-ON** 24 min., **TOTAL** 34 min.

KALE SALAD WITH HOT BACON DRESSING

Popular in Italian cuisine, lacinato kale has long, slightly bumpy dark blue-green leaves and a milder flavor than the curly variety.

- 2 bunches lacinato kale (about 1 lb.)
- 6 bacon slices
- 1 shallot, minced
- 1 garlic clove, minced
- 1/3 cup apple cider vinegar
- 1 Tbsp. honey
- 1/2 tsp. table salt
- 1/4 tsp. freshly ground black pepper
- 1/4 cup freshly grated Romano cheese
- 2 hard-cooked eggs, peeled and finely shredded

1. Trim and discard thick stems from kale; thinly slice leaves, and place in a large bowl.
2. Cook bacon in a large skillet over medium-high heat 8 minutes or until crisp; remove bacon, and drain on paper towels, reserving 3 Tbsp. drippings in skillet. Crumble bacon.
3. Sauté shallot and garlic in hot drippings 2 minutes or until just tender. Stir in vinegar, honey, salt, black pepper, and 1/4 cup water; bring to a boil. Remove from heat; stir in bacon.
4. Drizzle vinegar mixture over kale, and sprinkle with cheese; toss to coat. Top with shredded eggs.

MAKES 6 to 8 servings. **HANDS-ON** 20 min., **TOTAL** 20 min.

PECAN CORNBREAD

Serve this cornbread with pepper jelly and butter.

- 1 **cup finely chopped pecans**
- 1/3 **cup shortening**
- 2 **cups yellow self-rising cornmeal mix**
- 1 **Tbsp. sugar**
- 1/2 **tsp. ground red pepper**
- 2 **cups buttermilk**
- 1 **large egg**

1. Preheat oven to 350°. Bake pecans in a single layer in a shallow pan 5 to 8 minutes or until toasted and fragrant, stirring halfway through. Remove pecans from oven, and increase temperature to 425°.
2. Melt shortening in a 10-inch cast-iron skillet in oven 5 minutes. Stir together cornmeal mix, sugar, and ground red pepper in a medium bowl. Whisk together buttermilk and egg; add to cornmeal mixture, stirring just until moistened. Stir in pecans.
3. Remove skillet from oven; tilt skillet to coat, and pour shortening into batter, stirring until blended. Immediately pour batter into hot skillet.
4. Bake at 425° for 25 minutes or until edges are golden brown.

Note: If you have any cornbread left over and about 1 1/2 cups of Easy Hoppin' John (page 331) remaining, you can make a quick-and-easy cornbread salad. Crumble cornbread to measure 3 cups, and stir into leftover Easy Hoppin' John. Add 1/3 cup mayonnaise and 1/3 cup Ranch dressing, stirring until just moistened. Serve immediately, or cover and chill salad until ready to serve.

MAKES 6 to 8 servings. **HANDS-ON** 5 min., **TOTAL** 35 min.

FIG UPSIDE-DOWN CAKE

Port is a blend of a still wine, typically red, and brandy. Ruby port gets its name from the distinct ruby color created from the mix of grapes used to make this variety. Although the port is an exceptional addition to the whipped cream and pairs beautifully with the cake, it can be omitted.

- 1/2 **cup butter**
- 1 **cup firmly packed brown sugar**
- 12 **fresh Brown Turkey or Mission figs, halved**
- 1/2 **cup chopped walnuts**
- 2 **large eggs, separated**
- 1 **large egg yolk**
- 1 **cup granulated sugar**
- 1 **cup all-purpose flour**
- 1 **tsp. baking powder**
- 1 **tsp. chopped fresh rosemary**
- 1 **tsp. lemon zest**
- 1/4 **tsp. table salt**
- 1/4 **cup milk**
- 1/2 **tsp. vanilla extract**
- 3/4 **cup whipping cream**
- 2 **Tbsp. ruby port (optional)**

1. Preheat oven to 350°. Melt butter in a 10-inch cast-iron skillet over medium-low heat; sprinkle brown sugar over butter. Remove from heat. Arrange figs, cut sides down, over sugar mixture; sprinkle with walnuts.
2. Beat 3 egg yolks at high speed with an electric mixer until thick and pale; gradually add 1 cup granulated sugar, beating well. Stir together flour and next 4 ingredients; add to egg mixture alternately with milk, beginning and ending with flour mixture. Stir in vanilla.
3. Beat egg whites at high speed with an electric mixer until stiff peaks form; fold egg whites into batter. Pour batter over figs in skillet.
4. Bake at 350° for 38 to 40 minutes or until a wooden pick inserted in center comes out clean. Cool in skillet on a wire rack 10 minutes; invert cake onto a serving platter, scraping any syrup from bottom of skillet onto cake.
5. Beat whipping cream and port, if desired, until soft peaks form. Serve cake warm or at room temperature with whipped cream.

MAKES 6 to 8 servings. **HANDS-ON** 28 min.; **TOTAL** 1 hour, 8 min.

Fig Upside-Down Cake

Southern Baker

BAKE UP IMPRESSIVE RECIPES WITH A DECIDEDLY SOUTHERN TWIST

CHOCOLATE BREAKFAST WREATH

From Virginia to Atlanta, artisanal chocolate makers are popping up all over the South. Look to their products to infuse this holiday wreath with authentic regional flavor. Be sure to soften the butter until it's spreadable.

- 1/2 **cup warm milk (100° to 110°)**
- 2 **(1/4-oz.) envelopes active dry yeast**
- 1/3 **cup plus 1/2 cup sugar, divided**
- 4 1/2 **cups all-purpose flour, divided**
- 2 **tsp. kosher salt**
- 1 1/2 **cups soft butter, divided**
- 3 **large eggs, at room temperature**
- **Parchment paper**
- 1 **(4-oz.) bittersweet chocolate baking bar, finely chopped**
- **Easy Vanilla Glaze**

1. Combine milk, yeast, and 1/3 cup sugar in bowl of a heavy-duty electric stand mixer; let stand 5 minutes or until foamy. Gradually add 1 cup flour, beating at low speed until blended; scrape down sides. Add salt and 1 cup butter; beat at low speed until smooth. Add eggs, 1 at a time, beating until blended after each addition and scraping sides of bowl as needed. Gradually add remaining 3 1/2 cups flour, beating until blended. Increase speed to medium, and beat until dough forms a ball and begins to pull away from sides. Beat dough 2 more minutes or until smooth and elastic. Turn dough out onto a lightly floured surface, and knead 3 minutes.

2. Place dough in a greased large bowl, turning to grease top. Cover with plastic wrap, and let rise in a warm place (80° to 85°), free from drafts, 1 hour or until doubled in bulk. Punch dough down; turn out onto lightly floured parchment paper. Roll dough into an 18- x 12-inch rectangle.

3. Brush 6 Tbsp. soft butter over dough; sprinkle with chocolate and 1/2 cup sugar. Roll up dough, jelly-roll fashion, starting at 1 long side. Press edge to seal, and place dough, seam side down, on parchment paper.

4. Transfer parchment paper with dough onto a baking sheet. Shape rolled dough into a ring, pressing ends together to seal. Cut ring at 2-inch intervals, from outer edge up to (but not through) inside edge. Gently pull and twist cut pieces to show filling. Cover dough.

5. Let rise in a warm place (80° to 85°), free from drafts, 1 hour or until doubled in bulk. Preheat oven to 350°. Uncover dough. Melt remaining 2 Tbsp. butter; brush over dough. Bake at 350° for 30 to 40 minutes or until golden. Cool on pan 10 minutes. Drizzle Easy Vanilla Glaze over warm bread before serving.

MAKES 10 to 12 servings. **HANDS-ON** 25 min.; **TOTAL** 3 hours, 40 min., including glaze

Easy Vanilla Glaze

- 2 **cups powdered sugar**
- 3 to 4 **Tbsp. milk**
- 1/2 **tsp. vanilla extract**
- **Dash of table salt**

Whisk together powdered sugar, 3 Tbsp. milk, vanilla, and salt. Whisk in up to 1 Tbsp. milk, 1 tsp. at a time, to reach desired consistency.

MAKES 1 1/2 cups.

Cornbread Focaccia

CARAMELIZED ONION AND SWISS POPOVERS

A hot pan and oven temperature get popovers off to a high-rising start, while lowering the temperature midway through baking helps prevent an out-of-the-oven collapse.

- **1 cup all-purpose flour**
- **1 cup 2% reduced-fat milk, at room temperature**
- **3 large eggs, at room temperature**
- **2 Tbsp. butter, melted**
- **3/4 tsp. table salt**
- **1/4 cup (1 oz.) freshly shredded Swiss cheese**
- **1/4 cup Caramelized Sweet Onions, chopped**
- **1 Tbsp. chopped fresh chives**
- **4 tsp. canola oil**

1. Preheat oven to 425°. Heat a 12-cup muffin pan in oven 10 minutes.
2. Meanwhile, process first 5 ingredients in a blender or food processor 20 to 30 seconds or until smooth. Stir in cheese and next 2 ingredients.
3. Remove pan from oven; pour 1/2 tsp. oil into each of 8 muffin cups, filling center 6 cups and middle cup on each end. Heat in oven 5 minutes. Remove from oven. Divide batter among oiled muffin cups; return to oven immediately.
4. Bake at 425° for 15 to 20 minutes or until puffed and lightly browned around edges. Reduce oven temperature to 350°; bake 10 minutes or until tops are golden brown.
5. Transfer to a wire rack; cool 3 to 4 minutes before serving.

MAKES 8 servings. **HANDS-ON** 10 min.; **TOTAL** 1 hour, 35 min., including onions

Caramelized Sweet Onions

The trick is to cook the onions slowly, allowing the natural sugars to caramelize. Make several batches, and store in freezer.

- **12 cups sliced sweet onions (about 2 1/2 lb.)**
- **1 tsp. table salt**

Heat a large nonstick skillet over medium heat. Add sweet onions, and cook, stirring often, 30 minutes or until onions are caramel colored, sprinkling with salt halfway through.

Note: Store in refrigerator in an airtight container or a zip-top plastic freezer bag up to 1 week, or freeze up to 2 months.

MAKES about 3 cups.

CORNBREAD FOCACCIA

This Southern take on an Italian classic is packed with flavor—not just from a rainbow of toppings, but also thanks to a sprinkling of yeast in the batter that gives this cornbread a bread-like taste.

- **2 cups self-rising white cornmeal mix**
- **2 cups buttermilk**
- **1/2 cup all-purpose flour**
- **1 (1/4-oz.) envelope rapid-rise yeast**
- **2 large eggs, lightly beaten**
- **1/4 cup butter, melted**
- **2 Tbsp. sugar**
- **1 cup crumbled feta cheese**
- **1 cup coarsely chopped black olives**
- **3/4 cup grape tomatoes, cut in half**
- **1 Tbsp. coarsely chopped fresh rosemary**

Preheat oven to 375°. Heat a well-greased 12-inch cast-iron skillet in oven 5 minutes. Stir together cornmeal mix and next 6 ingredients just until moistened; pour into hot skillet. Sprinkle with feta cheese, olives, tomatoes, and rosemary. Bake at 375° for 30 minutes or until golden brown.

MAKES 8 to 10 servings. **HANDS-ON** 15 min., **TOTAL** 45 min.

Herbed Cornbread Focaccia: Prepare recipe as directed, omitting black olives and grape tomatoes; stir in 1 Tbsp. each chopped fresh basil and chopped fresh parsley.

SWEET POTATO BISCUITS

Cooked sweet potatoes make for a very moist and tender biscuit with a slightly orange hue. Serve these biscuits alongside a holiday feast or offer them as an upgrade to the classic breakfast spread.

- **5 cups self-rising flour**
- **1 Tbsp. sugar**
- **1 tsp. kosher salt**
- **1 cup cold butter, cut into small cubes**
- **1/4 cup cold shortening**
- **2 cups buttermilk**
- **1 cup mashed cooked sweet potato**
- **Parchment paper**
- **2 Tbsp. butter, melted**

1. Preheat oven to 425°. Stir together first 3 ingredients in a large bowl. Cut butter cubes and shortening into flour mixture with pastry blender or fork until crumbly. Cover and chill 10 minutes.
2. Whisk together buttermilk and sweet potato. Add to flour mixture, stirring just until dry ingredients are moistened.
3. Turn dough out onto a lightly floured surface, and knead lightly 3 or 4 times. Pat or roll dough to 3/4-inch thickness; cut with a 2-inch round cutter, reshaping scraps once. Place rounds on a parchment paper-lined baking sheet.
4. Bake at 425° for 18 to 20 minutes or until biscuits are golden brown. Remove from oven, and brush tops of biscuits with melted butter. Serve immediately.

MAKES 3 dozen. **HANDS-ON** 30 min.; **TOTAL** 1 hour, 20 min.

MINI BANANA-CRANBERRY-NUT BREAD LOAVES

These loaves may become your new holiday breakfast, or even gifting, tradition! Loaded up with cranberries and pecans with a citrus glaze, this is not your ordinary banana bread.

- **1 (8-oz.) package cream cheese, softened**
- **3/4 cup butter, softened**
- **2 cups sugar**
- **2 large eggs**
- **3 cups all-purpose flour**
- **1/2 tsp. baking powder**
- **1/2 tsp. baking soda**
- **1/2 tsp. table salt**
- **1 1/2 cups mashed ripe bananas**
- **3/4 cup chopped fresh cranberries**
- **1/2 tsp. vanilla extract**
- **3/4 cup chopped toasted pecans**
- **Orange Glaze**

1. Preheat oven to 350°. Beat cream cheese and butter at medium speed with an electric mixer until creamy. Gradually add sugar, beating until light and fluffy. Add eggs, 1 at a time, beating just until blended after each addition.
2. Combine flour and next 3 ingredients; gradually add to butter mixture; beat at low speed just until blended. Stir in bananas, next 2 ingredients, and pecans. Spoon about 1 1/2 cups batter into each of 5 greased and floured 5- x 3-inch miniature loaf pans.
3. Bake at 350° for 40 to 44 minutes or until a wooden pick inserted in center comes out clean and sides pull away from pans. Cool in pans 10 minutes. Transfer to wire racks. Prepare Orange Glaze. Drizzle over warm bread loaves, and cool 10 minutes.

MAKES 5 miniature loaves. **HANDS-ON** 25 min.; **TOTAL** 1 hour, 25 min.

Regular-Size Banana-Cranberry-Nut Bread Loaves: Spoon batter into 2 greased and floured 8- x 4-inch loaf pans. Bake at 350° for 1 hour and 10 minutes or until a wooden pick inserted in center comes out clean.

MAKES 2 loaves.

Mini Banana-Cranberry-Nut Bread Loaves

Orange Glaze

A drizzle of this glaze adds a citrusy, sweet topping to your favorite bread, sweet roll, or coffee cake.

- **1 cup powdered sugar**
- **1 tsp. loosely packed orange zest**
- **2 to 3 Tbsp. fresh orange juice**

Stir together powdered sugar and remaining ingredients in a small bowl until blended. Use immediately.

MAKES 1/2 cup.

STRAWBERRY LEMONADE MUFFINS

With bright springtime flavor thanks to fresh berries and lemons, these muffins would be perfect for a light breakfast or as an addition to a brunch spread. Make these extra special by serving alongside store-bought or homemade lemon curd.

- **2 1/2 cups self-rising flour**
- **1 1/4 cups sugar, divided**
- **1 (8-oz.) container sour cream**
- **1/2 cup butter, melted**
- **1 Tbsp. loosely packed lemon zest**
- **1/4 cup fresh lemon juice**
- **2 large eggs, lightly beaten**
- **1 1/2 cups diced fresh strawberries**

1. Preheat oven to 400°. Combine flour and 1 cup sugar in a large bowl; make a well in center of mixture.
2. Stir together sour cream and next 4 ingredients; add to flour mixture, stirring just until dry ingredients are moistened. Gently fold strawberries into batter. Spoon batter into lightly greased 12-cup muffin pans, filling three-fourths full. Sprinkle remaining 1/4 cup sugar over batter.
3. Bake at 400° for 16 to 18 minutes or until golden brown and a wooden pick inserted in center comes out clean. Cool in pans on a wire rack 1 minute; remove from pans to wire rack, and cool 10 minutes.

MAKES 15 muffins. **HANDS-ON** 15 min., **TOTAL** 42 min.

PUMPKIN CUPCAKES WITH BROWNED BUTTER FROSTING

While many pumpkin cakes feature a cream cheese frosting or glaze, we've updated this fall favorite with a rich frosting flavored with browned butter and vanilla paste. For a generous swirl of frosting on each cupcake, double the recipe for Browned Butter Frosting.

- **30 paper baking cups**
- **1 cup butter, softened**
- **2 cups sugar**
- **5 large eggs, separated**
- **1 1/2 tsp. pumpkin pie spice**
- **3 cups all-purpose flour, divided**
- **1 cup buttermilk**
- **1 tsp. baking soda**
- **1 cup chopped toasted pecans**
- **1 cup canned pumpkin**
- **Browned Butter Frosting**

1. Preheat oven to 350°. Place 30 paper baking cups in 3 (12-cup) standard-size muffin pans.
2. Beat butter at medium speed with a heavy-duty electric stand mixer until creamy. Gradually add sugar, beating until light and fluffy. Add egg yolks, 1 at a time, beating just until blended after each addition.
3. Stir together pumpkin pie spice and 2 3/4 cups flour in a medium bowl; stir together buttermilk and baking soda in a small bowl. Add flour mixture to butter mixture alternately with buttermilk mixture, beginning and ending with flour mixture. Beat at low speed just until blended after each addition.
4. Stir together pecans and remaining 1/4 cup flour. Fold pecan mixture and pumpkin into batter.
5. Beat egg whites at high speed with an electric mixer until stiff peaks form. Stir about one-third of egg whites into batter; fold in remaining egg whites. Spoon batter into muffin cups, filling about three-fourths full.
6. Bake at 350° for 18 to 22 minutes or until a wooden pick inserted in center comes out clean. Remove from pans to wire rack, and cool completely (about 20 minutes) before frosting.

MAKES 30 servings. **HANDS-ON** 30 min.; **TOTAL** 2 hours, 30 min., including frosting

Browned Butter Frosting

- **1 1/2 cups butter**
- **7 1/2 cups powdered sugar**
- **6 Tbsp. milk**
- **1 1/2 tsp. vanilla bean paste**

1. Cook 1 1/2 cups butter in a medium-size saucepan over medium heat, stirring constantly, 8 to 10 minutes or just until butter begins to turn golden brown; pour butter into a small bowl. Cover and chill 1 hour or until butter is cool and begins to solidify.
2. Beat browned butter at medium speed with an electric mixer until fluffy; gradually add powdered sugar alternately with milk, beginning and ending with sugar. Beat at low speed until well blended after each addition. Add in vanilla bean paste, and beat at medium speed for 5 minutes or until fluffy.

MAKES 3 cups.

HONEY-BALSAMIC-BLUEBERRY PIE

A touch of tangy-sweet balsamic vinegar, combined with honey, cinnamon, and a pinch of freshly ground black pepper, magnifies the sweetness of blueberries. We love this buttery crust, but if time is short, substitute a store-bought one and simply crimp the edges.

CRUST

- 3 **cups all-purpose flour**
- 3/4 **cup cold butter, sliced**
- 6 **Tbsp. cold vegetable shortening, sliced**
- 1 **tsp. kosher salt**
- 4 to 6 **Tbsp. ice-cold water**

FILLING

- 7 **cups fresh blueberries**
- 1/4 **cup cornstarch**
- 2 **Tbsp. balsamic vinegar**
- 1/2 **cup sugar**
- 1/3 **cup honey**
- 1 **tsp. vanilla extract**
- 1/4 **tsp. kosher salt**
- 1/4 **tsp. ground cinnamon**
- 1/8 **tsp. finely ground black pepper**
- **Butter**
- 2 **Tbsp. butter, cut into 1/4-inch cubes**
- 1 **large egg**

1. Prepare Crust: Process flour and next 3 ingredients in a food processor until mixture resembles coarse meal. With processor running, gradually add 4 Tbsp. ice-cold water, 1 Tbsp. at a time, and process until dough forms a ball and pulls away from sides of bowl, adding up to 2 Tbsp. more water, 1 Tbsp. at a time, if necessary. Divide dough in half, and flatten each half into a disk. Wrap each disk in plastic wrap, and chill 2 hours to 2 days.

2. Prepare Filling: Place 1 cup blueberries in a large bowl; crush with a wooden spoon. Stir cornstarch and vinegar into crushed berries until cornstarch dissolves. Stir sugar, next 5 ingredients, and remaining 6 cups blueberries into crushed berry mixture.

3. Unwrap 1 dough disk, and place on a lightly floured surface. Sprinkle with flour. Roll dough to 1/8-inch thickness. Fit dough into a greased (with butter) 9-inch deep-dish pie plate. Repeat rolling procedure with remaining dough disk; cut dough into 12 to 14 (1/2 inch-wide) strips. (You will have dough left over.)

4. Pour blueberry mixture into piecrust, and dot with butter cubes. Arrange piecrust strips in a lattice design over filling. Trim excess dough.

5. Reroll remaining dough, and cut into 6 (9- x 1/2-inch) strips. Twist together 2 strips at a time. Whisk together egg and 1 Tbsp. water. Brush a small amount of egg mixture around edge of pie. Arrange twisted strips around edge of pie, pressing lightly to adhere. Brush entire pie with remaining egg mixture. Freeze 20 minutes or until dough is firm.

6. Preheat oven to 425°. Bake pie on an aluminum foil-lined baking sheet at 425° for 20 minutes. Reduce oven temperature to 375°, and bake 20 more minutes. Cover pie with aluminum foil to prevent excessive browning; bake 25 to 30 more minutes (65 to 70 minutes total) or until crust is golden and filling bubbles in center. Remove from baking sheet to a wire rack; cool 1 hour before serving.

MAKES 8 servings. **HANDS-ON** 35 min.; **TOTAL** 5 hours, 5 min.

BANANAS FOSTER COFFEE CAKE WITH VANILLA-RUM SAUCE

You don't have to wait for Mardi Gras to bake up this rum-infused cake from New Orleans. Banana slices are first sautéed in a buttery rum glaze, then folded into a simple coffee cake batter and topped off with a crumble.

- **1 1/2 cups mashed ripe bananas**
- **7 Tbsp. light rum, divided**
- **2 cups firmly packed brown sugar, divided**
- **1 1/2 cups soft butter, divided**
- **2 tsp. vanilla extract, divided**
- **1 (8-oz.) package cream cheese, softened**
- **2 large eggs**
- **3 1/4 cups plus 3 Tbsp. all-purpose flour, divided**
- **5/8 tsp. table salt, divided**
- **1/2 tsp. baking powder**
- **1/2 tsp. baking soda**
- **1 1/2 cups chopped pecans**
- **1 tsp. ground cinnamon**
- **1 cup granulated sugar**
- **2 cups heavy cream**

1. Preheat oven to 350°. Cook bananas, 3 Tbsp. rum, 1/2 cup brown sugar, and 1/4 cup butter in a skillet until mixture is bubbly. Cool; stir in 1 tsp. vanilla.
2. Beat cream cheese and 1/2 cup butter at medium speed with an electric mixer until creamy. Add 1 cup brown sugar; beat until fluffy. Beat in eggs, 1 at a time.
3. Stir together 3 cups flour, 1/2 tsp. salt, and next 2 ingredients; add to cream cheese mixture. Beat at low speed until blended. Stir in banana mixture. Spoon into a greased and floured 13- x 9-inch pan.
4. Combine pecans, cinnamon, 1/2 cup brown sugar, and 1/4 cup flour. Melt 1/4 cup butter; stir into pecan mixture. Sprinkle over batter. Bake at 350° for 45 minutes or until a wooden pick inserted in center comes out clean. Cool in pan on a wire rack 10 minutes.
5. Combine granulated sugar, remaining 3 Tbsp. flour, and remaining 1/8 tsp. salt in a saucepan over medium heat. Add cream and remaining 1/2 cup butter; bring to a boil. Boil, whisking constantly, 2 minutes or until slightly thickened. Remove from heat; stir in remaining 1/4 cup rum and remaining 1 tsp. vanilla. Drizzle over warm cake.

MAKES 8 to 10 servings. **HANDS-ON** 20 min.; **TOTAL** 1 hour, 45 min.

DOUBLE CHOCOLATE CHIP COOKIES WITH BOURBON GANACHE

As if these cookies weren't decadent enough, we sandwiched Bourbon Ganache between them for more wow!

- **3/4 cup butter, softened**
- **3/4 cup granulated sugar**
- **3/4 cup firmly packed dark brown sugar**
- **2 large eggs**
- **1 1/2 tsp. vanilla extract**
- **2 1/2 cups all-purpose flour**
- **1 tsp. baking soda**
- **3/4 tsp. table salt**
- **1 (12-oz.) package semisweet chocolate morsels**
- **Parchment paper**
- **Bourbon Ganache**

1. Preheat oven to 350°. Beat butter and sugars at medium speed with a heavy-duty electric stand mixer until creamy. Add eggs, 1 at a time, beating just until blended after each addition. Add vanilla, beating until blended.
2. Combine flour and next 2 ingredients in a small bowl; gradually add to butter mixture, beating at low speed just until blended. Stir in morsels just until combined. Drop dough by level spoonfuls onto parchment paper-lined baking sheets, using a small cookie scoop (about 1 1/8 inches).
3. Bake at 350° for 12 minutes or until golden brown. Remove from baking sheets to wire racks, and cool completely (about 30 minutes).
4. Spread Bourbon Ganache on flat side of half of cookies (about 1 Tbsp. per cookie); top with remaining cookies. Cover and chill cookies 2 hours or until ganache is firm.

MAKES 2 1/2 dozen. **HANDS-ON** 45 min.; **TOTAL** 5 hours, 48 min.

Bourbon Ganache

- **1 (12-oz.) package semisweet chocolate morsels**
- **1/2 cup whipping cream**
- **3 Tbsp. bourbon**
- **3 Tbsp. softened butter**
- **1/2 tsp. vanilla extract**

Microwave semisweet chocolate morsels and whipping cream in a 2-qt. microwave-safe bowl at HIGH 1 1/2 to 2 minutes or until chocolate is melted and smooth, stirring at 30-second intervals. Whisk in bourbon, softened butter, and vanilla. Cover and chill, stirring occasionally, 1 hour and 30 minutes or until spreading consistency.

MAKES 2 1/2 cups.

Double Chocolate Chip Cookies with Bourbon Ganache

SALTED CARAMEL-CHOCOLATE PECAN TART

A cross between a chocolate tart and pecan pie, this is all the more stunning if you arrange autumn's new-crop Southern pecan halves from the center in an even spiral pattern.

TART SHELL

- **1 recipe Simple Piecrust, prepared through Step 2**

CHOCOLATE FILLING

- **1 1/2 cups sugar**
- **3/4 cup butter, melted**
- **1/3 cup all-purpose flour**
- **1/3 cup 100% cacao unsweetened cocoa**
- **1 Tbsp. light corn syrup**
- **1 tsp. vanilla extract**
- **3 large eggs**
- **1 cup chopped toasted pecans**

SALTED CARAMEL TOPPING

- **3/4 cup sugar**
- **1 Tbsp. fresh lemon juice**
- **1/3 cup heavy cream**
- **4 Tbsp. butter**
- **1/4 tsp. table salt**
- **2 cups toasted pecan halves**
- **1/2 tsp. sea salt**

1. Prepare Tart Shell: Preheat oven to 350°. Fit piecrust into a 10-inch tart pan with removable bottom.
2. Prepare Chocolate Filling: Stir together first 6 ingredients in a large bowl. Add eggs, stirring until well blended. Fold in chopped pecans. Pour mixture into tart shell.
3. Bake at 350° for 35 minutes. (Filling will be loose but will set as it cools.) Remove from oven to a wire rack.
4. Prepare Salted Caramel Topping: Bring 3/4 cup sugar, 1 Tbsp. lemon juice, and 1/4 cup water to a boil in a medium saucepan over high heat. (Do not stir.) Boil, swirling occasionally after sugar begins to change color, 8 minutes or until dark amber. (Do not walk away from the pan, as the sugar could burn quickly once it begins to change color.) Remove from heat; add cream and 4 Tbsp. butter. Stir constantly until bubbling stops and butter is incorporated (about 1 minute). Stir in table salt.
5. Arrange pecan halves on tart. Top with warm caramel. Cool 15 minutes; sprinkle with sea salt.

Note: We tested with Hershey's 100% Cacao Special Dark Cocoa and Maldon Sea Salt Flakes.

MAKES 8 servings. **HANDS-ON** 25 min.; **TOTAL** 1 hour, 20 min.

Simple Piecrust

- **1 1/4 cups all-purpose flour**
- **1/2 cup cold butter, cut into pieces**
- **1/4 tsp. table salt**
- **4 or 5 Tbsp. ice water**

1. Combine first 3 ingredients in a large bowl with a pastry blender until mixture resembles small peas. Sprinkle ice water, 1 Tbsp. at a time, over surface of mixture in bowl, and stir with a fork until dry ingredients are moistened. Shape into a ball; cover and chill 30 minutes.
2. Preheat oven to 425°. Roll dough into a 13-inch circle on a lightly floured surface. Fit into a 9-inch pie plate; fold edges under, and crimp. Line pastry with aluminum foil; fill with pie weights or dried beans.
3. Bake at 425° for 15 minutes. Remove weights and foil; bake 5 to 10 more minutes or until golden brown. Cool completely on a wire rack.

MAKES 1 (9-inch) piecrust

PEPPERMINT DIVINITY BARS

Make this recipe all the way through without stopping, spreading the warm divinity onto a still-warm cookie base. If the divinity is too cool, it will tear the cookie base as you spread it.

- **3 cups all-purpose flour**
- **1 Tbsp. baking powder**
- **1 tsp. kosher salt**
- **1 vanilla bean**
- **1 1/4 cups butter, softened**
- **2 cups sugar, divided**
- **Parchment paper**
- **1/4 cup light corn syrup**
- **2 large egg whites**
- **1 tsp. vanilla extract**
- **1/4 tsp. peppermint extract**
- **3/4 cup crushed hard peppermint candies, divided**

1. Preheat oven to 375°. Stir together first 3 ingredients.

2. Split vanilla bean; scrape seeds into bowl of a heavy-duty electric stand mixer; discard bean. Add butter and 1 cup sugar; beat at medium speed 2 minutes or until creamy. Add flour mixture; beat until blended.

3. Line bottom and sides of a 13- x 9-inch pan with parchment paper, allowing 2 to 3 inches to extend over sides; lightly grease parchment paper. Press dough into bottom of prepared pan. Bake at 375° for 20 minutes or until edges are golden brown.

4. Meanwhile, stir together corn syrup, 1/4 cup water, and remaining 1 cup sugar in a small saucepan over high heat, stirring just until sugar dissolves. Cook, without stirring, until a candy thermometer registers 250° (7 to 8 minutes).

5. While syrup cooks, beat egg whites at medium speed, using whisk attachment, until foamy.

6. When syrup reaches 250°, beat egg whites at medium-high speed until soft peaks form. While mixer is running, gradually add hot syrup to egg whites. Increase speed to high; beat until stiff peaks form. (Mixture should still be warm.) Add vanilla and peppermint extracts, and beat at medium speed just until combined. Fold in 1/2 cup peppermint candies.

7. Working quickly, spread mixture on warm cookie base, using a butter knife or offset spatula. Sprinkle with remaining 1/4 cup crushed peppermints, and cool.

8. Lift mixture from pan, using parchment paper sides as handles; cut into bars.

MAKES 32 servings. **HANDS-ON** 50 min.; **TOTAL** 2 hours, 10 min.

METRIC EQUIVALENTS

The recipes that appear in this cookbook use the standard United States method for measuring liquid and dry or solid ingredients (teaspoons, tablespoons, and cups). The information on this chart is provided to help cooks outside the U.S. successfully use these recipes. All equivalents are approximate.

METRIC EQUIVALENTS FOR DIFFERENT TYPES OF INGREDIENTS

A standard cup measure of a dry or solid ingredient will vary in weight depending on the type of ingredient. A standard cup of liquid is the same volume for any type of liquid. Use the following chart when converting standard cup measures to grams (weight) or milliliters (volume).

Standard Cup	Fine Powder (ex. flour)	Grain (ex. rice)	Granular (ex. sugar)	Liquid Solids (ex. butter)	Liquid (ex. milk)
1	140 g	150 g	190 g	200 g	240 ml
¾	105 g	113 g	143 g	150 g	180 ml
⅔	93 g	100 g	125 g	133 g	160 ml
½	70 g	75 g	95 g	100 g	120 ml
⅓	47 g	50 g	63 g	67 g	80 ml
¼	35 g	38 g	48 g	50 g	60 ml
⅛	18 g	19 g	24 g	25 g	30 ml

USEFUL EQUIVALENTS FOR LIQUID INGREDIENTS BY VOLUME

¼ tsp							=	1 ml		
½ tsp							=	2 ml		
1 tsp							=	5 ml		
3 tsp	=	1 Tbsp			=	½ fl oz	=	15 ml		
		2 Tbsp	=	⅛ cup	=	1 fl oz	=	30 ml		
		4 Tbsp	=	¼ cup	=	2 fl oz	=	60 ml		
		5⅓ Tbsp	=	⅓ cup	=	3 fl oz	=	80 ml		
		8 Tbsp	=	½ cup	=	4 fl oz	=	120 ml		
		10⅔ Tbsp	=	⅔ cup	=	5 fl oz	=	160 ml		
		12 Tbsp	=	¾ cup	=	6 fl oz	=	180 ml		
		16 Tbsp	=	1 cup	=	8 fl oz	=	240 ml		
		1 pt	=	2 cups	=	16 fl oz	=	480 ml		
		1 qt	=	4 cups	=	32 fl oz	=	960 ml		
						33 fl oz	=	1000 ml	=	1 l

USEFUL EQUIVALENTS FOR DRY INGREDIENTS BY WEIGHT

(To convert ounces to grams, multiply the number of ounces by 30.)

1 oz	=	1/16 lb	=	30 g
4 oz	=	¼ lb	=	120 g
8 oz	=	½ lb	=	240 g
12 oz	=	¾ lb	=	360 g
16 oz	=	1 lb	=	480 g

USEFUL EQUIVALENTS FOR LENGTH

(To convert inches to centimeters, multiply the number of inches by 2.5.)

1 in					=	2.5 cm		
6 in	=	½ ft			=	15 cm		
12 in	=	1 ft			=	30 cm		
36 in	=	3 ft	=	1 yd	=	90 cm		
40 in					=	100 cm	=	1 m

USEFUL EQUIVALENTS FOR COOKING/OVEN TEMPERATURES

	Fahrenheit	Celsius	Gas Mark
Freeze Water	32° F	0° C	
Room Temperature	68° F	20° C	
Boil Water	212° F	100° C	
Bake	325° F	160° C	3
	350° F	180° C	4
	375° F	190° C	5
	400° F	200° C	6
	425° F	220° C	7
	450° F	230° C	8
Broil			Grill

baking at high altitudes

Liquids boil at lower temperatures (below 212°), and moisture evaporates more quickly at high altitudes.Both of these factors significantly impact the quality of baked goods. Also, leavening gases (air, carbon dioxide, water vapor) expand faster. If you live at 3,000 feet or below, first try a recipe as is. Sometimes few, if any, changes are needed. But the higher you go, the more you'll have to adjust your ingredients and cooking times.

A Few Overall Tips

- Use shiny new baking pans. This seems to help mixtures rise, especially cake batters.
- Use butter, flour, and parchment paper to prep your baking pans for nonstick cooking. At high altitudes, baked goods tend to stick more to pans.
- Be exact in your measurements (once you've figured out what they should be). This is always important in baking, but especially so when you're up so high. Tiny variations in ingredients make a bigger difference at high altitudes than at sea level.
- Boost flavor. Seasonings and extracts tend to be more muted at higher altitudes, so increase them slightly.
- Have patience. You may have to bake your favorite sea-level recipe a few times, making slight adjustments each time, until it's worked out to suit your particular altitude.

ingredient/temperature adjustments

CHANGE	AT 3,000 FEET	AT 5,000 FEET	AT 7,000 FEET
Baking powder or baking soda	· Reduce each tsp. called for by up to ⅛ tsp.	· Reduce each tsp. called for by ⅛ to ¼ tsp.	· Reduce each tsp. called for by ¼ to ½ tsp.
Sugar	· Reduce each cup called for by up to 1 Tbsp.	· Reduce each cup called for by up to 2 Tbsp.	· Reduce each cup called for by 2 to 3 Tbsp.
Liquid	· Increase each cup called for by up to 2 Tbsp.	· Increase each cup called for by up to 2 to 4 Tbsp.	· Increase each cup called for by up to 3 to 4 Tbsp.
Oven temperature	· Increase 3° to 5°	· Increase 15°	· Increase 21° to 25°

Menu Index

This index lists every menu by suggested occasion.

Easiest Easter Ever

Serves 8
(page 84)

Sparkling Elderflower Lemonade
Roasted Leg of Lamb with Lemon-Herb Salt
Roasted Beets, Carrots and Sweet Onions
Shaved Vegetable Salad with Toasted Almonds
Cheese Grits Soufflé with Mushroom Gravy
Coconut-and-Pecan Strawberry Shortcakes

Derby Day

Serves 8 to 10
(page 100)

The *SL* Mint Julep
The Blush Lily
Jalapeño Cheese Straws
Chilled Sweet Pea Soup with Mint and Cream
Pork Tenderloin Sliders
Honey Custard with Berries

Simple Holiday Party

Serves 6 to 8
(page 326)

Milk Punch
Country Ham
Cheese-Grit-and-Chive Muffins

Make-Ahead Thanksgiving

Serves 8 to 10
(page 250)

Parsnip-Potato Soup
Dry-Brined-and-Marinated Smoked Turkey
Easy Butter Rolls with Sorghum Butter and Herb Butter
Roasted Vegetable Salad with Apple Cider Vinaigrette
Mashed Potatoes with Greens
Grandmother Carter's Cornbread Dressing
Slow-Cooker Sweet Potatoes with Bacon
Make-Ahead Gravy
Cherry-Spice Cake Trifle
Pumpkin-Espresso Tiramisù

The Baker's Brunch

Serves 8 to 10
(page 295)

Smoky Sausage-and-Grits Casserole
Mint Chocolate Chess Tarts
Christmas Morning Cinnamon Rolls
Apple-Pecan Streusel Muffins
Festive Fresh Fruit Salad with Mint-Lime Simple Syrup
Cream Cheese Pastries
Holiday Ham with Apricot Glaze

Holiday Tea

Serves 8
(page 310)

Hot Apple Tea
Chicken Salad Tomato Cups
Smoked Trout and Watercress Tea Sandwiches
Fig and Bacon Palmiers
Pumpernickel Tea Sandwiches
Hazelnut Chip Shortbread
Browned Butter Scones with Faux Clotted Cream
Petit Fours Presents

Get the Party Started

Serves 6 to 8
(page 316)

Cranberry Tea-Ini
Ruby Red Negroni
Turkey-Artichoke Cornucopias
Cranberry-Chutney Cheese Ball
Cheese Biscuit Barbecue Bites

Good Luck New Year

Serves 6 to 8
(page 331)

Easy Hoppin' John
Pork Chops with Bourbon-Rosemary-Mustard Sauce
Kale Salad with Hot Bacon Dressing
Pecan Cornbread
Fig Upside-Down Cake

Recipe Title Index

This index alphabetically lists every recipe by exact title.

A
African Chickpea Soup, 42
Alabama White Sauce, 133
Almond Poinsettia Cookies, 330
An Easy Lowcountry Oyster Roast, 25
Any Greens Sauté, 48
Apple Brandy-Caramel Sauce, 224
Apple-Cheddar Bread, 223
Apple Cider Syrup, 266
Apple Cider Vinaigrette, 255
Apple-Cream Cheese Bundt Cake, 268
Apple Fritter Pull-Apart Bread, 313
Apple-Pecan Streusel Muffins, 297
Apple-Spice Pancakes, 266
Avocado Ranch, 46

B
Bacon and Cheddar Belgian Waffles, 308
Bacon-and-Sweet Onion Jam, 102
Bacon Old-Fashioned, 115
Baked Beef and Fried Pepper Burritos, 308
Baked French Toast with Banana Brûlée, 98
Balsamic Green Beans with Tomatoes and Feta, 309
Banana-Nut Muffins, 223
Bananas Foster Coffee Cake with Vanilla-Rum Sauce, 338
Barbecue Deviled Eggs, 83
Barbecue Mac and Cheese, 225
BBQ Bacon, 62
Beef-and-Bean Chili, 30
Beef Rub, 80
Beer-Battered Pumpkin with Dipping Sauce, 238
Beer-Braised Pot Roast, 29
Belgian Spice Cookies, 291
Belle Chèvre Raspberry Tiramisù, 97
Best Ever Seafood Gumbo, 260
Biscuit Cinnamon Sweet Rolls, 224
Blackberry Chambord Jam, 220
Blackberry-Honey Mustard Sauce, 102
BLT Buttermilk Blue Slaw, 97
Blueberry-Lemon Zest Ice Cream, 201
Blueberry Yogurt Pops, 309
Blue Cheese Crostini with Balsamic-Roasted Grapes, 54
Blue Cheese-Pecan Popovers, 313
Blue Cheese Ranch, 46
Blue Cheese-Red Currant Cheese Ball, 274
Boozy Peach Shortcakes with Sweet Cream, 125
Bourbon-Butter-Salted Pecan Ice Cream, 201
Bourbon-Caramel Marshmallows, 290
Bourbon Ganache, 338
Bourbon-Spiked Maple-Bacon Mini Cakes, 315
Braised Beef-and-Mushroom Stew, 318
Brisket Tacos with Summer Salsa, 199
Broccolini with Pecans and Cane Syrup Vinaigrette, 281
Broiled Oysters, 26
Broiled Pork Chops with Basil Butter and Summer Squash, 194
Brown Butter Icing, 314
Browned Butter Frosting, 336
Browned Butter Scones with Faux Clotted Cream, 312
Brown Sugar-Brined Chicken, 137
Brown Sugar-Cream Cheese Frosting, 68
Brussels Sprouts and Sautéed Cabbage, 51
Brussels Sprouts with Parmesan Cream Sauce, 283
Bucatini, Ham, and Asparagus, 89
Burgers with Green Tomato Mayonnaise, 159
Butterflied and Barbecued Turkey, 324
Buttermilk-Glazed Cherry Sheet Cake, 155
Buttermilk-Lady Pea Soup with Bacon, 217
Buttermilk-Plum Ice Cream, 155
Buttermilk-Soaked, Bacon-Fried Chicken in Gravy, 234
Buttermilk Tandoori Marinade, 134
Buttermilk-Vegetable Curry, 46
Butternut-Pecan Sauté, 320
Buttery Nashville Hot Sauce, 133

C
Candied Carrot Curls, 67
Candy Cane Biscotti, 330
Cane Syrup Vinaigrette, 281
Caramel Apple Cake, 231
Caramelized Onion and Swiss Popovers, 334
Caramelized Sweet Onions, 334
Caramel-Peanut-Popcorn Snack Mix, 245
Carolina Burger Sauce, 139
Carolina Chicken Burgers with Creamy Ancho Slaw, 139
Carrot Soup with Brown Butter, Pecans, and Yogurt, 202
Chardonnay-Herb Marinade, 134
Charred Ambrosia with Toasted Coconut and Marshmallow Crème, 282
Charred Steak Salad with Spicy Dressing, 34
Cheddar-Horseradish-Walnut Cheese Ball, 274
Cheese Balls, Three Ways, 274
Cheese Biscuit Barbecue Bites, 317
Cheese-Grit-and-Chive Muffins, 300
Cheese Grits Soufflé with Mushroom Gravy, 85
Cheesy Bacon-and-Two-Onion Tart, 51
Cheesy Garlic French Bread, 265
Cheesy Pimiento Cornbread, 302
Cheesy Pimiento Corn Casserole, 321
Cherry-Plum Sauce, 279
Cherry-Rosemary Muffins, 70
Cherry-Spice Cake Trifle, 254
Chewy Ginger Cookies, 293
Chicken-and-Artichoke Penne, 88
Chicken-and-Black Bean Chimichangas, 226
Chicken-and-Prosciutto Tortelloni Soup, 265
Chicken, Artichoke, and Sun-Dried Tomato Pizza, 306
Chicken Cutlets with Herbed Mushroom Sauce, 50
Chicken, Farro, and Vegetable Salad with Lemon Vinaigrette, 64
Chicken Liver Mousse Crostini with Pepper Jelly, 272
Chicken Paillard with Citrus Salad and Couscous, 33
Chicken Salad Tomato Cups, 310
Chile-Rubbed Chops with Sweet Potatoes and Grilled Okra, 160
Chilled Sweet Pea Soup with Mint and Cream, 100
Chipotle Cream, 31
Chocolate-Bourbon-Butterscotch Icebox Cake, 122
Chocolate Breakfast Wreath, 333
Chocolate-Buttermilk Pudding, 45
Chocolate-Cappuccino Whoopie Pies, 329
Chocolate Chip-Toasted Pecan Pancakes, 266
Chocolate-Cream Cheese Frosting, 57
Chocolate-Mayonnaise Cake, 57
Chocolate Milk Granita, 219
Chocolate-Pecan Mousse Tart, 262
Chocolate-Peppermint Cheesecake, 289
Chocolate-Peppermint Crackle Cookies, 292
Chocolate Whipped Cream, 56
Chopped Salad with Steak, 196
Christmas Morning Cinnamon Rolls, 296
Christmas Salad, 283
Cinnamon Apple Tarts, 309
Citrus-Braised Chicken Thighs, 64
Citrus Chicken and Beet Salad, 305
Citrus Glaze, 263
Claire's Cream Cheese Swirl Brownies, 38
Classic Baked Macaroni & Cheese, 83
Classic Cocktail Sauce, 124
Classic Oyster Stew, 24
Classic Pimiento Cheese, 230
Classic Roasted Duck with Orange-Bourbon-Molasses Glaze, 278

Classic Succotash, 197
Classic Sweet Tea, 209
Coconut-Macadamia Nut Pound Cake, 289
Coconut-and-Pecan Strawberry Shortcakes, 86
Coconut Cream Cake, 93
Coconut-Pecan Frosting, 327
Coconut Snowballs, 291
Coffee Cake Muffins with Brown Butter Icing, 314
Coffee Liqueur Cookies-and-Cream Ice Cream, 201
Collard-and-Olive Pesto, 48
Cornbread Crumbles, 254
Cornbread Focaccia, 334
Cornbread-Vegetable Salad, 306
Cornmeal Thumbprint Cookies with Tomato Jam, 158
Cotton Candy Meringues, 58
Cowboy Brisket Sandwich, 81
Cranberry-Chutney Cheese Ball, 316
Cranberry Tea-Ini, 317
Cream Cheese Filling, 298
Cream Cheese-Honey Filling, 223
Cream Cheese Icing, 297
Cream Cheese Pastries, 298
Creamy Ancho Slaw, 139
Creamy Basil-Black Pepper Cucumbers, 73
Creamy Glaze, 224
Creamy Shrimp Dip with Crispy Wonton Chips, 272
Creamy Shrimp Piccata, 280
Creole Rice, 211
Creole Seafood Jambalaya, 240
Crispy Andouille Hush Puppies, 125
Crispy Eggplant, Tomato, and Provolone Stacks with Basil, 215
Crispy Salt-and-Pepper Grilled Chicken Thighs, 135
Crispy Wonton Chips, 272
Crown Pork Roast with Fig-Fennel Stuffing, 323
Crumb Crust, 118
Crunchy Jerk Tacos with Watermelon-Mango Salsa, 135
Crunchy Summer Salsa, 199
Cucumber-Basil Granita, 219
Cucumber-Mint Water, 144
Cucumbers with Ginger, Rice Vinegar, and Mint, 75

D

Decadent Cream Puffs with Praline Sauce and Toasted Pecans, 263
Diablo Farfalle, 211
Dirty Rice Risotto, 281
Dixie Chicken Salad with Grapes, Honey, Almonds, and Broccoli, 227
Double Chocolate Chip Cookies with Bourbon Ganache, 338
Dry-Brined-and-Marinated Smoked Turkey, 256

E

Easy Barbecue Shrimp, 248
Easy BBQ Ribs, 109
Easy Butter Rolls, 252
Easy Hoppin' John, 331
Easy Summer Green Beans, 193
Easy Vanilla Glaze, 333
Eggs in Fire Roasted Tomato Sauce, 307
El Sancho Barbecue Sauce, 80
Espresso Filling, 329

F

Fancy Barbecue Sauce, 109
Faux Clotted Cream, 312
Festive Fresh Fruit Salad with Mint-Lime Simple Syrup, 297
Feta-Olive-Fresh Herb Cheese Ball, 274
Fettuccine with Smoky Turnip Greens, Lemon, and Goat Cheese, 47
Fiery Green Beans, 42
Fig and Bacon Palmiers, 310
Fig-Fennel Stuffing, 323
Fig Upside-Down Cake, 332
5-Ingredient Sugar Cookies, 292
Fluffy Buttermilk Waffles, 45
Fluffy Coconut Frosting, 93
Fontina-Chive Yorkshire Puddings, 283
Fried Apple Pork Chops, 308
Fried Eggs, 60
Fruit Pickle Brine, 141

G

Game Day Venison Chili, 113
Garden Potato Salad, 74
Garlicky Beef-and-Bean Stir-Fry, 50
Garlic with Confit Oil, Thyme, and Black Pepper, 144
Garlic-Yogurt-Marinated Chicken Thighs, 65
Ginger "Beer Can" Chicken, 136
Gingerbread Baked Alaska, 287
Ginger-Peach Shortbread Cobbler, 153
Glazed Fruitcake Bars, 294
Gnocchi Mac and Cheese, 282
Goat Cheese Mashed Potatoes, 65
Goo Goo Cluster Brownies à la Mode, 115
Gouda Grits, 302
Grand Aïoli, 148
Grandmother Carter's Cornbread Dressing, 253
Grapefruit-Beet-Goat Cheese Flatbread, 273
Greek Baked Ziti, 242
Green Apple Mignonette, 26
Green Tomato Mayonnaise, 159
Green Tomato Relish, 108
Green Tomato Soup with Lump Crabmeat, 157
Grillades, 232
Grilled Chicken Caprese, 304
Grilled Corn and Tequila-Lime Butter, 114
Grilled Cornish Hens with Herb Brine, 279
Grilled Oysters, 26
Grilled Peach & Avocado Salad, 82
Grilled Pineapple Dessert Pizza, 206
Grilled Pork Loin Steaks with Cherry-Plum Sauce, 278
Grilled Salmon with Orange-Fennel Salad, 305
Grilled Teriyaki Shrimp Kabobs, 305
Grilled Tomato-Peach Pizza, 307
Grilled Triggerfish, 105
Grits Cakes with Poached Eggs and Country Gravy, 61

H

Ham-and-Corn Chowder, 319
Hazelnut Chip Shortbread, 311
Heartwarming Cider, 245
Herb Butter, 252
Herbed Breadcrumbs, 280
Herbed Buttermilk Ranch Dressing, 46
Herbed Cornbread Focaccia, 334
Herbed Cucumber-and-Tomato Salad, 243
Herbed Grits Dressing with Leeks and Mushrooms, 322
Herb Vinaigrette, 159
Holiday Ham with Apricot Glaze, 298
Holiday Spice Cake, 253
Homemade Pizza Dough, 207
Homemade Shake-and-Bake Pork Chops with Mustard Sauce, 193
Homemade Strawberry Milk, 94
Honey-Balsamic-Blueberry Pie, 337
Honey Custard with Berries, 102
Honey-Glazed Pork Tenderloin with Homemade Applesauce, 51
Honey Peanut Brittle, 42
Hoppin' John and Limpin' Susan, 38
Hoppin' John Cakes with Tomato-Jalapeño Gravy, 280
Hot Apple Tea, 310
Hot Chocolate, 290

I

Inside-Out Hot Brown Bites, 272
Italian-Style Salsa Verde, 101
Italian Turkey Meatloaves, 35

J

Jalapeño Biscuits, 248
Jalapeño Cheese Straws, 100
Judy's Bloody Mary Mix, 233

K

Kale-and-Blueberry Slaw with Buttermilk Dressing, 75
Kale Salad with Hot Bacon Dressing, 331
Key Lime-Buttermilk Icebox Pie with Baked Buttery Cracker Crust, 120

King Ranch Chicken Soup, 264
Korean Flank Steak, 304

L
Lady Pea-and-Corn Patties, 217
Lady Peas, 217
Layered Eggnog Blondies, 293
Layered Eggnog Cream with Puff Pastry, 290
Leek Fonduta, 202
Lemon and Shallot Brussels Sprouts, 309
Lemon-Blueberry Topping, 121
Lemon-Buttermilk Icebox Pie, 120
Lemon-Thyme Vinaigrette, 125
Lemon Vinaigrette, 65
Lemony Ranch, 46
Light Crab Cakes, 91
Lime-and-Mint Macerated Cherries, 153
Lime-Sriracha Aïoli, 148
Lime Vinaigrette, 212
Lime-Wasabi Sauce, 196
Loaded Potato Soup, 37
Lump Crabmeat Salad, 157

M
Make-Ahead Gravy, 251
Mango-Buttermilk Shakes, 44
Mango-Lemon-Buttermilk Icebox Pie with Baked Saltine Cracker Crust, 121
Maple-Bourbon Glaze, 315
Maple-Praline Cookies, 329
Maple-Praline Frosting, 329
Mashed Potatoes with Greens, 254
Meatball Sliders with Tomato Sauce, 28
Memphis Slaw, 83
Mexican Mac and Cheese, 225
Mexican Stew, 55
Milk Punch, 300
Mini Banana-Cranberry-Nut Bread Loaves, 335
Mini Chocolate Chess Tarts, 296
Mini Pumpkin-Honey-Beer Bread, 224
Mini Triple-Chocolate Buttermilk Pound Cakes, 44
Mint Simple Syrup, 100
Minty Lemonade, 74
Mixed Stone Fruit Pie, 153
Mocha Cream Filling, 327
Mocha-Hazelnut Dacquoise, 326
Molten Hazelnut Brownies, 309
Mushroom Gravy, 86
Mushroom Stroganoff, 226
Mustard-Cream Sauce, 279
Mustard Greens with Yogurt-Parmesan Dressing and Bacon Croutons, 48
My Go-To Old Fashioned, 103

N
Napa Cabbage-and-Sweet Pepper Slaw, 108
Nashville Farmers' Market Succotash, 115
Nectarine-Chile Ice Pops, 155
Nectarine Tarts with Honey, 155
Nutty Chocolate Thumbprints, 57

O
Old Timey Pickles, 220
Open-Faced Shrimp-and-Avocado Sandwiches, 214
Orange-and-Basil Macerated Cherries, 153
Orange Cream Granita, 219
Orange Glaze, 335
Orange-Infused Rum Gimlets, 149
Orzo Salad with Spicy Buttermilk Dressing, 200
Our Best Homemade Marshmallows, 290
Our Easiest Pumpkin Pie Ever, 237
Our Favorite Buttermilk Biscuit, 90
Oven-Roasted Bacon, 60
Overnight Slow-Cooker Grits, 60
Oyster-Bacon Pot Pie, 25

P
Pan-Seared Grouper with Balsamic Brown Butter Sauce, 105
Pan-Seared Skirt Steak and Chimichurri, 116
Parsnip-Potato Soup, 252
Pasta Shells with Spring Vegetables, 89
Pasta with Heirloom Tomatoes, Goat Cheese, and Basil, 159
PB&J Sammies, 58
Pea-and-Green Onion Soup, 51
Peach Divinity Icebox Pie, 121
Peach Pancakes, 308
Peanut Chicken Stew, 41
Peanut Sauce, 63
Peanutty Braised Chicken, 97
Peanut Whiskey and Cola, 275
Pear-and-Pumpkin Tart, 237
Peas and Kale Salad with Bacon Vinaigrette, 34
Pecan-and-Wild Rice Stuffing, 323
Pecan Cornbread, 332
Pecan-Cranberry Shortbread, 292
Pecan-Espresso Toffee, 262
Pecan Pie Bars, 262
Pecan Pound Cake, 263
Pecan-Rosemary Bacon, 62
Pecan Streusel, 297
Pecan Streusel Topping, 261
Peppered White Bean Mash, 94
Pepper Jelly Vinaigrette, 82
Peppermint Divinity Bars, 340
Peppermint Marshmallows, 290
Peppermint Red Velvet Bundt Cakes, 326
Perfect Chicken Kabobs, 133
Pesto Chicken Grill, 304
Pesto Egg Salad, 48
Pesto Mashers, 48
Pesto Mayo, 48
Petite Sweet Potato Biscuits with Pulled Pork and Slaw, 274
Petits Fours Presents, 312
Philly Cheesesteak Pizza, 306
Pickled Melon, 141
Pickled Peaches, 141
Pickled Strawberries, 141
Pistachio-Crusted Rack of Lamb, 325
Poached Eggs, 60
Pool Party Watermelon Punch, 95
Pork-and-Black Bean Chili, 31
Pork Chop Sandwiches with Gravy and Grits, 194
Pork Chops with Bourbon-Rosemary-Mustard Sauce, 331
Pork Tenderloin Sliders, 101
Potato Gratin with Bacon and Comté, 320
Powdered Sugar Glaze, 298, 326
Praline Cream-Beignet Tower, 286
Praline Frosting, 268
Praline Sauce, 263
Prosciutto and Tomato Sandwiches, 306
Pulled Pork Barbecue Nachos, 112
Pumpernickel Tea Sandwiches, 311
Pumpkin-and-Turnip Green Lasagna, 238
Pumpkin-Chocolate Brownies, 237
Pumpkin Cupcakes with Browned Butter Frosting, 336
Pumpkin-Espresso Tiramisù, 256
Pumpkin-Honey-Beer Bread, 224
Pumpkin-Honey-Beer Bread Pudding with Apple Brandy-Caramel Sauce, 224
Pumpkin-Pecan Streusel Pie, 261
Pumpkin Soup with Red Pepper Relish, 319

Q
Quick BBQ Chicken Pizzas, 110
Quick Cook Shrimp and Chorizo Paella, 113

R
Raspberry Limeade Granita, 219
Raspberry "Rhubars," 111
Red Pepper Relish, 319
Red Velvet Thumbprints, 291
Red Wine Marinade, 196
Regular-Size Banana-Cranberry-Nut Bread Loaves, 335
Retro Rum Punch, 275
Rib Rub, 109
Roasted Beets, Carrots, and Sweet Onions, 85
Roasted Brussels Sprouts with Sage Pesto, 321
Roasted Chicken Thighs with Herb Butter, 65
Roasted Cider-Brined Pork Loin with Green Tomato Chutney, 53
Roasted Fennel-and-Prosciutto Flatbread, 273
Roasted Garlic Duchess Potatoes, 282
Roasted Gulf Shrimp with Romesco Sauce, 105
Roasted Leg of Lamb with Lemon-Herb Salt, 84
Roasted Potato-and-Okra Salad, 124
Roasted Vegetable Salad with Apple Cider Vinaigrette, 255

Romaine Salad with Country Ham and Eggs, 34
Rosemary-Citrus Vinaigrette, 320
Rosemary Collins, 275
Rotini with Crumbled Turkey and Tomato Sauce, 36
Royal Icing, 328
Royal Red Icing, 330
Ruby Red Negroni, 317

S

Saffron Aïoli, 148
Sage-Brown Butter Squash Agnolotti, 307
Salmon Croquettes with Dill Sauce, 41
Salted Caramel-Chocolate Pecan Tart, 339
Sausage-and-Bean Stew, 69
Sautéed Sprouts, 35
Seared Flank Steak with Lime-Wasabi Sauce, 196
Sesame Chicken Tacos, 66
Sesame-Chicken Thigh Paillard with Peanut Sauce, 63
Shaved Vegetable Salad with Toasted Almonds, 85
Shepherd's Pie with Potato Crust, 242
Shrimp-and-New Potato Chowder, 264
Shrimp and Peas with Farfalle, 89
Shrimp and Pesto-Rice Salad, 33
Shrimp Boil with Green Olives and Lemon, 124
Shrimp 'n' Grits Pie, 61
Shrimp Spice Mix, 124
Simple Chicken Stock, 251
Simple Piecrust, 339
Simple Syrup, 209, 275
Skillet Pork Chops with Sautéed Peaches, 142
Skillet Steak and Wilted Kale, 226
Slow-Cooker Bolognese Sauce over Pappardelle Pasta, 247
Slow-Cooker Chicken Cacciatore with Spaghetti, 30
Slow-Cooker Chicken Lettuce Cups, 92
Slow-Cooker Sweet Potatoes with Bacon, 257
Slow-Cooker Tomato Sauce, 210
Smoked Beef Brisket Tostadas, 81
Smoked Chicken Pizza with White Barbecue Sauce, 207
Smoked Mojo Leg Quarters, 138
Smoked Sausage, Grilled Corn, and Sweet Onion Pizza, 207
Smoked Trout and Watercress Tea Sandwiches, 310
Smoked Trout Crostini with Radishes and Dill Cream, 273
Smoky Barbecue Popcorn, 245
Smoky Pork Stir-Fry, 229
Smoky Sausage-and-Grits Casserole, 295
Smoky Turkey-and-Sweet Potato Chili, 31
Snapper Baked in Parchment with Spring Vegetables, 104
Snowy Rum Glaze, 290
Sorghum Butter, 252
Southern Anadama Bread, 314
Southern Cheese Steak Sandwiches, 196
Southern Lemon Americano, 275
Southern Pimiento Mac and Cheese, 225
Southwest Cornmeal Bacon, 62
Southwest Pork Stew, 264
Sparkling Elderflower Lemonade, 86
Speakeasy Sparkler, 302
Speedy Homemade Mac and Cheese, 225
Spice Cake with Cranberry Filling, 286
Spiced Beef Kabobs with Herbed Cucumber-and-Tomato Salad, 243
Spiced Chocolate Stout Beer Cocktail, 275
Spiced Sorghum Snowflakes, 328
Spice-Rubbed Tenderloin with Mustard-Cream Sauce, 279
Spicy Andouille Spanish Tortilla, 95
Spicy Pumpkin Soup with Avocado Cream, 238
Spicy Simple Syrup, 275
Spicy-Sweet Pecans, 263
Spinach-and-Three-Herb Pesto, 73
Spinach-Feta Scones, 223
Stewed Green Beans and Tomatoes, 143
Sticky Sesame-Sorghum Marinade, 134
Stove-Top Grits, 60
Strawberry-Basil Ice Cream, 201
Strawberry-Blueberry Relish, 107
Strawberry Lemonade Muffins, 335
Strawberry-Lemon-Buttermilk Icebox Pie with Baked Gingersnap Crust, 120
Stuffed Peppers with Chèvre, Pecans, and Golden Raisins, 94
Sugar Cookie Stars, 329
Summer Corn-and-Golden Potato Chowder, 213
Summer Corn Soup with Corn Salsa, 96
Summer Pasta Salad with Lime Vinaigrette, 212
Sweet-and-Tangy Barbecue Sauce, 324
Sweet Corn Relish, 107
Sweetened Whipped Cream, 120
Sweet Mascarpone Cream, 86
Sweet Potato Biscuits, 275, 335
Sweet Potato Broth, 42
Sweet Potato Chips, 83

T

Tangy Fried Chicken with Dijon, 234
Tart Cherry-Chestnut Sausage Dressing, 322
Tex-Mex Meatballs in Red Chile Sauce, 242
The Blush Lily, 100
The Original "Pam-Cakes," 266
The *SL* Mint Julep, 100
The Ultimate Carrot Cake, 68
The Ultimate Chocolate Pie, 56
The Ultimate Classic Collards, 47
The Ultimate Fried Egg Sandwich with BBQ Bacon, 61
The Ultimate Grilled Cheese, 227
Three-Pepper Chowchow, 108
Tomato-and-Red Pepper Soup, 227
Tomato Jam, 158
Tomato-Peach Salad with Chicken, 137
Tomato, Watermelon, and Feta Skewers with Mint and Lime, 157
Triple Chocolate Brownie-Mousse Stacks, 288
Triple-Chocolate Buttermilk Pound Cake, 44
Triple Chocolate Cranberry Oatmeal Cookies, 70
True Smoked Beef Brisket, 80
Tuesday Night Hamburger Steak, 112
Turkey-Artichoke Cornucopias, 316
Turkey Dry Rub, 256
Turkey-Stuffed Peppers, 36
Tuscan Beef Tenderloin, 325
Twice-Baked Green Potatoes, 48

U

Ultimate Cheese Pizza, 207
Uncle Jack's Mac-and-Cheese, 29

V

Vanilla Glaze, 313
Veal Chops Milanese with Lemon and Herbs, 278
Vietnamese Peanut Sauce, 133
Vodka-Cream Sauce with Penne, 211

W

Waffled Bacon & Cheddar Grits, 61
Waffled Gruyère & Ham Grits, 61
Waffled Jalapeño & Monterey Jack Cheese Grits, 61
Watermelon, Arugula & Pecan Salad, 82
Whipped Honey-Rosemary Butter, 314
Whipped Peppermint Cream, 289
White Bean, Sausage, and Turnip Green Stew, 265
White Cocktail Sauce, 124
White Frosting, 330
White German Chocolate Cake, 327
Winter Citrus-Avocado Salad, 320
Wonder Wings, 133

Y

Yogurt-Tahini Sauce, 243

Z

Zucchini Stuffed with Lady Peas, 217

Month-by-Month Index

This index alphabetically lists every food article and accompanying recipes by month.

January

Community Cookbook, 38
Claire's Cream Cheese Swirl Brownies, 38
Hoppin' John and Limpin' Susan, 38
A Delicious New Diet, 37
Loaded Potato Soup, 37
5 Hearty Main-Dish Salads, 33
Charred Steak Salad with Spicy Dressing, 34
Chicken Paillard with Citrus Salad and Couscous, 33
Peas and Kale Salad with Bacon Vinaigrette, 34
Romaine Salad with Country Ham and Eggs, 34
Shrimp and Pesto-Rice Salad, 33
Mighty Good Meatloaf, 35
Italian Turkey Meatloaves, 35
Rotini with Crumbled Turkey and Tomato Sauce, 36
Sautéed Sprouts, 35
Turkey-Stuffed Peppers, 36
Slow-Cooked Perfection, 28
Beef-and-Bean Chili, 30
Beer-Braised Pot Roast, 29
Meatball Sliders with Tomato Sauce, 28
Pork-and-Black Bean Chili, 31
Slow-Cooker Chicken Cacciatore with Spaghetti, 30
Smoky Turkey-and-Sweet Potato Chili, 31
Uncle Jack's Mac-and-Cheese, 29
The *SL* Test Kitchen Academy, 32
Slow-Cooker Success, 32
The South Is Your Oyster, 24
Classic Oyster Stew, 24
An Easy Lowcountry Oyster Roast, 25
Green Apple Mignonette, 26
Grilled Oysters, 26
Oyster-Bacon Pot Pie, 25

February

Blue Cheese-Pecan Grapes, 54
Blue Cheese Crostini with Balsamic-Roasted Grapes, 54
Buttermilk Revival, 43
Buttermilk-Vegetable Curry, 46
Chocolate-Buttermilk Pudding, 45
Fluffy Buttermilk Waffles, 45
Herbed Buttermilk Ranch Dressing, 46
Mango-Buttermilk Shakes, 44
Triple-Chocolate Buttermilk Pound Cake, 44
Community Cookbook, 58
Cotton Candy Meringues, 58
PB&J Sammies, 58
Dark Magic, 56
Chocolate-Mayonnaise Cake, 57
Nutty Chocolate Thumbprints, 57
The Ultimate Chocolate Pie, 56
Eat Your Greens, 47
Any Greens Sauté, 48
Collard-and-Olive Pesto, 48
Fettuccine with Smoky Turnip Greens, Lemon, and Goat Cheese, 47
Mustard Greens with Yogurt-Parmesean Dressing and Bacon Croutons, 48
Twice-Baked Green Potatoes, 48
The Ultimate Classic Collards, 47
A Light Pork Supper, 53
Roasted Cider-Brined Pork Loin with Green Tomato Chutney, 53
The Next Generation of Soul Food, 40
African Chickpea Soup, 42
Fiery Green Beans, 42
Honey Peanut Brittle, 42
Peanut Chicken Stew, 41
Salmon Croquettes with Dill Sauce, 41
Sauté, Simmer, Savor, 55
Mexican Stew, 55
The *SL* Test Kitchen Academy, 52
Spectacular Fresh Butter & Buttermilk, 52
Ten-Dollar Dinners, 50
Brussels Sprouts and Sautéed Cabbage, 51
Cheesy Bacon-and-Two-Onion Tart, 51
Chicken Cutlets with Herbed Mushroom Sauce, 50
Garlicky Beef-and-Bean Stir-Fry, 50
Honey-Glazed Pork Tenderloin with Homemade Applesauce, 51
Pea-and-Green Onion Soup, 51

March

Community Cookbook, 70
Cherry-Rosemary Muffins, 70
Triple Chocolate Cranberry Oatmeal Cookies, 70
Crazy for Carrot Cake, 68
The Ultimate Carrot Cake, 68
A Fresh Fiesta, 66
Sesame Chicken Tacos, 66
The Saturday Morning Chef, 60
BBQ Bacon, 62
Fried Eggs, 60
Grits Cakes with Poached Eggs and Country Gravy, 61
Oven-Roasted Bacon, 60
Overnight Slow-Cooker Grits, 60
Pecan-Rosemary Bacon, 62
Poached Eggs, 60
Shrimp 'n' Grits Pie, 61
Southwest Cornmeal Bacon, 62
Stove-Top Grits, 60
The Ultimate Fried Egg Sandwich with BBQ Bacon, 61
Waffled Bacon & Cheddar Grits, 61
The *SL* Test Kitchen Academy, 67
Candied Carrot Curls, 67
Thigh and Mighty, 63
Chicken, Farro, and Vegetable Salad with Lemon Vinaigrette, 64
Citrus-Braised Chicken Thighs, 64
Garlic-Yogurt-Marinated Chicken Thighs, 65
Roasted Chicken Thighs with Herb Butter, 65
Sesame-Chicken Thigh Paillard with Peanut Sauce, 63
Try a New Stew, 69
Sausage-and-Bean Stew, 69

April

Community Cookbook, 94
Baked French Toast with Banana Brûlée, 98
Belle Chèvre Raspberry Tiramisù, 97
BLT Buttermilk Blue Slaw, 97
Homemade Strawberry Milk, 94
Peanutty Braised Chicken, 97
Peppered White Bean Mash, 94
Pool Party Watermelon Punch, 95
Spicy Andouille Spanish Tortilla, 95
Stuffed Peppers with Chèvre, Pecans, and Golden Raisins, 94
Summer Corn Soup with Corn Salsa, 96
Crab Takes the Cake, 91
Light Crab Cakes, 91
Easiest Easter Ever, 84
Cheese Grits Soufflé with Mushroom Gravy, 85
Coconut-and-Pecan Strawberry Shortcakes, 86
Roasted Beets, Carrots, and Sweet Onions, 85
Roasted Leg of Lamb with Lemon-Herb Salt, 84
Shaved Vegetable Salad with Toasted Almonds, 85
Fresh from Garden to Plate, 72
Creamy Basil-Black Pepper Cucumbers, 73
Cucumbers with Ginger, Rice Vinegar, and Mint, 75
Garden Potato Salad, 74

Kale-and-Blueberry Slaw with Buttermilk Dressing, 75
Minty Lemonade, 74
Spinach-and Three-Herb Pesto, 73
Our Best Coconut Cake, 93
Coconut Cream Cake, 93
The Pitmaster's Guide to Brisket, 78
Barbecue Deviled Eggs, 83
Classic Baked Macaroni & Cheese, 83
Cowboy Brisket Sandwich, 81
Grilled Peach & Avocado Salad, 82
Memphis Slaw, 83
Smoked Beef Brisket Tostadas, 81
Sweet Potato Chips, 83
True Smoked Beef Brisket, 80
Watermelon, Arugula & Pecan Salad, 82
The Pursuit of Perfect Biscuits, 90
Our Favorite Buttermilk Biscuit, 90
Saucy Chicken Lettuce Cups, 92
Slow-Cooker Chicken Lettuce Cups, 92
The *SL* Test Kitchen Academy, 87
Cooking Perfect Pasta, 87
30-Minute Spring Pastas, 88
Bucatini, Ham, and Asparagus, 89
Chicken-and-Artichoke Penne, 88
Pasta Shells with Spring Vegetables, 89
Shrimp and Peas with Farfalle, 89

May

Bake, Grill, Sear, or Roast: It's Fish Season, 104
Grilled Triggerfish, 105
Pan-Seared Grouper with Balsamic Brown Butter Sauce, 105
Roasted Gulf Shrimp with Romesco Sauce, 105
Snapper Baked in Parchment with Spring Vegetables, 104
Chowchows to Relish, 107
Green Tomato Relish, 108
Napa Cabbage-and-Sweet Pepper Slaw, 108
Strawberry-Blueberry Relish, 107
Sweet Corn Relish, 107
Three-Pepper Chowchow, 108
Community Cookbook, 112
Bacon Old-Fashioned, 115
Game Day Venison Chili, 113
Goo Goo Cluster Brownies à la Mode, 115
Grilled Corn and Tequila-Lime Butter, 114
Nashville Farmers' Market Succotash, 115
Pan-Seared Skirt Steak and Chimichurri, 116
Pulled Pork Barbecue Nachos, 112
Quick Cook Shrimp and Chorizo Paella, 113
Tuesday Night Hamburger Steak, 112
An Editor's Escape, 103
My Go-To Old Fashioned, 103
A Healthy Slice, 110
Quick BBQ Chicken Pizzas, 110
Host Derby Day, 100
Bacon-and-Sweet Onion Jam, 102
Blackberry-Honey Mustard Sauce, 102
The Blush Lily, 100
Chilled Sweet Pea Soup with Mint and Cream, 100
Honey Custard with Berries, 102
Italian-Style Salsa Verde, 101
Jalapeño Cheese Straws, 100
Mint Simple Syrup, 100
Pork Tenderloin Sliders, 101
The *SL* Mint Julep, 100
Raise the Bar, 111
Raspberry "Rhubars," 111
Slow-Cooker Ribs, 109
Easy BBQ Ribs, 109
The *SL* Test Kitchen Academy, 106
Buying Fresh Seafood, 106

June

Community Cookbook, 144
Cucumber-Mint Water, 144
Garlic with Confit Oil, Thyme, and Black Pepper, 144
Dinner in a Breeze, 142
Skillet Pork Chops with Sautéed Peaches, 142
Food Awards 2015, 126
The *SL* Test Kitchen's 24 favorite Southern-made foods, 127
Green Beans Like You've Never Had, 143
Stewed Green Beans and Tomatoes, 143
The Pies of Summer, 118
Chocolate-Bourbon-Butterscotch Icebox Cake, 122
Crumb Crust, 118
Key Lime-Buttermilk Icebox Pie with Baked Buttery Cracker Crust, 120
Lemon-Buttermilk Icebox Pie, 120
Mango-Lemon-Buttermilk Icebox Pie with Baked Saltine Cracker Crust, 121
Peach Divinity Icebox Pie, 121
Strawberry-Lemon-Buttermilk Icebox Pie with Baked Gingersnap Crust, 120
The *SL* Guide to Grilled Chicken, 132
Brown Sugar-Brined Chicken, 137
Carolina Chicken Burgers with Creamy Ancho Slaw, 139
Crispy Salt-and-Pepper Grilled Chicken Thighs, 135
Crunchy Jerk Tacos with Watermelon-Mango Salsa, 135
Ginger "Beer Can" Chicken, 136
Perfect Chicken Kabobs, 133
Smoked Mojo Leg Quarters, 138
Tomato-Peach Salad with Chicken, 137
Wonder Wings, 133
The *SL* Test Kitchen Academy, 141
Fruit Pickle Brine, 141
Pickled Melon, 141
Pickled Peaches, 141
Pickled Strawberries, 141
Taste of the Gulf, 123
Boozy Peach Shortcakes with Sweet Cream, 125
Crispy Andouille Hush Puppies, 125
Roasted Potato-and-Okra Salad, 124
Shrimp Boil with Green Olives and Lemon, 124

July

Chop, Chop!, 160
Broiled Pork Chops with Basil Butter and Summer Squash, 194
Chile-Rubbed Chops with Sweet Potatoes and Grilled Okra, 160
Easy Summer Green Beans, 193
Homemade Shake-and-Bake Pork Chops with Mustard Sauce, 193
Pork Chop Sandwiches with Gravy and Grits, 194
Community Cookbook, 202
Carrot Soup with Brown Butter, Pecans, and Yogurt, 202
Leek Fonduta, 202
Easy Ice Cream, 201
Blueberry-Lemon Zest Ice Cream, 201
Bourbon-Butter-Salted Pecan Ice Cream, 201
Coffee Liqueur Cookies-and-Cream Ice Cream, 201
Strawberry-Basil Ice Cream, 201
Light and Luscious Summer Pasta, 200
Orzo Salad with Spicy Buttermilk Dressing, 200
Loving Summer Tomatoes, 156
Burgers with Green Tomato Mayonnaise, 159
Cornmeal Thumbprint Cookies with Tomato Jam, 158
Green Tomato Soup with Lump Crabmeat, 157
Pasta with Heirloom Tomatoes, Goat Cheese, and Basil, 159
Tomato, Watermelon, and Feta Skewers with Mint and Lime, 157
Romancing the Stone, 150
Buttermilk-Glazed Cherry Sheet Cake, 155
Buttermilk-Plum Ice Cream, 155
Ginger-Peach Shortbread Cobbler, 153
Lime-and-Mint Macerated Cherries, 153
Mixed Stone Fruit Pie, 153
Nectarine-Chile Ice Pops, 155
Nectarine Tarts with Honey, 155
Orange-and-Basil Macerated Cherries, 153
The *SL* Test Kitchen Academy, 198
Summer Tomatoes, 198
Steak Nights, 195
Chopped Salad with Steak, 196
Seared Flank Steak with Lime-Wasabi Sauce, 196
Southern Cheese Steak Sandwiches, 196

Summer Succotash, 197
Classic Succotash, 197
Taco Night Reimagined, 199
Brisket Tacos with Summer Salsa, 199
The Ultimate Lazy Lunch, 146
Grand Aïoli, 148
Orange-Infused Rum Gimlets, 149

August
Backyard Pizzeria, 204
Grilled Pineapple Dessert Pizza, 206
Homemade Pizza Dough, 207
Smoked Chicken Pizza with White Barbecue Sauce, 207
Smoked Sausage, Grilled Corn, and Sweet Onion Pizza, 207
Ultimate Cheese Pizza, 207
Classic Sweet Tea, 209
Classic Sweet Tea, 209
Community Cookbook, 220
Blackberry Chambord Jam, 220
Old Timey Pickles, 220
Cool Down with Easy Granitas, 218
Chocolate Milk Granita, 219
Cucumber-Basil Granita, 219
Orange Cream Granita, 219
Raspberry Limeade Granita, 219
Fast, Fresh, and Filling, 212
Crispy Eggplant, Tomato, and Provolone Stacks with Basil, 215
Open-Faced Shrimp-and-Avocado Sandwiches, 214
Summer Corn-and-Golden Potato Chowder, 213
Summer Pasta Salad with Lime Vinaigrette, 212
Fresh, Versatile Tomato Sauce, 210
Creole Rice, 211
Diablo Farfalle, 211
Slow-Cooker Tomato Sauce, 210
Vodka-Cream Sauce with Penne, 211
One Pot, Three Ways, 216
Buttermilk-Lady Pea Soup with Bacon, 217
Lady Pea-and-Corn Patties, 217
Lady Peas, 217
Zucchini Stuffed with Lady Peas, 217
The *SL* Test Kitchen Academy, 208
Grilled Pizza, 208

September
Baked with Love, 222
Apple-Cheddar Bread, 223
Banana-Nut Muffins, 223
Biscuit Cinnamon Sweet Rolls, 224
Mini Pumpkin-Honey-Beer Bread, 224
Pumpkin-Honey-Beer Bread, 224
Pumpkin-Honey-Beer Bread Pudding with Apple Brandy-Caramel Sauce, 224
Spinach-Feta Scones, 223
Better Than Takeout, 229
Smoky Pork Stir-Fry, 229
Classic Pimiento Cheese, 230
Classic Pimiento Cheese, 230
Community Cookbook, 234
Buttermilk-Soaked, Bacon-Fried Chicken in Gravy, 234
Tangy Fried Chicken with Dijon, 234
Dinner in 20 Minutes Flat, 225
Chicken-and-Black Bean Chimichangas, 226
Dixie Chicken Salad with Grapes, Honey, Almonds, and Broccoli, 227
Mushroom Stroganoff, 226
Skillet Steak and Wilted Kale, 226
Speedy Homemade Mac and Cheese, 225
Tomato-and-Red Pepper Soup, 227
The Ultimate Grilled Cheese, 227
The *SL* Test Kitchen Academy, 228
Five Time-Saving Tools, 228
The Hunt Breakfast, 232
Grillades, 232
Judy's Bloody Mary Mix, 233
The No-Peel Apple Cake, 231
Caramel Apple Cake, 231

October
Beautiful Beef Dinners, 241
Greek Baked Ziti, 242
Shepherd's Pie with Potato Crust, 242
Spiced Beef Kabobs with Herbed Cucumber-and-Tomato Salad, 243
Tex-Mex Meatballs in Red Chile Sauce, 242
Bolognese Made Easy, 247
Slow-Cooker Bolognese Sauce over Pappardelle Pasta, 247
Community Cookbook, 248
Easy Barbecue Shrimp, 248
Jalapeño Biscuits, 248
Cooking for My Family, 240
Creole Seafood Jambalaya, 240
The Great Pumpkin Cookbook, 236
Beer-Battered Pumpkin with Dipping Sauce, 238
Our Easiest Pumpkin Pie Ever, 237
Pear-and-Pumpkin Tart, 237
Pumpkin-and-Turnip Green Lasagna, 238
Pumpkin-Chocolate Brownies, 237
Spicy Pumpkin Soup with Avocado Cream, 238
Make It a Movie Night, 244
Caramel-Peanut-Popcorn Snack Mix, 245
Heartwarming Cider, 245
Smoky Barbecue Popcorn, 245
The *SL* Test Kitchen Academy, 246
The *SL* Beef Primer, 246

November
The Best Pancakes Ever, 266
Apple-Spice Pancakes, 266
Chocolate Chip-Toasted Pecan Pancakes, 266
The Original "Pam-Cakes," 266
Community Cookbook, 268
Apple-Cream Cheese Bundt Cake, 268
Host a Gumbo Party, 259
Best Ever Seafood Gumbo, 260
Our Test Kitchen's Make-Ahead Thanksgiving, 250
Cherry-Spice Cake Trifle, 254
Dry-Brined-and-Marinated Smoked Turkey, 256
Easy Butter Rolls, 252
Grandmother Carter's Cornbread Dressing, 253
Holiday Spice Cake, 253
Make-Ahead Gravy, 251
Mashed Potatoes with Greens, 254
Parsnip-Potato Soup, 252
Pumpkin-Espresso Tiramisù, 256
Roasted Vegetable Salad with Apple Cider Vinaigrette, 255
Simple Chicken Stock, 251
Slow-Cooker Sweet Potatoes with Bacon, 257
Pecan Delights, 261
Chocolate-Pecan Mousse Tart, 262
Decadent Cream Puffs with Praline Sauce and Toasted Pecans, 263
Pecan-Espresso Toffee, 262
Pecan Pie Bars, 262
Pecan Pound Cake, 263
Pumpkin-Pecan Streusel Pie, 261
Spicy-Sweet Pecans, 263
Soulful Soups and Stews, 264
Cheesy Garlic French Bread, 265
Chicken-and-Prosciutto Tortelloni Soup, 265
King Ranch Chicken Soup, 264
Shrimp-and-New Potato Chowder, 264
Southwest Pork Stew, 264
White Bean, Sausage, and Turnip Green Stew, 265
The *SL* Test Kitchen Academy, 267
Our Secrets to the Best Pancakes, 267

December
The Baker's Brunch, 295
Mini Chocolate Chess Tarts, 296
Smoky Sausage-and-Grits Casserole, 295
Community Cookbook, 302
Cheesy Pimiento Cornbread, 302
Gouda Grits, 302
Speakeasy Sparkler, 302
5 Brunch Beauties, 296
Apple-Pecan Streusel Muffins, 297
Christmas Morning Cinnamon Rolls, 296
Cream Cheese Pastries, 298
Festive Fresh Fruit Salad with Mint-Lime Simple Syrup, 297
Holiday Ham with Apricot Glaze, 298
Ham for the Holidays, 299
Cheese-Grit-and-Chive Muffins, 300
Milk Punch, 300

Holiday 2015
Crowd Pleasing Cocktails, 275
Peanut Whiskey and Cola, 275
Retro Rum Punch, 275
Rosemary Collins, 275
Southern Lemon Americano, 275
Spiced Chocolate Stout Beer Cocktail, 275
8 Amazing Appetizers, 272
Cheese Balls, Three Ways, 274
Chicken Liver Mousse Crostini with Pepper Jelly, 272
Creamy Shrimp Dip with Crispy Wonton Chips, 272
Grapefruit-Beet-Goat Cheese Flatbread, 273
Inside-Out Hot Brown Bites, 272
Petite Sweet Potato Biscuits with Pulled Pork and Slaw, 274
Roasted Fennel-and-Prosciutto Flatbread, 273
Smoked Trout Crostini with Radishes and Dill Cream, 273
8 Spectacular Sides, 281
Broccolini with Pecans and Cane Syrup Vinaigrette, 281
Brussels Sprouts with Parmesan Cream Sauce, 283
Charred Ambrosia with Toasted Coconut and Marshmallow Crème, 282
Christmas Salad, 283
Dirty Rice Risotto, 281
Fontina-Chive Yorkshire Puddings, 283
Gnocchi Mac and Cheese, 282
Roasted Garlic Duchess Potatoes, 282
9 Dazzling Desserts, 286
Chocolate-Peppermint Cheesecake, 289
Coconut-Macadamia Nut Pound Cake, 289
Gingerbread Baked Alaska, 287
Hot Chocolate, 290
Layered Eggnog Cream with Puff Pastry, 290
Our Best Homemade Marshmallows, 290
Praline Cream-Beignet Tower, 286
Spice Cake with Cranberry Filling, 286
Triple Chocolate Brownie-Mousse Stacks, 288
9 Superfast Cookies, 291
Belgian Spice Cookies, 291
Chewy Ginger Cookies, 293
Chocolate-Peppermint Crackle Cookies, 292
Coconut Snowballs, 291
5-Ingredient Sugar Cookies, 292
Glazed Fruitcake Bars, 294
Layered Eggnog Blondies, 293
Pecan-Cranberry Shortbread, 292
Red Velvet Thumbprints, 291
7 Marvelous Mains, 278
Classic Roasted Duck with Orange-Bourbon-Molasses Glaze, 278
Creamy Shrimp Piccata, 280
Grilled Cornish Hens with Herb Brine, 279
Grilled Pork Loin Steaks with Cherry-Plum Sauce, 278
Hoppin' John Cakes with Tomato-Jalapeño Gravy, 280
Spice-Rubbed Tenderloin with Mustard-Cream Sauce, 279
Veal Chops Milanese with Lemon and Herbs, 278
The *SL* Test Kitchen Academy, 301
Smart Cookies, 301

Bonus Favorites

Baked Greats, 313
Apple Fritter Pull-Apart Bread, 313
Blue Cheese-Pecan Popovers, 313
Bourbon-Spiked Maple-Bacon Mini Cakes, 315
Coffee Cake Muffins with Brown Butter Icing, 314
Southern Anadama Bread, 314
A Big Batch of Cookie Favorites, 328
Almond Poinsettia Cookies, 330
Candy Cane Biscotti, 330
Chocolate-Cappuccino Whoopie Pies, 329
Maple-Praline Cookies, 329
Spiced Sorghum Snowflakes, 328
Sugar Cookie Stars, 329
Get the Party Started, 316
Cheese Biscuit Barbecue Bites, 317
Cranberry-Chutney Cheese Ball, 316
Cranberry Tea-Ini, 317
Ruby Red Negroni, 317
Turkey-Artichoke Cornucopias, 316
Good Luck New Year, 331
Easy Hoppin' John, 331
Fig Upside-Down Cake, 332
Kale Salad with Hot Bacon Dressing, 331
Pecan Cornbread, 332
Pork Chops with Bourbon-Rosemary-Mustard Sauce, 331
Holiday Sideboard, 320
Butternut-Pecan Sauté, 320
Cheesy Pimiento Corn Casserole, 321
Crown Pork Roast with Fig-Fennel Stuffing, 323
Herbed Grits Dressing with Leeks and Mushrooms, 322
Pecan-and-Wild Rice Stuffing, 323
Potato Gratin with Bacon and Comté, 320
Roasted Brussels Sprouts with Sage Pesto, 321
Tart Cherry-Chestnut Sausage Dressing, 322
Winter Citrus-Avocado Salad, 320
Holiday Tea, 310
Browned Butter Scones with Faux Clotted Cream, 312
Chicken Salad Tomato Cups, 310
Fig and Bacon Palmiers, 310
Hazelnut Chip Shortbread, 311
Hot Apple Tea, 310
Petits Fours Presents, 312
Pumpernickel Tea Sandwiches, 311
Smoked Trout and Watercress Tea Sandwiches, 310
Main Attractions, 324
Butterflied and Barbecued Turkey, 324
Pistachio-Crusted Rack of Lamb, 325
Tuscan Beef Tenderloin, 325
Quick-and-Easy, 304
Bacon and Cheddar Belgian Waffles, 308
Baked Beef and Fried Pepper Burritos, 308
Balsamic Green Beans with Tomatoes and Feta, 309
Blueberry Yogurt Pops, 309
Chicken, Artichoke, and Sun-Dried Tomato Pizza, 306
Cinnamon Apple Tarts, 309
Citrus Chicken and Beet Salad, 305
Cornbread-Vegetable Salad, 306
Eggs in Fire Roasted Tomato Sauce, 307
Fried Apple Pork Chops, 308
Grilled Chicken Caprese, 304
Grilled Salmon with Orange-Fennel Salad, 305
Grilled Teriyaki Shrimp Kabobs, 305
Grilled Tomato-Peach Pizza, 307
Korean Flank Steak, 304
Lemon and Shallot Brussels Sprouts, 309
Molten Hazelnut Brownies, 309
Peach Pancakes, 308
Pesto Chicken Grill, 304
Philly Cheesesteak Pizza, 306
Prosciutto and Tomato Sandwiches, 306
Sage-Brown Butter Squash Agnolotti, 307
Save Room, 326
Mocha-Hazelnut Dacquoise, 326
Peppermint Red Velvet Bundt Cakes, 326
White German Chocolate Cake, 327
Soups and Stews, 318
Braised Beef-and-Mushroom Stew, 318
Ham-and-Corn Chowder, 319
Pumpkin Soup with Red Pepper Relish, 319
Southern Baker, 333
Bananas Foster Coffee Cake with Vanilla-Rum Sauce, 338
Caramelized Onion and Swiss Popovers, 334
Chocolate Breakfast Wreath, 333
Cornbread Focaccia, 334
Double Chocolate Chip Cookies with Bourbon Ganache, 338
Honey-Balsamic-Blueberry Pie, 337
Mini Banana-Cranberry-Nut Bread Loaves, 335
Peppermint Divinity Bars, 340
Pumpkin Cupcakes with Browned Butter Frosting, 336
Salted Caramel-Chocolate Pecan Tart, 339
Simple Piecrust, 339
Strawberry Lemonade Muffins, 335
Sweet Potato Biscuits, 335

General Recipe Index

This index alphabetically lists every recipe by exact title.

A

AÏOLI
Grand Aïoli, 148
Lime-Sriracha Aïoli, 148
Saffron Aïoli, 148

ALMONDS
Chicken Salad with Grapes, Honey, Almonds, and Broccoli, Dixie, 227
Cookies, Almond Poinsettia, 330
Petits Fours Presents, 312
Toasted Almonds, Shaved Vegetable Salad with, 85

AMBROSIA
Charred Ambrosia with Toasted Coconut and Marshmallow Crème, 282

APPETIZERS. *SEE ALSO* SNACKS.
Biscuit Barbecue Bites, Cheese, 317
Biscuits with Pulled Pork and Slaw, Petite Sweet Potato, 274
Cheese
Ball, Blue Cheese-Red Currant Cheese, 274
Ball, Cheddar-Horseradish-Walnut Cheese, 274
Ball, Cranberry-Chutney Cheese, 316
Ball, Feta-Olive-Fresh Herb Cheese, 274
Balls, Three Ways, Cheese, 274
Straws, Jalapeño Cheese, 100
Chicken Salad Tomato Cups, 310
Chips, Crispy Wonton, 272
Crab Cakes, Light, 91
Crostini with Pepper Jelly, Chicken Liver Mousse, 272
Crostini with Radishes and Dill Cream, Smoked Trout, 273
Dips
Aïoli, Grand, 148
Aïoli, Lime-Sriracha, 148
Aïoli, Saffron, 148
Garlic with Confit Oil, Thyme, and Black Pepper, 144
Shrimp Dip with Crispy Wonton Chips, Creamy, 272
Flatbread, Grapefruit-Beet-Goat Cheese, 273
Flatbread, Roasted Fennel-and-Prosciutto, 273
Fonduta, Leek, 202
Hot Brown Bites, Inside-Out, 272
Palmiers, Fig and Bacon, 310
Pecans, Spicy-Sweet, 263
Petits Fours Presents, 312
Sandwiches, Pumpernickel Tea, 311
Sandwiches, Smoked Trout and Watercress Tea, 310
Sauces
Alabama White Sauce, 133
Buttery Nashville Hot Sauce, 133
Peanut Sauce, Vietnamese, 133
Scones with Faux Clotted Cream, Browned Butter, 312
Turkey-Artichoke Cornucopias, 316
Wings, Wonder, 133

APPLES
Bread, Apple-Cheddar, 223
Bread, Apple Fritter Pull-Apart, 313
Cake, Apple-Cream Cheese Bundt, 268
Cake, Caramel Apple, 231
Fried Apple Pork Chops, 308
Mignonette, Green Apple, 26
Muffins, Apple-Pecan Streusel, 297
Pancakes, Apple-Spice, 266
Sauce, Apple Brandy-Caramel, 224
Syrup, Apple Cider, 266
Tarts, Cinnamon Apple, 309
Tea, Hot Apple, 310
Vinaigrette, Apple Cider, 255

APPLESAUCE
Homemade Applesauce, Honey-Glazed Pork Tenderloin with, 51

APRICOTS
Glaze, Holiday Ham with Apricot, 298

ARTICHOKES
Cornucopias, Turkey-Artichoke, 316
Penne, Chicken-and-Artichoke, 88
Pizza, Chicken, Artichoke, and Sun-Dried Tomato, 306

ASPARAGUS
Bucatini, Ham, and Asparagus, 89

AVOCADOS
Ranch, Avocado, 46
Salad, Grilled Peach & Avocado, 82
Salad, Winter Citrus-Avocado, 320
Sandwiches, Open-Faced Shrimp-and-Avocado, 214

B

BACON
BBQ Bacon, 62
Belgian Waffles, Bacon and Cheddar, 308
Cakes, Bourbon-Spiked Maple-Bacon Mini, 315
Collards, The Ultimate Classic, 47
Croutons, Mustard Greens with Yogurt-Parmesan Dressing and Bacon, 48
Dressing, Kale Salad with Hot Bacon, 331
Grits, Waffled Bacon & Cheddar, 61
Hot Brown Bites, Inside-Out, 272
Jam, Bacon-and-Sweet Onion, 102
Oven-Roasted Bacon, 60
Palmiers, Fig and Bacon, 310
Pecan-Rosemary Bacon, 62
Pork Loin Steaks with Cherry-Plum Sauce, Grilled, 278
Potato Gratin with Bacon and Comté, 320
Pot Pie, Oyster-Bacon, 25
Slaw, BLT Buttermilk Blue, 97
Soup, Loaded Potato, 37
Soup with Bacon, Buttermilk-Lady Pea, 217
Southwest Cornmeal Bacon, 62
Sweet Potatoes with Bacon, Slow-Cooker, 257
Tart, Cheesy Bacon-and-Two-Onion, 51
Vinaigrette, Peas and Kale Salad with Bacon, 34

BANANAS
Bread Loaves, Mini Banana-Cranberry-Nut, 335
Bread Loaves, Regular-Size Banana-Cranberry-Nut, 335
Brûlée, Baked French Toast with Banana, 98
Coffee Cake with Vanilla-Rum Sauce, Bananas Foster, 338
Muffins, Banana-Nut, 223

BARBECUE. *SEE ALSO* GRILLED.
Bacon, BBQ, 62
Chicken Pizzas, Quick BBQ, 110
Deviled Eggs, Barbecue, 83
Mac and Cheese, Barbecue, 225
Popcorn, Smoky Barbecue, 245
Pork
Bites, Cheese Biscuit Barbecue, 317
Nachos, Pulled Pork Barbecue, 112
Ribs, Easy BBQ, 109
Sauces
El Sancho Barbecue Sauce, 80
Fancy Barbecue Sauce, 109
Sweet-and-Tangy Barbecue Sauce, 324
White Barbecue Sauce, Smoked Chicken Pizza with, 207
Shrimp, Easy Barbecue, 248
Turkey, Butterflied and Barbecued, 324

BEANS
Black
Chili, Pork-and-Black Bean, 31
Chimichangas, Chicken-and-Black Bean, 226

Soup with Avocado Cream, Spicy Pumpkin, 238
Stew, Southwest Pork, 264
Chickpea Soup, African, 42
Chili, Beef-and-Bean, 30
Green
Balsamic Green Beans with Tomatoes and Feta, 309
Easy Summer Green Beans, 193
Fiery Green Beans, 42
Stewed Green Beans and Tomatoes, 143
Stir-Fry, Garlicky Beef-and-Bean, 50
Stew, Sausage-and-Bean, 69
Succotash, Classic, 197
Succotash, Nashville Farmers' Market, 115
White Bean Mash, Peppered, 94
White Bean, Sausage, and Turnip Green Stew, 265
BEEF. *SEE ALSO* BEEF, GROUND; VEAL.
Brisket
Cowboy Brisket Sandwich, 81
Step by Step, Brisket, 79
Tacos with Summer Salsa, Brisket, 199
Smoked Beef Brisket Tostadas, 81
Smoked Beef Brisket, True, 80
Burritos, Baked Beef and Fried Pepper, 308
Pizza, Philly Cheesesteak, 306
Pot Roast, Beer-Braised, 29
Rub, Beef, 80
Steaks
Flank Steak, Korean, 304
Flank Steak with Lime-Wasabi Sauce, Seared, 196
Salad with Spicy Dressing, Charred Steak, 34
Salad with Steak, Chopped, 196
Sandwiches, Southern Cheese Steak, 196
Skillet Steak and Wilted Kale, 226
Skirt Steak and Chimichurri, Pan-Seared, 116
Stir-Fry, Garlicky Beef-and-Bean, 50
Stew, Braised Beef-and-Mushroom, 318
Tenderloin, Tuscan Beef, 325
Tenderloin with Mustard-Cream Sauce, Spice-Rubbed, 279
BEEF, GROUND
Bolognese Sauce over Pappardelle Pasta, Slow-Cooker, 247
Burgers with Green Tomato Mayonnaise, 159
Chili, Beef-and-Bean, 30
Hamburger Steak, Tuesday Night, 112
Kabobs with Herbed Cucumber-and-Tomato Salad, Spiced Beef, 243
Meatballs in Red Chile Sauce, Tex-Mex, 242
Meatball Sliders with Tomato Sauce, 28
Shepherd's Pie with Potato Crust, 242
Ziti, Greek Baked, 242
BEETS
Flatbread, Grapefruit-Beet-Goat Cheese, 273
Roasted Beets, Carrots, and Sweet Onions, 85
Salad, Christmas, 283
Salad, Citrus Chicken and Beet, 305
BEIGNETS
Tower, Praline Cream-Beignet, 286
BEVERAGES
Alcoholic
Americano, Southern Lemon, 275
Blush Lily, The, 100
Cocktail, Spiced Chocolate Stout Beer, 275
Cola, Peanut Whiskey and, 275
Collins, Rosemary, 275
Gimlets, Orange-Infused Rum, 149
Lemonade, Sparkling Elderflower, 86
Mint Julep, The *SL*, 100
Negroni, Ruby Red, 317
Old-Fashioned, Bacon, 115
Old Fashioned, My Go-To, 103
Sparkler, Speakeasy, 302
Tea-Ini, Cranberry, 317
Cider, Heartwarming, 245
Hot Chocolate, 290
Lemonade, Minty, 74
Milk, Homemade Strawberry, 94
Mix, Judy's Bloody Mary, 233
Punch
Milk Punch, 300
Rum Punch, Retro, 275
Watermelon Punch, Pool Party, 95
Shakes, Mango-Buttermilk, 44
Syrup, Simple, 209, 275
Syrup, Spicy Simple, 275
Tea, Classic Sweet, 209
Tea, Hot Apple, 310
Water, Cucumber-Mint, 144
BISCUITS
Buttermilk Biscuit, Our Favorite, 90
Cheese Biscuit Barbecue Bites, 317
Jalapeño Biscuits, 248
Rolls, Biscuit Cinnamon Sweet, 224
Sweet Potato Biscuits, 275, 335
Sweet Potato Biscuits with Pulled Pork and Slaw, Petite, 274
BLACKBERRIES
Jam, Blackberry Chambord, 220
Sauce, Blackberry-Honey Mustard, 102
BLUEBERRIES
Ice Cream, Blueberry-Lemon Zest, 201
Pops, Blueberry Yogurt, 309
Relish, Strawberry-Blueberry, 107
Slaw with Buttermilk Dressing, Kale-and-Blueberry, 75
Topping, Lemon-Blueberry, 121
BREADS. *SEE ALSO* BEIGNETS; BISCUITS; CORNBREADS; CROUTONS; FRENCH TOAST; HUSH PUPPIES; MUFFINS; PANCAKES; ROLLS; WAFFLES.
Apple-Cheddar Bread, 223
Banana Cranberry-Nut Bread Loaves, Mini, 335
Banana-Cranberry-Nut Bread Loaves, Regular-Size, 335
French Bread, Cheesy Garlic, 265
Popovers, Blue Cheese-Pecan, 313
Popovers, Caramelized Onion and Swiss, 334
Pudding with Apple Brandy-Caramel Sauce, Pumpkin-Honey-Beer Bread, 224
Pull-Apart Bread, Apple Fritter, 313
Pumpkin-Honey-Beer Bread, 224
Pumpkin-Honey-Beer Bread, Mini, 224
Scones, Spinach-Feta, 223
Scones with Faux Clotted Cream, Browned Butter, 312
Yeast
Anadama Bread, Southern, 314
Focaccia, Cornbread, 334
Focaccia, Herbed Cornbread, 334
Wreath, Chocolate Breakfast, 333
BROCCOLI
Chicken Salad with Grapes, Honey, Almonds, and Broccoli, Dixie, 227
BROCCOLINI
Pecans and Cane Syrup Vinaigrette, Broccolini with, 281
BRUSSELS SPROUTS
Lemon and Shallot Brussels Sprouts, 309
Parmesan Cream Sauce, Brussels Sprouts with, 283
Roasted Brussels Sprouts with Sage Pesto, 321
Sautéed Cabbage, Brussels Sprouts and, 51
Sautéed Sprouts, 35
BURRITOS
Beef and Fried Pepper Burritos, Baked, 308
BUTTER
Balsamic Brown Butter Sauce, Pan-Seared Grouper with, 105
Basil Butter and Summer Squash, Broiled Pork Chops with, 194
Brown Butter Icing, 314
Brown Butter, Pecans, and Yogurt, Carrot Soup with, 202
Browned Butter Frosting, 336
Browned Butter Scones with Faux Clotted Cream, 312
Herb Butter, 252
Herb Butter, Roasted Chicken Thighs with, 65
Sage-Brown Butter Squash Agnolotti, 307
Sorghum Butter, 252
Tequila-Lime Butter, Grilled Corn and, 114
Whipped Honey-Rosemary Butter, 314
BUTTERSCOTCH
Cake, Chocolate-Bourbon-Butterscotch Icebox, 122

C

CABBAGE. *SEE ALSO* SALADS/ SLAWS.
Sautéed Cabbage, Brussels Sprouts and, 51
CACCIATORE
Chicken Cacciatore with Spaghetti, Slow-Cooker, 30
CAKES
Apple-Cream Cheese Bundt Cake, 268
Caramel Apple Cake, 231
Carrot Cake, The Ultimate, 68
Cheesecake, Chocolate-Peppermint, 289
Cherry Sheet Cake, Buttermilk-Glazed, 155
Cherry-Spice Cake Trifle, 254
Chocolate-Bourbon-Butterscotch Icebox Cake, 122
Chocolate-Mayonnaise Cake, 57
Coconut Cream Cake, 93
Coffee Cake with Vanilla-Rum Sauce, Bananas Foster, 338
Cupcakes with Browned Butter Frosting, Pumpkin, 336
Mini Cakes, Bourbon-Spiked Maple-Bacon, 315
Peppermint Red Velvet Bundt Cakes, 326
Petits Fours Presents, 312
Pound
Coconut-Macadamia Nut Pound Cake, 289
Pecan Pound Cake, 263
Triple-Chocolate Buttermilk Pound Cake, 44
Triple-Chocolate Buttermilk Pound Cakes, Mini, 44
Shortcakes, Coconut-and-Pecan Strawberry, 86
Shortcakes with Sweet Cream, Boozy Peach, 125
Spice Cake, Holiday, 253
Spice Cake with Cranberry Filling, 286
Upside-Down Cake, Fig, 332
White German Chocolate Cake, 327
CANDIES
Peanut Brittle, Honey, 42
Toffee, Pecan-Espresso, 262
CARAMEL
Cake, Caramel Apple, 231
Marshmallows, Bourbon-Caramel, 290
Sauce, Apple Brandy-Caramel, 224
Snack Mix, Caramel-Peanut-Popcorn, 245
Tart, Salted Caramel-Chocolate Pecan, 339
CARROTS
Cake, The Ultimate Carrot, 68
Candied Carrot Curls, 67
Roasted Beets, Carrots, and Sweet Onions, 85
Soup with Brown Butter, Pecans, and Yogurt, Carrot, 202
CASSEROLES. *SEE ALSO* LASAGNA.
Corn Casserole, Cheesy Pimiento, 321
Mac and Cheese, Gnocchi, 282
Macaroni & Cheese, Classic Baked, 83
Potato Gratin with Bacon and Comté, 320
Sausage-and-Grits Casserole, Smoky, 295
Ziti, Greek Baked, 242
CHEESE. *SEE ALSO* APPETIZERS/ CHEESE.
Balsamic Green Beans with Tomatoes and Feta, 309
Breads
Apple-Cheddar Bread, 223
Belgian Waffles, Bacon and Cheddar, 308
Biscuit Barbecue Bites, Cheese, 317
Cornbread, Cheesy Pimiento, 302
French Bread, Cheesy Garlic, 265
Muffins, Cheese-Grit-and-Chive, 300
Popovers, Blue Cheese-Pecan, 313
Popovers, Caramelized Onion and Swiss, 334
Scones, Spinach-Feta, 223
Casseroles
Corn Casserole, Cheesy Pimiento, 321
Mac and Cheese, Gnocchi, 282
Macaroni & Cheese, Classic Baked, 83
Potato Gratin with Bacon and Comté, 320
Ziti, Greek Baked, 242
Desserts
Blondies, Layered Eggnog, 293
Cake, Apple-Cream Cheese Bundt, 268
Cream, Faux Clotted, 312
Cream, Sweet Mascarpone, 86
Filling, Cream Cheese, 298
Filling, Cream Cheese-Honey, 223
Frosting, Brown Sugar-Cream Cheese, 68
Frosting, Chocolate-Cream Cheese, 57
Icing, Cream Cheese, 297
Pastries, Cream Cheese, 298
Tiramisù, Belle Chèvre Raspberry, 97
Fettuccine with Smoky Turnip Greens, Lemon, and Goat Cheese, 47
Grapes, Blue Cheese-Pecan, 54
Grits, Gouda, 302
Grits Soufflé with Mushroom Gravy, Cheese, 85
Grits, Waffled Bacon & Cheddar, 61
Grits, Waffled Gruyère & Ham, 61
Grits, Waffled Jalapeño & Monterey Jack Cheese, 61
Mac and Cheese, Barbecue, 225
Mac and Cheese, Mexican, 225
Mac and Cheese, Southern Pimiento, 225
Mac and Cheese, Speedy Homemade, 225
Mac-and-Cheese, Uncle Jack's, 29
Pasta with Heirloom Tomatoes, Goat Cheese, and Basil, 159
Pimiento Cheese, Classic, 230
Pizza, Philly Cheesesteak, 306
Pizza, Ultimate Cheese, 207
Potatoes, Goat Cheese Mashed, 65
Potatoes, Twice-Baked Green, 48
Ranch, Blue Cheese, 46
Sandwiches
Grilled Cheese, The Ultimate, 227
Grilled Chicken Caprese, 304
Southern Cheese Steak Sandwiches, 196
Sauce, Brussels Sprouts with Parmesan Cream, 283
Soup, Loaded Potato, 37
Stacks with Basil, Crispy Eggplant, Tomato, and Provolone, 215
Straws, Jalapeño Cheese, 100
Stuffed Peppers, Turkey-, 36
Stuffed Peppers with Chèvre, Pecans, and Golden Raisins, 94
Tart, Cheesy Bacon-and-Two-Onion, 51
Yorkshire Puddings, Fontina-Chive, 283
CHERRIES
Cake, Buttermilk-Glazed Cherry Sheet, 155
Dressing, Tart Cherry-Chestnut Sausage, 322
Macerated Cherries, Lime-and-Mint, 153
Macerated Cherries, Orange-and-Basil, 153
Muffins, Cherry-Rosemary, 70
Sauce, Cherry-Plum, 279
Trifle, Cherry-Spice Cake, 254
CHICKEN
"Beer Can" Chicken, Ginger, 136
Brown Sugar-Brined Chicken, 137
Buttermilk-Soaked, Bacon-Fried Chicken in Gravy, 234
Cacciatore with Spaghetti, Slow-Cooker Chicken, 30
Chimichangas, Chicken-and-Black Bean, 226
Citrus-Braised Chicken Thighs, 64
Cutlets with Herbed Mushroom Sauce, Chicken, 50
Fried Chicken with Dijon, Tangy, 234
Kabobs, Perfect Chicken, 133
Lettuce Cups, Slow-Cooker Chicken, 92
Liver Mousse Crostini with Pepper Jelly, Chicken, 272
Marinated Chicken Thighs, Garlic-Yogurt-, 65
Paillard with Citrus Salad and Couscous, Chicken, 33
Peanutty Braised Chicken, 97
Penne, Chicken-and-Artichoke, 88
Pesto Chicken Grill, 304
Pizza, Chicken, Artichoke, and Sun-Dried Tomato, 306
Pizzas, Quick BBQ Chicken, 110
Pizza with White Barbecue Sauce, Smoked Chicken, 207
Roasted Chicken Thighs with Herb Butter, 65
Salads
Citrus Chicken and Beet Salad, 305
Dixie Chicken Salad with Grapes, Honey, Almonds, and Broccoli, 227
Farro, and Vegetable Salad with Lemon Vinaigrette, Chicken, 64

Tomato Cups, Chicken Salad, 310
Tomato-Peach Salad with Chicken, 137
Salt-and-Pepper Grilled Chicken Thighs, Crispy, 135
Sandwiches
Burgers with Creamy Ancho Slaw, Carolina Chicken, 139
Grilled Chicken Caprese, 304
Sesame-Chicken Thigh Paillard with Peanut Sauce, 63
Smoked Mojo Leg Quarters, 138
Soups
King Ranch Chicken Soup, 264
Stock, Simple Chicken, 251
Tortelloni Soup, Chicken-and-Prosciutto, 265
Stew, Peanut Chicken, 41
Tacos, Sesame Chicken, 66
Tacos with Watermelon-Mango Salsa, Crunchy Jerk, 135
Wings, Wonder, 133
CHILI
Beef-and-Bean Chili, 30
Pork-and-Black Bean Chili, 31
Turkey-and-Sweet Potato Chili, Smoky, 31
Venison Chili, Game Day, 113
CHIMICHANGAS
Chicken-and-Black Bean Chimichangas, 226
CHIMICHURRI
Skirt Steak and Chimichurri, Pan-Seared, 116
CHIPS
Sweet Potato Chips, 83
Wonton Chips, Crispy, 272
CHOCOLATE
Bars and Cookies
Brownies à la Mode, Goo Goo Cluster, 115
Brownies, Claire's Cream Cheese Swirl, 38
Brownies, Molten Hazelnut, 309
Brownies, Pumpkin-Chocolate, 237
Crackle Cookies, Chocolate-Peppermint, 292
Double Chocolate Chip Cookies with Bourbon Ganache, 338
Thumbprints, Nutty Chocolate, 57
Thumbprints, Red Velvet, 291
Triple Chocolate Cranberry Oatmeal Cookies, 70
Whoopie Pies, Chocolate-Cappuccino, 329
Beverages
Cocktail, Spiced Chocolate Stout Beer, 275
Hot Chocolate, 290
Breads
Pancakes, Chocolate Chip-Toasted Pecan, 266
Wreath, Chocolate Breakfast, 333
Cakes
Cheesecake, Chocolate-Peppermint, 289
Icebox Cake, Chocolate-Bourbon Butterscotch, 122
Mayonnaise Cake, Chocolate-, 57
Pound Cakes, Mini Triple-Chocolate Buttermilk, 44
Pound Cake, Triple-Chocolate Buttermilk, 44
White German Chocolate Cake, 327
Frosting, Filling, and Toppings
Bourbon Ganache, 338
Cream Cheese Frosting, Chocolate-, 57
Mocha Cream Filling, 327
Whipped Cream, Chocolate, 56
Granita, Chocolate Milk, 219
Mocha-Hazelnut Dacquoise, 326
Pie and Tarts
Chess Tarts, Mini Chocolate, 296
Mousse Tart, Chocolate-Pecan, 262
Salted Caramel-Chocolate Pecan Tart, 339
Ultimate Chocolate Pie, The, 56
Pudding, Chocolate-Buttermilk, 45
Stacks, Triple Chocolate Brownie Mousse, 288
Toffee, Pecan-Espresso, 262
CHOWCHOW
Three-Pepper Chowchow, 108
CHOWDERS
Corn-and-Golden Potato Chowder, Summer, 213
Ham-and-Corn Chowder, 319
Shrimp-and-New Potato Chowder, 264
CHUTNEY
Cheese Ball, Cranberry-Chutney, 316
Green Tomato Chutney, Roasted Cider-Brined Pork Loin with, 53
COCONUT
Cake, Coconut Cream, 93
Frosting, Coconut-Pecan, 327
Frosting, Fluffy Coconut, 93
Pound Cake, Coconut-Macadamia Nut, 289
Shortcakes, Coconut-and-Pecan Strawberry, 86
Snowballs, Coconut, 291
Toasted Coconut and Marshmallow Crème, Charred Ambrosia with, 282
COFFEE
Cappuccino Whoopie Pies, Chocolate-, 329
Espresso Filling, 329
Espresso Tiramisù, Pumpkin-, 256
Espresso Toffee, Pecan-, 262
Mocha Cream Filling, 327
Mocha-Hazelnut Dacquoise, 326
COLLARDS
Pesto, Collard-and-Olive, 48
Ultimate Classic Collards, The, 47
COMMUNITY COOKBOOK RECIPES
Appetizers
Garlic with Confit Oil, Thyme, and Black Pepper, 144
Leek Fonduta, 202
Beverages
Milk, Homemade Strawberry, 94
Old-Fashioned, Bacon, 115
Punch, Pool Party Watermelon, 95
Sparkler, Speakeasy, 302
Water, Cucumber-Mint, 144
Breads
Biscuits, Jalapeño, 248
Cornbread, Cheesy Pimiento, 302
French Toast with Banana Brûlée, Baked, 98
Muffins, Cherry-Rosemary, 70
Desserts
Brownies à la Mode, Goo Goo Cluster, 115
Brownies, Claire's Cream Cheese Swirl, 38
Cake, Apple-Cream Cheese Bundt, 268
Cookies, Triple Chocolate Cranberry Oatmeal, 70
Frosting, Praline, 268
Meringues, Cotton Candy, 58
Tiramisù, Belle Chèvre Raspberry, 97
Grits, Gouda, 302
Jam, Blackberry Chambord, 220
Main Dishes
Chicken in Gravy, Buttermilk-Soaked, Bacon-Fried, 234
Chicken, Peanutty Braised, 97
Chicken with Dijon, Tangy Fried, 234
Hamburger Steak, Tuesday Night, 112
Nachos, Pulled Pork Barbecue, 112
Paella, Quick Cook Shrimp and Chorizo, 113
Shrimp, Easy Barbecue, 248
Skirt Steak and Chimichurri, Pan-Seared, 116
Spanish Tortilla, Spicy Andouille, 95
Stuffed Peppers with Chèvre, Pecans, and Golden Raisins, 94
Pickles, Old Timey, 220
Slaw, BLT Buttermilk Blue, 97
Sammies, PB&J, 58
Side Dishes
Corn and Tequila-Lime Butter, Grilled, 114
Hoppin' John and Limpin' Susan, 38
Succotash, Nashville Farmers' Market, 115
White Bean Mash, Peppered, 94
Soups and Chili
Carrot Soup with Brown Butter, Pecans, and Yogurt, 202
Chili, Game Day Venison, 113
Corn Soup with Corn Salsa, Summer, 96
COOKIES
Bars and Squares
Blondies, Layered Eggnog, 293
Brownies à la Mode, Goo Goo Cluster, 115

Brownies, Claire's Cream Cheese Swirl, 38
Brownies, Molten Hazelnut, 309
Brownies, Pumpkin-Chocolate, 237
Fruitcake Bars, Glazed, 294
Pecan Pie Bars, 262
Peppermint Divinity Bars, 340
Raspberry "Rhubars," 111
Belgian Spice Cookies, 291
Biscotti, Candy Cane, 330
Crackle Cookies, Chocolate-Peppermint, 292
Drop
Chocolate-Cappuccino Whoopie Pies, 329
Coconut Snowballs, 291
Double Chocolate Chip Cookies with Bourbon Ganache, 338
Maple-Praline Cookies, 329
Triple Chocolate Cranberry Oatmeal Cookies, 70
Ginger Cookies, Chewy, 293
Rolled
Almond Poinsettia Cookies, 330
Shortbread, Hazelnut Chip, 311
Spiced Sorghum Snowflakes, 328
Sugar Cookie Stars, 329
Shortbread, Pecan-Cranberry, 292
Sugar Cookies, 5-Ingredient, 292
Thumbprint Cookies with Tomato Jam, Cornmeal, 158
Thumbprints, Nutty Chocolate, 57
Thumbprints, Red Velvet, 291
***COOKING LIGHT* RECIPES**
Crab Cakes, Light, 91
Main Dishes
Pizzas, Quick BBQ Chicken, 110
Pork Chops with Sautéed Peaches, Skillet, 142
Pork Loin with Green Tomato Chutney, Roasted Cider-Brined, 53
Pork Stir-Fry, Smoky, 229
Tacos, Sesame Chicken, 66
Salad with Spicy Buttermilk Dressing, Orzo, 200
Soup, Loaded Potato, 37
CORN
Casserole, Cheesy Pimiento Corn, 321
Chowder, Ham-and-Corn, 319
Chowder, Summer Corn-and-Golden Potato, 213
Grilled Corn, and Sweet Onion Pizza, Smoked Sausage, 207
Grilled Corn and Tequila-Lime Butter, 114
Patties, Lady Pea-and-Corn, 217
Relish, Sweet Corn, 107
Soup with Corn Salsa, Summer Corn, 96
Succotash, Classic, 197
Succotash, Nashville Farmers' Market, 115
CORNBREADS
Cheesy Pimiento Cornbread, 302
Crumbles, Cornbread, 254
Dressing, Grandmother Carter's Cornbread, 253
Focaccia, Cornbread, 334
Focaccia, Herbed Cornbread, 334
Pecan Cornbread, 332
Salad, Cornbread-Vegetable, 306
CORNISH HENS
Grilled Cornish Hens with Herb Brine, 279
COUSCOUS
Chicken Paillard with Citrus Salad and Couscous, 33
CRAB
Cakes, Light Crab, 91
Salad, Lump Crabmeat, 157
Soup with Lump Crabmeat, Green Tomato, 157
CRANBERRIES
Bread Loaves, Mini Banana-Cranberry-Nut, 335
Bread Loaves, Regular-Size Banana-Cranberry-Nut, 335
Cheese Ball, Cranberry-Chutney, 316
Cookies, Triple Chocolate Cranberry Oatmeal, 70
Filling and Apple Cider Frosting, Christmas Spice Layer Cake with Cranberry, 286
Shortbread, Pecan-Cranberry, 292
Tea-Ini, Cranberry, 317
CROSTINI
Chicken Liver Mousse Crostini with Pepper Jelly, 272
Smoked Trout Crostini with Radishes and Dill Cream, 273
CROUTONS
Bacon Croutons, Mustard Greens with Yogurt-Parmesan Dressing and, 48
CUCUMBERS
Basil-Black Pepper Cucumbers, Creamy, 73
Ginger, Rice Vinegar, and Mint, Cucumbers with, 75
Granita, Cucumber-Basil, 219
Pickles, Old Timey, 220
Salad, Herbed Cucumber-and-Tomato, 243
Water, Cucumber-Mint, 144
CURRY
Buttermilk-Vegetable Curry, 46
CUSTARD
Honey Custard with Berries, 102

D

DESSERTS. *SEE ALSO* CAKES; CANDIES; COOKIES; CUSTARD; FROSTINGS; ICE CREAM; MERINGUES; MOUSSE; PIES, PUFFS, AND PASTRIES; PUDDINGS.
Baked Alaska, Gingerbread, 287
Brownie-Mousse Stacks, Triple Chocolate, 288
Dacquoise, Mocha-Hazelnut, 326
Frozen
Granita, Chocolate Milk, 219
Granita, Orange Cream, 219
Granita, Raspberry Limeade, 219
Pops, Blueberry Yogurt, 309
Pops, Nectarine-Chile Ice, 155
Pizza, Grilled Pineapple Dessert, 206
Sauce, Apple Brandy-Caramel, 224
Sauce, Praline, 263
Tiramisù, Belle Chèvre Raspberry, 97
Tiramisù, Pumpkin-Espresso, 256
Tower, Praline Cream-Beignet, 286
Trifle, Cherry-Spice Cake, 254
DRESSINGS. *SEE ALSO* SALAD DRESSINGS; STUFFINGS.
Cherry-Chestnut Sausage Dressing, Tart, 322
Cornbread Dressing, Grandmother Carter's, 253
Grits Dressing with Leeks and Mushrooms, Herbed, 322
DUCK
Roasted Duck with Orange-Bourbon-Molasses Glaze, Classic, 278

E

EGGNOG
Blondies, Layered Eggnog, 293
Cream with Puff Pastry, Layered Eggnog, 290
EGGPLANT
Stacks with Basil, Crispy Eggplant, Tomato, and Provolone, 215
EGGS
Deviled Eggs, Barbecue, 83
Fire Roasted Tomato Sauce, Eggs in, 307
Fried Eggs, 60
Fried Egg Sandwich with BBQ Bacon, The Ultimate, 61
Poached Eggs, 60
Poached Eggs and Country Gravy, Grits Cakes with, 61
Salad, Pesto Egg, 48
Salad with Country Ham and Eggs, Romaine, 34
Sandwiches, Pumpernickel Tea, 311
Spanish Tortilla, Spicy Andouille, 95
ENTERTAINING WITH JULIA RECIPES
Gumbo, Best Ever Seafood, 260
Muffins, Cheese-Grit-and-Chive, 300
Punch, Milk, 300

F

FARRO
Salad with Lemon Vinaigrette, Chicken, Farro, and Vegetable, 64
FENNEL
Roasted Fennel-and-Prosciutto Flatbread, 273

Salad, Grilled Salmon with Orange-Fennel, 305
Stuffing, Fig-Fennel, 323

FETTUCCINE
Turnip Greens, Lemon, and Goat Cheese, Fettuccine with Smoky, 47

FIGS
Cake, Fig Upside-Down, 332
Palmiers, Fig and Bacon, 310
Stuffing, Fig-Fennel, 323

FILLINGS
Cranberry Filling, Spice Cake with, 286
Cream Cheese Filling, 298
Cream Cheese-Honey Filling, 223
Espresso Filling, 329
Mocha Cream Filling, 327
Tomato Jam, 158

FISH. *SEE ALSO* CRAB; OYSTERS; SALMON; SEAFOOD; SHRIMP.
Grouper with Balsamic Brown Butter Sauce, Pan-Seared, 105
Smoked Trout and Watercress Tea Sandwiches, 310
Smoked Trout Crostini with Radishes and Dill Cream, 273
Snapper Baked in Parchment with Spring Vegetables, 104
Triggerfish, Grilled, 105

FLATBREAD
Grapefruit-Beet-Goat Cheese Flatbread, 273
Roasted Fennel-and-Prosciutto Flatbread, 273

FONDUE
Leek Fonduta, 202

FRENCH TOAST
Baked French Toast with Banana Brûlée, 98

FROSTINGS
Brown Butter Icing, 314
Browned Butter Frosting, 336
Brown Sugar-Cream Cheese Frosting, 68
Chocolate-Cream Cheese Frosting, 57
Coconut Frosting, Fluffy, 93
Coconut-Pecan Frosting, 327
Cream Cheese Icing, 297
Maple-Praline Frosting, 329
Praline Frosting, 268
Royal Icing, 328
Royal Red Icing, 330
White Frosting, 330

FRUIT. *SEE ALSO* SPECIFIC TYPES.
Bars, Glazed Fruitcake, 294
Custard with Berries, Honey, 102
Pickle Brine, Fruit, 141
Pie, Mixed Stone Fruit, 153

Salads
Citrus Salad and Couscous, Chicken Paillard with, 33
Fresh Fruit Salad with Mint-Lime Simple Syrup, Festive, 297
Winter Citrus-Avocado Salad, 320

G

GARLIC
Aïoli, Grand, 148
Beef-and-Bean Stir-Fry, Garlicky, 50
Beef Tenderloin, Tuscan, 325
Chicken, Brown Sugar-Brined, 137
Chicken Lettuce Cups, Slow-Cooker, 92
Chicken Thighs, Garlic-Yogurt-Marinated, 65
Chili, Beef-and-Bean, 30
Confit Oil, Thyme, and Black Pepper, Garlic with, 144
Crown Pork Roast with Fig-Fennel Stuffing, 323
Leg Quarters, Smoked Mojo, 138
Marinade, Buttermilk Tandoori, 134
Pesto, Collard-and-Olive, 48
Roasted Garlic Duchess Potatoes, 282
Sauce, Vietnamese Peanut, 133
Stew, Mexican, 55
Stew, Sausage-and-Bean, 69
Vinaigrette, Herb, 159

GARNISH
Candied Carrot Curls, 67

GINGERBREAD
Baked Alaska, Gingerbread, 287

GLAZES
Apricot Glaze, Holiday Ham with, 298
Citrus Glaze, 263
Creamy Glaze, 224
Ginger Ale Glaze, 136
Maple-Bourbon Glaze, 315
Orange-Bourbon-Molasses Glaze, Classic Roasted Duck with, 278
Orange Glaze, 335
Powdered Sugar Glaze, 298, 326
Snowy Rum Glaze, 290
Vanilla Glaze, 313
Vanilla Glaze, Easy, 333

GNOCCHI
Mac and Cheese, Gnocchi, 282

GRANITAS
Chocolate Milk Granita, 219
Cucumber-Basil Granita, 219
Orange Cream Granita, 219
Raspberry Limeade Granita, 219

GRAPEFRUIT
Flatbread, Grapefruit-Beet-Goat Cheese, 273
Negroni, Ruby Red, 317

GRAPES
Blue Cheese-Pecan Grapes, 54
Chicken Salad with Grapes, Honey, Almonds, and Broccoli, Dixie, 227

GRAVIES
Country Gravy, Grits Cakes with Poached Eggs and, 61
Make-Ahead Gravy, 251
Mushroom Gravy, 86
Pork Chop Sandwiches with Gravy and Grits, 194
Tomato-Jalapeño Gravy, Hoppin' John Cakes with, 280

GREENS
Collard-and-Olive Pesto, 48
Collards, The Ultimate Classic, 47
Mashed Potatoes with Greens, 254
Mustard Greens with Yogurt-Parmesan Dressing and Bacon Croutons, 48
Potatoes, Twice-Baked Green, 48
Sauté, Any Greens, 48

Turnip
Fettuccine with Smoky Turnip Greens, Lemon, and Goat Cheese, 47
Hoppin' John Cakes with Tomato-Jalapeño Gravy, 280
Lasagna, Pumpkin-and-Turnip Green, 238
Stew, White Bean, Sausage, and Turnip Green, 265

GRILLED

Beef
Brisket Step by Step, 79
Brisket, True Smoked Beef, 80
Burgers with Green Tomato Mayonnaise, 159
Flank Steak, Korean, 304
Flank Steak with Lime-Wasabi Sauce, Seared, 196
Kabobs with Herbed Cucumber-and-Tomato Salad, Spiced Beef, 243
Steak Salad with Spicy Dressing, Charred, 34

Corn and Tequila-Lime Butter, Grilled, 114

Fish and Shellfish
Oyster Roast, An Easy Lowcountry, 25
Oysters, Grilled, 26
Salmon with Orange-Fennel Salad, Grilled, 305
Shrimp Kabobs, Grilled Teriyaki, 305
Triggerfish, Grilled, 105

Peach & Avocado Salad, Grilled, 82
Pizza, Grilled, 208
Pizza, Grilled Pineapple Dessert, 206
Pizza, Grilled Tomato-Peach, 307
Pizza, Smoked Sausage, Grilled Corn, and Sweet Onion, 207

Pork
Chops with Sweet Potatoes and Grilled Okra, Chile-Rubbed, 160
Loin Steaks with Cherry-Plum Sauce, Grilled Pork, 278

Poultry
Chicken, Brown Sugar-Brined, 137
Chicken Burgers with Creamy Ancho Slaw, Carolina, 139
Chicken Caprese, Grilled, 304
Chicken, Farro, and Vegetable Salad with Lemon Vinaigrette, 64
Chicken, Ginger "Beer Can," 136
Chicken Grill, Pesto, 304
Chicken Kabobs, Perfect, 133

Chicken Pizza with White Barbecue Sauce, Smoked, 207
Chicken Thighs, Crispy Salt-and-Pepper Grilled, 135
Cornish Hens with Herb Brine, Grilled, 279
Leg Quarters, Smoked Mojo, 138
Turkey, Dry-Brined-and-Marinated Smoked, 256
Wings, Wonder, 133

GRITS
Cakes with Poached Eggs and Country Gravy, Grits, 61
Casserole, Smoky Sausage-and-Grits, 295
Dressing with Leeks and Mushrooms, Herbed Grits, 322
Gouda Grits, 302
Muffins, Cheese-Grit-and-Chive, 300
Pie, Shrimp 'n' Grits, 61
Pork Chop Sandwiches with Gravy and Grits, 194
Slow-Cooker Grits, Overnight, 60
Soufflé with Mushroom Gravy, Cheese Grits, 85
Stove-Top Grits, 60
Waffled Bacon & Cheddar Grits, 61
Waffled Gruyère & Ham Grits, 61
Waffled Jalapeño & Monterey Jack Cheese Grits, 61

GUMBO
Seafood Gumbo, Best Ever, 260

H

HAM
Bucatini, Ham, and Asparagus, 89
Chowder, Ham-and-Corn, 319
Country Ham and Eggs, Romaine Salad with, 34
Grits, Waffled Gruyère & Ham, 61
Holiday Ham with Apricot Glaze, 298

Prosciutto
Flatbread, Roasted Fennel-and-Prosciutto, 273
Sandwiches, Prosciutto and Tomato, 306
Soup, Chicken-and-Prosciutto Tortelloni, 265

HONEY
Bread, Mini Pumpkin-Honey-Beer, 224
Bread Pudding with Apple Brandy-Caramel Sauce, Pumpkin-Honey-Beer, 224
Bread, Pumpkin-Honey-Beer, 224
Butter, Whipped Honey-Rosemary, 314
Chicken Salad with Grapes, Honey, Almonds, and Broccoli, Dixie, 227
Custard with Berries, Honey, 102
Filling, Cream Cheese-Honey, 223
Nectarine Tarts with Honey, 155
Peanut Brittle, Honey, 42
Pie, Honey-Balsamic-Blueberry, 337
Pork Tenderloin with Homemade Applesauce, Honey-Glazed, 51
Sauce, Blackberry-Honey Mustard, 102

HOPPIN' JOHN
Cakes with Tomato-Jalapeño Gravy, Hoppin' John, 280
Easy Hoppin' John, 331

HOSPITALITY RECIPES

Desserts
Cream, Sweet Mascarpone, 86
Pizza, Grilled Pineapple Dessert, 206
Shortcakes, Coconut-and-Pecan Strawberry, 86

French Bread, Cheesy Garlic, 265
Gravy, Mushroom, 86
Lemonade, Sparkling Elderflower, 86

Main Dishes
Leg of Lamb with Lemon-Herb Salt, Roasted, 84
Pizza, Smoked Sausage, Grilled Corn, and Sweet Onion, 207
Pizza, Ultimate Cheese, 207
Pizza with White Barbecue Sauce, Smoked Chicken, 207

Pizza Dough, Homemade, 207

Side Dishes
Cheese Grits Soufflé with Mushroom Gravy, 85
Roasted Beets, Carrots, and Sweet Onions, 85
Salad with Toasted Almonds, Shaved Vegetable, 85

Soups and Stews
Chicken-and-Prosciutto Tortelloni Soup, 265
Chicken Soup, King Ranch, 264
Pork Stew, Southwest, 264
Shrimp-and-New Potato Chowder, 264
White Bean, Sausage, and Turnip Green Stew, 265

HUSH PUPPIES
Andouille Hush Puppies, Crispy, 125

I

ICE CREAM
Blueberry-Lemon Zest Ice Cream, 201
Bourbon-Butter-Salted Pecan Ice Cream, 201
Buttermilk-Plum Ice Cream, 155
Coffee Liqueur Cookies-and-Cream Ice Cream, 201
Strawberry-Basil Ice Cream, 201

J

JAMBALAYA
Seafood Jambalaya, Creole, 240

JAMS
Bacon-and-Sweet Onion Jam, 102
Blackberry Chambord Jam, 220
Tomato Jam, 158

K

KABOBS
Beef Kabobs with Herbed Cucumber-and-Tomato Salad, Spiced, 243
Chicken Kabobs, Perfect, 133
Shrimp Kabobs, Grilled Teriyaki, 305
Tomato, Watermelon, and Feta Skewers with Mint and Lime, 157

KALE
Salad, Christmas, 283
Salad with Bacon Vinaigrette, Peas and Kale, 34
Salad with Hot Bacon Dressing, Kale, 331
Slaw with Buttermilk Dressing, Kale-and-Blueberry, 75
Wilted Kale, Skillet Steak and, 226

L

LAMB
Leg of Lamb with Lemon-Herb Salt, Roasted, 84
Rack of Lamb, Pistachio-Crusted, 325

LASAGNA
Pumpkin-and-Turnip Green Lasagna, 238

LEEKS
Dressing with Leeks and Mushrooms, Herbed Grits, 322
Fonduta, Leek, 202

LEMON

Beverages
Americano, Southern Lemon, 275
Lemonade, Minty, 74
Lemonade, Sparkling Elderflower, 86

Brussels Sprouts, Lemon and Shallot, 309

Desserts
Ice Cream, Blueberry-Lemon Zest, 201
Pie, Lemon-Buttermilk Icebox, 120
Pie with Baked Saltine Cracker Crust, Mango-Lemon-Buttermilk Icebox, 121
Pie with Baked Gingersnap Crust, Strawberry-Lemon-Buttermilk Icebox, 120
Topping, Lemon-Blueberry, 121

Fettuccine with Smoky Turnip Greens, Lemon, and Goat Cheese, 47
Muffins, Strawberry Lemonade, 335
Ranch, Lemony, 46
Salt, Roasted Leg of Lamb with Lemon-Herb, 84
Shrimp Boil with Green Olives and Lemon, 124
Shrimp Piccata, Creamy, 280
Veal Chops Milanese with Lemon and Herbs, 278
Vinaigrette, Lemon, 65
Vinaigrette, Lemon-Thyme, 125

LIME
Aïoli, Lime-Sriracha, 148
Butter, Grilled Corn and Tequila-Lime, 114
Granita, Raspberry Limeade, 219

Key Lime-Buttermilk Icebox Pie with Baked Buttery Cracker Crust, 120
Macerated Cherries, Lime-and-Mint, 153
Sauce, Lime-Wasabi, 196
Simple Syrup, Festive Fresh Fruit Salad with Mint-Lime, 297
Tomato, Watermelon, and Feta Skewers with Mint and Lime, 157
Vinaigrette, Lime, 212
LIVER
Chicken Liver Mousse Crostini with Pepper Jelly, 272
Dirty Rice Risotto, 281

M
MACADAMIA
Pound Cake, Coconut-Macadamia Nut, 289
MACARONI
Cheese
Barbecue Mac and Cheese, 225
Classic Baked Macaroni & Cheese, 83
Gnocchi Mac and Cheese, 282
Mexican Mac and Cheese, 225
Southern Pimiento Mac and Cheese, 225
Speedy Homemade Mac and Cheese, 225
Uncle Jack's Mac-and-Cheese, 29
MANGOES
Pie with Baked Saltine Cracker Crust, Mango-Lemon-Buttermilk Icebox, 121
Salsa, Crunchy Jerk Tacos with Watermelon-Mango, 135
Shakes, Mango-Buttermilk, 44
MARINADES
Buttermilk Tandoori Marinade, 134
Chardonnay-Herb Marinade, 134
Red Wine Marinade, 196
Sticky Sesame-Sorghum Marinade, 134
MARSHMALLOWS
Bourbon-Caramel Marshmallows, 290
Homemade Marshmallows, Our Best, 290
Peppermint Marshmallows, 290
MAYONNAISE
Cake, Chocolate-Mayonnaise, 57
Green Tomato Mayonnaise, 159
Pesto Mayo, 48
MEATBALLS
Sliders with Tomato Sauce, Meatball, 28
Tex-Mex Meatballs in Red Chile Sauce, 242
MEATLOAF
Turkey Meatloaves, Italian, 35
MELONS
Pickled Melon, 141
Watermelon
Punch, Pool Party Watermelon, 95
Salad, Watermelon, Arugula & Pecan, 82
Salsa, Crunchy Jerk Tacos with Watermelon-Mango, 135
Skewers with Mint and Lime, Tomato, Watermelon, and Feta, 157
MERINGUES
Cotton Candy Meringues, 58
Mocha-Hazelnut Dacquoise, 326
MICROWAVE
Desserts
Biscotti, Candy Cane, 330
Brownie-Mousse Stacks, Triple Chocolate, 288
Cakes, Mini Triple-Chocolate Buttermilk Pound, 44
Cake, Triple-Chocolate Buttermilk Pound, 44
Cake, White German Chocolate, 327
Cheesecake, Chocolate-Peppermint, 289
Cookies, Chocolate-Peppermint Crackle, 292
Cookies, Triple Chocolate Cranberry Oatmeal, 70
Ganache, Bourbon, 338
Pie, Mixed Stone Fruit, 153
Pie with Baked Gingersnap Crust, Strawberry-Lemon-Buttermilk Icebox, 120
Shortbread, Pecan-Cranberry, 292
Thumbprints, Nutty Chocolate, 57
Lemonade, Sparkling Elderflower, 86
Main Dishes
Chicken Tacos, Sesame, 66
Eggs in Fire Roasted Tomato Sauce, 307
Lasagna, Pumpkin-and-Turnip Green, 238
Shepherd's Pie with Potato Crust, 242
Rolls, Easy Butter, 252
Side Dishes
Butternut-Pecan Sauté, 320
Mashed Potatoes with Greens, 254
Soup, Loaded Potato, 37
Syrup, Apple Cider, 266
MOUSSE
Brownie-Mousse Stacks, Triple Chocolate, 288
Chicken Liver Mousse Crostini with Pepper Jelly, 272
Chocolate-Pecan Mousse Tart, 262
MUFFINS
Apple-Pecan Streusel Muffins, 297
Banana-Nut Muffins, 223
Cheese-Grit-and-Chive Muffins, 300
Cherry-Rosemary Muffins, 70
Coffee Cake Muffins with Brown Butter Icing, 314
Strawberry Lemonade Muffins, 335
MUSHROOMS
Beef Tenderloin, Tuscan, 325
Dressing with Leeks and Mushrooms, Herbed Grits, 322
Gravy, Mushroom, 86
Sauce, Chicken Cutlets with Herbed Mushroom, 50
Stew, Braised Beef-and-Mushroom, 318
Stroganoff, Mushroom, 226
MUSTARD
Sauce, Mustard-Cream, 279
Sauce, Pork Chops with Bourbon-Rosemary-Mustard, 331

N
NACHOS
Pulled Pork Barbecue Nachos, 112
NECTARINES
Ice Pops, Nectarine-Chile, 155
Tarts with Honey, Nectarine, 155
THE NEW SUNDAY SUPPER RECIPES
Main Dishes
Flank Steak with Lime-Wasabi Sauce, Seared, 196
Meatloaves, Italian Turkey, 35
Patties, Lady Pea-and-Corn, 217
Rotini with Crumbled Turkey and Tomato Sauce, 36
Salad with Steak, Chopped, 196
Sandwiches, Southern Cheese Steak, 196
Stuffed Peppers, Turkey-, 36
Marinade, Red Wine, 196
Sauce, Lime-Wasabi, 196
Side Dishes
Lady Peas, 217
Sautéed Sprouts, 35
Zucchini Stuffed with Lady Peas, 217
Soup with Bacon, Buttermilk-Lady Pea, 217

O
OATS
Cookies, Triple Chocolate Cranberry Oatmeal, 70
OKRA
Grilled Okra, Chile-Rubbed Chops with Sweet Potatoes and, 160
Hoppin' John and Limpin' Susan, 38
Salad, Roasted Potato-and-Okra, 124
OLIVES
Pesto, Collard-and-Olive, 48
Shrimp Boil with Green Olives and Lemon, 124
ONIONS
Caramelized Onion and Swiss Popovers, 334
Caramelized Sweet Onions, 334
Green Onion Soup, Pea-and-, 51
Jam, Bacon-and-Sweet Onion, 102
Pizza, Smoked Sausage, Grilled Corn, and Sweet Onion, 207
Roasted Beets, Carrots, and Sweet Onions, 85
Tart, Cheesy Bacon-and-Two-Onion, 51
ORANGE
Gimlets, Orange-Infused Rum, 149

Glaze, Classic Roasted Duck with Orange-Bourbon-Molasses, 278
Glaze, Orange, 335
Granita, Orange Cream, 219
Macerated Cherries, Orange-and-Basil, 153
Salad, Citrus Chicken and Beet, 305
Salad, Grilled Salmon with Orange-Fennel, 305

ORZO
Salad with Spicy Buttermilk Dressing, Orzo, 200

OYSTERS
Broiled Oysters, 26
Grilled Oysters, 26
Pot Pie, Oyster-Bacon, 25
Roast, An Easy Lowcountry Oyster, 25
Stew, Classic Oyster, 24

P

PAELLA
Shrimp and Chorizo Paella, Quick Cook, 113

PANCAKES
Apple-Spice Pancakes, 266
Chocolate Chip-Toasted Pecan Pancakes, 266
Original "Pam-Cakes," The, 266
Peach Pancakes, 308

PARSNIPS
Soup, Parsnip-Potato, 252

PASTA. *SEE ALSO* COUSCOUS; FETTUCCINE; GNOCCHI; LASAGNA; MACARONI; ORZO; SPAGHETTI.
Agnolotti, Sage-Brown Butter Squash, 307
Bucatini, Ham, and Asparagus, 89
Farfalle, Diablo, 211
Farfalle, Shrimp and Peas with, 89
Heirloom Tomatoes, Goat Cheese, and Basil, Pasta with, 159
Pappardelle Pasta, Slow-Cooker Bolognese Sauce over, 247
Penne, Chicken-and-Artichoke, 88
Penne, Vodka-Cream Sauce with, 211
Rotini with Crumbled Turkey and Tomato Sauce, 36
Salad with Lime Vinaigrette, Summer Pasta, 212
Shells with Spring Vegetables, Pasta, 89
Tortelloni Soup, Chicken-and-Prosciutto, 265
Ziti, Greek Baked, 242

PEACHES
Cobbler, Ginger-Peach Shortbread, 153
Pancakes, Peach, 308
Pickled Peaches, 141
Pie, Peach Divinity Icebox, 121
Pizza, Grilled Tomato-Peach, 307
Salad, Grilled Peach & Avocado, 82
Salad with Chicken, Tomato-Peach, 137
Salsa, Crunchy Summer, 199
Sautéed Peaches, Skillet Pork Chops with, 142
Shortcakes with Sweet Cream, Boozy Peach, 125

PEANUT BUTTER
Chicken, Peanutty Braised, 97
Sammies, PB&J, 58
Sauce, Peanut, 63
Stew, Peanut Chicken, 41
Thumbprints, Nutty Chocolate, 57

PEANUTS
Brittle, Honey Peanut, 42
Sauce, Vietnamese Peanut, 133
Snack Mix, Caramel-Peanut-Popcorn, 245
Whiskey and Cola, Peanut, 275

PEARS
Tart, Pear-and-Pumpkin, 237

PEAS
Black-eyed
Hoppin' John and Limpin' Susan, 38
Hoppin' John Cakes with Tomato-Jalapeño Gravy, 280
Hoppin' John, Easy, 331
Farfalle, Shrimp and Peas with, 89
Lady Pea-and-Corn Patties, 217
Lady Peas, 217
Lady Pea Soup with Bacon, Buttermilk-, 217
Lady Peas, Zucchini Stuffed with, 217
Salad with Bacon Vinaigrette, Peas and Kale, 34
Soup, Pea-and-Green Onion, 51
Soup with Mint and Cream, Chilled Sweet Pea, 100

PECANS. *SEE ALSO* PRALINE.
Bacon, Pecan-Rosemary, 62
Breads
Cornbread, Pecan, 332
Loaves, Mini Banana-Cranberry-Nut Bread, 335
Loaves, Regular-Size Banana-Cranberry-Nut Bread, 335
Muffins, Apple-Pecan Streusel, 297
Muffins, Banana-Nut, 223
Pancakes, Chocolate Chip-Toasted Pecan, 266
Popovers, Blue Cheese-Pecan, 313
Rolls, Christmas Morning Cinnamon, 296
Broccolini with Pecans and Cane Syrup Vinaigrette, 281
Desserts
Bars, Pecan Pie, 262
Cake, Pecan Pound, 263
Cookies, Maple-Praline, 329
Cream Puffs with Praline Sauce and Toasted Pecans, Decadent, 263
Frosting, Coconut-Pecan, 327
Ice Cream, Bourbon-Butter-Salted Pecan, 201
Pie, Pumpkin-Pecan Streusel, 261
Shortbread, Pecan-Cranberry, 292
Shortcakes, Coconut-and-Pecan Strawberry, 86
Tart, Chocolate-Pecan Mousse, 262
Tart, Salted Caramel-Chocolate Pecan, 339
Toffee, Pecan-Espresso, 262
Topping, Pecan Streusel, 261
Grapes, Blue Cheese-Pecan, 54
Salad, Watermelon, Arugula & Pecan, 82
Sauté, Butternut-Pecan, 320
Soup with Brown Butter, Pecans, and Yogurt, Carrot, 202
Spicy-Sweet Pecans, 263
Stuffed Peppers with Chèvre, Pecans, and Golden Raisins, 94
Stuffing, Pecan-and-Wild Rice, 323

PEPPERMINT
Bars, Peppermint Divinity, 340
Biscotti, Candy Cane, 330
Cakes, Peppermint Red Velvet Bundt, 326
Cheesecake, Chocolate-Peppermint, 289
Cookies, Chocolate-Peppermint Crackle, 292
Cream, Whipped Peppermint, 289
Marshmallows, Peppermint, 290

PEPPERS
Burritos, Baked Beef and Fried Pepper, 308
Chile Ice Pops, Nectarine-, 155
Chipotle Cream, 31
Chowchow, Three-Pepper, 108
Jalapeño
Biscuits, Jalapeño, 248
Cheese Straws, Jalapeño, 100
Gravy, Hoppin' John Cakes with Tomato-Jalapeño, 280
Grits, Waffled Jalapeño & Monterey Jack Cheese, 61
Red Pepper Relish, 319
Red Pepper Soup, Tomato-and-, 227
Salsa Verde, Italian-Style, 101
Slaw, Napa Cabbage-and-Sweet Pepper, 108
Stuffed Peppers, Turkey-, 36
Stuffed Peppers with Chèvre, Pecans, and Golden Raisins, 94

PESTO
Collard-and-Olive Pesto, 48
Egg Salad, Pesto, 48
Mashers, Pesto, 48
Mayo, Pesto, 48
Sage Pesto, Roasted Brussels Sprouts with, 321
Spinach-and-Three-Herb Pesto, 73

PICKLES
Fruit Pickle Brine, 141
Melon, Pickled, 141
Old Timey Pickles, 220
Peaches, Pickled, 141
Strawberries, Pickled, 141

PIES, PUFFS, AND PASTRIES
Chocolate Pie, The Ultimate, 56

Cobbler, Ginger-Peach Shortbread, 153
Cream Puffs with Praline Sauce and Toasted Pecans, Decadent, 263
Honey-Balsamic-Blueberry Pie, 337
Icebox
Key Lime-Buttermilk Icebox Pie with Baked Buttery Cracker Crust, 120
Lemon-Buttermilk Icebox Pie, 120
Mango-Lemon-Buttermilk Icebox Pie with Baked Saltine Cracker Crust, 121
Peach Divinity Icebox Pie, 121
Strawberry-Lemon-Buttermilk Icebox Pie with Baked Gingersnap Crust, 120
Main Dish
Oyster-Bacon Pot Pie, 25
Shepherd's Pie with Potato Crust, 242
Shrimp 'n' Grits Pie, 61
Mixed Stone Fruit Pie, 153
Pastries and Crusts
Cornucopias, Turkey-Artichoke, 316
Cream Cheese Pastries, 298
Crumb Crust, 118
Layered Eggnog Cream with Puff Pastry, 290
Palmiers, Fig and Bacon, 310
Simple Piecrust, 339
Pumpkin-Pecan Streusel Pie, 261
Pumpkin Pie Ever, Our Easiest, 237
Tarts
Bacon-and-Two-Onion Tart, Cheesy, 51
Chocolate Chess Tarts, Mini, 296
Chocolate-Pecan Mousse Tart, 262
Cinnamon Apple Tarts, 309
Nectarine Tarts with Honey, 155
Pear-and-Pumpkin Tart, 237
Salted Caramel-Chocolate Pecan Tart, 339
PINEAPPLE
Dessert Pizza, Grilled Pineapple, 206
PISTACHIO
Rack of Lamb, Pistachio-Crusted, 325
PIZZA. *SEE ALSO* FLATBREAD.
BBQ Chicken Pizzas, Quick, 110
Cheese Pizza, Ultimate, 207
Chicken, Artichoke, and Sun-Dried Tomato Pizza, 306
Dough, Homemade Pizza, 207
Grilled Pizza, 208
Philly Cheesesteak Pizza, 306
Smoked Chicken Pizza with White Barbecue Sauce, 207
Tomato-Peach Pizza, Grilled, 307
PLUMS
Ice Cream, Buttermilk-Plum, 155
Sauce, Cherry-Plum, 279
POPCORN
Smoky Barbecue Popcorn, 245
Snack Mix, Caramel-Peanut-Popcorn, 245
PORK. *SEE ALSO* BACON; HAM; SAUSAGE.
Chops
Bourbon-Rosemary-Mustard Sauce, Pork Chops with, 331
Broiled Pork Chops with Basil Butter and Summer Squash, 194
Chile-Rubbed Chops with Sweet Potatoes and Grilled Okra, 160
Fried Apple Pork Chops, 308
Sandwiches with Gravy and Grits, Pork Chop, 194
Shake-and-Bake Pork Chops with Mustard Sauce, Homemade, 193
Skillet Pork Chops with Sautéed Peaches, 142
Grillades, 232
Mac and Cheese, Barbecue, 225
Meatball Sliders with Tomato Sauce, 28
Pulled Pork and Slaw, Petite Sweet Potato Biscuits with, 274
Pulled Pork Barbecue Nachos, 112
Ribs
Chili, Pork-and-Black Bean, 31
Easy BBQ Ribs, 109
Roasts
Crown Pork Roast with Fig-Fennel Stuffing, 323
Loin Steaks with Cherry-Plum Sauce, Grilled Pork, 278
Roasted Cider-Brined Pork Loin with Green Tomato Chutney, 53
Stew, Mexican, 55
Stew, Southwest Pork, 264
Tenderloin
Honey-Glazed Pork Tenderloin with Homemade Applesauce, 51
Sliders, Pork Tenderloin, 101
Stir-Fry, Smoky Pork, 229
POTATOES. *SEE ALSO* SWEET POTATOES.
Chowder, Shrimp-and-New Potato, 264
Chowder, Summer Corn-and-Golden Potato, 213
Crust, Shepherd's Pie with Potato, 242
Gratin with Bacon and Comté, Potato, 320
Mashed
Goat Cheese Mashed Potatoes, 65
Greens, Mashed Potatoes with, 254
Pesto Mashers, 48
Roasted Garlic Duchess Potatoes, 282
Salad, Garden Potato, 74
Salad, Roasted Potato-and-Okra, 124
Soup, Loaded Potato, 37
Soup, Parsnip-Potato, 252
Spanish Tortilla, Spicy Andouille, 95
Twice-Baked Green Potatoes, 48
PRALINE
Cookies, Maple-Praline, 329
Cream-Beignet Tower, Praline, 286
Frosting, Maple-Praline, 329
Frosting, Praline, 268
Sauce and Toasted Pecans, Decadent Cream Puffs with Praline, 263
Sauce, Praline, 263
PUDDINGS
Bread Pudding with Apple Brandy-Caramel Sauce, Pumpkin-Honey-Beer, 224
Chocolate-Buttermilk Pudding, 45
Yorkshire Puddings, Fontina-Chive, 283
PUMPKIN
Beer-Battered Pumpkin with Dipping Sauce, 238
Bread, Mini Pumpkin-Honey-Beer, 224
Bread Pudding with Apple Brandy-Caramel Sauce, Pumpkin-Honey-Beer, 224
Bread, Pumpkin-Honey-Beer, 224
Brownies, Pumpkin-Chocolate, 237
Cupcakes with Browned Butter Frosting, Pumpkin, 336
Lasagna, Pumpkin-and-Turnip Green, 238
Pie Ever, Our Easiest Pumpkin, 237
Pie, Pumpkin-Pecan Streusel, 261
Soup with Avocado Cream, Spicy Pumpkin, 238
Soup with Red Pepper Relish, Pumpkin, 319
Tart, Pear-and-Pumpkin, 237
Tiramisù, Pumpkin-Espresso, 256

Q

QUICK-FIX SUPPERS RECIPES
Main Dishes
Bacon-and-Two-Onion Tart, Cheesy, 51
Bucatini, Ham, and Asparagus, 89
Chicken-and-Artichoke Penne, 88
Chicken Cutlets with Herbed Mushroom Sauce, 50
Chicken Paillard with Citrus Salad and Couscous, 33
Chimichangas, Chicken-and-Black Bean, 226
Chops with Sweet Potatoes and Grilled Okra, Chile-Rubbed, 160
Eggplant, Tomato, and Provolone Stacks with Basil, Crispy, 215
Grouper with Balsamic Brown Butter Sauce, Pan-Seared, 105
Mac and Cheese, Southern Pimiento, 225
Mac and Cheese, Speedy Homemade, 225
Mushroom Stroganoff, 226
Pasta Shells with Spring Vegetables, 89
Pork Chops with Basil Butter and Summer Squash, Broiled, 194
Pork Chops with Mustard Sauce, Homemade Shake-and-Bake, 193
Pork Tenderloin with Homemade Applesauce, Honey-Glazed, 51
Shrimp and Peas with Farfalle, 89
Shrimp with Romesco Sauce, Roasted Gulf, 105
Skillet Steak and Wilted Kale, 226
Snapper Baked in Parchment with Spring Vegetables, 104

Stir-Fry, Garlicky Beef-and-Bean, 50
Triggerfish, Grilled, 105
Salads and Salad Dressings
Chicken Salad with Grapes, Honey, Almonds, and Broccoli, Dixie, 227
Lime Vinaigrette, 212
Pasta Salad with Lime Vinaigrette, Summer, 212
Peas and Kale Salad with Bacon Vinaigrette, 34
Romaine Salad with Country Ham and Eggs, 34
Shrimp and Pesto-Rice Salad, 33
Steak Salad with Spicy Dressing, Charred, 34
Sandwiches
Grilled Cheese, The Ultimate, 227
Open-Faced Shrimp-and-Avocado Sandwiches, 214
Pork Chop Sandwiches with Gravy and Grits, 194
Side Dishes
Brussels Sprouts and Sautéed Cabbage, 51
Green Beans, Easy Summer, 193
Soups
Corn-and-Golden Potato Chowder, Summer, 213
Pea-and-Green Onion Soup, 51
Tomato-and-Red Pepper Soup, 227

R

RADISHES
Smoked Trout Crostini with Radishes and Dill Cream, 273
RASPBERRIES
Granita, Raspberry Limeade, 219
"Rhubars," Raspberry, 111
Tiramisù, Belle Chèvre Raspberry, 97
RED VELVET
Thumbprints, Red Velvet, 291
RELISHES. *SEE ALSO* CHOWCHOW; CHUTNEY; PESTO.
Corn Relish, Sweet, 107
Green Tomato Relish, 108
Red Pepper Relish, 319
Strawberry-Blueberry Relish, 107
RHUBARB
Raspberry "Rhubars," 111
RICE
Creole Rice, 211
Hoppin' John and Limpin' Susan, 38
Hoppin' John Cakes with Tomato-Jalapeño Gravy, 280
Hoppin' John, Easy, 331
Risotto, Dirty Rice, 281
Salad, Shrimp and Pesto-Rice, 33
Wild Rice Stuffing, Pecan-and-, 323
ROLLS
Biscuit Cinnamon Sweet Rolls, 224
Yeast
Cinnamon Rolls, Christmas Morning, 296
Easy Butter Rolls, 252

S

SALAD DRESSINGS
Avocado Ranch, 46
Blue Cheese Ranch, 46
Herbed Buttermilk Ranch Dressing, 46
Lemony Ranch, 46
Vinaigrette
Apple Cider Vinaigrette, 255
Lemon-Thyme Vinaigrette, 125
Lemon Vinaigrette, 65
Lime Vinaigrette, 212
Pepper Jelly Vinaigrette, 82
Rosemary-Citrus Vinaigrette, 320
SALADS
Chicken
Citrus Chicken and Beet Salad, 305
Dixie Chicken Salad with Grapes, Honey, Almonds, and Broccoli, 227
Farro, and Vegetable Salad with Lemon Vinaigrette, Chicken, 64
Tomato Cups, Chicken Salad, 310
Chopped Salad with Steak, 196
Christmas Salad, 283
Citrus Salad and Couscous, Chicken Paillard with, 33
Cornbread-Vegetable Salad, 306
Crabmeat Salad, Lump, 157
Cucumber-and-Tomato Salad, Herbed, 243
Egg Salad, Pesto, 48
Fruit Salad with Mint-Lime Simple Syrup, Festive Fresh, 297
Kale Salad with Hot Bacon Dressing, 331
Mustard Greens with Yogurt-Parmesan Dressing and Bacon Croutons, 48
Orange-Fennel Salad, Grilled Salmon with, 305
Orzo Salad with Spicy Buttermilk Dressing, 200
Pasta Salad with Lime Vinaigrette, Summer, 212
Peach & Avocado Salad, Grilled, 82
Peas and Kale Salad with Bacon Vinaigrette, 34
Potato-and-Okra Salad, Roasted, 124
Potato Salad, Garden, 74
Romaine Salad with Country Ham and Eggs, 34
Shrimp and Pesto-Rice Salad, 33
Slaws
Ancho Slaw, Creamy, 139
BLT Buttermilk Blue Slaw, 97
Kale-and-Blueberry Slaw with Buttermilk Dressing, 75
Memphis Slaw, 83
Napa Cabbage-and-Sweet Pepper Slaw, 108
Steak Salad with Spicy Dressing, Charred, 34
Tomato-Peach Salad with Chicken, 137
Tomato, Watermelon, and Feta Skewers with Mint and Lime, 157
Vegetable Salad with Apple Cider Vinaigrette, Roasted, 255
Vegetable Salad with Toasted Almonds, Shaved, 85
Watermelon, Arugula & Pecan Salad, 82
Winter Citrus-Avocado Salad, 320
SALMON
Croquettes with Dill Sauce, Salmon, 41
Grilled Salmon with Orange-Fennel Salad, 305
SALSAS. *SEE ALSO* PESTO; RELISHES; SAUCES; TOPPINGS.
Corn Salsa, Summer Corn Soup with, 96
Crunchy Summer Salsa, 199
Verde, Italian-Style Salsa, 101
Watermelon-Mango Salsa, Crunchy Jerk Tacos with, 135
SANDWICHES
Burgers with Creamy Ancho Slaw, Carolina Chicken, 139
Burgers with Green Tomato Mayonnaise, 159
Cheese Steak Sandwiches, Southern, 196
Cowboy Brisket Sandwich, 81
Fried Egg Sandwich with BBQ Bacon, The Ultimate, 61
Grilled Cheese, The Ultimate, 227
Grilled Chicken Caprese, 304
Open-Faced Shrimp-and-Avocado Sandwiches, 214
PB&J Sammies, 58
Pork Chop Sandwiches with Gravy and Grits, 194
Prosciutto and Tomato Sandwiches, 306
Sliders, Pork Tenderloin, 101
Sliders with Tomato Sauce, Meatball, 28
Tea Sandwiches, Pumpernickel, 311
Tea Sandwiches, Smoked Trout and Watercress, 310
SAUCES. *SEE ALSO* DESSERTS/ SAUCE; GLAZES; GRAVIES; PESTO; RELISHES; SALSAS; TOPPINGS.
Alabama White Sauce, 133
Balsamic Brown Butter Sauce, Pan-Seared Grouper with, 105
Barbecue Sauce, El Sancho, 80
Barbecue Sauce, Fancy, 109
Barbecue Sauce, Sweet-and-Tangy, 324
Blackberry-Honey Mustard Sauce, 102
Bolognese Sauce over Pappardelle Pasta, Slow-Cooker, 247
Bourbon-Rosemary-Mustard Sauce, Pork Chops with, 331
Buttery Nashville Hot Sauce, 133
Carolina Burger Sauce, 139
Cherry-Plum Sauce, 279
Cocktail Sauce, Classic, 124

Cocktail Sauce, White, 124
Crumbled Turkey and Tomato Sauce, Rotini with, 36
Dill Sauce, Salmon Croquettes with, 41
Dipping Sauce, Beer-Battered Pumpkin with, 238
Fire Roasted Tomato Sauce, Eggs in, 307
Herbed Mushroom Sauce, Chicken Cutlets with, 50
Lime-Wasabi Sauce, 196
Mignonette, Green Apple, 26
Mustard-Cream Sauce, 279
Mustard Sauce, Homemade Shake-and-Bake Pork Chops with, 193
Parmesan Cream Sauce, Brussels Sprouts with, 283
Peanut Sauce, 63
Peanut Sauce, Vietnamese, 133
Red Chile Sauce, Tex-Mex Meatballs in, 242
Romesco Sauce, Roasted Gulf Shrimp with, 105
Tomato Sauce, Meatball Sliders with, 28
Tomato Sauce, Slow-Cooker, 210
Vanilla-Rum Sauce, Bananas Foster Coffee Cake with, 338
Vodka-Cream Sauce with Penne, 211
White Barbecue Sauce, Smoked Chicken Pizza with, 207
Yogurt-Tahini Sauce, 243

SAUSAGE
Andouille Hush Puppies, Crispy, 125
Andouille Spanish Tortilla, Spicy, 95
Casserole, Smoky Sausage-and-Grits, 295
Chorizo Paella, Quick Cook Shrimp and, 113
Dressing, Tart Cherry-Chestnut Sausage, 322
Jambalaya, Creole Seafood, 240
Mac and Cheese, Mexican, 225
Smoked Sausage, Grilled Corn, and Sweet Onion Pizza, 207
Soup with Avocado Cream, Spicy Pumpkin, 238
Stew, Sausage-and-Bean, 69
Stew, White Bean, Sausage, and Turnip Green, 265

SAVE ROOM RECIPES
Cake, Caramel Apple, 231
Cake Chocolate-Mayonnaise, 57
Cake, Coconut Cream, 93
Cake, The Ultimate Carrot, 68
Chocolate Whipped Cream, 56
Frosting, Brown Sugar-Cream Cheese, 68
Frosting, Chocolate-Cream Cheese, 57
Frosting, Fluffy Coconut, 93
Granita, Chocolate Milk, 219
Granita, Cucumber-Basil, 219
Granita, Orange Cream, 219
Granita, Raspberry Limeade, 219
Ice Cream, Blueberry-Lemon Zest, 201
Ice Cream, Bourbon-Butter-Salted Pecan, 201
Ice Cream, Coffee Liqueur Cookies-and-Cream, 201
Ice Cream, Strawberry-Basil, 201
Pie, The Ultimate Chocolate, 56
Raspberry "Rhubars," 111

SEAFOOD. *SEE ALSO* CRAB; FISH; OYSTERS; SALMON; SHRIMP.
Gumbo, Best Ever Seafood, 260
Jambalaya, Creole Seafood, 240

SEASONINGS
Beef Rub, 80
Garlic with Confit Oil, Thyme, and Black Pepper, 144
Herb Brine, Grilled Cornish Hens with, 279
Lemon-Herb Salt, Roasted Leg of Lamb with, 84
Rib Rub, 109
Shrimp Spice Mix, 124
Turkey Dry Rub, 256

SHRIMP
Barbecue Shrimp, Easy, 248
Boil with Green Olives and Lemon, Shrimp, 124
Chowder, Shrimp-and-New Potato, 264
Dip with Crispy Wonton Chips, Creamy Shrimp, 272
Farfalle, Shrimp and Peas with, 89
Jambalaya, Creole Seafood, 240
Paella, Quick Cook Shrimp and Chorizo, 113
Piccata, Creamy Shrimp, 280
Pie, Shrimp 'n' Grits, 61
Roasted Gulf Shrimp with Romesco Sauce, 105
Salad, Shrimp and Pesto-Rice, 33
Sandwiches, Open-Faced Shrimp-and-Avocado, 214
Spice Mix, Shrimp, 124

SLOW COOKER
Chili and Stew
Beef-and-Bean Chili, 30
Pork-and-Black Bean Chili, 31
Smoky Turkey-and-Sweet Potato Chili, 31
Mexican Stew, 55
Grits, Overnight Slow-Cooker, 60
Mac-and-Cheese, Uncle Jack's, 29
Main Dishes
Bolognese Sauce over Pappardelle Pasta, Slow-Cooker, 247
Brisket Tacos with Summer Salsa, 199
Chicken Cacciatore with Spaghetti, Slow-Cooker, 30
Chicken Lettuce Cups, Slow-Cooker, 92
Pot Roast, Beer-Braised, 29
Ribs, Easy BBQ, 109
Sliders with Tomato Sauce, Meatball, 28
Sauce, Slow-Cooker, Tomato, 210
Side Dishes
Stewed Green Beans and Tomatoes, 143
Sweet Potatoes with Bacon, Slow-Cooker, 257

THE SLOW DOWN RECIPES
Creole Rice, 211
Main Dishes
Bolognese Sauce over Pappardelle Pasta, Slow-Cooker, 247
Brisket Tacos with Summer Salsa, 199
Chicken Lettuce Cups, Slow-Cooker, 92
Farfalle, Diablo, 211
Penne, Vodka-Cream Sauce with, 211
Ribs, Easy BBQ, 109
Stew, Mexican, 55
Stew, Sausage-and-Bean, 69
Rib Rub, 109
Salsa, Crunchy Summer, 199
Sauces
Barbecue Sauce, Fancy, 109
Tomato Sauce, Slow-Cooker, 210
Stewed Green Beans and Tomatoes, 143

THE *SL* TEST KITCHEN ACADEMY RECIPES
Candied Carrot Curls, 67
Cooking Perfect Pasta, 87
Fruit Pickle Brine, 141
Grilled Pizza, 208
Pickled Melon, 141
Pickled Peaches, 141
Pickled Strawberries, 141
Spectacular Fresh Butter & Buttermilk, 52

SNACKS
Chips, Crispy Wonton, 272
Chips, Sweet Potato, 83
Mix, Caramel-Peanut-Popcorn Snack, 245
Pecans, Spicy-Sweet, 263
Popcorn, Smoky Barbecue, 245
Pumpkin with Dipping Sauce, Beer-Battered, 238

SORGHUM
Snowflakes, Spiced Sorghum, 328
Marinade, Sticky Sesame-Sorghum, 134
Butter, Sorghum, 252

SOUFFLÉ
Cheese Grits Soufflé with Mushroom Gravy, 85

SOUPS. *SEE ALSO* CHILI; CHOWDERS; CURRY; GUMBO; JAMBALAYA; STEWS.
Broth, Sweet Potato, 42
Buttermilk-Lady Pea Soup with Bacon, 217
Carrot Soup with Brown Butter, Pecans, and Yogurt, 202
Chicken-and-Prosciutto Tortelloni Soup, 265
Chicken Soup, King Ranch, 264
Chickpea Soup, African, 42
Corn Soup with Corn Salsa, Summer, 96
Green Tomato Soup with Lump Crabmeat, 157
Parsnip-Potato Soup, 252
Pea-and-Green Onion Soup, 51

Pea Soup with Mint and Cream, Chilled Sweet, 100
Potato Soup, Loaded, 37
Pumpkin Soup with Avocado Cream, Spicy, 238
Pumpkin Soup with Red Pepper Relish, 319
Stock, Simple Chicken, 251
Tomato-and-Red Pepper Soup, 227

SOUTHERN CLASSIC RECIPES
Classic Succotash, 197
Pimiento Cheese, Classic, 230
Simple Syrup, 209
Classic Sweet Tea, 209

SOUTHERNERS RECIPE
Creole Seafood Jambalaya, 240

***SOUTHERN LIVING* RECIPE REVIVAL RECIPE**
Blue Cheese-Pecan Grapes, 54

SPAHGETTI
Chicken Cacciatore with Spaghetti, Slow-Cooker, 30

SPINACH
Beef Tenderloin, Tuscan, 325
Pesto, Spinach-and-Three-Herb, 73
Scones, Spinach-Feta, 223

SPREADS. *SEE ALSO* AÏOLI; BUTTER; JAM; MAYONNAISE.
Faux Clotted Cream, 312
Pimiento Cheese, Classic, 230

SQUASH. *SEE ALSO* ZUCCHINI.
Agnolotti, Sage-Brown Butter Squash, 307
Butternut-Pecan Sauté, 320
Summer Squash, Broiled Pork Chops with Basil Butter and, 194

SRIRACHA
Aïoli, Lime-Sriracha, 148

STEWS. *SEE ALSO* CHILI; CHOWDERS; CURRY; GUMBO; JAMBALAYA; SOUPS.
Beef-and-Mushroom Stew, Braised, 318
Mexican Stew, 55
Oyster Stew, Classic, 24
Peanut Chicken Stew, 41
Sausage-and-Bean Stew, 69
White Bean, Sausage, and Turnip Green Stew, 265

STRAWBERRIES
Ice Cream, Strawberry-Basil, 201
Milk, Homemade Strawberry, 94
Muffins, Strawberry Lemonade, 335
Pickled Strawberries, 141
Pie with Baked Gingersnap Crust, Strawberry-Lemon-Buttermilk Icebox, 120
Relish, Strawberry-Blueberry, 107
Shortcakes, Coconut-and-Pecan Strawberry, 86

STROGANOFF
Mushroom Stroganoff, 226

STUFFINGS
Fig-Fennel Stuffing, 323
Pecan-and-Wild Rice Stuffing, 323

SUCCOTASH
Classic Succotash, 197
Farmers' Market Succotash, Nashville, 115

SWEET POTATOES
Biscuits, Sweet Potato, 275, 335
Biscuits with Pulled Pork and Slaw, Petite Sweet Potato, 274
Broth, Sweet Potato, 42
Chili, Smoky Turkey-and-Sweet Potato, 31
Chips, Sweet Potato, 83
Chops with Sweet Potatoes and Grilled Okra, Chile-Rubbed, 160
Slow-Cooker Sweet Potatoes with Bacon, 257

SYRUPS
Apple Cider Syrup, 266
Cane Syrup Vinaigrette, 281
Simple Syrup, 209, 275
Simple Syrup, Festive Fresh Fruit Salad with Mint-Lime, 297
Simple Syrup, Spicy, 275

T

TACOS
Brisket Tacos with Summer Salsa, 199
Chicken Tacos, Sesame, 66
Crunchy Jerk Tacos with Watermelon-Mango Salsa, 135

TEA
Cranberry Tea-Ini, 317
Hot Apple Tea, 310

TERIYAKI
Shrimp Kabobs, Grilled Teriyaki, 305

TOMATOES
Balsamic Green Beans with Tomatoes and Feta, 309
Gravy, Hoppin' John Cakes with Tomato-Jalapeño, 280
Green Tomato Chutney, Roasted Cider-Brined Pork Loin with, 53
Green Tomato Mayonnaise, 159
Green Tomato Relish, 108
Heirloom Tomatoes, Goat Cheese, and Basil, Pasta with, 159
Jam, Tomato, 158
Pizza, Grilled Tomato-Peach, 307

Salads
Cups, Chicken Salad Tomato, 310
Herbed Cucumber-and-Tomato Salad, 243
Peach Salad with Chicken, Tomato-, 137
Slaw, BLT Buttermilk Blue, 97
Watermelon, and Feta Skewers with Mint and Lime, Tomato, 157

Sandwiches, Prosciutto and Tomato, 306

Sauces
Crumbled Turkey and Tomato Sauce, Rotini with, 36
Fire Roasted Tomato Sauce, Eggs in, 307
Slow-Cooker Tomato Sauce, 210
Tomato Sauce, Meatball Sliders with, 28

Soup, Tomato-and-Red Pepper, 227
Soup with Lump Crabmeat, Green Tomato, 157
Stacks with Basil, Crispy Eggplant, Tomato, and Provolone, 215
Stewed Green Beans and Tomatoes, 143
Sun-Dried Tomato Pizza, Chicken, Artichoke, and, 306

TOPPINGS. *SEE ALSO* FROSTINGS; GLAZES; GRAVIES; RELISHES; SALSAS; SAUCES.

Savory
Chipotle Cream, 31
Dill Cream, Smoked Trout Crostini with Radishes and, 273
Herbed Breadcrumbs, 280
Herb Vinaigrette, 159

Sweet
Bourbon Ganache, 338
Cane Syrup Vinaigrette, 281
Chocolate Whipped Cream, 56
Lemon-Blueberry Topping, 121
Lime-and-Mint Macerated Cherries, 153
Orange-and-Basil Macerated Cherries, 153
Pecan Streusel Topping, 261
Sweetened Whipped Cream, 120
Sweet Mascarpone Cream, 86
Whipped Peppermint Cream, 289

TORTILLA
Spanish Tortilla, Spicy Andouille, 95

TOSTADAS
Beef Brisket Tostadas, Smoked, 81

TURKEY
Butterflied and Barbecued Turkey, 324
Chili, Smoky Turkey-and-Sweet Potato, 31
Cornucopias, Turkey-Artichoke, 316
Dry-Brined-and-Marinated Smoked Turkey, 256
Dry Rub, Turkey, 256
Hot Brown Bites, Inside-Out, 272
Meatloaves, Italian Turkey, 35
Sauce, Rotini with Crumbled Turkey and Tomato, 36
Stuffed Peppers, Turkey-, 36

V

VANILLA
Glaze, Easy Vanilla, 333
Glaze, Vanilla, 313
Sauce, Bananas Foster Coffee Cake with Vanilla-Rum, 338

VEAL
Chops Milanese with Lemon and Herbs, Veal, 278

VEGETABLES. *SEE ALSO* SPECIFIC TYPES.
Curry, Buttermilk-Vegetable, 46

Pasta Shells with Spring Vegetables, 89
Salads
Chicken, Farro, and Vegetable Salad with Lemon Vinaigrette, 64
Cornbread-Vegetable Salad, 306
Roasted Vegetable Salad with Apple Cider Vinaigrette, 255
Shaved Vegetable Salad with Toasted Almonds, 85
Snapper Baked in Parchment with Spring Vegetables, 104
VENISON
Chili, Game Day Venison, 113

W

WAFFLES
Bacon and Cheddar Belgian Waffles, 308
Buttermilk Waffles, Fluffy, 45
WASABI
Sauce, Lime-Wasabi, 196
WHAT TO COOK NOW RECIPES
Breads
"Pam-Cakes," The Original, 266
Pancakes, Apple-Spice, 266
Pancakes, Chocolate Chip-Toasted Pecan, 266
Chowchow, Three-Pepper, 108
Main Dishes
Baked Ziti, Greek, 242
Beef Kabobs with Herbed Cucumber-and-Tomato Salad, Spiced, 243
Chicken Thighs, Citrus-Braised, 64
Chicken Thighs, Garlic-Yogurt-Marinated, 65
Chicken Thighs with Herb Butter, Roasted, 65
Meatballs in Red Chile Sauce, Tex-Mex, 242
Sesame-Chicken Thigh Paillard with Peanut Sauce, 63
Shepherd's Pie with Potato Crust, 242
Mashed Potatoes, Goat Cheese, 65
Relish, Green Tomato, 108
Relish, Strawberry-Blueberry, 107
Relish, Sweet Corn, 107
Salads and Salad Dressings
Chicken, Farro, and Vegetable Salad with Lemon Vinaigrette, 64
Herbed Cucumber-and-Tomato Salad, 243
Lemon Vinaigrette, 65
Slaw, Napa Cabbage-and-Sweet Pepper, 108
Sauce, Peanut, 63
Sauce, Yogurt-Tahini, 243
Syrup, Apple Cider, 266
WONTON
Wonton Chips, Crispy, 272

Y

YOGURT
Chicken Thighs, Garlic-Yogurt-Marinated, 65
Dressing and Bacon Croutons, Mustard Greens with Yogurt-Parmesan, 48
Pops, Blueberry Yogurt, 309
Sauce, Yogurt-Tahini, 243
Soup with Brown Butter, Pecans, and Yogurt, Carrot, 202

Z

ZUCCHINI
Stuffed with Lady Peas, Zucchini, 217

Published by Oxmoor House, an imprint of Time Inc. Books
1271 Avenue of the Americas, New York, NY 10020

Senior Editor: Katherine Cobbs
Editor: Susan Hernandez Ray
Editorial Assistant: April Smitherman
Assistant Project Editor: Lacie Pinyan
Art Director: Christopher Rhoads
Designers: Carol Damsky, Allison Sperando Potter
Junior Designer: AnnaMaria Jacob
Executive Photography Director: Iain Bagwell
Photo Editor: Kellie Lindsey
Senior Photo Stylist: Kay E. Clarke
Food Stylist: Catherine Crowell Steele
Test Kitchen Manager: Alyson Moreland Haynes
Senior Recipe Developer and Tester: Callie Nash
Recipe Developer and Tester: Karen Rankin
Senior Production Manager: Greg A. Amason
Assistant Production Director: Sue Chodakiewicz
Copy Editor: Donna Baldone
Proofreaders: Rebecca Brennan, Rebecca Henderson
Indexer: Mary Ann Laurens
Fellows: Jessica Baude, Rishon Hanners, Natalie Schumann, Mallory Short, Abigail Wilt

ISBN-13: 978-0-8487-4481-6
ISBN-10: 0-8487-4481-0
ISSN: 0272-2003

Printed in the United States of America

10 9 8 7 6 5 4 3 2 1

First Printing 2015

Cover: Spice Cake with Cranberry Filling, page 286
Page 1: Pumpkin-Pecan Streusel Pie, page 261

Southern Living®
Editor in Chief: Sid Evans
Creative Director: Robert Perino
Digital Managing Editor: Grant Dudley
Senior Executive Editor: Katy McColl
General Manager: Whitney Chen Wright
Executive Editor: Krissy Tiglias
Director of Photography: Jeanne Dozier Clayton
Style Director: Heather Chadduck Hillegas
Managing Editor: Mary Elizabeth McGinn Davis
Copy Chief: Susan Emack Alison
Art Director: Paul Carstensen
Associate Art Director: Tim Kilgore
Senior Designer: Betsy McCallen Lovell
Test Kitchen Director: Robby Melvin
Test Kitchen Professional: Pam Lolley
Recipe Editor: JoAnn Weatherly
Director of Food Styling: Erin Merhar
Editorial Assistants: Marian Cooper, Pat York
Photographers: Robbie Caponetto, Laurey W. Glenn, Alison Miksch, Hector Sanchez
Photo Editor: Paden Reich
Associate Photo Editor: Kate Phillips Robertson
Senior Photo Stylist: Buffy Hargett Miller
Prop Stylist: Caroline Murphy Cunningham
Assistant Copy Chief: Libby Montieth Minor
Assistant Production Manager: Rachel Ellis
Office Manager: Nellah Bailey McGough

To order additional publications, call 1-800-765-6400.

Favorite Recipes Journal

Jot down your family's and your favorite recipes for quick and handy reference. And don't forget to include the dishes that drew rave reviews when company came for dinner.

Recipe	Source/Page	Remarks

Recipe	Source/Page	Remarks